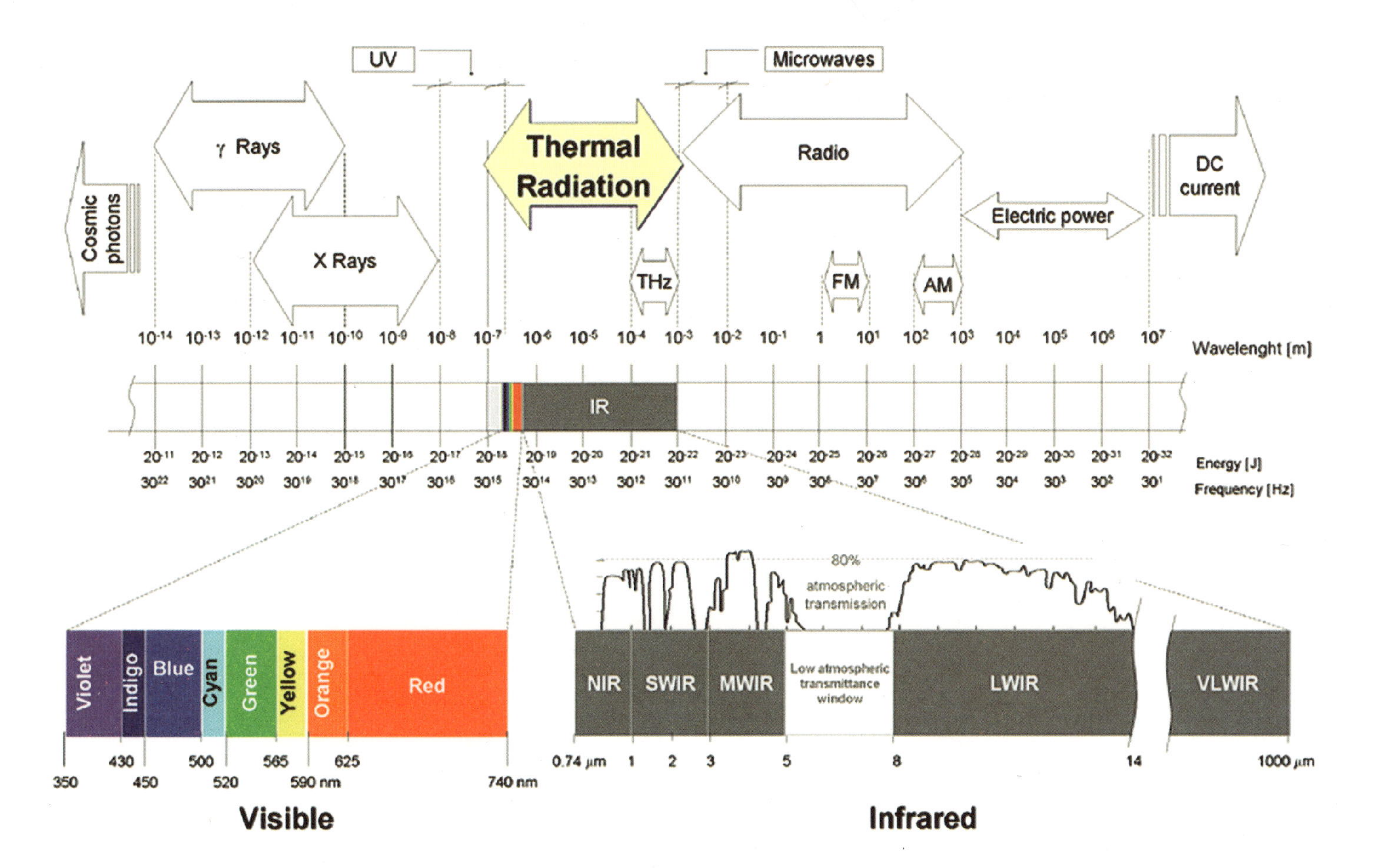
UV
Microwaves
Cosmic photons
γ Rays
X Rays
Thermal Radiation
Radio
DC current
Electric power
THz
FM
AM
Wavelenght [m]
IR
Energy [J]
Frequency [Hz]
80%
atmospheric transmission
Violet
Indigo
Blue
Cyan
Green
Yellow
Orange
Red
NIR
SWIR
MWIR
Low atmospheric transmittance window
LWIR
VLWIR
350
430
450
500
520
565
590 nm
625
740 nm
0.74 μm
1
2
3
5
8
14
1000 μm
Visible
Infrared

Third Edition

Fundamentals of MICROFABRICATION AND NANOTECHNOLOGY

VOLUME III

From MEMS to Bio-MEMS and Bio-NEMS

Manufacturing Techniques and Applications

Third Edition

Fundamentals of MICROFABRICATION AND NANOTECHNOLOGY

VOLUME III

From MEMS to Bio-MEMS and Bio-NEMS

Manufacturing Techniques and Applications

Marc J. Madou

CRC Press
Taylor & Francis Group
Boca Raton London New York

CRC Press is an imprint of the
Taylor & Francis Group, an **informa** business

CRC Press
Taylor & Francis Group
6000 Broken Sound Parkway NW, Suite 300
Boca Raton, FL 33487-2742

Printed and bound in India by Replika Press Pvt. Ltd.

International Standard Book Number: 978-1-4200-5516-0 (Hardback)

Visit the Taylor & Francis Web site at
http://www.taylorandfrancis.com

and the CRC Press Web site at
http://www.crcpress.com

I dedicate this third edition of Fundamentals of Microfabrication to my family in the US and in Belgium and to all MEMS and NEMS colleagues in labs in the US, Canada, India, Korea, Mexico, Malaysia, Switzerland, Sweden and Denmark that I have the pleasure to work with. The opportunity to carry out international research in MEMS and NEMS and writing a textbook about it has been rewarding in terms of research productivity but perhaps even more in cultural enrichment. Scientists have always been at the frontier of globalization because science is the biggest gift one country can give to another and perhaps the best road to a more peaceful world.

Contents

PART III
Miniaturization Application

Roadmap

From MEMS to Bio-MEMS: Manufacturing Techniques and Applications consists of ten chapters in three parts. In Part I, we compare available manufacturing options for microelectromechanical systems (MEMS) and nanoelectromechanical systems (NEMS), introduce inspection options for the produced parts (metrology), and summarize available modeling software for MEMS and NEMS. Nonlithography-based (traditional) manufacturing techniques are contrasted with lithography-based (nontraditional) methods in Chapter 1. In Chapter 2, we investigate nature as an engineering guide, and we contrast top-down and bottom-up approaches in Chapter 3. In Chapter 4 we learn about packaging, assembly, and self-assembly from integrated circuits to DNA and biological cells. In these chapters we aim to help the reader decide upon an optimized manufacturing option to tackle specific manufacturing problems. We then introduce some selected new MEMS and NEMS processes and materials in Chapter 5. Finally, we cover metrology techniques for MEMS and NEMS and summarize MEMS and NEMS modeling in Chapter 6.

Of the three chapters in Part II, the first deals with scaling laws, the second with actuators, and the third with issues surrounding power generation and the implementation of brains in miniaturized devices. In Chapter 7, on scaling laws, we look at scaling from both intuitive and mathematical points of view. We are especially interested in deviations from linear scaling, where downscaling reveals new unexpected physical and chemical phenomena. The scaling chapter also constitutes an introduction to the subsequent treatises on actuators in Chapter 8 and on power and brains in Chapter 9. An actuator—like a sensor—is a device that converts energy from one form to another. In the case of an actuator, one is interested in the ensuing action; in the case of a sensor, one is interested in the information garnered. Scaling enables one to compare various actuator mechanisms, such as those proposed to propel a rotor blade, bend a thin silicon membrane, or move fluids in fluidic channels. Power generation poses quite a challenge for miniaturization science; the smaller the power source, the less its total capacity. We will consider the powering of miniaturized equipment, as well as the miniaturization of power sources themselves, in Chapter 9. In the same chapter, we also discuss the different strategies for making micromachines smarter.

In Part III, Chapter 10, we review the MEMS and NEMS markets. Today, the MEMS field has made the transition out of the laboratory and into the marketplace with a plethora of new products, including a strong entry into the consumer electronics field (mobile phones, cameras, laptop computers, games, etc.). If MEMS is entering adolescence, the word *nano* still seems to derive from a verb that means "to seek and get venture capital funding" instead of from the Greek noun for dwarf. Today, the market dynamics for MEMS and NEMS commercial products are very different, and we treat MEMS and NEMS applications separately in this chapter.

As highlighted in Chapter 1, which compares nonlithography-based (traditional) and lithography-based (nontraditional) manufacturing, we hope that a better understanding of how to match different manufacturing options with a given application will guide the identification of additional killer applications for MEMS and NEMS and encourage more companies and research organizations to innovate faster based on their in-house manufacturing tools and know-how.

Note to the Reader: *From MEMS to Bio-NEMS: Manufacturing Techniques and Applications* was originally composed as part of a larger book that has since been broken up into three separate volumes. *From MEMS to*

Bio-NEMS: Manufacturing Techniques and Applications represents the third and final volume in this set. The other two volumes include *Solid-State Physics, Fluidics, and Analytical Techniques in Micro- and Nanotechnology* and *Manufacturing Techniques for Microfabrication and Nanotechnology.* Cross-references to these books appear throughout the text and will be referred to as Volume I and Volume II, respectively. The interested reader is encouraged to consult these volumes as necessary.

Author

Dr. Madou is the Chancellor's Professor in Mechanical and Aerospace Engineering (MEA) at the University of California, Irvine. He is also associated with UC Irvine's Department of Biomedical Engineering and the Department of Chemical Engineering and Materials Science. He is a Distinguished Honorary Professor at the Indian Institute of Technology Kanpur, India, and a World Class University Scholar (WCU) at UNIST in South Korea.

Dr. Madou was Vice President of Advanced Technology at Nanogen in San Diego, California. He specializes in the application of miniaturization technology to chemical and biological problems (bio-MEMS). He is the author of several books in this burgeoning field he helped pioneer both in academia and in industry. He founded several micromachining companies.

Many of his students became well known in their own right in academia and through successful MEMS start-ups. Dr. Madou was the founder of the SRI International's Microsensor Department, founder and president of Teknekron Sensor Development Corporation (TSDC), Visiting Miller Professor at UC Berkeley, and Endowed Chair at the Ohio State University (Professor in Chemistry and Materials Science and Engineering).

Some of Dr. Madou's recent research work involves artificial muscle for responsive drug delivery, carbon-MEMS (C-MEMS), a CD-based fluidic platform, solid-state pH electrodes, and integrating fluidics with DNA arrays, as well as label-free assays for the molecular diagnostics platform of the future.

To find out more about those recent research projects, visit http://www.biomems.net.

Acknowledgments

I thank all of the readers of the first and second editions of Fundamentals of Microfabrication as they made it worthwhile for me to finish this completely revised and very much expanded third edition. As in previous editions I had plenty of eager reviewers in my students and colleagues from all around the world. Students were especially helpful with the question and answer books that come with the three volumes that make up this third edition. I have acknowledged reviewers at the end of each chapter and students that worked on questions and answers are listed in the questions sections. The idea of treating MEMS and NEMS processes as some of a myriad of advanced manufacturing approaches came about while working on a WTEC report on International Assessment Of Research And Development In Micromanufacturing (http://www.wtec.org/micromfg/report/Micro-report.pdf). For that report we travelled around the US and abroad to visit the leading manufacturers of advanced technology products and quickly learned that innovation and advanced manufacturing are very much interlinked because new product demands stimulate the invention of new materials and processes. The loss of manufacturing in a country goes well beyond the loss of only one class of products. If a technical community is dissociated from manufacturing experience, such as making larger flat-panel displays or the latest mobile phones, such communities cannot invent and eventually can no longer teach engineering effectively. An equally sobering realization is that a country might still invent new technologies paid for by government grants, say in nanofabrication, but not be able to manufacture the products that incorporate them. It is naïve to believe that one can still design new products when disconnected from advanced manufacturing: for a good design one needs to know the latest manufacturing processes and newest materials. It is my sincerest hope that this third edition motivates some of the brightest students to start designing and making things again rather than joining financial institutions that produce nothing for society at large but rather break things.

Part I

From Traditional Manufacturing to Nanotechnology

In ten years the only manufacturing left in the United States will be 1) those facilities vital to the defense industry, 2) those industries that are uniquely high-tech, 3) those that cannot absorb long-distance freight charges, and 4) those industries that service "on the spot" instantaneous demand (although even that is questionable).

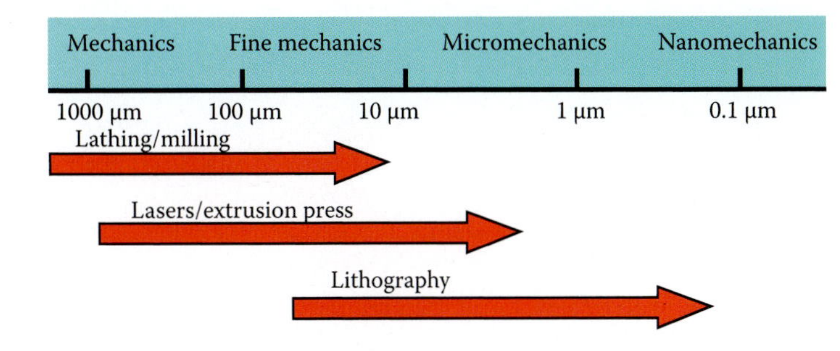

From macromachining to nanomachining.

Introduction to Part I

Introduction to Part I

In Part I, we compare available manufacturing options for microelectromechanical systems (MEMS) and nanoelectromechanical systems (NEMS), introduce inspection options for the produced parts (metrology), and summarize available modeling software for MEMS and NEMS. Nonlithography-based (traditional) manufacturing techniques are contrasted with lithography-based (nontraditional) methods in Chapter 1. In Chapter 2, we investigate nature as an engineering guide, and we contrast top-down and bottom-up approaches in Chapter 3. In Chapter 4 we learn about packaging, assembly, and self-assembly from integrated circuits (ICs) to DNA. In these chapters, we aim to help the reader decide upon an optimized manufacturing option to tackle specific manufacturing problems. We then introduce some selected MEMS and NEMS processes and materials in Chapter 5. Finally, we review metrology techniques for MEMS and NEMS and summarize MEMS and NEMS modeling in Chapter 6.

1

Nonlithography-Based (Traditional) and Lithography-Based (Nontraditional) Manufacturing Compared

Outline

1995–02, America lost 2 million industrial jobs, mostly to China. China lost 15 million of such jobs, mostly to machines.

Fortune

Despite the shrinking of America's industrial work force, our country's overall industrial output increased by 50% since 1992.

Economist

Charlie Chaplin's *Modern Times.*

Introduction

Micromachining, or microelectromechanical system/nanoelectromechanical system (MEMS/NEMS), is emerging as a set of new manufacturing tools to solve specific industrial problems rather than as a monolithic new industry with generic solutions for every manufacturing problem. It is important to evaluate the merit of using one certain MEMS/NEMS technique above all the other available micromanufacturing options [say, lithography-based LIGA* (i.e., nontraditional manufacturing) vs. nonlithography computer numerical control machining (CNC) (i.e., traditional manufacturing)] so as to find the technique that is optimal for the application at hand—in other words, one needs to zero-base the technological approach to the problem.[1] For example, micromachinists are often not aware of the capabilities of traditional machining (nonlithography) and use Si micromachining for parts that could have been made better with a more conventional manufacturing technology. By applying the right tool to the machining job at hand, we hope that micromachining will lead to many more successful commercial applications than there are today.

We start this chapter by listing reasons why one might want to miniaturize a given device at all. We follow this with a short treatise on the maturation of MEMS into a bona fide manufacturing technique that is complimentary today to traditional manufacturing methods. We go on contrasting serial, batch, and continuous manufacturing processes and distinguish between lithography-based (nontraditional) and nonlithography-based (traditional) manufacturing methods, capturing all of this in a large comparison table. In this same context, we explore the difference between truly three-dimensional (3D) manufacturing with equal versatility of machining along all three axes and the more constrained manufacturing methods, such as lithography-based techniques, where the capabilities in one dimension (say the height or *z*-axis) are very different from those in the other two dimensions (say the *x*, *y* plane). We then justify our recommendation to design MEMS from the package inward rather than starting from the MEMS itself. Finally, we introduce a decision tree to help determine the best manufacturing choice for a miniature device when given its detailed specifications. The utility of this decision tree is demonstrated using, as an example, the micromanufacture of a disposable glucose sensor with glucose meter.

Why Use Miniaturization Technology?

Over the last twenty years, many MEMS applications have proven their mettle in the marketplace, and consumer electronic applications that started emerging over the last five years (MEMS are now found in iPods, computers, cameras, GPSs, etc.) have catapulted MEMS once more in the public eye. A long list of reasons confirms why miniaturization presents so many opportunities for product innovation in so many different areas. Some of the most obvious reasons for miniaturization are summarized in Table 1.1. Usually, not all those reasons apply at once. For example, the small dimensions of micromachines might be crucial in medical and space applications but often lack importance in the automotive industry, where cost is the more important driver.

Table 1.1 Why Use Miniaturization Technologies?

- Minimizing of energy and materials consumption during manufacturing
- Redundancy and arrays
- Integration with electronics, simplifying systems (e.g., single-point vs. multipoint measurement)
- Reduction of power budget
- Taking advantage of scaling when scaling is working for us in the microdomain, e.g., faster devices, improved thermal management
- Increased selectivity and sensitivity
- Minimal invasiveness
- Wider dynamic range
- Exploitation of new effects through the breakdown of continuum theory in the microdomain
- Cost/performance advantages
- Improved reproducibility
- Improved accuracy and reliability
- Self-assembly and biomimetics with nanochemistry
- More intelligent materials with structures at the nanoscale

* LIGA is a German acronym for "Lithographie, Galvanoformung, Abformung," which in English is (x-ray) lithography, electroplating, and molding. It is a process in MEMS/NEMS that was developed in the early 1980s by a team under the leadership of E. Becker and W. Ehrfeld at the Institute for Nuclear Process Engineering [*Institut für Kernverfahrenstechnik* (IKVT)] at the Karlsruhe Nuclear Research Center. For details, Volume II, Chapter 10.

In Volume I, Chapter 1, when analyzing Moore's law describing the exponential growth in integrated circuit (IC) transistor density, we pointed out that similar exponential improvement in technology sophistication pertains to engineering skills in optics, genetics, and magnetic storage density. Humankind seems "hard-wired" to continue to innovate at an ever-increasing rate (see also Ray Kurzweil's Accelerating Returns, in Volume II, Chapter 2). Miniaturization is just one of the most important means by which humankind is shaping its own destiny rather than being at the mercy of evolution. Adaptation through evolution operates at a pace too slow to produce humans who thrive in the ever faster environmental changes they have created for themselves. In other words, we might have no choice but to adapt through ever-increasing technology sophistication rather than our DNA.

From Perception to Realization

MEMS Finally Succeeds in the Market

In the early 1980s, Si micromachining seemed often applied to show the world that University X also had a clean room or that research group Y could make a yet longer-lasting surface micromachined micromotor. The earlier predicament of Si-based micromachining as the "GaAs of the 1980s and early 1990s" [that is, a very good technology with little acceptance from an entrenched and highly standardized industry (think also Mac vs. PC)] had something to do with the intellectual/philosophical climate of silicon MEMS research and development in the early 1980s. Francis Bacon (1561–1626) explained how, in the advancement of knowledge, one is easily misled by what he calls "idolatry." Two of the "idols" he identifies point in the direction of a universal tendency to oversimplify, often manifested by the assumption of more order in a given body of phenomena than actually exists, and a tendency to be struck by novelty.[2]

Both idols apply to how Si micromachining became misrepresented. Little commonality exists between the many different microdevices made possible by micromachining. Actually, MEMS products represent discontinuous innovations—they are not simply incremental improvements on a previously existing technology. Each new product requires some new thinking on the part of the developers. The striking visual aspects of Si micromachined devices easily give a sense of novelty to any observer. This second idol drove most of the interest in micromachining by the popular press and the efforts of many academics. This climate contributed to overly optimistic expectations for very fast results, with a market size dwarfing the IC industry, and an overemphasis on Si as an answer to all miniaturization problems. In Chapter 10, on MEMS and NEMS applications, we suggest that a realistic number for the market of Si MEMS products today is less than 3% of the IC market (even the $6.3 billion MEMS sales in 2006, i.e., 3% of the $211 billion IC market for 2006, does include non-Si devices). Subsequently, government and industry funding sources experienced a hangover.

Bacon's idols aside, the very large academic involvement in the micromachining field had some other explanations. The IC industry technologically and financially outdistanced universities and forcibly pushed the latter to explore topics requiring lower startup expenses and areas where innovation still seemed likely. It was a perfect fit, as micromachining is an excellent topic for numerous PhD topics. In the 1980s, Si micromachining became a favored filler for clean room overcapacity, especially in Europe, where the IC industry suffered major setbacks against US and Japanese competition. By the end of the century, it was apparent that the Si micromachining industry was still very small compared with the IC industry (less than 3%) but had large numbers of people involved in its research and development. Middlehoek and Dauderstadt estimated, as early as 1994, that about 10,000 researchers worldwide were involved in Si sensor research and that $7.5 billion had been spent over the preceding twenty-five years.[3] Many MEMS researchers have now switched to nanotechnology (NEMS) research instead.

Another major shortcoming in Si MEMS, as practiced in academia, pertains to the lack of interdisciplinary teams. More than any other field, microengineering requires interdisciplinary teamwork rather than specialists. Multidisciplinary work from the design phase on is crucial in the development of a successful sensor product. Many of the participating university groups characterize themselves as multidisciplinary, but the opposite is often true; this lack of multidisciplinary focus continues to obstruct the production of more practical results. It reminds us of a statement by C.P. Snow (1905–1980), who remarked that the separate departments at a university had ceased to communicate with one another. The prefix *uni* in university, he said, lost its meaning as the institution, striving for more and more power as government funds replenished research dollars, had turned into a loose confederation of disconnected mini-states instead of an organization devoted to the joint search for knowledge and truth.

For micromachining to lose the stigma of a technology looking for an application, used in research backrooms and excess industrial clean rooms only, it became important for MEMS to realize some product successes. These industrial successes started appearing principally in the early to mid-1990s and include pressure sensors (on a small scale as early as 1972 at National Semiconductor), accelerometers (1992, from Analog Devices), projection displays (digital mirror devices from TI in 1995), ink-jet nozzles (thermal ink-jet from HP, 1984), electrokinetic platforms (Caliper and Agilent, 1999), fiber optic switches (NTT, 1995), and the first digital mirror device-based products (Texas Instruments, 1996). The first mechanical MEMS applications in industry were modest success stories only, with dollar evaluation in initial public offerings for mechanical MEMS companies typically only in the $5 million to $45 million range. Ever since biotechnology and information/communication industries started taking an interest in MEMS, huge stock market success stories began to appear, and large corporations bought small MEMS companies for billions of dollars. But today, many investors still cringe when they hear the terms *MEMS* or *NEMS*, because of the ill-fated optical MEMS bubble of the late 1990s and a BIOMEMS field that never seems to live up to its promises.

Fortunately, as we will learn in Chapter 10, on MEMS and NEMS applications, mechanical and optical MEMS in consumer electronics, started taking off over the last five years, and even MEMS for large optical switches in telecommunications is now regaining a foothold. The real MEMS breakthroughs continue to originate overwhelmingly from start-up companies (e.g., Illumina and Cepheid) and large corporations (e.g., ADI and Bosch) rather than academia. Micromachining in industry is more correctly seen as a diverse set of tools to solve practical problems in the crafting of subminiaturized 3D structures rather than as a goal by itself. The multidisciplinary approach and matrix organization of MEMS teams in industry further explains their more impressive MEMS accomplishments.

In academia, with its very vertical departmental structure and isolation from practical manufacturing and application understanding, miniaturization science is still often hyped beyond recognition in order to secure grants, tenure, or money from government agencies, large corporations, and venture capitalists. Describing academic MEMS efforts as unbiased fundamental work or fine-tuning of micromachining skills, rather than proclaiming each new result as the latest breakthrough in sensor technology or analytical chemistry, would often serve society better.

The new wave of nanomachining and nanochemistry efforts, mostly grounded in the fundamental research phase at this stage, holds the promise that university scholars might regain the miniaturization playing field in the near future. We can only hope that the hype surrounding nanotechnology settles down and that funding agencies and companies give researchers and scientists enough time to realize the stupendous opportunities that nanotechnology holds. To realize the economic potential of MEMS and NEMS, society will need a new generation of engineers who are comfortable with chemical and biological issues as well as many different types of micro- and nanomanufacturing considerations. Academia will have to become more matrix organized, if only to prepare the proper workforce for the coming age of nanomachining. From the fundamental point of view, micromachining and nanomachining are also of extreme interest, as these fields provide an excellent opportunity for operating experiments in a regime where continuum theory breaks down, holding the potential for discovery of important new chemical and physical phenomena. Unfortunately, mounting funding pressures in the United States are forcing scientists to sell research as development and development as manufacturing. By putting their work in the context of a practical solution too early, academics might have lost some of industry's trust in the capabilities of micromachining and nanomachining. A breakthrough in deep, cryogenic etching or yet thicker vertical photoresist walls should stand by itself and does not necessarily need to be illustrated with the fabrication of wheels for micromotors! A better practice in the future might be to compare these new academic results (say, on new deep UV photoresists) with other, more traditional approaches so that the work becomes of generic value to a potential user for his or her own specific application.

Beyond Si-Based MEMS

New Si-based MEMS products are only a small part of the total projected MEMS markets, and there is ample room for non-Si-based companies in the MEMS markets of the future (e.g., microplastic molders, microelectroplaters, microceramic parts producers, etc.). Market predictions from the 1980s and early 1990s about the Si MEMS revolution were very mistaken; it is only by broadening the definition that MEMS is living up to the numbers that were quoted at that time. In Chapter 10 of this volume on MEMS and NEMS markets, we will see that when including non-Si MEMS, which many market studies do include now, and especially when also counting nonlithography-based MEMS (e.g., precision engineered microproducts; see below) (which very few market studies do yet), MEMS might very well constitute 10% of all current IC sales.

The market does not care about the type of miniaturization method employed for a new product, and for some time the Si micromachining community, especially in the United States, did not appreciate the tremendous accomplishments of alternative miniaturization methods. The HP Kittyhawk Personal Storage Module, a 1.3-inch disk drive, a compact disc (CD) optical pickup, and IBM's thin-film magnetic read-write head shown in Figure 1.1 are triumphs of precision engineering, and none involve Si-MEMS. The Kittyhawk (Figure 1.1a), introduced by HP in 1992, never caught on and seems to have been introduced before its time (Sony has continued its efforts on a similar MiniDisc). Kittyhawk was discontinued in September 1994. Approximately 160,000 units were sold (compared with a projected two-year sales of 700,000 units). In 1996, largely because of Kittyhawk's failure, Hewlett Packard closed its Disk Memory Division and exited the disk drive business. (The story of HP Kittyhawk was described in a Jan 26, 2006 Harvard Business School business case, "Hewlett-Packard: The Flight of the Kittyhawk."[64]) CD optical pickups and CD technology in general are other marvels of precision engineering, funded by billions of dollars in research and development (R&D) at Philips and Sony (see Figure 1.1b and 1.1c). This technology has not been exploited nearly as well as it should outside its traditional application area. We envision many nontraditional applications for this platform, such as its use as a sample preparation station and a molecular diagnostic tool as explored in this chapter, in Volume I, Chapter 6, and in Chapter 5, this volume. In diagnostic applications, the same CD technology may be used to read both analytical data (e.g., as a color change of a chromophore) and information (in the form of pits on the CD), to serve as a very inexpensive microscope (see Volume I, Figures 6.93 and 6.94) or as a high-precision *x-y* stage, etc. In this respect, one could legitimately ask the question, why are optical pickup heads not included in MEMS market studies when hard disc drive read-write heads often are? (See Chapter 10, this volume.) The answer is that these choices were made quite arbitrarily, underscoring the need to identify the criteria/definitions of MEMS and the manufacturing means used to make the microcomponents covered in any MEMS marketing report. The thin-film magnetic heads, shown in Figure 1.1d, are an excellent example of the commercial success one can achieve when identifying the correct machining option and materials combinations for a miniaturization task at hand. At $10 billion in sales in 2007, the thin-film magnetic head is still the most successful MEMS product today. The manufacture of the thin-film head is principally based on photolithography and electroplating of the coils and does not involve any Si. In 2009, read-write (RW) heads are still projected to have a market the size of all other Si MEMS

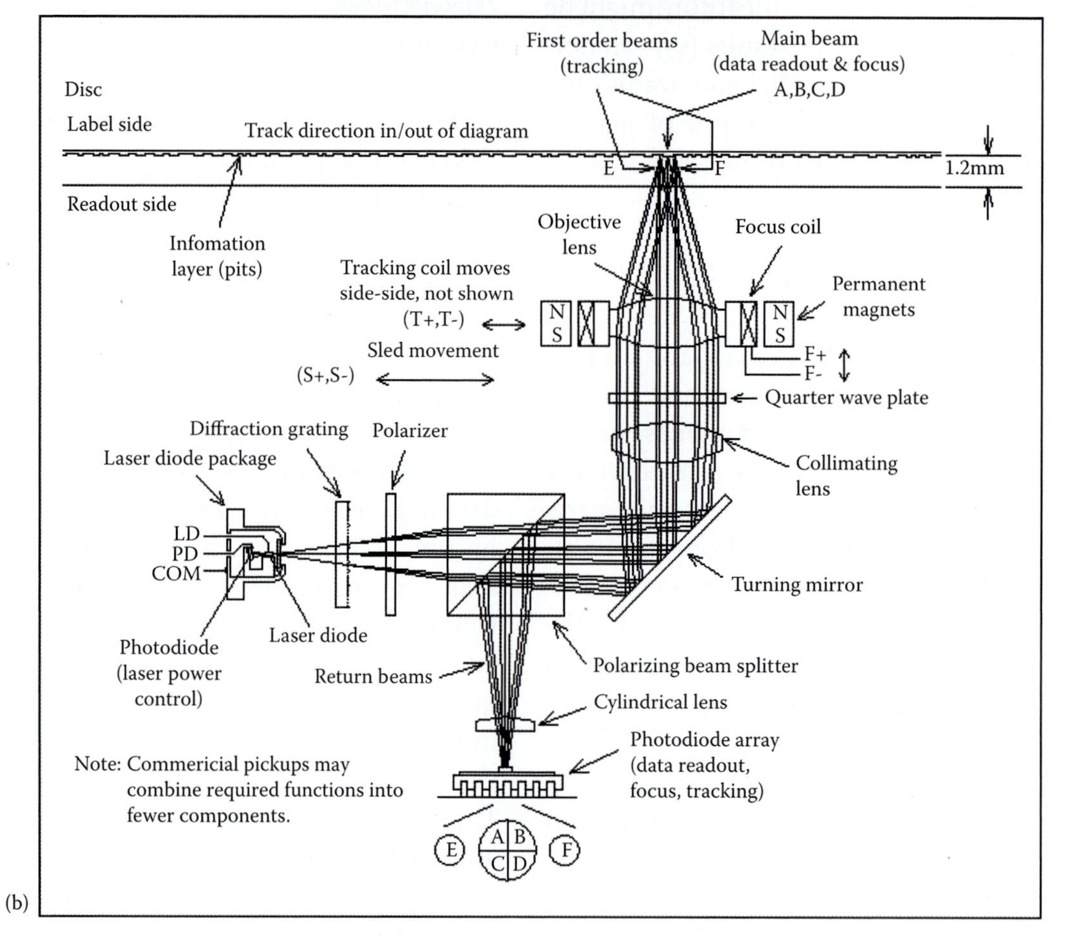

FIGURE 1.1 Triumphs of mechatronics. (a) HP's Kittyhawk. (From http://www.homebrewcpu.com/Pictures/kitty-hawk.jpg.) (b) CD optical pickup. General three-beam optical pickup organization. (From http://www.repairfaq.org/REPAIR/ F_cdfaq1.html.) (c) Physical realization embodied by the Sony KSS361A Optical Pickup (see http://www.repairfaq.org/REPAIR/kss110c.gif). (d) The evolution of magnetic read-write heads. From 1980 on, the magnetic coils were made by semiconductor-type technology (lithography and Cu plating) (http://dspace.cusat.ac.in/dspace/bitstream/123456789/2121/1/ GIANT%20MAGNETORESISTANCE%20EFFECT.pdf.)

applications put together (see RW in Figure 10.15, this volume).

Because there are few established engineering rules and practices available to decide upon an ideal manufacturing technique from the wealth of options available for a given microapplication, good matches of machining methodology and application are often missed or rediscovered. Figure 1.2 illustrates the evolution of the types of manufacturing methods attempted for the construction of microfluidic elements. In the 1960s, microfluidic elements were principally made from photoglass; after attempting

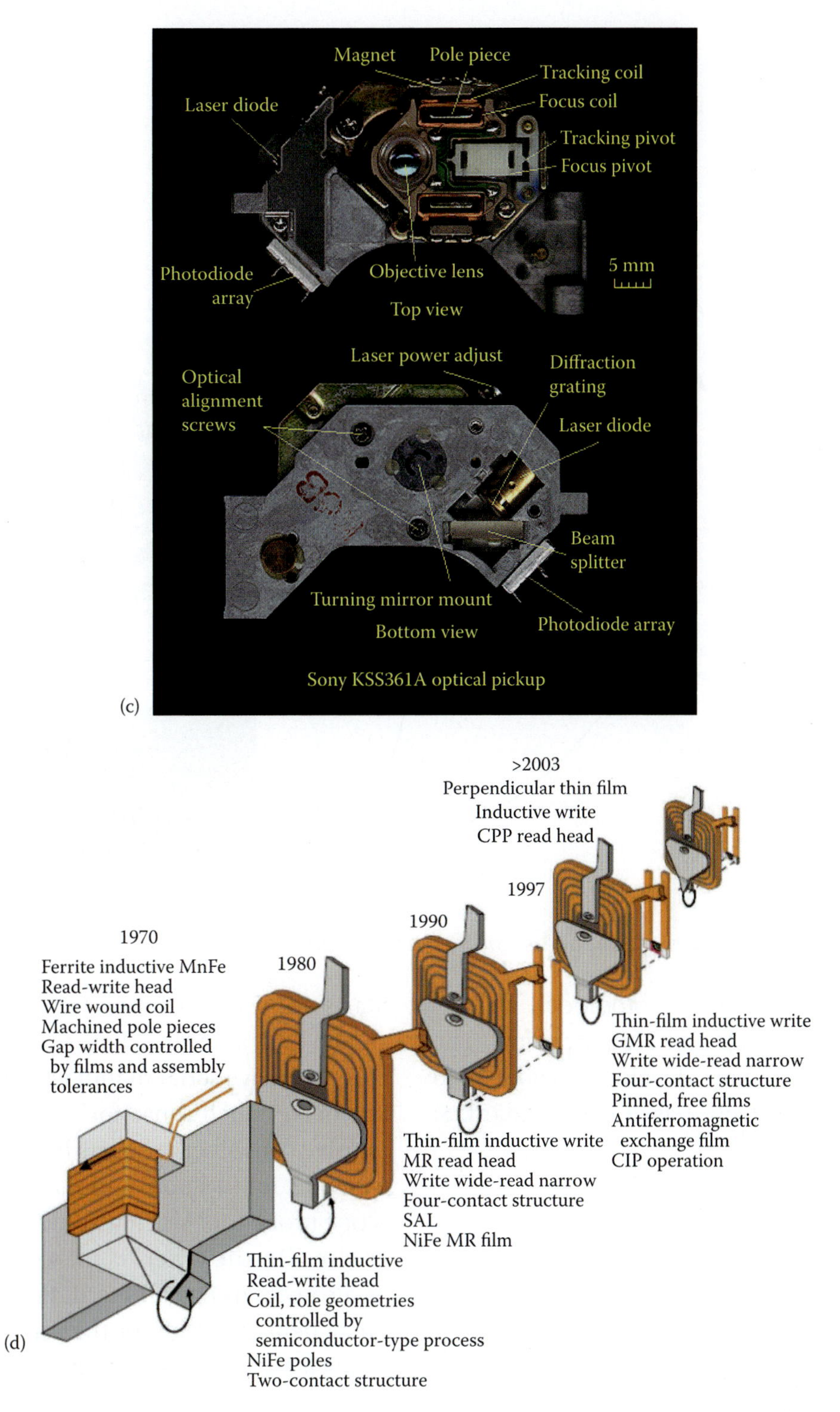

FIGURE 1.1 *(Continued)*

to craft the same devices by dry etching in Si in the 1980s at Stanford, LIGA was employed in the early 1990s. But by the mid-1990s, photoglass was rediscovered as the best machining option for some microfluidics applications. One might hope that by following a decision process flow researchers can avoid these costly loss and rediscovery cycles in the future.

As pointed out above, recent market studies cover Si and non-Si technology miniaturization applications, thereby achieving earlier projected large revenues that were expected to derive from Si micromachining alone. Some early market data on Si-based miniaturization were misleading in another way: they included big commercial successes

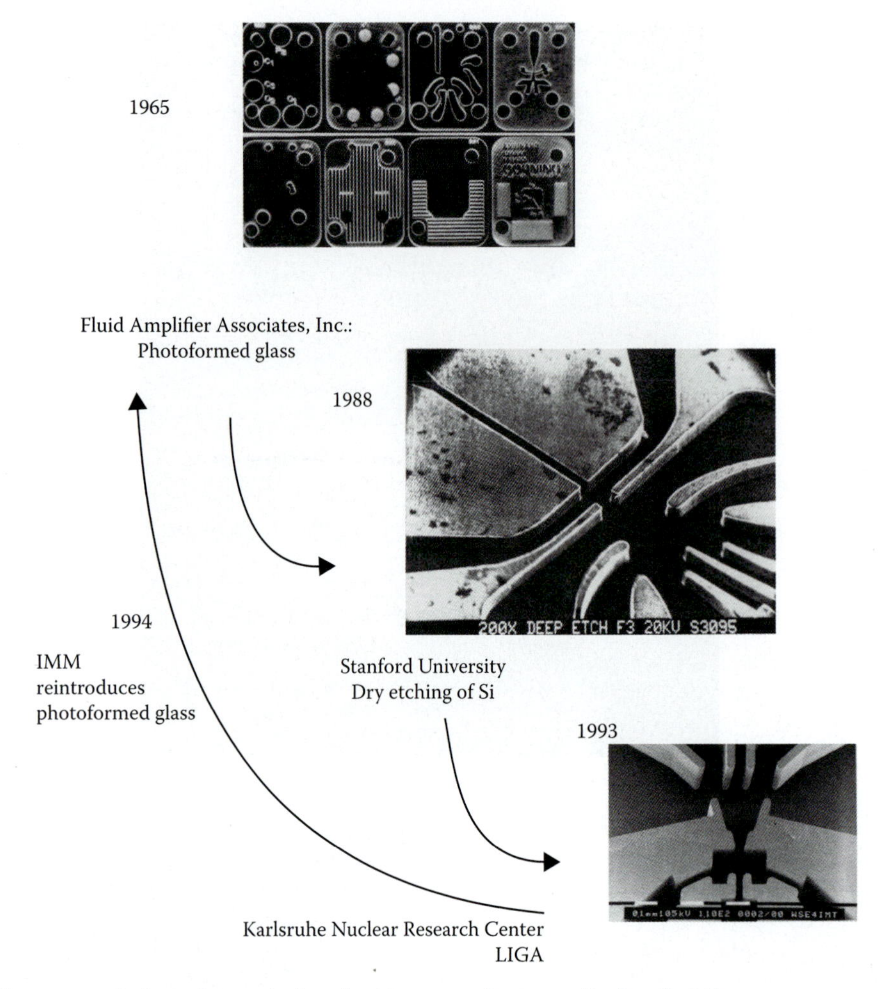

FIGURE 1.2 The rediscovery of photoformed glass for the manufacture of microfluidics.

in Si sensors such as simple photodiodes, charge-coupled detectors (CCDs), and Hall elements. The latter were, for the most part, developed before 1970 and are almost exclusively based on standard Si manufacturing techniques.

Serial, Batch, and Continuous Manufacturing Options

In serial manufacturing, a manufacturing station produces one product at the time. As an example for serial machining consider focused ion-beam milling, a mechanical removing technique detailed in Volume II, Chapter 6. Vasile et al., for example, describe a lathe mounted 20-keV Ga ion beam as the cutting tool for in-vacuum micromachining.[4,5] The ion beam they use is 0.3 μm in diameter, and they crafted tungsten needles, hooks, tuning forks, and specialized scanning probe tips. Two of their tungsten microstructures are shown in Volume II, Figure 6.33. Start-to-finish fabrication time for one of the objects shown here is about 2 hours! It should be noted that this expensive serial manufacturing technique might not make much sense for mass production of inexpensive consumer products but could be justified, for example, for the fabrication of an atomic force microscope probe tool set that could not be produced otherwise. On the other hand, to make a microshovel, as also shown in Volume II, Figure 6.33, is merely a fancy demonstration of the capabilities of this technique (call it a micro-"BS shovel"*).

In batch fabrication, a "group" or "batch" of identical products is produced in one single machining step. Batch fabrication techniques lend themselves to economies of scale that are unavailable with serial techniques. The use of batch processes is widespread throughout the manufacturing industry, producing a diverse range of products, including ICs,

* BS used to stand for *bullshit*; now it stands for *Bush Shit*.

pharmaceuticals, polymers, biochemicals, food products, and specialty chemicals. In the more advanced manufacturing techniques used for micromachining and nanomachining, batch fabrication typically involves some sort of a mask containing a desired repeated pattern that is positioned between the tool and work-piece. Then, that mask design is transferred to the work-piece using, for example, light (as in photolithography; Volume II, Chapter 1) or an abrasive slurry (as in abrasive jet machining; Volume II, Chapter 6). Alternatively, the tool itself contains an array of patterns that is transferred to the work-piece, for example, mechanically, through a slurry (as in ultrasonic machining; Volume II, Chapter 6), or thermomechanically (as in hot embossing; Volume II, Chapter 10). Most nontraditional micromachining carried out today, such as IC fabrication, relies on batch fabrication; that is, repetitive features are simultaneously defined on a work-piece. In the case of IC fabrication, an array pattern is photolithographically defined on a silicon wafer, and many wafers are then processed further to fabricate desired structures. We present as an example of batch fabrication, an array of finished CCD image sensors (to use, for example, in a digital camera), produced all in parallel on the same Si wafer, as shown in Figure 1.3. After production, these sensors need to be cut out of the Si wafer.

It is important to recognize that the use of a mask severely restricts the machining possibilities in the direction perpendicular to the mask. The resulting products, when using a mask, are projected shapes

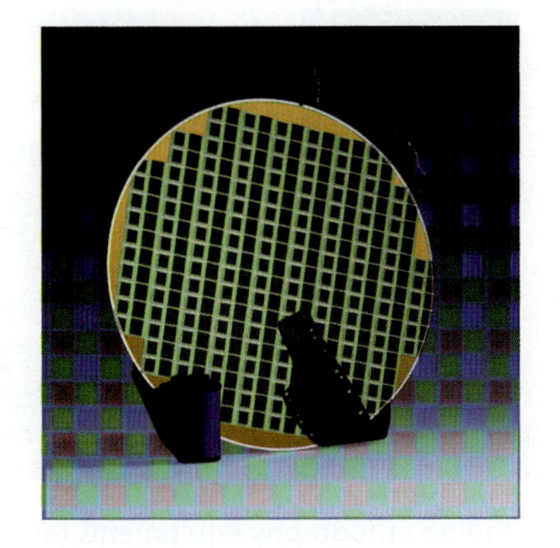

FIGURE 1.3 Charge-coupled detector (CCD) image sensors are formed on Si wafers and then cut apart. CCDs have replaced film as the image capture medium of preference. (Courtesy of IBM.)

of the image on the mask only, with limited opportunities to shape the sidewalls of the work-piece perpendicular to the image on the mask. To illustrate this, think about precision mechanical machining, which is truly 3D since a diamond tool can cut in any direction, and contrast this with using a photomask in photolithography or a patterned tool in ultrasonic machining, where material removal forces work only in one direction and the resulting shapes are principally 2D. In the latter case, the dimension perpendicular to the mask may have differing heights (and indeed, we could say that we are dealing with a 2½D structure), but the shape of the sidewall can be modified only in a separate machining step.

Beyond batch manufacturing, the ultimate in automation of manufacturing is continuous or web-based manufacturing, with a web moving from supply roll to take-up roll [also called roll-to-roll (R2R) manufacturing*], with visions of low-cost photovoltaics (PVs), thin-film transistors (TFTs), large-area organic light-emitting diode (OLED) lighting, flexible displays, low-cost sensors, radio frequency identification (RFID) tags, etc. (see Figure 1.4). In this case, the product is made continuously, small or large, on flexible substrates, as shown in Figure 1.5. The flexible substrates replacing glass or Si must possess the desirable optical properties, thermal stability, low coefficient of thermal expansion, surface smoothness, and moisture and solvent resistance. Today, polymer films and stainless steel are the two flexible substrates of choice. R2R manufacturing dramatically increases throughput, with a significant reduction of capital and device costs.

Gordon Moore has argued that the only industry "remotely comparable" in its rate of growth to the semiconductor industry is the printing industry. It is worth noting that newspaper printing constitutes a continuous process; here, we promote the idea of continuous (beyond batch) miniaturization processes that, like a printer connected to a personal computer, will print out miniature objects in a continuous mode. Many of the manufacturing techniques we encountered in Volume II are already or can be modified to a roll-to-roll operation. This includes older

* Not to be confused with "run-to-run" or "run-by-run" control in semiconductor manufacturing.

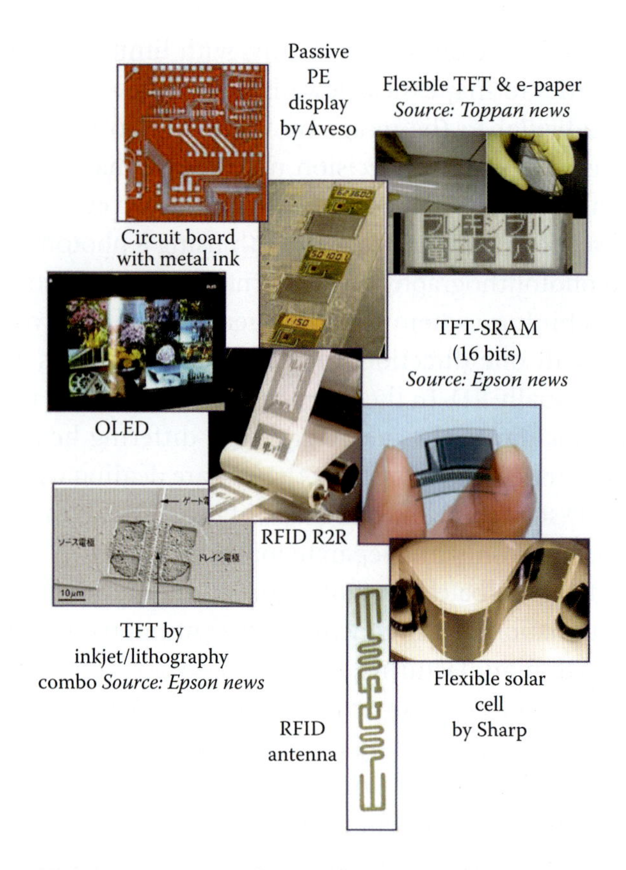

FIGURE 1.4 Examples of printed products on flexible substrates. For acronyms such as RFID, OLED, and TFT, please refer to the text.

technologies, such as doctor blade, ink-jet printing, laser machining, lamination, punching, drying and curing (in roll format ovens), picking and placing, hot-embossing, and sputtering (with cylindrical targets; see Volume II, Figure 7.12), and newer methods, such as imprint lithography [nanoimprint lithography (NIL) and step-and-flash imprint lithography (SFIL)], dip-pen lithography (DPL), etc. Soft lithography (patterning with flexible stamps; see Volume II, Figure 2.55) constitutes yet another possible means of transferring lithography patterns to a web-based process. In this case, the soft lithography pattern on the rubber stamp might be rolled up around a cylinder to ink a substrate continuously.

Processes, such as lithography, that need critical alignment and registration are the more challenging to incorporate in an R2R manufacturing process. But roll-to-roll or in-line equipment is also available for lithography. For illustration, in Figure 1.6, we show the lithography equipment procured by The Center for Advanced Microelectronics Manufacturing at Binghamton University in New York (http://camm.binghamton.edu). Established in 2005, this center is meant to spearhead development of next generation R2R electronics manufacturing capabilities. In Figure 1.6a we show a wet stripper/developer from Hollmuller-Siegmund that can handle a web of a width up to 15 in. The in-line lithography setup (Figure 1.6b) is from Azores Corporation and can handle webs that are 8 in. wide (expandable to 24 in.). This g-line machine comes with an accuracy of about 4 μm and can process photosensitive material at a speed of 230–760 mm/min. Hole-punch patterns are used for alignment. Many governments are taking notice of this important trend in electronics manufacturing. In Korea, for example, the Department of Printed Electronics Engineering in the World Class University (WCU) program of Sunchon National

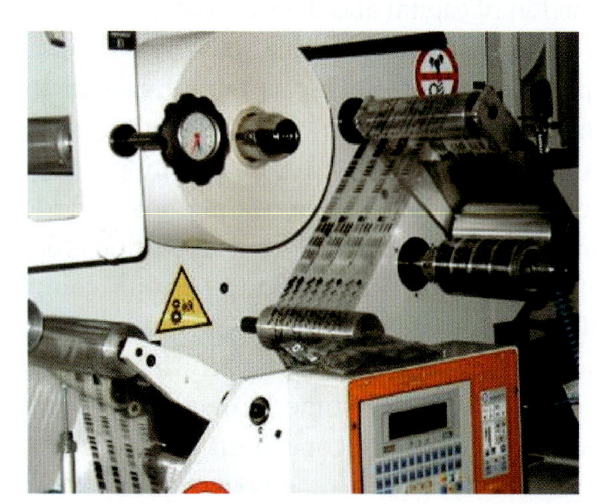

FIGURE 1.5 Roll-to-roll (R2R) manufacturing. Production in kilometers and use and sale in centimeters.

FIGURE 1.6 In-line lithography equipment. (a) wet stripper/developer from Hollmuller-Siegmund (can handle a web up to 15 in. wide). (b) in-line lithography setup from Azores Corporation can handle webs that are 8 in. wide (expandable to 24 in.).

University has been launched to lead in the fields of printed RFID and flexible lighting through Korean Government support that started in 2009.

Advances are being made continuously in "ink" technologies and printing processes and control; organic and printable electronics are expected to be a $35 billion industry by 2015 and reach over $300 billion in 2025, that is, almost twice the size of the Si industry today.

In Example 1.1, at the end of this chapter, we introduce a scenario for the continuous manufacture of polymer/metal-based biosensors—beyond batch.

Comparison Table

In Table 1.2, the most popular manufacturing techniques in crafting micromachines are compared. These tools are used to fabricate microstructures either in serial (Se), parallel (Ba for batch), or continuous (C) mode; the techniques can be organized in many different other ways, for example on the basis of the type of energy involved in the machining or as nonlithography-based (i.e., traditional—group 1) or lithography-based (i.e., nontraditional—group 2), i.e., truly 3D vs. truly 2½D. Table 1.2 is a listing of lithographic and nonlithographic miniaturization techniques with their respective resolutions, defined here as the smallest lateral features that have been generated. As a very rough rule of thumb, one might say that with dimensions below 100 μm one needs to switch from mechanical tooling (group 1) to optical tooling (group 2) to cut into a material.

Wet bulk micromachining, a very popular MEMS technique, had its genesis in the Si-based IC industry, but further development required the adoption and adaptation of several different processes and materials. Si-based wet bulk micromachining techniques involve anisotropic etching in the bulk of the material, often resulting in stunningly beautiful scanning electron microscope micrographs, are one of the most visual and "sexy" micromachining tools used—the MTV of miniaturization science. But in the trend toward miniaturization, almost a law of nature ever since the first oil crisis and the IC revolution, many other precision machining techniques on many different materials have also found commercial applications. Wet bulk Si micromachining, the most traditional form of Si micromachining (dominated by electrical engineers), has found acceptance in traditional precision machining circles (dominated by mechanical engineers) and vice versa: silicon micromachinists are starting to look at precision engineering techniques as a way to extend their tool kit. The most important characteristic for both disciplines is the ever-increasing quest for improved accuracies, higher aspect ratios, and reduced cost, and practitioners have recognized the complementary nature of their respective fields.

Whereas mechanical sensor manufacture is moving toward more integration of sensing functions with electronics, embodied in increased reliance on surface micromachining mostly involving the deposition and etching of thin films on top of a Si substrate, which is most compatible with IC fabrication, chemical and biosensors are moving away from integration and toward hybrid thick technology. This trend in chemical and biological sensors is caused by the need for modularity and the tremendous compatibility problems that arise when attempting to integrate chemical sensor materials with ICs.

From Table 1.2, Si micromachining then emerges as only one of the many options for precision machining. To emphasize the need for micromachinists to look beyond Si as the ultimate substrate and/or building material, we have presented many examples of non-Si micromachinery throughout this book. The need to incorporate new materials and processes is especially urgent for progress in chemical sensors and microinstrumentation, which rely on non-IC materials and often are relatively large.

The renewed interest in some of the non-Si-based machining methods stems principally from two major deficiencies of IC-based machining techniques: the difficulty of creating truly 3D microstructures and the problem of interfacing microstructures with the macroworld. A machining method with a range that covers macroscopic- to microscopic-scale devices is needed, as often micromachines cannot easily be handled. Laser beam machining (LBM), electrodischarge machining (EDM), ultraprecision machining, dry reactive ion etching (DRIE), LIGA, and pseudo-LIGA techniques are steps in that direction, forming so-called "handshake technologies," bridging the micro- and macroworlds. These

TABLE 1.2 MEMS Miniaturization Methods: Nonlithography-Based (Traditional) and Lithography-Based (Nontraditional)

Machining Method	Material/ Application	Typical Min/Max Size Feature	IC Compatible?	Tolerance	Important Reference on Technique	Aspect Ratio (Depth/ Width)	Initial Investment Cost/Access
Group 1: Traditional Techniques (Not Involving Photolithography-Defined Masks)							
Chemical milling (S), (Ba)	Almost all metals	From submillimeter to a few meters (*x*, *y*); max thickness (*z*) ±1 cm	Yes	Lateral tolerance 0.25–0.5 mm	Harris[6]	~1	Low/good
Electrochemical machining (S/A), (Ba)	Hard and soft metals, turbine blades, pistons, fuel-injection nozzles	Minimum-size devices larger than in chemical milling because of contacting need	Fair	Lateral tolerance <10 μm	(A) Romankiw,[7,8] (S) Phillips[9]	(S) 100	~$400,000/good
Electrodischarge machining (S), (Se)	Hard, brittle, conductive materials used for tools and dies	Minimum holes of 0.3 mm in 20-mm-thick plate	No	Lateral tolerance 5–20 μm	Kalpajian[10]	100	High/good: equipment with numerical control is common
Electrodischarge wire cutting (S), (Se)	Hard, brittle materials; many punch-and-die applications	Minimum rods 20 μm in diameter and 3 mm long	No	Lateral tolerance 1 μm	Saito[11]	>100	High/good: does not require special electrodes

Electron beam machining (S/A), (Se)	Hard-to-machine materials	(S) Most suited for large numbers of simple holes (<0.1 mm)	Fair	(S) ~10% of feature size (5 μm on a 50-μm hole)	Taniguchi[12]	(S) 10 is typical but 100 is possible	Very high/fair ~$100,000 when using a modified scanning electron microscope
Continuous deposition (A), (C); e.g., doctor's blade technology, tape casting	With all materials available in inks, e.g., glucose sensors	Most suited for inexpensive disposables, from 100 μm to a few millimeters	No	15 μm	Harper,[13] Fiori and De Portu[14]		Low/good
Focused ion-beam on a lathe (S/A), (Se)	Very pure IC materials	From submicrons to millimeters	Yes	(S) 50–100 nm	Vasile et al.[5]		High/poor
Hybrid thick film (A), (Ba)	Wide variety of materials available in inks	Minimum feature size 90 μm	Fair	12 μm	Harper[13]		~$30,000/good
Laser beam machining (S/A), (Se)	Complex profiles in hard materials	(S) Holes from 10 μm to 1.5 mm at all angles	Fair	1 μm	Helvajian[15]	(S) 50	~$50,000 but up to ~$400,000 for a five-axis system/good
Plasma-beam machining (S/A), (Se/Ba)	Very-high-temperature materials	(A) Only used for thick films >25 μm; (S) for very thick films >2.5 mm	No	(A) 20 μm for a 25-μm-thick film; (S) typical ± 3 mm but 0.8 mm is possible	Pfender[16]		~$600,000/fair

(Continued)

Table 1.2 MEMS Miniaturization Methods: Nonlithography-Based (Traditional) and Lithography-Based (Nontraditional) (*Continued*)

Machining Method	Material/ Application	Typical Min/Max Size Feature	IC Compatible?	Tolerance	Important Reference on Technique	Aspect Ratio (Depth/ Width)	Initial Investment Cost/Access
Stereolithography (A), (Se)	Polymeric photosensitive materials	Maximum 10 × 10 × 10 mm (*x*, *y*, *z*)	Yes	Minimum solidification 5, 5, 3 μm (x, *y*, *z*)	Ikuta,[64] stereo lithography[17]		$100,000–500,000/ good
Ultraprecision mechanical machining (S), (Se)	Form-stable materials	From submillimeters (e.g., 0.2-mm hole) to meters	No	1 nm	Boothroyd and Knight[18]		$400,000/good
Ultrasonic machining (S), (Se)	Hard and brittle materials	Holes from 50 μm to 75 mm	No	Lateral tolerance 10 μm	Bellows and Kohls[19]	2.5 μm for a 250-μm hole	$20,000/good
Microdispensing (A), (Se/Ba)	Wide variety of materials available in inks, especially suited to biomaterials	Minimum membrane volume: nanoliters to microliters	Yes	~3% volume	Li et al.[20]		Medium/poor
Injection molding, hot embossing (A), (Ba)	Plastics, ceramics, and metals	Minimum 250 μm; maximum several centimeters	No	±0.13% (e.g., ±2.6 μm for a 1.998-mm diameter)	Sakai,[21] Rak[22]	>5	Medium/good

Machining Method	Material/ Application	Typical Min/Max Size Feature	IC Compatible?	Tolerance	Important Reference on Technique	Aspect Ratio (Depth/Width)	Shape and Height/Depth	Initial Investment Cost/Access
Powder blasting (S), (Se)	Ceramics and hard materials >40 HRC	Minimum 50 µm up to several millimeters	No	1–2 µm	Belloy et al.[23]			Medium/poor
Abrasive water jet (S), (Se)	Metals, plastics, and ceramics	Min 0.8 mm typical, 80 µm possible	No	±50 µm	Waterjet[24]	>100		\$80,000–200,000/ poor
Group 2: Nontraditional Techniques (Involving Photolithography-Defined Masks)								
Photofabrication (S)	Plastic, glass (ceramic), e.g., fluidic elements	Max x, y = 40 × 40 cm and max z = 0.6 mm	Yes	Lateral tolerance 20 µm	Trotter[25]	~3 for photoplastics; ~20 for photoglass	x, y is free; z up to 6 mm	Medium/ poor to medium
Photochemical milling (S)	Printed circuit boards, lead frames, shadow masks	Max 60 × 60 cm, max thickness <0.5 mm	Yes	13 µm (printed circuits)	Allen[26]	~1	x, y is free; z up to 0.5 mm	Medium/ good
Wet etching of anisotropic materials (S)	Crystal Si, GaAs, quartz, SiC, InP	Max wafer size, min feature a few microns	Fair	1 µm	Kern[27]	100	x, y, z shape locked in by crystallography, z height of the wafer	Low/good
Dry etching (S)	Most solids	Max wafer size, min feature submicron	Good	0.1 µm	Manos and Flamm[28]	10	x, y shape free; z up to 200 µm	High/good
Poly-Si surface micromachining (S/A)	Poly-Si, Al, Ti, etc.	Max wafer size, min feature submicron	Good	0.5 µm	Johnstone and Parameswaran[29]		x, y free; z 0.1–10 µm, but preferably 1–2 µm	High/fair

(*Continued*)

TABLE 1.2 MEMS Miniaturization Methods: Nonlithography-Based (Traditional) and Lithography-Based (Nontraditional) (*Continued*)

Machining Method	Material/ Application	Typical Min/Max Size Feature	IC Compatible?	Tolerance	Important Reference on Technique	Aspect Ratio (Depth/Width)	Shape and Height/Depth	Initial Investment Cost/Access
Si on insulator (S)	Crystalline Si	Max wafer size, min feature submicron	Good	0.1 μm	Diem et al.[30]		*x*, *y* free; *z* height depending on the epilayer, e.g., 100 μm	High/poor
LIGA (S/A)	Ni, PMMA, Au, ceramic, etc.	Max 10 × 10 cm or more, min 0.2 μm	Fair	0.3 μm	Ehrfeld[31]	>100	*x*, *y* free, *z* up to 100 μm	High (>$35M)/ poor
UV-transparent resists (S/A)	Polyimide, SU-8, AZ-4000	Max size equals the wafer size	Good	0.5 μm	Ahn et al.,[32] SOTEC[33]	25	*x*, *y* free, *z* up to 1 mm	High/poor
Molded polysilicon HEXSIL (Keller) (S/A)	Poly-Si, Ni, etc.	Max several millimeters, min few microns	Good	0.5 μm	Keller and Ferrari[34]	10	*x*, *y* free; *z* up to 100 μm	High/poor
Erect polysilicon (Pister) (S/A)	Poly-Si	Max several millimeters, min submicron	Good	0.5 μm	Pister[35]	10	*x*, *y* free, *z* up to 1 mm	High/poor
Micromolding (A), (Ba)	Ceramics, plastics, metals	Max wafer size, min feature several microns	No		Bride et al.,[36] Zhao et al.,[37] Sander et al.[38]	3–20	*x*, *y* free, *z* up to 200 μm	Medium/ poor to fair

A, additive; Ba, batch; C, continuous; IC, integrated circuit; max, maximum; MEMS, microelectromechanical system; min, minimum; PMMA, poly(methylmethacrylate); S, subtractive; Se, serial. LIGA is a German acronym for "**L**ithographie, **G**alvanoformung, **A**bformung," which in English is (x-ray) lithography, electroplating, and molding. For a more in-depth study on manufacturing options, we refer to the literature: Harris,[6] Shaw,[39] DeVries,[40] Slocum,[41] and Evans.[42] Also visit the Louisiana Tech University's Institute for Micromanufacturing (http://www.latech.edu/ifm).

handshake technologies should be of importance beyond micromachines—for example, in ICs, where one is faced with the same dilemma of packaging, mounting, and wiring large-scale integrated chips (LSIs) and very large-scale integrated chips (VLSIs). Several of these new technologies are maskless and enable very fast prototyping. The merging of bulk micromachining with other new fabrication tools, such as surface micromachining and electroplating, and the adaptation of materials such as Ni and polyimides have fostered powerful new nontraditional precision engineering methods.

Table 1.2 comprises both additive (A) and subtractive (S) techniques. For additive techniques, a further comparison of thick-film versus thin-film deposition methods is presented in Table 1.3. The decision to go with a thin versus thick film has very important ramifications and has to be considered very deliberately. Resolution and minimum feature size for thin films are obviously superior. Also, the porosity, roughness, and purity of deposited metals are less reproducible with thick films. Finally, the geometric accuracy is poorer with thick films. On the other hand, the thick-film method displays versatility, which is often key in chemical sensor manufacture. Silk screening forms an excellent alternative when size does not matter but cost, in relatively small production volumes, does. For biomedical applications the size limitations, clear from Table 1.3, make thick-film sensors more appropriate for in vitro applications. For in vivo sensors, where size is more crucial, IC-based technologies might be more appropriate. Thin-film technology does not necessarily involve IC integration of the electronic functions. Table 1.4 provides a comparison of the economic and technical aspects in implementing thin-film technology, IC fabrication, thick film, and classic construction for sensors.[43] It should be mentioned that in sensor fabrication, 60–80% of the cost consists of packaging, an aspect not addressed in Table 1.4. In comparison with complementary metal-oxide-silicon (CMOS)-compatible thin-film sensors, the packaging of thick-film sensors usually is more straightforward. Since packaging expenses overshadow all other costs, this is a decisive criterion. In the case of chemical sensors, where the sides of the conductive Si substrate might shunt the sensing function through contact with the electrolyte, encapsulation is especially difficult compared with the packaging of an insulating plastic or ceramic substrate (see also Chapter 4, this volume, on packaging).

Table 1.3 Comparison of Thin- versus Thick-Film Technology

Property	Silicon/Thin Film	Hybrid/Thick Film
In-plane resolution	0.25 μm and better	12 μm
Minimum feature size	0.75 μm and better	90 μm
Temperature range	<125°C	>>125°C
Sensor size	Smaller	Small
Geometric accuracy	Very high	Poor
Deposition methods	Evaporation, sputtering, CVD*	Screen printing, stencil printing
Patterning methods	Etch through photomask; lift-off stencil	Screen photomask, etched stencil, machined stencil
Reliability nonencapsulated device	Low	High
Electronic compatibility	Good	Moderate
Versatility	Low	Very good
Roughness, purity, and porosity of deposited materials	Superior	Moderate
Energy consumption	Low	Moderate
Handling	Difficult	Easy
Approximate capital costs per unit	Very large, but very low in large volumes	Low; very low in moderate volumes

*CVD, chemical vapor deposition

Table 1.4 Comparison between Different Sensor Technologies: Economic and Technical Aspects

	Classic Construction	Thick-Film Technology	Thin-Film Technology	IC Technology*
Technology substrate	Wires and tubes	Screen printing Al_2O_3, plastic	Evaporation-sputtering Al_2O_3, glass, quartz	IC techniques silicon, GaAs
Initial investment	Very low	Moderate	High	High
Production line cost	>$10,000	>$100,000	>$400,000	>$800,000
Production	Manual production	Mass production	Mass production	Mass production
Units per year	1–1,000	1,000–1,000,000	10,000–10,000,000	100,000 and up
Prototype	Cheap	Cheap	Moderate	Expensive
Sensor price	Expensive sensor	Low cost per sensor	Low cost per sensor	Low cost per sensor
Use	Multiple use, in vitro, in vivo	Disposable, in vitro	Disposable, in vivo	Disposable, in vivo
Markets	Research, aerospace	Automotive, industrial	Industrial, medical	Medical, consumer
Dimensions	Large	Moderate	Small	Extreme miniaturization
Solidity	Fragile	Robust	Robust	Robust
Reproducibility	Low	Moderate	High	High
Maximum temperature	800°C	800°C	1,000°C	150°C (silicon)
Interfacing	External discrete devices	Smart sensors, surface mount	Smart sensors, surface mount	Smart sensors, CMOS,† bipolar

*IC, integrated circuit.
†CMOS, complementary metal-oxide-silicon.
Source: From Lambrechts, M., and W. Sansen. 1992. *Biosensors: Microelectrical devices*. Philadelphia: Institute of Physics Publishing. With permission.[43]

Selection criteria for the additive processes reviewed in Volume II depend on a variety of considerations, such as the following.

1. Limitations imposed by the substrate or the mask material: maximum temperature (T_{max}), surface morphology, and substrate structure and geometry
2. Apparatus requirement and availability
3. Limitations imposed by the material to be deposited: chemistry, purity, thickness, T_{max}, morphology, crystal structure, etc.
4. Rate of deposition to obtain the desired film quality
5. Adhesion of deposit to the substrate; necessity of adhesion layer or buffer layer
6. Total running time, including setup time and postcoating processes
7. Cost
8. Ease of automation
9. Safety and ecological considerations

Table 1.5 compares some of the additive processes important in microsensors and micromachining. This table and the above criteria complement the questions in the check-off list presented in Table 1.11. The latter is meant as a guide toward a more intelligent choice of an optimum machining substrate for the micromachining task at hand.

A subset of subtractive machining techniques from Table 1.2 is compared in Table 1.6. Here, metal removal rates, tolerance, surface finish, damage depth, and required power are listed.

The metal removal rate by electrochemical machining (ECM) and plasma arc machining (PAM) is much higher than that of the other machining processes. Compared with conventional mechanical CNC machining, the metal removal rates by ECM and PAM are, respectively, 0.3 and 1.25 times that of CNC, whereas others are only a small fraction of it. Power requirements for ECM and PAM are also comparatively high. The tolerance obtained

Table 1.5 Comparison of Additive Processes Important in Microsensors and Micromachining

	Species Deposit Ion Mechanism	Deposit Ion Rate	Depositing Species	Coverage of Complex-Shaped Objects	Coverage into Small Blind Holes	Metal Deposit Ion	Alloy Deposit Ion	Refractory Compound Deposit Ion	Energy of Depositing Species	Bombardment of Substrate/ Deposit by Inert Ions	Substrate Heating by External Means	Cost
Evaporation	Thermal energy	Very high (up to 750,000 Å/min)	Atoms and ions	Poor line-of-sight coverage	Poor	Yes	Yes	Yes	Low (0.1–0.5 eV)	No, normally	Yes, normally	Low
Sputtering	Momentum transfer	Low except for pure metals (e.g., Cu, 10,000 Å/ min)	Atoms and ions	Good, but non-uniform thickness distribution	Poor	Yes	Yes	Yes	Can be high (1–100 eV)	Depends on geometry	Not generally	High
CVD	Chemical reaction	Moderate (200–2500 Å/ min)	Atoms	Good	Limited	Yes	Yes	Yes	High in plasma-enhanced CVD (PECVD)	Possible	Yes	High
Electrodeposition	Precipitation of redox species	Low to high	Ions	Good	Limited	Yes, limited	Limited	Limited	Can be high	No	No	Low
Thermal spraying	Solidification of molten droplets	Very high	Droplets	No	Very limited	Yes	Yes	Yes	Can be high	Yes	No, normally	Very high

CVD, chemical vapor deposition.

Table 1.6 Machining Characteristics of Different Processes

	MRR (mm^3/min)	Tolerance Maintained (μm)	Surface Finish Required (μm)	Surface Damage Depth (μm)	Power Required for Machining (W)
USM	300	7.5	0.2–0.5	25	2,400
AJM	0.8	50	0.5–1.22	2.5	250
ECM	15,000	50	0.1–2.5	5	100,000
CHM	15	50	1.5–2.5	5	N/A
EDM	800	15	0.2–1.2	125	2,700
EBM	1.6	25	0.5–2.5	250	150–2,000
LBM	0.1	25	0.5–1.2	125	2–200
PAM	75,000	125	Rough	500	50,000
CNC	50,000	50	0.5–5	25	3,000

AJM, abrasive jet machining; CHM, chemical machining; CNC, computer numerical control machining; EBM, electron beam machining; ECM, electrochemical machining; EDM, electrodischarge machining; LBM, laser beam machining; MRR, metal removal rate; PAM, plasma arc machining; USM, ultrasonic machining.
Source: Based on http://www.cemr.wvu.edu.

by various processes except PAM is within the range of CNC machining, which means satisfactory dimensional accuracy can be maintained. Satisfactory surface finishes are obtained by all processes except PAM. Depth of surface damage is very small for abrasive jet machining (AJM), electrochemical machining (ECM), and chemical machining (CHM) processes, whereas it is very high in the case of electrodischarge machining (EDM), electron beam machining (EBM), PAM, and the like. [As we saw in Volume II, these techniques give rise to a heat-affected zone (HAZ).] For this reason, ECM can be employed for making dies and punches. AJM may be suitably employed for machining super alloys and refractory material; ECM has good applicability for machining steels and super alloys; CHM is very useful for aluminum and steel; EDM may well be adopted for machining steels, super alloys, titanium, and refractory material; and PAM can be suitably employed for machining aluminum, steels, and super alloys. Capital cost for ECM is very high compared with conventional CNC machining, whereas capital costs for AJM are PAM are comparatively low. EDM has a higher tooling cost than other machining processes. Power consumption cost is very low for PAM and laser beam machining (LBM) processes, whereas it is greater in case of ECM. Metal removal efficiency is higher for EBM and LBM than for other processes. Tool wear is very low for ECM, CHM, EBM, LBM, and PAM, whereas it is high for electrical discharge machining.

Given the tremendous variety of micromachining tools available, as evident from Tables 1.2–1.6, a truly multidisciplinary engineering education will be required to design miniature systems with the most appropriate building philosophy.

Design from the Package Inward

When approaching the development of a microdevice, we prefer using the terminology *miniaturization* rather than *MEMS*, as the latter term is still too loaded with the notion that Si technology represents the only solution. The choice of processes for manufacturing a 3D microdevice can best be made after studying the detailed requirements of the application. Since the package serves as the interface between the microstructure and the macroworld, and since it is the major contributor to cost and size, one should start the design with the package and work toward the best machining process for the micromachine inside. The package is often made with a traditional machining method such as turning and grinding or possibly with ultrasonic or wire electrodischarge machining.

For the micromachine inside the package, there may be many machining options, as both batch IC and traditional serial production techniques, perhaps in combination with micromolding, must be considered. Several new, batch Si micromachining techniques derived from the IC industry are gaining acceptance, but serial, precision machining techniques with specialized tools (e.g., ultrasonic machining or laser beam machining) are often the only commercially available methods to make intricate, small, 3D microcomponents.

Decision Tree for the Optimized Micromanufacturing Option

Introduction

In this section we develop a checklist on how to arrive at an optimum miniaturization strategy for a new MEMS/NEMS product, choosing between MEMS manufacturing options: from nonlithography-based to lithography-based, as captured in Table 1.2. To become a proficient MEMS/NEMS engineer one needs to first understand the application and market for the product very well. These aspects should be apparent from a very detailed specification list and discussions with the user/client. Once an appropriate sensor/actuator principle for the MEMS/NEMS has been chosen, on the basis of the specifications, through brainstorming sessions with the user and/or client and by making preliminary designs, one needs to develop a clear understanding on how this sensor/actuator principle scales into the microdomain. Then one must choose the optimum manufacturing approach and the optimized materials, including substrate material, for the most cost-effective practical implementation. These choices are tested through a new design phase and evaluation of critical process steps. A major challenge in this regard often concerns partitioning, i.e., how far to push the integration of electronics with the sensing function (hybrid vs. monolithic)? If the MEMS-based product compromises both a disposable and a fixed instrument the question becomes, what should be included with the MEMS disposable and what should be included in the fixed instrument? With partitioning decisions made, a computer-aided design (CAD) is followed by research prototyping, engineering prototyping, and finally α and β products for the user/client to test. With a satisfied customer, manufacturing of a new MEMS product may begin.

Throughout the rest of this chapter, we illustrate the use of our proposed decision tree with the search for a sensor to measure glucose in a drop of blood. Glucose sensing represents the single largest biosensor market, and a significant amount of all R&D in biosensors worldwide is geared toward improving this technology. Glucose self-testing by diabetics is one of the fastest-growing segments of the world diagnostics market. The World Health Organization estimates a worldwide diabetic population of more than 200 million (1–2% of the world population). Diabetes is the fifth leading cause of death in the United States, and 30% of our children will be diagnosed with diabetes in the future. The US healthcare costs related to diabetes are $132 billion per year. The specifications for a glucose self-test meter with disposable sensor include very low cost disposable (<15 cents), in vitro use, millions of devices manufactured per month, size of active area on the sensor of a few millimeters square (dictated by the size of the drop of blood), linear response with glucose concentration, and a simple handheld measuring device. Examples of glucose sensors/meters for diabetic care are shown in Figure 1.7. Our analysis will clarify why the current glucose sensors are constructed the way they are.

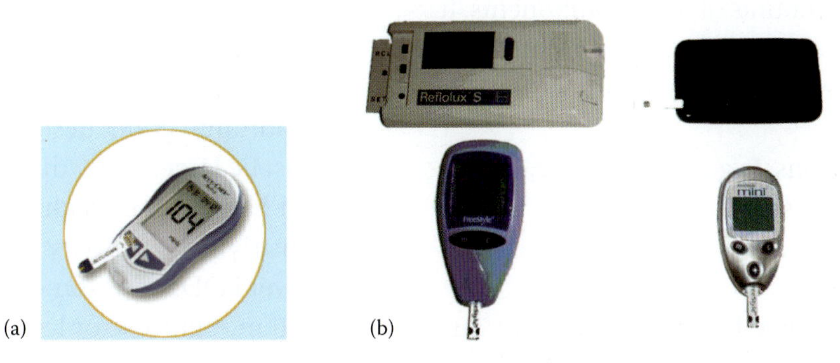

FIGURE 1.7 Blood glucose monitor. (a) Glucometer from Roche (Accu-Chek). (b) Other examples of blood glucose meters, 1993–2005.

Specifications

Applied miniaturization starts with a thorough understanding of the sensor or microsystem specification list, interviews with client/users, understanding of the application environment, and, equally important, a firm market appreciation. In choosing a sensor/actuator principle for a miniaturization application, the various possible sensing/actuating principles (see Table 8.2) need to be challenged with this detailed specification list. During discussions of the specifications required for measuring a certain measurand, jargon confusion is common; making face-to-face meetings with clients/user is mandatory. A specification list should include static and dynamic performance characteristics. Static specifications for a biomedical product with both a reader and a disposable might include the following.

- Package/interface with the world.
- Esthetics/design of package.
- Sensitivity. Change in signal (response) per change in concentration (or mass).
- Lower limit of detection (LOD). The minimum concentration (mass) that can be detected at a given confidence level. The detection limit is affected by the level of noise in the data. One can determine the level of noise by repetitively measuring the signal of a blank.
- Resolution. The smallest increment of change in the measured value that can be determined from the instrument's readout scale; e.g., for a pH electrode the accuracy is ±0.2 pH and the resolution is 0.1 pH.
- Selectivity/specificity. The ability to accurately quantify one species or component in the presence of variable amounts of other components. It has a great effect on the accuracy of the method.
- Stability (long-term and short-term).
- Response time.
- Ambient conditions (e.g., temperature, under or above water, pressure).
- Sample size.
- Operating life.
- Drift. Gradual change in signal that is not due to change in analyte content.
- Shelf life.
- Cost, size, weight of reader.
- Cost, size, weight of disposable.
- Number of devices needed.
- Accuracy.
- Precision.
- Output format/display.
- Operating voltage, current, power.
- Quality.
- Reliability, i.e., mean time between failures.
- Hysteresis.
- Threshold.
- Nonlinearity.

Typical dynamic specifications for such a biomedical product are as follows:

- Dynamic error response.
- Dynamic range. Typical sensor response reaches saturation because of either saturation of binding sites or instrument factors.
- Hysteresis.
- Instability and drift.
- Noise.

Given the very fragmented nature of the microsystems business, it is impossible to present a generic specification list, as a good list must also detail all application-related intricacies. As an example, consider a pH sensor for sensing pH in the secondary cooling flow loop of a nuclear reactor. The presence of radiation and water at high temperature and pressure will significantly reduce the potential sensor principles one may consider. The example also clarifies why a specification list can hardly ever be detailed enough.

The specifications for our example disposable glucose sensor and glucose meter are as follows.

- Package/interface. Handheld device with a slot that accepts disposable glucose sensor strips.
- Esthetics/design. Esthetics are more and more important: good design, easy to use. Letters must be large. Many diabetics are elderly people and have poor eyesight.
- Sensitivity. Sensitivity is about 100 nA/M.
- Lower LOD. Lowest measurable level should be 0.06 mM glucose or less.
- Dynamic range. Glucose assay range is from 1.1 to 33.3 mmol/L.

- Selectivity/specificity. Glucose test results are automatically adjusted for common interfering medications, vitamins, and endogenous substances, including up to 100 μg of acetaminophen, 20 mg/dL uric acid, 3 mg/dL ascorbic acid, 500 mg/dL cholesterol, and 3000 mg/dL triglycerides.
- Response time. The time it takes to read a glucose sensor may range from 5 to 60 seconds (modern meters are typically below 15 seconds).
- Ambient conditions. Sensor operating temperature range = 18–30°C.
- Sample size. Sample sizes vary from 30 to 0.3 μL.
- Operating life. Nonreplaceable battery life = 4000 tests.
- Shelf life. Six months.
- Cost, size, weight of reader. The glucose reader should be approximately the size of the palm of the hand, and it should be battery powered.
- Cost, size, weight of disposable. Very low cost (less than 15 cents). Strip is a few centimeters square in size, and the active spot is a few millimeters square. A consumable sensor element containing chemicals, which react with glucose in the drop of blood, is used for each measurement. Since sensors may vary from batch to batch, a provision might be required for the user to enter in a code that may be found on the batch of sensors. By entering the code into the glucose meter, the meter will be calibrated to that batch of glucose sensors.
- Number of devices needed. Each strip can be used only once and is then discarded. Millions of glucose sensor strips will be manufactured per month.
- Accuracy. The minimum acceptable accuracy for results produced by a glucose monitoring system shall be as follows: 95% of the individual glucose results shall fall within ±15 mg/dL (±0.83 mmol/L) of the results of the manufacturer's measurement procedure at glucose concentrations less than 75 mg/dL (4.2 mmol/L) and within ±20% at glucose concentrations greater than or equal to 75 mg/dL (4.2 mmol/L).
- Precision. Determined as a coefficient of variation on replicate tests on fresh blood in the 3.5–6% range.
- Output format/display. The glucose value in mg/dL or mmol/L displayed in a window. The preferred measurement unit varies by country. In the United States, mg/dL is preferred, and mmol/L is preferred in Canada and Europe. To convert mmol/L of glucose to mg/dL, multiply by 18.
- Reliability. Does not apply for glucose sensor: one-time use only.
- Hysteresis. Does not apply for glucose sensor: one-time use only.
- Nonlinearity. Linear glucose sensor response.

The specifications for a glucose meter listed above suggest a very low cost disposable (<15 cents) with an active area of a few millimeters square (dictated by the size of the drop of blood). These two data points alone preclude the use of Si technology, as this would represent a too-expensive solution. Since we desire a small, simple, and inexpensive measurement device, and a linear response to the glucose in blood, an amperometric approach stands out as the most attractive sensing option (see Volume I, Chapter 7). An amperometric sensing approach is very simple and inexpensive—one pair of thin-film metal electrodes on an insulating substrate—and the method produces a signal linear in glucose concentration. See, for example, the MediSense/Abbott sensor, the Precision QID, in Volume I, Figure 7.80. The linear response makes it more sensitive than a potentiometric device that features a log dependency of the voltage signal on concentration. As we learned in Volume I, Chapter 7, an optical approach tends to be more expensive than an electrochemical approach (see Volume I, Table 7.21).

Having established a sensing mechanism, we now need to know how the chosen effect (amperometry) scales into the microdomain.

Scaling Characteristics of Micromechanisms

To make the right choice from the different available sensing/actuating principles, it is also important to know how the phenomena scale in the microdomain.

Scaling effects are explored in detail in Chapter 7, this volume, and some intuitive guidance (rules of thumb) is provided in Table 1.7.

Scaling understanding is crucial early on when deciding on a sensing/transduction principle. For example, both absorption and amperometric detection approaches scale poorly compared with a luminescent or a potentiometric sensing approach.

We apply these scaling rules now to the glucose sensor example. From the above, we appreciate that an amperometric device scales more poorly than a potentiometric device (the smaller an amperometric device, the smaller the current, while the signal for a potentiometric device is scaling invariant), but an in vitro glucose sensor may be relatively large, and sensitivity is more important in this case. A potentiometric device would be as inexpensive as our chosen amperometric approach, but its sensitivity is not as good. Moreover, we saw in Volume I, Chapter 7, that by making an array of electrodes, the signal to noise ratio of a microamperometric glucose sensor may be enhanced because of increased mass-transport at the electrode edges. So we stick with the amperometric sensor as our choice for the proposed glucose meter.

Table 1.7 Some Important Effects of Scaling on Design of Miniaturized Devices

1. Scaling of forces: weak forces are preferred in the microdomain. Electrostatic forces [with a quadratic dependency on dimension (1^2)] under certain conditions outperform magnetic torque [with a cubic dependency on dimension (1^3)] since they become relatively stronger in small devices.
2. Increasing strength of materials in the microdomain. Single-crystal whiskers may be more than 1000 times stronger than the bulk material.
3. The dominance of surface effects in the microworld: for example, because of an increase in S/V (surface-to-volume ratio), better heat dissipation results (thermal isolation is difficult and cooling is excellent) and frictional forces increase.
4. Decrease of manufacturing accuracy of micromachines (see Volume II, Figure 6.3).
5. The need for error-insensitive design (fewer parts, flexible materials) and the use of nature as a guide in design philosophy for very small devices.
6. Amperometric and absorption measurements scale poorly into the microdomain compared with potentiometric and chemiluminescence (see Volume I, Chapter 7, and Chapter 7, this volume, for details). Our proposed glucose sensor is amperometric, so we need to look in some more detail into whether this averse scaling should influence our choice.
7. Mechanical, thermal, and electromagnetic response time constants and many other time constants shrink with advanced miniaturization, generally resulting in shorter response times. Microsystems, with their small inertial mass, are faster, leading to higher speed or frequency, but higher frequency of a cyclic process generally leads to larger losses and consequently to lower efficiency.
8. Power consumption is often dramatically reduced, especially if good thermal isolation is possible for elements with small heat capacity.
9. Energy sources scale very poorly (1^3), making onboard energy sources (e.g., batteries and fuel cells) less attractive for powering micromachines than radiating in energy (e.g., solar or laser light with micromachined photovoltaic converters or microwaves if extremely small receivers and converters were available).
10. Most actuators, which rely on power for inducing a movement, also scale poorly with reducing size, and it is often better to consider a "macroactuator," which can produce a micromotion, than a micromachined actuator with too small of a stroke.

Partitioning

Introduction

A major challenge in the manufacture of MEMS and NEMS revolves around correct partitioning, that is, how far to push integration of electronics with the MEMS or NEMS sensing function (hybrid vs. monolithic). What do we include with the MEMS disposable, and what can we put into the fixed reader instrument? Onboard or offboard fluidics? Battery or mains power? We first compare hybrid versus monolithic approaches and then illustrate the challenges of partitioning a MEMS instrument by looking at an example from the burgeoning field of microfluidics. Partitioning of electronics and instrument components obviously affects packaging needs, and we will come back to this point in Chapter 4, this volume, on packaging.

Monolithic versus Hybrid MEMS

Hybrid integration in the IC industry means combining thin-film ICs with thick-film technology. Viewed from the Si MEMS sensor angle, a hybrid sensor keeps the electronics separate from the sensor. The hybrid might then consist of two pieces of Si on the same substrate connected by a short wire bridge, or it might be a Si sensor mounted in a header plugged into an electronics board. In a monolithic MEMS approach, on the other hand, electronics and MEMS elements are cofabricated within one single (lengthy!) sequential Si process. Each step in such a sequence has a nonzero failure probability and, since probabilities multiply fast, the yield may be dramatically reduced for long process sequences. In this light, there are some positive aspects to a hybrid MEMS approach. In the latter, MEMS and electronics are fabricated in separate high-yield processes and joined within a common package/substrate, improving the overall yield. Microassembly in hybrid manufacture enables the mixture of components from separate optimal processes, including nonstandard processes and materials that are incompatible. For example, many optoelectronics require GaAs substrates that are incompatible with standard Si electronics (see also under microassembly in Chapter 4, this volume). In addition, in hybrid MEMS, *all* components needed for a complete MEMS function can be accommodated. Somewhat unexpectedly, parasitics are sometimes lower in a hybrid than in an integrated approach. The device components in a hybrid remain serviceable and, because the sensor package determines the final size, hybrids often are not much larger than their Si integrated counterparts. Hybrids usually are simpler to design and can be produced more economically in smaller-production volume runs. For chemical sensors and biosensors, they often represent the most viable option to implement micromachining technology. Prices for Si mechanical sensors given in market reports mostly refer to hybrid devices with the Si sensor mounted in a header and the signal conditioning kept on separate chips. As an example, Figure 1.8 represents an accelerometer built using a 3-inch Si wafer and involves a fusion bonding process to produce a proof mass, spring, and fixed capacitor plate. The interface circuits are designed to function with different sensors. The specific sensor configuration and calibration data are written into an EEPROM. It is ±2-g device with 2-mg resolution at 60 Hz. These accelerometers are used for aerospace navigation, in navigation in small commercial airplanes, and in helicopters.

From the precision engineering angle, a hybrid device is one in which parts produced by different technologies are joined and assembled. In the electronics industry thick film is often synonymous with hybrid manufacturing and thin film with front-end IC manufacturing. For more details on choosing between thin- and thick-film processes, consult Tables 1.3 and 1.4 above.

FIGURE 1.8 Commercially available Si accelerometer based on hybrid technology. CSEM 6100 accelerometer with interface circuitry in a T08 package. (From CSEM.)

For the proposed glucose meter, we will have all the electronics in the reader rather than in the disposable. It makes no sense to throw the electronics away with each used glucose sensor. Moreover, the glucose sensor will be constructed on an inexpensive nonsilicon substrate.

Many chemical sensors come with a pump and some fluidics, say for sample collection, calibration, rinsing, etc. In what follows, we decide what pumping mechanism to employ and how to partition the fluidics between sensor disposable and permanent instrument.

Partitioning in a Microfluidic Instrument

As we saw in Volume I, Chapter 6, microfluidic systems comprising nozzles, pumps, channels, reservoirs, columns, mixers, oscillators, diodes, amplifiers, and valves can be used for a variety of applications, including separation chemistry, drug dispensation, ink-jet printing, and general transport of liquids, gases, and liquid/gas mixtures. The advantages of these devices include lower fabrication costs, enhancement of analytical performance, lower power budget, and lower consumption of chemicals than traditional pumping and spraying systems. Ultimately, these developments will revolutionize applications where precise control of fluid flow is a necessity.

Generic technical challenges in making a microfluidic instrument—for a biotechnology application for example—include the partitioning of the various instrument functions such as power supply, polymerase chain reaction (PCR) for DNA sample amplification, mechanism for getting the sample into and out of the instrument, providing fresh viable reagents, and detection. Depending on the specific biotechnology application (say, clinical diagnostics or high-throughput drug screening), the electronics, the power for the propulsion mechanism, heaters (e.g., for PCR), etc., may be integrated with a disposable cassette holding the disposables or within the fixed instrument. For diagnostics, both the disposable and the reader instrument should be inexpensive; the disposable should thus involve as little complexity as possible—perhaps only reagents, plastic substrate, and fluidic conduits. The reader instrumentation should be inexpensive, small (perhaps hand-held), and very user friendly. For high-throughput screening, much more sophisticated fluid handling is typically incorporated into a tabletop, expensive research instrument. The disposable in the latter does not need to contain sealed fluids or dry stored chemicals, often making it less sophisticated than disposables in diagnostic application. In one extreme, one might envision a totally integrated option such as the micro-total analysis system[44] illustrated schematically in Figure 1.9a.[45] More realistically, one might look at a hybrid construct, as sketched in Figure 1.9b. For most

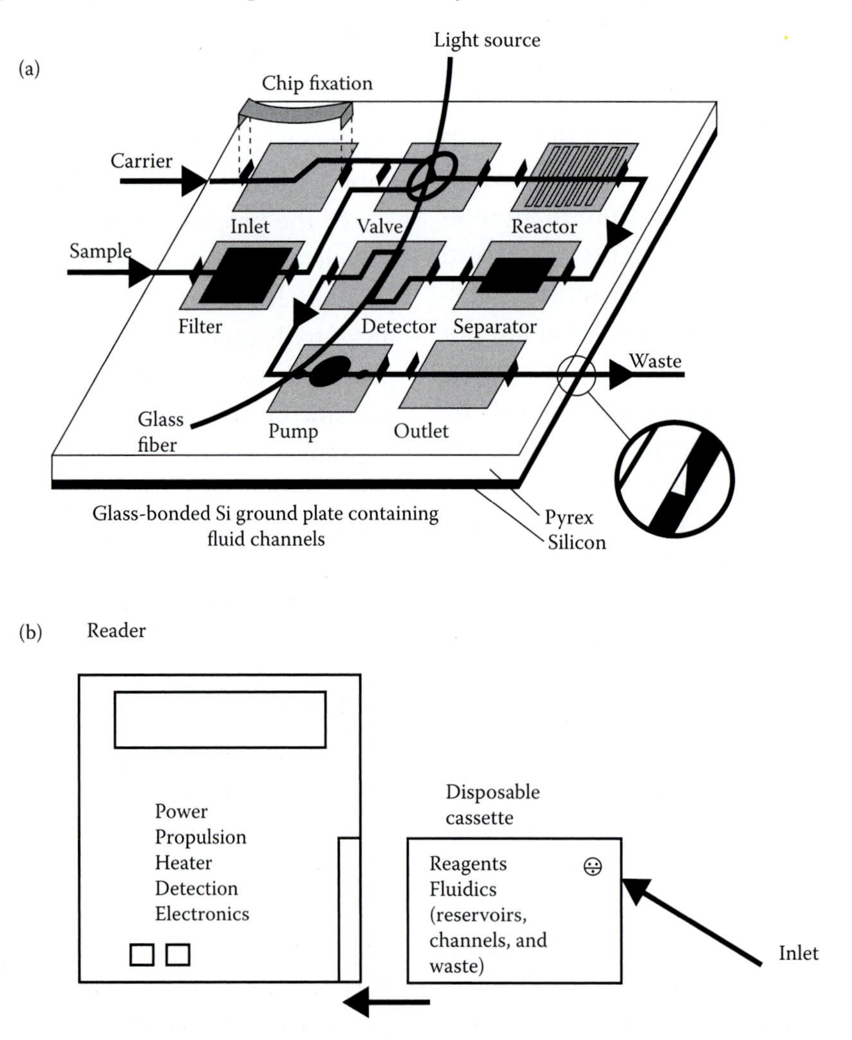

FIGURE 1.9 Partitioning of functions between instrument and a disposable. (a) Configuration for a micro-total analysis system (μTAS). (Based on Bergveld, P. 1995. In *Micro total analysis systems*, eds. A. van den Berg and P. Bergveld, 1–4. Twente, the Netherlands: Kluwer Academic Publishers. With permission.[45]) (b) Generic hybrid microinstrument, with inlet, valves, fluid propulsion, detection, waste, and thermal control. Various options exist for partitioning between cassette and reader.

diagnostic microinstruments, it indeed makes more sense to envision a disposable cassette incorporating the specific reagents needed for a certain test and a separate reader instrument. In a diagnostic application, a major challenge is how to store wet or dry chemicals in sealed reservoirs within the disposable package, making a microfluidic diagnostic instrument actually more challenging to develop than a high-throughput drug screening station.

Our disposable glucose sensor will only have a substrate, electrodes, a chemical cocktail and perhaps a very simple fluidic mechanism to make blood collection on the sensor surface easier. In the context of glucose sensing we will now briefly refer to systems that employ pumps to deliver insulin depending on the glucose sensor reading.

Fluid Propulsion Methods and MEMS Integration

Introduction

As we saw in Volume I, Chapter 6, there are various technologies for moving small quantities of fluids or suspended particles from reservoirs to mixing and reaction sites, to detectors, and eventually to waste or to a subsequent instrument. Methods to accomplish this include syringe and peristaltic pumps, electrochemical bubble generation, acoustics, magnetics, direct current (DC) and alternating current (AC) electrokinetics, and centrifuge. In Table 1.8 we compare some promising fluid propulsion means.[46] Only the simplest and least expensive pumping schemes should be integrated within the disposable itself.

Mechanical Pumps

From Volume I, Chapter 6, we remember that the pressure that mechanical pumps have to generate to propel fluids through capillaries will be higher the narrower the conduit (see Volume I, Equation 6.54). Piezoelectric (acoustic), electro-osmotic (DC electrokinetic), electrowetting (DC electrokinetic), and electrohydrodynamic pumping (AC electrokinetic) all scale more favorably in the microdomain and lend themselves better to pumping in MEMS devices. Despite this promise, almost all biotechnology equipment today is based on traditional external syringe or peristaltic pumps. Although integrated micromachined pumps based on two "one-way" valves may achieve a precise flow control on the order of 1 μl/min with fast response, high sensitivity, and negligible dead volume, these pumps require complicated fabrication processes, generate only modest flow rates and low pressures, and consume a large amount of chip area and considerable power. As a consequence of the low pumping pressure, they often become useless when device dimensions are reduced and resistance to fluid flow and required driving pressure increase. An example of a Si micromachined pump from Animas-Debiotech is shown in Figures 1.10 and 1.11. This belt-worn drug delivery pump features a piezoelectric design pioneered at the

Table 1.8 Comparison of Microfluidics Propulsion Techniques

	Fluid Propulsion Mechanism				
Property	Centrifuge (see Figure 1.14 and Volume I, Figure 6.89)	Pressure (e.g., blister pouch)	Acoustic	Electro-osmosis	Electrowetting
Valving solved	Yes, for liquids	Yes, for liquids and vapor	No	Yes, for liquids	Yes, for liquids
Maturity	Research and development	Products	Research	Products	Research
Propulsion force influenced by	Density and viscosity	Generic	Generic	pH, ionic strength	pH, ionic strength
Power source	Rotary motor	Mechanical roller (see Figure 1.12b)	5–40 V (peak to peak)	10 kV	1–10 kV
Materials	Plastics	Plastics	Piezoelectrics	Glass, plastics	Glass
Flow rate	From less than 1 nL to greater than 100 μl/s	20–500 μl/s	20 μl/s	0.001–1 μl/s	0.3 μl/s

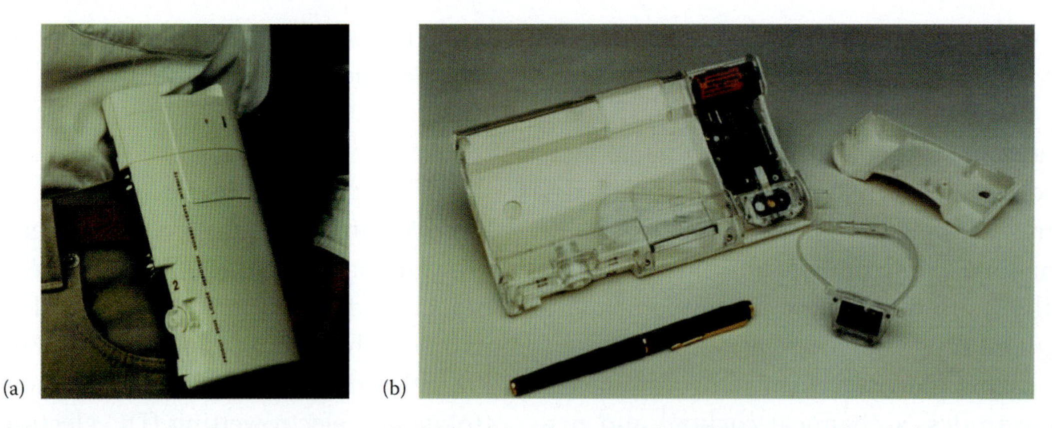

FIGURE 1.10 Micropump technology. (a) Animas-Debiotech's micropump is a Si micromachined piezopump meant to be a belt-worn drug delivery mechanism. The Oncojet is disposable and may deliver a drug at a rate of 10–50 mL/day. (b) The glass-Si sandwich pump is about 2 × 1 cm in size. (Courtesy of Ary Saaman, Animas-Debiotech.)

University of Twente, and a generic example of this MEMS pump is shown in Figure 1.11a.[47] A close-up of the actual Animas-Debiotech MEMS pump is shown in Figure 1.11b. The size of most Si micromachined pumps is quite large, and one wonders why Si was chosen as a substrate (http://www.diabetesnet.com/diabetes_technology/insulinpumps_debiotech.php). Reasonably, plastic molding technology is now being explored as a viable solution.[48,49]

The advantages of using larger peristaltic pumping today are that it relies on well-developed and commercially available components, that a wide range of flow rates is attainable, and that the pumping mechanism is not easily fouled.

Another example of a mechanical pump is embodied in the Johnson & Johnson blister pouch human immunodeficiency virus (HIV) test shown in Figure 1.12.[50] Automated sample preparation and amplification coupled with detection systems would revolutionize the medical diagnostic application of DNA amplification with PCR. An elegant example of progress in this direction comes from Johnson & Johnson (see also US patent 5422271).[50] This company's simple fluidic structure for the detection of the HIV is shown in Figure 1.12. The device was invented at Kodak and bought by Johnson & Johnson in 1994. The disposable pouch-based PCR instrument features a total amplification time of 43 min and a sensitivity and specificity adequate for the detection of HIV. An unknown sample and the PCR cocktail are injected into the amplification compartment through a fill port in the plastic pouch. Eight pouches are

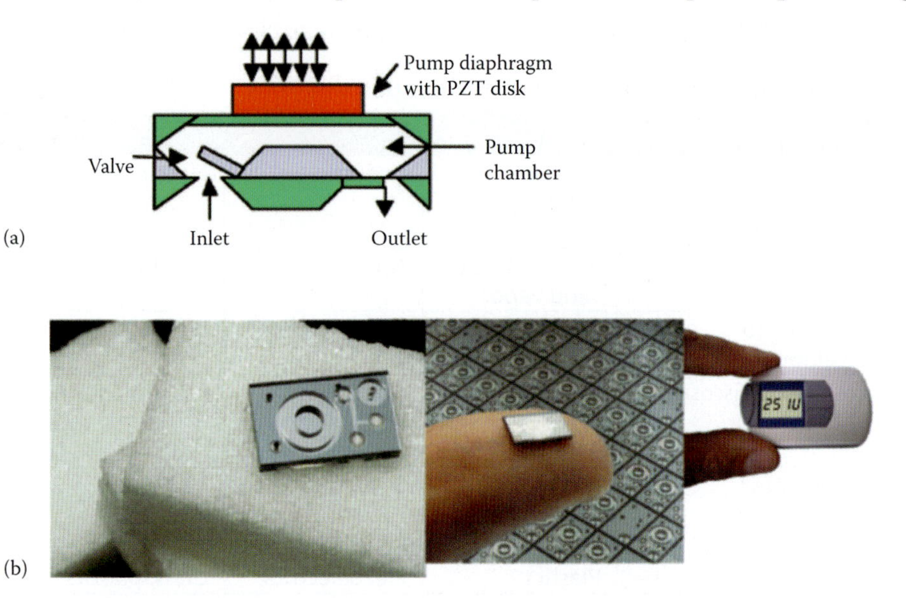

FIGURE 1.11 (a) Generic design of a piezopump.[47] (b) The Animas-Debiotech Si-based microelectromechanical system micropump (http://www.diabetesnet.com/diabetes_technology/insulinpumps_debiotech.php).

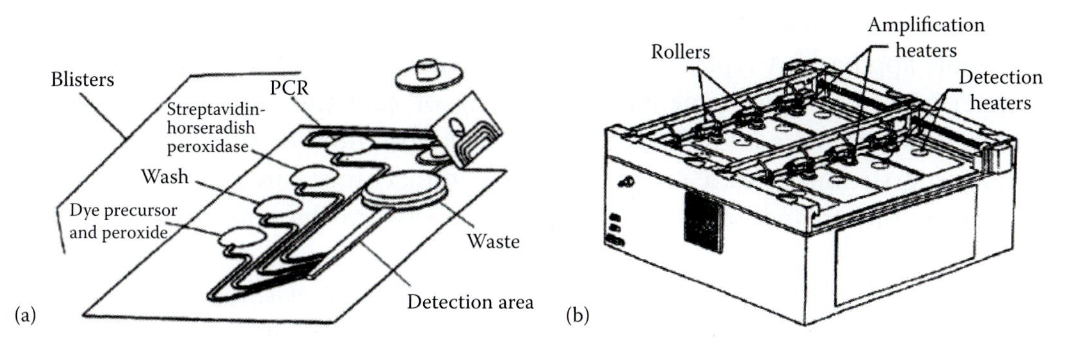

FIGURE 1.12 Johnson & Johnson blister pouch HIV test (a) and tabletop instrument with rollers, amplification, and detection heaters (b). See also reference 50 and US patent 5422271.

aligned on a fixed instrument with the amplification compartments on the upper amplification heaters, and the detector areas on the lower detection heaters. Upon completion of the amplification, the next analysis step is accomplished by an automatic roller that forces detection chemistry fluids and amplified samples from compartments through channels leading to a temperature-controlled detection area. The plastic channels in the pouch contain small nips that close off the channel and break open upon pressurization with a roller system, similar to the one used in credit card image transfer. These nips in the plastic form very inexpensive, single-use, normally closed valves. In the detection zone, biotinylated PCR products are captured via hybridization to probes immobilized on latex particles. Multiple probe sites can be accommodated in the detection areas. Streptavidin-horseradish peroxidase is mixed with those probes to allow for binding between the biotinylated probes and the streptavidin-horseradish peroxidase. Any material left unbound is then removed with the subsequent wash solution. In the presence of peroxidase, leuco dye is oxidized to a visible blue color, indicating a positive hybridization event. Replacing the relatively thick walls of traditional, low surface-to-volume ratio, polypropylene reaction tubes with the pouch "blisters," with their relatively high surface-to-volume ratio and very thin walls (100 μm), which are readily conformable to the shape of the instrument heaters, permits rapid heating and cooling.

The above PCR-based instrument is an example of optimum partitioning between disposable and fixed instrument for use in a medical diagnostic application. For this purpose, sample receptacle, reagents, and all fluidics, such as mixing, valving, and waste area, are integrated in an inexpensive disposable. Power supply, heaters, coolers, readout electronics, etc., are kept separate in an inexpensive disposable plastic pouch. This inexpensive and ingenious way to couple PCR and detection is difficult for a Si-based micromachining approach to compete with. Despite its promise, this HIV test was taken off the market, presumably because of manufacturing problems.

A disadvantage of the pouch approach is that it is rather difficult to further miniaturize because it is a pressure-driven system. The disposable cartridge in the i-STAT (now Abbott) blood analyzer device illustrated in Figure 10.42b of this volume also contains a blister pouch containing calibrant solution.

The proposed glucose meter of our example does not come with an insulin delivery pump, but, in addition to the Animas-Debiotech pump (Figures 1.10 and 1.11), many such insulin delivery pumps are commercially available. Some glucose meters communicate wirelessly to the pump, and some pumps include the meter.

Acoustic Streaming

Acoustic streaming is a constant (DC) fluid motion induced by an oscillating sound field at a solid-fluid boundary (see Volume I, Chapter 6, on fluidics). Transducers do not need to be integrated with the disposable; the cassette with capillary flow channels can simply be laid on top of the acoustic pump network in the reader instrument. Acoustic streaming is also well suited for mixing reagents.[51] The method is considerably more complex to implement than electro-osmosis (see below), but the insensitivity of acoustic streaming to the chemical nature of the fluids inside the fluidic channels and its ability to mix fluids very effectively make it a potentially viable approach. A typical flow rate measured for water in a small metal pipe lying on a piezoelectric plate is 0.02 mL/s at 40 V, peak to peak.[52] Today acoustic streaming as a propulsion mechanism remains in the research stage.

Electro-osmosis and Iontophoresis

Electro-osmosis is capable of delivering an appreciable DC electrokinetic flow rate in a capillary, does not involve any moving parts, and is easily implemented (see Volume I, Chapter 6, on fluidics). All that is needed is a metal electrode in some type of a reservoir at each end of the flow channel. Typical electro-osmotic flow velocities are on the order of 1 mm/s, with a 1200 V/cm applied electric field. For example, Jorgenson reported electro-osmotic flows of 1.7 mm/s.[53] This is fast enough for most analytical purposes. Some disadvantages of electro-osmosis are the required high voltage (1–30 kV power supply) and direct electrical-to-fluid contact with resulting sensitivity of flow rate to the charge of the capillary wall and to the ionic strength and pH of the solution. It is consequently more difficult to make it into a generic propulsion method. For example, liquids with high ionic strength cause excessive Joule heating; it is therefore difficult or impossible to pump biological fluids such as blood and urine. The sensitivity of the electrokinetic potential to the nature and strength of the electrolyte brings with it the insidious problem of pressure buildup at intersections where liquid columns of different ionic strengths meet. Micromachinists are starting to address the connection between the outside world and the capillaries in electrokinetic equipment. Gluing glass reservoirs for holding the platinum bias electrodes to the glass plate substrate with room temperature vulcanizing (RTV) silicone is a research solution but not an attractive manufacturing option.[54]

In the context of our glucose meter example, we need a blood collection/sampling mechanism. One method for interstitial fluid collection uses iontophoresis. The GlucoWatch® Biographer (cost, $599–799), shown in Figure 1.13, collects glucose through the skin, and no finger pricks are needed (Cygnus Corp.: http://www.glucowatch.com).* In this method, one draws out interstitial fluid by iontophoresis, and the charge and size exclusion properties of the reverse iontophoretic extraction lead to a very clean sample. Iontophoresis utilizes the passage of a constant, low-level electric current (~0.3 mA/cm^2) passed through the skin between two electrodes applied onto the surface of the skin. Electrolyte ions in the body act as the charge carriers for this current to iontophorese substances from within the body outward through the skin. Only small compounds pass through the skin,

(a)

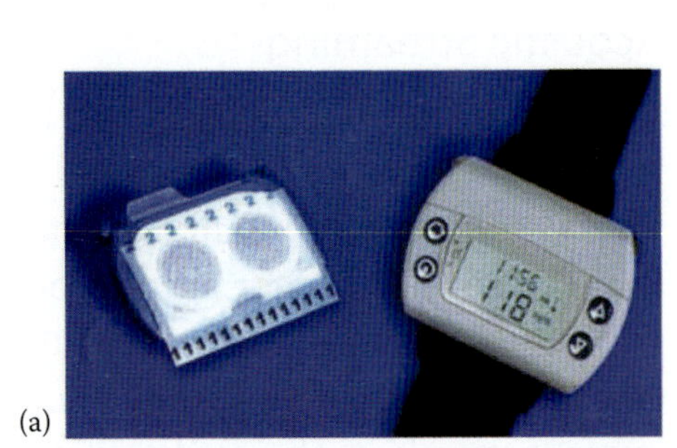

(b)

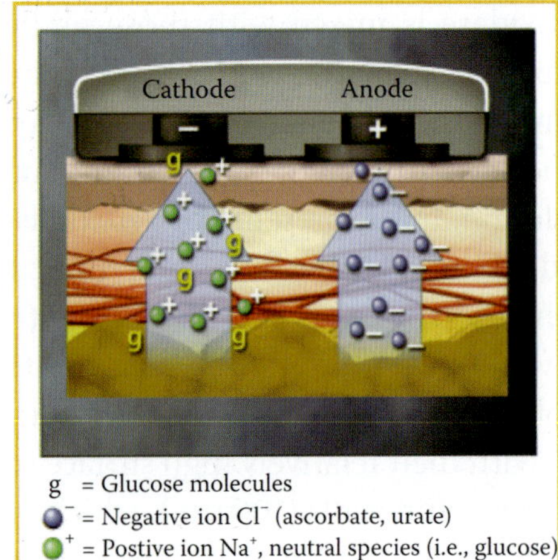

FIGURE 1.13 The GlucoWatch® G2™ Biographer (a) collects glucose through the skin via iontophoresis (b). (Cygnus Corp., http://www.glucowatch.com; http://www.animas.com.*)

* On March 1, 2006, Cygnus, who made the GlucoWatch Biographer, was closed down, and its assets were bought by Animas (who makes insulin pumps, see Figure). Animas is no longer making GlucoWatch Biographers. The U.S. Food and Drug Administration reports that at least 50% of users experienced mild or moderate skin irritation.

so there are no proteins (e.g., hemoglobin) in the extract. Because the skin has a net negative charge at physiological pH, positively charged sodium ions are the major current carriers. This migration of sodium ions toward the cathode induces an electro-osmotic flow toward this electrode. Uncharged molecules (e.g., glucose) are carried along by this electro-osmotic flow and thus preferentially extracted at the cathode. While glucose is collected at the cathode, interfering species (ascorbate and urate) collect at the anode. Micromolar concentrations of glucose are extracted by the process of iontophoresis compared with 5–10 mM concentrations normally seen in blood samples. To accurately measure this small amount of glucose, the GlucoWatch® Biographer utilizes an amperometric biosensor. Glucose oxidase enzyme located in the hydrogel discs of the AutoSensor reacts selectively with glucose to produce hydrogen peroxide (H_2O_2). The hydrogen peroxide is detected at a platinum-containing working electrode via an oxidation reaction producing an electric current that is measured by the circuitry of the Biographer. After a 2-hour warm-up period and a single-point calibration, the GlucoWatch® G2™ Biographer can provide glucose readings every 10 min for up to 13 hours. Obtaining a blood glucose value from a traditional finger stick glucose meter and then entering the value directly into the Biographer during the calibration phase constitute a single-point calibration. This calibration phase occurs at the end of the 2-hour warm-up cycle. The GlucoWatch provides information on trends and not necessarily in real time for treatment adjustments.

Concerns with the merit of trending only, calibration time, risk of infection and irritation,* and the irreproducibility of the watch eventually spelled the demise of this approach. So we will combine our glucose meter with a more conventional lancet or laser pricking (see below) to obtain a blood sample.

Centrifugal Pumping

Using a rotating disc, centrifugal pumping provides flow rates ranging from less than 10 nL/s to greater than 100 μL/s depending on disc geometry, rotational rate (revolutions per minute), and fluid properties (see Figure 1.14).[55] Pumping is relatively insensitive to physicochemical properties such as pH, ionic strength, or chemical composition (in contrast to AC and DC electrokinetic means of pumping). Aqueous solutions, solvents (e.g., dimethyl sulfoxide), surfactants, and biological fluids (blood, milk, urine) have all been pumped successfully. As explained in Volume I, Chapter 6, fluid gating may be accomplished using "capillary" valves in which capillary forces pin fluids at an enlargement in a channel until rotationally induced pressure is sufficient to overcome the capillary pressure (at the so-called "burst frequency"). Since the types and the amounts of fluids one can pump on a centrifugal platform span a greater dynamic range than for electrokinetic pumps, this approach seems more amenable to sample preparation tasks. The example fluidic platform based on a CD in Figure 1.14 is used by RotaPrep for cell lysis and DNA purification. This specific product uses magnetism, centrifugal force, and capillary action for system operation.

The disposable in this case is an inexpensive polycarbonate disc, containing only fluid manifolds, glass beads, and a metal disc. The permanent instrument has a motor, controls, and permanent magnets.

Vacuum Pressure Reservoir

Fluid transport can also be accomplished by employing chambers within the cassette or reader that are at a higher or lower pressure than the sample and reagent reservoirs. Valves are required to separate

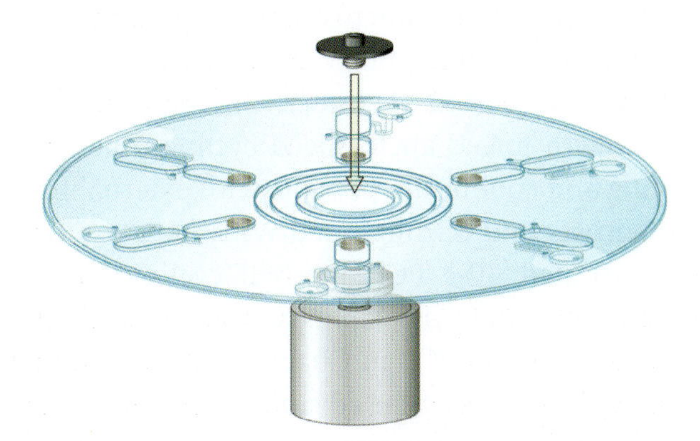

FIGURE 1.14 RotaPrep system for cell lysis (http://www.rotaprep.com). This specific product uses magnetism, centrifugal force, and capillary action for system operation. (Courtesy of Dr. Kido, CEO, RotaPrep.)

* On March 1, 2006, Cygnus, who made the GlucoWatch Biographer, was closed down, and its assets were bought by Animas (who makes insulin pumps). Animas is no longer making GlucoWatch Biographers. The U.S. Food and Drug Administration reports that at least 50% of users experienced mild or moderate skin irritation.

the evacuated or pressurized chambers from the rest of the fluid path until fluid motion is required. Opening a valve initiates the flow of fluid toward an evacuated chamber or away from a pressurized chamber. An advantage of this mechanism over those previously described is that it is a straightforward way to generate a large fluid-motive force. Its drawbacks are that it can only be actuated once, valves may be difficult to implement, and the reliable manufacture of pressurized or evacuated chambers may be challenging. Unless one uses osmotic pumping by implementing a salt reservoir, external actuation seems the safest implementation method for pressure-gradient actuation.

Heating and Cooling and MEMS Integration

Heating and cooling of fluid within a microinstrument, like fluid propulsion, can be accomplished with elements that are either contained within a disposable cassette or the reader. There are, superficially, some arguments to be made for integrating heating and cooling in the disposable. We learned in Volume II, Example 9.1, "Spray Pyrolysis Application: Taguchi Gas Sensor" that electrical current passing through resistors integrated on thin membranes can cause a very rapid temperature increase. This is because the resistor has very little thermal mass and is in intimate thermal contact with the membrane and the fluid to be heated. Integrated heaters are, in principle, simple to fabricate on a variety of materials. On the other hand, temperature control can also be accomplished with a heating and cooling system that is external to the microsystem. For example, the Johnson & Johnson pouch, surveyed earlier, uses two external heaters (Figure 1.12). The heaters are cooled with forced air to speed cycling times. Fast heating and cooling are possible with external heaters. Wittwer and Garling[56] and Wittwer et al.[57] have demonstrated 30-s PCR thermal cycling using a hot-air cycler and glass capillary PCR chambers (see the discussion on PCR in Chapter 10, this volume). External heaters offer the advantages of simpler (less expensive) microinstruments and no cassette-to-reader electrical connections. In general, however, external heaters are slower and require more electrical power than integrated heaters.

A third option is to heat fluid within the microinstrument using radiation generated in the reader. The radiation could be infrared, radio wave, or microwave; it could be adsorbed by the fluid or by absorptive structures (such as black ink) placed on the cassette. An advantage of radiative heating is that it avoids electrical contact from reader to disposable. Our proposed glucose sensor strip has no heater on board.

Sample Introduction

The exact mechanism for introducing samples into the microinstrument depends on the target sample. This is one of the most difficult areas—not only is pretreatment of the sample often required, but one also must transition from a relatively bulky external receptacle structure, which can be handled and accessed manually, to the microfluidic components inside without incurring too much dead space. Very few machining techniques cover the whole dynamic size range required. Because the sample introduction method may place important constraints on the manufacturing process, a few sample introduction options are considered here. Traditional sample injection systems, such as that for flow injection analysis, rely on some sort of commutating valve, either rotary or sliding, actuated manually or mechanically. Harrow and Janata[58] compared different commercially available sample injection systems, and Shoji et al.[59] were among the first to micromachine a sample injector consisting of two three-way microvalves and a channel. These valve systems today are too complex and expensive to consider their integration in a disposable. Perhaps the simplest way to introduce the sample into the microinstrument is by creating wells into which the sample is dropped. For example, most of the substrates with electro-osmotic pumps have plastic tubes glued to them to make sample reservoirs. Similarly, the Biotrack blood coagulation device requires that a drop of sample be placed in a depression on the instrument surface.[60] The sample is then wicked into an internal chamber by capillary action. Problems with sample wells include the large volume that they require and the need to accurately align the sample with the well if the sample well is small. Because the wells are open, one risks sample contamination or the inclusion of air bubbles in the sample. The latter are a major problem in miniature fluidic systems because of the associated surface tension, and debubblers, which are typically bulky, are often a necessity. Samples may also be injected directly into the microinstrument or disposable cassette. An example of this is the Johnson & Johnson pouch HIV test, where an injection septum is provided in the disposable pouch (Figure 1.12). This reduces the chance of sample contamination but again requires exact alignment of the injection needle with the sample injection port. Furthermore, the manufacture of the cassette pouch is complicated by the need for an injection septum. In an integrated sample, the collection device could be the cassette. For example, the cassette could be manufactured in the form of a blood collection tube (Vacutainer), a syringe, a pipette, or a swab. However, this would complicate the cassette manufacture considerably and may limit the cassette design to a specific type.

For the sample collection of our example glucose sensor, we came to understand earlier that iontophoresis still poses problems. Therefore, we will use a traditional lancet (Figure 1.15a). The Transdermal Biophotonic System from SpectRx (Figure 1.15b) collects interstitial fluid through micropores punctured in the outer layer of the skin by a laser (not FDA approved at this time), and glucose is measured in a patch that contains the glucose sensor. Measuring in interstitial fluids rather than blood, this collection method is similar to the GlucoWatch (see above) in that it provides information only on trends and not necessarily real time for treatment adjustments.

Manufacturing Technique of the Disposable Cassette

The choice of how to partition fluid transport and heating techniques must go hand in hand with the choice of disposable/cassette manufacturing technique. The principal advantages of nontraditional methods for making microinstruments are that very intricate structures with very fine features can be

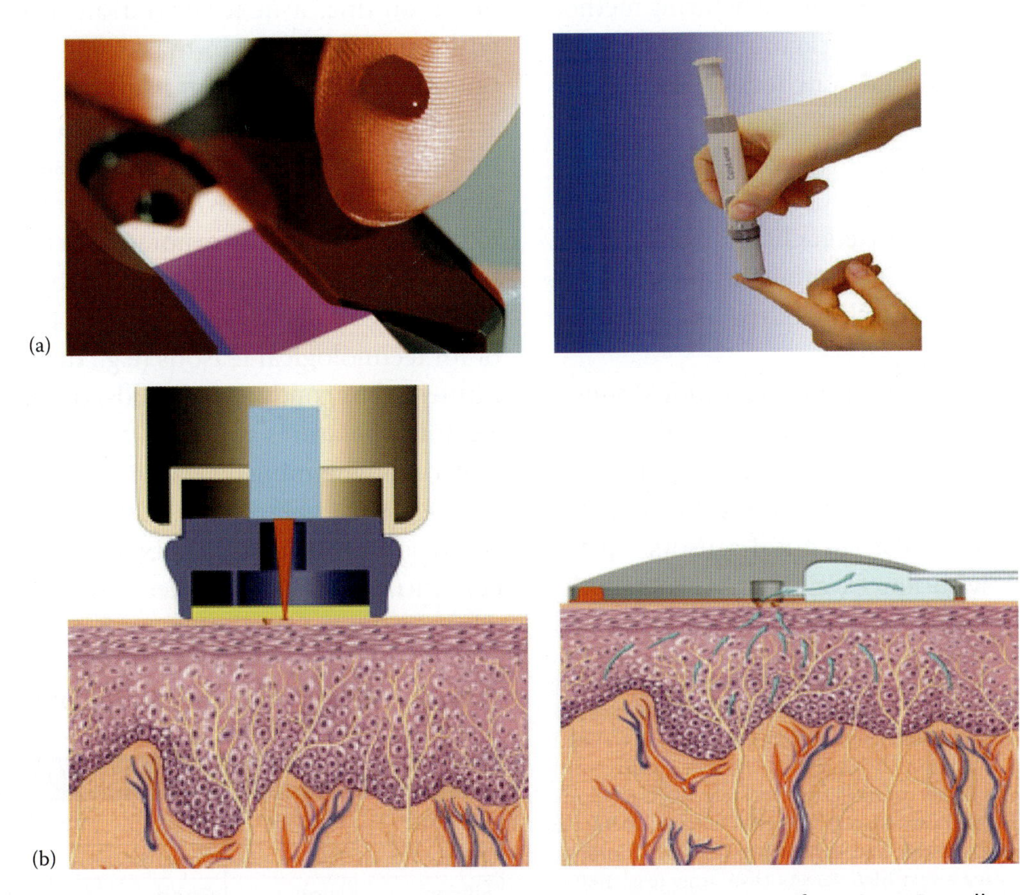

FIGURE 1.15 (a) Lancet for blood collection. (b) The Transdermal Biophotonic System from SpectRx collects interstitial fluid through micropores punctured in the outer layer of skin by a laser (http://www.spectrx.com).

produced, and heaters, electro-osmosis electrodes, and acoustic streaming pumps can be readily realized. On the other hand, nontraditional microfabrication is optimized for making devices that are considerably smaller than most biochemical reaction sample volumes. A 100-μl reagent volume is the same size as a Si chip; that is, 14 mm on a side—huge by microelectronics standards. It is certainly possible to make Si micromachined parts that are as large as the starting wafer (typically 4 in. in diameter); however, the costs of the wafer and wafer processing are likely to be much higher than for other manufacturing techniques. In addition, variation from a previously developed microfabrication sequence can require an expensive and time-consuming process development effort. In the meantime, traditional methods have provided answers to most of the miniaturization needs in diagnostic applications. Where MEMS contributes is in realizing some crucial components of the instrument, for example, the detector, the fluidic conduits, and the actuators. Most solutions rely on plastic rather than Si. We will settle on a choice for our glucose sensor's substrate and manufacturing method after we introduce some background on substrate choices and have launched our check-off list.

Substrate Choice

Here we compare substrate choices such as Si, ceramic, glass, plastics, and others in terms of traditional mechanical machinability, metallization ease, and cost.

For many mechanical sensor applications, single-crystal Si, based on its intrinsic mechanical stability and the feasibility of integrating sensing and electronics on the same substrate, presents an excellent substrate choice. For chemical sensors, on the other hand, Si, with few exceptions,* is merely the substrate and as such it is not necessarily the most attractive option.

In Table 1.9, we show a performance comparison of substrate materials in terms of cost, metallization ease, and machinability. Both ceramic and glass substrates are difficult to machine, and plastic substrates are not readily amenable to metallization. Silicon has the highest material cost per unit area, but this cost can often be offset by the small feature sizes possible in a Si implementation. Silicon, with or without passivating layers, due to its extreme flatness, relative low cost, and well-established coating procedures, is often the preferred substrate—especially for thin films. A lot of thin-film deposition equipment is built to accommodate Si wafers, and as other substrates are harder to accommodate, this lends Si a convenience advantage. There is also a greater flexibility in design and manufacturing with Si technology compared with other substrates. In addition, the initial capital equipment investment, although much more expensive, is not product-specific. Once a first product is on-line, a next generation of new products will require changes in masks and process steps—but not in the equipment itself.

The disadvantages of using Si are usually most pronounced with increasing device size and low production volumes and when electronics do not need to be, or cannot be, incorporated on the same Si substrate. The latter could be either for cost reasons (e.g., in the case of disposables, such as glucose sensors) or for technological reasons (e.g., the devices will be immersed in conductive liquids, or they must operate at temperatures above 150°C).

An overwhelming determining factor for substrate choice is the final package of the device. A chemical sensor on an insulating substrate is almost always easier to package than a piece of Si with conductive edges in need of insulation.

Table 1.9 Performance Comparison of Substrate Materials

Substrate	Cost	Metallization	Machinability
Ceramic	Medium	Fair	Poor
Plastic	Low	Poor	Fair
Silicon	High	Good	Very good
Glass	Low	Good	Poor

* The ion-sensitive field effect transistor (ISFET) is one example where chemistry and electronics are intimately integrated.[61] The difficulties encountered while integrating chemistry with electronics in an ISFET were highlighted in Volume I, Chapter 7. This device, thirty years after its invention, is finally commercially available. A second example is Nanogen's high-density DNA array (see Volume II, Chapter 8, on chemical-, photochemical-, and electrochemical-based forming). The problem of wiring a very high-density DNA array (e.g., 10,000 sites) makes on-site electronics attractive in this case, outweighing both packaging and cost issues.

As we pointed out earlier, packaging is so important in sensors that, as a rule, sensor design should start from the package rather than from the sensor. In this context, an easier-to-package substrate has a huge advantage. The latter is the most important reason why chemical sensor development in the industry retrenched from an all-out move toward integration on Si in the 1970s and early 1980s to a hybrid thick film on ceramic approach in the late 1980s and early 1990s. In US academic circles, chemical sensor integration with electronics continued until the late 1980s, stopped for a while, and then re-emerged with the successful introduction of a commercial ion-sensitive field effect transistor (ISFET)[61] and DNA arrays with integrated electronics;[62] in Europe and Japan, such efforts were never completely abandoned.

A good engineering guideline to determine whether Si is suitable for a given mechanical application is the following: there should be at least two benefits arising from the use of Si over other substrates (aside from the possibility of integrating the electronics on the same substrate). For example, in the case of a torsional micromirror, as shown in Figure 1.16, fabricated based on a silicon-on-insulator (SOI) approach, the two Si-derived benefits are 1) excellent mirror-like surface and 2) superior torsional behavior of the Si mirror hinges. On the basis of these two important properties, Si has become a formidable contender for making arrays of micromirrors for optical switching. Note that the mirror in Figure 1.16 is not optimized for a high-density mirror switch array. For such an application, the hinges should be very short or hidden behind the mirror, as the mirror surface exposed to the incoming light should be maximized. The design as shown could be used, for example, for a linear scanner.

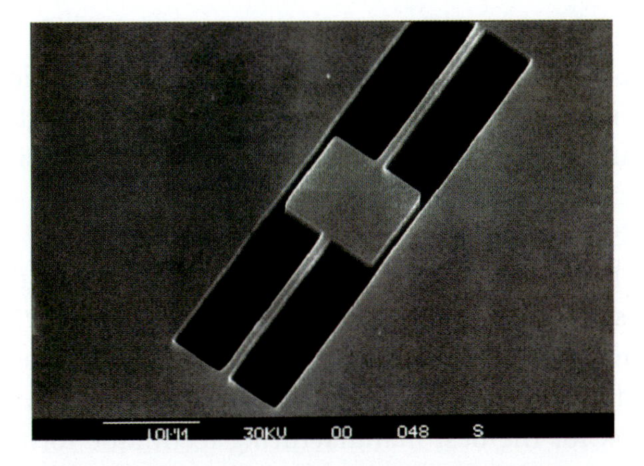

FIGURE 1.16 Silicon-on-insulator (SOI) micromirror.

In Table 1.10 we compare some of the material properties for single-crystal SiC, Si, GaAs, and diamond with relevance to their role as substrates in electronics and MEMS. Not enough good data are available yet to assemble a similar table for the polycrystalline version of the listed semiconductors. From a comparison of Si and GaAs for micromachining applications in this table, we conclude that GaAs is a better material for thermal isolation and for higher-temperature operation. Single-crystal GaAs lends itself as a material for micromechanics because of its attractive optoelectronic and thermal properties. The material is less attractive for mechanical devices, with a yield load smaller by a factor of 2 compared with Si. The many heterostructures possible with GaAs make a wide variety of optical components, such as lasers and optical waveguides, feasible. The piezoelectric effect enables piezoelectric transducers. On the basis of the high-electron mobility, the material is also ideal for the measurement of a magnetic field through the Hall effect. "Macro" pressure, temperature, and vibration sensors, making use of the influence of external pressure and temperature on the band gap, have been built.

Micromachining is proving to be a good method to incorporate micromirrors in resonant optical cavities for tunable lasers. Moreover, micromachining using gallium arsenide and group III–V compound semiconductors is a practical way to integrate radio frequency switches, antennas, and other custom high-frequency components with ultrahigh-speed electronic devices for wireless communications (see Chapter 10, this volume).

Checklist

Some twenty years ago, Stephen Senturia (Massachusetts Institute of Technology) posed a rhetorical question about building micromachines: "Can we design microrobotic devices without knowing the mechanical properties of the materials involved?" In subsequent years, characterizing mechanical properties of MEMS materials became popular. However, an equally important question—"Would

TABLE 1.10 Material Properties of Four Important MEMS Materials at 300 K

Property	3C-SiC	GaAs	Si	Diamond
Melting point (°C)	2,830 (pressure is 35 bar; decomposes)	1,238	1,415	4,000 (phase change occurs)
Maximum operating temperature (°C)	873	460	300	1,100
Thermal conductivity (W/cm °C)	4.9	0.5	1.57; comparable to metals such as carbon steel (0.97) and Al (2.36)	20
Linear thermal expansion coefficient (× 10–6 °C–1)	4.7	5.9	2.35; the low expansion coefficient of Si is closer to that of quartz (7.1) than that of a metal (e.g., 25 for Al), making it insensitive to thermal shock	0.08
Young's modulus (GPa)	448	75	190 (111); the elastic modulus is similar to that of steel (206–235)	1,035
Physical stability	Excellent	Fair; sublimation of As is a problem	Good	Excellent
Energy gap (eV)	2.39	1.42, direct	1.12, indirect	5.5
Chemical resistance	Very good	Poor	Good	Excellent
Electron mobility (cm2/V s)	1,000	8,500	1,500	2,200
Hole mobility (cm2/V s)	50	400	600	1,600
Density	3.2	5.3	2.32; lower density than Al (2.7); thus, it has a high stiffness-to-weight ratio	3.5
Yield strength (GPa)	21 (for 6H-SiC)	2.0	7 (steel is 2.1); IC-grade Si is stronger than steel	53
Breakdown voltage (× 106 V/cm)	2	0.4	0.3	10
Dielectric constant	9.7	13.1	11.9	5.5
Lattice constant (Å)	4.36	5.65	5.43	3.57
Knoop hardness (kg/mm2)	3,980	600	1,000 (stainless steel is 660)	10,000
Saturated electron drift velocity (× 107 cm/s)	2.2	2	1.1	2.7

you build a microstructure without knowing all the available construction methods?"—often remains unaddressed. Choices of materials and machining processes are intertwined, and both must be confronted early on in the preliminary design phase.

In Table 1.11, we introduce a checklist to help make the decision of optimized manufacturing option and substrate material easier. For the checklist, which serves as a control gate, we start from the assumption that the market and specifications list are well understood and that the detailed specifications list, scaling appreciation, and market understanding agree best with one particular sensing approach (in the example of the glucose sensor, we decided on an amperometric approach).

The checklist starts with a question about the design of the interface of the micromachine with the real world—in other words, with its package. Once the package design is established, one can look for the best manufacturing technique for the micromachined part inside. This strategy dictates

Table 1.11 Checklist to Determine Substrate and Machining Approach for a New Micromachining Application

1. What will the package of the sensor or system be, and how does it interface with the real world? The package and the interface with the environment determine size and cost of the total product and the nature of the microdevice inside, as well as the answers to most of the following questions.
2. Is there a need for integration of electronic functions in the micromachine? Such need (e.g., because the sensor has a very high impedance) necessitates a semiconductor substrate. Above, we proposed a rule of thumb to justify Si as a substrate; there must be at least two compelling reasons besides the possibility of integrated electronics [for the Si on insulator (SOI) micromirror in Figure 1.16, e.g., the reasons are the optical quality mirror surface of silicon and its excellent torsional behavior].
3. How many of the microdevices will be made (production volume or number of units) and what is the unit complexity (number of devices in a unit)? The number might suggest a serial (small number), hybrid (large number), batch (very large number), or even a continuous (moving web-based) approach.
4. Cost per device? The cost may suggest a serial (high cost, >$40) or Si batch (low cost, <$2) approach. For very low cost (glucose sensor strips, <$0.20), a very large batch or continuous fabrication process becomes mandatory.
5. Does the substrate merely function as a support? If so, glass, ceramic, or even plastic and cardboard all become options. If the substrate has a mechanical function, Si is an excellent candidate. If the substrate must have good optical properties, materials such as GaAs and poly(methylmethacrylate) are candidates.
6. Is there a need for modularity? Modularity is important with chemical sensor arrays, in which integration often is counterproductive because of the incompatibility issues faced when depositing different chemical sensor coatings on an integrated chip.
7. What are the expected relative tolerances on the lateral dimensions, and what is the required depth-to-width aspect ratio of features built into the substrate? Very small relative tolerances on lateral dimensions cannot be achieved yet; 1% lateral tolerance on a 100-μm line, in optical lithography, is considered good, but 10^{-3} on 1 cm in diamond milling is very poor (see Volume II, Figure 6.3). Large aspect ratios (say, >20) might necessitate wet anisotropic etching of a single-crystalline material, room temperature and cryogenic dry etching, or LIGA and pseudo-LIGA processes.
8. To what environment (air, water, or other) will the device be exposed? Sensors exposed to aqueous environments such as blood present more packaging problems and make integration of electronics difficult.
9. Which substrate makes packaging requirements less stringent? For a sensor in aqueous solutions, for example, a ceramic substrate requires no protection of the sides of the ceramic strip; a Si sensor, on the other hand, is difficult to insulate and package since a conductive medium might electrically short out the chemical sensor signal via the conductive Si sidewalls.
10. Is the desired microstructure a truly 3D part (e.g., a contact lens), or is it a projected form? A truly 3D part suggests traditional precision machining, for example, diamond turning or a molding process; projected forms allow for a lithography process.
11. What are the thermal requirements?
    - Maximum temperature of exposure (temperatures >150°C may make integrated Si functions impossible). If integration and high temperature are needed, specially processed Si, SOI, or GaAs become candidates.
    - Is thermal conductivity needed?
    - Is a thermal match with other materials important?
12. What are the flatness requirements (often in connection with the optical properties of the substrate)?
    - Average roughness, Ra?
    - One or both sides polished?
13. What are the optical requirements?
    - Transparency in certain wavelength regions?
    - Index of refraction?
    - Reflectivity?
14. What are the electrical and magnetic requirements?
    - Conductor versus insulator?
    - Dielectric constant?
    - Magnetic properties?
15. What is the process compatibility?
    - Is the substrate part of the process?
    - Chemical compatibility?
    - Ease of metallization?
    - Machinability?
16. What are the strain-dependent properties?
    - Piezoresistivity?
    - Piezoelectricity?
    - Fracture behavior?
    - Young's modulus?

itself, because the packaging cost contributes largely to the overall cost of any micromachined product and often overwhelms size specifications as well. The reason for expensive packaging can be ascribed to the many serial labor processes involved. Also, each device may require individual attention. The size of the package is dictated by the need to interface to a bigger structure, by limitations of traditional machining methods, or by the need for manual handling of the structure. Micromachining techniques such as fusion bonding, anodic bonding, and other integrated encapsulation techniques, commonly used in bulk and surface micromachining, might help transform more and more of the serial packaging steps into parallel batch steps even for ICs, which will prove advantageous in keeping costs of the total structure down. In the author's opinion, micromachined batch packaging solutions represent one of the most unmined micromachining opportunities.

Next in the checklist, one needs to establish whether to integrate active electronics on the substrate and confirm the expected temperature range of operation (if too high, the latter might void the integration option). A further consideration concerns number and cost of the devices and which substrate will make packaging easier. When faced with a new micromachining problem, one should consider all available precision manufacturing options—that is, zero-base the approach, be willing to forget all preconceived notions about how the part should be manufactured.

Different markets dictate different machining approaches. For example, chemical sensors, serving a fragmented market, are better constructed in a modular hybrid thick-film approach, but automotive accelerometers, serving a huge and uniform market, are better built in an integrated polysilicon design. The choice of materials for miniaturization of sensors, actuators, and power sources will depend also on the particular sensing principle employed.

We now return once more to the glucose meter example summarized in Figure 1.17. In this exercise, the first round of brainstorming (based on the specification list, and the scaling input) for a disposable glucose sensor resulted in the suggestion of an amperometric approach. We are ready now to face the first design iteration by looking at material and machining options for the proposed electrochemical glucose sensor. For implementation of the preferred sensor materials and manufacturing technology, we challenge the electrochemical conversion option with the list of manufacturing options presented in Tables 1.2 to 1.6 and substrate materials options (Tables 1.9 and 1.10). In the decision process, we use the checklist in Table 1.11 as the control gate. For the glucose sensor this results in thick-film sensor electrodes on disposable plastic (polyvinyl chloride) as the best suggested approach. Screen printing is a well-known technique for making patterns on various substrates (i.e., paper, plastic, wood, glass, ceramics, etc.). This coincides with the principal solution pursued by industry. See, for example, the MediSense/Abbott sensor, the Precision QID, which was shown in Volume I, Figure 7.80. This sensor consists basically of a two-electrode system with a carbon working electrode (carbon ink) and counter—and reference—electrode combined in a Ag/AgCl electrode (Ag/AgCl ink) (Figure 1.18). The working electrode is covered with another ink for the enzyme/mediator system.

Example 1.1: Proposed Scenario for Continuous Manufacture of Polymer/Metal-Based Biosensors

While the number of new tools available for microfabrication has grown dramatically, few methods have been gainfully applied to disposable biosensor construction for clinical applications. In contrast to mechanical sensors, where the excellent mechanical properties of single-crystalline Si often tend to favor Si technology, the choice of the optimum manufacturing technology for chemical sensors is far less evident. For chemical sensors, Si in the sensor often serves no other role than that of a passive substrate. This past decade the following trend emerged: mechanical sensors (pressure, acceleration, temperature, etc.) moved toward more integration (for example, in accelerometers with CMOS-compatible surface micromachining), while chemical sensors and biosensors moved away from integration and from Si as a substrate. For the latter, hybrid technology on plastic or ceramic substrates with silk screening

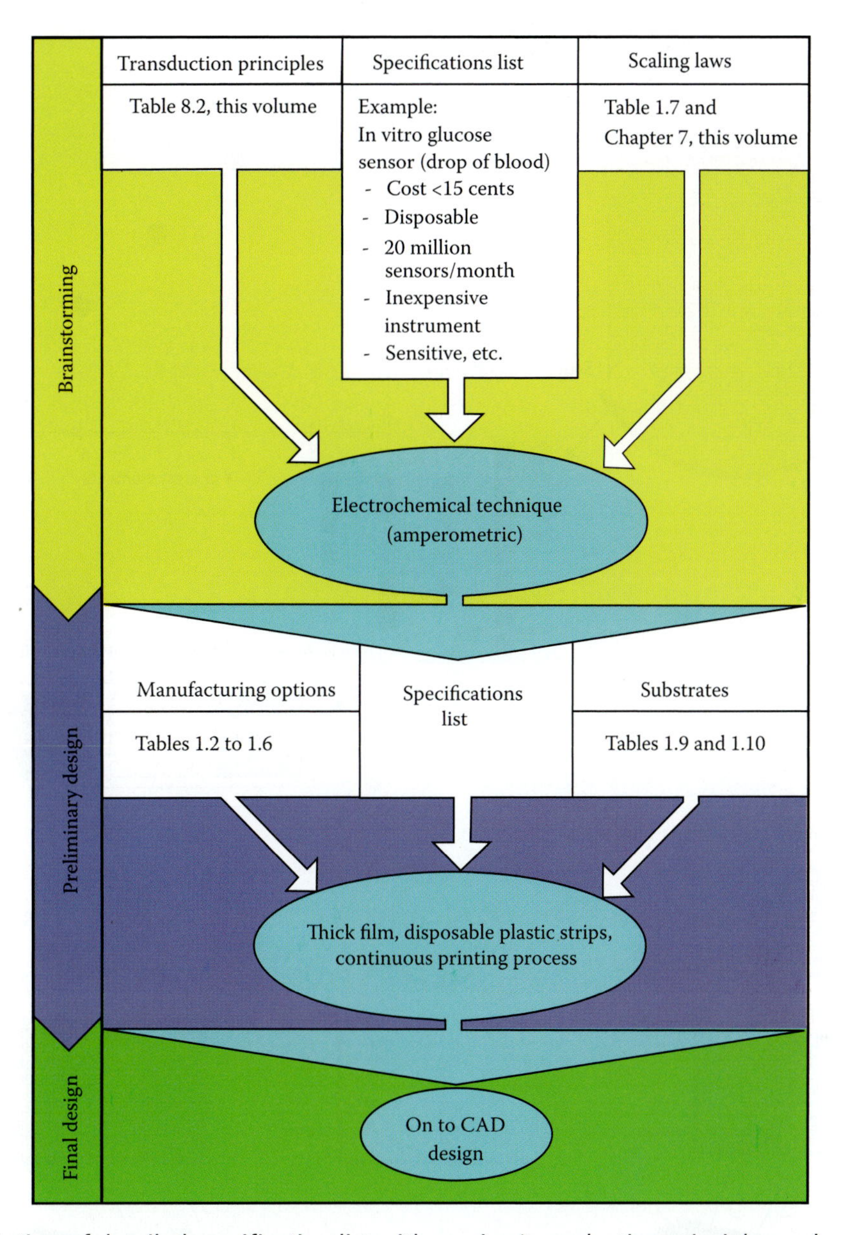

FIGURE 1.17 Confrontation of detailed specification list with sensing/transduction principles and scaling considerations in the brainstorming phase and subsequent searching of manufacturing options by going over a check-off list in the preliminary design phase in the process of developing a disposable glucose sensor (see text).

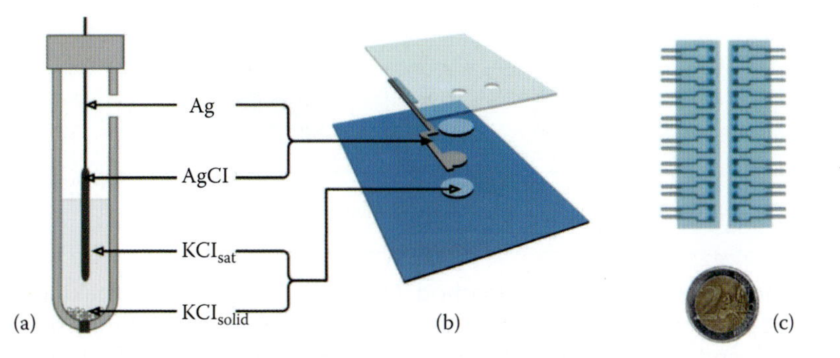

FIGURE 1.18 Structure of a conventional reference electrode of Ag/AgCl/KCl type (a) and design of a planar reference electrode fabricated by means of screen printing (b and c).

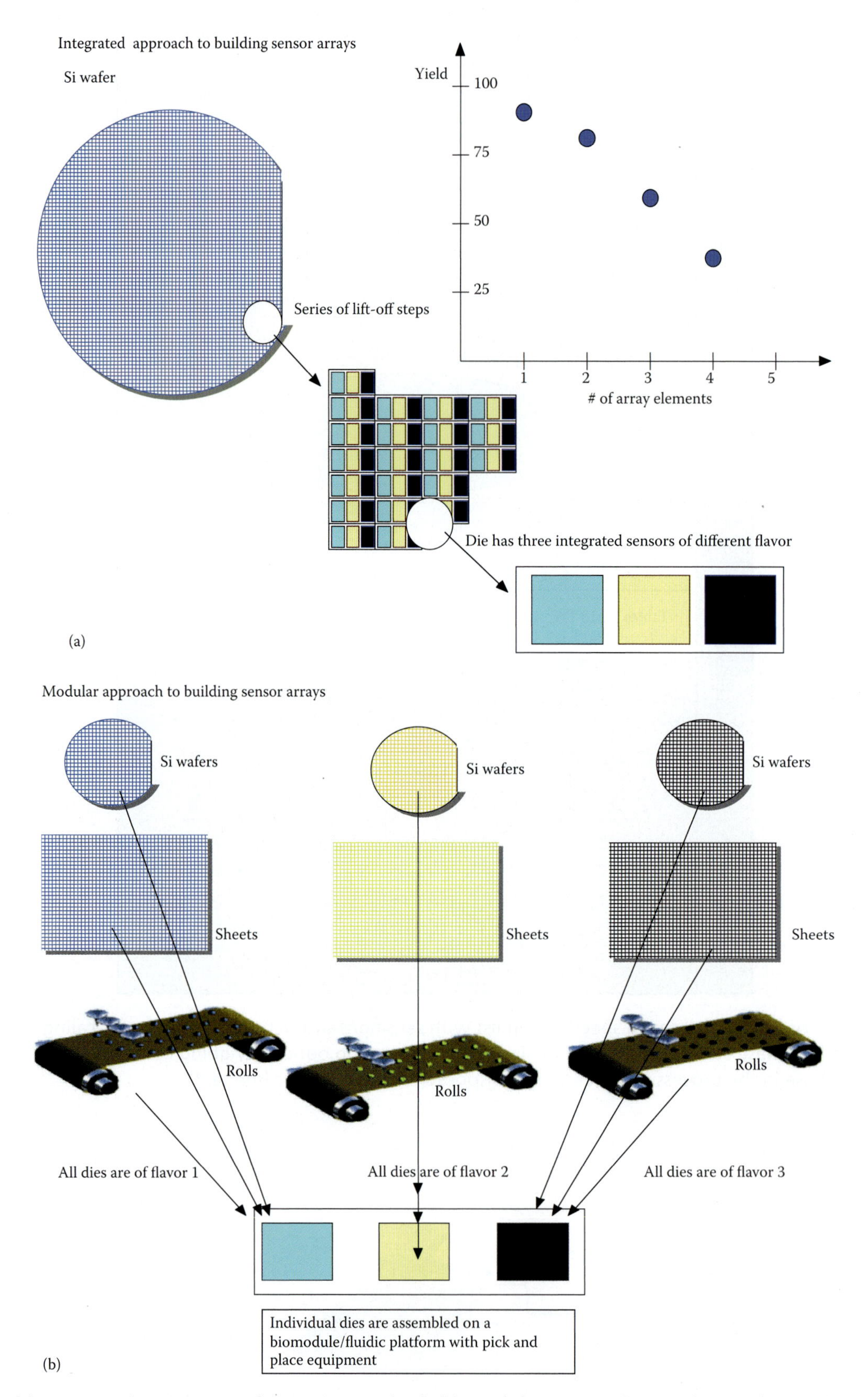

FIGURE 1.19 (a) Integrated sensor array fabrication method. (b) Modular approach to making a chemical sensor array. Wafers, sheets, or rolls of substrate with sensors of one type only are optimized separately to ensure optimum yield. They are then cut out and with pick-and-place equipment put into a biomodule/fluidic platform. Different sensor panels can be put together easily, as there are no compatibility issues to deal with.[63]

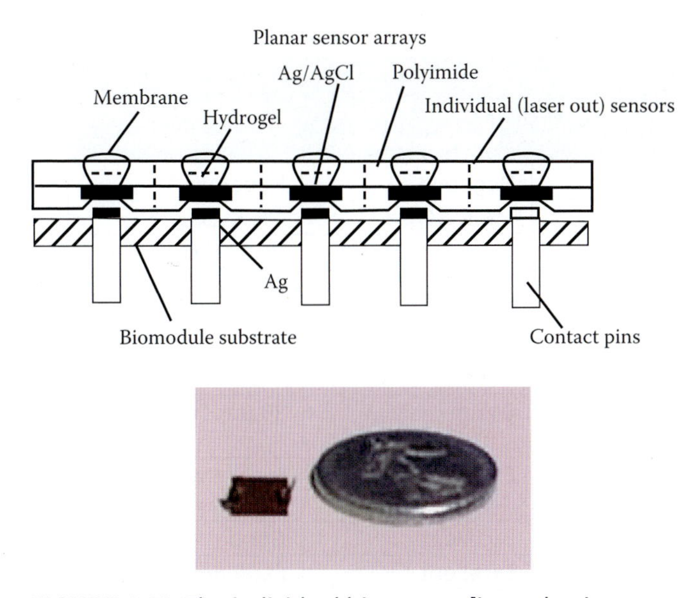

FIGURE 1.20 The individual biosensors [ion-selective electrodes (ISEs)] are combined in a so-called "biomodule" with pick-and-place equipment.[63]

or drop delivery systems for the application of the organic layers became more popular. Microfabrication of Si-based electrochemical sensors, using IC technology, has been especially challenging, mainly due to process incompatibility issues, packaging problems, failure to incorporate a true reference electrode, and the difficulties involved in patterning relatively thick organic layers such as ion-selective membranes and hydrogels [see also i-STAT (Abbott Point of Care, http://abbottpointofcare.com) in Chapter 10 of this volume]. The low cost requirements and the tremendous variety and fragmentation of biomedical sensor applications necessitates a nonsilicon, modular approach. Having addressed materials choice and modularity, various competing manufacturing processes must be compared. Today, in fabricating somewhat larger disposable biosensors, continuous manufacturing processes are used that yield much less expensive devices than silicon batch processes can afford. As an example, consider the mass production of amperometric glucose sensors where the cost per sensor target is 10 cents. Using a Si batch approach, it is very difficult to make a glucose sensor for less than $1 (or any disposable biosensor for that matter). The current process to mass produce glucose sensors involves such proven technology as doctor's blade on a continuous moving web, making a 10-cent cost per sensor quite possible. It is our belief that for disposable microsensors to become an economic reality such continuous or semicontinuous manufacturing processes will have to be further developed and adapted to microsensor construction. In Figure 1.19, we contrast an integrated approach to chemical sensor array fabrication with a more modular manufacturing approach.[63] The integrated sensor array with three types of sensing functions on a silicon substrate, depending on process compatibility, often has a rapidly decreasing yield. The more modular approach shows how one type of sensor might be made on a silicon wafer, a plastic sheet, or a continuous roll. The biosensors fabricated here are electrochemical sensors that can be combined in "biomodules" with pick-and-place equipment after they have been laser-cut from their individual processed sheets (see Figure 1.20). By making sensors of only one type on the same substrate, this approach allows for the fabrication of a sensor array composed of sensors that may otherwise have fabrication incompatibilities, thus increasing the yield of the final array. Since any panel could be put together in a factory upon demand without reconsidering process incompatibilities, more products could be introduced into the market all at once.

The ultimate goal is a manufacturing process for disposable sensors that merges traditional machining options with IC-based manufacturing options and is more akin to packaging processes (e.g., drop delivery, pick-and-place) than to the

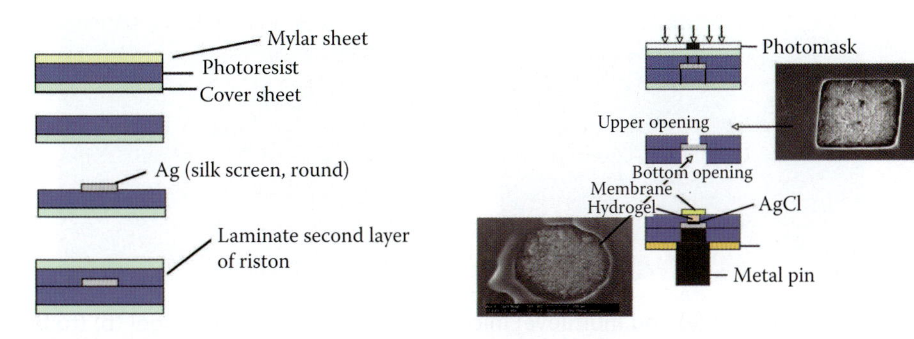

FIGURE 1.21 Schematic illustration of the procedure for fabrication of an individual array element.[63]

Table 1.12 Basic Properties of the Different Photoresist Materials Used to Build Fluidic Elements[63]

Photoresist	XP SU-8	Pyralin PI2721 (Photosensitive Polyimide)	Pyralux PC1025 Dry Resist Film	Riston 4600 Dry Resist Film
Resist tone	Negative	Negative	Negative	Negative
Developer	PGMEA*	DE6180	1% K_2CO_3	1% K_2CO_3
Etching profile	Vertical sidewall	Undercut	Vertical sidewall	Vertical sidewall
Smallest feature	2 μm	2 μm	20 μm	20 μm
Compatibility with plastic substrate in developing	No (surface coating and back cover are needed)	No (surface coating and back cover are needed)	Yes	Yes
Thickness of single layer	Any thickness up to 1 mm	~20 μm	64 μm	30 μm
Bonding to dry resist film	Excellent	Excellent	Excellent	Excellent
Adhesion to substrate	Excellent	Excellent	Excellent	Good
Uniformity	Good	Fair	Excellent	Excellent
Flexibility	Not applicable	Not applicable	Excellent	Brittle
Developing	Vibration needed	Vibration needed	Washing needed	Easy and quick
Multicoating availability	Yes	Yes	Multilayers by lamination	Multilayers by lamination

*PGMEA, propylene glycol methyl ether acetate.

front end part of the IC industry (e.g., lift-off, integration).

As a concrete example, the manufacturing procedure of an individual ion selective electrode array element, say, for sensing potassium, is illustrated in Figure 1.21. These electrochemical planar sensors may be made on a variety of polymer substrates; currently we use 5 × 5 in dry negative photoresist sheets (e.g., Pyralux). Pyralux comes in rolls, so it will eventually be possible to make these types of sensors in a continuous process (as depicted in Figure 1.19). An important simplification in the sensor manufacture depicted in Figure 1.21 is the self-aligned step. By exposing the dry resist from the top through the photomask, both cavities in the photoresist bilayer are made at the same time. The top cavity results from the pattern on the mask while the bottom cavity originates from the silk-screened Ag pattern, which acts as a mask for the lower part of the resist. No expensive double exposure system is required.

Fluid handling for calibration or washing steps may be implemented on the basis of the same dry photoresist technology. Fluidic structures have been made directly in all of the various resist systems listed in Table 1.12. The use of two or more photoresists in combination has enabled the fabrication of photopatternable multilayer microfluidic structures. For example, Madou et al.[63] have sealed photopatterned microchannels of SU-8 with sheets of the flexible photoresist, Pyralux, as shown in Figure 1.22a. Subsequent patterning of the Pyralux allows for the creation of more complex three-dimensional fluidic structures, as seen from Figure 1.22b. The

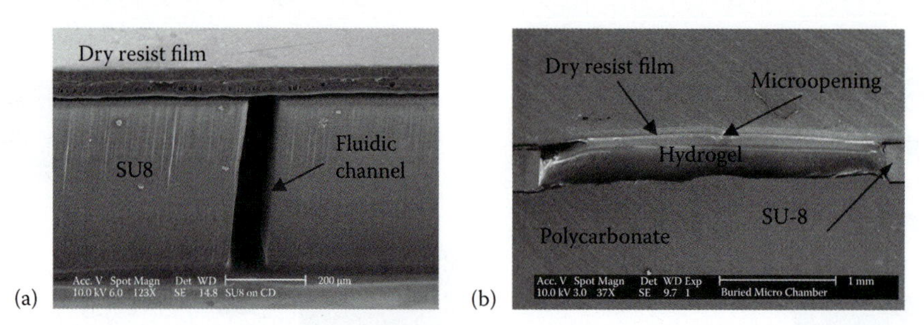

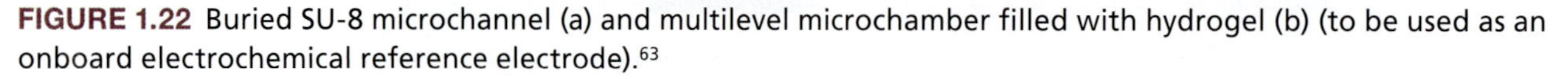
FIGURE 1.22 Buried SU-8 microchannel (a) and multilevel microchamber filled with hydrogel (b) (to be used as an onboard electrochemical reference electrode).[63]

latter structure can be used, for example, to seal liquids or hydrogels such as for an onboard electrochemical reference electrode.

The above "beyond batch" fabrication sequence is generic and applies equally well if optical sensor probes have been selected and will make it possible to fabricate biosensors in an affordable manner. This will lead to a wide variety of additional biomedical products currently impossible to fabricate using silicon technology.

Questions

1.1: List three different methods one can utilize to deposit a thin-metal film on a MEMS structure.

1.2: What do the following acronyms stand for? MEMS, VLSI, DUV, DOF, SEM, NGL, NA, MTF, OAI, OPC, RIE, and CMOS.

1.3: Compare laser-machining (LM) with e-beam machining (EBM).

1.4: The following bulleted list includes examples of micromaching tools:

(a) Classify them according to the applied energy appearance (W: wet chemical and electrochemical machining; M: mechanical machining; E: electrothermal machining) at the workpiece.

(b) Classify them according to machining methods (S: subtractive; A: additive; S/A: subtractive and additive).

- Photofabrication
- Laser beam machining
- Diamond milling
- Chemical and electrochemical milling
- Electron beam machining
- Photochemical milling
- Dry etching
- Ultrasonic machining
- Plasma beam machining
- Abrasive jet machining
- Electroplating and electroless plating
- Stereolithography
- Electrodischarge machining

1.5: Provide a comparison of the economic and technical aspects in the construction of sensors with thin-film technology, IC fabrication (semiconductor substrates), thick film, and classic construction.

1.6: Why use miniaturization technology?

1.7: Get calibrated: What is a mil or a thou? How many microns in a mil? How many thousandths of an inch (thou or mil) are there in a millimeter? Approximately how thick is a human hair? Approximately how big is a virus? A bacterium? A red blood cell? How thick is a standard sheet of paper?

1.8: How small is small? There are many manufacturing processes other than silicon micromachining for making small parts. Do some research to find specifications for various manufacturing processes, small devices, and tools. Search the Internet, call companies, use the Yellow Pages, use a Thomas's Register, ask a machinist, ask your instructor, etc. We're looking for the lower limits on traditional machining or tools that are currently commercially available at a local job shop or a tool distributor. The objective is to get a feel for the boundary between MEMS and traditional machining.

1.9: Reconstruct the decision tree to reach an optimized micromanufacturing option for a given miniaturization problem.

1.10: Explain the statement that MEMS engineers should design from the MEMS package inwards.

1.11: Make a comparison of additive thin- versus thick-film technology.

1.12: Explain the progress from serial to batch to beyond batch manufacturing (R2R).

1.13: Characterize objects made by laser machining, mechanical machining, E-beam machining, and plastic molding using the following criteria:

- Projected versus 3D
- Serial, batch, or continuous manufacturing
- Top-down or bottom-up machining

References

1. Block, B. Personal communication.
2. Van Doren, C. 1991. *A history of knowledge.* New York: Ballantine Books.

3. Middlehoek, S., and U. Dauderstadt. 1994. Haben Mikrosensoren aus Silizium eine Zukunft? *Technische Rundschau* July:102–05.
4. Vasile, M. J., C. Biddick, and S. A. Schwalm. 1993. Microfabrication by ion milling: The lathe technique. *Micromechanical systems*. A. P. Pisano, J. Jara-Almonte, and W. Trimmer, Eds. New York: ASME Press, p. 81.
5. Vasile, M. J., C. Biddick, and S. A. Schwalm. 1994. Microfabrication by ion milling: The lathe technique. *J Vac Sci Technol* B12:2388–93.
6. Harris, T. W. 1976. *Chemical milling*. Oxford, United Kingdom: Clarendon Press.
7. Romankiw, L. T. 1997. Electroforming of electronic devices. *J Plating & Surf Finish* 84:10–16.
8. Romankiw, L. T. 1997. Evolution of the plating through lithographic mask technology. *Electrochim Acta* 42:2985–3005.
9. Phillips, R. E. 1986. What is electrochemical grinding and how does it work? *Carbide Tool* 18:12–14.
10. Kalpajian, S. 1984. *Manufacturing processes for engineering materials*. Reading, MA: Addison-Wesley.
11. Saito, N. 1964. Mechanism of electric discharge machining. *Bull Jpn Soc Precis Eng* 1:39.
12. Taniguchi, N. 1984. Research on and development of energy beam processing of materials in Japan. *Bull Jpn Soc Precis Eng* 18:117–25.
13. Harper, C. A., ed. 1982. *Handbook of thick film hybrid microelectronics*. New York: McGraw-Hill.
14. Fiori, C., and G. De Portu. 1986. Tape casting: a technique for preparing and studying new materials. *Br Ceram Proc* 38:213–25.
15. Helvajian, H., ed. 1995. *Microengineering technology for space systems*. El Segundo, CA: Aerospace Corporation.
16. Pfender, E. 1988. Fundamental studies associated with the plasma spray process. *Surf Coat Technol* 34:1–14.
17. UMN. http://www.nrri.umn.edu/NLTC/SLA.htm.
18. Boothroyd, G., and W. A. Knight. 1989. *Fundamentals of machining and machine tools*. New York: Marcel Dekker.
19. Bellows, G., and J. B. Kohls. 1982. Drilling without drills. *Am Mach Special Report* 743:187.
20. Li, B., P. A. Clark, and K. H. Church. 2007. Robust direct-write dispensing tool and solutions for micro/meso-scale manufacturing and packaging. *Proceedings of the 2007 international manufacturing science and engineering conference*. Atlanta, GA.
21. Sakai, T. 1992. State of the art injection moulding of high performance ceramics. *Adv Polym Technol* 11:53–67.
22. Rak, Z. S. 1998. From technical ceramics to high-performance ceramics. *Ceramic Forum Int* 75: 19–24.
23. Belloy, E., P. Q. Pham, A. Sayah, and M. A. M. Gijs. 2000. In *Eurosensors XIV*, eds. R. de Reus and S. Bouwstra, 255–56. Copenhagen, Denmark: MIC-Mikroelektronik Centret.
24. Waterjet. http://www.waterjets.org.
25. Trotter, D. M. 1991. Photochromic and photosensitive glass. *Sci Am* 124:124–29.
26. Allen, D. M. 1986. *The principles and practice of photochemical machining and photoetching*. Bristol, United Kingdom: Adam Hilger.
27. Kern, W. 1978. Chemical etching of silicon, germanium, gallium arsenide, and gallium phosphide. *RCA Rev* 39:278–308.
28. Manos, D. M., and D. L. Flamm, eds. 1989. *Plasma etching: An introduction*. Boston: Academic Press.
29. Johnstone, R. W., and M. Parameswaran. 2004. *An introduction to surface-micromachining*. New York: Springer.
30. Diem, B., M. T. Delaye, F. Michel, S. Renard, and G. Delapierre. 1993. SOI (SIMOX) as a substrate for surface micromachining of single crystalline silicon sensors and actuators. *Transducers '93*. Yokohama, Japan.
31. Ehrfeld, W. 1990. The LIGA process for microsystems. *Proceedings: Micro system technologies '90*. Berlin, pp. 521–28.
32. Ahn, C. H., Y. J. Kim, and M. G. Allen. 1993. A planar variable reluctance magnetic micromotor with fully integrated stator and wrapped coils. *Proceedings: IEEE micro electro mechanical systems (MEMS '93)*. Fort Lauderdale, FL, February 7–10, pp. 1–6.
33. SOTEC. http://dmtsun.epfl.ch/~lguerin/sotec.html.
34. Keller, C., and M. Ferrari. 1994. In *Technical digest: 1994 solid state sensor and actuator workshop*, 132–37. Hilton Head Island, SC.
35. Pister, K. S. J. 1992. In *Technical digest: 1992 solid state sensor and actuator workshop*, 136–39. Hilton Head Island, SC.
36. Bride, J. A., S. Baskaran, N. Taylor, J. W. Halloran, W. H. Juan, S. W. Pang, and M. O'Donnell. 1993. Photolithographic micromolding of ceramics using plasma etched polyimide patterns. *Appl Phys Lett* 63:3379–81.
37. Zhao, X.-M., Y. Xia, and G. M. Whitesides. 1996. Fabrication of three-dimensional microstructures: microtransfer molding. *Adv Mater* 8:837–40.
38. Sander, D., R. Hoffmann, V. Reiling, and J. Muller. 1995. Fabrication of metallic microstructures by electroplating using deep-etched silicon molds. *J Microelectromech Syst* 4:81–86.
39. Shaw, M. C. 1984. *Metal cutting principles*. Oxford, United Kingdom: Clarendon Press.
40. DeVries, W. R. 1992. *Analysis of material removal processes*. New York: Springer-Verlag.
41. Slocum, A. H. 1992. *Precision machine design*. Englewood Cliffs, NJ: Prentice Hall.
42. Evans, C. 1989. *Precision engineering: An evolutionary view*. Bedford, United Kingdom: Cranfield Press.
43. Lambrechts, M., and W. Sansen. 1992. *Biosensors: Microelectrical devices*. Philadelphia: Institute of Physics Publishing.
44. Lebbink, G. K. 1994. Microsystem technology: Een niet te missen ontwikkeling. *Mikroniek* 3:70–73.
45. Bergveld, P. 1995. In *Micro total analysis systems*, eds. A. van den Berg and P. Bergveld, 1–4. Twente, the Netherlands: Kluwer Academic Publishers.
46. Madou, M. J., and G. J. Kellogg. 1998. *Progress in biomedical optics*. Eds. G. E. Cohn and A. Katzir, 80–93. San Jose, CA: SPIE.
47. van Lintel, H. T. G., F. C. M. van der Pol, and S. Bouwstra. 1988. A piezoelectric micropump based on micromachining of silicon. *Sensors Actuators* 15:153–67.
48. Schomburg, W. K., J. Fahrenberg, D. Maas, and R. Rapp. 1993. Active valves and pumps for microfluidics. *J Micromech Microeng* 3:216–18.
49. Nguyen, N. T., X. Y. Huang, and K. C. Toh. 2002. MEMS—micropumps: A review. *J Fluid Eng* 124:384–92.
50. Findlay, J. B., S. M. Atwood, L. Bergmeyer, J. Chemelli, K. Christy, T. Cummins, W. Donish, T. Ekeze, J. Falvo, and D. Patterson. 1993. Automated closed-vessel system for in vitro diagnostics based on polymerase chain reaction. *Clin Chem* 39:1927–33. See also United States Patent 5422271.

51. Tsao, T. R., R. M. Moroney, B. A. Martin, and R. M. White. 1991. Electrochemical detection of localized mixing produced by ultrasonic flexural waves. *IEEE 1991 ultrasonics symposium proceedings*. Orlando, FL.
52. Miyazaki, S., T. Kawai, and M. Araragi. 1991. A piezo-electric pump driven by a flexural progressive wave. *Proceedings: IEEE micro electro mechanical systems (MEMS '91)*, 283–88. Nara, Japan.
53. Jorgenson, J. W., and E. J. Guthrie. 1983. Liquid chromatography in open-tubular columns. *J Chrom* 255: 335–48.
54. Jacobson, S. C., L. B. Kounty, R. Hergenroder, A. W. Moore, and J. M. Ramsey. 1994: Microchip capillary electrophoresis with an integrated postcolumn reactor. *Anal Chem* 66:3472–76.
55. Kellogg, G. J., T. E. Arnold, B. L. Carvalho, D. C. Duffy, and N. F. Sheppard. 2000. Centrifugal microfluidics: applications. *Micro total analysis systems*. Eds. A. van den Berg, W. Olthuis, and P. Bergveld. Enschede, the Netherlands: Kluwer Academic Publishers, pp. 239–42.
56. Wittwer, C. T., and D. J. Garling. 1991. Rapid cycle DNA amplification: time and temperature optimization. *BioTechniques* 10:76–83.
57. Wittwer, C. T., G. C. Fillmore, and D. J. Garling. 1990. Minimizing the time required for DNA amplifications by efficient heat transfer to small samples. *Anal Biochem* 186:328–31.
58. Harrow, J. J., and J. Janata. 1983. Comparison of sample injection systems for flow injection analysis. *Anal Chem* 55:2461–63.
59. Shoji, S., S. Nakagawa, and M. Esashi. 1990. Micropump and sample-injector for integrated chemical analyzing systems. *Sensors Actuators A* A21:189–92.
60. Hillman, R. S., M. E. Cobb, J. D. Allen, I. Gibbons, V. E. Ostoich, and L. J. Winfrey. 1990. Capillary flow device. US Patent 4963498.
61. Sandifer, J. R., and J. J. Voycheck. 1999. A review of biosensor and industrial applications of pH-ISFETs and an evaluation of Honeywell's "DuraFET." *Mikrochim Acta* 131:91–98.
62. Thewes, R. 2002. Sensor arrays for fully-electronic DNA detection on CMOS. *IEEE international solid-state circuits conference (ISSCC)*. San Francisco, CA.
63. Madou, M., and J. Florkey. 2000. From batch to continuous manufacturing of microbiomedical devices. *Chem Rev* 100:2679–91.
64. Christensen, C. M. 2006. Hewlett–Packard: The Flight of the Kittyhawk(A). http://cb.hbsp.harvard.edu/cb/web/product_detail.seam?R=606088-pdf-ENG&ConversationId=51866.
65. Ikuta, K., K. Hirowatari, and T. Ogata. 1994. Three dimensional integrated fluid systems (MIFS) fabricated by Stereo Lithography. In *IEEE International Workshop on Micro Electro Mechanical Systems (MEMS '94)*, pp. 1–6. Oiso, Japan.

2

Nature as an Engineering Guide: Biomimetics

Outline

What then is the vocation of the whole man? So far as I can make out, his vocation is to be a creator: and if you ask me, creator of what, I answer—creator of real values ... And if you ask me what motive can be appealed to, what driving power can be relied on, to bring out the creative element in men and women, there is only one answer I can give; but I give it without hesitation—the love of beauty, innate in everybody, but suppressed, smothered, thwarted in most of us ...

L. P. Jacks
The Education of the Whole Man

Fall of Icarus. Mimicking nature—biomimetics. The Flemish artist Peter Paul Rubens (1636–1638) depicts Daedalus looking on as his son Icarus is falling from the sky. The picture is sometimes used to illustrate human arrogance. In the current context, the painting illustrates an early attempt at biomimetics: the fixing of bird wings onto human arms with beeswax. When Icarus flew too close to the sun, the wax melted.

Watson and Crick with DNA model.

Introduction

In biomimetics, one studies how nature, building atom by atom, i.e., through bottom-up manufacturing, through eons of evolution of life, developed materials, structures, processes, and intelligence to inspire and improve the engineering and design of artificial materials, human-made structures and processes (e.g., software). Human manufacturing technology works in the opposite direction, i.e., it builds top-down; in most current manufacturing we tend to start with larger building blocks and use stiff materials (e.g., Si or stainless steel), whereas nature prefers small building blocks and mostly soft, low-Young's modulus materials (e.g., muscle or skin). Throughout history, biomimetics has been attempted but often with less than satisfactory results. Bird flight, for example, did not lead to aircraft, but mathematical expressions from aerodynamics did. As a consequence, from the middle of the eighteenth century to about thirty years ago, engineers were tempted to engineer around nature's obstacles rather than be inspired by nature itself. Today though, in fields ranging from artificial intelligence to microelectromechanical systems (MEMS), nanoelectromechanical systems (NEMS), and smart materials, the perceived advantages of bottom-up designs and manufacturing are convincing many scientists to research natural, biomimetic approaches and manufacturing methods. In this chapter, we first introduce the most important engineering feats of nature and then explain how these are applied in today's biotechnology. It can be argued that molecular scientists and genetic engineers were practicing nanotechnology long before the name became popular with electrical and mechanical engineers. Actually molecular biology or "wet nanotechnology" has been called by some "nanotechnology that works." To be fair, molecular biology discoveries were rarely seen in the light of improved manufacturing techniques, as the discoverers were usually aiming for improved diagnostics and therapeutics. Within this context we introduce the use of natural polymers and their mutants not only for optimized sensing and realizing new drugs but also for the fabrication of actuators and new manufacturing techniques in general.

We start this chapter on a philosophical note, inquiring into the wisdom of biomimetics. We then detail a long list of nature's engineering feats. We try to define life on Earth and other planets, list the many ideas about the emergence of life, introduce DNA and RNA as the software of life and proteins as its hardware, and explain the build of a biological cell and the genetic code. We dwell on how nature generates power, while biomimetic approaches to packaging of nanocomponents are covered in Chapter 4, this volume, on packaging, assembly, and self-assembly, and on implementation of artificial intelligence in Chapter 9, on power and brains in miniature devices.

After introducing a definition of biotechnology we present examples of the use of MEMS and NEMS in biotechnology. We finish this chapter with a comparison of human- and nature-engineered structures.

This chapter is also meant as a bridge from *dry* engineering, in air and with rigid materials (mechanical and electrical engineering), to *wet* engineering, with flexible materials in aqueous solutions (biology and molecular engineering). Miniaturization science in 2011 is still practiced predominantly by electrical and mechanical engineers, but as applications are becoming more and more biological in nature, biologists, materials scientists, chemists, and physicists must work together with electrical and mechanical engineers to develop new miniaturization solutions. We hope that this chapter will further stimulate such collaborations.

Icarus Revisited?

Biomimetic-oriented manufacturing is a hot topic today, but some caution might be in order. As the legend of Icarus reminds us, at least some previous attempts at biomimetics have resulted in failure. The chapter-opening *Fall of Icarus* illustrates an early attempt at biomimetics. Icarus and his father Daedalus fixed bird wings onto their arms with beeswax. When Icarus flew too close to the sun, the wax melted and he fell to his death in the Mediterranean.

Smallness is the ancestral condition of life. Large organisms are built bottom-up from cells rather than divided into cells. Cells are, in turn, fabricated from yet smaller entities. A single *Escherichia coli* bacterium, for example, is shaped like a cylinder about 1 μm across and 3 μm long. It contains at least 3000 different molecular parts and will divide every 20 min when nutrients are available. No wonder scientists are intrigued by the miniaturization skills displayed by nature! But zoologist Steven Vogel, in his book *Cats' Paws and Catapults*,[1] asks a question very relevant to this chapter: "Who then is the better technologist? Mother Nature or the human engineer?" Natural design and biomimetics are very popular today and, according to Vogel, in danger of being overused or misinterpreted. Vogel presents a balanced view on the relative merits of natural and human mechanical design. He points out that although nature often exemplifies desired features, mimicking it has proved useful on surprisingly few occasions (as the Icarus legend attests). Vogel's explanation is that human and natural mechanical designs are two individually well-integrated technologies, but they are within very separate contexts. Each might be uniquely integrated by its own elements of internal harmony and consistency, but an impressive aspect of one may have little relevance to the other. A prominent example is metallurgy. In human technology, metals enable stamping, forging, casting, grounding, slicing, and sawing. No known organism uses pieces of pure metal for any mechanical purpose; they make stiff materials into artifacts by internal growth and surface deposition. So humanity's diverse array of metal manufacturing techniques is of no value in nature. Nature, on the other hand, does not use steel or favor the production of flat surfaces and sharp corners—all very useful in human manufacturing. Nature builds with proteins and produces mostly curved surfaces and rounded corners, resulting in such masterfully engineered objects as biological cells. Both natural and human manufacturing approaches have their merits within their own proper frame of reference. As a consequence, Vogel maintains, most attempts at mimicking nature have failed; we do not fly in aircraft with flapping wings or use wood to build skyscrapers. Vogel's comparison of human and nature's engineering feats principally deals with large designs, where he finds them competitive.

However, it is at the nanoscale level that nature truly excels, and perhaps Vogel's contention might not hold true in nanoengineering. In the nanoworld, nature is way ahead of human engineering, as it works with much smaller, more versatile building blocks and has mastered the self-assembly and multiparallel construction of those building blocks to create large, hierarchically arranged organisms that are much more complex than even the most complex human machines. Today, using human engineering techniques such as MEMS and NEMS, we are not able to build a single micromachine displaying anywhere near the complexity of even the simplest biological cell.

With biomimetic bottom-up manufacturing techniques, we are able to synthesize natural and mutant nucleic acids and proteins. Using well-established approaches such as recombinant DNA, gene shuffling, and knockout mutations, we can alter the genome of a cell, but there has been little progress made assembling an artificial cell or even the simplest organelle. But all this might be changing soon: at the end of 2007, a team of 20 scientists constructed a synthetic chromosome that is 381 genes long and contains 582,970 base pairs of genetic code, using lab-made chemicals at the J. Craig Venter Institute.[2] Prior to this work, the largest synthesized DNA contained only 32,000 base pairs. The J. Craig Venter team achieved its technical breakthrough by chemically making DNA fragments in the lab and developing new methods for their assembly and reproduction. After several years of work perfecting chemical assembly, the team members found that they could use homologous recombination (a process that cells use to repair damage to their chromosomes) in the yeast *Saccharomyces cerevisiae* to rapidly construct the entire artificial bacterial chromosome from large synthesized subassemblies. This team then started working toward the ultimate goal of inserting a synthetic chromosome into a cell and booting it up to create the first synthetic organism. The artificial chromosome was to be transplanted into a living bacterial cell, and in the final stage of the process, it was anticipated to take control of the cell and in effect become a new life form. The new life form would still depend for its ability to replicate itself and metabolize on the molecular machinery of the cell into which it has been introduced, and in that sense, it would not be a wholly synthetic life form, rather one more in tune with a parasitic life form. In May 2010, Venter's team did indeed create such a "synthetic cell," i.e., a normal bacterial cell controlled by DNA made in the lab. The Venter team first decoded the genome of the bacterium *Mycoplasma mycoides*. Then they purchased about 1000 bits of DNA about 1000 letters long that make up that bacterial genome. These bits were stitched together in the right order by putting them into yeast cells. The completed synthetic genome, containing almost 1000 genes, was then put into a different kind of bacterial cell, *Mycoplasma capricolum*, where it replaced the native DNA. The man-made DNA booted up the host cell, which began to replicate, making millions of *M. mycoides* cells.[70] Possible applications of this technology are unlimited. For example, in medicine, imagine that synthetic genes could be introduced into individuals suffering from a genetic disease, where they are missing genes or have damaged ones. The ability to correct the original faulty gene could potentially constitute a cure for a number of genetic diseases. Since building large genomes is now feasible and scalable, other important applications, such as biofuels, can be envisioned. Along this line, Venter's team is working already with Exxon Mobil to develop algae that soak up carbon dioxide and convert it into hydrocarbon fuel, and on new vaccines with Novartis. The hope is that many other complex, engineered microbes will be developed to clean up water, produce new forms of food, textiles, pharmaceuticals and industrial chemicals in ways that naturally occurring microbes cannot achieve.

The reason that nature is leading in nanotechnology may be sought in the fact that evolution has worked much longer on developing a single biological cell (3.5 billion years for the first life-forms to evolve) than on the larger life-forms (*Homo sapiens sapiens* first appeared only about 120,000 years ago). In other words, many designs were experimented with over a much longer period of time to come to the optimally designed single-cell life forms than for the larger living creatures such as trees and humans. It will perhaps be through nanochemistry that humans can aspire to reach the same level of sophistication in the manufacture of nanostructures as nature does. Moreover, the flight of birds did inspire the Wright brothers' first successful fixed-wing aircraft. But, as we are cautioned to do, the Wright brothers did not blindly mimic nature: they used fixed-wing aircraft, not the flapping bird wings of Icarus. That is, one should learn from nature's manufacturing techniques and adapt them to human/societal needs rather than blindly mimic king nature's approaches. After all, nature's manufacturing techniques are geared toward maximizing procreation, not toward maximizing speed and efficiency of product manufacturing techniques in a modern human societal context.

As humans learn to build with the same construction set as nature does, we are bound to challenge nature in nanoengineering. Because it uses relatively large building blocks, human manufacturing is rapid, and the expectation is that nature, because it uses much smaller building blocks (for example, atoms with a diameter of 0.3 nm), must be very slow. To offset the time it takes to work with small, basic building blocks, nature, in growing an organism, relies on an additive process featuring massive parallelism and self-assembly, the latter of which is introduced as one of the key characteristics of bottom-up manufacturing, detailed in Chapter 3, this volume.

Nature's Engineering Feats

Life?

The question of how exactly to define life has been a constant dilemma throughout humankind's history. Scientists of the stature of Schrödinger, Fermi, Arrhenius, Hoyle, von Helmholtz, Eigen, Haldane, von Neumann, Urey, Oparin, Monod, Dyson, de Duve, and Crick, to name a few, have been preoccupied with this question. From the plethora of ideas and the many attempts to characterize life that emerged from these thinkers, we select a few that promise the longest staying power. We refer to "ideas" and "characterizations" of life rather than "definitions," as no definition has yet been presented that is accurate enough.[3]

One of the better ideas comes from the book *Life Itself: Its Origin and Nature* by Francis Crick, who describes life as a process that is able to directly replicate its own instructions and indirectly replicate any machinery needed to execute those instructions.[4] Others consider something to be alive if it exhibits a minimal set of features that include,

Table 2.1 Characteristics Associated with Living Things

Characteristic	Remarks
Typical chemistry	Proteins, nucleic acids, fatty acids, sugars and water. Several of these chemicals can also be found in nonliving matter.
Very high degree of complexity	The incredible complexity guarantees the unpredictability of organisms. The weather is also very complex though and few would consider it alive.
Self-organization	Perhaps more important than complexity is organized complexity. Then again, there are also self-assembled monolayers and micelles that we would not call alive.
Chirality	Life on Earth almost exclusively uses left-handed amino acids and right-handed DNA. Experiments show that proteins cannot form if both right- and left-handed DNA are present.
Self-repair	The object is capable of rectifying errors in either its metabolic or genetic machinery.
Autonomy or self-determination	Spontaneity, inner freedom—perhaps the most enigmatic characteristic of life to explain, and one that sets life apart from nonlife.
Decrease in entropy	The increasing complexity of a living organism goes hand in hand with a decreasing entropy. Unfortunately, such a definition also characterizes a refrigerator.
Metabolism and homeostasis	Taking up substances (nutrition) to garner energy and excrete others reaching an equilibrium body shape. But what does this characterization make of a dormant microorganism?
Self-recognition	Distinguish foreign from self and reject foreign.
Sensitivity	Response to environmental stimuli.
Information content	The information needed to replicate an organism is encoded in the genes.
Replication	Identical offspring are generated. Then again, many molecules and crystals form templates for replication.
Evolution (natural selection)	Populations change over time.
Reproduction	Offspring is more than a facsimile; it also must incorporate a copy of the replication mechanism. Slight differences in the offspring are implied. Clays can replicate as well as reproduce (see text).
Growth and development	Growth and gradual evolutionary development, often from a single cell, to more complex structures. Crystals also grow, but there is no actual evolutionary adaptation.
Heredity	Transmit information from generation to generation with some slight changes embedded.

for example, metabolism, self-repair, and replication. A more complete set of characteristics associated with living "things" is summarized in Table 2.1. Not all organisms exhibit all these features all at once, but a combination of some of these characteristics can describe almost anything that can be defined as *alive* on Earth.

From Table 2.1, we learn that thermodynamics offers one of the many descriptions of life. In a living system, materials and free energy (ΔG) flow into the organism, and waste and heat (ΔH) flow out of it. Living things, from the lowly bacteria to the more sophisticated, intelligent humans, all are highly ordered compared with their constituent molecules and seem to violate the law of increasing entropy (second law of thermodynamics). The second law of thermodynamics states that in any spontaneous process, there always is an increase in entropy. Living creatures have spontaneously developed into more ordered and sophisticated systems through evolution with an accompanying decrease in entropy (ΔS). This apparent violation of the second law of thermodynamics is easily explained if one recognizes that the evolutionary process takes place within the broader context of a universe in which chaos keeps on increasing. In more rigorous thermodynamic terms, the argument goes as follows: the entropy of the universe may be expressed as the sum of the system under study (life) and its surroundings, as follows.

$$\Delta S_{universe} = \Delta S_{system} + \Delta S_{surroundings} \tag{2.1}$$

In life, large molecules assemble from simple, small ones into those that are increasingly complex; that is, ΔS_{system} is negative. To reconcile this with Equation 2.1, we must remember that $\Delta S_{universe}$ must be positive, which can be achieved only if $\Delta S_{surroundings}$ is large and positive.[5*]

To illustrate how weak some of these "definitions" of life are, just consider a thermodynamic definition, often presented in the literature: a living system is one that can maintain itself in a state of low entropy. However, a definition in which a system reducing its entropy through interaction with its environment is considered alive would also qualify your refrigerator—or the weather, for that matter.

Computer science (through modeling of artificial life), mathematics (through complexity theory), and information theory all contribute their own definitions. The following are some of the basic tenets. In artificial life, or a-life, computer-virus-like entities are created in an artificial computer world (e.g., Tom Ray's *Tierra* and *Network Tierra*);[6] these creatures reproduce, mutate, evolve, and die. Some scientists actually have considered these digital entities alive. Critics of a-life are skeptical because of the many simplifications and the arbitrariness of its rules. Independent of the issue of a-life being alive or not, important information about what life is and how it evolved might be gathered in days rather than over millions or billions of years. Indeed, because of the simulation power of today's computers, Darwinian evolution scenarios, involving creatures of varying complexity and different degrees of environmental pressures, can be played out over short periods of time.

The continuing study of complex systems may result in a better definition of *life*. Complexity scientists suggest that life is an "emergent" or "self-organizational" property that arises spontaneously when a broth of different chemicals reaches a certain level of complexity.[3] Others are adamant that it is the principle of information economy in self-developing systems that provided the guiding principle of biological evolution. Artificial life and complexity theory is touched upon in Chapter 9, on power and brains in miniature devices, and in "Emergence of Life: Biogenesis" below. Some artificial life Web sites can be found at http://alife .org and http://www.alcyone.com/max/links/alife .html. It is perhaps best at this point to admit that, today at least, because of its complexity, we cannot define what life really is and that describing life obviously is easier than defining it.

Life on Other Planets?

To garner a better understanding of what life is, we might look for it on other planets. The search for extraterrestrial intelligence (SETI) (http://setiathome.ssl.berkeley.edu) is privately funded and kept somewhat at arms' length by NASA out of fear of criticism by overzealous politicians, who would rather spend tax dollars finding weapons of mass destruction. Use of miniaturized systems figures prominently in this quest, and NASA is an important funding source for miniaturization science. It is beyond the scope of this book to dwell on SETI, but a brief introduction of the Drake equation, which predicts the probability of finding intelligent life on other planets, might motivate interested readers to explore the topic further. The Drake equation has been a starting point for almost all theoretical speculation on the existence of extraterrestrial intelligence since 1960.[7] The equation is a product of a large number of elements, most of which are only vaguely known:

$$\underbrace{N = R^{*} \times f_p \times n_e}_{\text{Physical}} \times \underbrace{f_l \times f_i}_{\text{Biological}} \times \underbrace{f_c \times L}_{\text{Cultural}} \tag{2.2}$$

The elements forming the Drake equation are as follows.

Physical Elements

R^* = the rate at which stars are formed in our galaxy per year; f_p = the fraction of stars, once formed, that will have a planetary system; n_e = the number of planets in each planetary system with the right environment suitable for life.

* But where did the order of the universe as a whole come from? The second law of thermodynamics must have been violated during creation! Pointing out that the expanding universe allows increasing room for order to form could deflate this point, but then one may further ask where did the energy and mass of the universe originate? The first law of thermodynamics (conservation of energy) must have been violated at creation. This point may perhaps be refuted as well since the positive energy of the universe is exactly balanced by the negative gravitational potential energy: $E = k(\text{inetic}) + mc^2$ (matter/energy) $- V_g$ (gravitational) $= 0$. So perhaps no energy was required to produce the universe?

Biological Elements

f_l = the probability that life will develop on a suitable planet; f_i = the probability that life will develop to an intelligent state.

Cultural Elements

f_c = the probability that intelligent life will develop a culture capable of communicating over interstellar distances; L = the time (in years) that such a culture will spend trying to communicate.

We know that N is no less than 1, as manifested by life on Earth. Beyond that, the widest range of N values has been proposed—from 1 all the way up to >100 million! *Paradigms Lost* by John Casti (1989)[7] presents an excellent treatise on the subject. Casti revisited the topic in 2000[8] and points out that, since his *Paradigms Lost* of 1989, the evidence for planetary systems outside our own solar system has become overwhelming, increasing n_e in Equation 2.2.

Most of the "life" sensors that NASA has developed so far are based on the assumption that life on other planets is similar to that on Earth and are based on sensors for flow of typical Earth-type feed materials (e.g., oxygen) and Earth-type waste materials (e.g., carbon dioxide), or they may be heat flow detectors. A few such "life" detectors were tested on Mars. Oyama, for example, in a Mars *Viking* experiment, wetted a soil sample with amino acids and other nutrients and then monitored common metabolic gases, such as carbon dioxide, oxygen, and hydrogen.[9] Perhaps too many experiments designed to find life on other planets are based on the expectation of Earth-like life and a specific molecular structure rather than on an abstract system featuring properties such as adaptability, flexibility, reproduction, and self-organization. Life elsewhere might well involve molecules other than DNA.

Crick, in his 1981 *Life Itself: Its Origins and Nature,*[4] details the Panspermia theory, quite controversial at the time, in which life on Earth came from somewhere else in the universe. This, of course, puts the question about what life is, or how it emerged, on another planet and does not answer the underlying question. Interestingly, the Panspermia theory has found new supporters. In August 1996, it was announced that features that were explained as life forms more primitive than those on Earth were found in a 4-pound meteorite retrieved on Earth but originating from Mars* named ALH84001[8] (Figure 2.1). A mineralogical study concluded that the temperature of the rock never rose much above that of boiling water, and a living microbe or spore deeper inside would have survived the voyage. There have also been new suggestions that extremely sturdy life forms might reach Earth unscathed via dust or small objects such as comets.[10,11] We do, indeed, know now that some types of meteorites contain carbon and organic chemicals, including amino acids.[11] In this context, the amino acid glycine was found inside the galactic cloud Sagittarius B2 in 1994.[8] An issue that the cosmic seed theory does address is how life on Earth could get off to such a quick start; it might have arrived nearly fully developed!

FIGURE 2.1 ALH84001. A rock from Mars retrieved on Earth.

Emergence of Life: Biogenesis

Introduction

Religious debates aside (Genesis with creationism, vitalism, reincarnation, etc.), and ignoring earlier suggested life forces such as psyche, air, lifeblood, phlogiston, electricity, and radioactivity, the question about the "emergence" of life has been even more controversial than the question about what life itself is. The two questions, as suggested in the preceding section, are intertwined: if we knew how

* The 1996 NASA announcement of potential proof for life on Mars is still mired in controversy relating to life originating from inorganic or organic sources.

life evolved, we might be able to define it better. There are several schools of thought on the issue. We briefly review the most prominent ones.

DNA, RNA, and Proteins: Who Was on First?

Four billion years ago, there was so much cratering on the moon and on Earth's surface that no life could have taken hold. The oldest fossilized bacteria date the start of life on Earth to 3.8 billion years ago. The starting point of "life," in a "narrow" window between 3.8 and 4 billion years ago, might have been a simple molecule with the ability to catalyze its own formation from raw materials found in its immediate surroundings—*replication through autocatalysis*. The period of "competing chemistries" before terrestrial life developed is known as the *prebiotic* phase. One or several self-replicating "mother molecules" must have led to the molecules of life we have come to know: deoxyribonucleic acid (DNA), ribonucleic acid (RNA), and proteins. In Table 2.2 we summarize some of the most common terms/definitions used in the context of the emergence of life (biogenesis) and genetics.

Many of the debates have focused around the primacy of DNA versus RNA versus proteins. We discuss life's functional polymers—DNA, RNA, and proteins—and the genetic code in more detail below. For now, it will suffice to expose the purported roles of DNA, RNA, and proteins in the emergence of life.

1. *DNA first.* DNA is the repository of genetic information and would seem like a logical candidate to be the first molecule of terrestrial life to follow one or more of the mother molecules of prebiotic times. However, this is not so. It is widely accepted that the carrier of the genetic code could not have been at the cradle of life, as it cannot replicate without the presence of certain proteins. To be specific, it needs enzymes unzipping its double helix structure before it can replicate. Some theories have RNA emerging first, others proteins, and yet others have both of them arising separately and developing a symbiotic relationship. In all instances DNA would have emerged later as a more reliable repository for genetic information.
2. *RNA first.* RNA is involved in the copying of the DNA genetic code and the assembly of amino acids into proteins on the basis of the copied DNA blueprint. As told by Casti, Eigen et al., in 1974, showed that from a broth containing the four RNA bases and an enzyme, RNA could self-assemble.[8] Thomas Cech (1989 Nobel Prize winner for Chemistry for the codiscovery of ribozymes, and now director of the Howard Hughes Medical Institute), when at the University of Colorado, demonstrated that RNA exhibits enzymatic properties and can catalyze the replication of other RNA molecules (i.e., ribozymes).[8] According to Matt Ridley, RNA was Greece to DNA's Rome, Homer to its Virgil.[12]
3. *Proteins first.* Proteins are the basis of a wide variety of functions (e.g., catalysis of reactions and immunoreactions to combat disease) and structural elements (e.g., in skin and hair). There is also evidence for the protein-first model. Reza Ghadiri, for example, showed that in homogeneous solutions, certain small peptides can catalyze the formation of other peptides, form symbiotic relationships, and spontaneously develop error-correcting mechanisms that prevent mutant forms from replicating.[8] In 1997, Lee et al.[13] published results in *Nature* on a helical peptide that is able to replicate itself, so the debate of what came first, RNA or protein, remains unsettled.
4. There also are serious problems with the pre-DNA models. The RNA-first model, for example, is thought to be too prone to errors. Coding errors are induced by mutations, and they are quite frequent in RNA, as there is only one strand of the molecule, and there would have been no "proofreading" enzymes available. As we will see below, DNA features two strands and special enzymes that repair coding errors before the code is translated into proteins. The RNA error rate explains why nature eventually switched to the more secure DNA molecule in order to protect its software. In the case of the proteins-first model, it is not easily understood how proteins could have passed on their heritage. An example of a symbiotic relationship theory is the one postulated by

Table 2.2 Common Terms Used in Biogenesis and Genetics

Term	Definition
Allele	Slight variant of the same gene in a given species.
Archaea	Archaeans are single-celled creatures that join bacteria to make up a category of life called the prokaryotes (pronounced "pro-carry-oats").
Chromosomes	A complex structured set of DNA with proteins inside the cell nucleus. It stores genetic information.
DNA	Biological chain molecule built from four kinds of building blocks, the nucleotides. DNA, normally found as a double helix spiral of two complementary strands, is the carrier of genetic information in all cellular life forms and many viruses.
Eukaryotes	Creatures whose cells have a nucleus, which contains the genetic information (pronounced "ew-carry-oats").
Gene	A segment of DNA containing the necessary information to produce a particular protein.
Genome	All of the DNA possessed by an individual member of a species.
Genotype	The particular collection of various possible gene alleles present in the genome of a given individual.
Germ cells	The cells of the body involved in reproduction; i.e., the sperm of the male and the eggs of the female are formed from germ cells.
Heterozygous	If there is more than one allele of a given gene within the population, an individual might have two different alleles for a given gene, in which case that individual is said to be heterozygous for that gene.
Homologous chromosomes	Chromosomes containing the same gene sequence.
Homozygous	Having the same alleles at a particular gene locus on homologous chromosomes.
Mutation	Modification of a DNA sequence extending from a single base change to the addition or elimination of an entire DNA fragment.
Phenotype	The physically discernible features resulting from a particular genotype.
Polymorphic genes	Genes that have allelic forms.
Prokaryotes	Prokaryotes' genetic material, or DNA, is not enclosed in a central cellular compartment called the nucleus. Bacteria and archaea are the only organisms of this class.
Protein	Molecules made up of long chains of amino acids. They build tissues, organs, etc., and carry out many critical functions in the body.
RNA	A nucleotide chain that differs from DNA in having the sugar ribose instead of deoxyribose and having the base uracil instead of thymine. RNA helps translate the instructions encoded by DNA in order to synthesize proteins.
Single-nucleotide polymorphisms	Differences (polymorphism) of individual bases within a genome from different individuals.
Somatic cells	All of the cells in the body except the germinating cells, such as those in the eggs and sperm.

Freeman Dyson. In this attractive model, a vesicle or cell (the hardware) metabolizes proteins but has no genome (the software) until a primitive nucleic acid replicating molecule invades the vesicle and finds it advantageous to replicate not only itself but also all the helpful proteins it finds in the cell.[8]

Made from Clay?

A rival theory to the DNA-, RNA-, or proteins-first models has life originating from inorganic materials such as clays. (Life originating from clay! Where have we heard this before?) This concept was first advanced in 1982 by Cairns-Smith,[8,10] a chemist at the University of Glasgow, Scotland. In a test tube, small clay plates (platelets), less fragile than nucleic acids, given the right feed stock, will replicate—that is, identical offspring are generated (see Figure 2.2a). In a natural, more challenging setting in which a small change might induce the platelets to change shape (evolution), they may even reproduce—that is, their offspring become somewhat different, and these offspring may in turn replicate (identical offspring) or reproduce (offspring changed in shape). Microcrystals of clay, typically flat platelets with two large plane surfaces exposed to the surrounding medium, consist of a silicate lattice with a regular array of ionic sites but an irregular distribution of metal ions (e.g., Mg^{2+} and Al^{3+}) occupying these sites. The precise arrangement of metal ions in the clay may be capable of catalyzing specific reactions of organic molecules on the clay surface. These organic molecules in turn can bestow certain benefits onto the clay crystals: perhaps a more advantageous platelet shape or cementing the crystal planes together better—anything that protects them against the environment and speeds up (catalyzes) their replication. During crystal growth, silicate and metal ions from the surrounding medium will tend to carry the same pattern of ionic charges as the clay layer below it. In this way, the metal ion patterns, the carriers of information, are replicated and selected by environmental pressures. Replication and reproduction of clay are illustrated in Figure 2.2a. In a clay such as kaolinite, many cross-sectional shapes of crystallites are possible. The electron micrograph of the structure of kaolinite in Figure 2.2b illustrates the basic idea. The distribution of metal ions on these platelets, in association with organic molecules from the surrounding medium, helps favor one platelet out-replicating another. Physical aspects of the crystal act as the phenotype (what an organism—in this case a crystal—looks like as a consequence of its genotype) and cause selection to choose among the genotypes (the genes that an organism possesses).

Cairns-Smith suggests that clays came first, enzymes second, cells third, and genes fourth. After natural clay *learned* to direct synthesis of organic molecules, clay and organics *learned* to make a cell membrane. These cells containing clay as the hereditary unit eventually *discovered* RNA, which "piggybacked" at first on clays as they fitted snuggly into clay lattices. Clays might then have found a way of fabricating these self-replicating organics and, consequently, they would have quickly out-replicated other clays. By piggybacking on clays, RNA could have existed for a long time before becoming self-replicating. Once the organic molecules on the clay started replicating faster and faster by absorbing organic material from their surroundings, "organic life" might have taken over from "inorganic life." The latter is called a genetic takeover. For Cairns-Smith's bibliography and work, see http://originoflife.net/cairns_smith/index.html. There was not much experimental evidence for the Cairns-Smith's model until 1996, when Ertem and Ferris discovered that long strands of RNA, given the right feedstock, form spontaneously on the surface of montmorillonite (a type of clay).[14]

Life's Origin beneath the Earth's Surface?

The Archaea (also Archaeabacteria) constitute the earliest and most basic life form: relatively simple single-celled organisms that are chemotrophic; i.e., they do not need light, as they make biomass via chemical energy rather than light. Today, these organisms are still often are found in "extremophiles" living in harsh, toxic environments. Archaea use

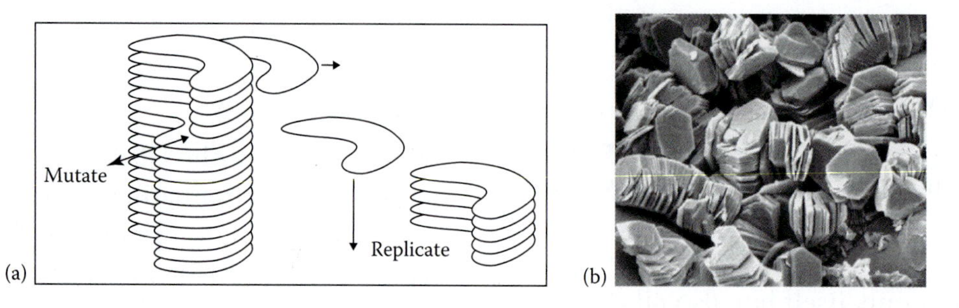

FIGURE 2.2 Replication and reproduction. (a) Replication and reproduction in a stack of clay platelets. (b) Electron micrographs of the structure of kaolinite clay.

a remarkable diversity of metabolic tricks and adaptations to survive under apparently impossible conditions. A lot of information has been gathered on these "superbugs" in the last twenty years; they have been found in oil wells, the Dead Sea, the Antarctic, near volcanic deep-sea vents, and deep down below the Earth's surface. From the Archaea two quite different groups developed, namely, the Eubacteria (bacteria) and the Eukarya (eukaryotes). Archaeans and Eubacteria (bacteria) make up a category of life called the prokaryotes. In prokaryotes, the genetic material, or DNA, is not enclosed in a nucleus, a central cellular compartment. The Eukarya include all organisms made up of cells with nuclei; that is, their DNA is walled off in a separate compartment of the cell (see also below under "Cells," Figure 2.20). It should be noted that extremophiles also include bacteria, and these hardy organisms can be grouped as acidophiles (acid loving), halophiles (salt loving), thermophiles (heat loving), hyperthermophiles (the present record is 121°C), psychrophiles (cold loving), alkaliphiles (thriving at high pH), barophiles (surviving under high-pressure levels, especially in deep sea vents), osmophiles (surviving in high-sugar environments), xerophiles (surviving in hot deserts where water is scarce), anaerobes (surviving in habitats lacking oxygen), microaerophiles (surviving under low-oxygen conditions), endoliths (dwelling in rocks and caves), and toxitolerants (able to withstand high levels of damaging agents, for example, living in water saturated with benzene or in the water core of a nuclear reactor).

Carl Woese, from the University of Illinois, in 1977, had a hard time convincing the world about the existence of Archaeans.[15] Woese found that Archaeans clustered together as a group well away from the usual bacteria and the eukaryotes. Because of their vast differences in genetic makeup, he proposed that life be divided into three domains: Eukaryota, Eubacteria, and Archaebacteria (Figure 2.3). Craig Venter, in 1996, completed the genome sequence of the first Archaean, *Methanococcus jannaschii*. The genome data of this hardy bacterium, thriving at a balmy 85°C and 200 atm of pressure,

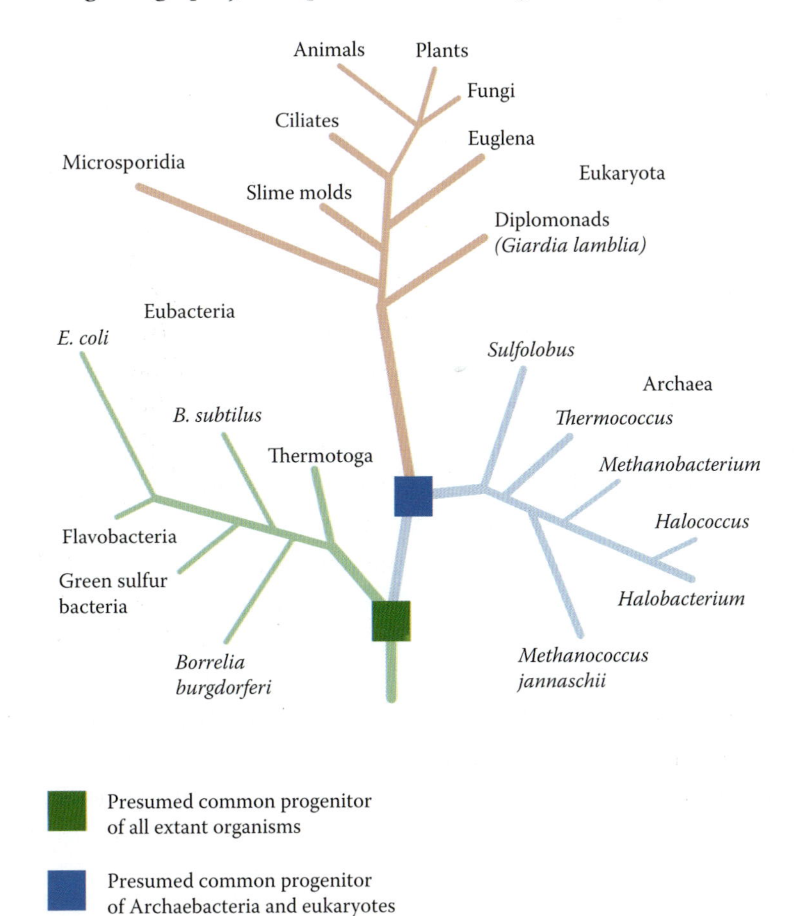

FIGURE 2.3 The tree of life.

includes the sequence of more than 1700 extremely heat-stable proteins never before sequenced.

The idea that chemicals beneath the Earth's surface produced these prebiotic replicators is attractive to many researchers. Early Earth, with high concentrations of $CO_2/H_2S/H_2$ etc., low oxygen levels, and high temperatures, was largely inhospitable to life as we know it today. Life forms that could evolve in such an environment needed to adapt to these extreme conditions. Many hyperthermophiles and Archaea are H_2 oxidizers. Thus, it is widely held now that extremophiles represent the earliest forms of life, with nonextreme forms evolving after cyanobacteria had accumulated enough O_2 in the atmosphere.

The recognition of life in those harsh environments on Earth makes the prospect of life on a harsh planet like Mars a bit more likely, and Archaea are favorites of the adherents of the Panspermia theory. A Web site with good information on extremophiles is http://www.astrobiology.com/extreme.html.

Extremophiles present a tremendous opportunity for more robust biotechnology processes and nanochemistry. Ordinary natural macromolecules cannot withstand extreme environments, making the prospect of widely used bottom-up, biomimetic devices less bright unless proteins from these superbugs are used to implement more robust nanochemistry. A well-known example of this approach is that of the hyperthermophile *Thermus aquaticus* used as the DNA polymerase in the very useful polymerase chain reaction (PCR) technique, which amplifies DNA samples by heat cycling the DNA sample together with the required chemistry cocktail (see "Polymerase Chain Reaction"). In Appendix 2A, we list other applications of extremophiles.

Complexity Theory

A groundswell of opinion has it that the difference between life and nonlife is only one of degree of organization, not of the type of components involved, and that life is how things behave, rather than of what they are made. Along this line, complexity scientists suggest that life is an "emergent" or "self-organizational" property that arises spontaneously when a large set of chemicals reaches a certain level of complexity (see also Chapter 3, "Self-Organization"). The background here is quite profound: humans are reasonably capable of calculating behavior of inorganic materials and the structures built from them, but we have had much less success with the more complex organic systems. This is where the science of complexity comes in—just like scaling and artificial life, complexity is a subdiscipline of biomathematics and represents a general attempt to elucidate fundamental biological principles on the basis of mathematics.

The Belgian Nobelist Ilya Prigogine (1917–2003) models complex systems that are far removed from thermodynamic equilibrium and describes their self-organizing* properties. These self-organizing properties have a tendency to reach critical "bifurcation" or "indecision points," where their behavior may leap into chaos or greater complexity and stability.[16] The latter scenario, Prigogine envisions, is how life might have started, that is, on the precipice between order and chaos.

An outstanding example of work in complexity is the theoretical work by Kauffman, from the Santa Fe Institute, who showed that an autocatalytic set of reactions leads to "emergent" and "self-organizational" behavior when a certain level of interconnectedness is reached between the reactions.[17] Kaufmann's Web site can be found at http://wn.com/Stuart_Kauffman. There is plenty of experimental evidence gathering for his theory in the laboratory. We described Ghadiri's protein work already (see "DNA, RNA, and Proteins: Who Was on First?"). In similar work, Rebek has demonstrated that synthetic chemicals can also reproduce, mutate, evolve, and combine with other reproducing chemicals.[18] Skeptics not only point to the simplifications and the arbitrariness of the rules of complexity theory but also point out that life is actually not an example of self-organization but one of genetically directed or specified organization. In other words, in life there is a software code directing the organization that emerges from complexity; in the complexity theory, on the other hand, there is no underlying software. In complexity theory, the physical boundaries of the system lead to the

* In Chapter 3, we clarify the distinction between self-assembly (near equilibrium) and self-organization (far removed from equilibrium) (see "Self-Assembly and Self-Organization").

emergence of organization. Despite the skeptics, this discipline is proving itself. We need to know how the system's software came about before we can understand how life emerged.

A Model from Information and Communication Theory

Yet another way of looking at the emergence of life comes from the theory of information and communication. Considering that DNA stores the information needed to construct and operate an organism, noise, a form of disorder, may then be regarded as the encroaching entropy. One then has to explain the creation of information (negative entropy) from noise, which again seems to violate the second law of thermodynamics until we realize that we are dealing with an open system (see above). Mutations in this model generate the noise, and it is the environment, via natural selection, that selects which information out of that noise ends up in the genetic message. The ultimate question in this model is, from where did the information content of the universe come?[10] The above considerations can be expressed in some simple mathematical formulations. There is a direct connection between information content (I) in a system and its entropy (S); an increase in entropy implies a decrease of information, and vice versa. Boltzmann linked entropy, S, to the computed probability of sorting objects into bins—a set of N into subsets of sizes n_i as follows:

$$S = -k\sum_i p_i \ln p_i \tag{2.3}$$

where p_i is n_i/N, and k is the Boltzmann constant ($3.2983\ 10^{-24}$ calories/°C). On Boltzmann's tombstone, Equation 2.3 is written out as follows:

$$S = k \log W \tag{2.4}$$

where W is the probability of a given state. One of the classic equations of information, derived by Claude Shannon in the 1940s, is formally similar to Equation 2.3.

$$I = \sum_i p_i \log_2 p_i \tag{2.5}$$

Information, I, is a dimensionless entity, but its partner entity, entropy, has a dimension; comparing Equations 2.3 and 2.5, one notices that S and I have opposite signs and otherwise differ only in their scaling factors, and to convert from one to the other expression, one may write $S = -(k \ln 2)I$, or an entropy unit equals $-k\ln 2$ bits. Information is thus a concept equivalent to entropy, and "life" can be described in terms of one or the other.* Lowenstein promotes the idea that the first molecular mutants that came about were those that made the most efficient information loops, an information loop where the product promoted its own production.[19] As with digital computers, it is in the resetting of the memory register where the unavoidable thermodynamic price is paid. Each time a molecule has been produced on its template, the molecule needs to be reset for spinning out the next "clone."

For further reading on the origins of life in general, good starting points are http://www.seti.org/seti and http://www.panspermia.org/.

DNA and RNA: The Software of Life

In all known life forms on Earth, heredity, or genotype, is encoded by deoxyribonucleic acid (DNA). This molecule first appeared on Earth more than 3.5 billion years ago. Sections of DNA, called genes, constitute the chemical recipe expressed in the actual form of an organism or its phenotype. DNA—a master "read-only" memory—guides, records, and passes on every evolutionary development of all carbon-based life on Earth. Its building blocks are sugar (deoxyribose), phosphate, and four different nitrogen-containing bases, namely guanine (G), cytosine (C), thymine (T), and adenine (A). A unit of a base, sugar, and phosphate is a *nucleotide*, and since there are four different bases in DNA, there are also four different nucleotides in DNA. A model of a DNA molecule is shown in Figure 2.4b.

In 1953, James Watson and Francis Crick (see photo on chapter-opening page), building on experimental results by Rosalind Franklin (Figure 2.5), Maurice Wilkins, and many others, elucidated the

* William Dembski, a proponent of intelligent design, has claimed that natural causes are incapable of generating information. He calls this the *law of conservation of information*. From Equations 2.3, 2.4, and 2.5, information gain is equivalent to loss of entropy, and Dembski is confusing entropy and energy: entropy is not a conserved quantity like energy.

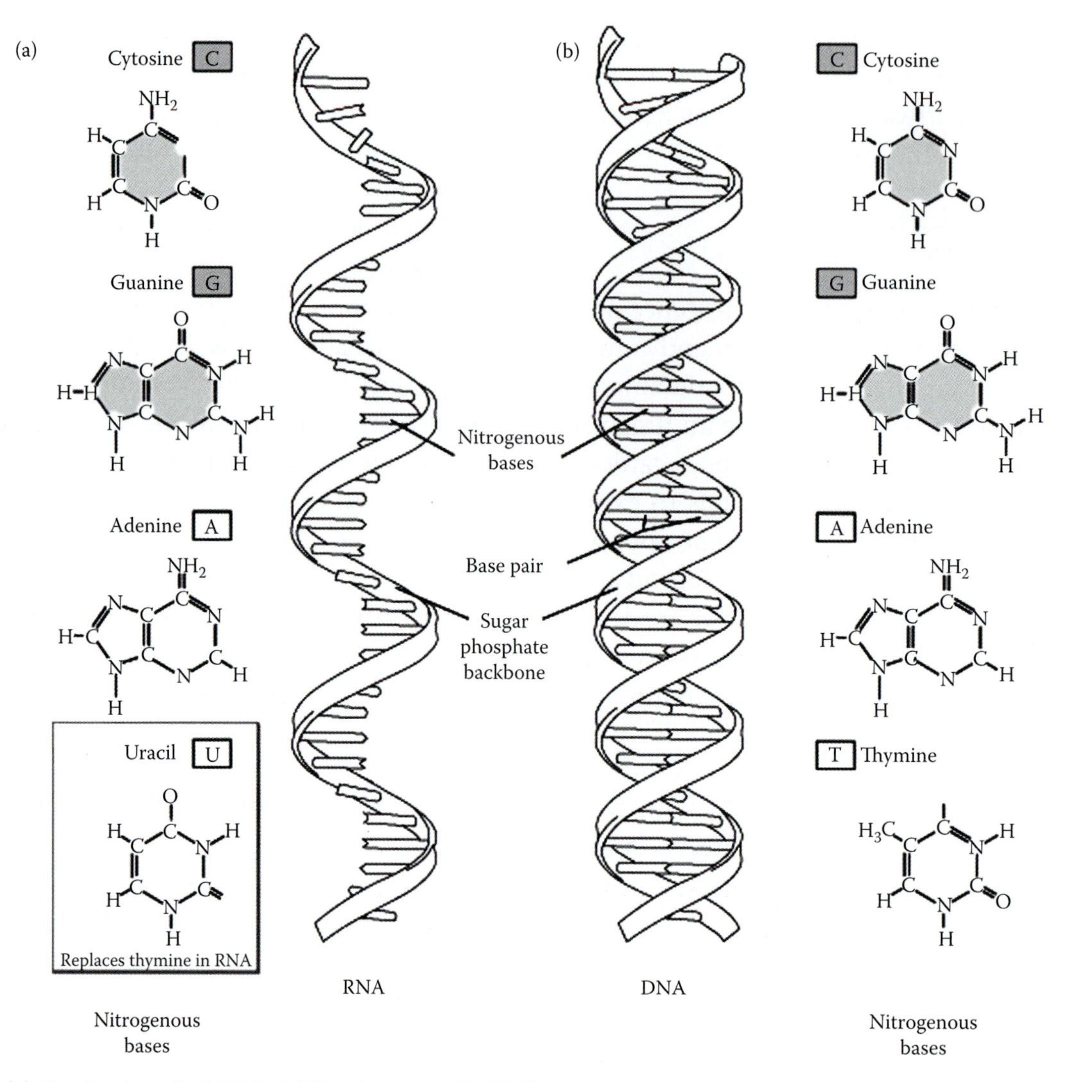

FIGURE 2.4 (a) Single-stranded RNA. (b) Double-stranded DNA.

double helix motif of DNA. Within a few years, it became clear how DNA, with the help of some other more or less complex molecules, is able to store and pass on hereditary information. This insight, among others, earned a Nobel Prize for Watson, Crick, and Wilkins in 1962. A fascinating historical, scientific, and ethical perspective on the discovery of the human blueprint is that of Robert Shapiro, *The Human Blueprint: The Race to Unlock the Secrets of Our Genetic Script.*[20] While scientific, Shapiro's book manages an entertaining account of the personalities involved in this greatest of scientific discoveries. Watson's autobiography *Double Helix* reads like a detective story and has become a classic bestseller.[21] Contrasting it with Crick's autobiography, *What Mad Pursuit*,[22] reveals how differently these two Nobel laureates measured their own contributions and that of others leading to that same monumental discovery (Crick passed away in July 2004).

The double helix DNA molecule consists of a right-handed spiral coil of two linked strands and is about 20 Å in diameter. Watson, during a dinner talk, perhaps after a bit too much wine, could muster only, "It is beautiful."* A model of the molecule as shown in Figure 2.4 is a bit misleading, since DNA is in reality a very narrow but extremely long molecule. From an

* It has been noted that the height of one unit of the DNA helix is related to its width in a proportion equal to the *Golden Ratio*. Throughout history, the ratio for length to width of rectangles of 1.61803398874989484820 has been considered the most pleasing to the eye. The Greeks named this ratio the Golden Ratio. In the world of mathematics, the numeric value is called "phi," named for the Greek sculptor Phidias. Many artists who lived after Phidias have used this proportion. Leonardo Da Vinci called it the "divine proportion" and featured it in many of his paintings. Long before Phidias, the Great Pyramids were proportioned based on the Golden Ratio.

FIGURE 2.5 Rosalind Franklin.

engineering point of view, DNA is a highly asymmetric molecule, with a width on the atomic scale and a length on the macroscopic scale, that is, from millimeters to centimeters and up to meters for a human.

This asymmetry makes DNA quite susceptible to breakage by shearing forces. The winding staircase of the DNA molecule in Figure 2.4 has railings made up by alternating molecules of sugar and phosphate; the steps are made up by pairs of bases. The four bases guanine (G), cytosine (C), thymine (T), and adenine (A) form only two types of pairs. An A base always joins to a T base via two hydrogen bonds (in shorthand, A = T), and a G always joins a C via three hydrogen bonds (G ≡ C). Walking up the staircase of the double helix, staying close to one side of the steps, the climber might read at his or her feet a base sequence such as AGCTAT; coming back down, on the other side of the steps, the descending person will read TCGATA. DNA often operates in conjunction with another nucleic acid, namely, ribonucleic acid (RNA). RNA is involved in the copying of the DNA genetic code and the assembly of amino acids into proteins based on the copied DNA blueprint. RNA uses a base called uracil (U) in place of thymine (T) and also employs a slightly different sugar ring (ribose instead of deoxyribose) (Figure 2.4a). Apart from these two differences, the components of RNA are identical to those of DNA.

The joining of two single-stranded DNA (ssDNA) strands into one double-stranded DNA (dsDNA) is called *hybridization*. The Watson-Crick pairing or hybridizing of DNA bases is shown in Figure 2.6.

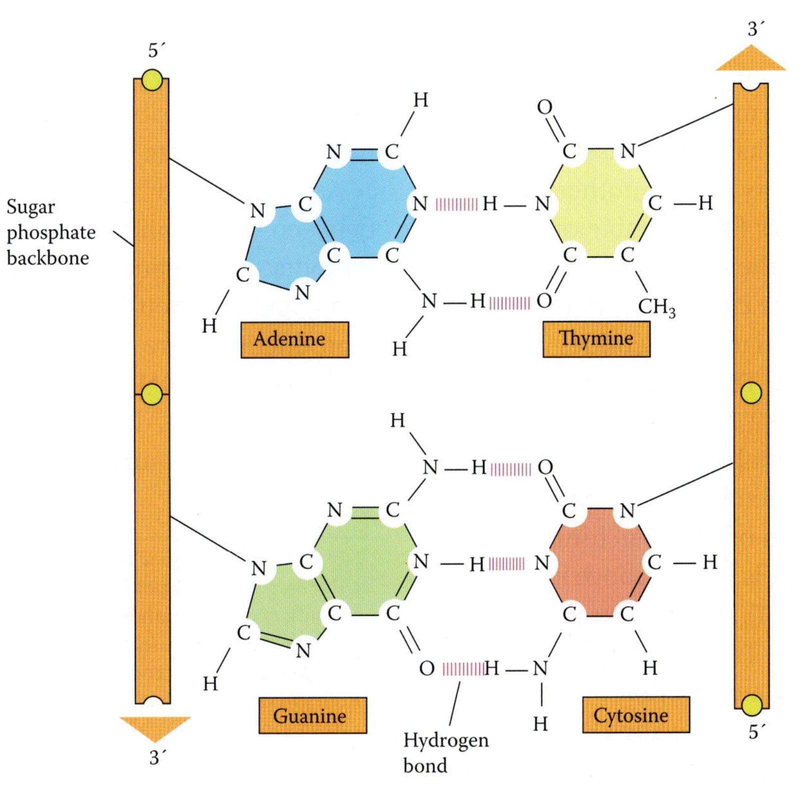

FIGURE 2.6 Base pairing in DNA.

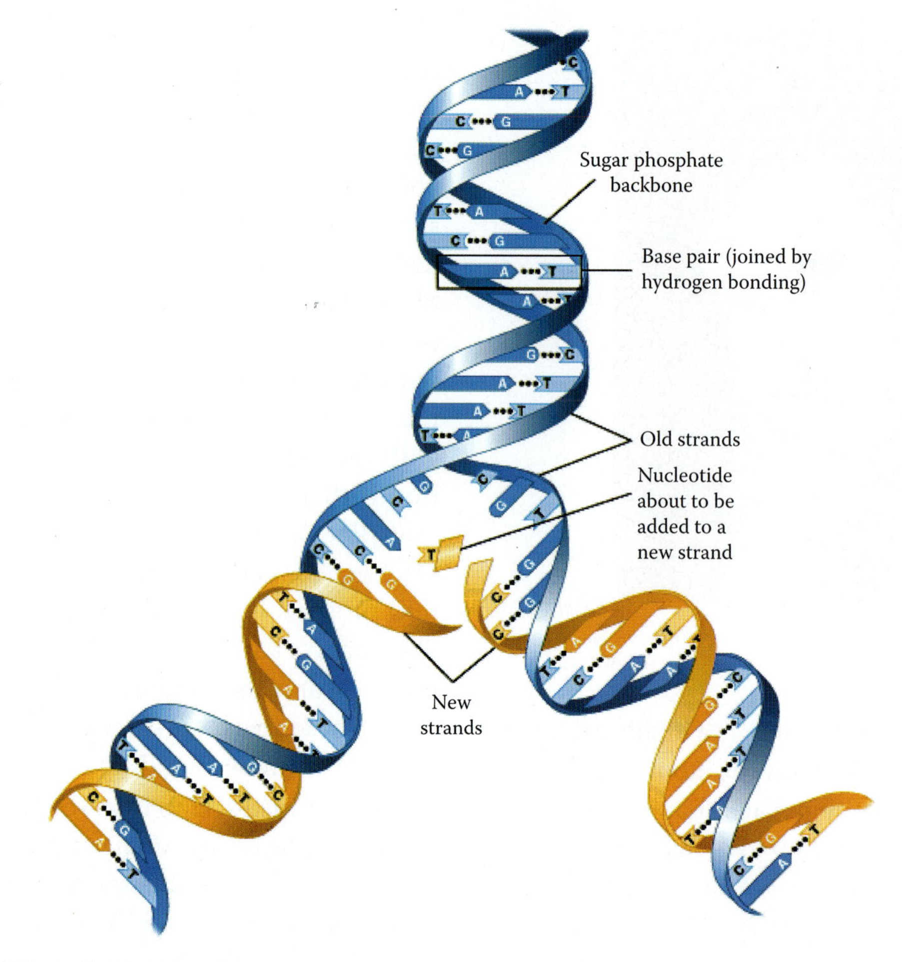

FIGURE 2.7 The DNA replication process.

To pass on information, a DNA molecule simply splits apart down the middle and then rebuilds matching parts to each half. The two strands in the ladder are complementary to one another, and, if the strands separate, each is fully capable of recreating the base sequences of the other with the help of a specialized enzyme. The enzyme, DNA polymerase, links up the appropriate nucleotides to make two daughter strands. The replication process is demonstrated in Figure 2.7. From this figure, we see that once the original double helix "unzips," the two separated strands act as templates (molecular molds), directing the production of new matching strands. If a double-stranded DNA is heated above its so-called "melting temperature," T_m, the two strands separate into two single strands. The melting temperature is a function of the ion concentration of the solution and the G-C content in the specific DNA sequence. When the temperature is reduced again, the two strands will eventually pair again by diffusion, and consequently "rehybridize" or "renature." The process just described is the one taken advantage of in PCR to replicate DNA (see above).

As mentioned earlier, one must distinguish between DNA replication as shown in Figure 2.7 and reproduction. Reproduction involves making modified copies of itself that can reproduce in return. If reproduction were flawless, evolution would never have occurred. DNA not only replicates and allows for reproduction, it also is the blueprint* for the synthesis of proteins. Further below, we introduce

* Not everyone sanctions the blueprint analogy. Ian Steward in *Life's Other Secret* correctly points out that genes are actually more like recipes in a cookbook rather than engineering blueprints; they tell us what ingredients to use, in what order, and in what quantities, but they do not provide a complete, accurate plan of the final result. Matt Ridley in *Genome* prefers the analogy of an immense book, a recipe of extravagant length; he also considers the term *blueprint* a poor analogy for a one-dimensional digital code. Then again, Robert Shapiro titled his book *The Human Blueprint: The Race to Unlock the Secrets of Our Genetic Script*.

some important characteristics of proteins, but we must first learn how a reading of the DNA blueprint results in their synthesis or production; i.e., we need to understand the mechanism of the genetic code.

Proteins: Hardware of Life and Nature's Robots*

Introduction

Virtually every process and product in an organism depends on proteins. In addition to their roles as enzymes (catalysts), messengers between cells, and defenders against infection, they are the principal constituents of cell membranes, tendons, muscles, skin, blood, bone, silk, hair, and other structural materials. The human body alone contains more than 100,000 distinct types of proteins, and all are synthesized in the same basic way. Just like nucleic acids, they are macromolecules, long polymeric chains. In the case of proteins, the chains are composed of just twenty different types of amino acids that are always joined in the simplest possible manner: linearly, in long chains (see Appendix 2B for a listing of their names and chemical formulas). There are many more amino acids, but nature only uses the twenty listed.

Biomolecules can process electrical and/or optical signals, but what proteins excel in is at recognizing and reacting to three-dimensional shapes. Proteins have twenty letters in their alphabet rather than the four of DNA, and they have the ability to adapt an endless number of three-dimensional (3D) shapes, making them much more versatile molecules than DNA. Information in biology, we shall see, is stored in the 3D shape of molecules.

Making Proteins

Amino acids are the basic components that make up all proteins. They are left-handed chiral molecules (except for the simplest, glycine) containing both a carboxylic acid group and an amine group. Chiral molecules are molecules that come in two forms, enantiomers, which are structurally identical except that one is the mirror image of the other. Like left- and right-hand gloves, these molecules cannot be superimposed (Figure 2.8). All chiral amino acids, forming proteins, in nature are "left-handed," and nucleotides, forming the genetic material, are "right-handed." It is one of the great unifying principles of biochemistry that the key molecules of life have the same hand in all organisms. The twenty common amino acids can be combined in an almost infinite number of ways to produce proteins. The reaction involved is the formation of a peptide bond (amide bond) in a so-called "condensation reaction" between the amine group of one amino acid and the carboxylic acid of another amino acid. In this reaction, a water molecule is released, as shown in Figure 2.9.

Each type of protein differs in its sequence and number of amino acids; therefore, it is the sequence of the chemically different side chains that makes each protein distinct. The two ends of a polypeptide chain are chemically different: the end carrying the free amino group (NH_3^+ or NH_2, depending on pH) is the amino, or N, terminus, and the one carrying the free carboxyl group (COO^- or COOH, depending on pH) is the carboxyl, or C, terminus. The amino acid sequence of a protein is always presented in the N to C direction, reading from left to right.

When only a few amino acids are linked, the product is called a *peptide*, the difference between a peptide and a protein being one of size. A protein with a molecular weight of 50,000 is also referred to as a protein of 50,000 daltons or 50 kilodaltons (kD). One dalton is equal to one atomic mass unit.

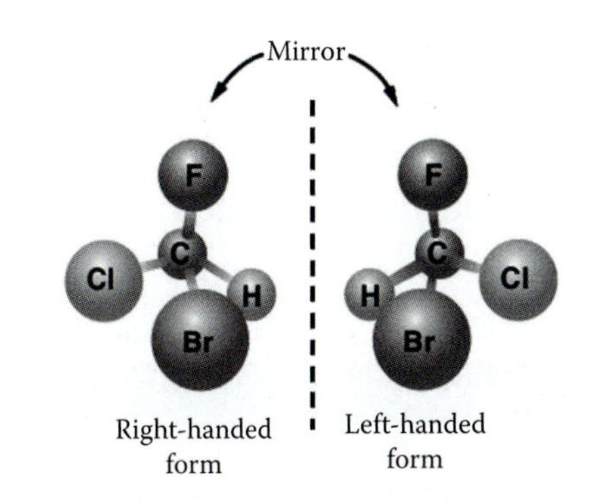

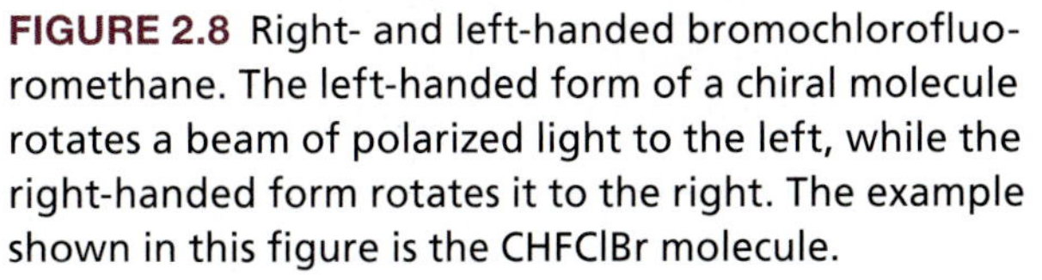

FIGURE 2.8 Right- and left-handed bromochlorofluoromethane. The left-handed form of a chiral molecule rotates a beam of polarized light to the left, while the right-handed form rotates it to the right. The example shown in this figure is the CHFClBr molecule.

* From the title of the book *Nature's Robots: A History of Proteins* by Charles Tanford and Jacqueline Reynolds.

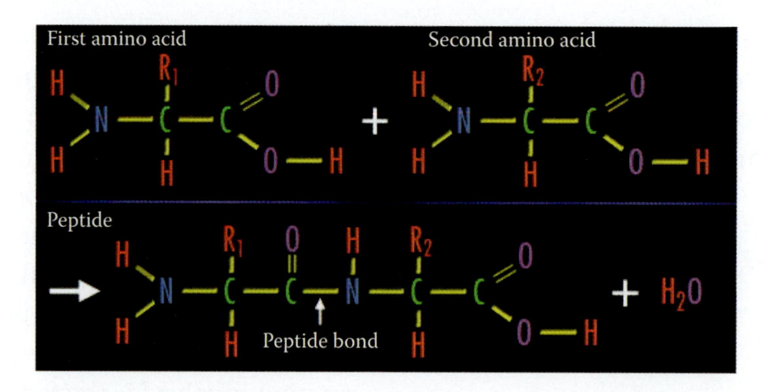

FIGURE 2.9 Water release in protein formation: proteins consist of a polypeptide backbone with attached side chains.

In Table 2.3, example proteins and their function are reviewed.

The Odds against the Emergence of Proteins

All models of the emergence of life eventually rely on the theory that a chance combination of chemicals leads to increasingly complex molecules and eventually to "life." Some simple organic molecules were already present in Earth's primeval atmosphere, as first demonstrated in 1953 in the now famous Miller-Urey experiments.[7] Miller, inspired by a suggestion of his advisor, Urey, reproduced a potential primeval atmosphere in laboratory glassware. When heating methane (carbon source), ammonia, and hydrogen with water, and passing an electrical discharge, to simulate lightning, through the mixture, he produced significant quantities of organic compounds, including some amino acids (e.g., glycine and alanine). He obtained this result after running the experiment for one week. Earth's early atmosphere was at first believed to have been reducing. Although Miller used hydrogen, ammonia, and methane, today it is believed that the atmosphere was more of a neutral mixture, consisting mainly of carbon dioxide and nitrogen and only small amounts of methane and ammonia. Since 1953, the Miller-Urey experiments have been reproduced with different simulated environments, such as the electric spark being replaced by a furnace, an ultraviolet lamp, shock waves, and energized chemicals. It turns out that amino acid synthesis is rather easy; as we saw before, they actually have been found even in meteorites.[3,7,10] The issue is not so much how to make individual amino acids but how to string them into the three-dimensional complicated proteins that life needs. The odds are stacked against making even the smallest protein this way. Take a small protein of 100 amino acids; with 20 different amino acids, the number of differently arranged proteins corresponds to 20 multiplied by itself 100 times (which corresponds to 1.26×10^{130}). Even the total number of stars in the observable universe, on the order of 10^{21}, is trivially small compared with the gigantic odds of stitching together a simple protein properly.[10] The latter is the major reason why scientists such as Crick and Davies came to believe that life on Earth was seeded from interstellar space, as there was not enough time to reach that degree of complexity on Earth in such a short amount of time.[23]

TABLE 2.3 Examples of Proteins

Structural	Functional
Collagen (found in bone and skin)	Hormones (control body functions)
Keratin (principal component of hair and nails)	Antibodies (fight infections)
Fibrin (helps blood clot)	Enzymes (catalyze chemical reactions)
Elastin (major part of ligaments)	Hemoglobin (carries oxygen and carbon dioxide in the blood)
	Myoglobin (transports oxygen in muscles)

Protein Structure

Proteins begin their life as linear chains of linked amino acids; however, most of them do not stay that way long, as is clear from Figure 2.10. The floppy protein chains quickly fold to minimize free energy. This minimization involves hydrophilic groups (groups that like water) orienting themselves toward the aqueous environment and hydrophobic groups (groups that dislike water) orienting themselves inward, that is, away from the aqueous or hydrophilic environment. Similarly, there are groups attracted to one another and groups that repel one another. The sequence of amino acids constitutes the primary structure of the protein. The secondary protein structure can follow several motifs—an α-helix (a right-handed spiral), a β-pleated sheet, or a ribbon. The tertiary protein structure describes how secondary structural motifs are packed together to give the molecule its three-dimensional shape. The tertiary structure of an enzyme molecule creates cavities and grooves on the molecule surface into or onto which only selected partner or ligand molecules fit. Those select ligand molecules, while bound to the enzyme, may in turn react as they present the right functional chemical groups to the immediate environment. The quaternary structure refers to the grouping of individual proteins into protein clusters that carry out a task collectively. The final assembly of any protein or protein group is all programmed by the original amino acid sequence in the linear primary structure. The seamless differentiation of proteins into different components results in one smooth and uniform manufacturing process. Every living body is thus a unique molecular machine in which function at the nanoscale dictates the cellular and whole organ behavior at the macroscale.[24] The various protein structural organizational levels are illustrated in Figure 2.11.

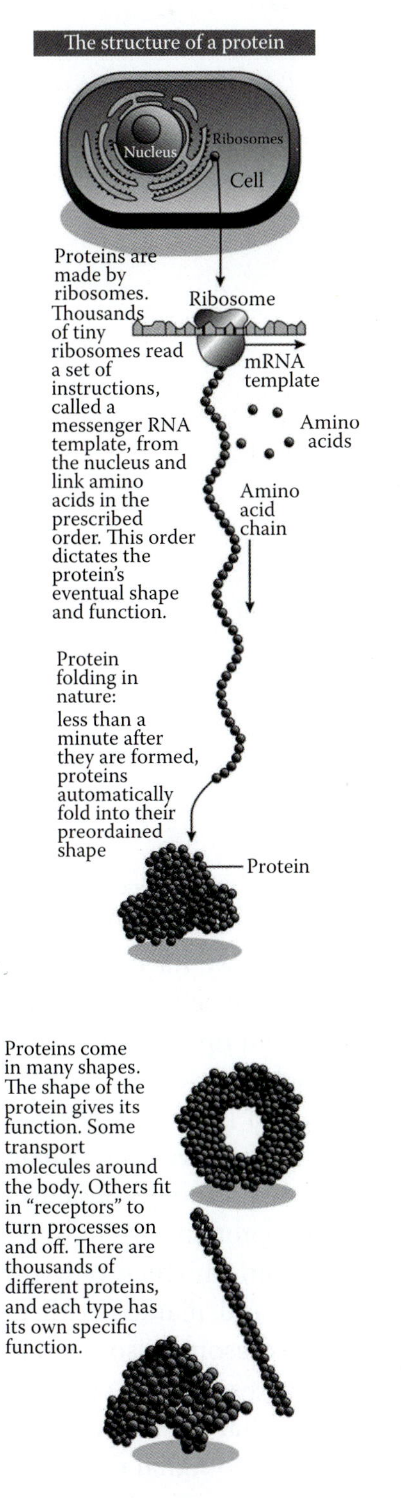

FIGURE 2.10 A protein emerges from a ribosome.

Today's computers can barely simulate a fraction of the fascinatingly complex protein-folding process. With that task in mind, IBM developed a computer called Blue Gene. Blue Gene is 500 times faster than other supercomputers. It was the first petaflop computer (10^{15}), and it is able to simulate interactions of certain proteins with their natural ligands and potential drug candidates and thereby potentially aid in the design of more efficient and potent drugs.

Protein Self-Repair and Recycling

Proteins and derived structures (e.g., tissues such as skin) are of use only in a small operational range of temperature and pH. Despite this narrow dynamic

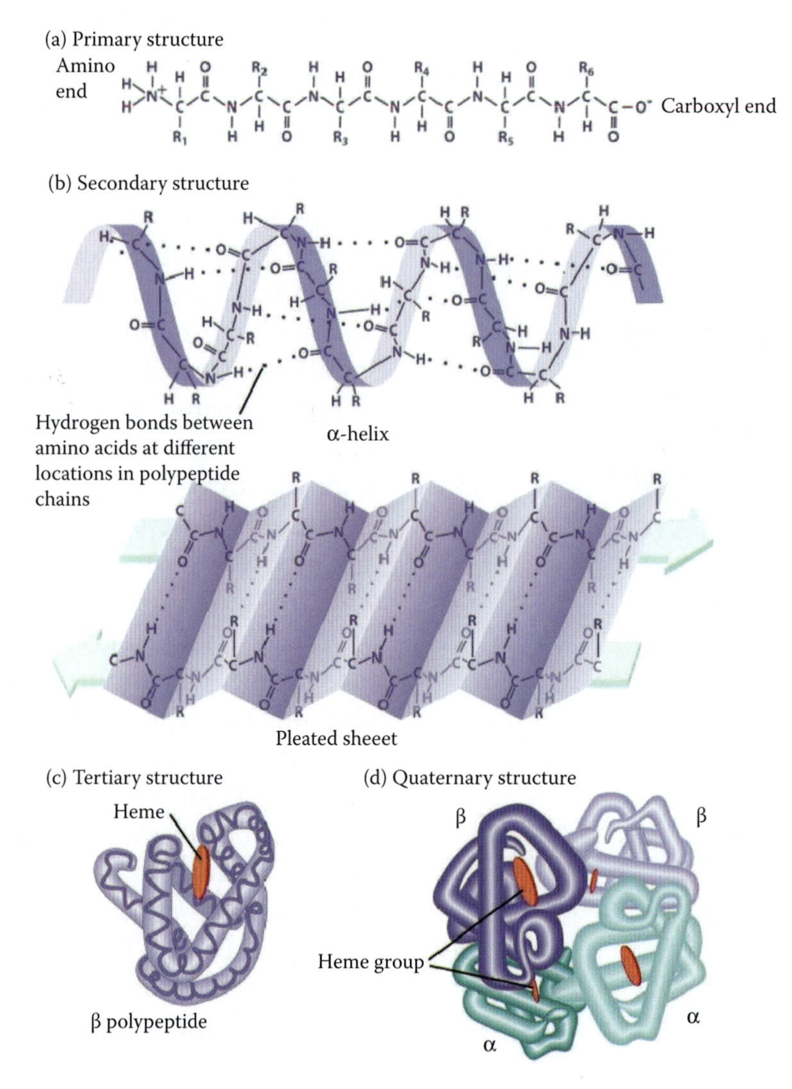

FIGURE 2.11 The various protein structural organizational levels.

range, the synthesis of natural organisms—starting with atoms and molecules, through intermediate structures such as fibers and grains, to components such as skin, bone, and teeth, has led to marvels of enduring engineering such as trees, birds, and humans. Humans usually use stronger (higher Young's modulus) building materials than proteins (say steel or aluminum), but nature has another major advantage, namely, self-repair. The external physical observable form, function, and behavior of living organisms (i.e., the phenotype) are not permanent constructs; they derive from the entire genetic information (i.e., the genotype). Some structures, such as skin and brain, are in a state of constant regeneration; proteins are continually synthesized de novo. Other structures, such as the eyes, are made up by proteins with some of the slowest turnover rates. This type of self-repair compensates for many of the deficiencies of life's construction materials. In human engineering, we have nothing of this kind yet, and together with self-assembly, self-repair is a must for the bottom-up approach in human engineering to succeed.

Cells of nearly all life forms possess systems for recycling proteins that have been damaged or are no longer needed. Ubiquitin tags proteins for degradation in a complex protein system, the proteasome, a molecular abattoir, in which only those proteins that are unfolded and marked with ubiquitin are destroyed. The proteasome also plays an important role in the immune system as it cuts foreign proteins down into chunks that the immune cells then present on their surfaces to signal the start of the attack by antibodies.[25]

The Genetic Code

In this section, we address the question of how nature, using DNA as a template, builds a plethora of different proteins from just twenty different amino acids. From our description of DNA, we learned that there are four different bases and four different nucleotides in DNA. All cells contain the same DNA, but different genes are expressed depending on the local need for an enzyme, a muscle cell, a neuron, or a bone. In this process, DNA is the repository for the construction blueprint, but it is not responsible for building the proteins. This hard work is done by another nucleic acid, namely, RNA. RNA uses a base called uracil (U) in place of thymine (T), and it also employs a slightly different sugar ring (ribose instead of deoxyribose). Apart from these two differences, the components of RNA are identical to those of DNA.

The details of the protein-building process are sketched in Figure 2.12. The process of protein manufacture (e.g., of insulin) may be compared to that of the manufacture of a car (such comparison was first made by Casti).[7] It all starts with copying the car's blueprint at headquarters, corresponding to transcription or copying of DNA in the cell nucleus. Next is a quality control step in which the copy of

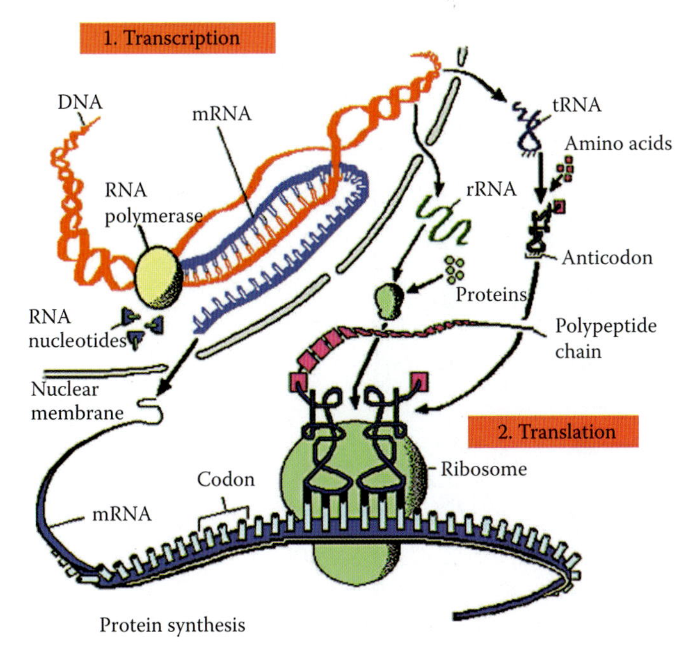

FIGURE 2.12 Protein manufacture: transcription and translation.

the blueprint of the car is inspected, a process that has its analog in the editing, or snipping out, of introns in protein manufacture. This step is followed by mailing of the copies to the various manufacturing plants—the ribosomes in the case of a cell. After a new car, or protein, is finished, it is packaged and shipped, and, finally, if the car or protein is no longer needed, it is disassembled and recycled.

A bit more slowly, and in more detail, the story unfolds as follows. In protein manufacture, an enzyme first unzips a section of DNA, and the gene (DNA segment) that codes for the protein (in this case insulin) is freed for expression. All genes use the ATG codon to serve as the marker to start protein synthesis and TAA to stop it. Expression of a specific gene is switched on and off by proteins physically attaching themselves to promoter and enhancer sequences near the start of the gene text. In other words, the blueprint is unfolded for reading and copying one specific part of the original only (one car or protein is copied). Messenger RNA (mRNA) interprets the freed construction blueprint from the DNA inside the nucleus of the cell in a process called *transcription*. In transcription, an mRNA strand is synthesized by complementary base pairing between the nucleotide bases of the DNA and the mRNA. In Detroit lingo, this translates to the photocopying of the blueprint for a certain type of car. In eukaryotic cells, the mRNA now leaves the nucleus and acts as a template for protein synthesis in ribosomes. In our car analogy, copies of the car's blueprint are being distributed to different factories in Ohio and Indiana. The code to make proteins on the mRNA template is a triplet code—a sequence of three bases, or a *codon*, describes one amino acid.

If each individual base is coded for only one amino acid, then the four bases would only code for four amino acids. A pair of bases could code for sixteen different amino acids (4^2), and a triplet could handle sixty-four (4^3). It is the latter code, the one with the largest bandwidth available to call on the twenty amino acids, that nature has chosen to build proteins (see Appendix 2C for a list of the twenty amino acids). Actually, an amino acid may be encoded by between one and six codons, several different triplets may code for the same amino acid, and three triplets are assigned to the instruction for a stop. In 1967, Har Gobind Khorana and Marshall Nirenberg were able to correlate each one of the twenty amino acids with the sixty-four possible codons. The resulting universal decoder chart is presented in Appendix 2C.

In many cases, stretches of DNA coding for polypeptide chains are not continuous but are interrupted by large stretches of "nonsense" sequences. Even within genes, there are stretches of nucleotides, called *introns*, that don't code for anything; Casti calls them *commercial breaks*.[8] Exons within the gene are the only segments where protein coding occurs (see Figure 2.13). In humans, there is only about one gene per 100,000 base pairs, and the mean gene extends over 27 kilobases, whereas the mean transcript is 1.3 kilobases.

The validity of the mRNA copy is checked by verifying the integrity of all base pairs (also called *editing*). A team of enzymes snips out the introns and reconnects the parts that contain useful information called *exons*, which are the expressed genetic sequence. This editing process reduces the error rate to about one in 1 billion base pair replications. Although most editing is done by enzymes, some RNA molecules can do it unassisted. These are called *ribozymes*, reflecting the

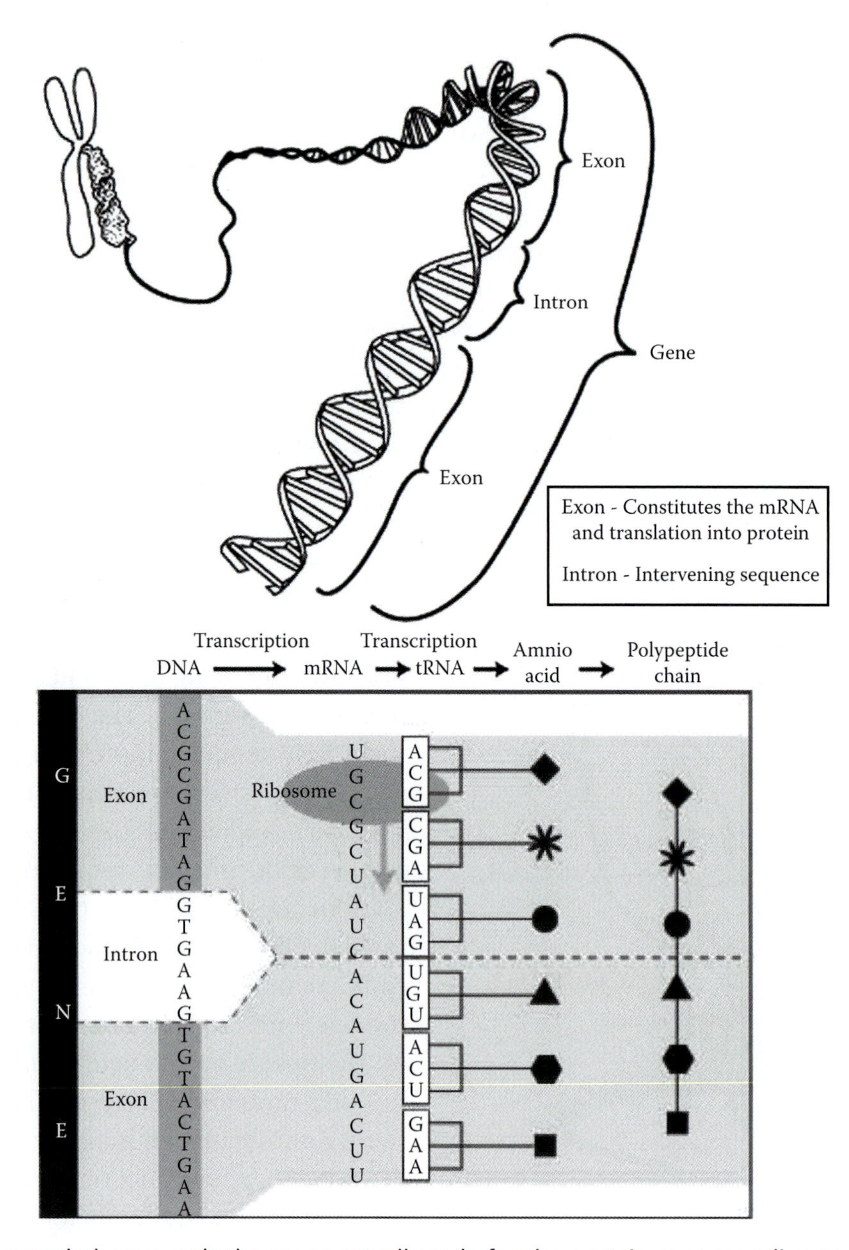

FIGURE 2.13 In the expanded gene, only the exons actually code for the protein corresponding to the gene.

fact that they show -enzyme-like tendencies. The nonsense DNA—perhaps up to 95% of the human genome—has been called *junk* DNA in the past but more recently has been upgraded to *selfish* DNA, as scientists recognized that it might play some role in heredity after all. Crick speculated that this selfish DNA might be involved in the rapid evolution of some of the complex genetic control mechanisms essential for higher organisms (lower organisms have none or much less of this selfish DNA). Crick believed selfish DNA might be there for its own sake rather than that of the organism itself. The same claim is made by Richard Dawkins[26] in his book *The Selfish Gene.**

We continue now our story about the genetic code and car manufacture and study the process that follows transcription and editing—that is, translation of the blueprint into proteins, or how to go from a blueprint copy to a car.

Living organisms feature armies of three-dimensional protein construction sites called *ribosomes* (see also Figure 2.12) in which any conceivable protein can be manufactured. Small RNA molecules called *transport RNA* (tRNA) transport the proper amino acids to the ribosomes, which are cell organelles located in the cytosol outside the nucleus, to build new peptide chains on the mRNA template in the so-called *translation* process. Each of the twenty amino acids has its own tRNAs equipped with a codon, complementary to the triplet of bases that encodes the tRNA's amino acid on the mRNA template. In the car analogy, individual car components are being shipped to the car assembly plant. The ribosomes bind together the amino acids, docked in place by the tRNAs, in a growing protein chain at a rate of five amino acids per second. Many ribosomes can be lined up on one mRNA strand so that a full copy of a new protein—for example, insulin—can be delivered every few seconds. The insulin protein just fabricated in the ribosome contains a short sequence specifying the part of the cell to which it should be sent. This is the equivalent of a bar code on a new car. When this signal sequence (the bar code) has emerged from the ribosome, the cell's transport mechanism is activated, the bar code is read by a signal recognition particle, and in the case of an emerging insulin it is, together with its ribosome, directed toward the membrane of the cell's shipping dock, the endoplasmic reticulum (ER). The ribosome attaches to the ER membrane, and the protein is threaded through a pore in the ER. Once inside the ER, the signal sequence or bar code is removed, the protein—assisted by molecular chaperones—folds into its native shape and is packaged into a small sphere (vesicle) that buds out of the ER membrane. The vesicle then transports the insulin to the Golgi apparatus. The Golgi apparatus, similar in appearance to and perhaps continuous with the ER, is a region of smooth, stacked membranous sacs. Cell biologists have shown that the Golgi apparatus modifies proteins, which it receives from the ER in vesicles, and prepares them, in a series of steps, for excretion to the exterior. Just as a letter is taken to the post office to be mailed to another location, proteins are transported to the Golgi apparatus for shipment to other locations in the body. Within the Golgi apparatus, the proteins and lipids are labeled with sequences of molecules (an address), which tells the body where the product should be delivered.

The French biochemists Francois Jacob and Jacques Monod showed in the 1960s that genes regulate one another, switching each other on and off via the proteins they encode. For example, some genes that encode proteins used in the cell (called *structural genes*) are partnered with regulatory genes that encode repressor genes. With the regulatory gene switched on, the repressor protein is synthesized and binds to the structural gene, preventing it from being *expressed* (transcribed and translated into its protein). Jacob and Monod called these regulated stretches of DNA *operons*.

The fact that information flows from DNA to RNA to protein is called the *central dogma* of molecular biology. This "dogma" also implied that there was only one protein per gene. A first serious attack on the first part of this central dogma came in 1970 when David Baltimore and Howard Temin discovered an enzyme that does just the opposite: it transforms the RNA genome of a group of RNA viruses, now called *retroviruses*, into a DNA copy that can be

* *The Selfish Gene* is an elegant account of neo-Darwinism in which Dawkins argues the view that DNA rules.

inserted into a genome. The family of retroviruses includes HIV. The enzyme in question is a reverse transcriptase. In comparison with DNA polymerase that replicates DNA chains, reverse transcriptase is very error-prone. The latter is what makes it so difficult to make an HIV vaccine. Baltimore and Temin received a Nobel Prize in 1975 for this discovery.

Gene Copying Errors: The Error Catastrophe

A higher organism has about thirty thousand genes capable of storing about one hundred million bits of information, each of which may be subject to copying errors. Too many copying errors can cause the reproduction machine itself to come to a halt. The German biochemist Eigen called this the *error catastrophe*.[27] Eigen, who believed in an RNA-first world, introduced a simple mathematical formulation for characterizing this error catastrophe. He assumed that a self-replicating system is specified by N bits of information and that each time a single bit is replicated from parent to offspring the probability of making and error is given by ε. He also accepted that natural selection penalizes copying errors by a selection factor S. In other words, an error-free replication has a selective advantage of S over a system with one error, and so on. Then according to Eigen the criterion for survival of a replicating organism is:

$$N\varepsilon < \log S \tag{2.6}$$

If the above condition does not hold, the species will die out, but with the inequality satisfied, the selective advantage of the error-free system is great enough to maintain a population alive with few errors. To understand the deeper meaning of Equation 2.6, we may look back to the information theory of life (see above) for some further insights. The left side of the inequality represents the information loss through copying errors in each generation. The right is the number of bits supplied by the selective action of the environment. If the information supplied is less than the information lost in each generation, a progressive degeneration is inevitable. The selective advantage of an error-free system cannot be too large, so that log S is probably not much greater than 1 (life always survives on the edge of chaos). The greater the number of genes an organism possesses, the lower the error rate must be to avoid error catastrophe. Human beings accumulate about 100 mutations per generation, with N of the order of 10^8, to satisfy Equation 2.6, ε must be of the order of 10^{-8}. In other words to avoid error catastrophe, the copying error in human DNA should be less than one in a hundred million per replication. In human cells, an editing process weeds out errors, and the remaining error rate is actually one in a billion. Bacteria with considerably fewer genes than humans can get away with many more errors (they can tolerate up to one in a million), and a virus can survive an even greater number of errors. The latter is the reason why drug-resistant mutations emerge with such rapidity, always keeping researchers one drug behind. This will continue to challenge chemists and the pharmaceutical industry, keeping the latter happy and rich for a long time to come.

Genes and Chromosomes

A gene, as we learned above, is a segment of DNA with a unique sequence of nucleotides that encodes information—the genetic code—for assembling amino acids into a particular protein. Genes are located on chromosomes and humans have twenty-three chromosome pairs (Figure 2.14). Genes are isolated from chromosomes through purification of RNA molecules. Since RNA is not very stable, this purified material is first converted back into DNA known as *complementary DNA* (cDNA). To store and make copies of genes, the cDNA is stitched into a small circular bacterial chromosome. Whenever more of the gene is needed, the bacterial colony containing the gene of interest is grown, and the required amount of cDNA is isolated.

The biggest surprise of the rough analysis of the first sequencing of the human genome (see below) was the low number of genes it contains. For years scientists had been predicting that human DNA would contain somewhere between 100,000 and 140,000 genes. It turns out we may have as few as 26,000—a genome about the size of a rat or a corn plant, with roughly a third more genes than the fruit fly. We are coming to the somewhat uncomfortable conclusion that there is no relation between the complexity of an organism and the number of chromosomes or genes or the length of the DNA strand. *Lysandra atlantica* (a butterfly) has 250 chromosomes

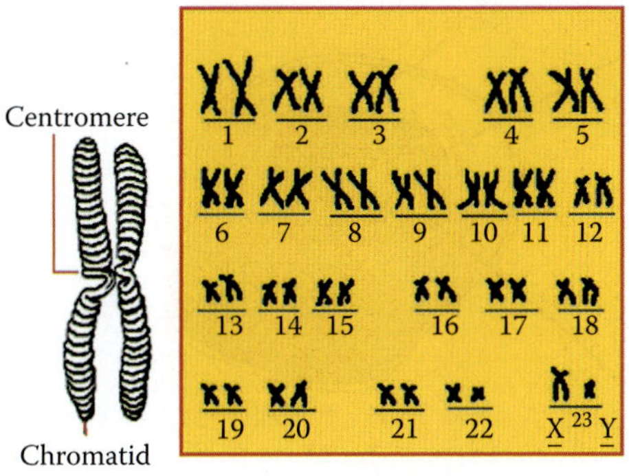

Organism	Genome Size (Bases)	Estimated Genes
Human (*Homo sapiens*)	3 billion	30,000
Laboratory mouse (*M. musculus*)	2.6 billion	30,000
Mustard weed (*A. thaliana*)	100 million	25,000
Roundworm (*C. elegans*)	97 million	19,000
Fruit fly (*D. melanogaster*)	137 million	13,000
Yeast (*S. cerevisiae*)	12.1 million	6,000
Bacterium (*E. coli*)	4.6 million	3,200
Human immunodeficiency virus (HIV)	9,700	9

FIGURE 2.14 Human chromosomes and genome size (bases) and estimated number of genes for some different organisms.

versus our meager 46. Nor is there a correlation with the number of genes; the fruit fly has 13,700, whereas *Caenorhabditis elegans* (a 1-mm nematode worm) has 20,500. Despite the fact that the eye of the fruit fly is composed of more cells than are found in the entire *Caenorhabditis* worm (profoundly more complex in structure and behavior than the microscopic roundworm), the fruit fly has 5000 fewer genes! Finally, there is also no correlation with the length of the DNA, with, for example, the *Amoeba dubia* at 670,000,000,000 bases and *Homo sapiens* at 3,000,000,000. Obviously, the complexity of organisms is not reflected in the complexity of its genes. Some people are referring to this dilemma, the lack of correlation between the genome and the complexity of the organism, as the cosmic joke. The cosmic joke violates a second tenet of the central dogma, i.e., one gene to one protein. Ron Evans, Michael Rosenfeld and others clearly demonstrated that the same RNA gene transcript can be spliced in many ways to encode multiple proteins.[28] This discovery shed much-needed light into the low number of genes–large number of proteins paradox.

Only 40% of a chromosome consists of DNA. The rest is made up of packaging protein materials. Figure 2.15 illustrates an exploded view of a chromosome as well as some other important cell components. The total human genome, that is, the full set of chromosomes, comprises about 3 billion DNA base pairs. One strand of human DNA stretched out would measure up to 5 cm. Since there are forty-six chromosomes in a human cell, this means we have about 230 cm, or 6 feet, of DNA for each cell. Within the chromosome, the DNA is obviously very well coiled and packed. This remarkable feat of packaging is described in more detail in Chapter 4 on packaging (see Figure 4.1). In this packed state, the DNA is so tightly bundled that it is entirely inaccessible to gene-activating enzymes. Only as stretches of chromosome move out of this condensed state are genes exposed and available for expression. Inside the nucleus of each cell there are two complete sets of the human genome. Egg and sperm cells contain only one copy, and red blood cells do not have any DNA.

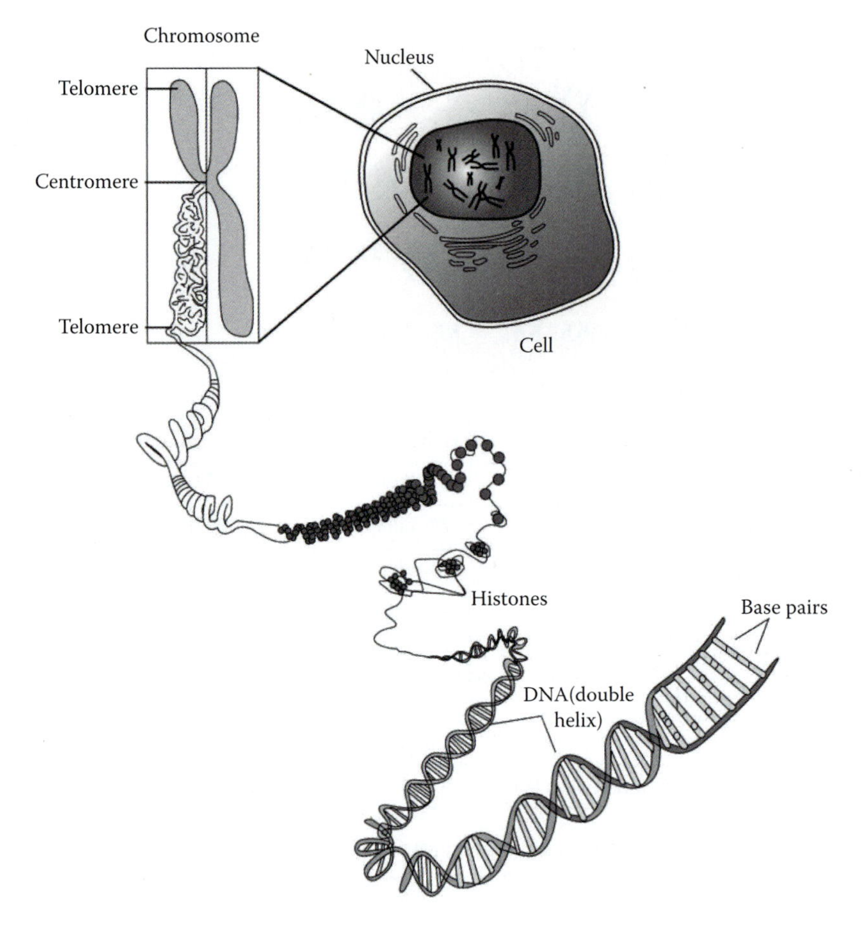

FIGURE 2.15 From chromosome in the nucleus to DNA.

Evolution

The driving force for the sequence of events that led to the plethora of life forms we know today comes from the ever-changing challenges presented by the environment to living organisms in a process called *evolution* or *natural selection*. Evolution theory tells us that over long periods of time, life forms change. Perfect replication of DNA would only have led to a planet with very little diversity among its life forms. Errors during DNA replication do occur, and when errors occur, they can manifest themselves as mutations in the organisms that inherit them. Mutations not only arise from copying errors but may also be caused by physical damage to the DNA in the genes, such as that caused by radiation and certain chemicals. If an alteration of a gene results in an observable change in some aspect of an individual's phenotype, and if this change is passed on to his or her offspring, then we say that a mutation has occurred or a new allele of a gene was created. Only mutations in germ cells are passed on to the offspring: mutations in somatic cells (all the cells of an organism except for the germ cells) do of course occur, but they are not passed on. An allele is defined as one of the two alternative states of a gene that occupy the same locus on homologous chromosomes (chromosomes containing the same gene sequence) and that are responsible for alternative traits. Not all genes have alleles; many genes within a given species are invariant among individuals. If there is more than one allele of a given gene within a population, an individual might inherit two different alleles for a given gene (one from the father and one from the mother), in which case that individual is said to be heterozygous for that gene. Some alleles are dominant over others, and with a heterozygote the dominant allele will be the one that is expressed in the phenotype. With two identical alleles, the individual is called monozygotic. Random errors in the DNA coding are mostly disadvantageous, sometimes neutral, and very rarely beneficial, conferring an advantage on the mutant. Over long periods of time, favorable traits will be transmitted more efficiently to the offspring, and unfavorable ones will occur with decreasing frequency, until they are eliminated. For any two unrelated individuals, there is roughly one alteration every 1000 bases, or 3 million sequence variations altogether. These sites of sequence variations are called single-nucleotide polymorphisms (SNPs, pronounced *snips*). Those "snips" are critically important in the development of new molecular diagnostics (see applications in Chapter 10).

The concepts of evolution and inheritance were introduced in the second half of the nineteenth century with Charles Darwin's *On the Origin of Species* (1859). Gregor Mendel published in *Transactions of the Brunn Natural History Society* on heredity of peas in 1865, but his work was not widely known until 1900, when it was rediscovered. Darwin had predicted a continuous variation within a species with gradual change over long periods of time. The perpetuation of any life form requires a fine balance between fidelity of replication and variation from generation to generation. A slight imbalance, and extermination certainly follows. Among Mendel's pea plants, a more discrete variation manifested itself, with the plants exhibiting either white or purple flowers with no continuous variation in between. In the early twentieth century, evidence pointed toward gene mutations having a big step effect rather than a continuous effect. Only later did it become apparent that although the units of heredity—genes—are discrete, many mutations do have small-scale effects, and that combinations of genes can produce continuous variations, such as height in humans. The mathematical foundation of Darwin's theory was on solid footing by the 1930s. Mendel's very discrete heredity results were found to be coincidental, because the traits he studied were controlled by single genes only. If this had not been the case, he would not have obtained such clear-cut results.

There is some controversy as to whether natural selection explains all features of evolution. One extreme position is that of Dawkins, who, in his books *The Selfish Gene*[26] and *The Blind Watchmaker,*[29] contends that everything is shaped by natural selection alone. The opposite extreme, much more controversial, is represented by, for example, Micheal J. Behe with his intelligent design who claims that Darwin's mechanism—natural selection working on variations—can explain small changes but does not explain how complex molecular systems, such as vision, might have been gradually produced (for an introduction visit http://www.talkorigins.org/faqs/behe/review.html). Intelligent design is a criticism of Darwinism, not a theory, as it does not logically follow from experimental evidence. In between, there are scientists such as the late Stephen Jay Gould, who believed that there must be other factors involved and that selection alone is not capable of explaining all patterning forces in the history of life (http://www.stephenjaygould.org/library/gould_structure.html). This view supports complexity theory, which sees evolution as a complex system, not to be studied by its individual parts alone. This implies that self-organizational properties typical of complex systems would go hand in hand with selection. Artificial life simulations also have the potential to solve some of the major problems that evolutionary biologists have struggled with for decades. The first one has to do with the concept of predeterminism in evolution. Evolution is not supposed to have purpose or a predetermined goal in mind—mutations are random, which makes it difficult to explain how life organized itself into increasingly complicated organisms. However, computer simulations of artificial life do suggest a tendency for systems to organize themselves into increasingly complex forms for purely mathematical reasons. A second problem is that of mass extinctions, which computer simulations suggest are the norm, not the exception.[6]

Transposomes

It is estimated that less than 5% of the DNA in a human contains gene sequences that code for proteins. Noncoding DNA is assumed to consist of structural elements and repetitive sequences such as transposons. Transposons, also called *jumping genes,* are stretches of DNA that jump from site to site within one genome or "catch a ride" with a virus and move from species to species in so-called horizontal transmission.

Barbara McClintock won the 1983 Nobel Prize for her work on transposons in Indian corn; she postulated transposons as far back as the 1940s. It is now widely accepted that genes operate as parts of networks almost like those formed by neurons in the brain, and individual genes may either inhibit the action of another gene or cause it to be more likely to express itself (translate into proteins). Because of this additional complexity, it seems increasingly likely that a traditional reductionist approach will not be able to solve all of the issues involved in the expression of the 30,000 or so human genes. Answers may come from the science of complex systems instead, and, as a result, nonlinearity and complexity mathematics may become the most important contributors to the understanding of life and evolution. Along this line of thinking, Roy Britten at Caltech and many others believe that evolution cannot be explained by single-point mutations only.[30] Britten's position is that transposons are a much larger source of natural variation. To make his argument, he has focused on the Alu transposons that are unique to primates. Alus make up more than 5% of our DNA and could conceivably be responsible for the diversity of primates. Alus, for some unknown reason, seem to have started spreading widely around 30 to 50 million years ago. Wanda Reynolds discovered that Alus have the power to alter which sets of genes are expressed, by which hormone, and when they are triggered, thereby elevating or reducing gene expression in certain tissue so that the evolution of a species can gradually be changed (see also http://www.ncbi.nlm.nih.gov/pmc/articles/PMC38434/). This process might have been an important factor in creating diversity in primates. Despite the fact that the genetic distance separating us from pygmy chimps is only 1.6%, it is possible that not all our differences are the result of this (the chimp-human split occurred not much more than 10 million years ago, possibly even less than 5 million). The proximity of Alus might have regulated which sets of genes were triggered, enabling *Homo sapiens* to appear around 40,000 years ago. Transposons can be regarded as an internal environmental challenge to the evolution of life forms rather than the more familiar external environmental factors. Transposons, by posing a continuing challenge to increasingly complex genomes, have helped evolution along and might even be the reason why more complicated organisms such as humans could emerge at all. The reason for the dense packing of chromosomes in more complex living systems is probably to keep transposons better hidden within them. As a consequence, only 0.2% of all spontaneous human mutations are caused by transposons, although they make up a large portion of our genome. By contrast, fruit fly transposons constitute only 10–20% of the genome, yet they are responsible for as much as 85% of spontaneous fruit fly mutations. The chromosomes of the fruit fly hide the transposons much less than our chromosomes do.

In 1998, Schatz and coworkers at Yale University identified yet another major role that transposons might have played in evolution, namely, the development of a powerful immune system in vertebrates.[31] A powerful immune system is created by mixing gene fragments to create a vast array of antibody genes in millions of B cells.* Schatz's group has demonstrated that the mechanism for the creation of these antibody cocktails involves the RAG1 and RAG2 genes, which produce the proteins used to "cut and paste" the gene fragments into new combinations. RAGs have lost their ability to reinsert themselves at random in our genome, and as a result, they are not only harmless but actually beneficial. Transposons also might be dangerous, as their presence near or on a functional gene may damage or destroy it; they have been associated with diseases such as leukemia, hemophilia, and breast cancer.

Human Genome Project

The aim of the $3 billion Human Genome Project, started in the mid-1980s, was to decode the details of the software running the genetic code machinery depicted in Figure 2.12 by 2005.[32] Scientists working on the project located, mapped, and identified the ~30,000 individual genes carried on our twenty-three pairs of chromosomes by 2001. If the

* B cells develop from stem cells in the bone marrow. Each B cell is programmed to make one specific antibody. For example, one B cell will make an antibody to block a virus that causes the common cold, while another produces an antibody to attack a bacterium that causes pneumonia.

human genome is compared to a book, the twenty-three chromosomes may be considered the chapters, with each chapter containing many stories, or genes. Each story contains many paragraphs, called *exons*, interrupted by advertisements called *introns*. The words are *codons*, and each word is written in letters, or *bases*.[12] One of the project's approaches involved cloning fragments of DNA, inserting them into *E. coli* bacteria, and looking for shared sequences of DNA that could be identified chemically. The Human Genome Project stressed free access, including via the Internet, to all the information resulting from this multinational project (http://genomics.energy.gov). On the other hand, Celera, a competing private company, had already applied for patents on 6500 different gene sequences. In elucidating the genome, Celera used an approach in which the genetic material was shredded into tiny pieces and, after analysis, a supercomputer tried to put the pieces back together. In September 1999, Celera succeeded in mapping the genome of the fruit fly (*Drosophila melanogaster*), which contains "only" 180 million base pairs. The Human Genome Project announced the elucidation of the map of human chromosome 22, the second smallest, on December 2, 1999.[33] Although this chromosome holds only 1.6–1.8% of the human genetic code, it is believed to contain genes that play a role in at least thirty-five diseases, including congenital heart disease, mental retardation, leukemia, and schizophrenia. By June 26, 2000, Bill Clinton, president of the United States and Tony Blair, prime minister of the United Kingdom, simultaneously announced that the rough draft of the Human Genome was complete. We can now only hope that humankind has it in its genes to exploit the results for the benefit of humanity rather than for making a very few very rich individuals yet richer at the cost of their own country and especially to the detriment of third world nations.

Identification of the DNA sequence is only the beginning; the real payoff is still to come; molecular biologists worldwide that wrote down the 3.5-billion-bit genetic code are now trying to link specific genes and the proteins for which they code. It is becoming apparent that the expression of genes and the folding of proteins might better be solved by complexity theory than with a reductionist approach because of the extreme interconnectedness of the subcomponents.

Death and Aging

With the aging of the population in the industrialized countries, research funding for the study of death and aging is rapidly increasing. Some things about aging are known, but more needs to be learned, and plenty of controversy surrounds the topic. Here are some generally accepted facts. The simpler the life form, the easier it is to prolong its life or to regenerate parts of itself. Bacteria, for instance, continue to grow and produce progeny as long as they have food and a nontoxic environment. They do not die because of aging. Aging becomes increasingly important as a species acquires more sophisticated, specialized cells and organs. In these more complex life forms, there are two types of cells, namely germ cells and somatic cells. Germ cells are potentially immortal (like bacteria, they can reproduce the whole living entity forever); somatic cells are there for the greater efficiency of particular functions but cannot reproduce the whole organism. If a worm is cut in half, the tail, which contains germ cells, can reproduce a whole worm; the head, which only contains somatic cells, dies. The higher the ratio of somatic cells to germ cells, the more prone an organism is to aging. Mutating the genes that control how worms respond to hormones resembling insulin enable worms to live two to five times longer.[34,35]

Chromosomes are capped and protected by special DNA pieces called *telomeres*, in the same way that shoelaces are prevented from becoming frayed by aglets. These telomeres in mammals consist of thousands of TTAGGG repeats and act somewhat as an internal clock for the aging of cells: each time a cell divides, the telomeres on the chromosomes get shorter, and when they are short enough, they signal the cell to stop dividing. By adding telomerase, an enzyme that rebuilds telomeres, cells cultured in a dish become immortal (the patent rights for telomerase are owned by Geron in California). Cancer cells keep on dividing in part because they reactivate their telomerase! It is not yet clear how telomeres relate to death and aging in humans, since so many other factors play a role. It is reasonable to expect, though, that some cells and tissue—especially those that divide quickly—become crippled by shortened telomeres. Short telomeres in old people might weaken the immune system and the capability of regenerating bone and skin, all of which contain rapidly dividing cells. The more times DNA is replicated, the more likely it is that errors are introduced. This link between age and mutation frequency is clear from parental-age-dependent genetic disorders such as Down's syndrome. Dolly, the cloned sheep, died in 2003, at age seven, very young for a sheep (see Figure 2.18). It has been speculated that Dolly started off life with a telomeric "clock" set already at her mother's age. This has put somewhat of a damper on cloning enthusiasm.

Many factors are claimed to influence average lifetime. Free oxygen radicals decrease the average life span (Linus Pauling's vitamin C is an antioxidant, and he claimed that it increased lifetime!).[36] Mutant forms of the enzyme catalase decrease the average life span; food restriction and less sexual activity (or more sex later in life—figure that one out!) will make you live longer. Cooler temperatures and an efficient glucose metabolism also prolong life (goodbye Florida). Although we can expect to live longer and healthier in the new millennium, Hayflick discovered in 1961 that normal human cells, when grown in a culture dish, divide a limited number of times (mean value is 61) and then die. This ultimate ceiling for life has been dubbed the Hayflick limit. Hayflick believes that our maximum life span is fixed by our genes at about 125 years.[37–40] Evolutionarily, there is no reason for human life to go on after the reproductive years are over (just look at the Rolling Stones).

Cloning

Introduction

The last decade of the twentieth century brought one of the most important milestones in molecular biology: the creation of the cloned sheep Dolly in 1996,[41] an event that made unprecedented headlines, both in scientific journals and in the popular press. The result was that suddenly the entire world was introduced to cloning. Cloning has become a household term, but most people are still unacquainted with the scientific processes involved in cloning and how cloning is actually executed. In this section, the cloning process and its implications from a molecular biology point of view is introduced. As shown below, there are three different types of cloning: 1) DNA cloning or recombinant DNA technology, 2) reproductive cloning, and 3) therapeutic cloning.

DNA Cloning, or Recombinant DNA Technology

First, we should mention that the terms cloning, gene cloning, recombinant DNA technology, and molecular cloning are used interchangeably to denote the same genetic engineering method. The term *cloning* denotes the process by which genetic material from one species is transferred to a self-replicating genetic element, such as a bacterial plasmid or a virus. A plasmid is an extrachromosomal circular piece of DNA distinct from the normal bacterial genome that contains genes and that is capable of producing multiples copies of itself when introduced into a host organism (i.e., bacteria, yeast, mammalian, or plant cells). Thus, plasmids replicate desired genes; in nature, these genes are from the plasmid's same organism (bacteria, yeast, etc.), but in recombinant DNA technology, scientists introduce foreign genes and use the plasmid to copy said genes. For example, in genetically modified foods, a gene encoding for a natural pesticide can be incorporated into a corn cell, giving rise to corn that now can produce its own pesticide.

Of all the terms used to denote cloning, *recombinant DNA technology* is perhaps the one that best defines what the procedure entails. As explained, the method involves taking a gene that encodes for a desired protein and introducing it within a bacterial plasmid, thereby recombining this foreign DNA with that of the plasmid. More specifically, the gene to be cloned first is isolated or cut out from chromosomal DNA by employing restriction enzymes, a type of enzymes that cleave DNA when they encounter a specific sequence of DNA bases that they recognize, leaving a distinct sequence of bases at both ends. The resulting DNA fragment containing the gene is then isolated and purified from the mixture of chromosomal DNA, so it is ready to be inserted into the plasmid. Likewise, the plasmid is also cut open, or cleaved, with the same restriction enzyme, leaving the same sequence of bases at each end of the open plasmid as those present in the isolated gene. In that manner, when the gene fragment and the open plasmid are mixed together in the presence of another enzyme—ligase, whose function is to bind or "glue" together DNA sequences—they recombine like two pieces in a puzzle. After the plasmid is constructed, it is ready to be inserted into a host—most commonly a bacterium—where it will take advantage of the host's cell machinery to replicate in large amounts. The plasmid then expresses the gene that was inserted, thus producing the protein encoded by the desired gene. Figure 2.16 depicts a generic cloning procedure performed in bacteria.

Reproductive Cloning

The process of reproductive cloning is different from DNA cloning as described above. Its goal is to create an animal from the nuclear DNA of an existing one. The principle of this method involves the removal of the nucleus of an egg containing all the DNA of the cell and replacing it with that from another cell. In the laboratory, the newly created egg is then treated with chemicals or electrical stimulation so it can divide. Once the embryo reaches a certain cellular stage, it is inserted into the uterus of a female host, or "surrogate mother," where it grows and develops until birth (see Figure 2.17). It is important to note that cloned animals are not

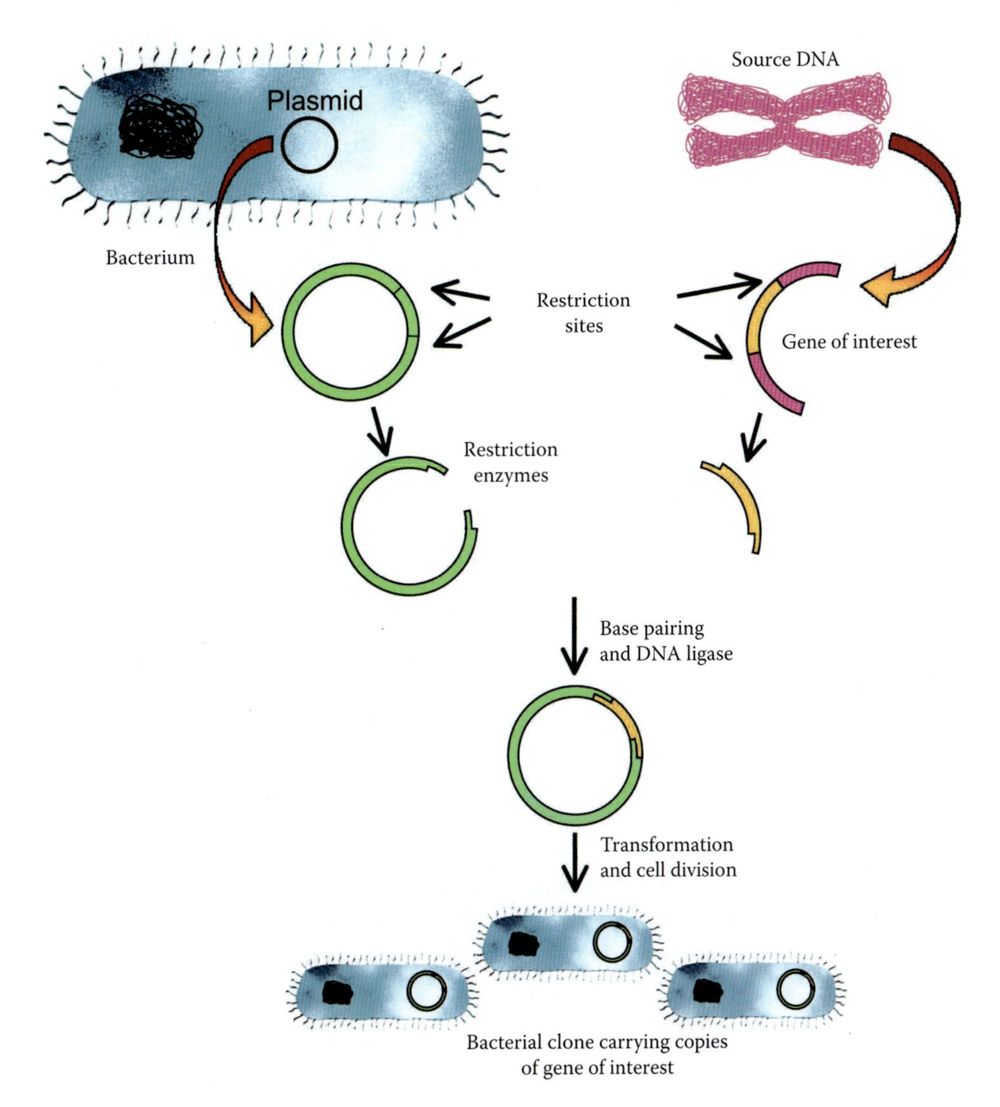

FIGURE 2.16 A generic cloning procedure performed in bacteria.

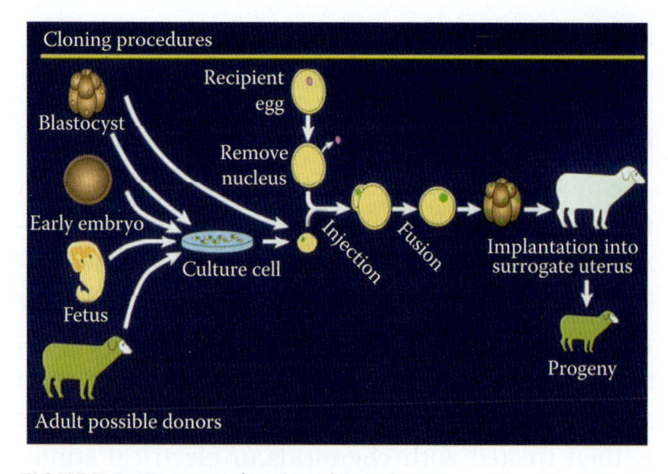

FIGURE 2.17 Reproductive cloning.

FIGURE 2.18 Dolly with her firstborn lamb, Bonnie. All of Dolly's offspring were bred the old-fashioned way. (Roslin Institute Image Library, http://www.roslin.ac.uk.)

exact replicas of their progenitors because the clones, in addition to the nuclear DNA from the progenitor, have mitochondrial DNA that is derived from the original egg. Mitochondrial DNA plays a role in the aging process, an important consideration when choosing the egg donor.

Reproductive cloning is not as effective as DNA cloning, where virtually any gene can be cloned into a suitable host, with the exception of a few incompatibilities. Reproductive cloning methods, although continuously improving, still are inefficient, with more than 90% of the embryos failing to produce viable offspring. To date, not enough data are available to allow one to predict the long-term effects of cloning, partly because a large number of the cloned animals produced are dying at a premature age. The first example of reproductive cloning, performed in 1952, resulted in a tadpole.[42]

Although other examples of cloned amphibians and other animals exist,[43] the most notable example of this type of cloning involves the creation of Dolly the sheep (named after the country singer Dolly Parton). Dolly (Figure 2.18) was cloned by a team of Scottish scientists led by Dr. Ian Wilmut,[41] currently at the Queen's Medical Research Institute at the University of Edinburgh. Dolly's cloning was exemplary in that she was the first large mammal to be cloned, thus bringing to reality the possibility of cloning other large animals, including humans. Unfortunately, Dolly's life was not easy; she suffered from crippling arthritis and lung cancer in addition to having to deal with paparazzi following her at all times—she was a Dolly Parton kind of celebrity! As a result, Dolly had to be euthanized when she was only six years old, a rather short life span compared to the eleven to twelve years of life of regular sheep.

Therapeutic Cloning, or Biomedical Cloning

The goal of this type of cloning is to produce an artificial organ or tissue that can be transplanted back into the individual who supplied the original DNA. The procedure, in its first stages, is almost identical to that of reproductive cloning, and it involves first removing the nucleus of an egg and then replacing the genetic material by inserting the nuclear DNA of the individual for whom a new organ or tissue is needed. The egg is allowed to develop until it has produced a large number of stem cells. After that, the stem cells are harvested from the embryo. This procedure results in the destruction of the embryo. The stem cells are allowed to reproduce in vitro and are chemically treated so as to develop in different types of stem cells that will give rise to the desired organ or tissue type. Once the artificial tissue or organ has been produced, it can be implanted into the patient (see Figure 2.19). The fact that the embryo is being destroyed during the cloning process brings up a number of ethical issues (see discussion in the section below). Nevertheless, the potential benefits of therapeutic cloning are tremendous, and researchers believe that these cells will be invaluable in the treatment of all sorts of diseases, among others, those of neurological origin

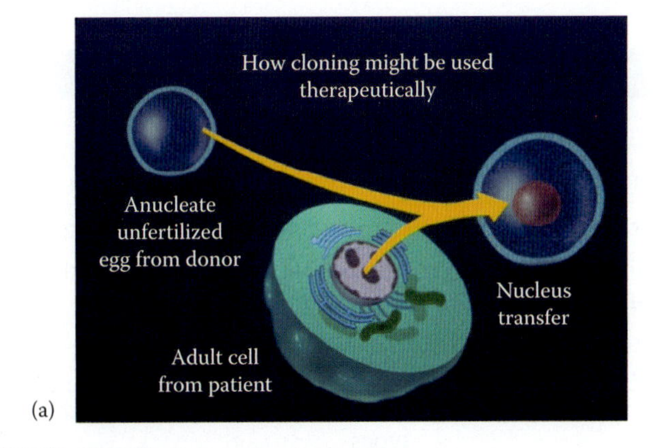

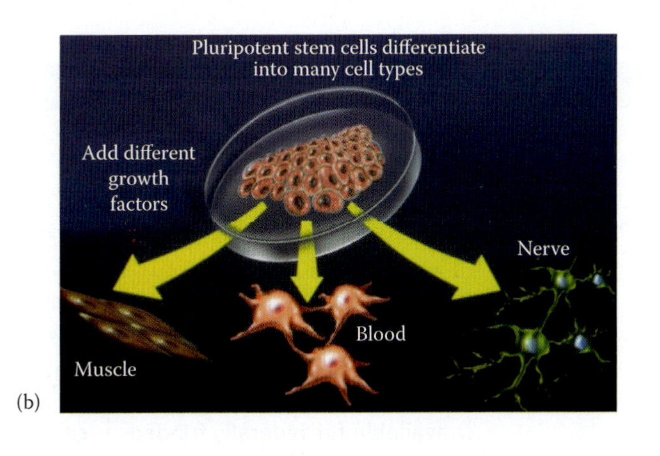

FIGURE 2.19 Therapeutic cloning.

(e.g., Parkinson's, Alzheimer's), cardiac diseases, and autoimmune diseases.

Stem Cells

The human body has more than 200 different types of cells, all derived from the same original stem cells. These stem cells are capable of reproducing themselves and transforming into different kinds, depending on what type of signal they receive while growing. This results in the production of a large number of diverse specialized cells responsible for organ and tissue formation in animals. Much research has been devoted to understanding the signaling events and pathways that lead to the transformation of a pool of identical unspecialized stem cells into this distinct very specialized cells originators of organs and tissues. Embryonic stem cells can develop into all different types of cells, whereas adult stem cells (present in fully developed adult individuals) are limited to the production of only certain types of specialized cells. In addition to embryonic and adult stem cells, rather recently, stem cells have been identified in placenta and in umbilical cord blood. The latter two types can develop into various types of blood cells. The characteristics and differences between embryonic and adult stem cells are shown in Table 2.4, below.

There is much to be learned about stem cells, but undoubtedly, the knowledge gained thus far already shows the vast number of applications and benefits that their manipulation and production will have in medicine, agriculture, and in the environment. A clear example of the potential of embryonic stem cells in medicine involves the preparation of active functioning nerve cells in mouse brains. In order to realize all the benefits of stem cells, protocols and methods will need to constantly be improved and simplified. In that regard, in 2006 researchers demonstrated the feasibility of producing embryonic stem cells from the morula (i.e., solid mass of cells that develops about six days after fertilization of an egg cell) of a mouse, thus showing that stem cells can be made without animal products in a culture. In November 2007, two separate teams of researchers were able to reprogram skin cells and turn them into stem cells.[44,45] The teams, led by James Thomson of the University of Wisconsin-Madison and by Shinya Yamanaka of Kyoto University in Japan, first identified four genes that confer the ability to reprogram certain types of cells. They then introduced these genes into skin cells from different sources and individuals by employing a retrovirus to essentially infect the skin cells and carry these genes into the nucleus of the cells. Once the new genes were part of the skin cells, they reprogrammed them, converting the skin cells into pluripotent stem cells. Because the original cells came from different types of tissue (i.e., facial skin, connective tissue, fetal skin, and baby foreskin) and from a variety of individuals (i.e., a 36-year-old woman, a 69-year-old man, a fetus, and a baby), the researchers are confident that the method will be able to be generalized and employable in most individuals. While the efficiency of the method was not the best—the Japanese team reported that 1 in 5,000 cells treated turned into a stem cell, and the US team, 1 in 10,000—both methods still result

Table 2.4 Stem Cell Overview

Characteristics of Stem Cells			
	Embryonic Stem Cells		
	In Vitro Fertilization	Nuclear Transfer	Adult Stem Cells: Adult Tissues
Properties	Produce all cell types Easy to manipulate and grow in the laboratory Can obtain embryos from in vitro fertilization clinics	Produce all cell types Easy to manipulate and grow in the laboratory Stem cells are genetically matched to patient	Successful in certain treatments Stem cells are genetically matched to patient
Limitations	Limited number of cells available for federally funded research Risk of creating tumors from transplanting undifferentiated cells	Not achieved with human cells yet Risk of creating tumors from transplanting undifferentiated cells	Results in limited numbers of cell types Not found in all tissues or organs Difficult to manipulate and grow in the laboratory
Ethical issues	Destruction of human embryos Require informed consent from embryo donor	Destruction of human embryos Require informed consent from embryo donor Possible misapplication in reproductive cloning	No ethical issues have been raised

Source: Adapted from the National Academies of Sciences, http://dels.nas.edu/dels/rpt_briefs/Understanding_Stem_Cells.pdf.

in a good number of cell lines that are viable. The implications of this discovery are vast, as this is the first real demonstration of conversion of cells from mature adult individuals into stem cells. The major downfall of these cell-reprogramming experiments is that they use a retrovirus to carry the reprogramming genes into the skin cells to be transformed. Retroviruses are known to cause tumors in human cells, and thus, the risk of inducing tumors cannot be trivialized. However, at this moment the possible positive outcomes of this new breakthrough technology outweigh the negative. Perhaps the most interesting application stemming from the work of these groups is the possibility of regeneration of organs and tissues of a patient from his/her own cells. Moreover, since the methods developed bypass the use of embryos or oocytes, the ethical dilemmas and controversy associated with their use will be avoided.

Finally, it is important to note that stem cells are also useful in studying human development, for testing new drugs and screening of toxins, and for testing gene therapy methods.

Reproductive Cloning and Stem Cell Research: Implications in Bioethics

Dolly's creation and life led to controversy, and as a consequence, reports regarding the ethics of cloning were drafted in the European Union (EU), the United States, and a number of other countries, including Canada and Australia. Specifically, in 1997, after Dolly's creation, President Clinton responded quickly to the prospect of human cloning, banning the use of federal funds for human cloning in the United States. Having considered the scientific and moral implications of said research, he also asked privately funded researchers to implement a temporary moratorium on human cloning research until the Bioethics Advisory Committee set up guidelines regarding this matter. Moreover, in 2001, then President George W. Bush created The President's Council on Bioethics (http://www.bioethics.gov/topics/cloning_index.html), whose mission is to advise the President on bioethical issues that may emerge because of advances in biomedical science and technology. Currently, there are no federal laws in the United States that prohibit human reproductive cloning. To date, federal regulations on most human biotechnologies are not adequate, and more comprehensive regulatory policies are needed. Regarding stem cell research, in 2001, President Bush imposed a limitation on federal funding earmarked to embryonic stem cell research. This policy decision has been a source of controversy and debate, with many members of Bush's own Republican Party regretting this limitation and siding with the more progressive Democratic members of Congress. Also noteworthy are some of the most outspoken supporters of stem cell research, Republican public figures Nancy Reagan, former First Lady, and Arnold Schwarzenegger, the California governor. In that regard, then Governor Schwarzenegger led the country by proposing a bill, Proposition 71, in his state to fund stem cell research. Californians voted to pass the bill in 2004, which authorized up to $3 billion over 10 years for embryonic stem cell research in California. With this funding, the California Institute for Regenerative Medicine was created to foster stem cell research and appropriate the funds provided by the state of California. Other states have followed suit with the creation of stem cell research programs, albeit not as ambitious or comprehensive as the California one.

In 2005, the National Academies of Sciences of the United States published the Guidelines for Human Embryonic Stem Cell Research, which provide a common set of scientific and ethical guidelines aimed at aiding scientists from both the public and private sectors to establish new protocols in stem cell research. These guidelines state the need for the monitoring of all human embryonic stem cell research and outline specific methods for the derivation of new stem cell lines; for example, experimenting on human embryos by inserting stem cells into them is not allowed. Furthermore, the rapid advances in stem cell research prompted the National Academies to create a panel composed of experts in the field to monitor and review scientific developments; to change ethical, legal, and policy issues; and to prepare periodic reports to update the guidelines. More information on this can be found at http://www.nationalacademies.org/stemcells.

Within the EU, the European Convention on Human Rights and Biomedicine has prohibited human cloning in one of its protocols; however, only a few countries—namely Greece, Spain, and Portugal—have ratified this protocol. Furthermore, the ambiguity continues in the EU with regard to any kind of legal standing on reproductive human cloning despite the fact that the Charter of Fundamental Rights of the European Union has prohibited its practice. The United Nations decided to introduce debate on this issue among the country members of its General Assembly; as a result, in March 2005, the United Nations Declaration on Human Cloning was adopted.

For information on National Polices Governing New Technologies of Human Genetic Modification in a large number of different nations, see http://geneticsand society.org/article.php?id=304. For readers interested in the topic of bioethics, information can be found at http://ethics.sandiego.edu/Applied/Bioethics/index.asp. In addition, a good source of information and interesting reading can be found on the Web site of Professor Lee Silver of Princeton University, an expert in bioethics, http://www.leemsilver.net/challenging/index.htm. Dr. Silver's views are provocative as well as entertaining, and interested readers will enjoy his written work and audiovisual pieces.

Biological Cells and Their Machines

Prokaryotes and Eukaryotes

Earlier, we learned how cells might have emerged from a soup of chemicals, and their more detailed construction and operation will be instructive for the future of miniaturization science. Living cells and their organelles have dimensions in the micrometer and submicrometer range and constitute the first truly autonomous micromachines. There is much more membrane area (99%) surrounding the internal organelles than there is outer membrane surface (1%).

Microscopic sizes are the normal working size for the biological world, a biological cell of about 5–10 μm being one of the larger components. The human body contains approximately 100 trillion of these cells. Some of the simplest organisms, such as bacteria and algae, have no cell nucleus, and their genetic material is mixed in with the aqueous cytoplasm. These life forms are known as *prokaryotes,* a designation used to convey their relatively simple status as protocells, and they are the forerunners to multicellular organisms. Most multicellular organisms have their genetic material in a separate chamber within the cell, the *nucleus,* and are called *eukaryotes,* literally, "true nucleus." These two types of cells are compared in Figure 2.20.

Prokaryotes (from bacteria to blue-green algae) took about 500 million years to evolve, but the evolution of the first eukaryotic cells took 2500 million years. Clearly the development of this more elaborate construction was a crucial step in the evolution toward more complex life forms; early eukaryotic cells began the evolutionary path that led to complex multicellular animals and plants. The more complex a multicellular organism, the more different types of cells it has; a blue whale, for example, has approximately 120 types of cells, whereas a mushroom has fewer than 10. Eukaryotic cells come in a very wide variety of shapes and sizes, but on average a cell has a volume of about 3.4×10^{-9} ml and weighs roughly 3.5×10^{-9} grams. Cell size, on the upper end (400 μm^3), is limited by the ability of diffusion to bring nutrients to the appropriate parts inside the cell and to dispose of waste products. The smallest size for a free-living organism (0.02 μm^3) most likely is set by the catalytic efficiency of enzyme and the protein synthetic chemistry machinery.

Organelles

Introduction

Organelles, visible in the eukaryotic cell in Figure 2.20b, are specialized structures. The functions of several of these cell organelles are described in the following sections. As depicted in Figure 2.21, a cell might be compared to a city, with the corresponding functions listed below the images. The high degree of spatial/temporal organization of molecules and organelles within cells is made possible by protein machines that transport components to various destinations within the cytoplasm. These different tiny motors are specialized for specific tasks, such as cell division, cell movement, crawling, maintenance of cell shape, movement of internal organelles, synthesis of ATP, RNA synthesis, DNA synthesis, etc.

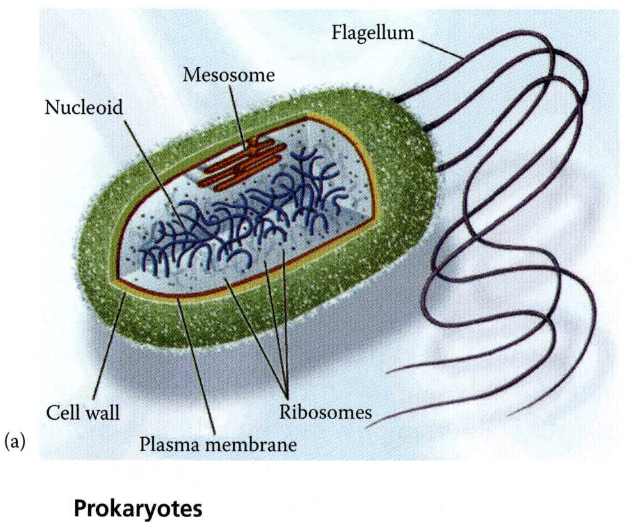

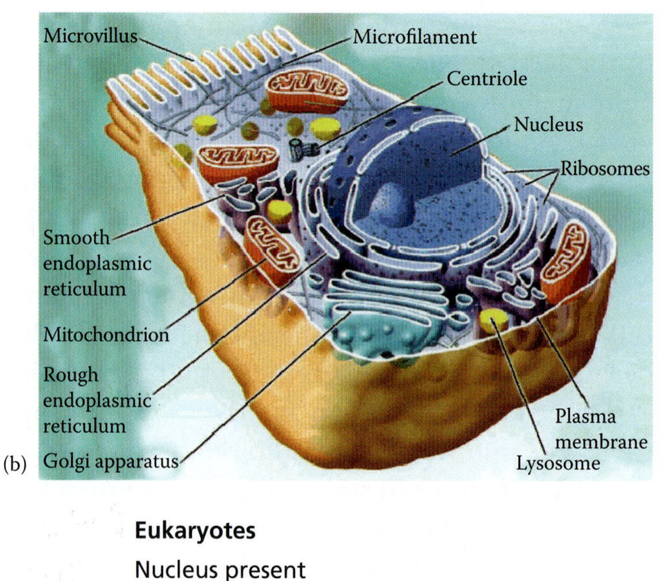

Prokaryotes

No nucleus—have a nuclear "area"

No membrane-bound organelles

DNA present usually as 1 large chromosome

Primitive cells usually quite small

Archebacteria, Eubacteria

Eukaryotes

Nucleus present

Have membrane-bound organelles

DNA in nucleus

"Modern" cells, can be quite large

Plants, Animals, Fungi, Protists

FIGURE 2.20 Comparison of a prokaryotic cell (a) and an eukaryotic cell (b).

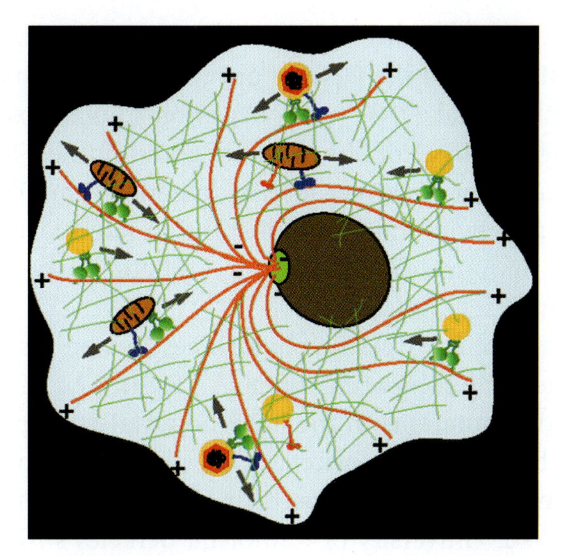

Workers/machines	Proteins
Roads	Actin filaments
Highways	Microtubules
Trucks	Molecular motors
Power plants	Mitochondria
Factories	Ribosomes
Library	Nucleus with DNA
Post office	Golgi apparatus
Police	Chaperones
Gates, keys, passes	Ion channels
Electric fences	Membranes
Train tracks	Actin filament network
Motors, generators	ATP synthases
Vehicles	Hemoglobin
Train control center	Centrosome
Copy machines	Polymerases
Chain couplers	Ligases
Bulldozers, destroyers	Proteases, proteasomes
Internet nodes	Neuron synapses
Mail sorting machines	Protein sorting machines

FIGURE 2.21 The cell is like a city.

Mitochondria and Phylogenetic Trees

The nucleus contains the genetic material (library), and the mitochondria (power plants), through utilization of pyruvate and oxygen, produce ATP (adenosine triphosphate), a compound that provides energy for cellular activities. There is convincing evidence that these mitochondria used to be independent structures with a life of their own that became incorporated into eukaryotic cells over time—very much like bacteria that permanently colonize more complex cells.[46] Mitochondria, for example, contain DNA that uses a slightly different code for amino acids than does the same DNA in the nucleus of the cell. As in bacteria, the length of DNA is closed into a loop and the mitochondria reproduced by splitting in two. These are amazing findings, as the DNA code for amino acids is otherwise the same for species as different as humans and fish (see under "The Genetic Code"). While each of us necessarily has two parents, our mitochondria and mitochondrial DNA come from the ovum and hence from our mothers. Our mothers, in turn, got their mitochondrial DNA from their mothers, and so on. Thus, while our nuclear DNA is a mish-mash of the DNA of our four grandparents, mitochondrial DNA (mtDNA) is an almost exact copy of the DNA of our maternal grandmother. The match is not exact because of mutations: mtDNA has a pointwise mutation substitution rate 10 times faster than nuclear DNA—so it is a better record of evolutionary change.

Evolutionary trees or phylogenies can be reconstructed by comparing mtDNA (Figure 2.22). In fact, the mutations in the mitochondrial DNA provide the molecular clock that allows us to determine how much time has elapsed since the *mitochondrial* Eve lived. University of California Berkeley's Allan Wilson, in 1987, compared mtDNA of many racial groups.[47] Knowing the average mutation rate, he figured out that we all derive from a single version of mtDNA that existed 200,000 years ago in the cells of an African woman—the common ancestor of all of humanity. This work is now seen as overly simplistic. Construction of an exact phylogenic tree is indeed still controversial: for example, it is not necessarily true that the molecular clock is constant; i.e., mutation rates may change, and environmental pressure may change (http://www.sciencemag.org/feature/data/tol). Moreover, the mathematical descriptions and algorithms that may lead to a historically correct phylogenetic tree remain to be developed (http://tolweb.org/tree).

Mitochondria and Power

Here we dwell for a moment as we consider the operation of the cells' power stations—the mitochondria—in more detail (Figure 2.23). We study how cells derive their power from the sun by relying on mitochondria. Energy is required for mechanical work, transportation, and chemical work. When considering building artificial cells, powering the micro- or nanounits is one of the first priorities. This might mean, in human engineering terms, capturing radiant energy (e.g., sunlight) with a solar cell

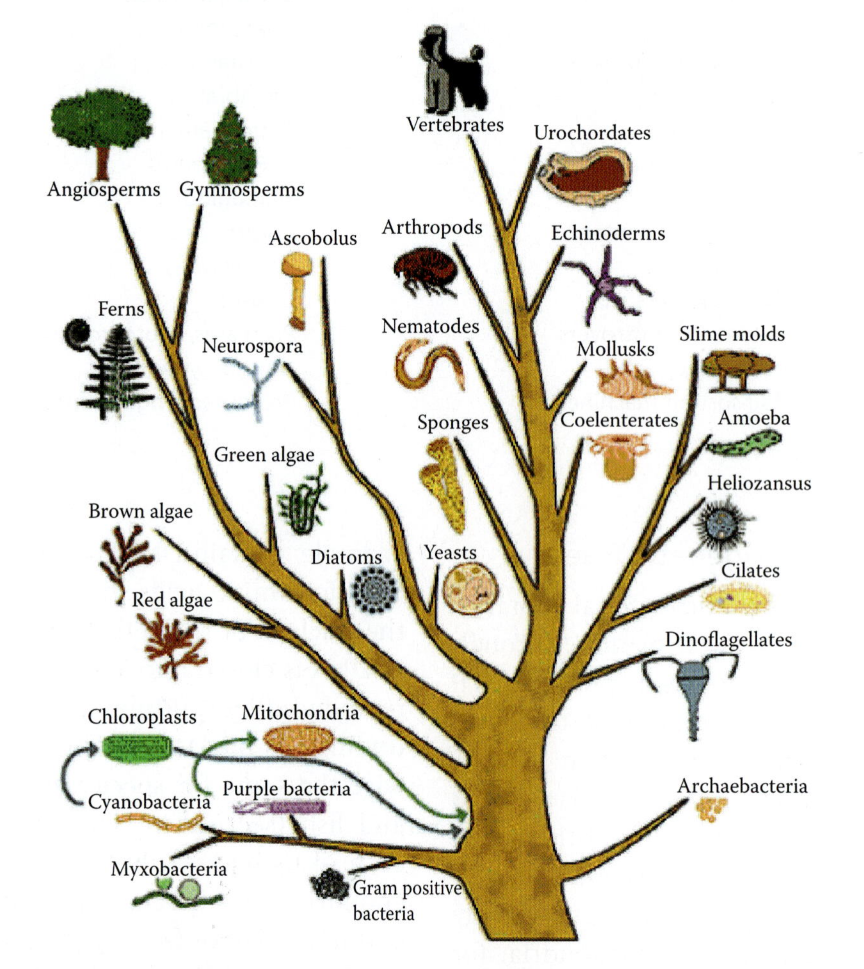

FIGURE 2.22 Evolutionary trees or phylogenies can be reconstructed by comparing mitochondrial DNA.

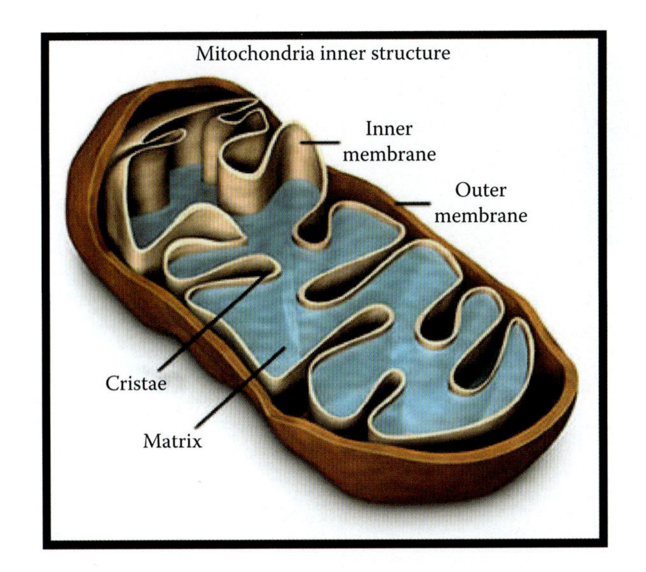

FIGURE 2.23 Mitochondria.

and converting it to a fuel that can be stored for later use (e.g., hydrogen through water electrolysis).

In Figure 2.24a we show the molecular structure of the energy currency of a biological cell, i.e., adenosine triphosphate or ATP, the immediate source of energy in all cells of all living organisms on Earth. The structure of an ATP molecule contains three molecules of phosphate linked to each other; at the pH of normal biological fluids, they have a negative charge. When energy is required, ATP is hydrolyzed to adenosine diphosphate (ADP) and inorganic phosphate (P_i). Back in the mitochondria, ADP is recycled back to ATP almost immediately, so that the energy is again available to do needed work (recharging of the batteries) (Figure 2.24b). The production of ATP is due to three different cell processes, namely photosynthesis, fermentation, and respiration. In photosynthesis, nature uses stacks of thylakoids in a chloroplast, a plant organelle as shown in Figure 2.25. We will detail the photosynthesis process, the conversion of carbon dioxide and water into glucose with the help of sunlight, shortly.

For now it suffices to say that through photosynthesis, plants, algae, and some bacteria store solar energy in the form of the high-energy chemical bonds of glucose. In that process carbon dioxide and water are consumed, and some leftover oxygen is released in the air. This energy consuming process is said to be anabolic; i.e., large molecules are synthesized from smaller ones in a scenario schematically illustrated in Figure 2.26a ($\Delta G > 0$). Most living organisms (including plants and animals) then reverse photosynthesis by taking glucose apart and releasing energy to power their own growth and development. In this second process, oxygen is consumed, and water and carbon dioxide, the original ingredients consumed by the plants, algae or bacteria, are released. This energy producing process is said to be catabolic, i.e., breaking down complex

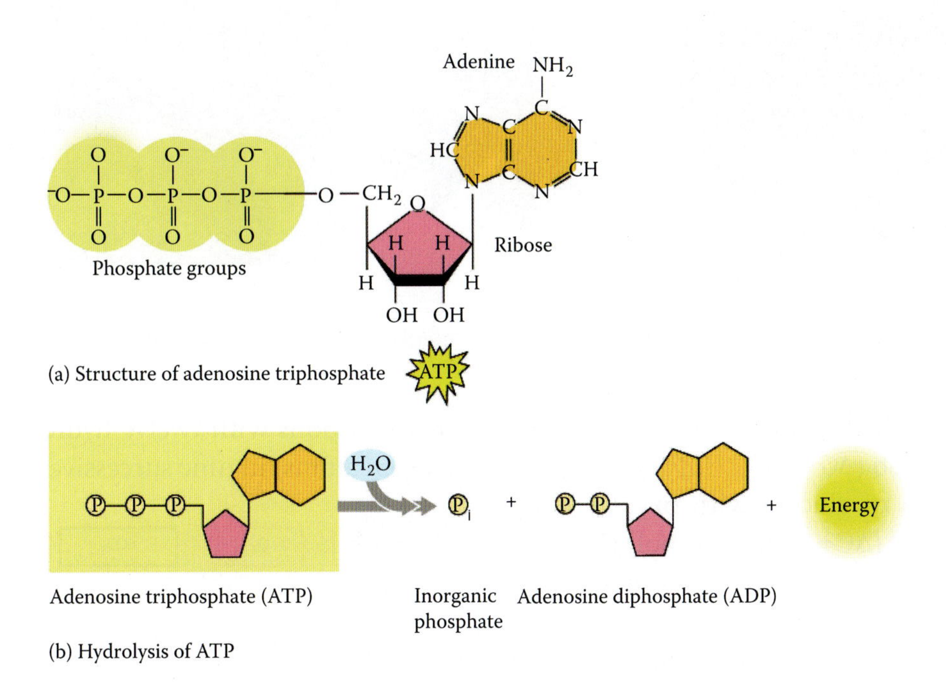

FIGURE 2.24 (a) Molecular structure of adenosine triphosphate (ATP). (b) Hydrolysis of ATP to ADP.

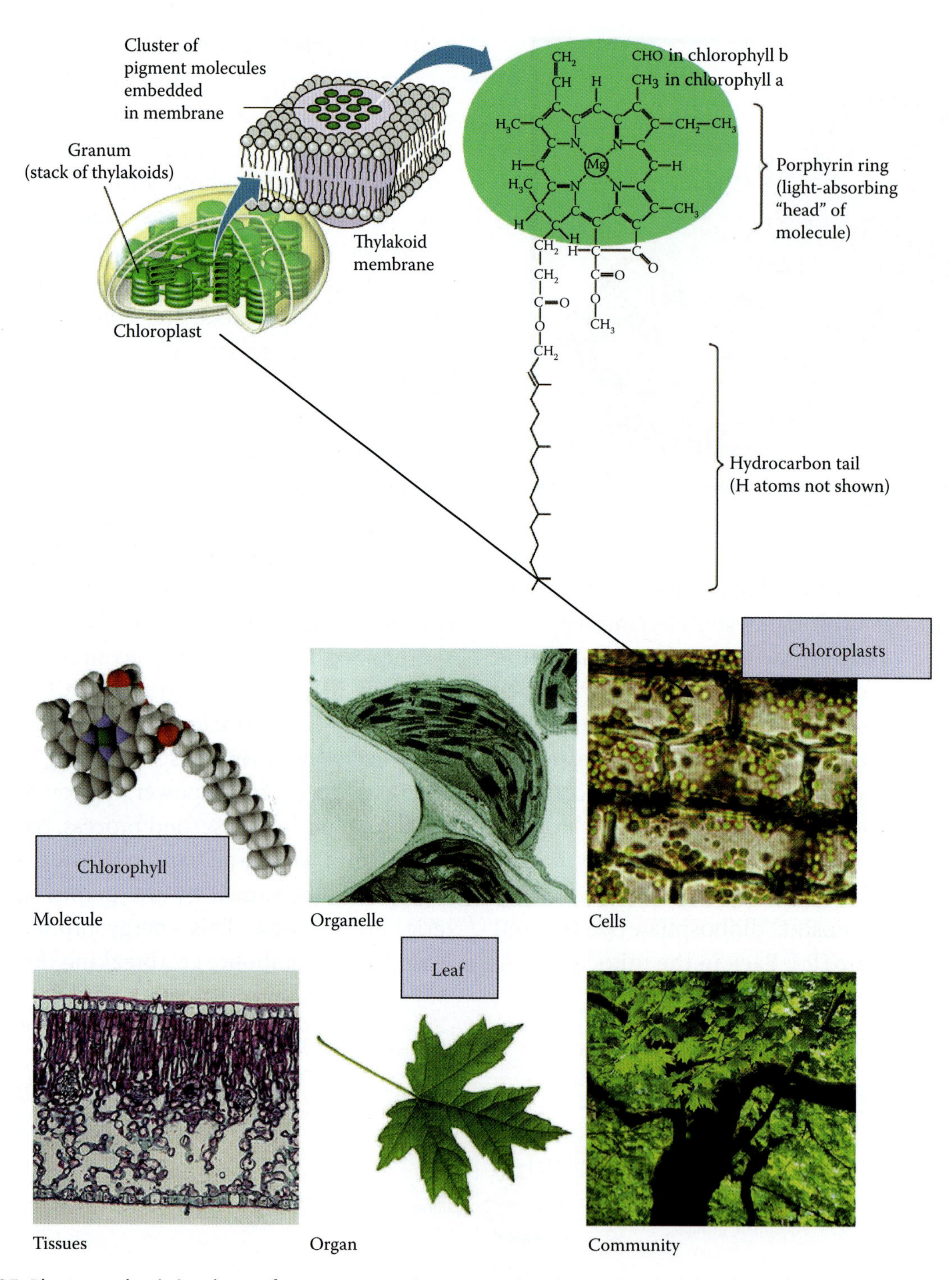

FIGURE 2.25 Photosynthesis in plants: from trees to pigment molecules in the thylakoids in chloroplasts.

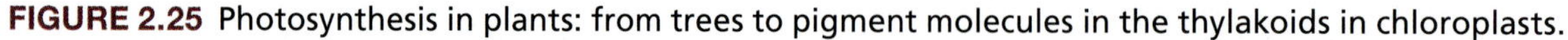

molecules into smaller, simpler ones. This is illustrated in Figure 2.26b ($\Delta G < 0$).

Direct oxidation of glucose with oxygen would release 686 kcal/mol ($\Delta G = -686$ kcal/mol) (Reaction 2.1), but cells cannot use the energy stored in glucose directly.

In the cytoplasm of a cell, first, nine different enzymes must help convert the glucose (a chain of six carbons with twelve hydrogen atoms and six oxygen atoms) in nine successive chemical reactions

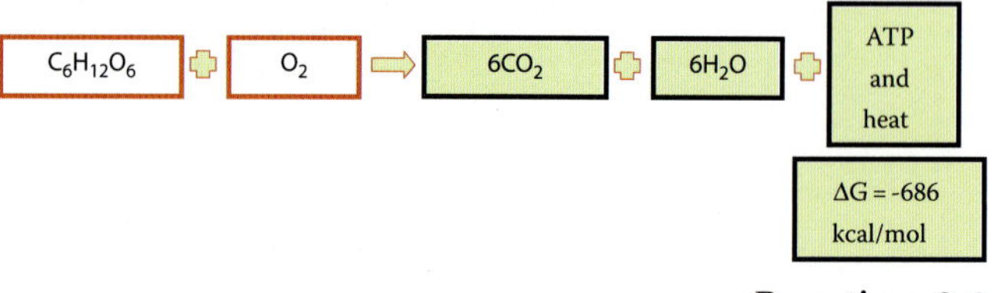

Reaction 2.1

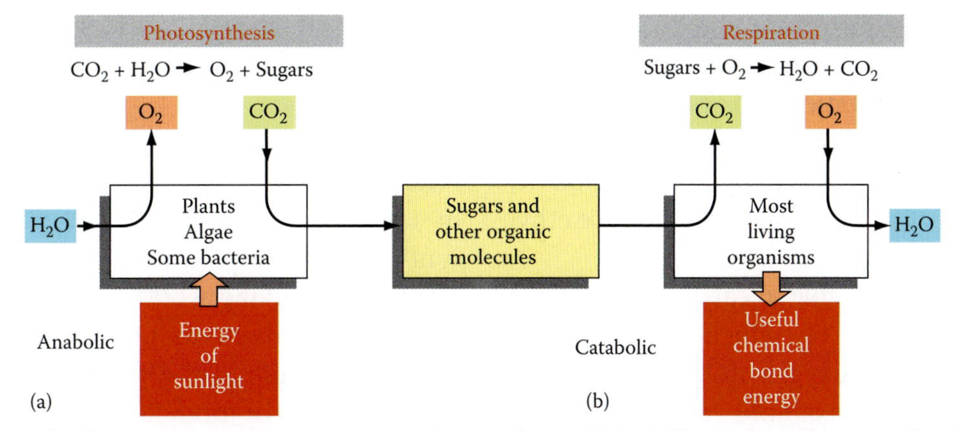

FIGURE 2.26 (a) Anabolic process: use of energy to conduct unfavorable/uphill reactions (from small to large molecules). (b) Catabolic processes: fuel breakdown to deliver energy (from large molecules to small).

to two molecules of pyruvate (a three-carbon chain with three hydrogens and three oxygens). This complicated process is called *glycolysis*. After glycolysis, it is the pyruvate that carries on as the energy source originally produced with the solar energy input, and transports it to the mitochondria. The mitochondria takes the energy garnered during photosynthesis (in the case of plants) or the digested food (in the case of animals) and stored as pyruvate and converts it to ATP. This conversion in the mitchondria occurs at a very high surface inner membrane in a complex set of reactions, including the famous Krebs cycle. The Krebs cycle is an aerobic process (i.e., it involves the use of oxygen) that is responsible for cellular respiration. Hydrolysis of the ATP result in the release of one of three P_i groups, along with 7 kcal/mol, and the formation of ADP (Figure 2.24b). ATP thus packs a wallop of concentrated solar power. These tiny solar batteries are released by the mitochondria into the cell to power all types of uphill reactions ($\Delta G > 0$). In releasing their power, ATP battery molecules shed instantly one phosphate (P_i) and are discharged to become ADP (ATP minus one P_i). In biochemistry, phosphorylation is the addition of a P_i group to a protein or a small molecule. Beyond doubt, protein phosphorylation is the most important regulatory event in eukaryotic cells. Many enzymes and receptors are turned on or off by phosphorylation and dephosphorylation. Phosphorylation is catalyzed by various specific protein kinases, whereas phosphatases dephosphorylate. The discharged ADP solar battery, in need of recharging now, diffuses until it finds a mitochondria power station, where it will regain its lost phosphate. Cells obviously recycle the ATP batteries they use for work.

To make ATP from pyruvate, as we show in Figure 2.27, the most prevalent and efficient pathway is cellular aerobic respiration. In cellular aerobic respiration, the conversion of 1 mol of glucose yields up to 36 mol of ATP; therefore, the energy efficiency is 36 × 7 kcal/mol divided by 686 kcal/mol = 252/686 = 36.7% (compared with about 3% efficiency in the burning of oil or gasoline). In the absence

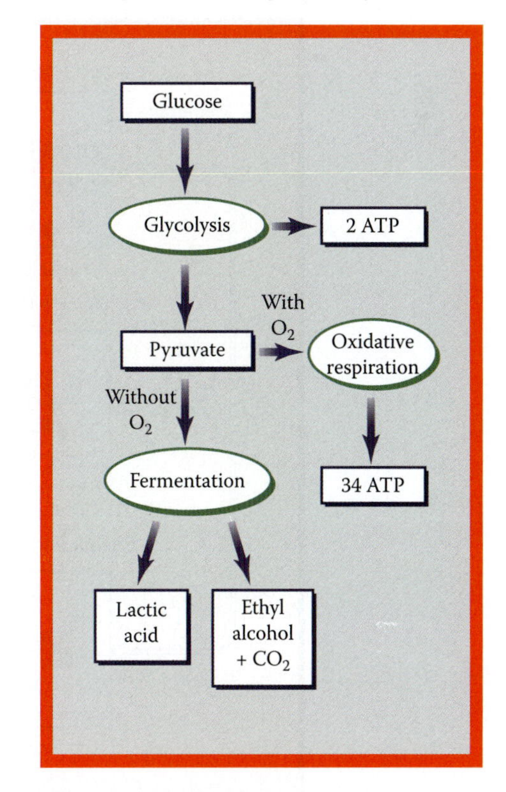

FIGURE 2.27 Various pathways to making of pyruvate, the food source for mitochondria.

of oxygen, certain cells are still able to make ATP, albeit in a less efficient matter. This can be accomplished through fermentation and anaerobic respiration. Fermentation is the breakdown of sugars in the absence of oxygen to release energy, carbon dioxide, and alcohol (or lactic acid). Yeast cells produce ethyl alcohol by fermentation.

Some prokaryotes are able to carry out anaerobic respiration, respiration in which an inorganic molecule other than oxygen (O_2) is the final electron acceptor. For example, certain bacteria known as sulfate reducers can transfer electrons to sulfate (SO_4^{2-}), reducing it to H_2S. Other bacteria, called nitrate reducers, can transfer electrons to nitrate (NO_3^-), reducing it to nitrite (NO_2^-). Other nitrate reducers can reduce nitrate even further, to nitrous oxide (NO) or nitrogen gas (N_2). The total energy yield per glucose oxidized is less than with aerobic respiration, with a theoretical maximum yield of 36 ATP molecules or fewer.

Cellular respiration, fermentation, and anaerobic respiration are catabolic, energy-yielding pathways. The totality of an organism's chemical reactions is called *metabolism*. Metabolism manages the material and energy resources of the cell. In Figure 2.28 the various processes that ATP energizes are summarized.

Since animals cannot make food directly from water and CO_2, chloroplasts and mitchondria work in tandem, as illustrated in Figure 2.29. Photosynthesis and respiration are complementary processes in the living world. Photosynthesis uses the energy of sunlight to produce sugars and other organic molecules. These molecules, in turn, serve as energy sources or food. Respiration is a process that uses O_2 and forms CO_2 from the same carbon atoms that had been taken up as CO_2 and converted into sugars by photosynthesis. In respiration, organisms obtain the energy that they need to survive. It is the cycling of molecules between chloroplasts and mitochondria that allows a flow of energy from the sun through all living things. Photosynthesis preceded respiration on the Earth by, probably, billions of years before enough O_2 was released to create an atmosphere rich in oxygen. Ultimately, the process of photosynthesis (Reaction 2.2) can be considered one of the most, if not *the* most, important chemical reaction on Earth.

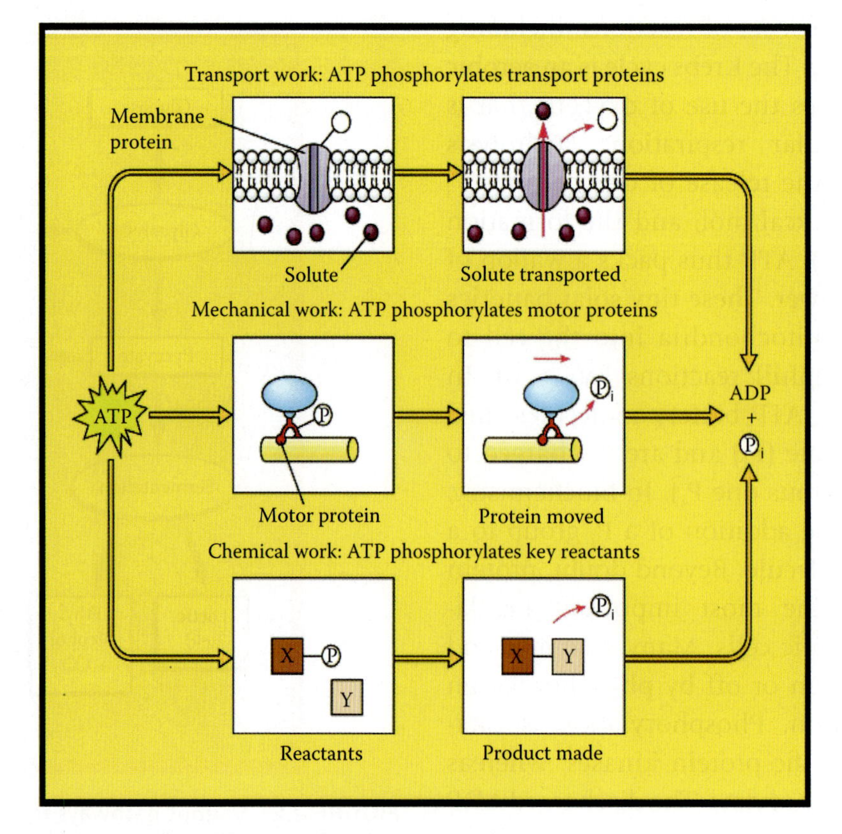

FIGURE 2.28 The various processes that ATP energizes: ATP at work.

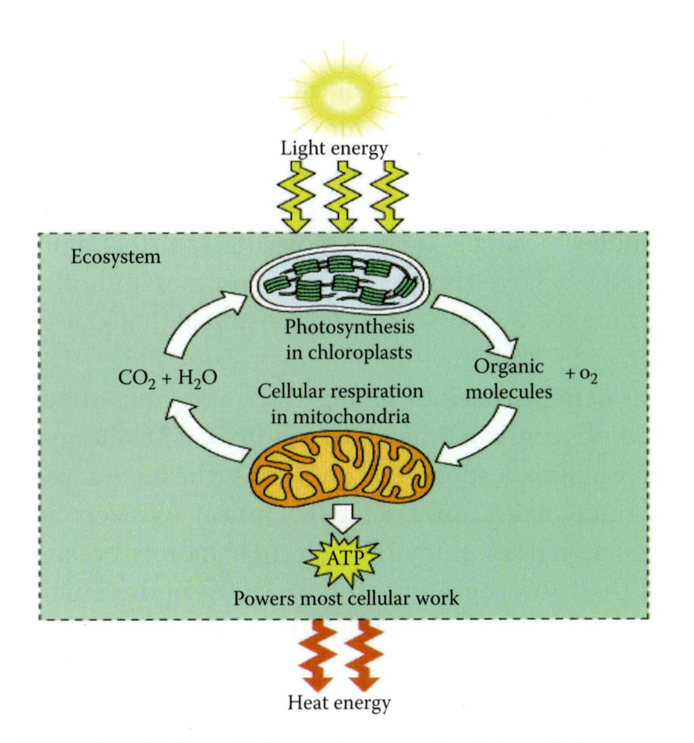

FIGURE 2.29 From light to photosynthesis to cellular respiration and heat energy.

$$6CO_2 + 12H_2O + \text{light energy} \Rightarrow C_6H_{12}O_6 + 6O_2 + 6H_2O \qquad \text{Reaction 2.2}$$

For building artificial cells, we would be better off building plant like-cells, as the source for fuel is in the atmosphere (water and/or CO_2) and harvesting energy from the sunlight is relatively simple. So let us take a more detailed look at how the chloroplasts in plant leaves do it (Figure 2.25).

Pigments in the leaves of plants absorb sunlight, principally through chlorophyll in the chloroplasts. Chloroplasts are found mainly in mesophyll cells forming the tissues in the interior of the leaf (Figure 2.25). Chlorophyll comes in two slightly different forms, called chl a and chl b. Most leaves contain at least two additional types of pigments, carotenes and xanthophylls, that are capable of absorbing light of different wavelengths. The more abundant chlorophylls usually mask the latter two types of pigments. There are about half a million chloroplasts per square millimeter of leaf surface. The color of a leaf comes from chlorophyll, the green pigment in the chloroplasts. Oxygen exits and carbon dioxide enters the leaf through microscopic pores, stomata, in the leaf. Veins deliver water from the roots, and carry off sugar from mesophyll cells to other plant areas (Figure 2.30).

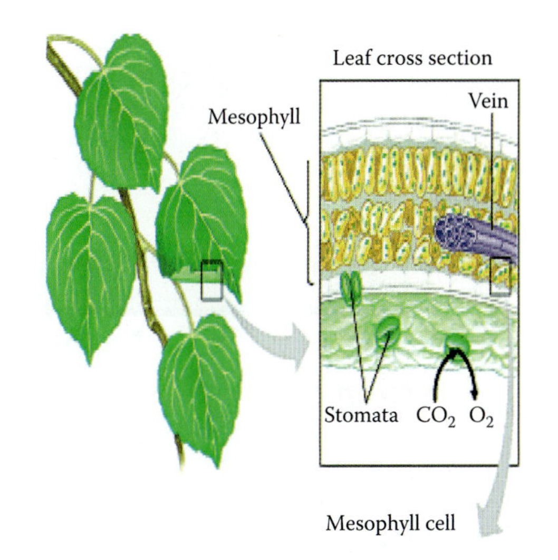

FIGURE 2.30 Stomata and veins in a green leaf.

Only chl a can participate directly in the light reactions. Other pigments in the thylakoid membrane (e.g., chl b, carotenoids) can absorb light and transfer the energy to chl a. These pigments thus broaden the spectrum of solar light (wavelengths) that can drive photosynthesis. Plants only use a narrow range of frequencies from sunlight between 400 and 700 nm. Light reactions on the thylakoid membrane convert solar energy and produce oxygen, and dark reactions fix CO_2 in the Calvin-Benson cycle to make sugar, as shown in Figure 2.31.

Molecular Motors

Other subcomponents of biological cells exhibit equally fascinating feats of nanoengineering. For illustration, consider molecular motors, composed of a discrete number of atoms that produce molecular or supramolecular motions. Molecular motors derive energy from ATP that is present in the cytoplasm and nucleoplasm of every cell and is the universal energy currency for all biological processes (see Figure 2.24). Living things can use ATP like a battery. Molecular motors are in a sense nanomachines that do mechanical work through the hydrolysis of ATP, the energy source of the cell. During hydrolysis of ATP, a shape change occurs within the motor protein (a mechanochemical process), and mechanical work is performed. This mechanical work is used to move the motor along a track in order to perform a load transport function or to apply forces to the filament for cell motility and cell

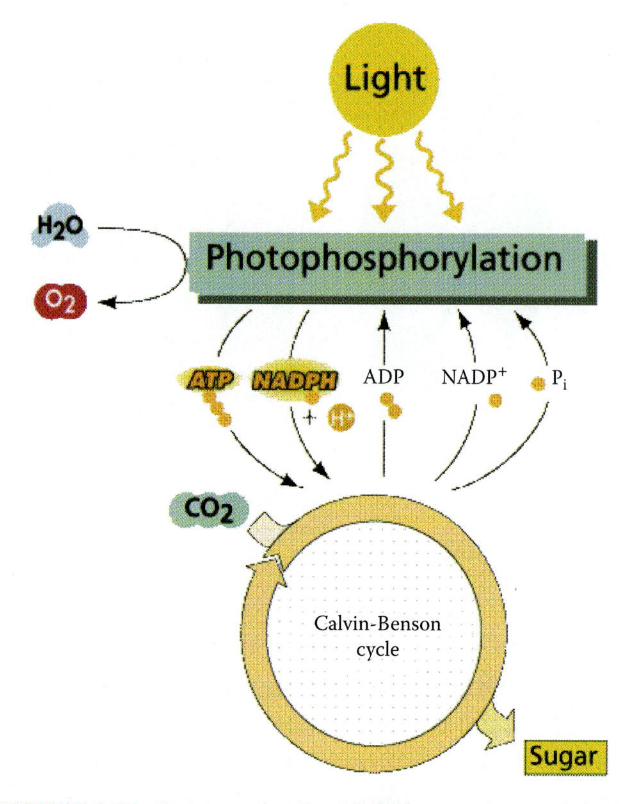

FIGURE 2.31 Photosynthesis with light reaction and dark reaction in the Calvin-Benson cycle.

division. Three types of motors can be distinguished, i.e., linear motors (such as myosin, dynein, and kinesin), rotary motors (F_0-F_1 ATP synthase) and polymerization motors (RNA, ssDNA); their tracks and functions are summarized in Table 2.5. Many of proteins can be considered nanoactuators, not only rotary or linear motor proteins. In some cases, even a simple binding of a protein and a specific target molecule—a chemical reaction—results in definite shape change—a mechanical effect. A conformational change might be put to good use in genetically engineered nanoactuators or biosensors (see calmodulin example, below).

Kinesin is a linear motor protein that steps along the microtubuli, the tubelike fibrils of the cell's skeleton, and is responsible for carrying organelles over a long distance within the cell (Figure 2.32). Most kinesins "walk" toward the plus ends of microtubules. A single kinesin molecule can drag along relatively large things like chromosomes and other organelles. Svoboda et al. demonstrated that kinesin can even work in vitro on a microscope slide.[48] This was demonstrated by gluing kinesin on microtubuli and observing it drag along a little ball of silica gel. Its strength is about 5 piconewtons (the force a laser pointer makes on a screen). Optical tweezers are well suited for studying molecular motors because of their low spring constants. Using optical traps, this same group determined that kinesin moves in individual steps that are 8 nm (80 Å) in length, a size consistent with its dimensions (Figure 2.33).[48] Microtubules can also move themselves on a glass slide surface containing bound kinesin (a single molecule bound to glass can move a microtubule). For the latest on kinesin research, visit the kinesin homepage at http://www.proweb.org/kinesin.

Noji et al. discovered that the γ-subunit of the hydrophilic F_1 portion of ATPase rotates in response to the synthesis/hydrolysis of ATP (Figure 2.34).[49] The moving part of an ATPase is a central protein shaft (or rotor, in electric-motor terms) that rotates in response to electrochemical reactions with each of the molecule's three proton channels (comparable to the electromagnets in the stator coil of an electric motor). The ubiquitous ATPase was the first rotary motor enzyme ever found, and the force generated by this motor protein (>100 pN) is among the greatest of any known molecular motor. The enzyme puts out a very large torque, comparable

TABLE 2.5 Molecular Motors and Their Tracks and Functions

Type	Track	Function
Myosin	F-actin, filament	Cell crawling, cell division, muscle contraction
Dynein	Microtubules	Eukaryotic cell (produces the axonemal beating of cilia and flagella and also transports cargo along microtubules toward the cell nucleus)
Kinesin	Microtubules	Organelle transport, chromosome segregation, moves cargo inside cells away from the nucleus along microtubules
Polymerase	RNA, single-stranded	Nuclear processes (transcribes RNA from a DNA template)

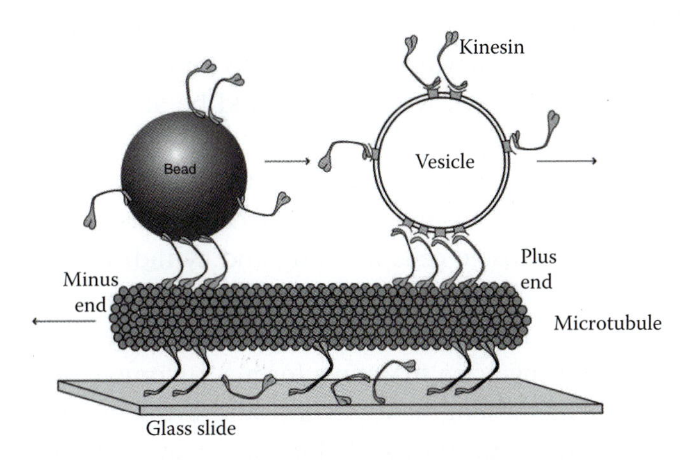

FIGURE 2.32 Kinesin motor. Schematic diagram showing movements mediated by kinesin, an ATP-driven motor, on microtubule tracks. Vesicles and beads containing kinesin molecules on their surface move toward the plus end of microtubules. Microtubules can also move themselves along a glass slide containing bound kinesin.[48]

to a person rotating a 150-m rod. The enzyme can spin such a long filament because it automatically ratchets down the rotation rate. A no-load rotational velocity of 1020 rpm was calculated, and the motor has a diameter of less than 12 nm. Montemagno et al.,[50] then at Cornell University, integrated these genetically engineered motor proteins with NEMS, for example, by attaching the protein nanomotors to arrays of nickel posts and attaching nickel rotor blades to the protein molecules.[50] The "handle" for attaching the ATPase motor to the nanofabricated metallic substrates is a synthetic peptide composed of histidine and other amino acids. The histidine peptide allows the molecular motors to adhere to nanofabricated patterns of gold, copper, or nickel—the three standard contact materials in integrated circuits.

The researchers attached fluorescent microspheres (~1 μm in diameter) to the ATPase molecule's rotor and observed the microsphere movement with a differential interferometer and with a charge-coupled device camera.

The authors envision that F_1-ATPase motors will be used to pump fluids and open and close valves in nano- and microfluidic devices and provide mechanical drives for a new class of nanomechanical devices.[50] This work was cited by *Discover* magazine as one of the most promising new technologies of 1999. Today, twelve years later, that promise remains unmet.

Bacteria swim in viscous liquid environments by rotating helical propellers known as flagella, shown in Figure 2.35 (see also Volume I, Chapter 6, "Mixing, Stirring, and Diffusion in Low Reynolds Number Fluids"). The bacterial flagellum is a nanomachine made of about twenty-five different proteins, each of them in multiple copies ranging from a few to tens of thousands. It is constructed by self-assembly of these large numbers of proteins, each into a different

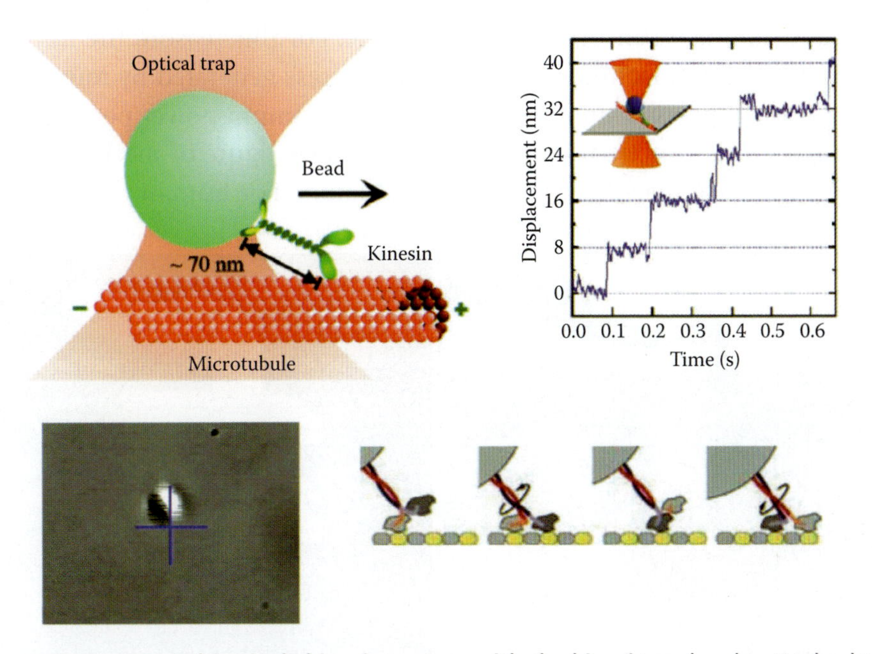

FIGURE 2.33 Laser tweezers were used to study kinesin motors with the kinesin molecule attached to a bead traveling over a microtubule. (From http://www.stanford.edu/group/blocklab.)

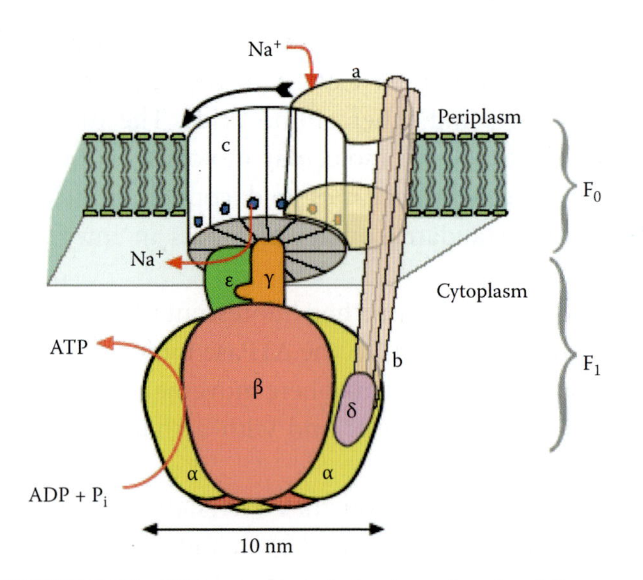

FIGURE 2.34 ATPase motors. A single molecule of F_0 F_1-ATPase acts as a rotary motor—the smallest known. A central rotor of radius ~1 nm, formed by its γ-subunit, turns in a stator barrel of radius ~5 nm formed by three α- and three β-subunits.

part that exerts a different function, such as a rotary motor, bushing, drive shaft, rotation-switch regulator, universal joint, helical propeller, and rotary promoter for self-assembly. The rotary motor, with a diameter of only 30–40 nm, drives the rotation of the flagellum at around 300 Hz, at a power level of 10–16 W, with energy conversion efficiency close to 100%. These protein motors appear to be fueled by a flow of protons across a pH gradient and can push an average-sized cell at 30,000 nm/s or 15 body-lengths per second—and can also run in reverse![51] In comparison, the surface micromachined motors of Volume II, Chapter 7, are hopelessly impractical.

The structural designs and functional mechanisms revealed in the complex machinery of the bacterial flagellum could provide many novel technologies that may become the basis for future nanotechnology, from which we should be able to find many useful applications.

Mammalian muscle, nature's ubiquitous actuator for larger organisms, is another chemomechanical actuator. Muscle cells are long and cylindrical and made up of thick myosin filaments surrounded by thin actin filaments. A central myosin filament and surrounding actin filaments form a sarcomere unit, which has a length of 2.5 μm and a thickness of 10–20 nm. The thin and thick filaments are interdigitated, and upon muscle contraction, the thick filaments are pulled along the thin filament in a ratchet-like manner (Figure 2.36). The linear motion is effected by the myosin head, which is attached to the thick filament via a flexible hinge and binds to the actin filament (Figure 2.37). This linear protein motor again uses chemical energy released in the conversion of ATP to ADP and a phosphate ion (P_i). The hydrolysis of 1 mol of ATP releases about 10 kcal of energy (i.e., 6.96×10^{-20} J per molecule).

The maximum static muscle force generated per unit cross-sectional area (i.e., tension, stress) is about 0.350 MPa, a constant number for all vertebrate muscle fibers. In vertebrate muscle, the maximum force generated can be held only for short periods of time because of muscle fatigue. The maximum sustainable force usually is about 30% of the peak value. For this reason, the maximum static sustainable stress generated by muscle is about 0.100 MPa. The maximum power per unit mass (in W/kg) is an important figure of merit for robotic and prosthetic actuators. For human muscle, it typically measures

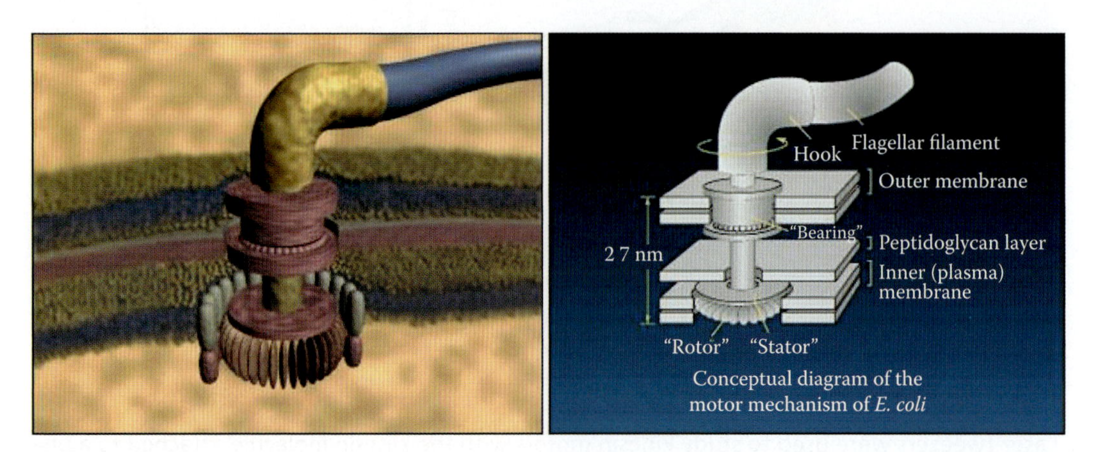

FIGURE 2.35 Bacteria swim in viscous liquid environments by rotating helical propellers called flagella.

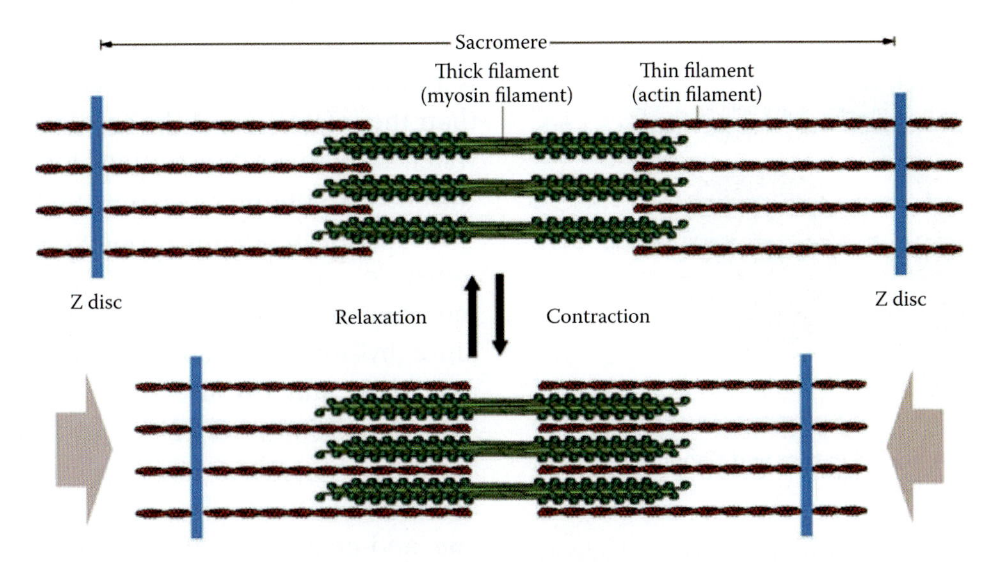

FIGURE 2.36 Muscle cells: the linear motion of muscles, which results in muscle contraction, is driven by chemical energy as ATP is hydrolyzed.

about 50 W kg^{-1} but can be as high as 200 W kg^{-1} for some muscles for a brief period of time. To illustrate the excellent cycle lifetime of a muscle, we need only look at cardiac muscle. The heart beats more than 3×10^9 times in the lifetime of an average person; an excellent lifetime compared with any artificial actuator. When trying to mimic a muscle with an artificial actuator, one of the most difficult properties to emulate is the extreme change in stiffness occurring between resting muscle and maximally activated muscle. Stiffness can increase as much as 5 times from rest to a 100% contraction.[52] On the basis of these observations, researchers preparing artificial muscles are now shifting from using stiff inorganic materials with a high Young's modulus, such as Si, to low-Young's modulus organic materials, such as hydrogels and redox polymers.

The beating motion of hair-like cilia on cells and of flagella depends on an interplay of a pair of proteins, dynein and tubulin (see Figure 2.38). Dyneins walk toward the minus ends (the microtubule part with less expand and shrink activity) of microtubules, causing sliding of microtubules, which leads to a beating motion. This protein pair is equivalent to the myosin and actin pair in muscle tissue (see Figure 2.37).

Unlike the cells of plants and fungi, animal cells do not have a rigid cell wall. This feature was lost in the distant past by the single-celled organisms that

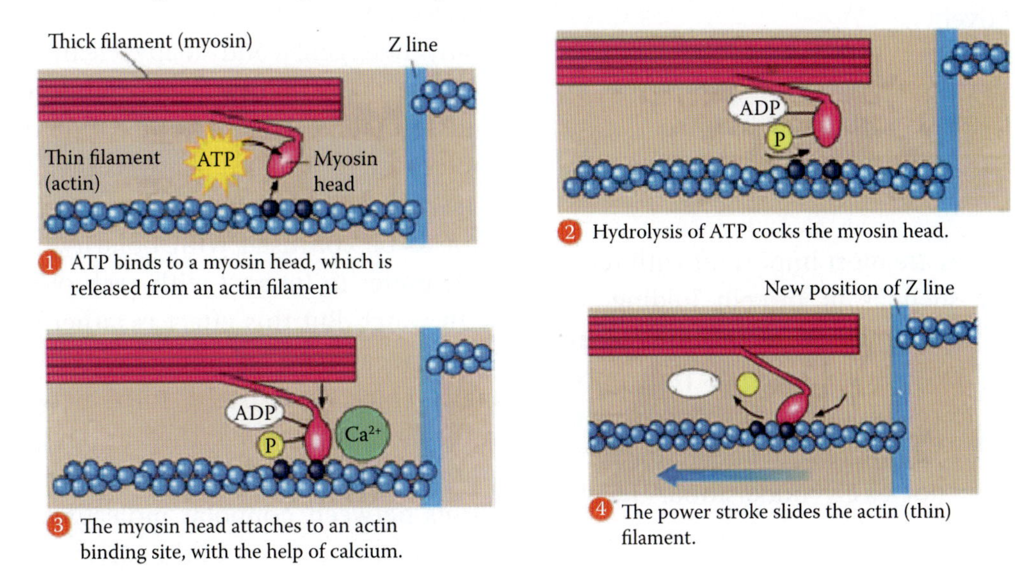

FIGURE 2.37 The linear motion is effected by the myosin head, which is attached to the thick filament via a flexible hinge, and binds to the actin filament.

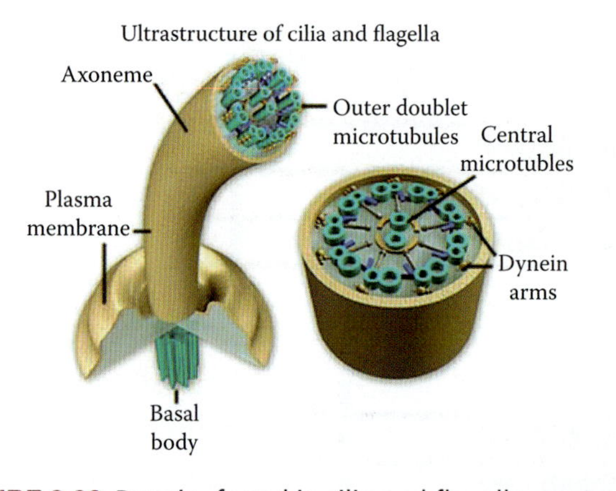

FIGURE 2.38 Dynein, found in cilia and flagella, causes a beating motion.

gave rise to the kingdom Animalia. Most eukaryotic cells, both animal and plant, range in size between 1 and 100 μm. The lack of a rigid cell wall allowed animals to develop a greater diversity of cell types, tissues, and organs. Specialized cells that formed nerves and muscles—tissues impossible for plants to evolve—gave these organisms mobility. The ability to move about by the use of specialized muscle tissues is a hallmark of the animal world, though a few animals, primarily sponges, do not possess differentiated tissues. Notably, protozoans locomote, but only via nonmuscular means, in effect, using cilia, flagella, and pseudopodia.

Different types of cells, such as neuronal cells and brain cells will be reviewed in more detail in the context of equipping micro- or nanomachines with a muscle and brains in Chapters 8 and 9, this volume, respectively.

Hydrophobic and Hydrophilic Forces Drive Self-Assembly in Nature

Introduction

Hydrophobic forces are most important with regard to self-assembly such as in protein folding, cell wall formation and formation of macromolecular assemblies (e.g., DNA hybridization). The concept of hydrophobic forces only was introduced in the 1950s and is hard to define, as the underlying physical principles are not easily understood.* Jencks's definition is one of the better ones: "hydrophobicity is an interaction between molecules that is stronger than the interaction of the separate molecules with water, and can not be accounted for by covalent, electrostatic, H-bonding or charge transfer processes."

As we saw in Volume I, Chapter 7, a hydrophobic molecule, also termed a nonpolar or lipophilic molecule, is a water-fearing or -repelling molecule. Since hydrophobic molecules are not electrically polarized, and because they are unable to form hydrogen bonds, water repels them in favor of bonding with other water molecules. Examples of hydrophobic molecules include the alkanes, oils, fats, and greasy substances in general. Hydrophilic or polar molecules is the term used for water-loving molecules. Examples include polar organic groups such as amino, carboxyl, and hydroxyl groups, such as those in amines, carboxylic acids, and alcohols. Amphiphilic or amphipathic molecules are molecules that have both a hydrophobic and a hydrophilic component, which makes them at the same time water-loving and water-repelling. They are compounds that exhibit both hydrophilic and hydrophobic properties simultaneously, for example, fatty acids and detergents. One type of an amphiphilic molecule has a water-soluble carboxylic group head and a water-insoluble hydrocarbon tail. Amphiphilic molecules of this sort are called surfactants. A surfactant is a soluble chemical amphiphilic compound that reduces the surface tension between two liquids. It is used in many detergents and soapy cleaning compounds because of its ability to surround grease particles and solubilize them. Another type of amphiphilic molecules, introduced below, is the phospholipids.

Hydrocarbon groups have greater polarizability (not to be confused with polarity) than water, so they prefer to interact with each other rather than with water. But this effect is rather small, so there must be a more dominant effect at work, especially since even if the force between nonpolar molecules is repulsive they are aggregating when put together in a polar medium. The complex that forms between nonpolar molecules has less unfavorable interaction with the solvent than the sum of the uncomplexed molecules (the lesser of two evils so to speak) (Figure 2.39). It is this latter effect that

* "The magnitude, range and origin of the hydrophobic interaction have been a mystery ever since the pioneering work by Kauzman and Tanford" (J. Israelachvili, 2005).[53]

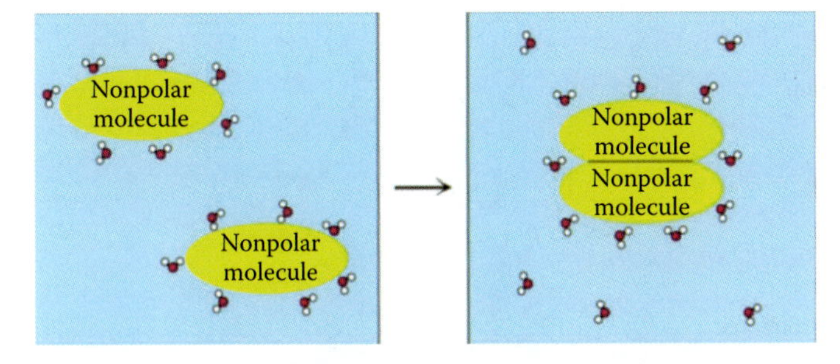

FIGURE 2.39 Aggregation reduces the hydrophobic surface area that requires ordering of water molecules for optimized solvation.

causes the hydrophobic interaction—which in itself is thus incorrectly named since the energetic driving force comes from the hydrophilic molecules.

The hydrophobic effect is illustrated in more detail in Figure 2.40 and in the sequence of events in Figure 2.41. Both figures clarify what happens when we dissolve lipid molecules in water. In Figure 2.40, we illustrate how hydrophobic compounds upset the structure of water and lead to the formation of a clathrate. A clathrate, clathrate compound, or cage compound is a chemical substance consisting of a lattice of one type of molecule trapping and containing a second type of molecule. The ordering of the shell of water molecules around the hydrocarbon solute in a clathrate compound (so as to minimize "dangling" H-bonds) causes a significant decrease in the entropy of the water ($\Delta S < 0$). Typically, the total change in entropy, ΔS, in dissolving small hydrocarbon molecules in water (at 298 K) is about −100 J K^{-1} mol^{-1}. This decrease in entropy (associated with the ordering of molecules) makes it unfavorable to mix "hydrophobic molecules" in water, or solvent entropy favors the removal of nonpolar solutes from solution. The overall enthalpy change of interaction of a nonpolar solute with water, on the other hand,

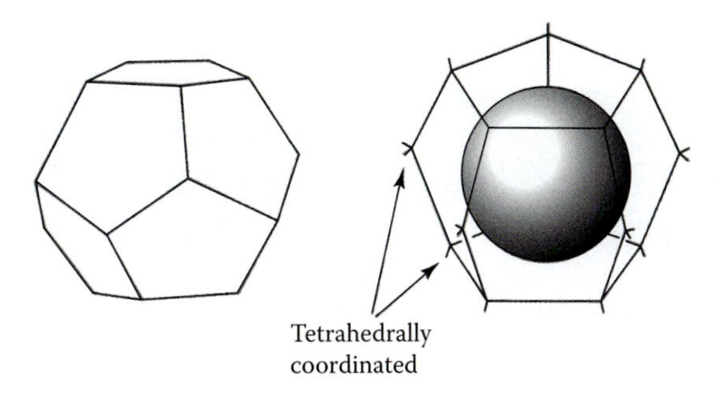

FIGURE 2.40 A water "cage" around another molecule or a clathrate.

is not particularly unfavorable ($\Delta H < 0$) because the nonpolar molecules induce cage-like ordering of the first shell of water molecules (clathrate), strengthening their H-bonding. Comparing the numbers, it is clear that the origin of the hydrophobic effect is mostly entropic.

The latter is also clear from the following example. When a hydrophobic molecule such as n-butane is added to water at 25°C, the Gibbs free energy changes as follows.

$$\Delta G = \Delta H - T\Delta S = -4.3 + 28.7 = +24.5 \text{ kJ/mol} \quad (2.7)$$

Generalizing, ΔH is related to bond formation/breaking, and ΔS is related to configurational freedom and water ordering. For a favorable reaction, we want the equilibrium constant to be $K > 1$, $\Delta G < 0$, $\Delta H < 0$, and $\Delta S > 0$, where $K = \exp(-\Delta G^{\circ}/RT)$ with $R = 8.3145\ JK^{-1}mol^{-1}$. In the case of ΔH – negative, we have bond-making gain, and with ΔS – negative we lose entropy because of increased ordering. It is thus enthalpically favorable to place n-butane in water, but it is unfavorable from an entropic perspective (increasing order).

The sequence of events depicted in Figure 2.41 illustrates what happens when amphiphilic or amphipathic molecules are placed in water to form a micelle. A micelle is an aggregate of amphipathic molecules in water, with the nonpolar portions in the interior and the polar portions at the exterior surface, exposed to water. In Figure 2.41a, the water molecules around one amphiphilic molecule reduce the entropy of the water through ordering (forming of a clathrate). In Figure 2.41b, we show a dispersion of lipid molecules, where each molecule forces surrounding water molecules to become highly ordered. In Figure 2.41c, only lipids at the edge of the cluster force the ordering of water. Fewer water molecules are ordered, and the entropy is thus increased. Finally, in Figure 2.41d, all hydrophobic groups are sequestered from water; the ordered shell of water is minimized, the entropy is further increased, and a micelle is formed.

The two immiscible phases (hydrophilic and hydrophobic) change so that their corresponding interfacial area is minimized in the micelle construct. This effect constitutes a phase separation. Depending on their concentration and ionic strength in solution,

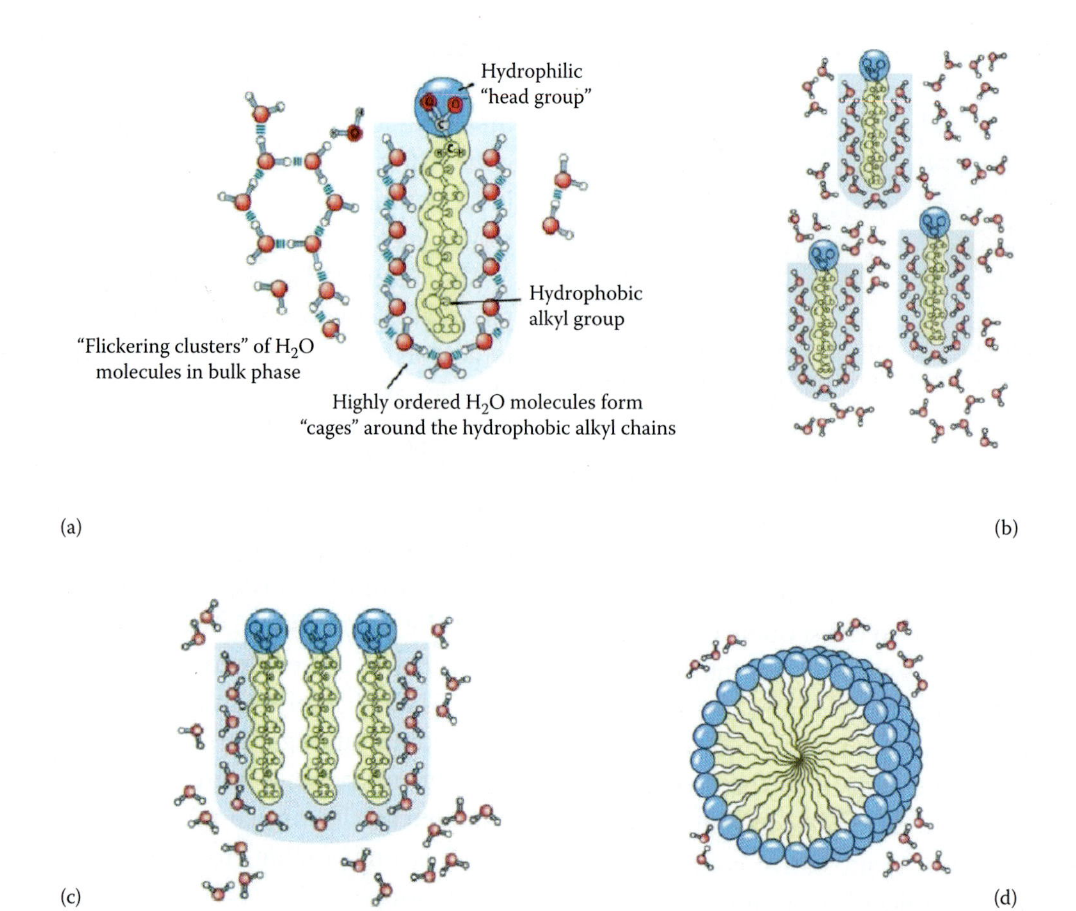

FIGURE 2.41 Steps involved in making a micelle. See text for explanation. (a) The water molecules around one amphiphilic molecule reduce the entropy of the water through ordering (forming of a clathrate). (b) Dispersion of lipid molecules, where each molecule forces surrounding water molecules to become highly ordered. (c) Only lipids at the edge of the cluster force the ordering of water. Fewer water molecules are ordered, and the entropy is thus increased. (d) All hydrophobic groups are sequestered from water; the ordered shell of water is minimized, the entropy is further increased, and a micelle is formed.

the temperature, and other factors, surfactants can assemble into spherical micelles, cylindrical micelles, or bilayers (membranes) or even saddle surfaces in bicontinuous structures (Figure 2.42). A bicontinuous surface, as shown in Figure 2.43, separates the oil phase and water phase. The oil and water are completely separated, but both are continuous across the system. Bilayer sheets are bent into "cubic phases," in which sheets are bent into a complex, ordered network of channels (see also http://www.nationalacademies.org/bpa/reports/bmm/bmm.html).

In a low-dielectric-constant solvent, micelles turn inside out, shielding their hydrophilic heads and forming a reverse micelle. Micelles form spontaneously—i.e., $\Delta G < 0$. When water is added to dried phospholipids, multilamellar vesicles (MLVs) form,

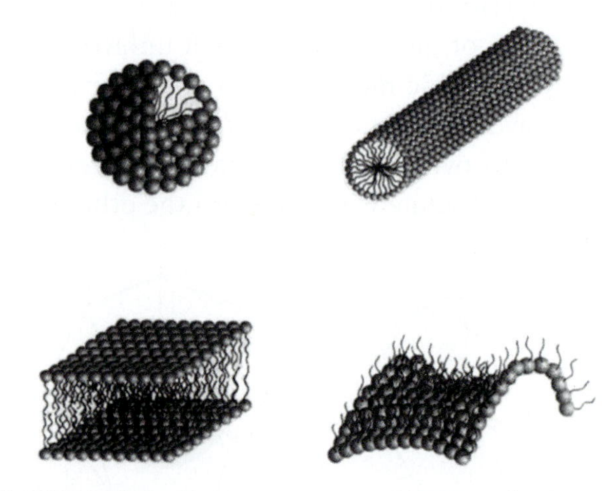

FIGURE 2.42 Surfactants can assemble into spherical micelles, cylindrical micelles, bilayers (membranes), or saddle surfaces in bicontinuous structures. (From I. W. Hamley. 2000. *Introduction to soft matter.* (2nd edition), J. Wiley, Chichester. With permission.)

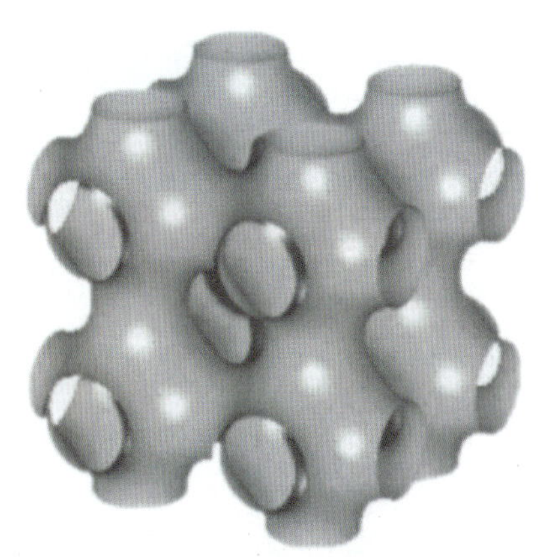

FIGURE 2.43 Surfactants can create a bicontinuous surface to separate an oil phase and a water phase. (From R. A. L. Jones. 2002. *Soft condensed matter.* Oxford University Press, Oxford. With permission.)

with water trapped between the successive polar head groups of the multiple bilayers. Sonicating these produces small unilamellar vesicles (SUVs).

From Micelles to Liposomes and Biological Cells

Introduction

It appears that the chemistry of life cannot take place in anything much simpler than a cell. A virus, a prototype life form, needs to invade a cell and assume control of the cell's chemical machinery by reprogramming it to produce more viruses. For life to have started, it seems essential that the important chemicals were somehow concentrated in order to increase their chance of meeting and reacting. This could have occurred when water evaporated in a small pond, or it could have involved reactions on a clay surface (meeting on a two-dimensional plane is more probable than meeting in three-dimensional space). Another possibility is that the chemicals became grouped in some kind of a vesicle. Thus, they could not diffuse away into the primordial broth. One of the many unanswered questions is, when did this cell-wall precursor take hold, before, during, or after the prebiotic chemical steps?

Micelles, Liposomes, and Biological Cells

Another type of an amphiphilic molecule (besides the detergents encountered above) consists of a phospholipid with an ionic phosphate head group and two hydrocarbon chains attached to it. A glycerol molecule links the two hydrocarbon chains together, and the phosphate group often has other hydrophilic molecules, such as choline and serine, attached to it. When a choline attaches, the molecule is called *phosphatidylcholine,* one of the most abundant phospholipids in all the cell membranes (Figure 2.44A). To form a micelle, water-soluble heads of amphiphilic molecules maintain contact with water, while the water insoluble ends are tucked inside.

A spherical micelle is the most stable configuration of amphipathic lipids that have a conical shape, such as fatty acids (Figure 2.44Ba). A bilayer is the most stable configuration of amphipathic lipids with a cylindrical shape, such as phospholipids (Figure 2.44Bb). Lipid bilayers are back-to-back arrangements of monolayers, and they typically form when there is more than one fatty acid chain connected to the polar head group. Double-chain fatty acids do not pack as well as single-chain fatty acids in a micelle configuration. Unilamellar vesicles, as shown in Figure 2.44Bc, are known as liposomes; multilamellar vesicles containing one or more bilayers arranged around the initial unilamellar vesicle. These types of vesicles are currently used as vehicles to deliver drugs in treatment of various diseases. For example, technologies have been developed to enclose chemotherapy and other drugs in unilamellar liposome bubbles.

The hydrophobic effect is essential for life. It forces the spontaneous clustering of lipids, forming biological membranes. Early in the development of life on Earth, chemicals organized themselves into cells, just like micelles. In humans, they are principally composed of phospholipids, formed from fatty acids, with some specialized proteins embedded in them, as shown in Figure 2.45. A wide variety of amphiphilic molecules can be accepted into a cell membrane. About half of a membrane surface is covered with proteins and sugars performing a wide variety of roles.

The bilayer that constitutes the cell surface—the plasma membrane—is permeable to some molecules, either passive or mediated by transmembrane proteins (pumps), impermeable to others (most water soluble molecules), and very flexible; call it a

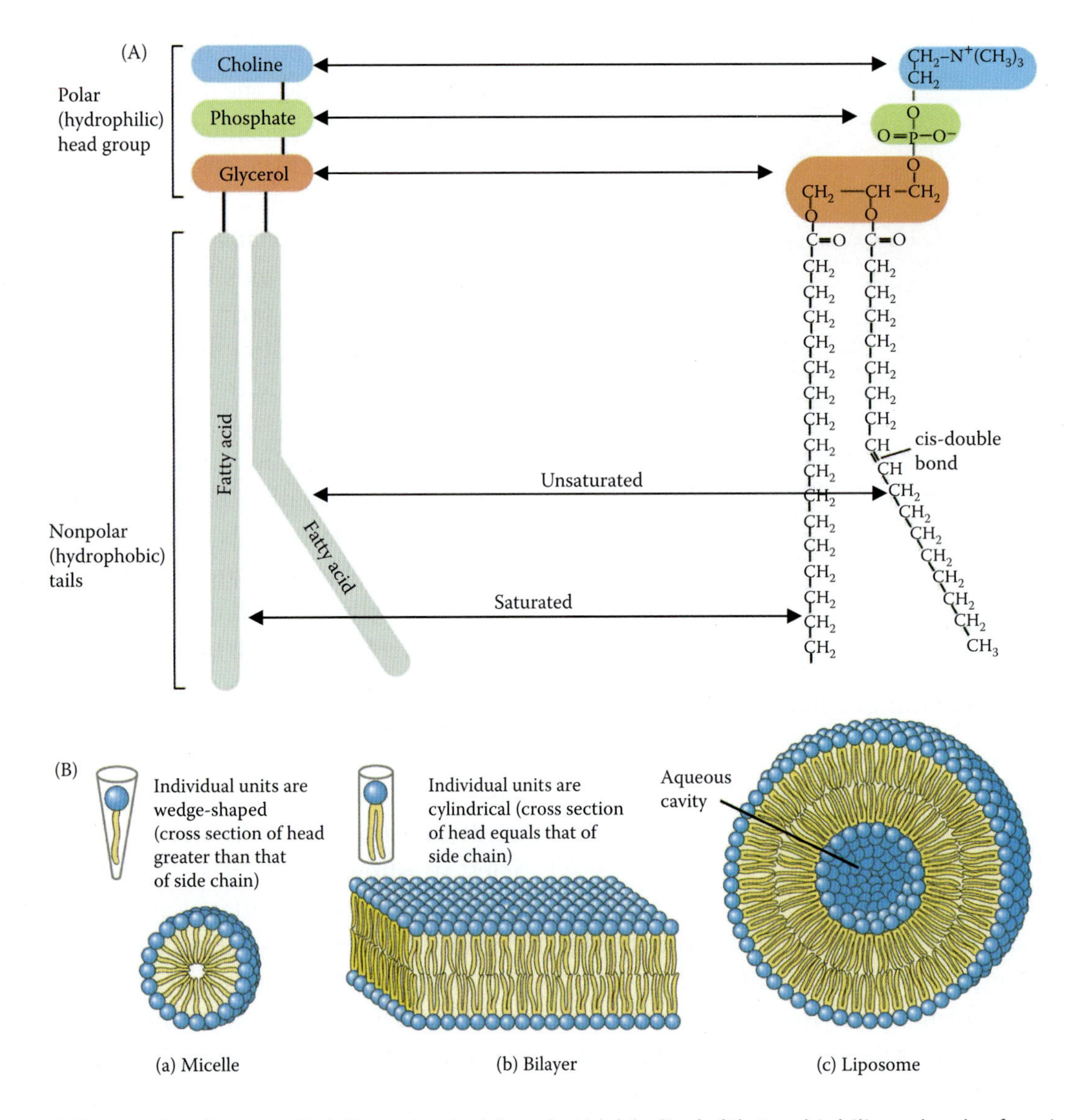

FIGURE 2.44 (A) Example of an amphiphilic molecule (phosphatidylcholine). (B) Amphiphilic molecules forming micelle (a), lipid membranes (b), and unilamellar liposome (c).

dynamic two-dimensional fluid in which lipids and proteins can move laterally. The membrane is also a very good electrical insulator, typically holding 70 mV over 45 Å (i.e., a field strength of 1.6 10^7 V/m), has no corners, and self-assembles.

As widely accepted as the idea of a plasma membrane with embedded proteins as ion channels and ion pumps has become, there are still valid objections to this model, as Gerald Pollack lucidly explains in *Cells, Gels and the Engines of Life.*[54] Pollack presents plenty of evidence that a cell is not enveloped by an ion-impermeant barrier. In this elegantly written book, he explains most cell functions by invoking the gel nature of the cytoplasm.

Natural Biopolymers

Even the most complex self-assembled structures of natural biopolymers, such as nucleic acids and proteins, are all driven by very simple balances of coil entropy and bond enthalpy; the coil entropy wants to expand, but the folded structure is held in place by enthalpic bonds such as hydrogen bonds, ionic bonds, and hydrophobic bonds. Being individually weak, these noncovalent interactions nevertheless can sum to an impressive total, often several hundreds of kilojoules, when many are present within a given macromolecule or between macromolecules. This explains the stability of the macromolecular structures. But being individually

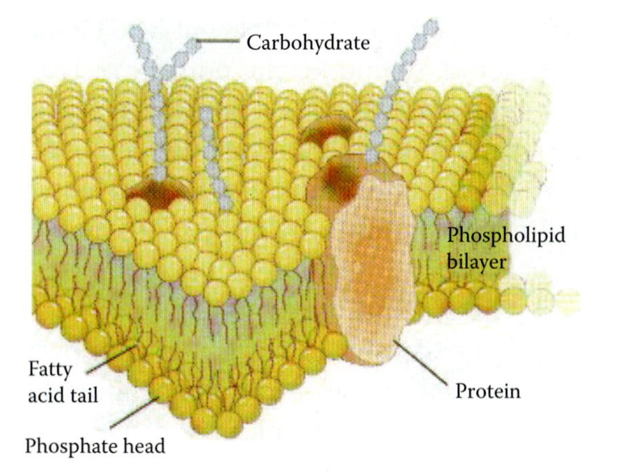

FIGURE 2.45 The cell wall of an animal cell is a phospholipid bilayer with embedded proteins regulating a series of functions, e.g., transport, recognition.

weak, individual noncovalent bonds in macromolecular structures can be broken and reformed rather easily. The latter in turn explains the flexibility in macromolecular structures. All biochemical processes have thus enthalpic and entropic contributions that combine to dictate spontaneity. DNA is a straightforward example of this, with a balance of mostly H-bonds and coil entropy (Figure 2.46). DNA structure arises from a favorable enthalpic term (enthalpy-dominated state) compensating for an unfavorable entropic term ($\Delta H < 0$ and $\Delta S < 0$). At a critical temperature $T_{critical}$, this reverses and entropy wins out. This *melting temperature* can be derived by solving $\Delta G = \Delta H - T\Delta S$, with ΔH from H-bonds balanced by ΔS from the amount of disorder, as follows.

$$T_{critical} = \frac{\Delta H}{\Delta S} \tag{2.8}$$

Because there are three hydrogen bonds between cytosine and guanine and only two between adenine and thymine, a higher C, G content makes for a higher melting temperature. The latter is illustrated in Figure 2.47.

In the 1980s, Nadrian Seeman (http://seemanlab4.chem.nyu.edu) from New York University took advantage of the intrinsic self-assembling properties of DNA molecules, as described here, to construct DNA molecular complexes. By synthesizing branched DNA and exploiting this behavior, Seeman and coworkers managed to assemble increasingly complex structures that do not occur in the natural environment (DNA-mediated assembly is described in Chapter 4, Figure 4.49). The use of stable branched DNA molecules permits one to make stick figures as depicted in Figure 4.49a, where we show a stable branched DNA molecule, and in Figure 4.49b, where we show such a molecule with sticky ends; four of these sticky-ended molecules are shown assembled into a quadrilateral. More exploits from the world of DNA nanotechnology, such as DNA origami, can be found in Chapter 3, this volume, on bottom-up manufacturing.

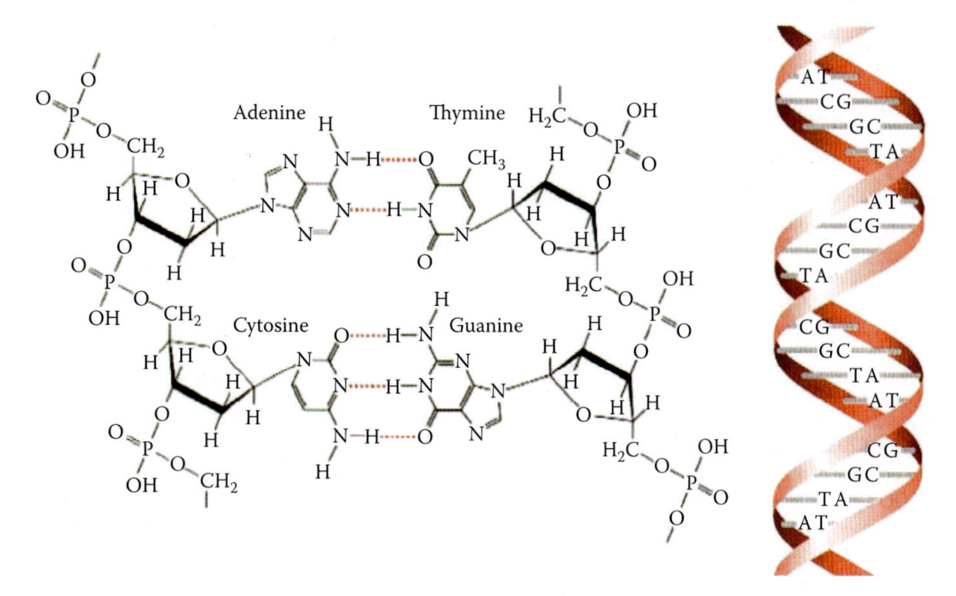

FIGURE 2.46 DNA: a question of entropy of coiling versus enthalpy of the hydrogen bonds (two between adenine and thymine and three between cytosine and guanine).

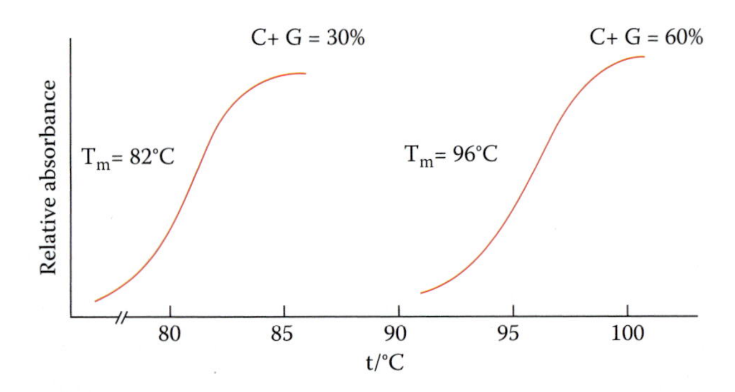

FIGURE 2.47 Melting temperature or critical temperature of DNA is reached at the moment that the entropy wins out. The higher the CG content, the higher the melting temperature.

Proteins are polypeptides prepared from 20 different amino acids, with a few different $\Delta H < 0$ bonds holding them together, to form a vast number of folded structures that govern biological function and biomaterials. As in micelle and DNA assembly, entropy and enthalpy balances are the basis of protein self-assembling behavior, such as the formation of a β-sheet, illustrated in the beautifully intricate "β-barrel" structure in green fluorescent protein (GFP) (see Volume I, Figure 7.108). In self-assembled, native proteins, the enthalpy wins over entropy (*H* dominated) and an increase in the temperature helps the entropy term till we observe denaturing or unfolding, as illustrated in Figure 2.48. Denaturing of a protein is the conversion of a biologically functional molecule into a nonfunctional form. There are many denatured states but only one native state, so entropically, the denatured protein state is by far the most favored condition for the protein to be in. Proteins can regenerate to their native state but only slowly. Denatured proteins have a greater tendency to aggregate. The transition from native to denatured happens at a critical temperature given by Equation 2.8. A typical coiling/ordering ΔS penalty for a complex protein such as GFP may be of the order of 40,000 kJ, the number of H-bonds needed to balance this, at ~50 kJ each, is about 800 (ignoring any hydrophobic bond help). Most of the associative effects are enthalpic in origin, except for the hydrophobic effect, which is essentially entropic in origin.

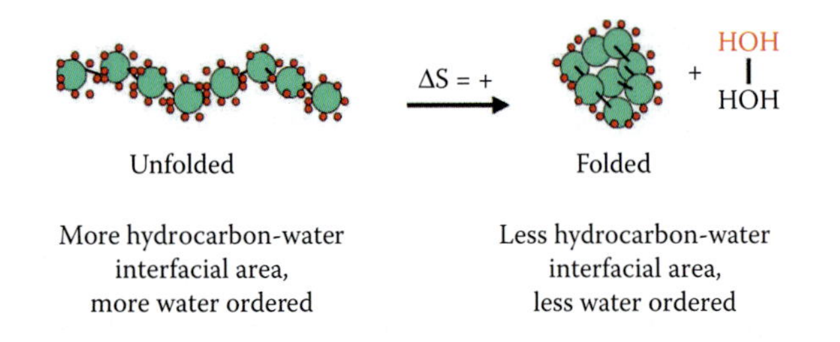

FIGURE 2.48 Protein folding and unfolding (denaturing).

In the exceptional case of a protein with an entropy dominated balance, such as with pyruvate carboxylase, a decrease in the temperature lowers the entropy term, and we observe denaturation at lower temperatures.

Biotechnology and Biomimetics

Introduction

Today, biotechnology is a major application target for miniaturization science, as engineered structures and nature's building tools have become similar in size, and financial rewards promise to be much larger for these life science ventures than they ever were for mechanical MEMS applications (see Chapter 10). After putting forward a definition for biotechnology and presenting examples of the use of miniaturization science in biotechnology, i.e., bio-MEMS, we make some concluding remarks on biomimetic approaches and finish this chapter by comparing some human and natural engineered structures. As examples of micromachining and nanomachining in biotechnology (bio-MEMS) we survey three examples here: genetically engineered proteins, miniaturized PCR and DNA chips. Other bio-MEMS and bio-NEMS examples, including electrochemical and optical immunosensors, microphysiometers, responsive drug delivery pills, glucose sensors, blood gas and blood electrolyte sensors, lab on a chips (LOCs), etc., can be found throughout the book. This section constitutes an attempt at sketching possible avenues for merging of biotechnology with nanochemistry and nanofabrication.

Biotechnology: Definition

Fermentation processes known from ancient times (the oldest known written recipe for beer brewing is written on a clay cuneiform tablet, part of an epic poem devoted to Ninkasi, the Sumerian goddess of beer) evolved eventually into a larger endeavor called biotechnology, employing either cells or cell extracts (e.g., enzymes) to produce foods, drugs, insecticides, etc. Generally, modern biotechnology is an umbrella term that covers various techniques, applied mainly to cells or molecules, which take advantage of biological processes in very precise ways. The discoveries of three different enzymes in the late 1960s, especially, heralded its advent. The first of these discoveries was reverse transcriptase enzyme, which converts RNA into the corresponding DNA; the second, an enzyme that cuts DNA at predefined places; and the third, an enzyme that joins separate pieces of DNA. Using these gene technology tools, the pooled content of a cell can thus be used to manufacture proteins that can be used for human consumption (e.g., as medicine). Because the genetic code is universal, almost any cell in any organism can "read" a gene and translate it into a desired protein. Bacteria that have been genetically engineered to carry the human insulin gene, for example, can produce human insulin. Cells might also be reengineered and substituted in living things to correct for inherited deficiency or performance. "Gene therapy" has made great progress, especially in plant engineering.

Genetically Engineered Proteins

Binding chemistry of a target to a specific protein at the nanoscale translates into a mechanical movement/motion of the protein, known as a change in conformation. This conformational change of proteins upon binding with a target molecule has been exploited in the design of very selective and sensitive biosensors. This is accomplished by labeling the protein, say with a fluorescent probe, in such a way that the probe does not hinder the binding of the protein with its target, while the change in signal reported by the probe upon binding is maximized.

Two important examples of genetically engineered proteins: green fluorescent protein (GFP) and calmodulin. GFP, from the molecular system of the bioluminescent jellyfish *Aequorea victoria,* we saw in Volume I, Chapter 7, are making quite an impact on biotechnology today as fluorescent markers. Here we will explore genetic manipulation of the calmodulin protein and its uses.

Through judicious genetic manipulation of a protein such as calmodulin [a 148-amino acid (17-kD) molecule, shown in Figure 2.49], a rationally designed molecular biosensor can thus be developed. The key is to always select the right location for attachment of the probe in the protein structure. Calmodulin is a "regulatory" protein, involved in protein-protein interactions and in signaling pathways that involve stimulating or inhibiting the activities of many enzymes and proteins. Calmodulin is a notably tough molecule: it withstands a very low pHs and even survives boiling water. One of its binding partners is Ca^{2+} ion, which matches the versatility of the protein. Calcium ions are abundant in all biological systems, and its importance derives from its role as a universal messenger ion. The ion conveys signals received at the cell surface to the inside of the cell (the concentration of calcium inside the cell is 1,000- to 10,000-fold less than in the extracellular fluid). Calcium ions are involved in such processes as the regulation of muscle contraction; the secretion of hormones, digestive enzymes, and neurotransmitters; the transport of salt and water across the intestinal lining; and the control of glycogen metabolism in the liver. The binding of Ca^{2+} to multiple sites of a calmodulin molecule induces a major conformational change—a hinge-type motion—that converts it from an inactive to an active form, allowing for interactions with other proteins and peptides and thus regulating the function of

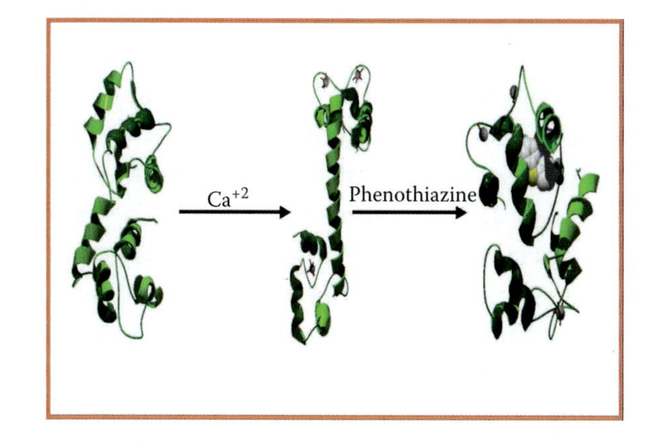

FIGURE 2.49 Model showing the exposure of drug binding site in calmodulin following binding of Ca^{2+}. Upon binding to calcium, calmodulin undergoes a change in conformation, which exposes two hydrophobic pockets located in the N- and C-domains. Certain hydrophobic peptides and the antipsychotic phenothiazine class of drugs interact with these exposed hydrophobic pockets.

the various enzymes and proteins. The calmodulin crystal structure has been well studied using x-ray crystallography and nuclear magnetic resonance (NMR) techniques, and it has been found to consist of two domains, the N- and C-domains.* Two high-affinity calcium-binding sites are located in the C-domain, and the other two calcium-binding sites (the low-affinity sites) are located in the N domain. Upon binding to calcium, calmodulin undergoes a change in conformation to an active form, which exposes two hydrophobic pockets located in the N- and C-domains. Certain hydrophobic peptides and the antipsychotic phenothiazine class of drugs interact with these exposed hydrophobic pockets, and in the presence of calcium ions, calmodulin also becomes a sensor for those peptides and drugs (Figure 2.49). Schauer-Vukasinovic et al.,[55] through site-directed mutagenesis of calmodulin, produced several mutants that varied from one another by a single cysteine residue at various positions in the protein skeleton. N-[2-(1-maleimidyl) ethyl]- 7-(diethylamino)coumarin-3-carboxamide (MDCC) (Invitrogen, CA, USA-prod #D-10253), a commercially available, environmentally sensitive, thiol-reactive fluorophore label was attached to the cysteine residue in each of the mutants, and the change in fluorescence signal upon binding the calcium ions was measured. Mutation sites were chosen on the basis of both the crystal and NMR structures of free calmodulin and calmodulin complexed with calcium. The MDCC-labeled calmodulin mutant showed a change in the fluorescence signal in the presence of calcium ions, the magnitude of which was dependent on the concentration of calcium.[55] By comparing the enhancement of the fluorescent response of the different mutants, a Ca^{2+} sensor with a detection limit as low as 2×10^{-9} M was built. The biggest change was noted for the mutant with a single cysteine at amino acid residue 109.[55] This sensing system is highly selective over other metals, including magnesium, europium, and barium. A calibration plot for calcium obtained with the MDCC-labeled calmodulin (CaM) mutant is shown in Figure 2.50.

* The N terminus of the protein denotes the first amino acid of its sequence, while the C terminus corresponds to the last one.

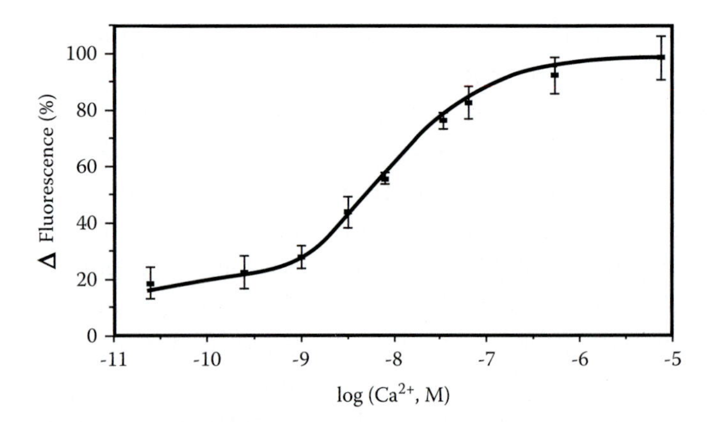

FIGURE 2.50 Calibration plot for calcium obtained with the MDCC [N-[2-(1-maleimidyl)ethyl]- 7-(diethylamino)coumarin-3-carboxamide]-labeled calmodulin (CaM) mutant. The *x*-axis denotes the concentrations of free calcium.[56]

In another study by the same team, when phenothiazine was added to calcium-bound fluorophore-labeled CaM109, a maximum of 100% quenching in the fluorescence was observed. This team then developed an compact disc based fluidic platform (see also Volume I, Chapter 6 on CD fluidics) for the sensing of phenothiazine-class of drugs based on this drug-dependent change in fluorescence signal.[56] This biosensing system also has been proposed as a model for the high-throughput screening of drugs that selectively interact with binding proteins.[57]

Polymerase Chain Reaction

The polymerase chain reaction (PCR) is to genes what Gutenberg's printing press was to the written word, but it took the 1995 O. J. Simpson trial to make PCR a household word. PCR perhaps is the most important tool of biotechnology today; it was conceived by Kary Mullis in 1983 (Figure 2.51). Mullis was granted a patent in 1989, and he received the 1993 Nobel Prize in Chemistry for his invention.[58] The invention of PCR, amplifying the amount of DNA in a sample, is as important for the biotechnology field as the transistor, amplifying electrons, is for electronics.

Mullis, a maverick, unbending to authority, never got the financial rewards or academic credit for his invention, a victim, to some degree, of commercial greed and academic jealousy. Mullis' technique enables the replication of any stretch of DNA as long as the arrangement of nucleotides at the two ends is known. Through PCR, a desired DNA sequence

FIGURE 2.51 Kary Mullis, inventor of polymerase chain reaction.

of viral, bacterial, plant, or human origin can be amplified hundreds of millions of times in a matter of hours. By amplifying minute amounts of DNA to analyzable quantities and thus circumventing the need for lengthy cell-culture methods, the technique has had major impact on clinical medicine, genetic disease diagnostics, forensic science, and evolutionary biology. In addition, the ability to isolate genes from organisms by essentially copying them directly from their genome has allowed for the rapid cloning of a myriad of genes whose function was unknown. This has also benefited the fields of genomics and proteomics and thus had an impact on not only science but medicine and biotechnology as well.

Through miniaturization, PCR is becoming faster and applicable to ever smaller samples. The original PCR process is based on the heat-stable DNA polymerase enzyme from *Thermus aquaticus* (*Taq*) (see Figure 2.52), a thermophilic bacterium that can withstand high temperatures and that can synthesize a complementary strand to a given DNA strand in a mixture containing the four DNA bases and two primer DNA fragments (each about 20 bases long) flanking the target sequence.

The mixture is heated to separate (denature or melt) strands of double-helix DNA containing the target sequence and then cooled to allow 1) the primers to find and bind (anneal) to their complementary sequences on the separate strands, and 2) the *Taq* polymerase to extend the primers into new complementary strands. The *Taq* polymerase needs about 1 min to synthesize 1 kb pair, so the synthesis time depends on the length of the product. Repeated heating and cooling cycles multiply the target DNA exponentially, since each new double strand separates to become two templates for further synthesis. In Figure 2.53a, we illustrate how each PCR cycle doubles the amount of DNA.

A typical temperature profile (Figure 2.53b) for PCR is as follows:

1. Denature at 93°C for 15–30 s
2. Anneal primer at 55°C for 15–30 s
3. Extend primers at 72°C for 30–60 s

The primer extension step has to be increased by roughly 60 s/kb to generate products longer than a few hundred bases. The above are typical instrument times; in fact, the denaturing and annealing steps occur almost instantly, but temperature rates in commercial instruments (Figure 2.53c) usually are less than 1°C/s when metal blocks or water are used for thermal equilibration and samples are contained in plastic microcentrifuge tubes. Instantaneous temperature changes are not possible because of sample,

FIGURE 2.52 The bacterium *Thermus aquaticus* was first discovered in several springs in the Great Fountain area of the Lower Geyser Basin at Yellowstone National Park.

(a)

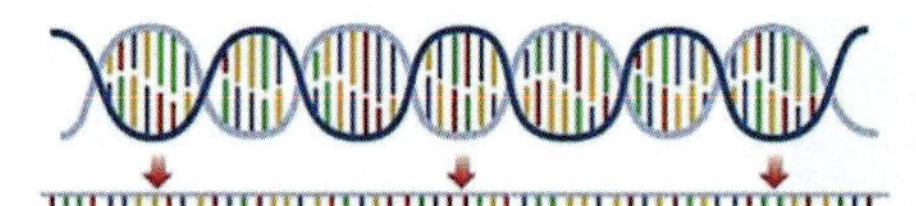

Step 1 – Denaturation (optimal temperature is 94°C). By heating the DNA, the double strand melts and opens to single-stranded DNA.

Step 2 – Annealing (optimal temperature is 60°C). The single-stranded primers bind to their complementary single-stranded bases on the denaturated DNA.

Step 3 – Extension: 72°C is the ideal temperature for the *Taq* polymerase to attach and start copying the template. The result is two new helixes in place of the first.

Step 4 – By applying this cycle several times, the quantity of DNA obtained is quickly enough to perform any analysis. Starting with one DNA molecule, after just 20 cycles there will be a million copies, and after 30 cycles there will be a billion copies.

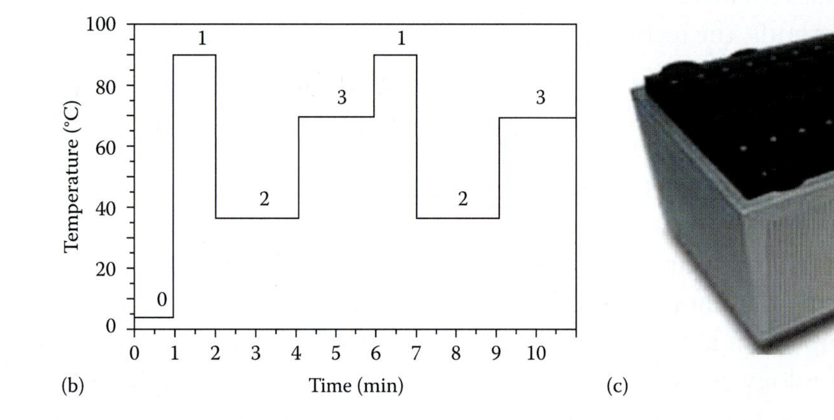

FIGURE 2.53 Polymerase chain reaction (PCR). (a) How PCR cycle doubles DNA. (b) A typical PCR temperature profile (1, denature; 2, anneal primers; 3, extend primers; see text). (c) Commercial PCR instrument from Cepheid: the SmartCycler II System.

container, and cycler heat capacities, and extended amplification times of 2–6 h result. During the periods where sample temperature is making a transition from one temperature to another, extraneous, undesirable reactions occur that consume important reagents and create unwanted and interfering compounds.[59] Micromachining has improved PCR in several respects.

By micromachining thermally isolated, low-thermal-mass PCR chambers, one is able to mass-produce a faster, less expensive, more energy efficient, and more specific PCR instrument. To appreciate better the fast thermal cycling of a smaller heater, see Chapter 8 on miniaturization of thermal elements. Given the PCR chambers' small size, smaller amounts of sample and expensive reagents are required. Moreover, rapid transitions from one temperature to another ensure that the sample spends a minimum of time at undesirable intermediate temperatures, so that the amplified DNA has optimum fidelity and purity.[59] Northrup et al.[60] fashioned Si microchambers with integrated thermally isolated heater elements and found that thermal cycling resulted in amplification without the need for active cooling. This is a direct result of the more effective cooling of these miniaturized Si structures (see Chapter 8). Northrup et al.[60] found that they could heat the PCR reactants as quickly as 35°C/s. Further work showed that integrated heaters on both sides of the reactant volume are required for optimal temperature uniformity. Later, it became evident that to make an economic miniaturized PCR instrument for diagnostic applications, the heater/cooler should be made part of a fixed reader instrument to enable a less expensive disposable unit. Along this line, Wilding et al.[61] performed a less integrated experiment with Si/glass microchambers of 4.4 to 8.9 μL inserted in a commercially available heater/cooler structure. The PCR total cycling time was reduced, but specificity was found to be inferior to that of commercial cyclers. This inferior performance may be linked to the fact that Si is poisonous to the *Taq* polymerase enzyme. The specificity problem can be avoided by derivatizing the Si surface or by machining PCR chambers from a more enzyme-compatible material. Another glaring problem with the two described Si-micromachined PCR approaches is that they do not provide an easy way to apply and remove the DNA sample. This is a major problem with miniaturized PCR and fluidic microsystems in general: the fabrication of connectors, receptacles, and packages that link the microstructure to the macroworld is still in its infancy (see Chapter 4). In principle, to avoid a hybrid connecting/packaging approach, this requires a manufacturing method with a very wide dynamic range, i.e., from microns to several millimeters or even centimeters. Also, for this reason using Si is not necessarily the best approach. The next three PCR examples illustrate nonsilicon alternatives.

Wittwer and Garling[62] and Wittwer et al.[63] optimized PCR temperature and cycle times, using a thermal cycler based on recirculating hot air and 10-μL samples in thin glass capillary tubes, with the sample temperature monitored by a miniature thermocouple. They found that product specificity could be improved significantly while decreasing the required amplification time by an order of magnitude. For a 536-base pair β-globin fragment from human DNA, optimal denaturation at 92–94°C took place in less than 1 s. Annealing for 1 s or less at 54–56°C gave the best product specificity and yield. Only elongation, at 75–79°C, required longer times. Denaturation yield actually decreased with times greater than 30 s, and nonspecific amplification was minimized with a rapid denaturation to annealing temperature transition. Wittwer's device has been commercialized by Idaho Technology; the RapidCycler from this company can perform 30 PCR cycles in less than 10 min.

The Si PCR work of Northrup et al.[60] eventually led to a commercial portable PCR instrument, i.e., the SmartCycler system (http://www.smartcycler.com) from Cepheid (see Figure 2.53c). This instrument incorporates a nonsilicon, rather large but thin PCR chamber with high thermal conductivity ceramic heater plates and integrated temperature sensors. At Cepheid it was understood early on that a major negative consequence of miniaturizing analysis equipment, in extremely small sample sizes, presents numerous chemicals and organisms that are present at levels below the typical detection limits. The topic of scaling of sample size was reviewed in Volume I, Chapter 6, on fluidics. There we saw that biological chemicals associated with clinical chemistry assays (between 10^{14} and 10^{20} copies/mL) and immunoassays (between 10^{7} and 10^{18} copies/mL) are readily assayed with very small sample volumes, in the range of picoliters and nanoliters. However, as evident from Figure 10.49, the minimum sample volume required for accurate DNA assays (from 10^{7} copies/mL to fewer than 100 copies/mL) is 100 μL, but a volume several milliliters of sample is more typical. As a consequence, an important Cepheid focus is the design of microfluidic diagnostic systems with sample preparation of larger input volumes. Along this line, the company fabricated a DNA capture device in which a continuous flow of biological sample results in a gradual accumulation of nucleic acid on the large silicon dioxide surface area of the chip.[59] It is not clear to this author why the latter device should be in Si, as there are many very large surface, porous materials available that are much less expensive.

DNA Chips

Introduction

With the rough draft of the Human Genome Project completed, the next most important question to solve is how genes relate to human health. Why do some people get certain diseases while others remain disease-free? A vast amount of expanding data supports the fact that it is rare for a single defect in one location on one gene to be responsible for a specific disease. The trend is toward the study of the biochemical and molecular functioning of a set of genes and how this complicated interaction leads to a diseased state. One way to tackle this study, we saw in Volume II, Chapter 8, is by using DNA microarrays. A DNA microarray, or DNA chip, consists of DNA molecules ranging from 20 to 1000 or more base pairs concentrated into specific areas on a solid support such as a glass chip.[64] The arrays are used as a capturing device of specific complementary gene products. Thousands of the same oligonucleotides can be attached in a specific location so that gene expression may be observed by counting the amount of oligos that bind. In this way, specific gene expression can be observed in particular cells, during specific stages of development, in diseased versus normal individuals, after administration of a drug, or in malignant cells, for example (see Volume II, Figure 8.23). Alternatively, one can incorporate many different oligonucleotides representing different genes on the same chip so that screening for the expression of many genes simultaneously is possible. The uses of this technology are broad and include disease diagnosis, individual response to drug therapy, and understanding the difference of gene expression between healthy and normal individuals.

Almost all current genetic testing either uses sequencing or hybridization. Sequencing involves hydrolyzing (digesting) a piece of DNA under controlled conditions, derivatizing the pieces as they come off the DNA strands, and analyzing them by chromatography or electrophoresis. In hybridization, pieces of genes bind to other pieces containing complementary sequences. Of the two techniques, sequencing is the more comprehensive and versatile, but hybridization is faster and, for small pieces of genes, more efficient. Rapid identification of genes or gene sequencing has become one of the main aims of medical diagnostics.

The field of DNA arrays involves microfluidics, MEMS machining, and self-assembly of DNA probes, which puts it at the forefront of mixing top-down and bottom-up manufacturing techniques. For the various microfabrication methodologies of DNA chips see Volume II, Chapter 8. In what follows we briefly review the major applications of microfabricated DNA chips.

Applications of DNA Arrays

In a typical cDNA array experiment, the genes of interest are first amplified using PCR (see above). After purification, the PCR products are printed on microscope glass slides by using a robot (see Volume II, Figure 8.22). For example, a DNA microarray, containing all of the genes in yeast (6400 of them) can be used to look at specific changes of gene expression in metabolism. Using in vitro reverse transcription, both the test and reference total RNA are labeled using two different fluorescent dyes. Laser excitation causes a specific emission that generates fingerprint-like spectra, which are measured using a scanning confocal laser microscope. The data are basically viewed as a change in the color of the two fluorescent readings at a specific point on the slide (a gene) when compared with a reference sample.

In an alternative DNA array approach, Fodor et al.[65] at Affymax (http://www.affymax.com) developed an elegant method combining on-chip combinatorial chemistry with established hybridization techniques and advanced fluorescence detection to make arrays of oligonucleotides (see Volume II, Figure 8.16). After synthesizing oligonucleotide arrays onto a glass slide, the glass slide is incubated with fluorescent-labeled DNA probes. After the DNA probes are hybridized to the oligonucleotide array, the complementary sequences are determined by fluorescent scanning. The oligonucleotide array format is used to detect mutations, locate target regions within genes, investigate gene expression, and identify gene function.[65] In this approach, the chemical synthesis and analysis of large numbers of potentially useful compounds can be carried out on a single small chip (say, a centimeter square); therefore, it becomes orders of magnitude quicker and cheaper to identify potential ligands to build new drugs or analyze DNA sequences.[66] A key advantage of this approach is that it enables chips to be manufactured directly from sequence databases, thereby removing the uncertain and burdensome aspects of sample handling and tracking. Another advantage of this technology is that the use of synthetic reagents minimizes chip to chip variation by ensuring a high degree of precision in each coupling cycle. One disadvantage of this approach, however, is the need for photomasks, which are expensive and time-consuming to design and build. In Volume II, Chapter 8, we described how the more recent maskless array synthesizer might get around the latter obstacle (Figure 8.19). Also, the hybridization can occur only through DNA fragments finding each other through diffusion, which is slow.[67] The Nanogen DNA chip described in Example 8.2 (Volume II, Chapter 8) is much faster, as DNA fragments reach the chip hybridization sites through electrophoresis (Nanogen, a San Diego–based company, went out of business in early 2010). A more detailed list of pros and cons of the various DNA array approaches is presented in Volume II, Chapter 8. The integration of DNA arrays within a sample preparation platform (e.g., a microfluidic CD) was discussed in Volume I, Chapter 6, on fluidics.

Biomimetics: Some Concluding Remarks

Early examples of successful biomimetics include artificial materials for implants, tissue engineering, an artificial retina based on the architecture of the human eye, and neural network software based on the human brain. The devices mentioned are relatively large, however, and are fabricated using top-down methodologies. The neural network software example is only a very crude approximation of a human brain. Human-made devices crafted with bottom-up methods will open up a much more rewarding paradigm in human manufacturing; nanochemistry holds the promise of the versatility of design offered by nature itself, and molecular self-assembly and replication add to the tremendous appeal of this type of nanotechnology. Obviously, there is something very attractive about the small size of nature's preferred building blocks such as proteins. In biological systems, the energy efficiency is approximately proportional to the 2/3 power of the linear dimension. This is because metabolism is proportional to the second power of the linear dimension (surface of the organism, l^2), and energy uptake (feeding) is proportional to the body volume (l^3), so the smaller organism is, the higher its efficiency (see the discussions of *scaling laws* in Chapter 7). Nature provides excellent examples in the design of efficient micro- and nanosystems as it optimizes scaling laws in the microdomain and even exploits the quantum-size effects of its components. These biosystems have been tested and selected by eons of evolution, and at a time when we find that micromachining tools are limiting the fabrication of ever smaller devices, exploration of some of nature's tools and building blocks promises to be very rewarding. As one of the many motivating examples of the promise of biomimetics, consider enzymes. There is a plethora of enzymes that outperform synthetic catalysts by several orders of magnitude. For instance, there is no artificial catalyst for producing ammonia from its elements at ambient temperature and pressure, as the nitrogenase of nodule-forming bacteria does (the industrial Haber-Bosch synthesis of ammonia from nitrogen and hydrogen at high temperatures and pressures pales in comparison).

Most miniaturization, especially in integrated circuits (ICs), over the next ten years will be based on top-down nanofabrication, and Moore's law will continue to run its course. In parallel, we will continue to adapt bottom-up nanochemistry principles and combine them with nanofabrication. There is a significant gap between the scale of individual molecular structures of nanochemistry and the sub-micrometer structures of nanofabrication. It is exactly in that gap, from about one nanometer to several hundred nanometers, where fundamental materials properties are defined. It is also on the nanometer scale that quantum effects become significant.

There is little doubt that biotechnology analysis tools will keep on improving at an increasingly faster rate. Desktop DNA sequencers and 3D protein readers will be a reality in the not-so-distant future. Genetic engineering will have the most profound impact, though, on how humankind looks at manufacturing. Since it is already possible to synthesize a virus bottom-up, given the sequence of the bases in its genes, it seems quite likely that we will be able to manufacture synthetic viruses designed to enter a cell and carry out diagnostic and therapeutic tasks. Venter, we saw earlier, is using bottom-up synthesis methods to create a minimal gene setup required to sustain life in a test tube. Photovoltaic solar cells today have a high conversion efficiency of about 10–15% but are expensive to deploy and maintain. Crops grown for energy are also expensive, involve harvesting, and are only 1% efficient. It is conceivable that genetic engineering will enable the production of energy crops that convert sunlight into fuel at a 10% efficiency.[68] Through nanochemistry, the current digital information technology revolution might well be followed by a new analog manufacturing revolution. Today, computers let us shape our digital environment, but by giving computers the means to manipulate the analog world of atoms as easily as they manipulate bits, the same kind of personalization may be brought to our physical three-dimensional environment. In this context Gershenfeld from the Massachusetts Institute of Technology Media Laboratory envisions a personal fabricator akin to the personal computer.[69]

A human society based on nanomachining could perhaps be a much more balanced one, with a manufacturing approach based on how the species itself is made. Products will be based on a fundamental understanding of the assembly of their ultimate components, atoms, molecules, and proteins, and on how to induce self-assembly into useful objects. Materials will be degradable, flexible, and fully reusable. The smaller building blocks used in manufacture will enable products of more variety and intelligence. There will be less emphasis on the traditional engineering materials, such as steel, wood, stone, composites, and carbon, and proteins will become much more important.

The transition toward a nanosociety will require a major shift in workforce skill level as manipulation of data and applying knowledge of bioengineering will be part of a manufacturing worker's daily duties. In academia, less hyperspecialization and better grounding in all the sciences and engineering will be a must as the traditional "dry" engineering and sciences, such as electrical engineering and mechanical engineering, will merge with the "wet" sciences and engineering, such as biology and bioengineering.

Appendix 2A: Uses of Extremophiles

Source	Uses
Hyperthermophiles	
DNA polymerases	DNA amplification by PCR
Alkaline phosphatase	Diagnostics
Proteases and lipases	Dairy products
Lipases, pullulanases, and proteases	Detergents
Proteases	Baking, brewing, and amino acid production from keratin
Amylases, α-glucosidase, pullulanase, and xylose/glucose isomerases	Baking, brewing, and amino acid production from keratin
Alcohol dehydrogenase	Chemical synthesis
Xylanases	Paper bleaching
Lenthionin	Pharmaceutical
S-layer proteins and lipids	Molecular sieves
Oil degrading microorganisms	Surfactants for oil recovery
Sulfur oxidizing microorganisms	Bioleaching, coal and waste gas desulfurization
Hyperthermophilic consortia	Waste treatment and methane production

(Continued)

Appendix 2A: (*Continued*)

Source	Uses
Psychrophiles	
Alkaline phosphatase	Molecular biology
Proteases, lipases, cellulases, and amylases	Detergents
Lipases and proteases	Cheese manufacture and dairy production
Proteases	Contact-lens cleaning solutions, meat tenderizing
Polyunsaturated fatty acids	Food additives, dietary supplements
Various enzymes	Modifying flavors
ß-Galactosidase	Lactose hydrolysis in milk products
Ice nucleating proteins	Artificial snow, ice cream, other freezing applications in the food industry
Ice minus microorganisms	Frost protectants for sensitive plants
Various enzymes (e.g., dehydrogenases)	Biotransformations
Various enzymes (e.g., oxidases)	Bioremediation, environmental biosensors
Methanogens	Methane production
Halophiles	
Bacteriorhodopsin	Optical switches and photocurrent generators in bioelectronics
Polyhydroxyalkanoates	Medical plastics
Rheological polymers	Oil recovery
Eukaryotic homologs (e.g., *myc* oncogene product)	Cancer detection, screening antitumor drugs
Lipids	Liposomes for drug delivery and cosmetic packaging
Lipids	Heating oil
Compatible solutes	Protein and cell protectants in variety of industrial uses, e.g., freezing, heating
Various enzymes, e.g., nucleases, amylases, proteases	Various industrial uses, e.g., flavoring agents
γ-Linoleic acid, ß-carotene, and cell extracts (e.g., *Spirulina* and *Dunaliella*)	Health foods, dietary supplements, food coloring, and feedstock
Microorganisms	Fermenting fish sauces and modifying food textures and flavors
Microorganisms	Waste transformation and degradation, e.g., hypersaline waste brines contaminated with a wide range of organics
Membranes	Surfactants for pharmaceuticals
Alkaliphiles	
Proteases, cellulases, xylanases, lipases, and pullulanases	Detergents
Proteases	Gelatin removal on x-ray film
Elastases, keratinases	Hide dehairing
Cyclodextrins	Foodstuffs, chemicals, and pharmaceuticals
Xylanases and proteases	Pulp bleaching
Pectinases	Fine papers, waste treatment, and degumming
Alkaliphilic halophiles	Oil recovery
Various microorganisms	Antibiotics
Acidophiles	
Sulfur-oxidizing microorganisms	Recovery of metals and desulfurication of coal
Microorganisms	Organic acids and solvents

Appendix 2B: The Twenty Amino Acids Found in Biological Systems

Alanine, arginine, asparagine, aspartic acid, cysteine, glutamic acid, glutamine, glycine, histidine, isoleucine, leucine, lysine, methionine, phenylalanine, proline, serine, threonine, tryptophan, tyrosine, and valine.

All amino acids have the same general formula:

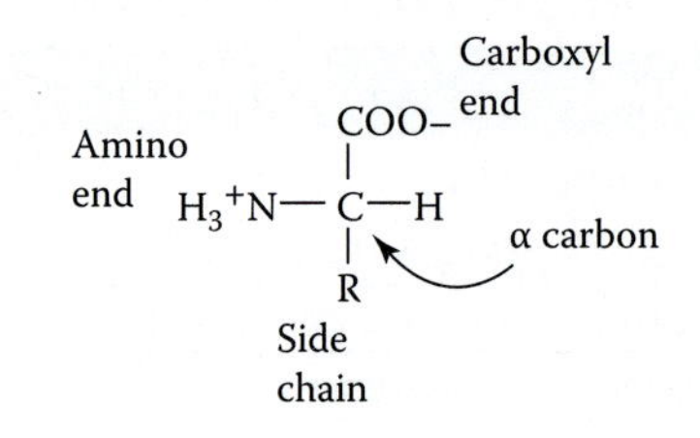

Amino acids with hydrophobic side groups

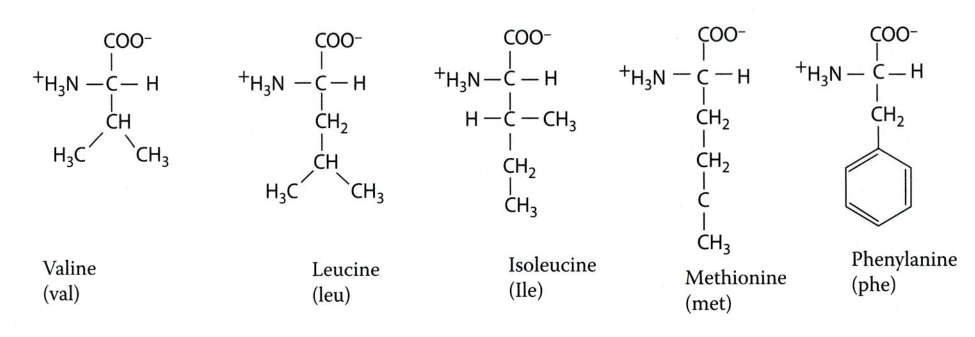

Amino acids with hydrophilic side groups

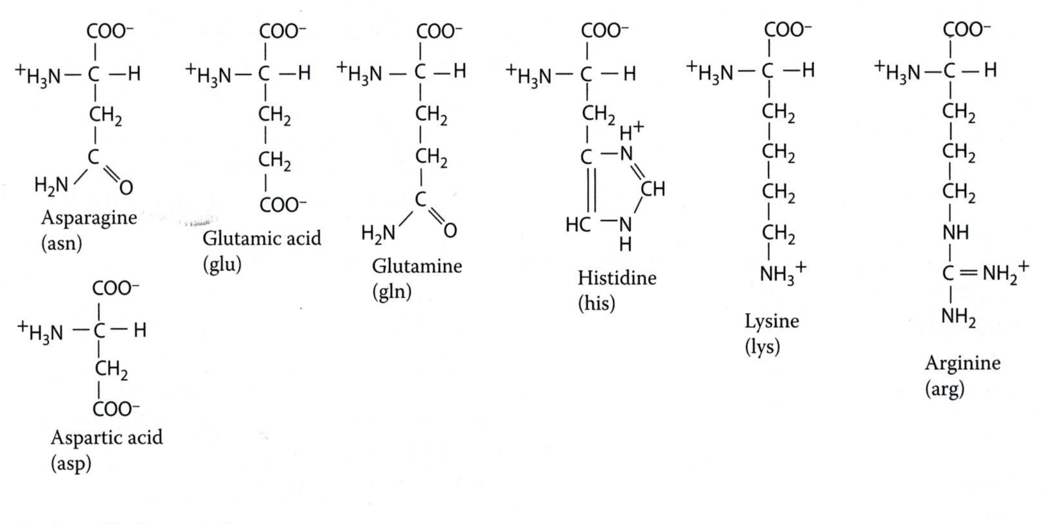

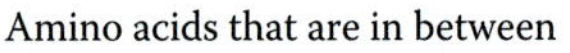
Amino acids that are in between

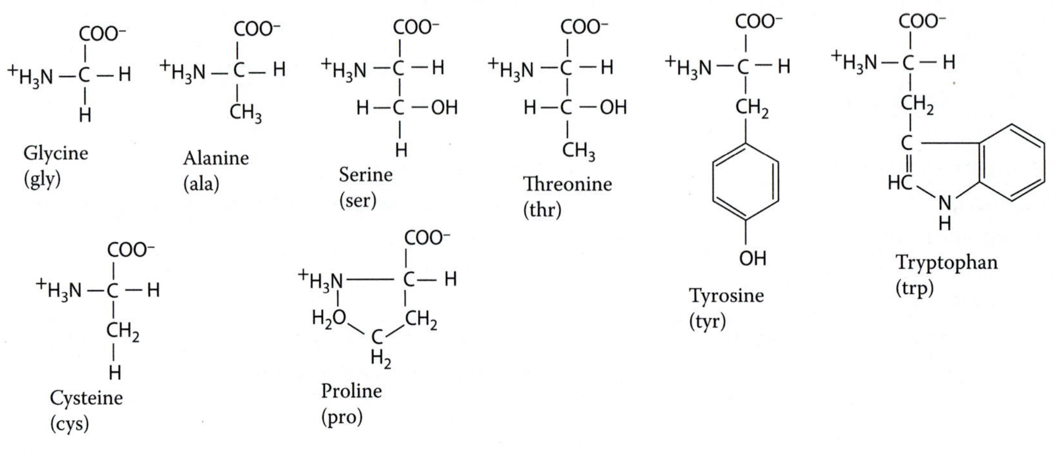

Appendix 2C: The Universal Code

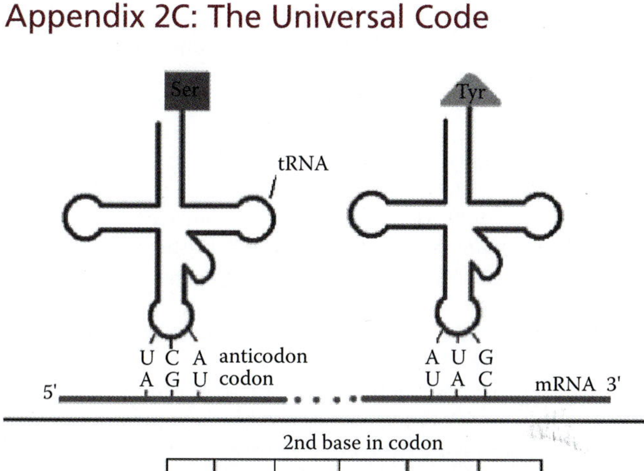

2nd base in codon

1st base in codon	U		A	G	3rd base in codon
U	Phe	Ser	Tyr	Cys	U
	Phe	Ser	Tyr	Cys	C
	Leu	Ser	STOP	STOP	A
	Leu	Ser	STOP	Trp	G
C	Leu	Pro	His	Arg	U
	Leu	Pro	His	Arg	C
	Leu	Pro	Gln	Arg	A
	Leu	Pro	Gln	Arg	G
A	Ile	Thr	Asn	Ser	U
	Ile	Thr	Asn	Ser	C
	Ile	Thr	Lys	Arg	A
	Met	Thr	Lys	Arg	G
G	Val	Ala	Asp	Gly	U
	Val	Ala	Asp	Gly	C
	Val	Ala	Glu	Gly	A
	Val	Ala	Glu	Gly	G

FIGURE A2C.1 The genetic code.

Questions

2.1: Immunoassays are based on the specific binding of antibodies with antigens.
   (a) Describe an ELISA.
   (b) Explain how nonspecific binding can produce unwanted signals in both immunoassays and in DNA assays. (*Nonspecific binding* refers to the binding of adsorption of molecules onto the surface without any preference or selectivity.)

2.2: With the aid of diagrams, describe:
   (a) A competitive immunoassay
   (b) A sandwich immunoassay

2.3: Why are living systems, at first glance, violating the second law of thermodynamics?

2.4: Briefly discuss the "who was on first" dilemma: DNA, RNA, or proteins?

2.5: What models are complexity and communication sciences contributing to the biogenesis question?

2.6: Draw a diagram to illustrate how proteins are manufactured.

2.7: List examples of emergent properties of complex systems.

2.8: Find some examples of man-made materials that feature self-repair.

2.9: Which of the following statements regarding GFP is True/False?
   ________ The excitation maxima of GFP are at 395 and 470 nm.
   ________ GFP is extracted from *A. victoria* jellyfish.
   ________ GFP has a highly unstable molecular structure.
   ________ GFP emits light at high intensity but with a very low quantum efficiency.
   ________ GFP gets activated by red or infrared light.

2.10: What are DNA microarrays? What are their applications?

2.11: Define what an antigen and an antibody are. What class of proteins are antibodies? Sketch an antibody, and illustrate where the antigen binds to the antibody. Explain what forces contribute to the antigen-antibody interaction.

2.12: Explain the error catastrophe in gene copying.

2.13: Explain the difference between DNA cloning, reproductive cloning, and biomedical or therapeutic cloning.

2.14: PCR is a very important analytical methodology to amplify genetic material. Briefly describe how it works and the merits/problems miniaturization of PCR entails.

2.15: Contrast human manufacturing techniques with nature's methods.

2.16: What does the term "binding constant" refer to in antibody chemistry? What is the value of a typical binding constant between an antigen and an antibody? What does the term "cross-reactivity" mean?

2.17: What is the difference between polyclonal and monoclonal antibodies? Briefly discuss the advantages and/or disadvantages of monoclonal vs. polyclonal antibodies.

2.18: The antigen-antibody interaction is used in a good number of bioanalytical techniques. One such example is affinity chromatography.

Describe the principle of affinity chromatography and give one example where you would need to use affinity chromatography.

2.19: Match the following terms with the appropriate definitions:

(1) Fluorophore	(a) Crystals composed of materials from periodic groups II-VI, III-V, or IV-VI and can be conjugated to biomolecules
(2) Stokes shift	(b) Method used to measure the ability of a molecule to move around in a cell over time
(3) FRAP	(c) Intense fluorescent probe used in many biological studies
(4) Quantum dots	(d) Difference in wavelength between fluorescence excitation maximum and the fluorescence emission maximum
(5) Green fluorescent protein	(e) Molecule that can be excited by light to emit fluorescence

Thanks to Mr. Youssef Farhat, UC Irvine.

2.20: What is a biosensor? Draw a schematic of a typical biosensor with its components and explain its operation with an example.

References

1. Vogel, S. 1998. *Cats' paws and catapults.* New York: W. W. Norton.
2. Gibson, D. G., G. A. Benders, C. Andrews-Pfannkoch, E. A. Denisova, H. Baden-Tillson, J. Zaveri, T. B. Stockwell, A. Brownley, D. W. Thomas, M. A. Algire, C. Merryman, L. Young, V. N. Noskov, J. I. Glass, J. C. Venter, C. A. Hutchison III, and H. O. Smith. 2008. Complete chemical synthesis, assembly, and cloning of a mycoplasma genitalium genome. *Science* 319:1215–20.
3. Morris, R. 1999. *Artificial worlds: Computers, complexity, and the riddle of life.* New York: Plenum.
4. Crick, F. 1981. *Life itself: Its origin and nature.* New York: Simon & Schuster.
5. Haynie, D. T. 2001. *Biological thermodynamics.* Cambridge, United Kingdom: Cambridge University Press.
6. Stewart, I. 1998. *Life's other secret: The new mathematics of the living world.* New York: Wiley.
7. Casti, J. L. 1989. *Paradigms lost.* New York: William Morrow.
8. Casti, J. L. 2000. *Paradigms regained.* New York: William Morrow.
9. Poundstone, W. 1999. *Carl Sagan: A life in the cosmos.* New York: Henry Holt.
10. Davies, P. 1999. *The fifth miracle: The search for the origin and meaning of life.* New York: Simon & Schuster.
11. de Duve, C. 1995. *Vital dust: The origin and evolution of life on earth.* New York: Basic Books.
12. Ridley, M. 2000. *Genome: The autobiography of a species in 23 chapters.* New York: Perennial.
13. Lee, D. H., K. Severin, Y. Yokobayashi, and M. R. Ghadriri. 1997. Emergence of symbiosis in peptide self-replication through a hypercyclic network. *Nature* 390: 591–94.
14. Ertem, G., and J. P. Ferris. 1996. Synthesis of RNA oligomers on heterogeneous templates. *Nature* 379:238–40.
15. Woese, C. R., and G. E. Fox. 1977. Phylogenetic structure of the prokaryotic domain: The primary kingdoms. *Proc Natl Acad Sci USA* 74:5088–90.
16. Capra, F. 1996. *The web of life.* New York: Doubleday.
17. Kauffman, S. 2000. *Investigations.* Oxford, United Kingdom: Oxford University Press.
18. Rebek, J. 1994. Synthetic self-replicating molecules. *Sci Am* 271:48–55.
19. Loewenstein, W. R. 1999. *The touchstone of life.* Oxford, United Kingdom: Oxford University Press.
20. Shapiro, R. 1991. *The human blueprint: The race to unlock the secrets of our genetic script.* New York: St. Martin's Press.
21. Watson, J. D. 1968. *The double helix.* New York: New American Library.
22. Crick, F. 1988. *What mad pursuit: A personal view of scientific discovery.* New York: Basic Books.
23. Maddox, J. 1988. *What remains to be discovered.* New York: Simon & Schuster.
24. Robinson, D. W. 1994. Medicine, molecules, and microengineering. *Nanobiology* 3:147–48.
25. Goldberg, A. L., S. J. Elledge, and J. W. Harper. 2001. The cellular chamber of doom. *Sci Am* 284:68–73.
26. Dawkins, R. 1989. *The selfish gene.* Oxford, United Kingdom: Oxford University Press.
27. Eigen, M., and P. Schuster. 1979. *The hypercycle: A principle of natural self-organization.* New York: Springer-Verlag.
28. Edwards, S. A. 2006. *The nanotech pioneers: Where are they taking us?* Weinheim, Berlin: Wiley-VCH Verlag GmbH & Co.
29. Dawkins, R. 1986. *The blind watchmaker.* New York: W. W. Norton.
30. Britten, R. J., and E. H. Davidson. 1969. Gene regulation for higher cells: a theory. *Science* 165:349–57.
31. Agrawal, A., Q. M. Eastman, and D. G. Schatz. 1998. Transposition mediated by RAG1 and RAG2 and its implications for the evolution of the immune system. *Nature* 394:744–51.
32. Cantor, C. R. 1990. Orchestrating the Humane Genome Project. *Science* 248:49–51.
33. Dunham, I., N. Shimizu, B. A. Roe, S. Chissoe, A. R. Hunt, J. E. Collins, et al. 1999. The DNA sequence of human chromosome 22. *Nature* 402:489–95.
34. Kenyon, C. 1996. Ponce d'Elegans: genetic quest for the fountain of youth. *Cell* 84:501–4.

35. Kenyon, C. 1996. Environmental factors and gene activities that influence lifespan. *C. elegans*. Ed. D. L. Riddle. Cold Spring Harbor Monograph Series no. 33. Cold Spring Harbor, NY: Cold Spring Harbor Laboratory Press.
36. Passwater, R. A. 1985. *The antioxidants.* New Canaan, CT: Keats Publishing.
37. Hayflick, L., and P. S. Moorhead. 1961. The serial cultivation of human diploid cell strains. *Exp Cell Res* 25:585–621.
38. Hayflick, L. 1985. The cell biology of aging. *Clin Geriatr Med* 1:15–27.
39. Hayflick, L. 1994. *How and why we age.* New York: Ballantine Books.
40. Hayflick, L. 2000. New approaches to old age. *Nature* 403:365.
41. Wilmut, I., A. E. Schnieke, J. McWhir, A. J. Kind, and K. H. S. Campbell. 1997. Viable offspring derived from fetal and adult mammalian cells. *Nature* 385:810–13.
42. Briggs, R., and T. J. King. 1952. Transplantation of living nuclei from blastula cells into enucleated frogs' eggs. *Proc Natl Acad Sci USA* 38, 455–63.
43. Di Berardino, M. A., R. G. McKinnell, and D. P. Wolf. 2003. The golden anniversary of cloning: A celebratory essay. *Differentiation* 71:398–401.
44. Takahashi, K., K. Tanabe, M. Ohnuki, M. Narita, T. Ichisaka, K. Tomoda, and S. Yamanaka. 2007. Induction of pluripotent stem cells from adult human fibroblasts by defined factors. *Cell* 131:861–72.
45. Yu, J., M. A. Vodyanik, K. Smuga-Otto, J. Antosiewicz-Bourget, J. L. Frane, S. Tian, J. Nie, G. A. Jonsdottir, V. Ruotti, R. Stewart, I. I. Slukvin, and J. A. Thomson. 2007. Induced pluripotent stem cell lines derived from human somatic cells. *Science* 5858:1917–20.
46. Margulis, L., and D. Sagan. 1986. *Microcosmos: Four billion years of evolution from our microbial ancestors.* New York: Simon & Schuster.
47. Cann, R. L., M. Stoneking, and A. C. Wilson. 1987. Mitochondrial DNA and human evolution. *Nature* 325:31–36.
48. Svoboda, K., C. F. Schmidt, B. J. Schnapp, and S. M. Block. 1993. Direct observation of kinesin stepping by optical trapping interferometry. *Nature* 365:721–27.
49. Noji, H., R. Yasuda, M. Yoshida, and K. Kinoshita. 1997. Direct observation of the rotation F1-ATPase. *Nature* 386:299–302.
50. Montemagno, C., G. Bachand, S. Stelick, and M. Bachand. 1999. Constructing biological motor powered nanomechanical devices. (Special conference issue of *Nanotechnology*: Sixth Foresight Conference on Molecular Nanotechnology.) *Nanotechnology* 10:225–31.
51. Kudo, S., Y. Magariyama, and S. Aizawa. 1990. Abrupt changes in flagellar rotation observed by laser dark-field microscopy. *Nature* 346:677–79.
52. Hunter, I. W., and S. Lafontaine. 1992. A comparison of muscle with artificial actuators. *Technical Digest: 1992 Solid State Sensor and Actuator Workshop*. Hilton Head Island, SC.
53. Meyer, E. E., K. J. Rosenberg, and J. Israelachvili. 2006. Recent progress in understanding hydrophobic interactions. *Proc Natl Acad Sci USA* 103:15739–46.
54. Pollack, G. H. 2001. *Cells, gels and the engines of life.* Seattle, WA: Ebner and Sons Publishers.
55. Schauer-Vukasinovic, V., L. Cullen, and S. Daunert. 1997. Rational design of a calcium sensing system based on induced conformational changes of calmodulin. *J Am Chem Soc* 119:11102–103.
56. Douglass, P. M., B. R. Wenner, R. D. Johnson, G. Barrett, J. C. Ball, P.-H. Hsu, J. Lee, L. G. Bachas, M. Madou, and S. Daunert. 2000. Biomimetic detection schemes on the lab CD. *Proceedings, NanoSpace 2000.* Houston, TX: The Institute for Advanced Interdisciplinary Research.
57. Wenner, B. R., P. M. Douglass, S. Shrestha, B. V. Sharma, S. Lai, M. Madou, and S. Duanert. 2001. Genetically designed biosensing systems for high-throughput screening of pharmaceuticals, clinical diagnostics, and environmental monitoring. *Progress in biomedical optics and imaging.* SPIE.
58. Mullis, K. 1998. *Dancing naked in the minefield.* New York: Vintage Books.
59. Petersen, K., W. McMillan, G. Kovacs, A. Northrup, L. Christel, and F. Pourahmadi. 1998. The promise of miniaturized clinical diagnostic systems. *IVD Technology* July [online].
60. Northrup, M. A., M. T. Ching, R. M. White, and R. T. Watson. 1993. DNA amplification with a microfabricated reaction chamber. *7th International Conference on Solid-State Sensors and Actuators (Transducers '93)*. Yokohama, Japan.
61. Wilding, P., M. A. Shoffner, and L. J. Kricka. 1994. PCR in a silicon microstructure. *Clin Chem* 40:1815–18.
62. Wittwer, C. T., and D. J. Garling. 1991. Rapid cycle DNA amplification: Time and temperature optimization. *BioTechniques* 10:76–83.
63. Wittwer, C. T., G. C. Fillmore, and D. J. Garling. 1990. Minimizing the time required for DNA amplifications by efficient heat transfer to small samples. *Anal Biochem* 186:328–31.
64. Schena, M., R. A. Heller, T. P. Theriault, K. Konrad, E. Lachenmeier, and R. W. Davis. 1998. Microarrays: Biotechnology's discovery platform for functional genomics. *TIB Tech* 16:301–06.
65. Fodor, S. P. A., J. L. Read, M. C. Pirrung, L. Stryer, A. T. Lu, and D. Solas. 1991. Light-directed, spatially addressable parallel chemical synthesis. *Science* 251:767–73.
66. Eggers, M., M. Hogan, R. K. Reich, J. Lamture, D. Ehrlich, M. Hollis, B. Kosicki, T. Powdrill, K. Beattie, S. Smith, et al. 1994. A microchip for quantitative detection of molecules utilizing luminescent and radioisotope reporter groups. *Biotechniques* 17(3):516–25.
67. Friend, S. H. 1999. DNA microarrays and expression profiling in clinical practice. *BMJ* 319:1306.
68. Dyson, F. J. 1999. *The sun, the genome, and the internet—Tools of scientific revolutions.* Oxford, United Kingdom: Oxford University Press.
69. Gershenfeld, N. 1999. *When things start to think.* New York: Henry Holt and Company.
70. Gibson, D. G., J. I. Glass, C. Lartigue, V. N. Noskov, R.-Y. Chuang, M. A. Algire, G. A. Benders, M. G. Montague, L. Ma, M. M. Moodie, C. Merryman, S. Vashee, R. Krishnakumar, N. Assad-Garcia, C. Andrews-Pfannkoch, E. A. Denisova, L. Young, Z.-Q Qi, T. H. Segall-Shapiro, C. H. Calvey, P. P. Parmar, C. A. Hutchison III, H. O. Smith, and J. C. Venter. 2010. Creation of a bacterial cell controlled by a chemically synthesized genome. *Science* 329(5987):52–56.

3

Nanotechnology: Top-Down and Bottom-Up Manufacturing Approaches Compared

Outline

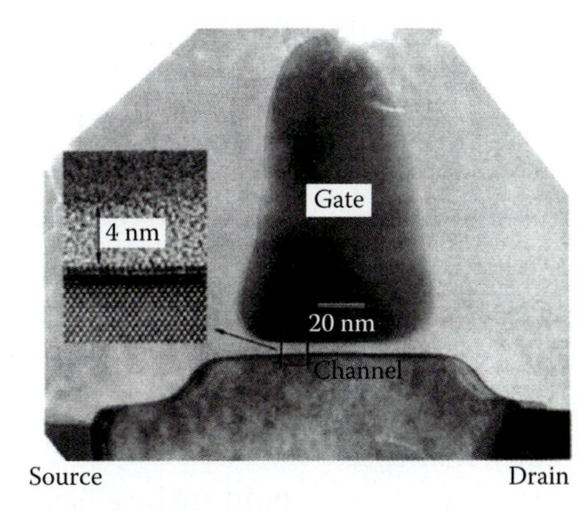

Top-down nanotechnology. Electron micrograph of an SiO_2/Si interface in a Si N-channel MOS transistor fabricated using top-down methods.

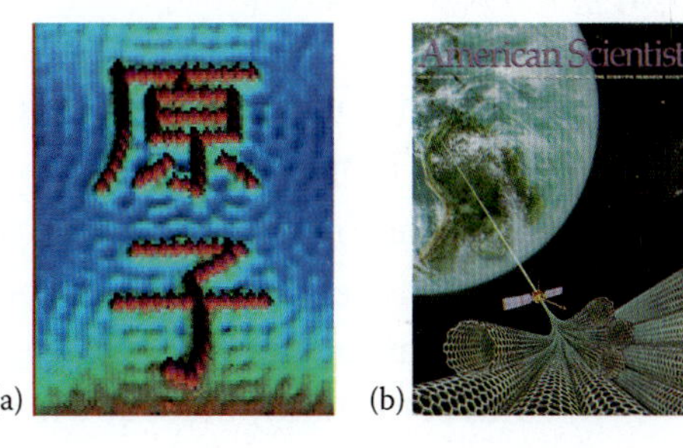

(a) Kanji character for "atom," by Lutz and Eigler, IBM, Almaden. An example of manipulation/mechanosynthesis of iron atoms on a copper surface with a scanning tunneling microscope. (Courtesy of D. Eigler.) Mechanical manipulation of atoms is a bottom-up technique. (b) Cover of *American Scientist*: the carbon nanotube launch station. Carbon nanotubes are usually made with bottom-up techniques.

Our ability to arrange atoms lies at the foundation of technology.

K. Eric Drexel
1986

Introduction

Nario Taniguchi coined the word *nanotechnology* in 1974 to describe machining with tolerances of less than 1 μm (see Volume II, Figure 6.3). Humans are still best at building structures we can grasp with our hands and see with our eyes, and we have only relatively recently started to intentionally craft objects at the nanoscale with nanotechnology. Over the last two decades, nanotechnology has come to mean two quite different things. In the broadest sense it is any technology dealing with building structures between 1 and 100 nm in size, in a narrower sense, as promoted by K. Eric Drexel, it involves designing and building machines in which every atom and chemical bond is precisely specified.[1]

About the first definition, one commentator quipped that some companies are willing to say their product is based on nanotechnology if it is made of atoms![1] Clearly, some further refinement is in order for a working definition. Our criteria for nanotechnology are that the miniaturized structures have at least one dimension smaller than 100 nm, are crafted with a novel technique, and have been intentionally designed with a specific nanofeature in mind. Medieval church stained glass, with embedded gold nanoparticles, does not qualify, since the reason for the color dependence on the size of the particles was not yet appreciated in the Middle Ages (Figure 3.1). Finally, nanotechnology attempts to control or manipulate matter at the atomic scale. Our definition dovetails nicely with the criteria adopted by the National Nanotechnology Institute (http://www.nano.gov/):

1. Nanotechnology involves structures in the 1–100-nm range
2. Nanotechnology creates and uses structures that have novel size-based properties
3. Nanotechnology takes advantage of the ability to control or manipulate at the atomic scale.

FIGURE 3.1 Stained glass: *Labors of the Months* (Norwich, England, ca. 1480). The ruby color is due to embedded gold nanoparticles.

The manufacture of devices with dimensions between 1 and 100 nm is based on either top-down manufacturing methods (starting from bigger building blocks, say a whole Si wafer, and chiseling them down into smaller and smaller pieces by cutting, etching, and slicing) or bottom-up manufacturing methods (in which small particles such as atoms, molecules, and atom clusters are added for the construction of bigger functional units). The top-down approach to nanotechnology we call *nanofabrication* or *nanomachining*—an extension of the micromachining approaches [microelectromechanical system (MEMS)] introduced in previous chapters. The bottom-up approach we call *nanochemistry*. Nanochemistry is also called *molecular nanotechnology*; it is the arranging and building of more complex organic and inorganic structures at molecular level by chemistry, molecular biology, or mechanical positional assembly. An example of this second approach is the self-assembly of a monolayer from individual molecules on a gold surface (see Volume II, Figure 8.14). Bottom-up methods are nature's way of growing things. In biomimetics, one studies how nature, through eons of time, developed manufacturing methods, materials, structures, and intelligence. These biomimetic studies, expanded upon in Chapter 2, this volume, are now inspiring design, materials, and engineering of human-made miniature objects.

Both nanomachining and nanochemistry are considered nanotechnology. We start this chapter off by contrasting these two opposite methods to obtain nanostructures. Along the way we highlight the Drexel-Smalley debate about the "correct path" toward nanotechnology. We then list all the properties that make nanodevices special, i.e., physical and chemical behavior that depend on particle size and allow one to engineer their properties. In one extreme, we have atoms and molecules, generally less than a nanometer in size, studied in quantum chemistry, and in the other extreme we have bulk condensed matter physics that deals with solids with an infinite array of bound atoms. Nanoscience deals with the in-between mesoworld where neither quantum chemistry nor the classic laws of physics apply. The listing of nanoparticle properties is followed by a section on preparation of nanostructures in general: from high-energy ball milling to photolithography to gas phase and liquid phase processes. Finally, as more specific examples, we detail properties and highlight bottom-up manufacturing approaches of carbon nanotubes, graphene sheets, and nanowires.

Top-Down versus Bottom-Up Manufacturing

General Characteristics of Top-Down and Bottom-Up Manufacturing

Top-down nanofabrication is principally based on lithography and traditional mechanical machining, etching, and grinding, whereas bottom-up nanochemistry is based on chemical synthesis, self-assembly, and positional assembly. The distinction is clarified in Figure 3.2. These complementary developments constitute different aspects of nanotechnology; both aim to create and use structures, devices, and systems in the size range of about 0.1–100 nm (covering the atomic, molecular, and macromolecular length scales).

There are some important distinctions to be made between these two approaches to making very small things. Whereas top-down methods essentially impose a structure or pattern on a substrate, in self-assembly and self-organization-based bottom-up methods the assembly or organization of atomic and molecular constituents into larger structures is inherent through processes in the manipulated system. Self-assembly and self-organization are driven by thermodynamics and hold the intriguing potential of delegating more and more mechanical or logical operations to materials instead of to manufacturing machines. Assembly of atoms or molecules into larger aggregates in chemical synthesis is driven by chemical reactions with the Gibbs free energy changes dominated by the enthalpy changes associated with breaking and creating bonds ($\Delta G = \Delta H - T\Delta S$). In self-assembly the driving forces are

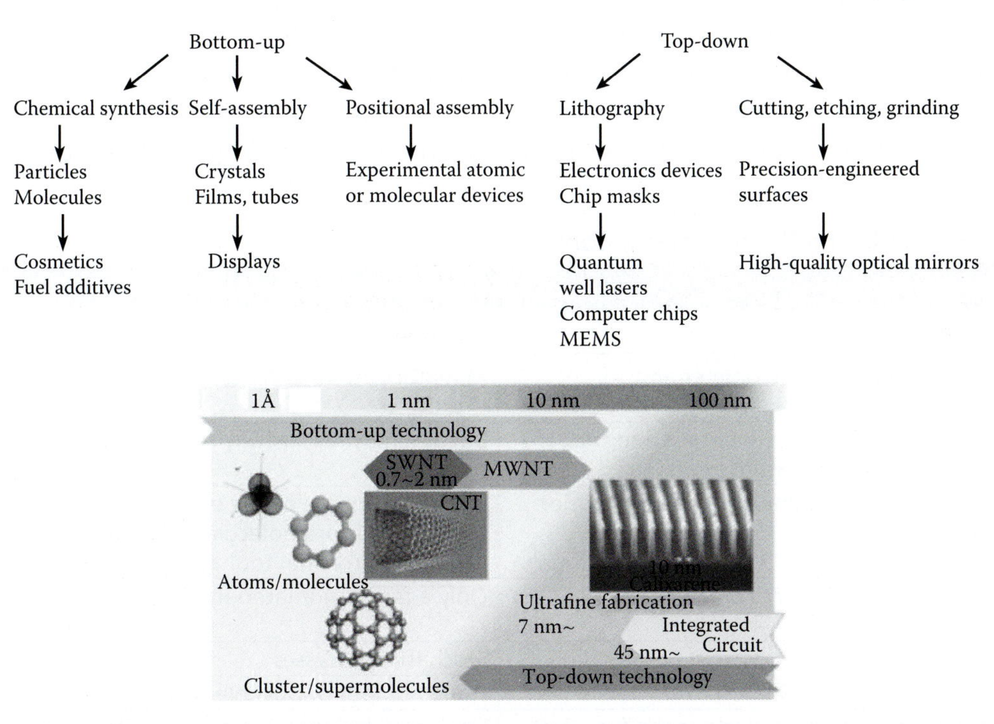

FIGURE 3.2 Contrasting bottom-up nanofabrication with top-down nanochemistry.[2] (SWNT: single wall nanotube; MWNT: multiwall nanotube; CNT: carbon nanotube.)

weak interactions (hydrogen bonds, van der Waals forces, hydrophobic/hydrophilic interactions, etc.; see Volume I, Chapter 7) and are usually driven by entropy changes only. Both chemical synthesis and self-assembly-based bottom-up processes are based on growth phenomena, and any growth scenario is governed by the competition between kinetics and thermodynamics. Positional assembly is the least mature bottom-up technique and uses mechanical means, e.g., an atomic force microscope (AFM), to assemble nanostructures by forcing atoms or molecules in close enough proximity and with the right orientation to form or break bonds (see Volume II, Figures 2.43 to 2.45).

A top-down manufacturing process (for example, the writing of an e-beam pattern) is deterministic, but a bottom-up process (for example, the chemical synthesis of a quantum dot from a colloidal solution) is stochastic. In the latter case, instead of dealing with only one possible "reality" of how the process might evolve under time, in a stochastic or random process there is some indeterminacy in its future evolution, described by probability distributions. This means that even if the initial condition is known, there are many possibilities to which the process might go, but some paths are more probable and others less (statistics take over). Current top-down manufacturing techniques are in a sense crude and wasteful compared with the bottom-up manufacturing techniques we survey in this chapter. Top-down, making a smooth edge requires grinding away and discarding trillions of atoms (i.e., removing excess material). The hope for bottom-up manufacturing is to build with atomic precision without wasting material.

Applying modified integrated circuit (IC) methods, MEMS led to major advances in manufacturing strategies, but the method still presents a limited choice of materials and works with building blocks that are large and crude compared with the nanochemistry arsenal. As humankind learns to build with the same construction set nature uses, we are bound to challenge nature in nanoengineering. Because it uses relatively large building blocks, human manufacturing is rapid. The expectation is that nature, because it uses much smaller building blocks, for example, atoms with a diameter of 0.3 nm, must be very slow. To offset the time it takes to work with small, basic building blocks, nature, in growing an organism, relies on additive processes featuring massive parallelism and self-assembly. Massive parallelism and self-assembly are two areas where human technology still is very backward compared with nature's methods. In Table 3.1 we compare some of the key characteristics of bottom-up manufacturing (nanochemistry) with MEMS technology.

In Figure 3.3 we illustrate the convergence over time of top-down and bottom-up methodologies. The lines representing top-down manufacturing in Figure 3.3 are a simplified representation of a 1974 diagram by Norio Taniguchi (see Volume II, Figure 6.3). The higher of the two lines, marking

Table 3.1 MEMS and Nanochemistry Compared

MEMS Technology	Nanochemistry
Top-down	Bottom-up
Miniaturization with micrometer and submicrometer tolerance	Miniaturization with atomic accuracy
Builds mostly on solid-state physics and the band model from quantum mechanics	Builds almost exclusively on quantum mechanics
Evolved from the integrated circuit fabrication technology	Evolved from molecular engineering, colloid chemistry, supramolecular chemistry, and scanning proximal tool assembly
Many MEMS products are commercially successful	Only a few nanomaterials are on the market
Deterministic processes	Stochastic processes
Limited set of building materials	Huge number of possible building materials
Relatively large building blocks make for a fast process	Very small building blocks necessitate massive parallelism and self-assembly

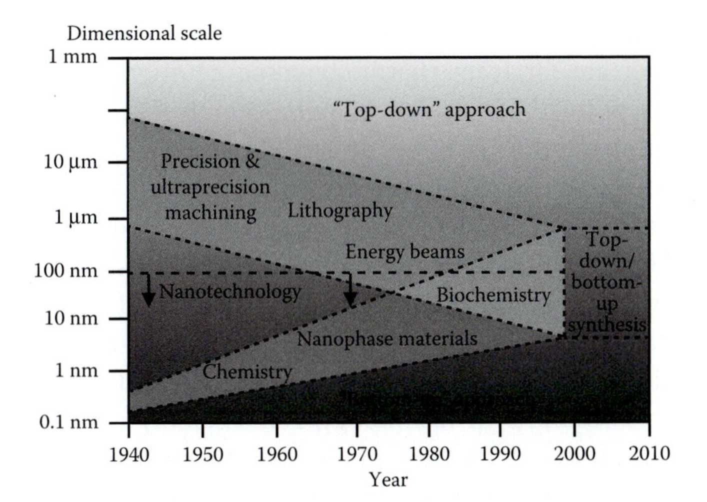

FIGURE 3.3 Convergence of bottom-up nanochemistry and top-down nanofabrication. The nanotechnology line is placed at 100 nm^2.

progress of absolute tolerance in top-down machining, corresponds to precision machining. The lower line represents absolute tolerance in ultraprecision machining, lithography, and energy beam techniques. Some caution is in order about the use of the term *absolute tolerance* in nanotechnology: a more rigorous definition of nanotechnology should differentiate between absolute tolerances in lateral and vertical nanostructures. Deposition of nanometer-thick films is relatively easy with all types of "traditional" chemical vapor deposition (CVD) and physical vapor deposition (PVD) techniques, but patterning of a thin film laterally, with critical dimensions (CDs) that extend from about 100 nm down to atomic orders of magnitude, is still quite a challenge when employing top-down methods. Bottom-up methods hold the potential for nanometric control in all three dimensions. This explains why we tend to make quantum wells with lithography-based top-down methods, since only the thickness direction requires tight control, but quantum wires and quantum dots are mostly made with bottom-up methods, since for those quantum structures we also need control in the lateral directions. Vertically layered structures, i.e., nanoscaled thin films, are obviously much easier to achieve than laterally patterned features. When people discuss breakthroughs in top-down nanotechnology, they mostly refer to better lateral patterning techniques. In the same context, in Volume II, Chapter 6, we explain how the excellent absolute tolerances for precision and ultraprecision mechanical machining only hold true for surface finishes, i.e., in the thickness direction, while in the lateral directions absolute tolerances are again worse. Mechanical machining lateral tolerances are worse than the lateral tolerances achievable with lithography, and those in turn are worse than what we can achieve with nanochemistry techniques: using electron beam machines we can cut shapes with a precision of 2.5 nm, about 1,000 times better than what is achievable in a standard machine shop (2.5 μm is standard, and 1 μm may be achieved in specialty shops), and precision in nanochemistry is about 1 Å, 25 times better than what e-beam offers today. The two lines in Figure 3.3 representing the bottom-up manufacturing progress in achievable absolute tolerances show how bottom-up processes evolved to enable, ever-larger structures through progress in nanochemistry. It is seen from this figure that from the mid-1970s on, feature sizes that could be achieved by nanofabrication (top-down) and nanochemistry (bottom-up) started to overlap. Over the next ten years, the size overlap of objects fabricated with either approach will continue, and molecular engineers and supramolecular chemists, jointly with IC, MEMS, and nanoelectromechanical system (NEMS) specialists, now making devices with 45-nm feature sizes and below, will find new ways to combine top-down and bottom-up manufacturing methods.

In Volume II, Chapter 6, we commented that the reason for the slowdown in electronic device miniaturization hinges on securing more mechanical stability in the positioning stages, but even more so on manufacturing cost and the fact that transistors are becoming too leaky below the 32-nm node, where quantum effects start playing a role. To further extend manufacturing accuracy in electronics one has to look beyond current manufacturing methodologies and consider methods such as molecular engineering and nanofabrication of quantum devices. Micromachining of special lithography tools will play a pivotal role in those new machining approaches. For example, the scanning tunneling microscope (STM), based on micromachining ultrasharp tips on Si cantilevers, creates machining opportunities with accuracies on the atomic level, well beyond the predictions of the

Taniguchi and Moore curves in Volume II, Figure 6.3. For mechanical NEMS structures, operation in the micro- and nanodomains imposes barriers other than those described for electronic devices. New design philosophies, producing results that are more error-insensitive with simpler mechanisms, smaller numbers of parts, and, wherever possible, flexible members rather than rigid ones must be adopted. Most of this book deals with top-down machining; in this chapter we deal in some more detail with bottom-up manufacturing.

Important Concepts in Bottom-Up Manufacturing

Introduction

Before reviewing nanostructure properties and listing methods for bottom-up nanotechnology manufacturing, some important contributing concepts and their definitions are in order. What follows are clarifications and definitions for concepts that are often utilized in bottom-up manufacturing but are not always well understood, such as self-assembly, self-organization, biomimetics, supramolecular chemistry, self-replication, and molecular assemblers.

Self-Assembly and Self-Organization

Self-Assembly

Self-assembly and self-organization are important concepts in both nanotechnology and biology. According to Maasden and Kaiser, about 10% of papers dealing with nanotechnology published between 1990 and 2005 address the concept of self-assembly, and among those, 3% deal with self-organization.[3] The distinction between self-assembly and self-organization is not always readily apparent, and this causes considerable confusion, especially when self-replication (see below) is thrown into the discussion.

Thermodynamics represents the simplest way to differentiate between self-assembly and self-organization. Self-assembly concerns equilibrium situations and is purely driven by the tendency of systems to minimize their free energy in accordance with the second law of thermodynamics. Self-organization, on the other hand, occurs in systems that are open and are driven by a constant input of energy and that are far from thermodynamic equilibrium. As we pointed out earlier, in self-assembly free energy minimization is often driven by entropy changes, ΔS (Equation 2.7). For natural examples of self-assembly, consider phospholipids with hydrophobic and hydrophilic ends placed in aqueous solution spontaneously forming stable structures (Figures 2.41–2.43, this volume) or the intricate three-dimensional (3D) patterns in protein folding (Figure 2.11, this volume). In technology, self-assembly is taken advantage of, for example, in sensors such as DNA probes and immunosensors, but is not used much in human manufacturing yet. We use self-assembly in manufacturing, for example, when growing crystals, in building self-assembled monolayers (see Volume II, Figure 8.14), and in colloid chemistry. In colloidal chemistry one can assemble inorganic solids, such as CdSe crystals, with astonishing regularity in size and properties, comparable to what can be accomplished with nanolithography, and this method has become an important route for building quantum dots (see Figure 3.23).

There is an upswing in self-assembly work that is bringing a renewed excitement to the nanotechnology field. In Chapter 4, this volume, on packaging, for example, we detail the differences between deterministic and stochastic parallel microassembly and clarify the need and tremendous opportunities for self-assembly in packaging of nanocomponents. Recent examples of work in self-assembly based packaging, clarified in Chapter 4, include DNA-mediated self-assembly of metallic and semiconductor components and micro-origami assembly of optical components and of small polymer sheets (see also the example that follows).

Example of Self-Assembly: Scaffolded DNA Origami

Weight

The reproducible generation of atomically precise patterns and shapes is at the heart of nanotechnology. Although research has shown improvement at creating nanosized shapes, it is still a tremendous challenge to create desired nanostructures that always do have the same size and shape with

FIGURE 3.4 Origami art. (From Origami.com, http://www.origami.com.)

atomic accuracy. Here we review a recently developed method that holds that potential. It involves a technique for folding long single strands of DNA (ssDNA) into arbitrary two-dimensional shapes using synthesized DNA staple strands in so-called "scaffolded DNA origami." Scaffolded DNA origami builds on Seeman's DNA nanotechnology, discussed in Chapters 2 and 4, this volume (e.g., Figure 4.49) and pushes it quite a few steps further.

Origami is the art of folding paper; it originated in China as "Zhe Zhi" and was adapted by Japanese monks in the sixth century AD (see an example of origami in Figure 3.4). Ancient origami is not self-assembly, but the scaffolded DNA origami reviewed here is. DNA by its nature does exhibit self-assembly (see base pairing in Chapter 2, Figure 2.6). The same characteristics that make DNA so effective genetically also make it a good choice for programmed self-assembly in nanotechnology:

- DNA can be manipulated to create exact copies that are extremely accurate (atomic accuracy).
- DNA is predictable and programmable.
- DNA also has the ability to store enormous amounts of information.

The most important part of controlling the way a DNA strand folds is knowing its genetic nucleotide structure. The scaffold strand that has been used most often is circular genomic DNA from the virus M13mp18 (natural M13mp18 is 7249 bases long). Synthesized DNA staple strands then bind to complementary base pairs on the scaffold M13mp18 viral genome as illustrated in Figure 3.5. Both the DNA strand and the staple strand must be known to avoid unwanted pairing and to ensure that the stapling takes place in desired locations only, as illustrated in Figure 3.6.

The design for a desired shape is made by raster filling the shape with the ~7-kilobase single-stranded scaffold molecule and by choosing more than 200 short oligonucleotide staple strands to hold the scaffold in place. Once synthesized, a large excess of staples (100×) is mixed with scaffold and annealed in a polymerase chain reaction (PCR) machine (<2 h) (for details on how PCR works, see Chapter 2, this volume, Figure 2.53). Synthesized DNA staples self-assemble, with the viral strands resulting in folded area shapes (squares, disks, five-pointed-star, etc.) approximately 100 nm in diameter, with a spatial resolution of 6 nm. A nucleotide may serve as a 6-nm pixel in these designs, so any type of word or image can be programmed. Individual DNA structures can also be made to form larger assemblies of two-dimensional (2D) periodic lattices. Sample DNA origami are deposited on mica and imaged with an AFM in tapping mode. Experimentally verified by the creation of a dozen shapes and patterns, the method is easy, has a high yield, and lends itself well to automated design and manufacture. So far, computer-aided design (CAD) tools for scaffolded DNA origami are simple, require hand-design of the folding path only, and are restricted to two-dimensional designs. If the method gains wide acceptance, better CAD tools will be required.[4]

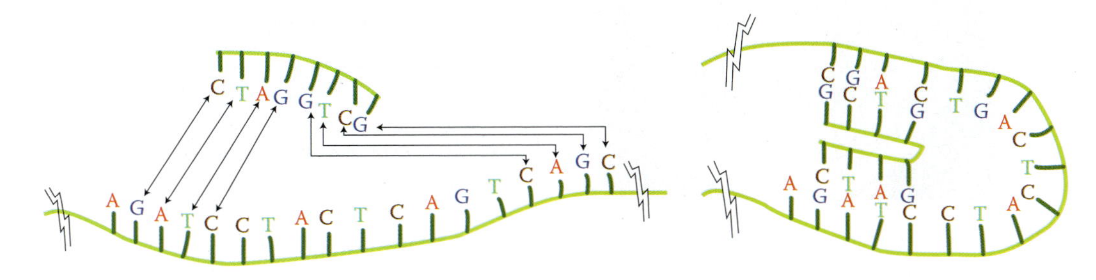

FIGURE 3.5 DNA staple concept. The sequence of DNA provides a unique address for each location. Staple strands bind locations together according to the design.

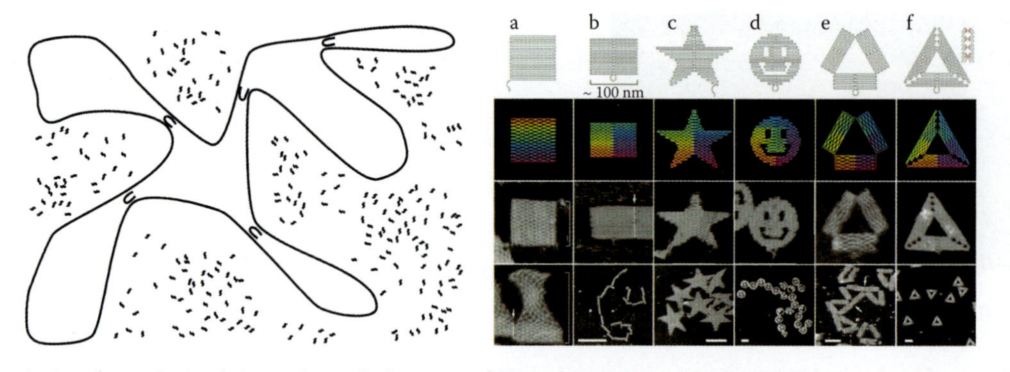

FIGURE 3.6 The design for a desired shape is made by raster filling the shape with the 7-kilobase single-stranded scaffold molecule and by choosing more than two hundred short oligonucleotide staple strands to hold the scaffold in place.[4]

Self-Organization

Self-organization is a term for those types of pattern forming systems that are open and thus driven by a constant input of energy and that are far from thermodynamic equilibrium. The Rayleigh-Bénard instability, or, more simply, Bénard heat convection, is regarded as a classic case of self-organization. This type of heat convection is named after French chemist Henri Bénard, who first recorded the effect in 1900. Because the British physicist Lord Rayleigh further investigated Bénard's observation, the conditions that produce it came to be called the *Rayleigh-Bénard instability*. Anyone can observe the formation of Bénard cells by heating a thin layer of oil in a pan, as illustrated in Figure 3.7. With the oil uniformly heated from below, a constant heat flux is established, moving from the bottom to the top of the oil layer in the pan. The oil itself remains at arest, and when the temperature difference between top and bottom of the oil layer reaches a certain critical value, thermal conduction is replaced by heat convection, in which the coherent motion of large number of molecules transfers the heat. At this point a very strikingly ordered pattern of hexagonal cells (a "honeycomb") appears, in which hot oil rises through the center of the cells, while cooler liquid descends to the bottom along the cell walls. The same physics underlies a weather phenomenon that leaves hexagonal patterns behind in sand or snow. The sun heats up the desert sand or a snowfield, and warm air from the surface rises, while cold air descends from the upper layers of the atmosphere. The conflict is resolved again by the emergence of orderly patterns of fluid flow under the action of an imposed temperature gradient, resulting in hexagonal circulation vortices that leave their imprints on sand dunes in the desert and on arctic snowfields.

During the 1960s, Ilya Prigogine (1917–2003), winner of the Nobel Prize in Chemistry, 1977, developed a new, nonlinear, irreversible thermodynamics, that mathematically describes the self-organization phenomenon described here.[5] It is

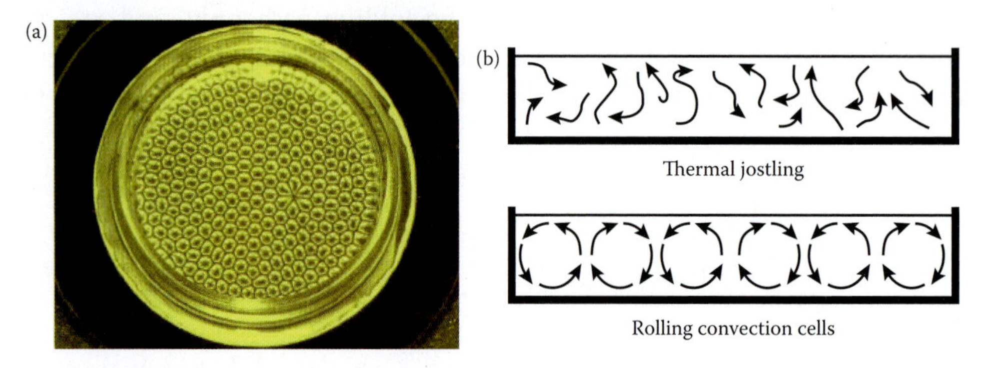

FIGURE 3.7 (a) Patterns of hexagonal Bénard cells viewed from above. The most interesting stable patterns appear right at the edge of chaos. This is happening far from equilibrium, thanks to the energy flowing through an open system and being dissipated. (b) The hexagonal cells—the simple global order—is more efficient at dissipating heat than the simple local motions.

quite unlikely that anyone would have deduced these phenomena from the Navier-Stokes equations (see Volume I, Chapter 6). On the basis of Prigogine's work, Bénard cells are now understood to arise from the properties of the liquid state at a bifurcation point in which a self-organizing structure emerges. The huge number of degrees of freedom of individual molecules in the fluid at this critical point suddenly organize into a coherent and simple flow pattern that dissipates heat more effectively than simple thermal conduction. In this new emergent behavior, convection (transport of heat by mass movement) in self-organized spatial structures becomes the dominant process. This change is initiated by an instability (noise) in the system, and the result obeys the second law of thermodynamics, since this mechanism is a more efficient way of disposing of the system's heat. More disorder must generally prevail (increase in ΔS), but one can still have a local increase in order, such as in the hexagonal Bénard cells, as long as overall they contribute to the wider disordering of the surroundings, i.e., the dissipation of energy as entropy or waste heat.

What is happening at a molecular level in the frying pan is that when the pan is only gently warmed, each heated molecule moves by itself, and the molecules all chaotically barge about and only eventually emerge at the surface, where they shed their excess heat. But at a certain critical temperature, a phase transition occurs. The jostling molecules are becoming entrained into a coherent flow, the rolling convection current that moves them smoothly from the heated bottom to the cooler surface. The hexagonal cells—the simple global order—is more efficient at dissipating heat than the simple local motions (see Figure 3.7b). But when one heats a Bénard cell just a little more, its orderly hexagonal cells soon break up and instead turbulence results and disorder appears again. The order turns out to exist only for a very finely controlled balance between heat input and heat dissipation.

In self-organization what differs from what happens in self-assembly is that internal variations, called *fluctuations* or *noise*, trigger self-organization. The increase in organization corresponds to a decrease of entropy equivalent to an increase in redundancy, information, or constraint. In other words, after the self-organization process, less ambiguity about which state the system is in results. Prigogine called systems that continuously export entropy in order to maintain their organization *dissipative structures*, i.e., systems that exist far from thermodynamic equilibrium, dissipate the heat generated to sustain them, and have the capacity to change to higher levels of orderliness or self-organization. In classic thermodynamics the dissipation of energy in heat transfer, friction, and the like was always regarded as waste. Dissipative structures introduced a radical change in this view by showing that in open systems, dissipation becomes a source of order. According to Prigogine, dissipative systems contain subsystems that continuously fluctuate. At times, a single such fluctuation or a combination of them may become so magnified by possible feedback that it shatters the preexisting organization in what is called an emergent new property. At such moments, called *bifurcation points*, it is impossible to determine in advance whether the system will disintegrate into "chaos" or leap to a new, more differentiated, higher level of "order." While dissipative structures receive their energy from outside, the instabilities and jumps to new forms of organization are the result of internal fluctuations amplified by positive feedback loops. Thus, amplifying "runaway" feedback, which was traditionally seen as destructive, now appears as a source of new order and complexity in this model.

Self-assembly and self-organization are fundamentally different from a thermodynamics point of view: while self-assembly stems from minimization of the free energy, self-organization requires an open system and a constant energy input.* Whereas self-assembly is quite well understood and is clearly important at small length scales in biology (protein folding, for example), self-organization is not as well understood, and the length scale at which it comes into play in biology is still under debate. For the specific case of cell membrane assembly, it is believed that both processes eventually lead to quite similar patterns, but that their evolution occurs on very different length and time scales. Self-assembly produces periodic protein patterns on a spatial scale below 0.1 μm in a few seconds followed by extremely slow coarsening,

* Unfortunately, important players in this field, such as Whitesides, seem to maintain their own definitions of self-assembly, and J.M. Lehn uses both terms more or less interchangeably to characterize the synthesis of supramolecular structures.

whereas self-organization results in a pattern wavelength comparable to the typical cell size of 100 μm within a few minutes, suggesting different biological functions for the two processes.[6] One thing that everyone agrees on is that living organisms cannot arise wholly from self-assembly, because in the absence of a continuous supply of energy they die. In summary, crudely said: viruses self-assemble, but elephants (perhaps) self-organize (http://www.softmachines.org/wordpress/?p=178). Another parallel to living systems here is that such dynamical structures are not devised to exhibit this behavior: they develop spontaneously from random initial conditions (we emphasize: not from special initial conditions).

The behavior of interest, such as that of the Bénard cells, is often found in the transition between order and chaos—at the edge of chaos—and is classified as a kind of organized complexity. The same type of behavior is involved in adaptive path minimization by ants, wasp and termite nest building, army ant raiding, fish schooling and bird flocking, coordinated cooperation in slime molds, synchronized firefly flashing, evolution by natural selection, game theory, the evolution of cooperation, etc. When a flock of birds is taking off from a lake (see Figure 3.8), birds compete for position, try to get free of the maelstrom of their fellow birds and go up into the air. They want to be part of the flock, yet at the same time try to avoid collisions with each other. Computer models show that each bird's attempt to keep a minimum and maximum distance from others causes flight paths to couple into feedback loops of attraction and repulsion. Positive and negative feedback balance so that the individual birds appear transformed into a single organism. The late Prigogine wrote about the reception his irreversible thermodynamics received at first:

> It is difficult today to give an account of the hostility that such an approach was to meet. For example, I remember that towards the end of 1946, at the Brussels IUPAP meeting, after a presentation of the thermodynamics of irreversible processes, a specialist of great repute said to me, in substance: "I am surprised that you give more attention to irreversible phenomena, which are essentially transitory, than to the final result of their evolution, equilibrium."
>
> **Ilya Prigogine**
> *1980*

FIGURE 3.8 Flock formation as an example of emergent, spontaneous organization. (From http://www.cs.vu.nl/~schut/bird12.jpg.)

Biomimetic Nanomanufacturing

Introduction

Biomimetics is the application of designs, materials, and systems found in nature to engineering and technology. As we saw in Chapter 2, this volume, where we cover biomimetics in much more detail, engineers and scientists tend to increasingly turn toward nature for inspiration on designs, materials, and systems, spawning a number of technology innovations. There is an obvious reason behind this trend: the rigorous competition of natural selection means that waste and inefficiency are not tolerated in natural systems, unlike many of the technologies developed by humans. For this same reason, many researchers now give natural, biomimetic manufacturing methods another look. Within this context, the guiding principles involved in the construction of living organisms are as follows.[7,8]

- The use of composites (rather than monolithic materials in human engineering)
- Several successive levels of organization (hierarchy), often in the form of two prevalent natural design motives: fractals (trees, lungs, the brain, the cardiovascular system, etc.) or layered structures (tree rings, onions, etc.) and combinations thereof
- Very wide range of manufacturing sizes: from atoms to several meters, i.e., 10^{12} orders of magnitude

- The use of carbon instead of Si
- Ambient temperature versus the high temperatures used in human engineering
- Low energy (ATP) versus high energy
- Mobility of components (Brownian motion) versus rigidity with no or few fluctuations
- Operation in aqueous solution rather air (wet vs. dry engineering)
- Compartmentalization in cells
- Plasticity: soft, flexible materials that change conformation in response to the environment versus rigid components
- Self-assembly (see above), self-organization (see above), and self-replication all through weak bonds
- Order out of noise versus noise as a nuisance (see above)
- The use of templates (e.g., the reading of genes by messenger RNA to build proteins in the cells' ribosomes, as shown in Figure 2.12)
- Variable number of components versus fixed number of components
- Adhesive surfaces (van der Waals) versus separated surfaces
- Robustness through stochastics versus through redundancy

These principles provide inspiration for the implementation of human nanotechnology. As noted in Chapter 2, this volume, biomimetics should not be taken too far beyond inspiration from nature, because the ultimate goal in nature is spreading genetic information to the next generation, not to generate the most efficient manufacturing technique. We add a few comments now for several of the attributes of a natural growing process listed above.

Nanocomposites

Nanocomposite materials are widespread in biological systems, such as bone, cell walls, and soft body tissue. The nacre (mother-of-pearl) of abalone shell, an oriented coating composed of alternating layers of aragonite ($CaCO_3$) and a biopolymer, is an oft-studied example (Figure 3.9). The laminated structure provides strength, hardness, and toughness. Nacre is twice as hard and 1000 times as tough as its constituent phases. Humankind has accumulated plenty of experience with composite materials, but attempts to make "biomimetic" nanocomposite assemblies are quite recent. Along this line, Sellinger et al. reported on an efficient self-assembly process for preparing nanolaminated organic-inorganic coatings that mimic nacre.[9] New biomimetic materials such as scaffolds for regrowth of tendons tend to be composites too, although not on the nanoscale; in the case of tendon scaffolds, composites are made from carbon fibers and polylactic acid.[7]

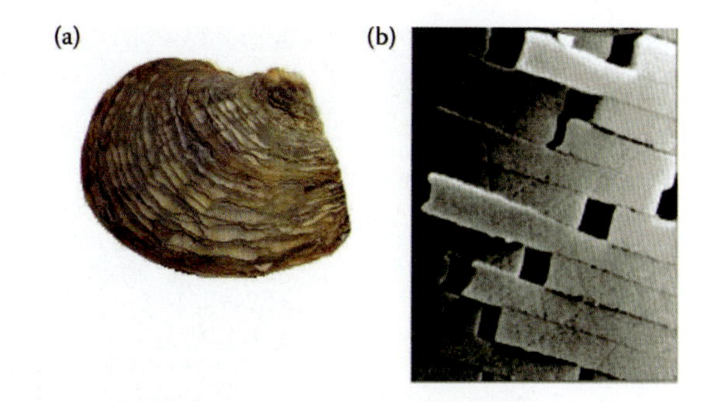

FIGURE 3.9 Abalone shell nanocomposite structure: (a) abalone shell and (b) scanning electron microscope image of the aragonite composite structure.

Hierarchical Organization and Range of Manufacturing Sizes

Proteins, bone, and tissue fibers have a hierarchical organization (see Figure 2.11 for the various protein structural levels) much more complex than that of materials of the nonliving world (e.g., crystals). This organization bestows many benefits on living organisms, such as a high strength-to-density ratio and the ability to serve several functions simultaneously. The ability of humans to mimic this feat is still limited and will improve only if we become more adept at manipulating a wider variety of materials at the nanoscale, use smaller building blocks, and build in three dimensions with more dexterity. Traditional lithography is mostly limited to 2D structures and is impractical for printing feature sizes below 0.1 μm. While 3D lithography techniques, as described in Volume II, Chapter 9, are limited in feature size, they do offer a possible route to 3D microstructures.

Two ubiquitous design patterns of hierarchical structures in nature are fractals (trees, brains, river

FIGURE 3.10 Examples of fractals. (a) Romanesco cauliflower. (b) The fractal coastline of eastern Greenland, with its many fjords.

deltas, fjords, snowflakes, blood circulation system, etc.; see Figure 3.10) and layers (onions, tree rings, etc.) and combinations thereof. Consider a tree where the trunk forks into branches and the branches into smaller twigs. Twigs have leaves, which themselves repeat the dendrite pattern in their veins. The tree also has a layered build-up radially in its stem, branches, and twigs. An object is called a *fractal* if it has structures at a large number of levels. If they are identical, the fractal is said to be *self-similar* (see Chapter 7, this volume, on scaling, for more details on the similarity concept). Self-similarity can also be called *scale invariance* because we always find the same pattern upon zooming into at different length-scales.

In Chapter 9, this volume, we show that the underlying reason for fractal hierarchical structures is simply that such a shape maximizes surface area while minimizing the energy involved in reaching each point on that surface. A fractal network can both absorb and excrete at a maximal rate, whether it is water and electricity, oxygen, carbon dioxide, bloodstream nutrients, or even neural control information. We calculate in Chapter 9 that for a fractal the resistance is minimized and the surface-to-volume (S/V) scale is invariant. We also demonstrate that in a zero-order approximation, a layered structure attains the same goals as a fractal, with both resistance and S/V scaling, in the same way a fractal does when miniaturized.

Using these ubiquitous hierarchical design patterns (fractals and layers) and combinations thereof, nature covers sizes of biological life forms from the atom to several meters (e.g., a 10-m-tall tree), i.e., it covers at least twelve orders of magnitude in size. Human manufacturing techniques, on the other hand, typically are limited to much smaller ranges of size capability (e.g., lithography from 90 nm to 1 cm, i.e., hardly three orders of magnitude). In Figure 3.11, we sketch the challenge involved in making an artificial fractal, say a carbon tree. As indicated in this drawing, as the size ranges we work with approach the atomic and molecular size, stochastic manufacturing techniques (1 Å–100 nm), perhaps from a solution in a beaker, must take over from the larger scale deterministic manufacturing means, such as lithography (100 nm–1 cm) and computer numerical controlled (CNC) machining (1 cm–10 m). In other words, we are forced to switch to different manufacturing options as we move across different length-scales.

The Use of Carbon Instead of Si

Although silicon, human's foremost technological material, is abundant in silicate minerals, carbon is the key to life. In Chapter 5, this volume we explore how in nature, carbon occupies a very special place as the essential building block of life, and we notice how also in human enterprise carbon in its various forms has acquired an enviable position. To name a few, diamonds are made of carbon, and graphite, coke, diamond-like carbon, and glassy carbon are all forms of carbon. So are the more recently discovered buckyballs, carbon fibers, and carbon

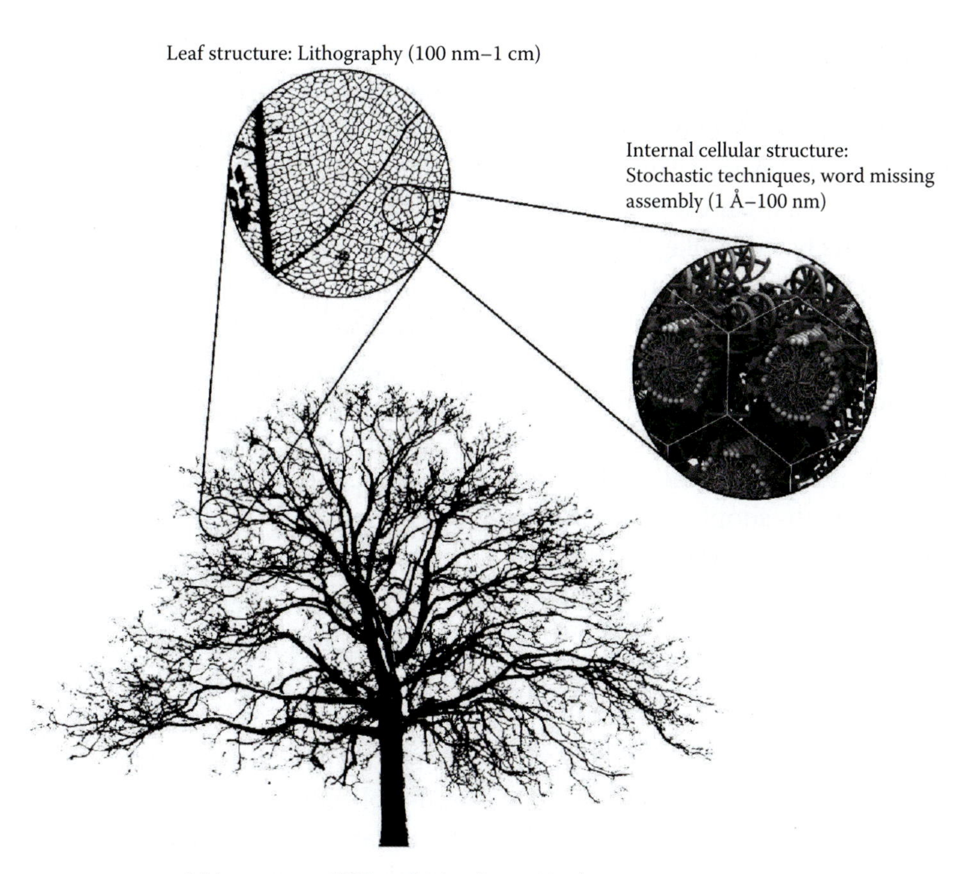

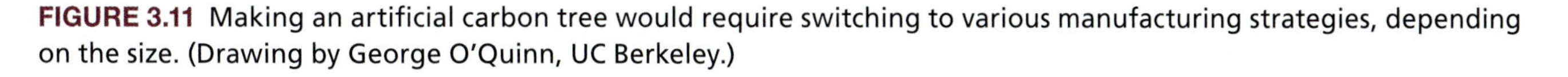

FIGURE 3.11 Making an artificial carbon tree would require switching to various manufacturing strategies, depending on the size. (Drawing by George O'Quinn, UC Berkeley.)

nanotubes (CNTs). It can rightfully be argued that carbon is now also becoming a more and more important material in human technology. Some key advantages of carbon compared with Si are summarized in Table 5.9. As we see from this table, carbon polymerizes better than Si (making organic chemistry a field all on its own); it comes in all types of forms (amorphous, graphite, nanotubes, etc.); it has a wide electrochemical stability window; and it is biocompatible, low cost, chemically inert, and easy to functionalize.

In Chapter 5, this volume, we further introduce the reader to a vision of a carbon-based MEMS and carbon-based NEMS world.

Ambient Temperature, ATP, Soft Materials, Compartimentalization (Cells), and Water Management

Of the many motivating examples of the promise of biomimetics we encountered in Chapter 2, this volume, consider enzymes. There are a plethora of enzymes that outperform synthetic catalysts by several orders of magnitude. For instance, there is no human-made catalyst for producing ammonia from its elements at ambient temperature and pressure, as the nitrogenase of nodule-forming bacteria does (the industrial Haber-Bosch synthesis of ammonia from nitrogen and hydrogen at high temperatures and pressures pales in comparison).

Molecular motors, also covered in Chapter 2, this volume, are in a sense nanomachines that do mechanical work through the hydrolysis of ATP, the energy source of the cell (Figure 2.24). During hydrolysis of ATP, a shape change occurs within the motor protein (a mechanochemical process), and mechanical work is performed. The force generated by a motor protein is as large as 100 pN, a very large torque, comparable to a person rotating a 150-m rod. Such a rotary motor, with a diameter of only 30–40 nm, drives the rotation of the flagellum at around 300 Hz at a power level of 10–16 W, with energy conversion efficiency close to 100%. These protein motors appear to be fuelled by a flow of protons across a pH gradient and can push an average-sized cell at 30,000 nm/s, or 15 body-lengths per second—and can also run in reverse! The structural designs and functional mechanisms revealed in the complex machinery of the bacterial flagellum could provide many novel technologies that may become the basis for future nanotechnology, from which we should be able to find many useful applications.

In comparison, the surface micromachined motors of Volume II, Chapter 7 (see Figure 7.64) are hopelessly impractical. The biological motors are more error-insensitive, with simpler mechanisms and smaller numbers of parts, and wherever possible, flexible members rather than rigid ones are adopted.

In biology, it is important to note that the compartmentalization in cells constitutes an efficient water management system. Since neither proteins nor DNA functions in a dry environment, water needs to be actively controlled by the cell wall. It is no coincidence that more and more of the nanotechnology is focusing on nanochemistry (wet) instead of nanofabrication (dry). In human technology, when working with small droplets the evaporation rate is very high because of the high surface-to-volume ratio, requiring a water management system. Similarly, when working with natural polymers, results can only be obtained when working in an aqueous environment.

Self-Replication and the Use of Templates

We defined self-assembly and self-organization earlier in this chapter: self-replication is yet different. Self-replication is a process in which an object makes a copy of itself. Biological cells, we saw in Chapter 2, this volume, will, given a suitable environment, reproduce by cell division. During cell division, DNA is replicated and is transmitted to offspring during reproduction. Biological viruses can reproduce, but only by hijacking the reproductive machinery of cells. Computer viruses reproduce using the hardware and software already present on computers. Memes reproduce using the human mind and culture as their reproductive machinery. In self-replication, we saw in Chapter 2, this volume, molecules such as DNA encode the genetic information for cells and the resulting organisms, thus allowing the information to be passed on from cell to cell and from generation to generation.[10] PCR and research results with RNA and certain peptides, described in Chapter 2, this volume, are manifestations of this self-replication capability of biological molecules.

Example of Self-Replication in Human Engineering: Self-Replicating Robots

Self-replicating machines have been the subject of theoretical discussion since the early days of computing and robotics. Only three human made mechanical devices that can replicate themselves have been reported so far. One uses Lego parts assembled in a two-dimensional pattern by moving them along tracks, and another uses an arrangement of wooden tiles that tumble into a new arrangement when given a push. The third and most recent self-replicating machine is the most fascinating one. Hod Lipson, from Cornell Laboratory, made the first robots that self-replicate.[11] His robots are made up of a set of modular cubes called *molecubes;* each cube contains identical machinery and a computer program for replication. These machines are simple compared with biological replication, but they demonstrate that mechanical replication is possible and is not limited to biology. Electromagnets on the face of the cubes allow them to selectively attach to and detach from one another, and a complete robot consists of several cubes linked together. Each cube is divided in half along a long diagonal, which allows a robot composed of many cubes to bend, reconfigure, and manipulate other cubes (Figure 3.12).

A tower of cubes can thus bend itself over at a right angle to pick up another cube (Figure 3.13). To start

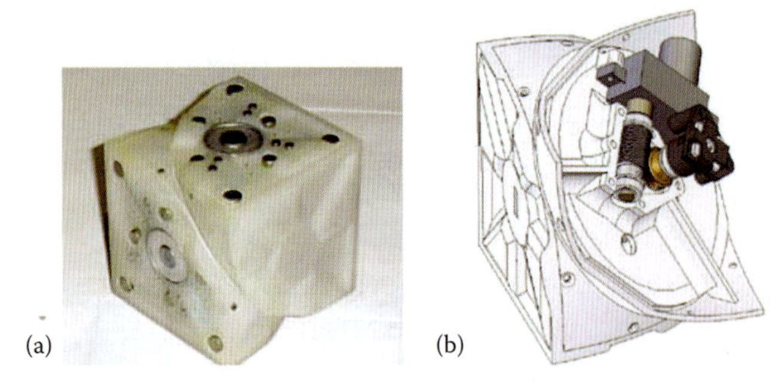

FIGURE 3.12 (a) Each module of the self-replicating robot is a cube about 4 in. on a side, able to swivel along a diagonal. (b) A cutaway drawing shows the motor mechanism.[11]

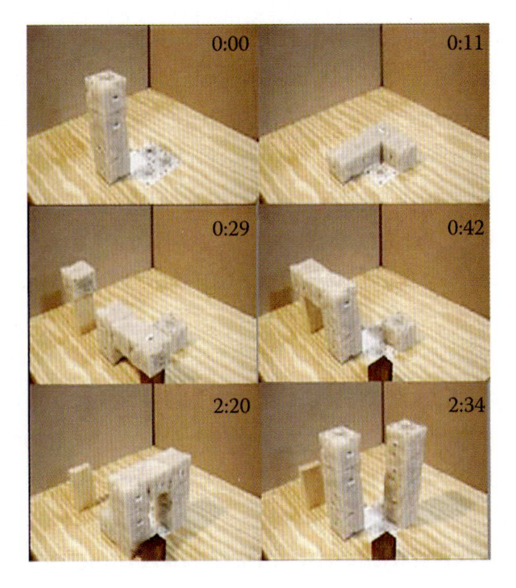

FIGURE 3.13 Frames from a video show the replication process. A robot consisting of a stack of four cubes begins by bending over and depositing one of its cubes on the table. The remaining three cubes pick up additional cubes from "feeding stations" and transfer them to the new robot, which assists in the process by standing itself up. Video is available online at http://www.news.cornell.edu/stories/may05/selfrep.ws.html.

the replication process, a stack of cubes bends over and sets its top cube on the table. Then it bends to one side or another to pick up a new cube and deposit it on top of the first cube on the table. Repeating this process, one robot made up of a stack of cubes creates another just like itself. One robot cannot reach across another robot of the same height. The robot being built assists in completing its own construction. These replicating robots are very dependent on their environment; they draw power through contacts on the surface of the table and cannot replicate unless the experimenters "feed" them by supplying additional modules.

Supramolecular Chemistry

Chemistry beyond the molecule gave rise to supramolecular chemistry—a new field of chemistry pioneered by researchers such as Nobel laureate Jean-Marie Lehn from the University of Strasbourg.[12,13] Self-assembly leads to large supramolecular architectures: no chemist needs to specify the individual atom or molecule positions in these systems.[14] For example, after disintegration of ribosomes (see Figure 2.12 for a ribosome in action) into their protein and RNA building blocks, the molecules will reorganize into functional units in a test tube given the right conditions.

An early suggestion of supramolecular chemistry came from Nobel laureate Hermann Emil Fischer who, in 1890, suggested that enzyme-substrate interactions take the form of a "lock-and-key,"* implying the concepts of molecular recognition and host-guest chemistry. The same lock-and-key assembly is involved in neurotransmitters assembling with molecular recognition sites on proteins that elicit chemical responses, for example, acetylcholine recognition at nerve-muscle junctions, resulting in muscle contractions.[15] The fundamental concepts underpinning supramolecular science are processes in which atoms, molecules, aggregates of molecules, and components arrange themselves into ordered, functioning entities without human intervention.[16] In the early twentieth century noncovalent bonds were understood in gradually more detail. The weak interactions typically involved are as follows (see Volume I, Chapter 7, for more details):

- Hydrogen bonds (described by Latimer and Rodebush in 1920)
- Electrostatic attractions
- Van der Waals forces (the existence of intermolecular forces was first postulated by Johannes Diderik van der Waals in 1873)
- Hydrophobic interactions
- Weak covalent bonds

Through supramolecular chemistry with weak ionic and hydrogen bonds and van der Waals interactions (0.1–5 kcal/mol), structures with dimensions of 1–100 nm are possible. For chemical synthesis, the largest of these dimensions often is too large, and for microfabrication, it often is too small. The types of interactions listed above allowed scientists to finally interpret protein and DNA behavior. Chemists started to study synthetic structures based on no covalent interactions, such as micelles and microemulsions. Eventually, these concepts, derived from

* The lock-and-key concept of biochemical interactions has, over the years, been modified in favor of a "hand-in-a-glove" concept, because some reorganization (conformational change) is required to make two biomolecules complementary to each other.

biological systems, were applied to synthetic chemical systems. The first big breakthrough along this line came in the 1960s with the synthesis of the crown ethers by Charles J. Pedersen.[17] Donald J. Cram, Jean-Marie Lehn, and Fritz Vögtle became active in the synthesis of shape- and ion-selective receptors, and throughout the 1980s research in the area gathered a rapid pace, with concepts such as mechanically interlocked molecular architectures emerging. The emerging science of nanotechnology also had a strong influence on the subject, with building blocks such as fullerenes, nanoparticles, and dendrimers becoming involved in synthetic systems. By the 1990s, supramolecular chemistry had become an established avenue toward nanoelectronic and nanomechanical systems, with researchers such as Fraser Stoddart developing molecular machinery and highly complex self-assembled structures[18] and researchers such as Itamar Willner developing sensors and methods of electronic and biological interfacing.[19] Stoddart pioneered the development of the use of molecular recognition and self-assembly processes in template-directed protocols for the syntheses of two-state mechanically interlocked compounds (bistable catenanes and rotaxanes) that have been employed as molecular switches and as motor-molecules in the fabrication of nanoelectronic devices and NEMS (see Figure 3.14 for a rotaxane switch).[20] Rotaxanes can act as molecular switches because they can oscillate back and forth between two stable states.

Yet more recently, electrochemical and photochemical motifs have become integrated into supramolecular systems in order to increase functionality, research into synthetic self-replicating system began, and work on molecular information processing devices began. For example, the Stoddart and Zink groups have developed donor-chromophore-acceptor-based molecular triads that can transduce light into electrical energy by mimicking the photosynthesis pathway.[21] A nanoscale power supply in the form of a light-harvesting molecular triad has been developed (Figure 3.15). Donor-chromophore-acceptor-based molecular triads that can convert light to electrical energy are some of the most effective molecules that can serve the purpose of energy transduction. Yet more impressively, it has been demonstrated that the photocurrent generated by the triad can be utilized to drive a supramolecular machine in the form of a pseudorotaxane.[22] The molecular triad is based on tetrathiafulvalene-porphyrin-fullerene (TTF–P–C60), which generates electrical current by harnessing light energy when self-assembled onto gold electrodes. The triad is composed of three unique electroactive components: 1) an electron donating TTF unit, 2) a chromophoric porphyrin unit, and 3) an electron-accepting C60 unit. A disulfide-based anchoring group was tagged to the TTF end of the molecule in order to allow self-assembly onto gold surfaces. In a closed electrical circuit, the triad-functionalized working electrode generates a switchable photocurrent of ~1.5 $\mu A/cm^2$, when irradiated with a 413 nm krypton-ion laser. The electrical energy generated by the triad is then exploited to drive a supramolecular machine in the form of a [2]pseudorotaxane composed of a π-electron-deficient tetracationic cyclobis(paraquat-

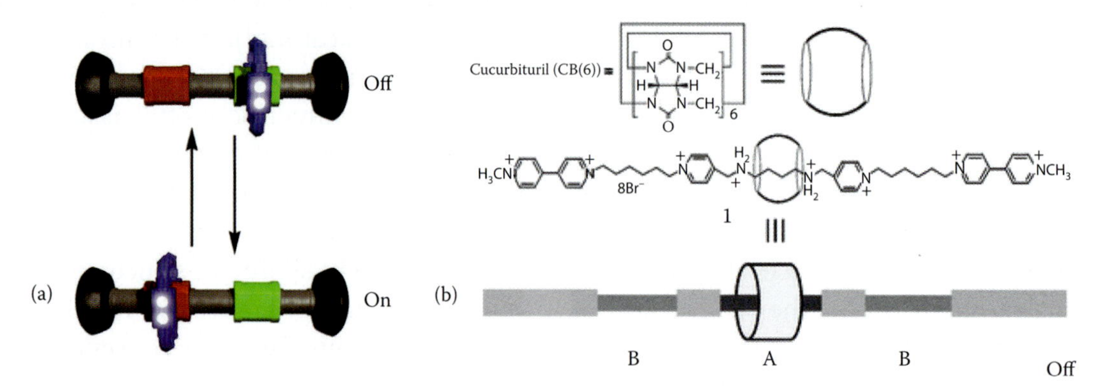

FIGURE 3.14 Rotaxene switch. (a) Model; (b) molecular structure and another model. A **cucurbituril** is a macrocyclic molecule consisting of several glycoluril [=$C_4H_2N_4O_2$=] repeat units, each joined to the next one by two methylene [-CH_2-] bridges to form a closed band. Cucurbiturils are commonly written as **cucurbit[*n*]uril**, where *n* is the number of repeat units. A common abbreviation is **CB[*n*]**. Adapted from Stoddart and Nepogodiev.[20]

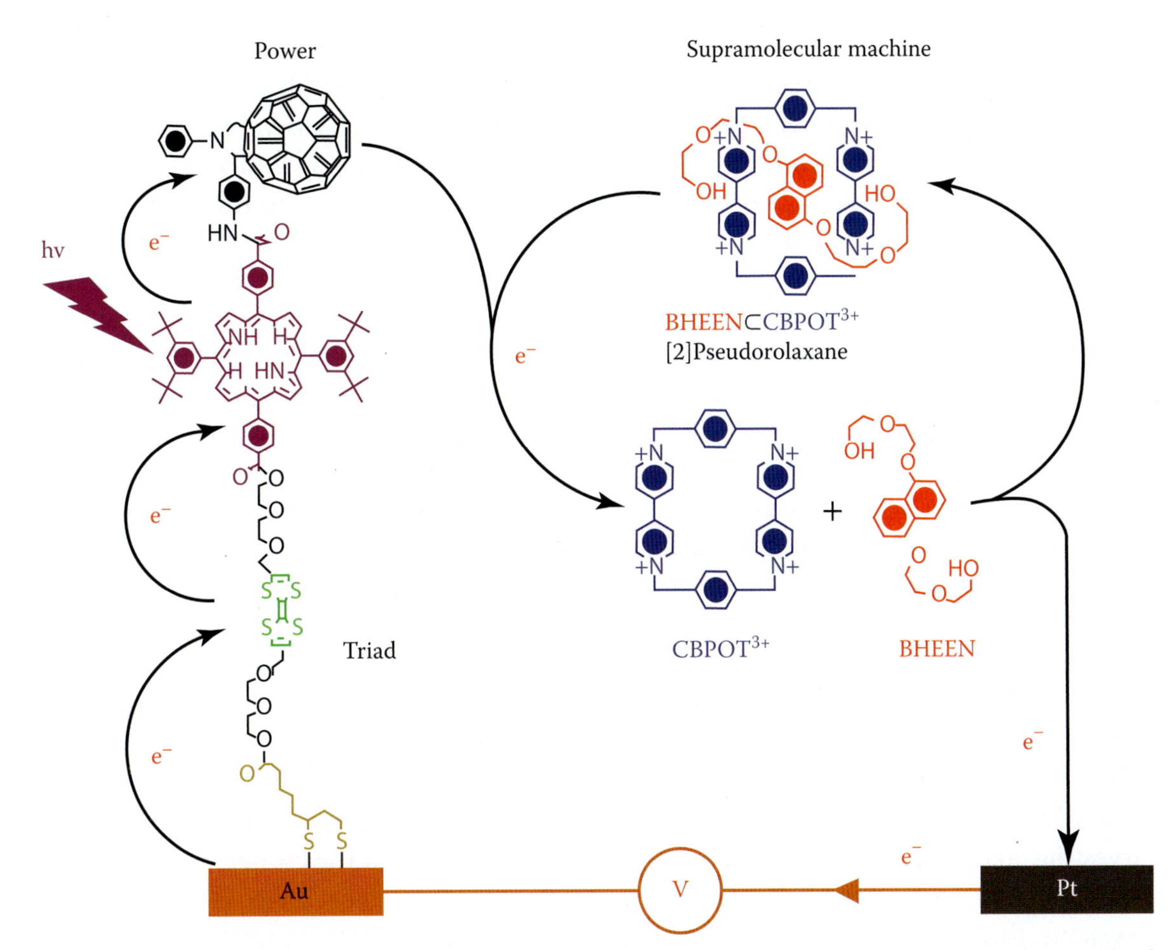

FIGURE 3.15 The photocurrent generated by the triad can be utilized to drive a supramolecular machine in the form of a pseudorotaxane. The [2]pseudorotaxane is composed of a π-electron-deficient tetracationic cyclobis (paraquat-*p*-phenylene) (CBPQT4+) cyclophane and a π-electron-rich 1,5-bis[(2-hydroxy-ethoxy)-ethoxy] naphthalene (BHEEN) thread. Adapted from Saha et al. 2005. Powering a supramolecular machine with a photoactive molecular triad. *Small* 1:87–90.[22]

p-phenylene) (CBPQT4+) cyclophane and a π-electron-rich 1,5-bis[(2-hydroxy-ethoxy)-ethoxy] naphthalene (BHEEN) thread (Figure 3.15). The dethreading of CBPQT4+ cyclophane from the BHEEN thread was monitored by measuring the increase in the fluorescence intensity of the BHEEN unit. A gradual increase in the fluorescence intensity of the BHEEN upon illumination confirmed that the dethreading process is driven by the photocurrent generated by the triad.

Dendrimers form a special set of molecules in supramolecular chemistry. They are globular polymer-based macromolecules of monodisperse size in which all bonds emerge treelike from a central core in a regular branching pattern and may be as small as 3 nm across (the size of a typical protein). These fractal-like macromolecules, introduced by Tomalia et al.[23] and Newkome et al.[24] in 1985 have attracted a lot of attention because of their novel structure and intriguing properties.[25,26] Dendrimer synthesis allows for nearly complete control over the critical molecular design parameters, including size, shape, reaction site distribution, surface/interior chemistry (e.g., number and size of internal voids), flexibility, and topology. The process involves a stepwise, repetitive sequence that guarantees completed reactions for each generation of reaction products, resulting in monodisperse final products. There are three major strategies involved in effective dendrimer synthesis: the starburst divergent strategy, the convergent growth strategy, and the self-assembly strategy. In the divergent approach, growth starts at the core and proceeds radially outward toward the periphery, and in the convergent approach, growth starts at what will become the periphery of the dendrimer and proceeds inward. Although the majority of dendrimers prepared to date have been constructed of covalent bonds[25,27] using convergent or divergent strategies, several noncovalent dendrimers[25,28] have been prepared by a variety of self-assembly processes involving, for example, hydrogen bonding[29] or supramolecular coordination chemistry.[30]

Balzani and coworkers[31,32] described the use of metal branching coordination centers, such as ruthenium and osmium, and multidentate ligands for the construction of metallodendrimers centers more than a decade ago. Since this work, a large number of self-assembled dendritic structures have been prepared by using metal ligand complexes (Figure 3.16a), hydrogen bonding moieties (Figure 3.16b), or ionic interactions (Figure 3.16c) for the assembly.

Figure 3.16a shows a small bimetallic dendrimer obtained in a flexible synthesis strategy that is used to prepare many larger dendrimers with a variety of arrangements of the internal and peripheral metal centers.[32] These polynuclear complexes may be redox-active and/or luminescent, can be used as electrochemical and/or photochemical molecular devices, and may prove useful in areas such as energy harvesting and conversion, signaling, and diagnostics. Among the numerous self-assembling dendritic systems that have been reported, those involving assembly into columnar or spherical structures are particularly intriguing. Percec and coworkers[33,34] used van der Waals interactions and H-bonding to self-assemble functionalized polyether dendrons into cylindrical columnar or spherical assemblies that mimic the shape and architecture of certain types of viruses. Similarly, Aida and coworkers[35] have used weak metal-metal interactions to effect the hierarchical self-assembly of small dendrimers into luminescent superhelical fibers. Given their sizes, optical properties, and electronic properties, such assemblies could well find display, storage, or other applications in micro- and nanoelectronics.

As summarized in Figure 3.17, dendrimers can be designed to aggregate to form cylinders or spheres depending upon the nature of the fundamental building unit. Pie-shaped dendrimers assemble into cylinder segments (donuts) that then further aggregate into cylinders (Figure 3.17a). Cone-type dendrimers organize into spheres (two to twelve per sphere depending on generation), which then pack into the unusual Pm3-n (A-15)-type cubic crystal structure having spheres at the corners and body center of the cube plus two on each face (Figure 3.17b).

The layered architecture of dendrimers, their globular shape, highly controlled monodisperse size,

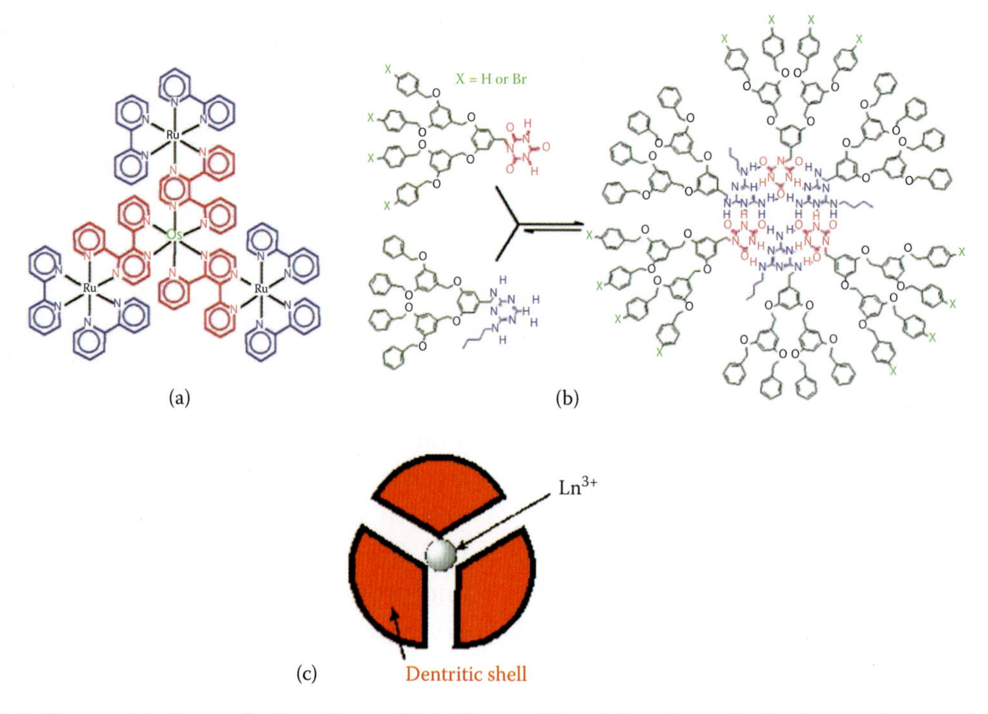

FIGURE 3.16 (a) Self-assembled metallodendrimer. (b) Self-assembly of dendrimer by hydrogen bonding. (c) Schematic representation of self-assembled ionic dendrimer. (From Denti, S., et al. 1992. *J Am Chem Soc* 114:2944–50; Balzani, V., et al. 1996. *Chem Rev* 96:759–833; Fréchet, J.M.J. 2002. *Proc Natl Acad Sci USA* 99:4782–87. [31,32,36])

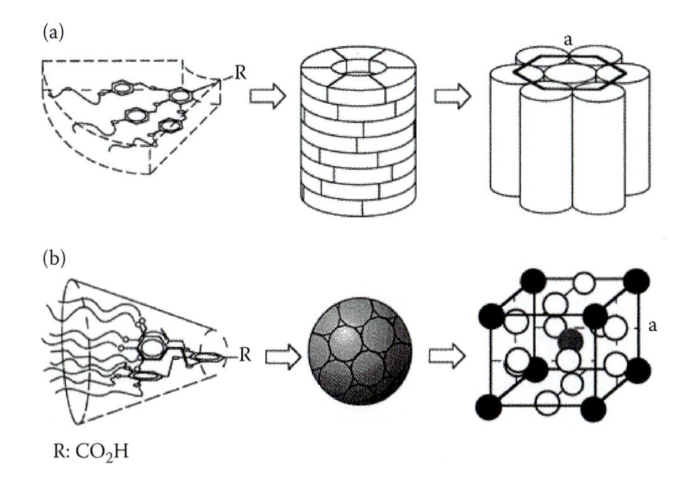

FIGURE 3.17 Assembling cylinders (a) and spheres (b) from dendritic molecules.[37]

radially controlled chemical makeup, multivalent periphery, variable inner volume, and controlled intramolecular dynamics endow dendrimers with unique features such as the ability to morph into a variety of supramolecular arrangements in response to external stimuli. On the basis of these properties, dendrimers may provide the key to reliable and economical fabrication of functional nanoscale materials with unique electronic, optical, optoelectronic, magnetic, chemical, or biological properties. Some of these applications are already being researched. Specially designed dendrimers with a hydrophobic interior and a hydrophilic[38] periphery, so-called "unimolecular micelles," are capable of molecular inclusion and solubilize hydrophobic molecules, such as pyrene in aqueous solution, and are at least as good at it than traditional micelles; moreover, they do not display a critical micelle concentration.[39,40] Classic micelles are thermodynamic aggregates of amphiphilic molecules and therefore dynamic assemblies of small molecules, while "unimolecular micelles" are static and retain their cohesion regardless of concentration. Also, dendrimer monolayers are finding a number of applications from surface-confined chemical sensor arrays[41,42] or affinity biosensors[43,44] to resists for nanolithography.[38,45]

Molecular Assembler

A molecular assembler was defined by K. Eric Drexler as a "device able to guide chemical reactions by positioning reactive molecules with atomic precision."[46] Some biological molecules, such as ribosomes, constructing a wide variety of proteins from their amino acid constituents, fit this definition (see Figure 2.12). However, the term *molecular assembler*, as defined by Drexel, refers to artificial mechanical machines, envisioned to work in air rather than in an aqueous solution. Molecular assemblers are broader than self-replicating machines (see above and Figures 3.12 and 3.13), as in theory they can both self-replicate and make other products. Drexel logically pointed out that if self-replication of molecular assemblers is not restrained, it might lead to competition with naturally occurring organisms. This is referred to as the "gray goo" scenario.[47] One method to construct molecular assemblers in solutions is to mimic natural evolutionary processes. As a result, production of complex molecular assemblers might evolve from simpler systems.

One of the most outspoken critics of the mechanical "molecular assemblers" was Richard Smalley (1943–2005). Smalley argued that such assemblers were not physically possible and introduced two major objections to them, termed the "fat fingers problem" and the "sticky fingers problem."[48] He believed that these two problems would exclude the possibility of molecular assemblers that worked by precision picking and placing of individual atoms.

In the fat fingers problem, Smalley argues that in an ordinary chemical reaction, five to fifteen atoms near the reaction site engage in an intricate three-dimensional waltz in a very cramped region of space measuring no more than a nanometer on each side, so there is not enough room in the nanometer-size reaction region to accommodate all the fingers of the mechanical manipulators necessary to exert complete control of the chemistry. Drexel's counterargument is that two reactants can be brought together with controlled trajectories if one reactant is bound to a substrate and the second reactant is positioned and moved by a single finger. He correctly points out that this has already been done experimentally by using the tip of an STM.[49] For example, Ho and Lee[50] physically bound a CO molecule to an iron atom on a silver substrate using an STM (see Volume II, Figures 2.44 and 2.45, for more complex STM molecular assembly). Drexel also points out that if steric constraints near the tool tip do make it too difficult to manipulate particular individual atoms or small molecules with sufficient reliability, a simple alternative is to use conventional solution or gas phase chemistry for the bulk synthesis of somewhat larger nanoparts, consisting of 10–100 atoms. These larger nanoparts could then be bound to a positional mechanical device and assembled into larger structures without further significant steric constraints. This is the approach taken by the ribosome in the synthesis of proteins, where individual amino acids are sequentially assembled into an atomically precise polypeptide without the need to manipulate individual atoms. *Atomically precise* is a description of the precision of the final product, not a description of the manufacturing method. It must be argued though that once we are dealing with a cluster of atoms the original objective of atomic scale control has been abandoned and a ribosome still recognizes the individual base pairs to stitch the protein together from its amino acid components.

Smalley also advances the sticky fingers problem, which is the claim that "the atoms of the manipulator tool will adhere to the atom that is being moved. So it will often be impossible to release this minuscule building block in precisely the right spot."[48] Drexel's counterargument is that the sticky fingers problem is not a fundamental barrier to building mechanical assemblers since reactions exist that allow the synthesis of a useful range of precise molecular structures. As an example, he mentions that the application of a voltage between a manipulator tool and the workpiece that cause the target atom or moiety to move to the desired position. He points again to work by Ho and Lee[50] for experimental proof.

A final criticism by Smalley was that these mechanical assemblers would be far too slow to be ever practical.[48] Here, a counterargument could be the possible use of arrays of assemblers, as we saw in the IBM Millipede (see Volume II, Figure 2.4b).

All in all, the Smalley-Drexel debate is about how to make molecular assemblers, not whether they are possible or not. To push either extreme, i.e., to claim that only biology/chemistry can do it (Smalley) or to put too much emphasis on mechanical assembly (Drexel), is to ignore that the solution might well be one that incorporates a combination of both. Drexel is indeed pointing at chemical synthesis (up to a nanocluster) and then mechanical positioning for further assembling of these clusters as a possible solution.

Nanostructure Properties

Introduction

In this section we introduce all the different properties that nanostructures bring to the table that do differ from microstructure properties. On the broadest level the differences may be summarized as shown in Table 3.2. Nanostructures occupy very high surface area and have a vastly increased surface area per unit mass, e.g., upwards of 1000 m^2 per gram. In terms of suface-to-volume (S/V) ratio, a 3-nm iron particle has 50% of its atoms on the surface, a 10-nm particle has 20%, and a 30 nm particle only 5% (see Volume I, Figure 2.49). Nanoparticles are quantum mechanical, wave-like, with discrete energies and quantized currents. Quantum size effects result in unique mechanical, electronic, photonic, and magnetic properties. The chemical reactivity of nanoscale materials greatly differs from that of more macroscopic forms. Microfabricated microstructures, on the other hand, exhibit relatively small S/V, behave mostly classically (semiclassically) and

TABLE 3.2 Comparison of Microstructures and Nanostructures

	Microstructure	Nanostructure
Physics	Semiclassic	Quantum mechanical
Electron nature	Particle-like	Wave-like
E- or k-space	Continuous	Discrete
Current	Continuous	Quantized
Decision	Deterministic	Probabilistic
Fabrication method	Microfabrication	Nanofabrication
Surface area/volume	Small	Very large
Packing density	Low	Very high

are particle-like, with continuous energy and currents. All the properties of nanoparticles we mention below are ultimately a consequence of the more fundamental ones listed in Table 3.2.

Melting Temperature, Thermal Transport, and Heat Capacity

In Chapter 7, this volume, on scaling, we learn that the surface-to-volume (S/V) ratio of particles scales as $1/r$, where r is the characteristic dimension of the particles. The smaller a particle, the more of its atoms find themselves at it surface. A bulk solid material will typically have less than 1% of its atoms on its surface, whereas 10-nm particles have about 15% of surface atoms. The high S/V ratio of nanoparticles makes them more reactive as catalysts in chemical reactions and lowers their melting temperature (T_m). There is an inverse linear relationship between melting temperature and the surface-to-volume ratio. This makes sense since atoms on a surface are more easily accessed and rearranged than atoms in the bulk. As a consequence, the melting temperature of particles is always lower than the bulk. As observed from Figure 3.18 the melting point of gold particles decreases dramatically as the particle size gets below 5 nm. For nanoparticles embedded in a matrix, melting point may be lower or higher, depending on the strength of the interaction between the particles and the matrix.

Macroscopic thermal transport theory, with Fourier's law (Volume I, Equation 3.53) and the diffusion equation (Volume I, Equation 3.54), breaks down for applications involving thermal transport in NEMS. Whereas for bulk materials, interface effects (surface or grain boundaries) dominate only at low temperatures, in nanometer-scale thin films and particles, the room temperature thermal conductivity can be significantly modified. One can think about the reduction in thermal conductivity in these structures as corresponding to a lower mean free path for the energy flux. In the case of conductors, electrons dominate the thermal conductivity and the mean free path depends on electron scattering phenomena. At lower temperatures, the electron mean free path (λ_e) is constant and controlled by defect scattering, whereas at higher temperatures, phonon scattering takes over and the mean free electron path falls off, as shown in Volume I, Figure 3.161, where we plot mean free electron path (λ_e) versus temperature. From Volume I, Figure 3.162, plotting electronic thermal conductivity (κ_e) versus temperature for Cu and Al, we see that thermal conductivity increases with temperature in the defect scattering regime and then goes down with temperature increase when phonon conduction takes over at higher temperatures. Phonons dominate the thermal conductivity in insulators. Phonons, like electrons, scatter at boundaries, defects, and other phonons (phonon-phonon scattering). For nanodevices the device boundary might take over from defect control. Scattering of heat flow at any of these interfaces significantly reduces thermal conductivity. The size effect on the thermal conductivity can exceed two orders of magnitude for layers of thickness near 1 μm at $T < 10$ K. Figure 3.19 illustrates this for three different thin-film Si layers (thickness d_s) versus bulk Si.[52]

Heat capacity (C_v) of a material is related to its atomic structure or its vibrational and configurational

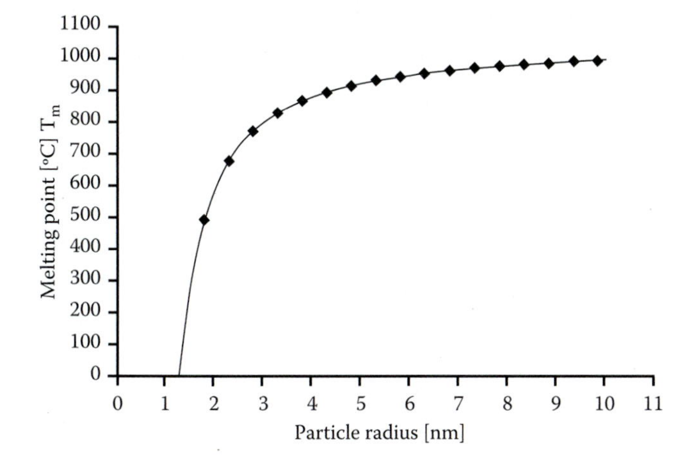

FIGURE 3.18 Melting temperature (T_m) of gold particles as a function of their radius.[51]

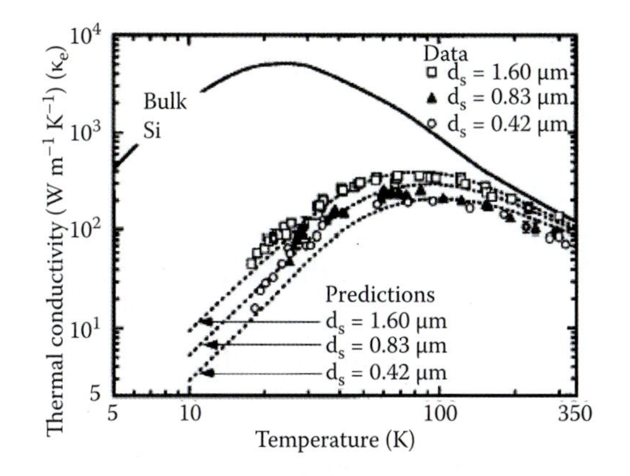

FIGURE 3.19 Thermal conductivity (κ_e) of thin Si films versus bulk Si. d_s is the Si layer thickness.[52]

entropy, which is controlled by all the possible nearest-neighbor configurations. Nanocrystalline materials feature ultrafine crystalline grains. A large fraction of atoms are located in the metastable grain boundaries in which the nearest-neighbor configurations are much different from those in the crystallites. The grain-boundary possesses an excess volume with respect to the perfect crystal lattice, therefore, heat capacities (C_v) of the nanocrystalline materials are expected to be higher than those of the corresponding coarse-grained polycrystalline counterparts.

For nanocrystalline materials, the C_v values, in practice, are indeed observed to be higher than their bulk counterparts. For example:

Pd: 48% up from 25 to 37 J/mol·K at 250 K for 6 nm crystalline
Cu: 8.3% up from 24 to 26 J/mol·K at 250 K for 8 nm
Ru: 22% up from 23 to 28 J/mol·K at 250 K for 6 nm

The phonon contribution to the molar specific heat capacity (C_v) of these polycrystalline materials, we saw in Volume I, Chapter 3, is given by the Dulong-Petit law (Volume I, Figure 3.15), with values for different materials approaching $C_v = 3R$ or 25 J/mol·K at high temperatures. As calculated by Debye, a constant value for C_v is reached at $T >> \theta_D$ and for $T << \theta_D$ (where θ_D is the Debye temperature) he found:

$$C_v = \frac{12}{5}\pi^4 R\left(\frac{T}{\theta_D}\right)^3$$

(Volume I, Equation 3.331)

A T^3 temperature dependency for the specific heat stems from a T^4 dependency of the internal energy U ($C_v = \frac{dU}{dT}$). Since the density of phonon states, [$G(E)$], depends on the dimensionality of the structure, the specific heat of the lattice also depends on the dimensionality of the structure (3D, 2D, 1D, or 0D). Generalizing for different dimensions (D), one derives:

$$U \propto T^{D+1} \text{ and } C_{v,lat} \propto T^D$$

(Volume I, Equation 3.338)

Besides phonon contributions, if the material is an electronic conductor, the electronic contribution to the heat capacity must be counted as well. The electronic specific heat capacity for a 3D structure we introduced in Volume I, Chapter 3 as follows.

$$C_{v,el} = \frac{dU}{dT} = \frac{\pi^2}{3} k_B^2 T G(E = E_F)_{3D} = \frac{\pi^2}{2}\left(\frac{k_B T}{E_F}\right) n_{3D} k_B$$

(Volume I, Equation 3.307)

This expression must be modified again for different dimensionalities since the internal energy (U) depends on the density of electronic states $G(E)_{\#D}$ and $C_{v,el} = \frac{dU}{dT}$.

Thermoelectric Power

In the 1960s, semiconductor thermoelectrics were considered for replacement of mechanical refrigerators and diesel generators. The idea was that if transistors could replace vacuum tubes for electronics, then surely thermoelectrics could replace those outdated machines. The efficiency of cooling and heating is controlled by a dimensionless parameter Z given by the following equation:

$$ZT = \frac{S^2 \sigma T}{\kappa} \tag{8.78}$$

where T is the absolute temperature; σ and κ are the electrical and total thermal conductivities (lattice and electrons) of the material, respectively; and S is the Seebeck coefficient. Consequently, requirements for a good thermoelectric material are a high figure of merit ZT; this in turn requires a high Seebeck coefficient (S), a high electrical conductivity (σ), and a low thermal conductivity (κ). For the best thermoelectric materials, one has a ZT of ~0.9, which remained the upper limit for many years. The biggest problem was that ZT had to be improved from ~0.9 to 3 or higher to become commercially attractive [ZT needs to improve to >1.3 at 300 K for a major impact in electronics cooling and to around 2.5 for a revolutionary impact in air-conditioning and power from waste heat (energy scavenging)]. Minimizing thermal conductivity and maximizing electrical conductivity has been the biggest dilemma for the last forty years. Dresselhaus et al. show that lower dimensional structures such as nanowires based on Bi, BiSb, Bi_2Te_3, and SiGe nanotechnology give additional controls and can lead to a higher *ZT* (see Chapter 8, this volume).[53] Combining a superlattice with a nanowire structure enhances thermopower because of the sharper density of states than ordinary 1D nanowires (enhances *S*) and reduction of the lattice thermal conductivity by increasing the phonon scattering at the segment interfaces (decreases κ). The reason is that the higher density of electron states at the Fermi level increases σ, the increased phonon-boundary scattering lowers κ, and both factors contribute to making *ZT* higher. Figure 8.32 shows the thermoelectric figure of merit *ZT* of a Bi quantum well and Bi quantum wire as a function of size.

In 2008, two teams, one at Caltech and one at the University of California, Berkeley, separately reported that they could increase Si's ability to convert heat into electric current by as much as 100 times[54,55] (see also http://www.spectrum.ieee.org/jan08/5879). Bulk Si used to be a poor material for thermoelectric conversion, because its thermal conductivity is too high; heat travels across it too well so that it is difficult to create the necessary temperature gradient. Bulk Si at room temperature has a *ZT* of only 0.01, but the Berkeley team increased that to 0.4, and the Caltech team increased it to 0.6. This puts Si nanowires almost on par with bismuth telluride. Making thermoelectric devices out of Si, which is much more abundant, cheaper, and easier to handle than bismuth telluride, could help create a large market for these devices.

By fashioning Si into nanowires with diameters of 10–100 nm and introducing defects in the wires that slow the flow of phonons, both teams have found that they can decrease silicon's thermal conductivity without losing too much of its electronic conductivity, and in doing so increase *ZT*. Defects at three different length scales were engineered into the nanowires. First, with silicon in the shape of nanowires, the material is small enough compared with the phonons that the size of the wires themselves affects how the phonons move. Secondly, the surface of the nanowires is made rough, introducing a set of defects at a smaller scale. Finally, the silicon nanowires are doped with boron to introduce defects at the atomic level. Heath at Caltech induced a greater drop in thermal conductivity by making his nanowires even smaller than Yang's group at Berkeley—only 10–20 nm in diameter.[55] Normally, a wire can carry two types of phonons: one that causes the wire's diameter to expand or contract, and one that causes it to lengthen or shorten. But when the nanowires get small enough, these two vibration modes merge into a single type of phonon, which slows down the heat transport even more. Unfortunately, using wires only 10 nm wide may give the best results for thermal conductivity, but the electrical conductivity crucial to thermoelectric conversion also drops.

One of the applications for these silicon nanowires is for recycling waste heat from computer chips into electricity, i.e., energy scavenging (see Figure 3.20).

Mechanical Properties

Small grains, over a certain size range, lead to improved strength and hardness in lightweight nanocomposites and nanomaterials, altering bending and compression properties. This mechanical change is explained as follows. Dislocations are blocked by grain boundaries, so slip is blocked, and therefore, since the smaller the grain size the larger the surface of the grain boundaries, a larger elastic limit is obtained. The latter is expressed in the empirical Hall-Petch equation representing the maximum elastic yield strength (σ_Y = stress at which the material permanently deforms) of a polycrystalline material as a function of grain size. As we learned

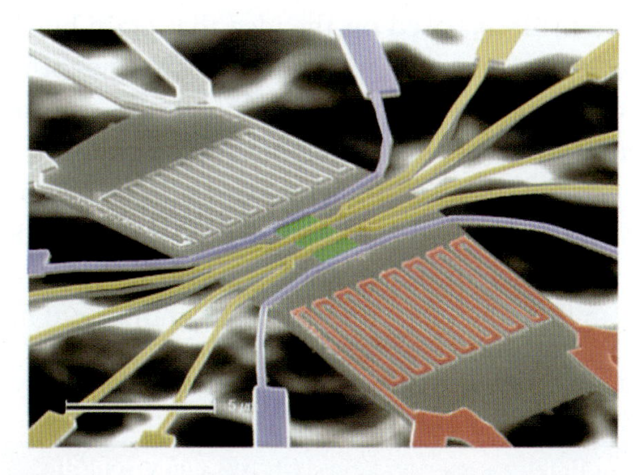

FIGURE 3.20 A nanofurnace: an array of nanowires (green) convert heat from the temperature difference between two slivers of a microchip. Current flowing in through a heater (red) causes the temperature difference. (From James Heath, http://www.spectrum.ieee.org/jan08/5879.)

in Volume I, Chapter 2, Hall and Petch, working independently, found that the yield strength of a polycrystalline material follows the following dependence:

$$\sigma_Y = \sigma_0 + \frac{K}{\sqrt{d}}$$

(Volume I, Equation 2.70)

where d is the average diameter of the grains in micrometers, with σ_0 the frictionless stress (N/m^2) and K a constant. The strength of a material thus depends on grain size (see Table 3.3). The strength of a material thus depends on grain size. In a small grain, a dislocation gets to the boundary and stops, i.e., slip stops. In a large grain, the dislocation can travel farther. So small grain size equates to more strength. For example, the elastic limit of copper doubles when the grain size falls from 100 μm to 25 μm. Instead of yield strength, one might also plot the hardness (H) of the material as a function of grain size, which creates a similar relationship.

The Hall-Petch relation holds for conventional grain sizes (from 0.1 to 100 μm) but for grains smaller than 20 nm, this relation breaks down (see Figure 3.21). For nanometer grains, a reverse Hall-Petch effect has actually been observed, i.e., the strength and hardness of materials from a small grain size on starts to decrease with decreasing grain size. In other words, an optimal grain size (d_c) exists, as suggested by the plot in Figure 3.21 (see also Volume II, Figure 7.54). The classic Hall-Petch relationship is based on the idea that grain boundaries act as obstacles to dislocation movement and since dislocations are carriers of plastic deformation, this manifests itself macroscopically as an increase in material strength. The Hall-Petch behavior breaks down when the smallest dislocation loop no longer fits inside a grain. The deformation mechanism for materials with very small grains (<20 nm) is different and it has been suggested that plastic deformation in this case is no longer dominated by dislocation motion and it is believed that individual atoms migrate, diffuse, and slide along grain boundaries and through triple junctions (Y-shaped grain boundary intersections).

Table 3.3 The Strength of a Material Depends on Grain Size

Material	Yield Strength
Cold worked copper	393 MPa
400-nm copper	443 MPa
100-nm nanograin copper	900 MPa
10-nm nanograin copper	2.9 GPa

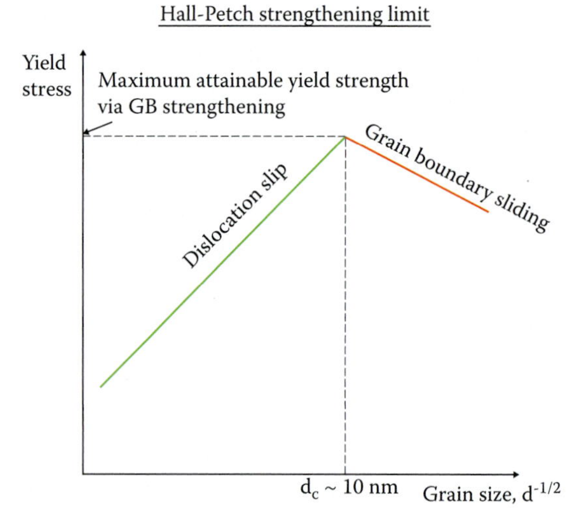

FIGURE 3.21 Hall-Petch reversal.

The increase of the surface-to-volume (S/V) ratio in nanoparticles leads to an increasing dominance of the behavior of the surface atoms over those in the interior of the particle. This results in more interactions between the intermixed components in ceramic nanocomposites; those nanophase powders compact more easily in the sintering process, and successful sintering enhances their hardness.

Chemical Effects

Surface chemistry effects that are barely noticeable in large particle systems become overwhelming in nanoparticle systems that demonstrate enhanced reactivity due to their unusual adsorptive properties. In bulk solids, all molecules are surrounded by and bound to neighboring atoms, and forces are in balance. Surface atoms on the other hand are bound only on one side, leaving unbalanced atomic and molecular forces on the surface. These forces attract gases and molecules through the van der Waals force, and at low temperatures this leads to physical adsorption or physisorption. At high temperatures, electron sharing or valence bonding with gas atoms, resulting in chemical adsorption or chemisorptions, may satisfy those same unbalanced surface forces. Chemisorption is a much stronger interaction than physisorption, with a heat of adsorption up to 800 KJ/mol. Adsorbing gas or vapor molecule splits into atoms, radicals, or ions, which form a chemical bond with the adsorption site. These processes are the basis for heterogeneous catalysis (key to production of fertilizers, pharmaceuticals, synthetic fibers, solvents, surfactants, gasolines, other fuels, and automobile catalytic converters). Nanostructured materials offer many advantages for this type of catalysis applications including the following:

- Huge surface area, high proportion of atoms on the surface
- Enhanced intrinsic chemical activity as size gets smaller, for example, likely due to changes in crystal shape
- When the shape changes from cubic to polyhedral, the number of edges and corner sites goes up significantly
- As crystal size gets smaller, anion/cation vacancies can increase, thus affecting surface energy; also, surface atoms can be distorted in their bonding patterns
- Enhanced solubility, sintering at lower *T*, more adsorptive capacity

All the properties discussed so far, unlike the quantum effects that we turn our attention to next, change gradually as the dimensions of the particles change.

Blue Shift in Semiconductor Nanoparticles

In Volume I, Chapter 3, we saw that whenever electrons are confined to a small space their behavior as tiny quantum wave packets takes precedence over their particle behavior. For a particle confined to a region of the *x*-axis of length Δx, the uncertainty in momentum is given by $\Delta p_x \sim \hbar/\Delta x$ (see Equation 3.106). If the particle has a mass *m*, the confinement gives it an additional kinetic energy of magnitude $E_c = (\Delta p_x)^2/2m \sim \hbar^2/2m(\Delta x)^2$ (Volume I, Equation 3.162). The confinement energy becomes significant if it is comparable to or greater than the kinetic energy of the particle due to its thermal motion, i.e., $E_c \sim \hbar^2/2m(\Delta x)^2 > \frac{1}{2}k_BT$, and the quantum size effect will be important if $\Delta x \sim \hbar/(mk_BT)^{1/2} \sim \hbar/p_x \sim \lambda_{DB}$ (Where λ_{DB} is the thermal de Broglie wavelength, i.e., roughly the average de Broglie wavelength of the particles in an ideal gas at the specified temperature, see Volume I, Equation 3.216). Thus, if the particle motion is confined in a region of the order of the particle de Broglie wavelength, one will observe effects of size quantization.

When the de Broglie wavelength of an electron is comparable to the size of the system in one or more directions, since only an integer number of half-waves of the electrons can be put in any finite system, one obtains a set of discrete electron states and energy levels, each of which is characterized by the corresponding number of half-wavelengths. This situation is summarized for 3D, 2D, 1D, and 0D degrees of freedom of electrons in Figure 3.22 (see also Volume I, Chapter 3).

A semiconductor quantum dot (QD) can be regarded as an artificial solid-state atom with delta-function-like density of states. Electrons in a quantum dot arrange themselves as if they were part of an atom, even though there is no nucleus for them to surround. The atom these assembled electrons emulate depends on the number of electrons and the exact geometry of the confining nanoprison. Here we calculate, as an example, the spacing between the lowest two energy levels, $\Delta E = E_{n+1} - E_n$, for a spherical

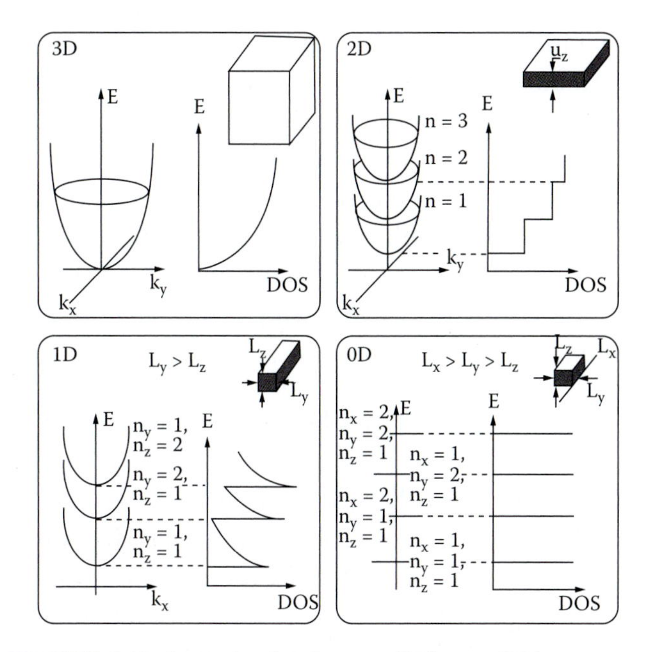

FIGURE 3.22 Quantization in zero (3D), one (2D), two (1D), and three (0D) directions.

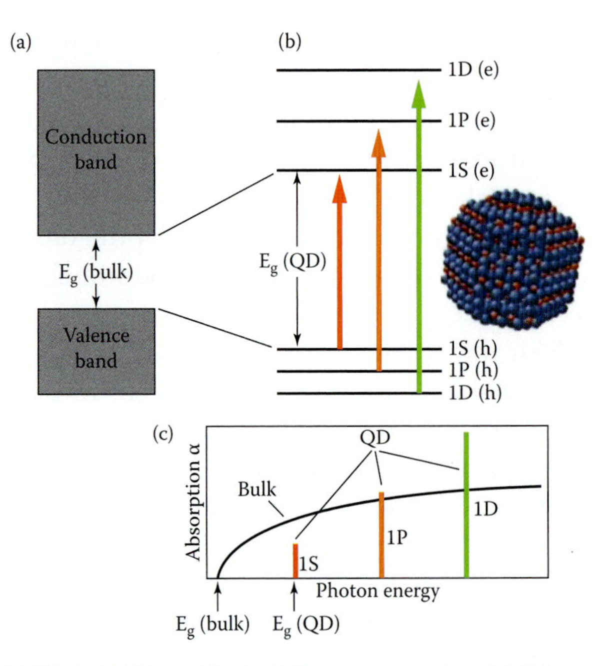

FIGURE 3.23 Blue shift in CdSe quantum dot. (a) CdSe bulk semiconductor. (b) CdSe quantum dot (QD). (c) Quantum dot photon energy absorption spectrum.

CdSe semiconductor quantum dot (see Figure 3.23). Such a spherical nanocrystal may be made from solution using colloid chemistry instead of the epitaxial methods used to make the cubic nanoparticle considered in Volume I, Figure 3.141. The dimensions of this nanocrystal are smaller than the electron de Broglie wavelength or the exciton Bohr radii (see also Volume I, Chapter 3). Semiconductor QDs possess high radiative efficiency, are compatible with semiconductor device technology, and are considered promising elements for implementing the coherent control of the quantum state, which is essential for quantum information processing.

In the case of an electron in a spherical quantum well, the Hamiltonian separates into angular and radial parts, and because of the spherical symmetry this results in the following eigen-energies and eigen-states:

$$E_{n,l,m} = E_{n,l} \quad \text{and} \quad \Psi(r,\theta,\phi) = Y_{l,m}(\theta,\phi)R_{n,l}(r) \tag{3.1}$$

with $Y_{l,m}(\theta,\phi)$ the spherical harmonics and $R_{n,l}(r)$ the radial wave functions. The energy levels and wave functions do of course depend on the details of the confining potential. If that potential is an infinite spherical well, where $V=0$ for $r<R$ and $V=\infty$ for $r>R$, then the solutions are the atom-like spherical Bessel functions from Volume I, Figure 3.81, described in the following expressions:

$$E_{n,1} = \frac{\hbar^2 \beta_{n,1}^2}{2m_e^* R^2} \quad \text{and} \tag{3.2}$$

$$R_{n,1} = j_1\left(\frac{\beta_{n,1} r}{R}\right) \quad \text{for } r < R \tag{3.3}$$

where $j_l(x)$ is the lth spherical Bessel function and the coefficient $\beta_{n,l}$ is the nth zero crossing of $j_l(x)$, with example values $\beta_{0,0} = \pi$ (**1S**), $\beta_{0,1} = 4.5$ (**1P**), $\beta_{0,2} = 5.8$ (**1D**), $\beta_{1,0} = 2\pi$ (**2S**), and $\beta_{1,1} = 7.7$ (**2P**). In parentheses and boldface, we list the atomic notations for the orbitals involved. This outcome explains why these spherical nanocrystals are often called *artificial atoms*.

The spacing between the lowest two energy levels $\Delta E = E_{0,1} - E_{0,0}$, also called the *1S transition*, for a CdSe nanocrystal with a conduction band effective mass m_e^* of 0.13 m_e and a radius R of 2 nm is calculated, using Equation 3.2, as follows:

$$E_{n,1} = \frac{2.9\ \text{eV}}{R^2}\left(\frac{\beta_{n,1}}{\beta_{0,0}}\right)^2 \tag{3.4}$$

or $\Delta E = E_{0,1} - E_{0,0} = 0.76$ eV, which is much larger than the thermal energy of the carriers (kT). This result is schematically represented in Figure 3.23, where in panel c we compare the absorption α of light by bulk CdSe with that by a nanocrystal of CdSe, clearly indicating that the bandgap is wider for the nanoparticle, which results in a blue shift. The probability of optical transitions between two quantum states is controlled by Fermi's Golden Rule (see Volume I, Chapter 3), which predicts that optical absorption must be proportional to the density of states, because a photon cannot be absorbed if there is no final state available for the electronic transition to go to. Bulk semiconductor and nanocrystal have the same overall absorption per unit volume when integrated over all frequencies. But in the case of a nanocrystal, the absorption spectrum consists of a series of discrete transitions with very high absorption transitions (sharp peaks), whereas a bulk semiconductor has a continuous absorption spectrum. This spectral concentration of a quantum dot enhances all resonant effects and increases the energy selectivity. The sharpening of the density of states (DOS) at specific energies induced by quantum confinement is at the origin of many improvements in the properties of nanostructured materials compared with bulk materials.

The 1S (*e*) electron level shown in Figure 3.23b increases in energy with a decreasing *R*, while the 1S (*h*) energy level decreases for a decreasing *R*. As a consequence, the bandgap of CdSe grows over a wide range as one varies *R*, resulting in a dramatic shift in the optical spectra with changing particle size. This is called the *blue shift*, as the energies shift to higher values or shorter wavelengths for a decreasing particle size. The particle size sensitivity of optical spectra enables one to tune continuously those spectra from blue to infrared, for both absorption and emission. This can lead to applications ranging from new more intense light-emitting diodes to more stable fluorescent labels. As an important example, the very strong optical transitions at particular frequencies have motivated researchers to develop quantum dot lasers (the laser principle, which is explained in Volume I, Chapter 5). Arakawa and Sakaki, in 1982, predicted significant increases in the gain and decreases in the threshold current of semiconductor lasers utilizing quantum dots.[56] These predictions set off an intense effort worldwide to fabricate such structures.

In Volume I, Chapter 5, we compared the energy separation calculated above for a semiconductor particle to that of a metal nanoparticle of the same size (2 nm). We found that quantum effects are much more readily observed for semiconductor particles than for metal particles (see "Optical Properties of Metals").

Plasmonic Effects in Metal Nanoparticles

Surface plasmon (SP) waves are coherent electron oscillations at the interface between any two materials where the real part of the dielectric function changes sign across the interface (say a metal-dielectric interface). In the case of metal nanoparticles, we saw in Volume I, Chapter 5, that these surface plasmon waves are localized and are readily observed. These "localized plasmons" may be excited directly by light absorption. For metallic nanoparticles significantly smaller than the wavelength of light, light absorption then concentrates to a narrow wavelength range and results in brilliant colors (see Volume I, Figure 5.94). With changing particle size and shape, those colors change because of changes in the surface plasmon resonance frequency ω_p, the frequency at which conduction electrons oscillate in response to the alternating electric field of the incident electromagnetic radiation (see Equation 5.216). For a gold spherical particle, the frequency is about 0.58 of the bulk plasma frequency. Although the bulk plasma frequency for bulk gold is in the UV range, the SP frequency is in the visible range (close to 520 nm). The Au plasmon effects are illustrated by the historic stained glass which adorn medieval cathedrals (see Figure 3.1). Only materials with free electrons (essentially Au, Ag, and Cu and the alkali metals) possess plasmon resonances in the visible spectrum. Au and Ag often are used in nanophotonics because of their excellent chemical stability and very good conductivity. For nonspherical particles, such as rods, the resonance wavelength depends also on the orientation of the electric field. Confinement and quantization phenomena are visible in semiconductors already at dimensions of 200 nm, whereas in metals they typically are seen only at 1–2 nm. Metallic nanostructures exhibit major changes in their optical spectra due to light absorption by plasmonics waves at the surface of these structures. The plasmonic effects of metal nanoparticles often mask the metal nanoparticle quantum effects.

Transparent Ceramic Nanoparticles

Previously, we dealt with the fascinating quantum properties of semiconductor quantum dots and the plasmonic behavior of metal nanoparticles. Here we clarify the properties of insulating ceramic nanoparticles. Ceramics can be defined as inorganic, nonmetallic materials. They are typically crystalline in nature and are compounds formed between metallic and nonmetallic elements, such as aluminum and oxygen (alumina: Al_2O_3), calcium and oxygen (calcia: CaO), and silicon and nitrogen (silicon nitride: Si_3N_4). The two main categories of ceramic nanoparticles are metal oxide ceramics, such as titanium, zinc, aluminum, calcium, and iron oxides, to name just a few, and silicate nanoparticles (silicates, or silicon oxides), generally in the form of nanoscale flakes of clays. It is the width of the forbidden bandgap (E_g) that determines whether these ceramics act as insulators ($E_g > 2$ eV) or as semiconductors ($E_g <$ 2 eV). The fact that the dimensions of these ceramic nanoparticles are below the critical wavelength of light renders them transparent, which makes for a convenient property for their use in packaging, cosmetics, and coatings. Ceramic nanoparticles, like metal and semiconductor nanoparticles, also can be formed into coatings and bulk materials at lower temperatures than their non-nanocounterparts, reducing manufacturing costs.

FIGURE 3.24 Water-repellant wood: BASF's lotus-effect aerosol spray combines nanoparticles with hydrophobic polymers such as polypropylene, polyethylene, and waxes. As it dries, the coating develops a nanostructure through self-assembly. BASF says that the spray particularly suits rough surfaces such as paper, leather, textiles, and masonry.

Superhydrophobicity and the Lotus Effect

In Volume I, Chapter 6, we saw how nanofeatures on a surface improves the wettability for hydrophilic surfaces ($\theta < 90°$). A drop on such surface will seem to sink into the hydrophilic surface (see Volume I, Figure 6.45a). This phenomenon is called *superhydrophilicity*. For hydrophobic surfaces ($\theta > 90°$) the wettability decreases for a nanotextured surface compared with a polished surface of the same material. It is energetically unfavorable to wet a rough hydrophobic surface and this result in an increased water repellency or superhydrophobicity (Volume I, Figure 6.45b). Superhydrophic surfaces, like a lotus leaf, are also self-cleaning. As we saw in Volume I, Figure 6.46, a droplet on an inclined superhydrophobic surface does not slide off; it rolls off. When droplets roll over contaminant particles, the particles are removed from the surface if the force of absorption of the particles by the droplet is higher than the stiction force between the particles and the surface. Usually the force needed to remove particles is very low because of the minimized contact area between the contaminants and the surface. As a result, the droplet cleans the leaf by rolling off the surface. The combination of the chemistry, the nanostructures, and the adherence properties of dirt and water to the surface, is what Barthlott and Neinhuis named the lotus effect.[57] In the lotus plant, the effect arises because lotus leaves have a very fine surface structure and are coated with hydrophobic wax crystals of around 1 nm in diameter. In the lotus plant, the actual contact area is only 2–3% of the droplet-covered surface.

Several surface treatments have been developed for roof tiles, fabrics, wood, paper, etc., that can stay dry and clean themselves in the same way as the lotus leaf (see Figure 3.24). Water does not adhere to the wood surface shown in Figure 3.24 but rolls off the paint, picking up and washing away debris in the process. By remaining dry, the coating also resists better mold, mildew, and algae. The lotus effect can be achieved, for example, by using special fluorochemical or silicone treatments on structured surfaces or with compositions containing microscale and nanoscale particulates. In one method, by Guo et al.,[58] an aluminum surface is made superhydrophobic by immersing it in sodium hydroxide for several hours (which roughens the surface) followed by spin coating a layer of perfluorononane to a thickness of 2 nm. This procedure increases the water contact angle from 67° to 168°. Electron microscope images show that the aluminum surface resembles that of a lotus surface, with a porous microstructure containing trapped air. Any surface treatment that creates micro- or nanoscale roughness with features having a height to width ratio greater than 1 that is subsequently coated with a thin hydrophobic coating will exhibit the lotus effect. Clothing that repels water has already been developed and marketed by brands such as Gap and Dockers; it uses a fabric named Nano-Care http://www.usato day.com/tech/columnist/cckev060.htm.

The BASF's lotus-effect aerosol spray, demonstrated in Figure 3.24, combines nanoparticles with hydrophobic polymers such as polypropylene, polyethylene, and waxes. It also includes a propellant gas. As it dries, the coating develops a nanostructure through self-assembly.

Magnetic Properties of Nanoparticles

Magnetic behavior of nanostructured materials was investigated from the early days of nanotechnology. One of the key findings so far is that the coercivity H_c, the intensity of the magnetic field needed to reduce the magnetization of a ferromagnetic material to zero after it has reached saturation, as a function of particle size goes through a maximum (see Figure 8.52).[59,60] The coercivity first increases linearly with decreasing grain size. The reason for the coercivity H_c increasing proportional to $1/d$ is that as particles decrease in size the magnetic domain size starts to coincide more and more with the grain size. As an example, for Fe the $1/d$ scaling is observed from 100 μm down to 40 nm. In the single-domain region a change of magnetization can only be produced by rotation of the magnetization of a single grain or crystallite as a whole, which requires strong magnetic fields, depending on the shape and anisotropy of the particle. As the size of the magnetic elements is further scaled down the coercivity goes over a peak and then decreases very fast (with a d^6 dependence). In this regime, a transformation from ferromagnetic to superparamagnetic behavior takes place. Typical particle sizes for the ferro- to superparamagnetic phase transformation are between 10 and 20 nm for oxides and 1–3 nm for metals. In the superparamagnetic state of the material, the room temperature thermal energy, kT, overcomes the magnetostatic energy of the domain or particle. The resulting fluctuations in the direction of magnetization cause the magnetic field to average to zero. Thus, the material behaves in a manner similar to paramagnetism, except that instead of each individual atom being independently influenced by an external magnetic field, the magnetic moment of the entire crystallite tends to align with the magnetic field. Superparamagnetic particles are attracted to a magnetic field but retain no residual magnetism after the field is removed (no hysteresis). These superparamagnetic nanoparticles, feature a high saturation magnetization, high relative permeability, low coercivity, high resistivity, and low hysteresis loss, and all of this benefits a large number of applications,* including in ferrofluids (see Figure 8.53).

The quest for a high-throughput manufacturing method for nanoparticles—be they semiconductors, metals, or insulators—is driven by their many extraordinary properties, some of which are summarized in Table 3.4.

Nanoparticle Manufacture

Introduction

Generally, nanoparticle manufacturing techniques sort out in either "bottom-up" methods or "top down" methods. In bottom-up methods, as we saw above, the nanoparticles are synthesized directly to the desired shape and size. One such bottom-up method is chemical synthesis in a liquid. An example of a top-down technique is a mechanical method such as ball milling (also called *mechanical attrition*). A further way of categorizing nanoparticle production techniques is based on the phase in which the process takes place, i.e., gas phase, liquid phase, and solid phase methods. The solid phase method is limited in terms of producing nanometer scale particles. The liquid phase method and the gas phase method are more suitable. Yet another way of classifying nanoparticle manufacturing techniques, the method we adapt in this chapter, is as follows:

1. In situ fabrication: These techniques incorporate lithography, vacuum coating [e.g., molecular beam epitaxial growth (MBE) and metallorganic chemical vapor deposition (MOCVD) epitaxial growth] and spray coating on solid surfaces.
2. Mechanical manufacturing: This is a top-down method that reduces the size of particles by attrition, for example, ball milling or planetary grinding.
3. Liquid phase synthesis (wet chemistry): This is fundamentally a bottom-up technique; i.e., it starts with ions or molecules and builds up into larger structures. These nanoparticle

* On the other hand, it is also because of superparamagnetism that hard drives are expected to stop growing once they reach a density of 150 gigabits per square inch.

Table 3.4 Semiconductor, Metal, and Insulator Nanoparticle Properties

Semiconductor
Quantum confinement at dimensions ~10–100 nm. Blue shift when the nanoparticle is of the size of the exciton's Bohr radius.
Light absorption is determined by the band-edge transition, which is strongly size- and shape-dependent, making the absorption highly tunable.
Density of states: electronic energy levels around the Fermi level are strongly affected for sizes in the 10–100-nm range.
Luminescence is very intense and can be tuned.
The I-V behavior is characterized by the Coulomb blockade.
Quantum efficiency (~0.1–0.5), spectral range of emission (400–2000 nm), and cross-section ~10^{-19} m^2.
A very high S/V ratio leads to very high catalytic properties and lowers the melting temperature T_m.
Metal
Quantum confinement at smaller dimensions ~2–3 nm. Electron motion is limited by the size of nanoparticle.
Light absorption is determined by the surface plasmon resonance, which is size- and shape-dependent.
Density of states: electronic energy levels around the Fermi level are strongly affected for sizes in the 1–2-nm range.
Luminescence is very weak.
The I-V behavior is characterized by the Coulomb blockade.
Quantum efficiency (~10^{-4}), spectral range of emission (500–9000 nm), and cross-section ~10^{-13} m^2.
A very high S/V ratio promotes very high catalytic properties and lowers melting temperature T_m.
Superparamagnetic nanoparticles feature a high saturation magnetization, high relative permeability, low coercivity, high resistivity, and low hysteresis loss, and all of this benefits ferrofluids.
Insulator
Improved sintering and hardness. Nanophase powder compacts more easily in the sintering process. Successful sintering enhances hardness of materials. Nanophase powders compact more easily than their analogous submicrometer particles.
Strength and hardness. Better for materials with smaller crystal grains, as dictated by the Hall-Petch relation (Volume I, Equation 2.70).
The specific heat capacity (C_v) of nanocrystalline materials is higher than that of their bulk counterparts.
A very high S/V ratio promotes very high catalytic properties and lowers melting temperature T_m.

manufacturing techniques historically can be placed under the title of colloid chemistry and involve classic "sol-gel" processes or other aggregation processes.

4. Gas phase synthesis: These methods include plasma vaporization, chemical vapor synthesis, and laser ablation.

Terminology one encounters frequently when discussing manufacture of nanoparticles includes:

- Cluster: A collection of units (atoms or reactive molecules) of up to about fifty units
- Colloids: A stable liquid phase containing particles in the 1–1000 nm range
- Nanoparticle: A solid particle in the 1–100 nm range that could be noncrystalline, an aggregate of crystals, or a single crystal
- Nanocrystal: A solid particle that is a single crystal in the nanometer range

In Situ Fabrication

Lithography Techniques

In a lithographic top-down approach, electron beam or ion beam lithography may be used to etch in situ a pattern of columns in a quantum well heterostructure [e.g., a GaAs/AlGaAs quantum well (Figure 3.25)]. The quantum well heterostructure provides

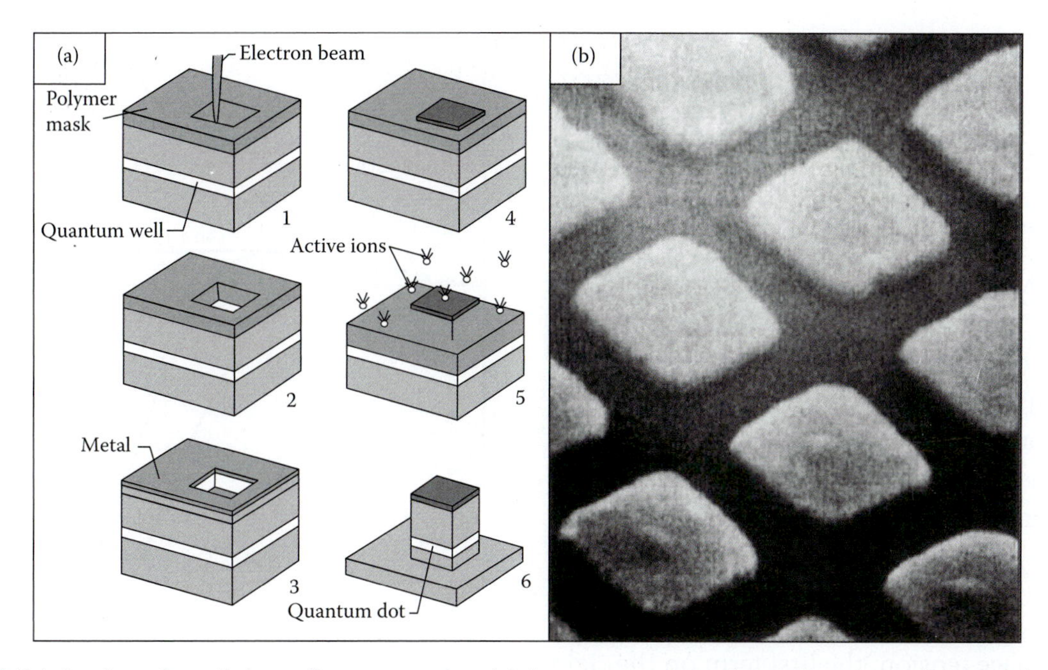

FIGURE 3.25 Fabrication of small, laterally patterned multiple quantum dot columns by e-beam lithography in a GaAs/AlGaAs. (a) Process of quantum dot etching. (b) Etched quantum dots in a Gas/AlGaAs well (scanning electron microscope picture).[61]

1D electron confinement in a plane because of the mismatch of the bandgaps of the two composing materials resulting in the formation of a potential energy well at their interfaces (see also Volume I, Figures 3.122 and 3.128). In the example in Figure 3.25, a positive resist is used as electron or ion beam mask and in the exposed areas the mask is removed. Next, the entire surface is covered with a thin metal layer. After a lift-off development, the polymer film and the protective metal layer are removed, and a clean surface of the sample is obtained, except for the previously exposed areas, where the metal layer remains (4). Next, by reactive ion etching, the areas not protected by the metal mask, slim pillars are created, containing the cut-out fragments of quantum wells (5–6). In this way, the motion of electrons, which initially is confined in the plane of the quantum well, is further restricted to a small pillar with a diameter of the order of 10–100 nm. Owing to the simplicity of the production of thin, homogeneous quantum wells, GaAs is the most commonly used material for creating dots by means of etching. Disadvantages of this technique are that it is slow, low density, and prone to defect formation.

As pointed out in Volume I, Chapter 3, confinement in lateral directions may also be implemented using electrostatic confinement with metal electrodes, lithographically patterned on the surface of a quantum well structure (see Volume I, Figure 3.138, for a quantum wire and Figure 3.144 for a quantum dot).

Self-Organized Epitaxial Growth

Self-organized epitaxial growth is another example of in situ nanoparticle manufacture. Oriented growth is referred to as epitaxy and, as we saw in Volume II, Chapter 7, it is subdivided into homoepitaxy and heteroepitaxy. Homoepitaxy is epitaxy performed with only one material: a crystalline film is grown on a substrate or film of the same material. Heteroepitaxy is epitaxy performed with materials that are different from each other: a crystalline film grows on a crystalline substrate or film of another material. This technology may be applied for growing crystalline films of materials of which single crystals cannot be obtained and for the integration of crystalline layers of different materials. Examples include gallium nitride (GaN) on sapphire or aluminum gallium indium phosphide (AlGaInP) on gallium arsenide (GaAs). Here, we will use the heteroepitaxy technique to form islands (quantum dots) of one material on the surface of another. To understand this method better, we first gather some

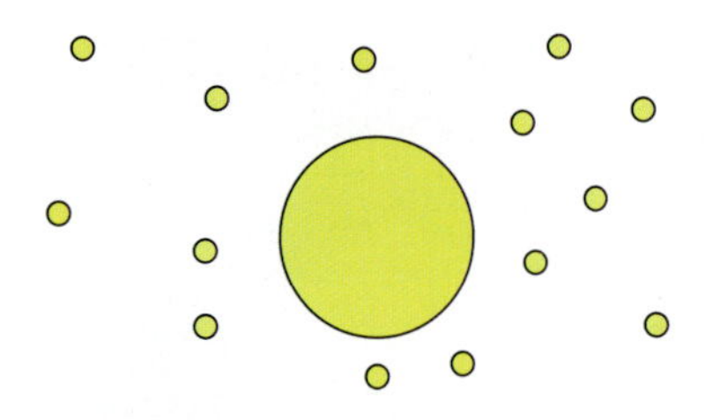

FIGURE 3.26 Nanoparticle growth in a vapor.

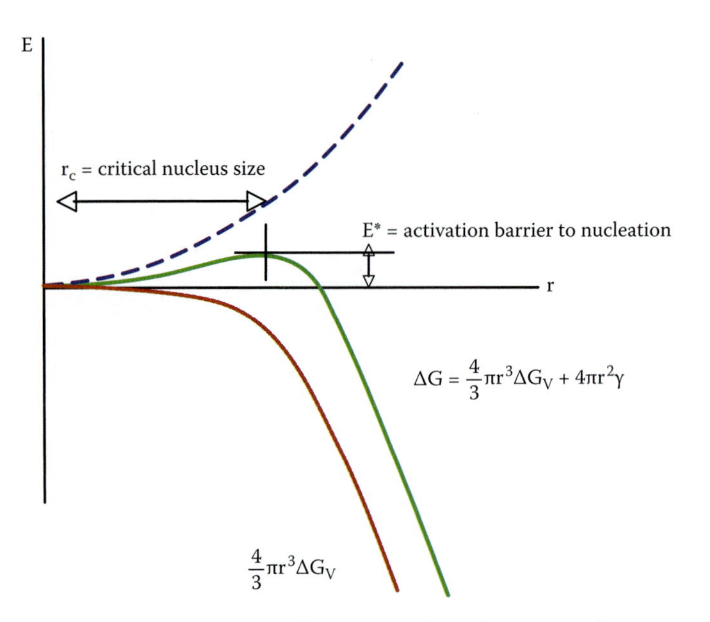

FIGURE 3.27 The Gibbs free energy as a function of particle radius.

background on nucleation and growth of particles in air and on a solid surface.

Consider a droplet of radius r in contact with a vapor (Figure 3.26). The Gibbs free energy change ΔG for this system is given by the following expression:

$$\Delta G = \frac{4}{3}\pi r^3 \Delta G_V + 4\pi r^2 \gamma \tag{3.5}$$

where γ = surface tension, the first term on the right is the free energy change per unit volume, and the second term on the right is the free energy change per unit area of surface. The first term can be calculated as follows:

$$\Delta G_V = \frac{kT}{\Omega}\ln\left(\frac{P_V}{P_S}\right) = \frac{kT}{\Omega}\ln(1+S) \tag{3.6}$$

with S defined as supersatuation $S = \frac{P_v - P_s}{P_s}$, Ω = the atomic volume (volume per atom), P_v the actual pressure in the gas phase, and P_s the saturated vapor pressure at equilibrium. When $P > P_s$, one defines supersaturation, $P = P_s$ means equilibrium is attained, and at $P < P_s$ the gas phase is undersaturated. The presence of surface tension, as illustrated in Figure 3.27, always produces an activation barrier to the nucleation of condensed phases. This is true for solids and for liquids (see, for example, the case of etching a single crystal in a solution as illustrated in Volume II, Figure 4.60).

The critical nucleus size, marked r_c in Figure 3.27, can be calculated from the following:

$$\frac{\partial \Delta G}{\partial r} = 4\pi r^2 \Delta G_V + 8\pi r \gamma \tag{3.7}$$

The maximum of the curve is at $\frac{\partial \Delta G}{\partial r} = 0$, so that $r_{critical} = \frac{2-\gamma}{\Delta G_v}$ and, with $\Delta G_v = \frac{kT}{\Omega}\ln\left(\frac{P_v}{P_s}\right) = \frac{kT}{\Omega}(1+S)$ (Equation 3.6) we obtain:

$$r_{critical} = \frac{-2\gamma\Omega}{kT\ln(1+S)} \tag{3.8}$$

and also:

$$\Delta G^*_{critical} = \frac{16\pi\gamma^3}{3(\Delta G_v)^2} \tag{3.9}$$

From Equation 3.8, as supersaturation S increases, the critical nucleus size decreases, so that for very small particles one needs high super-saturation and rapid nucleation.

The deposition rate of atoms on the nanoparticle is then given as follows:

$$\frac{dN}{dt} = hn_s A^* \exp^{\frac{-\Delta G^*_{critical}}{kT}} \tag{3.10}$$

where h = the Arrhenius pre-exponential
n_s = the number density of atoms in the gas phase
A* = the area of the critical nucleus

The Schwöbel barrier ($\Delta G^*_{critical}$) determines the kinetics (dN/dt) of the system.

For condensation of atoms on a solid surface (Figure 3.28) instead of on a spherical particle, Equation 3.5 is modified to the following expression:

$$\Delta G = l_1 l_2 h \Delta G_V + \underbrace{l_1 l_2 (\gamma)}_{\text{Surface tension}} + \underbrace{l_1\beta_1 + l_2\beta_2}_{\text{Line tension}} \tag{3.11}$$

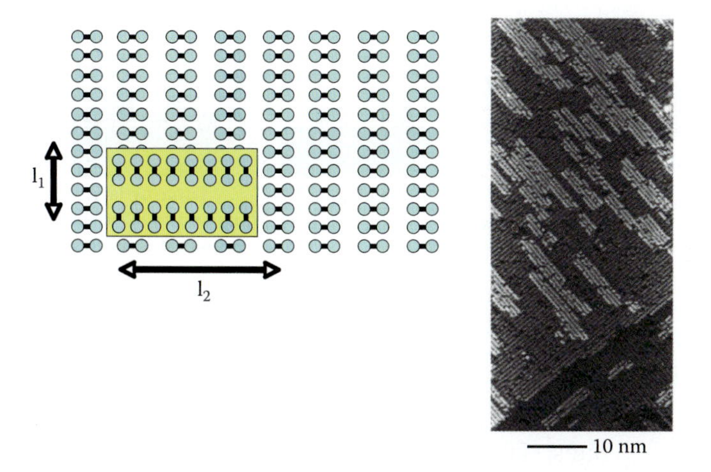

FIGURE 3.28 Island formation on flat solid surfaces.[62]

To identify l_1 and l_2 see Figure 3.28. The product l_1l_2 corresponds to the area of the adsorbate coverage. Where γ is the surface tension and β_1 and β_2 are line tensions. A line tension represents an additional contribution to the surface tension due to the presence of steps. The step line tension may be calculated as the difference between the surface tension of stepped and flat surfaces per step and step length. When an adsorbate coverage as shown in Figure 3.28 exceeds one monolayer, one speaks of thin-film growth. Thin-film growth is controlled by the interplay of thermodynamics and kinetics, and the following elementary processes take part: adsorption, surface diffusion, re-evaporation, capturing by defects, and combination with other adatoms to form clusters (nucleation). Small clusters are metastable, but they become stable at a critical island size, determined by the energy gain for condensation and the energy cost to form new surfaces. Depending on the surface diffusion of atoms, different growth modes result, governed by the bond strength between the atoms in the layer and the atom-substrate bonds. The occurrence of these different growth modes can be understood qualitatively in terms of the surface tension, γ (surface energy) as depicted in Figure 3.29.

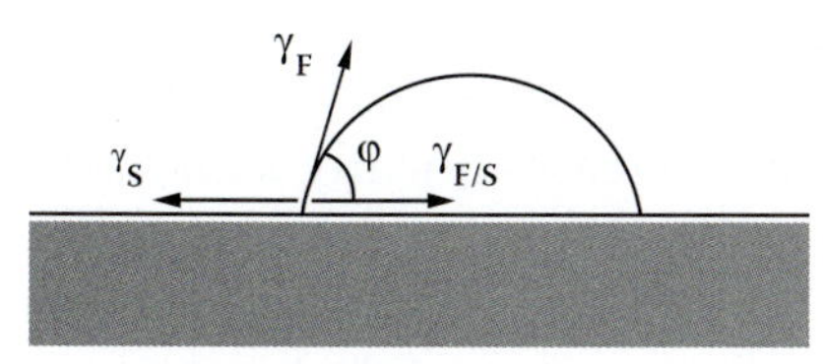

FIGURE 3.29 Surface tension of a film forming on a solid surface (see also Volume I, Figure 6.43).

The surface tension, represented by the symbol γ (see Volume I, Equation 6.88), is the force along a line of unit length, where the force is parallel to the surface but perpendicular to the line. Surface tension is therefore measured in force per unit length. Its SI unit is Newton per meter, but the cgs unit of dynes per centimeter is most commonly used. The wetting angle φ of a drop-like island, illustrated in Figure 3.29, is determined by the surface tensions of substrate (γ_S), film (γ_F), and film-substrate interface (γ_{FS}). Balance of these forces changes during growth. Flat surfaces exhibit three canonical growth modes, depending on the surface and interface energies. Beyond that, one finds more exotic growth modes at stepped surfaces.

In the case of $\gamma_S > \gamma_F + \gamma_{FS}$, typically, the first layer wets the surface but subsequent layers do not. The first atomic layer tries to coat the whole surface to provide optimal free energy reduction but with subsequent layers the situation changes: the first atomic layer reduces the surface energy of the substrate and new atoms do not experience the same interface energy as the first atomic layer. In addition, the added layers have to absorb the misfit strain energy, which grows with increasing film thickness (see below). Therefore, one can have either a continued layer-by-layer growth mode (Frank-van der Merwe mode), which is relatively rare, or the formation of islands on top of a flat first layer (Stranski-Krastanov growth mode). If the energy balance at the interface is tipped the opposite way, i.e., $\gamma_S < \gamma_F + \gamma_{FS}$, the overlayer has a tendency to nucleate three-dimensional islands right away and leave the low-energy substrate exposed (Volmer-Weber growth). This situation usually occurs when growing more reactive materials on top of an inert substrate, such as in the case of transition metals on a noble metal or an oxide.

The Stranski-Krastanov growth (SK growth; also spelled Stransky-Krastanov or Stranski-Krastanow) is one of the three primary modes by which thin films grow epitaxially at a crystal surface or interface. Also known as *layer-plus-island growth*, the SK growth mechanism was first noted by Ivan Stranski and L. Von Krastanov in 1939. It was not until 1958, however, in a seminal work by Ernst Bauer published in *Zeitschrift für Kristallographie*, that the

SK, Volmer-Weber, and Frank-van der Merwe mechanisms were systematically classified as the primary thin-film growth processes. Since then, SK growth has been the subject of intense investigation, not only to better understand the complex thermodynamics and kinetics at the core of thin-film formation, but also as a route to fabricating novel nanostructures for application in the microelectronics industry.

The range of possible growth modes becomes richer when going down in dimensionality. For example, when growing one-dimensional stripes on a two-dimensional, stepped surface. In addition to an extension of the three growth modes described above, one finds additional phenomena that are related to the interaction with the one-dimensional steps. As we saw in Equation 3.11 when stepped surfaces are involved line tensions due to the presence of steps contribute to the surface tension. Layer-by-layer growth turns into row-by-row growth, which is often called *step flow growth* because the step edge propagates downhill. There are other modes as well, such as the reverse step flow, where the material accumulates at the top of the step edge and propagates in the opposite direction. For manipulating these different growth modes, it is important to understand and control the surface and interface energies. Broadly speaking, one distinguishes three classes of solids where three different terms dominate the surface energy. Metals try to achieve the highest possible number of neighbors. Therefore, the surface with the lowest free energy is close-packed. Ionic solids, such as alkali halides, have the constraint that a stable surface needs to be charge neutral. Otherwise the Coulomb energy of a sheet of charge diverges. Covalent solids, such as semiconductors, have very directional bonds that rearrange themselves extensively in order to minimize the density of broken bonds at the surface. Bond counting is a first approximation, but the intricate rearrangement of atoms at semiconductor surfaces often defies our expectations. The absolute values of the surface energy are typically highest for transition metals and rare earths (2–3 Jm^{-2}); intermediate for covalent semiconductors, noble metals, and simple metals (1–2 Jm^{-2}); and lowest for ionic insulators (0.3–1 Jm^{-2}). The three growth modes are summarized in Figure 3.30.

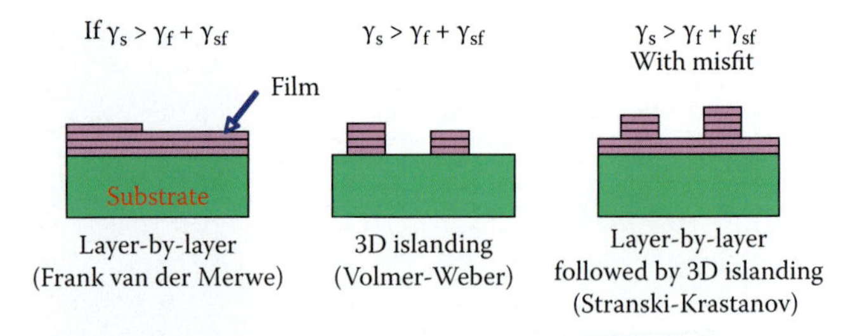

FIGURE 3.30 Summary of the three film growth mechanisms.

1. $\gamma_S > \gamma_F + \gamma_{FS}$. At the highest surface diffusion rates, layer-by-layer or Frank-van der Merwe growth mode is observed. The growing layer reduces the substrate surface very fast, i.e., it wets the surface.
2. $\gamma_S < \gamma_F + \gamma_{FS}$. High rate of incoming atoms and high Ehrlich-Schwöbel barrier, leads to island growth or Volmer-Weber growth. The growing layer wants to minimize interface energy and its own surface energy and the layer "balls up" on the surface.
3. $\gamma_S > \gamma_F + \gamma_{FS}$ with misfit. Layer plus island or SK growth. SK growth is possible in a heteroepitaxial system (misfit). This is the case of most interest for nanomanufacturing.

If the crystal structure (lattice constants) is different for substrate and film, strain appears at the interface because of this misfit. The strain depends on the size of this misfit: $\varepsilon = (b - a)/a$, with *a* and *b* the lattice constants for substrate and film (see also Vegard's law, Volume I, Equation 4.74). The elastic strain of a lattice mismatch can be accommodated either by a pseudomorphic growth or by misfit dislocations, as illustrated in Figure 3.31.

In Figure 3.32, this is illustrated for the case of Ge on Si, where we see how the island structure forms (see also Volume I, Figure 4.59). In this case, since the Ge lattice is 4% bigger than Si, bonding of Ge to Si compresses the Ge lattice by 4%. As the deposited film gets thicker, atoms continue to be forced to adapt to a different lattice constant, and the total strain energy increases linearly with the thickness of the growing film:

$$E_{strain} = \lambda\left(\frac{\Delta a}{a}\right)^2 At \tag{3.12}$$

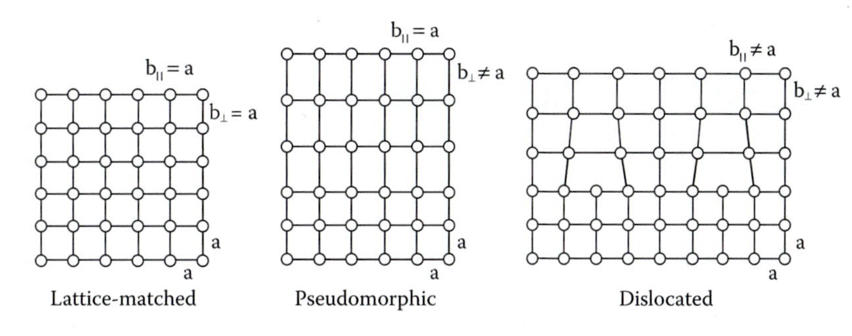

FIGURE 3.31 The deposited atoms try to adopt to the lattice constant. A lattice mismatch can be accommodated by pseudomorphic growth or by misfit dislocations.

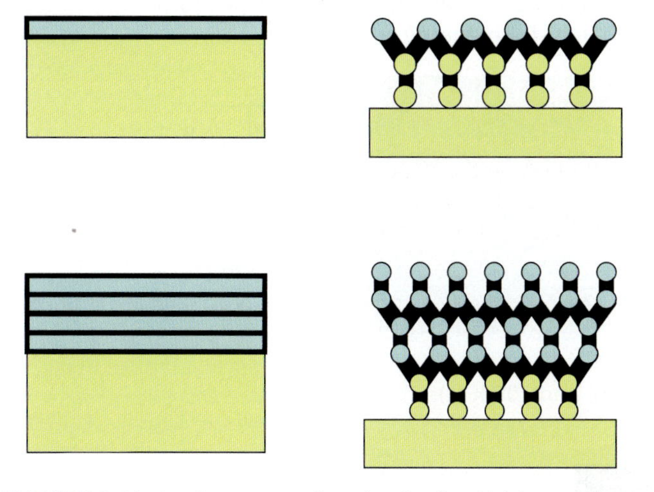

FIGURE 3.32 As the germanium lattice is 4% bigger than that of Si, bonding of Ge to Si compresses the Ge lattice by 4%. As the deposited film gets thicker, atoms continue to be forced to adapt to a different lattice constant, and the total strain energy increases linearly with the thickness of the growing film.

where

λ = elastic modulus
Δa = forced change in lattice constant
a = "free" lattice constant
A = area
t = thickness

In lattice-mismatched systems, self-organized quantum dots can be achieved by utilizing the Stranski-Krastanov growth mode (in an MBE or MOCVD reactor). In this mode, isolated islands form spontaneously above a certain critical thickness to relieve the mismatch strain energy. This is illustrated in Figure 3.33.

Depending on the growth conditions, one can get compact islands or ramified (fractal) islands. The compactness of the islands depends on the diffusion probability along its edges: this in turn depends on the surface temperature. In the extreme case of hit-and-stick, at low temperatures, where surface diffusion is limited, a diffusion limited aggregation model holds, which predicts fractal islands. This is demonstrated for the example of Pt growth on Pt(111) in Figure 3.34a (left-most image). The equilibrium shapes are hexagons (at 700 K) (Figure 3.34 panel c). The size distribution of the islands is affected by critical island size, coverage, and substrate temperature. According to scaling theory, one can derive a scaling relation for the size distribution, which depends only on the critical size.[63]

Certain surfactants may change the growth modes. Adding surfactants, a modification of the kinetics takes place rather than a change of the energetics. For example, the adatom diffusivity on the terraces might be influenced, the number of nucleation centers can be changed, the Schwöbel barrier may be changed, etc. The classic example is the growth of Ge on Si(111). On the bare surface there is a Stranski-Krastanov (SK) growth. With Sb as a surfactant, the film grows in a layer-by-layer fashion. Another example is the influence of H on Si(111) on the growth of Ag.

A successful method for quantum dot laser fabrication is based on the self-organized SK growth described here. SK growth occurs, we just saw, when a few monolayers of QD material (e.g., InGaAs) are deposited on a lattice mismatched material (e.g., GaAs). First, a quantum well-like wetting layer forms, from which quantum dots emerge. Strain, surface energy, and interaction with surrounding dots determine the dot shape, size, and distribution. Self-organized quantum dots are distributed on a semiconductor surface in a random way, and their size is subject to variations of a few percent (see

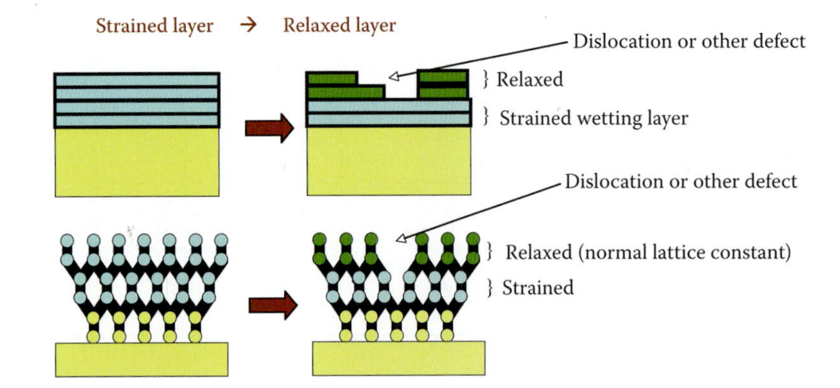

FIGURE 3.33 Stranski-Krastanov growth mode (in an MBE growth or MOCVD reactor): isolated islands form spontaneously above a certain critical thickness to relieve the mismatch strain energy.

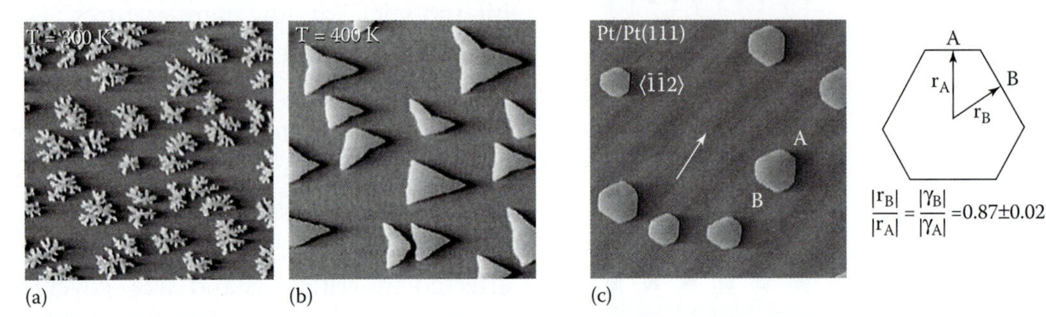

FIGURE 3.34 STM image of Pt growth on Pt(111). At 300 (a), 400 (b), and 700 K (c).

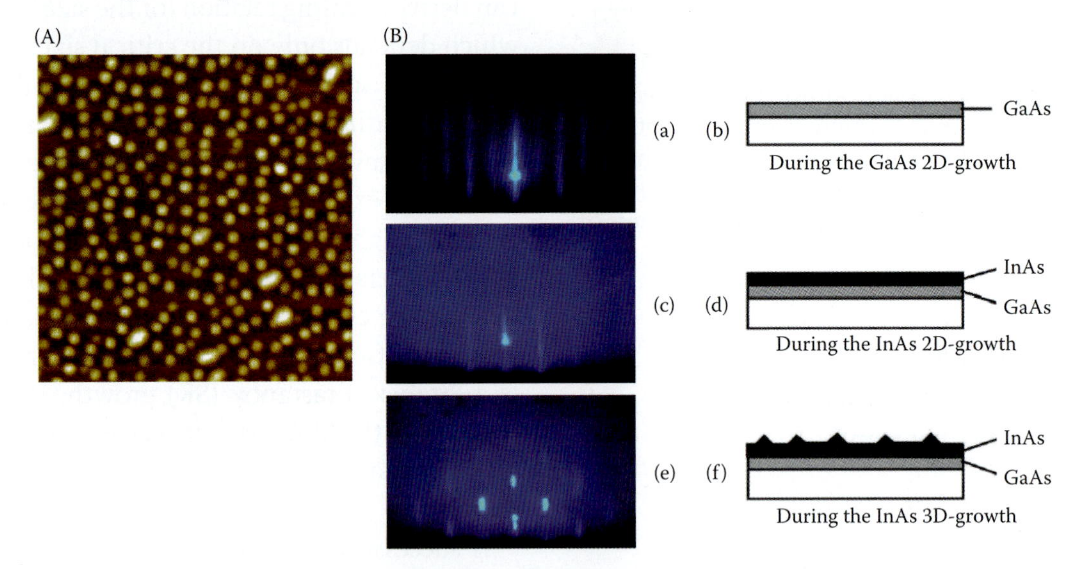

FIGURE 3.35 InAs quantum dots on a GaAs surface. (A) Atomic force microscope (AFM) picture of InAs/GaAs quantum dots. (B) Reflection high-energy electron diffraction (RHEED) patterns (a, c, e) during the two-dimensional (2D) deposition of GaAs (b) and the 2D deposition of InAs (d), and the 3D deposition of InAs (f) on a (100)GaAs buffer layer.[64]

Figure 3.35 with InAs quantum dots on a GaAs surface). While single quantum dots have sharp, atom-like emission lines, ensembles of quantum dots show an inhomogeneously broadened spectrum. Size, distribution, composition, and shape of the quantum dots can be chosen by different growth conditions such as temperature and precursor pressure. Larger quantum dots emit longer wavelengths, and smaller dots emit shorter wavelengths. The main advantages of self-organized growth are its integration into a single growth run, the low defect density, and the high dot density. The main limitations of this method are the cost of fabrication and the lack of control over positioning of individual dots. Some claim that Stranski-Krastanov methods for growing self-assembly quantum dots has rendered the lithographic approach to semiconductor quantum dot fabrication virtually obsolete.

Reflection high-energy electron diffraction (RHEED) is very sensitive to surface changes, because of either structural changes or adsorption, and is a good method by which to monitor island growth. In RHEED, a high-energy electron beam (3–100 keV) is directed at the sample surface at a grazing angle. The distance between the streaks in a RHEED pattern (see Figure 3.35) gives us the surface lattice unit cell size. For an atomically flat surface, sharp RHEED patterns result and with rougher surfaces one gets diffused RHEED patterns. In layer-by-layer growth mode, RHEED oscillations are observed (Figure 3.35Ba). These are a direct measure of growth rates in MBE: the oscillation frequency corresponds to the monolayer growth rate. The incident angle dependence of the oscillations suggest that interference between electrons scattering from the underlying layer and the partially grown layer contribute to these oscillations. The magnitude of the RHEED oscillations damps out (see Figure 3.35Be) because as the growth progresses, islands nucleate before the previous layer is finished.

Epitaxial growth on patterned substrates can organize the quantum dots on a surface into a regular pattern, as illustrated in Figure 3.36 where the growth of AlGaAs in pyramid-shaped recesses is shown. The recesses are formed by selective ion etching. The main disadvantage of this technique is that the density of QDs is limited by the mask pattern.

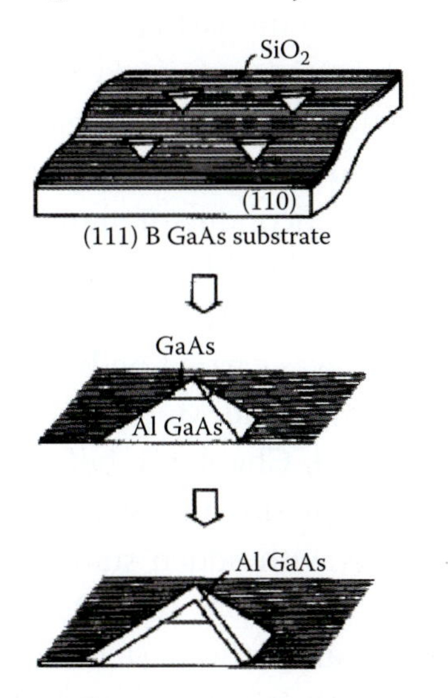

FIGURE 3.36 Growth of quantum dots in pyramid-shaped recesses on a GaAs substrate.[65]

Solid-State Processes

Solid-state processes, involving top-down grinding or milling to produce ceramic nanoparticles, are also referred to as mechanical attrition methods. In mechanical attrition, macro- or microscale particles are ground in a ball mill or in other size-reducing mechanisms. The resulting particles are air classified to recover the nanoparticles. Ball milling involves using fixed and moving elements to reduce larger particles into micron-sized and nanosized particles. The material to be ground is reduced in size because of the interaction of the fixed elements, the moving elements, and the material itself. The milling materials, the milling time, and the atmosphere all affect the nanoparticle properties. The kinetic energy transfer from balls to the powder in the milling bowl accomplishes the reduction in grain size. Since the kinetic energy of the balls is a function of their mass and velocity, dense materials such as steel or tungsten carbide are preferred. Other materials used as balls are agate, sintered corundum, zirconium dioxide, Teflon, chrome nickel, silicon nitride, etc. In order to prevent excessive abrasion, the hardness of the grinding bowl used and of the grinding balls must be higher than that of the materials that are being ground. Normally, grinding bowls and grinding balls should be made of the same material.

In the initial stage of milling, a fast decrease of grain size occurs that slows down after extended milling. Initially the kinetic energy transfer leads to the production of a myriad of dislocations in the powder grains resulting in atomic level strains. At a certain strain level, the dislocations annihilate and recombine to form small-angle grain boundaries, separating the individual grains. This way, subgrains with reduced grain size are formed. During further milling, this process extends throughout the entire sample. To maintain this reduction in size, the material must experience very high stresses. However, extended traditional ball milling cannot maintain the required high stresses, and hence reduction of grain size is limited. The other two parameters also causing a limit to grain size reduction in traditional ball milling are the local temperature developed due to ball collisions and the overall temperature in the bowl. A rise in temperature also results from

Table 3.5 Traditional Ball Milling Compared with High-Energy Ball Milling

	Ball Milling	High-Energy Ball Milling
Milling time	<1 h	20–200 h
Impact energy (W/g per ball)	0.001	0.2
Particle size	μm	μm clusters
Structural changes	No	Yes
Chemical changes	No	Yes
Atmosphere	No control	Vacuum, inert gas
Temperature	No control	Liquid nitrogen, −700°C

collisions between balls and balls, balls and powder, and balls and wall. Traditional ball milling thus produces only fine powders. In addition, traditional ball milling requires an increasing amount of energy and time to produce incrementally smaller particles. To produce nanomaterials, high-energy ball milling must be used (see Table 3.5).

High-energy ball milling uses steel balls to transfer mechanical energy into materials and creates structure changes and chemical reactions at ambient temperature (mechanochemistry). The impact energy is considerably higher, longer milling times are employed, and the process is carried out at liquid nitrogen temperatures in vacuum or in an inert atmosphere. High-energy ball milling can produce nanosized oxides and nitrides (<100 nm) via mechanical chemistry and has been recognized as one important synthesis method for nanomaterials, including nanoparticle, nanotube, nanocrystalline, nanoporous, and nanocomposite materials.

High-energy ball milling (see Figure 3.37 for a typical high-energy ball mill) is used for the commercial manufacture of magnetic, catalytic, and structural nanoparticles. Unfortunately, ball milling usually produces nanoparticles that are nonuniform in size and shape. The attrition approach is used to produce nanoparticles from materials that do not readily lend themselves to gas or liquid phase preparation.

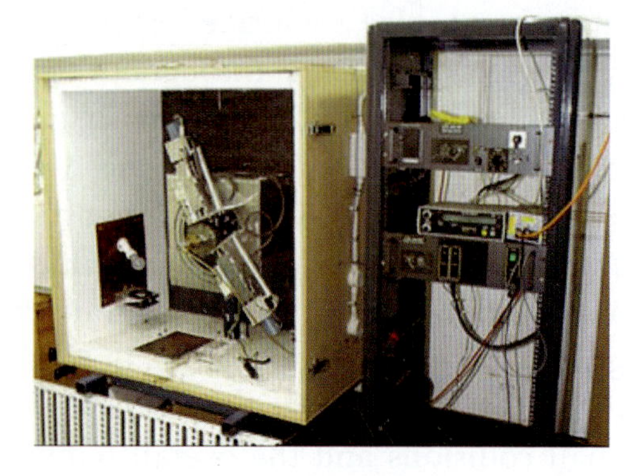

FIGURE 3.37 High-energy ball milling in a rotating steel ball mill.

Liquid Phase Synthesis

Several liquid phase nanoparticle synthesis methods were reviewed in Volumes I and II. Here, we briefly review the most salient points about those different liquid phase synthesis methods.

An important example of liquid phase nanoparticle synthesis is colloidal processing as detailed in Volume I, Chapter 7, where the Derjaguin, Landau, Verwey, and Overbeek theory (the DLVO theory) was used to describe the stability of a colloid (see "Intermolecular Forces"). A colloidal system, we remember, consists of two separate phases: a dispersed phase (or internal phase) and a continuous phase (or dispersion medium) that can be a solid, a liquid, or a gas. The colloids of interest here are sols, in which solid particles are dispersed in a liquid. Chemical reactions can be fine-tuned into forming dispersed nanoparticles, and capping the dispersed solid particles, formed by these chemical reactions with appropriate ligands, stabilizes them. This method is widely used for preparation of quantum dots (semiconductor nanoparticles) and was pioneered by Columbia University's Louis Brus (Volume I, Figure 3.68) in the late 1970s when he was at Bell Labs and it was carried on by two of his post docs, Paul Alivisatos and Moungi Bawendi. As we learned in Volume I, Chapter 3, before 1993, quantum dots (QDs) were often prepared from colloidal aqueous solutions with added stabilizing agents to avoid colloid precipitation. In 1993, Bawendi and coworkers synthesized the best luminescent CdSe quantum dots made up to that time by using a high-temperature organometallic procedure instead.[66,67]

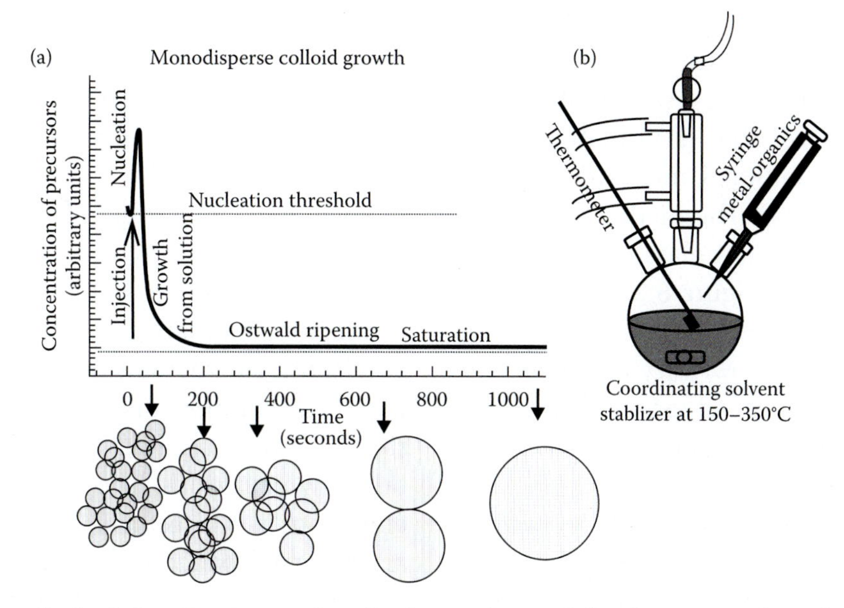

FIGURE 3.38 (a) Schematic depicting the stages of nucleation and growth for the preparation of monodisperse quantum dots (QDs) in the framework of the La Mer and Dinegar model. As nanocrystals grow over time, a size series of QDs may be isolated by periodically removing aliquots from the reaction vessel. (b) Representation of a typical apparatus used in monodisperse QD synthesis.[68]

Crystallites from ~12 Å to ~115 Å in diameter with consistent crystal structure, surface derivatization, and a high degree of monodispersity were prepared in a single reaction based on the pyrolysis of organometallic reagents by their injection into a hot coordinating nonaqueous solvent (see Figure 3.38b and Volume I, Figures 7.102 and 7.103).

A schematic to explain the formation of uniform nanoparticles in solution is shown in Figure 3.38a. In homogeneous precipitation, a short single burst of nucleation occurs when the concentration of constituent species reaches critical supersaturation. LaMer and Dinegar (1950) applied classic nucleation theory to qualitatively describe the kinetics of this type of burst nucleation. They proposed that from a strong initial supersaturation, a rapid nucleation or burst of particles occurs, followed by the slower absorption of diffusing atomic matter onto these nucleated particles. Their key observation was that the reduction in supersaturation after the nucleation burst strongly depletes the local availability (concentration) of constituent species, thus decreasing the rate of further nucleation, leading ultimately to a narrow size distribution of the nucleated, diffusively growing nanocrystals. Ostwald ripening may broaden the resulting distribution of nanocrystals. In Figure 3.38a this process starts at about 300s after injection. In Ostwald ripening, many small crystals initially form in a system, but they slowly disappear, except for a few that grow larger at the expense of the small crystals. The smaller crystals act as "nutrients" for the bigger crystals. This is a spontaneous process that occurs because larger crystals are more energetically favored than smaller crystals. While the formation of many small crystals is kinetically favored (i.e., they nucleate more easily), large crystals are thermodynamically favored. Thus, from a standpoint of kinetics, it is easier to nucleate many small crystals. However, small crystals have a larger ratio of surface area to volume than large crystals. Molecules on the surface are energetically less stable than the ones already well ordered and packed in the interior. Large crystals, with their greater volume to surface area ratio, represent a lower energy state.

Colloidal synthesis is by far the cheapest method and has the advantage of being carried out in benchtop conditions. For an example of a typical set-up see the apparatus in Figure 3.38b. This method is simple and can fabricate large quantities of quantum dots.

In Volume II, Chapter 8, we introduced sol-gel processing, which produces particles (typically metal oxides) by a series of hydrolysis and condensation reactions to form a sol of particles suspended in a liquid phase (Volume II, Figure 8.2). The sol-gel technique constitutes a chemical forming process for making very small particles—say, 20–40 nm in diameter—a particle size that is virtually impossible to make by conventional mechanical grinding (Volume II, Figure 8.1). The method is used to synthesize glass, ceramic, and glass-ceramic nanoparticles. From Volume II, Figure 8.2, ultrafine and uniform ceramic powders may be formed either by grinding of a xerogel or by slowly drying the sol. The disadvantages of sol-gel processing include the toxic chemicals required for the reactions and the lengthy preparation time.

Additional liquid phase nanoparticle synthesis techniques include sonochemical processing, cavitation processing, and microemulsion processing. In sonochemistry, an acoustic cavitation process generates a transient localized hot zone with extremely high temperature and pressure gradients. Such sudden changes in temperature and pressure assist the destruction of the sonochemical precursor (e.g., an organometallic compound) and the formation of nanoparticles. The technique can be used to produce large volumes of materials for industrial applications.

In hydrodynamic cavitation, nanoparticles are generated through creation and release of gas bubbles inside a sol-gel solution.[69,70] The sol-gel solution is mixed by rapidly pressurizing it in a supercritical drying chamber and exposing it to cavitational disturbance and high-temperature heating. The erupted hydrodynamic bubbles are responsible for nucleation, growth, and quenching of the nanoparticles. Acoustic cavitation in liquids actually leads to two different types of major effects, i.e., physical phenomena such as streaming, turbulence, microjet, shear, etc. and chemical phenomena such as radical production. The collapse of cavities can indeed generate highly reactive radicals, which may induce specific chemical reactions. One easily recognizes how in the case of gas-solid reactions, the dissolution of solids is enhanced due to the turbulent mixing generated by hydrodynamic cavitation. The vigorous mixing enhances the transport of gas molecules to the solid surface resulting in an increase in the mass transfer and hence the overall reaction rate. The particle size can be controlled by adjusting the pressure and the solution retention time in the cavitation chamber. Gedanken[69] and Gogate[70] reviewed the use of sonochemistry for the fabrication of inorganic nanomaterials of various shapes, size, structure, and phases.

Microemulsions (liquid in a liquid) and micelles have been used for synthesis of metallic, semiconductor, silica, barium sulfate, magnetic, and superconductor nanoparticles.[71] In contrast to ordinary emulsions, microemulsions form upon simple mixing of the components and do not require the high shear conditions generally used in the formation of ordinary emulsions. The two basic types of microemulsions are direct [oil dispersed in water (o/w)] and reversed [water dispersed in oil (w/o)]. The technique is useful for large-scale production of nanoparticles using relatively simple and inexpensive hardware.

The diagram in Figure 3.39 explains the basic scheme for microemulsion nanoparticle synthesis. Basically, each microemulsion type (A and B) contains a reactant of the synthesis, and when they are forced together (in a reactor) exchange occurs causing the reactants to combine. Because there is limited reactant in each droplet, the size of the nanoparticle is controllable. The ability to synthesize nanoparticles in a water/supercritical fluid microemulsion was also realized.[72–75] This method promises to be a highly useful route to controlled nanoparticle synthesis because of the added control variables afforded by the tunability of the solvent quality (density) through pressure and temperature.

Liquid phase synthesis techniques in general are better than gas phase techniques (see next) for controlling the final shape and size of the nanoparticles. The ultimate size is controlled, as in the case of gas

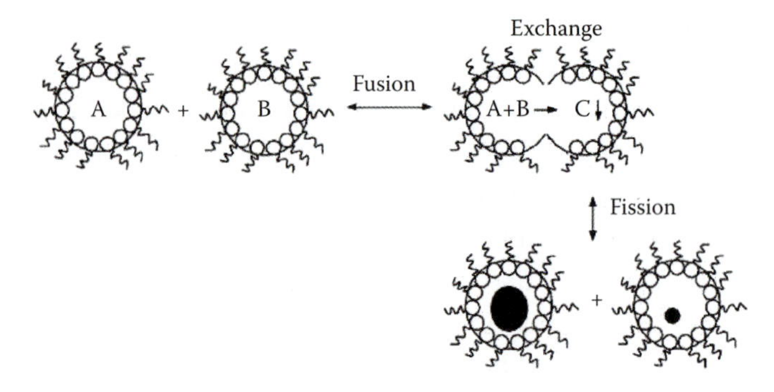

FIGURE 3.39 Microemulsion nanoparticle synthesis.

phase synthesis, by either stopping the process when the desired size is reached or by choosing chemicals that form particles that are stable, and stop growing at a certain size. These approaches are relatively low cost and high throughput, but contamination from precursor chemicals may be a problem.

Gas Phase Synthesis

Introduction

In gas phase nanoparticle synthesis a gas phase mixture is rendered thermodynamically unstable relative to the formation of the desired nanoparticles. Supersaturation is achieved by vaporizing material into a background gas and condensation to form particles is induced by cooling, e.g., by mixing with a cool gas or expansion through a nozzle. Once nucleation occurs, the remaining supersaturation is relieved by additional condensation on virgin substrate areas, or reaction of vapor phase molecules continues on deposited particles. This initiates the particle growth phase and rapid quenching after nucleation prevents particles that are too large.

In Table 3.6 we list several types of gas phase nanoparticle synthesis techniques grouped into those that use solid precursors and those that employ liquid or vapor precursors. The latter category is called *chemical vapor synthesis* (CVS), a modified chemical vapor deposition (CVD) method where the process parameters are adjusted to form nanoparticles instead of a thin film.

Gas Phase Synthesis with Solid-State Precursors

Inert-gas aggregation frequently is used to make nanoparticles from solid metals with low melting points (e.g., Bi). The metal is vaporized in a vacuum chamber and then supercooled with an inert gas stream. The supercooled metal vapor condenses into nanometer-sized particles that can be entrained in the inert gas stream and deposited on a substrate or studied in situ. Inert gases are used to avoid oxidation when creating metal nanoparticles, whereas a reactive oxygen atmosphere is used to produce metal oxide ceramic particles. The main advantage of this approach is the low contamination level. Final particle size is controlled by varying temperature, gas environment, and evaporation rate.

Pulsed laser ablation, like ion cluster beam deposition, is a nanoparticle gas phase synthesis method that uses solid precursors. Both techniques were reviewed in Volume II, Chapter 7. Pulsed laser ablation is used to vaporize a plume of material. Generally, the method can only produce small amounts of nanoparticles, but it can vaporize materials that are not easily evaporated, e.g., in the synthesis of Si, MgO, titania, and hydrogenated-silicon nanoparticles. There is a strong dependence of particle formation dynamics on the background gas. The use of pulsed laser ablation shortens the reaction time and allows for the preparation of very small particles. In 1995, at Rice University, it was reported that synthesis of carbon nanotubes could be accomplished by laser ablation.[76] A pulsed or continuous laser was used to vaporize a graphite target placed in an oven at 1200°C (Volume II, Figure 7.18). Contact with a cooled copper substrate caused the carbon atoms to be deposited in the form of nanotubes. This method has a carbon nanotube yield of around 70% and produces primarily single-walled carbon nanotubes with a controllable diameter determined by the reaction temperature. However, it is more

Table 3.6 Gas Phase Synthesis Methods for Nanoparticles

Methods Using Solid Precursors	Methods Using Liquid or Vapor Precursors
Inert and reactive gas condensation	Spray pyrolysis using different types of atomizers (e.g., pressure electrostatic, ultrasonic) and salt-assisted spray pyrolysis (SASP)
Pulsed laser ablation	Laser pyrolysis/photochemical synthesis
Ion cluster beam deposition	Low-temperature reactive synthesis
Spark discharge deposition	Thermal plasma synthesis
	Flame decomposition
	Flame spray pyrolysis

expensive than either arc discharge or chemical vapor synthesis.

Ion cluster beam technology also qualifies as a gas phase nanoparticle synthesis method. In ion cluster beam technology the cooling occurs when the vapor exits a special evaporation cell. As shown in Volume II, Figure 7.14, the heating of the evaporant in an evaporation cell with a small opening causes an adiabatic expansion from more than 100 to 10^{-5} or 10^{-7} mbar, upon exiting that cell. The expansion causes a sudden cooling, inducing the formation of atom clusters.

In spark discharge deposition, electrodes made of the metal to be vaporized in the presence of an inert background gas are polarized until the breakdown voltage is reached and an arc is formed across the electrodes that vaporizes a small amount of metal (e.g., Ni). This is the very technique that led to the discovery of carbon nanotubes in 1991 by NEC's Sumio Iijima (Volume I, Figure 3.64). Carbon nanotubes were found in the carbon soot of graphite electrodes during an arc discharge, by using a current of 100 amps. During this process, the carbon contained in the negative electrode sublimates because of the high temperatures caused by the discharge. Because nanotubes initially were discovered using this technique, it has been the most widely used method of nanotube synthesis. The yield for this method is up to 30% by weight, and it produces both single-walled and multiwalled nanotubes with lengths of up to 50 μm.

Gas Phase Synthesis with Liquid and Vapor Precursors

The entire range of reaction regimes and corresponding microstructures (epitaxial, polycrystalline, columnar, granular films, and aerogel coatings, as well as nanopowders) in going from chemical vapor deposition (CVD) to chemical vapor synthesis (CVS) of nanoparticles are illustrated in Figure 3.40.

In chemical vapor synthesis (CVS as opposed to CVD), vapor phase precursors are brought into a hot-wall reactor under nucleating condition. In CVS, vapor phase nucleation of particles is favored over film deposition characteristic for CVD. CVS constitutes a very flexible technique that can produce a wide range of materials and takes advantage of the huge database of precursor chemistries developed for CVD processes (see Volume II, Chapter 7). Examples include oxide-coated Si nanoparticles for high-density nonvolatile memory devices and tungsten (W) nanoparticles by decomposition of tungsten hexacarbonyl. The method allows for the formation of doped or multicomponent nanoparticles by the use of multiple precursors. For example, nanocrystalline europium-doped yttria is synthesized from organometallic yttrium and europium precursors. Other multicomponent particles that have been made by CVS include erbium in Si nanoparticles and zirconia particles doped with alumina.

Both in CVD and CVS, precursors are metalorganics, carbonyls, hydrides, chlorides, and other volatile compounds. The energy for the conversion

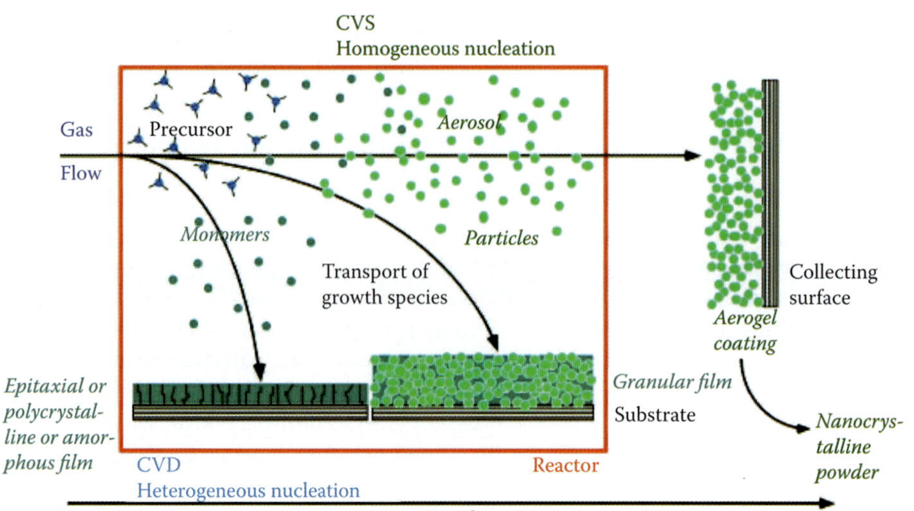

FIGURE 3.40 From chemical vapor deposition (CVD) to chemical vapor synthesis (CVS).

of the reactants into nanoparticles is supplied in hot wall (external furnace), flame (reaction enthalpy), plasma (microwave or radiofrequency), and laser (photolysis or pyrolysis) reactors.

The most important process parameters determining the quality and usability of the nanopowders are the total pressure (typical range from 100 to 100,000 Pa), the precursor material (decomposition kinetics and ligands determining the impurity level), the partial pressure of the precursor (determining the production rate and particle size), the temperature or power of the energy source, the carrier gas (mass flow determining the residence time), and the reactor geometry. The nanoparticles are extracted from the aerosol by means of filters, thermophoretic collectors, electrostatic precipitators, or scrubbing in a liquid. A typical laboratory reactor consists of a precursor delivery system, a reaction zone, a particle collector, and a pumping system. Modifications of the precursor delivery system and the reaction zone allow the synthesis of pure oxides, doped oxides, coated nanoparticles, functionalized nanoparticles, and granular films. Several CVS methods, including spray pyrolysis, salt-assisted spray pyrolysis (SASP), flame spray and electrospray pyrolysis for nanoparticle synthesis were detailed in Volume II, Chapter 9. For further reading on nanoparticle fabrication, a good resource is Jacak et al.[77] You may also want to consult Green and O'Brien.[78]

Carbon Nanostructures: Fullerenes

Introduction

Carbon, as we learn in Chapter 5, this volume, comes in an amazing number of varieties that may be grouped as crystalline, amorphous, nanoparticle, and engineered. The chemical genius of carbon is that it can combine with itself in different ways to create structures with entirely different properties. Two prominent examples of carbon structures are diamond and graphite. In pure sp^3 hybridization, four valence electrons are equally shared between neighboring carbon atoms to yield hard diamond, whereas in pure sp^2 hybridization, three in-plane covalent bonds leave a fourth delocalized electron to produce slippery graphite (see Figure 3.41 and Chapter 5, this volume: the carbon phase diagram in Table 5.9, Inset 2). In Volume I, Chapter 3, using the semiconductor band model, we explained how this same material acts as an insulator (diamond), a metal (graphite), or a superconductor (buckminsterfullerenes).

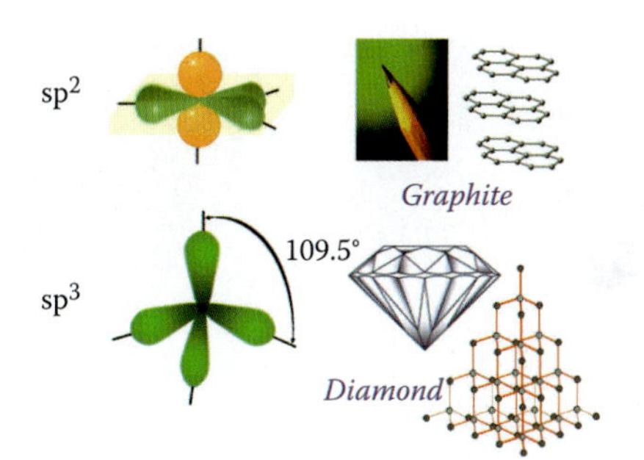

FIGURE 3.41 From graphite to diamond with carbon.[79]

In this section we only focus on fullerenes, carbon nanotubes, and graphene. Fullerenes are large, closed-cage, carbon atom clusters (Figure 3.42 and Volume I, Figure 1.29) and have several special properties that, until their discovery, were not found in any other material. Before the first synthesis and detection of the smaller fullerenes C60 and C70, it generally was assumed that these large spherical molecules were unstable. However, some Russian scientists already calculated that C60 in the gas phase should be stable and comes with a relatively large band gap. As is the case of many, important scientific breakthroughs, fullerenes were discovered accidentally. In 1985, Curl, Kroto, Smalley, and coworkers found strange results in mass spectra of evaporated carbon samples.[80] Herewith, the stability of C60 in the gas phase was proven, starting off the search for other fullerenes.

Since their discovery in 1991 by NEC's Sumio Iijima (Volume I, Figure 3.64), multiwalled carbon nanotubes (MWNTs) (Figure 3.43 and Volume I,

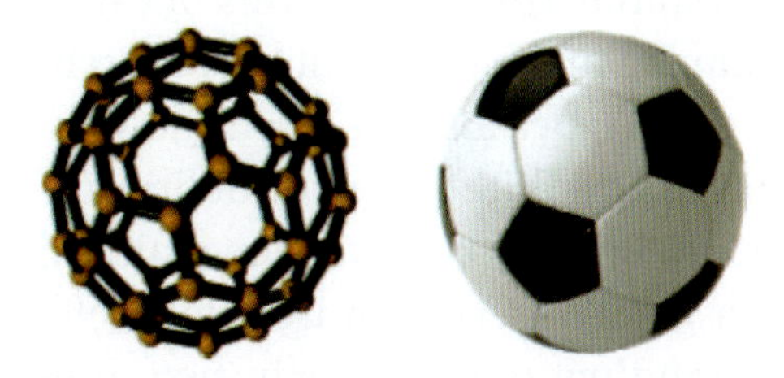

FIGURE 3.42 Sixty carbon atoms in the shape of a sphere—a buckyball. Every molecule combines twenty hexagons with twelve pentagons, resulting in a sphere shaped like the Buckminster Fuller geodesic sphere.

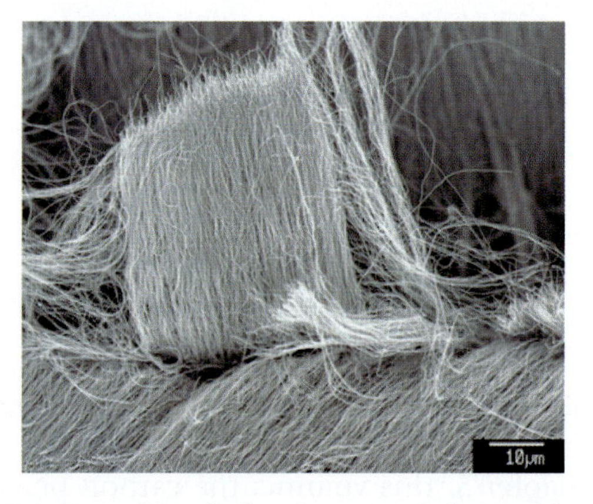

FIGURE 3.43 Multiwalled carbon nanotubes produced by the chemical vapor deposition (CVD) method. (Courtesy of the Department of Materials Science and Metallurgy, University of Cambridge.)

Figure 1.30) have been investigated by researchers all over the world.[81] With their large length (up to several millimeters) and small diameter (a few nanometers), they may be regarded as nearly one-dimensional fullerenes. Therefore, these materials possess additional interesting electronic, mechanical, and molecular properties. Carbon nanotubes, like fullerenes and graphene, profit from the intrusion of quantum phenomena and can be formed into a cylinder one atom thick in so-called "single-walled carbon nanotubes" (SWNTs) to give a wire that is perfect and without defects, much like the single-crystal whiskers we discussed in Volume I, Chapter 2.

Graphene is a flat monolayer of carbon atoms in a 2D honeycomb lattice. It is the basic building block for graphite materials of all other dimensionalities: 0D (buckyballs, C60), 1D (nanotubes), 2D (graphene), and 3D (graphite) (see Figure 3.44). Graphene is the two-dimensional crystalline form of carbon: a single layer of carbon atoms arranged in hexagons, like a sheet of chicken wire with an atom at each nexus. The familiar pencil-lead form of carbon, graphite, consists of layers of carbon atoms tightly bonded in the plane but only loosely bonded between planes; because the layers move easily over one another, graphite is a good lubricant. In fact, these graphite layers are graphene, although they had never been observed in isolation before 2004. As free-standing objects, such two-dimensional crystals were actually believed to be impossible to create—even to exist—until Russian-born physicists Andre Geim and Konstantin Novoselev at the University of Manchester actually made free-standing graphene sheets in 2004.[82] Graphene sheets actually are easy to make but hard to see (Figure 3.45). It is now presumed that tiny fragments of graphene sheets are produced (along with quantities of other debris) whenever graphite is abraded, such as when a line is drawn with a pencil (Figure 3.45).

FIGURE 3.44 Graphene as the building block for graphite materials of all other dimensionalities.[94]

A summary of the many fullerene discoveries made since the discovery of C60 is presented in Table 3.7.

Buckyballs: 0D

A buckyball (short for *buckminsterfullerene*) is a 60-carbon atom molecule named for its resemblance to the geodesic spheres of architect Buckminster Fuller (Figure 3.42). Electrons in such a structure are confined in 3 dimensions and such a zero dimensional (0D) molecule or quantum dot comes with a de Broglie wavelength $\lambda_{DB} > d$ (the diameter of the particle) (see also Volume I, Table 3.11).

FIGURE 3.45 Easy-to-make, hard-to-see graphene sheets.[79]

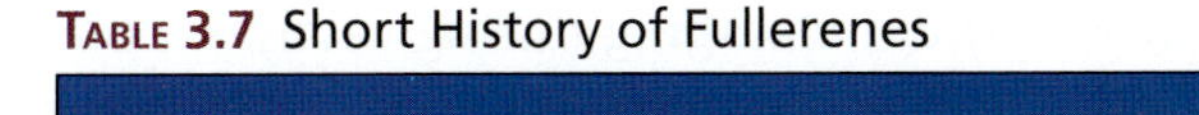

Table 3.7 Short History of Fullerenes

1985: Finding of evidence for C60 in a mass spectrometry analysis of evaporated carbon samples by Kroto et al.[80]
1991: Discovery of multiwalled carbon nanotubes by Iijima[81]
1992: Conductivity of carbon nanotubes established by Mintmire et al.[83]
1993: Structural rigidity of carbon nanotubes measured by Overney et al.[84]
1993: Synthesis of single-walled nanotubes by Iijima and Ichihashi[85]
1995: Nanotubes used as field emitters by Rinzler et al.[86]
1997: Hydrogen storage in nanotubes demonstrated by Dillon et al.[87]
1998: Demonstration of the first carbon nanotube transistor by Dekker et al.[88]
1998: Synthesis of nanotube peapods by Smith et al.[89,90]
2000: Thermal conductivity of nanotubes measured by Tománek et al.[91]
2001: Integration of carbon nanotubes for logic circuits by Collins et al.[92]
2001: Intrinsic superconductivity of carbon nanotubes demonstrated by Kociak et al.[93]
2004: Discovery of stable graphene sheets by Novoselov et al.[82]

This soccer ball-shaped molecule measures just under a nanometer across and was discovered by Robert F. Curl, Jr. (Rice University), Harold W. Kroto (University of Sussex), the late Richard E. Smalley (Rice University), and some graduate students in August of 1985.[80] The professors (but not the graduate students—surprise!) received the 1996 Nobel Prize in Chemistry for this discovery. Spherical molecules like the one depicted in Figure 3.42, called *fullerenes*, come in various sizes and are identified by the letter C followed by the number of carbon atoms, for example, C60, C70, and C80.

Smalley made his early C60 buckyballs by vaporizing carbon with a laser and allowing the carbon to condense in a cold zone (see laser ablation in Figure 3.56, and Volume II, Figure 7.18).

Until C60 could be created in somewhat larger quantities, scientists were stumped in their attempts to study the new material. In 1990, Dr. Donald R. Huffman, a physicist at the University of Arizona, and Dr. Wolfgang Kratschmer of the Max Planck Institute in Heidelberg, Germany, solved this production problem with the development of the so-called "Huffman-Kratschmer reactor."[95,96] By adjusting the gas pressure in an apparatus basically consisting of an arc welder with carbon rods, they made large quantities of soot rich in buckyballs. The arc discharge technique in general was detailed in Volume II, Chapter 9 (see Volume II, Figure 9.25), and a carbon-arc discharge setup is shown in Figure 3.51 below. A graduate student of Dr. Kroto, Jonathan Hare, succeeded in separating buckyballs from the soot produced in the carbon arc by dissolving it in benzene; ordinary graphite does not dissolve, but buckyballs form a red solution. After the benzene is evaporated, the orange-brown crystals that remain are pure fullerene carbon.

Ever since their discovery, expectations have been high that these hollow, pure carbon nanostructures could be used as drug delivery agents, as environmental tracers, as antioxidants, in medical imaging, to make stronger polymers, for more efficient fuel cell membranes, as light detectors, in superconductors, and in diamond-hard coatings for hard disks. As an antioxidant in health and medicine, a buckyball can supply an electron from one of its carbon bonds and neutralize a radical's unpaired electron. In the polymer area, DuPont and Exxon Mobil are using buckyballs to develop stronger polymers. Sony is developing a more efficient fuel-cell membrane. German industrial giant Siemens developed a buckyball-based light detector, and Seagate patented buckyballs to develop diamond-hard coatings for computer disc drives. Of the many tantalizing applications involving buckyballs, imagine buckyballs filled with a specific atom or a molecule, analogous to the ferretin liver protein caging iron oxide molecules. Because of the extreme stability of these molecules, it was difficult at first to open and close buckyballs. However, by rendering one of the thirty double bonds making up a C60 cage more reactive, Wudl at the University of California, Santa Barbara, succeeded in opening and closing the cage, contributing to a widening class of container molecules (endohedrals) tailored to fit a variety of target molecules.[97-99]

At MIT a fullerene mass production method was developed (together with a startup company, Nano-C, Inc.) in which hydrocarbons and oxygen are mixed and the hydrocarbons are burned at a low pressure. Nano-C has a prototype burner in its headquarters that can produce 1 ton of fullerenes a year. The standard atmosphere in the burner is a low-pressure mixture of oxygen and benzene, heated to over 3140°F. By manipulating conditions inside the reactor, different types of fullerenes can be produced. Nano-C claims that they obtain a yield in fullerenes of 95% obviating the usual separation step of buckyballs from soot (http://www.nano-c.com/fullerenes.html). C60 fullerenes are the least expensive fullerene and used to fetch $25 per gram. Larger fullerenes can cost thousands of dollars per gram. Nano-C says they can now make C60 fullerenes for about $4 or $5 per gram. They expect the cost to tumble below $1 with larger-scale equipment. The Frontier Carbon Corporation (FCC), headquartered in Tokyo, Japan, is the first company in the world to mass-produce fullerenes at a commercial scale, and it increased its fullerene production capacity to 40 tons/yr in April 2003 (http://www.f-carbon.com). FCC now sells fullerenes at a price one-tenth that of the existing market and has plans to decrease its price even further.

Because of the decrease in price and unique characteristics of fullerenes, some face creams and antiaging creams based on them have already entered the market (http://www.buckyusa.com). Other near-term projected industrial applications include batteries, lubricants, grinding and polishing material, polymer additives, and rubber additives.

Very few buckyball applications are financially lucrative yet. There also might be an increase in environmental concerns as, although fullerenes are found to be bactericidal, they also seem to cause brain damage in fish.[100]

Nanotubes: 1D

Carbon nanotubes (CNTs) had been observed before their "invention" in 1991. Carbon nanofibers (diameter range, 3–100 nm; length range, 0.1–1000 mm) were known for a long time as a nuisance often emerging during catalytic conversion of carbon-containing gases. In 1959, Roger Bacon produced images of carbon nanotubes,[101] and in the 1980s, Howard Tennant even applied for a patent on a method to produce them. But it was only in 1991 that Sumio Iijima, at NEC's Fundamental Research Lab, found that through carbon self-organization it is possible to deliberately fabricate those tiny graphite tubes, perhaps the smallest electrical wires ever made.[81,85] These nanotubes are called *buckytubes*, because they are structurally similar to the carbon buckyballs. CNTs can be described as a sheet of graphite, i.e., graphene rolled into a cylinder, constructed from hexagonal rings of carbon and with caps at the ends, making the shorter ones look like pills. It is possible to open those caps through oxidation: the pentagons in the end caps of nanotubes are more reactive than the sidewall, so during oxidation, the caps easily are removed while the sidewall stays intact. The first buckytubes discovered consisted of several nested, concentric cylinders with nanometer-scale diameters, called *multiwalled nanotubes* (MWNTs) (Figure 3.46). In the nested configuration MWNTs, inner and outer tubes slide and rotate easily against each other. Like Russian nested dolls, these type of structures typically feature about ten nested tubes. By 1993, Iijima and Donald Bethune at IBM independently created single-walled nanotubes consisting of just one layer of carbon atoms (SWNT) (Figure 3.47).[102] A typical

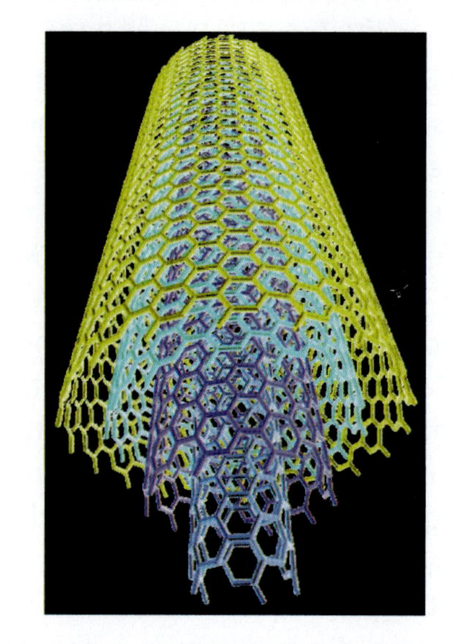

FIGURE 3.46 Artistic rendition of multiwalled nanotubes. (From A. Rochefort, Nano-CERCA, University of Montreal.)

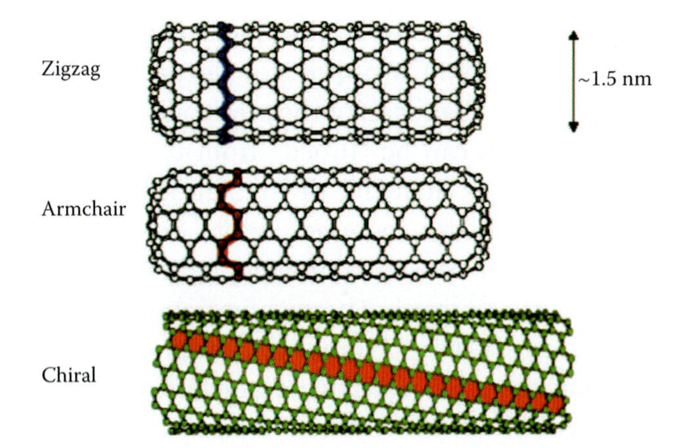

FIGURE 3.47 A single-walled carbon nanotube diameter is ~1.5 nm. Three types of lattice orientations can be identified: zigzag, armchair, and chiral nanotubes.

SWNT diameter is 1–1.5 nm, and tube lengths of several millimeters have been achieved.

Although carbon nanotubes are well ordered, STM images rarely show the sixfold symmetry of their graphene lattice. The images instead reveal the backscattering of electron waves from defects on the walls.

Nanotubes are very stiff and very stable and can be built with their length exceeding their thickness thousands of times. They fold and buckle but do not snap, and their hollow interior enables one to put materials inside.

Linear nanotubes may behave as metals or as semiconductors, depending on their diameter and the exact details of the 2D lattice orientation. Three types of lattice orientations can be identified. In armchair nanotubes, there is a line of hexagons parallel to the axis of the nanotube; in zigzag nanotubes, there is a line of carbon bonds down the center; and chiral nanotubes exhibit a twist or spiral around the nanotube (see Figure 3.47). It is the amount of twist of the lattice that determines whether a nanotube is metallic or semiconducting. Armchair nanotubes are all metallic, whereas about two-thirds of zigzag and chiral nanotubes are semiconducting (the remaining one-third are, again, metallic).

A bit more quantitative the geometry of a CNT is determined by its chiral vector (**c**) defined as follows (see Figure 3.48):

$$\mathbf{c} = n_1\mathbf{a}_1 + n_2\mathbf{a}_2 \tag{3.13}$$

where $\mathbf{a}_1$ and $\mathbf{a}_2$ are primitive lattice vectors of graphene and n_1 and n_2 are integers (chiral numbers). The basis of this 2D lattice contains two atoms and

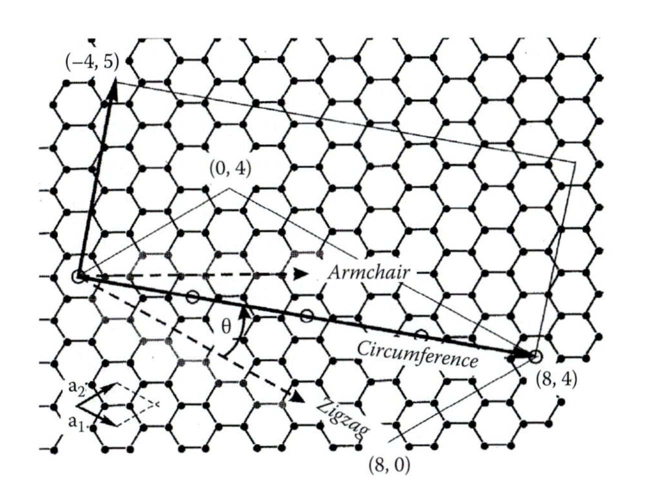

FIGURE 3.48 Graphene sheet and chiral vector.

the side length d of a hexagon in the lattice is 0.14 nm long. The chiral angle is the angle θ between the chiral vector **c** and the unit vector $\mathbf{a}_1$. If the chiral angle is 0°, $n_2 = 0$, and the nanotube is classified as zigzag. If the chiral angle is 30°, $n_1 = n_2$, and the nanotube is classified as armchair. If **c** is not along a mirror line (see Figure 3.48) the structure is chiral and these tubes, with angles between 0° and 30°, are referred to as chiral nanotubes. Chiral structures have no mirror symmetry and sort out into "left-handed" and "right-handed" versions. Due to the symmetry and unique electronic properties of graphene layers, the structure and dimensions of a nanotube strongly determines its electrical properties. For a given (n_1, n_2) nanotube, if $2n_1 + n_2 = 3q$ (where q is an integer), then the nanotube is metallic, otherwise the nanotube is a semiconductor. Thus all armchair ($n_1 = n_2$) nanotubes are metallic, and nanotubes such as (5,0), (6,4), (9,1), are all semiconducting. An alternative representation of this condition is that if $(n_1 - n_2)/3 = q$, then the SWNT is metallic. In theory, metallic nanotubes can have an electrical current density more than 1,000 times stronger than metals such as copper. The energy gap for the semiconducting fibers decreases with increasing fiber diameter d and in the limit of d_∞, we obtain the 2D case of a zero-gap semiconductor. The chirality of a SWNT can be determined by Raman spectroscopy and STM and transmission electron microscopy (TEM); STM and AFM can determine its diameter.

Also from Figure 3.48, if n is the greatest common divisor (4) of the chiral numbers n_1 (8) and n_2 (4), then the number of lattice points (open circles) along the chiral vector is $n + 1$ (there are five open circles in Figure 3.48).

In an ordinary wire, the charge transport is diffusive, and the diffusive length of the electrons, l, is much smaller than the wire length ($l \ll L$) (see Figure 3.49a). The unique electronic properties of carbon nanotubes are due to the quantum confinement of electrons perpendicular to the axis of the tube. In that direction, the monolayer thick graphene sheet confines electrons. Around the circumference of the nanotube, periodic boundary conditions hold. For example, if a nanotube has twelve hexagons around its circumference, the thirteenth hexagonal will coincide with the first. In other words, going around the cylindrical tube just once introduces a phase difference of 2π. Because of this quantum confinement, electrons propagate along the nanotube axis, and their wave vectors point in that direction. The actual number of one-dimensional conduction and valence bands of a nanotube depends on the number of standing waves that are set up around the circumference of the nanotube. The electrons in an ideal nanotube then behave like waves traveling down a smooth channel with no atoms to bump into, and they do not undergo scattering. This allows the nanotube to pass current without resistance along its length and permits very high current densities. This type of quantum movement of an electron within a nanotube is called *ballistic transport* ($l \ggg L$) (see Figure 3.49b). The contacts are part of the device and all the scattering occurs at the contacts; hence, all the voltage drops also at the contacts. The resistance, as we saw in Volume I, Chapter 3, is quantized in amounts of R = 25.8 kΩ (see also the Landauer equation: Volume I, Equation 3.274).

In metallic nanotubes, electrons can easily move from the valence to the conduction band. There is no gap between the valence and conduction band in armchair nanotubes, but as we just saw, an

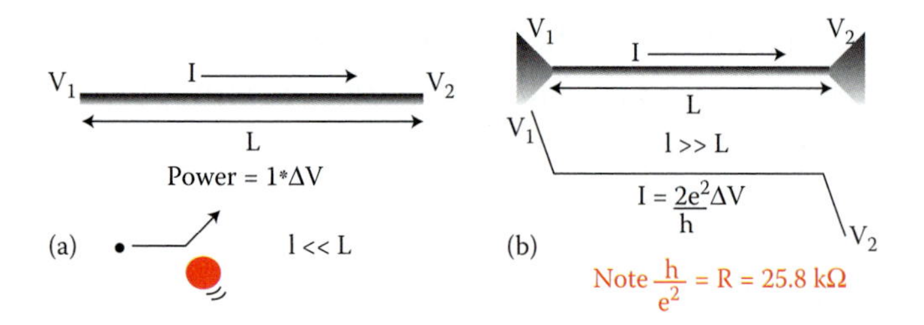

FIGURE 3.49 (a) Diffusive transport in an ordinary wire. (b) Ballistic transport in a quantum wire.

energy gap exists between the valence and conduction bands in about two-thirds of zigzag and chiral nanotubes. The bandgap of the semiconducting carbon nanotubes can be "tuned" from 0.5 to 1.5 eV by adjusting the diameter of the nanotube; as the nanotube diameter increases, more wave vectors are allowed in the circumferential direction and the band gap narrows. The carbon nanotube band gap is inversely proportional to the tube diameter and approaches zero at large diameters, just as for a graphene sheet (at a nanotube diameter of about 3 nm, the band gap becomes comparable to thermal energies at room temperature). The fact that the electronic structure of a carbon nanotube depends only on its physical geometry, without any doping, is unique to solid-state physics and is the basis for many of its proposed applications in electronics.

Nanotubes, we explained earlier, are made up of graphene sheets, and the unusual electronic properties of graphene arise from the fact that the carbon atom has four electrons, three of which are tied up in bonding with its neighbors. But the unbound fourth electrons are in orbitals extending vertically above and below the plane, and the hybridization of these spreads across the whole graphene sheet (see also the section below on graphene). The energy vs wavevector (**E-k**) surfaces and band diagrams of a metallic and a semiconducting carbon nanotube are compared in Figure 3.50A.

As we saw above carbon nanotubes can be viewed as a rolled up graphene sheet (Figure 3.50Aa and b). The periodic boundary condition only allows quantized wave vectors around the circumferential direction, which generates one-dimensional bands for carbon nanotubes. The nanotube can be either metallic (Figure 3.50Ac) or semiconducting (Figure 3.50Ad). If $(n_1 - n_2)/3 = q$ (with q an integer), then the SWNT is metallic. So with $n_1 - n_2 = 0$ we are dealing with a metallic tube and the bandgap is zero. And if $n_1 - n_2/3$ is not an integer we have a semiconducting tube that exhibits a forbidden band gap in the **E-k** diagram.

As we learned in Volume I, Chapter 3, carbon nanotubes have peaks in their density of states (DOS) function that are called *criticalities* or *Van Hove singularities*. These singularities in the DOS function lead to sharp peaks in optical spectra (see Figure 3.50B) and can be observed directly with scanning tunneling microscopy

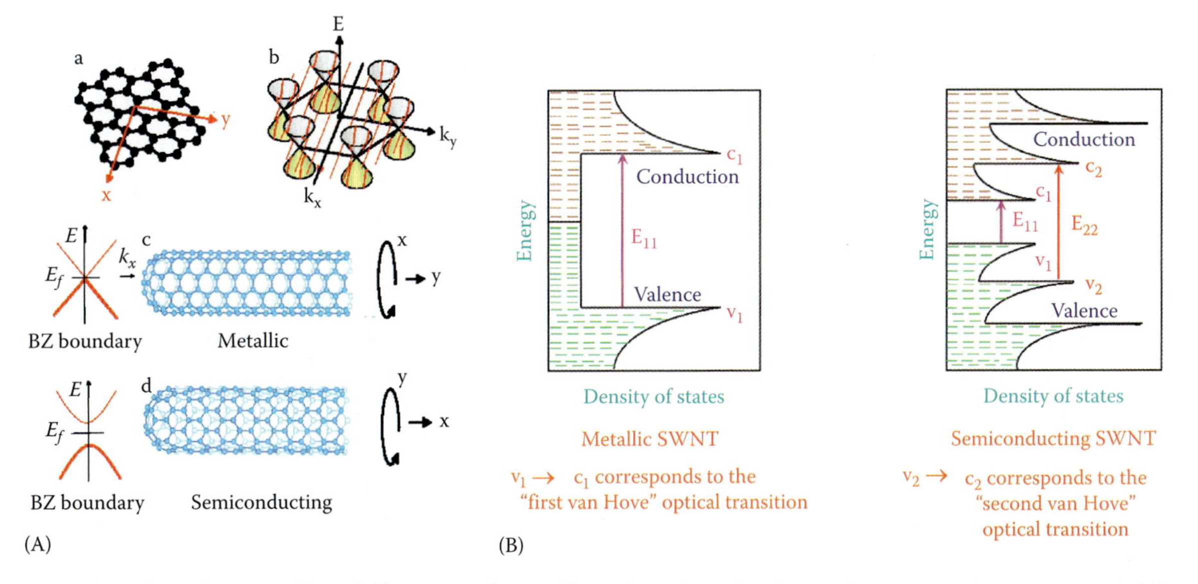

FIGURE 3.50 The E-k surfaces and band diagrams of metallic and semiconducting carbon nanotubes compared. (A) Energy vs wave vector plot. Carbon nanotubes can be viewed as rolled up graphene sheets (a,b). The periodic boundary condition only allows quantized wave vectors around the circumferential direction, which generates one-dimensional bands for carbon nanotubes. The nanotube can be either metallic (c) or semiconducting (d), depending on whether (n_1 – n_2) is the multiple of 3. (B) Sharp optical transitions due to Van Hove singularities. Optical transitions between the v_1 – c_1, v_2 – c_2, etc., states of semiconducting or metallic nanotubes are traditionally labeled as S_{11}, S_{22}, M_{11}, etc. The energies between the Van Hove singularities depend on the nanotube structure. Thus by varying this structure, one can tune the optoelectronic properties of carbon nanotubes. Optical transitions are sharp (~10 meV) and strong. Consequently, it is relatively easy to selectively excite nanotubes having certain (n_1, n_2) indices, as well as to detect optical signals from individual nanotubes.

(STM) (see Volume I, Figure 3.135). At low temperatures, the derivative of the STM current with respect to voltage (*dI*/*dV*) is proportional to the density of states at the tunneling electron energy, weighted by the density of those states at the STM tip position. Since dI/dV ~ local density of states (LDOS), the LDOS is directly observed from an STM scan. In practice, during scanning tunneling spectroscopy measurements, a constant tip-sample separation is first established by fixing the set-point current at a given bias voltage. Next, the feedback of the STM is turned off and a small sinusoidal modulation is added to the DC bias voltage. The resulting current modulation (*dI*/*dV*) is then read off from a lock-in amplifier, into which the tunneling current signal from the current preamplifier is fed. The output signal from the lock-in amplifier is plotted as a function of the DC bias voltage and the tip position to generate a tunneling spectrum *dI*/*dV* (*V*, *r*), and hence the LDOS. By taking *dI*/*dV* along a line or by taking a two-dimensional *dI*/*dV* map, changes in LDOS around a impurity, or among nanotubes with different crystalline orientations can be revealed and compared with theoretical predictions. The density of states (DOS) functions for a metallic and a semiconducting carbon nanotube are illustrated in Figure 3.50B. The peaks in these DOS functions are the Van Hove singularities. In Volume I, Figure 3.135, the same DOS functions are shown in the top two plates, and the differential conductance is illustrated in the bottom two plates. Metallic carbon tubes have a constant DOS around E = 0, whereas for semiconducting ones, there is a gap at E = 0 (DOS = 0).

Carbon Nanotube Manufacture

Like fullerenes, nanotubes may be formed in macroscopic quantities in an arc discharge apparatus as described in Volume II, Chapter 9 (see Figure 9.25). In the case of a carbon arc, one refers to such a setup as a Huffman-Kratschmer reactor.[95] In a carbon arc as illustrated in Figure 3.51, two carbon rods are placed end to end with a separation distance of about 1 mm and are used as electrodes. In an enclosed volume of an inert gas (He, Ar) at a low pressure (50–70 mbar), a current of 50–100 A, driven by 20 V, creates a high-temperature arc between the electrodes. The discharge vaporizes one rod forming a deposit on the other. Producing nanotubes in high yield depends upon the uniformity in the plasma arc and also upon

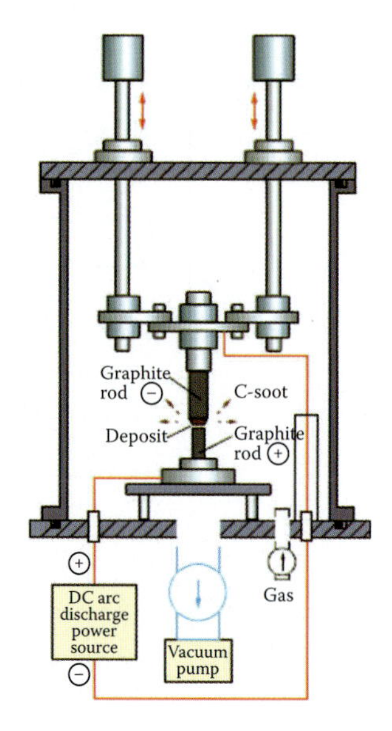

FIGURE 3.51 In the case of a carbon arc, one refers to such a setup as a Huffman-Kratschmer reactor.[95]

the temperature of the deposit on the second carbon electrode. At the extremely high temperatures of the arc plasma, a wide range of carbon materials is formed. This may include amorphous carbon, fullerenes, nanotubes, endohedrals (carbon structures with enclosures), nanocrystallites, and graphite crystals. A typical yield for nanotubes is around 30% by weight. It is the most common and arguably the easiest way to produce carbon nanotubes; however, as pointed out, it produces a mixture of items such as "soot" and catalytic metals in the end product, and the tubes are deposited in random sizes and directions.

There are a few ways of improving the process of arc discharge for producing either SWNT or MWNT (see Figure 3.52). In order to favor the production of SWNT, the anode in the apparatus is doped with a metal catalyst. The most important metals to catalyze the growth of graphitic carbon nanofibers are iron, cobalt, and nickel; chromium, vanadium, and molybdenum have also been studied. The metals have been used both as bulk particles (size typically 100 nm) and as supported particles (10–50 nm). MWNTs do not need any catalysts to grow.

By controlling the mixture of helium and argon gas, one can also control the diameter distribution of the produced nanotubes. Supposedly, this is due to the fact that the different mixtures have different diffusion coefficients and thermal conductivities.

Metal-filled carbon nanotubes (endohedrals) also have been produced with the arc-discharge method, for example, by doping a 99.4% graphite anode with a transition metal such as Cr or Ni, a rare earth such as Yb or Dy, or a covalent element such as S or Ge. A growth mechanism based on a catalytic process involving three elements, i.e., carbon, a metal, and sulfur has been proposed. The excellent protective nature of the outer graphitic cages against oxidation of the inner materials was demonstrated.[103] These carbon-coated nanoparticles are of potential importance in the study of nanowire protection against oxidation, magnetic data storage, magnetic toner xerography, magnetic inks, and ferrofluids.

To favor multiwalled carbon nanotube (MWNT) production, both electrodes in the arc discharge apparatus should be graphite (Figure 3.52). Synthesis

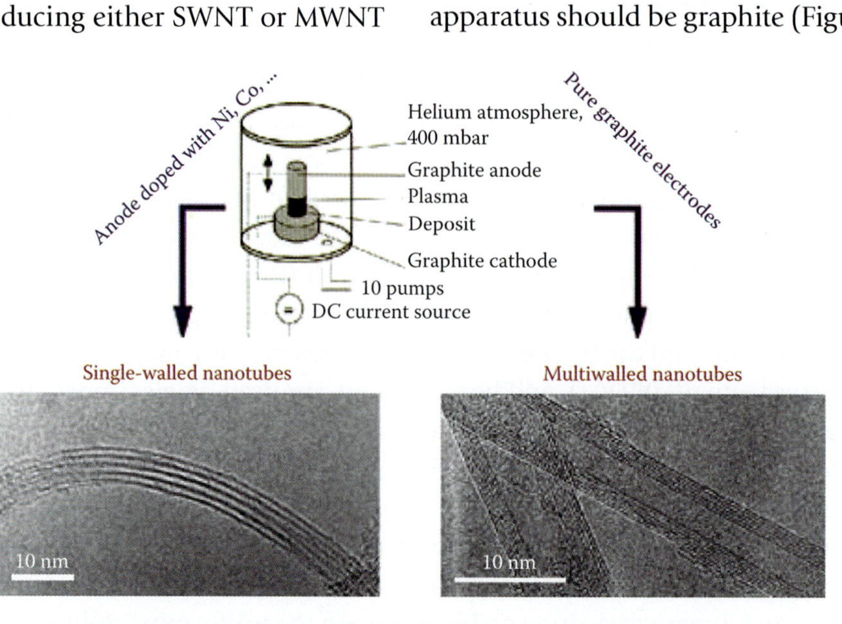

FIGURE 3.52 A carbon arc can be used to obtain single-walled and multiwalled nanotubes.

in a magnetic field and with a rotating arc discharge also improves the MWNT yield. Magnets are placed around electrodes and help align nanotubes during formation. Purification is needed in this case, which may lead to loss of structure and can cause disorder in the nanotube walls. Purification of carbon nanotubes is quite difficult because they are insoluble in organic solvents and metals are often encased in a carbon coating. Possible solutions entail acid oxidation, gas or water oxidation, filtration, and chromatography.

Besides carbon plasma arcing, other established methods to produce nanotubes are chemical vapor deposition (CVD) and laser ablation. Morinubo Endo of Shinshu University was the first to make nanotubes using CVD.[104] Pushing the field one step further, Hongjie Dai et al. at Stanford University devised a novel method for growing vertical single-walled carbon nanotubes (SWNTs) on a large scale. This group modified a plasma-enhanced chemical vapor deposition station and achieved ultrahigh-yield growth of SWNTs, thus increasing their application into commercial products.[105] By patterning a catalyst on a planar substrate they were able to control where the tubes form, for example, to make carbon nanotube arrays on a silicon wafer.[106] Typically, methane, carbon monoxide, synthesis gas (H_2/CO), ethyne, and ethene in the temperature range 700–1200 K are employed to provide the carbon atoms in the process. Dai and his colleagues discovered that the key to attaining SWNTs is to add oxygen to the CVD reaction. While carbon atoms try to form the nanotube's planar structure, hydrogen radicals are eating away at the carbon tube. Adding oxygen remedies the problem by scavenging the hydrogen radicals, creating a carbon-rich and hydrogen-deficient environment. These conditions jumpstart growth and spawn a vertical forest of nanotubes. (Dai and his colleagues were able to create 4-in. wafers blanketed with SWNTs. In addition, they devised a method for lifting the nanotubes off the original growth substrate and transferring them onto a variety of more desirable substrates such as plastics and metals—materials incompatible with the high temperatures required for nanotube growth.)

A typical CVD setup, we saw in Volume II, Chapter 7, consists of a target substrate held in a quartz tube placed inside a furnace (see Figure 3.53). The yield of this process ranges from 20% to 100%, the process is easy to scale up, and nanotubes of great length can be produced. For example, with fast-heating CVD, CNTs that are millimeters long can be grown while maintaining control over their orientation. Fast heating minimizes interaction of the growing tubes with the substrate. Using this process, Huang et al. grew a square 2D network of SWNTs guided by flow of carrier gas (Fe/Mo catalyst and CO_2 gas) (see Volume II, Figure 7.41).[108] The longest carbon nanotubes, as long as 18.5 cm, were reported in 2009. They were grown on Si substrates using an improved chemical vapor deposition (CVD).[109]

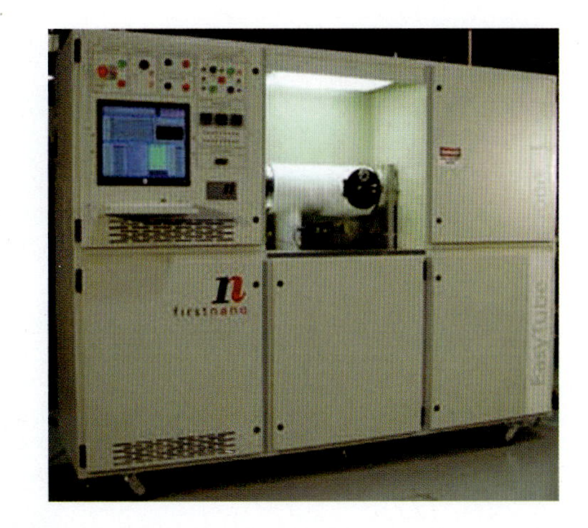

FIGURE 3.53 Plasma-enhanced chemical vapor deposition setup for nanotubes: Firstnano EasyTube 3000.

The general picture that has emerged for nanofiber formation is shown in Figure 3.54, where we contrast base growth and tip growth. In base growth, the catalyst particle can be found at the base of the nanotube, whereas in tip growth the nanoparticle is seen at the very tip of the tube. The latter seems to be the most frequent growth mode.

It is not exactly known yet how the carbon atoms in any of the manufacturingtechniques detailed here condense into tubes. Hoogenraad has proposed the following model for nucleation and growth in the case of carbon nanowires grown on Ni catalyst particles (Figure 3.55).[107] The process starts when methane decomposes into carbon and hydrogen atoms at the nickel catalyst surface (1–2). Next, H_2 molecules desorb, and carbon dissolves and forms a (substoichiometric) nickel carbide (3). This nickel carbide is metastable with respect to nickel metal and graphite so that after a short time (e.g., 10 min), the carbide phase decomposes into metallic nickel

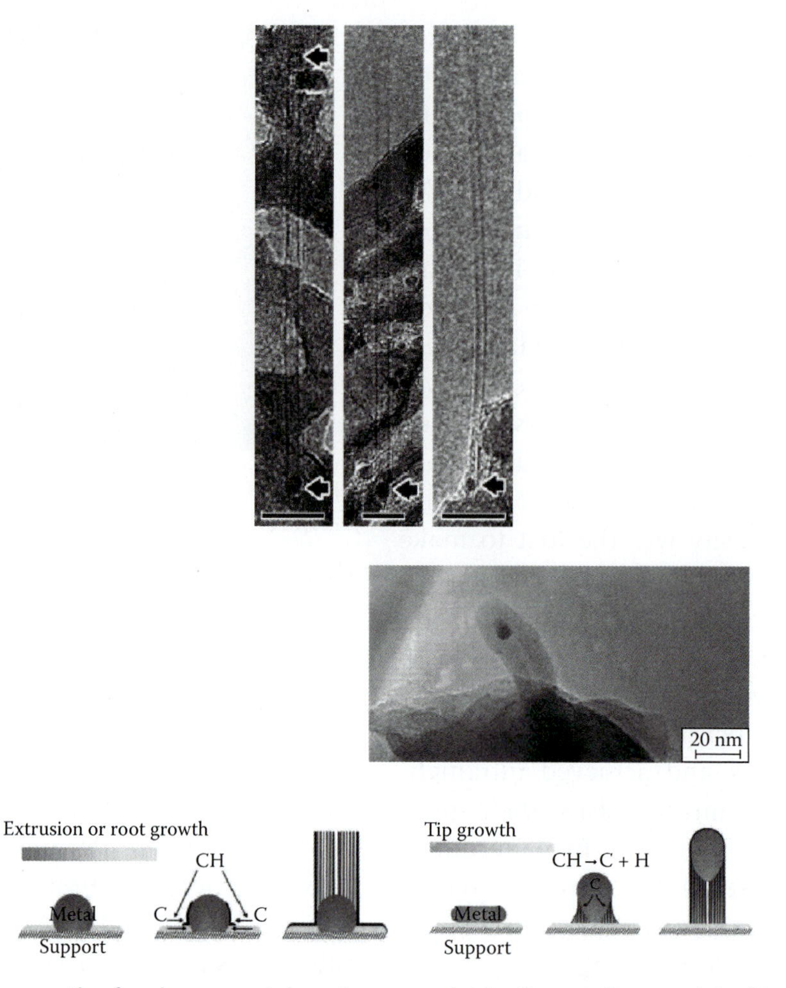

FIGURE 3.54 Base and tip growth of carbon nanotubes. Arrows point to the catalyst particle. (Hoogenraad, M.S. 1995. Utrecht University, PhD thesis. Utrecht, the Netherlands.[107])

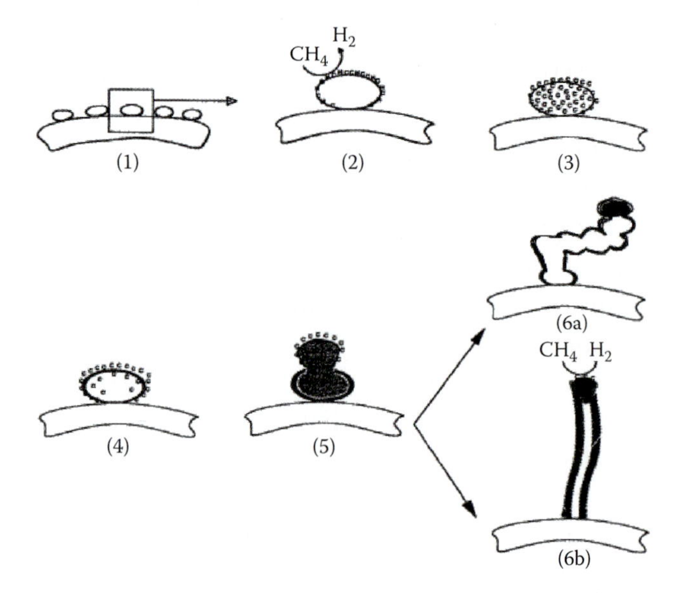

FIGURE 3.55 Mechanism for the nucleation and growth of a carbon nanofiber from methane catalyzed by a supported metal particle. (Based on the work of Hoogenraad, M.S. 1995. Utrecht University, PhD thesis. Utrecht, the Netherlands.[107])

and graphite with the graphite encapsulating the nickel particle in question (4). Hoogenraad now speculates that the metal particle is squeezed out of its encapsulation because of pressure buildup due to the formation of graphite layers at the internal surface of the graphite envelope, on the one hand, and liquid-like behavior of the metal under these conditions, on the other (5). As soon as the metal is pushed out, a fresh Ni surface is exposed to the methane and growth continues. Finally, a steady-state process occurs with either pulsed growth (6a) or smooth growth of a straight fiber (6b). This model explains why, more often than not, metal particles are found at the tip of the carbon fiber: the graphite fiber pushes the metal particle from the support and continues to grow at the back of the particle.

This type of mechanism corresponds to the vapor-liquid-solid (VLS) model as originally proposed in 1964 by R.S. Wagner and W.C. Ellis[110,140]

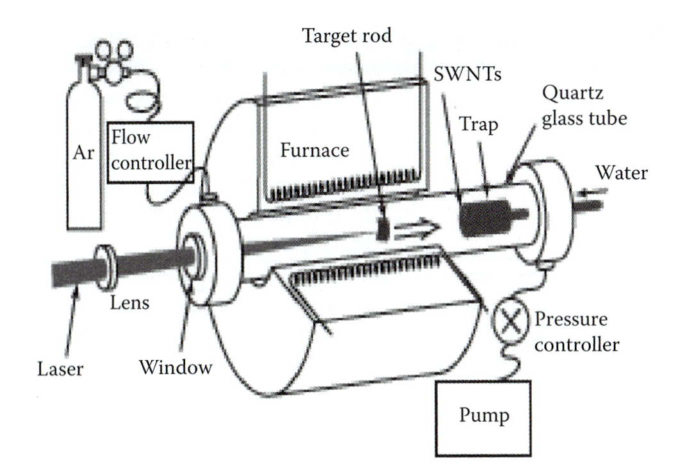

FIGURE 3.56 Laser ablation station for carbon nanotube production.[112] (With kind permission from Elsevier.)

for the Au-catalyzed Si whisker growth (see Figures 3.60 and 3.61).

Smalley made his early buckyballs by laser ablation in a furnace, as shown in Figure 3.56 (see also Volume II, Figure 7.18). The same setup can be utilized to make carbon nanotubes.

Smalley and coworkers at Rice University pioneered the use of intense laser pulses to bombard graphite rods to generate the hot carbon gas from which nanotubes settle on a copper collector. Typical yield is 70%, and the method tends to produce single-walled nanotubes. The method is quite expensive because of the use of the lasers.

In Table 3.8 we compare the pros and cons of carbon-arc discharge with chemical vapor deposition and pulse laser vaporization for the fabrication of carbon nanotubes.

Carbon Nanotube Applications

Electronic Properties

Some of the amazing properties of nanotubes are summarized in Table 3.9 below. These properties could well make carbon nanotubes (CNTs) the "Si" of this century.

It is possible for the carbon nanotube to supply the basic building blocks for an integrated circuit, as

Table 3.8 Comparison of the Pros and Cons of Carbon Nanotube Synthesis by Carbon-Arc Discharge with Chemical Vapor Deposition and Pulse Laser Vaporization

Method:	Carbon-arc discharge	Chemical vapor deposition	Pulse laser vaporization
Who:	Ebbesen and Ajayan, NEC, Japan 1992[113]	Endo, Shinshu University, Nagano, Japan[114]	Smalley, Rice, 1995[76]
How:	Connect two graphite rods to a power supply, place them a few millimeters apart, and throw the switch; at 100 amps, carbon vaporizes and forms a hot plasma	Place substrate in oven, heat to 600°C, and slowly add a carbon-bearing gas such as methane; as gas decomposes it frees up carbon atoms, which recombine in the form of NTs	Blast graphite with intense laser pulses; use the laser pulses rather than electricity to generate carbon gas from which the NTs form try various conditions until hitting upon one that produces prodigious amounts of SWNTs
Yield:	30–90%	20–100%	Up to 70%
SWNT:	Short tubes with diameters of 0.6–1.4 nm	Long tubes with diameters ranging from 0.6–4 nm	Long bundles of tubes (5–20 μm), with individual diameter from 1–2 nm
MWNT:	Short tubes with inner diameter of 1–3 nm and outer diameter of approximately 10 nm	Long tubes with diameters ranging from 10–240 nm	Not very much interest in this technique because it is too expensive, but MWNT synthesis is possible
Pros:	Can easily produce SWNT and MWNTs; SWNTs have few structural defects; MWNTs without catalyst, not too expensive, open-air synthesis possible	Easiest to scale up to industrial production; long length, simple process; SWNT diameter controllable, quite pure	Primarily SWNTs, with good diameter control and few defects; reaction product quite pure
Cons:	Tubes tend to be short with random sizes and directions; often needs a lot of purification	NTs are usually MWNTs and often riddled with defects	Costly technique because it requires expensive lasers and high-power equipment, but improving

TABLE 3.9 Some Physical Properties of Carbon Nanotubes

	Property	Remark/Comparison	Applications
Size	The average diameter of an SWNT is 1.2–1.4 nm	E-beam lithography produces lines 30 nm wide	Nanotube-tipped AFMs can trace an individual strand of DNA. The hollow structures may store hydrogen, lithium ions, or even drug molecules for slow release devices
High specific surface	>1000 m^2/g	For comparison carbon aerosol gels have a specific surface area (200–350 m^2/g)	Materials with large specific surface areas are used for energy storage as in capacitors, batteries, and fuel cells. They are also used for storage, purification, and separation of chemicals
Density	1.33–1.40 gcm^{-3} Lowest for armchair, highest for chiral	Al is 2.7 g cm^{-3}	The applications for low density materials are far ranging from consumer products such as food containers, to packaging materials, to crash protection energy absorbers
Tensile strength	Estimated to be 150 GPa	High-strength carbon steel alloys, 1 GPa. Carbon nanotubes have the greatest tensile strength of any material. It is also the stiffest known material, with a tremendously high elastic modulus (see Young's modulus)	More rugged tips for AFM and STM, superstrong composites
Resilience	Can be bent at large angle without damage	Metals and carbon fibers fracture at grain boundaries	Mechanical memory switches and nanotweezers, superstrong materials
Young's modulus	1–6 TPa (SWNT) 0.27–0.95 Tpa (MWNT)	The Young's modulus of the best nanotubes can be higher than 1000 GPa which is approximately 5x higher than steel. Reported numbers vary widely	This high Young's modulus combined with an excellent tensile strength and the lightness of carbon nanotubes, gives them great potential in applications such as aerospace (see "space elevator" in the heading of this chapter)
Current carrying capacity	Estimated at 1×10^{13} Am^{-2}	Cu wires will carry no more than 1×10^{6} A cm^{-2}	Potential for terahertz FETs
Thermal conductivity and phonon mean free path	Thermal conductivity for SWNT is as high as 3000 W/mK along the nanotube axis. The phonon mean free path length is ~100 nm	Compare this to copper, a metal well-known for its good thermal conductivity, which transmits 385 W/mK. The thermal conductivity of carbon nanotubes is dependent on the temperature and the large phonon mean free paths	Efficient heat sinking in electronics and optoelectronics
Electrical resistivity	10^{-4} ohm.cm for ropes of SWNTs at room temperature		Wiring in electronic circuitry
Field emission	Activates phosphors at 1–2 V with electrodes spaced 1 μm apart	Mo tips (Spindt cathodes, see Volume II, Figure 2.42) require fields of 50–100 V/μm	Flat-panel displays, nanovacuum tubes
Temperature stability	In vacuum up to 2800°C and 750°C in air	Metal in ICs melts between 600°C and 1000°C	High-temperature environments

TABLE 3.9 Some Physical Properties of Carbon Nanotubes (*Continued*)

	Property	Remark/Comparison	Applications
Quantum effects	For every nanotube circumference there is a unique band gap	Easy to manipulate electronic properties	New generation of quantum devices
Chemical sensitivity	Oxygen, halogens, and alkalis cause a drastic resistance change at room temperature	Most solid-state gas sensors only work at higher temperatures (e.g., >350°C)	Supersensitive, room temperature chemical sensors (selectivity might be an issue)
High aspect ratio	Easily 1000:1		The high aspect ratio of CNTs imparts electrical conductivity to polymers at lower loadings, compared to conventional additive materials, such as carbon black, chopped carbon fiber, or stainless steel fiber
Cost	~$100/g	Au, ~$10/g	High cost will make many of the above applications impossible until prices come down
Band gap	0 for metallic and 0.5 eV for semiconductors		Semiconductive type can be used as a transistor and metallic as an interconnect

For a good introductory reading on fullerenes, we suggest *The Chemistry of the Fullerenes* by Hirsch.[119] For a good introduction to nanotubes, we refer to the book *Carbon Nanotubes* by Dresselhaus et al.[120] and the special section "Carbon Nanotubes" in *Physics World*.[121]

the metallic nanotubes can act as highly conductive wires and the semiconducting nanotubes can be fabricated into transistors. Cees Dekker in Delft,[88] Paul McEuen at Cornell,[112] Phaedron Avouris at IBM,[115] and Charles Lieber at Harvard[116] all have demonstrated that single nanotubes can act as transistors. A field-effect transistor—a three-terminal switching device—can be constructed of only one semiconducting SWNT. By applying a voltage to a gate electrode, the nanotube can be switched from a conducting to an insulating state. In Volume I, Figure 1.30b, we show a single semiconducting nanotube contacted by two electrodes. The Si substrate, which is covered by a layer of SiO_2 300 nm thick, acts as a back gate. In 2002, researchers built the world's first array of transistors out of carbon nanotubes.[117]

Another carbon nanotube electronic application involves field emission. When a solid is subjected to a sufficiently high electric field, electrons near the Fermi level can be extracted from the solid by tunneling through the surface potential barrier (see Volume II, Figure 2.17). This emission current depends on both the strength of the local electric field at the surface and the material's work function (which denotes the energy necessary to extract an electron from its highest bounded state to the vacuum level). The applied electric field must be very high in order to extract an electron, a condition easily fulfilled for carbon nanotubes, because their elongated shape ensures a very large electric field concentration. For technological applications, the emissive material should also have a low threshold emission field and remain stable at very high current densities. Furthermore, an ideal emitter should be nanometer size, have high structural integrity and high electrical conductivity, exhibit a small energy spread, and demonstrate excellent chemical stability. Carbon nanotubes possess all these properties. The growth of aligned arrays of CNTs inside the pores of self-organized nanoporous alumina templates is an especially intriguing application in this regard (see Chapter 5, this volume). Govyadinov et al.[118] synthesized arrays of CNTs (deposited with catalytic CVD) in alumina pores with diameters in the sub-100-nm range and densities from 10^8 to 10^{11} tips per cm^2. These prototype CNT arrays have been used as field emitters (carbon has a low work function) and shown to exhibit field emission at a low threshold of 3–5 V/μm and emission current densities up to 100 mA/cm^2 in low-vacuum conditions.

However, a bottleneck in the use of nanotubes for field emitter application is the dependence of the conductivity and emission stability of the nanotubes on the fabrication process and synthesis conditions. Examples of applications for nanotubes as field emitting devices are flat-panel displays (see Volume II, Figure 7.90), gas-discharge tubes in telecom networks, electron guns for electron microscopes, AFM tips, and microwave amplifiers.

Mechanical Properties

Estimates vary as to precisely how strong a single-carbon nanotube can be, but experimental results have shown their tensile strength to be in excess of 60 times stronger than high-grade steel. Depending upon the specific paper at hand, the reported numerical values for the Young's modulus range from 1–6 TPa. This scatter is known as "Yakobson's Paradox" after Boris Yakobson (Rice University) who first discussed this.[122] In a recent paper, Young Huang and co-workers address this issue and provide a resolution.[123] According to Huang et al. the errors originate in the wide scatter of assumptions about "*h*" the thickness of the single-walled nanotube or the graphene sheet. Theoretically, SWNTs could have a Young's modulus of between 1 and 6 TPa. MWNTs are less strong because the individual cylinders slide with respect to each other. Ropes of SWNTs are also less strong. The individual tubes can pull out by shearing, and eventually the whole rope will break. This happens at stresses far below the tensile strength of individual nanotubes. Nanotubes sustain large strains in tension without showing signs of fracture, and in other directions, nanotubes are highly flexible. Because of their flexibility, nanotubes are used, for example, in scanning probe instruments. CNTs are not nearly as strong under compression than under tension. Because of their hollow structure and high aspect ratio, they tend to undergo bucking when placed under compressive, torsional, or bending stress. Since MWNT tips are conducting, they can be used in STM and AFM instruments (Volume II, Figure 2.40). Advantages are the improved resolution in comparison with conventional Si or metal tips, while the tips do not suffer from crashes with the surfaces because of their high elasticity. However, nanotube vibration due to the large length remains an important issue, until shorter nanotubes can be grown controllably. Because of the stiffness of carbon nanotubes, they should be ideal candidates for structural applications. For example, they may be used as reinforcements in high-strength, low-weight, and high-performance composites (see below).

Energy Storage and Conversion

See Chapter 9, this volume.

Chemical Sensors and Nanoreactors

Nanotube tips can be modified chemically through attachment of functional groups. Because of this, nanotubes can be used as molecular probes, with potential applications in chemistry and biology. Because of the small channels, strong capillary forces exist in nanotubes. These forces are strong enough to hold gases and fluids in nanotubes. In this way, it may be possible to fill the cavities of the nanotubes to create nanowires. The critical issue here is the wetting characteristic of nanotubes. Because of their smaller pore sizes, filling SWNTs is more difficult than filling MWNTs. If it becomes possible to keep fluids inside nanotubes, it could also be possible to perform chemical reactions inside their cavities. In particular, organic solvents wet nanotubes easily. In this case we could speak of a nanoreactor. Table 3.10 summarizes some chemical and biological sensor applications of carbon nanotubes.

Carbon Nanotubes in Composites

One of the most important applications of carbon nanotubes is projected to be as reinforcement in composite materials. However, there have not been many successful experiments that show that nanotubes are better fillers than the traditionally used carbon fibers. The main problem is to create a good interface between the nanotubes and the polymer matrix, as nanotubes are very smooth and have a small diameter that is nearly the same as that of a polymer chain. Secondly, nanotube aggregates, which are very common, behave differently in response to loads than individual nanotubes do. Limiting factors for good load transfer could be sliding of cylinders in MWNTs and shearing of tubes in SWNT ropes. To solve this problem, the

TABLE 3.10 Chemical and Biological Applications of Carbon Nanotubes

System	Target Species	Salient Feature
SWNT	NH_3 and NO_2	Sensitive to 200 ppm of NO_2 and 1% of NH_3.[124]
SWNT	N_2, He, O_2, and Ar	Gas concentrations as low as 100 ppm can be detected.[125]
MWCNT	NH_3	Gas concentrations as low as 10 ppm can be detected. The sensor showed a reversible response of few minutes.[126]
Poly(*o*-anisidine)-coated CNT	HCl	Nine times increase in sensitivity compared with uncoated CNT.[127]
MWNT-SiO_2	CO_2, O_2, and NH_3	Sensor response times are approximately 45 s, 4 min, and 2 min for CO_2, O_2, and NH_3, respectively. The sensor response is reversible for O_2 and CO_2 but irreversible for NH_3.[128]
SWNT	D-Glucose	Senses D-glucose in solution phase by two distinct mechanisms of signal transduction: fluorescence and charge transfer.[129]
Poly(vinylferrocene)-derivatized MWCNT	Glucose	Glucose concentration in real blood sample can be determined.[130]

aggregates need to be broken up and dispersed or cross-linked to prevent slippage. A main advantage of using nanotubes for structural polymer composites is that nanotube reinforcements will increase the toughness of the composites by absorbing energy during their highly flexible elastic behavior. Other advantages are the low density of the nanotubes, an increased electrical conduction, and better performance during compressive load. Another possibility, which is an example of a nonstructural application, is filling of photoactive polymers with nanotubes. Poly-*p*-phenylenevinylene filled with MWNTs and SWNTs is a composite that has been used for several experiments. These composites show a large increase in conductivity with only a small loss in photoluminescence and electroluminescence yields. Another benefit is that the composite is more robust than the pure polymer. Of course, nanotube-polymer composites could be used also in other areas. For instance, they could be used in the biochemical field as membranes for molecular separations or for osteointegration (growth of bone cells). However, these areas are less well explored. The most important thing we have to know about nanotubes for efficient use of nanotubes as reinforcing fibers is knowledge of how to manipulate the surfaces chemically to enhance interfacial behavior between the individual nanotubes and the matrix material.

Conclusion

Carbon nanotubes continue to hold a tremendous promise as a structural and electronic material. Some types of nanotubes might present a danger similar to that posed by asbestos, which has tiny fibers that can get into lung tissues and cause cancerous lesions.[131] Researchers are suggesting more study and caution in handling the material.

The Rise of Graphene:*[138] 2D

Graphene is the name given to a monolayer of carbon atoms arranged into benzene rings; it is used to describe properties of many carbon-based

* Based on the paper of the same title by Geim and Novoselov[119] (*Nature Materials* 2007;6:183–91).

materials, including graphite, fullerenes, and nanotubes (see Figures 3.44 and 3.45). The graphene theory was first developed by Philip Wallace in 1947 as an approximation to help understand the electronic properties of more complex, three-dimensional graphite.[132] Landau and Peierls, 70 years ago, claimed that 2D crystals are thermodynamically unstable and that divergent contribution of thermal fluctuations in 2D crystals lattices would lead to displacements of atoms comparable to the interatomic distances and thus destroy the 2D layer. They also concluded that the melting temperature of these thin films would rapidly decrease with decreasing thickness. Furthermore, in quantum field theory and statistical mechanics, the Mermin-Wagner theorem (also known as *Mermin-Wagner-Hohenberg theorem* or *Coleman theorem*) says that crystalline order is impossible in two-dimensional structures at non non-zero temperatures, and this has been interpreted as saying that a freestanding, two-dimensional crystal would be disrupted by thermodynamic forces (hence "2D materials were presumed not to exist").[133] It was believed that because of instabilities, a 2D graphene structure would fold up, or crumple, to form a more stable 3D structure. In summary, planar graphene was presumed not to exist in the free state, being unstable with respect to the formation of curved structures such as soot, fullerenes, and nanotubes.

However, in 2004, Novoselov et al. obtained relatively large graphene sheets (eventually, up to 100 μm in size) by mechanical exfoliation (repeated peeling) of 3D graphite crystals.[82] Their motivation allegedly was to study the electrical properties of thin graphite films. As purely two-dimensional crystals were unknown before and presumed not to exist, their discovery of individual planes of graphite may be viewed as accidental. The Manchester experimental discovery of 2D crystal matter was at first put in doubt until 2005, when the groups of Andre Geim and Philip Kim of Columbia University proved that the obtained graphitic layers exhibited the expected theoretical electronic properties of graphene.[134,135] On October 5, 2010, the Royal Swedish Academy of Sciences awarded the Nobel Prize in Physics to Andre Geim and Konstantin Novoselov University of Manchester, UK, "for groundbreaking experiments regarding the two-dimensional material graphene," (http://nobelprize.org/nobel_prizes/physics/laureates/2010/press.html). Single layers of graphite were previously (starting from the 1970s) grown epitaxially on top of other materials. Epitaxial graphene also consists of a single-atom-thick hexagonal lattice of sp^2-bonded carbon atoms, as in free-standing graphene, but there is significant charge transfer from the substrate to the epitaxial graphene, and, in some cases, hybridization between the *d* orbitals of the substrate atoms and π orbitals of the graphene layer. This interaction significantly alters the electronic structure of the epitaxial graphene. The Manchester free-standing graphene on the other hand is a single-atom-thick crystal of a macroscopic size that is either suspended or interacting only very weakly with a substrate.

The electronic structure of graphene approaches the 3D limit of graphite at ten layers of graphene. Today, graphene sheets are prepared by chemical exfoliation (solution-based), epitaxial growth by CVD (same as nanotubes), epitaxy on catalytic surfaces (Ni, Pt), and mechanical cleavage. Single graphene layers are not that difficult to make (see pencil in Figure 3.45), but they are more difficult to identify. With AFM it is possible, but the throughput is very low on large areas, and SEM does not reveal the number of atomic layers. A solution involves the placement of graphene on top of a Si wafer with a carefully chosen thickness of SiO_2; in this way graphene becomes visible even in an optical microscope. An interference-like contrast appears with an SiO_2 thickness of 300 or 90 nm (see Figure 3.57).

FIGURE 3.57 Graphene on SiO_2 observed in an optical microscope.[117]

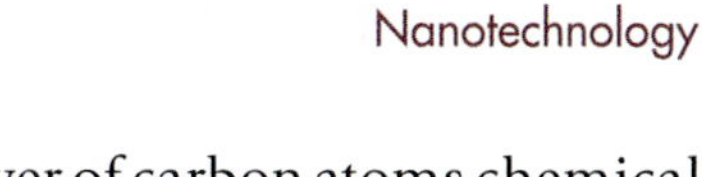

FIGURE 3.58 Graphene sheet with ripples.[119]

The atomic structure of isolated, single-layer graphene has been studied by transmission electron microscopy (TEM) on sheets of graphene suspended between bars of a metallic grid.[136] These electron diffraction patterns showed the expected hexagonal lattice of graphene but also revealed "rippling" of the flat sheet, with an amplitude of about 1 nm (see Figure 3.58). These ripples may be intrinsic to graphene as a result of the instability of two-dimensional crystals referred to above, or they may be extrinsic, originating perhaps from the ubiquitous dirt seen in all TEM images of graphene. While an infinitely large single layer of graphene would be in direct contradiction to the Mermin-Wagner theorem (see above), a finite-size 2D crystal of graphene could be stable. The observed ripples in suspended layers of graphene, it has been proposed, are caused by thermal fluctuations in the material; graphene adjusts to the thermal fluctuations, which could threaten to destroy the structure, by adjusting its bond length to accommodate the fluctuations. Within this context, it is actually debatable whether graphene is truly 2D or not, because of this natural tendency to ripple.[137]

Because graphene is somewhat tricky to make and even trickier to see and handle, most experiments have been performed not on free-standing monolayers but on two types of graphene samples: exfoliated graphene, deposited on a silicon-oxide/silicon substrate (see Figure 3.57), and epitaxial graphene, a layer of carbon atoms chemically deposited on (and chemically bonded to) a substrate of silicon carbide.

The surge of interest in the electronic properties of graphene has largely been motivated by the discovery that electron mobility in graphene is ten times higher than in commercial-grade silicon (up to 100,000 $cm^2\ V^{-1}\ s^{-1}$, limited by impurity scattering, and weakly dependent on temperature), ballistic transport is observed on the submicron scale (up to 0.3 μm at 300 K), and the quantum Hall effect can be observed at room temperature (extending the previous temperature range by a factor of 10). All these properties raise the possibility of high-efficiency, low-power, carbon-based electronics. The surprising electronic properties of graphene are ascribed to the presence of charge carriers that behave as if they are massless, "relativistic" quasiparticles called *Dirac fermions*.

A 2D graphene crystal is very different from a 3D silicon single crystal such as Si. In a semiconductor such as Si, electrons and holes interact with the periodic field of the atomic lattice to form quasiparticles. For example, in Volume I, Chapters 3 and 4, we saw how an electron moving through a conventional solid is described as having an effective mass (m_e^*) that takes into account the drag on its momentum from the surrounding crystal lattice as well as from interactions with other particles. The energy of quasiparticles in a solid depends on their momentum, a relationship described by the energy bands. In a typical 3D semiconductor, the energy bands are "parabolic."

Between them is the forbidden band gap, representing the amount of energy it takes to promote an electron from the valence band to the conduction band. The energy (E) of an electron in such a crystal depends quadratically on its momentum (p), as given by the following equation.

$$E = \frac{\hbar^2 \mathbf{k}^2}{2m_e^*}\left(= \frac{p^2}{2m_e^*}\right)$$

(Volume I, Equation 3.213)

While electron transport in most solids is accurately described by the nonrelativistic Schrödinger

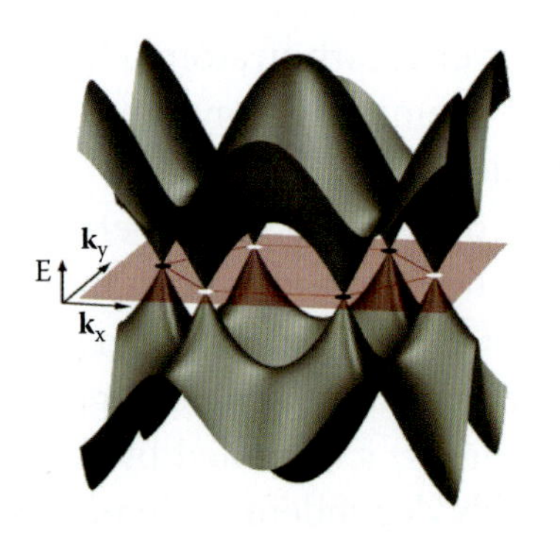

FIGURE 3.59 *E*-**k** surface of graphene. Intrinsic graphene is a semimetal or zero-gap semiconductor.

equation, graphene is quite different from most conventional three-dimensional materials. Unlike in ordinary semiconductors, graphs of the valence band and the conduction band in an *E*-**k** diagram for graphene are smooth-sided cones that meet at six points, called the *Dirac points* (see Figure 3.59). One interesting consequence of this unique band structure is that the electrons in graphene are "sort of free." The *E*-**k** relation near the Dirac points is linear for low energies, leading to zero effective mass for electrons and holes. Charge carriers in this energy range thus behave like relativistic particles (they behave like photons) described by the Dirac equation for spin 1/2 particles. The linear dispersion relationship in the low energy region is simply given by the following equation:

$$E = v_F \mathbf{k} \tag{3.14}$$

where the wave number (**k**) represents momentum and the Fermi velocity (v_F) stands in for the speed of light. Dirac fermions* travel at a constant speed—a small fraction of the speed of light.[138] Graphene has all the ideal properties to become an almost ideal component for integrated circuits. It has a high carrier mobility, as well as low noise, allowing it to be utilized as the channel in a FET. In 2008, the smallest transistor so far, one atom thick, ten atoms wide, was made of graphene.[139] IBM researchers are reporting progress in circuits based on narrow ribbons of planar graphene. The narrow graphene ribbons confine electrons in the narrow, single-planed graphene ribbons, which open up the possibility of building quantum wires and quantum wells. Narrow ribbons of graphene behave much like nanotubes and may be more amenable to manufacturing techniques such as etching. The current technology limits how thin graphene ribbons can be made. It does not allow them to be made thin enough to push to room temperature, so the work is carried out at lower temperatures.

Besides its electronic properties, graphene also is researched for its mechanical, thermal, and optical properties.

Despite all the excitement, don't expect "graphenium" microprocessors just yet. Think about carbon nanotubes, observed in 1991: they have created tremendous excitement, but today, there are no commercial CNT-based breakthrough products on the market—although they supposedly make your golf balls go straighter (a breakthrough for some golfers at least). Moreover, graphene currently is one of the most expensive materials on Earth, with a sample that can be placed on the cross-section of a human hair costing more than $1000 (as of April 2008). The price may fall dramatically, though, if and when commercial production methods are developed in the future.

Noncarbon Nanotubes and Nanowires

A nanotube, strictly speaking, is any tube with nanoscale dimensions, but generally the term is used to refer to carbon nanotubes. With a thin shell that is only one atom wide, nanotubes are hollow, giving them characteristics significantly different from those of solid nanowires made of silicon or other materials. Nanowires can be defined as structures that have a lateral size constrained to tens of nanometers or less and an unconstrained longitudinal size. At these scales, quantum mechanical effects are important. Such wires are also known as *quantum wires*. Many different types of nanowires exist, including metallic (e.g., Ni, Pt, Au), semiconducting (e.g., Si, InP, GaN, etc.), and insulating (e.g., SiO_2, TiO_2). Molecular nanowires are composed of repeating molecular units

* Dirac fermions have also been invoked to explain the quantum Hall effect in graphene, the magnetic-field-driven metal–insulator-like transition in graphite, superfluidity in ^3He, and the exotic pseudogap phase of high-temperature superconductors.

either organic (e.g., DNA and protein) or inorganic (e.g., $Mo_6S_{9-x}I_x$).

A major drawback to the utility of carbon nanotubes is the limited control over their chirality, which we saw above directly affects their electronic properties. At present no synthesis methods capable of controlling this feature are available, and selective synthesis of carbon nanotubes having only semiconducting characteristics remains elusive. A commonly studied noncarbon nanotube variety is made of boron nitride (BN). BN is the III-V homolog to graphene, and in this 2D lattice B and N occupy different sublattices—this lowers the symmetry and leads to new physical effects. BN nanotubes are all insulating, with a band gap of about 5 eV. Unlike carbon nanotubes, the conductive properties of BN nanotubes are more or less independent of chirality, diameter, and even the number of walls in the tube. Theoretically, the existence of silicon nanotubes with an sp^2 atomic arrangement is also predicted: creation of this type of nanotube is considered an important challenge for the future.

By making clever choices of seed crystals and growing conditions, it is possible to cause crystals to assume unusual shapes. As originally proposed by R.S. Wagner and W.C. Ellis, from Bell Labs, for the Au-catalyzed Si whisker growth, a vapor-liquid-solid (VLS) mechanism still is mostly used (Figure 3.60).[140]

As illustrated in Figure 3.61, (111)-oriented Si "whiskers" grow from a small Au particle island on a Si(111) surface when heated at 950°C and exposed

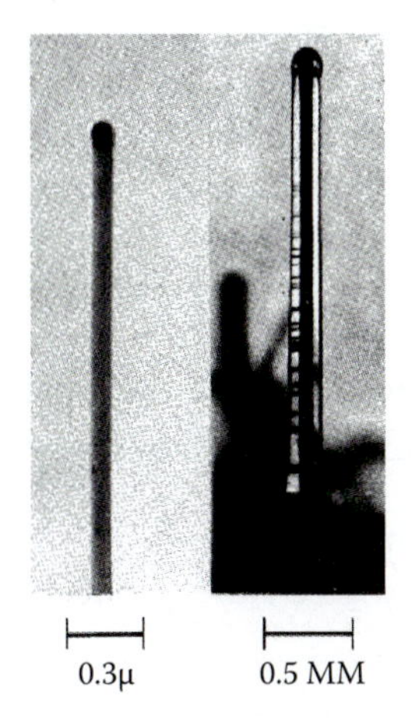

FIGURE 3.60 Si whisker growth: Au-catalyzed Si whisker growth based on a vapor-liquid-solid (VLS) mechanism.

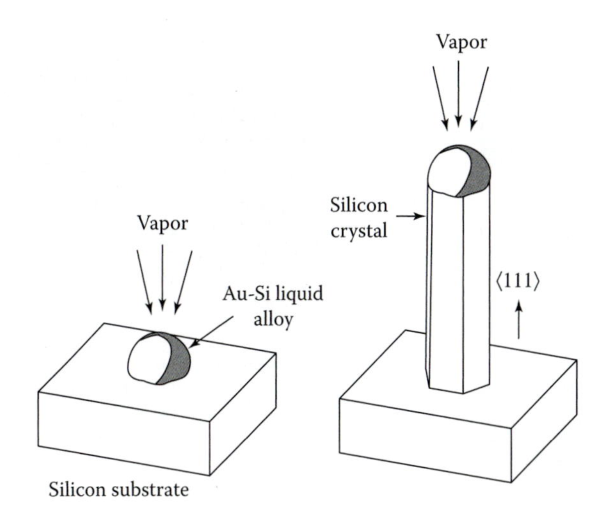

FIGURE 3.61 The vapor-liquid-solid mechanism for single-crystal Si growth.

to a flow of $SiCl_4$ and H_2 in a CVD reactor. Similar results are obtained with Pt, Ag, Pd, Cu, and Ni. When the substrate is heated up, a gold monolayer de-wets (unbinds) from the substrate in liquid form. From there, the liquid gold coalesces into nanoclusters and naturally self-assembles into gold nanoclusters. The gold acts as the catalyst in the VLS model. The liquid droplet is the preferred site for CVD deposition, and once the percentage of source material has supersaturated the nanoclusters, it begins to solidify and grow outward in a perfect single-crystal nanowire. In this growth mechanism the "impurity" is essential and remains in evidence from a small "globule" that is present at the tip of the whiskers during the growth (see Figure 3.61). The role of the impurity is to form a liquid alloy droplet at relatively low *T*. The selection of the impurity is important, as the diameter of the nanowire roughly is controlled by the diameter of the nanoclusters. The length of a nanowire is controlled by the amount of time the source is left on.

The VLS model corresponds to the model for the nucleation and growth of a carbon nanofiber from methane catalyzed by a supported metal particle shown in Figure 3.55 (based on the work of Hoogenraad[107]).

Nanoscale catalysts have been used to seed long, wire-like single crystals of compounds such as indium phosphide or gallium arsenide. The nanowire field got a shot in the arm (a rebirth, so to speak) with efforts by Charles Lieber (Harvard) (Volume I, Figure 3.65), Peidong Yang (http://www.cchem.berkeley.edu/

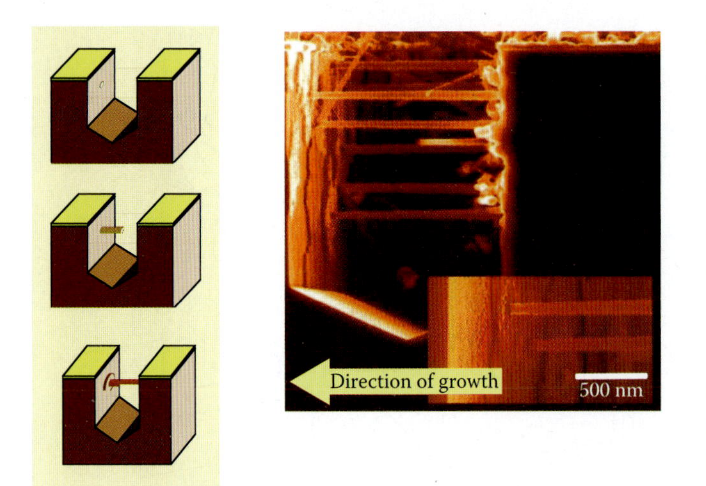

FIGURE 3.62 Lateral growth of nanowires between Si(111) surfaces. (From M. Saif Islam, S. Sharma, T.I. Kamins, and R.S. Williams, *Nanotechnology* 15;L5–L8, 2004. With permission.[141])

pdygrp/main.html), James Heath (http://www.its.caltech.edu/~heathgrp), and Hongkun Park (http://www.people.fas.harvard.edu/~hpark). Lieber's group at Harvard (http://cmliris.harvard.edu) reported arranging indium phosphide semiconducting nanowires into a simple configuration that resembled the lines in a tic-tac-toe board. The team used electron beam lithography to place electrical contacts at the ends of the nanowires in order to show that the array was electronically active.

Saif Islam et al. came up with the idea of connecting vertical Si(111) walls by growing Si nanowires laterally. Starting with a good catalyst on one wall, if there is another vertical surface in its close vicinity, the nanowire hits that surface and makes a bridge.[141] This is illustrated in Figure 3.62.

Electrochemical deposition of nanowires into a porous medium is another popular technique for making metallic nanowires. The porous medium may be porous silica, alumina, polycarbonate membranes, and even polyaniline nanotubules. Nanowire diameters as low as ~10 nm have been reported, and wire length may range from ~1 μm to >20 μm depending on the wire's diameter. Using nanoporous alumina as a template for the deposition of nanostructures of a material of choice, such as metals, semiconductors, and polymers, will be explained in Chapter 5, this volume (in Figure 5.10, the electrochemical deposition of the nanowire thermoelectric material Bi_2Te_3 is illustrated). Dresselhaus et al. show that lower-dimensional structures, such as nanowires based on Bi, BiSb, Bi_2Te_3, and SiGe nanotechnology, give additional controls that could lead to a higher ZT (see Chapter 8, this volume).[54]

TABLE 3.11 Nanowires and Their Applications

Material	Application
Si	Electronics, sensors
Cobalt	Magnetic
Germanium	Electronics, infrared detectors
Tin oxide	Chemical sensors
Indium oxide	Chemical sensors, biosensors
Indium tin oxide	Transparent conductive film in display electrodes, solar cells, organic light-emitting diodes
ZnO	UV lasers, field emission devices, chemical sensors
Copper oxide	Field emission device
Wide bandgap nitrides (GaN)	High temperature electronics, UV detectors and lasers, automotive electronics and sensors
Boron nitride	Insulator
Indium phosphide	Electronics, optoelectronics
Zinc selenide	Photonics (Q-switch, blue-green laser diode, blue UV photodetector)
Bi_2Te_3	Thermoelectric
Copper, tungsten	Electrical interconnects

Table 3.12 Nanowire Chemical Sensor Applications

Metal Oxide	Target Species	Salient Features
V_2O_5 nanofibers	1-Butylamine, toluene, propanol	Extremely high sensitivity was measured for 1-butylamine (below 30 ppb) and moderate sensitivity for ammonia. In contrast, only very little sensitivity was observed for toluene and 1-propanol vapors.[142]
SnO_2 nanobelts	CO, NO_2, ethanol	Sensitivity at the level of a few parts per billion.[143]
In_2O_3 nanowires	NH_3, NO_2	The response times have been determined to be 5 s for 100-ppm NO_2 and 10 s for 1% NH_3, and the lowest detectable concentrations are 0.5 ppm for NO_2 and 0.02% for NH_3.[144]
ZnO nanowires	Ethanol	Sensitive to ethanol concentration is in the range of 1–100 ppm. Sensitivity increases sharply as the temperature is raised from 200 to 300°C.[145]
MoO_3 nanorods	Ethanol and CO	The detection limit for ethanol and CO is lower than 30 ppm.[146]
Cd-doped ZnO nanowire	Relative humidity	Cd-doped ZnO nanowires show a clear positive temperature coefficient of resistance effect, which is quite abnormal compared with pure ZnO nanowires.[147]
SnO_2	Dimethyl methylphosphonate (DMMP)	Sensitive to 53 ppb DMMP; can be improved via doping nanobelts with catalytic additives.[148]

Advanced lithographic techniques (electron beam) are also used to pattern catalyst seeds for nanowire growth. In Table 3.11 we list some typical nanowires and their applications, and in Table 3.12, we review nanowire chemical sensor applications.

Questions

Questions by Dr. Marc Madou assisted by Mr. George O'Quinn, UC Berkeley

3.1: Explain the "fat fingers" and "sticky fingers" controversy, and suggest some possible ways that the problems associated with fat fingers and sticky fingers may be overcome.

3.2: Explain the difference between self-assembly, self-organization, and self-replication.

3.3: Why are fractal structures used consistently in nature? What benefits do these structures entail?

3.4: Make a listing of material properties that change in the nanoscale and explain why they change.

3.5: Explain the process of Stransky-Krastanov (SK) growth and how it can be used.

3.6: What effects do the chiral, zigzag, and armchair structures have on the properties of carbon nanotubes?

3.7: What are some important distinctions between top-down and bottom-up manufacturing (nanochemistry)?

3.8: Give some examples of bottom-up manufacturing methods.

3.9: Characterize the guiding principles involved in the growth of living organisms.

3.10: What does irreversible thermodynamics have to do with self-organization?

3.11: In the last few years, applications of quantum dots have become increasingly popular among analytical chemists. Quantum dots are now commercially available for a variety of applications, and many examples of their uses can be found in the literature. Briefly describe what quantum dots are and describe some of their applications.

3.12: Nanotechnology: hype or reality? Write one page.

References

1. Hall, J. S. 2005. *Nanofuture: What's next for nanotechnology.* Amherst, NY: Prometheus Books.
2. Royal Society. 2004. The Royal Society and The Royal Academy of Engineering.
3. Maasden, S. 2006. The assembled self of nanotechnology: the career of self-assembly as a metaphor. *EASST Conference.* Lausanne, Switzerland.
4. Rothemund, P. 2006. Folding DNA to create nanoscale shapes and patterns. *Nature* 440:297–302.
5. Prigogine, I. 1961. *Thermodynamics of irreversible processes.* New York: Wiley.
6. John, K., and M. Bar. 2005. Alternative mechanisms of structuring biomembranes: self-assembly versus self-organization. *Phys Rev Lett* 95:198101.
7. Ball, P. 1997. *Made to measure.* Princeton, NJ: Princeton University Press.
8. Jones, R. 2007. *Soft machines: Nanotechnology and life.* Oxford, United Kingdom: Oxford University Press.
9. Sellinger, A., P. M. Weiss, A. Nguygen, Y. Lu, R. A. Assink, W. Gong, and C. J. Brinker. 1998. Continuous self-assembly of organic-inorganic nanocomposite coatings that mimic nacre. *Nature* 394:256–59.
10. Achilles, T., and G. Von Kiedrowski. 1993. A self-replicating system from three starting materials. *Agnew Chem Int Ed Engl* 32:1198–201.
11. Lipson, H. 2005. Robotics: Self-reproducing machines. *Nature* 435:163–64.
12. Lehn, J. M. 1988. Supramolecular chemistry: Scope and perspectives (Nobel lecture). *Agnew Chem Int Ed Engl* 27:89–112.
13. Lehn, J. M. 1993. Supramolecular chemistry. *Nature* 260:1762–63.
14. Ringsdorf, H., B. Schlarb, and J. Venzmer. 1988. Molecular architecture and function of polymeric oriented systems: Models for the study of organization, surface recognition, and dynamics. *Agnew Chem Int Ed Engl* 27:113–58.
15. Whitesides, G. M., J. P. Mathias, and C. T. Seto. 1991. Molecular self-assembly and nanochemistry: A chemical strategy for the synthesis of nanostructures. *Science* 254:1312–18.
16. Whitesides, G. M. 1995. Self-assembling materials. *Sci Am* 146–49.
17. Pederson, C. J. 1988. The discovery of crown ethers. *Science* 241:536–40.
18. Stoddart, J. F., and D. B. Amabilimo. 1995. Interlocked and intertwined structures and superstructures. *Chem Rev* 95:2725–28.
19. Willner, I., and S. Rubin. 1996. Control of the structure and functions of biomaterials by light. *Agnew Chem Int Ed Engl* 35:367–385.
20. Stoddart, J. F., and S. A. Nepogodiev. 1998. Cyclodextrin-based catenanes and rotaxanes. *Chem Rev* 98:1959–76.
21. Stoddart, J. F., D. S. Hecht, R. J. A. Ramirez, M. Briman, E. Artukovic, K. S. Chichak, and G. Gruner. 2006. Bioinspired detection of light using a porphyrin-sensitized single-wall nanotube field effect transistor. *Nano Lett* 6:2031–36.
22. Saha, S., L. E. Johansson, A. H. Flood, H.-R. Tseng, J. I. Zink, and J. F. Stoddart. 2005. Powering a supramolecular machine with a photoactive molecular triad. *Small* 1:87–90.
23. Tomalia, D. A., H. Baker, J. Dewald, M. Hall, G. Kallos, S. Martin, J. Roeck, J. Ryder, and P. Smith. 1985. A new class of polymers: starburst-dendritic macromolecules. *Polym J* 17:117–32.
24. Newkome, G. R., Z. Yao, H. Baker, and V. K. Gupta. 1985. Cascade molecules: A new approach to micelles. *J Org Chem* 50:2003–04.
25. Newkome, G. R., C. N. Moorefield, and F. Vögtle. 1996. *Dentric molecules: Concepts, syntheses, perspectives.* Weinheim, Germany: VCH.
26. Fréchet, J. M. J., and Donald A. Tomalia. 2001. *Dendrimers and other dendritic polymers.* Weinheim, Germany: Wiley-VCH.
27. Grayson, S. M., and J. M. J. Fréchet. 2001. Convergent dendrons and dendrimers: From synthesis to applications. *Chem Rev* 101:3819–68.
28. Zeng, F., and S. C. Zimmerman. 1997. Dendrimers in supramolecular chemistry: From molecular recognition to self-assembly. *Chem Rev* 97:1681–712.
29. Zimmerman, S. C., F. W. Zeng, D. E. Reichert, and S. V. Kolotuchin. 1996. Self-assembling dendrimers. *Science* 271:1095–98.
30. Ward, M. D. 2000. Supramolecular coordination chemistry. *R Soc Chem Annu Rep A* 96:345–85.
31. Denti, S., S. Campagna, S. Serroni, M. Ciano, and V. Balzani. 1992. Decanuclear homo- and heterometallic polypyridine complexes. Synthesis, absorption spectra, luminescence, electrochemical oxidation, intercomponent energy transfer. *J Am Chem Soc* 114:2944–50.
32. Balzani, V., A. Juris, M. Venturi, S. Campagna, and S. Serroni. 1996. Luminescent and redox-active polynuclear transition-metal complexes. *Chem Rev* 96:759–833.
33. Percec, V., W. D. Cho, P. E. Mosier, G. Ungar, and D. J. P. Yeardley. 1998. Structural analysis of cylindrical and spherical supramolecular dendrimers quantifies the concept of monodendritic shape control by generation number. *J Am Chem Soc* 120:11061–70.
34. Hudson, S. D., H.-T. Jung, V. Percec, W.-D. Cho, G. Johansson, G. Ungar, and V. S. K. Balagurusamy. 1997. Direct visualization of individual cylindrical and spherical supramolecular dendrimers. *Science* 278:449–52.
35. Enomoto, M., A. Kishimura, and T. Aida. 2001. Coordination metallacycles of an achiral dendron self-assemble via metal-metal interaction to form luminescent superhelical fibers. *J Am Chem Soc* 123:5608–09.
36. Fréchet, J. M. J. 2002. Supramolecular chemistry and self-assembly special feature dendrimers and supramolecular chemistry. *Proc Natl Acad Sci USA* 99(8):4782–87.
37. Cagin, T., G. Wang, R. Martin, and W. A. Goddard, III. 1999. Molecular modeling of dendrimers for nanoscale applications. *Seventh Foresight Conference on Molecular Nanotechnology.* Santa Clara, CA.
38. Tully, D. C., K. Wilder, J. M. J. Fréchet, A. R. Trimble, and C. F. Quate. 1999. Dendrimer-based self assembled monolayers as resists for scanning probe lithography. *Adv Mater* 11:314–18.
39. Hawker, C. J., K. L. Wooley, and J. M. J. Fréchet. 1993. Unimolecular micelles and globular amphiphiles: Dendritic macromolecules as novel recyclable solubilization agents. *J Chem Soc Perkin Trans* 1:1287–97.
40. Kono, K., M. Liu, and J. M. J. Fréchet. 1999. Design of dendritic macromolecules containing folate or methotrexate residues. *Bioconjugate Chem* 10:1115–21.

41. Wells, M., and R. M. Crooks. 1996. Interactions between organized, surface-confined monolayers and vapor-phase probe molecules. 10. Preparation and properties of chemically sensitive dendrimer surfaces. *J Am Chem Soc* 118:3988–89.
42. Crooks, R. M., and A. J. Ricco. 1998. New organic materials suitable for use in chemical sensor arrays. *Acc Chem Res* 31:219–27.
43. Yoon, H. C., M. Y. Hong, and H. S. Kim. 2000. Affinity biosensor for avidin using a double-functionalized dendrimer monolayer on gold electrode. *Anal Biochem* 282:121–28.
44. Yoon, H. C., M. Y. Hong, and H. S. Kim. 2001. Reversible association/dissociation reaction of avidin on the dendrimer monolayer functionalized with a biotin-analog for a reagentless affinity-sensing surface. *Langmuir* 17:1234–39.
45. Tully, D. C., A. R. Trimble, J. M. J. Fréchet, K. Wilder, and C. F. Quate. 1999. Synthesis and preparation of ionically bound dendrimer monolayers and application towards scanning probe lithography. *Chem Mater* 11:2892–98.
46. Drexler, K. E. 1981. Molecular engineering: An approach to the development of general capabilities for molecular manipulation. *Proc Natl Acad Sci USA* 78:5275–78.
47. Drexler, K. E. 1986. *Engines of creation: The coming era of nanotechnology.* New York: Doubleday.
48. Smalley, R. E. 2001. Of chemistry, love and nanobots. *Sci Am* 285:76–77.
49. Drexler, K. E., D. Forrest, R. A. Freitas, J. S. Hall, N. Jacobstein, T. McKendree, R. Merkle, and C. Peterson. 2001. *On physics, fundamentals, and nanorobots: a rebuttal to Smalley's assertion that self-replicating mechanical nanorobots are simply not possible.* Palo Alto, CA: Institute for Molecular Manufacturing.
50. Ho, W., and H. Lee. 1999. Single bond formation and characterization with a scanning tunneling microscope. *Science* 286:1719–22.
51. Klabunde, K. J. 2001. *Nanoscale materials in chemistry.* New York: Wiley-Interscience.
52. Asheghi, M., Y. S. Ju, and K. E. Goodson. 1997. Thermal conductivity of SOI device layers. *Proceedings of the IEEE International SOI conference, Tenaya Lodge at Yosemite.* Fish Camp, CA.
53. Dresselhaus, M. S., X. Sun, and Y. Lin. 2000. Theoretical investigation of thermoelectric transport properties of cylindrical bi nanowires. *Phys Rev B* 62:4610–23.
54. Yang, P., A. I. Hochbaum, R. Chen, R. D. Delgado, W. Liang, E. C. Garnett, M. Najarian, and A. Majumdar. 2008. Rough silicon nanowires as high performance thermoelectric materials. *Nature* 451:163–67.
55. Heath, J. R., A. Boukai, Y. Bunimovich, J. Tahir-Kheli, J. Yu, and W. A. Goddard. 2008. Silicon nanowires as efficient thermoelectric materials. *Nature* 451:168–71.
56. Arakawa, Y., and H. Sakaki. 1982. Multidimensional quantum well laser and temperature dependence of its threshold current. *Appl Phys Lett* 40:939.
57. Barthlott, W., and C. Neinhuis. 1997. The purity of sacred lotus or escape from contamination in biological surfaces. *Planta* 202:1–8.
58. Guo, Z., F. Zhou, J. Hao, and W. Liu. 2005. Stable biomimetic super-hydrophobic engineering materials. *J Am Chem Soc* 127:15670–15671.
59. Herzer, G. 1997. Nanocrystalline soft magnetic alloys. *Handbook of magnetic materials,* ed. K. H. J. Buschow, 417–61. Amsterdam: Elsevier Science B.V.
60. Herzer, G., and H. Warlimont. 1992. *Nanostruct Mater* 1:292.
61. Jacak, L. 2000. Semiconductor quantum dots: Towards a new generation of semiconductor devices. *Eur J Phys* 21:487–97.
62. Hashizume, T., R. J. Hamers, J. E. Demuth, and K. Markert. 1990. Initial stage deposition of Ag on the Si(001)-(2X1) surface studied by STM/STS. *J Vac Sci Tech A* 8:249.
63. Blackman, J. A., and A. Wilding. 1991. Scaling theory of island growth in thin films. *Europhys Lett* 16:115–20.
64. Franchi, S., G. Trevisi, L. Seravalli, and P. Frigeri. 2003. Quantum dot nanostructures and molecular beam epitaxy. progress in crystal growth and characterization of materials. *Progress in Crystal Growth and Characterization of Materials* 47:166.
65. Fukui, T. E. A. 1991. GaAs tetrahedral quantum dot structures fabricated using selective area metalorganic chemical vapor deposition. *Appl Phys Lett* 58:2018–20.
66. Murray, C. B., C. R. Kagan, and M. G. Bawendi. 1995. Self-organization of CdSe nanocrystallites into three-dimensional quantum dot superlattices. *Science* 270:1335–38.
67. Murray, C. B., D. J. Norris, and M. G. Bawendi. 1993. Synthesis and characterization of nearly monodisperse CdE (E=S, Se, Te) semiconductor nanocrystallites. *J Am Chem Soc* 115:8706–15.
68. Murray, C. B., C. R. Kagan, and M. G. Bawendi. 2000. Synthesis and characterization of monodisperse nanocrystals and close-packed nanocrystal assemblies. *Annu Rev Mater* 30:545–610.
69. Gedanken, A. 2004. Using sonochemistry for the fabrication of nanomaterials. *Ultrason Sonochem* 11(2):47–55.
70. Gogate, P. R. 2008. Cavitational reactors for process intensification of chemical processing applications: A critical review. *Chem Eng Process* 47(4):515–27.
71. Arturo Lopez-Quintela, M. 2003. Synthesis of nanomaterials in microemulsions: Formation mechanisms and growth control. *Curr Opin Coll Int Sci* 8:137–44.
72. Eckert, C. A., B. L. Knutson, and P. G. Debenedetti. 1996. Supercritical fluids as solvents for chemical and materials processing. *Nature* 383:313.
73. Reverchon, E., and R. Adami. 2006. Nanomaterials and supercritical fluids. *J Supercrit Fluids* 37:1.
74. Liu, J. C., Y. Ikushima, and Z. Shervani. 2003. Environmentally benign preparation of metal nano-particles by using water-in-CO_2 microemulsions technology. *Curr Opin Solid State Mater* 7:255.
75. Eastoe, J., A. Dupont, and D. C. Steytler. 2003. Fluorinated surfactants in supercritical CO_2. *Curr Opin Coll Int Sci* 8:267.
76. Smalley, R. E., D. T. Colbert, A. Thess, P. Nikolaev, and T. Guo. 1995. Catalytic growth of single-walled nanotubes by laser vaporization. *Chem Phys Lett* 243:49–54.
77. Jacak, L., A. Wojs, and P. Hawrylak. 1998. *Quantum dots.* Berlin: Springer Verlag.
78. Green, M., and P. O'Brien. 1999. Recent advances in the preparation of semiconductors as isolated nanometric particles: New routes to quantum dots. *Chem Comm* 22:2235–41.
79. Castro Neto, A. H., and E. A. Kim. 2008. Graphene as an electronic membrane. *Europhys Lett* 84:57 007.
80. Kroto, H. W., R. E. Smalley, R. F. Curl, J. R. Heath, and S. C. O'Brien. 1985. C60: Buckminsterfullerene. *Nature* 318:162–63.
81. Iijima, S. 1991. Helical microtubules of graphitic carbon. *Nature* 354:56–58.

82. Novoselov, K. S., A. K. Geim, S. V. Morozov, D. Jiang, Y. Zhang, S. V. Dubonos, I. V. Grigorieva, and A. A. Firsov. 2004. Electric field effect in atomically thin carbon films. *Science* 306:666–69.
83. Mintmire, J. W., B. I. Dunlap, and C. T. White. 1992. Are fullerene tubules metallic? *Phys Rev Lett* 68:631–34.
84. Overney, G., W. Zhong, and D. Tomanek. 1993. Structural rigidity and low frequency vibrational modes of long carbon tubules. *Zeitschrift für Physik D Atoms, Molecules and Clusters* 27:93–96.
85. Iijima, S., and T. Ichihashi. 1993. Single-shell carbon nanotubes of 1-nm diameter. *Nature* 363:603–05.
86. Rinzler, A. G., J. H. Hafner, P. Nikolaev, P. Nordlander, D. T. Colbert, R. E. Smalley, L. Lou, S. G. Kim, and D. Tománek. 1995. Unraveling nanotubes: Field emission from an atomic wire. *Science* 269:1550–53.
87. Dillon, A. C., K. M. Jones, T. A. Bekkedahl, C. H. Kiang, D. S. Bethune, and M. J. Heben. 1997. Storage of hydrogen in single-walled carbon nanotubes. *Nature* 386: 377–79.
88. Dekker, C., S. J. Tans, and A. R. M. Verschueren. 1998. Room-temperature transistor based on a single carbon nanotube. *Nature* 393:49–52.
89. Smith, B. W., M. Monthioux, and D. E. Luzzi. 1998. Encapsulated C60 in carbon nanotubes. *Nature* 396:323–24.
90. Smith, B. W., D. E. Luzzi, M. Monthioux, B. Burteaux, A. Claye, and J. E. Fischer. 1999. Abundance of encapsulated C60 in single-wall carbon nanotubes. *Chem Phys Lett* 310:21–24.
91. Tománek, D., S. Berber, and Y. Kwon. 2000. Unusually high thermal conductivity of carbon nanotubes. *Phys Rev Lett* 84:4613–16.
92. Collins, P. G., M. S. Arnold, and P. Avouris. 2001. Engineering carbon nanotubes and nanotube circuits using electrical breakdown. *Science* 292:706–09.
93. Kociak, M., A. Y. Kasumov, S. Guéron, B. Reulet, I. I. Khodos, Y. B. Gorbatov, V. T. Volkov, L. Vaccarini, and H. Bouchiat. 2001. Superconductivity in ropes of single-walled carbon nanotubes. *Phys Rev Lett* 86:2416–19.
94. Schedin, F., A. K. Geim, S. V. Morozov, E. W. Hill, P. Blake, M. I. Katsnelson, and K. S. Novoselov. 2007. Detection of individual gas molecules adsorbed on graphene. *Nat Mater* 6:183–91.
95. Saito, Y. 1994. Synthesis and characterization of carbon nanocapsules encaging metal and carbide crystallites. San Francisco: Fullerenes.
96. Kratschmer, W., L. D. Lamb, K. Fostiropoulos, and D. R. Huffman. 1990. Solid C_{60}: A new form of carbon. *Nature* 347:354.
97. Hummelen, J. C., M. Prato, and F. Wudl. 1995. There is a hole in my Bucky. *J Am Chem Soc* 117:7003.
98. Nemecek, S. 1995. A tight fit. *Sci Am* 34–36.
99. Akasaka, T., F. Wudl, and S. Nagase. 2010. *Chemistry of Nanocarbons*. Chichester: Wiley.
100. Holmes, B. 2004. Buckyballs cause brain damage in fish. *New Scientist,* March 29. http://www.newscientist.com/news/news.jsp?id=ns99994825.
101. Bacon, R. 1960. Growth, structure, and properties of graphite whiskers. *J Appl Phys* 31:283–90.
102. Bethune, D. S., C. H. Klang, M. S. D. Vries, G. Gorman, R. Savoy, J. Vazquez, and R. Beyers. 1993. Cobalt-catalysed growth of carbon nanotubes with single-atomic-layer walls. *Nature* 363:605–07.
103. Seraphin, S. 1995. Single-walled tubes and encapsulation of nanocrystals into carbon clusters. *Electrochem Soc* 142:290–97.
104. Endo, M., K. Takeuchi, S. Igarashi, K. Kobori, M. Shiraishi, and H. W. Kroto. 1993. The production and structure of pyrolytic carbon nanotubes (PCNTs). *J Phys Chem Solids* 54:1841–48.
105. Dai, H., Y. Li, D. Mann, M. Rolandi, W. Kim, A. Ural, S. Hung, A. Javey, J. Cao, D. Wang, E. Yenilmez, Q. Wang, J. F. Gibbons, and Y. Nishi. 2004. Preferential growth of semiconducting single-walled carbon nanotubes by a plasma enhanced CVD method. *Nano Lett* 4:317–21.
106. Dai, H., J. Kong, H. T. Soh, A. M. Cassell, and C. F. Quate. 1998. Synthesis of individual single-walled carbon nanotubes on patterned silicon wafers. *Nature* 395: 878–81.
107. Hoogenraad, M. S. 1995. Growth and utilization of carbon fibrils. PhD thesis, Utrecht University, Utrecht, the Netherlands.
108. Huang, S., M. Woodson, R. E. Smalley, and J. Liu. 2004. Growth mechanism of oriented long single walled carbon nanotubes using "fast-heating" chemical vapor deposition process. *Nano Lett* 4:1025–28.
109. Wang, X., Q. Li, J. Xie, Z. Jin, J. Wang, Y. Li, K. Jiang, and S. Fan. 2009. Fabrication of ultralong and electrically uniform single-walled carbon nanotubes on clean substrates. *Nano Lett* 9(9):3137–41.
110. Wagner, R. S. 1970. VLS mechanisms of crystal growth. In *Whisker technology*, ed. A. P. Levitt, 47–119. New York: Wiley.
111. Ando, Y., X. Zhao, T. Sugai, and M. Kumar. 2004. Growing carbon nanotubes. *Materials Today* 7:22–29.
112. McEuen, P. L., A. Zettl, J. Hone, M. Bockrath, A. G. Rinzler, and R. E. Smalley. 2000. Chemical doping of individual semiconducting carbon-nanotube ropes. *Phys Rev B* 61, R10606–08.
113. Ebbesen, T. W., and P. M. Ajayan. 1992. Large-scale synthesis of carbon nanotubes. *Nature* 358:220.
114. Endo, M., K. Takeuchi, S. Igarashi, K. Kobori, M. Shiraishi, and H. W. Kroto. 1993. The production and structure of pyrolytic carbon nanotubes (PCNTs). *J Phys Chem Solids* 54:1841–48.
115. Avouris, P., T. Hertel, R. Martel, T. Schmidt, and H. R. Shea. 1998. Single- and multi-wall carbon nanotube field-effect transistors. *Appl Phys Lett* 73:2447–49.
116. Lieber, C. M., P. Yang, M. Ouyang, and J. Hu. 1999. Controlled growth and electrical properties of heterojunctions of carbon nanotubes and silicon nanowires. *Nature* 399:48–51.
117. Dai, H., A. Ural, Y. Li, A. Javey, and Q. Wang. 2002. Carbon nanotube transistor arrays for multistage complementary logic and ring oscillators. *Nano Lett* 2:929–32.
118. Govyadinov, A., P. Mardilovich, and D. Routkevitch. 2000. Field emission cathode from aligned arrays of carbon nanotubes. *Electrochemical Society Abstracts* 553: 2000–28.
119. Hirsch, A. 1994. *The chemistry of fullerenes.* Stuttgart, Germany: George Thieme Verlag.
120. Dresselhaus, M. S., G. Dresselhaus, and P. Avouris. 2000. *Carbon nanotubes.* Heidelberg, Germany: Springer-Verlag.
121. Special section: Carbon nanotubes. 2000. *Phys World* 13:29–53.
122. Yakobson, B. I., C. J. Brabec, and J. Bernholc. 1996. Nanomechanics of carbon tubes: Instabilities beyond linear response, *Phys Rev Lett* 76:2511–14.
123. Peng, J., J. Wu, K. C. Hwang, J. Song, and Y. Huang. 2008. Can a single-wall carbon nanotube be modeled as a thin shell? *J Mech Phys Solids* 56(6):2213–24.

124. Zettl, A., P. G. Collins, K. Bradley, and M. Ishigami. 2000. Extreme oxygen sensitivity of electronic properties of carbon nanotubes. *Science* 287:1801–04.
125. Rao, A. M., N. Gothard, K. McGuire, S. Chopra, and A. Pham. 2003. Selective gas detection using a carbon nanotube sensor. *Appl Phys Lett* 83:2280–82.
126. Hara, M., G. Zhou, and J. Suehiro. 2003. Fabrication of a carbon nanotube-based gas sensor using dielectrophoresis and its application for ammonia detection by impedance spectroscopy. *J Phys D Appl Phys* 36, L109–14.
127. Kenny, J. M., L. Valentini, V. Bavastrello, E. Stura, I. Amentano, and C. Nicolini. 2004. Sensors for inorganic vapor detection based on carbon nanotubes and poly(o-anisidine) nanocomposite material. *Chem Phys Lett* 383:617–22.
128. Grimes, C. A., K. G. Ong, and K. Zeng. 2002. A wireless, passive carbon nanotube-based gas sensor. *Sensors J* 2:82–88.
129. Strano, M. S., D. A. Heller, S. Baik, and P. W. Barone. 2005. Near-infrared optical sensors based on single-walled carbon nanotubes. *Nat Mater* 4:86–92.
130. Compton, R. G., B. Ljuki, C. E. Banks, C. Salter, and A. Crossley. 2006. Electrochemically polymerised composites of multi-walled carbon nanotubes and poly(vinylferrocene) and their use as modified electrodes: Application to glucose sensing. *Analyst* 131:670–77.
131. Lison, D., J. Mullera, F. Huauxa, N. Moreaub, P. Missona, J. Heiliera, M. Delosc, M. Arrasa, A. Fonsecab, and J. B. Nagyb. 2005. Respiratory toxicity of multi-wall carbon nanotubes. *Toxicol Appl Pharmacol* 207:221–31.
132. Wallace, P. R. 1947. The band theory of graphite. *Phys Rev* 71:622–34.
133. Wagner, H., and N. D. Mermin. 1966. Absence of ferromagnetism or antiferromagnetism in one- or two-dimensional isotropic Heisenberg models. *Phys Rev Lett* 17: 1133–36.
134. Novoselov, K. S., A. K. Geim, D. Jiang, F. Schedin, T. J. Booth, V. V. Khotkevich, and S. V. Morozov. 2005. Two-dimensional atomic crystals. *Proc Natl Acad Sci USA* 105:10451–53.
135. http://nobelprize.org/nobel_prizes/physics/laureates/2010/press.html
136. Geim, A. K., and A. H. MacDonald. 2007. Graphene: Exploring carbon flatland. *Phys Today* 60:35–41.
137. Novoselov, K. S., A. K. Geim, J. C. Meyer, M. I. Katsnelson, T. J. Booth, and S. Roth. 2007. The structure of suspended graphene sheets. *Nature* 446:60–63.
138. Geim, A. K., and K. S. Novoselov. 2007. The rise of graphene. *Nat Mater* 6:183–91.
139. Novoselov, K. S., A. K. Geim, A. Das, S. Pisana, B. Chakraborty, S. Piscanec, S. K. Saha, U. V. Waghmare, H. R. Krishnamurthy, A. C. Ferrari, and A. K. Sood. 2008. Monitoring dopants by raman scattering in an electrochemically top-gated graphene transistor. *Nat Nanotechnol* 3:210–15.
140. Wagner, R. S., and W. C. Ellis. 1964. Vapor-liquid-solid mechanism of single crystal growth. *Appl Phys Lett* 4:89–90.
141. Saif Islam, M., S. Sharma, T. I. Kamins, and R. S. Williams. 2004. Ultrahigh-density silicon nanobridges formed between two vertical silicon surfaces. *Nanotechnology* 15:L5–L8.
142. Raible, I., M. Burghard, U. Schlecht, A. Yasuda, and T. Vossmeyer. 2005. V2O5 nanofibres: Novel gas sensors with extremely high sensitivity and selectivity to amines. *Sensors Actuators B Chem* 106:730–35.
143. Comini, E., G. Faglia, G. Sberveglieri, Z. Pan, and Z. L. Wang. 2002. Stable and highly sensitive gas sensors based on semiconducting oxide nanobelts. *Appl Phys Lett* 81:1869–71.
144. Li, C., D. Zhang, X. Liu, S. Han, T. Tang, J. Han, and C. Zhou. 2003. In2O3 nanowires as chemical sensors. *Appl Phys Lett* 82:1613–15.
145. Wan, Q., Q. H. Li, Y. J. Chen, T. H. Wang, X. L. He, J. P. Li, and C. L. Lin. 2004. Fabrication and ethanol sensing characteristics of ZnO nanowire gas sensors. *Appl Phys Lett* 84:3654–56.
146. Comini, E., L. Yubao, Y. Brando, and G. Sberveglieri. 2005. Gas sensing properties of MoO3 nanorods to CO and CH_3OH. *Chem Phys Lett* 407:368–71.
147. Wan, Q., Q. H. Li, Y. J. Chen, T. H. Wang, X. L. He, X. G. Gao, and J. P. Li. 2004. Positive temperature coefficient resistance and humidity sensing properties of Cd-doped ZnO nanowires. *Appl Phys Lett* 84:3085–87.
148. Yu, C., Q. Hao, S. Saha, L. Shi, X. Kong, and Z. L. Wang. 2005. Integration of metal oxide nanobelts with microsystems for nerve agent detection. *Appl Phys Lett* 86:063101.

4

Packaging, Assembly, and Self-Assembly

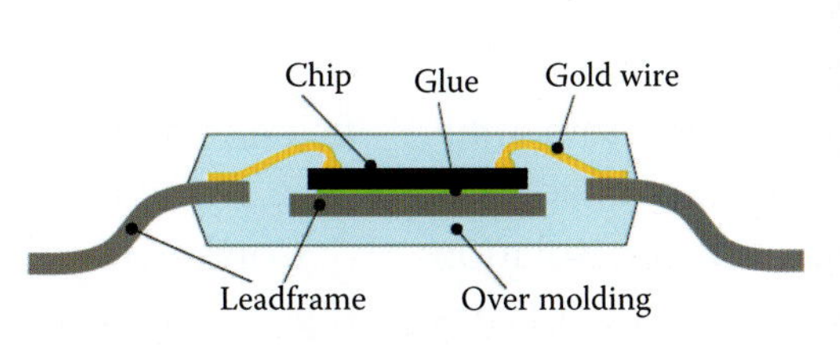

The quad flat pack (QFP) is a package for chips with 100 and more leads.

Programmable self-assembly: constructing global shapes using biologically inspired local interactions and origami mathematics. (Courtesy of Radhika Nagpal, Massachusetts Institute of Technology.)

Outline

> Once the components were expensive and the wires cheap.
> Now the components are cheap and the wires expensive.
>
> **Len Schaper**
> *Arkansas University*

> The aboriginal people of Australia believed that the world began only when their Ancestor sang it into existence.
>
> **Bruce Chatwin**
> *The Songlines*

In the end the supremacy of organic biomaterials is tied in with the question of scale. Organic machinery can be made much smaller. Such clever things become possible as sockets, which can recognize, hold and manipulate other molecules. In any competition to do with molecular control the system with the smallest fingers will win.

Graham Cairns-Smith
1985

What is the Heart, but a Spring; and the Nerves, but so many Springs; and the Joynts, but so many Wheeles, given motion to the whole Body, such as was intended by the Artificer.

Thomas Hobbes

Introduction

This chapter addresses packaging, assembly, and self-assembly. Under packaging of micromachines, we consider issues that differ the most from integrated circuit (IC) packaging; with micromachines one faces the paradox of needing hermetic isolation from the environment that must be sensed. This has proven to be a challenge for mechanical sensors, such as for pressure and, to a lesser extent, for accelerometers, but it is the Achilles' heel of chemical and biological sensors. We distinguish between wafer-level (batch) packaging steps and serial packaging processes and learn that micromachining itself plays an important role in transforming costly serial packaging steps into less costly, wafer-level processes.

During packaging, miniature components are put together into a housing structure in a microassembly step. This task becomes more cumbersome with smaller building blocks and will increasingly necessitate the use of parallel self-assembly techniques. Self-assembly is imperative when components are nanoscale.

Packaging

Introduction

In electronic device and mechanical micromachine fabrication, processes can be divided into three major groups.

1. First, fabrication with additive and subtractive processes, as described in Volume II. These also are called *front-end processes*.
2. Packaging, involving processes such as bonding, wafer scribing, lead attachment, and encapsulation in a protective body.
3. Testing, including package leak tests, electrical integrity, and sensor functionality.

The last two process groups incorporate the most costly steps and are called *back-end processes*. In the case of a micromachine, which often interfaces with a hostile environment and where testing might involve chemical and/or mechanical parameters, packaging and testing are yet more difficult and expensive than in the case of ICs. More than 70% of the sensor cost may be determined by its package, and the physical size of a micromachined sensor often is dwarfed by the size of its package. Most conventional packaging approaches are space inefficient, with volumetric efficiencies, even in the case of ICs, often less than 1%.[1]

Contrasting the inefficiency of human packaging methods with nature's excellence at packaging is striking. As an example, consider how DNA is packed in the nucleus of a cell. Biomolecules in a solution coil up with a gyration radius r_g given by the de Gennes formula, $r = (Lb)^{1/2}$, where L is the molecule's contour length and b is determined by the degree to which the molecule resists thermal motion.[2] The total length of all DNA in a human cell is 2 m, which is a million times the cell nucleus diameter. For b of DNA, a value of 100 nm has been determined. The resulting radius of the 2-m DNA molecule is calculated at 0.45 mm, 1000 times the nucleus diameter. How does nature pack all this into the nucleus of a cell? The answer is efficient packaging. Like a thread on a spool, DNA is wound around a set of nuclear proteins—the histones. The molecule is wound twice around one "spool" and then passes onto the next, and so on. A necklace of histones is neatly packed into a chromosome. This packaging scenario is illustrated in Figure 4.1. This way, a 0.45-mm polymer coil is squeezed into a nucleus of less than a micrometer.[3]

Packaging issues of ICs and mechanical micromachines are somewhat similar, whereas packaging of devices that are used in fluids (e.g., level sensor) or that contain liquids (e.g., some electrochemical sensors) are totally different. The latter require very

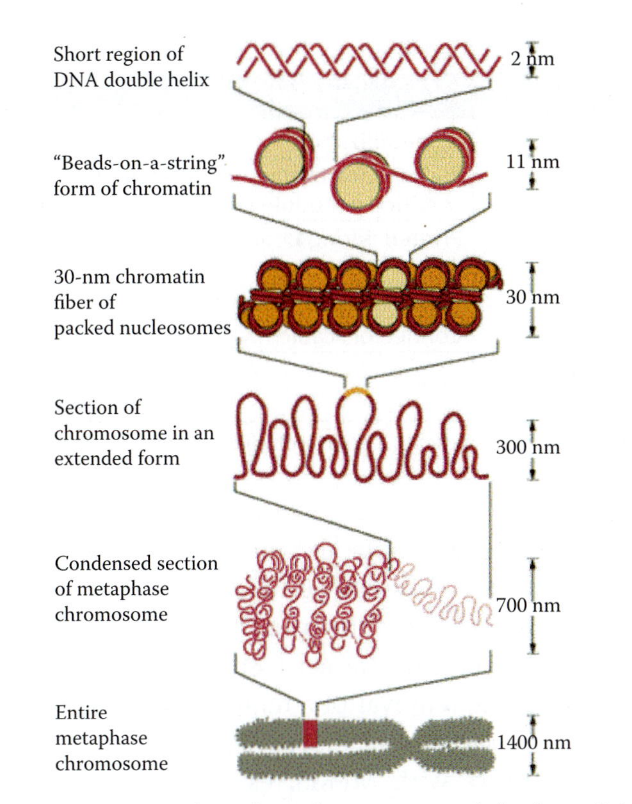

FIGURE 4.1 DNA is packaged very compactly in the cell. In eukaryotic cells, it is assembled into chromosomes, which have a hierarchical structure. The double-stranded DNA is wound twice around disc-like protein "beads" made from eight separate protein units called *histones*. The DNA-histone assembly is a nucleosome. The DNA chain, with nucleosomes all along its length, is then wound into a helix, which is itself packaged into a compact bundle in the chromosome.

different partitioning solutions and often must face biocompatibility issues as well.

Packaging in ICs and MEMS

Packaging is an essential bridge between components and their environment. In ICs, packaging provides at least four functions:

1. Signal redistribution
2. Mechanical support
3. Power distribution
4. Thermal management

In chemical and biological MEMS, the package might have several additional functions, for example:

1. Biocompatibility for in vivo devices
2. Fluidic manifold to guide liquids to different parts of the MEMS structure
3. Window/opening for analyte to interact with the chemical sensitive part of the MEMS structure

The signal redistribution is easy to understand: the electrical contacts of an IC are too closely spaced to accommodate the interconnection capacity of a traditional printed wiring board (PWB). Packaging redistributes contacts over a larger, more manageable surface—it fans out the electrical path. The mechanical support provides rigidity, stress release, and protection from the environment (e.g., against electromagnetic interference). Power distribution is similar to signal distribution except that power delivery systems are more robust than signal delivery systems, and the thermal management function is there to provide adequate thermal transport to sustain operation for the product lifetime.[1]

The various levels of packaging and their interconnections in the IC industry that we adapt here also for micromachines are summarized in Table 4.1[1] and illustrated in Figure 4.2a.[4,5] The lowest level in the IC electronics (L0 in the hierarchy) is single IC features interconnected on a single die with IC metallization lines into a level 1 IC or discrete component. Single-chip packages and multichip modules are at level 2. Levels 3 and 4 are the PWB (L3) and chassis or box (L4), respectively. Level 5 is the system itself. The hierarchy for packaging a micromachine such as a pressure sensor is illustrated in Figure 4.2b and 4.2c. For more information on advanced packaging schemes in ICs, we refer to the excellent reviews by Lyke,[1] Jensen,[4] and Lau.[6]

Packages for microelectronic systems have evolved over five decades to accommodate devices that largely have no need to interact with the environment. Packages for micromachines have all the same functions as listed for ICs but are considerably more complex, as they serve to protect from the environment while, somewhat in contradiction, enabling interaction with that environment so as to measure or affect the desired physical or chemical parameters. With the advent of integrated sensors, the protection problem became more complicated. Silicon circuitry is sensitive to temperature, moisture, magnetic field,

Table 4.1 Connection between Packaging Elements at Various Levels in the Hierarchy for ICs and Micromachines

Level	Element	Interconnected by
Level 0	Transistor within IC or resonator in MEMS	IC metallization
Level 1	ICs or discrete components such as a MEMS silicon/glass pressure sensor	Package leadframes (single-chip) or multichip module interconnection system
Level 2	Single- and multichip IC packages or a pressure sensor in a TO header in MEMS	Printed wiring boards
Level 3	Printed wiring boards	Connectors/backplanes (busses)
Level 4	Chassis or box	Connectors/cable harnesses
Level 5	System itself, e.g., a computer (IC-based) or a gas alarm system (MEMS-based)	

Source: Adapted from Lyke, J. C. 1995. In *Microengineering technologies for space systems,* ed. H. Helvajian, 131–80. El Segundo, CA: Aerospace Corporation.[1]

electromagnetic interference, and light, to name just a few items. The package must protect the onboard circuitry while simultaneously exposing the sensor to the effect it measures. For example, in an in vivo integrated sensor, a true hermetic seal is necessary to protect the circuitry from the effects of blood, and the housing needs to be biocompatible as well. The housing also might be equipped with a fluidic manifold, for example, to guide the analyte to the sensor surface. Sometimes the circuit itself can be used to reduce packaging concerns, for example, to transmit data about effects which cannot be screened out, as in the removal of a temperature effect from a pressure signal. But more frequently, integration of electronics with MEMS, especially when dealing with measuring chemical or biological signals,

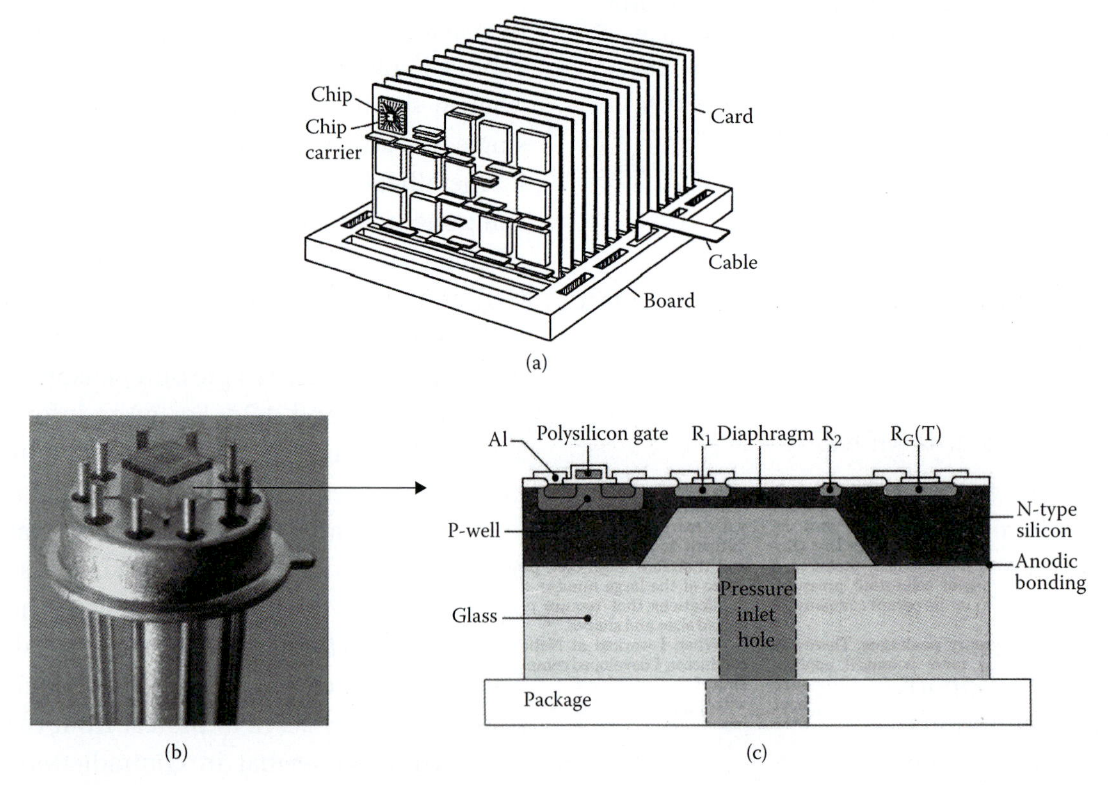

FIGURE 4.2 The packaging hierarchy in the integrated circuit (IC) world and in micromachining. (a) A large electronic unit with ICs. Transistors on the IC are level L0. The IC itself is level L1. The chip carrier is level L2, and the printed wiring board (card) is level L3. The chassis (or box) is level L4. Level 5 (L5), not shown, is the system itself, e.g., a computer. (From Jensen, J. R. 1989. In *Microelectronics processing,* ed. D. W. Hess and K. F. Jensen, 441–504. Washington, DC: American Chemical Society. With permission.[4]) (b) and (c) A micromachined unit: piezoresistive pressure sensor from NEC. L0, diaphragm on die; L1, Si/glass piece; L2, sensor mounted in TO header.

causes more problems than it solves. A package must also provide communication links, remove heat, and enable handling and testing. The material used must be one that will afford physical protection against normal process handling during and after assembly, testing, and prescribed mechanical shock. Chemical protection is also required during assembly and in practical usage. For example, during the process of installing dice on substrates, the package undergoes a series of cleanings and, under normal usage, the package may be exposed to oxygen, moisture, oil, gasoline, and salt water. The package must also be capable of providing an interior environment compatible with device performance and reliability; for example, a high-Q* resonator might need a good vacuum. Physical sensors usually have a physical barrier layer or fluid; for example, a pressure transmission fluid interposed between the sensing Si elements and the environment outside, making the encapsulation problem less severe.

The sensor community is still searching for the perfect way to protect sensors from their environment while at the same time probing that environment. The packaging problem is the least severe for a physical sensor such as an accelerometer, which can be sealed and protected from all chemical environment effects while probing the inertial effects it measures. The problem is most severe for chemical and biological sensors, which must be exposed directly to an unfriendly world. Generalizing, chemical and biological sensors have been orders of magnitude more difficult to commercialize than physical sensors, in no small part because of packaging problems. In the latter case, problems are especially severe with electronic components and sensors integrated on the same side of a single Si substrate. For that reason, chemical and biological sensors are better made with hybrid MEMS. In hybrid MEMS, MEMS and electronics are fabricated in separate processes and put together afterward within a common package. Making a hybrid device or putting the electronics on the other side of the wafer often provides better protection of the electronics from the environment. The challenges in packaging chemical and biological MEMS are further compounded by the fact that these devices usually are less expensive than ICs or physical MEMS sensors and that they are often disposable.

Front-End and Back-End Processes

During the front-end processing phase, an IC or MEMS is created on a silicon substrate (referred to as a wafer). During the back-end processing phase, the IC chips created during the front-end processing phase are encapsulated into packages and thoroughly inspected before becoming completed products. The front-end and back-end processes are preceded by a design phase and by the mask creation in which the function of the IC is defined, the electric circuit is designed, and a mask for IC manufacturing is created on the basis of the design.

In this chapter we are primarily concerned with the more expensive, serial back-end processes. The back-end process used to start after the last of the IC fabrication steps, i.e., to make the interconnections within an underlying IC (from level 0 to level 1 in Table 4.1). Unfortunately, things are not that simple anymore; lately making internal IC connections has become so complicated that even within that process one talks about front end and back end of the line steps. As advanced ICs contain more than a 100 million transistors, the importance of internal IC connections is steadily increasing. As shown in Figure 4.3a, eight metal layers already are quite common. The old standard for making these metal lines was using insulating layers made of silicon dioxide, via formation by plasma etching of holes, covering the whole surface with tungsten (W), polishing the tungsten down to the silicon dioxide, leaving the vias filled with tungsten plugs, and finally sputtering an aluminum layer and selective etching it to leave the connecting Al lines. As we saw in Volume II, Chapter 8, the more advanced process for constructing interconnections in ICs uses the "dual damascene" technique (see Figure 4.3b). This process, as we glean

* As we will learn in Chapter 8, this volume, there are two different common definitions of the Q factor of a resonator (see "Quality Factor" in Chapter 8). One is via the energy storage where the Q factor is given as 2π times the ratio of the stored energy to the energy dissipated per oscillation cycle, or equivalently the ratio of the stored energy to the energy dissipated per radian of the oscillation. The other definition is via the resonance bandwidth in which the Q factor is the ratio of the resonance frequency f_0 and the full width at half-maximum (FWHM) bandwidth Δf of the resonance. Both definitions are equivalent only in the limit of weakly damped oscillations, i.e., for high Q values.

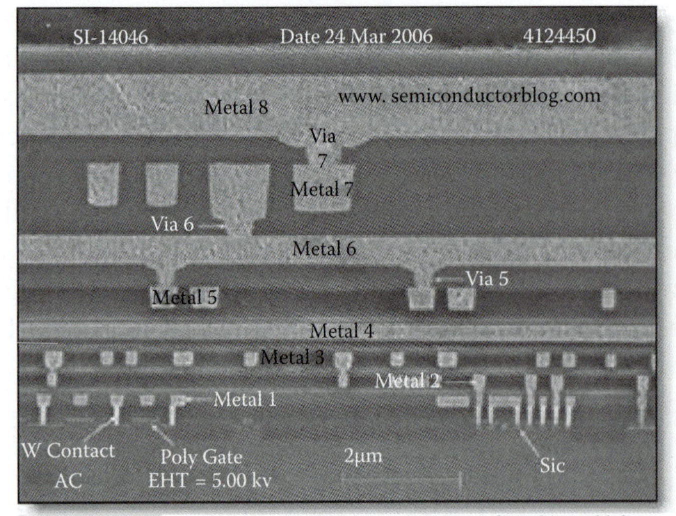

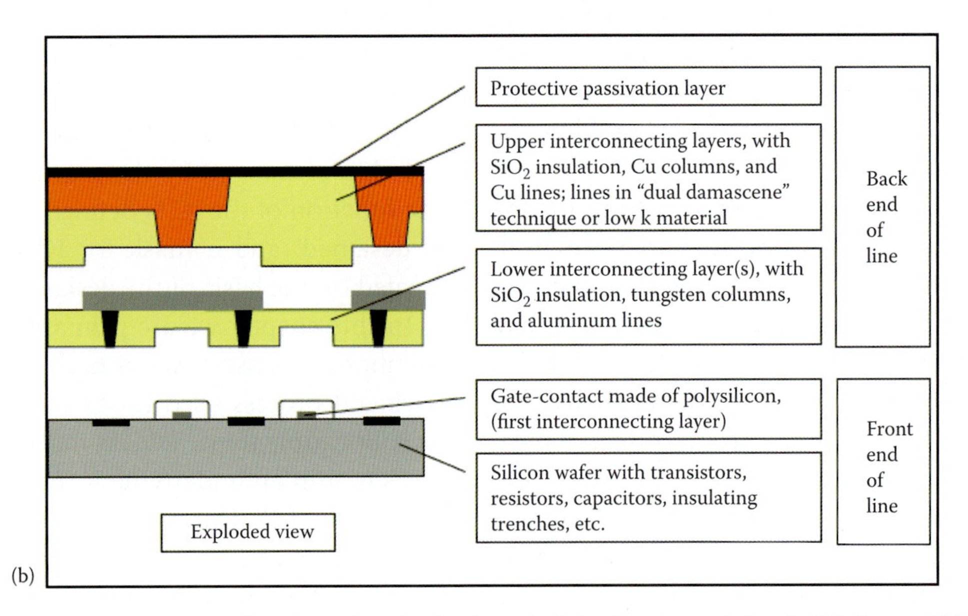

FIGURE 4.3 The back-end process usually takes of at the back end of the integrated circuit (IC) line, which makes the interconnections within an IC (from level L0 to L1). In the case of very complex interconnecting schemes such as the dual damascene, one also talks about back-end and front-end processes with the interconnecting scheme. (a) Eight-metal complementary metal-oxide-silicon chip. (b) Dual damascene technique is used to connect upper interconnecting layers with lower connecting layers (back end of line) and transistors on the wafer (front end of line).

from Figure 4.3b, is itself split up in a back-end and a front end of the line. In the dual damascene process, the last step in the front end of the line is a first interconnecting layer made up of poly-Si for the gate contacts. The back end of the IC interconnection process then starts with an insulating layer, either again made of silicon dioxide or from inorganic or organic materials with lower dielectric constants (so-called *low-k materials*). Next comes the creation of beds for lines and vias in two subsequent photoimaging process steps followed by plasma etching of the insulating layer. In a subsequent process step, the whole surface is covered with copper, and chemical mechanical polishing (CMP) removes the copper down to the top side of the insulating layer (see Figure 4.3b and Volume II, Figure 8.41).

CMP has developed into a very important step in the process chain of wafer fabrication (see Volume II, Chapter 1, Figure 1.66 for a CMP setup). The CMP slurry contains 1) chemistry to modify, e.g., oxidize, the top layer, and 2) abrasive particles to remove the modified material. CMP serves to planarize insulating layers, to remove tungsten or copper, to define vias and lines, and to back grind the wafer to make thin flexible chips.

The resulting flat chip surface now features input/output (I/O) pads that are of rectangular shape and made from the metal of the top layer, i.e., from aluminum or copper (in GaAs chips, gold is used for contacts). These exposed metal surfaces are slightly alloyed to increase hardness and longtime stability. The chip surface, except the I/O pads and the dicing lines, are covered by a passivation made from silicon oxide and silicon nitride. Historically, this passivation was a crucial step in the development of plastic packages. Plastic packaged chips are not as well protected from humidity as chips in hermetically sealed housings.

Testing the outcome of a wafer process depends on hundreds of parameters that all have to be tightly controlled during production. All materials and parameter settings are recorded to allow the back tracing of deviations. Processed chips still on a wafer are difficult to test, because measurements at high frequencies with hundreds of contacts are difficult and expensive. Therefore, only a so-called *static test* at low frequencies is performed to do a first wafer screening, perhaps in a manual probe station with needle tips and a microscope, as shown in Figure 4.4.

The packaged chip—with its built-in (shielded) fan out of the contacts—is the best location in the fabrication chain to do a full testing of the device

FIGURE 4.4 Optical microscope with needle-tip manipulators.

(L2; see below). But before we can mount the chip in its package we need to dice it.

Dicing

One of the final steps in front-end processing of an IC or three-dimensional (3D) microstructure fabrication and the first in back-end packaging is sawing the finished Si wafer into individual "dice." After the wafer is probed (see above; Figure 4.4), it is mounted onto sticky tape and put onto the dicing saw (Figure 4.5). A typical saw blade consists of a thick hub with a thin rim impregnated with diamond or carbide grit. Rotating at a speed of several thousand revolutions per minute, the blade encounters the Si wafer at a feed rate on the order of a centimeter per second, sawing partially or completely through the wafer. After dicing, the elastic plastic sheet can be stretched to distance the chips in case the saw kerf went all the way through the Si wafer.

Except when cutting a silicon-glass bonded wafer combination, standard IC cutting practices regarding surfactants, cleanliness, and blade width/depth ratio apply to MEMS wafers as well. Typical kerf widths are 50–200 μm, while typical roughness along the kerf edge is 10–50 μm. The restrictions of mechanical sawing include this roughness, which precludes the fabrication of smooth structures along the outside edge of the die. In addition, the vibration inherent in the sawing process makes it difficult to securely hold down dice during sawing if these dice are less than 0.5 mm on a side. For smaller dice, micromachining is necessary for edge definition and separation of individual devices. In the latter case, anisotropically etched V-grooves extending almost through the depth of the wafer separate the individual dice, which might be broken apart by applying a small mechanical force. Alternatively, the dice

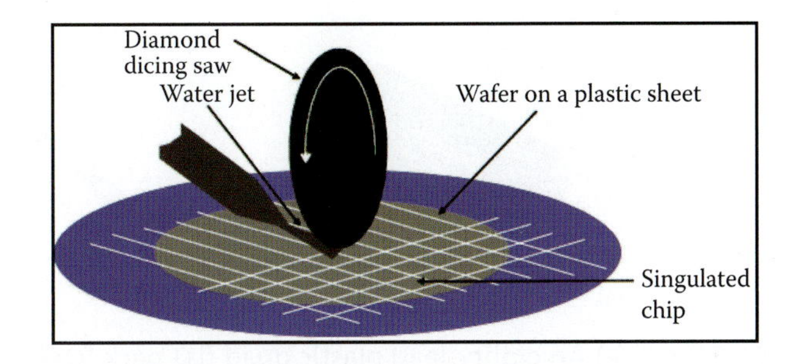

FIGURE 4.5 Dicing operation.

are etched completely free at the moment critical components (e.g., a pressure-sensitive membrane) have reached the specified thickness (see V-groove Thickness Control in Volume II, Figure 4.44).

Most mechanical sensors and actuators are equipped with a bonded cap or cover (Si, glass, etc.), protecting them during dicing. In the case of surface micromachined devices, a final sacrificial release etch (see Volume II, Chapter 7) is performed after the dicing is complete. This procedure ensures that there are no free mechanical structures during the dicing, but it implies also that the microstructures must be freed on individual dice, making for a more costly device. The TI DMD™ arrays described in Volume II, Example 7.2 are handled this way: they are diced first, and then the organic sacrificial layer is etched in an oxygen plasma. The cap approach is not applicable in the case of flow sensors, and chemical and biological sensors or other devices that require direct contact with their surrounding environments.

Level 1 Packaging (L1) and Level 2 Packaging (L2)

Introduction

Level 1 devices are packaged with leadframes (single-chip), transistor outline (TO)-headers, or multichip module interconnection systems, and the resulting devices are then packed onto printed wiring boards (see Table 4.1). Micromachining deviates most dramatically from ICs in the nonstandard nature of packages from level 2 up. In the case of some sensors, standard TO-5 and TO-8 headers still might be useful, but in most cases the package dictated by the specific application is of a much more complex nature.

Single-Device Packaging

A typical package for single chips with 100 and more leads is the quad flat pack (QFP) illustrated in Figure 4.6. The process steps necessary to build such a QFP are the stamping or etching of a leadframe, gluing the chip to the leadframe (also named die bonding), wire bonding the chip to the leadframe, overmolding everything except the outer leads and cutting and bending the leads to gull-wing shape. In the cross-section of the QFP, in Figure 4.6b, the different elements of a QFP are visible: the chip, the leadframe and the glue, the gold wires, the overmolding, and the final form of the leads. This package is now ready to be soldered to a printed wiring board.

Leadframes are stamped or etched strips of metal sheets (Figure 4.7). Chips can be mounted in one or several rows on the large central pads. The outside dimensions of the strips—width, length, and the holes at the edge—are standardized. The basic material is an iron-nickel-alloy. The inner leads, the lands for the gold wire, have to be gold or silver plated. The outer leads are pretinned for the later soldering to the printed wiring board. An auxiliary rectangular line circumventing the main package is needed to seal the overmolding form. This line is stamped out after overmolding.

The die is glued to the large central pad of the lead frame with a conductive epoxy resin. Precise mounting of the chips is required to facilitate the subsequent wire bonding. The glue has to be cured before wire bonding starts. Because standard glues cure in tens of minutes, rather than in seconds, the curing is quite a disturbing factor in a production line. Batches of leadframes have to be placed in a curing oven.

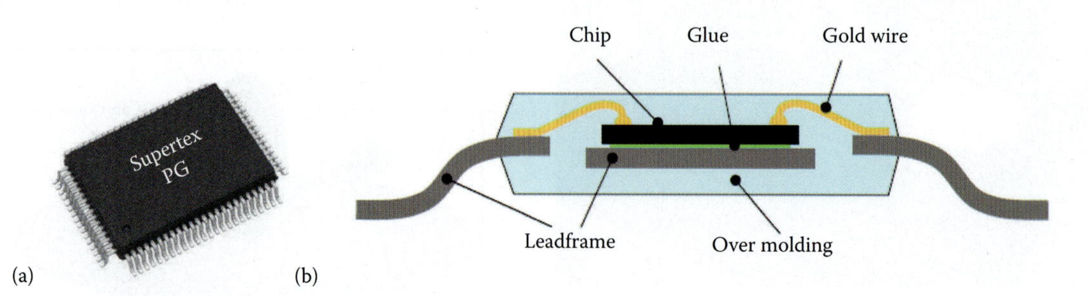

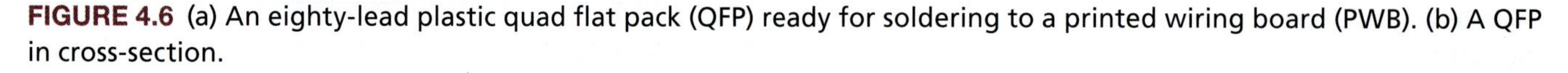

FIGURE 4.6 (a) An eighty-lead plastic quad flat pack (QFP) ready for soldering to a printed wiring board (PWB). (b) A QFP in cross-section.

FIGURE 4.7 Leadframes are stamped or etched strips of metal sheets.

The QFP shown in Figure 4.6 is a relatively old packaging scheme introduced in the 1980s. The dual inline package (DIP), what most people envision when they think of an IC "chip," was developed even earlier (in the 1960s) (Figure 4.8a). In a DIP, the leads are perpendicular to the body of the package, making them ideally suited for insertion mounting onto printed wiring boards (PWBs) using plated through-holes.

A QFP is an example of surface mount technology (SMT). In SMT, the leads do not penetrate the PWB surface, and ICs are mounted, by reflow solder technology, on both sides of the PWB, resulting in higher-density electronics. In the industry, SMT has largely replaced the through-hole technology construction method of fitting components with wire leads into holes in the circuit board. An SMT component usually

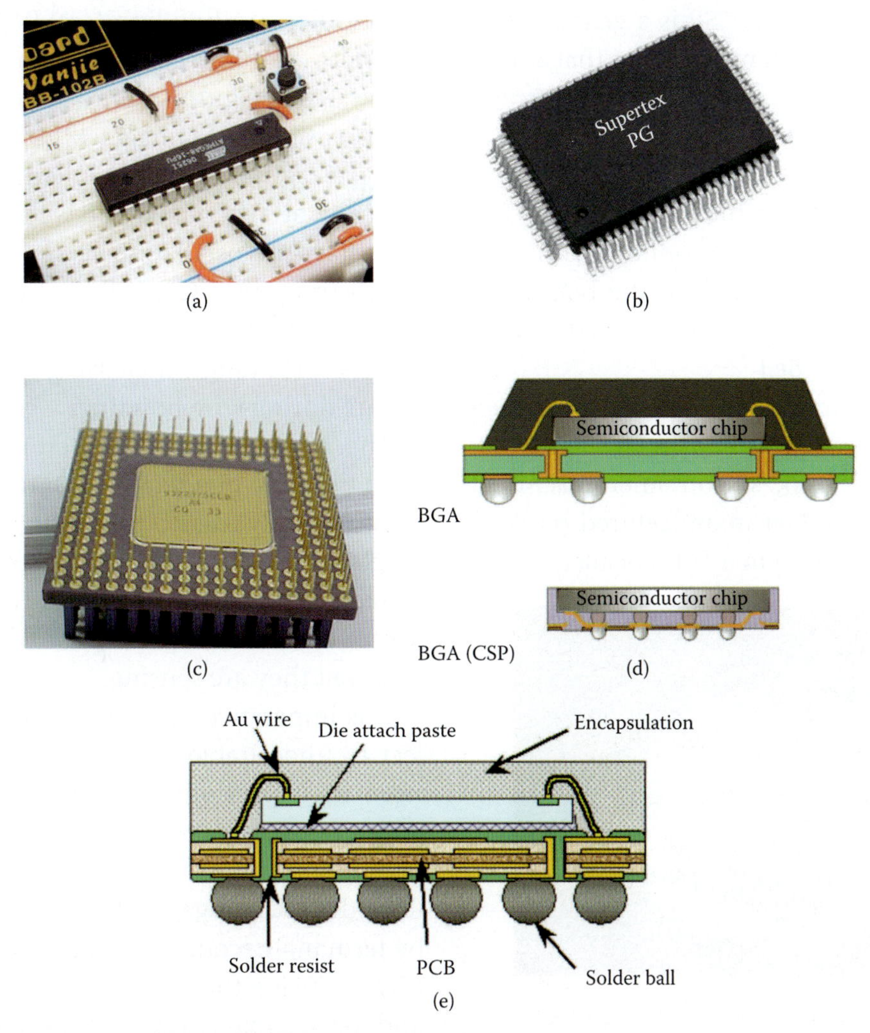

FIGURE 4.8 Packaging options. (a) The dual inline package (DIP), developed in the sixties. (b) The quad flat pack (QFP) was developed in the 1980s. It is a first example of surface mount technology. (c) Pin grid arrays (PGAs) were also developed in the 1980s. (d) Ball grid array (BGA), developed in the early 1990s as a replacement for QFPs, including a BGA implemented as chip scale packaging (CSP). (e) Detail of chip scale package (CSP).

is smaller than its through-hole counterpart because it has either smaller leads or no leads at all. The QFPs (see above and Figure 4.8b) and pin grid arrays (PGAs; Figure 4.8c) were developed in the 1980s. There are leadless or leaded ones and come in plastic or ceramic bodies. With contacts on all four sides, up to ~200 I/Os are possible with QFPs, and PGAs can have up to ~600 I/Os. Ball grid arrays (BGAs) (Figure 4.8d) were developed in the early 1990s as a replacement for QFPs. They feature solder bumps in a matrix on the bottom of the package. They have a higher density, smaller footprint, and shorter electrical paths (i.e., faster signal propagation) but are more expensive than QFPs. BGAs are the natural choice for multichip module (MCM) packaging (see below). Chip scale packaging (CSP) was developed in the late 1990s (see bottom drawing labeled "BGA (CSP)" in Figure 4.8d and Figure 4.8e). CSP is a general term for all semiconductor packaging schemes that aim to be almost the same size as the chip itself (typically about 1.2 times the bare chip). In CSP, the package is an extension of the bare IC and permits easier handling, testing, and assembly. In one version of CSP, the BGA packaging scheme is modified with electrodes that are directly installed on the bottom of the IC. [See bottom drawing labeled "BGA (CSP)" in Figure 4.8d and Figure 4.8e.]

A simple single die mounting method, often used in MEMS, is in a header, as shown in Figure 4.9, where we show the bulk silicon micromachined piezoresistive accelerometer manufactured by TRW NovaSensor (Fremont, CA) in a TO-5 header.

FIGURE 4.9 Commercially available Si accelerometer based on hybrid technology; bulk-micromachined 50-g accelerometer bonded into a TO-5 header. Signal conditioning electronics are on a separate chip. (Courtesy of TRW NovaSensor.)

After dicing a MEMS sensor die, the die attachment step to the header follows. In a typical header, gold-plated pins are hermetically sealed to the header base with glass eyelets, which provide a hermetic feed through with both high electrical and mechanical integrity. To protect the mounted sensor, a metal can is hermetically resistance-welded to the header. Since the thermal expansion coefficient of the package material to which the die is attached typically differs substantially from that of Si and glass, the die attachment must compensate for this. The three methods used the most for die attach are eutectic (Au-Si alloy), epoxy, and silicone rubber, in order of highest to lowest stress.[1,4] The next step in the process consists of the electrical connection to the Si sensor. Gold or aluminum wire is bonded by thermosonic, thermocompression, or wedge-wedge ultrasonic wire bonding (see "Wire Bonding", this chapter). The fine gold wires in the TO-5 package illustrated in Figure 4.9 are barely visible. Wire bonding occasionally damages MEMS components during packaging, as the applied ultrasonic energy, typically at a frequency between 50 and 100 kHz, may start the oscillation of suspended mechanical microstructures. Coincidentally, many micromachined structures have resonant frequencies in the 50–100 kHz range, increasing the risk of structural failure during wire bonding.

The TO-5 packaged accelerometer in Figure 4.9 is an example of a relatively simple packaging solution; typically MEMS packaging is a lot more challenging. Indeed, an accelerometer can be completely shielded from its environment, whereas many other MEMS sensors must be exposed directly to the environment they are sensing.

It is imperative that sensor elements are not subject to undesirable mechanical stresses originating from their packaging structure. Slow creep in the adhesive or epoxy that bonds the silicon die to the package housing, for example, may lead to long-term drift. Corrugated structural members, as used by Jerman,[7] decouple the sensor from its encapsulation, reducing the influence of temperature changes and packaging stress. Corrugated members have been implemented, for example, in single-crystal silicon, polysilicon, and polyimide. The technique was described in Volume II, Chapter 7, as one of the

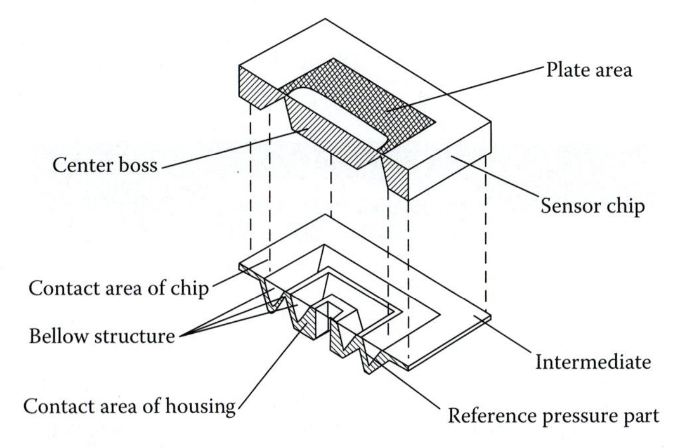

FIGURE 4.10 Stress release bellows. A corrugated intermediate Si piece decouples the sensor from its housing. (From Offereins, H. L., H. Sandmaier, B. Folkmer, U. Steger, and W. Lang. 1991. In *6th International Conference on Solid-State Sensors and Actuators [Transducers '91]*, 986–89. San Francisco. With permission.[10])

methods to control stress in micromachined structures. The stress-decoupling, corrugated bellows either surround the sensor structure itself[8,9] or the sensor chip is mounted via an intermediate with a bellow structure to the housing.[10] The latter implementation is sketched in Figure 4.10.

Thermal management in MEMS involves cooling heat-dissipating devices, especially thermal actuators, but also includes understanding and controlling the sources of temperature fluctuations that may adversely affect the performance of the device. Thermal modeling must include both the die and its package. To facilitate heat flow from die to package, a die-attach scheme must be used that exhibits high thermal conductivity.

The last assembly step, after die attachment and electrical contacting, involves the protection of the silicon die and electrical leads. This is especially important for sensors and actuators that are to be used in intimate contact with harsh environments. Various methods are available, including the following:[1,4]

1. Vapor-deposited organics for mildly aggressive environments {e.g., 2–3-μm thick low-pressure chemical vapor-deposited [LPCVD] parylene [poly(*p*-xylene)]}. A negative aspect of this form of passivation is that the added layer often makes the die too stiff for accurate sensing or actuating functions.
2. Silicone oil over the die for medium-isolated pressure sensors.
3. Coating of the die surface with soft substances such as silicone gel. Hardening of the gel over time is a major problem. Silicones do not form a moisture barrier, and the exact mechanism by which they protect the die as a surface coat is not understood.[11]
4. SiC (e.g., deposited with plasma-enhanced chemical vapor deposition) for harsh environments.[12]
5. A plastic or ceramic cap for particle and handling protection.
6. A welded-on nickel cap with pressure port.

Multichip Packaging

A multichip module (MCM) is a chip package that contains two or more bare chips mounted close together on a common substrate. Shorter tracks between the individual chips increase performance and eliminate much of the noise that external tracks between individual chip packages pick up. There are four main categories of MCMs based on the substrate used:

1. Laminated (MCM-L): Constructed of plastic laminate-based dielectrics and copper conductors using PWB technology. These are the lowest cost MCMs.
2. Ceramic (MCM-C): Constructed on cofired ceramic or glass-ceramic substrates using thick-film (screen printing) technologies.
3. Deposited (MCM-D): Uses a dielectric layer over a ceramic, glass, or metal substrate, and thin film interconnects are created on the dielectric layer. These are the highest-performance MCMs.
4. Si-based (MCM-Si): Uses a silicon substrate with tracks created in the silicon like regular ICs. Transistors can also be formed in the substrate.

ICs and micromachined chips may be packed laterally or they may be stacked on top of one another. In the latter case, individual dice, blocks of dice, and entire wafers may be stacked with electrical interconnections either made by wires (see Figure 4.11)

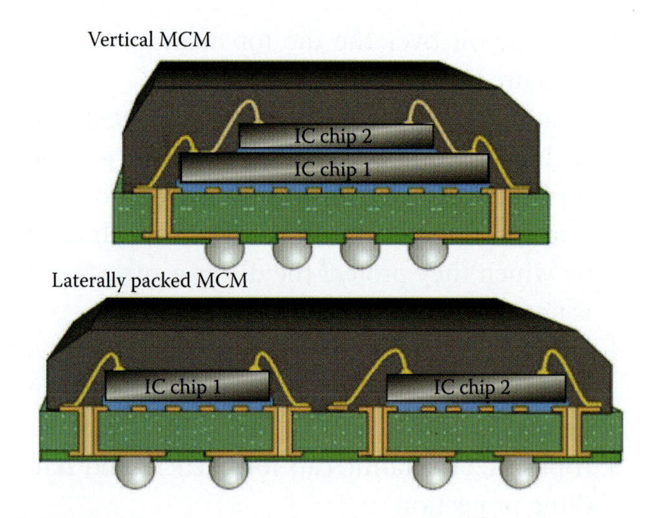

FIGURE 4.11 Schematic of a two-chip MCM.

or interconnects running vertically from plane to plane.[13] A commercial example of a four-chip MCM is shown in Figure 4.12.

MCMs increase the interconnect density on printed circuit boards and speed up the processing time. Multichip modules are an important supporting technology for MEMS, as most MEMS are better partitioned into separate sensing and electronic parts rather than being built monolithically in Si. MCM also allows complementary metal-oxide-silicon (CMOS) and bipolar technologies to be used in the same package.

Instead of the wire bonds shown in the MCMs in Figure 4.11, flip-chip bonding often is used to make contact to the chip. In flip-chip bonding, as its name implies, the die is bonded, facedown, on its substrate without the use of connecting wires (see "Electrical Connection Methods" and Figure 4.13). Electrical contacts are made directly between plated solder bumps on bond pads on the die and metal pads on the packaging substrate. The intimate spacing between the die and the substrate (<200 μm) presents two major advantages. With wire bonding, the pads on the die must always be positioned on the periphery to avoid crossing wires, but with flip-chip bonding, the bond pads may be placed over the entire die. This results in a significant increase in density of possible I/O connections (up to 700 simultaneously I/Os). Second, the effective inductance of each interconnect is minuscule because of the short height of the solder bump compared with that of a gold wire. These are the reasons why the integrated circuit industry has adopted flip chip for high-density, fast electronic circuits. Flip-chip mounting is equally attractive to the MEMS industry. Again, the ability to closely package a number of distinct dice on a single-package substrate with multiple levels of embedded electrical connections represents a major advantage. This approach may be used, for example, for bonding to electrically connect and package three dice, including an accelerometer die, a yaw-rate sensing die, and an electronic application-specific integrated circuit (ASIC) onto one ceramic substrate to build a fully self-contained navigation system. A similar unit could be built with wire bonding, but it would be significantly larger and its reliability would be questionable given the large number of "loose" gold wires within the package.[14] In the next section we explain how a typical flip-chip process works.

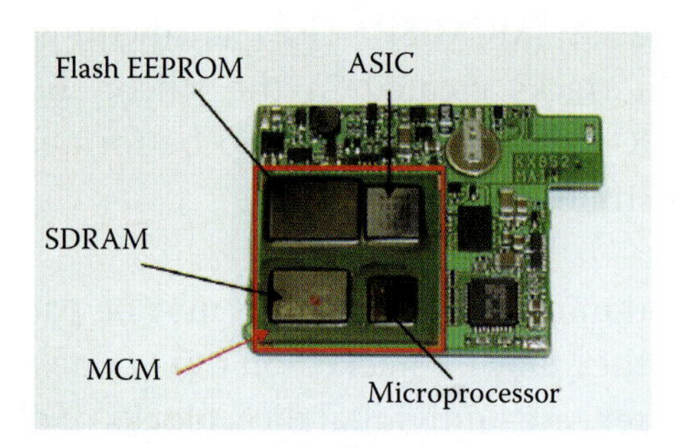

FIGURE 4.12 A 2-mm-thick MCM holding four ICs used in Casio's Exilim digital camera.

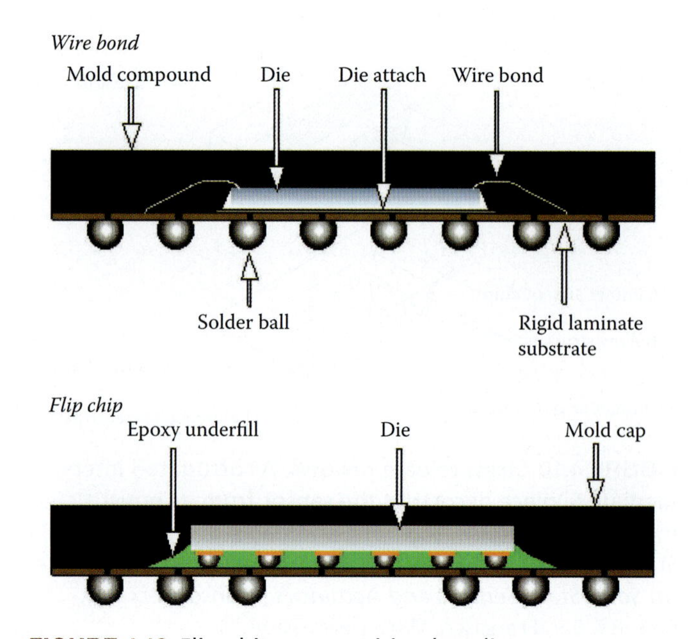

FIGURE 4.13 Flip-chip versus wiring bonding.

Electrical Connection Methods

Introduction

The three most used electrical connection methods between a chip and a substrate are illustrated in Figure 4.14. They are wire bonding, tape automated bonding (TAB), and flip-chip bonding.

Wire Bonding

The pads on a chip and the pads on a substrate (e.g., a leadframe, but also a printed wiring board or ceramic substrate) often are interconnected by wire bonding. A gold, copper, or aluminum wire is bonded between each bonding pad of the chip and the corresponding pad on the substrate. There are two major types of wire bonding: ball-wedge and wedge-wedge bonding. In gold ball-wedge bonding, a gold wire is fed through a ceramic capillary, as shown in Figure 4.15. The protruding wire end is melted by an electrical spark to form a ball, which is bonded to the first pad. This operation requires both ultrasonic excitation of the capillary and heat application to the substrate (200°C). The capillary guides the gold wire now to the second pad, where it forms a wedge and cuts the

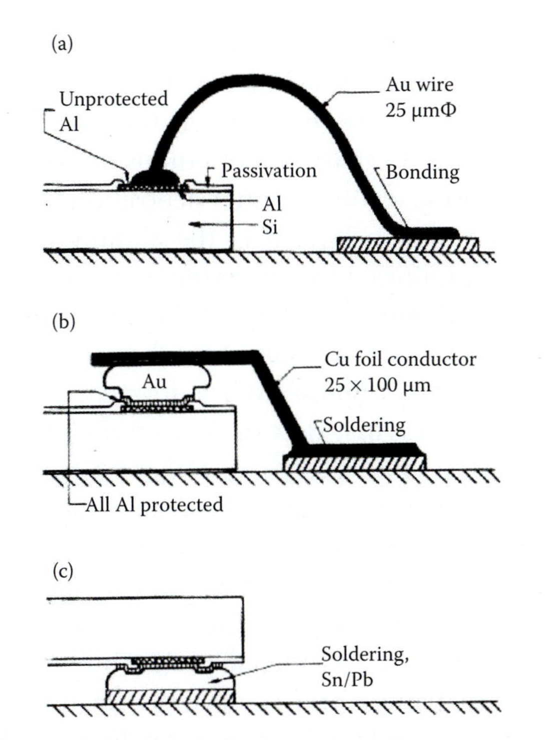

FIGURE 4.14 The three most used electrical connection methods between a chip and a substrate: (a) wire bonding; (b) tape automated bonding (TAB); (c) flip-chip bonding.

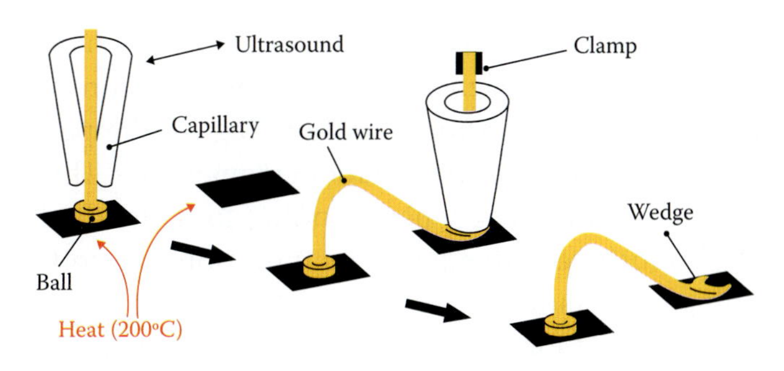

FIGURE 4.15 Gold ball-wedge bonding.

wire in a single step. The wire has to be clamped to cut it, and ultrasound action and heat application also are necessary to achieve good adhesion of the wedge to the second pad. Copper ball-wedge bonding is also possible when the copper wire is flushed by an inert gas during ball formation. Copper wires can be used instead of gold wires. They fit well to the copper pads of the most recent ICs, and their electrical conductivity is higher. Copper wires also are stiffer than gold wires and exhibit less wire sweep during overmolding. After the mounting and bonding, process steps to establish corrosion protection and mechanical protection of the chip are needed, either by hermetic sealing of the assembled board or locally by "glob top" sealing with a droplet of sealing material, e.g., epoxy. Wire bonding of bare chips is used in hybrids, most often with a ceramic substrate. Chip on board is also used with glass/epoxy laminates in some consumer applications (pocket calculators, watches, smart cards, etc.), and occasionally in professional applications.

An aluminum ball cannot be formed as easily in air as it can with gold because aluminum oxidizes immediately in air (much faster even than copper). Another technique, wedge-wedge bonding, has to be used as illustrated in Figure 4.16. Aluminum is

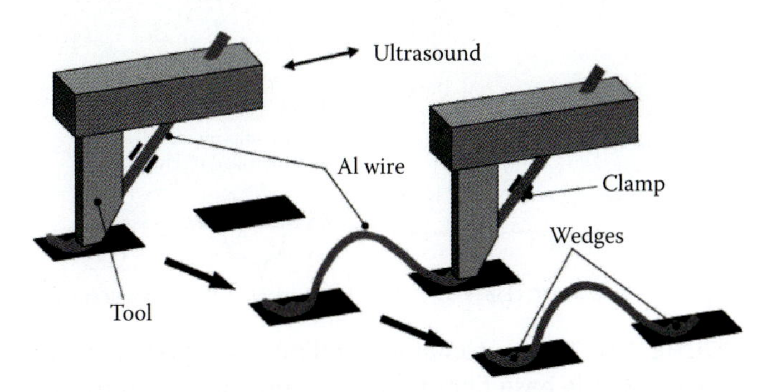

FIGURE 4.16 Wedge-wedge bonding of aluminum wire.

a soft metal, but its oxide at the surface is very hard and has to be broken by the ultrasound action of the wedging tool. Aluminum bonding can be performed with the substrate at room temperature. This is a huge advantage when bonding to a standard PWB substrate with a low glass transition, i.e., softening temperature (T_g) of around 150°C. The wedging tool used for aluminum bonding is directional, and it has to be rotated for different bonds in different directions. It is also larger than the capillary in ball-wedge bonding, so it needs more room on its back side and cannot reach into deep holes between components.

Tape automated bonding (TAB) is illustrated in Figures 4.14b and 4.17. It uses a leadframe made from a fine film of copper conductor on a dielectric film such as polyimide for interconnections between chip and substrate. The conductor patterns in the leadframe are imaged using a photolithography process from a continuous copper layer on the dielectric carrier film. The TAB leads connecting the die and the tape are the inner lead bonds, while those that connect the tape to the package or to external circuits are known as the outer lead bonds (see Figure 4.17). The mounting is done such that the bonding sites on the die, usually in the form of bumps or balls made of gold or solder, are connected to fine conductors on the tape, which provide the means of connecting the die to the package or directly to external circuits (Figure 4.17). To facilitate the connection of the die bumps or balls to their corresponding leads on the TAB circuit, holes are punched on the tape where the die bumps will be positioned. The conductor traces of the tape are then cantilevered over the punched holes to meet the bumps of the die. The leads are soldered to the substrate with a specialized tool called *thermode* ("impulse soldering"). Finally, the bare chip is encapsulated with epoxy or plastic (glob topped). The main advantage of TAB is the tight pitch of the circuit. Line pitches of 45 μm are achievable (22.5-μm lines/spaces). This allows for high-density circuits. TAB is in widespread use in liquid crystal displays, electronic watches, and other high-volume consumer products. Other TAB applications are VLSI circuits with a high lead count and other high-end needs.

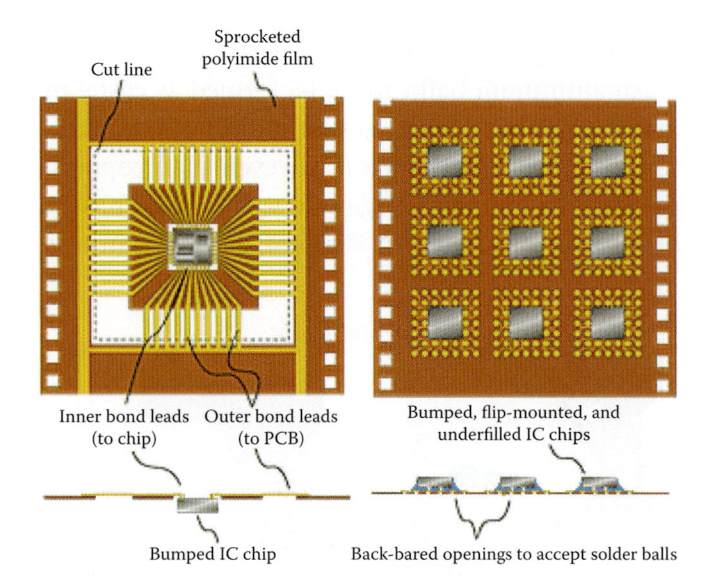

FIGURE 4.17 Tape automated bonding (TAB) takes advantage of flexible base films to make interconnections from the chip to a printed wiring board.

Flip-Chip Bonding

The concept of the flip-chip bonding process, with the semiconductor chip assembled facedown onto a substrate, is ideal for limited size applications because there is no extra area needed for contacting on the sides of the component (see Figure 4.13). Flip chip is carried out by mounting the chip upside down on the substrate, after having deposited solder bumps on all interconnection pads of the chip. The chip is then soldered directly onto the substrate, which requires close thermal matching of the chip and substrate. To make solder bumps on the die, a typical process involves sputter deposition of a thin titanium adhesion layer over the aluminum bond pad metal, followed by sputtering of a copper seed layer. The titanium and copper are patterned with photoresist to form a pedestal for the solder bump. Before the solder bump itself is formed, a thicker layer of copper is first electroplated on the patterned pedestal. The solder bump, typically a tin-lead alloy, is electroplated over the electroplated copper. On the substrate side, solder paste is screen-printed in patterns matching the landing sites of the solder bumps on the die. Pick-and-place robots position the die, facedown, and align the solder bumps to the solder-paste pattern on the substrate. Heating in an oven or under infrared radiation melts the solder, and the bond is created. Often an underfill step is required to fill the void space between the die and the package substrate with epoxy. The underfill not only adds mechanical strength to the assembly but also protects the connections from environmental hazards. Underfill adhesive may be dispensed along one or more edges of the die and is drawn into the space under the die through capillary action. Heat-curing the underfill adhesive completes the assembly process.

Conductive adhesives are a viable alternative to tin-lead solders for flip chip joining. Conductive or nonconductive adhesives may attach a gold stud bumped chip to a substrate. The conductive adhesives used are either isotropic, conducting in all directions, or they are anisotropic, conducting in a preferred direction only. In Figure 4.18 we show a schematic drawing of flip chip bonding with isotropically conductive adhesives (ICAs) and anisotropically conductive adhesives (ACAs). In either case, the adhesives act as the electrical and mechanical joining agents.

Isotropic conductive adhesives consist of an adhesive binder filled with conductive particles that are normally in contact with each other and provide a low electrical resistance in all directions. Generally, the polymer resin is an epoxy and the conducting particles are silver. The adhesive may be dispensed by stencil printing onto the substrate bond pads, or the bumped die may be dipped into a thin layer of the adhesive, coating only the bumps. Stencil printing of the conductive adhesive limits minimum pad pitch to about 90 μm. Since dipping places adhesive only on the bump surface, the minimum bump spacing is smaller than for stenciling, to pad pitches of 60 μm or less. After the isotropic conductive adhesive is heat cured, a nonconducting underfill adhesive is applied to completely fill the under-chip space.

Anisotropically conductive adhesives are pastes or films that are filled with metal particles or metal coated polymer spheres to a concentration that ensures electrical insulation in all directions before bonding, but after bonding the adhesive becomes electrically conductive in the *z*-direction. Anisotropically conductive adhesives have the ability to connect finer-pitch devices.

Also, nonconductive adhesives can be used for flip chip bonding, in this case the joint surfaces are forced into intimate contact by the adhesive between the component and substrate.

In general, the pad pitch of the die is finer with adhesive flip-chip than possible with solder flip-chip. The cleaning step is not as rigorous as with solder-based systems; it eliminates process steps such as coating the interconnect pads with solder and eliminates the use of lead. On the downside, the electrical conductivity of ICAs and ACAs is inferior to that of tin-lead solders. ICAs have about 1

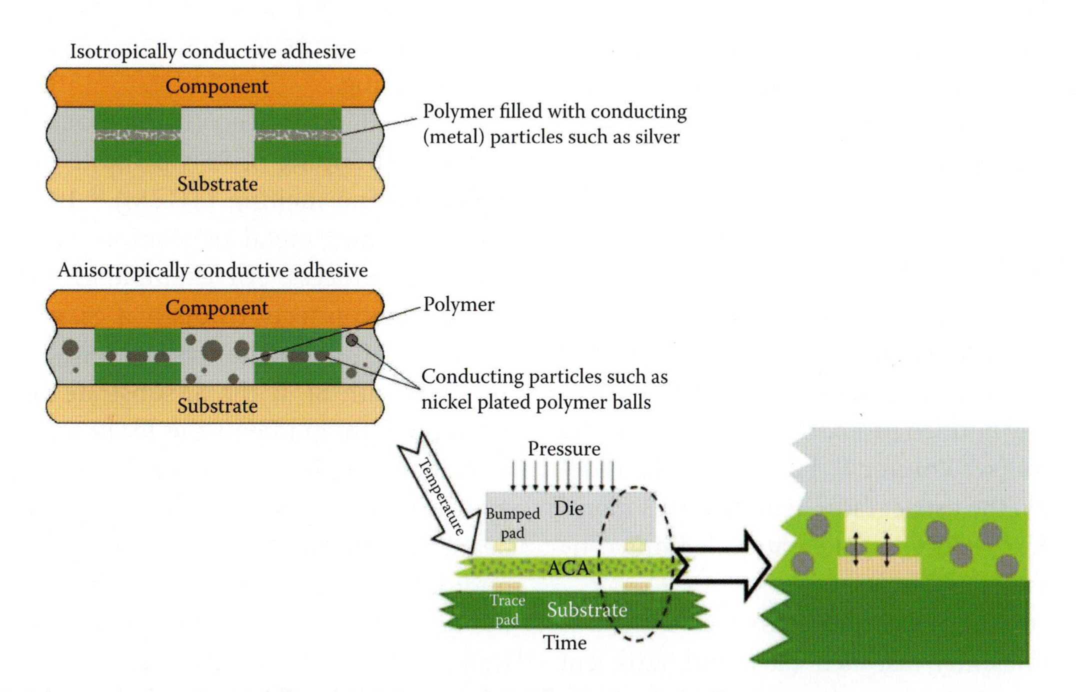

FIGURE 4.18 Schematic drawings of flip-chip joints made with conductive adhesives.

magnitude greater volume resistivity than eutectic tin-lead solder.

Flip-chip technology gives the highest packaging density of the different interconnection methods described here. It gives excellent electrical characteristics and high reliability when properly used, but it is a very demanding technology both to establish and operate. Initial costs are high, favoring high-volume or high-end applications. The performance in high-frequency applications is superior to other interconnection methods, because the length of the connection path is minimized. Also, reliability is better than with packaged components because of the decreased number of connections. Flip-chip technology is potentially cheaper than wire bonding because bonding of all connections takes place simultaneously, whereas with wire bonding one bond is made at a time. In practice, however, this price benefit is not always achieved because of immature processes, e.g., the cost of die bumping with current processes can be significant, especially in low volumes. Today flip-chips are widely used for watches, mobile phones, portable communicators, disc drives, hearing aids, LCD displays, automotive engine controllers, and mainframe computers.

Flip-chip mounting is not compatible with all MEMS devices. For example, there is a risk of damaging the thin diaphragm of a pressure sensor during a flip-chip process.

In what follows we review the different options for through-wafer electrical interconnects. Most methods work for Si only, but others may also be used to form vias in polymers, glass, and ceramics.

Making Vias through Silicon

At first glance, chemical etching is the simplest way to form holes (vias) through Si. The author applied this method for the fabrication of an electrochemical sensor array bonded to a bottom Si chip carrying the electronics (see Volume II, Figures 4.92 to 4.96). Unfortunately, the aspect ratio, length versus diameter, is low, less than 1, so integration is limited by the space between the holes.

With dry etching of vias, problems used to include a lack of sufficiently fast etch rates and sufficient selectivity over the mask. However, advances in high-density plasma sources now allow for etch rates up to 4 μm/min, a selectivity of Si versus SiO_2 of over 150 (70:1 for resists), and aspect ratios of 30:1.[15,16] Along this line, Chow et al. present a through-wafer electrical interconnect fabrication process compatible with standard CMOS processing.[17] In this pre-CMOS process step, a high-density SF_6 plasma (Bosch process; see Volume II, Chapter 3) is used to etch circular holes of 18–25 μm in a 400-μm-thick wafer. By etching from each side, the aspect ratio capability required of the etching process is halved.

Another via alternative is laser drilling. With laser drilling, an aspect ratio of 1:50 can be reached, and relatively high drilling speeds (e.g., 10 holes per second) are possible at a power density of 10^{11} W/cm^2.

Ultrasonic drilling makes for cleaner vias than laser machining, which, unless there is a reactive gas involved or a femtosecond laser, leaves debris on via rims. The ultrasonic technique is limited to "large" diameter holes (100 μm and up). Laser-machined holes are compared with ultrasonically drilled holes in Volume II, Figure 6.21. In all the techniques for making a through-wafer electrical interconnect thus far described, after the hole is created, a metal or highly doped polysilicon must be deposited for electrical connection. In the case of a conductive substrate, the walls of the hole must be passivated first, for example, by growing a thermal oxide. Most methods described to deposit the metal in the vias combine sputtering and electroless or electroplating.

Localized, very deep (throughout the thickness of a wafer) doping of Si is possible by temperature gradient zone melting (TGZM).[18] In this case, via formation and metal deposition become one and the same process. An example is the fabrication of Al interconnects through a Si wafer. At sufficiently high temperatures, aluminum will form an alloy with Si. If the Si substrate is subjected to a temperature gradient, the molten alloy zone will migrate to the hotter side of the wafer. In practice, Al thermomigration in Si starts with electron beam deposition of a thick layer of Al (>5 μm) and photolithographically defining the aluminum into the desired pattern on one side of the Si wafer. The wafer is then radiatively heated from the other side to temperatures considerably higher (1000–1200°C) than the eutectic temperature to form the molten zone. The one-sided heating imposes a thermal gradient of up to 50°C/cm across the molten zone, resulting in a highly aluminum doped (2×10^{19} cm^{-3}) zone of single-crystal Si. Because of the speed (the zone moves through the silicon on the order of 10 μm/min), wafers can be doped through their thickness in minutes. With TGZM, the concomitant side diffusion is a few micrometers compared with millimeters for solid-state diffusion.

In principle, this technique offers very exciting opportunities for novel 3D structures, as connections can be made front to back on a Si wafer. However, the lines through the silicon, being active junctions, are light sensitive, and the Al on the exit side can turn very rough. Also, because the process is so extreme, it needs to precede the implementation of any other structure on the wafer.

Bonding of Layers in MEMS

Introduction

Wafer bonding is a technology that has found widespread use in IC as well as MEMS fabrication. It allows the permanent bonding of two wafers. Applications include packaging, fabricating 3D structures, and forming multilayer devices. In MEMS, more often than with ICs, one needs to bond a variety of materials together, including Si to a glass pedestal (e.g., in pressure sensors), Si to Si (in many multistack MEMS devices), and polymer to polymer (e.g., in fluidic devices).

Assisted Thermal Bonding or Anodic Bonding

Field-assisted thermal bonding, also known as anodic bonding, electrostatic bonding, or the Mallory process, is commonly used for joining glass to silicon. The main utility of the process stems from the relatively low process temperature (the temperature limit for IC processed Si substrates is about 450°C). Since the glass and Si remain rigid during anodic bonding, it is possible to attach glass to Si surfaces, preserving etched features in either the glass or the silicon. This method is mostly applicable to wafer scale die bonding (L1).

A bond can be established between a sodium-rich glass, say Corning #7740 (Pyrex), and virtually any metal.[19] Besides Pyrex, Corning #7070, soda lime #0080, potash soda lead #0120, and aluminosilicate #1720 are suitable as well.[20] In the case of Si, Pyrex is most commonly used. Bonding can be accomplished on a hot plate in atmosphere or vacuum at temperatures between 180°C and 500°C. Typical voltages, depending on the thickness of the glass and the temperature, range from 200 to 1000 V. The operating temperatures are near the glass-softening point but well below its melting point, as well as below the sintering temperature of standard AlSi metallization. At the most elevated temperatures, the wafers are bonded in 5–10 min, depending on voltage and bonded area.[20] Compared with Si fusion bonding (see below), anodic bonding has the advantage of being a lower-temperature process with a lower residual stress and less stringent requirements for the surface quality of the wafers. Figure 4.19 represents a schematic of an anodic bonding setup and an image of the actual operation. Generally, one places a glass plate on top of the Si wafer and makes a pinpoint contact to the uppermost surface of the glass piece, which is held at a constant negative bias with respect to the electrically grounded silicon. The bonding is easy to follow: looking through the glass, the bonded region moves from the contact cathode pinpoint outward and can be detected visually through the glass by the disappearance of the interference fringes. When the whole area displays a dark gray color, the bonding is complete.

A constant current, instead of constant voltage, could also be used but is avoided, since dielectric breakdown may occur after the bonding is complete and the interface becomes an insulator (see bonding mechanism below). The contacting surfaces need to be flat (surface roughness $R_a < 1$ μm and warp/bow < 5 μm) and dust-free for a good bond to form. The native or thermal oxide layer on the Si must be thinner than 200 nm. The height of metal lines crossing under the bonded area should be less than 600 Å, or 1000 Å for hermetic sealing. The thermal expansion coefficients of the bonded materials must match in the range of bonding. In Figure 4.20, we show the thermal expansion coefficient of Si, a borosilicate

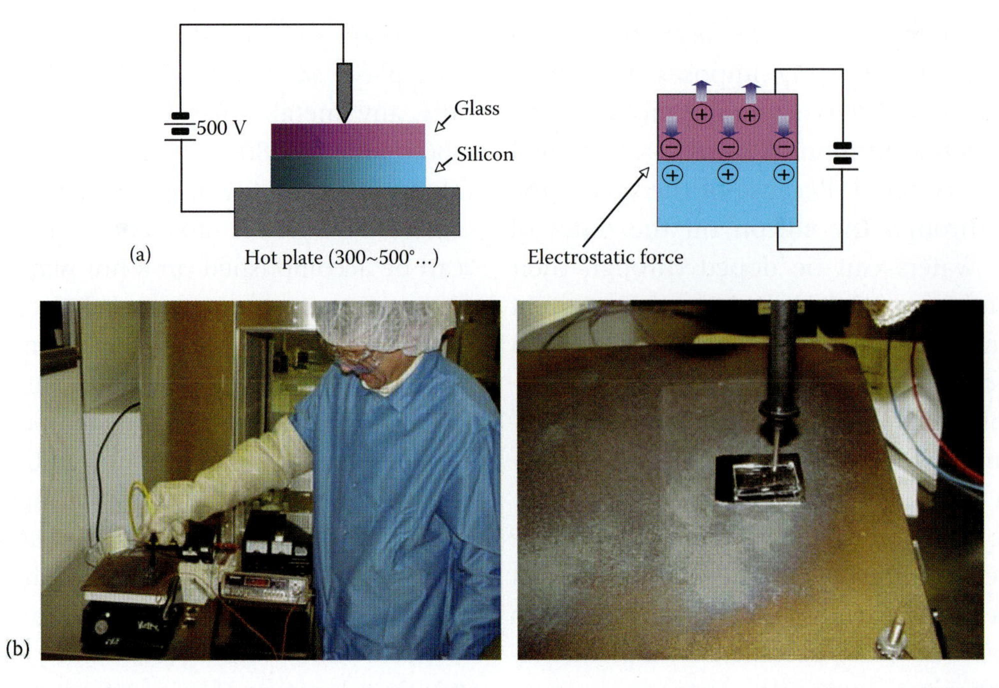

FIGURE 4.19 (a) Principle sketch of anodic glass-to-Si bonding. Control parameters are temperature (300–400°C), bias voltage (700–1200 V), time (~2 min), and materials (glasses, Si, SiO_2). (b) Actual anodic bonding process on a simple hot plate.

glass, and a Pyrex-like glass (HOYA's SD-2) as a function of temperature.[21] The curves of borosilicate glass and silicon single-crystal wafer cross at about 240°C. When anodic bonding is performed at 400°C, the difference of the expansions at high temperature creates residual stress in the Si chip during cooling to room temperature.

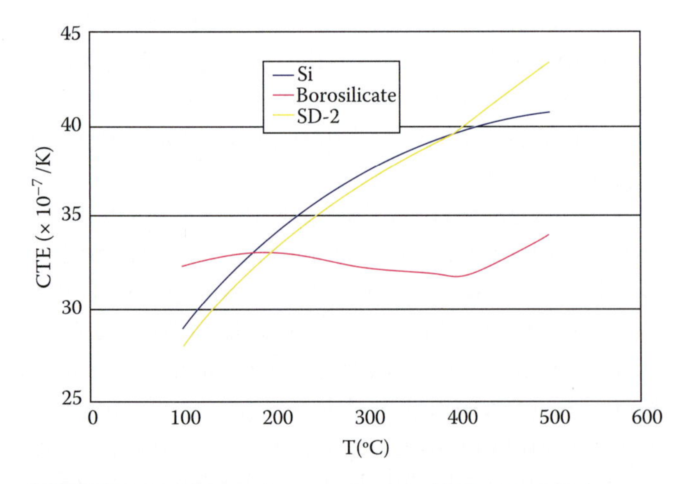

FIGURE 4.20 Thermal expansion coefficients of Si, borosilicate, and a Pyrex-like glass, HOYA's SD-2, a substrate engineered to minimize the distortion or bowing effect caused by the thermal mismatch between the two wafers. CTE, coefficient of thermal expansion. (From Peeters, E. 1994. *Process development for 3D silicon microstructures, with application to mechanical sensor design*. PhD thesis, Catholic University of Louvain, Belgium.[21])

Above 450°C, the thermal properties of the materials begin to deviate seriously; therefore, the process should be limited to 450°C. One also expects that Si would be under compression for seal temperatures below 280°C and under tension for temperatures in excess of 280°C.[21] Wafer curvature measurements indicate, however, that the transition from concave 7440 glass/Si sandwiches (Si under compression) to convex sandwiches (Si under tension) lies around a seal temperature of 315°C.[21,22] This indicates that other non-negligible, stress-inducing effects add an additional compressive component. As we learned before, for most applications tensile stress is preferred over compressive stress, and a considerable safety margin toward higher bonding temperatures must be respected to avoid buckled Si membranes and bridges.

The anodic bonding mechanism is not yet completely understood. Electrochemical, electrostatic, and thermal mechanisms and combinations thereof have been suggested to explain bond formation, but the dominant mechanism has not been clearly defined. It is suggested that at elevated temperatures, the glass becomes a conductive solid electrolyte, and the bonding results through the migration of sodium (Pyrex contains approximately 3–5% sodium) toward the cathode (see Figure 4.19). As it moves, it leaves a space charge (bound negative charges) in the region of the glass/silicon interface. Most of the applied voltage drop occurs across this space charge region, and the high electrical field between the glass and Si results in an electrostatic force, which pulls the glass and Si into intimate contact. During this charging process, the electric field is high enough to allow a drift of oxygen to the positive electrode (Si) reacting with silicon and creating Si-O bonds. The elevated temperatures and high electric field thus result in covalent bonds forming between the surface atoms of the glass and the silicon. A good quantitative discussion on the many important effects in anodic bonding is by Anthony.[23]

Field-assisted bonding has also been applied to bond GaAs to glass. Corning #0211 is used at 360°C, and a bias of 800 V is applied for 30 min to complete the bonding process. It is well known that GaAs forms very poorly adhering oxides, leading to poor anodic bonding; prebaking of the glass at 400°C for 15 h in a reducing atmosphere (H_2 and N_2) is reported to lead to better bonding.[24] Von Arx et al. bonded glass capsules to a smooth polysilicon surface to form a hermetically sealed cavity large enough to contain hybrid circuitry of a biocompatible implant.[25]

The high electrical field and the migration of sodium make anodic bonding of glass plates to Si a rather difficult technology. The mismatch in thermal expansion coefficient between the glass and the Si causes both thermally induced and built-in mechanical stress. In addition, the viscous behavior of the glass results in degraded long-term stability of the components. As a result of these problems, several modifications of the basic technology have been introduced (see below). A typical commercial instrument for anodic bonding is from Electronic Visions Co.'s 500 Series-Wafer Bonding Systems (e.g., the EV560).

Field-Assisted Thermal Bonding Modifications

The pinpoint method for anodic bonding, as illustrated in Figure 4.19, requires a very high bias voltage and a long period of time to bond areas far removed from the cathode point, since the electrical field in the glass substrate diminishes fast as the distance from the pinpoint cathode increases. At NEC, a Ti mesh bias electrode is deposited over the whole glass wafer to accomplish faster bonding. Because of the mesh assistance, the whole wafer may be Si bonded at 400°C and 600 V in less than 5 min, compared with over an hour at the same temperature and voltage without the mesh.[26]

Another modification of anodic bonding by Sander[27] involves deposition of intermediate layers of Si dioxide and aluminum to screen the underlying Si from harm from the high electrical fields. First, Si dioxide is thermally grown on the Si surface. Then, a layer of aluminum is deposited on the oxide surface. Finally, a piece of glass is bonded to the aluminum. This technique produces a good hermetic seal, but the soft aluminum may be expected to creep after bonding, producing drift in the sensor output. In addition, the aluminum is not corrosion resistant for in vivo applications, so that the bond area may corrode rapidly. A similar modification of glass-to-silicon bonding also uses a sandwich structure, but a layer of polycrystalline silicon (polysilicon) is deposited on the oxide surface instead of aluminum, and a piece of glass is bonded to the polysilicon (see also Von Arx et al.[25]).

It is also possible to anodically bond two Si wafers with a thin intermediate sputtered[28,29] or evaporated[30] borosilicate glass layer (4–7 μm thick). Because two Si wafers are used, there are fewer problems associated with mismatch in thermal expansion coefficient. In addition, since the glass layer is very thin, the voltage necessary for the bond is usually no more than about 50–100 V. Using a thin intermediate film makes the devices akin to a monolithic structure. Residual stresses and thermal expansion coefficient mismatches only have minor effects on device performance. The surfaces of the Si to be bonded should be polished, and one of the two wafers should be coated with the thin glass layer. This intermediate layer may be sputtered from a Pyrex target, resulting in a rather slow process. Hanneborg et al., for example, report a growth rate of 100 nm/h at a pressure of 5 mTorr argon and sputtering power of 1.6 W/cm^2.[29] The films, as deposited, are highly stressed and must be annealed before bonding at the annealing point of 565°C in a wet oxygen atmosphere.[20] With a 4-μm-thick sputtered glass layer, 50 V and a temperature of 450–550°C suffices to create a hermetic bond. With sputtered layers below 4 μm thick, a thermally grown SiO_2 must be included between the glass and the silicon to avoid electrical breakdown during anodic bonding. The bond strength with Pyrex as an intermediate layer, measured by pulling the wafers apart, ranges from 2 to 3 MPa. As the Pyrex thin films etch very slowly in KOH, they can be used as a mask material for anisotropic etching.[31] Esashi et al.[32] report room temperature anodic bonding using a 0.5–4-μm-thick film sputtered from a Corning glass #7570 target. This glass has a softening point of 440°C compared with 821°C for Pyrex, and bonding is completed within 10 min with 30–60 V applied. A pull test reveals a resulting bond strength in excess of 1.5 MPa. Application of pressure up to 160 kPa on the wafer sandwich apparently supports the bonding process; at a pressure of 160 kPa, the minimum voltage to achieve bonding drops by a factor of 2. The slow deposition rate of glass sputtering presents a major disadvantage for this technique, but evaporated glass layers can be deposited with much higher rates up to 4 μm/min and could provide a solution; unfortunately, in this case pinholes tend to reduce the breakdown voltage.[30]

Field-assisted bonding between two Si wafers, both provided with a thermally grown oxide film, is also reported to be successful. Both Si wafers must be covered with 1 μm of oxide. Bonding of a bare Si wafer to a second wafer with oxide failed because of oxide breakdown under very small applied bias. Temperatures range from 850°C to 950°C, and a voltage of 30 V must be applied for 45 min at the chosen bonding temperature.[33]

Silicon Fusion Bonding

The ability to bond two Si wafers directly without intermediate layers or applying an electric field simplifies the fabrication of many devices. This direct bonding of silicon to silicon, Si fusion bonding (SFB), is based on a chemical reaction between OH groups present at the surface of the native or grown oxides covering the wafers. The bonding relies on the tendency for both very smooth and flat surfaces to adhere, and heat annealing to increase bond strength. The method is of great interest for the fabrication of silicon-on-insulator (SOI) structures. The processes involved in making these bonds are simpler than other currently employed bonding techniques, the yields are higher, the costs are lower, and the mechanical sensors built according to this principle exhibit an improved performance. Finally, as demonstrated in Figure 4.21, the size of an SFB-type device can be almost 50% smaller than that of a conventional anodically bonded chip.

Wafer bonding may be achieved by placing the surfaces of two wafers in close contact and

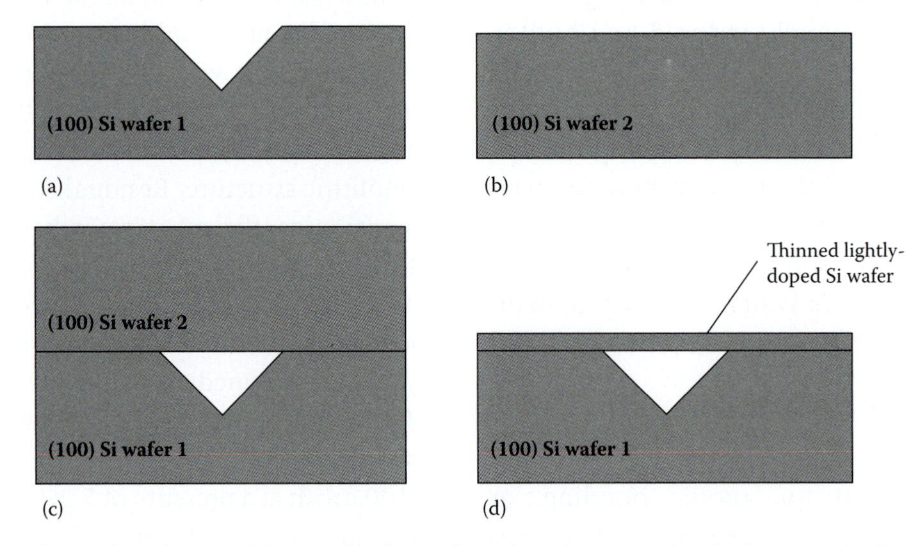

FIGURE 4.21 Silicon fusion bonding enables smaller-footprint microelectromechanical system devices to be fabricated. (a) Etch a silicon wafer to create a square pit (inverted pyramid). (b) A plain second silicon wafer is cleaned and ready for fusion bonding to the first wafer. (c) Bond the two wafer using fusion bonding. (d) The second wafer is then thinned down from the back side using either mechanical-chemical polishing or etching.

inserting them in an oxidizing ambient at temperatures greater than 800°C. It may be noted that the term *silicon fusion bonding* is somewhat misleading; the fusion point of Si at atmospheric pressure is 1410°C, well below the relevant process temperatures.[31] The quality of the bond depends critically on temperature and the roughness of the surface. Flatness requirements are much more stringent than for anodic bondings, with a microroughness R_a < 10 or 40 Å and bow < 5 μm versus 1 μm R_a and bow < 5 μm in anodic bonding. Because of the high temperature involved, active electronics cannot be incorporated before the bonding takes place (the temperature limit for bonding IC-processed standard Si substrates is about 420–450°C; see "IC Compatibility" in Volume II, Chapter 7). Higher temperature (above 1000°C) is usually required to get voidless and high-strength bonding.[34] Bond strength up to 20 MPa has been reported. The application of a small pressure during the bonding process further increases the final bond strength.[35] The bonding can be carried out successfully with one oxidized Si wafer to another bare Si wafer, two oxidized wafers, or two bare Si wafers, and even between one wafer with a thin layer of nitride (100–200 nm) and one bare wafer or two wafers with a thin layer of nitride. The same fusion-bonding technique can be applied for bonding quartz wafers, GaAs to Si, and Si to glass. Provided that the surfaces are mirror smooth and can be hydrated, the bonding process proceeds in an identical fashion as for Si-Si bonding (see Schmidt[36] and references therein).

According to most references, before fusion bonding, the oxidized Si surfaces must undergo hydration. Hydration is usually accomplished by soaking the wafers in a H_2O_2-H_2SO_4 mixture, diluted H_2SO_4, or boiling nitric acid. After this treatment, a hydrophilic top layer consisting of –OH bonds is formed on the oxide surface. An additional treatment in an oxygen plasma greatly enhances the number of –OH groups at the surface.[37] Then, the wafers are rinsed in deionized water and dried. Contacting the mirrored surfaces at room temperature in a clean-air environment forms self-bonding throughout the wafer surface, with considerable bonding forces.[38] In a transmission infrared microscope, a so-called *bonding wave* can be seen to propagate over the whole wafer in a matter of seconds. The bonded pair of wafers can be handled without danger of the sandwich falling apart during transportation. A subsequent high-temperature anneal increases the bond strength by more than an order of magnitude. The self-bonding is the same phenomenon described under stiction of surface micromachined features (see Volume II, Chapter 7).

At present, the fusion bonding mechanism is not completely clear. However, the polymerization of silanol bonds is believed to be the main bonding reaction:

$$\text{Si–OH} + \text{HO–Si} \rightarrow H_2O + \text{Si–O–Si} \qquad \text{Reaction 4.1}$$

Figure 4.22 shows the suggested bonding mechanism, including the transformation from silanol bonds to siloxane bonds. Silanol groups give rise to hydrogen bonding, which takes place spontaneously even at room temperature and without applying pressure, as long as the two wafers are extremely clean and smooth.[38]

Measurements of bond strength as a function of anneal temperature indicate three distinct bond strength regions, according to Schmidt.[36] For anneal temperatures below 300°C, the bond strength remains the same as the spontaneous bond strength measured before the anneal. For self-bonded

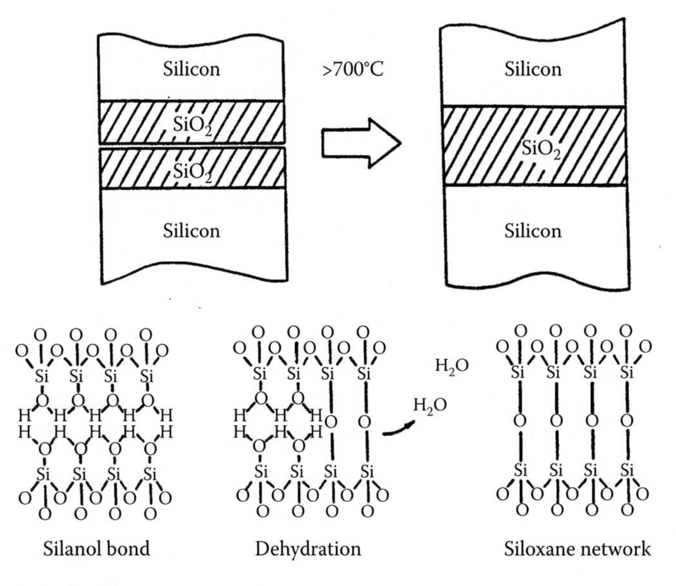

FIGURE 4.22 Schematic illustration of the silicon-to-silicon bonding by fusion and a proposed mechanism.

wafers with hydrogen bonding between the silanol groups of opposite surfaces, Tong et al.[39] measured an activation energy of 0.07 eV for temperatures <110°C—fairly close to the theoretical value of around 0.05 eV for hydrogen bonding. At about 300°C, according to Schmidt, the –OH groups form water molecules, and the voids that are sometimes observed at this temperature are believed to be due to water vapor formed in the process.[40] Stengl et al. report that the formation of voids can be avoided by first contacting the wafers at a temperature of 50°C.[41] The voids tend to disappear also at temperatures above 300°C, while the bond strength increases and then levels out. It is assumed that in this regime, Si-O-Si bonds (siloxane) start forming between the surfaces (Reaction 4.1). Tong et al.[39] found a pronounced increase of interface energy of room-temperature, self-bonded hydrophilic Si/Si, Si/SiO_2, and SiO_2/SiO_2 wafers after storage in air at room temperature and up to 150°C for periods between 10 and 400 h. The interfacial energy increase is ascribed to the generation of additional –OH groups due to a reaction with water and the strained oxide and/or Si below 100°C, as well as the formation of stronger siloxane bonds, which, this group claims, already forms from temperatures below 150°C. They find that the siloxane groups have a much higher bond energy with an activation energy of 1.8–2.1 eV.[39] After prolonged storage, interface bubbles were observed that seem to contain hydrogen and hydrocarbons. At the highest temperatures (>800°C), Schmidt finds the bond strength starts increasing again, and at 1000°C, the bond is reaching the fracture strength of single-crystalline silicon: 10–20 MPa. In this third region, it has been suggested that surfaces can deform more easily (oxide flow), bringing them into better contact.[36] At these temperatures, oxygen also diffuses into the Si lattice. Multiple Si wafers can be fusion bonded together. In some applications, up to a total of seven wafers have been bonded together.

Recall that for fusion bonding, surfaces have to be chemically activated to prepare for the bond. This is quite an undesirable step because the chemicals needed for surface preparation often are not compatible with other micromachining materials. Another way to activate the bonding surface and to make it hydrophilic is to use a plasma activation step. The plasma slightly roughens the surface and increases the number of –OH groups, in a manner similar to hydrophilic chemical activation. The plasma activation is a dry process that is much more compatible with many micromachining process flows. Also, because it is a dry process, it is compatible with released devices (no stiction problems). The bond still needs to be annealed, just as in regular fusion bonding, but the anneal temperatures and times required for a strong bond are much less (400°C vs. 1100°C), possibly because of the slightly roughened surface.

It should be noted that spontaneous bonding of Si wafers also has been claimed with hydrophobic surfaces. It was suggested that such low-temperature bonding is due to van der Waals forces, whereas hydrogen bridging is involved in hydrophilic surfaces. The bonding energy obtained with hydrophobic wafers was as low as 26 mJ/m^2, but after annealing at 600°C, the bonding energy reached a value of 2.5 J/m^2, only attainable at 900°C with hydrophilic wafers.[42] This is in contradiction with most other research done, which concludes that the wafers must be hydrophilic for SFB.

Early on, it was reported that Si_3N_4 was a non-bondable surface.[43] Based on Reaction 4.1, one would assume that the low-temperature direct bonding of Si_3N_4 could take place only if oxidation of the nitride introduced silanol groups. Indeed, an oxidized surface of nitride was reported to bond at high temperatures consistent with Si fusion bonding.[44] Unexpectedly, Bower et al.[45] demonstrated extremely strong bonds between Si wafers coated with a smooth, clean layer of LPCVD Si_3N_4 at temperatures ranging between 90°C and 300°C. A 500-Å-thick nitride was first deposited in the LPCVD reactor and then, without breaking the vacuum, activated in an ammonia stream for several minutes. This group speculates that a chemical reaction such as the following is responsible for bonding where the released hydrogen diffuses from the bonding interface:

$$\equiv Si-(N_xH_y) + (N_xH_y)-Si \equiv \rightarrow \equiv Si-(2N_x)-Si \equiv + yH_2$$

Reaction 4.2

Fracture occurs at about 2 MPa. In a further extension of this work, it was found that an NH_3 plasma treatment of Si, SiO_2, and TiN for 5 min or longer also provided direct bondable surfaces. Using Si_3N_4 as an intermediate layer, Bower and coworkers bonded Si to Si, Si to SiO_2, and Si to GaAs.[46] This low-activation energy bonding allows bonding of optoelectronic components such as GaAs with completed electronic circuitry on Si with Al metallization in place.

Thermal Bonding with Intermediate Layers

When incorporating an intermediate layer between two substrates, many thermal bonding techniques are feasible. If the intermediate layers can be patterned by lithography, both L0 and L1 packaging becomes possible. Anodic bonding occurs uniformly, but the high fields are detrimental to field-effect devices. Some intermediate layer materials with a low melting point [e.g., phosphosilicate glass (PSG)] do not need electric field for bonding, but the thickness and uniformity are hard to control. An ideal method would require neither an electrical field nor high temperatures and would be very uniform. With films that become soft at low temperatures, all features on the wafer may be covered smoothly as long as they are not a significant fraction of the intermediate film thickness itself.

LPCVD PSG has been used for Si-to-Si wafer bonding. Fusion of two Si wafers coated with 1–2-μm-thick PSG layers is fast. Excellent bonding results, provided the wafers are clean and reasonably flat. Unfortunately, the bonding process occurs at a high temperature of 1100°C for 30 min.[47]

Several suppliers of glass materials offer low-temperature sealing glasses known as glass frits or glass solders. Corning Glass Works, for example, has introduced a series of glass frits (#75xx) with sealing temperatures ranging from 415°C to 650°C.[48] These glasses are applied by a variety of methods, including spraying, screen-printing, extrusion, and sedimentation. After the glass is deposited, it needs to be preglazed to remove the organic residues produced by vehicle and binder decomposition (see "Types of Inks/Pastes and Applications" in Volume II, Chapter 9). The substrate-glass-substrate sandwich is then heated to the sealing temperature while a slight pressure is applied (>1 psi). Devitrifying glasses and thermoplastic, vitreous glasses are offered. The devitrifying glasses are thermosetting, and they crystallize during the sealing procedure, thereby changing their mechanical properties. The vitreous glasses are thermoplastic, melt at a lower temperature than devitrifying glasses, and flow during sealing, but their material properties do not change after sealing is completed. An important problem today remains the uncertainty regarding the mechanical and chemical behavior of these glasses.[27] Moreover, almost all of the low-melting-point formulations of glass frits contain some lead, which is a major drawback because the European Union and Japan are forbidding the use of lead in all electronics manufacturing in the coming years.

Ko et al.[20] used RF sputtering from a target made from a Corning 7593 glass frit to obtain a thinner (8000 Å) and more uniform intermediate glass film for bonding. The sputtered glass did not need glazing. Annealing at 650°C in an oxygen atmosphere resulted in an excellent bond strength of two Si wafers coated with these films.

Legtenberg et al.[49] used a thin film of atmospheric pressure chemical vapor-deposited (APCVD) boron oxide with a softening temperature of 450°C. However, boron oxide is hygroscopic and does not present a viable solution. Field et al.[50] used boron-doped Si dioxide, which also becomes soft at 450°C. The doping is performed in a solid source drive-in furnace. The bond seal is hermetic, and a crack propagates right through the material rather than along the bonded surface. A problem with this method is the sensitivity of the film to phosphorous contamination.

Spin-on-glass (SOG) was reported by Yamada et al.[51] and Quenzer and Benecke[52] In Yamada et al.'s effort, SOG [$Si(OH)_x$ (with $2 < x < 4$)] was coated on the wafer surface to be bonded. After the 50-nm-thick film was baked at a temperature of 250°C for 10 min, the wafers were contacted and pressurized in vacuum at 250°C for 1 h. The bonding was already strong after this step, but wafers were further annealed for 1 h at 1150°C in air to sinter the SOG layer and to improve the breakdown voltage of the layer. Quenzer et al. used sodium silicate as an intermediate layer. The wafers were made hydrophilic in an RCA1 cleaning step, and the glass was spun on from a dilute sodium silicate solution. The film thickness was between 3 and 100 nm, and annealing at 200°C produced good bonds between bare Si wafers, Si wafers with native oxide, thermal oxide, and Si nitride. No need for further annealing is mentioned in this work.

To bond the unpolished back side of a Si die to another Si part, an aqua-gel based on a hydrophilic pyrogenic silica powder (grain diameter smaller than 7 nm) and polyvinyl alcohol (PVA) as a binder may be used. In this scheme, the adhesion depends on hydrogen bonding between the –OH groups of the PVA, the SiO_2 grains, and the Si surfaces. No special surface treatment is required, and the bonding can be achieved in 15 min at room temperature. Bond strength is in excess of 1 MPa. This new technique works for bonding Si to Si, Si to SiO_2, SiO_2 to SiO_2, and glass to Si or SiO_2. Surprisingly, this technique was also shown to work for bonding III-V materials. This type of bonding is proposed for the fashioning of multichip-on-silicon packaging (L2 level).[53]

Silicon microstructures can be sealed together by eutectic bonding. Eutectic films are alloys of two materials and act as the "glue" or solder to bond two wafers. Solder is a form of eutectic alloy that is commonly used in building printed circuit boards. Solder has been used since antiquity to bond metal surfaces to one another. It is an alloy with a low melting point, typically containing tin or indium, used to join electrical components to a circuit board or to join metal objects together. In Au-Si eutectic bonding, one applies a contact force and increases the temperature beyond the Au-Si eutectic temperature (363°C). This results in diffusion of silicon into the gold (not diffusion of gold into silicon) until the Au-Si eutectic composition is reached (19 atomic percent Si). The Au-Si eutectic bonding takes place at a temperature of 363°C, which is well below the critical temperature for Al metallized components, and Au is a quite commonly used thin-film material. These are good reasons to select the Au-Si eutectic, but other material combinations are possible. It is possible to bond bare Si against Au-covered Si, or Au-covered Si against Au-covered Si. Another alternative is to use Au-Si preform with a composition close to the eutectic concentration. Eutectic bonding in MEMS packaging is demonstrated below, in Figure 4.37, where we illustrate a wafer-to-wafer transfer of an encapulsation package. Tiensuu et al. have demonstrated a mean fracture strength of the Au-Si bond of 148 MPa.[54] This compares well with fusion bonding with a typical bond strength of 5–15 MPa.

There are some considerable disadvantages associated with Au-Si eutectic bonding. It is difficult to obtain complete bonding over large areas, and even native oxides prevent the bonding to take place and eutectic preform (e.g., with Au-Sn) bonding is reported to introduce substantial mounting stress in piezoresistive sensors, causing long-term drift due to relaxation of the built-in stress.[20]

Bonding with Organic Layers

Two silicon wafers also can be bonded using an intermediate polymer film that forms the "glue" between the two. Polymers such as polyimide, or "epoxy" can be used; if the glue can be deposited in a uniform film and has good adhesion to Si, it can be used to bond the two wafers together. One of the issues with polymers is that as they are baked to harden them, they produce vapors of solvents within them, and these vapors create voids, which tend to push the two wafers apart. Therefore, polymers that tend to outgas very little are desired. One of these materials is benzo-cyclobutene (BCB), which is used as an insulator in building integrated circuits. BCB can be spin deposited and then patterned like photoresist. Thus, it is possible to create fine features on one wafer that can then be used to bond to a companion wafer. BCB patterns retain their shape after bonding, and the bond strength is very high (one needs to destroy the BCB to break the bonded wafers). It is also possible to use another polymer, such as Parylene, which is simply a plastic. When heated, the plastic joins the two wafers together. Parylene is a very inert, tough, and easy to deposit polymer. It is used extensively in implantable biomedical systems. It can be evaporated at room temperature (0.1–10 μm range) and when deposited it is conformal, uniform, and pinhole-free.

A most interesting option for packaging is through lithographic patterning of thick resist layers. Prominent new candidates for packaging micromachines are thick ultraviolet photoresists, such as AZ-4000 and SU-8, reviewed in Volume II, Chapter 1, as well as LIGA resists such as PMMA.

Using polymers, very low bonding temperatures are possible. The bond strength can be quite high, no metal ions are present, and the elastic properties can reduce stress. One can envision that with thick enough polymer layers, packaging as part of the front-end processing will become possible (up to level 2). Disadvantages include impossibility of hermetic seals (see Figure 4.29), high vapor pressure, and in cases where a high Young's modulus and low hysteresis is required poor mechanical properties.

An example of photopatterned bonding is provided by the flexible polysiloxane interconnection between two substrates demonstrated by Arquint et al.[55] UV-sensitive cross-linking of polysiloxane layers with a thickness of several hundred micrometers is used to pattern the polymer onto the first substrate, and a condensation reaction is used to form a chemical bond to the second surface. The procedure is illustrated in Figure 4.23.[55]

Good bonding results with photopatternable resists also have been obtained by den Besten et al.,[56] who used negative photoresist at a bonding temperature of 130°C. In an important potential application, UV-curable encapsulant resins have been employed for wafer-level packaging of ion-sensitive field effect transistors (ISFETs).[57]

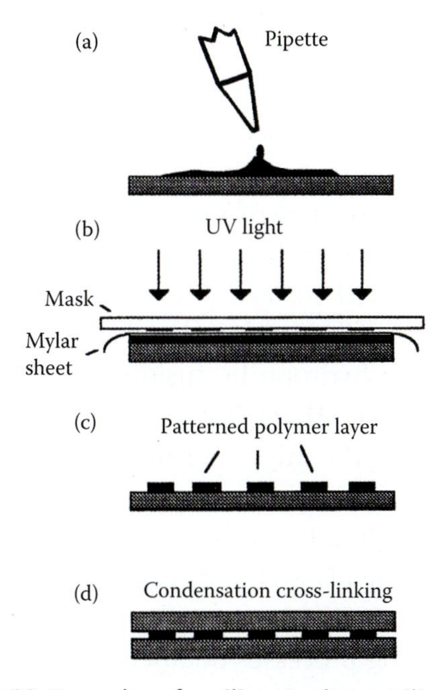

FIGURE 4.23 Procedure for silicon/polymer/silicon bonding: (a) Monomer solution is deposited either by spin coating or casting; (b) exposure to ultraviolet light for photopolymerization; (c) polymeric pattern is developed in xylene; (d) second wafer is pressed onto the polymeric pattern and left for humidity-induced polymerization in ambient air (10 h). (From Arquint, P., P. D. van der Wal, B. H. van der Schoot, and N. F. de Rooij. 1995. In *8th International Conference on Solid-State Sensors and Actuators (Transducers '95)*, 263–64. Stockholm, Sweden. With permission.[55])

Bonding of Plastic to Plastic

A major concern of polymer-based MEMS, especially in microfluidic devices, is how to bond plastic parts together. This bonding or application of plastic lids to plastic microparts can be achieved using adhesives or tapes, plastic welding (e.g., hot plate and ultrasonic welding), and selected organic solvents (i.e., partially dissolving the bonding surfaces and evaporating the solvent). In our laboratory, several of these techniques have been tried, including vacuum-assisted heat bonding, bonding with double-sided adhesive tape, and water-soluble polymer-assisted bonding. Figure 4.24 shows a cross-section of microchannels bonded by these three methods. For large features (several hundred micrometers to millimeters), conventional plastic welding techniques such as hot plate, ultrasonic, and infrared (IR) welding can be used. Applying vacuum before welding may reduce voids at the welded interface, and consequently, lower pressure and temperature can be used in welding. This minimizes the molten polymer from flowing into the bonded channels (see Figure 4.24a). Ideally, small overflow traps should be designed along the edge of the channels (as shown in Figure 4.25) to completely prevent the channels from being blocked by the molten polymer during welding. For smaller channel sizes (less than 100 μm), the design and fabrication of overflow traps become difficult. Double-sided adhesive tape can be a useful way for microbonding, as shown in Figure 4.24b. The main disadvantage is that the presence of an adhesive tape tends to change the channel size and channel wall properties. This is the technique we often used in the assembly of the compact disc (CD) fluidic platform (see Chapter 5, Figure 5.28). For very small channels (less than 1 μm), prefilling the channels with a water-soluble polymer (serving as a sacrificial material), drying the polymer, sealing the plate with channels to a cover lid by either welding or adhesive bonding, and then washing out the water-soluble polymer turns out to be a viable way to

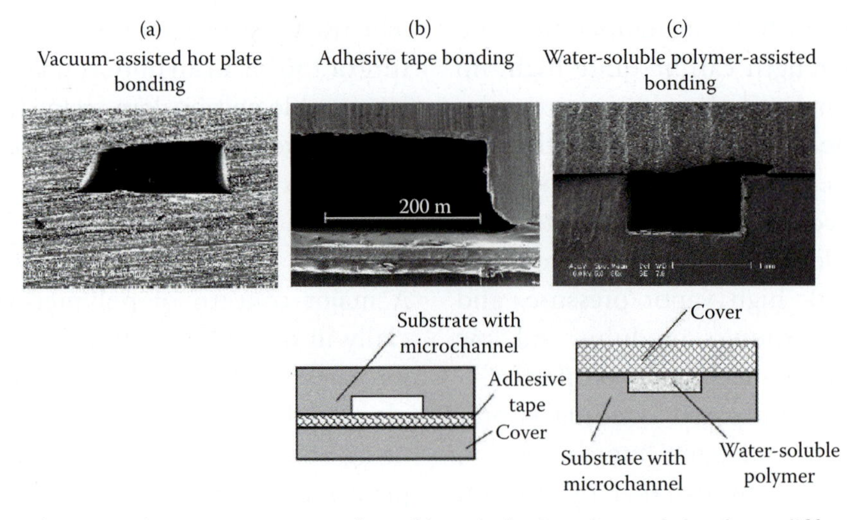

FIGURE 4.24 Scanning electron microscope cross-section of bonded microchannels by three different methods (see text). (Courtesy of Dr. Marc Madou and Dr. James Lee, Ohio State University.)

prevent channel blockage during bonding (see Figure 4.24c). This method, however, is more tedious and may not always be applicable. Development of low-cost, high-speed, and reliable bonding techniques for microfluidic devices remains an important and challenging issue in bio-MEMS fabrication.

Polydimethylsiloxane (PDMS), introduced in Volume II, Chapter 2, is the most common example of a group of polymers known as silicones. Silicones are used as a component in silicone grease and other silicone-based lubricants, as well as in defoaming agents, damping fluids, heat transfer fluids, cosmetics, hair conditioner, and other applications; they are also the primary component in Silly Putty. PDMS is used extensively in microfluidics research to create channels, valves, pumps, and other components that make up microfluidic systems. As we saw in Volume II, Chapter 2, the molecular structure of PDMS is made up of a backbone of siloxane with two methyl groups attached to each silicon atom. The backbone is terminated with either a methyl group or hydroxyl group. PDMS can be reversibly and irreversibly bonded to a variety of substrates. Without pretreatment, PDMS bonds reversibly to glass structures with a bond strength that can survive fluidic pressures of ~4 psi, enough for many microfluidics applications. Activating the PDMS surfaces with a plasma treatment before bonding can create an irreversible bond. The plasma treatment temporarily replaces one of the methyl groups of the PDMS with a silanol group. This activation typically lasts for tens of minutes. If the surface of an activated PDMS layer comes in contact with another activated surface, the silanol groups combine to create a siloxane bond and a water molecule, similar to the hydrophilic fusion bonding process described above. There is no need to anneal this bond because the high permeability of the surrounding polymer allows the water molecule to diffuse away easily (although heat speeds the process up). Irreversible PDMS bonds can stand pressures of up to 70 psi.

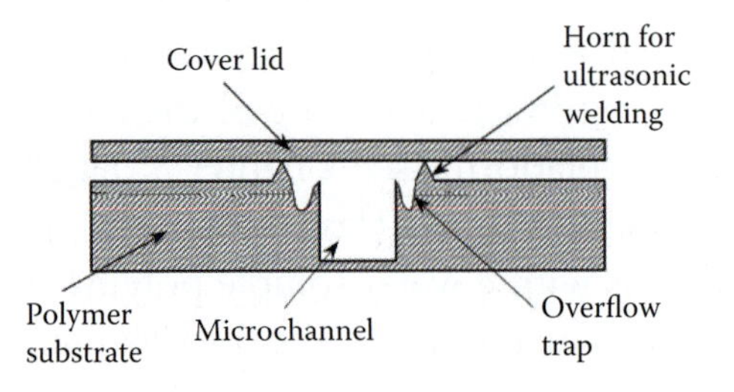

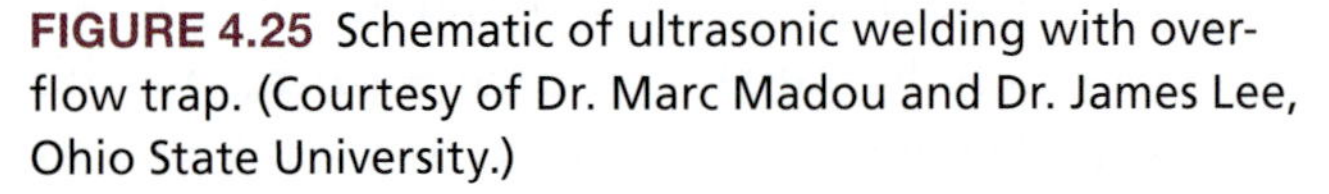

FIGURE 4.25 Schematic of ultrasonic welding with overflow trap. (Courtesy of Dr. Marc Madou and Dr. James Lee, Ohio State University.)

The bonding processes detailed here are summarized in Table 4.2. A good review of polymer micromolding, including polymer bonding, is that of Becker and Gärtner.[58]

Alignment during Bonding

Obtaining good alignment between the device wafer and support substrate (e.g., glass or another Si wafer) during bonding poses some extra challenges, because a transfer from aligner to annealing furnace

Table 4.2 Comparison of Bonding Techniques

	Anodic	Fusion	Glass Frit	Eutectic	Solder	Parylene	PDMS
Bond strength	Very strong	Excellent	Strong	Strong	Strong	OK	Weak
Hermeticity	Good	Excellent	Good	Good	Good	Poor/OK	Poor
Temperature							
Formation	250–400°C	200–1000°C	400–500°C	>363°C	57–400°C	230°C	RT–90°C
Service	>400°C	>1000°C	<400°C	<363°C	<<formation	<<230°C	<90°C
TCE mismatch	OK	Good	OK	OK	Poor	OK	OK
CMOS-compatible	OK	Poor	OK	OK	Good	Good	Good
Process complexity	Low	Some	Some	Some	Some	Some	Some
Planarization capability	Poor	Very poor	Good	Good	Good	Good	Good
Maturity	Decades	Several years	Several years	Several years	Years	Newer	Years
Cost	$	$$$	$$	$$$	$$	$$	$

generally is required after alignment. One technique used for bonding alignment is to generate holes in both the device wafer and the substrate, followed by putting them in a specially designed fixture to perform the bonding. The accuracy achievable with this option is only around 50 μm, though. Higher alignment accuracy (~2.5 μm) can be achieved by using a bonding machine equipped with an in situ optical alignment setup. Aligned wafer bonding has been worked on extensively by Bower et al.[59] They used an infrared aligner modified to hold two imprinted wafers face to face while projecting an infrared image of the surfaces to a viewing screen (see Figure 4.26).[59] An array of V-grooves etched into the surface of the Si wafers was then precisely aligned, and the wafers were brought in contact for initial bonding. The

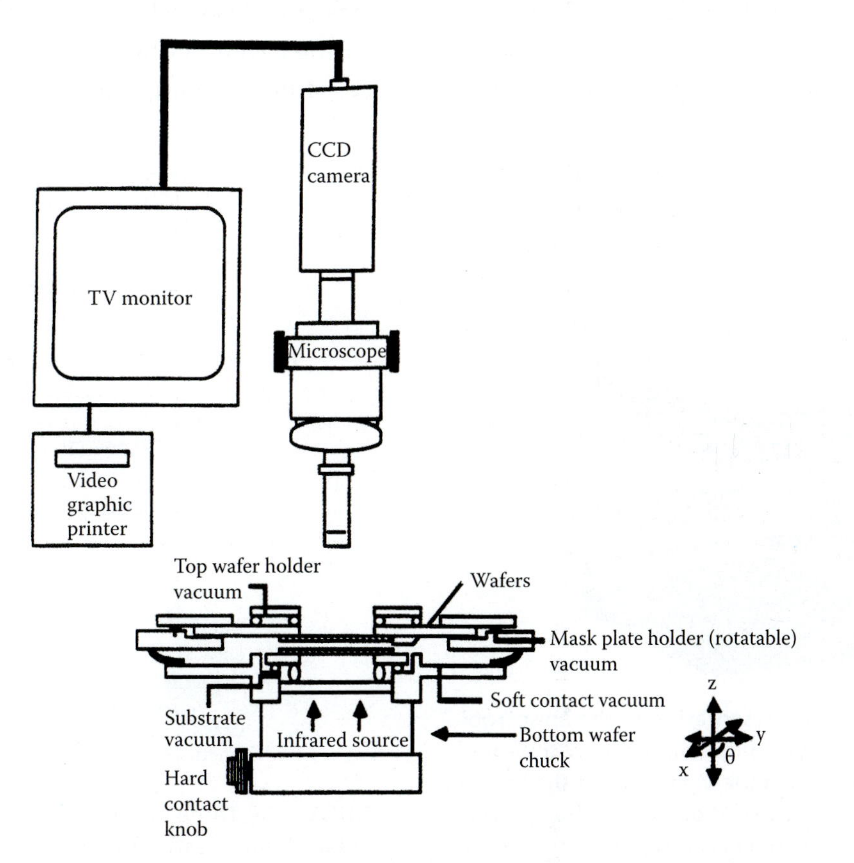

FIGURE 4.26 Alignment fixture for bonding. The schematic diagram shows the infrared aligner system used for wafer bonding. Wafer holders are designed with a large opening at the center to allow maximum infrared energy to pass through to the camera. (Courtesy of Dr. R. Bower, University of California, Davis.)

(111) planes that define the walls of the V-groove create a shadow image without the necessity of metal features commonly used for two-sided backside alignment. The initial bonding is nothing more than the normal soft contact provided by the aligner itself. Hydrogen bonds are formed between the two surfaces in intimate contact. These relatively weak bonds of about 0.05 eV are sufficient to hold the two pieces firmly together without relative displacement during the transport from the aligner to the annealing furnace. Subsequent high-temperature annealing (950°C for 30 min) was used to strengthen and complete the chemical bonding. The aligned-wafer technique also can be used when dealing with dissimilar substrates.

Shoaf and Feinerman[60] developed an alignment technique to assist in precise Au-Si eutectic bonding of Si structures. A (100)Si wafer is anisotropically etched to create V-grooves around the periphery of the structure to be bonded. Gold is deposited onto one of the wafers prior to dicing into individual dice. Optical fibers are placed into the orthogonal V-grooves, as demonstrated in Figure 4.27, and used as precision location keys for assembly prior to bonding. The entire structure is placed on a hot chuck at 400°C for bonding, and the fibers are removed after bonding. Results have shown a maximum misalignment of 5 μm for a 1 × 1 cm die. This technique allows a sensor die to be precisely bonded to an electronics die without the aid of a microscope or micropositioners.

FIGURE 4.27 Schematic representation of the anisotropically etched V-groove/optical fiber alignment technique. Optical fibers are placed into the V-grooves of the bottom Si die. A second die with etched V-grooves is then placed on top of the first die. The fibers act as precision locating keys and align the two Si dice. (From Shoaf, S. E., and A. D. Feinerman. 1994. Aligned Au-Si eutectic bonding of silicon structures. *J Vac Sci Technol* A12:19–22.[60])

Imaging and Bond Strength and Package Hermeticity Tests

Methods for imaging a bonded pair of Si wafers include infrared transmission (voids larger than 20–30 μm can be seen this way), ultrasound (qualitative information about the bond quality), and x-ray topography. The most common mechanical techniques to characterize bond strength are illustrated in Figure 4.28.[36] Both the burst test (Figure 4.28a) and the tensile/shear test (Figure 4.28b) yield important engineering insights for sensor construction but do not yield information about the detailed nature of the bond because of the complicated loading of the interface.[36] In the Maszara method,[61] a thin blade is inserted between the bonded wafers and a crack is introduced (Figure 4.28c). The length of the crack, measured by infrared imaging, gives a value for the surface energy inferred through knowledge of the sample and blade thickness and the elastic properties of the wafer. This method has the advantage of creating a well-defined loading on the bonded interface. Unfortunately, the surface energy is a fourth

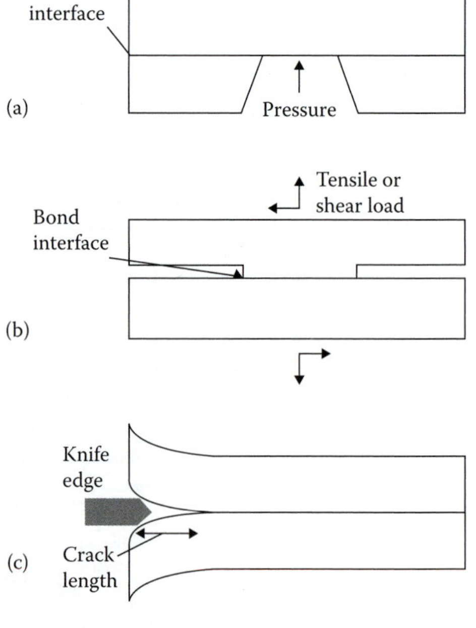

FIGURE 4.28 Three methods for bond strength measurements. (a) Burst test. (b) Tensile and shear test. (c) Maszera test. (From Schmidt, M. A. 1994. In *Technical Digest: 1994 Solid State Sensor and Actuator Workshop*, 127–30. Hilton Head Island, SC. With permission.[36])

power of the crack length and uncertainties in measuring that length make for large errors.

The hermeticity of sealed cavities in packaging is important for physical protection purposes and, in some cases, for the performance of the sensor inside. The word *hermetic* means airtight, so a good hermetic seal does not let gasses, vapors, or liquids pass across it. Some materials are better than others at creating hermetic seals because their molecules (or atoms) are more closely spaced together. Hermeticity is an important parameter for micropackaging because certain MEMS devices need vacuum to operate and ICs need a dry environment to maintain their reliability. For example, the quality factor, Q, of a resonator, the vacuum reference of an absolute pressure sensor, and the cavity of a pneumatic infrared sensor (i.e., a Golay cell) all are critically dependent on a good hermetic package. Testing the hermeticity of electronic devices usually is carried out by helium leak detection. While this technique, with a minimum detectable leak rate of 5×10^{-11} to 5×10^{-10} Torr L/s, is appropriate for testing of relatively large packages, it is unacceptable for nearly every Si sensor application (it is not valid for package sizes under 1000 nL). Nese et al. have introduced Fourier transform infrared (FTIR) spectroscopy for measuring the gas concentration inside a sealed Si cavity using N_2O as a tracer gas.[62] Although the sensitivity for leakage is about the same as with a helium leak detector, the method is accumulative, so by prolonging the exposure times, the sensitivity can be increased, which is not possible with the dynamic helium leakage testing method.

Although anodic bonding may be carried out in a vacuum chamber, the pressure inside a Si cavity differs from the chamber pressure because of gas generation, probably oxygen, during bonding. To control cavity pressure for critical damping of packaged micromechanical devices, Minami et al.[63] used nonevaporable getters. Getters are agents that counteract harmful contaminants within a sealed package; they may include solids, liquids, gases, and combinations thereof. The solid nonevaporable getter used by Minami et al.[62] is a Ni/Cr ribbon covered with a mixture of porous Ti and Zr-V-Fe alloy that absorbs gases after activation at 400°C and is built into the microdevice. These authors monitored the effectiveness of their method to control cavity pressure by measuring deflection of a thin membrane covering the cavity.

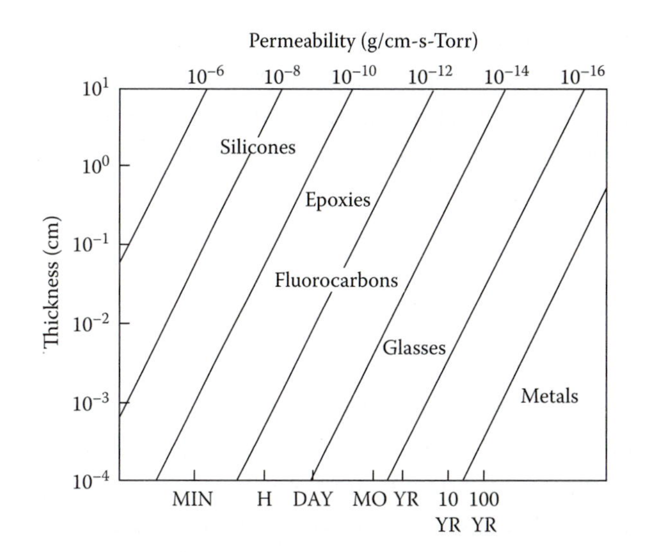

FIGURE 4.29 The calculated time for moisture to permeate various sealant materials (to 50% of the exterior humidity) in one defined geometry. Organics are orders of magnitude more permeable than materials typically used for hermetic seals. (From Striny, K. M. 1988. In *VLSI Technology,* ed. S. M. Sze, 566–611. New York: McGraw-Hill. With permission.[64])

Figure 4.29 shows the relative capabilities of several materials to exclude moisture from the encapsulated components over long periods. Organic materials are not good candidates for hermetic packages; for almost all high-reliability applications, the hermetic seal is made of glass or metal. Plastic packaging for integrated circuits (ICs) is governed by several industry standards that have evolved over the years; plastic packaging for MEMS, on the other hand, is not yet governed by any standard. For the simplest mechanical sensors, standard or slightly modified IC plastic packages are used. New plastic packaging technologies and their standardization for MEMS will begin only once the volume of MEMS products approaches a larger fraction of that of ICs.

For measuring moisture penetration in a package, temperature-accelerated soak tests may be performed. Moisture penetration can be followed, for example, with an integrated on-chip dew-point sensor.[25] Striny[64] points out that the cost or difficulty of obtaining a truly hermetic seal for a particular package often is prohibitive. In that case, in addition to polymer lid sealing, surface die coats, for example with silicones, are applied. From Figure 4.29,[64] we can conclude that silicones do not act as a moisture barrier; the exact mechanism by which they protect the die when applied as a surface coat is not yet well understood.[65]

Examples of MEMS Packages

Introduction

Pressure sensors and accelerometers typically fit in standard packages, but few other MEMS devices do. Examples of nonstandard packages that we have already discussed include the catheter-based electrochemical sensor shown in Volume II, Figures 4.92 to 4.96, where a dual-lumen catheter forms the basis of the packaging scheme, and the thin-film magnetic head illustrated in Chapter 1, Figure 1.1d, this volume. The very fragmented nature of micromachining applications and the nonstandardness of the package necessitate a design approach starting from the package.

Cap-on-Carrier and Cap-on-Chip

In Figure 4.30, two MEMS packaging schemes are compared. The cap-on-carrier packaging method (L1) (Figure 4.30a) is suitable for accelerometers, gyroscopes, and other motion detectors that can be hermetically sealed. It is the prevalent method in industry today. The cap-on chip is an on-chip packaging method (Figure 4.30b), also called *chip scale packaging* (CSP), that is the preferred approach, as the packaging can be made part of the front-end batch manufacturing process and may be much smaller. The two approaches are contrasted in Figure 4.31.

One reoccurring problem in MEMS packaging is how to protect sensitive movable mechanical structures during dicing. The dicing process is very harsh, and any loose structure might break off. Fortunately, in most MEMS process sequences, the step where movable structures are released is the final so-called *release step* (typically, a sacrificial material is etched away to free the movable part; see Volume II, Chapter 7). By carrying this release step out after dicing, breakage of sensitive MEMS structures is avoided.

But individual die-level release is expensive and slow. Released dice, as shown in Figure 4.31, are packaged in ceramic or metal cavity packages. These packages often are large and expensive. Alternately, one does the MEMS release at the wafer level and then wafer-level packages the devices before dicing. Wafer-level packaging must follow the wafer-level release, again to avoid damaging the MEMS. The process is called *CSP* with caps-on-chip. These caps may be made of glass or another Si piece and typically are still quite bulky (millimeters rather than tens or hundreds of micrometers). These bulk wafer caps are

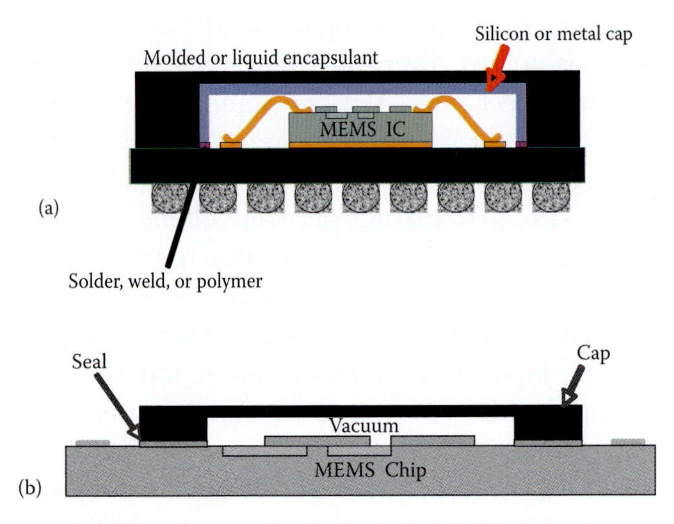

FIGURE 4.30 Cap-on-carrier (a) and cap-on-chip (b) compared.

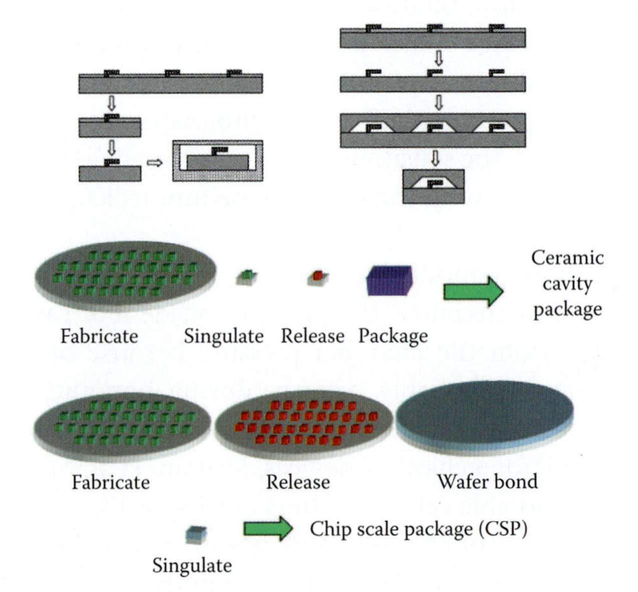

FIGURE 4.31 Microelectromechanical systems (MEMS) are either diced, then released and finally packaged in ceramic or metal cavity packages. Or one does the MEMS release at the wafer level and then wafer-level package the devices before dicing. The latter process is called *chip scale packaging* (CSP).

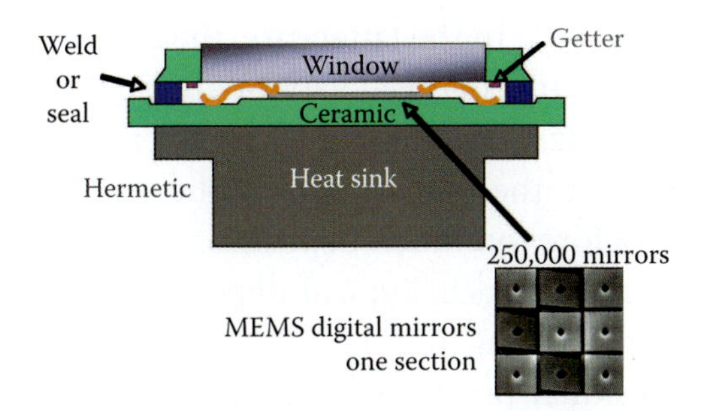

FIGURE 4.32 Packaged Digital Mirror Device™ from Texas Instruments utilizes a vacuum hermetic package with a transparent optical window. The package also includes a getter to adsorb any residual moisture.

the current industry standard, used, for example, in Motorola accelerometers, Bosch gyroscopes, Clarisay SAW filters, and Radant MEMS switches. The bulk wafer caps are robust, hermetic, and implemented on a wafer-level, but they require a large on-chip area for the seal ring. Below, we review a number of alternative wafer-level cavity sealings, i.e., cap-on-chip techniques such as the microassembled HEXSIL caps, techniques that are today mostly in the research and development phase. They all have a considerable smaller footprint than the bulk caps discussed here.

The Digital Mirror Device™ from Texas Instruments, discussed in Volume II, Example 7.2, utilizes a vacuum hermetic package with a transparent optical window as shown in Figure 4.32. The package also includes a getter to adsorb any residual moisture. This package is an example of a special type of cap-on-carrier package.

Next-Generation Cap-on-Chip Cavity Sealing

Introduction

Many microdevices have been fabricated making use of some type of cavity sealing or cap-on-chip technique. Cavity sealing can serve as a batch-compatible zero- and first-level packaging technique by encapsulating a die feature (L0) or a whole die at a time (L1). In bulk micromachining (Volume II, Chapter 4) and Si fusion-bonded (SFB) surface micromachining (Volume II, Chapter 7), cavities are fabricated by bonding, respectively, a glass plate (anodic bonding) or Si wafer (fusion bonding) over etched cavities in a bottom Si wafer. The Si and glass layers bonded typically are rather thick and do not lend themselves well to die feature level packaging. In polysilicon and selective epitaxy surface micromachining (see Volume II, Chapter 7), on the other hand, sealing cavities often is a more integral part of the overall fabrication process, and both die and die features might be packaged this way. These and other lithography-defined packages, such as those involving ultraviolet patternable polymers, lend themselves to inexpensive batch solutions and represent a new technology area where micromachining could provide solutions for the MEMS and IC industries.

Sealing of Polysilicon and Silicon Nitride Cavities

In Figure 4.33a, we illustrate how a polysilicon vacuum shell encapsulates a polysilicon resonator element. The micromachined surface package (microshell) illustrated is much smaller than typical bulk-micromachined packages. Microshells can be made by defining thin gaps (~100 nm) between the substrate and the perimeter of the structural elements by etching away a sacrificial layer sandwiched

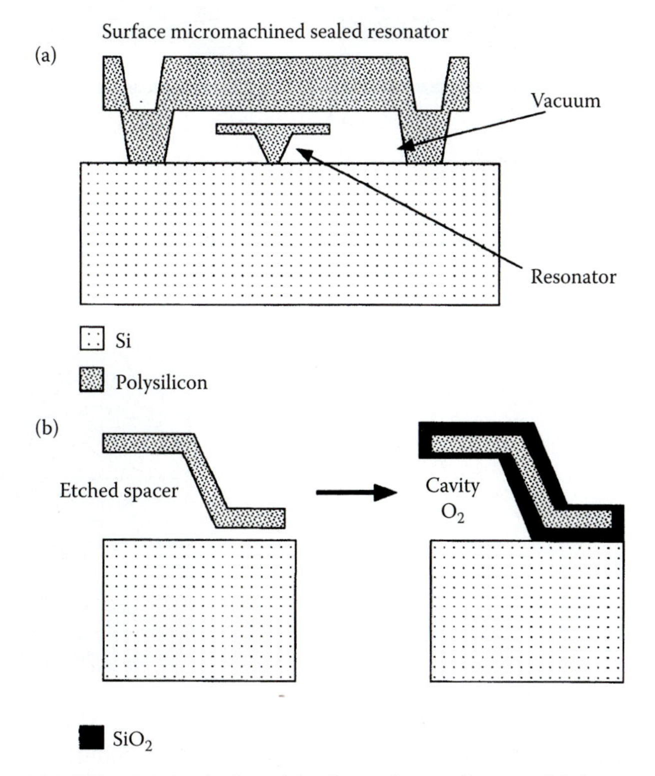

FIGURE 4.33 Sealed cavities in surface micromachining. (a) Typical sealed cavity with a resonator structure inside. (b) Reactive sealing.

between the two and then sealing the resulting gaps. In so-called *reactive sealing*, demonstrated in the schematic fabrication sequence in Figure 4.33b, thermal oxidation of the polysilicon and Si substrate at 1000°C seals the narrow openings left after removal of the spacer phosphosilicate glass (PSG).[66–68] Reactive sealing is possible even in vacuum because of the reaction of oxygen, trapped inside the cavity.

Alternatively, sealant films, such as oxides and nitrides, can be deposited over small etchant holes,[67,69] as illustrated in the fabrication sequence in Figure 4.34A. In the latter case, the cavity is sealed by low-pressure chemical vapor-deposited (LPCVD) silicon nitride. The excellent coverage of this method ensures that the nitride closes the etch channel quickly before too much deposition in the chamber itself can take place. Typical deposition conditions during the sealing step are 850°C and 250 mTorr, so the residual pressure in the microshell, at room temperature (assuming ideal gases) should be about 67 mTorr. Some researchers have reported residual pressures of 200–300 mTorr.[70] Eaton and Smith[71]

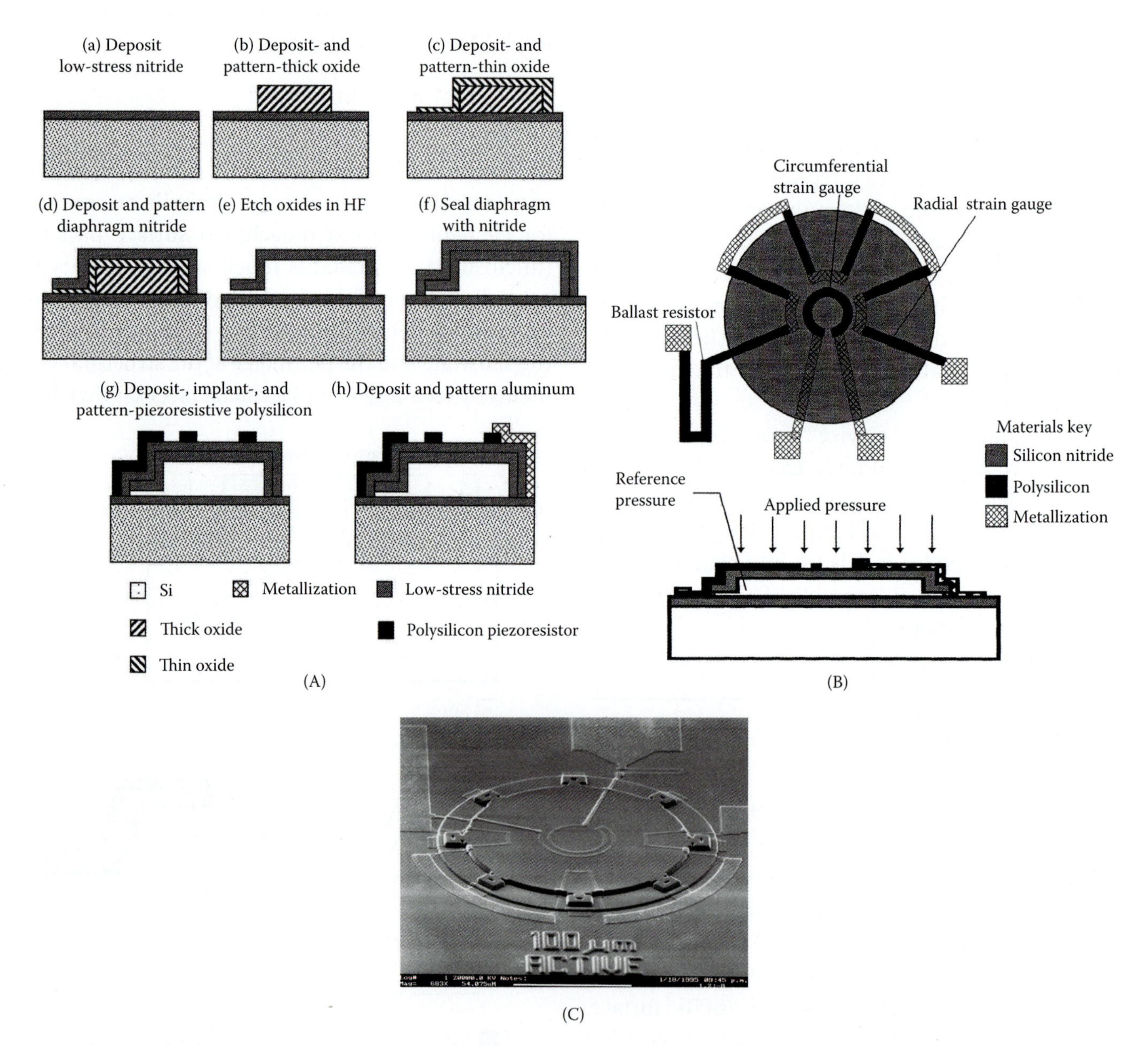

FIGURE 4.34 Absolute pressure sensor.[70] (A) Schematic fabrication sequence of vacuum shell with low-pressure chemical vapor-deposited Si nitride sealing. The vacuum shell constitutes the reference chamber of an absolute pressure sensor. (B) Schematic of absolute pressure sensor with Si nitride membrane. (C) Scanning electron microscope microphotograph of a finished device. (Courtesy of Dr. J. H. Smith, Sandia National Laboratories.)

established that the resulting cavity pressure is stable for a given membrane, but the residual pressure is variable and nonrepeatable across a substrate and even less from substrate to substrate.

The fabrication time of packaging shells often makes for a long process. The sacrificial material inside the shell is removed very slowly given the poor access via narrow etch holes to it by the etchant. Lebouitz et al. at the University of California, Berkeley, introduced an interesting means to speed up the fabrication of microshells.[72] The process is outlined in Figure 4.35. Permeable polysilicon windows (see Volume II, Chapter 7) are used as an etch access for removing the underlying sacrificial PSG. Using concentrated HF, shells 3 μm high and as wide as 1 mm have been cleared of PSG in less than 120 s. Subsequent low-pressure hermetic sealing using low-stress Si nitride leads to deposition of less than 100 Å inside the package.

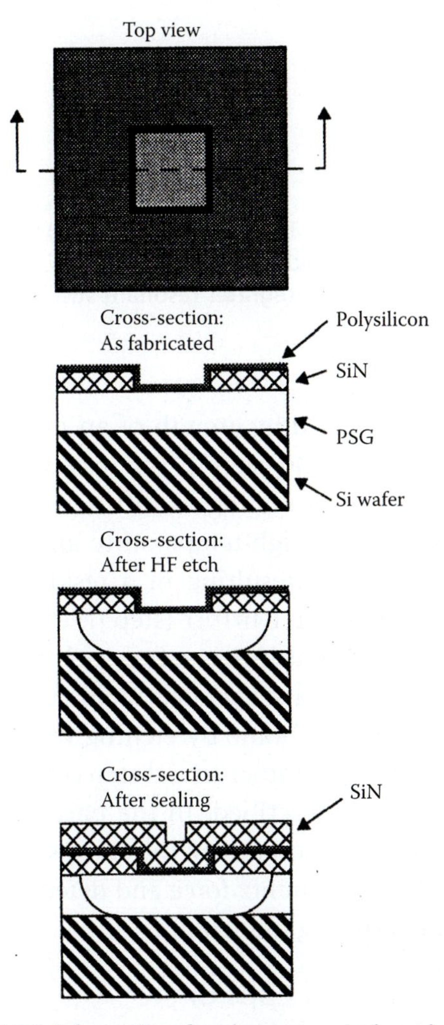

FIGURE 4.35 Schematic of etch access window design, operation, and sealing. [From Lebouitz, K. S., R. T. Howe, and A. P. Pisano. 1995. In *8th international conference on solid-state sensors and actuators (Transducers '95)*, 224–27. Stockholm, Sweden. With permission.[72]]

Sealed cavities, as shown here, lend themselves well to pressure sensing applications. Depending on the atmosphere to which the chip is exposed to during the sealing process, gauge, vacuum, or absolute pressure sensors can be created.[73,74] The process outline in Figure 4.34A actually illustrates the case of an absolute pressure sensor with a Si_3N_4 membrane.[71] The sensor consists of a circular Si nitride diaphragm, which forms the top of a sealed vacuum cavity, providing the pressure reference. Polysilicon strain gauges are fabricated on top of the diaphragm and the measured resistance changes are, to first order, directly proportional to the applied pressure. In Figure 4.34B, we show a schematic of the sensor, and in Figure 4.34C, we feature an scanning electron microscope (SEM) microphotograph of a finished device.

Of reactive sealing and sealant films, the reactive sealing process is by far the most elegant and highest in performance. The first commercial absolute polysilicon pressure sensor, incorporating a reactively sealed vacuum shell, was introduced for automotive applications in 1995 by SSI Technologies, but it is not a product offering anymore.[75] Another application of the above type of sealed surface shells is the vacuum packaging of lateral surface resonators. Most resonator applications share a need for resonance quality factors from 100 to 10,000. However, the operation of comb-drive microstructures, as shown in Volume II, Figure 7.76, in ambient atmosphere results in low-quality factors of less than 100 due to air damping above and below the moving microstructure.[76] Vacuum encapsulation is thus essential for high-Q applications. It also has been shown that lateral comb-drive microresonators,[77] besides their use in commercial accelerometers, have potential applications in areas such as mechanical filters for signal processing,[78] noncontact electrostatic voltage sensing,[79] and rotation-rate sensing.[80]

Epitaxial Deposition Cavity Sealing

Deposition of a set of epitaxial Si layers with varying doping has shown to afford the formation of a hermetically sealed surface cavity.[81] The cavity, in this

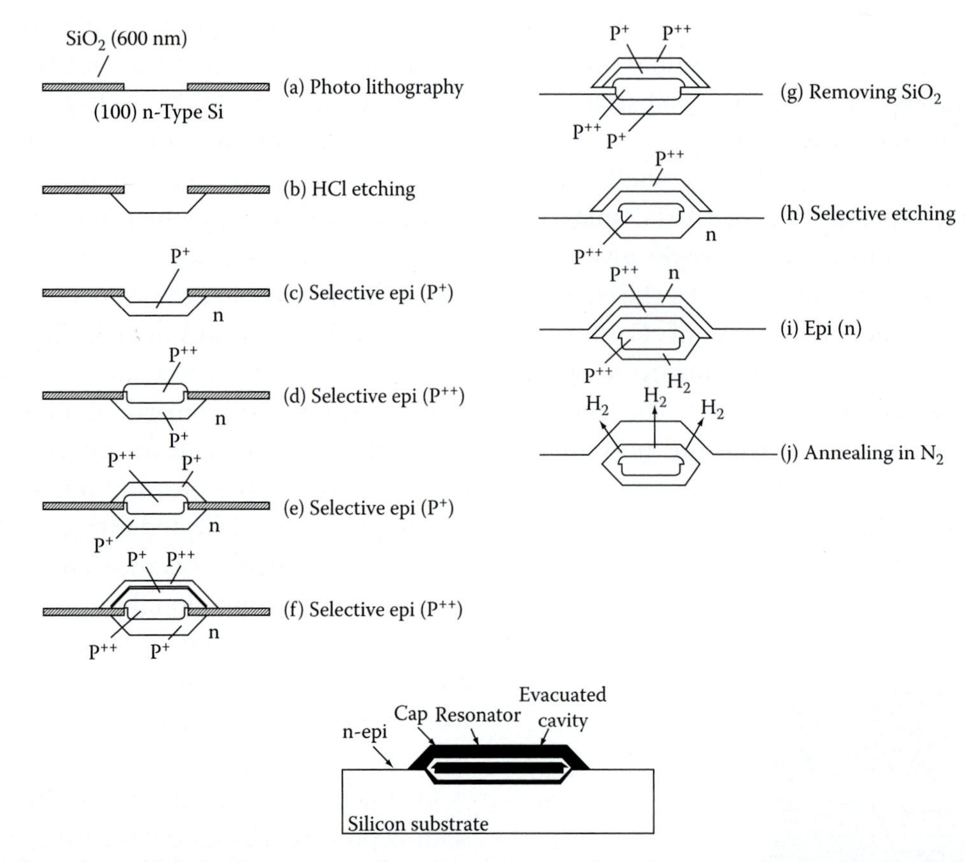

FIGURE 4.36 Epilayer-based fabrication process of a vacuum-encapsulated resonant beam. Process from (b) to (f) is carried out in one batch epitaxial process. For process step details see text. (From Ikeda, K., H. Kuwayama, T. Kobayashi, T. Watanabe, T. Nishikawa, T. Yoshida, and K. Harada. 1990. Silicon pressure sensor integrates resonant strain gauge on diaphragm. *Sensors Actuators A* A21:146–50. With permission.[82])

example of selective epitaxy surface micromachining, is formed by selective etching of p^+ epitaxial Si over more heavily doped Si p^{++} layers. The process is illustrated in Figure 4.36. The fabrication starts with a HCl dry etch at 1050°C in an epireactor through a hole in an oxide mask (Figure 4.36, steps a and b). Then follows a selective epitaxial growth of silicon in the following sequence: doped p^+, p^{++}, and p^+, p^{++} as indicated in Figure 4.36, steps c, d, e, and f. These steps are all carried out in the same epireactor, simply by changing the concentration of the B_2H_6 dopant. The next step consists of stripping of the oxide in HF (Figure 4.36, step g), followed by selective electrochemical etching in hydrazine (step h). In this electrochemical etching step, the n-substrate is passivated against the etching by the imposed potential and the p^{++} structures by the boron etch-stop mechanism. At this juncture, one ends up with a microbridge covered by a cap, both made out of single-crystalline, heavily boron-doped Si (p^{++}). The cap finally is sealed by growth of an n-type epitaxial layer (Figure 4.36, step i). The residual hydrogen in the cavity, after sealing, is diffused through the epitaxial layer by high-temperature annealing in a nitrogen ambient, resulting in a residual pressure inside the cavity of 1 mTorr (step j).[81,82] To make a pressure sensor that incorporates such an encapsulated resonator in a suspended membrane, one has to fabricate a membrane by etching from the back side of the wafer; a rather trivial step compared with the process just described. In the particular device shown here, the resonance of the encapsulated beam is activated by a Lorentz force and detected by measuring the resulting inductance.

Microassembled Caps or HEXSIL Cavity Sealing

Researchers at the University of California, Berkeley (UCB) developed another vacuum encapsulation method suitable for L0 and L1 level packaging involving the wafer-to-wafer transfer of micromachined caps as demonstrated in Figure 4.37a.[83] Reactive sealing requires a temperature of 1000°C, and thick SiN sealing requires 850°C. Moreover, in the latter technique some sealant gas deposits on the encapsulated microdevices.

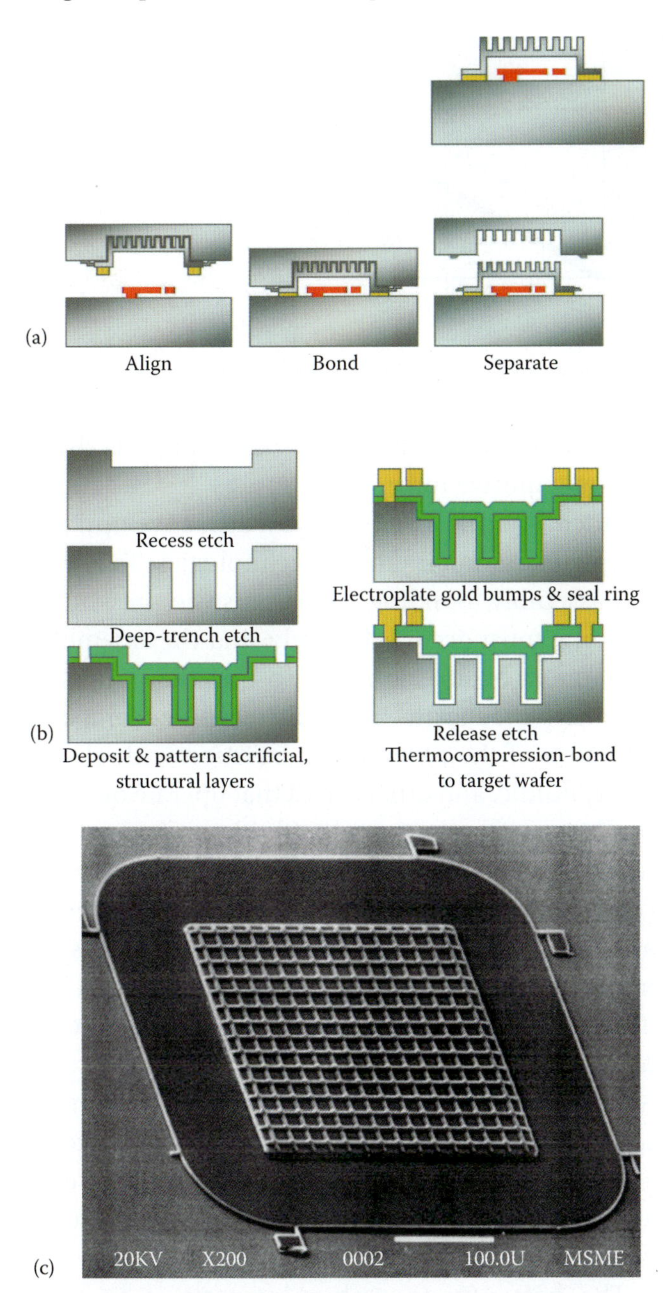

FIGURE 4.37 Wafer-to-wafer transfer of encapsulation structures. (From Hok, B., C. Dubon, and C. Ovren. 1983. Anodic bonding of galium arsenide to glass. *Appl Phys Lett* 43:267–69. With permission.[24]) (a) Principle of the wafer-to-wafer transfer method of encapsulation caps. (b) Fabrication of the tethered caps in a HEXSIL process. (c) Scanning electron microscope micrograph of a transferred cap.[82] (Courtesy of Dr. M. Cohen, University of California, Berkeley.)

In the alternative UCB packaging method, tethered cap structures are sealed down to the substrate employing a low-temperature Au-Si eutectic bond (the gold-silicon eutectic at 363°C is safely below that for aluminum-silicon). The encapsulation caps are made in the HEXSIL* process, as demonstrated in Figure 4.37b (see Volume II, Chapter 7, for details on the HEXSIL process). The caps have the general shape of a top hat and are suspended by breakaway polysilicon flexures in the handle wafer. The brim of the hat is coated with 0.7 μm of gold. The HEXSIL wafer with the embedded caps is positioned ($\Delta x <$ 10 μm) over the Si wafer with the active devices to be encapsulated. In situ annealing in vacuum by infrared heating forms the eutectic bond between the Si of the bottom wafer and the gold rim of the cap. The transfer occurs at a chamber pressure of 10^{-5} Torr, with 10 psi of mechanical pressure applied to the wafer sandwich. When the handle wafer is withdrawn, the polysilicon flexures break, leaving the caps sealed to the substrate wafer. In some limited test runs in a laboratory setting, a transfer yield of 100% was obtained for arrays of thirty caps. The HEXSIL mold wafer is reusable. An example of a transferred encapsulation cap is shown in the SEM micrograph in Figure 4.37c. To summarize, in the microassembled caps process, microcaps are fabricated on a donor wafer and transferred to a target wafer by wafer bonding and separation. The thin seal ring on the cap requires very little real estate (~1% of a bulk cap). The process potentially might be much less expensive than wafer-bonded caps as well.

Many of the newfound ways of hermetically sealing and bonding different layers and making contacts between them increasingly interest IC manufacturers. An important opportunity for micromachinists is to transfer the developed 3D machining technologies to the newest generations of 3D ICs.

Assembly and Self-Assembly

Introduction

Assembly is used to build complex products from relatively simple parts and to integrate incompatible

* HEXSIL is the acronym for hexagonal honeycomb poly-Si.

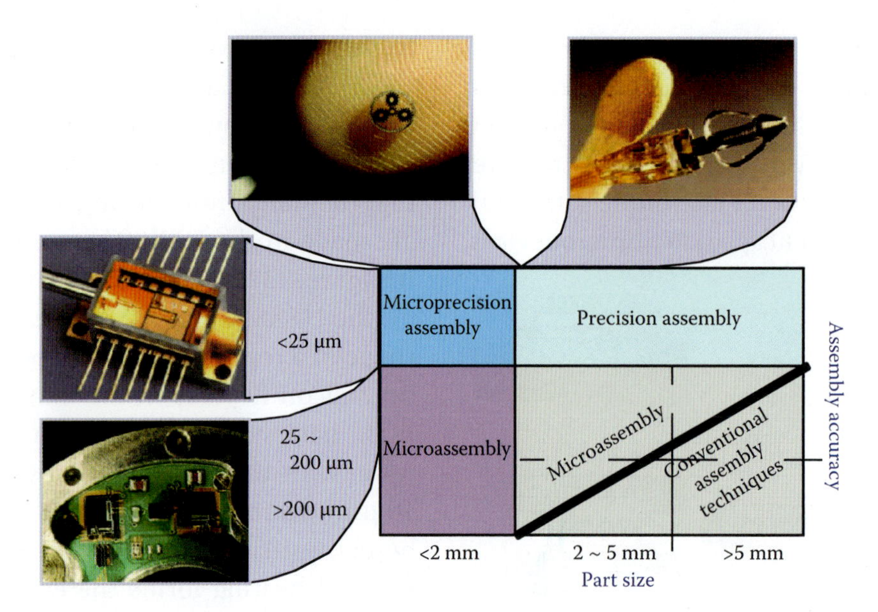

FIGURE 4.38 Assembly processes: part size assembled and assembly accuracy. (Courtesy of Kory Ehmann.)

manufacturing processes. Today, IC and MEMS assembly is mostly carried out by humans with tweezers and microscopes or with pick-and-place robots.[84] Micro- and mesoscale assembly processes handle objects from several micrometers to several millimeters in size. In Figure 4.38, we consider part size and assembly accuracy.

The current challenge is to assemble smaller and smaller components from different technologies with micro- and nanoprecision. The application field of micro- and nanoprecision assembly lies well beyond conventional or microscale assembly, with part dimensions larger than 2 mm (see Figure 4.38). The various micro- and nanoassembly approaches, divided into serial and parallel processes, are summarized in Table 4.3. In serial micro- and nanoassembly, parts are picked up and placed together on a substrate or workpiece one by one. Table 4.3 covers many methods enabling serial manipulation of very small parts/particles, including individual atoms for nanoassembly. Unless one can manipulate many parts/particles in parallel, most of these techniques will remain research tools rather than manufacturing options. In parallel micro- and nanoassembly, where parts are assembled simultaneously, we distinguish between deterministic and stochastic assembly.[84] In the former, the relationship between the part and its destination on a substrate is known in advance. In the latter, this relationship is unknown or random; parts "self-assemble" during a stochastic process driven by some motive force.

Stochastic self-assembly is an attractive option to integrate dissimilar process technologies. To achieve self-assembly, bonding forces must be present, the bonding must occur selectively, and the assembling parts must be able to move so that they can come together. Stochastic assembly often encompasses vibration of the parts in combination with electrostatic, fluidic, and other forces that operate on singulated parts in various media (fluids, air, or vacuum). After reviewing scaling issues of the assembly process, several examples of micro- and nanoassembly are presented. The smaller the building blocks, the higher the need for stochastic self-assembly, as handling the individual components becomes more and more of a mechanical challenge, and the process would be too slow. Nature, building on the nanoscale, does rely heavily on self-assembly (Chapter 2), and DNA-assisted self-assembly is an early attempt at co-opting nature for human manufacturing endeavors.

The best way to grab microparts is to use a microgripper, where grasping forces and jaw size match the requirements. Many kinds of microgrippers have been built. In Volume II, Figure 7.88, we show an example of a microgripper made of high-aspect-ratio molded polycrystalline silicon. The white bar at the bottom of the picture represents 100 μm. The actuator is an electrically heated

TABLE 4.3 Micro- and Nanoassembly Approaches

Serial Microassembly		Parallel Microassembly			
Type	**Refs.**	**Type**	**Refs.**	**Type**	**Refs.**
Deterministic		Deterministic		Stochastic	
Examples		**Examples**		**Examples**	
Manual assembly with tweezers and microscopes		Flip-chip wafer-to-wafer see, e.g., Figure 4.13	87, 88	Fluidic agitation and mating part shapes	89–91
Visually based and teleoperated microassembly	85, 86	Microgripper arrays for massively parallel pick-and-place (see Volume II, Figure 7.88 for an individual gripper)	94	Vibratory agitation and electrostatic force fields	95–99
High-precision macroscopic robots with submicron resolution	92, 93			Vibratory agitation and mating part shapes	103
Microtweezers	94, 100–102			Mating patterns of self-assembling monolayers	105
Optical trapping, see, e.g., Figure 2.33 this volume	104			Magnetic field-mediated separation	107
Microcapillary/pipette-based positioning, see, e.g., Figure 4.43	106			Dielectrophoretic trapping	109
AFM/STM atom manipulation, see, e.g., Volume II, Figures 2.43 to 2.45	108			Ultrasonic levitation and transportation in standing acoustic waves	110
				Hydrodynamic/electro-osmotic object manipulation in microchannels	111, 112

Source: Based on Böhringer, K. F., R. S. Fearing, and K. Y. Goldberg. 1998. In *Handbook of industrial robotics,* ed. S. Nof. New York: Wiley.[84]

thermal expansion beam, which causes the compound lever linkage to move the tips. Such grippers have been used to seize various microscopic objects, including 2.7-μm-diameter polystyrene spheres, dried red blood cells, and various protozoa.

Scaling of the Assembly Process

Scaling effects in microassembly represent a daunting task. The newest pick-and-place robots with submicrometer resolution must be used, as conventional robots that pick, transport, and place small parts have a control error of 100 μm at best, which would translate in relative errors of 100% on micrometer-sized parts. It might be difficult to pick up large objects with a robot gripper, but small parts (<1 mm and masses less than 10^{-6} g) are difficult to release, as adhesion forces between gripper and object are significant compared with gravitational forces at that scale.[84] In Chapter 7, this volume, on scaling, we will indeed learn that in the microdomain, surface forces such as adhesion dominate over volume forces such as gravity or inertia. Adhesive forces involved in pick-and-place include surface tension due to moisture on the part and gripper, van der Waals forces (significant for gaps below 100 nm), and electrostatic attractions (e.g., induced by triboelectrification).

In Figure 4.39, we compare the magnitude of attractive surface forces (electrostatic, van der Waals, and surface tension) with the gravitational force as a function of part size, assuming a sphere (part)-plane (gripper) interaction. It is observed that in the microworld, surface forces dominate over gravitational influence. At 1 μm the gravitational

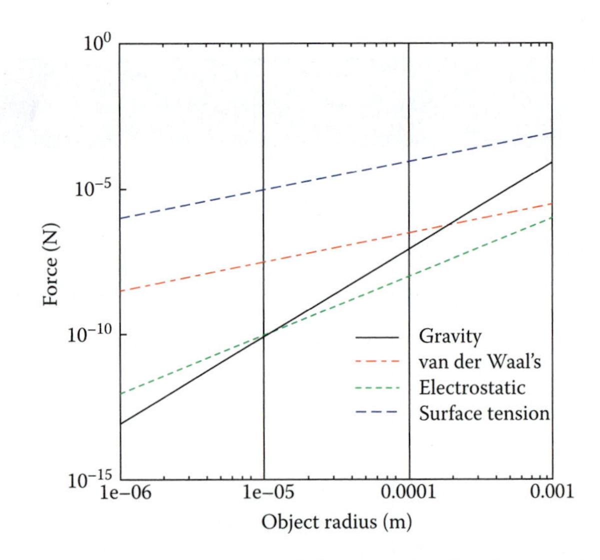

FIGURE 4.39 Comparison of the magnitude of attractive surface forces, i.e., electrostatic, van der Waals, and surface tension, with the gravitational force as a function of part size, assuming an interaction between sphere (part) and plane (gripper).

force is the least important force of the four, and at 10 μm the gravitational force equals the electrostatic force.

The electrostatic force is caused by charge attraction. Between a charged spherical object and a conductive plane it is approximated as follows:

$$F_{el} = \frac{\pi}{4\varepsilon_0}\frac{\varepsilon-\varepsilon_0}{\varepsilon+\varepsilon_0}d^2\sigma^2 \tag{4.1}$$

where

- d = the diameter of the particle
- σ = the charge density on the surface
- ε_0 and ε = the air's and the plane's dielectric constants, respectively

The distance z between the sphere and the plane is assumed to be very small (Figure 4.40).

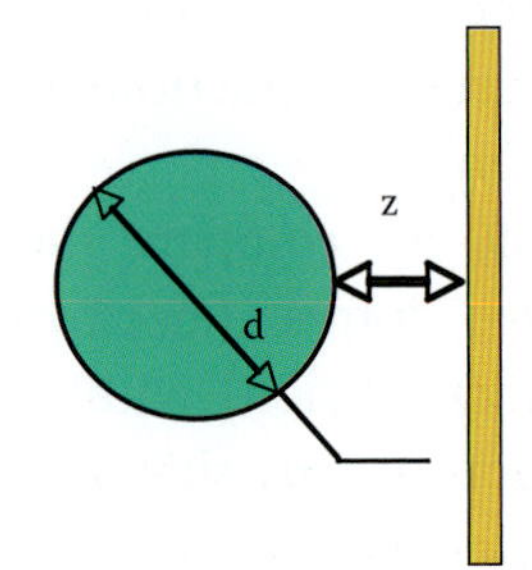

FIGURE 4.40 The electrostatic force between a spherical particle and a wall.

It is difficult to ensure that parts and grippers are electrically neutral because of the friction force and the difference in contact potentials between part and gripper. The gripper is typically coated with conductive coatings and grounded (the electrostatic force between two good insulators at 1 μm is 10^6 Pa!).

We already know that van der Waals forces may cause sticking together of MEMS parts and that this stiction may become permanent (see Volume II, Chapter 7, on stiction in surface micromaching). The van der Waals force is an atomic force caused by momentary movements of electrons, i.e., dipoles (see Volume I, Chapter 7). The sphere-plane van der Waals force is calculated as follows (see also Volume I, Table 7.8):

$$F_{vdw} = \frac{Hd}{12\,z^2} \tag{4.2}$$

with the Hamaker constant H (see Volume I, Chapter 7, Table 7.6 for some typical values), the object diameter d, and the object-plane distance z (it is again assumed that $z \ll d$). This force becomes significant at less than 100 nm and increases with contact area (remember the popular study of Gecko Setae, with a force as high as 60 μN on a single seta; see Volume I, Chapter 7). Rough surfaces increase the effective distance z and therefore reduce the van der Waals force. Making the gripper surface rough might compromise ultimate positioning precision though. Also, gripper coatings can reduce the van der Waals force.

The presence of water on surfaces increases the normal force between the contacting surfaces as a result of capillary effects. The source of the liquid film between a gripper and part can be either a preexisting film of liquid and/or capillary condensates of water vapor from the environment. The surface tension force can be attractive or repulsive depending upon the configuration and the materials involved. The degree of hydrophobicity/hydrophilicity affects the contact angle (θ) and thus the surface tension force. The capillary tension force between a ball and a plate, with reference to Figure 4.41, is given by the following:

$$F_{tens} = \pi R_2^2\gamma\left(\frac{1}{R_1}+\frac{1}{R_2}\right) \tag{4.3}$$

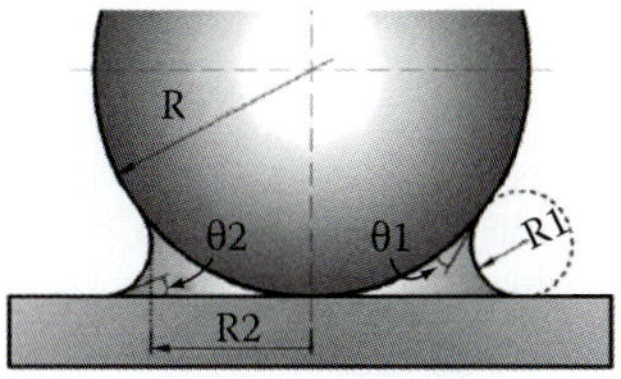

FIGURE 4.41 The capillary tension force between a ball and a plate.

where $\gamma = 72 \times 10^{-3}$ N/m is the water interfacial energy, and R_1 and R_2 are the axial and the tangential radius of the meniscus. To a good approximation, $2R_1$ corresponds to the thickness of the water film between the ball and the plane at the level of the meniscus, and both radii R_1 and R_2 can be expressed as function of ball radius R.

For small values of $2R_1$ the capillary force between the ball and the plane is given by the following equation:

$$F_{tens} = 2\pi R\gamma\left(2 + \sqrt{\frac{R_1}{R}}\right) \tag{4.4}$$

For $R_1 << R$, the capillary force F_{tens} does not depend on the meniscus radius, and this relation simplifies to the following:

$$F_{tens} = 4\pi R\gamma = 2\pi d\gamma \tag{4.5}$$

where R is the ball radius and d its diameter. When humidity is high, or with hydrophilic surfaces, a liquid film between a spherical object and a planar surface contributes a large capillary force. This effect can be reduced, for example, by working in a controlled environment or in a vacuum.

Surface forces make the release of objects after picking them up very difficult. To overcome these adhesion problems, some simple mechanisms for small-scale manufacture have been developed. They include electrostatic probes and vacuum probes. Also, the smallest components often are manipulated in a liquid medium (e.g., with laser trapping or dielectrophoresis). Microassembly obviously requires new handling techniques and electrostatic contact-less handling appears to be quite promising. Using a desktop factory (DTF) (see Volume II, Chapter 6), Nakao et al. used an electrostatic probe tip to assemble the microhouse shown in Figure 4.42.[113] The electrically charged tool can pick up and place down micro-objects, and a semiconductor laser bonds both metal and nonmetal materials.

FIGURE 4.42 Scanning electron microscope of a microhouse. Using a desktop factory and an electrostatic probe tip, a microhouse was assembled.[113]

Vacuum pipettes, commonly used in microbiology for handling living cells, are another microassembly option. A vacuum gripper consists of a glass pipette and a vacuum control unit, and the actuating force is generated by the pressure difference on both sides of microcomponents. They are easy to use and are capable of picking, holding, and placing, and to overcome adhesion during placement they may use stripping off, pushing, or blowing away (see Figure 4.43).

On a large scale, vibration is used extensively in industrial part feeders that singulate, position, and orient parts before they are transferred to an assembly station. Sony's Advanced Parts Orientation System (APOS) part feeder, for example, uses an array of nests (silhouette traps) cut into a vibrating plate.[114] The traps and the vibratory motion are designed so that parts can remain in the nest only in one particular orientation. The traps in the vibrating plate are filled by tilting the plate and letting the parts flow across it. The APOS system with its controlled vibration to provide stochastic motion, combined with gravity as a motive force for nonprehensile manipulation, is a good model for microassembly. Again, scaling effects must be taken into account; vibration relies on the inertia of the parts and becomes less efficient with decreasing part size, so that higher vibration amplitudes or frequencies are required for smaller parts.

In the microscale, objects can be transported, for example, using acoustic traveling waves generated on a surface (see Chapter 8, this volume, Figure 8.28). Two types of waves are used, Lamb and Rayleigh waves. Lamb waves cause bending of the plate over its whole thickness (the substrate thickness is small

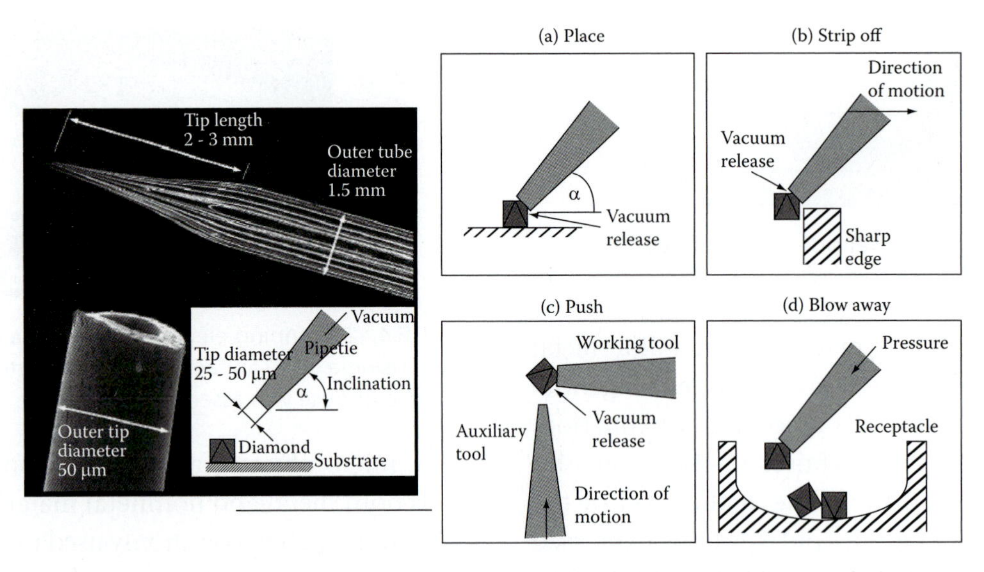

FIGURE 4.43 A glass pipette and a vacuum control unit are capable of picking, holding, and placing, and to overcome adhesion during placement they may use stripping off, pushing, or blowing away.

compared with the wavelength), while Rayleigh waves propagate only in the surface layer of the plate. Objects placed on the surface move in the opposite direction of the traveling wave (see Figure 8.28). Separation of particles may be effected by the competition between the acoustic force and the electrostatic force. The displacement of particles from pressure nodes in acoustic standing waves depends on the effective charge, radius, and stiffness of the particles. Yasuda has used lead zirconate titanate (PZT) transducers in a quartz cell to continuously concentrate, separate, and mix cells (e.g., blood cells).[115]

Another option is the use of distributed micromotion systems using arrays of tiny simple actuators that cooperate to move objects over relatively large distances and possibly in different directions and orientations. Several MEMS research groups have built such actuator arrays for micromanipulation; these arrays typically consist of a regular grid of "motion pixels" (mechanical members such as cilia).[116–123] These actuators are called *cilia* by analogy to their biological counterparts (see Chapter 8, this volume, Figure 8.11, and Chapter 5, this volume, Figure 5.13). Cilia arrays can move small objects placed on them by synchronous vibrations of the tiny hairs. To move objects in the x direction as well as the y direction, four cilia, one for each direction and sense, are grouped in motion pixels. The motion system consists of an array of these motion pixels. Appropriate control allows motion in any direction.

Table 4.4 summarizes some of the assembly issues one faces when manipulating parts in the mesoscale to the microscale and then into the nanoscale.

Serial Microassembly

In Volume II, Chapter 6, we saw how a 280-μm-high Japanese temple (Gojyunoto) on a polyimide platform was erected with the help of a microrobot manipulator. Many of the serial manipulation methods listed in Table 4.3 are used for manipulation of even smaller components, smaller than the typical MEMS device, such as biological cells, subcellular organelles, and even atoms.[125]

A small-enough object that is relatively transparent to laser light of a particular frequency refracts an incident laser beam, bending the light. As a result of this refraction, momentum is transferred from the light to the particle (see laser tweezers in Figure 2.33 this volume and in Volume I, Figure 5.34). When the geometry of light beams and particle is correct, the imparted momentum pulls the particle in the direction of the incident beam. The beam (or multiple beams) can then hold the particle in place, or by moving the beam(s), the laser operator can move the particle from place to place. In other words, light has momentum* and can be used to catch and manipulate objects in a size range from nanometers to micrometers in so-called *optical tweezers*.

* As far back as 1619, Johannes Kepler explained that the tails of comets are always directed away from the sun by the pressure of sunlight.

TABLE 4.4 Assembly Issues as a Function of Part Size[124]

Assembly Attribute	Assembly Scale		
	Mesoscale	Microscale	Nanoscale
Positioning	Easy	Difficult	Very difficult
Velocity	cm/s or m/s not unusual	Slow (μm/s or mm/s), vibration suppression	Very slow (nm/s or μm/s)
Force sensing and control	Easy, necessary to avoid part damage and improve manipulability	Difficult; range of forces could be as low as μN	Difficult; AFM used to measure force
Dominant forces	Gravity, friction	Surface forces (stiction, friction, electrostatic, Van der Waals)	Molecular/atomic forces
Throughput	Serial assembly provides adequate throughput	Serial assembly usually not sufficient; parallel manipulation methods preferred	Parallel manipulation methods or self-assembly necessary
Gripper	Mechanical	Micromechanical, gripper-free manipulation preferred	Other methods needed: optical, proximity forces, etc.
Fixturing	Mechanical	Micromechanical fixturing must be used	Chemical
Compliance	Gripper compliance not necessary if force measured	Gripper compliance usually necessary	Mechanical compliance not applicable
Vision	Easy	Difficult (expensive optics)	Impossible with visible wavelengths; SEM and TEM are used

The laser wavelengths in optical tweezers typically are between 0.7 and 1.06 μm, and 25 to 500 mW of power is generated in a focal spot between 0.5 and 1.0 mm in diameter. The first work with optical tweezers was performed by Ashkin et al. at AT&T Bell Laboratories. In pioneering works in 1985 and 1986, they successfully demonstrated that small particles could be trapped by focused low-power laser light (<1 W).[126,127] The first demonstration on trapping of living biological objects was also by Ashkin et al. in 1987.[104] The authors not only trapped and manipulated single cells, they demonstrated that the trap was harmless to living biological species. For example, reproduction of yeast cells was achieved in a trap capable of manipulating them at velocities of 100 m/s. Combined with other techniques (such as a laser scalpel), a number of important studies within microbiology have been carried out and many more can be visualized in the future. The difference between tweezers and scalpels is that the latter employ short pulses of high irradiance, whereas tweezers use continuous, low-irradiance beams. The laser scalpel can be used for cutting biological objects inside cells (e.g., DNA segments). Steven Chu,* a 1997 Nobelist in Physics (see Volume I, Figure 1.28) showed while at Stanford University that laser tweezers can be used to stretch molecules such as DNA: with colloidal spheres attached to the DNA strands, the spheres were trapped and the DNA was stretched by moving in fluid flows[128] (see also Chapter 2, Figure 2.33). The most common form of optical tweezers in biology is the "single-beam gradient trap," immobilizing transparent dielectric particles near the focal point of a tightly focused continuous-wave laser with a force proportional to the electric field gradient:[125,129,130]

$$F_{grad} = -n\left(\frac{1}{2}\alpha \nabla E^2\right) \quad (4.6)$$

where

n = the index of refraction of the solvent/medium

α = the polarizability of the particle

Such a beam generates forces in the piconewton range, more than sufficient for trapping cells and moving organelles inside and outside the cells. Usually, the technique cannot be used to lift objects much larger than the wavelength of the light. Links to pages related to optical tweezers can be found at http://www.phys.umu.se/exphys/OpticalTweezers.

Two different serial manipulation approaches can be used together. For example, cages for biological

* Dr. Steven Chu is currently the US Secretary of Energy.

cells have been made by combining optical trapping and dielectrophoresis.[131] Such a combination enables manipulation of two bioparticles in close proximity, which would be impossible using two laser beams or two electrical fields, as those fields would interfere. Dielectrophoresis is an AC electrokinetic effect (electro-osmosis and electrophoresis are direct current electrokinetic effects) discussed in detail in Volume I, Chapter 7. It is a powerful tool for separating particles of the size of biological cells but is more difficult to apply for particles below 1 μm.

In mechanosynthesis, an STM is used to manipulate individual atoms, as demonstrated in Volume II, Figure 2.43, where Pt atoms have been arranged on a Pt(111) surface. The Kanji character for *atom* in Figure 6.26 was also made this way. As pointed out in Chapter 1, serial manufacturing processes that use building blocks that are too small are not viable manufacturing candidates unless an army of them can work together on the same construction project.

Parallel Microassembly

The concept of self-assembly is inspired by biological processes in which complementary pairs such as antigens bind to antibodies on the basis of shape recognition. In the nonbiological world, objects can be similarly assembled on the basis of geometry. The advantage of self-assembly is that it is a batch process, in which hundreds or thousands of objects or even more are assembled at a time. When the interface is well designed, the objects are also self-aligned, and different types of objects can be assembled simultaneously by using different interfaces.

An example of deterministic parallel assembly by Cohn et al. for wafer-to-wafer transfer of MEMS and other microstructures was shown already, in Figure 4.37.[83,87] In this HEXSIL flip-chip process, the finished microstructures are suspended on breakaway tethers on the substrate. The target wafer with indium solder bumps is precisely aligned and pressed against the substrate, such that the microstructures are cold welded onto the target wafer. In another example of deterministic parallel microassembly, Keller et al. demonstrated microgrippers (see Volume II, Figure 7.88) and proposed gripper arrays for parallel transfer of palletized microparts.[94,102]

Many stochastic approaches for microassembly are under investigation. Yando, for example, employs an array of magnetic sites to assemble magnetically coated semiconductor parts in a square array. The components are vibrated at a gradually attenuated amplitude to "place" the array elements.[132] Similarly, Yeh et al.[90,91] trap semiconductor light-emitting diodes (LEDs), suspended in a liquid, in micromachined wells on a wafer by solvent-surface forces. As illustrated in Figure 4.44, carrier fluid containing the GaAs parts is dispensed over the host Si wafer with etched holes to assemble the GaAs blocks. Because of the trapezoidal design, the blocks fit preferentially into the holes in the design orientation. Random mechanical vibration of the microstructures enables large numbers of microstructures to be positioned into precise registration with the sites. More than 90% of the holes etched in Si were correctly filled by the blocks before the carrier fluid evaporated. However, during evaporation, surface tension pulls some objects out of the holes, reducing the yield to 30–70% locally. This problem can be addressed by using liquids with lower surface tension or by using supercritical drying methods.

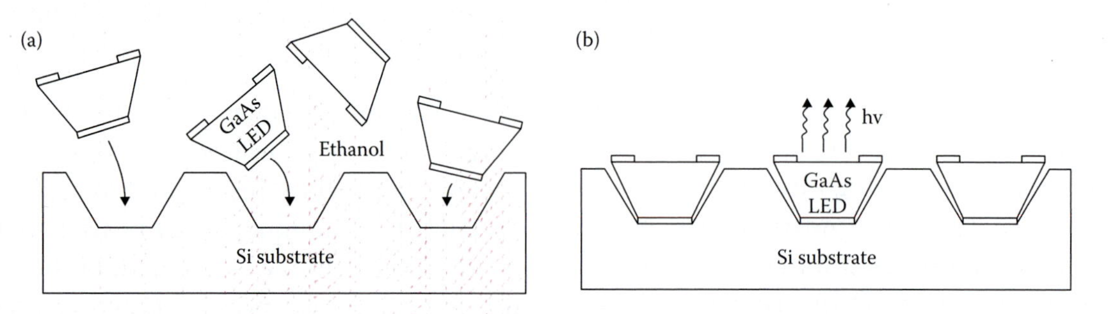

FIGURE 4.44 Fluid self-assembly of GaAs light-emitting diodes (LEDs) on Si micromachined substrate. (a) Solution containing the GaAs blocks is dispensed over the patterned silicon substrate. (b) Silicon substrate with GaAs LEDs integrated by fluidic self-assembly. [From Yeh, H.-J., and J. S. Smith. 1994. In *Proceedings of the IEEE International Workshop on Micro Electro Mechanical Systems (MEMS '94)*, 279–84. Oiso, Japan. With permission.[90]]

In 1991, Cohn et al.[96] first reported on stochastic assembly experiments that use vibration and gravitational forces to assemble periodic lattices of up to 1000 silicon chiplets.[133] Later, the same group[96] demonstrated an interesting variation on the work of Yando[132] and Yeh et al.[90,91] by using alignment capabilities over electrostatic traps. The process is illustrated in Figure 4.45. A critical problem in attempting to vibrate microstructures into specific sites from a liquid is that when structures adhere to the substrate or to each other, their progress toward optimal, precise registration is halted. This is caused by the same stiction phenomena we described for surface micromachined structures in Volume II, Chapter 7. Cohn et al. addressed this problem by levitating the microstructures a short distance above the target electrostatic trap site before letting them settle. As a result, random sticking is prevented, and critical positioning and orientation of the microstructure with respect to the trap become possible with a relatively low applied field (10 V/μm). The levitation over the trap site is accomplished by creating a short-range repulsive force between the microstructure and the target site. Microstructures with a relatively high permittivity (ε = 10) are suspended in a low-permittivity solvent (hexane), resulting in a net attraction to the charged electrodes making up the sites. However, a small amount of a more polar solvent (acetone) is added, which segregates into the high-field region between the two attracted parts. This prevents contact between the site and the suspended microstructure. Critical positioning and orientation become possible, in the manner of a compass needle resting on a low-friction bearing.[96] Once the sites are populated, the polar solvent is titrated out, gradually lowering the trapped devices to the target sites and pinning them onto the surface. A subsequent sintering step could effect permanent attachment—say, by eutectic bonding. Böhringer et al. subsequently demonstrated the ability to break surface forces using ultralow-amplitude vibration in vacuum, which holds promise as an extremely sensitive technique for positioning parts as well as discriminating part orientation, shape, and other physical properties.[97] Hosokawa et al.[134] modeled the kinetics of self-assembly and, using planar parts of various simple geometries floating on water, demonstrated two-dimensional (2D) self-assembly. The units, consisting of a polyimide layer on a polysilicon layer (assembly units were 400 μm in size), selectively bonded to each other because of surface tension (a dominant force in the microscale; see above). The floating parts that were at equal height attracted each other. Sharp corners induced yet larger selective attractive forces, and parts located at different heights were mutually repulsed. To disturb the self-assembly system, magnetic or fluid forces were used. The floating units tended to align, with their sharpest features pointing toward each other.

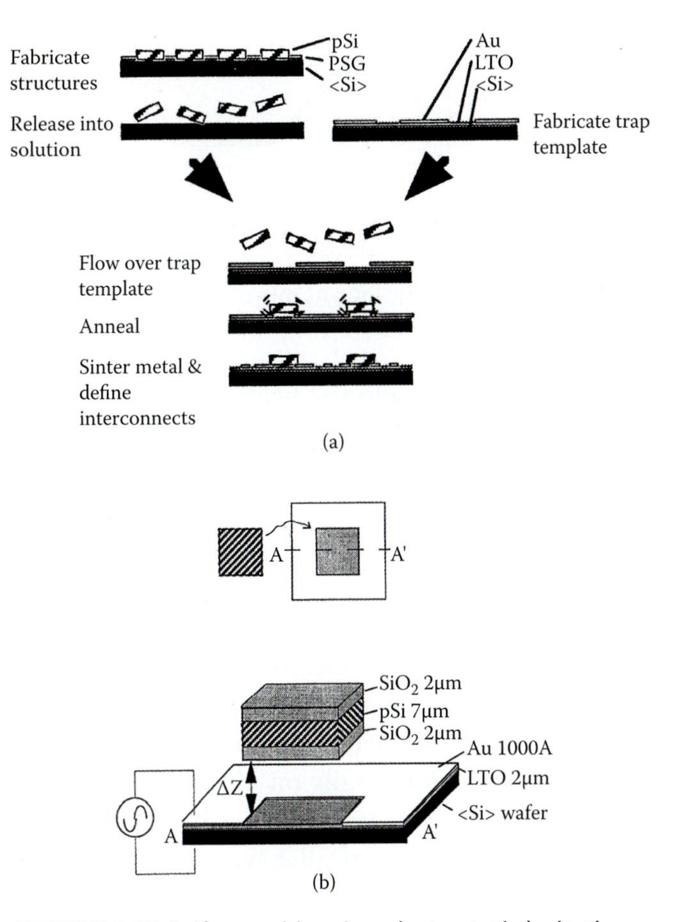

FIGURE 4.45 Self-assembly using electrostatic levitation. (a) Batch assembly of microstructures into binding sites on a substrate. Structures may include microelectromechanical system, complementary metal-oxide-silicon, optoelectronics, magnets, etc. (b) Detail of an electrostatic binding site. The section view shows position of a levitated microstructure. The levitation height (Δz) is in the range of 0–100 mm. The microstructure, a SiO_2-pSi-SiO_2 sandwich, has an average dielectric constant (ε) of 10, between that for hexane and acetone. The aperture in the gold film may range from ~2 to 100 mm in diameter, depending on the size of the microstructure to be trapped. (Courtesy of Dr. M. Cohen, University of California, Berkeley.)

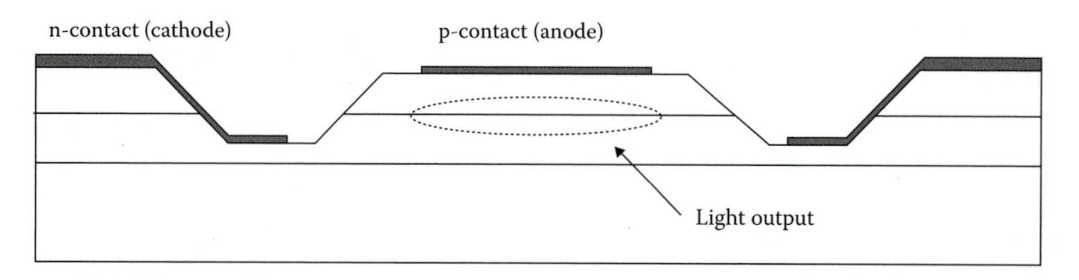

FIGURE 4.46 Schematic InGaAs microlight-emitting diodes have contacts on one side of the device as shown in the cross-section. (Courtesy of Dr. M. Heller, Nanogen.)

At Nanogen, Inc., electric fields are used to transport and control the placement of proteins, RNA, and DNA, as well as semiconductor components on a substrate equipped with an array of electrodes in a low-conductivity solution.[99,135] The electrophoretic nano assembly of DNA on the Nanogen platform is reviewed in Volume II, Chapter 8, Example 8.2. Here we illustrate the use of electric fields for precision placement of microlight-emitting diodes (microLEDs) (20 μm in diameter) onto a Si substrate using electro-osmosis (for details on electrophoresis and electro-osmosis, see Volume I, Chapter 7). The InGaAs microLEDs used all have contacts on one side of the device, as shown in the cross-section schematic in Figure 4.46. The corresponding substrate site has two sets of electrodes designed for the mounting process. As seen in Figure 4.47, a first pair of larger horseshoe electrodes brackets an inner set of electrodes. The inner set of electrodes is designed to mate with and bond to an LED and is made of standard tin/lead, electroplated to a height 50 nm above the surrounding surface. The linear structures seen in the pictures connect the exposed electrodes and are insulated to avoid shunting through the overlaying aqueous solution. Activation of the outer electrodes with respect to a counterelectrode (not shown in the figure) causes an electro-osmotic flow (see Volume I, Chapter 6) along the surface of the chip and moves the LED placed in the solution. Once the LED is on the electrode, the electro-osmotic flow is briefly reversed, and the reversal of the fluid flow drives the LED onto the awaiting bonding points

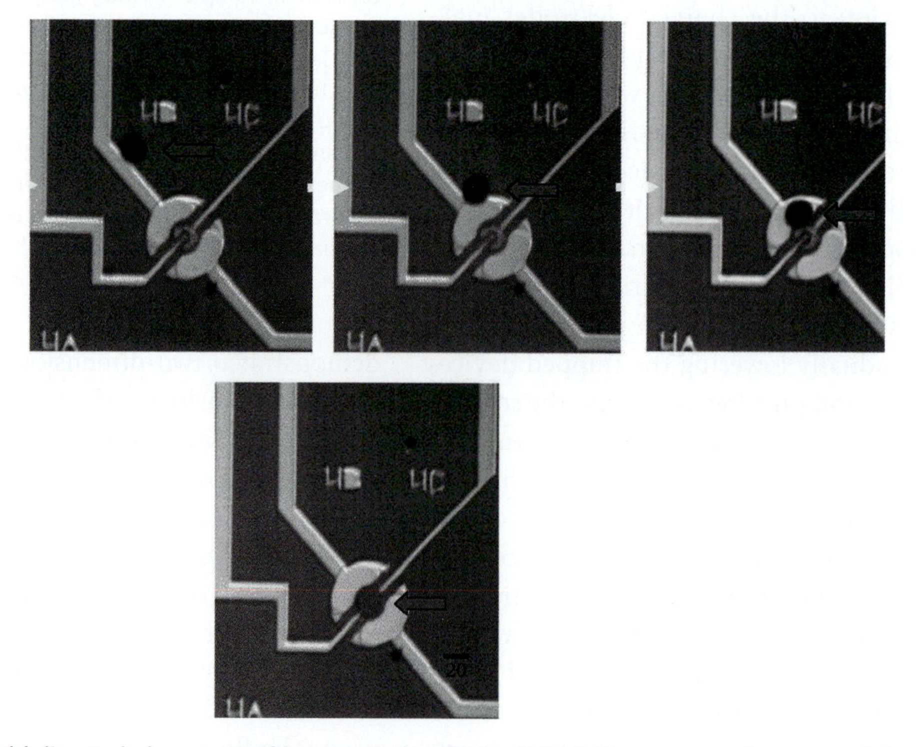

FIGURE 4.47 Electric field directed placement of light-emitting diodes (LEDs). The arrow points toward the moving LED. (Courtesy of Dr. M. Heller, Nanogen.)

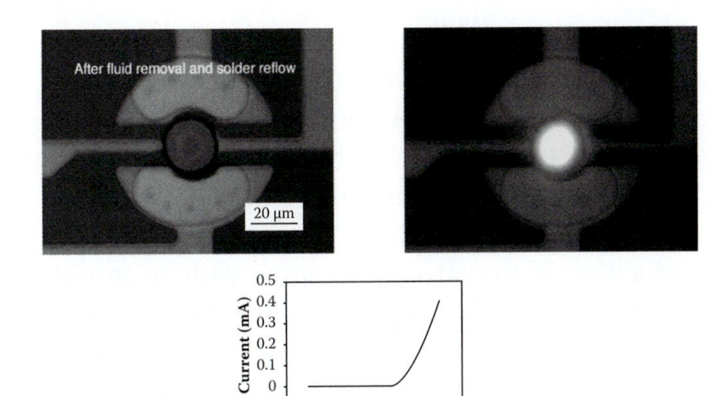

FIGURE 4.48 Circuit integrity and light-emitting diode (LED) activation demonstrated by forward biasing the LED. (Courtesy of Dr. M. Heller, Nanogen.)

with microscale precision. After the fluid is removed, the solder is reflowed under forming gas at 250°C. Circuit integrity and LED activation are demonstrated by forward-biasing the LED using the silicon circuit's contacts, as demonstrated in Figure 4.48.

The Nanogen authors point out that the same process could be used to assemble a large variety of components simultaneously and selectively on a circuit by the parallel activation of sets of electrodes while maintaining a number of components present in the low-conductivity solution. They speculate that the process might be aided by the use of a self-sorting mechanism by attaching short sections of complementary DNA on components and target sites on the circuit (see next section).[135]

Self-Assembly Examples

DNA-Mediated Assembly

While work continues on self-assembly of MEMS components with top-down methods, many researchers have recognized that the exquisite molecular recognition of various complementary biological molecules may be used to assist in the assembly of micro- or nanocomponents into complex networks. In this approach, DNA nanostructures and DNA mediated assembly of nano- and microstructures have received most attention in the literature. In the DNA nanostructures approach, pioneered by N. C. Seeman, the 2-nm-wide DNA molecules themselves are used to form 2D and 3D structures.[139–143] As shown in Figure 4.49a, DNA four-armed branched junctions are made by properly choosing the sequence of complementary strands. These DNA scaffolds can be linked together to make periodic crystals as shown in Figure 4.49b.

To use DNA for assembling nano- or microparts, one first needs to attach DNA molecules to these parts. The most widely used method is the covalent bond between sulfur and gold, that is, a metal thiolate (~44 kcal/mol). DNA molecules can be functionalized with thiol (S-H) or disulfide (S-S) groups, and upon immersion of microparts with clean gold surfaces in solutions of such thiol or disulfide derivatized oligonucleotides, the sulfur adsorbs on the gold surfaces, forming a single layer of molecules. Using this methodology, Mirkin et al. described a method of assembling colloidal gold

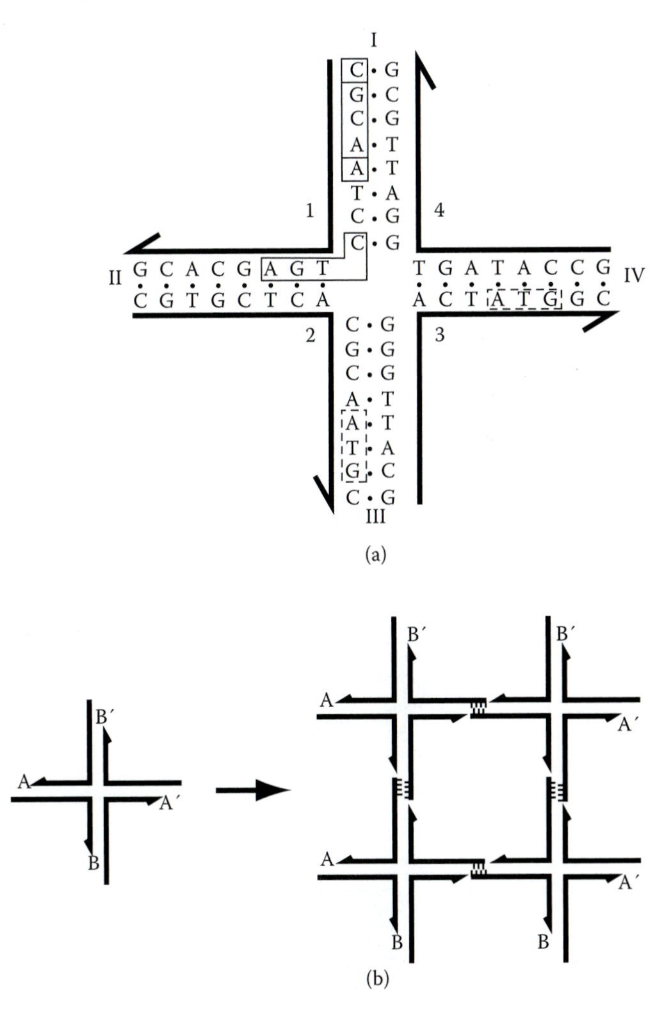

FIGURE 4.49 (a) Four-armed stable branched junction made from DNA molecules. (b) Use of branched junctions to form periodic crystals. (Based on Seeman, N. C. 1991. *Nanotechnology* 149.[140])

nanoparticles into macroscopic aggregates using DNA as linking element.[144] Noncomplementary DNA oligonucleotides were attached to the surfaces of two batches of 13-nm gold particles. An oligonucleotide duplex with ends complementary to the grafted sequences was then introduced into the particle mixture, and the nanoparticles self-assembled into aggregates. The process is illustrated in Figure 4.50. It can be reversed by increasing the temperature, which melts the double-stranded DNA linkers. Closely packed assemblies of aggregates with uniform particle spacings of about 60 Å were demonstrated this way.

The concept of DNA-mediated self-assembly has been extended to metallic nanowires/rods.[145] The idea behind this is to fabricate gold and/or platinum wires, functionalize them with single-stranded DNA (ssDNA), and assemble a wiring network on a substrate that has the complementary ssDNA molecules attached at specific sites. Thus, self-assembly of interconnects and wires is rendered possible. DNA has also been used as a template for the fabrication of nanowires. Braun et al.[136] formed a DNA bridge between two gold electrodes, again using thiol attachment. Once a DNA bridge is formed between the 12–16-μm spacing of the electrodes, a chemical deposition process is used to deposit silver ions along the DNA through Ag^+/Na^+ ion exchange and reduction of the silver complexes with hydroquinone. The result is a silver nanowire that is formed using DNA as a template or skeleton. The most ambitious goal is the DNA-assisted assembly (molecular "Velcro") of semiconductor active devices, such as transistors, laser diodes, etc., proposed by Edman et al. at Nanogen.[135] For a review of this area, check out Bashir[137] and Lee et al.[138]

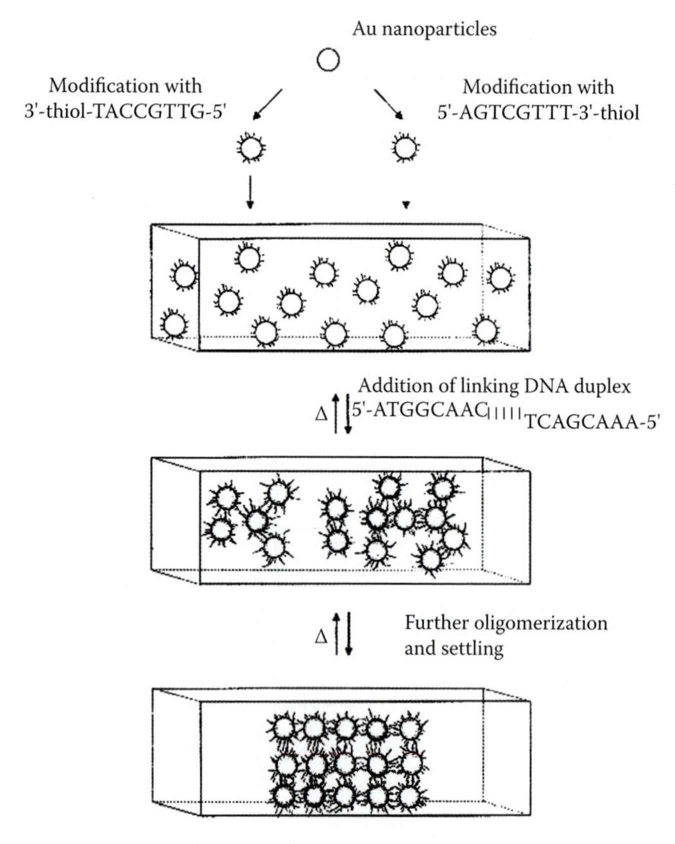

FIGURE 4.50 Fabrication process for the aggregated assembly of DNA conjugated gold nanoparticles. (Based on Mirkin, C. A., R. L. Letsinger, R. C. Mucic, and J. J. Storhoff. 1996. *Nature* 382:607.[144])

Micro-Origami

Often MEMS includes hinged parts that must be set in place before operation. This can lead to challenging and time consuming manual manipulation of components at small length scales. Researchers from the ATR Adaptive Communications Research Laboratories in Japan (http://www.atr.co.jp) have developed a technique they call micro-origami to fabricate MEMS devices that automatically move parts into position.[146] Micro-origami makes use of the energy stored as strain between lattice-mismatched semiconductor epitaxial layers to create self-positioning, three-dimensional structures with simple and robust hinges. The key to the micro-origami technique is to manufacture hinges out of a pair of material layers with slightly different atomic spacings. The lattice mismatch causes a stress that in turn bends a hinge, as shown generically in Figure 4.51. In the case shown, a distributed Bragg reflector* (DBR) or micromirror is moved out of the plane of the substrate. Micro-origami structures are manufactured by using epitaxial growth techniques to construct multilayered films designed with folding hinges etched into their surfaces in the manner of

* A distributed Bragg reflector is a structure formed from multiple layers of alternating materials with various refractive indexes. Each layer boundary causes a partial reflection of an optical wave. For waves whose wavelength is close to four times the optical thickness of the layers, the many reflections combine with constructive interference, and the layers act as a high-quality reflector. The range of wavelengths that are reflected is called the *photonic stopband*. Within this range of wavelengths, light is "forbidden" to propagate in the structure.

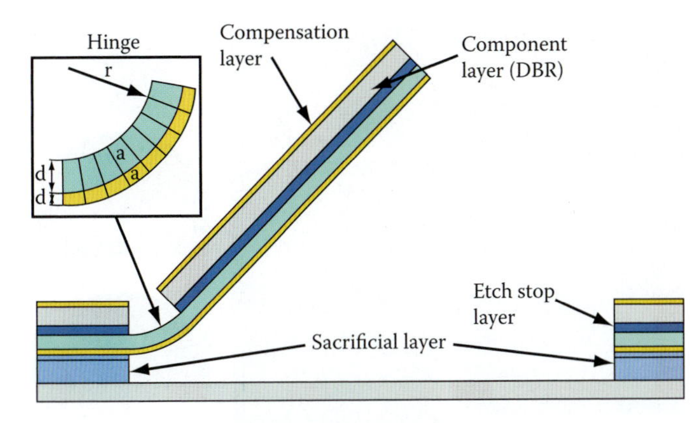

FIGURE 4.51 Schematic of the origami process. (From http://microfanatic.com/origami.html.)

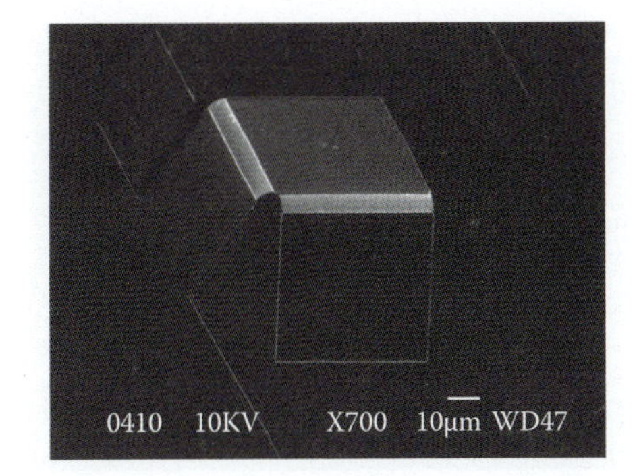

FIGURE 4.52 GaAs corner plates assembled by the strain in two hinge regions. (From http://microfanatic.com/origami.html.)

the creased folds of paper. Adjusting the thickness and the length of the hinge that connects the plates to each other controls the angle between different plates. When the film is removed from its substrate by etching away a sacrificial layer, the microstructure takes on its designed shape because of the released stress in the hinge areas. In this way, this fabrication method reflects the design principles of origami, the traditional Japanese art of sculpting objects from folded paper. Because it is possible to fabricate three-dimensional structures on a semiconductor substrate, these structures can be integrated with lasers and photodetectors.

This fabrication method of simple and robust hinges for movable parts can be applied to any pair of lattice-mismatched epitaxial layers in semiconductors or metals. The first work by the ATR team involved a multilayer structure, including AlGaAs/GaAs component layers and an InGaAs strained layer, which was grown by molecular beam epitaxy on a GaAs substrate. After definition of the hinge and mirror's shape by photolithography, the micromirrors were released from the substrate by selective etching. They moved to their final position powered by the strain release in the InGaAs layer. The curvature radius of the hinges is determined by the thickness ratio of the bilayer and the indium composition in the strained layer. Depending on the growth structure, the hinge can fold up or down, making the fabrication of complex 3D semiconductor structures possible. For example, in Figure 4.52, we show GaAs corner plates assembled by the strain in two hinge regions, and in Figure 4.53 we illustrate a microstage fabricated on a GaAs substrate using hinges with different bending directions called *tani-ori* (valley fold) and *yama-ori* (mountain fold) in origami. In the latter case, a multilayered structure including InGaAs strained layers was grown by molecular-beam epitaxy on a GaAs substrate. After the multilayered structure from the substrate is released by the selective etching of a sacrificial layer, the microstage moved into its final position powered by the strain release in the InGaAs layers, and the stage was kept parallel to the substrate owing to the combination of the two types of hinges.[147]

The ATR team also fabricated SiGe/Si free-standing micro-objects on an SOI substrate using SiO_2 as sacrificial layer material. SiGe/Si strained films were grown by molecular beam epitaxy on silicon-on-insulator (SOI) substrates and were released from the substrate by selective etching of the SiO_2 layer. The released films rolled up because of the elastic strain in the SiGe layer. Microtubes and microspirals with a diameter of about 40 μm were obtained as shown in Figure 4.54.[148]

FIGURE 4.53 A self-assembled microstage fabricated using hinges with different bending directions. (From http://microfanatic.com/origami.html.)

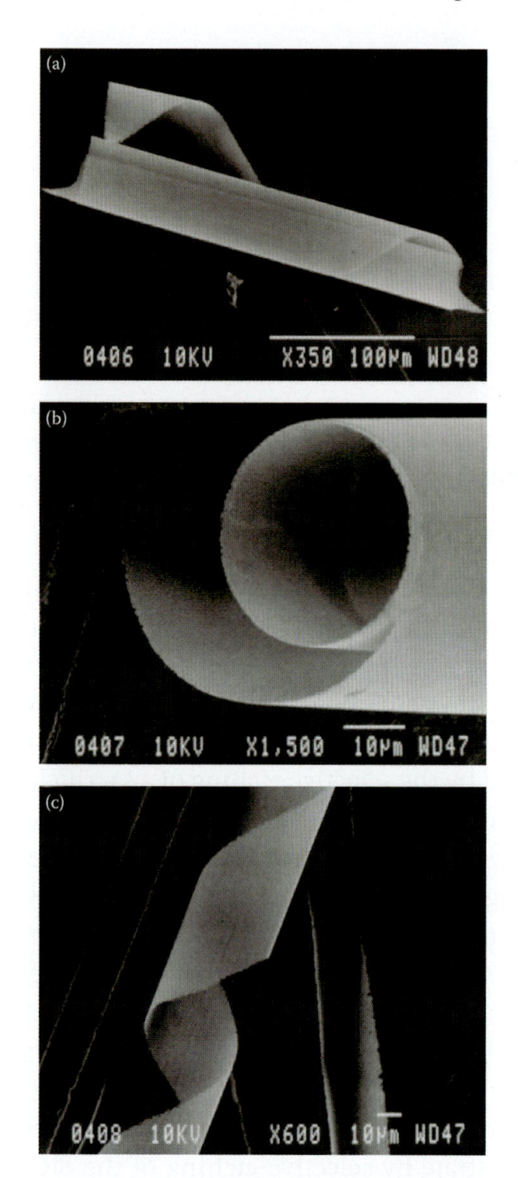

FIGURE 4.54 Free standing micro-objects on SOI substrate. (a) Top view of released SiGe/Si film scrolled in a microtube. (b) Rolled film at large magnification. (c) SiGe/Si microspiral. Curvature radius is about 20 μm.[148]

On a somewhat larger scale, researchers at the École Supérieure de Physique et de Chimie Industrielles in Paris, together with a team from the Paris Institute of Technology, have shown that water droplets can be used to make flat sheets of plastic (40–80 μm thick), a couple of millimeters across fold into 3D objects such as tiny pyramids, boxes, and spheres.[149] A drop of water, large enough to reach all corners, is placed on top of a cut-out shape. As the liquid evaporates, its surface tension tugs the membrane closer around the water's decreasing volume, until the remaining liquid is completely encapsulated inside a sphere, tetrahedron,

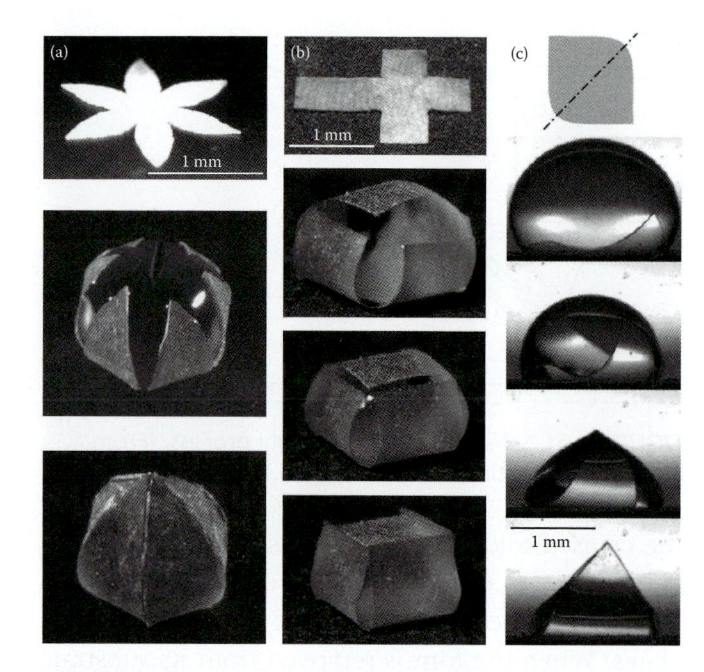

FIGURE 4.55 It is the interaction between elasticity and capillarity that is used to produce three-dimensional structures through the wrapping of a liquid droplet by a planar sheet. Tuning of the initial flat shape to obtain (a) a spherical encapsulation, (b) a cubic encapsulation, or (c) a triangular mode-2 fold.[149]

or some other shape. Tailoring the initial geometry of the thin membrane, as demonstrated with some experimental results in Figure 4.55, controls the final encapsulated 3D shape. For example, a flower-like pattern produces a sphere; a triangle becomes a tetrahedron, and a flat cross shape folds into a cube. Varying the thickness of the sheet controls how much a structure folds up, and as expected, the effect of surface tension becomes stronger as the size is decreased.

In Chapter 3, this volume, we saw several examples of DNA origami (see Figures 3.5 and 3.6).

The intersection between the subjects of origami and mathematics is rich with interesting results and applications. For details, consult http://kahuna.merrimack.edu/~thull/origamimath.html and http://www.paperfolding.com/math.

Questions

4.1: Explain front-end and back-end processes in IC or MEMS manufacturing. To which hierarchical level of manufacturing is each one related?

4.2: What are some of the advantages of a silicon-silicon bonding process over anodic bonding? Are there any disadvantages?

4.3: What is the role of packaging in ICs and in chemical and biological MEMS?

4.4: Which is easier to package, an absolute pressure sensor or an accelerometer? A pressure sensor or a pH sensor?

4.5: Describe and compare different wafer bonding techniques.

4.6: Describe the process of silicon-to-glass anodic bonding. Make a sketch of the setup and describe typical conditions required. What is the process that takes place on the molecular level?

4.7: Suggest a method for preventing an accelerometer sensor element from being subjected to undesirable mechanical stresses (from temperature changes or packing stress).

4.8: One of the modifications of anodic bonding is to deposit intermediate layers of silicon dioxide and aluminium. What are the advantages and disadvantages of this method?

4.9: Surface forces make the release of objects after picking them up in microassembly very difficult. Suggest several methods to make this process easier.

4.10: What are the three major electrical connection methods in chip-mounting?

4.11: Describe at least four techniques one could use to make a via through an Si wafer.

4.12: Describe as many methods that you know to vacuum seal a MEMS device on-chip.

4.13: Describe the difference between serial microassembly and parallel microassembly.

4.14: How do gravity, van der Waal's, electrostatic, and surface tension scale as a function of part size in microassembly, assuming a sphere (part) and a plane (gripper) interaction?

4.15: What is likely to be a major concern with DNA-assisted assembly? How could you improve the situation?

References

1. Lyke, J. C. 1995. Packaging technologies for space-based microsystems and their elements. *Microengineering technologies for space systems*. Ed. H. Helvajian, 131–80. El Segundo, CA: Aerospace Corporation.
2. de Gennes, P. G. 1979. *Scaling concepts in polymer physics*. Ithaca, NY: Cornell University Press.
3. Frank-Kamenetskii, M. D. 1997. *Unraveling DNA: The most important molecule of life*. Reading, MA: Perseus Books.
4. Jensen, J. R. 1989. *Microelectronics processing*. Eds. D. W. Hess and K. F. Jensen, 441–504. Washington, DC: American Chemical Society.
5. Allen, R. 1984. Sensors in silicon. *High Technology* September:43–81.
6. Lau, J. H. 1998. *Electronic packaging: Design, materials, process, and reliability*. New York: McGraw-Hill.
7. Jerman, J. H. 1990. The fabrication and use of micromachined corrugated silicon diaphragms. *Sensors Actuators A* A23:988–92.
8. Spiering, V. L., S. Bouwstra, J. Burger, and M. Elwenspoek. 1993. Membranes fabricated with a deep singel corrugation for package stress reduction and residual stress relief. *Fourth European Workshop on Micromechanics (MME '93)*. Neuchatel, Switzerland.
9. Spiering, V. L., S. Bouwstra, R. M. E. J. Spiering, and M. Elwenspoek. 1991. On-chip decoupling zone for package-stress reduction. *Sixth International Conference on Solid-State Sensors and Actuators (Transducers '91)*. San Francisco.
10. Offereins, H. L., H. Sandmaier, B. Folkmer, U. Steger, and W. Lang. 1991. Stress free assembly technique for a silicon based pressure sensor. *Sixth International Conference on Solid-State Sensors and Actuators (Transducers '91)*. San Francisco.
11. Tanger, R. K. 1976. Hermeticity of polymeric lid sealants. *26th Electronic Components Conference*. San Francisco.
12. Flannery, A. F., N. J. Mourlas, C. W. Storment, S. Tsai, S. H. Tan, J. Heck, D. Monk, T. Kim, B. Gogoi, and G. T. Kovacs. 1998. PECVD silicon carbide as a chemically resistant material for micromachined transducers. *Sensors Actuators A* 70:48–55.
13. Linder, S., H. Baltes, F. Gnaedinger, and E. Doering. 1994. Fabrication technology for wafer through-hole interconnections and three-dimensional stacks of chips and wafers. *Proceedings: IEEE Micro Electro Mechanical Systems (MEMS '94)*. Oiso, Japan.
14. Maluf, N. 2000. *An introduction to microelectromechanical systems engineering*. Boston: Artech House.
15. Bhardwaj, J. K., and H. Ashraf. 1995. Advanced Si etching using high density plasmas. *Micromachining and microfabrication process technology (proceedings of the SPIE)*. Austin, TX.
16. Craven, D., K. Yu, and T. Pandhumsoporn. 1995. Etching technology and applications for "through-the-wafer" silicon etching. *Micromachining and microfabrication process technology (proceedings of the SPIE)*. Austin, TX.
17. Chow, E. M., A. Partridge, C. F. Quate, and T. W. Kenny. 2000. Through-wafer electrical interconnects compatible with Stanford semiconductor processing. *Solid-State Sensor and Actuator Workshop*. Hilton Head Island, SC.
18. Lischner, D. J., H. Basseches, and F. A. D'Altroy. 1985. Observations of the temperature gradient zone melting process for isolating small devices. *J Electrochem Soc* 132:2991–96.
19. Wallis, G., and D. I. Pomerantz. 1969. Field assisted glass-metal sealing. *J Appl Phys* 40:3946–49.
20. Ko, W. H., J. T. Suminto, and G. J. Yeh. 1985. Bonding techniques for microsensors. *Micromachining and micropackaging of transducers. Eds.* C. D. Fung, P. W. Cheung, W. H. Ko, and D. G. Fleming. Amsterdam: Elsevier.

21. Peeters, E. 1994. *Process development for 3D silicon microstructures, with application to mechanical sensor design.* PhD thesis, Catholic University of Louvain, Belgium.
22. Puers, B., A. Cozma, and E. Van De Weyer. 1993. *Intermediate project report.* KU Leuven.
23. Anthony, T. R. 1983. Anodic bonding of imperfect surfaces. *J Appl Phys* 54:2419–28.
24. Hok, B., C. Dubon, and C. Ovren. 1983. Anodic bonding of galium arsenide to glass. *Appl Phys Lett* 43:267–69.
25. Von Arx, J., B. Ziaie, M. Dokmeci, and K. Najafi. 1995. Hermeticity testing of glass silicon packages with high-density on-chip feedthroughs. *Eighth International Conference on Solid-State Sensors and Actuators (Transducers '95).* Stockholm, Sweden.
26. Ito, N., K. Yamad, H. Okada, M. Nishimura, and T. Kuriyama. 1995. A rapid selective anodic bonding method. *Eighth International Conference on Solid-State Sensors and Actuators (Transducers '95).* Stockholm, Sweden.
27. Sander, C. S. 1980. A bipolar-compatible monolithic capacitive pressure sensor. PhD thesis, #G558-10, Stanford University, December 1980.
28. Brooks, A. D., and R. P. Donovan. 1972. Low temperature electrostatic Si-to-Si seals using sputtered borosilicate glass. *J Electrochem Soc* 119:545–46.
29. Hanneborg, A., M. Nese, H. Jacobsen, and R. Holm. 1992. Silicon-to-thin-film anodic bonding. *J Micromech Microeng* 2:117–21.
30. Krause, P., M. Sporys, E. Obermeier, K. Lange, and S. Grigull. 1995. Silicon to silicon anodic bonding using evaporated glass. *Eighth International Conference on Solid-State Sensors and Actuators (Transducers '95).* Stockholm, Sweden.
31. Elwenspoek, M., H. Gardeniers, M. de Boer, and A. Prak. 1994. *Micromechanics. Report no. 122830.* Twente, The Netherlands: University of Twente.
32. Esashi, M., A. Nakano, S. Shoji, and H. Hebiguchi. 1990. Low-temperature silicon-to-silicon bonding with intermediate low melting point glass. *Sensors Actuators A* A23:931–34.
33. Anthony, T. R. 1985. Dielectric isolation of silicon by anodic bonding. *J Appl Phys* 58:1240–47.
34. Ohashi, H., J. Ohura, T. Tsukakoshi, and M. Shimbo. 1986. Improved dielectrically isolated device integration by silicon-wafer direct bonding (SDB) technique. *Technical Digest: IEEE International Electron Devices Meeting (IEDM '86).* Los Angeles, pp. 211–13.
35. Kissinger, G., and W. Kissinger. 1993. Void-free silicon-wafer-bond strengthening in the 200–400°C range. *Sensors Actuators A* A36:149–56.
36. Schmidt, M. A. 1994. Silicon wafer bonding technology for micromechanical devices. *Technical digest: 1994 Solid State Sensor and Actuator Workshop.* Hilton Head Island, SC.
37. Sun, L. G., J. Zhan, Q. Y. Tong, S. J. Xie, Y. M. Caim, and S. J. Lu. 1988. Cool plasma activated surface in silicon wafer direct bonding technology. *J Physique Colloq C* 49:79–82.
38. Shimbo, M., K. Furukawa, and K. Tanzawa. 1986. Silicon-to-silicon direct bonding method. *J Appl Phys* 60:2987–89.
39. Tong, Q.-Y., G. Cha, R. Gafiteanu, and U. Gosele. 1994. Low temperature wafer direct bonding. *J Micromechan Syst* 3:29–35.
40. Barth, P. W. 1990. Silicon fusion bonding for fabrication of sensors, actuators and microstructures. *Sensors Actuators A* A23:919–26.
41. Stengl, R., T. Tan, and U. Gosele. 1989. A model for the silicon wafer bonding process. *Jpn J Appl Phys Part I* 28:1735–41.
42. Backlund, Y., K. Ljungberg, and A. Soderbarg. 1992. A suggested mechanism for silicon direct bonding from studying hydrophilic and hydrophobic surfaces. *J Micromech Microeng* 2:158–60.
43. Lasky, J. B. 1986. Wafer bonding for silicon-on-insulator technologies. *Appl Phys Lett* 48:78–80.
44. Ismail, M. S., R. W. Bower, J. L. Veteran, and O. J. Marsh. 1990. Silicon nitride direct bonding. *Electron Lett* 26:1045–46.
45. Bower, R. W., M. S. Ismail, and B. E. Roberds. 1993. Low temperature Si_3N_4 direct bonding. *Appl Phys Lett* 62:3485–97.
46. Ismail, M. S., R. W. Bower, B. E. Roberds, and S. N. Farrens. 1993. *Seventh International Conference on Solid-State Sensors and Actuators (Transducers '93).* Yokohama, Japan.
47. Anacker, W., E. Bassous, F. F. Fang, R. E. Mundie, and H. N. Yu. 1976. Fabrication of multiprobe miniature electrical connector. *IBM Tech Bull* 19:372–74.
48. Editorial. 1981. Sealing glass. *Corning Technical Publication.* Corning Glass Works.
49. Legtenberg, R., S. Bouwstra, and M. Elwenspoek. 1991. Low-temperature glass bonding for sensor applications using boron oxide thin films. *J Micromech Microeng* 1:157–60.
50. Field, L. A., and R. Muller. 1990. Fusing silicon wafers with low melting temperature glass. *Sensors Actuators A* A23:935–38.
51. Yamada, A., T. Kawasaki, and M. Kawashima. 1987. SOI wafer bonding with spin-on glass as adhesive. *Electronic Lett* 23:39–40.
52. Quenzer, H. J., and W. Benecke. 1992. Low-temperature silicon wafer bonding. *Sensors Actuators A* A32:340–44.
53. Guerin, L., M. A. Schaer, R. Sachot, and M. Dutoit. 1995. Proposal for new multichip-on-silicon packaging scheme. *Eighth International Conference on Solid-State Sensors and Actuators (Transducers '95).* Stockholm, Sweden.
54. Tiensuu, A.-L., J.-Å. Schweitz, and S. Johansson. 1995. In situ investigation of precise high strength micro assembly using Au-Si eutectic bonding. *Eighth International Conference on Solid-State Sensors and Actuators (Transducers '95).* Stockholm, Sweden.
55. Arquint, P., P. D. van der Wal, B. H. van der Schoot, and N. F. de Rooij. 1995. Flexible polysiloxane interconnection between two substrates for microsystem assembly. *Eighth International Conference on Solid-State Sensors and Actuators (Transducers '95).* Stockholm, Sweden.
56. den Besten, C., R. E. G. van Hal, J. Munoz, and P. Bergveld. 1992. Polymer bonding of micro-machined silicon structures. *Proceedings: IEEE Micro Electro Mechanical Systems (MEMS '92).* Travemunde, Germany.
57. Munoz, J., A. Bratov, R. Mas, N. Abramova, C. Dominguez, and J. Bartroli. 1995. Packaging of ISFETs at the wafer level by photopatternable encapsulant resins. *Eighth International Conference on Solid-State Sensors and Actuators (Transducers '95).* Stockholm, Sweden.
58. Becker, H., and C. Gärtner. 2000. Polymer microfabrication methods for microfluidic analytical applications. *Electrophoresis* 21:12–26.
59. Bower, R. W., M. S. Ismail, and S. N. Farrens. 1991. Aligned wafer bonding: A key to three dimensional microstructures. *J Electron Mater* 20:383–87.
60. Shoaf, S. E., and A. D. Feinerman. 1994. Aligned Au-Si eutectic bonding of silicon structures. *J Vac Sci Technol* A12:19–22.
61. Maszara, W. P., G. Goetz, A. Caviglia, and J. B. McKitterick. 1988. Bonding of silicon wafers for silicon-on-insulator. *J Appl Phys* 64:4943–50.

62. Nese, M., R. W. Bernstein, I.-R. Johansen, and R. Spooren. 1995. New method for testing hermeticity of silicon sensor structures. *Eighth International Conference on Solid-State Sensors and Actuators (Transducers '95)*. Stockholm, Sweden.
63. Minami, K., T. Moriuchi, and M. Esashi. 1995. Cavity pressure control for critical damping of packaged micro mechanical devices. *Eighth International Conference on Solid-State Sensors and Actuators (Transducers '95)*. Stockholm, Sweden.
64. Striny, K. M. 1988. Assembly techniques and packaging of *VLSI* devices. *VLSI Technology*. Ed. S. M. Sze, 566–611. New York: McGraw-Hill.
65. Traeger, R. K. 1976. Hermeticity of polymeric lid sealants. *26th Electronic Components Conference*. San Francisco.
66. Guckel, H., and D. W. Burns. 1984. Planar processed polysilicon sealed cavities for pressure transducer arrays. *Technical Digest: IEEE International Electron Devices Meeting (IEDM '84)*. San Francisco.
67. Guckel, H., and D. W. Burns. 1986. Fabrication techniques for integrated sensor microstructures. *IEEE International '86*. Los Angeles.
68. Guckel, H., and D. W. Burns. 1985. A technology for integrated transducers. *International Conference on Solid-State Sensors and Actuators.* Philadelphia.
69. Guckel, H., D. W. Burns, C. K. Nesler, and C. R. Rutigliano. 1987. Fine grained polysilicon and its application to planer pressure transducers. *Fourth International Conference on Solid-State Sensors and Actuators (Transducers '87)*. Tokyo, Japan.
70. Lin, L., K. M. McNair, R. T. Howe, and A. P. Pisano. 1993. Vacuum-encapsulated lateral microresonators. *Seventh International Conference on Solid-State Sensors and Actuators (Transducers '93)*. Yokohama, Japan.
71. Eaton, W. P., and J. H. Smith. 1995. A *CMOS-compatible, surface-micromachined pressure sensor* for aqueous ultrasonic application. *Smart Structures and Materials 1995: Smart Electronics (Proceedings of the SPIE)*. San Diego, CA.
72. Lebouitz, K. S., R. T. Howe, and A. P. Pisano. 1995. Permeable polysilicon etch-access windows for microshell fabrication. *Eighth International Conference on Solid-State Sensors and Actuators (Transducers '95)*. Stockholm, Sweden.
73. Guckel, H., D. W. Burns, and C. R. Rutigliano. 1986. In *Technical Digest: 1986 Solid State Sensor and Actuator Workshop.* Hilton Head Island, SC.
74. Erskine, J. C. 1983. Polycrystalline silicon-on-metal strain gauge transducers. *IEEE Trans Electron Devices* 30: 796–801.
75. *SSI Technologies.* 1995. Janesville, WI: SSI Technologies.
76. Cho, Y.-H., B. M. Kwak, A. P. Pisano, and R. T. Howe. 1993. Viscous energy dissipation in laterally oscillating planar microstructures: a theoretical and experimental study. *Proceedings: IEEE Micro Electro Mechanical Systems (MEMS '93)*. Fort Lauderdale, FL.
77. Tang, W. C., T. H. Nguyen, and R. T. Howe. 1989. Laterally driven polysilicon resonant microstructures. *Sensors Actuators* 20:25–32.
78. Lin, L., T. C. Nguyen, R. T. Howe, and A. P. Pisano. 1992. Microelectromechanical filters for signal processing. *Proceedings: IEEE Micro Electro Mechanical Systems (MEMS '92)*. Travemunde, Germany.
79. Hsu, C. H., and R. S. Muller. 1991. Micromechanical electrostatic voltmeter. *Sixth international conference on solid-state sensors and actuators (Transducers '91)*. San Francisco.
80. Bernstein, J., S. Cho, A. T. King, A. Kourepenis, P. Maciel, and M. Weinberg. 1993. A micromachined comb-drive tuning fork rate gyroscope. *Proceedings: Micro Electro Mechanical Systems (MEMS '93)*. Fort Lauderdale, FL.
81. Ikeda, K., H. Kuwayama, T. Kobayashi, T. Watanabe, T. Nishikawa, T. Yoshida, and K. Harada. 1990. Three-dimensional micromachining of silicon pressure sensor integrating resonant strain gauge on diaphragm. *Sensors Actuators A* A23:1007–10.
82. Ikeda, K., H. Kuwayama, T. Kobayashi, T. Watanabe, T. Nishikawa, T. Yoshida, and K. Harada. 1990. Silicon pressure sensor integrates resonant strain gauge on diaphragm. *Sensors Actuators A* A21:146–50.
83. Cohn, M. B., Y. Liang, R. T. Howe, and A. P. Pisano. 1996. Wafer-to-wafer transfer of microstructures for vacuum packaging. *Technical Digest:1996 Solid State Sensor and Actuator Workshop*. Hilton Head Island, SC.
84. Böhringer, K. F., R. S. Fearing, and K. Y. Goldberg. 1998. Microassembly. *Handbook of industrial robotics.* Ed. S. Nof. New York: Wiley, pp. 1045–66.
85. Nelson, B., and B. Vikramaditya. 1997. Visually guided microassembly using optical microscopes and active vision techniques. *Proceedings: IEEE Int. Conf. on Robotics and Automation (ICRA)*. Albuquerque, NM: IEEE.
86. Feddema, J. T., and R. W. Simon. 1998. CAD-driven microassembly and visual servoing. *Proceedings: IEEE Int. Conf. on Robotics and Automation (ICRA)*. Leuven, Belgium: IEEE.
87. Cohn, M. B., and R. T. Howe. 1997. Wafer-to-wafer transfer of microstructures using breakaway tethers. *IEEE solid-state sensor and actuator workshop.* Hilton Head, SC.
88. Singh, A., D. A. Horsley, M. B. Cohn, A. P. Pisano, and R. T. Howe. 1997. Batch transfer of microstructures using flip-chip solder bump bonding. *Digest: Int. Conf. on Solid State Sensors and Actuators*. Chicago: Transducers Research Foundation.
89. Yeh, H.-J., and J. S. Smith. 1994. Fluidic self-assembly for the integration of GaAs light-emitting diodes on Si substrates. *IEEE Photonics Tech Lett* 6:706–708.
90. Yeh, H.-J., and J. S. Smith. 1994. Fluidic self-assembly of microstructures and its application to the integration of GaAs on Si. *Proceedings of the IEEE International Workshop on Micro Electro Mechanical Systems (MEMS '94)*. Oiso, Japan.
91. Yeh, H.-J., and J. S. Smith. 1994. Integration of GaAs vertical-cavity surface-emitting laser on Si by substrate removal. *Appl Phys Lett* 64:1466–68.
92. Quaid, A. E., and R. L. Hollis. 1996. Cooperative 2-dof robots for precision assembly. *Proceedings: IEEE Int. Conf. on Robotics and Automation (ICRA)*. Minneapolis, MN: IEEE.
93. Zesch, W. 1997. Multi-degree-of-freedom micropositioning using stepping principles. PhD thesis, Swiss Federal Institute of Technology, Zurich, Switzerland.
94. Keller, C. G., and R. T. Howe. 1997. Hexsil tweezers for teleoperated micro-assembly. *Proceedings: IEEE Workshop on Micro Electro Mechanical Systems (MEMS)*. Nagoya, Japan.
95. Cohn, M. B. 1992. *Self-assembly of microfabricated devices.* US Patent 5,355,577.
96. Cohn, M. B., R. T. Howe, and A. P. Pisano. 1995. Self-assembly of microsystems using non-contact electrostatic traps. *Proceedings of the ASME International Congress and Exposition, Symposium on Micromechanical Systems (IC '95)*. San Francisco.

97. Böhringer, K.-F., M. B. Cohn, K. Goldberg, R. Howe, and A. Pisano. 1997. Electrostatic self-assembly aided by ultrasonic vibration. *AVS 44th National Symposium*. San Jose, CA.
98. Böhringer, K.-F., K. Goldberg, M. B. Cohn, R. Howe, and A. Pisano. 1998. Parallel microassembly with electrostatic force fields. *Proceedings: IEEE Int. Conf. on Robotics and Automation (ICRA)*. Leuven, Belgium.
99. Edman, C. F., C. Gurtner, R. E. Formosa, J. J. Coleman, and M. J. Heller. 2000. Electric-field-directed pick-and-place assembly. *HDI* October:30–35.
100. Kim, C.-J., A. P. Pisano, and R. S. Muller. 1992. Silicon-processed overhanging microgripper. *J Microelectromech Syst* 1:31–36.
101. Pister, K. S. J., M. W. Judy, S. R. Burgett, and R. S. Fearing. 1992. Microfabricated hinges. *Sensors Actuators A* 33:249–56.
102. Keller, C., and R. T. Howe. 1995. Hexsil bimorphs for vertical actuation. *Digest: Int. Conf. on Solid-State Sensors and Actuators*. Stockholm, Sweden.
103. Hosokawa, K., I. Shimoyama, and H. Miura. 1995. Dynamics of self-assembling systems: Analogy with chemical kinetics. *Artificial Life* 1:413–27.
104. Ashkin, A., J. M. Dziedzic, and T. Yamane. 1987. Optical trapping and manipulation of single cells using infrared laser beams. *Nature* 330:769–71.
105. Srinivasan, U., and R. Howe. 1997.
106. Brown, K. T., and D. G. Flaming. 1992. Advanced micropipette techniques for cell physiology. *IBRO handbook: Methods in neurosciences*. Chichester United Kingdom: Wiley.
107. Miltenyi, S., W. Mueller, W. Weichel, and A. Radbruch. 1990. High-gradient magnetic cell separation with MACS. *Cytometry* 11:231–38.
108. Stroscio, J. A., and D. M. Eigler. 1991. Atomic and molecular manipulation with the scanning tunneling microscope. *Science* 254:1319–26.
109. Schnelle, T., R. Hagedorn, G. Fuhr, S. Fiedler, and T. Müller. 1993. Three-dimensional electric field traps for manipulation of cells—calculation and experimental verification. *Biochim Biophys Acta* 1157:127–40.
110. Coackley, W. T. 1997. Ultrasonic separations in analytical biotechnology. *TIBTECH* 5:506–11.
111. Giddings, J. C. 1993. Field-flow fractionation: analysis of macromolecular, colloidal and particulate materials. *Science* 260:1456–64.
112. Fu, A., C. Spence, A. Scherer, F. H. Arnold, and S. R. Quake. 1999. A microfabricated fluorescence-activated cell sorter. *Nat Biotech* 17:1109–11.
113. Nakao, M., K. Tsuchiya, K. Matsumoto, and Y. Hatamura. 2001. Micro handling with rotational needle-type tools under real time observation. *CIRP Annals Manufact Tech* 50:9–12.
114. Hitakawa, H. 1988. Advanced parts orientation system has wide application. *Assembly Automation* 8:147–50.
115. Yasuda, K. 2000. Non-destructive mixing, concentration, fractionation and separation of μm-sized particles in liquid by ultrasound. *Micro total analysis systems 2000*. Eds. A. van den Berg, W. Olthuis, and P. Bergveld, 343–46. Enschede, The Netherlands: Kluwer Academic Publishers.
116. Darling, R. B., J. W. Suh, and G. T. Kovacs. 1998. Ciliary microactuator array for scanning electron microscope positioning stage. *J Vac Sci Technol A* 16:1998–2002.
117. Pister, K. S. J., R. Fearing, and R. Howe. 1990. A planar air levitated electrostatic actuator system. *Proceedings: IEEE Workshop on Micro Electro Mechanical Systems (MEMS)*. Napa Valley, CA.
118. Fujita, H. 1993. Group work of microactuators. *International Advanced Robot Program Workshop on Micromachine Technologies and Systems*. Tokyo, Japan.
119. Böhringer, K.-F., B. R. Donald, R. Mihailovich, and N. C. MacDonald. 1994. A theory of manipulation and control for microfabricated actuator arrays. *Proceedings: IEEE Workshop on Micro Electro Mechanical Systems (MEMS)*. Oiso, Japan.
120. Storment, C. W., D. A. Borkholder, V. Westerlind, J. W. Suh, N. I. Maluf, and G. T. A. Kovacs. 1994. Flexible, dry-released process for aluminum electrostatic actuators. *J Microelectromech Syst* 3:90–96.
121. Suh, J. W., S. F. Glander, R. B. Darling, C. W. Storment, and G. T. A. Kovacs. 1996. Combined organic thermal and electrostatic omnidirectional ciliary microactuator array for object positioning and inspection. *Proceedings: Solid-State Sensor and Actuator Workshop*. Hilton Head Island,SC.
122. Liu, C., T. Tsao, P. Will, Y. Tai, and W. Liu. 1995. In *Digest: Int. Conf. on Solid-State Sensors and Actuators*. Stockholm, Sweden: Transducers Research Foundation.
123. Liu, W., and P. Will. 1995. Parts manipulation on an intelligent motion surface. *IEEE/RSJ Int. Workshop on Intelligent Robots, and Systems (IROS)*. Pittsburgh, PA.
124. Popa, D. O., and H. E. Stephanou. 2004. Micro and meso scale robotic as sembly. *Proceedings of WTEC Workshop: Review of US Research in Robotics*. WTEC.
125. Wheeler, A. R., K. Morishima, D. W. Arnold, A. B. Rossi, and R. N. Zare. 2000. Single organelle analysis with integrated chip electrophoresis and optical tweezers. *Micro total analysis systems 2000*. Eds. A. van den Berg, W. Olthuis, and P. Bergveld, 25–28. Enschede, The Netherlands: Kluwer Academic Publishers.
126. Ashkin, A., and J. M. Dziedzic. 1985. Observation of radiation-pressure trapping of particles by alternating light beams. *Phys Rev Lett* 54:1245–48.
127. Ashkin, A., J. M. Dziedzic, J. E. Bjorkholm, and S. Chu. 1986. Observation of a single-beam gradient force optical trap for dielectrical particles. *Optics Letters* 11:228–90.
128. Perkins, T. T., D. E. Smith, R. G. Larson, and S. Chu. 1995. Stretching of a single tethered polymer in a uniform flow. *Science* 268:83–87.
129. Ashkin, A., and J. M. Dziedzic. 1987. Optical trapping and manipulation of viruses and bacteria. *Science* 235:1517–20.
130. Svoboda, K., and S. M. Block. 1994. Biological applications of optical forces. *Ann Rev Biophys Biomol Struc* 23:247–85.
131. Fuhr, G. R., and C. Reichle. 2000. Living cells in opto-electrical cages: characterisation, manipulation and force measurements. *Micro total analysis system*. Eds. A. van den Berg, W. Olthuis, and P. Bergveld. Enschede, The Netherlands: Kluwer Academic Publishers.
132. Yando, S. 1969. Method and apparatus for fabricating an array of discrete elements. US Patent 3439416.
133. Cohn, M. B., C. J. Kim, and A. P. Pisano. 1991. Self-assembling electrical networks: an application of micromachining technology. *Digest: Int. Conf. Solid-State Sensors and Actuators*. San Francisco: Transducers Research Foundation.
134. Hosokawa, K., I. Shimoyama, and H. Miura. 1996. Two-dimensional micro-self-assembly using the surface tension of water. *Ninth Annual International Workshop on Micro Electro Mechanical Systems (MEMS '96)*. San Diego, CA.

135. Edman, C. F., R. B. Swint, C. Gurtner, R. E. Formosa, S. D. Roh, K. E. Lee, P. D. Swanson, D. E. Ackley, J. J. Coleman, and M. J. Heller. 2000. Electric field directed assembly of an InGaAs LED onto silicon circuitry. *IEEE Photonics Tech Lett* 12:1198–200.
136. Braun, E., Y. Eichen, U. Sivan, and G. B. Yoseph. 1998. DNA-templated assembly and electrode attachment of a conducting silver wire. *Nature* 391:775.
137. Bashir, R. 2001. DNA nanobiostructures. *Materials Today* 4:30–9.
138. Lee, S. W., W. J. Chang, and R. Bashir, "Bottom-up" approach for implementing nano/microstructure using biological and chemical interactions. *Biotech Bioprocess Eng* 12:185–99.
139. Seeman, N. C. J. 1982. Nucleic acid junctions and lattices. *Theor Biol* 99:237–47.
140. Seeman, N. C. 1991. The use of branched DNA for nanoscale fabrication. *Nanotechnology* 2:149–59.
141. Seeman, N. C., Y. Zhang, and J. Chen. 1994. *J Vac Sci Technol* A12:1895.
142. Seeman, N. C. 1998. DNA nanotechnology: novel DNA constructions. *Annu Rev Biophys Biomol Struc* 27:225–48.
143. Winfree, E., F. Liu, L. Wenzler, and N. C. Seeman. 1998. Design and self-assembly of two-dimensional DNA crystals. *Nature* 394:539–44.
144. Mirkin, C. A., R. L. Letsinger, R. C. Mucic, and J. J. Storhoff. 1996. A DNA-based method for rationally assembling nanoparticles into macroscopic materials. *Nature* 382: 607–09.
145. Martin, B. R., D. J. Dermody, B. D. Reiss, M. Fang, L. A. Lyon, M. J. Natan, and T. E. Mallouk. 1999. Orthogonal self-assembly of colloidal gold-platinum nanorods. *Adv Materials* 11:1021–25.
146. Vorob'ev, A., P. Vaccaro, K. Kubota, S. Saravanan, and T. Aida. 2003. Array of micromachined components fabricated using "micro-origami" method. *Jpn J Appl Phys* 42:4024–26.
147. Kubota, K., T. Fleischmann, S. Saravanan, P. O. Vaccaro, and T. Aida. 2003. Self-Assembly of Microstage Using Micro-Origami on GaAs. *Jpn J Appl Phys* 42:4079–83.
148. Vorob'ev, A., P. O. Vaccaro, K. Kubota, T. Aida, T. Tokuda, T. Hayashi, Y. Sakano, J. Ohta, and M. Nunoshita. 2003. SiGe/Si microtubes fabricated on a silicon-on-insulator substrate. *J Phys D Appl Phys* 36:L71–L73.
149. Charlotte, P., P. Reverdy, L. Doppler, J. Bico, B. Roman, and C. N. Baroud. 2007. Capillary origami: Spontaneous wrapping of a droplet with an elastic sheet. *Phys Rev Lett* 98:156103.

5

Selected Materials and Processes for MEMS and NEMS

Outline

Micro glass blowing, a process developed by Andrei Shkel and others at the University of California, Irvine. The diameter of the glass spheres on a Si wafer is approximately 900 μm. (Courtesy of Dr. Andrei Shkel.)

Carbon microflowers, made in a process developed by Ashutosh Sharma and others at IIT Kanpur, India. An example of a carbon-based microelectromechanical system (C-MEMS) based on pyrolysis of resorcinol-formaldehyde gels (RF-gels). (Courtesy of Dr. Ashutosh Sharma, IIT Kanpur, India.)

Microtable made with carbon MEMS at UCI (Courtesy of Dr. Marc Madou).

Introduction

Silicon is one of the most commonly used micromachining materials not only because of its excellent electronic and mechanical properties, the incredible degree of control one can exert over doping this semiconductor material, and the ease of insulating its surface with SiO_2 but also because it has been the dominant material in the integrated circuit (IC) industry ever since the early 1950s. However, silicon is not perfect for all applications, nor can it function on its own: the IC arsenal features a small group of additional materials, such as the metals Al, Cu, W, B, and P and the insulators SiO_2 and Si_3N_4. There are plenty of new demands that are challenging the continued dominance of silicon and will require the introduction of new materials that work in conjunction with silicon or necessitate the switch to alternative semiconductors altogether: higher speeds; better heat dissipation; merging of sensors and actuators with active electronics, optoelectronics, and quantum devices; lower power consumption; more environmentally friendly manufacturing processes, etc. In microelectromechanical systems (MEMS) and nanoelectromechanical systems (NEMS), and in some application-specific ICs, alternative semiconductor materials already are being used, including gallium arsenide (GaAs), $Al_xGa_{1-x}As$ compounds, SiGe, CdSe, SiC, TiO_2, ZnO, and carbon nanotubes. A wide variety of polymers, ceramics and metals, foreign to the standard IC process also have been adopted, especially in bio-MEMS. In this chapter, we review some selected materials and processes that have affected and will continue to affect the direction MEMS and NEMS take on next. The chapter is headed by a description of possible substrate materials for MEMS and NEMS: quartz, GaAs, diamond, SiC, ceramics (with emphasis on Al_2O_3), polyimides, silicon nitride, and amorphous and hydrogenated amorphous silicon. For exciting new MEMS/NEMS techniques, we survey carbon-based MEMS (C-MEMS), bulk metallic glasses (BMGs), fast microfluidic compact disc (CD) prototyping, and micro glass blowing. Finally, we dedicate a section to biocompatibility and the various chemistries for surface derivation of solids to make them biocompatible. Studies of in vivo MEMS devices are regaining popularity, making a deeper understanding of biocompatibility a necessity.

Selected MEMS and NEMS Materials

Introduction

Miniaturization—a mantra in electrical engineering for a long time—has become a widely accepted discipline in the other engineering branches. This has led to a wider and wider array of materials used in miniaturization science. Here we dwell upon the importance of the substrate material, i.e., the principal material on which or in which to fabricate the MEMS or NEMS. In Chapter 1, this volume, we presented a section on substrate choice, including a comparison of important substrate materials for MEMS and NEMS (see Tables 1.9 and 1.10). These tables include materials that are merely "passive" substrates (e.g., plastics or cardboard support structures) and others that are "active" and may constitute an active member of a micromachine (e.g., GaAs or quartz).[1] If the substrate to be machined must combine good electronic properties with excellent mechanical behavior, single-crystal Si is still the favored material. The electronic qualifications of Si are well known through the resounding successes scored by the IC industry in this and the previous century (see Volume I, Chapter 4). But, when comparing the yield strength, density, specific strength (i.e., ratio of yield strength to density), Knoop hardness, Young's modulus, thermal conductivity, and thermal expansion coefficient for some of the most important technological materials (Volume I, Tables 4.15 and 4.16), one realizes that Si also is an outstanding mechanical sensor/actuator substrate. Although a large percentage of MEMS and NEMS relies on single-crystal Si, and the IC industry is almost exclusively based on it, the need for all types of miniaturized devices with a wide variety of new functions is rapidly changing this picture. This is particularly true for chemical and biological sensors, actuators, harsh environment devices, and microfluidic structures. But even for mechanical devices, Si is far from the only choice, with quartz, for example, being an excellent contender. For biomedical applications and microfluidics, glass and polymers are becoming an increasingly favored miniaturization substrate. Fast ICs, radio frequency (RF)-MEMS, optoelectronics, and NEMS are pushing GaAs, $Al_xGa_{1-x}As$ compounds, and SiGe to the foreground. For harsh environments, ceramic micromachining is emerging as a new opportunity. A check-off list to determine the optimum substrate for a given micromachining task was introduced in Chapter 1, Table 1.11. Many MEMS/NEMS substrates were covered earlier in the book; some additional details on quartz, GaAs, diamond, SiC, ceramics (with emphasis on Al_2O_3), polyimides, silicon nitride, and amorphous and hydrogenated amorphous silicon, which were not treated in sufficient detail elsewhere, follow.

Quartz

Quartz Properties and Quartz Machining

Quartz is the only material known that possesses the following combination of properties:

- Piezoelectric
- Zero-temperature expansion coefficient cuts
- Stress-compensated cut exists
- Low loss [i.e., high-quality factor (Q)]
- Easy to process: low solubility in everything, under "normal" conditions, except fluoride and hot alkali etchants
- Hard but not brittle
- Abundant in nature; easy to grow in large quantities, at low cost, and with relatively high purity and perfection

Some quartz properties are summarized in Table 5.1, and in Figure 5.1, a quartz angular rate sensor (tuning fork-based), developed by Jan Söderkvist (Colibri Pro Development AB, Sweden) for automotive and other applications, is shown. The tines of the tuning fork sensor element shown are 2.5 mm long. The sensor is well suited, for example, to space applications because of its small size and ability to withstand harsh conditions.

The properties of quartz vary greatly with crystallographic direction: when a quartz sphere is etched deeply in an HF bath, the sphere takes on a triangular shape when viewed along the *z*-axis and a lens shape when viewed along the *y*-axis. The etching rate is more than 100 times faster along the fastest etching rate direction (the *z*-direction) than along the slowest direction (the slow-*x*-direction). Quartz, in some respects, presents us with an even more ideal sensor material than Si because of its nearly temperature-independent thermal expansion

Table 5.1 Some Important Physical Properties of Quartz

Property	Value ‖ Z	Value ⊥ Z	Temperature Dependence
Thermal conductivity [cal/(cm/s/°C)]	29×10^{-3}	16×10^{-3}	Decreases with T
Dielectric constant	4.63 (ε_1)	4.52 (ε_3)	Decreases with T
Thermal expansion coefficient (1/°C)	7.1×10^{-6}	13.2×10^{-6}	Increases with T
Electrical resistivity (Ω·cm)	0.1×10^{15} (ionic)	20×10^{15} (electronic)	Decreases with T
Piezoelectric constant, d (pC/N)	2.31 (d_{11})		
	0.73 (d_{14})		
Young's modulus (N/m²)	9.7×10^{10}	7.6×10^{10}	Decreases with T
Density (kg/m³)	2.66×10^3		
Maximum coupling coefficient, k	0.1		
Curie temperature (°C)	573		
Young's modulus E(GPa)	107		

‖ Z, parallel with Z; ⊥ Z, perpendicular to Z. T, temperature. See also Chapter 8, this volume, Table 8.7.

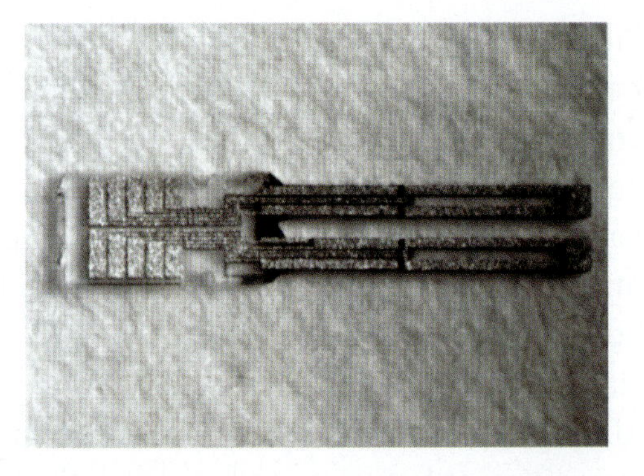

FIGURE 5.1 Quartz angular rate sensor (tuning fork) developed by Jan Söderkvist, Colibri Pro Development AB (Sweden), for automotive and other applications. The sensor is well suited to space applications because of its small size and ability to withstand harsh conditions. The tines of the tuning fork sensor element shown are 2.5 mm long.

coefficient. The thermal expansion coefficient of quartz is $7.8 \times 10^{-6}/°C$ along the z-direction and $14.3 \times 10^{-6}/°C$ perpendicular to the z-direction; the temperature coefficient of density therefore is $-36.4 \times 10^{-6}/°C$. The temperature coefficients of the elastic constants range from $-3300 \times 10^{-6}/°C$ (for C_{12}) to $+164 \times 10^{-6}/°C$ (for C_{66}). For the proper angles of cut, the sum of the different temperature coefficients may cancel, i.e., temperature-compensated cuts exist in quartz. Micromachined quartz is chemically and mechanically robust and may be used in high-temperature, high-shock applications where Si would fail. A variety of quartz micromachined piezoelectric devices also are on the market, such as electronic filters, resonators, and wristwatch tuning forks. Quartz exhibits very good temperature and long-term stability and therefore is ideal for high-precision oscillators. The frequency-accuracy, including all the environmental effects, after one year of aging, is of the order of few parts per million to tens of parts per million.

Prior to about 1956, the material used for quartz resonators was natural quartz, i.e., mined quartz. Natural quartz has been replaced by quartz grown in large single crystals in factories: "cultured quartz." Wafers up to 3 in. in diameter are commercially available as starting material. Although this quartz often is referred to as "synthetic quartz," nobody has yet found a way to synthesize single-crystal quartz directly from silicon and oxygen. Large quartz bars (typically ~15 cm long) of uniform size and shape are grown from small, irregularly shaped pieces of quartz. So, strictly speaking, the quartz is cultured quartz. Quartz is a common material in the earth's crust (e.g., sand is mostly quartz); however, the high-purity crystals needed for quartz growing are not as common. Most of the nutrient materials used by quartz growers are mined in Brazil and the United States.

Machining methods for quartz include mechanical machining, such as diamond saw cutting, grinding, lapping, polishing, and, relatively more recently, ultrasonic machining and wet and dry chemical etching. Lapping and polishing can be used for fashioning quartz plates as thin as 100 μm. These techniques do not lend themselves to making small, complex shapes, though. For the latter applications, ultrasonic machining has been applied to make, for example, high-precision bulk wave oscillators and complex-shaped piezoelectric resonator sensors in the millimeter size range, but the technique is not suitable for mass production. Wet chemical etching of quartz using HF/NH_4F on z-cut quartz can be used for submillimeter fashioning of complex quartz shapes and currently is used for mass production of wristwatch tuning fork resonators.[2] Etching rates depend strongly on the crystallographic orientation, with the z-cut exhibiting the highest etching rates: 1 mm/min at 50°C, with typical etch ratios in the range of 50:1 to 500:1 with respect to x-cut and y-cut plates.[3,4] Conventional photolithography is used to pattern shapes on quartz with chromium/gold for masking layers. High repeatability and automatic batch processing yield lower cost devices. Similar technologies are used for the fabrication of force sensors[5] and accelerometers. The tolerances of these quartz devices for mechanical sensing applications are, as expected from the temperature-independent thermal expansion coefficient, better than those of Si devices. Quartz also allows a greater number of different geometrical shapes than Si, and there is no need for deposition of an insulator between conductor and substrates. Dufour et al.[6] presented a balanced comparison between micromachined pressure sensors using quartz versus silicon vibrating beams and concluded that both materials are strong candidates for mass-produced micromachines. Companies such as Systron Donner[7] that have been selling traditional precision

Table 5.2 Accuracies of Various Oscillators

Oscillator Type[a]	Accuracy[b]	Typical Applications
Crystal oscillator (XO)	10^{-5} to 10^{-4}	Computer timing
Temperature-compensated crystal oscillator (TXCO)	10^{-6}	Frequency control in tactical radios
Microcomputer-compensated crystal oscillator (MCXO)	10^{-8} to 10^{-7}	Spread spectrum system clock
Oven-controlled crystal oscillator (OCXO)	10^{-8} (with 10^{-10} per g option)	Navigation system clock and frequency standard, MTI
Small atomic frequency standard (Rb, RbXO)	10^{-9}	C^3 satellite terminals, bistatic, and multistatic radar
High-peformance atomic standard (Cs)	10^{-12} to 10^{-11}	Strategic C^3, EW

[a] Sizes range from <5 cm³ for clock oscillators to >30 L for cesium standards. Costs range from <$5 for clock oscillators to >$50,000 for cesium standards.
[b] Including environmental effects (e.g., –40°C to +75°C) and one year of aging.
MTI, moving-target indicator radar; C^3, command, control, and communications radar; EW, early-warning radar.

quartz products such as high-temperature, high-shock-resistance accelerometers are appreciating the potential competition from a Si technology and are developing silicon technology in parallel.

Hjort et al.[8] developed a most interesting, novel anisotropic dry/wet etching scheme for single-crystalline quartz. By bombarding quartz with fast heavy ions [e.g., ^{197}Au at 11.6 MeV/a and ^{129}Xe at 11.4 MeV/a specific energy (where a is the atomic mass unit)], deep nanosized amorphous cylindrical latent tracks (±10 mm wide) are formed. The bombarded wafers are then Au-Cr masked and etched in 20 M KOH at 143°C. By masking parts of the wafer prior to etching, microstructures can be generated along the ion track direction. The masked zone remains latent, and the tracks can be eliminated in an annealing step. When the ion tracks are etched in the mask windows, the latent pores are opened, and they widen until they overlap, creating the desired microfeatures. By irradiating the sample off its normal, slanted microstructures can be generated. The aspect ratio of the cylindrical nanotracks reaches a whopping 10^4, and after etching, aspect ratios larger than 100 result. Features as large as 80 µm were demonstrated in this fascinating new micromachining approach.

As mentioned earlier, quartz resonators constitute important quartz micromachining applications. From Table 5.2, accuracies of various oscillators range from 10^{-4} to 10^{-11}, and quartz can deliver even the highest accuracies required. Moreover, quartz comes with an excellent Q (see Chapter 8, this volume, Equation 8.28). Q is proportional to the decay time and is inversely proportional to the line width of the resonance peak. The higher the Q, the higher the frequency stability and accuracy capability of a resonator; i.e., high Q is a necessary but not a sufficient condition for high accuracy.*

Silicon MEMS (Si-MEMS) startup companies Discera, Silicon Clocks, and SiTime are currently targeting the quartz oscillator market. They are building packaged Si-MEMS oscillators, consisting of a MEMS resonator and the electrical circuits for the oscillator function and frequency and temperature compensation. The micromechanical resonator is a Si micromachined structure that vibrates at a specific frequency because of an external excitation (e.g., electrostatic). Today, the main technologies for oscillator clock products are quartz, ceramic and complementary metal-oxide-silicon (CMOS) silicon clocks. Si-MEMS is just barely emerging as a potential competitor. The market is still mostly served by quartz (70–80% of the value), with Japanese companies such as Kyocera, Rakon, and Epson Toyocom in control of this market. Silicon timing or CMOS clocks (i.e. non-MEMS, simply electronic) are available from, for example, Maxim or Linear Technologies and gradually are replacing

* If, e.g., $Q = 10^6$, then 10^{-10} accuracy requires the ability to determine the center of the resonance curve to 0.01% of the line width, and stability (for some averaging time) of 10^{-12} requires the ability to stay near the peak of the resonance curve to 10^{-6} of line width.

ceramic devices in applications with more relaxed specifications. Somewhere between CMOS clocks and quartz oscillators, Si-MEMS clocks now are looking for their piece of this multibillion-dollar pie (see Chapter 10, this volume).

The ~5-MHz resonant frequency resonator from SiTime is illustrated in Figure 5.2 and is made of single-crystal silicon. The resonator is ~200 μm on a side by ~10 μm thick. The trench gap is 0.4 μm and the *Q* is ~75 k at room temperature. The resonator is integrated into standard silicon CMOS chips, which makes the oscillator inexpensive to produce. All oscillator frequencies are derived from the same 5-MHz resonator. The oscillator is compensated by measuring the temperature with a bandgap thermometer on the same CMOS die. The temperature resolution is ~0.05°C, giving a frequency resolution of ~1.5 ppm.

An analysis of the competition between silicon and quartz for MEMS oscillators constitutes a good exercise to acquire a better insight into all of the arguments that need consideration in such a comparison (it is also another opportunity to test the decision tree presented in Chapter 1, "Decision Tree for the Optimized Micromanufacturing Option"). Si-MEMS oscillators are not new; the first publications mentioning MEMS resonators for oscillator applications appeared in the 1980s. Until recently, however, the poor temperature stability of silicon and the need for expensive ceramic or metal vacuum packaging prevented Si-MEMS oscillators from becoming a serious alternative to quartz. But, progress in processes, packaging and integration of the circuitry over the last five years has gone quite a way to address those issues.

Si-MEMS technology, in theory, offers major advantages over quartz in terms of size (especially for mobile applications), power consumption, integration cost, and ability to provide multiple operation frequencies. Let us look at each of these criteria in somewhat more detail, starting with the size argument (http://www.memsinvestorjournal.com/2006/10/mems_oscillator.html).

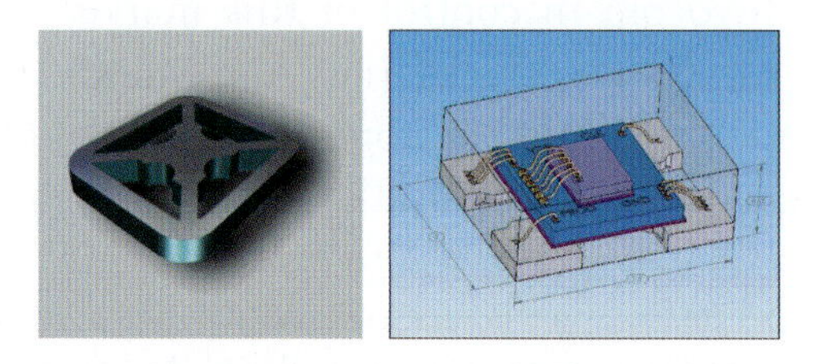

FIGURE 5.2 The ~5 MHz resonant frequency resonator from SiTime is made of single-crystal silicon.

Size Argument

Si-MEMS resonator elements have traditionally been much smaller than their quartz MEMS counterparts. This used to be the key selling argument in the mobile communication industry. A tiny MEMS oscillator often is pictured on top of a quartz oscillator that is 100 times its size to seal this argument. However, as is clear from the above discussed quartz micromachining efforts, the latest quartz MEMS products are very small as well. A new generation of quartz MEMS is emerging, with dimensions similar to those of the first Si-MEMS oscillators (e.g., 2 × 2.5 × 0.8 mm for a quartz device from NDK compared with 2 × 2.5 × 0.85 mm for the Si-MEMS structure from SiTime). The first generation Si-MEMS oscillators are pin-to-pin compatible with quartz products, allowing an easy one-on-one replacement. Since the size of the package will be the same, the main advantage of Si-MEMS oscillators in the "small oscillator segment" will not be size but price. The manufacturing and encapsulating of small quartz devices is expensive compared with Si-MEMS (see next). Printed circuit board (PCB) space also can be saved with Si-MEMS oscillators, which do not require additional bypass capacitors and resistors.

Low Cost Argument

Si-MEMS resonators can be manufactured at a lower cost than quartz MEMS oscillators since Si-MEMS leverages IC batch manufacturing techniques much better. The first commercial Si-MEMS oscillators are hybrid parts, though; e.g., SiTime's Si-MEMS resonators are wire bonded to an ASIC. SiTime uses an innovative wafer-scale packaging solution that is licensed from Bosch and is compatible with CMOS processing. It involves placing resonators on top of an SOI wafer and encapsulating them with an epitaxial layer of silicon, facilitating sealing at high temperatures and creating a barrier to moisture (see Chapter 4, this volume, on packaging, assembly, and self-assembly). The Si resonator is inexpensive in volume and is estimated to be less than $0.10. The typical price of a Si-MEMS oscillator will start at $0.50 and may reach a few dollars in some instances.

Multifrequency Argument

Several RF-MEMS resonators with different frequencies can be manufactured on the same die, allowing for multifrequency devices while retaining small size and low cost. Crystals, on the other hand, are single-frequency devices. If multiple frequencies are needed such as in a cell phone, e.g., a 32 kHz oscillator for the real-time clock and several MHz oscillators for transmission, reception, and processing functions, several discrete-packaged crystals must be used.

High Q Argument

Si-MEMS devices exhibit a higher *Q* than CMOS clocks. However, Si-MEMS has not yet reached the quality factor of quartz (100,000–200,000 for quartz compared with 75,000 for Si-MEMS). The related phase noise of MEMS also is an issue that must be overcome to reach the performance required for temperature-compensated crystal oscillators (TCXOs).

Integration Argument

Contrary to the case with quartz, Si-based RF-MEMS oscillators are manufactured with CMOS-compatible processes. This could pave the way to true single-chip solution with built-in frequency references. Advantages in terms of power consumption (lower parasitics from the wires) and size are obvious. However, as stated above, the first commercially available Si-MEMS oscillators are actually hybrid parts.

Low Power Argument

The low power consumption of MEMS resonators is a key advantage in portable applications. However, MEMS solutions only reduce power consumption when monolithically integrated, because all of the parasitics associated with an off-chip component (i.e. wire bonding quartz or discrete MEMS) are eliminated. This advantage also is diminished in a hybrid part because of wire-bond parasitics.

Summary

Of the human-grown single crystals, quartz, at ~3000 tons per year, is second only to silicon grown in quantity (three to four times as much Si was grown annually as of 1997). Despite the mechanical/thermal advantages of quartz, quartz micromachining lags considerably behind that of Si. This is mainly due to the lack of plasma processes to structure quartz with the versatility achieved in the case of Si. Another reason that quartz is not competing better with Si as a mechanical MEMS substrate is the lack of large quartz substrates and its cost: a quartz wafer, 76.2 mm in diameter, 0.45 mm thick, and polished on one side costs about $200, while the average selling prices for 200- and 300-mm silicon wafers are about $45 and $200, respectively.

GaAs

GaAs Properties

The first gallium arsenide (GaAs) metal semiconductor field effect transistor (MESFET) was proposed in 1966 and the first GaAs high-electron mobility transistor (HEMT) was demonstrated in 1978 at the Bell Laboratories. However, GaAs is only now rising from its position as the technology that remains the technology of the future to true technological wonder. The main economic driver for widespread use of GaAs semiconductors was the emergence of the digital cellular phone, which needs faster IC technology than Si can provide, and ever since 1994, the GaAs industry has grown at least 25% per year. As the micromachining industry largely evolves in lockstep with the IC industry, we can expect GaAs micromachines to succeed as GaAs use penetrates the IC industry further. The growing importance of quantum mechanical devices also is pushing GaAs toward becoming a more important MEMS/NEMS substrate contender. As we pointed out in Volume I, Chapter 3, an important feature of the III-V technologies in general, and of GaAs in particular, is the possibility of forming compatible ternary and quaternary compounds. Using GaAs as the substrate material, the formation of $Al_xGa_{1-x}As$ compounds is especially attractive, since their lattice constants are nearly equal to those of GaAs, and aluminum and gallium atoms are easily substituted in the lattice without causing too much strain. Today, epitaxy techniques have matured, and both high-quality MOCVD and molecular beam epitaxy (MBE) epitaxy are commercially available. Layers deposited with these techniques have very little induced strain, which allows them to be grown almost arbitrarily thick. In Volume I, Chapter 3, we describe GaAs-based quantum wells; Volume I, Chapter 5 covers double-barrier resonant tunneling diodes and quantum cascade lasers, all based on layers of gallium aluminum arsenide sandwiched between gallium arsenide layers. With its direct bandgap, GaAs can be used to emit light efficiently, whereas silicon has an indirect bandgap and is very poor at emitting light (see Volume I, Chapter 3). Nonetheless, recent advances in nanotechnology, as we described in Volume I, Chapters 3 and 5, may make silicon light-emitting diodes and lasers possible after all. Another important application of GaAs (and III-V compounds in general) is for high-efficiency solar cells. In 1970, Zhores Alferov (see Volume I, Figure 3.60) and his team in the USSR created the first GaAs heterostructure solar cells. Today, the combination of GaAs with germanium and indium gallium phosphide is the basis of a triple-junction solar cell that holds the record for solar light conversion efficiency of over 32% and can operate also with light as concentrated as 2000 suns.* This type of solar cell powers the robots *Spirit* and *Opportunity*, which are exploring the Mars surface.

In Chapter 1, this volume, Table 1.10, we compare some of the material properties of single-crystal SiC, Si, GaAs, and diamond with relevance to electronics and MEMS/NEMS. Gallium arsenide, it emerges, is a very promising MEMS/NEMS substrate for a variety of reasons. These include well-known optoelectronic properties such as a direct bandgap transition (see above) and high-mobility electrons. The higher saturated electron velocity and higher electron mobility of GaAs compared with those of Si allow it to function at frequencies in excess of 250 GHz. GaAs HEMTs are a typical high-electron mobility application, and the same property also renders the material ideal for the measurement of magnetic fields through the Hall effect (see Volume I, Equation 3.64). Also, GaAs devices generate less noise than silicon devices when operated at those high frequencies. Because of its high switching speed, GaAs, it would seem, also is more ideal for computer applications than Si. In the 1980s it often was suggested that the IC market might switch from Si to GaAs, which it never really did (from there its unfortunate designation as "the technology that remains the technology of the future"). Cray even built a GaAs-based computer in the early 1990s, the Cray-3, but the company filed for bankruptcy in 1995. Because it has a larger bandgap than Si, GaAs is considered to be a better material for high-temperature electronics. GaAs devices can be operated at higher power levels than equivalent silicon devices because GaAs has a higher breakdown voltage. Pressure, temperature, and vibration sensors, making use of the influence of external pressure and temperature on the bandgap, have already been built.

GaAs has piezoelectric properties comparable to those of quartz, and various physical effects give higher piezoresistive values than those of silicon. The piezoelectric effect enables GaAs-based piezoelectric transducers, and one may excite any vibration mode used in quartz if the proper crystal orientation is observed. Overall, the material is less attractive for purely mechanical MEMS devices, though, since the yield load is smaller by a factor of two compared with silicon.[9] We summarize some of the most important GaAs properties in Table 5.3.

GaAs Micromachining

Besides wet[9] and dry[10] bulk micromachining of GaAs, some GaAs surface micromachining work has been reported.[11] Because of the zinc blend structure of GaAs (same as Si), the wet etching of {111} crystal planes proceeds more slowly than all others. In wet bulk micromachining of GaAs, a unique triangular prism-shaped bridge can be obtained through anisotropic etching by appropriate placement of the etch mask (see Figure 5.3). This unique profile, not possible in silicon, is formed because of the

* Where one sun equals 1000 W/m^2 at the point of arrival. One sun is a unit of power flux.

TABLE 5.3 GaAs Properties

Property	GaAs
Atomic or molecular weight	144.63
Atoms or molecules/cm³	4.42×10^{22}
Breakdown field, V/cm>	$\sim 4 \times 10^{5}$
Chemical resistance	Poor
Crystal structure	Zinc blend
De Broglie electron wavelength (Å)	240
Debye temperature	360
Density, g/cm³	5.32
Dielectric constant (high frequency)	10.89
Dielectric constant (static)	12.9
Effective density of states: valence band N_v (cm^{-3})	7.0×10^{18}
Effective density of states: conduction band N_c (cm^{-3})	4.7×10^{17}
Effective mass: electrons	$0.063\ m_0$
Effective mass: holes	$0.051\ m_0$
Electron affinity (eV)	4.07
Energy gap (eV)	1.42 direct
Energy gap (eV)	1.43
Intrinsic carrier concentration, n_i (cm^3)	1.8×10^{6}
Intrinsic mobility: Electron, cm²/V-s	8600
Intrinsic mobility: Hole, cm²/V-s	400
Knoop hardness (kg/mm²)	600
Lattice constant, nm	0.5653
Maximum operating temperature (°C)	460
Melting point (°C)	1238
Number of atoms in 1 cm³	4.42×10^{22}
Optical phonon energy	0.035
Physical stability	Fair; sublimation of arsenic is a problem
Saturated electron drift velocity (× 10⁷ cm/s)	2
Symmetry group	T_d^2– F43 m
Temperature coefficient of expansion (°C⁻¹)	6.8×10^{-6}
Thermal conductivity (W/cm-°C)	0.46
Yield strength (GPa)	2.0
Young's modulus (GPa)	75

See also http://www.ioffe.rssi.ru/SVA/NSM/Semicond/GaAs/index.html.

crystallographic polarity in some axial directions, and the differences between the etch rates of the A{111} planes and the other low-index planes.[12]

Although slower etching planes exist, GaAs does not present real etch stop planes, as observed in silicon when using, for example, KOH-based solutions (see Volume II, Chapter 4). In the latter etchants, the {111} Si planes present etch rates of up to 400 times lower than the other planes. Surface micromachined sensor and actuator structures made of metal on a GaAs substrate have been demonstrated in schemes compatible with normal GaAs IC processing. The most frequently used sacrificial wet and dry etch systems for III-V crystals are presented in the review papers of K. Hjort[14] and S. D. Collins[15] and in the work by Karam et al.[16] It was shown that to craft surface micromachined structures in epitaxial GaAs, $Al_xGa_{1-x}As$ ($x \geq 0.5$) may act as a sacrificial

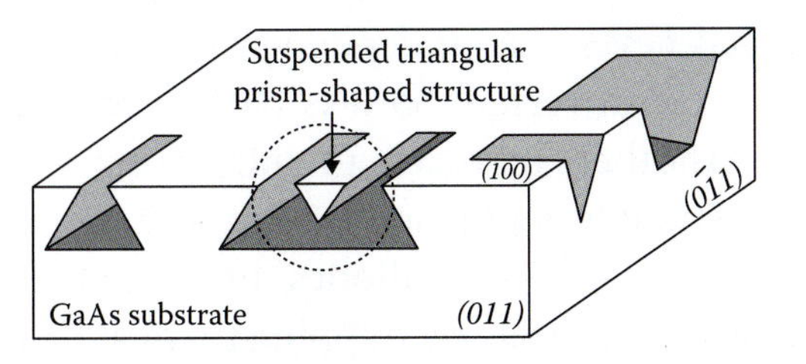

FIGURE 5.3 Free-standing triangular bridge in GaAs.[13]

layer. By using several steps of MOCVD epitaxial layer regrowth, structures of polysilicon-like complexity were built.[17,18] Note that not only different material compositions but also different dopant concentrations and damaged regions of GaAs can act as stop and sacrificial layers.

In Volume II, Chapter 3, we detail the GaAs single-crystal reactive etching and metallization process, which includes chemically assisted ion beam etching (CAIBE) and reactive ion etching (RIE) for vertical and undercut dry etches, respectively. Suspended and movable GaAs structures are produced with a 25:1 aspect ratio of vertical depth (10 μm) to lateral width (400 nm) (Volume II, Figure 3.43).

Ribas from TIMA developed a front-side bulk GaAs micromachining approach compatible with standard GaAs microelectronics technologies for the fabrication of integrated, low-cost, high-volume microsystem applications (http://cmp.imag.fr/products/mems/thesis.pdf).[13] Dr. Ribas also developed a set of computer-aided design tools for GaAs MEMS layout level design, such as cross-section and three-dimensional layout viewers, layout generators, and a bulk GaAs etching simulator within the Mentor Graphics environment. In the same context of GaAs IC compatibility, Karam et al.[16] are investigating gallium arsenide micromachining techniques using high-electron mobility transfer (HEMT) and metal Schottky field effect transistor (MESFET) foundry processes.

In an interesting example of self-assembly, two-dimensional arrays of self-positioning, free-standing GaAs structures were fabricated using a new method of surface micromachining called *micro-origami*. This technique is based on epitaxial growth of III-V multilayers on GaAs substrates followed by photolithography and wet etching, as we described in Chapter 4, this volume, on packaging, assembly, and self-assembly.

Progress in processes to deposit GaAs epitaxially on Si could benefit both IC and micromachining industries. GaAs grown epitaxially on Si substrates has great potential to lower the costs, improve the manufacturability, and improve the thermal characteristics of GaAs-based ICs. GaAs/Si wafers are larger and stronger and lighter than GaAs wafers.[19] This combination opens the possibility of monolithic integration of GaAs-based ICs with Si-based ICs. GaAs-on-Si has significant market potential, both as a substitution technology for manufacturing traditional GaAs devices and as a new technology for monolithic integration of GaAs devices and silicon integrated circuits. As a substitution technology, GaAs-on-Si holds the potential to fundamentally change the economics of manufacturing GaAs integrated circuits and GaAs optical components. As a technology for monolithic integration of GaAs devices and silicon integrated circuits, GaAs-on-Si technology is even more interesting. In fiber optics, integration of GaAs lasers and photodetectors with silicon control circuitry would become feasible. Because of these obvious advantages, many attempts have been made over the past thirty years to grow GaAs on Si substrates. The GaAs lattice constant (5.65 Å) is about 4% larger than silicon's (5.43 Å), and this difference makes it difficult to grow high-quality GaAs directly on silicon. Instead, Motorola researchers place a thin layer (about 50 Å) of strontium titanate ($SrTiO_3$) between the two semiconductors.[20] Strontium titanate's lattice constant falls roughly halfway between GaAs and Si. Moreover, an amorphous layer (10–20 Å thick) forms between Si and $SrTiO_3$. This layer, it is hypothesized, absorbs the lattice mismatch strain between the two materials, allowing the crystalline $SrTiO_3$ to form a normal lattice without distortion from the underlying Si. Because the $SrTiO_3$ layer is very thin, the amorphous interlayer also absorbs the mismatch strain between the $SrTiO_3$ and GaAs layers.

In 2001, scientists from Motorola reported on supposedly device-quality GaAs epitaxial layers with low dislocation densities, close to that of bulk GaAs crystals (~$10^4/cm^2$), grown onto Si by molecular beam epitaxy (MBE) (Figure 5.4).[21] In spite of the initial worldwide publicity, Motorola decided to stop its activity in this area in 2003.[22] The presence of a low dislocation density in the epitaxial GaAs layers has never been independently confirmed and remains controversial. So it is back to the drawing board. Another approach to integrating GaAs on a Si substrate was pursued by Yeh and Smith, who assembled GaAs laser diodes in micromachined wells in a Si substrate by trapping them as discussed in Chapter 4, this volume (see Figure 4.44).[23,24]

FIGURE 5.4 Motorola Labs created the world's first 8-in. GaAs on silicon wafer and worked with epitaxial wafer manufacturer IQE to create the world's first 12-in. GaAs on silicon wafer.

In the micromachining world, possible applications of GaAs MEMS are optical shutters or choppers, actuators in monolithic microwave integrated circuits (MMIC), sensors using piezo- or optoelectrical properties of GaAs, or applications favoring integration of micromechanical devices with electronic circuitry for fast signal processing, high operating temperature, or high radiation tolerance. GaAs micromachining might constitute a good method by which to incorporate micromirrors in resonant optical cavities for tunable lasers. Also, micromachining using gallium arsenide and group III-V compound semiconductors is a practical way to integrate RF switches, antennas, and other custom high-frequency components with ultrahigh-speed electronic devices for wireless communications (see Chapter 10, this volume).

Summary

Overall, it remains hard for GaAs to compete with Si as a semiconductor substrate, but GaAs did start making major commercial inroads for high-frequency applications with the introduction of mobile phones. Silicon presents at least three major advantages over GaAs, explaining why it is so hard for GaAs to compete. To begin, silicon is abundant and cheap to process. Silicon's greater physical strength allows working with larger wafers (maximum of ~300 mm compared with ~150 mm diameter for GaAs). Silicon is abundantly available in the Earth's crust in the form of all types of silicate minerals, and the economy of scale available to the silicon IC industry discourages a more rapid adaptation of GaAs. The second major advantage of Si over GaAs is the existence of silicon dioxide—one of the best and most stable insulators that may be created with a very low defect concentration at interface Si/SiO_2. Silicon dioxide can easily be incorporated onto silicon circuits, and such layers adhere very well to the underlying Si. GaAs does not form such a stable adherent insulating layer (its native oxides are typically leaky). The third, perhaps most important advantage of silicon over GaAs is that it possesses a much higher hole mobility. This high hole mobility allows for the fabrication of higher-speed p-channel field effect transistors required for CMOS logic. Because they lack a fast CMOS structure, GaAs logic circuits have much higher power consumption, which has made them unable to compete with silicon logic circuits.

As suggested before, the micromachining industry largely evolves with and piggybacks on the IC industry, so we can expect GaAs micromachines to succeed as GaAs use penetrates the IC industry further. The most effective approach to the cost reduction of any device is to increase the diameter of GaAs wafers. Along this line, Anadigics Inc., which manufactures RF ICs for the wireless handset and broadband communications markets, is building a 6-in. GaAs fab in China, a certain sign that GaAs has arrived (cost per 6-in. GaAs wafer is projected at $300).

The SiGe devices market has grown more rapidly than some anticipated, and III-V producers have some reason to be concerned about the transition from Si to SiGe instead of using GaAs in a number of key applications. With the talk of SiGe (see Volume I, Figure 4.59) matching performance with GaAs and beating it on cost, GaAs still edges out SiGe in terms of efficiency, so for those applications where this is less important—e.g., the shorter-range wireless networks, such as WiFi and Bluetooth—SiGe is going head to head with RF-CMOS.

Diamond

Introduction

Diamond, the hardest of all natural materials, is a wide bandgap semiconductor material with excellent thermal, tribological, and mechanical properties. The thermal conductivity of diamond is higher than that of any other material near room temperature, making it an ideal material for heat sinking. Diamond components may be as much as 10,000 times more wear-resistant than those made from silicon and come with a much higher thermal and chemical stability. Diamond is also a biocompatible material, so that it could be used in the body as a drug-dispensing unit without initiating an allergic reaction. This is because carbon (the elemental ingredient of diamond) is chemically benign (see also below, under C-MEMS). If it could be micromachined easily, the exceptional physical and chemical properties of diamond would promise to expand the range of applications for microdevices. It is an ideal candidate for MEMS applications, particularly in harsh environments. In addition, diamond reduces stiction compared with silicon (diamond's low coefficient of friction is comparable to that of Teflon), perhaps making diamond-surface micromachined devices easier to manufacture. Diamond can be doped to change it from an insulator to a semiconductor: p-doping works fine, but it remains difficult to make an n-type diamond. We compare the properties of diamond with those of silicon, GaAs, and SiC in Table 5.4.

Diamond Deposition and Diamond Micromachining

Chemical vapor deposition (CVD) of single-crystalline diamond has been the subject of a great deal of research since the early 1980s. Very high-quality diamond thin films with negligible levels of defects and impurities (even superior to natural diamonds and high-pressure, high-temperature diamonds) are achievable when using microwave plasma chemical vapor deposition. However, high- quality homoepitaxial single-crystal CVD diamonds are still very expensive and are available only on small areas (3×3 or 4×4 cm^2) because of the need of single-crystalline diamond substrates.[25,26]

Alternatively, polycrystalline diamond film, deposited by CVD, is one potentially viable approach

Table 5.4 Properties of Diamond and Other Semiconductors at 293 K

	Diamond	Silicon	GaAs	SiC
Lattice constant (Å)	3.567	5.431	5.653	a = 3.086; c = 15.117
Hardness (kgf/mm²)	7,000–10,000	1,000		1,875–3,980
Density (g/cm³)	3.52	2.30	5.32	3.10
Refractive index	2.4195	3.448	3.4	2.65–2.69
Thermal conductivity (W/cm·K)	20	1.41	0.455	4
Specific heat (J/g·K)	0.52	0.70	0.35	0.65
Optical phonon energy (meV)	163	63	35	100
Longitudinal phonon velocity (cm/s)	1.8×10^6	9×10^5	5.2×10^5	1.3×10^6
Coefficient of thermal expansion	0.8×10^{-6}	2.5×10^{-6}	5.9×10^{-6}	2.9×10^{-6}
Debye temperature (K)	1,860	645	344	1,200
Electron saturated velocity (cm/s)	1.5×10^7	1×10^7	0.8×10^7	2×10^7
Hole saturated velocity (cm/s)	1.05×10^7	9×10^6	1×10^7	1×10^7
Electron mobility (cm²/V·s)	2,000 (500)	1,420	8,800	600
Hole mobility (cm²/V·s)	1,500 (500)	470	400	650
Effective mass				
Electron	0.57	1.1	0.068	0.45
Hole	1.2	0.8	0.5	1.0
Dielectric constant	5.7	11.9	12.5	10
Intrinsic resistivity (Ω·cm)	$>10^{15}$	10^5	10^8	$>10^{15}$
Breakdown field (V/cm)	$1–20 \times 10^6$	3×10^5	3.5×10^5	$1–5 \times 10^6$
Bandgap (eV)	5.47	1.12	1.42	2.2
Effective density of states				
Conduction band (cm⁻¹)	1×10^{19}	2.8×10^{19}	4.7×10^{17}	7×10^{18}
Valence band (cm⁻¹)	3×10^{19}	1.04×10^{19}	7×10^{18}	2.5×10^{19}

to high-temperature, harsh-environment MEMS[27] (see also http://www.nasatech.com/Briefs/Feb99/NPO20529.html). Chemical vapor deposition of diamond on Si substrates is compatible with the integration of Si-based electronics. The most common polycrystalline diamond CVD process involves a flowing mixture of methane and hydrogen, typically at a total pressure of 45 Torr (6 kPa) and a substrate temperature of 800–950°C. Diamond heteroepitaxially deposited on Si is polycrystalline (Figure 5.5) with a large number of grain boundaries, a high intrinsic stress, and a rough surface, making it difficult to use in MEMS. The grain size of typical diamonds grown from a methane and hydrogen gas source by microwave plasma CVD on Si is around 30 nm to 10 μm.[28] Problems with polycrystalline diamond films also include difficulties in growing good-quality diamond films and making reliable ohmic contacts to the material, as well as poor reproducibility.[29] This leaves us with ultrananocrystalline diamond (2–5 nm grain size)[30] and amorphous diamond[31] as the two most suitable diamond candidates for MEMS applications.

At Argonne National Laboratories, "ultrananocrystalline" diamond films have been deposited by a novel CVD process that may overcome some of the problems mentioned above (http://www.anl.gov/techtransfer/Available_Technologies/Chemistry/uncd_flc.html). Argonne's patented CVD method was developed at first using fullerenes as the carbon source. Fullerene powder was vaporized and introduced into an argon plasma, causing the fullerenes to fragment into two-atom carbon molecules (dimers). Silicon or other substrate materials are "primed" with fine diamond powder. When the carbon dimers settle out of the plasma onto the substrate, they arrange into a film of small diamond crystals about 3–5 nm in diameter. Subsequent work showed that the same result can be achieved by introducing methane into an argon plasma as long as little or no additional hydrogen is present. The Argonne method differs in two major respects from other CVD methods for making diamond film. First, the molecular building block is different (carbon dimers rather than methyl radicals), and second, little or no hydrogen is present in the plasma (~1%), whereas in other methods the plasma contains 97–99.5% hydrogen along with methane. The ultrananocrystalline films produced by Argonne's method are completely free of intergranular flaws and nondiamond secondary phases that degrade the properties of conventionally produced diamond films, have crystals about 50–200 times smaller than those in other films, and are 10–20 times smoother than conventional high-purity CVD diamond films.

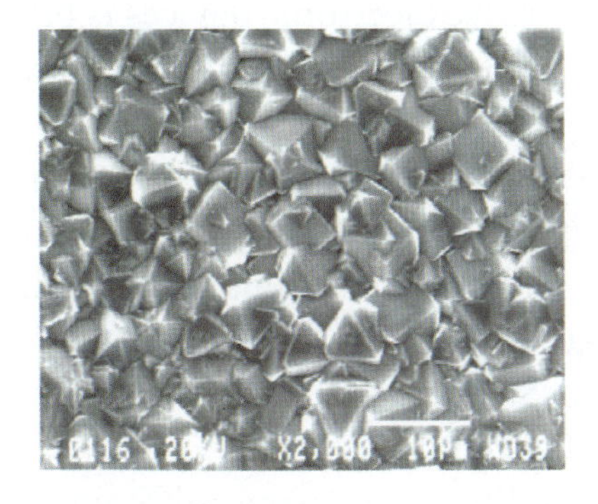

FIGURE 5.5 A typical SEM image of a CVD diamond surface on a Si substrate. (Courtesy of Dr. Chunlei Wang, Florida International University.)

For surface micromachining, a "starter" (nucleation) layer of nanocrystalline diamond powder on a silicon substrate is patterned, and the diamond film forms only in the areas where the nucleation layer remains. Flat, free-standing diamond films can also be grown on silicon dioxide. The diamond layer may be released from the substrate by etching away the silicon dioxide. The result is free-standing diamond structures as little as 300 nm thick with features as small as 100 nm and friction coefficients as low as 0.01. Argonne is also developing deposition methods that permit integrated fabrication of complete devices, such as micrometer-sized pinwheels, gears, turbines, and micromotors, without manual assembly. Also, hollow, three-dimensional (3D) structures can be fabricated in diamond surface micromachining. In this case, silicon is used as the sacrificial material (e.g., a disc or pin) for the desired diamond component. The diamond film is grown on the Si structure and conforms to its shape. The silicon is etched away, leaving a diamond cylinder or tube. It has been shown that this method is capable of making convex hollow structures (e.g., a hollow pyramid)—shapes that cannot be produced by conventional silicon lithographic methods.

Sandia researchers have explored amorphous diamond for micromachining. Amorphous diamond is the second hardest substance known, after crystalline diamond (http://www.sandia.gov/media/NewsRel/NR2000/diamond.htm). They use pulsed laser deposition, and a simple proof-of-principle device—a diamond interdigitated comb—takes about 3 h to fabricate. Subsequent annealing of the diamond film device down to zero stress (to prevent warpage) takes a few minutes. Before this work, amorphous diamond itself had been impractical because its tremendous internal stresses—hundreds of atmospheres—made it impossible for the material to stand alone or to coat thickly on any but the strongest substrates.

Diamond-based MEMS, such as cantilevers, membranes, and microturbines, can be fabricated using two strategies: lithographic patterning and selective deposition.[31,32] In the selective deposition method, diamond microstructures are fabricated by selective diamond growth on Si using a patterned SiO_2 mask. With this selective growth approach, piezoresistors and thermal resistor structures have been patterned.[32]

In lithographic diamond patterning, diamond films are synthesized on an appropriate substrate such as Si. This is followed by diamond surface micromachining processes similar to the ones used in the fabrication of poly-Si-MEMS. Diamond layers can be masked by photoresist or metals, followed by dry etching (RIE) in oxygen-containing plasmas.

Diamond films also can be directly patterned using focused ion beam (FIB) milling without the use of a mask.[33] Figure 5.6 shows CVD diamond microscale p-i-p electronic devices fabricated on homoepitaxially grown single-crystalline layers by FIB etching; the latter devices were used to investigate carrier transport properties of diamond at high electric field strengths (more than10^7 V/cm).[33]

A number of other diamond-based devices have been investigated; these include basic electronics elements, sensors, and actuators.[34–36] Although their cost, performance, and processibility are still not compatible with Si-based MEMS, diamond as a complementary MEMS material remains very promising, especially for use in extreme environments.

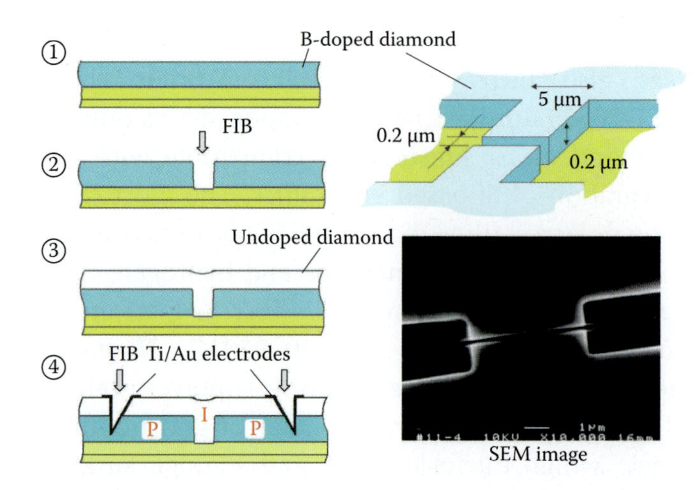

FIGURE 5.6 Schematic fabrication processes for a diamond p-i-p structure consisting of a narrow *intrinsic* (undoped) region sandwiched between two p-type (boron-doped) regions. Top view schematic drawing and a scanning electron microscope picture show narrow (0.2-μm) grooves formed in the 0.2-μm-thick p layer (at step 2). These grooves are filled with i diamond (at step 3).[33]

SiC

Like diamond, silicon carbide (SiC) is well known for its mechanical hardness, chemical inertness, high thermal conductivity, and electrical stability at temperatures well above 300°C, making it another excellent candidate for high-temperature MEMS. Both exhibit piezoresistive properties. In comparison with diamond, other SiC features are that it can be doped both p-type and n-type fairly easily and that it allows a natural oxide to be grown on its surface, which make SiC an attractive candidate for alternative semiconductor MEMS material. Because of its large bandgap and good carrier mobility, it can be employed in high-temperature and high-power applications. In addition, the fabrication technology for SiC active electronic devices is based mostly on processes established in the Si microelectronics industry. One property that makes SiC films particularly attractive for micromachining is that these films easily can be patterned by dry etching using Al masks. Patterned SiC films actually can be used as passivation layers in the micromachining of the underlying Si substrate (SiC can withstand both KOH and HF etching).[37]

SiC crystallizes in many different polytypes, which differ from one another in the stacking sequence of a repeat unit consisting of two planes of close-packed Si and C atoms. The two most common SiC polytypes are 3C-SiC and 6H-SiC. The 3C polytype, also known as β-SiC, is the only polytype with a cubic structure. 3C-SiC crystallizes in a ZnS-type structure; hence, it can be deposited on Si. Historically, most research has focused on developing 6H-SiC as a semiconductor material for high-temperature and high-power electronics. However, the small wafer size (<2 in.) and the inability to grow epitaxial layers on substrates other than 6H-SiC have hindered the development of the 6H polytype. Over the last seven years we have witnessed a growing interest in 3C-SiC as a MEMS material. As we saw above, unlike 6H-SiC, the 3C-SiC polytype can be epitaxially grown on single-crystal silicon substrates. Large-area substrates enable low-cost batch processing, essential in making SiC MEMS devices viable for mass production applications (e.g., automotive). It was at Case Western Reserve University (CWRU) that the first successful depositions of spatially uniform, single-crystal, 3C-SiC films on 4-in. (100) silicon wafers were made. These epitaxial 3C-SiC films on Si were used to craft structures such as diaphragms and cantilever beams. Since SiC films are highly resistant to KOH and EDP etching, dry etching and Al masks are employed for patterning. Because these diamond films must be grown directly on single-crystal silicon, surface micromachining with a sacrificial layer was not possible at first. The solution was again invented at CWRU and involves growing polycrystalline cubic silicon carbide (poly-SiC) films of about 2 μm thickness by an APCVD process on 4-in. polysilicon-coated (100) silicon wafers.[38,39] Surface micromachining is achieved by using the underlying polysilicon film as the sacrificial layer. The poly-SiC deposition process is carried out at 1280°C in a cold-wall, RF-induction heated, vertical APCVD reactor. After deposition, the SiC film is polished to reduce the surface roughness (Ra) from ~400 Å in the as-grown film to <40 Å. The poly-SiC lateral resonant device shown in Figure 5.7 was made by this surface micromachining approach, and the finished device exhibits Qs as high as 215,000 at pressures below 10^{-5} Torr (see also "Quality Factor" in Chapter 8, this volume, on actuators). Resonant frequency drifts of less than 18 ppm/h have been observed, and device operation at elevated temperatures as high as 950°C has been achieved. The poly-SiC surface micromachining steps are outlined in Figure 5.8.

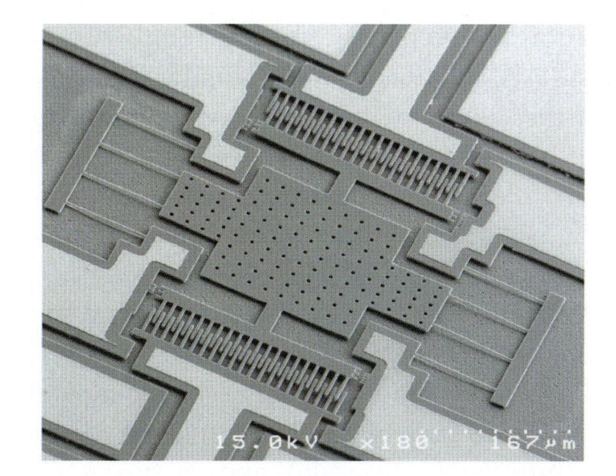

FIGURE 5.7 Scanning electron microscope (SEM) micrograph of a released poly-SiC lateral resonant device. The suspension beam length and width are nominally 100 and 2.5 μm, respectively. Exposed poly-SiC shows up as dark gray, while Ni metallization appears light gray. (Courtesy of Dr. M. Mehregany, Case Western Reserve University.)

In Chapter 1, this volume, Table 1.10, we compared some of the material properties of single-crystal SiC with Si, GaAs, and diamond with relevance to electronics and MEMS/NEMS. For an update on this

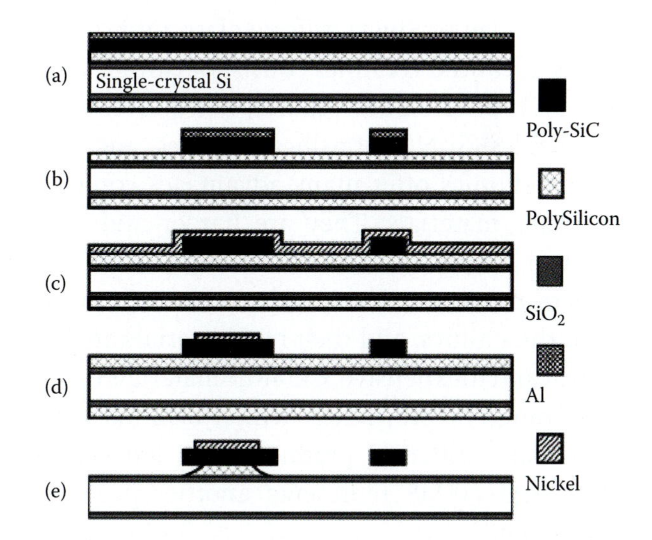

FIGURE 5.8 Schematic description of the poly-SiC resonator fabrication process showing cross-sections after a 5000-Å-thick layer of aluminum is sputter deposited over the polished poly-SiC (a); the Al layer is patterned using lithography and anisotropic $CHF_3/O_2/He$ plasma etching to transfer the Al pattern into the poly-SiC (b); the Al layer is stripped and a 7500-Å-thick layer of Ni is deposited by sputtering (c); the Ni film is patterned for contacts using photolithography and wet etching (d); and poly-SiC is released in 40 wt% KOH at 40°C (e).

table for SiC properties, consult http://www.ioffe.rssi.ru/SVA/NSM/Semicond/SiC/index.html.

Further advances in SiC deposition and patterning techniques have led to the development of single and multilevel implementations similar to the polysilicon MUMPs (multiuser MEMS process). The multiuser silicon carbide (MUSiC) is an eight-mask, four polycrystalline SiC (poly-SiC) layer surface micromachining process.

The State of Ohio has funded high-temperature MEMS efforts both at Case Western Reserve University (http://mems.case.edu) and at NASA Glenn (http://www.grc.nasa.gov/WWW/SiC/SiC.html) and made these two centers leaders in the field.

Bulk and Surface Microstructuring of Ceramics

Ceramics encompass a vast array of materials, and a concise definition of ceramics is almost impossible. A workable definition of ceramics is that of a class of refractory, inorganic, and nonmetallic materials. This class of materials can be divided into traditional and advanced ceramics. Traditional ceramics include clay products, silicate glass, and cement, whereas advanced ceramics consist of carbides (SiC; see above), pure oxides (quartz; see above: Al_2O_3, etc.), nitrides (Si_3N_4), nonsilicate glasses, and many others. Ceramics offer many advantages compared with other materials. They are harder and stiffer than steel, more heat and corrosion resistant than metals or polymers, and less dense than most metals and their alloys, and their raw materials are both plentiful and inexpensive. Ceramic materials display a wide range of properties, which facilitates their use in many different product areas and ceramic MEMS, and NEMS represents another promising development.

The inherent chemical resistance of refractory materials makes their anisotropic micromachining very difficult and expensive.[40,41] In Volume II, Chapter 9, we promoted the use of plasma spray deposition for making oxygen-sensitive ZrO_2 membranes in batch mode (see Volume II, Figure 9.24). Another potential solution for working with these poorly machinable materials is by casting or by gas phase deposition into molds. MIT's micromachined gas turbines, for example, were made of thick (10–200 μm) SiC or silicon nitride, deposited by CVD into Si molds (http://web.mit.edu/aeroastro/faculty/index.html). Sol-gel-type solidification of ceramic liquid precursors in molds, as covered in Volume II, Chapter 8, is another available option—less expensive but yielding low- to medium-resolution microstructures only. Deep UV resists may be used as alternative, sacrificial molds. In most cases a high-temperature sintering step will significantly shrink the dimensions of the high-temperature part, reducing achievable accuracy. In Volume II, Chapter 10, on LIGA, we described how UV and LIGA techniques provide some of the highest-resolution ceramic parts. Finally, in Volume II, Chapter 7, we introduced atomic layer deposition by which precursor gases or vapors alternately are pulsed onto the substrate surface to make Al_2O_3 (see Volume II, Figures 7.47 and 7.48).

Here we introduce Al_2O_3 and TiO_2 micromachining work for high-accuracy ceramic MEMS with the potential for a plethora of novel micro- and nanomachining applications.

Micromachining of alumina ceramic appears particularly appealing and might have advantages over other high-temperature MEMS substrates, such as quartz, 3C-SiC, diamond, and GaAs (see Chapter 1, this volume, Table 1.10), in terms of cost, manufacturability, and performance.

Routkevitch et al.[42] and Li et al.[43] have made fascinating microstructures from nanoporous anodic alumina films (also known as porous anodic alumina). Anodic alumina is a self-organized material with unique pore morphology, composition, and properties. Regular porous films, from less than 100 nm thick to more than 300 μm, are formed when aluminum metal is anodized in solutions of phosphoric, oxalic, or sulfuric acid (in low concentrations—from 2 wt% to 10 wt%). Pore diameter is tunable from a few nm to several hundred nm, and pore densities vary from 10^9 to 10^{11} cm^{-2}. For typical anodization voltages (50–100 V), the pore diameter ranges from 0.15 to 0.3 μm, and by sintering above 830°C in air, the amorphous anodic alumina may be converted into crystalline material, thereby increasing its chemical and thermal

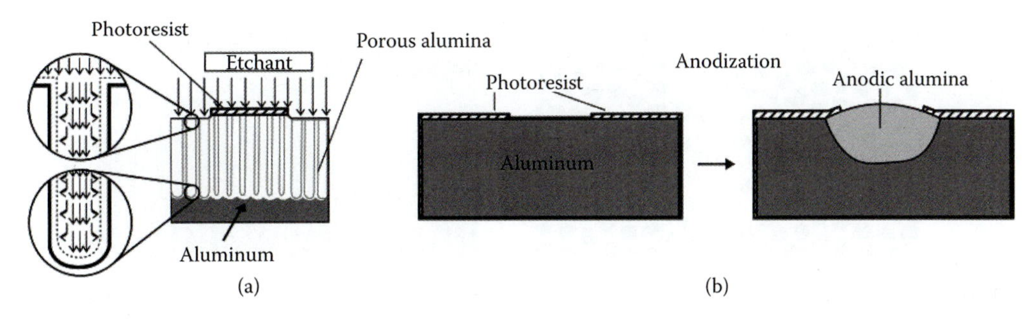

FIGURE 5.9 Schematic of two approaches for micromachining of anodic alumina. (a) High-resolution micromachining. (b) Lower-resolution microstructure prepared by localized anodization of aluminum. (Based on Routkevitch, D., A. Govyadinov, and P. Mardilovich. 2000. High aspect ratio, high resolution ceramic MEMS. *2000 International Mechanical Engineering Congress and Exposition*, 1–6.[42])

stability dramatically. The schematic in Figure 5.9 summarizes two approaches for micromachining of porous anodic alumina. In the higher-accuracy micromachining method (Figure 5.9a), the alumina film is anisotropically etched after patterning with conventional lithography. As prepared, anodic alumina dissolves readily at a pH lower than 4.2 and at a pH higher than 9.9, and the dissolution process is highly anisotropic. The walls of the porous pipes through the alumina film (see Figure 5.10) consist of an anion-contaminated outer layer and a purer inner alumina layer. The outer anion contaminated layer etches 25 times as fast as the purer inner layer. This etching rate difference and the anisotropic nature of the pores themselves, which provides access for the etchant throughout the entire thickness of the film, results in microstructures with aspect ratios as high as 50 (Tan et al.[41] projected the possibility of a 1000:1 aspect ratio) and a resolution better than 2 μm. The lateral underetching of the photoresist mask stops once the purer alumina at the wall of the first pore closed off at the top by the mask is reached by the etchant. Obviously, the smaller the pore diameter, the better the expected resolution of the technique. Routkevitch et al.[42] developed a second type of alumina micromachining process based on local anodization of aluminum metal. This process is simpler but has a somewhat poorer resolution limit (~20 μm) as pictured schematically in Figure 5.9b. In this case, the sidewalls do not remain vertical, since aluminum under the mask gets oxidized, and the alumina expands as compared with the initial volume of aluminum. Either the aluminum metal or the aluminum oxide may be etched selectively to reveal textured surfaces, with one being the exact imprint of the other. For selective etching of anodic alumina, a mixture of chromic and phosphoric acids is used, and aluminum may be selectively etched away in a solution of $CuCl_2$ in HCl. This work presents a whole new approach to microstructuring ceramic and metal surfaces.

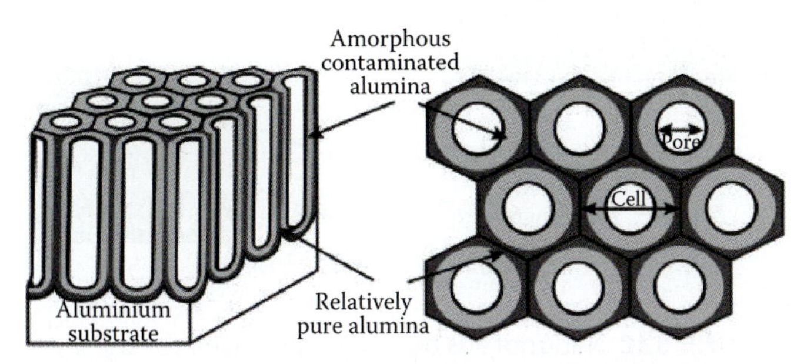

FIGURE 5.10 Schematic morphology and chemical composition of amorphous alumina. (Based on Routkevitch, D., A. Govyadinov, and P. Mardilovich. 2000. High aspect ratio, high resolution ceramic MEMS. *2000 International Mechanical Engineering Congress and Exposition*, 1–6.[42])

The nanoporous morphology of this ceramic MEMS material presents a controllable, very high surface area (up to 100 m^2/g) with numerous opportunities for applications such as a mechanical filter or as a support for catalysts and gas sensors, the latter of which require a high number of triple points (i.e., sites where gas reactants, catalyst, and support can meet). Recently, the method has been used in template-assisted nanowire growth as well.

Mechanical filters based on porous alumina are called *Anodisc™ Membrane Filters*. These aluminum oxide membrane filters are available in three nominal sizes (47, 25, and 13 mm diameter) and three pore sizes (0.2, 0.1, and 0.02 μm). Inorganic mechanical

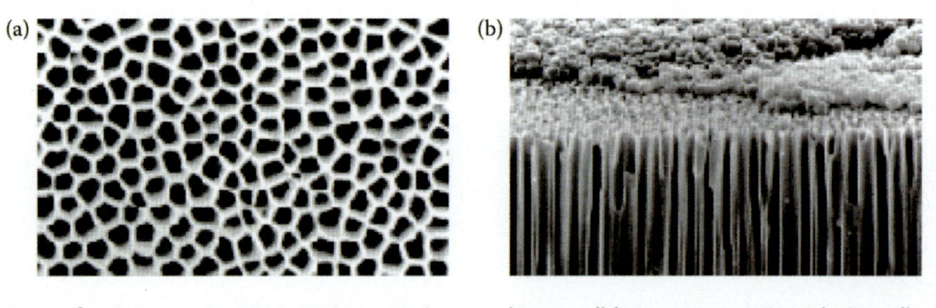

FIGURE 5.11 (a) Top view of a 0.2-μm Anopore™ inorganic membrane. (b) A cross-sectional view (below right) showing 0.23-μm latex microspheres retained on the membrane's surface. Membrane thickness is approximately 60 μm. Note the extraordinarily high pore density, which makes very high flow rates possible—unusual for membrane filters with pore sizes this small.

filters have several advantages over polymer membranes (e.g., polycarbonate sheets, about 10 μm thick, with pore sizes in the range of 15–12,000 nm), including higher flow rates, sieving of particles at the surface rather than trapping at an uncertain depth within the membrane and temperature and electron beam radiation resistant (see Figure 5.11) For more information, see http://www.2spi.com/catalog/spec_prep/filter2.shtml.

Porous alumina-based humidity sensors have been demonstrated.[44] The insulating material also makes for an excellent low-power and very-high-temperature microheater substrate with demonstrated temperatures up to 1200°C, power consumption as low as 50 μW/°C, and a very fast response.[44] The latter application represents an important alternative to Si as a microheater substrate in Taguchi sensors (see Volume II, Figures 9.11 and 9.12); Si is limited in temperature range and is a rather good heat conductor, so that it needs to be passivated for this application (it is also considerably more expensive).

The nanopores can also serve as a template for the deposition of nanostructures of a material of choice, such as metals, semiconductors, and polymers.[45] This is demonstrated in Figure 5.12 for the electrochemical deposition of nanowire thermoelectric material Bi_2Te_3. The process allows the oxide barrier layer to be removed/dissolved so that pores are in contact with the substrate. Filling rates of up to 90% have been achieved (T. Sands, HEMI group, http://www.mse.berkeley.edu/groups/Sands/HEMI/nanoTE.html).

The growth of aligned arrays of carbon nanotubes (CNTs) inside the pores of the self-organized nanoporous alumina templates is an especially intriguing application. Govyadinov et al.[46] synthesized arrays of CNTs (deposited with catalytic CVD) in alumina pores with diameters in the sub-100-nm range and densities from 10^8 to 10^{11} tips per cm^2. These prototype CNT arrays have been used as field emitters (carbon has a low work function) and shown to exhibit field emission at a low threshold of 3–5 V/μm and emission current densities up to 100 mA/cm^2 in low-vacuum conditions.

Other applications of porous alumina include microchannel plates, mesoscopic engines, filters, membranes, and nozzles and apertures.

Nanostructures similar to the ones shown in porous alumina have been produced by photoelectrochemical etching of polycrystalline n-type semiconductors, such as TiO_2, CdSe, CdS, and ZnO.[47] The most detailed work was performed on TiO_2. Results demonstrate an underlying anisotropic etching mechanism substantially different from the one

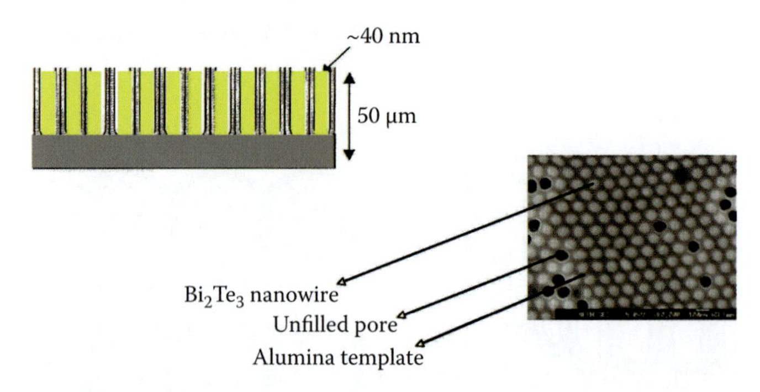

FIGURE 5.12 Nanoporous alumina works well for the electrochemical deposition of thermoelectric materials (Bi_2Te_3) and for metals. The process allows for the removal/dissolution of the oxide barrier layer so that pores are in contact with substrate. Filling rates of the pores of up to 90% have been achieved. (From T. Sands/HEMI group.)

at play in anodized alumina films.[48] Anodization of polycrystalline TiO_2 in a sulfuric acid solution under UV irradiation causes photoelectrochemical corrosion or photoetching. After a prolonged photoetching procedure, regularly ordered submicrometer porous structures (called *nanohoneycombs*) are observed at the TiO_2 surface. The quadrangle cells of the revealed nanohoneycomb are a few hundred nanometers wide and several micrometers deep. The thin walls (10–20 nm) of the pattern consist of {110} crystal faces. Etching selectively dissolves the grains, while the grain boundaries remain. The explanation for this selective etching behavior is readily understood from the fact that the holes created in the valence band of the semiconductor by the UV light readily corrode the bulk of the grains but recombine with electrons at the grain boundaries, leaving those intact. Besides their use in a dye-sensitized solar-cell and as a photocatalyst, fewer applications have been suggested for these new nanostructures than for porous alumina.

Conceptually, the alumina and TiO_2 micromachining approaches described here are somewhat similar to the quartz micromachining approach discussed earlier in which high-aspect ratio micromachining is achieved in single-crystal quartz by etching along the nuclear tracks left after bombardment with heavy ions. Obviously, alumina micromachining is a lot less complicated and costly, and so is the TiO_2 etching procedure, although the latter entails a rather difficult to control photoetching step and results in much less regular patterns than porous alumina.

Polymer MEMS-Polyimides

Introduction

Quartz and ceramic micromachining, discussed above, principally address harsh environment MEMS. Polymer materials also have some distinct advantages over more traditional engineering materials used in MEMS. Incentives for using polymers include increased fracture strength, low Young's modulus, high elongation, and reduced material costs. Furthermore, polymers are often inert and biocompatible, making them attractive for use in biological and chemical applications. As we have seen, in Volume I, Chapter 6, on fluidics, polymers are the most-used substrate for all types of fluidic platforms. We saw plenty of other examples of polymer MEMS in the section on soft lithography and dry photoresists in Volume II, Chapters 1 and 2, on lithography, in Volume II, Chapter 10, on replication and LIGA, and in Chapter 4, this volume, on packaging, assembly, and self-assembly. We conclude this section on new MEMS and NEMS substrate materials by considering the many applications of polyimides as an example of polymer MEMS.

Polyimide MEMS

Polyimides are part of a group of high-temperature polymers that have been in commercial use since the 1970s. Because of their structure, involving very strong carbon ring bonds, they do not melt and flow as most thermosets and thermoplasts do. Polyimides are synthesized from two monomers, a dianhydride and a diamine. Commercial products are supplied as soluble polyamic acid (PAA) intermediates, which undergo a thermal imidization with the evolution of water to form the insoluble polyimide. Polyimide films feature excellent thermal stability (up to ~450°C for short periods of time, $T_g > 300°C$), low dielectric properties ($\varepsilon = 3.3$) and a resistivity of ~10^{16} ohm-cm, superior chemical resistance, toughness, wear resistance, and flame retardance. Because of their flexibility they also exhibit interesting mechanical properties, and the material is easy to process. However, its tendency to absorb moisture causes the polyimide layers to swell and their dielectric constants to increase significantly. Another negative aspect of polyimides is their short shelf-life stability, which still needs to be addressed by organic chemists. Various polyimide analogs have been developed to avoid some of these inherent undesirable properties.

Current macro applications of nonphotosensitive polyimides include ball-bearing separators and mechanical seals. Polyimides in the microworld are used as passivating and interlayer dielectrics, planarizing compounds, reactive ion masks, alpha-particle barriers, humidity-sensitive materials,[49,50] liquid crystal display, color filters[51] (by incorporating dyes), and other optical elements, such as waveguides. In planarization, polyimides smooth the undulation caused by topographic features on the wafer, so that the top-imaging layer has a much smaller thickness variation (see Volume II, Chapter 1, on photolithography). A typical polyimide, designed for MEMS applications, packaging dielectric and as substrate material is PI 2611, a low-moisture, low-stress (coefficient of thermal expansion (CTE) = 5 ppm), and high modulus of elasticity (8.5 GPa) polyimide. This polyimide can be patterned using dry etching techniques as well as TMAH etching (for example, using the developers MF 319 and AZ 300). Film thicknesses in the range 4–8 µm can be obtained in a single coat with good uniformity by varying spin speeds in the range 5000–1000 rpm. The polyimide substrate can withstand metal sputtering such as Cr, Au, and Al without substantial damage.

An organic ciliary array of thin-film polyimide bimorph microactuators exploiting combined thermal and electrostatic control was employed to implement sensorless manipulation strategies for small objects. The tasks of parts translation, rotation, orientation, and centering were demonstrated using small integrated circuit (IC) dice by Kovacs et al. (see Figure 5.13).[52,53] To induce motion on a part that is placed on an array of cilia, the cilia are actuated in a cyclic, gait-like fashion. In each cycle, the part is moved in a certain direction by the motion of the actuators that are in contact with it. The speed of the moving part depends on the (horizontal) displacement of the actuators per cycle, as well as the frequency

FIGURE 5.13 Polyimide cilia motion pixel (scanning electron microscope micrograph). Four actuators in a common center configuration make up a motion pixel. Each cilium is 430 µm long and bends up to 120 µm out of the plane. (Figure by John Suh.[52])

of cycle repetition. It also depends on the surface properties and weight of the moving part.

In our own work, we have used polyimide sheets as precursor material to design carbon structures in C-MEMS (see below), and in building flat panel displays at SRI International, Capp Spindt used polyimide posts to separate the glass plate from the Si field emitters (see Volume II, Figure 7.90).

Photosensitive polyimides have found many applications in both integrated circuitry and in MEMS and NEMS. Photosensitive polyimide is typically a negative tone resist and in use as photoresist, polyimide precursors, called polyamic acids (*PAAs*), are spun on the wafer; upon exposure to UV light, cross-linking results (more recently introduced positive polyimide resists are discussed in Volume II, Chapter 1, on photolithography; see Volume II, Figure 1.28). During development, the unexposed regions are dissolved, and final curing by further heat treatment leads to a chemical transformation (known as *imidization*) of the remaining cross-linked material, which yields polyimide. At a temperature of 275°C, more than 99% of the polyimide precursor is converted to polyimide. As with other negative resists, an oxygen effect is noted in the exposure. The actinic sensitivity of the resist usually is confined to 365 nm.

Polyimide photoresists often are used as permanent photoresists; that is, as resists that after exposure and curing can be left behind as a structural and/or functional component. For example, they have been used as a dielectric, especially when the uniquely low dielectric constant of the polyimide gives devices a decreased capacitance, resulting in increased speed for electronic applications.[54] In multiple spin-coats, thick polyimide films can be deposited (>100 µm). Since the films are so transparent, UV lithography permits the fabrication of high aspect ratio features (see LIGA-like processes in Volume II, Chapters 1 and 10). An important commercial application of both photosensitive and nonphotosensitive polyimides is in multiple chip modules.[55] As we explained in Volume II, Chapter 1 on lithography in the multiple chip module application, the use of photosensitive polyimides dramatically simplifies the manufacturing procedure as compared with nonphotosensitive option, as illustrated in Figure 1.28.[55]

In a finished device, up to five layers of metallization are insulated by 15–20-μm-thick polyimide films on a ceramic wiring board.

Photosensitive polyimides also come as dry-film photoresists. Dry photoresists were introduced by DuPont as far back as 1968. These dry films come in rolls and consist of a photopolymerizable layer sandwiched between a polyester support film and a separator sheet (see Volume II, Figure 1.33). They do not need to be spin coated on, as they can, after removal of the separator sheet, be laminated onto a substrate. Also, the polyester cover sheet protects the resist film from oxygen diffusion, which would inhibit the cross-linking reaction. Both chlorinated and aqueous-base-developable dry film resist systems are available. The current resolution capabilities of dry film resists are less than 75 μm. The fact that the laminated dry film can bridge or "tent" over holes and cavities not only benefits the manufacture of printed wiring boards (PWBs) but can be exploited in the manufacture of biosensors as well, as we illustrate in Chapter 1, this volume, Example 1.1. For other photosensitive dry resist systems, see Table 1.12.

Exploration of IC and MEMS applications of modified polyimides also has started. A common approach to reducing the dielectric constant of polyimides is the inclusion of organofluorine components, in the form of pendant perfluoroalkyl groups. The polyamic acid precursor for fluorinated polyimides is commonly based on hexafluorodianhydrideoxydianiline. The inclusion of fluorinated monomers in the polyimide backbone has been found to reduce their moisture absorption and dielectric constant. Unfortunately, these polyimides generally show an increased susceptibility to chemical attack, making their use in multilayer fabrication questionable. Recently, however, newer formulations have improved the chemical resistance of the polyimide and have demonstrated a unique wet etch capability for vias with aspect ratios approaching 1.2:1.

Special perfluorinated polyimides were synthesized at the NTT laboratories in Japan. These remarkable materials are resistant to soldering temperatures (260°C) and are highly transparent at the wavelengths of optical communications (1.0–1.77 mm). In addition, their low dielectric constants and refractive indices match the conventional fluorinated polyimides, whereas their birefringence is lower.[56] As we saw in Volume II, Example 3.4, dry plasma etching of these modified polyimides results in very high aspect ratio microstructures with interesting optical properties.

Silicon Nitride

Introduction

Silicon nitride (Si_xN_y) is a commonly used material in microcircuit and microsensor fabrication because of its many superior chemical, electrical, optical, and mechanical properties. The material provides an efficient passivation barrier to the diffusion of water and to mobile ions, particularly Na^+. It also oxidizes slowly (about 30 times less than silicon) and has highly selective etch rates over SiO_2 and Si in many etchants. Some applications of silicon nitride are optical waveguides (nitride/oxide), encapsulant (diffusion barrier to water and ions), insulator (10^{16} Ω-cm, high dielectric strength, field breakdown limit of 10^7 V/cm), mechanical protection layer, etch mask, oxidation barrier, and ion implant mask (density is 1.4 times that of SiO_2). Silicon nitride also is hard, with a Young's modulus higher than that of Si, and can be used, for example, as a bearing material in micromotors.[57]

Silicon nitride can be deposited by a wide variety of the CVD techniques reviewed in Volume II, Chapter 7—APCVD, low-pressure chemical vapor deposition (LPCVD), and plasma-enhanced chemical vapor deposition (PECVD)—and its intrinsic stresses can be controlled by the specifics of the deposition process. Silicon nitride and silicon oxide deposited with these techniques usually exhibit too much residual stress, which hampers their use as mechanical components. However, CVD of mixed silicon oxynitride can produce substantially stress-free components.

Nitride often is deposited from SiH_4 or other Si containing gases and NH_3 in a reaction such as the following.

$$3SiCl_2 + 4NH_3 \rightarrow Si_3N_4 + 6HCl + 6H_2$$

Reaction 5.1

In this CVD process, the stoichiometry of the resulting nitride can be moved toward a silicon-rich composition by providing excess silane or dichlorosilane compared with ammonia.

PECVD Nitride

Plasma-deposited silicon nitride, also plasma nitride or SiN, is used as the encapsulating material for the final passivation of devices. The plasma-deposited nitride provides excellent scratch protection, serves as a moisture barrier, and prevents sodium diffusion. Because of the low deposition temperature, 300–350°C, the nitride can be deposited over the final device metallization. Plasma-deposited nitride and oxide both act as insulators between metallization levels, which is particularly useful when the bottom metal level is aluminum or gold. The silicon nitride that results from PECVD in the gas mixture of Reaction 5.1 has two shortcomings: high hydrogen content (in the range of 20–30 atomic percent) and high stress. The high compressive stress (up to 5×10^9 dyn/cm^2) can cause wafer warping and voiding and cracking of underlying aluminum lines.[58] The hydrogen in the nitride also leads to degraded MOSFET lifetimes. To avoid hydrogen incorporation, source gases containing little or no hydrogen, such as nitrogen, may be employed instead of ammonia as the nitrogen source. Also, a reduced flow of SiH_4 results in less Si-H in the film. The hydrogen content and the amount of stress in the film are closely linked. Compressive stress, for example, changes to tensile stress upon annealing to 490°C in proportion to the Si-H bond concentration. By adding N_2O to the nitride deposition chemistry an oxynitride forms with lower stress characteristics; however, oxynitrides are somewhat less effective as moisture and ion barriers than nitrides.

One key advantage of PECVD nitride, besides the low temperature aspect, is the ability to control stress during deposition. Silicon nitride deposited at the typical 13.56 MHz plasma excitation frequency exhibits tensile stress of about 400 MPa. When depositing the film at 50 kHz, a compressive stress of 200 MPa is typical. We discussed the effect of RF frequency on nitride stress, hydrogen content, density, and the wet etch rate in more detail in Volume II, Chapter 7. We concluded that low frequency (high energy bombardment) results in films with low compressive stress, lower etch rates, and higher density. Therefore, the effect of ion bombardment is more pronounced at lower pressures, and better quality CVD films ensue, characterized as films with a low wet etch rate (high film density) and low compressive stress. Experimental results show that as the reactor pressure is lowered, film stress goes from tensile to compressive (Volume II, Figure 7.59) and wet etch rates decrease. Often, stress dynamically changes when the film is exposed to the atmosphere and subsequent heating; the higher the ion bombardment during PECVD, the higher the stress. Stress also is affected by moisture exposure and temperature cycling (for a good review, see Wu and Rosler[59]).

We compare properties of silicon nitride formed by LPCVD and PECVD in Table 5.5, and Table 5.6 highlights typical PECVD process parameters. The refractive index of the deposited silicon nitride films is a measure of impurity content and overall quality. Its value ranges from 1.8 to 2.5 for PECVD films as compared with 2.01 for stoichiometric LPCVD (see below). The higher numbers indicate excess silicon, and lower numbers represent excess oxygen.

LPCVD Nitride

In the IC industry, stoichiometric silicon nitride (Si_3N_4) is LPCVD deposited at 700–900°C and at 200–500 mTorr, and it functions as an oxidation mask and as a gate dielectric in combination with thermally grown SiO_2. In micromachining, LPCVD nitride serves as an important mechanical membrane material and isolation/buffer layer. The standard source gases are SiH_2Cl_2 and ammonia, and at the above specified temperatures and pressures, this leads to a deposition rate of about 30 Å/min. The deposition of the amorphous silicon nitride is reaction-limited, so it deposits on both sides of the wafers with equal thickness. Dichlorosilane is used instead of silane because it results in more uniform film thickness, and the wafers can be spaced closer together for larger loads. For stoichiometric silicon nitride, typical gas flows are in a 10:1 ratio of ammonia to dichlorosilane. Stoichiometric silicon nitride has a large residual tensile stress of about 10^{10} dynes/cm^2, and as a consequence, film thickness

TABLE 5.5 Properties of Silicon Nitride

Deposition	LPCVD	Plasma-Enhanced CVD
Temperature (°C)	700–800	250–350
Density (g/cm^3)	2.9–3.2	2.4–2.8
Pinholes	No	Yes
Throughput	High	Low
Step coverage	Conformable	Poor
Particles	Few	Many
Film quality	Excellent	Poor
Dielectric constant	6–7	6–9
Resistivity (ohm-cm)	10^{16}	10^{6}–10^{15}
Refractive index	2.01	1.8–2.5
Atom % H	4–8	20–25
Energy gap	5 eV	4–5 eV
Dielectric strength (10^6 V/cm)	10	5
Etch rate in conc. HF	200 Å/min	
Etch rate in BHF	5–10 Å/min	
Residual stress (10^9 dyne/cm^2)	1 T	2 C–5 T
Poisson ratio	0.27	
Young's modulus	270 GPa	
TCE	1.6×10^{-6}/°C	

Sources: Based on 1) Adams, A. C. 1988. In *VLSI technology*, ed. S. M. Sze. New York: McGraw-Hill[60]; 2) Sinha, A. K., and T. E. Smith. 1978. Thermal stresses and cracking resistance of dielectric films. *J Appl Phys* 49:2423–26[61]; and 3) Retajczyk, T. F. J., and A. K. Sinha. 1980. Elastic stiffness and thermal expansion coefficients of various refractory silicides and silicon nitride films. *Thin Solid Films* 70:241–47.[62]
Note: C = compressive; T = tensile.

often is limited to a few thousands angstroms. By increasing the Si content in silicon nitride, the tensile film stress reduces (even to compressive), the film turns more transparent, and the HF etch rate lowers. Such films result by increasing the dichlorosilane to ammonia ratio.[63] At an ammonia to dichlorosilane ratio of approximately 1 to 6, the films are nearly stress free, for deposition temperatures of 850°C and pressures of 500 mTorr.[64] Figure 5.14 illustrates the effect of gas flow ratio and deposition temperature on stress and the corresponding refractive index and HF etch rate.[63]

TABLE 5.6 Silicon Nitride Plasma-Enhanced Chemical Vapor Deposition Process Conditions

Flow (sccm)	SiH_4	190–270
	NH_3	1900
	N_2	1000
Temperature (°C)	T.C.	350 or 400
	Wafer	~330 or 380
Pressure (Torr)		2.9
RF power (watts)	1100	
Deposition rate (Å/min)		1200–1700
Refractive index		2.0

Source: Wu, T. H. T., and R. S. Rosler. 1992. Stress in PSG and nitride films as related to film properties and annealing. *Solid State Technol* May, 65–71. With permission.[59]

Silicon-rich or low-stress nitride has emerged as an important micromechanical material. Low residual stress means that relatively thick films can be deposited and patterned without fracture. Low etch rate in HF means that films of silicon-rich nitride survive release etches better than stoichiometric silicon nitride. The etch characteristics of LPCVD Si_xN_y are summarized in Table 5.7. The properties of a LPCVD Si_xN_y film, deposited by the reaction of $SiCl_2H_2$ and NH_3 (5:1 by volume) at 850°C, were already summarized in Table 5.5.

Amorphous Silicon and Hydrogenated Amorphous Silicon

Amorphous silicon behaves quite different from the coarse- and the fine-grained polysilicon discussed in Volume II, Chapter 7. Amorphous Si can be stress annealed at temperatures as low as 400°C.[65] This low-temperature anneal makes the material

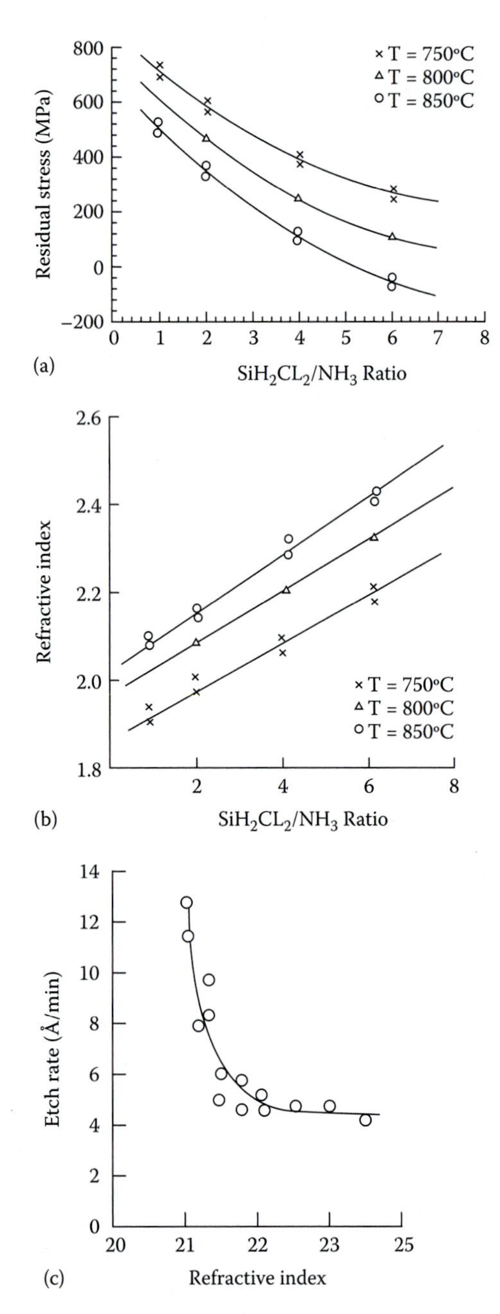

FIGURE 5.14 Silicon nitride low-pressure chemical vapor deposition parameters. (a) Effect of gas-flow ratio and deposition temperature on stress in nitride films. (b) The corresponding index of refraction. (c) The corresponding HF etch rate. (From Sakimoto, M., H. Yoshihara, and T. Ohkubo. 1982. Silicon nitride single-layer x-ray mask. *J Vac Sci Technol* 21:1017–21. With permission.[63])

compatible with almost any active electronic component. Unfortunately, little is known about the mechanical properties of amorphous Si.

Hydrogenated amorphous Si (α-Si:H), with its interesting electronic properties, is even less well understood in terms of its mechanical properties. If the mechanical properties of hydrogenated amorphous Si were found to be as good as those of polysilicon, the material might make a better choice than polysilicon as a MEMS material, given its better electronic characteristics.

TABLE 5.7 Etching Behavior of Low-Pressure Chemical Vapor-Deposited Si_3N_4

Etchant	Temperature (°C)	Etch Rate (Å/min)	Selectivity of Si_3N_4:SiO_2:Si
H_3PO_4	180	100	10:1:0.3
CF_4–4% O_2 plasma		250	3:2.5:17
BHF	25	5–10	1:200:±0
HF (40%)	25	200	1:>100:0.1

The amorphous polysilicon material produces a high breakdown strength (7–9 MV/cm) oxide with low leakage currents (vs. a low breakdown voltage and large leakage currents for polycrystalline Si oxides). Amorphous polysilicon also attains a broad maximum in its dielectric function without the characteristic sharp structures near 295 and 365 nm (4.2 and 3.4 eV) of crystalline polysilicon. Approximate refractive index values at a wavelength of 600 nm are 4.1 for crystalline polysilicon and 4.5 for amorphous material.[60] As deposited, the material is under compression, but an anneal at temperatures as low as 400°C reduces the stress significantly, even leading to tensile behavior.[65]

Like polysilicon, amorphous silicon exhibits a strong piezoresistive effect. The gauge factor is about five times smaller than that of single-crystal Si, but the temperature coefficient of the resistance (TCR) is lower than that of Si.

Hydrogenated amorphous silicon enables the fabrication of active semiconductor devices on foreign substrates at temperatures between 200°C and 300°C. The technology, first primarily applied to the manufacturing of photovoltaic panels, now quickly is expanding into the field of large-area microelectronics, such as active matrix liquid crystal displays (AMLCD). It is somehow surprising that micromachinists have not taken more advantage of this material to either power surface micromachines or implement electronics cheaply on nonsilicon substrates. In Figure 5.15 we compare polycrystalline-based solar cells with ultrathin amorphous Si on glass and ultrathin amorphous Si on flexible substrates.

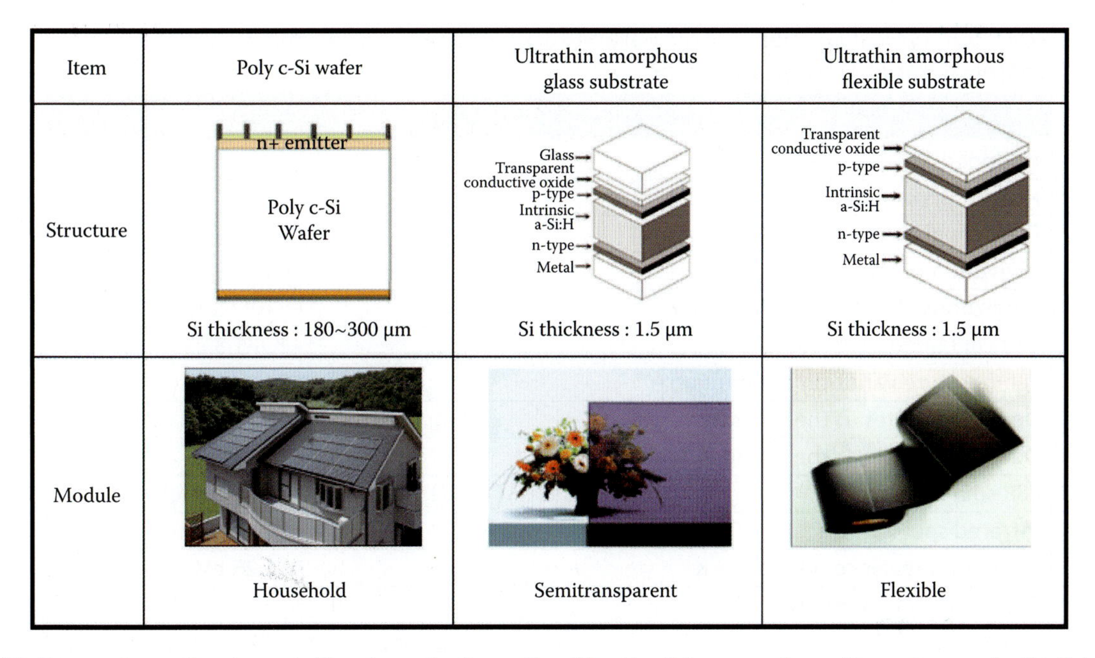

FIGURE 5.15 Comparison of polycrystalline-based solar cells with ultrathin amorphous Si on glass and ultrathin amorphous Si on flexible substrates. (From http://www.displaybank.com.)

Spear and Le Comber[66] showed that, in contrast to α-Si, α-Si:H could be doped both n- and p-type. Singly bonded hydrogen, incorporated at the silicon dangling bonds, reduces the electronic defect density from $\sim 10^{19}/cm^3$ to $\sim 10^{16}/cm^3$ (typical H concentrations are 5–10 atomic percent—several orders of magnitude higher than needed to passivate all the Si dangling bonds). The lower defect density results in a Fermi level that is free to move, unlike in ordinary amorphous Si, where it is pinned. Other interesting electronic properties are associated with α-Si:H—exposure of α-Si:H to light increases photoconductivity by four to six orders of magnitude, and its relatively high-electron mobility ($\sim 1\ cm^2/V\ s^{-1}$) enables fabrication of useful thin-film transistors. Lee et al.[67] noted that hydrogenated amorphous silicon solar cells are an attractive means to realize an onboard power supply for integrated micromechanical systems. They point out that the absorption coefficient of α-Si:H is more than an order of magnitude larger than that of single-crystal Si near the maximum solar photon energy region of 500 nm. Accordingly, the optimum thickness of the active layer in an α-Si:H solar cell can measure 1 μm, much smaller than that of single-crystal Si solar cells. By interconnecting 100 individual solar cells in series, the measured open circuit potential reaches as high as 150 V under AM 1.5 conditions, a voltage high enough to drive onboard electrostatic actuators.

Hydrogenated amorphous silicon is manufactured by plasma-enhanced chemical vapor deposition (PECVD) from silane. Usually, planar RF-driven diode sources using SiH_4 or SiH_4/H_2 mixtures are used. Typical pressures of 75 mTorr and temperatures between 200°C and 300°C allow silane decomposition with Si deposition as the dominant reaction. Decomposition occurs by electron impact ionization, producing many different neutral and ionic species.[68] Deposition rates for usable device quality α-Si:H generally do not exceed ~2–5 Å/s, due to the effects of temperature, pressure, and discharge power. Table 5.8 gives state-of-the-art parameters for α-Si:H prepared by PECVD. Although its semiconducting properties are inferior to those of single-crystal Si, the material is finding more and more applications. Some examples are thin film transistor (TFT) switches for picture elements in AMLCDs,[69] page-wide TFT-addressed document scanners, and high-voltage TFTs capable of switching up to 500 V.[70] An excellent source for further information on amorphous silicon is the book *Plasma Deposition of Amorphous Silicon-Based Materials*.[71]

Table 5.8 Typical Optoelectronic Parameters Obtained for Plasma-Enhanced Chemical Vapor Deposition α-Si:H

	Symbol	Parameter
Undoped		
Hydrogen content		~10%
Dark conductivity at 300 K	σ_D	~10^{-10} (Ω-cm)$^{-1}$
Activation energy	E_σ	0.8–0.9 eV
Pre-exponent conductivity factor	σ_0	>10^3 (Ω-cm)$^{-1}$
Optical bandgap at 300 K	E_g	1.7–1.8 eV
Temperature variation of bandgap	$E_g(T)$	2–4 × 10^{-4} eV/K
Density of states at the minimum	g_{min}	>10^{15}–10^{17} cm^3/eV
Density of states at the conduction band edge		~10^{15}/cm^3
ESR spin density	N_s	~10^{21}/cm^3-eV
Infrared spectra		2000/640 cm^{-1}
Photoluminescence peak at 77 K		~1.25 eV
Extended state mobility		
Electrons	μ_n or μ_e	>10 cm^2/V-s
Holes	μ_p or μ_h	~1 cm^2/V-s
Drift mobility		
Electrons	μ_n or μ_e	~1 cm^2/V-s
Holes	μ_p or μ_h	~10^{-2} cm^2/V-s
Conduction band tail slope		25 meV
Valence band tail slope		40 meV
Hole diffusion length		~1 µm
Doped amorphous		
n-type[a]	σ_D	10^{-2} (Ω-cm)$^{-1}$
	E_g	~0.2 eV
p-type[b]	σ_D	10^{-3} (Ω-cm)$^{-1}$
	E_g	~0.3 eV
Doped microcrystalline		
n-type[c]	σ_D	≥1 (Ω-cm)$^{-1}$
	E_g	≤0.05 eV
p-type[d]	σ_D	≥1 (Ω-cm)$^{-1}$
	E_g	≤0.05 eV

Source: Crowley, J. L. 1992. Plasma enhanced CVD for flat panel displays. *Solid State Technol* February, 94–98. With permission.[68]

[a] 1% PH_3 added to gas phase.

[b] 1% B_2H_6 added to gas phase.

[c] 1% PH_3 added to dilute SiH_4/H_2, or 500 vppm PH_3 added to SiF_4/H_2 (8:1) gas mixtures. Relatively high powers are involved.

[d] 1% B_2H_6 added to dilute SiH_4/H_2.

Selected MEMS and NEMS Processes

Introduction

In this section we review selected MEMS/NEMS processes that might affect the direction that MEMS/NEMS takes next. Many relatively new and important processes, such as soft lithography (see Volume II, Chapter 1) and atomic layer deposition (see Volume II, Chapter 7), are covered elsewhere in this work. The following additional selection is rather arbitrary and complements the new materials/processes covered elsewhere. For exciting new MEMS/NEMS techniques, we survey carbon-based MEMS (C-MEMS), bulk metallic glasses (BMG), fast microfluidic CD prototyping, and micro glass blowing.

C-MEMS

Introduction

In nature, carbon occupies a very special place as the essential building block of life. In human enterprise, too, carbon in its various forms has acquired an enviable position. To just name a few, diamonds are made of carbon, and graphite, coke, diamond-like carbon, and glassy carbon are all forms of carbon. So are the more recently discovered buckyballs, carbon fibers, and carbon nanotubes (CNTs).[72,73] Carboneous materials have been used in various application fields because their widely different crystalline structures and morphologies enable different physical, chemical, mechanical, thermal, and electrical uses. For example, glassy carbon electrodes often are used because of their wide electrochemical stability window, low background currents and low cost. Graphite and hard carbons are well known for their utility in Li ion battery applications because of their Li intercalation/deintercalation capacity. Other properties, such as biocompatibility, chemically inertness, and ease of functionality have made carbon very attractive for biosensor and implantable device applications. Moreover, CNTs are of tremendous current interest in both fundamental research and for nanoelectronics applications, including flat panel displays, chemical sensors, and the next generation of transistors (see Volume I, Figure 1.30). Recently, new microfabrication methods for carboneous materials have received a lot of attention, especially with regard to their use as microelectrodes in electrochemical sensors and miniaturized energy storage/energy conversion devices. Some key advantages of carbon compared with Si are summarized in Table 5.9. Although silicon, human's foremost technological material, is abundant in silicate minerals, carbon is the key to life. It can rightfully be argued that carbon is becoming a more and more important material in human technology. Besides the common use of carbons in sensors and batteries, carbon nanotubes might form the basis for a future carbon-electronics industry. We speculate that C-MEMS, described here, might be the material to use to contact those carbon-nanotube-based devices.

Additional important C-MEMS properties are included in Table 5.9: the material has a very wide

TABLE 5.9 Carbon: The Next Si?

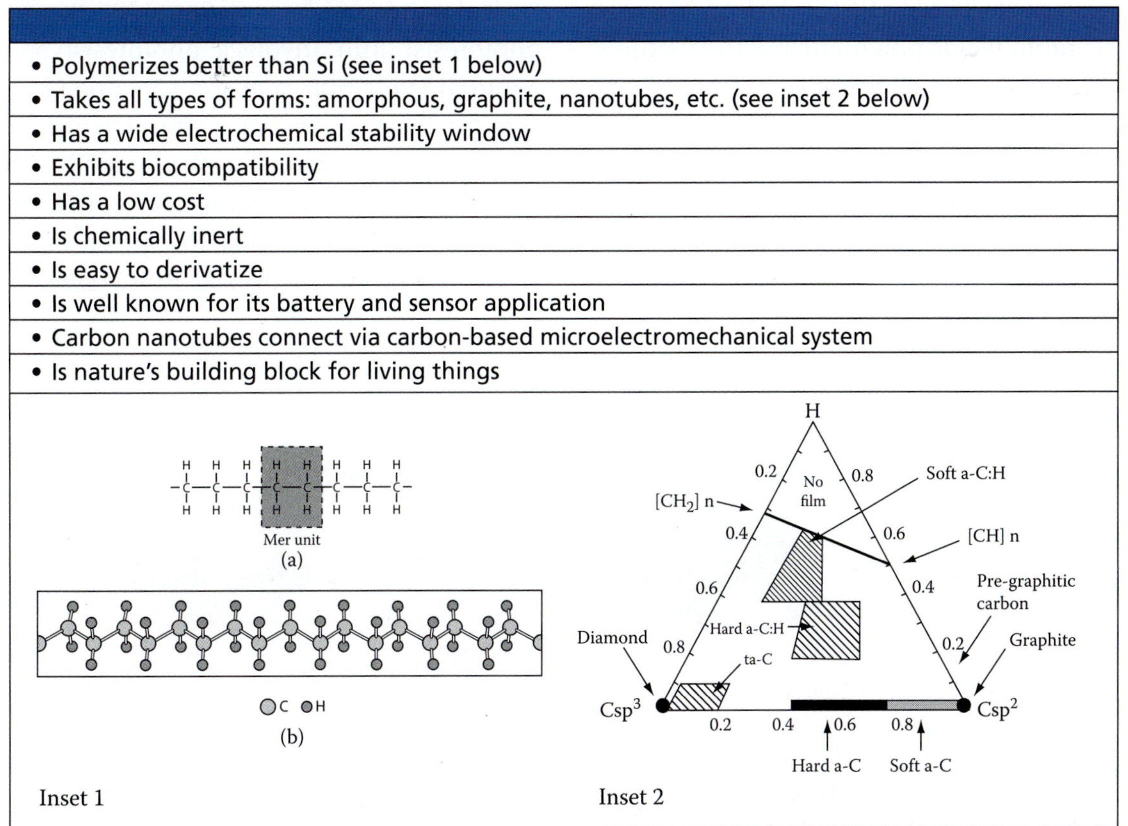

• Polymerizes better than Si (see inset 1 below)
• Takes all types of forms: amorphous, graphite, nanotubes, etc. (see inset 2 below)
• Has a wide electrochemical stability window
• Exhibits biocompatibility
• Has a low cost
• Is chemically inert
• Is easy to derivatize
• Is well known for its battery and sensor application
• Carbon nanotubes connect via carbon-based microelectromechanical system
• Is nature's building block for living things

electrochemical stability window, exhibits excellent biocompatibility, and is low cost, and the surface of this very chemically inert material is easy to functionalize. The material has particular importance in bio-MEMS applications, including electronic DNA arrays (replacing the platinum shown in Nanogen's DNA array; see Figure 8.51 this Volume), glucose sensors, and microbatteries (see below).

Microfabrication of Carbon: C-MEMS Process

For microfabrication of carbon structures, IC processing technologies, such as focused ion beam (FIB) and reactive ion etching (RIE), are time consuming and expensive because of the need of high-vacuum systems. Low feature resolution and poor repeatability of the carbon composition, as well as the widely varying properties of the resulting devices, limit the application of screen printing with commercial carbon inks for carbon microfabrication.[73] Recently, more nontraditional carbon microfabrication methods were reported. Whitesides and co-workers described the use of soft lithography (see Volume II, Chapter 2) to fabricate glassy carbon microstructures.[74] In this approach, micromolding of a resin such as poly(furfuryl alcohol) in an elastomeric mold yields polymeric microstructures, the microstructures are converted to free-standing carbon by heat treatment (500–1100 °C) in an inert atmosphere (see Volume II, Figure 2.55).

The carbon microfabrication technique introduced by this author is based on the pyrolysis of shaped/patterned polymers (e.g., resists patterned with photolithography) at different temperatures and different ambient atmospheres.[75,76] Carbon, because it is brittle and hard, is a difficult material to machine. Polymers, on the other hand, can be machined easily in a wide variety of machine tools. The underlying principle of C-MEMS is to choose an easy to work with polymer precursor, machine, or pattern this precursor material and then convert it to carbon by pyrolysis.

In one embodiment of the C-MEMS process, a photoresist is patterned by photolithography and is subsequently pyrolyzed at high temperatures in an oxygen-free environment. By changing the lithography conditions, soft and hard baking times and temperatures, additives to the resist, pyrolysis time, temperature, and environment, C-MEMS permits a wide variety of interesting new MEMS and NEMS applications that employ structures having a wide variety of shapes, resistivities, and mechanical properties. The process is detailed in Figure 5.16.[73]

The photolithography to make C-MEMS structures, illustrated in Figure 5.16, includes the usual spin coating, soft bake (not shown), near-UV exposure, postbake (not shown), and development. A typical patterning process for a 200-μm-thick SU-8 photoresist film involves photoresist spinning at approximately 500 rpm for 12 s and then at 1400 rpm for 30 s, followed by a 10-min soft-bake at 65°C and finally a soft-bake for 80 min at 95°C. Exposure is performed in a Karl Suss MJB3 contact aligner for about 100 s. The postbake is carried out for 2 min at 65°C and for 30 min at 95°C. Development is

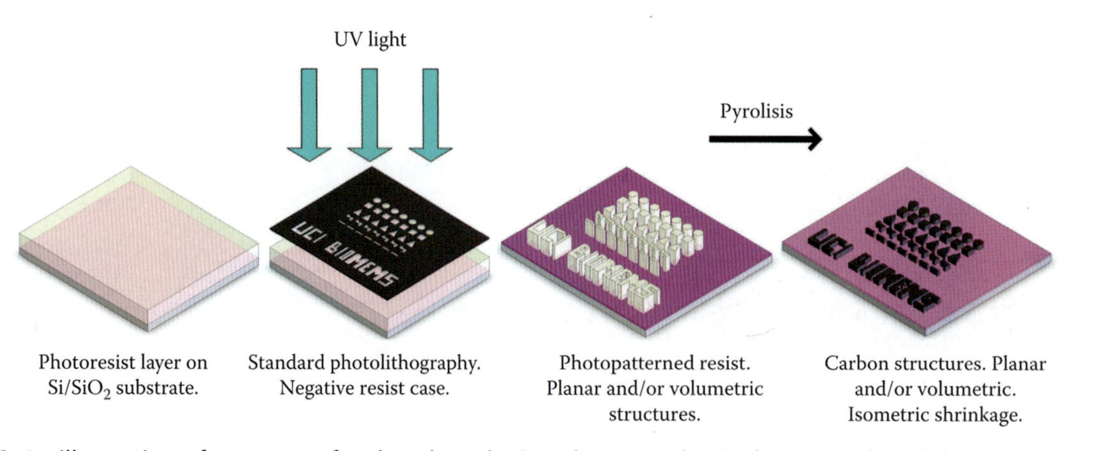

FIGURE 5.16 An illustration of one type of carbon-based microelectromechanical system: photolithography patterning of a polymer precursor and pyrolysis (see text for details).

carried out using a SU-8 developer from MicroChem (NANO™ SU-8 developer).

Photoresist-derived C-MEMS architectures are obtained in a two-step pyrolysis process in an open-ended quartz-tube furnace, in which samples are postbaked in an N_2 atmosphere at 300°C for about 40 min first and then heated in a N_2 atmosphere (2000 sccm) up to 900°C. At this point, the N_2 gas is shut off and forming gas [H_2 (5%)/N_2] is introduced (2000 sccm) for 1 h, and then the heater is turned off and the samples are cooled down again in an N_2 atmosphere to room temperature. The heating rate is about 10°C/min, and the total cooling time is about 10 h.

The resulting carbon material was found to be amorphous and glass carbon-like in electrochemical behavior.[77] The average height of the SU-8 posts shown in two typical scanning electron microscope (SEM) pictures (Figure 5.17a and 5.17b), is around 300 μm.

As shown in Figure 5.17c and 5.17d, after pyrolysis, the overall shape of the cylindrical posts largely is retained, and a typical aspect ratio of the carbon post achieved is around 10:1. It was found that the C-MEMS posts shrink much less during the pyrolysis process near the base of the structures than at the midsection because of the good adhesion of SU-8 to the substrate.

The processes described above easily can be extended to two-level or multilevel C-MEMS structures. This feature was used for making Li-ion-based C-MEMS batteries.[78] In the C-MEMS battery depicted in Figure 5.18, one layer of carbon constitutes the current collector for the rows of anode and cathode posts made in a second C-MEMS layer.[78]

In Figure 5.18, the current collector is 20 μm thick and the posts are 250 μm high. C-MEMS Li ion batteries, as shown here, hold the promise of changing the battery world in three significant ways (see also Chapter 9, this volume, on power and brains in miniature devices): 1) Smarter batteries: C-MEMS batteries contain a multitude of anodes and cathodes (we call them *baxels*, by analogy to *pixels*) that can be interconnected to provide the optimal current or voltage depending on the application need. In other words, we can perform in-battery smart load

FIGURE 5.17 Typical scanning electron microscope photos. (a) and (b) SU-8 post arrays before pyrolysis. (c) and (d) Carbon post arrays after pyrolysis.

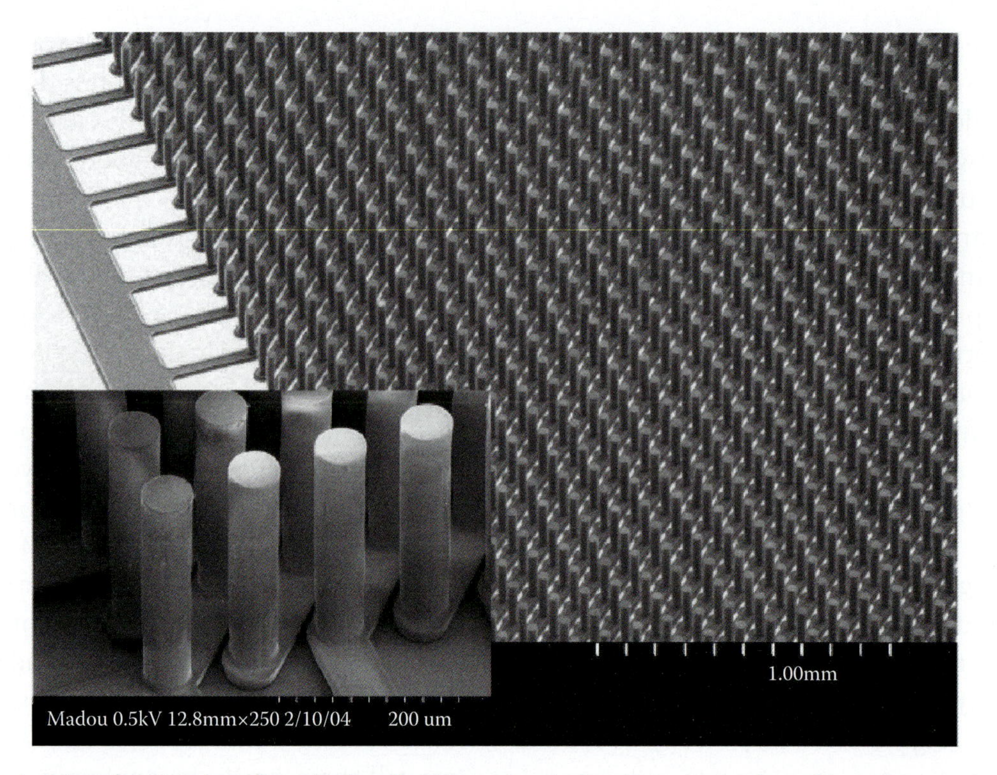

FIGURE 5.18 Typical two-level carbon-based microelectromechanical system electrodes with carbon contacts underneath. The inset image shows enlarged SU-8 two-level structure.[78]

leveling; 2) Faster charging batteries: in charging the C-MEMS battery, Li ions only need to intercalate into the thin individual carbon posts rather than intercalating into one big chunk of carbon; 3) Theoretically, the capacity of the C-MEMS batteries, depending on the aspect ratio of the posts and the thickness of the electrolyte, can be significantly higher than that of current Li ion batteries.[79]

In order to build yet higher aspect ratio C-MEMS structures (up to 40:1), we developed a three-level C-MEMS process. In a multilayer C-MEMS process, each layer of SU-8 photoresist is exposed and baked separately, but the whole assembly is developed all at the same time in the last step. Typical SEM photos of three-level C-MEMS devices are shown in Figure 5.19, with a three-level C-MEMS structure in which the first two are in good alignment but the third is out of alignment.

By carefully controlling the lithography processing parameters, a variety of complex 3D C-MEMS structures, such as suspended carbon wires, bridges, plates, and self-organized posts (carbon flowers) and ribbons (networks), were also microfabricated by Madou and coworkers.[72] These intricate 3D structures are fabricated using nontraditional process recipes, such as overexposure, underdevelopment, photoresist additives, directed flow of the developer, and exploitation of surface tension in the developing photoresist patterns. Suspended carbon fibers are shown in Figure 5.20, and a C-MEMS table was shown in the inset in the heading of this chapter. The diameter of the suspended carbon fibers in Figure 5.20 is submicrometer. The wire resistance between two posts was measured at room temperature and in air using two point probes, and a typical resistivity

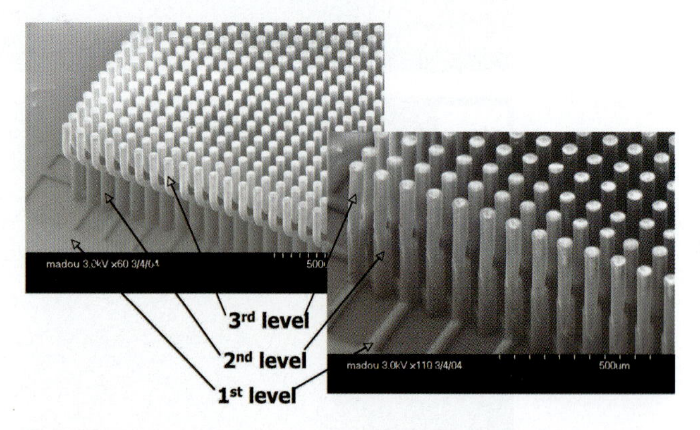

FIGURE 5.19 Typical scanning electron microscope photos of three-level carbon-based microelectromechanical system (C-MEMS) devices: a three-level C-MEMS structure with two levels in good alignment but with a third level out of alignment.

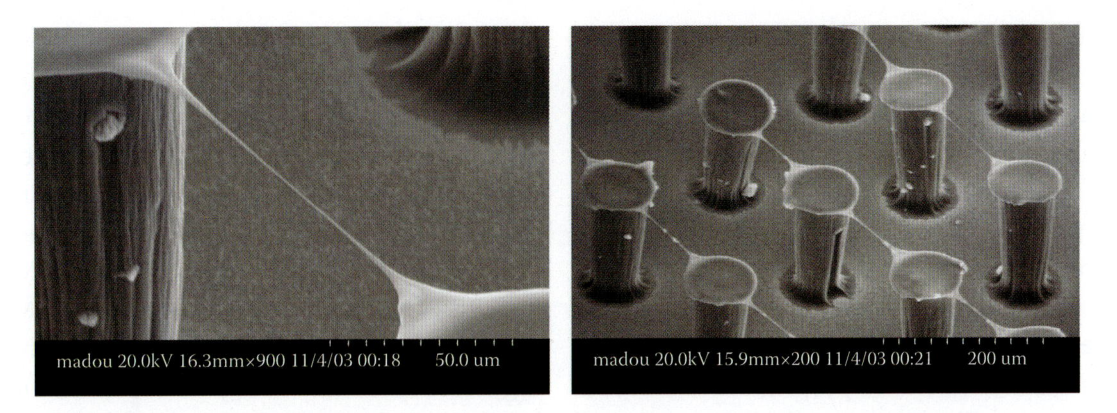

FIGURE 5.20 Carbon-based microelectromechanical system suspended carbon wires. Washcloth lines.

of 1.8×10^{-2} Ωcm was obtained. The Madou team is currently working on measuring the influence of chemical reactions on the suspended wire's resistivity. The aim is to turn these amorphous carbon "washcloth lines" into "washcloth line sensors" by derivatizing the C-MEMS surface with a monolayer of a sensor molecules (e.g., proteins, DNA). It is anticipated that the fact that these nanowires are removed from the supporting substrate and are thus suspended in air or liquid will lead to significant advantages over other nanowire-based sensors typically embedded on a solid substrate. This arrangement exposes the sensor to its sensing environment in all directions and avoids deleterious substrate influences, such as contamination and charge shunting.

In Figure 5.21, we illustrate self-organized groups of carbon posts (bunched posts)—a feature most readily observed when starting with high-aspect ratio SU-8 arrays with posts higher than 300 μm. It should be noted that when bunching of the SU-8 posts occurs, the bowed posts in each bunch remain structurally intact when converted to carbon. This means that the aggregation/bunching takes place after UV exposure, i.e., during postbake, development, or the drying after development. Bunching could in principle occur at the postbake temperature, as that temperature (65°C) is above the glass transition temperature of unexposed SU-8 (50–55°C); at those temperatures, the unexposed areas reflow, which could enable the posts to move toward each other. From SEM observations, after the postbake, the mask patterns are transferred with high fidelity. Clearly no aggregation or bunching is happening during postbake. The opportunity to control and optimize self-organized two-dimensional textures obviously presents itself only during the development process itself. We speculate at this point that the dominant cause for the bunching of resist posts is surface tension. When the developer solution is removed, it gently pulls posts that are tall and close enough together into symmetric patterns.

The electrochemical characteristics of carbon produced by pyrolysis of photoresists have been extensively studied.[76,77] It is found that electrochemical reactions on pyrolyzed photoresist exhibit reaction kinetics very similar to those on glassy carbon. We also found that C-MEMS electrodes can

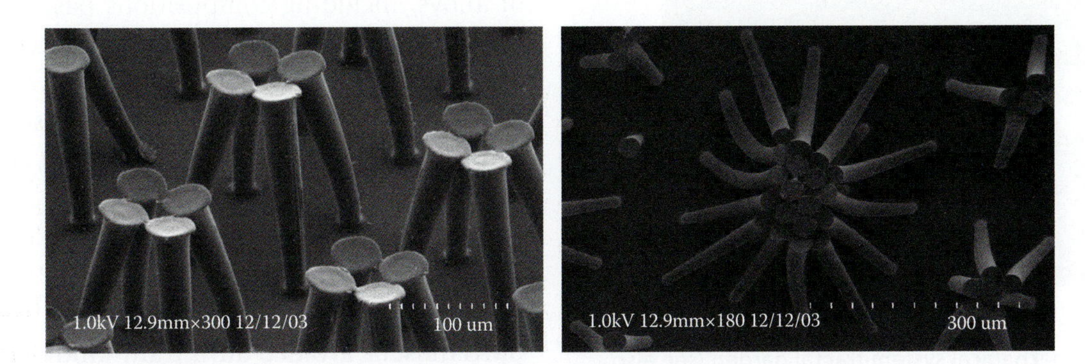

FIGURE 5.21 Carbon flowers. Self-assembly of polymer structures converted into carbon.

be reversibly charged and discharged with Li ions,[78] which led to the construction of C-MEMS Li ion batteries, as shown in Figure 5.18. The advantage of using photoresists as the starting material for the microfabrication of various carbon structures is that the photoresists can be very finely patterned by lithography techniques, and hence, a wide variety of repeatable shapes are possible. Moreover, different temperature treatments result in different resistivities and mechanical properties. Molecular rectification and conductance switching in pyrolyzed carbon-based molecular junctions were studied by McCreery et al.[80] They found that the current/voltage behavior showed strong and reproducible rectification and strong dependence on temperature and scan rate. Pyrolyzed photoresist carbon films were demonstrated to be useful in electroanalytical and microfluidic applications; for example, pyrolyzed carbon has been used in conjunction with a microchip capillary electrophoretic device.[81]

Madou and co-workers, as explained in Chapter 9, this volume, have presented theoretical arguments demonstrating that making fractal-like carbon electrodes will lead to better sensors and more energy efficient batteries and fuel cells.[82] In ongoing work, our team is attempting to make what we call a fractal garden. An artistic rendition of this concept is shown in Figure 5.22. In this fractal garden, each element in the array of fractal electrodes can be addressed individually. These electrodes maximize surface area while minimizing internal resistance and can be implemented gainfully for the sensors, batteries, and fuel cells of the future.

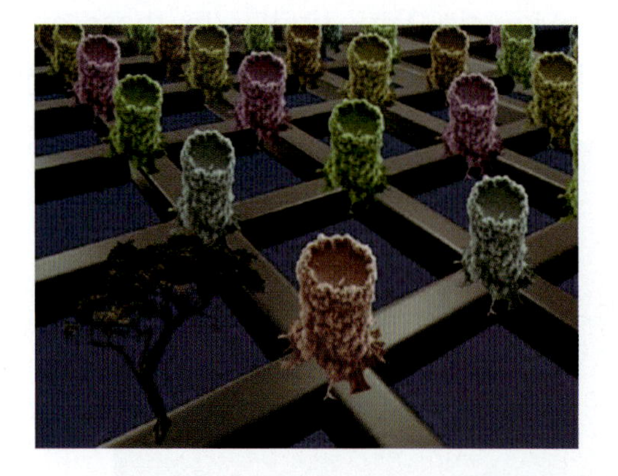

FIGURE 5.22 An artistic rendition of a carbon-based microelectromechanical system fractal garden (see also Chapter 9, this volume).

In Volume I, Chapter 6, on fluidics, we introduced 3D C-MEMS dielectrophoresis (see Volume I, Example 6.2). The main concept behind using 3D carbon electrodes here was to reduce the mean distance of any targeted particle contained in a channel or chamber to the closest electrode surface.[83]

The author believes that C-MEMS/NEMS technology can find applications in several MEMS fields as a substitute for silicon-based devices. This technology can be revolutionary not only in energy applications, such as biofuel cells, microbatteries, and supercapacitors, but also in biological and chemical sensing applications. The C-MEMS/NEMS approach gives the development engineers unprecedented freedom in designing and manufacturing high-surface area conductive structures through the use of new MEMS materials and innovative fabrication techniques. The development of novel biofuel cells based on high-aspect ratio 3D carbon structures from millimeter to nanometer size, integrating "top-down" and "bottom-up" processing approaches and combining soft biological components with hard MEMS/NEMS structures, could have the effect of making the bio-MEMS world less Si-centric and start a trend of lithographically patterning materials other than Si. Beyond the impact on the MEMS/NEMS/bio-MEMS community, one expects a broader anticipated impact on industry and on the technology end user.

Bulk Metallic Glasses

Bulk metallic glasses (BMGs) are multicomponent alloys for which crystallization is avoided during cooling. Sluggish crystallization kinetics and a corresponding ease of solidification into the amorphous state have been observed for a wide range of alloys, including compositions based on Zr, Cu, Ni, Ti, Fe, Mg, Pd, Pt, and Au.[84–94] Because of their random amorphous structure and lack of dislocation formation during deformation, these materials exhibit very high strength and elasticity, outstanding resilience, good corrosion resistance, and a near-perfect elastic behavior. For example, a typical BMG based on zirconium yields at 2000 MPa after it has elastically deformed by 2%.[95] Crystallization of a BMG must be avoided during any kind of processing, since it degrades the properties of the alloy. One consequence of the sluggish crystallization kinetics in BMG forming alloys is the presence of a supercooled liquid region (SCLR). In this temperature region, the amorphous BMG first relaxes into a viscous liquid before it eventually crystallizes. This behavior results in a large window of material processing opportunities, where the BMG, in its supercooled liquid state, can be processed under pressures and temperatures similar to those used for plastics,[96] with several minutes of processing time allowed before crystallization sets in and starts degrading the alloy properties. Thus, thermoplastic forming (TPF) is carried out by reheating the BMG into its supercooled liquid region, and a relatively small forming pressure is sufficient to fill even the smallest and most complicated features while avoiding crystallization.[97] In other words, we can process the metal as if it were a polymer, as demonstrated in Figure 5.23.

Thermoplastic forming with BMGs has a strong potential as a technology for MEMS and NEMS fabrication.[99–100] The amorphous structure of a BMG makes it an ideal material for small geometries, since it lacks an intrinsic limitation like the grain size in crystalline materials or the large molecular weight of polymers. The BMG is homogeneous and isotropic down to the atomic scale. The absence of crystallization during solidification and the low processing temperature during TPF dramatically reduce solidification shrinkage to levels below 0.2% for most BMGs of technical relevance.[97] This low shrinkage, together with the fact that fast cooling is not required after TPF, allows one to precisely replicate mold features.

The supercooled liquid state of the BMG can be utilized for other processes, one of which is hot separation.[98] Thermoplastic forming into a mold cavity results in precise replication of micro features, which, however, are connected to a typically large reservoir of amorphous material. For MEMS fabrication and also for three-dimensional microparts, this reservoir must be separated from the part. Hot separation, where a scraper separates reservoir from part as shown in Figure 5.23, is carried out in the SCLR. This leaves a plain surface on which further fabrication steps can be carried out or a three-dimensional part after the mold is etched. The low-viscosity supercooled liquid state of the BMG can also be used to erase features in the BMG,[101] as shown in Figure 5.24. This is accomplished through surface tension forces alone. Such a process has potential for use as a rewrite mechanism for data storage, as well as to smooth the surface of microparts (where the finish is limited by the surface roughness of the mold).

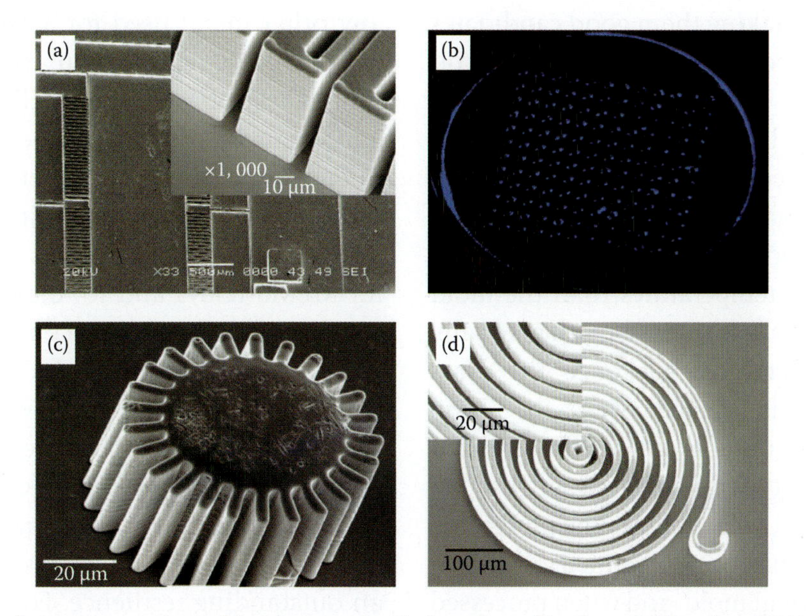

FIGURE 5.23 Examples of structures created by thermoplastic forming of bulk metallic glass (BMG). The top images depict precise replication of typical microelectromechanical system features with zirconium-based BMG (a) and a microporous membrane (b). Separation of the microparts from the BMG reservoir using a hot-separation technique results in three-dimensional microparts (c and d).[98] (Courtesy of Dr. Jan Schroers, Yale University.)

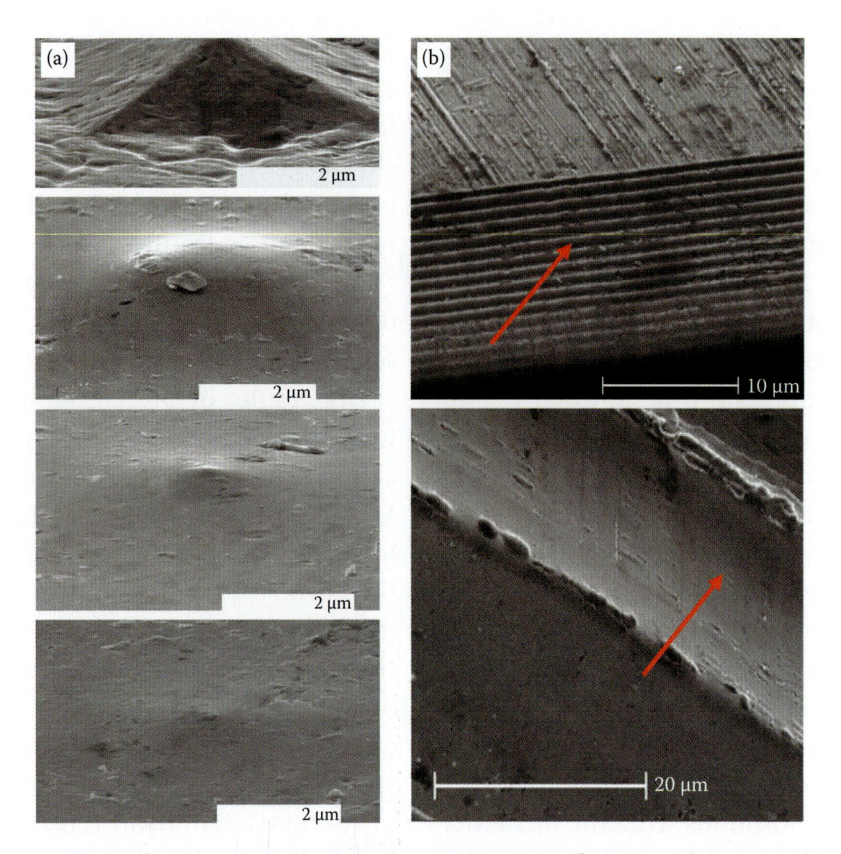

FIGURE 5.24 Forces exerted by surface tension alone can be used to erase features (a) or smoothing of the surface (b) in bulk metallic glass while processing it in the supercooled liquid region.[101]

Because of favorable chemical selectivity in BMGs, sacrificial etching and patterning can be performed without damaging the amorphous material.[98] For example, Zr- and Pt-based BMGs are resistant to most acids and bases, making them good candidates for postprocessing (selective wet-etching) of silicon after the TPF process.

Not only can BMGs be used to replicate a mold, but they can also be used as the mold material itself from which various materials are replicated. Even other BMGs can be replicated on a BMG mold, as long as they soften at lower temperatures than the mold material. BMGs typically lose their strength gradually with increasing temperature, but once the glass transition temperature is reached, where the glassy BMG relaxes into a metastable liquid, the strength significantly drops. Consequently, clever use of BMGs with decreasing T_g values allows the processing sequence depicted in Figure 5.25, where a BMG is used to replicate a conventional mold, and when processed at a lower temperature, as a mold itself.[98]

MEMS and microstructure applications are dictated by the BMG properties, processability, and process compatibility. Mechanical properties and processing temperature ranges for a few selected BMGs are summarized in Table 5.10. Their very high yield strength values, which by far exceed those of any other metal used for MEMS, are beneficial for almost every application. In particular, medical probes, skin-based drug delivery patches, sensors and actuators in high-shock environments, and microthrusters can all benefit from high strength. The high yield strength also implies a high hardness. This leads to less wear and is particularly interesting for MEMS relays and RF dielectric switches, which require robust performance after 10+ million contact cycles. The absence of dislocations also results in a very large elastic strain limit. Within the elastic region the material approaches perfect elastic behavior, reflected in low losses upon cyclic loading, which is particularly useful for resonators. Together, the high yield strength and large elastic strain limit suggest an outstanding resilience. For example, the 17-MPa resilience of a Zr-based BMG is about 1000 times higher than the resilience of nickel. Even the best conventional metallic spring material, high-carbon

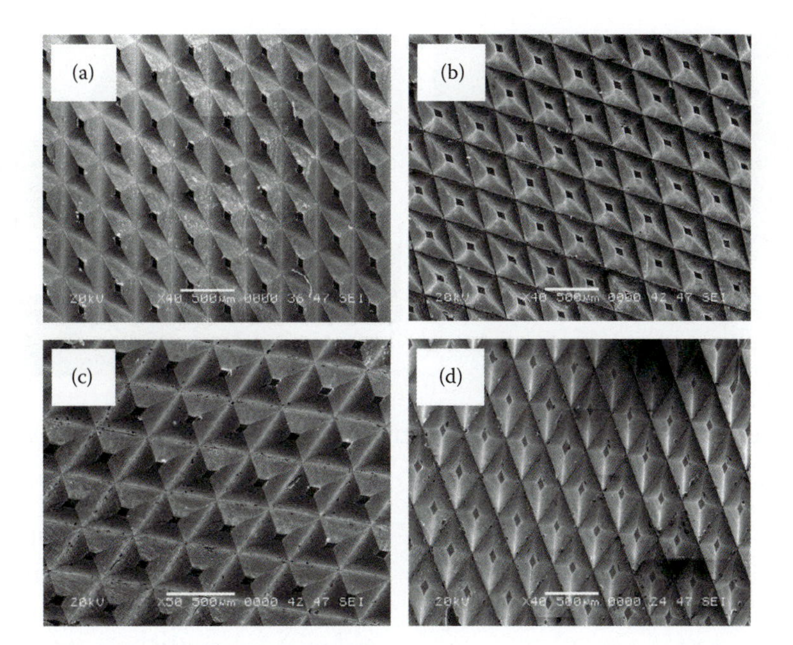

FIGURE 5.25 Sequential use of various bulk metallic glasses (BMGs) with different softening behavior as mold and molded material. The nickel master (a) is replicated at 450°C with a Zr-based BMG (b). This structure is then used as a mold for a Pt-based BMG formed at 280°C (c). A Au-based BMG with an even lower glass transition temperature of 130°C (softening behavior) is used to replicate the Pt-based BMG structure. The Au-based BMG is processed at 160°C, and perfect replication of the Pt-based BMG mold can be achieved (d).[102]

spring steel, possesses a resilience below 2 MPa. Obviously, their outstanding resilience makes BMGs excellent materials for any spring application involving energy storage and harvest, as well as for valve applications.

BMGs are available in a wide range of densities. The high density of $Pt_{57.5}Cu_{14.7}Ni_{5.3}P_{22.5}$, 15.3 g/cm^3, makes it an attractive candidate for improving the sensitivity of accelerometers.

The ability to use BMGs as a material to replicate parts on a mold and as a mold material itself suggests an alternative to the LIGA process (see Volume II, Chapter 10) for the mass production of microparts. BMG molds could supplant disposable molds.

Summarizing, BMGs are an entirely new class of material best described as an ultrahigh-strength material that can be processed like a plastic. There are many reasons why this class of materials might become an important MEMS/NEMS material:

- No intrinsic size limitation (like grain boundaries in crystalline materials)
- Perfectly isotropic
- Very large strength and strain
- Perfectly elastic (high Q factor)
- Some are very inert in most environments
- Very low processing temperature (60–400°C for various BMGs)

Table 5.10 Selected Properties and Processing Temperatures for BMGs with Potential for Microelectromechanical System and Microstructural Applications

BMG Material	Density (g/cm^3)	Yield Strength (MPa)	ε_{el} (%)	Y (GPa)	G (GPa)	B (GPa)	E_{el} (MPa)	ν	Thermoplastic Forming Temperature (°C)
$Au_{49}Ag_{5.5}Pd_{2.3}Cu_{26.9}Si_{16.3}$	11	1200	1.5	66.38	23.43	132.3	7.5	0.41	150–190
$Pt_{57.5}Cu_{14.7}Ni_{5.3}P_{22.5}$	15.3	1470	1.3	94.8	33.3	243	8	0.42	250–300
$Zr_{44}Ti_{11}Cu_{10}Ni_{10}Be_{25}$	6.1	1860	1.9	96.7	35.6	114	17	0.36	380–470

Source: Jan Schroers, Yale University.

FIGURE 5.26 A computer numerically controlled milling machine. Quick Circuit 5000 machine cuts fluidic design in a polycarbonate disc.

- Very low flow stress
- Features as small as 30 nm can be replicated
- Molding process reduces internal stresses and porosity to an undetectable level

Fast Microfluidic CD Prototyping

In Volume I, Chapter 6, we analyzed CD-based microfluidics.[103] Here we detail some simple methods to implement fast prototyping of microfluidic CDs in a research laboratory setting.

Many microfluidic applications utilize relatively large volumes (on the order of 20 μL to several milliliters) that are not always compatible with typical microfabrication techniques. This holds especially true for biological applications, including sample preparation and nucleic acid analysis (see also Chapter 10, this volume, under sample preparation). Centrifugal platforms in which microfluidic CDs are used for biological analysis require these larger volumes and may feature small conduits and DNA analysis chambers, which means that a combination of macro- and microfabrication techniques must be used to create these devices.

Computer numerical control (CNC) machining can be used to create most macro features (>100 μm) on a microfluidic CD. Inexpensive plastics, such as polycarbonate (PC) or ABS, are used as substrates in which channels down to ~1 mm in width can be machined using standard endmill and drill bits (Figure 5.26, CNC machine, Quick Circuit 5000). The polymer substrate thicknesses can be as small as 200 μm. Using a layering approach, different sheets of thin machined plastic CDs can be laminated together using pressure-sensitive adhesive (PSA) layers (PSA-FLEX mount DFM 200 clear V-95). The PSA layers may be as thin as 100 μm and are cut using a CNC cutter-plotter (Figure 5.27, CNC plotter, Graphtec CE-2000-60). The use of PSA layers allows for channels as narrow as 250 μm. When sandwiched between two plastic sheets, these channels in the PSA connect the larger CNC machined reservoirs in the PC discs.

Even the simplest standard microfluidic CD consists of no fewer than five layers, as illustrated in Figure 5.28. Usually the fluidic compact discs made in our lab are composed of three optically clear polycarbonate disks held together by two pressure-

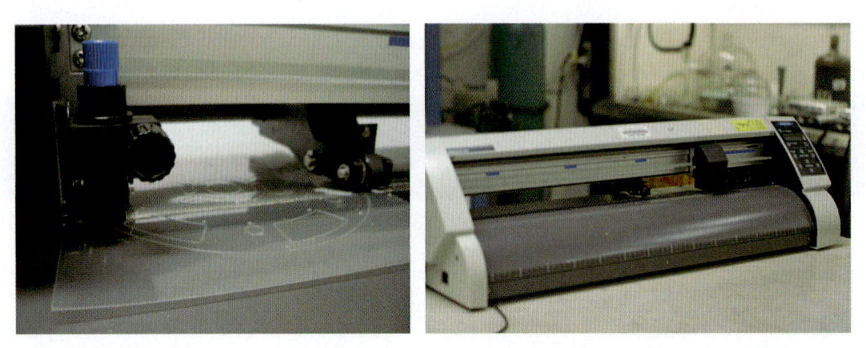

FIGURE 5.27 A roll-feed cutter plotter cuts out the pressure-sensitive adhesive layer designs. Graphtec CE-2000–60.

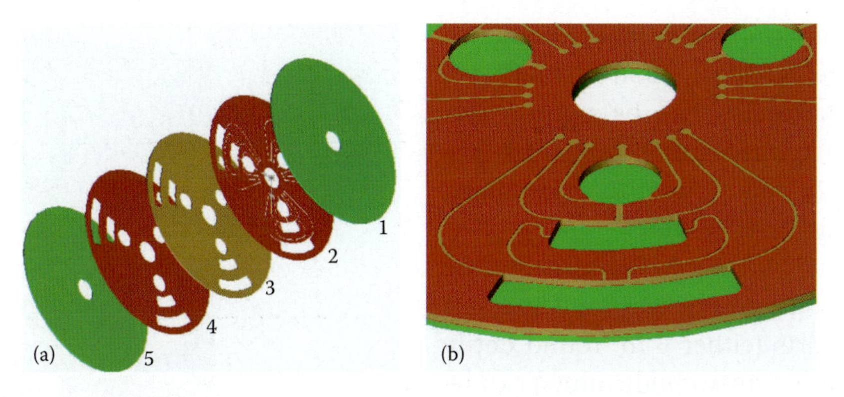

FIGURE 5.28 Computer numerical control (CNC) machined compact disc (CD) assembly. (a) Illustration showing five layers fabricated by a CNC machine and vinyl cutter: 1, top plate (1.2 mm); 2, pressure-sensitive adhesive layer (100 μm); 3, middle plate (1.2 mm); 4, pressure-sensitive adhesive layer (100 μm); 5, bottom disc (1.2 mm). (b) Schematic illustration of assembled CD structure.

sensitive adhesive layers. The top plate (including venting holes), the middle plate (containing all fluidic chambers), and the bottom plate contain the alignment holes at the same positions. Specifically, the five layers shown in Figure 5.28 are as follows: 1) top polycarbonate CD with CNC-machined sample loading, sample removal, and air venting holes; 2) pressure-sensitive adhesive with channel features cut using a plotter; 3) middle polycarbonate CD with machined channel features; 4) pressure-sensitive adhesive with channel features cut using a plotter; 5) solid bottom polycarbonate CD to seal off the channels. Microfluidic CD platforms can involve more layers to accommodate more complex fluidics. Moreover, different devices and substances can be placed inside the CD during fabrication, such as beads, lyophilized reagents, or filters. The CDs also can be exposed to O_2 plasma treatment or functionalized with bovine serum albumin to create hydrophilic and hydrophobic surfaces, respectively. Once the appropriate pieces have been designed and machined, they are aligned centrally and radially and laminated together using the PSA layers. The fabrication process ends with running the CDs through an industrial press to ensure excellent adhesion and sealing between all CD layers. After the assembled layers are pressed together, they are fed through a laminator to press the layers together tightly and ensure a good seal (Figure 5.29).

While the majority of CD platforms developed in our laboratory utilize standard macromachining processes as described above, microfabrication is easily integrated onto the CD platform. Integration usually takes the form of creating microfluidic polydimethylsiloxane (PDMS) molds on 6-in. Si wafers using multilevel, thick-resist lithography (Figure 5.30). Briefly, thick photoresist is spun on a Si wafer, prebaked to remove solvents, exposed to ultraviolet (UV) to catalyze cross-linking, postbaked to finish cross-linking, and developed to remove unexposed

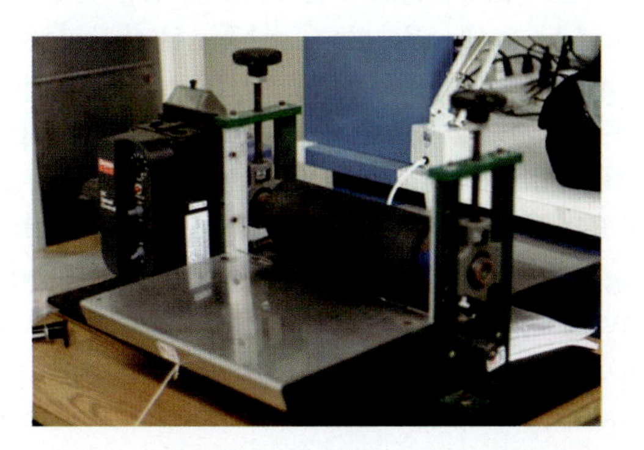

FIGURE 5.29 A laminator is used to press CNC machined discs together with pressure-sensitive adhesive layers.

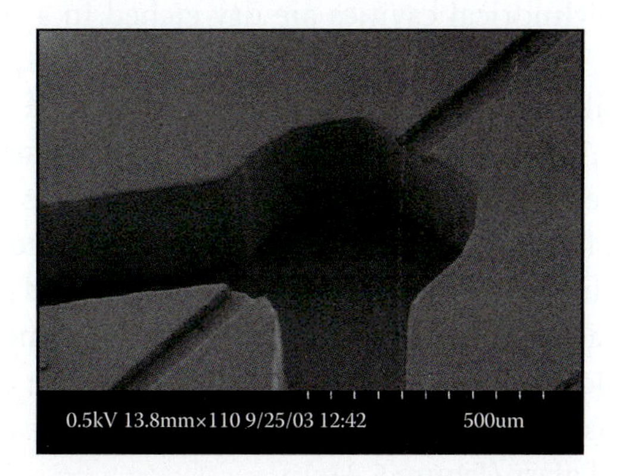

FIGURE 5.30 A scanning electron microscope image of a typical two-level polydimethylsiloxane flow cell structure. The microchannel is 25 μm deep, and the chamber is 250 μm deep.

photoresist. This creates the master mold (a negative of the desired channel features), upon which PDMS (a biocompatible elastomer) can be poured and cured. This process, termed soft lithography (see Volume II, Chapter 2), can be used to make many PDMS parts from a single Si mold.

Sample loading and venting holes can then be punched using small-gauge syringes, and the soft lithography PDMS parts (either 6-in. round entire PDMS CDs or individual microfluidic units) can be placed onto a polycarbonate CD, or in some cases bound to glass slides and secured with a specially designed slide holder CD. Most often, the passive binding properties of PDMS on glass/plastic are enough to prevent leakage during use. However, an O_2 plasma treatment process can be used to create a permanent bond with PDMS. Almost all processes and materials used to create the microfluidic CD platforms are relatively inexpensive, with the exception of the initial Si mold needed when incorporating microfabrication.

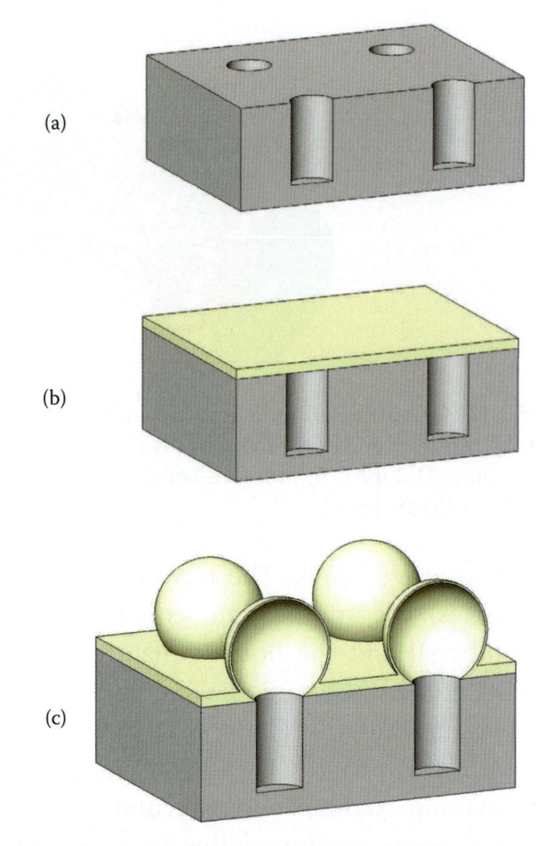

FIGURE 5.31 Micro glass blowing process sequence. (a) Step 1: etching of deep cavities in a silicon wafer. (b) Step 2: anodic bonding of glass wafer to the silicon wafer. (c) Step 3: blowing of glass spheres by heating sample inside furnace.

Micro Glass Blowing

Eklund and Shkel developed a wafer-level glass blowing process for the formation of three-dimensional glass microstructures.[104] In contrast to conventional glass blowing, where relatively large glass structures are shaped one at a time, the micro glass blowing process allows for parallel fabrication of thousands of microscopic glass parts.

The micro glass blowing fabrication process is illustrated in Figure 5.31. In a first step (Figure 5.31a), deep cylindrical cavities are dry etched in a silicon wafer. The etch mask (not shown) is removed, and a borosilicate glass wafer is anodically bonded to the silicon wafer (Figure 5.31b). Next, the parts are rapidly heated inside a furnace to a temperature above the softening point of the glass. This elevated temperature causes the pressure of the gas trapped inside the sealed cavities to increase. The glass deforms into spherical shapes because of the combination of the increased gas pressure and the low viscosity of the heated glass (Figure 5.31c).

The size of the blown glass spheres depends on the dimensions of the etched cavity, the furnace temperature, and the pressure and temperature at which the cavities are sealed. The total volume of the etched cavity and the blown glass sphere is proportional to the volume of the etched cavity, with the proportionality constant being the ratio between the final and initial temperatures. For example, if a cavity is sealed in atmospheric pressure and at room temperature and the furnace temperature is 900°C, the final volume is approximately four times as great as the volume of the etched cavity (1173 K vs. 295 K). The volume enclosed by only the glass sphere is therefore three times as great as the volume of the etched cavity.

In Figure 5.32 we show fabricated glass spheres, all with a diameter of approximately 900 μm. A layer of AZ P4620 photoresist was used as the etch mask during the deep reactive ion etching of the cylindrical cavities. For the sample shown in Figure 5.32a, a silicon wafer with a thickness of 450 μm was used, and the etched cavities were 350 μm deep and 750 μm in diameter. For the parts in Figure 5.32b and 5.32c, cavities with a depth of 750 μm and a

FIGURE 5.32 Fabricated glass structures. (a) Diced chip on US quarter coin. (b) Close-up of fabricated glass sphere. (c) Array of fabricated glass spheres.

diameter of 500 μm were etched in silicon wafers with a thickness of 1 mm. Pyrex 7740 glass wafers with an initial thickness of 100 μm were used for the anodic bonding of all samples, and the spheres were formed inside a furnace set to 850°C. This technology may be utilized for several different types of MEMS devices, including microfabricated glass lenses, three-dimensional microfluidic networks, and spherical gas confinement chambers for atomic MEMS.[105]

Biocompatibility and Surface Modification

Definition

Before new materials or devices can be used in conjunction with a biological system, whether cells or organisms, their effect on that system must be tested. In the simplest terms, a substance or device that has no effect on the time scale tested is said to be *biocompatible.* This can be verified at first by observing whether the material or device under study forms a suitable substrate for the growth of cultured cells. Such preliminary screening eliminates materials that have direct, harmful effects. To fully test the biocompatibility of a new material or device, one needs to test it on a wide range of different types of cells; for applications in humans, it also has to be tested on animals for any less direct effects. Biocompatibility is obviously a complex phenomenon. Some materials and devices are immediately toxic to all cells or organisms; some are toxic only to specific animals; others are harmful in the long term (e.g., carcinogens); others, while not toxic, may cause marked changes in the normal behavior of the system.[106] The definition of biocompatibility of a material or device also varies depending on how the material or device is to be used. If a device is to be implanted into a tissue, one may want the cells of the tissue to treat it as part of the tissue to stick to it and grow on it (not an advisable strategy for a chemical-type sensor where membranes might become blocked). But if a device is injected into the bloodstream, it is vital that the blood cells not stick to it, as this results in a clot, which could be fatal to the biological system. In the case of a medical nanorobot (perhaps a future version of a pharmacy-on-a-chip shown in Volume II, Figure 4.100) floating or swimming in the circulation, cell and tissue adhesion is again undesirable. The term *biocompatibility* evolved over time, and today an accepted working definition is the ability of a material or device to perform with an appropriate host response in a specific application.

History

Biocompatibility work has a long history. Two thousand years ago, gold was already used in dental applications; glass eyes and wooden teeth were introduced next, and plastics were developed in the early 1900s. After World War II, parachute cloth was used as vascular prosthesis. Finally, over the last two decades, active implants (implants providing more than a mechanical support function), such as biosensors and drug delivery systems, have further challenged our understanding of biocompatibility issues. Indeed, in many of these active devices an efficient transport of chemicals to and from the device is essential, something that was not an issue for more traditional implants. Clogging of the pores or membranes in such active devices will render the device useless, though the device may remain perfectly harmless to the functioning of the biological system itself. Over the last fifteen years, considerable work has been directed to create nano- and microtextured surfaces that are cell-friendly and promote controlled cell growth. This promises to become another important application area for MEMS. One envisions substrates for cell culturing, the growth of nerves for neural implants, textures that encourage neural regeneration and the acceptance of prostheses, and microsubstrates for bioartificial organs and as a medium for the regeneration of tissues such as bone and tendon. Typical materials used or proposed, with or without nano- or microtexturing, include titanium, niobium, carbon, some ceramics, polyimide, some polyurethanes, Teflon, Parylene, Makrolon, polystyrene, PMMA, PEEK, polycarbonate, polypropylene, and silicone. Often, there are conflicting views (remember the breast implant controversy), and in vivo sensor applications remain so elusive that many companies have abandoned implantable devices all together and are focusing on ex vivo systems instead.[107] The field where perhaps the most progress has been made in terms of biocompatible materials evaluation is in the area of biomedical materials for drug delivery systems (see, e.g., Chapter 8, this volume, Table 8.20).[108–110] The biocompatibility of silicon was discussed in Volume I, Chapter 4.

Mechanism

A foreign material implanted in a host tissue causes a cascade of events to occur at the tissue/material interface. It starts with a noncovalent adsorption of plasma protein from blood onto the surface. This protein adsorption is quite a bit faster than the transport of host cells to the foreign surface. By the late 1960s, several researchers already knew that within 10 s of exposure to blood or plasma, a uniform ~6-nm layer of fibrinogen forms on surfaces such as Si, Pt, Ge, and Ta. Fibrinogen is a 340-kilodalton soluble plasma glycoprotein about 47.5 nm in length. It is the major surface protein. The protein-coated surface, in turn, mediates the types of cells that may adhere to the surface, which ultimately determines the type of tissue that forms in the vicinity of the implant. The type and state of adsorbed proteins, including their conformational changes, initiate coagulation via platelet adhesion, and subsequent collagen deposition leads to the encapsulation of the microdevice by fibrous tissue. Excessive fibrosis may cause tissue necrosis, granulomas, or tumor genesis.[111] The adsorption and denaturation of adsorbed fibrinogen molecules actually is commonly used as a biocompatibility indicator; the amounts of denatured fibrinogen accumulated on a surface correlates closely to the extent of biomaterial-mediated inflammation. Adsorbed amounts of denatured fibrinogen on surgical implants that are regarded "biocompatible" (such as titanium and stainless steel) are about ~2.1 mg/m^2 (the same as for CVD diamond).

There are two important viewpoints from which to approach biocompatibility:

1. The implant perspective: the protection of implanted devices against the corrosive effects of body fluids and tissue, their sterilizability, and continued operation.
2. The biological system perspective: the avoidance of blood clotting and long-term cytotoxicity, and the continued health of the host.

The former has been named *sensocompatibility*.[107] An important aspect of senscompatibility is biofouling, the adhesion of proteins and other organic matter on the sensor surface. As pointed out above, biofouling is a special concern for active implants such as biosensors (see, e.g., Volume II, Figure 4.92 for an in vivo pH, CO_2, and O_2 sensor), where an efficient transport of chemicals to and from the device is essential. Immune rejection during organ transplants may be prevented by pharmacologically depressing the immune system by the administration of cyclosporine. Such a strategy is, of course, not applicable for the long-term biocompatibility of implants, whether they are biosensors or cell-transplant-based therapeutics—for example, Langerhans or pancreatic islets for type I diabetes,[112] cochlear implants, or heart pacemakers. Perhaps a pharmacy-on-the-chip, as shown in Volume II, Figure 4.100, could also monitor for biofouling and release anti-clotting chemicals such as heparin or herudin as required to maintain the function of the sensor. It should be kept in mind that biofouling occurs both for invasive and noninvasive sensors, but obviously intervention is easier with the latter. In Figure 5.33, we show a schematic of an amperometric invasive glucose sensor, perhaps the most widely studied and most "successful" implantable sensor to date.[107] It is assumed that the needle-type glucose sensor is embedded in tissue and in contact with blood (for details on the operation of amperometric glucose sensors, see Chapter 1, this volume).[113] Failure modes (many packaging related) include the following:

- Lead detachment (could be avoided with a telemetric device)
- Electrical short
- Membrane delamination
- Membrane biofouling
- Electrode passivation
- Enzyme degradation
- Membrane biodegradation
- Fibrous encapsulation
- Electrode fouling

Surface Modifications

Surface Modifications for Biocompatibility

Most success today for control of protein and cell adhesion on surfaces has been achieved by surface immobilization of polyethylene oxide or polyethylene glycol (PEG; one refers to a "pegylated surface") forming a water-soluble, nontoxic, and nonimmunogenic polymer film.[114,115] Zhang et al. modified silicon surfaces by covalent attachment of a self-assembled PEG film and demonstrated an effective suppression of both plasma protein adsorption and cell attachment to such modified silicon surfaces.[116] A pegylated surface may cut the amount of protein adsorbed down by a factor of ten. Glow-discharge plasma-deposited tetraethylene glycol dimethyl ether can reduce the fibrinogen adsorption to ~0.2 mg/m^2 (~350 molecules/μm^2) on many different types of substrates. A specific strategy to fabricate biocompatible ion-selective electrodes of the type shown in Volume I, Figure 7.45, implemented by Brooks et al.,[117] involves covalent attachment of heparin to the surface of derivatized cellulose triacetate membranes, which are subsequently impregnated with the potassium-selective ionophore valinomycin. The resulting ion-selective electrodes respond to potassium and have selectivity coefficients on the same order of magnitude as those of conventional poly(vinyl chloride)-based electrodes.

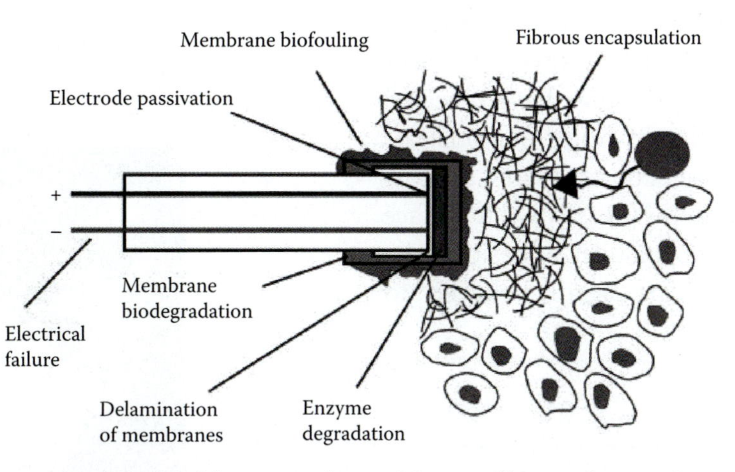

FIGURE 5.33 Schematic illustration of a blood-borne analyte exiting capillary and traversing to a needle-type glucose biosensor embedded in tissue. In addition to normal component failure—electrical failure, enzyme degradation, and membrane delamination—the sensor can fail for several physiologically related reasons, such as membrane biodegradation, electrode passivation, and reduction of analyte access due to fibrous encapsulation and membrane biofouling. (Based on Wisniewski, N., F. Moussy, and W. M. Reichert. 2000. Characterization of implantable biosensor membrane biofouling. *Fresenius J Anal Chem* 366:611–21.[107])

It was found that the heparin layer does not alter significantly the response characteristics of the electrodes. The biological activity of the immobilized heparin was measured in terms of its inactivation of blood coagulation factor Xa; the covalently anchored heparin was able to inactivate factor Xa.

Other approaches that have worked well to avoid protein adsorption are based on biomimetics. This involves the development of membranes that mimic the surface of biological membranes, that is, the design of "stealth" materials. The external region of a cell membrane, the glycocalyx, is studded with glycosylated molecules, which direct specific interactions such as cell-cell recognition and contribute to the steric repulsion that prevents undesirable nonspecific binding of other molecules or cells. By modifying a pyrolitic graphite surface with oligosaccharide surfactant polymers, which, like glycocalyx, provide a dense and confluent layer that mimics the nonadhesive properties of a cell layer, spontaneous adsorption on diverse hydrophobic surfaces is suppressed by at least ~90%.[118] The highly hydrated dextran in these layers protrudes into the aqueous phase and suppresses nonspecific adsorption of plasma proteins. Coating surfaces with phosphorylcholine groups to mimic red blood cell surfaces also leads to diminished protein adsorption. One such material is poly(methacryloyloxyethyl phosphorylcholine-co-*n*-butylmethacrylate) (see Figure 5.34). This polymer, which behaves as a hydrogel, shows very low protein and cell adsorption. Studies have revealed that sensors coated with this hydrogel demonstrate excellent sensor response characteristics, while scanning electron microscopy (SEM) of the surface of the sensors indicates no platelet adhesion. The antifouling characteristics are again believed to be due to the extreme hydrophilic nature of the surface layer; proteins have difficulty adsorbing on the surface because of the layer of bound water.[119] For a further review of in vivo biosensors, consult Frasier;[120] for surface modification strategies, see Wisniewski and Reichert.[121]

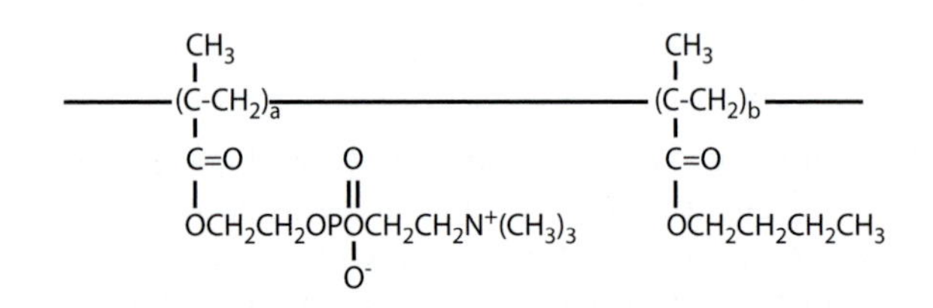

FIGURE 5.34 Structure of poly(methacryloyloxyethyl phosphorylcholine-co-*n*-butylmethacrylate).

Questions

Questions by Dr. Marc Madou assisted by Mr. Omid Rohani, UC Irvine

5.1: What properties make silicon a good mechanical sensor/actuator substrate?

5.2: What has prevented Si-MEMS oscillators from becoming a serious alternative to quartz so far?

5.3: What are the possible applications of GaAs in general and GaAs MEMS in particular?

5.4: What are the two most suitable diamond variations for MEMS application? Explain briefly.

5.5: What method would you suggest for patterning convex hollow structures?

5.6: What are the main advantages of ceramics MEMS compared to other materials?

5.7: NASA's space shuttle has a long manipulator robot arm, also known as the Shuttle Remote Manipulator System or SRMS (see figure below), that permits astronauts to launch and retrieve satellites. It is also used to view and monitor the outside of the space shuttle using a mounted video camera. Select a suitable material for the arm.

5.8: List some of the applications of nonphotosensitive and photosensitive polyimides in microtechnology.

5.9: How does an increase in Si content in silicon nitride change the following parameters?
(a) Tensile film stress
(b) Film transparency
(c) HF etch rate

5.10: In the micro glass blowing method, what are the parameters on which the sizes of blown glass spheres depend?

5.11: Describe the merits and problems associated with micromachining in GaAs, Si, diamond, and 3C-SiC.

5.12: (a) Detail the C-MEMS process.
(b) What advantages might a C-MEMS world have compared to a Si-world?

5.13: What are bulk metallic glasses (BMGs)? What are their properties? How would you use them in MEMS and NEMS?

5.14: Present a definition of biocompatibility of a MEMS structure.

5.15: A thermistor is a device used to measure temperature by taking advantage of the change in electrical conductivity when the temperature changes. Select a material that might serve as a thermistor in the 500 to 1000°C temperature range.

5.16: Describe the ionic bonding between magnesium and chlorine. What are the consequences for the projected strength of the compound that forms and the electronic/ionic conductivity of such a compound?

5.17: Silica (SiO_2) is often given as an example of a covalently bonded material. In reality, silica exhibits both ionic and covalent bonding. What fraction of the bonding is covalent? Give some example applications in which silica is used. To answer this question first find a figure that plots electronegativity vs. atom type.

5.18: Silica is used for making long lengths of optical fibers (see figure below). Being a covalently and ionically bonded material, the strength of the Si-O bonds is expected to be high. Other factors such as susceptibility of silica surfaces to react with water vapor in atmosphere have a deleterious effect on the strength of silica fibers. Given this, what design strategies can you think of such that silica fibers could still be bent to a considerable degree without breaking?

5.19: Define each of the following terms: stress, strain, elastic modulus, Young's modulus of typical materials (ceramics, glasses, semiconductors, metals, and polymers), Poisson's ratio, and yield strength.

5.20: Why are p-type piezoresistors most commonly used in Si pressure sensors?

5.21: In the context of using piezoresistivity to sense strain, what are the advantages/disadvantages of using silicon versus silicon nitride for pressure sensor diaphragms? How does residual stress affect the sensitivity?

References

1. Senturia, S. D., and R. T. Howe. 1990. Mechanical properties and CAD *Lecture notes.* Boston: Massachusetts Institute of Technology.
2. Studer, B., and W. Zingg. 1990. In *4th European Frequency and Time Forum*. Neuchatel, Switzerland, pp. 635–58.
3. Tellier, C. R., and F. Jouffroy. 1983. Orientation effects in chemical etching of quartz plates. *J Mater Sci* 18:3621–32.
4. Ueda, T., F. Kohsaka, T. Lino, and D. Yamazoki. 1987. Theory to predict etching shapes in quartz and applications to design devices. *Trans Soc Inst Control Eng* 23: 1–6.
5. Chuang, S. S. 1983. In *Proceedings: 37th Annual Frequency Control Symposium 1983*, 248–54. Philadelphia: IEEE.
6. Dufour, M., M. T. Delaye, F. Michel, J. S. Danel, B. Diem, and G. Delapierre. 1991. A comparison between micromachined pressure sensors using quartz or silicon vibrating beams. *6th International Conference on Solid-State Sensors and Actuators (Transducers '91)*. San Francisco, June 24–27, 1991, pp. 668–71.
7. Editorial. 1994. Concord, CA: BEI Systron Donner.
8. Hjort, K., G. Thornell, R. Spohr, and J.-Å. Schweitz. 1996. Heavy ion induced etch anisotropy in single crystalline quartz. *Ninth Annual International Workshop on Micro Electro Mechanical Systems*. San Diego, February 11–15, 1996, pp. 267–71.

9. Ericson, F., S. Johansson, and J.-Å. Schweitz. 1988. Hardness and fracture toughness of semiconducting materials studied by indentation and erosion techniques. *J Mater Sci Eng* A105/106:131–41.
10. Takebe, T., T. Yamamoto, M. Fujii, and K. Kobayashi. 1993. Fundamental selective etching characteristics of $HF+H_2O_2+H_2O$ mixtures for GaAs. *J Electrochem Soc* 140:1169–80.
11. Zhang, Z. L., and N. C. MacDonald. 1993. Fabrication of submicron high-aspect-ratio GaAs actuators. *J Microelectromech Syst* 2:66–73.
12. Tarui, Y., Y. Komiya, and Y. Harada. 1971. Preferential etching and etched profile of GaAs. *J Electrochem Soc* 118:119–122.
13. Ribas, R. P. 1998. *Maskless front-side bulk micromachining compatible to standard GaAs IC technology*. Grenoble, France: TIMA Laboratory.
14. Hjort, K. 1996. Sacrificial etching of III-V compounds for micromechanical devices. *J Micromechanics Microeng* 6:370–375.
15. Collins, S. D. 1997. Etch stop techniques for micromachining. *J Electrochem Soc* 144:2242–2262.
16. Karam, J. M., B. Courtois, M. Holjo, J. L. Leclercq, and P. Viktorotovitch. 1996. Collective fabrication of gallium arsenide based microsystems. *SPIE: Micromachining and Microfabrication Process Technology II*. Austin, TX, October 14–15, 1996, pp. 315–26.
17. Hjort, K., J.-Å. Schweitz, and B. Hok. 1990. In *Proceedings: IEEE Micro Electro Mechanical Systems (MEMS '90)*. Napa Valley, CA, pp. 73–76.
18. Hjort, K., J.-Å. Schweitz, S. Andersson, O. Kordina, and E. Janzen. 1992. In *Proceedings: IEEE Micro Electro Mechanical Systems (MEMS '92)*. Travemunde, Germany, February 4–7, 1992, 83–86.
19. Morkoc, H., H. Unlu, H. Zabel, and N. Otsuka. 1988. Gallium arsenide on silicon: A review. *Solid State Technol* March: 71–76.
20. Ramdani, J., R. Droopad, Z. Yu, J. A. Curless, C. D. Overgaard, J. Finder, K. Eisenbeiser, J. A. Hallmark, W. J. Ooms, V. Kaushik, P. Alluri, and S. Pietambaram. 2000. Interface characterization of high-quality SrTiO3 thin films on Si(100) substrates grown by molecular beam epitaxy. *Applied Surface Science* 159–160:127–133.
21. Singer, P. 2001. GaAs-on-silicon, finally! *Semiconductor International* October.
22. LaPedus, M., and P. Clarke. 2003. Motorola closes GaAs-on-silicon wafer subsidiary. *EETimes*.
23. Yeh, H. J., and J. S. Smith. 1994. Fluidic self-assembly of microstructures and its application to the integration of GaAs on Si. *IEEE International Workshop on Micro Electro Mechanical Systems (MEMS '94)*. Oiso, Japan, January 25–28, 1994, pp. 279–84.
24. Yeh, H. J., and J. S. Smith. 1994. Integration of GaAs vertical-cavity surface-emitting laser on Si by substrate removal. *Appl Phys Lett* 64:1466–68.
25. Teraji, T., S. Mitani, C. Wang, and T. Ito. 2002. High-rate deposition of high-quality homoepitaxial diamond films (review). *New Diamond Front Carbon Tech* 12:355.
26. Wang, C., M. Irie, K. Kimura, T. Teraji, and T. Ito. 2001. Boron-doped diamond film homoepitaxially grown on high-quality chemical-vapor deposited diamond (100). *Jpn J Appl Phys* 40:4145.
27. Herb, J. A., M. G. Peters, S. C. Terry, and J. H. Jerman. 1990. PECVD diamond films for use in silicon microstructures. *Sensors Actuators A* A23:982–87.
28. Spear, K. E., and J. P. Dismukes. 1994. *Synthetic diamond: Emerging CVD science and technology.* New York: Wiley-Interscience.
29. Obermeier, E. 1995. In *8th International Conference on Solid-State Sensors and Actuators (Transducers '95)*. Stockholm, Sweden, June 25–29, 1995, pp. 178–81.
30. Auciello, O., J. Birrell, J. A. Carlisle, J. E. Gerbi, X. Xiao, B. Peng, and H. D. Espinosa. 2004. Materials science and fabrication processes for a new MEMS technology based on ultrananocrystalline diamond thin films. *J Phys Condens Matter* 16:R539–52.
31. Webster, J. R., C. W. Dyck, J. P. Sullivan, T. A. Friedmann, and A. J. Carton. 2004. Performance of amorphous diamond RF MEMS capacitive switch. *Electron Lett* 40:43–44.
32. Kohn, E., P. Gluche, and M. Adamschik. 1999. Diamond MEMS: A new emerging technology. *Diamond Relat Mater* 8:934–40.
33. Irie, M., S. Endo, C. L. Wang, and T. Ito. 2003. Fabrication and properties of lateral p-i-p structures using single-crystalline CVD diamond layers for high electric field applications. *Diamond Relat Mater* 12:1563–68.
34. Lari, T., A. Oh, N. Wermes, H. Kagan, M. Keil, and W. Trischuk. 2005. Characterization and modeling of non-uniform charge collection in CVD diamond pixel detectors. *Nuclear Instruments Methods Phys Res Sec A Accel Spectrom Detect Associated Equip* 537:581–93.
35. Muller, R., M. Adamschik, D. Steidl, E. Kohn, S. Thamasett, S. Stiller, H. Hanke, and V. Hombach. 2004. Application of CVD-diamond for catheter ablation in the heart. *Diamond Relat Mater* 13:1080–83.
36. Achard, J., F. Silva, H. Scheider, R. S. Sussmann, A. Tallaire, A. Gicquel, and M. C. Castex. 2004. The use of CVD diamond for high-power switching using electron beam excitation. *Diamond Relat Mater* 13:876–80.
37. Krotz, G., W. Legner, C. Wagner, H. Moller, H. Sonntag, and G. Muller. 1995. Silicon carbide as a mechanical material. *8th International Conference on Solid-State Sensors and Actuators (Transducers '95)*. Stockholm, Sweden, June 25–29, 1995, 186–89.
38. Fleischman, A. J., S. Roy, C. A. Zorman, M. Mehregany, and L. G. Matus. 1996. Polycrystalline silicon carbide for surface micromachining. *9th Annual International Workshop on Micro Electro Mechanical Systems*. San Diego, CA, February 11–15, 1996, pp. 234–38.
39. Roy, S., A. K. McIlwain, R. G. DeAnna, A. J. Fleischman, R. K. Burla, C. A. Zorman, and M. Mehregany. 2000. In *Solid-State Sensor and Actuator Workshop 22–25*. Hilton Head Island, SC: Transducers Research Foundation.
40. Mardilovich, P., D. Routkevitch, and A. Govyadinov. 2000. Hybrid micromachining and surface microstructuring of alumina ceramic. *Electrochem Soc Proc* 200-19:33–42.
41. Tan, S., M. Reed, H. Han, and R. Boudreu. 1995. High aspect ratio microstructures on porous anodic aluminum oxide. *Proceedings of the Eighth International Workshop on Micro Electro Mechanical Systems (MEMS-95)*. Amsterdam, January 1995, pp. 267–72.
42. Routkevitch, D., A. Govyadinov, and P. Mardilovich. 2000. High aspect ratio, high resolution ceramic MEMS. *2000 International Mechanical Engineering Congress and Exposition*. Orlando, FL, November 5–10, 2000, pp. 1–6.
43. Li, A., F. Muller, A. Birner, K. Nielsch, and U. Gosele. 1999. Fabrication and microstructuring of hexagonally ordered two-dimensional nanopore arrays in anodic alumina. *Adv Materials* 11:483–87.

44. Govyadinov, A., P. Mardilovich, K. Novogradecz, S. Hooker, and D. Routkevitch. 2000. Anodic alumina MEMS: Applications and devices. In *2000 International Mechanical Engineering Congress and Exposition*, Orlando, FL, November 5–10, 2000, pp. 1–6.
45. Mardilovich, P., D. Routkevitch, and A. Govyadinov. 2000. New approach for surface microstructuring. In *2000 International Mechanical Engineering Congress and Exposition*. Orlando, FL, November 5–10, 2000, pp. 1–5.
46. Govyadinov, A., P. Mardilovich, and D. Routkevitch. 2000. Field emission cathode from aligned arrays of carbon nanotubes. *Electrochem Soc Abstracts* 2000-2:Abstract 553.
47. Sugiura, T., T. Yoshida, and H. Minoura. 1998. Designing a TiO_2 nano-rhoneycomb structure using photoelectrochemical etching. *Electrochem Solid-State Lett* 1:175–77.
48. Sugiura, T., S. Itoh, T. Ooi, T. Yoshida, K. Kuroda, and H. Minoura. 1999. Evolution of a skeleton structures TiO_2 surface consisting of grain boundaries. *Electroanal Chem* 473:204–08.
49. Schubert, P. J., and J. H. Nevin. 1985. A polyimide-based capacitive humidity sensor. *IEEE Trans Electron Devices* ED-32:1220–24.
50. Ralston, A. R. K., C. F. Klein, P. E. Thoma, and D. D. Denton. 1995. A model for the relative environmental stability of a series of polyimide capacitance humidity sensors. *8th International Conference on Solid-State Sensors and Actuators (Transducers '95)*. Stockholm, Sweden, June 25–29, 1995, pp. 821–24.
51. Latham, W. J., and D. W. Hawley. 1988. Color filters from dyed polyimides. *Solid State Technol* 31:223–26.
52. Suh, J. W., S. F. Glander, R. B. Darling, C. W. Storment, and G. T. A. Kovacs. 1996. Combined organic thermal and electrostatic omnidirectional ciliary microactuator array for object positioning and inspection. *Proceedings: Solid-State Sensor and Actuator Workshop*. Hilton Head, SC, June 2–6, 1996, pp. 168–73.
53. Darling, R. B., J. W. Suh, and G. T. Kovacs. 1998. Ciliary microactuator array for scanning electron microscope positioning stage. *J Vac Sci Technol A* 16:1998–2002.
54. Makino, D. 1994. In *Polymers for microelectronics: Resists and dialectrics* (eds. Thompson, L. F., C. G. Willson, and S. Tagawa) Washington, DC: American Chemical Society.
55. Studt, T. 1992. Polyimides: Hot stuff for the '90s. *Res Dev* August: 30–31.
56. Ando, S., T. Matsuura, and S. Sasaki. 1994. Polymers for microelectronics: Resists and dielectrics, ed. L. F. Thompson, C. G. Willson, and S. Tagawa. *ACS Symposium Series* 537:304–22.
57. Pool, R. 1988. Microscopic motor is a first step. *Res News* October: 379–80.
58. Rosler, R. S. 1991. The evolution of commercial plasma enhanced CVD systems. *Solid State Technol* June: 67–71.
59. Wu, T. H. T., and R. S. Rosler. 1992. Stress in PSG and nitride films as related to film properties and annealing. *Solid State Technol* May: 65–71.
60. Adams, A. C. 1988. In *VLSI technology*, ed. S. M. Sze. New York: McGraw-Hill, pp. 259–60.
61. Sinha, A. K., and T. E. Smith. 1978. Thermal stresses and cracking resistance of dielectric films. *J Appl Phys* 49:2423–26.
62. Retajczyk, T. F. J., and A. K. Sinha. 1980. Elastic stiffness and thermal expansion coefficients of various refractory silicides and silicon nitride films. *Thin Solid Films* 70:241–47.
63. Sakimoto, M., H. Yoshihara, and T. Ohkubo. 1982. Silicon nitride single-layer x-ray mask. *J Vac Sci Technol* 21:1017–21.
64. Ibid.
65. Chang, S., W. Eaton, J. Fulmer, C. Gonzalez, B. Underwood, J. Wong, and R. L. Smith. 1991. Micromechanical structures in amorphous silicon. *6th International Conference on Solid-State Sensors and Actuators (Transducers '91)*. San Francisco, June 24–27, 1991, pp. 751–54.
66. Spear, W. E., and P. G. Le Comber. 1975. Substitutional doping of amorphous silicon. *Solid State Commun* 17:1193–96.
67. Lee, J. B., Z. Chen, M. G. Allen, A. Rohatgi, and R. Arya. 1995. A miniaturized high-voltage solar cell array as an electrostatic MEMS power supply. *J Microelectromech Syst* 4:102–08.
68. Crowley, J. L. 1992. Plasma enhanced CVD for flat panel displays. *Solid State Technol* February: 94–98.
69. Holbrook, D. S., and J. D. McKibben. 1992. Microlithography for large area flat panel display substrates. *Solid State Technol* May: 166–72.
70. Bohm, M. 1988. Advances in amorphous silicon based thin film microelectronics. *Solid State Technol* September.
71. Bruno, G., P. Capezzuto, and A. Madan, Eds. 1995. *Plasma deposition of amorphous silicon-based materials.* Boston: Academic Press.
72. Wang, C., and M. Madou. 2005. From MEMS to NEMs with carbon. *Biosensors and Bioelectronics* 20:2181–87.
73. Wang, C., G. Jia, L. H. Taherabadi, and M. J. Madou. 2005. A novel method for the fabrication of high-aspect ratio C-MEMS structures. *J MEMS* 14:348–58.
74. Schueller, O. J. A., S. T. Brittain, C. Marzolin, and G. M. Whitesides. 1997. Fabrication and characterization of glassy carbon MEMS. *Chem Mater* 9:125–39.
75. Madou, M., A. Lai, G. Schmidt, X. Song, K. Kinoshita, M. Fendorf, A. Zetl, and R. White. 1997. Carbon micromachining (CMEMS). In *Chemical and biological sensors and analytical electrochemical methods*, ed. A. Ricco, M. B., P. Vanysck, G. Horvai, and A. Silva. Pennington, NJ: Electrochemical Society Proceedings Series, pp. 61.
76. Kim, J., X. Song, K. Kinoshita, M. Madou, and R. White. 1998. Electrochemical studies of carbon films for pyrolyzed photoresist. *J Electrochem Soc* 145:2314–19.
77. Ranganathan, S., R. McCreery, S. M. Majji, and M. Madou. 2000. Photoresist derived carbon for microelectromechanical systems and electrochemical applications. *J Electrochem Soc* 147:277–82.
78. Wang, C., L. Taherabadi, G. Jia, M. Madou, Y. Yeh, and B. Dunn. 2004. C-MEMS for the manufacture of 3D microbatteries. *Electrochem Solid-State Lett* 7:A435–38.
79. Hart, R. W., H. S. White, B. Dunn, and D. R. Rolison. 2003. 3D microbatteries. *Electrochem Commun* 5:120–23.
80. McCreery, R., J. Dieringer, A. O. Solak, B. Snyder, A. M. Nowak, W. R. McGovern, and S. Duvall. 2003. Molecular rectification and conductance switching in carbon-based molecular junctions by structural rearrangement accompanying electron injection. *JACS* 125:10748–58.
81. Hebert, N. E., B. Snyder, R. L. McCreery, W. G. Kuhr, and S. A. Brazill. 2003. Performance of pyrolyzed photoresist carbon films in a microchip capillary electrophoresis device with sinusoidal voltammetric detection. *Anal Chem* 75:4265–71.
82. Park, B. Y., R. Zaouk, C. Wang, and M. J. Madou. 2007. A case for fractal electrodes in electrochemical applications. *J Electrochem Soc* 154:1–5.

83. Park, B. Y., and M. J. Madou. 2005. 3-D electrode design for flow-through dielectrophoretic systems. *Electrophoresis* 26:3745.
84. Inoue, A., T. Zhang, and T. Masumoto. 1990. Zr-Al-Ni amorphous-alloys with high glass-transition temperature and significant supercooled liquid region. *Materials Transactions Jim* 31:177–83.
85. Peker, A., and W. L. Johnson. 1993. A highly processable metallic-glass: Zr41.2ti13.8cu12.5ni10.0be22.5. *Appl Phys Lett* 63:2342–44.
86. Ponnambalam, V., S. J. Poon, and G. J. Shiflet. 2004. Fe-Mn-Cr-Mo-(Y,Ln)-C-B (Ln = lanthanides) bulk metallic glasses as formable amorphous steel alloys. *J Mater Res* 19:3046–52.
87. Lu, Z. P., C. T. Liu, J. R. Thompson, and W. D. Porter. 2004. Structural amorphous steels. *Phys Rev Lett* 92:245503.
88. Zhang, Q. S., W. Zhang, and A. Inoue. 2006. New Cu-Zr-based bulk metallic glasses with large diameters of up to 1.5 cm. *Scripta Materialia* 55:711–13.
89. Xu, D. H., G. Duan, W. L. Johnson, and C. Garland. 2004. Formation and properties of new Ni-based amorphous alloys with critical casting thickness up to 5 mm. *Acta Materialia* 52:3493–97.
90. Lin, X. H., and W. L. Johnson. 1995. Formation of Ti-Zr-Cu-Ni bulk metallic glasses. *J Appl Phys* 78:6514–19.
91. Inoue, A., A. Kato, T. Zhang, S. G. Kim, and T. Masumoto. 1991. Mg-Cu-Y amorphous-alloys with high mechanical strengths produced by a metallic mold casting method. *Materials Transactions Jim* 32:609–16.
92. Nishiyama, N., and A. Inoue. 1999. Supercooling investigation and critical cooling rate for glass formation in P-Cu-Ni-P alloy. *Acta Materialia* 47:1487–95.
93. Schroers, J., and W. L. Johnson. 2004. Highly processable bulk metallic glass-forming alloys in the Pt-Co-Ni-Cu-P system. *Appl Phys Lett* 84:3666–68.
94. Schroers, J., B. Lohwongwatana, W. L. Johnson, and A. Peker. 2005. Gold based bulk metallic glass. *Appl Phys Lett* 87:61912.
95. Johnson, W. L. 1999. Bulk glass-forming metallic alloys: Science and technology. *MRS Bull* 24:42–56.
96. Schroers, J., and N. Paton. 2006. Amorphous metal alloys form like plastics. *Adv Materials Processes* 164:61–63.
97. Schroers, J. 2005. The superplastic forming of bulk metallic glasses. *JOM* 57:35–39.
98. Schroers, J., Q. Pham, and A. Desai. 2007. Thermoplastic forming of bulk metallic glass—A technology for MEMS and microstructure fabrication. *J MEMS* 16:240–47.
99. Saotome, Y., K. Imai, S. Shioda, S. Shimizu, T. Zhang, and A. Inoue. 2002. The micro-nanoformability of Pt-based metallic glass and the nanoforming of three-dimensional structures. *Intermetallics* 10:1241–47.
100. Jeong, H. W., S. Hata, and A. Shimokohbe. 2003. Microforming of three-dimensional microstructures from thin-film metallic glass. *J MEMS* 12:42–52.
101. Kumar, G., and J. Schroers. 2008. Write and erase mechanism on bulk metallic glass. *Appl Phys Lett* 92:031901.
102. Schroers, J., Q. Pham, and A. Desai. 2007. Thermoplastic forming of bulk metallic glass—A technology for MEMS and microstructure fabrication. *J MEMS* 16:240–47.
103. Madou, M., J. Zoval, G. Jia, H. Kido, J. Kim, and N. Kim. 2008. Lab on a CD. *Annu Rev Biomed Eng* 8:601–28.
104. Eklund, E. J., A. M. Shkel, S. Knappe, E. Donley, and J. Kitching. 2008. Glass-blown spherical microcells for chip-scale atomic devices. *Sensors Actuators A* 143:175–80.
105. Eklund, E. J., and A. M. Shkel. 2007. Glass blowing on a wafer level. *J MEMS* 16:232–39.
106. Von Recum, A., ed. 1998. *Handbook of biomaterials evaluation: Scientific, technical, and clinical testing of implant materials.* New York: Hemisphere.
107. Wisniewski, N., F. Moussy, and W. M. Reichert. 2000. Characterization of implantable biosensor membrane biofouling. *Fresenius J Anal Chem* 366:611–21.
108. Kaetsu, I. 1995. Biocompatible and biofunctional membranes by means of radiation techniques. *Nucl Instrum Methods Phys Res B* 105:294–301.
109. Schlosser, M., and M. Ziegler. 1997. Biocompatibility of active implantable devices. In *Biosensors in the body: Continuous in vivo monitoring,* ed. D. M. Frasier. New York: Wiley.
110. Reichert, W. M., and A. A. Sharkawy. 1999. Biosensors. In Handbook of biomaterials evaluation (ed. von Recum, A.) 439–60. Philadelphia: Taylor and Francis.
111. Park, J. B., and R. S. Lakes. 1993. *Biomaterials: An introduction.* New York: Springer.
112. Desai, T. A., D. Hansford, and M. Ferrari. 1999. Characterization of micromachined silicon membranes for immunoisolation and bioseparation applications. *J Membr Sci* 159:221–31.
113. Madou, M. J., and S. R. Morrison. 1989. *Chemical sensing with solid state devices.* New York: Academic Press.
114. Mrksich, M., and G. M. Whitesides. 1996. Using self-assembled monolayers to understand the interactions of man-made surfaces with proteins and cells. *Annu Rev Biophys Biomol Struct* 25:55–78.
115. Ratner, B. D., and A. S. Hoffman. 1996. Thin films, grafts, and coatings. In *Biomaterials science: An introduction to materials in medicine,* eds. B. D. Ratner, A. S. Hoffman, F. J. Schoen, and J. E. Lemons. San Diego, CA: Academic Press, p. 105.
116. Zhang, M., T. Desai, and M. Ferrari. 1998. Proteins and cells on PEG immobilized silicon surfaces. *Biomaterials* 19:953–60.
117. Brooks, K. A., J. R. Allen, P. W. Feldhoff, and L. G. Bachas. 1996. Effect of surface-attached heparin on the response of potassium-selective electrodes. *Anal Chem* 68:1439–43.
118. Holland, N. B., Y. Qiu, M. Ruegsegger, and R. E. Marchant. 1998. Biomimetic engineering of non-adhesive glycocalyx-like surfaces using oligosaccharide surfactant polymers. *Nature* 392:799–801.
119. Ishihara, K., N. Nakabayashi, M. Sakakida, N. Kenro, and M. Shichiri. 1998. In *American Chemical Society Annual Meeting.* Orlando, FL.
120. Frasier, D. M., ed. 1997. *Biosensors in the body: Continuous in vivo monitoring.* New York: Wiley.
121. Wisniewski, N., and W. Reichert. 2000. Methods for reducing biosensor membrane biofouling. *Colloids Surf B: Biointerfaces* 18:197–21.

6

Metrology and MEMS/NEMS Modeling

Outline

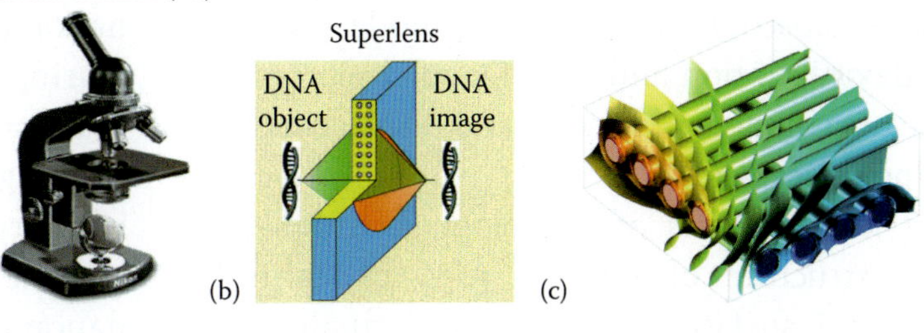

From (a) microscopes to (b) superlenses. (c) Model of microelectromechanical system heat exchanger. (From http://www.comsol.com/showroom/gallery/252.php.) In a new field such as nanotechnology it is often the consultants and conference organizers making the only money. Another group often cashing in early are measuring instrument or metrology developers. Edwards compared this to the hardware stores and whorehouses making more money in the gold rush of 1849 than the actual gold miners. (Steven A. Edwards, in *The Nanotech Pioneers,* 2006.)

Steven A. Edwards

Introduction

This chapter is divided into two parts; one deals with metrology of microelectromechanical systems (MEMS) and nanoelectromechanical systems (NEMS), the other with their modeling. Metrology is the science that deals with measurements. Tools for MEMS and NEMS metrology are either analytical techniques derived from analytical chemistry or size measuring techniques borrowed from the semiconductor industry. We start the metrology section by exploring techniques to visualize MEMS and NEMS devices and to measure sizes such as film thickness and critical dimensions (CDs), i.e., dimensional metrology. Dimensional metrology tools we cover are light microscopes (including dark-field, bright-field, phase contrast, and fluorescence), scanning confocal and two-photon microscopes, white light interferometers, ellipsometers, scanning electron microscopes (SEMs), and a set of scanning proximal probe microscopes (SPMs) including scanning near-field optical microscopes (NSOMs or SNOMs), scanning tunneling microscopes (STMs), and atomic force microscopes (AFMs).

In the category of material property measuring equipment, we start with a review of specialized SPMs for material property determination. SPMs are used not only to visualize structures (see SNOM, STM, and AFM, above) but also to analyze a wide variety of material properties such as magnetic, thermal, and mechanical behavior. We then consider tools used to measure conductivity and carrier type of a material. MEMS devices often are mechanically dynamic by nature, hence the need to measure dynamic motion of MEMS devices. We study mechanical properties using interferometry with a stroboscopic light-emitting diode (LED) to freeze device motion and laser Doppler vibrometry to measure contactless velocity and displacement of vibrating structures.

Next are examples of metrology techniques to analyze structure of samples (e.g., crystallinity, defects, and surface roughness) and finally analytical techniques for chemical composition. In the structural metrology category, we review stylus profilometry, which can be used for surface roughness and transmission electron microscopy (TEM) with utility for crystallinity determination. As examples of elemental analysis of surfaces, we discuss energy dispersive analysis of x-rays (EDAX), a standard procedure for identifying and quantifying elemental composition of specimen surfaces to shallow depths, and scanning Auger microscopy (SAM), which can identify every atom from lithium on up.

We close off the metrology section by looking at MEMS approaches to metrology. This includes on-chip MEMS devices to measure physical properties such as stress, Young's modulus, Poisson ratio, natural frequency, etc.

Modeling is the use of mathematical equations to simulate and predict real events and processes. In the modeling section, we introduce finite element analysis and review software packages helpful in the design phase of miniaturized devices, that is, software for computer-aided design (CAD) in MEMS. The emphasis in this modeling section is on MEMS modeling. NEMS and BIONEMS require molecular and quantum effect modeling. These techniques are used to construct, display, manipulate, simulate, and analyze molecular structures and to calculate their properties. The topic is beyond the scope of the current work, and only some useful software packages will be referred to.

Metrology

Introduction

The history of the last two industrial revolutions—first the machine revolution (1800–1920) and second the semiconductor revolution (1950–2010)—reveals that a nanotechnology revolution might not happen until the considerable lag of new nanometrology technologies has been remedied. During the machine revolution, the milli-inch (1 mil = 25 μm) was the measuring stick, and it could be gauged by vernier calipers. The semiconductor industry today has as a dominant measure 50 nm, accessible through electron imaging and SPMs. The nanometer revolution is centered at the 1 nm length-scale, for which we do not yet have the appropriate high-throughput tools. Today, high-throughput imaging or measuring sub-30-nm features, using a conventional SEM or SPMs (say, an AFM or an STM), results in fuzzy, noisy images. Fast, sharp, low-noise images at the nanoscale will take Herculean efforts. This does not bode well for the swift arrival of so-called *nanomanufacturing*, where rapid, accurate, high-resolution measurements are essential for profitability. On the positive side, in research settings, measurements, images, and motion can be controlled to within 10 picometers (10^{-11} m).

MEMS and NEMS metrology includes CD measurements of metal lines on an accelerometer, establishing surface roughness of a pressure sensitive membrane, determining compositional profiles of a pn-junction, and measuring magnetic, mechanical, electrical, and optical properties (see Figure 6.1). Accordingly, metrology equipment is divided into dimensional metrology (I), material property studies (electrical, optical, magnetic, mechanical, etc.) (II), structure identification (surface roughness, defects, crystallinity) (III), and chemical composition

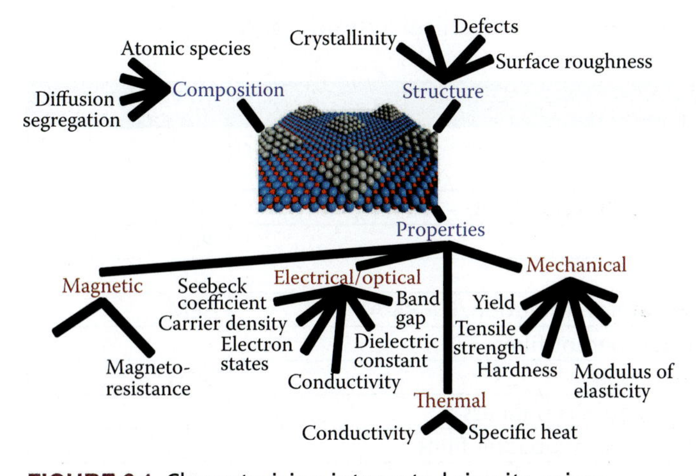

FIGURE 6.1 Characterizing integrated circuits, microelectromechanical systems, and nanoelectromechanical systems.

analysis (IV). In Table 6.1, we group metrology techniques in these four categories and list examples of typical tools.

Several metrology tools are used for different types of metrology. For example, scanning electron microscopy (SEM) can be used for dimensional metrology [e.g., to measure the CD of an integrated circuit (IC)] as well as in structural metrology (e.g., to measure defects in an IC). If the SEM comes equipped with EDAX, which it often does, chemical composition information can be collected as well. Similarly, SPMs may image MEMS, measure their size, or establish important material properties.

The measurement tools in Table 6.1, when used in IC, MEMS, or NEMS manufacturing, are further described as being stand-alone or integrated with the manufacturing line.[1] A stand-alone tool is a separate piece of equipment that is either off-line, at-line, or in-line. Off-line is an instrument that is available only outside the fab (usually involves destructive or contaminating testing). An at-line piece of equipment is one that is available in the fab itself, and in-line tools are used during production to measure, for example, processed Si wafers. An integrated tool is one that is combined with a manufacturing line and is either on-line or in situ. An on-line tool is available at the process workstation to measure patterned wafers but is not able to measure during wafer processing. In the sections below, we select some of these techniques for further scrutiny. In Table 6.2 additional characteristics of the most important metrology techniques from Table 6.1 are listed, and in Figure 6.2 we compare this most important set of analytical tools in terms of detection limit, sampling depth, and spot size.

Tools to See and Size: Dimensional Metrology

Introduction

An important category in metrology equipment comprises seeing and sizing tools, for example, to image an etched profile in a MEMS structure, to measure film thickness, or to determine a critical dimension (CD) on an IC. From Table 6.1, seeing tools range from optical microscopy (OM), to scanning electron microscopy (SEM) and scanning proximal probe microscopes (SPMs), including scanning tunneling microscopy (STM), atomic force microscopy (AFM), and scanning near-field optical microscopy (NSOM or SNOM). Film thickness is accessible, for example, through ellipsometry, photoacoustic measurements, x-ray fluorescence (XRF), reflection spectroscopy, and interferometry. CD in ICs and MEMS and NEMS are typically measured using a scanning electron microscope (CD-SEM) or an atomic force microscope (CD-AFM) with subnanometer resolution and accuracy. In the sub-100-nm IC generations, CDs are becoming harder and harder to control. With 65-nm features, for example, one typically has an error budget of only ±5 nm and current CD-SEM metrology techniques offer about ±2 nm precision. An AFM, in theory, should provide better resolution, but unfortunately AFM scanning is slow and suffers from probe tip wear. Another entry in CD measurements is the use of scatterometry that involves measurements on IC wafers by diffraction from grating test patterns. Scatterometry is a faster and more economic way to characterize periodic patterns on a wafer to relieve/bypass the heavy workload of CD-SEM and CD-AFM in a typical fab. Scatterometry involves flood exposure of a field, the measurement of changes in the polarization state of the reflected light, and calculation of IC-structure characteristics via real-time or library-based regression software algorithms. With visible light, scatterometry affords a resolution, R, of about 0.15 nm and can deliver multiple measurements, including

Table 6.1 Metrology Equipment for Integrated Circuits, Microelectromechanical Systems, and Nanoelectromechanical Systems

	Examples	Abbreviation
Dimensional metrology (e.g., film thickness, CD) (I)		
	High-resolution microscopy (R = diffraction limit, ~150 nm)	HM
	Phase contrast microscope	PCM
	Scanning tunneling microscope	STM
	Scanning electron microscope (lateral resolution, R = 35 Å)	SEM
	Ellipsometer for thickness of transparent films	
	Scanning proximal probe microscope	SPM
	Stylus profilometer for sample surface roughness	
	Reflection spectroscopy for thickness of opaque films	
	X-ray fluorescence for film thickness	XRF
	Photoacoustic technology for film thickness	
	Scatterometer for analysis of periodic structures. CD measurement	
	Vernier alignment scales (on-chip)	
Metrology for material properties (II)		
Mechanical properties		
	Micro/nanoindentation with scanning proximal probe	SPM
	Residual stress measurements with interferometry	
	Vibrational analysis with laser Doppler	
Electrical properties		
	Capacitance	
	Resistance with a four-point probe	
	Electron energy loss spectroscopy	EELS
	Carrier type with hot probe	
	Dopant concentration with thermal wave system	
Optical properties		
	Ellipsometer	
	Photoelectron spectroscopy	XPS, UPS
	Ultraviolet/visible spectroscopy	UV/VIS
	Scanning near-field optical microscopy	NSOM/SNOM
Magnetic properties		
	Secondary electron microscopy with polarization analysis	SEMPA
	Scanning magnetic force microscopes	SFM
Metrology for structural properties (III)		
	Scanning electron microscope, R = 35 Å (lateral)	SEM
	Transmission electron microscope, R = 2 Å	TEM
	Scanning probe microscopies	SPMs such as STM and AFM
	High-resolution microscopy (R = diffraction limit, ~150 nm)	HM
	Field ion microscope	FIM
	Phase contrast microscope	PCM
	X-ray microscope (Synchrotron radiation and x-ray optics; R = ~1 Å)	
	Low energy electron emission microscopy (low energy e-beam; R = 20 nm)	LEEM
	Focused ion beam	FIB
	X-ray diffraction spectroscopy	XRD
	Scanning x-ray photoelectron microscopy	SXPEM

Table 6.1 Metrology Equipment for Integrated Circuits, Microelectromechanical Systems, and Nanoelectromechanical Systems (*Continued*)

	Examples	Abbreviation
Metrology for chemical composition (IV)		
	Auger electron spectroscopy	AES
	Energy dispersive analysis of x-rays	EDAX or EDS
	X-ray photoelectron spectroscopy	XPS
	Photoemission electron microscopy: lateral R = ~100 nm	PEEM
	Secondary ion mass spectroscopy	SIMS
	Mass spectrometry	MS
	Rutherford back scattering	RBS
	X-ray fluorescence	XRF
	Nuclear magnetic resonance spectroscopy	NMR
	Photoluminescence spectroscopy	PL
	Cathodoluminescence spectroscopy	CL
	Raman spectroscopy	RAMAN
	Infrared spectroscopy	IR, FTIR
	Nuclear magnetic resonance spectroscopy	NMR
	Photoluminescence spectroscopy	PL
	Neutron activation analysis	NAA
	Ultraviolet/visible spectroscopy	UV/VIS

R, resolution.

CDs, feature shapes (i.e., CD profiles), overlay, and film thickness on patterned structures, including stacked films. Because scatterometry tools are diffraction based, the target must be a periodic structure that consists of a single feature repeated with a well-defined pitch. The technique cannot measure isolated single lines or more complex three-dimensional (3D) shapes.

There is a need not only for sizing metrology with faster throughput but also for more equipment that provides 3D information, not only in ICs but particularly for the newer areas of MEMS and NEMS, which add a more prominent *z*-dimension to parts. These three-dimensional parts require metrology tools that can measure in three dimensions. Most available tools are two-dimensional (2D) in nature, and measurements are made on the top or bottom of the microcomponent and use an algorithm to provide a pseudo-3D image (2.5D in reality), as often no sidewall information is available.

We cannot detail all tools that are available for dimensional metrology and limit ourselves to a review of light microscopes (including dark-field, light-field, phase contrast, and fluorescence), scanning confocal and two-photon microscopy, white light interferometry, scanning electron microscopy (SEM), NSOM, STMs, AFMs, and ellipsometry.

Light Microscopy

Since the resolution of the human eye is limited to about 100 micrometers, microscopes have always been the engines for new discoveries. Optical microscopy (OM) is usually what one thinks of when studying microscopy; this is also where we are starting our discussion. It was in 1590 that the Dutch spectacle-makers Hans Janssen and his son Zacharias Janssen invented a compound microscope with a magnification of ×3 to ×9. Robert Hooke, in 1665, published *Micrographia*, a collection of biological micrographs and he coined the word *cell* for the structures he observed in cork bark. In 1674 Anton van Leeuwenhoek improved on a simple microscope for viewing biological specimens and discovered bacteria (from tooth scrapings), protozoans (from pond water—he called them *animalcules*), sperm (his own, we presume), and blood cells (magnification of ×270). In Table 6.3, a brief history of microscopy is provided. The table includes light microscopy as well as electron microscopy, proximal probe microscopy, and near field superlenses.

Table 6.2 Metrology for Integrated Circuits (ICs), Microelectromechanical Systems (MEMS), and Nanoelectromechanical Systems (NEMS)

	Spot Size or Probe Area	Resolution/ Magnification	Thickness Range	Applications	Remarks (n = Refractive Index)
Reflection spectroscopy (visible: 400–900 nm)	3.5–50 μm		40 Å to 5 μm	Nonabsorbing films.	Limit measures below 100 Å, n measures only above 700 Å. Fast.
Reflection spectroscopy (UV: 200–400 nm)	10 μm		25 Å to 2 μm	Transparent very thin films on metals.	Limit measures below 50 Å, n measures only above 100 Å.
Ellipsometry (632.8 nm standard), 790, 830, 1,300, and 1,500 nm (optional)	12–100 μm	1–2 μm	10–1,000 Å	Nonabsorbing films.	Automated, highest accuracy. Thickness and n measurements.
Optical microscopes		Magnification 10–1,000	0.3–0 5 μm	Universal tool for all dimensions > 0.29 μm.	Poor depth of field.
Scanning electron microscope (SEM)	Spots down to 10–30 Å	Lateral: –35 Å (standard SEM); –9 Å (field-emission SEM); –150 Å (low-voltage SEM); magnification is 20–150,000		SEM pictures, CD measures below 0.8 μm.	Good for surface topography. Large depth of focus. Needs vacuum. Can be destructive.
Transmission electron microscope (TEM)	10 Å	1–2 Å (resolving power 10^4 times better than optical microscope). Magnification is 500–500,000	10 Å to 1 μm	Atomic structure defect analysis nanoprobe.	Requires thin sample preparation.
Scanning acoustic microscopy	1.3 mm (50 MH_z)	Lateral: 10 nm to 2 mm Depth: 10 nm to 5 mm	Signal pene trates 1λ	Inspection inside a ceramic packaged IC. Interface inspection, wide range of samples.	Nondestructive. Subsurface information. Needs liquid medium.
Stylus profilometer	Tip size limits bandwidth	Resolution of better than 1 Å and lateral resolution of ~0.03 mm		Measures sharp steps from a few angstroms to several micrometers. Materials and process R&D, low volume QC.	Measures surface profile using a contact stylus. May damage soft samples.
Nomarski microscope (differential interference contrast)	0.3 μm			Separation of images of different layers. Sample roughness inspection.	Pseudo-three-dimensional image.
Confocal laser scanning microscope (CLSM)	0.3 μm	0.3 μm	<1 μm	Cross-sectioning. Independent layer imaging.	Nondestructive. No vacuum. Diffraction limit resolution.
Atomic force microscope (AFM)	Radius of curvature of probe tip: 50–100 Å	Atomic imaging (one atom)	<1 Å	Surface roughness. Tribology.	Slow scan. Probe tip wear. Non-destructive.

TABLE 6.2 Metrology for Integrated Circuits (ICs), Microelectromechanical Systems (MEMS), and Nanoelectromechanical Systems (NEMS) (*Continued*)

	Spot Size or Probe Area	Resolution/ Magnification	Thickness Range	Applications	Remarks (n = Refractive Index)
Total integrated scattering		With visible light 0.15 μm is possible		Lithography control, etch structure monitoring and metal grain size determination.	Nondestructive, rapid quantitative process monitor.
Scanning tunneling microscope	50–100 Å	Lateral: 0.1–1 nm Depth: 0.1 nm		Atomic resolution. measurement of chemical properties.	Sample must be conduct. Slow scan. Probe wear. Cannot measure surfaces larger than a few μm².
Optical interference microscope		Lateral: 0.2–2 μm Depth: 400 μm to 2 cm		Cross-sectioning.	Lateral resolution limited by diffraction.
Near-field scanning optical microscope	0.02–4 μm Depth: 10 nm	Lateral: 1–200 nm		General imaging. Lithography.	No vacuum.

See also http://www.mwrn.com/product/product.htmt

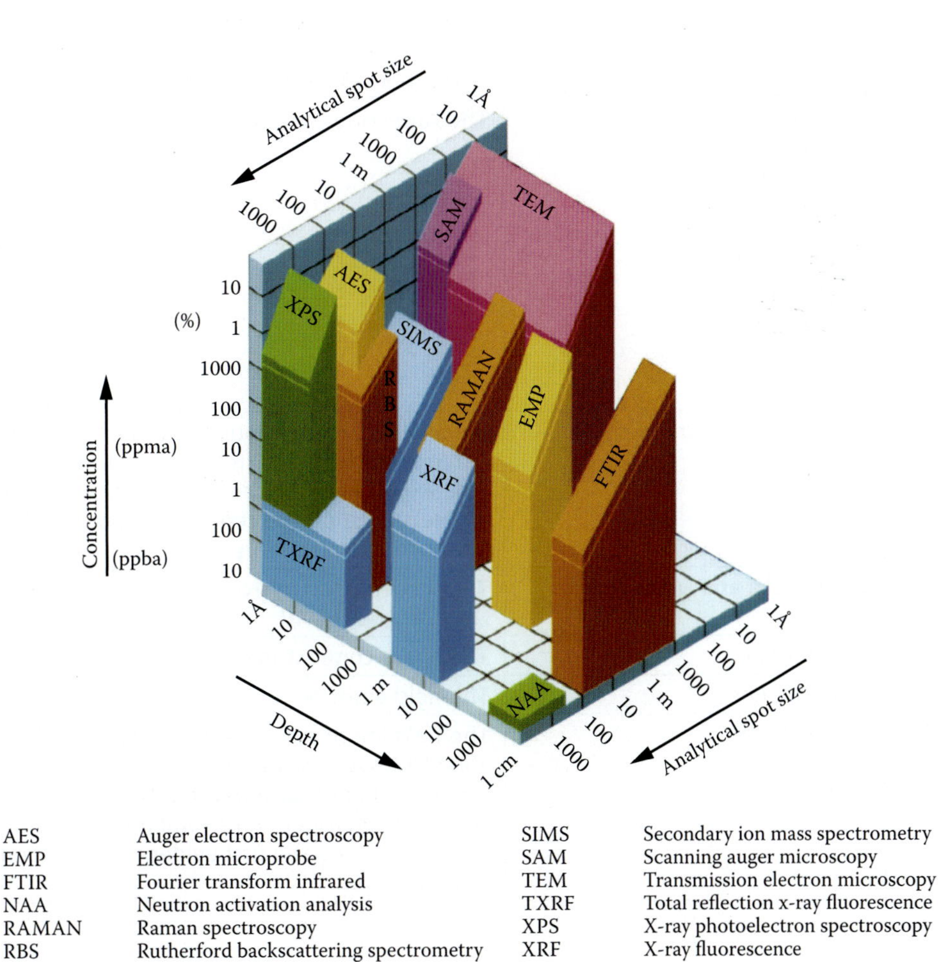

AES	Auger electron spectroscopy	SIMS	Secondary ion mass spectrometry
EMP	Electron microprobe	SAM	Scanning auger microscopy
FTIR	Fourier transform infrared	TEM	Transmission electron microscopy
NAA	Neutron activation analysis	TXRF	Total reflection x-ray fluorescence
RAMAN	Raman spectroscopy	XPS	X-ray photoelectron spectroscopy
RBS	Rutherford backscattering spectrometry	XRF	X-ray fluorescence

FIGURE 6.2 Comparison of analytical techniques: detection limits, sampling depth, and spot size. (Courtesy of Dr. Shaffner, National Institute of Standards and Technology.)

Table 6.3 Seeing Better and Better

Year	Researcher	Discovery/Instrument
1590	Hans and Zacharias Janssen	First compound microscope, magnification × 3 and × 9
1590–1609	Galileo	One of the earliest microscopists; came up with the term *microscope*
1660	Marcello Malpighi	Father of embryology and early histology-observed blood capillaries
1665	Robert Hooke	Publishes the book *Micrographia* with the study of thin slices of cork
1674	Anton van Leeuwenhook	Microscope with ×275 magnification Location of specimen Lens
1827	Robert Brown	While examining pollen grains and the spores of mosses suspended in water under a microscope, Brown observed minute particles within vacuoles in the pollen grains executing a continuous jittery motion.
1860s	Ernst Abbe (1840-1905)	The Abbe sine condition: intrinsic diffraction limit
1926	Jean Baptiste Perrin	After Albert Einstein published his theoretical explanation of Brownian motion in terms of atoms (1905), Perrin did the experimental work to test Einstein's predictions, thereby settling the century-long dispute about John Dalton's atomic theory.
1931	Ernst Ruska and Max Knoll	The first electron microscope [a transmission electron microscope (TEM)]
1935	Frits Zernike	The phase contrast microscope
1936	Erwin Müller	The field emission microscope (FEM)
1951	Erwin Müller	The field ion microscope (FIM). Müller published his historic first FIM paper describing the significant improvement in contrast and resolution brought about by imaging with positive (hydrogen) ions compared with imaging by FEM and presenting the first evidence that atomic resolution was achieved.

Table 6.3 Seeing Better and Better (*Continued*)

Year	Researcher	Discovery/Instrument
1957	Marvin Minsky	Confocal microscope
1967	Erwin Müller (the same Müller)	The first atom probe, also known as the *atom probe field ion microscope,* incorporates FIM capability to give an atomic map of the specimen surface, by means of which the user selects atoms for chemical identification by time-of-flight mass spectrometry.
1981	G. Binnig, H. Rohrer	The scanning tunneling microscope (STM)
1984	Pohl	Near-field optical microscope
1986	Binnig, Quate, Gerber	The atomic force microscope (AFM)
1988	A. Cerezo, T. Godfrey, G. Smith	Added a position-sensitive detector to the atom probe → 3D
1988	Kingo Itaya	The electrochemical scanning tunneling microscope
1991	M. Nonnenmacher, M. P. O'Boyle, and H. K. Wickramasinghe	Kelvin probe force microscope is the Kelvin mode of scanning probe microscopy; it measures contact potential difference between the probe and the sample.
2003	Zhang	Silver near-field superlens

The first lenses were made in about 1000 AD, and spectacles came about around 1200 AD. From Table 6.3, it is obvious that optical microscopes have come a long way since van Leeuwenhoek's animalcules (1674), Brown's observation of moving pollen grains (Brownian motion) (1827), and Jean Perrin's demonstration of Einstein's proof that atoms really do exist (1926 Nobel Prize for Physics) by observing the distribution of particles undergoing Brownian motion. Ernst Ruska and Max Knoll invented the electron microscope in 1931, interestingly; at first they were working with the electron as a particle concept. But as de Broglie showed, the wavelength of an electron is 100,000 times smaller than the smallest wavelength of light, small enough to allow for 2.2-Å resolution. That type of resolution was only achieved forty years after Ruska and Knoll's breakthrough, and by the 1970s, researchers were able to visualize individual proteins and DNA molecules.

Microscopy helps to describe the size and structure of samples. This information is often critical for understanding the mechanisms of forming the sample or of the observed performance of the material. A very important question to ask regarding a microscope is about its ultimate resolution: what is the smallest feature size that it can resolve? Resolution is usually different when comparing lateral resolution (resolution in the plane of the sample) and vertical resolution. One also needs to know the dynamic range of the instrument. Techniques that have a high resolution often are not able to image large objects or feature sizes that vary over several orders of magnitude (they have a small field of view).

Roughly speaking, the lateral resolution of an optical microscope is limited to Abbe's limit, about 1/2 of the wavelength of the light being diffracted through the optical lensing elements of the microscope [i.e., Abbe's formula: $R = \lambda/2NA$, with $NA = n \sin \theta_{max}$ (see Volume I, Equations 5.6 and 5.3)]. In theory, the R using visible light (shortest wavelength is 400 nm) and the best numerical aperture (NA) (1.4) is about 200 nm, but in practice it is challenging to get good optical resolution down to the 1-μm range. Based on Abbe's formula, the lateral R of a microscope can be improved by working at shorter wavelengths, λ, by using a higher aperture lens (Figure 6.3) and

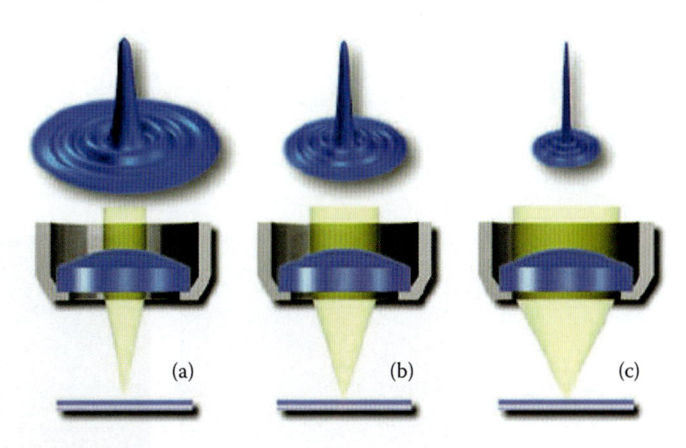

FIGURE 6.3 Increasing numerical aperture (NA) of the lens from (a) to (c). As NA increases, the acceptance angle increases and the Airy disc decreases.

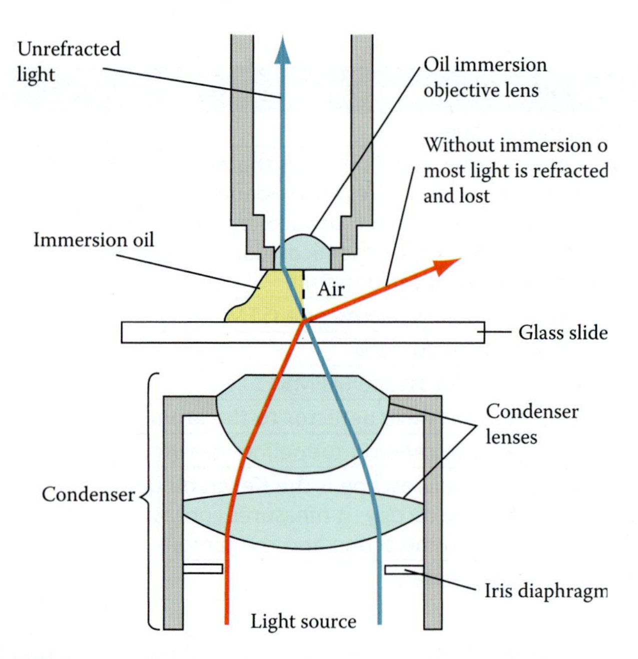

FIGURE 6.4 Oil immersion microscopy: increasing the refractive index of the medium (*n*).

in a medium with a higher index of refraction (*n*) (Figure 6.4). A larger NA means that the lens has a larger acceptance angle, so it can collect higher orders of diffracted light enhancing the resolution of the image. Working with oil between sample and objective lens, in so-called *oil immersion microscopy*, ensures a better capture of all diffracted wavelengths. The refractive index n (1.0 in air, 1.33 in water, and 1.55 in immersion oil) is the light-bending ability of a medium. The light coming from a sample under the microscope may bend in air so much that it misses the small, high-magnification lens. Immersion oil is used to keep light from bending too far.

The Abbe imaging resolution is a theoretical limit. Light passing through a circular aperture interferes with itself, creating ring-shaped diffraction patterns, known as *Airy patterns*, that blur the image (Volume I, Figures 5.13 and 5.14), and the Rayleigh criterion, which takes these Airy patterns into account, yields a minimum spatial resolution as follows:

$$R = \frac{0.61\,\lambda}{NA} \qquad \text{(Volume I, Equation 5.13)}$$

Objective lenses in a microscope are designed to correct for aberrations: achromats are the lowest level of correction, and apochromats, are the highest and most sophisticated. Eyepieces can have a reticle (graticle) etched inside to impose a scale on an image for measurement purposes. Light microscopy often is divided up into bright-field, dark-field, phase contrast, and fluorescence microscopy. In Figure 6.5 we illustrate bright-field (Figure 6.5a), dark-field (Figure 6.5b), and phase-contrast microscopy (Figure 6.5c). In bright-field microscopy, dark objects are

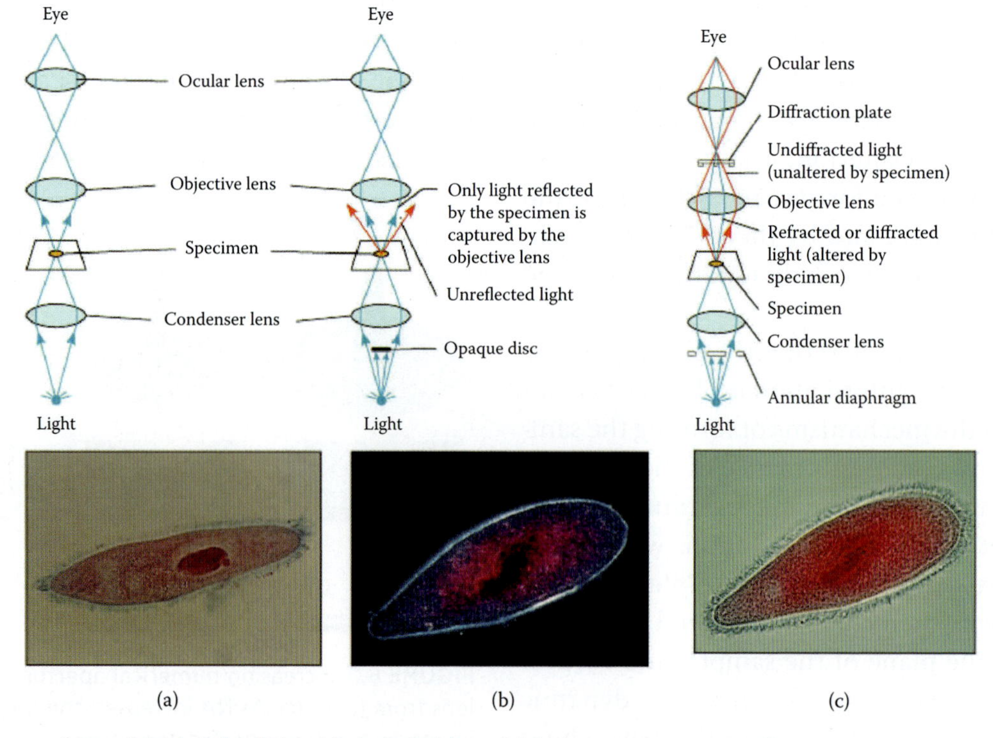

FIGURE 6.5 Comparison of (a) bright-field, (b) dark-field, and (c) phase-microscopy of a protozoa.

made visible against a bright background, and in dark-field microscopy, light objects are made visible against a dark background. In bright-field illumination, light reflected off the sample does not enter the objective lens, whereas in dark field it does.

The human eye does not see phase changes in a light beam directly. This is unfortunate, as phase changes carry a large amount of additional information about a sample. To make a phase change visible, one needs to combine the light passing through the sample with a reference so that the resulting interference reveals the phase information of the sample. In phase-contrast microscopy, one uses an annular transparent phase plate* so that the change in phase of light that passes through the sample is increased by shifting back ¼ wavelength relative to the background, while background light is shifted forward ¼ wavelength. When the waves recombine they produce a ½-wavelength shift, resulting in excellent contrast. The phase contrast microscope converts these phase changes to changes in brightness that the eye can see. A differential interference contrast (DIC) microscope, also known as a *Nomarski microscope*, is similar to a phase contrast microscope in that imaging again produces amplitude difference from phase differences in the specimen. The Nomarski microscope uses polarized light and emphasis differences in the refractive index of transparent samples. This gives relief to a sample corresponding to the variation of optical density. In Figure 6.6, we show a standard light microscope (Olympus BH2).

In Table 6.4, we compare the resolution of various types of optical microscopes. For a discussion of NSOM, see below.

Vibration is a fundamental problem in all microscopy experiments. Buildings vibrate because of motors, air conditioners, passing vehicles, etc. Vibrations blur images since the excitation, detection, and sample regions of the microscopes may move with respect to each other during the course of the experiment. Most buildings have vibrational resonances in the 1–100 Hz range and feature amplitudes in the micrometer range. The effect of vibrations is mitigated by constructing the microscope on stages connected to each other either very rigidly or very loosely. The rigid connections permit the transmission of only high-frequency vibrations, while the loose connections permit only low-frequency vibrations. Compare this with the implementation of high-pass and low-pass filters in optics or electronics. Often, this filtering is achieved by building the microscope to be as small and rigid as possible and then mounting the entire system on pneumatically supported legs that couple only loosely to the building. Sometimes an additional stage using soft springs is included. While these procedures reduce mechanical vibration, acoustical vibrations may still couple to the microscope and circumvent these vibration isolation stages. Including a surrounding of acoustic-absorbing material is often necessary for the best microscopy work.

* Rings in a phase plate can include attenuating layers (absorption but no phase shift), phase-shifting layers (phase shift but no absorption), or any combination of the two.

Light Microscopy Summary

Optical microscopes have the great advantages of being fast, in situ,† and nondestructive. Test parts usually do not have to be modified [e.g., coated with a conductive material, as in scanning electron microscopy (SEM)] from their original form/condition. Quantitation in light microscopy may be achieved with a reticle etched into one of the eyepieces. Images can be registered on film or, better, recorded electronically to a charge-coupled device (CCD) camera. Optical microscopes tend to be repeatable for features as small as 0.25 μm. Optical metrology hardware has a resolution that is diffraction limited, and it has a poor ability to produce images with clear intensity changes to accurately detect edges. Often locating the edge of a part is quite difficult, as location varies with lighting condition, noise, and assumptions made in the edge position algorithm.[2] Other significant errors of optical techniques typically stem from interference, resonance, shadowing, secondary reflections, and lens distortions. Perhaps the most important limitation of optical

† In situ experiment: an experiment performed on a sample while it is still located in its native environment. With an in situ measurement, there is less risk of altering the sample's true properties. Ex situ experiment: an experiment performed on a sample after it has been removed from the location wherein it was formed. A wider range of experimental techniques are available.

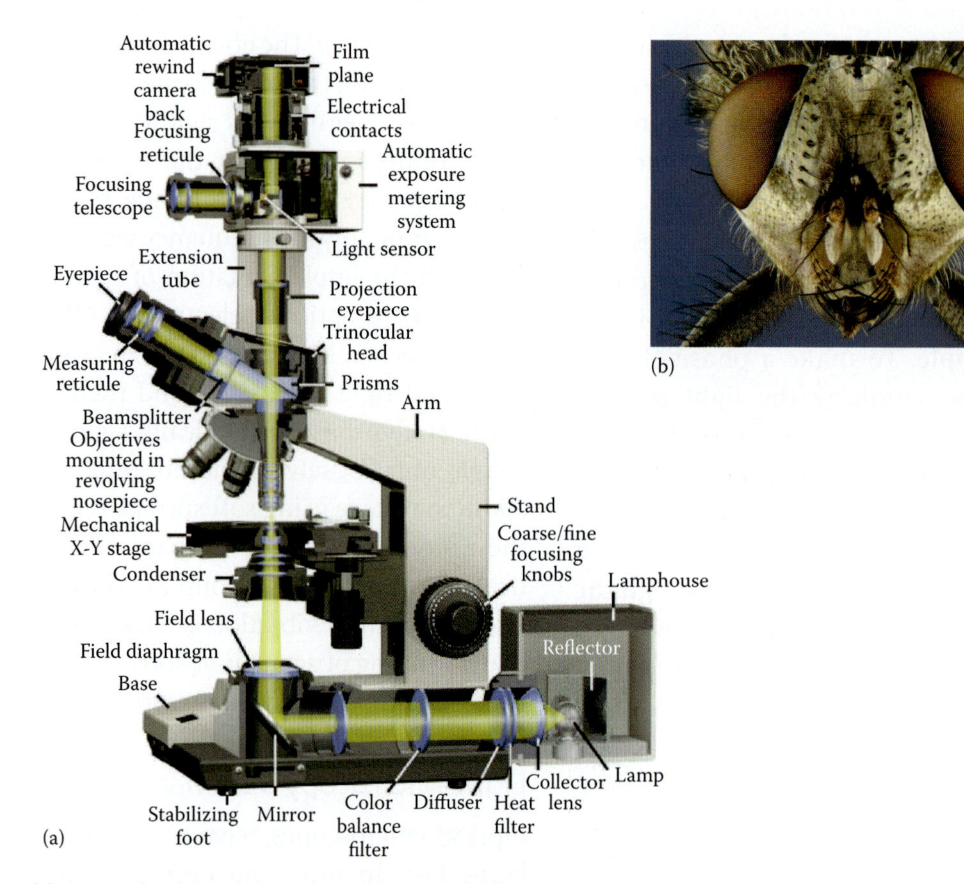

FIGURE 6.6 (a) A standard light microscope (Olympus BH2 research microscope). (b) First place winner, Nikon's Small World 2005 competition: Charles B. Krebs, muscoid fly (house fly) (× 6.25), reflected light.

microscopes for MEMS inspection is the inability to acquire true three-dimensional data. Some optical microscopes are integrated with software that uses image-processing techniques to determine the *z*-height at which the image is taking place. The current state-of-the-art software uses a projected grid (a Ronchi grid) to determine the height at which the microscope is focused in one region of the image. If the region selected has multiple focus points (i.e., the region selected is not all on one plane), the algorithm assigns the average value for the *z*-height. Further edge detection algorithms are run to extract *x* and *y* data from the microscope image. This technique, in theory, produces three-dimensional data from an image; however, the algorithms used after finding the *z*-height in one location of the image assume that all of the data are on the same plane. Thus, the data acquired from vision systems such as these can be characterized as 2.5D data sets.[2]

Table 6.4 Resolution of Various Types of Optical Microscopes

Objective (Magnification/Numerical Aperture)	Illumination	Resolution (μm)
×10/0.25	White light	1.34
×20/0.40	White light	0.84
×100/0.95	White light	0.35
×100/0.95	488 nm, confocal	0.22
×100/1.35 (with glycerin immersion)	248 nm (UV)	<0.16
Optical fiber (scanning near-field optical microscopy; values down to 10 nm have been reported)	488 nm	0.05

A good Web site for an introduction to optical microscopes can be found at http://www.olympusmicro.com/primer/opticalmicroscopy.html. Also visit http://micro.magnet.fsu.edu/index.html and go through the tutorials—they are excellent!

Fluorescence Microscopy

In BIOMEMS, fluorescence microscopy is a most important tool by which to study cells, DNA arrays, tissue, etc. It is used to study properties of samples

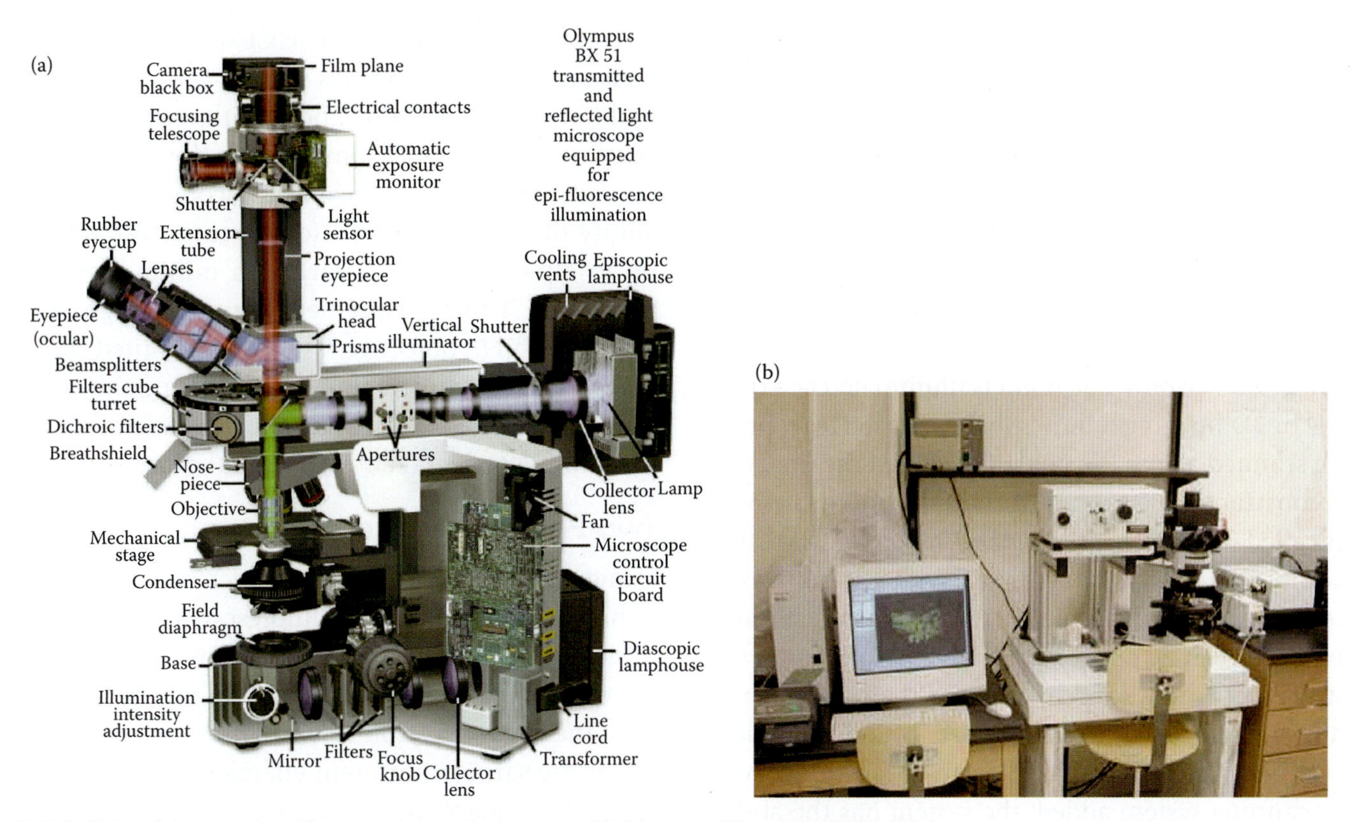

FIGURE 6.7 (a) Typical epifluorescence microscope. (b) Olympus Fluoview 300 LSM.

using the phenomena of fluorescence and phosphorescence (see Volume I, Chapter 7). Samples are labeled with dyes that emit visible light when bombarded with shorter ultraviolet (UV) rays (see Volume I, Figure 7.93). A fluorescence microscope is an optical microscope with an ultraviolet radiation source and a filter that protects the viewer's eye. The filter that the microscope uses is a special dichroic mirror (or more precisely, a dichromatic mirror). This mirror reflects light shorter than a certain wavelength and passes light longer than that wavelength. Thus the observer sees only the emitted red light from the fluorescent dye, rather than seeing scattered purple light. Most fluorescence microscopes are epifluorescence, i.e., excitation and observation of the fluorescence is from above (epi). In reflected light fluorescence, the intensity of the image, I, varies directly as the fourth power of the numerical aperture (NA) of the objective in use, as well as inversely as the square of the magnification, M:

$$\mathrm{I} \propto \frac{(\mathrm{NA})^4}{(\mathrm{M})^2} \tag{6.1}$$

e.g., a ×40 objective with an NA of 1 will yield images more than a five times brighter than a ×40 objective with a NA of 0.65. Electronic sensors (e.g., a CCD) are typically used to measure the intensity, as they give a much higher sensitivity than film. To avoid fading of the fluorophore labels, a phenomenon known as *photobleaching*, the UV light intensity and duration need to be carefully chosen. In Figure 6.7 we show a typical fluorescence microscope.

Confocal Microscopy

Normally, in fluorescence microscopy, the sample is completely illuminated by the excitation light, so that the whole sample is fluorescing at the same time. The highest intensity of the excitation light is at the focal point of the lens, but nonetheless, the other parts of the sample do get some of this light and they do fluoresce as well.* This contributes to a background haze in the resulting fluorescence image. To get rid of this haze and make for a sharper image, one resorts to confocal microscopy. In such a microscope, patented in 1957 by Marvin Minsky (see inset in Figure 6.9), one places a small pinhole in a conjugate optical plane in front of the detector to eliminate out-

* From introductory optics, we remember the formula $1/s + 1/s' = 1/f$ for locating the image formed by a lens: points do not need to be at the focal point of the lens in order for the lens to form an image.

of-focus light (Figure 6.9). Because the focal point of the objective lens of the microscope forms an image where the pinhole is, these two points are known as *conjugate points*. The pinhole is conjugate to the focal point of the lens; thus, it is a confocal pinhole (or alternatively, the sample plane and the pinhole/screen are conjugate planes). Only light from the plane of focus will pass through the pinhole, so that the image quality is much better than that of wide-field images. Since only one point is illuminated at a time, 2D or 3D imaging requires that a focused laser beam be scanned across the field of view and the image be reconstructed electronically. This type of microscopy is called *confocal laser scanning microscopy* (CLSM), as it combines a confocal microscope with a scanning system in order to image an entire specimen. Although scanning can be performed in several different ways, it is most often done by moving the beam, which alleviates focus problems caused by objective lens scanning and is faster than specimen scanning. With a scanning system added, the system has the ability to scan multiple times on different imaging planes, resulting in a three-dimensional data set.

In confocal microscopy, the background rejection is thus substantially improved and the method allows for optical sectioning; moreover, the lateral and axial extent of the point spread function (see Volume I, Figure 5.12) is reduced by about 30% compared with that in a wide-field microscope. As a consequence, focal plane lateral resolution is improved by a factor of 1.5 over ordinary microscopy for incoherent imaging (Rayleigh limit or $R = 0.61\ \lambda/NA$; Volume I, Equation 5.13), as follows:

$$R_{xy} = 0.4\frac{\lambda_{emission}}{NA} \tag{6.2}$$

The vertical depth resolution of a confocal microscope enables optical sectioning and is poorer than the lateral resolution:

$$R_z = 1.4\frac{\lambda_{emission} n}{NA^2} \tag{6.3}$$

For typical objectives ($NA \leq 1.4$) and visible light ($\lambda \geq 400$ nm), the best lateral resolution is $R_{xy} = 0.13\ \mu m$ and the vertical resolution is $R_z = 0.43\ \mu m$, so that the depth discrimination is not comparable to the focal plane resolution.

Work has been done to use CLSM to perform 3D analysis of microstructures such as micro end mills and hot embossing tools, both with overall dimensions on the order of 1 mm^3. One of the most important advantages of confocal microscopy found is the ability of this microscope to measure steep slopes, up to almost 90° on a part with minimal surface roughness. This measurement requires a high-resolution, high-NA objective, which has a limited lateral measuring field unsuitable for measuring the entire sample. Because of this, a stitching procedure is used to combine scans taken with several objectives.

Two-Photon Microscopy

Using two-photon confocal microscopy further enhances the resolution of fluorescence microscopy. Multiphoton fluorescence microscopy is a relatively novel imaging technique that relies on the quasisimultaneous absorption of two or more photons (of either the same or different energies) by a molecule (see also quantum lithography in Volume II, Chapter 2). Two-photon microscopy was pioneered by Winfried Denk in the laboratory of Watt Webb at Cornell University.[4] He combined the idea of two-photon absorption with the use of a laser scanner.

The use of two-photon absorption has several important advantages. First, the probability of two-photon absorption increases quadratically with illumination intensity, essentially limiting excitation to a tiny focal volume (see Figure 6.8). This limits automatically out-of-focus background excitation and provides innate confocality, since only fluorophores in the plane of the focused beam are excited. One obvious benefit of this is to obviate the need for the confocal aperture in front of the detector used in one-photon confocal microscopy, maximizing the amount of light detected. Second, the use of longer wavelength excitation light slows the rate of photobleaching of the fluorescent labels on the sample, allowing the use of higher magnification objectives. Third, longer wavelength light is scattered less as it passes through a sample such as biological tissue, resulting in a greater penetration depth into the sample. These factors increase both the resolution and duration of measurement available with this technique. Fourth, the fact that two-photon excitation microscopy does not require a pinhole to obtain

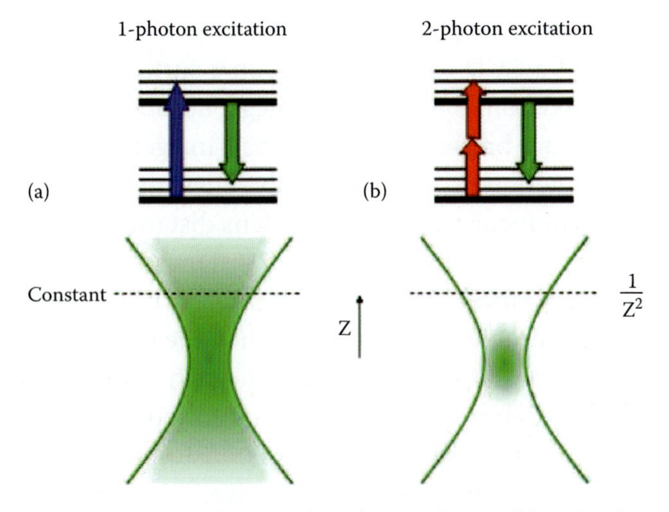

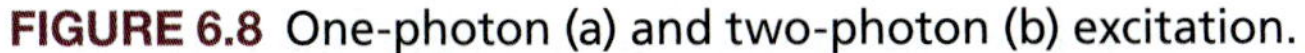

FIGURE 6.8 One-photon (a) and two-photon (b) excitation.

three-dimensional resolution allows more flexibility in detection geometry. Two-photon microscopy is thus a novel form of laser-based scanning microscopy that allows three-dimensional imaging and avoids many of the problems inherent in confocal microscopy.

In Figure 6.9 we compare confocal microscopy with laser scanning two-photon microscopy. Resolution limits of quantum lithography were analyzed in Volume II, Chapter 2. The results derived there directly apply to two-photon microscopy discussed here.

Interferometry

Another powerful method for MEMS part inspection is scanning white light interferometry (SWLI), also called *noncontact profilometry*. Interferometry, a method to determine a point in space on the basis of phase differences between a reflected probe beam and an isolated reference beam, has long been used by various industries as one of the most accurate ways to profile a surface. In Figure 6.10 and the example below, we illustrate how a typical interferometer can be used to measure distances in a number of wavelengths or to measure wavelengths. A Michelson interferometer, the particular two-path interferometer shown here, is constructed using a beam-splitter, i.e., a partially silvered mirror that transmits and reflects half of the incident intensity. This is useful because the reflected and transmitted beams are coherent, i.e., they have a definite common phase, even when the incident light has a time-varying random phase. This interferometer allows a

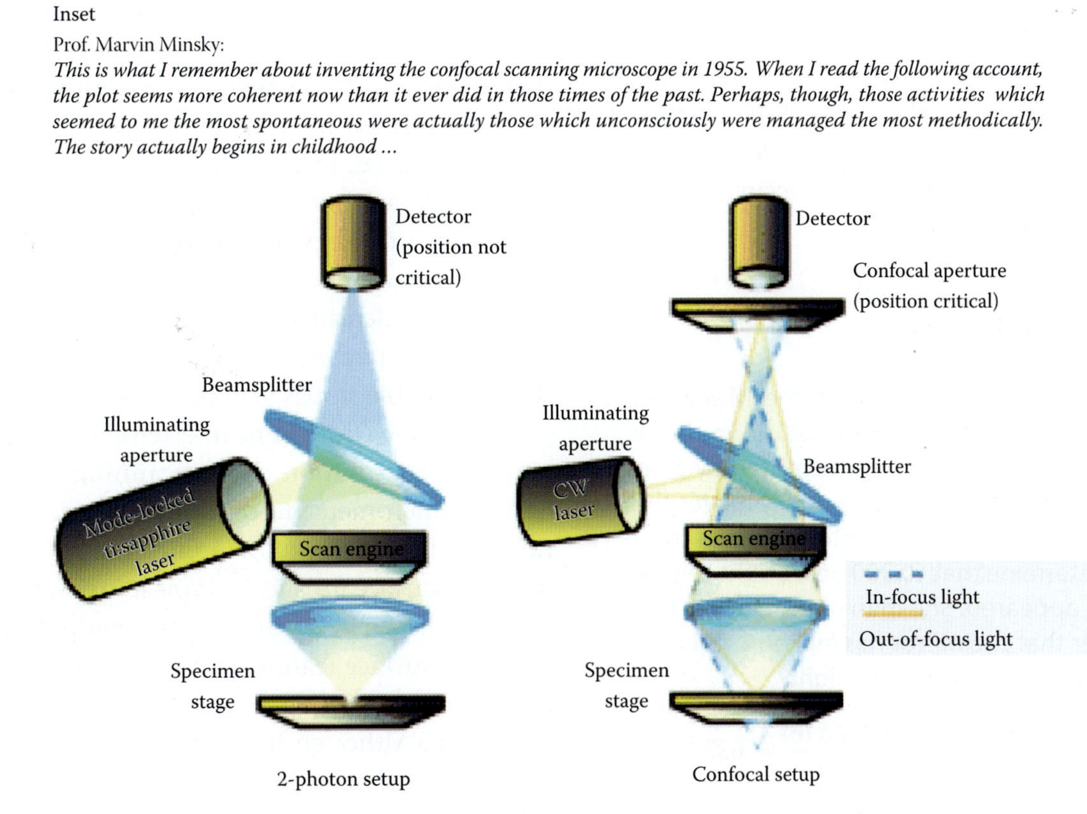

FIGURE 6.9 Comparison of a confocal microscope with a two-photon microscope. Inset, Marvin Minsky remembering the confocal microscope.

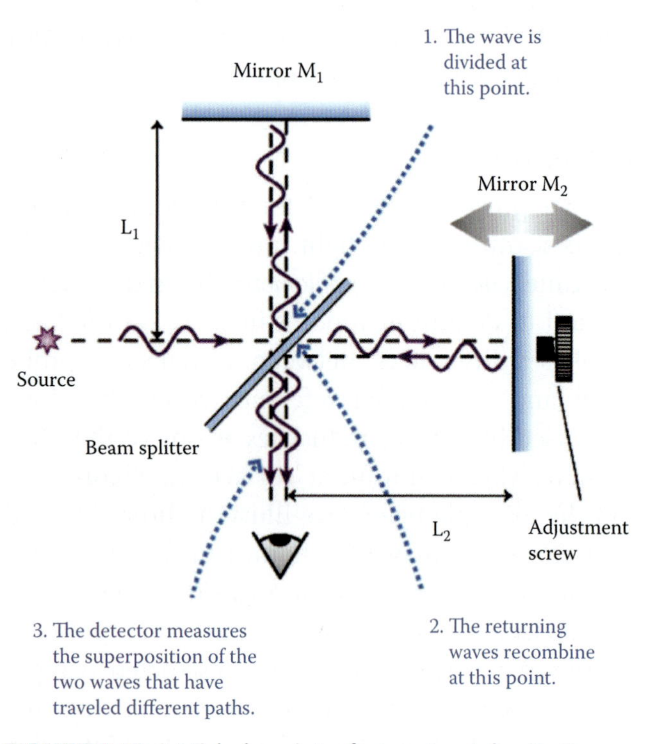

FIGURE 6.10 A Michelson interferometer. John G. Cramer, Professor of Physics, University of Washington.

beam of light to be sent along two paths, reflected, and recombined so that it can interfere. Reflection from a surface produces a "phase-flip" of 180° in the reflected light. However, when light is reflected at an angle other than 180°, the phase change in the reflected light is the same as the reflection angle. Thus, light reflected at 90° has a 90° change in phase. If the interferometer is set for an interference maximum, a new maximum will occur each time one of the arms is moved by $\lambda/2$.

Example 6.1 (Dr. J. G. Cramer): Measuring Wavelengths

Question: An experimenter uses the Michelson interferometer shown in Figure 6.10 to measure one of the wavelengths of light emitted by electrically excited neon atoms. She slowly moves the mirror M2 and uses a photo-detector and a computer to determine that 10,000 new bright central spots have appeared. She then determines with a micrometer that the mirror has moved 3.164 mm. What is the wavelength of the light?

Answer: $\lambda = \frac{2\Delta L_2}{\Delta m} = \frac{2(0.003164 \text{ m})}{(10{,}000)} = 632.8 \text{ nm}$

If the wavelength is known, then of course a distance can be determined.

In SWLI, recombined light is directed to image plane of a CCD camera. Points on a surface that are separated from the lens by an integer number of wavelengths are bright and those a half-integer apart are dark. Interference is strong only when the reflected light is in focus i.e., the sample-lens distance is at the focal position. The sample-lens distance is scanned around the focal length and ach pixel of the CCD will show a strong intensity change when the lens reaches the focal position corresponding to each point on the surface. High-resolution position information comes from a linear variable differential transformer (LVDT) connected to the lens scanning drive.

SWLI has the ability to quickly measure step heights changes and deflections (see also mechanical property measurements, below): it is possible to resolve step height changes greater than $\lambda/4$. Using white light increases the dynamic range of the measurement and allows for higher resolution by comparing data from multiple wavelengths. In addition, when integrated with an image processing system, SWLI can provide lateral dimensions. By using image processing, line widths, diameters, and feature positions can be measured as well. However, the lateral resolution of commercially available systems is poor, except when equipped with high-power objectives that severely limit the field of view. Moreover, these tools are limited in their ability to measure sloped surfaces. Indeed, interference can occur only if light is reflected back into the objective lens. If the surface angle is inclined beyond the acceptance angle of the lens, no interference is observed. The largest slope that can be identified is typically around 30° with a × 100 objective. As the objective power decreases, the identifiable slopes also decrease.

In practice, white light interferometers have subnanometer resolution in the scanning direction, at best submicrometer resolution in the lateral directions, and can be used on a multitude of parts with different surface finishes. A typical instrument, from Wyko, is shown in Figure 6.11. The equipment shown measures surface features with 0.45–11.8-μm lateral resolution and an approximately 0.10-nm vertical resolution. Although initially developed mostly for surface characterization, such as surface roughness, SWLI is currently used to make a wide variety of other measurements on MEMS parts.

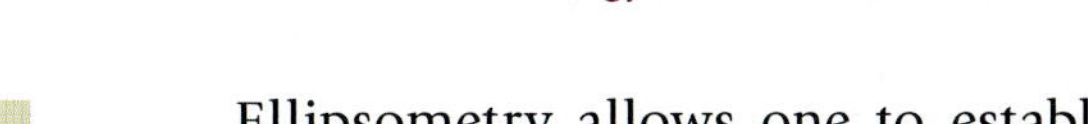

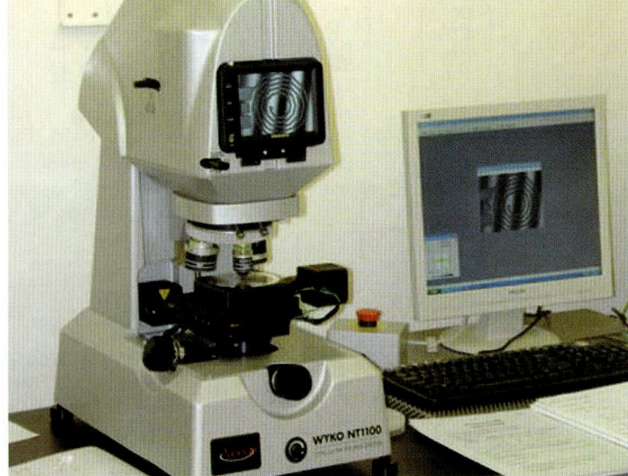

FIGURE 6.11 Optical Wyko white-light profilometers with dynamic microelectromechanical system capability. This capability enables both in-plane and out-of-plane resonant frequency measurements.

Ellipsometry

Introduction

Fresnel derived the equations that determine the reflection/transmission coefficients of light at interfaces in the early nineteenth century, and Paul Drude invented ellipsometry, based on these equations, in 1887. Ellipsometry is a noncontact optical technique used for analysis and metrology and became very important in the 1960s, as it was demonstrated to have the sensitivity required to measure nanometer-scale thin films used in microelectronics.

Ellipsometry allows one to establish very accurately and with high reproducibility the complex dielectric function of a given material. In Volume I, Chapter 3, we saw how the dielectric function is dependent on both frequency (ω) and wave-vector **k**: ε (ω, **k**). In ellipsometry, one determines the change in polarization state of light reflected from a sample. This polarization change is represented as an amplitude ratio, Ψ, and a phase difference, Δ. Since ellipsometry is measuring the ratio (or difference) of two values (rather than the absolute value of either), it is very robust, accurate, and reproducible. For instance, it is relatively insensitive to scatter and fluctuations and requires no standard sample or reference beam. The measured response depends on optical properties and thickness of individual materials. Transparent films from subnanometers up to several micrometers can be measured, and the surface upon which the film is measured can be a semiconductor, dielectric or metal. The film can be transparent or absorbing on a transparent or opaque substrate as long as there is enough contrast between substrate and film. It is a model-based technique that, besides being used to determine film thickness and optical constants, can be employed to characterize composition, crystallinity, roughness, and doping level—in short, all material properties that can change the optical characteristics. In Figure 6.12 we illustrate a typical ellipsometer setup.

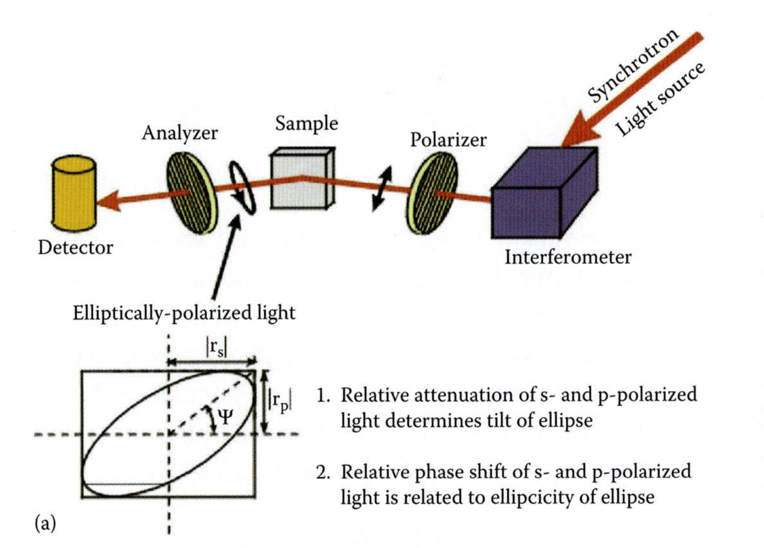

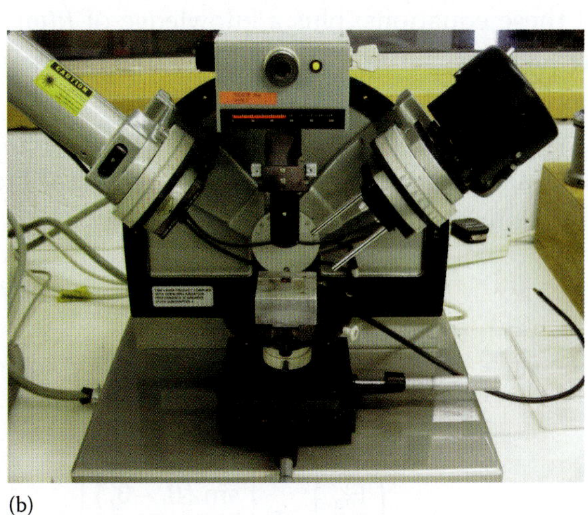

FIGURE 6.12 Schematic (a) and photograph (b) of an ellipsometer. The ellipsometer measures polarization changes in terms of two parameters: Δ (sometimes cosΔ) and Ψ.

From the schematic in Figure 6.12a, we see that a first polarizer linearly polarizes incident light with finite field components in the directions parallel (*p*-component) and perpendicular (*s*-component) to the plane of incidence of the light. The plane of incidence of light is the plane containing the surface that is normal to the sample and the incident direction of light. The wavelength of the measuring polarized light can range from the ultraviolet to the far-infrared. (In Figure 6.12a, a synchrotron light source is used in order to obtain a high-enough intensity of far-infrared light.) Upon reflection, the *s*- and *p*-components of the light undergo different attenuations and phase shifts, resulting in elliptically polarized reflected light, giving the technique its name. The ellipse of polarization of the reflected light is measured with a second polarizer (called the *analyzer*), as illustrated in Figure 6.12a.

Theory

When light interacts with a material, the Maxwell equations remain satisfied by imposing the proper boundary conditions at the interface (see Volume I, Table 5.2). At an interface a portion of light reflects, and the remainder transmits at the refracted angle (see Volume I, Figure 5.36). It is relatively easy to measure reflectance, *R*, of a material, and from this the more difficult-to-access complex refractive index, n_c, can be derived. The values for the refractive index, *n*, and extinction coefficient, κ, which make up the complex refractive index n_c, are not independent but are linked by the Kramers-Kronig relations. Either of these equations, plus a knowledge of $R(\omega)$ at all frequencies, permits one to disentangle the separate values of $n(\omega)$ and $\kappa(\omega)$.

With reference to Volume I, Figure 5.38, we obtained the four amplitude coefficients (the Fresnel equations) as follows:

$$r(\omega)_{\parallel} = \left(\frac{E^r_m}{E^i_m}\right)_{\parallel} = \left[\frac{\tan(\theta_i - \theta_t)}{\tan(\theta_i + \theta_t)}\right]_{\parallel}$$

(Volume I, Equation 5.140)

$$r(\omega)_{\perp} = \left(\frac{E^r_m}{E^i_m}\right)_{\perp} = \left[\frac{\sin(\theta_i - \theta_t)}{\sin(\theta_i + \theta_t)}\right]_{\perp}$$

(Volume I, Equation 5.141)

$$t(\omega)_{\parallel} = \left(\frac{E^t_m}{E^i_m}\right)_{\parallel} = \left[\frac{2\sin\theta_t\cos\theta_i}{\sin(\theta_i + \theta_t)\cos(\theta_i - \theta_t)}\right]_{\parallel}$$

(Volume I, Equation 5.142)

$$t(\omega)_{\perp} = \left(\frac{E^t_m}{E^i_m}\right)_{\perp} = \left[\frac{2\sin\theta_t\cos\theta_i}{\sin(\theta_i + \theta_t)}\right]_{\perp}$$

(Volume I, Equation 5.143)

where $r(\omega)$ is the dimensionless reflectivity coefficient defined as the ratio of the reflected electric field to the incident electric field, and $t(\omega)$ is the dimensionless transmission coefficient.

Reflection and transmission, we saw, are related by the following equation.

$$t(\omega) = 1 + r(\omega)$$

(Volume I, Equation 5.144)

A complex $r(\omega)$ [or $t(\omega)$] means that a phase shift is introduced at the interface upon reflection (or transmission). The link between reflectivity at normal incidence and the complex refractive index was calculated in Volume I, Chapter 5 as follows:

$$r(\omega) = \frac{E^r_m}{E^i_m} = \left(\frac{n + i\kappa - 1}{n + i\kappa + 1}\right) = \frac{n_c - 1}{n_c + 1}$$

(Volume I, Equation 5.147)

The quantity that is actually measured in experiments is not reflectivity, *r*, but reflectance, *R*, defined as the ratio of the reflected intensity to the incident intensity, as follows:

$$R(\omega) = \frac{|E^r_m|^2}{|E^i_m|^2} = \left|\frac{n_c - 1}{n_c + 1}\right|^2 = \frac{(n-1)^2 + \kappa^2}{(n+1)^2 + \kappa^2}$$

(Volume I, Equation 5.148)

The phase, $\theta(\omega)$, of the reflected wave, it can be shown, may be calculated from the measured reflectance, $R(\omega)$, if known over all frequencies (Kramers-Kronig). With both $R(\omega)$ and $\theta(\omega)$ known, we can use Volume I, Equation 5.147 to calculate $n(\omega)$ and $\kappa(\omega)$; as we saw earlier, $n(\omega)$ and $\kappa(\omega)$ are related to the complex dielectric function, ε_r, as follows:

$$n_c(\omega) = n(\omega) + i\kappa(\omega) = \sqrt{\varepsilon_r'(\omega) + i\varepsilon_r''(\omega)}$$

(Volume I, Equation 5.115)

Plugging the values for $n(\omega)$ and $\kappa(\omega)$ into Equation 5.115, we thus obtain the complex dielectric function.

We now need to explain how the Fresnel equations are linked to the experimental parameters obtained from ellipsometry. The reflectivity coefficients $r(\omega)_{||}$ and $r(\omega)_{\perp}$ in what follows are symbolized more simply as r_p (p for parallel) and r_s (s for senkrecht: German for normal or perpendicular). Using Snell's law (Volume I, Equation 5.19), the expressions for r_s and r_p (Volume I, Equations 5.140 and 5.141) can be rewritten as follows:

$$r_s = \left(\frac{E_m^r}{E_m^i}\right)_{\perp} = \left[\frac{n_t \cos\theta_i - n_t \cos\theta_t}{n_i \cos\theta_t + n_t \cos\theta_i}\right]_{\perp} \quad (6.4)$$

$$r_p = \left(\frac{E_m^r}{E_m^i}\right)_{\|} = \left[\frac{n_t \cos\theta_i - n_i \cos\theta_t}{n_i \cos\theta_t + n_t \cos\theta_i}\right]_{\|} \quad (6.5)$$

In these expressions we now see the explicit appearance of the refractive index of the two media (n_i and n_t; see Volume I, Figure 5.38). Thin-film and multilayer structures involve multiple interfaces, with Fresnel reflection and transmission coefficients applicable at each. The superposition of multiple waves introduces interference that depends on the relative phase of each light wave. To track the relative phase of each light component so as to correctly determine the phase of the overall reflected or transmitted beam, a film phase thickness is introduced as follows (see Figure 6.13).

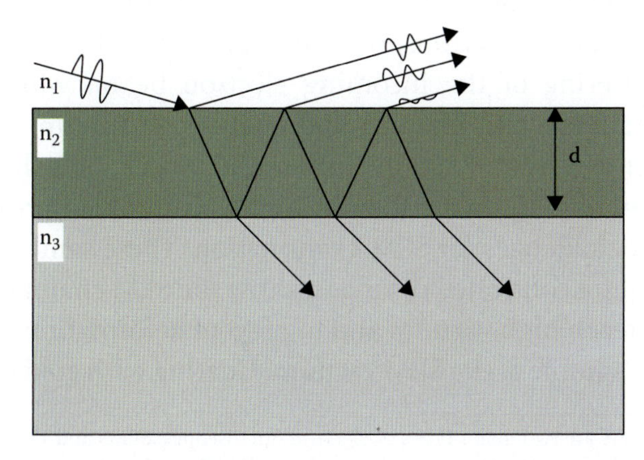

FIGURE 6.13 Light reflects and refracts at each interface, which leads to multiple beams in a thin film. Interference between beams depends on relative phase and amplitude of the electric fields. Fresnel: reflection and transmission can be used to calculate the response from each contributing beam.

$$\beta = 2\pi\left(\frac{d}{\lambda}\right) n_2 \cos\theta_2 \quad (6.6)$$

The reflectivity coefficients r_p and r_s for this case are given as the following.

$$r_p = \frac{r_{p(1,2)} + r_{p(2,3)} \exp(-i2\beta)}{1 + r_{p(1,2)} r_{p(2,3)} \exp(-i2\beta)} \quad (6.7)$$

$$r_s = \frac{r_{s(1,2)} + r_{s(2,3)} \exp(-i2\beta)}{1 + r_{s(1,2)} r_{s(2,3)} \exp(-i2\beta)} \quad (6.8)$$

Ellipsometry determines how p- and s-components of the light change upon reflection or transmission in relation to each other. A known polarization is reflected or transmitted from the sample, and the output polarization is measured. The ellipsometer measures this polarization change of the light reflected from the sample in terms of two parameters: Δ (sometimes $\cos\Delta$) and Ψ (see Figure 6.12a). These values are related to the ratio of the Fresnel amplitude reflection coefficients for p- and s-polarized light as follows:

$$\frac{r(\omega)_{\|}}{r(\omega)_{\perp}} = \frac{r_p}{r_s} = \tan\Psi \exp(i\Delta) \quad (6.9)$$

This ratio is complex, with $\tan\Psi$ measuring the relative attenuation of s- and p-polarized light (which determines the tilt of the ellipse in Figure 6.12a) and Δ measuring the phase difference between p- and s-polarized reflected light (related to the ellipticity). In the case of an ideal bulk material, the result in Equation 6.9 can be directly converted to give the complex dielectric constant and complex refractive index as:

$$\varepsilon_r = n_c^{\,2} = \sin^2(\phi)\left[1 + \tan^2(\phi)\left(\frac{1 - \frac{r_p}{r_s}}{1 + \frac{r_p}{r_s}}\right)\right] \quad (6.10)$$

This equation is an oversimplification implying that there are no surface layers of any type. A realistic bulk material comes with a surface oxide and at least some roughness, and Equation 6.10 would include these as part of the bulk optical constants. As a consequence, ellipsometry is used as an indirect method/model based technique, i.e., in general the measured Ψ and Δ cannot be converted directly into the optical constants of the sample. From Ψ, Δ, and an optical model, the desired information can be

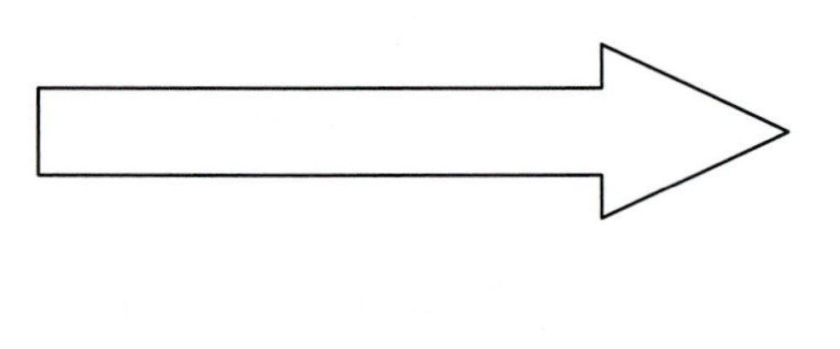

FIGURE 6.14 From Ψ, Δ, and an optical model the desired information can be extracted.

extracted, as illustrated in Figure 6.14. After a sample is measured, a model is constructed and is used to calculate the predicted response from the Fresnel's equations, which describe each material layer in the sample with its own thickness and optical constants. If these values are not all known, an estimate is made for a preliminary calculation. The calculated values are then compared with the experimental data, and any unknown material property can be varied to improve the match between experiment and calculation. Finding the best match between the model and the experiment is typically achieved through regression. An estimator, such as the mean squared error (MSE), is used to quantify the difference between curves. The unknown parameters are allowed to vary until the minimum MSE is reached.

Ellipsometers may be single-wavelength or spectroscopic and may be have single or multiple angles of incidence. They may also operate with a rotating element or as nulling ellipsometer, and they may be single-point or imaging. Spectroscopic ellipsometer allows the accurate characterization of a range of properties, including the layer thickness, optical constants, composition, crystallinity, anisotropy, and uniformity. Thickness determinations ranging from a few angstroms to tens of micrometers are possible for single layers and complex multilayer stacks.

Ellipsometry measures the ratio of two values, so it is highly accurate and reproducible, it does not need a reference sample, and it is not very susceptible to light source fluctuation. Since it measures phase, it is highly sensitive to the presence of thin films (down to submonolayer coverage). The sampling region ranges from 30 μm to 10 mm, and the depth of sampling, which depends on the absorption of the light in the sample, is typically 100 Å or more. For further reading we refer to the *Handbook of Ellipsometry*.[5]

Scanning Electron Microscopy

In ICs, MEMS, and NEMS, scanning electron microscopes (SEMs) like optical microscopes, are fabrication line workhorses. Scanning electron microscopy is one of the primary tools used for analysis of MEMS and NEMS devices, one that was used long before the development of scanning probe techniques. Ernst Ruska, the inventor of the electron microscope (1931),* received his Nobel in 1986, at age of 80, and shared the prize with Gerd Binnig and Heinrich Rohrer for the scanning tunneling microscope (STM) (see below), invented in 1981.

In an SEM, the surface or microvolume to be examined is irradiated with a finely focused electron beam; the beam may be swept in a raster† across the surface of the specimen, which is contained in a vacuum, to form images or can be static to obtain an analysis at just one position. The electrons are produced in an electron gun that accelerates them at the operating voltage. The operating voltage ranges from 0 to between 30 and 60 kV, depending upon the type of SEM.

Scanning electron microscopy requires a vacuum system so that the electrons can make the trip from the source in the electron gun, e.g., a thermal field emitter to the sample and to the detector (an electron multiplier). Insulating samples have traditionally required pretreatment in the form of coating with a conductive film, typically gold or carbon, before SEM examination. Nonconductive samples are subject to a buildup of electrons on their surface. This buildup of electrons, or charging, eventually causes scattering of the incoming electron beam, which interferes with imaging and analysis. Furthermore, samples that contain substantial water or other materials that volatilize in high vacuum also present challenges for SEM examination. These samples require controlled drying to allow the SEM chamber to reach high vacuum and to prevent deformation of the sample at the SEM vacuum. Coating with gold or

* There are two major types of electron microscopes: SEMs and TEMs. In the former, electrons scattered from a thick analyzed sample; in the latter, electrons transmitted through a very thin specimen are utilized. The first electron microscopes that were developed were actually TEMs (see "Transmission Electron Microscopy").

† The beam is scanned over the specimen in a series of lines and frames called a *raster*, just like the (much weaker) electron beam in an ordinary television.

carbon and drying are not needed if an environmental scanning electron microscope (ESEM) is used (see below). The focusing of electrons in an SEM is achieved magnetically instead of with a polished and shaped lens. Electron beams (probes) of sizes down to ~6 nm are attainable with conventional thermionic emission sources, although smaller probes, ~2 nm, can be achieved using field emission sources (for a listing of different electron sources, see Volume II, Figure 2.17). The SEM also does not have lenses to magnify the image as in the optical microscope; instead, magnification results from the ratio of the area scanned on the specimen to the area of the television screen [cathode ray tube (CRT)] (× 25 to × 250,000 or more). Increasing the magnification in an SEM is therefore achieved quite simply by scanning the electron beam over a smaller area of the specimen. The raster movement is accomplished by means of small coils of wire carrying the controlling current (the scan coils), and magnification is therefore controlled by the current supplied to the *x*, *y* scanning coils and not by objective lens power. A schematic drawing of an electron microscope is shown, together with a photograph of a commercial machine in Figure 6.15.

We remember that the resolution, *R*, of a light microscope can be calculated as $0.61\lambda/\text{NA}$ (Raleigh criterion; Volume I, Equation 5.13). In this expression, $\text{NA} = n\sin\theta_{max} = \frac{D}{2f} = \frac{1}{2F\#}$, with *F#* being the effective *F*-number, which equals *f*/*D*, or focal length over lens diameter. The assumption here is a large *F#*, which implies $\theta \lll 1$ and $\text{NA} = \sin\theta \sim \tan\theta \sim \theta$; this is called the *paraxial approximation* (see Volume I, Chapter 5). A good light microscope objective has an *f*/*D* (effective F-number) of about 2, so with λ = ~500 nm, a typical optical microscope *R* is about 1 μm. This resolution can be much improved by using a scanning electron microscope. The matter wave of an electron (accelerated at 1,000,000 V) is 0.001 nm (see de Broglie wavelength; Volume I, Equation 3.96) versus 450 nm for a light microscope; in other words, electron waves are 100,000 times shorter than the waves of visible light. So electrons have tremendous power to resolve minute structures because the resolving power is a function of wavelength. Moreover this wavelength depends on the applied voltage and increasing the voltage leads to faster electrons with yet shorter de Broglie

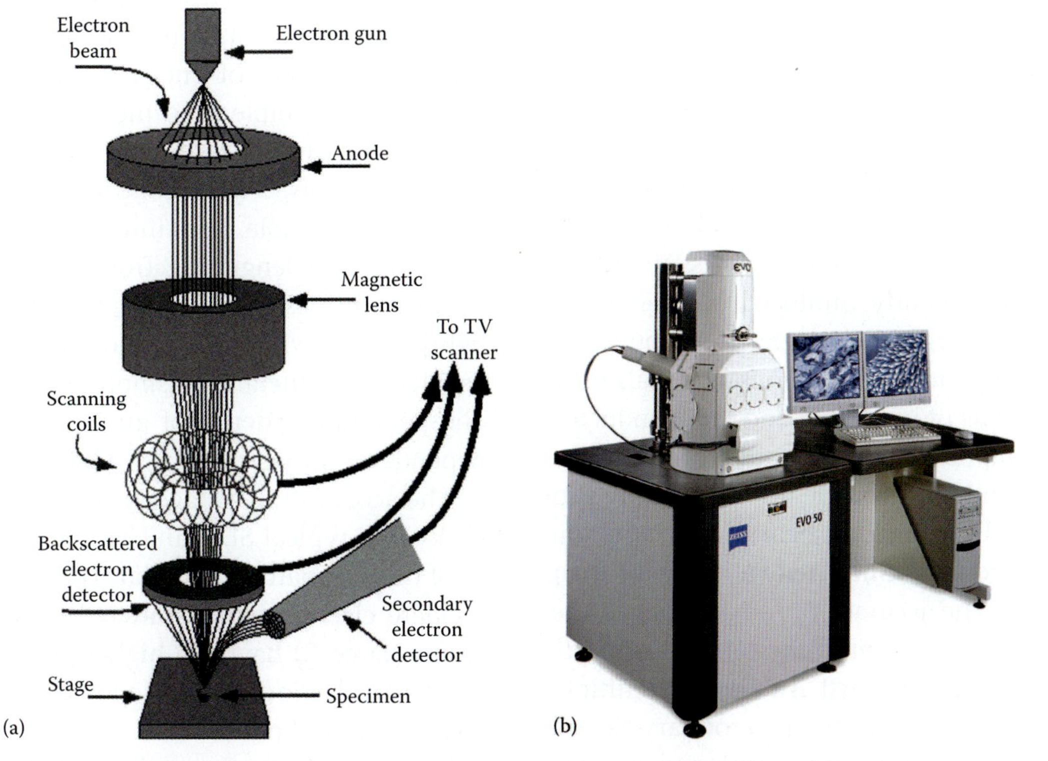

FIGURE 6.15 (a) Schematic of scanning electron microscope (SEM). (b) Carl Zeiss SEM. Scanning electron microscopy provides detailed three-dimensional view.

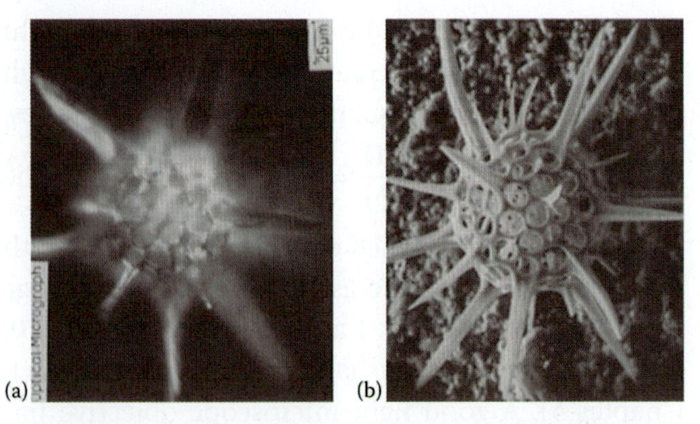

FIGURE 6.16 *Trochodiscus longispinus* in optical microscopy (a) and scanning electron microscopy (b). Note improved depth of field and resolving capability of the scanning electron microscopy experiment. (Taken from Goldstein, J., D. Newbury, D. Joy, C. Lyman, P. Echlin, E. Lifshin, L. Sawyer, and J. Michael. 2003. *Scanning electron microscopy and x-ray microanalysis.* New York: Kluwer Academic/Plenum Publishers.[6])

wavelengths. The effective F-number (also *F#* or *f/D*) for an electron microscope, on the other hand, is about 100 (quite large). The much shorter wavelength despite the poorer *f/D* gives an SEM a resolution 1000-fold better than that of the best light microscope. This improved resolution is dramatically illustrated in Figure 6.16, where we compare an image from an optical microscope (OM) with that of a SEM of the same *Trochodiscus longispinus*.

As the rastered focused electron beam irradiates the sample, secondary electrons are ejected from the sample. Secondary electrons are specimen electrons that obtain energy by inelastic collisions with beam electrons. They are defined as electrons emitted from the specimen with energy less than 50 eV and are predominantly produced by the interactions between energetic beam electrons and weakly bonded conduction-band electrons in metals or the valence electrons of insulators and semiconductors. With a flux amplitude that depends on the nature of the material and on the angle that the feature's surface makes with the incident beam, these secondary electrons are counted and used to create an image of the sample. The accuracy of the images captured is highly dependent on machine capability and the specific part being examined. An electron multiplier is used to detect and amplify the secondary electron intensity as a function of beam position to create the SEM image (see more on SEM detectors on page 308).

The output generated from the electron detector is displayed on a CRT rastered in synchronization with the electron beam. The final result is a two-dimensional image on a screen. Since no coordinate data are directly output from the SEM, performing any analysis other than line width measurements directly with the SEM software becomes difficult. Thus, SEMs are ideal for visualizing MEMS parts but are inadequate tools for quantitative analysis.[2]

The type of particles produced from the interaction of an electron beam with a solid sample is not limited to secondary electrons but may include backscattered electrons (BSEs), elastically scattered electrons, characteristic x-rays, and other photons of various energies. One of the main features of the SEM is that, in principle, any radiation emitted from the specimen or any measurable change in the specimen can be used to provide the signal to modulate the CRT and thus provide the contrast.

From Figure 6.17, it is obvious that several processes occur when electrons bombard a solid surface and that different emission mechanisms arise from different depths within the sample. The two basic types of electron scattering events that take place in an SEM are elastic scattering, which causes backscattering of electrons without loss or little loss of kinetic energy, and inelastic scattering, which results in transfer of energy from the beam electrons to the sample. The inelastic scattering leads to the generation of secondary electrons but also to Auger electrons, x-rays (continuous and characteristic), electron-hole pairs in conductors and insulators, long wavelength electromagnetic radiation or cathodoluminescence [visible, UV, and infrared (IR) light], lattice vibrations (phonons), and electron oscillations in metals (plasmons, specimen current). The penetration depth of an electron into a solid depends on the beam energy, which in turn depends on the accelerating voltage, usually between 30 and 60 kV in a typical SEM and on the atomic number, *Z*, of the bombarded material. The probability of an incident electron being scattered varies, with larger atoms (large *Z*) having a higher probability of producing an elastic collision, so low-*Z* materials have a greater depth of penetration.

When a 20 KV e-beam is used, say for Ni (Z = 28), Auger electrons originate from very close to

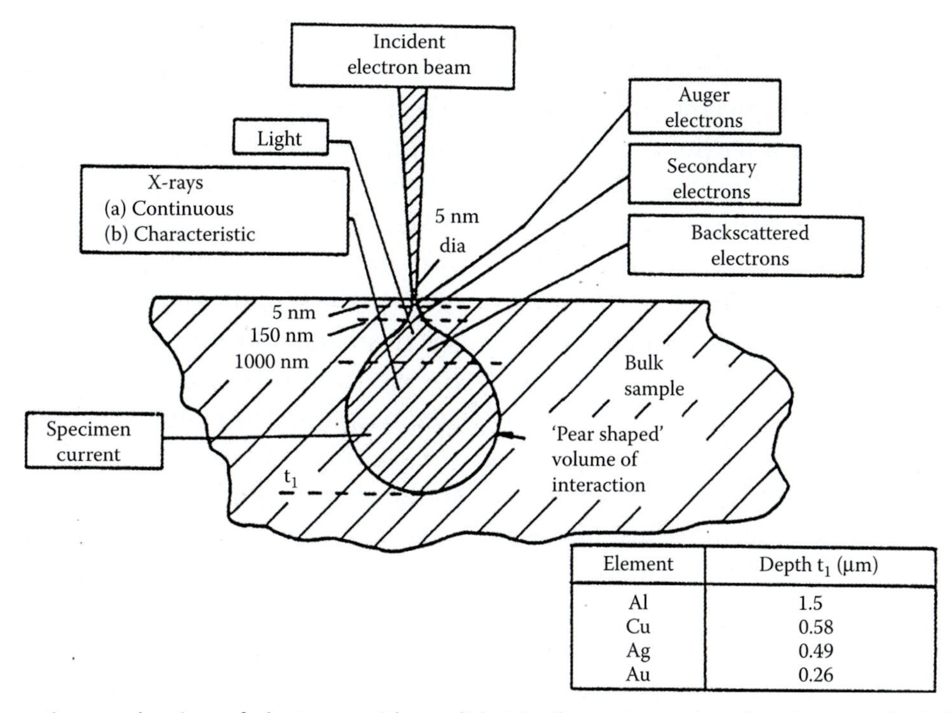

Element	Depth t_1 (μm)
Al	1.5
Cu	0.58
Ag	0.49
Au	0.26

FIGURE 6.17 Interaction mechanism of electrons with a solid. A bulk specimen showing the spread of the beam and the activated volume, giving rise to backscattered, secondary, and Auger electrons (i), characteristic x-rays (ii), and electron current (iii). Typical values given for the depth of electron penetration and the excited volume for a 30-keV, 100-nm-diameter electron probe.

the surface (10–30 Å), secondary electrons from 100 Å deep, backscattering electrons from 1–2 μm deep, and x-rays from 5 μm deep (the depth to which the incoming electrons may penetrate). Cathodoluminescence (CL), which arises from the recombination of electron-hole pairs generated by the incident beam, consists of visible photons (light). These can escape the sample only when emitted close to the surface, making cathodoluminescence very surface specific.

The interaction volume of the e-beam with the sample (pear shape shown in Figure 6.17) can be made visible in photoresist materials such as PMMA. Molecular bond damage (scission) during electron bombardment renders these materials sensitive to etching in a suitable solvent revealing the exposed volume. This phenomenon is the basis for e-beam lithography as detailed in Volume II, Chapter 2. The interaction volume as a function of beam energy can be estimated from Monte Carlo calculations (see also Volume II, Figure 2.15).

In Figure 6.18 the relative intensity versus abundance of the various emission phenomena under e-beam radiation are indicated. Whereas surface topography is obtained when low energy secondary electrons are collected, atomic number or orientation information is obtained from higher-energy backscattered electrons and elastically scattered electrons. The latter may be used to differentiate between surface roughness, porosity, granular deposits, stress-related gross microcracks (often used in conjunction with microsectioning), and the observation of grain boundaries in unetched samples.

Elastically scattered electrons have the same energy as the incident electron beam and retain orientation and phase information. These electrons are used, for example, in diffraction imaging experiments, such as low-energy electron diffraction (LEED). In surface

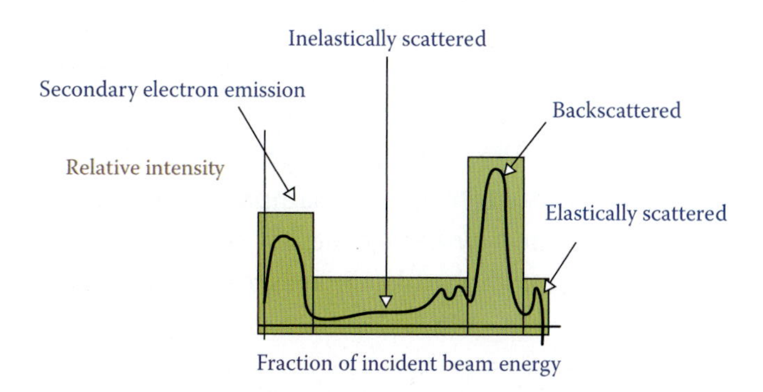

FIGURE 6.18 Relative intensity versus abundance of various emission phenomena under e-beam radiation.

science, well-ordered single-crystal surfaces are often investigated this way. Electrons penetrate (energy 50–300 eV) only a few monolayers into the sample; therefore, they are surface sensitive (for comparison, x-rays are more bulk sensitive). The observed diffraction pattern is the reciprocal lattice of the geometric surface lattice. Well-ordered surfaces exhibit sharp bright spots and low background intensity. The presence of surface defects and crystallographic imperfection results in broadening and weakening of the spots and increased background.

The backscattered electrons (BSEs) have an energy of about 80% of the incident beam. They are emitted more or less specularly. These electrons, just like secondary electrons, can be used to form topographic SEM images, but they have a higher energy than secondary electrons and also yield compositional information. Larger atoms (large atomic number, Z) have a higher probability of producing an elastic collision because of their greater cross-sectional area. Consequently, the number of BSEs reaching a BSE detector is proportional to the mean atomic number of the sample. Thus, a "brighter" BSE intensity correlates with greater average Z in the sample, and "dark" areas have lower average Z. BSE images are very helpful for obtaining compositional maps of a sample and for quickly distinguishing different phases. BSE images are limited to a grayscale range, though, because they only record one variable, average Z (a combination of all the elements in a sample). Thus, they do not convey as much information as can be obtained by elemental composition mapping using EDS* x-ray detectors (see "EDAX (EDX or EDS)").

SEM electron detectors can be based on different principles and oriented and operated differently to preferentially detect either secondary or backscattered electrons. A scintillator-photomultiplier system known as the *Everhart-Thorny* (E-T) *detector* detects the secondary electron signal, which is much more dependent on the sample topography than other electrons emitted from the sample. Topographic images obtained with secondary electrons actually look remarkably like images of solid objects viewed with light. A typical E-T detector consists of a Faraday cage in front of a scintillator, which in turn is coupled to a light pipe leading to a photomultiplier tube. The Faraday cage is kept at a positive potential of a few hundred volts so as to efficiently collect most of the secondary electrons emitted from the sample (remember they have an energy less than 50 eV). The scintillator typically has a positive voltage of several kilovolts applied to its surface, so that the electrons that pass the Faraday cage are accelerated onto the scintillator. When the electrons strike the scintillator they produce light, which is directed to the photomultiplier by a light pipe.

Using BSEs to construct topographical images either relies on a simplified scintillator–light pipe–photomultiplier (E-T), as used for secondary electrons, or uses solid state detectors. This stripped-down version of the E-T detector consists of scintillator coupled to a photomultiplier by a light pipe. Because there is no collection potential, the collection efficiency of these detectors is directly related to the scintillator size and its proximity to the sample. The simplest solid-state detector is a p-n junction with electrodes on the front and back of the sample. When hit by electrons, the holes tend to migrate to one electrode while the electrons migrate to the other, producing a current, the total of which is dependent on the electron flux and the electron energy. Response time of the detector is, unfortunately, slow but can be improved by putting a potential across the diode but this comes at the expense of increased noise in the signal. Often a set of solid-state detectors for simultaneous collection of backscattered electrons in different directions is used.

Detectors above the sample collect electrons scattered as a function of sample composition, whereas detectors placed to the side collect electrons scattered as a function of surface topography. BSE images can be obtained nearly instantly, depending on scan rate, and at any magnification within the instrument range.

The composition of samples is sensitive to the backscattered coefficient (η), which varies monotonically with Z. It has been shown that the coefficient can be expressed as follows.

$$\eta = -0.254 + 0.016Z - 1.86 \times 10^{-4} Z^2 + 8.3 \times 10^{-7} Z^3 \quad (6.11)$$

* EDAX (EDX or EDS): Energy dispersive analysis of x-rays (EDAX) is also called *EDX* and, more and more frequently, EDS. See "EDAX (EDX or EDS)".

The magnitude of the compositional or atomic number contrast, C, from two phases of backscattered coefficients, η_1 and η_2, is readily calculated as follows.

$$C = \frac{\eta_1 - \eta_2}{\eta_1} \tag{6.12}$$

This method is not recommended for phases with similar atomic number, as the resolution may be poor.

The inelastically scattered electrons have interacted strongly with the substrate electrons, and in the process they have lost all information of their initial direction and phase. These include Auger electrons, photoelectrons, and e-h pair recombination processes. These have mainly spectroscopic value but can also be used for imaging (see SAM, "Auger Electron Spectroscopy and Scanning Auger Microscopy").

In general, a high vacuum is required in SEM, as the sample environment and the samples have to be clean, dry, and electrically conductive. An environmental SEM (ESEM) offers high-resolution secondary electron imaging in a gaseous environment of practical any composition, at pressures as high as 50 Torr and temperatures as high as 1500°C. In this case wet, oily, dirty, nonconductive samples may be examined in their natural states without modification or preparation. An immediate concern is that the gas molecules will scatter the electrons and degrade the beam. To avoid this, instead of using a single pressure-limiting aperture as used in conventional SEM, the ESEM uses multiple pressure limiting apertures to separate the sample chamber from the vacuum. This leaves the column still at very high vacuum, but the sample chamber may sustain pressures as high as 50 Torr.

An ESEM uses a special type of detector that can function in a nonvacuum environment. The detector uses the principle of gas ionization. A positive potential of a few hundred volts is applied to the detector, which attracts the secondary electrons emitted by the sample. As the electrons accelerate in the detector field, they collide with gas molecules, creating additional electrons, amplifying the original secondary electron signal, and generating more positive ions. The detector collects the amplified secondary electron signal and passes it on to an electron amplifier.

In nonconductive samples, the positive ions created in the gas ionization process are attracted to the sample surface, and they effectively suppress charging artifacts.

A short review of environmental scanning electron microscopy by Krisada Kimseng and Marcel Meissel can be found at http://www.calce.umd.edu/TSFA/ESEM.pdf.

Scanning Proximal Microscopes

Introduction

A scanning proximal microscope (SPM) is a mechanical imaging instrument in which a small (<1 μm) probe is scanned over a sample surface. All SPMs owe their ultimate existence to the scanning tunneling microscope (STM), invented by IBM research scientists Gerd Binnig and Heinrich Rohrer in the early 1980s.

By monitoring the motion of a tiny probe tip, the surface topography and/or images of surface physical properties are measured with an SPM. Because close proximity or contact between the specimen and probe is a general requirement, SPMs need a feedback system that precisely controls the physical separation between probe and sample. Several methods are available to monitor the *z*-position of the probe tip at all times:

- Interferometric measurement of the tip amplitude
- Electron tunneling (limited to conductive specimens)
- Detection of the light emitted through the probe tip
- Constant force (atomic force feedback) is the most common method; it can be further subdivided as follows:
  - Diffraction of a separate light source by the tip
  - Mechanical sensor attached to the tip (e.g., a quartz tuning fork)
- Capacitance detection

The two most commonly used mechanisms of tip positioning are optical methods that monitor the tip vibration amplitude (usually interferometric), and nonoptical tuning fork techniques.

In addition to the *z*-axis feed-back control, an *x*-*y*-*z* scanner (usually driven piezoelectrically) controls the movement of the probe over the sample surface. The x-y-z scanner moves either the sample or the proximal probe. With the sample attached to the scanner, the sample moves under the fixed probe tip in a raster pattern, and an image is generated from the interaction of the tip and the sample. The size of the area that can be imaged depends on the maximum displacement of the scanner used. The typical size scale of features measured ranges from the atomic level (less than 1 nm) to more than 100 μm. A computer monitors the probe position and gathers data from the feedback system to control both the scanning of the tip (or sample) and the separation of the tip from the sample surface. The information generated by sensing the interaction between the probe and sample surface is collected and recorded by the computer point-by-point during the raster movement, and the computer renders the data into two-dimensional data sets (lines). The interactions between probe and sample used for imaging in scanning probe microscopes include electron tunneling, magnetic forces, electrical forces, electrochemical interactions, mechanical interactions, capacitance, ion conductance, Hall coefficient, thermal properties, and optical properties. SPMs offer an alternative to noncontact techniques and are characterized by their potential for very high resolution (subangstrom), limited only by the probe size and the property probed.

As illustrated in Table 6.5, SPMs are used in a wide range of applications. The three most used SPMs are the near-field scanning optical microscope (NSOM), the STM, and the atomic force microscope (AFM). In this section, we analyze the potential of these three SPMs for geometrical metrology. The utility of other specialized SPMs in property metrology is surveyed below (see "Tools to Measure Materials Property: Property Metrology").

NSOM or SNOM

To attain resolution beyond Abbe's diffraction limit of 200 nm, the textbook knowledge used to be that a complicated and costly SEM (0.1 nm resolution) or STM was required. Although these instruments do have a better resolution, they sacrifice many of the advantages of an optical microscope, such as nondestructiveness, low cost, high speed, reliability, versatility, accessibility, and ease of use. Thus, it would be very attractive to miniaturize a light microscope and have it operate below Abbe's diffraction limit. This objective is met by combining a scanning proximal probe technique with optical microscopy (OM) in so-called *scanning near-field OM*, a technique developed in the mid-1980s. We encountered NSOM first in Volume I, Chapter 5; here we give some additional background supplementing the information presented there.

The idea of using a subwavelength aperture to improve optical resolution was first proposed by E. H. Synge[7] in a letter to Einstein in 1928 and again, independently, in 1956 by John A. O'Keefe.[8] These ideas were realized only much later. First results were

Table 6.5 Scanning Proximal Microscopes and Their Applications

Scanning Proximal Microscopes	Examples of Applications
Scanning tunneling microscopy	Topography, density of states
Atomic force microscopy	Topography, force measurement
Lateral force microscopy	Friction
Magnetic force microscopy	Magnetism
Electrostatic force microscopy	Charge distribution
Scanning near-field optical microscopy	Optical properties
Scanning capacitance microscopy	Dielectric constant, doping
Scanning thermal microscopy	Temperature
Ballistic electron emission microscopy	Interface structure
Spin-polarized scanning tunneling microscopy	Spin structure
Kelvin probe force microscope is the Kelvin mode of scanning probe microscopy	Measures contact potential difference between the probe and the sample
Scanning electrochemical microscopy	Electrochemistry
Photon emission scanning tunneling microscopy	Chemical identification

achieved in 1972, by Ash and Nicholls,[9] in the microwave region of the spectrum, and in 1984, in the visible, by D. W. Pohl et al.[10] During image acquisition in NSOM, the point light source must be scanned over the surface, without touching it, and the optical signal from the surface must be collected and detected. To realize a constant probe-sample distance, a feedback mechanism that is unrelated to the NSOM/SNOM signal is utilized. Harootunian et al.[11] were the first to demonstrate SNOM microscopy in fluorescence mode. This result led eventually to single-molecule detection experiments by several groups.

During image acquisition in NSOM, a point light source is scanned very close to the surface, without touching it, and the optical signal from the surface is detected. In order to achieve an optical resolution better than Abbe's diffraction limit (the resolution limit of conventional OM), the probe tip must be brought within the near-field region, with the illumination source closer to the specimen than the wavelength of the illuminating radiation. For NSOM, the separation distance between probe and specimen surface is typically on the order of a few nanometers. Radiation near the source is highly collimated within the near-field region, but after propagation of a few wavelengths' distance from the specimen, the radiation experiences significant diffraction and enters the far-field regime. To make a point light source, laser light is fed to a microscopic aperture via an optical fiber. The aperture can be a tapered fiber with its shaft coated with a metal, a microfabricated probe, or a tapered pipette (see example aperture SNOM implementations in Volume I, Chapter 5). The aperture-based resolution is limited to about $\lambda/10$. With careful probe design and sample preparation a resolution of ~30 nm has been achieved.

One challenge for aperture based SNOM is the small aperture resulting in a very low light throughput. Scattering SNOM or apertureless SNOM, developed in the 1990s, is an alternative that avoids this low light intensity problem. Apertureless near-field optics, as we saw in Volume I, Chapter 5, uses a sharp, metallic subwavelength tip instead of an optical point source. The sharp metal tip acts as a nanoscopic light scatterer when placed in near-field distance to the sample. Oblique incident light is scattered at the tip, and the resulting strongly enhanced local plasmonic field is used for excitation of the sample. The resulting optical response signal is collected via an objective in far field and can be used for imaging and chemical analysis. The tip radius in this case is the only factor determining resolution. The best resolution attained today is 10 nm, and it is expected that this can be reduced to atomic level resolution. A significant problem with the scattering SNOM approach is that the light scattered by the tip carrying the information on tip-sample interaction is easily overwhelmed by background light. Therefore the method is particularly suitable for light-matter interaction processes where the light detected has a different wavelength from the excitation light (e.g., Raman scattering, fluorescence, second harmonic generation). Consider tip-enhanced Raman spectroscopy as an example. In tip-enhanced enhanced Raman spectroscopy (TERS), a silver-coated tip is scanned across the illuminated area, acting as a Raman signal enhancer. As we saw in Volume I, Chapter 5, Raman signals of molecules deposited on silver or in close proximity to silver nanoparticles can be increased by a factor of 10^{14}. It is believed that this enhancement results principally from the resonant enhancement of the surface plasmon mode localized at the apex of the sharp metallic tip. This technique represents an elegant combination of two Nobel Prize-winning discoveries: the Raman effect and the scanning tunneling microscope. TERS has created a new tool that achieves a long-awaited goal in surface science, namely the correlation of topographic and chemical information of the same sample region at nanometer scale. Although the enhancement processes allowing for TERS are still not fundamentally understood, the use of an illuminated scanning probe tip to greatly enhance Raman scattering from the sample underneath the tip is one of the most exciting developments in optical spectroscopy.

In the rest of this discussion, we focus on aperture-based SNOM only.

In an aperture-scanning NSOM instrument, the quantitative point-spread function in the near field can be approximated by a Gaussian profile whose 1/e intensity value is of the same order as the radius of the aperture at the tip of the NSOM probe. The mode of light propagation is primarily evanescent (and parallel to the specimen surface) when the

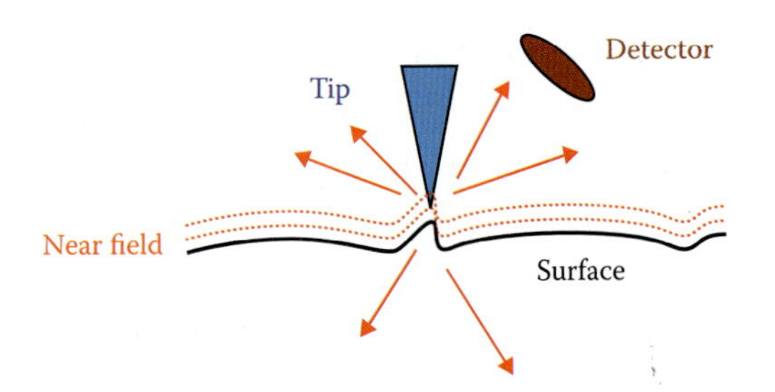

FIGURE 6.19 Light source is held in near-field to the surface.

radius of the illuminating source is less than one-third of the imaging light wavelength as shown in Figure 6.19. The NSOM configuration illustrated here positions the objective in the far field, in the conventional manner, for collection of the image-forming optical signal.

A schematic illustrating the control and information flow of an inverted optical microscope-based NSOM system is presented in Figure 6.20a. A laser source is coupled to a fiber optic probe for near-field specimen illumination. In the NSOM setup shown, the probe tip movements are controlled through an optical feedback loop incorporating a second laser that is focused on the tip. Two-dimensional data sets gathered by the NSOM instrument are registered and displayed as a 3D reconstruction on a computer monitor. Figure 6.20b is a photograph of an actual near-field scanning optical microscope set-up.

To improve signal-to-noise ratios for the feedback signal, the NSOM tip is oscillated at its resonance frequency. This allows for more sensitive lock-in detection techniques, eliminating positional detection problems associated with low-frequency noise and drift. As the oscillating tip approaches the sample surface, forces between the tip and specimen change both the peak resonance and the quality factor. These changes can be monitored by several techniques, classifiable in two broad groups: 1) shear-force mode and 2) tapping mode. In the shear-force mode, to control the tip-specimen gap during imaging, one utilizes lateral oscillation shear forces generated between a straight optical fiber and the sample (Figure 6.20a). The fiber is drawn to a sharp point and coated on the sides with an opaque material (usually Al). A small (~50-nm) hole is formed in the coating at the apex of the fiber and the evanescent field associated with this aperture is used to image the sample. In the optical feedback approach, illustrated in Figure 6.20a, a second laser is tightly focused as close to the end of the NSOM probe as possible. With the laser feedback established, the probe is then vibrated, at a known frequency, utilizing a dither piezo. A split photodiode collects the laser light, and the difference between the signals from each side of the detector is determined. The main problem associated with such an optical type of feedback mechanism is that the light source, which is used to detect the tip vibration frequency, phase, and amplitude, becomes a

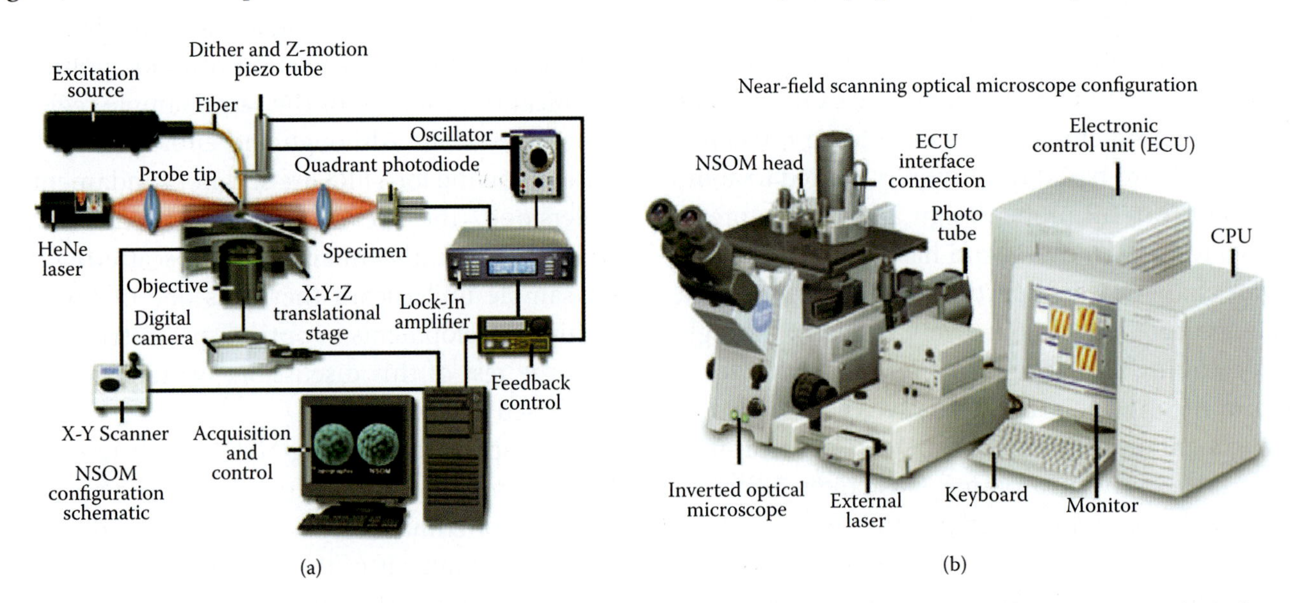

FIGURE 6.20 A schematic (a) illustrating the control and information flow of an inverted optical microscope-based scanning near-field optical microscopy system. (b) Actual photo of an Olympus NSOM system (http://www.olympusmicro.com/primer/techniques/nearfield/ nearfieldintro.html).

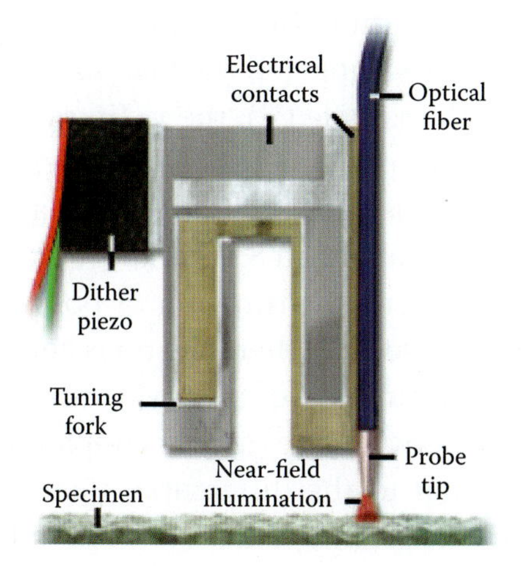

FIGURE 6.21 Tuning fork feedback (http://www.olympusmicro.com/primer/techniques/nearfield/nearfieldintro.html).

potential source of stray photons that can interfere with the detection of the NSOM signal. To realize a constant probe sample distance, a feedback mechanism that is unrelated to the NSOM signal is obviously preferred. This is the primary reason why the nonoptical feedback method based on the tuning-fork technique, illustrated in Figure 6.21, has become increasingly popular. Tuning forks are incorporated into the NSOM to serve as inexpensive and simple, nonoptical excitation and detection devices. When quartz tuning forks are utilized for regulation in a feedback loop, their very high mechanical quality factor, Q (as high as approximately 10,000), and corresponding high gain provide the system with high sensitivity to small forces, typically on the order of piconewtons. The fiber is typically fixed to one leg of a tuning fork, which is caused to oscillate. Interaction between the tip and the sample (due to shear forces—hence the name shear-force mode) leads to damping of this oscillation, and the height is adjusted to maintain a constant oscillation frequency.

In the tapping mode, to generate the feedback signal, one relies on atomic forces modulating the oscillation of a cantilever probe. In this case, a bent optical fiber, effectively functioning as an AFM cantilever probe, may be used as the light source. The nonoptical feedback method based on the tuning-fork technique, illustrated in Figure 6.21, is also applicable here. It is also possible to use an AFM-type cantilever probe, in which the cantilever has an aperture at its apex; a lens focuses light into the aperture and an AFM-type feedback mechanism is used.

For the best image quality, shear-force feedback techniques are restricted to specimens that have low surface relief, and longer scan times are required compared with operation in tapping mode. Moreover, the interactions that underlie shear force imaging are not very well understood. However, the straight probes employed in shear-force feedback techniques are easier to fabricate, are lower cost, and have a higher light throughput than their bent probe counterparts.

Previously developed high-resolution scanning techniques, such as scanning electron microscopy (SEM), transmission electron microscopy (TEM), scanning tunneling microscopy (STM), and atomic force microscopy (AFM), do not benefit from the wide array of contrast mechanisms available to an OM such as SNOM and, in most cases, are limited to the study of specimen surfaces only. Aside from the available contrast-enhancing techniques of staining, fluorescence, polarization, phase contrast, and differential interference contrast, optical methods have inherent spectroscopic and temporal resolution capabilities.

Advantages and disadvantages of SNOM and a list of applications are provided in Volume I, Chapter 5, "NSOM Applications."

Scanning Tunneling Microscope

The scanning tunneling microscope (STM) was the first and remains the most famous of the scanning probe microscope techniques. As detailed in Volume I, Chapter 3, scanning tunneling microscopy measures topography of surface electronic states using a tunneling current that is dependent on the separation between the probe tip and a conductive sample surface (metals or semiconductors). The violation of classical physics in tunneling is allowed by the quantum mechanical uncertainty principle. On the basis of this principle, a particle can violate classical physics by tunneling inside and even through an energy barrier, ΔE, for a short time, $\Delta t \sim \hbar/\Delta E$ (see Volume I, Equation 3.107). The exponential decay of the wave function inside a rectangular energy barrier that stretches from $x = 0$ to $x = L$ and has a height of V_0 is given as follows:

$$\psi(x) = Ae^{-\alpha x}$$

(Volume I, Equation 3.185)

with $\alpha^2 = \frac{2m(V_0 - E)}{\hbar^2}$ (Volume I, Equation 3.165). If the barrier is narrow enough (L is small), there will be a finite probability, P, of finding the particle on the other side of the barrier, as follows:

$$P = |\psi(x)|^2 = A^2 e^{-2\alpha L}$$
(Volume I, Equation 3.186)

where A is a function of the energy, E, of the particle and the barrier height, V_0. The probability of finding an electron on the other side of a barrier of width L can be probed with a fine needle tip from a STM. The tunneling current, picked up by the sharp needle point of an STM, based on Volume I, Equation 3.186, is given by the following equation:

$$I = f_w(E) A^2 e^{-2\alpha L}$$
(Volume I, Equation 3.187)

where $f_w(E)$ is the Fermi-Dirac function, which contains a weighted joint local density of electronic states (DOS) in the solid surface that is being probed and those states in the needle point (see Figure 6.22). Tunneling occurs between states of the same energy and the electron's energy does not change during the tunneling event. From Volume I, Equation 3.187, when L changes by 1 Å, the current changes by a factor of about 10. Obviously, the tunneling current is very sensitive to the gap distance. The size of the gap in STM experiments is kept on the order of a couple of angstroms. The DOS can vary as a function of position on the sample, which means that one can define a local density of states (LDOS). LDOS depends on both energy and on position. An intuitive way to think about LDOS is that it gives the density of electrons of a certain energy at that particular spatial location [LDOS (x, y, E)]. An STM measures and controls the current that flows between the tip and the sample (I), the bias voltage between the tip and the sample (V), the xy (in sample plane) position of the tip, and the z (perpendicular to sample plane) distance between the tip and sample. Using these variables, an STM is able to measure the LDOS of a material as function of position on the surface (controlled by where the tip is above the surface) and as a function of energy (controlled by the bias voltage between the tip and sample). The LDOS is proportional to the differential increase in tunneling current given a differential increase in bias voltage; in other words, one can measure the LDOS by measuring dI/dV. This was illustrated in Volume I, Figure 3.135, for a metallic and semiconducting carbon nanotube where singularities (sharp peaks) in the density of states function are observed directly in scanning tunneling microscopy images.

An STM may be viewed as an electron microscope that uses a single atom on a tip to attain atomic resolution. In a bit less lofty way, it may be described as a glorified phonograph needle, as illustrated in Figure 6.23. Resolution is determined by manufacturing a probe whose tip is of the same dimension we seek to resolve. If one can get a single atom at the tip, the vast majority of the current will run through it and thus give us atomic resolution. Tungsten is commonly used for the tip because one can use electrochemical-etching techniques to create very sharp tips like the one shown in Figure 6.24. This tip is brought to within a few angstroms of the surface that we wish to probe, and a small bias voltage (10 mV to 2 V) is applied between the tip and the sample while an electrical

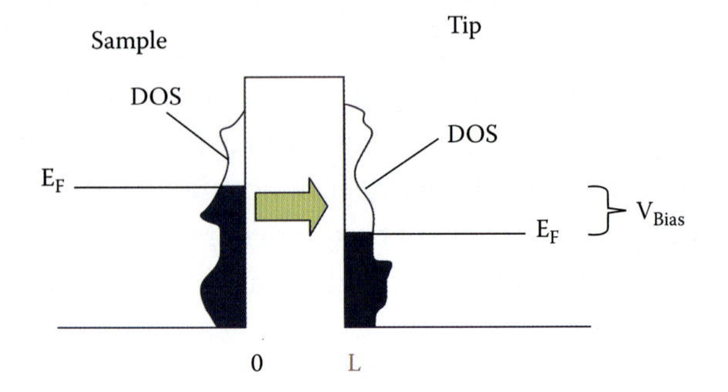

FIGURE 6.22 Density of state functions in sample and tip.

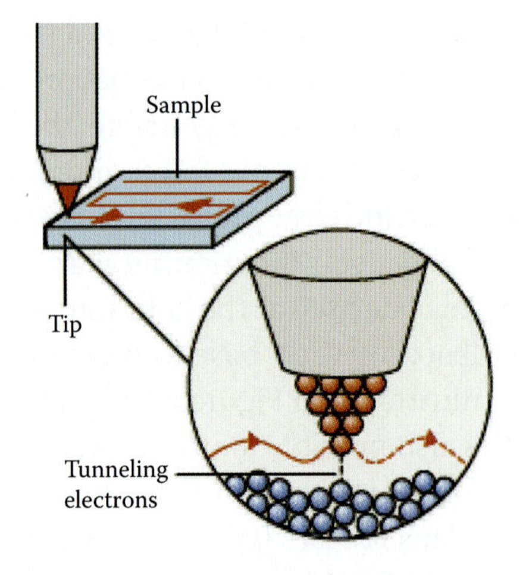

FIGURE 6.23 Scanning tunneling microscope.

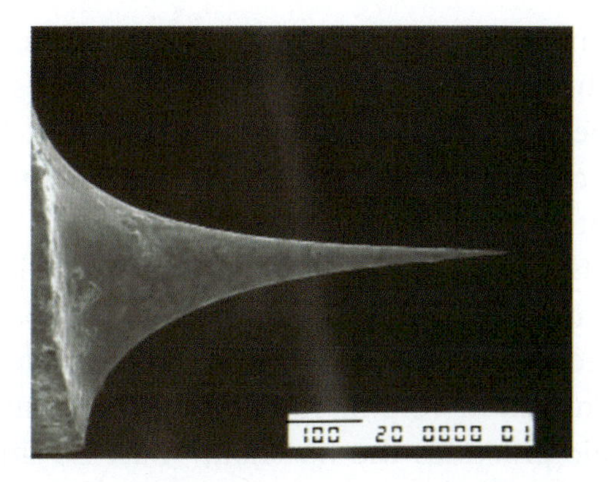

FIGURE 6.24 Tungsten scanning tunneling microscope tip; magnification ×150.

current flows by electrons tunneling across the gap between the two conductors. As we saw above, this tunneling current magnitude depends upon the gap voltage, the gap distance, and the local density of electron states (Volume I, Equation 3.187).

At its heart, a scanning probe microscope has a piezoelectric material (e.g., quartz) that expands or contracts slightly when a voltage is applied to it. These voltages can be carefully controlled and cause one end of the crystal to move with better precision than the width of a single atom. The needle or tip is brought very close to the surface and moved over it piezoelectrically and a tunneling current, resulting from the overlap of the tip's electron wave functions with the conducting surface is measured. This tunneling current is generated, even though there is a break in the circuit. Using a feedback servo loop, the tip slowly scans across the surface and is raised and lowered in order to keep the signal constant and maintain a distance of a couple of angstroms, where the tunneling current is at a measurable level.

An STM does not measure nuclear position directly; rather, it measures the electron density clouds on the surface of the sample. In some cases, the electron clouds represent the atom locations pretty well, but not always. The trace from an STM (Figure 6.25A) can be interpreted as a grid (Figure 6.25Ba) that can be shown as a grayscale picture (Figure 6.25Bb). The grayscale picture can be interpreted as a contour map (Figure 6.25Bc), which can then be averaged out to make it smooth (Figure 6.25Bd) and finally can be colored (Figure 6.25C). This explains the color of the Kanji character, made by manipulation of individual iron atoms on a copper surface with an STM, in Figure 6.26.

The vertical resolution of an STM is remarkably good because of the exponential dependence of the current on vertical height (L in Volume I, Equation 3.187). This means that 1-Å change in height corresponds to almost 1 order of magnitude difference in current. The noise floor is in the tens of femtometer

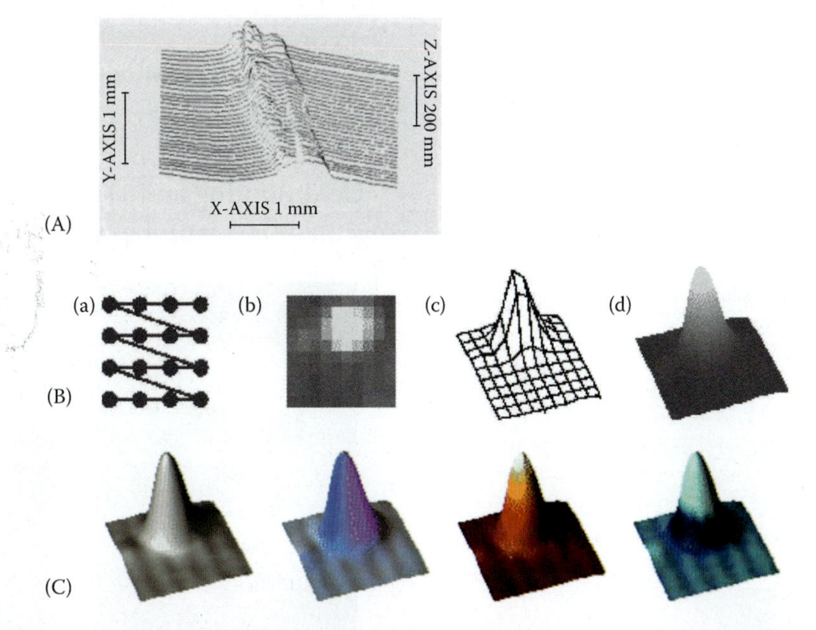

FIGURE 6.25 (A) Trace of a scanning tunneling microscope (STM). Shown is a trace of an individual turn mark on a diamond-turned aluminum substrate to be used for subsequent magnetic film deposition for a high-capacity hard disc drive. (B) The trace from an STM can be interpreted as a grid (a), which can be shown as a grayscale picture (b). The grayscale picture can be interpreted as a contour map (c) that can then be averaged out to make smooth (d) and finally colored (C).

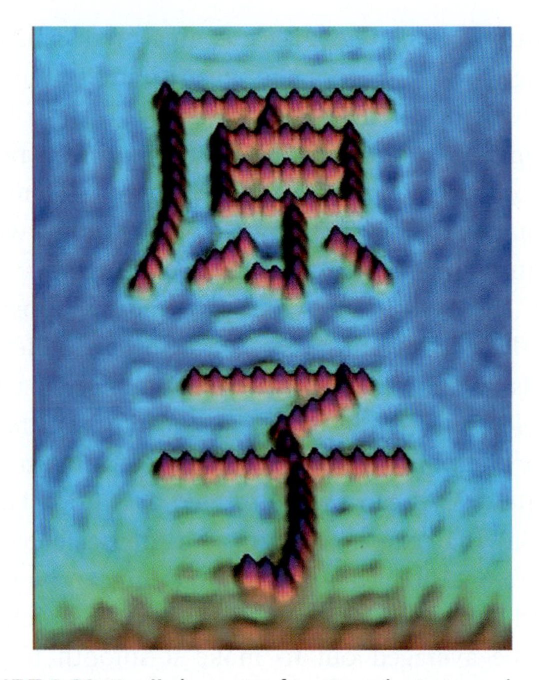

FIGURE 6.26 Kanji character for *atom*, by Lutz and Eigler (IBM Almaden). An example of manipulation/mechanosynthesis of iron atoms on a Cu surface with a scanning tunneling microscope. (Courtesy of D. Eigler.)

range. The lateral resolution depends on the tip radius as clarified in Figure 6.27. A tip with radius R = 1000 Å actually focuses 90% of the current in a circle of radius (Δx in Figure 6.27) of 45 Å. A 100-Å tip focuses 90% of the current in a circle of radius 14 Å. In practice, a 2-Å lateral resolution is routine. These numbers can be ascertained by rewriting $I = f_w(E)A^2e^{-2\alpha L}$ (Volume I, Equation 3.187) as:

$$I(\Delta x) = f_w(E)A^2e^{-2\alpha\frac{\Delta x^2}{2R}} \tag{6.13}$$

There are two major modes of operation for an STM: constant current and constant height. In constant current mode one maintains a constant tunneling current by adjusting the height (separation between tip and sample). In constant height mode one, maintains the height and measures the current.

STMs can be used in solution, as there still is a vacuum gap, even in water. This mode of STM operation is called *scanning electrochemical microscopy*, which we encountered in Volume II, Chapter 4. To minimize faradaic processes the tip shank must be shielded with melted wax or plastic to expose only the last few nanometers of metal. The tunneling current to the exposed metal tip must be large compared with faradaic current to the tip shank.

STMs are limited to imaging parts with conductive surfaces, and electronic inhomogeneities can have significant effects on the acquired topographical images. Vibrations in the probing mechanism limits gap width stability, which, in turn, can affect the fidelity of the measurements. All SPMs are limited, in the same sense as white light interferometers, to a maximum measurable slope. When features with perpendicular sidewalls are scanned, the data typically exhibit a slope that is actually not present. The height of measurable features is also limited to the probe tip length, which is typically <10 μm in commercial systems. This limitation severely prohibits the inspection of high-aspect ratio parts with dimensions on the order of millimeters. Though these tools have extremely high resolution, it is unfeasible to collect scans that cover all of the surfaces of a part, given the limited scan range of the tools.

Gerd Binning and Heinrich Rohrer (Figure 6.28) invented the scanning tunneling microscope at IBM

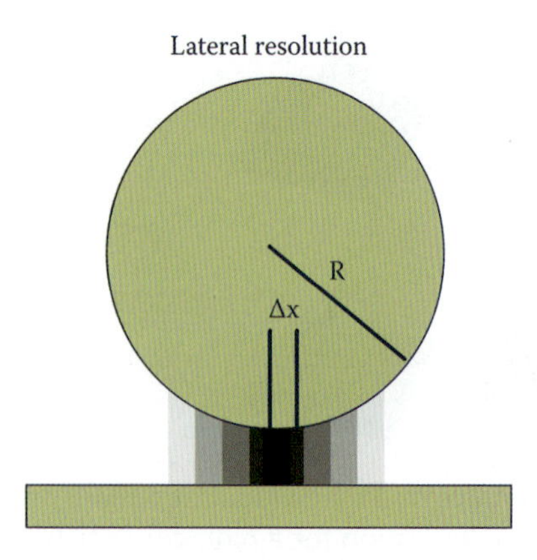

FIGURE 6.27 Lateral resolution of a scanning tunneling microscope.

FIGURE 6.28 The inventors of the scanning tunneling microscope. Nobel laureates in Physics, 1986.

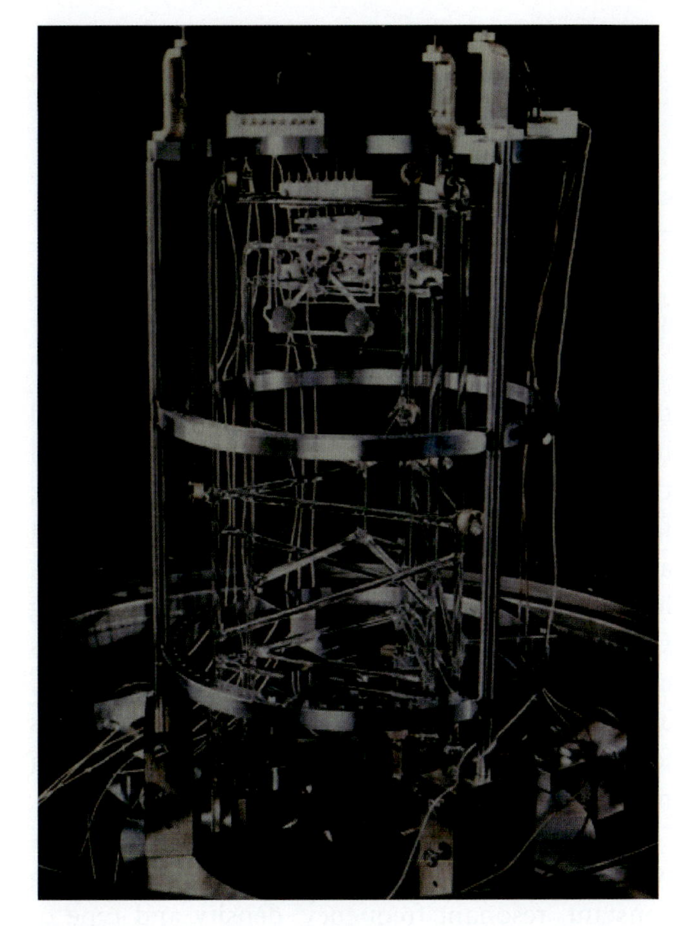

FIGURE 6.29 Binning and Rohrer's original scanning tunneling microscope.

Zürich in 1981. They shared the Nobel Prize for Physics in 1986 for their microscope (Figure 6.29).

Atomic Force Microscopy

Scanning force microscopy, also called *atomic force microscopy* (AFM), constitutes a major extension of the STM technique. The technique, first described in 1986, depends on forces (repulsive or attractive) between atoms of a sample and a probe (Figure 6.30).

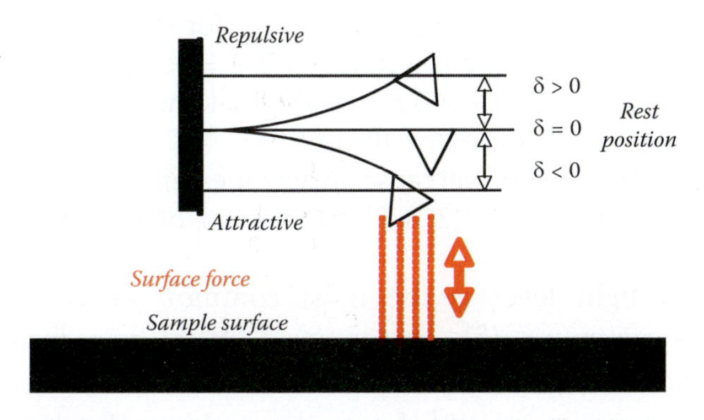

FIGURE 6.30 An atomic force microscope measures surface forces.

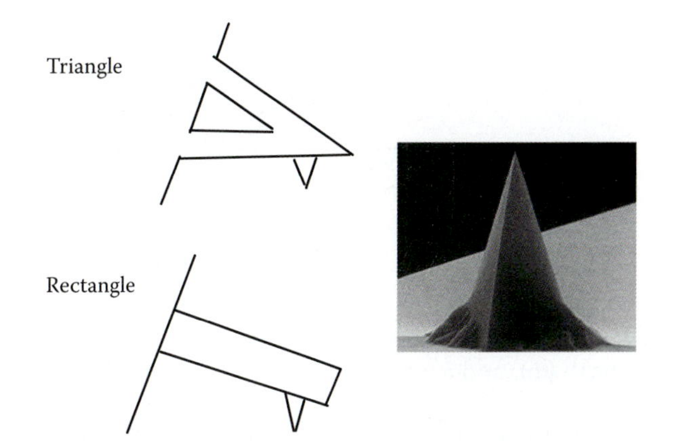

FIGURE 6.31 Cantilevers are either triangular or rectangular. Material can be silicon on silicon nitride (plastic in development). Probes shapes can be pyramidal or conical and may be sharpened.

No current flows between sample and AFM tip, so the sample need not be conducting.

In a typical AFM, a cantilever with a tip (the probe; see Figure 6.31) is oscillated above the sample surface. The oscillation amplitude is less than10 nm, and in the so-called *noncontact mode*, the tip does not touch the sample. Constant oscillation amplitude is maintained and surface forces, which extend from 1 to 10 nm above the surface, decrease the resonant frequency of the cantilever. This in turn changes the amplitude of oscillation. Typical surface forces range from the piconewton to the nanonewton [quantum mechanical (covalent, metallic bonds): 1–3 nN; Coulomb (dipole, ionic): 0.1–5 nN; polarization induced (dipoles): 0.02–0.1 nN].

A cantilever must now be designed to convert these very small mechanical forces into measurable deflection, δ.

A microfabricated cantilever beam and probe tip, as shown in Figure 6.32, deflects in response to an applied force and can be represented as a linear elastic, Hookean spring:

$$F = -k\delta \tag{6.14}$$

where δ = displacement at the end of cantilever (in meters) (this is what we actually measure in a force spectroscopy experiment)

F = external force applied to cantilever (in newtons) (this we can calculate from δ)

k = cantilever spring constant

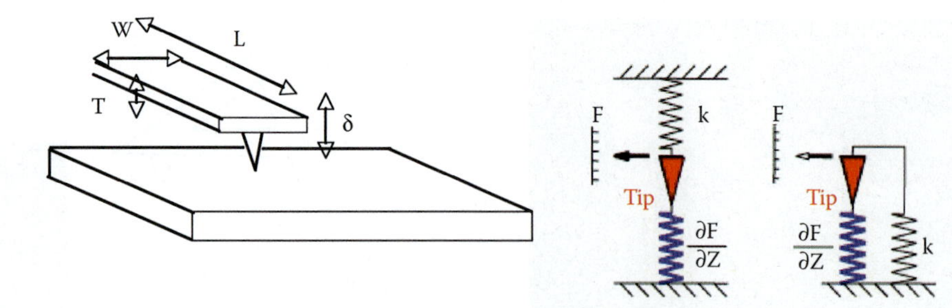

FIGURE 6.32 Atomic force microscope cantilever model. T = cantilever thickness (m), W = cantilever width (m), L = cantilever length (m), and δ = deflection of the end of the cantilever.

For the cantilever in Figure 6.32 the spring constant (in N/m) is given as follows:

$$k = \frac{EW}{4}\left(\frac{T}{L}\right)^3 \quad (6.15)$$

where T = cantilever thickness (in meters)
W = cantilever width (in meters)
L = cantilever length (in meters)
E = Young's (elastic) modulus of cantilever material (in pascals or N/m^2)

For Si, $E = 11 \times 10^{10}\ N/m^2$.

We can calculate k independently, and typical values are 0.01–100 N/m. The resonance frequency f, of the cantilever is given as follows:

$$f = 0.162\sqrt{\frac{E}{\rho}\frac{T}{L^2}} \quad (6.16)$$

where ρ is the density of the cantilever material. Typical values for f are 7–500 kHz.

The interaction of surface forces with this cantilever can now be modeled as a spring in series with the cantilever spring, as pictured in Figure 6.32 or, remembering that $\omega = \sqrt{\frac{k}{m}}$:

$$\omega = \sqrt{\frac{k - \frac{\partial F}{\partial z}}{m}} \quad (6.17)$$

where $\partial F/\partial z$ is the force gradient (F'). For small amplitudes, a Taylor expansion of the force can be carried out, and a frequency shift proportional to the force gradient is found, as follows:

$$\frac{\Delta\omega}{\omega} = -\frac{1}{2k}\frac{\partial F}{\partial z} \quad (6.18)$$

In practice one needs to compromise between a cantilever with a small k and a large resonance frequency f. The resonance frequency (f) should indeed be higher than the data acquisition rate and the noise level (kHz range). To increase the resonance frequency, the cantilever should be stiff (large k), short and thin. On the other hand, to maximize sensitivity, a soft cantilever (small k) is preferred. But making k too small, one might have a problem of cantilevers crashing into the sample. With a k value of ~1 N/m, an AFM can measure forces of pN, and even fN is feasible. The force detection limit is the Brownian motion of the cantilever (thermal noise limited).

In Table 6.6 we list some example AFM cantilevers with their force sensitivity, Young's modulus, spring constant, resonant frequency, density and type of material, and shape. For an excellent discussion and references on cantilever design, consult http://www.asylumresearch.com/springconstant.asp.

Example 6.2

Exercise: In atomic force microscopy, an ultrasharp tip on the end of a silicon cantilever beam is used to probe a surface at the nanoscale. By how much is the beam deflected by thermal motion? The cantilever has a spring constant of ~10 N/m.

Answer: The energy required for deflection of the beam by a distance δ is $E = ½ k\delta^2$. At a temperature of 300 K, the thermal energy is on the order of $kT = 6 \times 10^{-21}$ J.

This energy will cause an average deflection of the beam by $\delta = (2E/k)^{0.5} = 1 \times 10^{-7}$ m, or 100 nm.

A light lever is the most common deflection measurement (δ) sensor used in atomic force microscopes. In this approach, when the cantilever moves up and down, the position of a laser spot on a photodetector moves up and down. As a

Table 6.6 Example Atomic Force Microscope Cantilevers

Force Sensitivity (N)	k (N/m)	ω	ρ (g/cm^{-3})	Module Young E (10^{10} N/m^2)	Dimensions (mm)	Material
~10^{-8}	5000	22 kHz	8.9	22	4 × 0.25 diameter	Ni wire
~10^{-11}	2.5	33 MHz	2.3	11	2 × 0.5 × 0.1 μm	Si bar
~10^{-14}	0.004	8 kHz	3.1	32	0.2 × 0.036 × 0.3 μm	Si_3N_4 (V-shape)
10^{-12}	0.4	14 kHz	2.2	7	0.25 × 5 μm diameter	Quartz fiber

position-sensitive photodetector, a four-quadrant photodiode (current in each quadrant changes with light intensity) is used, as illustrated in Figure 6.33.

In an alternative approach a piezoresistive sensor is embedded in the cantilever. This approach was discussed in Volume I, Example 4.1.

There are three main AFM techniques in use: contact, noncontact, and tapping mode. In the "repulsive" contact AFM, the tip is in perpetual contact with the sample as it slides across the surface. The tip is attached to the end of a cantilever with a low spring constant, *k*, while the feedback keeps the vertical deflection of the cantilever constant. In contact mode, the AFM measures hard-sphere repulsion forces between the tip and sample. Forces range from nano- to micronewtons in ambient conditions and even lower (0.1 nN or less) in liquids. This method provides the highest resolution but may damage the surface. It is the only AFM technique that can provide atomic-resolution images.

In noncontact mode, the AFM derives topographic images from measurements of the van der Waals attractive forces. The tip does not touch the sample;

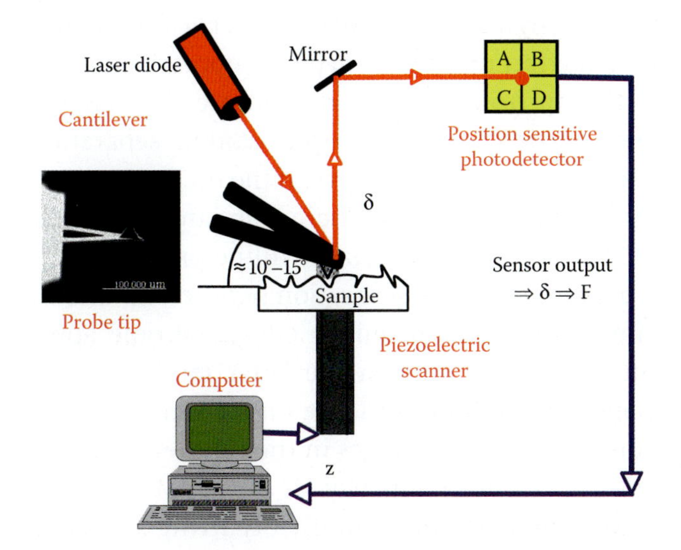

FIGURE 6.33 A light lever atomic force microscope setup.

instead, the cantilever with the tip oscillates (amplitude <10 nm) above the surface. The resonant frequency of the cantilever is decreased by the attractive surface force, which extends from 1 to 10 nm above the surface (see Volume I, Chapter 7, "Calculating Interaction Forces between Surfaces and Particles"). The noncontact mode has the lowest resolution of the three main AFM approaches.

Tapping mode AFM, the most commonly used of all AFM modes, is a patented technique (Veeco Instruments) that maps topography by lightly tapping the surface with an oscillating probe tip. This eliminates the shear forces potentially damaging to soft samples. The AFM tip taps the sample surface during the closest point of approach in each oscillation cycle, the amplitude of which is kept unchanged by the feedback loop. In this intermittent tip contact, force is measured with improved resolution over noncontact AFM. Surface forces again cause an oscillation frequency shift, but the tap also causes the cantilever/tip to lose energy, giving additional information about the surface. Another major advantage of tapping mode is related to limitations that can arise because of the thin layer of liquid that forms on most sample surfaces in an ambient imaging environment, i.e., in air or some other gas. The amplitude of the cantilever oscillation in tapping mode is typically on the order of 20–100 of nanometers, which ensures that the tip does not get stuck in this liquid layer. The amplitude used in noncontact AFM is much smaller (<10 nm; see above), and as a result, the noncontact tip often gets stuck in the liquid layer. In general, tapping mode is much more effective than noncontact AFM for imaging larger scan sizes that may include large variations in sample topography. Tapping mode can be performed in gases, liquids, and some vacuum environments.

The presence of a feedback loop is one of the subtler differences between AFMs and older

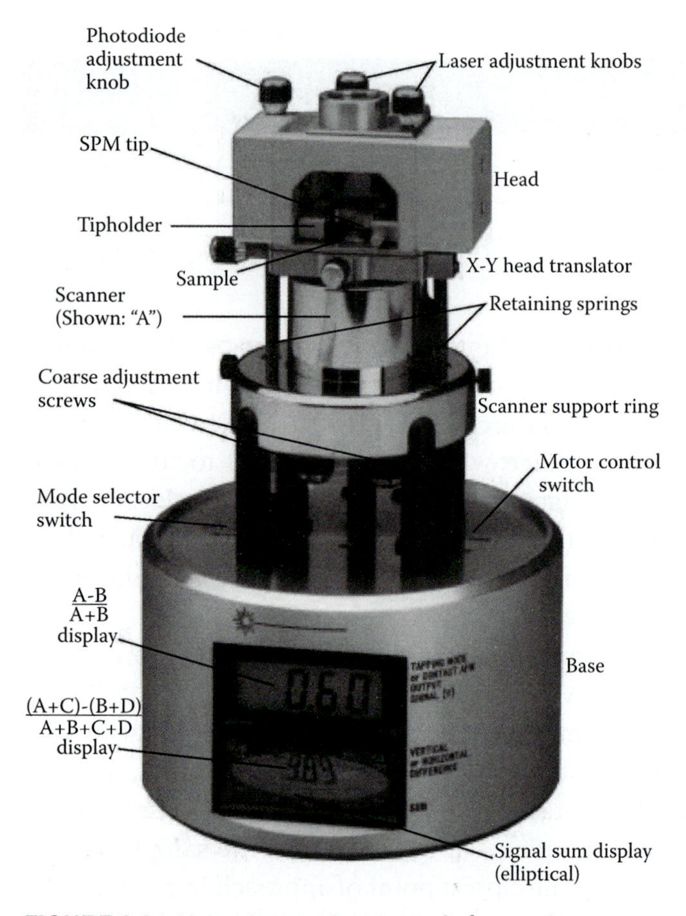

FIGURE 6.34 Veeco Instruments atomic force microscope.

stylus-based instruments such as record players and stylus profilometers (see Figure 6.40). The AFM not only measures the force on the sample but also regulates it, allowing acquisition of images at very low forces. One point of interest: the faster the feedback loop can correct deviations of the cantilever deflection, the faster the AFM can acquire images; therefore, a well-constructed feedback loop is essential to microscope performance. AFM feedback loops tend to have a bandwidth of about 10 kHz, resulting in image acquisition times of about 1 min.

In Figure 6.34 we show a photograph of a typical AFM.

Tools to Measure Materials Properties: Property Metrology

Introduction

While most MEMS metrology techniques focus on dimensional metrology and some on dynamic metrology, few systems exist for measuring other properties of microsystems, such as electrical, magnetic, thermal, and mechanical characteristics. In the category of material property measuring equipment, we start with a review of scanning proximal probe microscopes (SPMs) for determination of material properties, such as magnetic, thermal, and mechanical behavior. We then consider tools used to measure conductivity and carrier type of a material. MEMS devices are often mechanically dynamic by nature. Therefore, there is also a need to measure dynamic motion of MEMS devices. We cover two mechanical property metrology techniques: using interferometry with a stroboscopic LED to freeze device motion and laser Doppler vibrometry to measure contactless velocity and displacement of vibrating structures.

Material Property Measuring Scanning Proximal Probe Microscopes

The scanning proximal microscopes introduced above also can be utilized for a variety of material property metrology tasks (see Table 6.5). We briefly describe a number of the best-known techniques.

Magnetic force microscopy (MFM) is an imaging mode derived from atomic force microscopy (AFM) used in the noncontact mode that maps magnetic force gradients above a sample surface with a resolution of 10–25 nm. An image acquired with a MFM probe tip coated with a magnetized material (e.g., CoCr or NiFe) contains information about both the topography and the magnetic properties of a surface. Which effect dominates depends upon the distance of the tip from the surface, because the magnetic force persists for greater tip-to-sample separations than the van der Waals force. If the tip is close to the surface, in the region where standard noncontact AFM is operated, the image will be predominantly topographic. As the separation between tip and the sample increases, magnetic effects become apparent. While scanning in the latter mode, it is the magnetic field's dependence on tip-sample separation that induces changes in the cantilever's amplitude, resonance frequency, or phase. MFM can be used to image both naturally occurring and deliberately written magnetic domains. A MFM typically measures magnetic properties and topography of a surface in a two-pass fashion. The first pass is a standard AFM trace that maps out the surface topography by gently tapping the tip along the surface. A second pass then samples the magnetic stray field by scanning at a constant height above the surface (see Figure 6.35).

Indentation (hardness) testing is very common for bulk materials where the direct relationship between bulk hardness and yield strength is well known. It can be measured by pressing a hard, specially shaped point into the surface and observing indentation. This type of measurement is of little use for measuring thin films below 5×10^4 Å. Consequently, very little is known about the hardness of thin films. Nanoindenting and scratching are new AFM-based methods to characterize material mechanical properties on a very small scale. Features less than 100 nm across, as well as thin films less than 5 nm thick, can be evaluated. Test methods include indentation for comparative and quantitative hardness determination and scratching for evaluation of wear resistance and thin-film adhesion. During nanoindentation, a hard tip of known geometry makes an impression in a sample, with a penetration depth and force dynamically measured at resolutions of nanometers and micronewtons, respectively. The resulting force-depth curve can be analyzed to give bulk properties such as modulus and hardness, as well as to study deformation mechanisms. Empirical relations have correlated hardness with Young's modulus and with uniaxial strength of thin films. Hardness calculations must include both plastic and long-distance elastic deformation. If the indentation is deeper than 10% of the film, corrections for elastic hardness contribution of the substrate must also be included.[12]

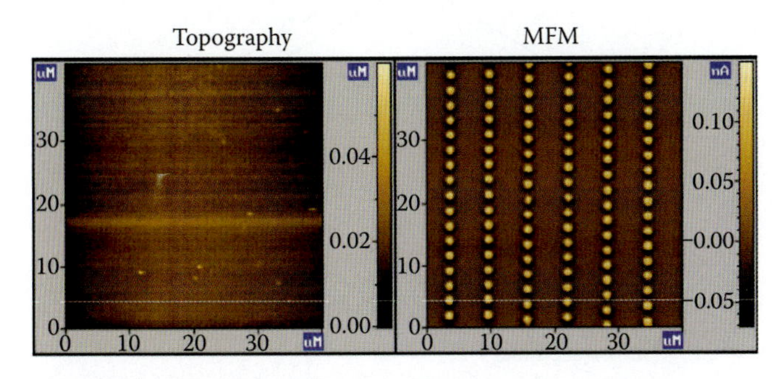

FIGURE 6.35 Magnetic force microscopy of a magneto-optical disc. (Veeco Instruments.)

Since in this new specialized instrument, load and displacement data are collected while the indentation is being introduced to the thin film; this eliminates the errors associated with later measurement of indentation size and provides continuous monitoring of load/displacement data similar to a standard tensile test. Load resolution may be 0.25 μN, displacement resolution 0.2–0.4 nm, and x-y sample position accuracy 0.5 μm. Diamond and sapphire are the primary materials of nanoindenting tips, but other hard materials also can be used, such as quartz, silicon, tungsten, steel, tungsten carbide, and almost any other hard metal or ceramic.

Bushan provides an excellent introduction to this field in the *Handbook of Micro/Nanotribology*.[13] In Figure 6.36 an indentation and scratch on polyurethane film is shown (17-μm scan).

In electrostatic force microscopy (EFM) a grounded tip first acquires the surface topography using the tapping mode. A voltage between the tip and the sample is applied in the second scan to collect electrostatic data (50–100 nm above the surface). EFM measures electric field gradient and distribution above the sample surface. EFM is used to monitor continuity and electric field patterns on samples such as semiconductor devices and composite conductors, as well as for basic research on electric fields on the microscopic scale. For thermal microscopy, there are quite a few options, as listed in Table 6.7, but scanning thermal microscopy (SThM) provides by far the best spatial resolution.

FIGURE 6.36 Indentation and scratch on polyurethane film (17-μm scan). (Asylum Research, http://www.asylumresearch.com/Products/NanoI/NanoIndenter.pdf.)

Table 6.7 Thermal Microscopy for Micro-Nano Devices

Techniques	Spatial Resolution
Infrared thermometry	1–10 μm*
Laser surface reflectance	1 μm*
Raman spectroscopy	1 μm*
Liquid crystals	1 μm*
Near-field optical thermometry	<1 μm
Scanning thermal microscopy	<100 nm

* Diffraction limit for far-field optics.

When an ultraminiaturized resistance is used as the scanning probe in an AFM, the image contrast reflects variations in thermal properties across the sample. The probe can be used as a thermometer or as a localized DC or AC heat source. When used as a thermometer, the image contrast represents variations in temperature across the surface. Hot spots, and also defects causing hot spots, in integrated circuits (ICs) and semiconductor devices may be detected this way. When used as a highly localized DC heat source, thermal conductivity/diffusivity variations across the surface can be mapped. In the AC heat source mode, thermal waves can be generated. Because of the evanescent nature of these waves, one can control the maximum depth of the sample that is imaged, according to the depth of penetration of the wave, which varies with the frequency chosen. SThM therefore looks below the surface in a controlled manner.

Conductivity and Carrier Type of a Material

In this section we cover how to go about measuring the conductivity and carrier type of thin films, such as thin-film resistors. In ICs and MEMS/NEMS devices, resistors are either thin-film metals or semiconductors. Metal resistors are deposited by evaporation, sputtering, or plating, whereas doping a semiconductor substrate creates semiconductor resistors. Metal resistors are patterned by wet etch, dry etch, or lift off and come in typical thicknesses of 0.1–10 μm and widths between 1 and 100 μm. Typical resistivity values range from 3 μΩcm for Al to 100 μΩcm for Nichrome. Metal films have nearly linear resistance-temperature behavior over wide temperature ranges, with a typical resistivity coefficient of 10^{-3}/°C. Issues that arise with metal resistors are morphological variations such as grain size changes and surface passivation. Semiconductor resistors, created by doping a semiconductor substrate, have a typical resistivity from 10^{-3} to 10^{7} Ωcm. The resistance-temperature behavior of semiconductor is much more dependent on temperature and impurities than that of metals and feature a typical resistivity coefficient per °C of −0.05. The biggest issues with semiconductor resistors are dopant concentration uniformity and deleterious effects of high-temperature processing.

In Volume I, Chapter 3, we saw that the Hall effect enables the measurement of the free-carrier type and concentration in semiconductors. Here we review two simpler techniques to determine these parameters, i.e., hot-point probe and four-point probe.

In Figure 6.37 we consider a very thin conductor film, of length *l*, width *W*, and thickness *t*, so that the current path area A = tW. The resistance, *R*, of this thin conductive layer is proportional to the resistivity, ρ, and inversely proportional to the thickness, *t*. It is convenient to define a sheet resistance, R_s, that is equal to ρ/t. Sheet resistance may be thought of as a material property for conductors that are essentially two-dimensional. For the resistance of a thin deposited conductor layer of length *L*, thickness *t*, and width *W*, we can write the following equation:

$$R = \frac{\rho}{t}\frac{L}{W} = R_s \frac{L}{W} \qquad (6.19)$$

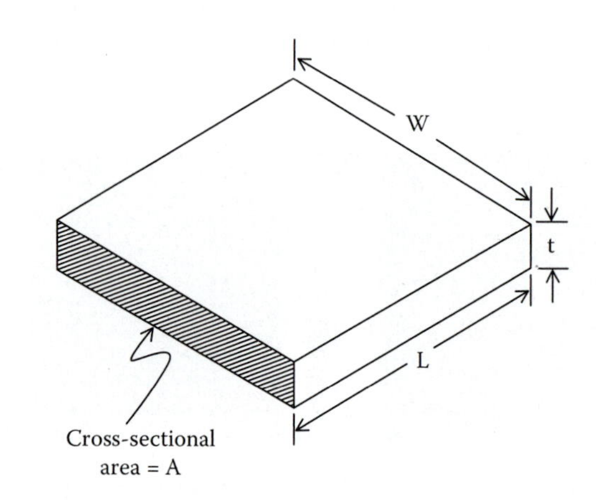

FIGURE 6.37 Sample geometry for a conductor thin film.

The ratio L/W is referred to as the number of squares (~) and is dimensionless. The sheet resistance has the units of ohms, but it is convenient to refer to it as ohms per square (Ω/~). The resistance of a rectangular thin layer is therefore the sheet resistance times the number of squares (*L*/*W*). The thickness and the number of squares, which can be accurately controlled with lithography, primarily determine the sheet resistance of a thin film.

The latter can be accurately controlled with lithography, whereas the sheet resistance primarily is determined by the film thickness carrier type and conductivity. Sheet resistance of a newly deposited thin film is typically measured with a four-point probe, an important tool in any IC or MEMS laboratory. For measuring resistance of a thin film, one could just measure the current that flows for a given applied voltage between two electrodes contacting the sample. However, in that case, contact resistances associated with the probes and current spreading problems around the probes can be significant and are not easily accounted for. Using four metal probes, as shown in Figure 6.38, allows one to force the current through the two outer probes using a constant current source and measure the voltage between the two inner probes. There still might be contact resistance and current spreading problems around the two outer probes, but since the voltage drop is measured between the two inner probes, problems with probe contacts are eliminated since no current flows through these inner contacts. The measured resistance (*V*/*I*) is converted to resistivity using the following relationship:

$$\rho = \frac{V}{I}.2\pi SF \tag{6.20}$$

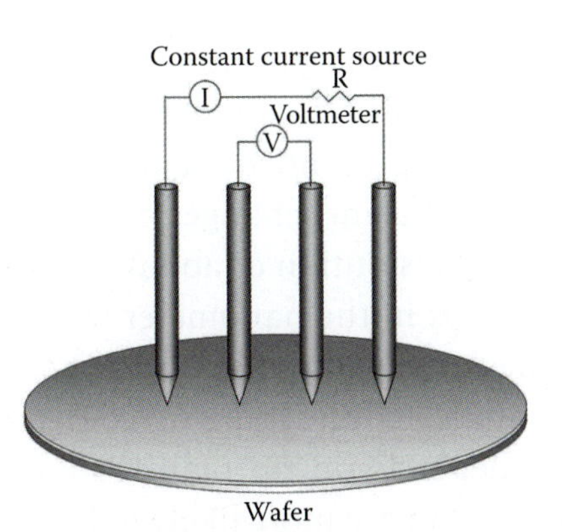

FIGURE 6.38 Four-point probe measurement arrangement.

with *S* the distance in centimeters between the metal probe needles and *F* a correction term. In the case the thickness, *t*, of the wafer or a measured film is much larger than *S* ($t >>>> S$), the correction factor *F* in Equation 6.18 equals 1. For thin films with $S >>> t$, Equation 6.20 must be replaced by the following:

$$\rho = \frac{\pi t}{\ln 2}\frac{V}{I} \tag{6.21}$$

and since $R_s = \rho/t$, we can calculate the sheet resistance as follows:

$$R_s = \frac{\pi}{\ln 2}\frac{V}{I} = 4.53\frac{V}{I} \tag{6.22}$$

The latter expression is most handy and easy to memorize.

The technique is more powerful than the sheet resistance method described above because it can determine the material type, carrier concentration, and carrier mobility separately. To determine whether a wafer is n- or p-type, a hot-point probe is a simple and reliable tool. The basic operation of this probe is illustrated in Figure 6.39. Two metal probes make ohmic contact with the semiconductor surface, and one probe is heated 25–100°C higher than the other. A voltmeter measures a potential difference across the probes, and the voltage polarity indicates whether the material is n- or p-type. In the case of an n-type sample, the majority carriers are electrons. At the hot probe, the thermal energy of the electrons is higher than at the cold probe, so the electrons will tend to diffuse away from the hot probe driven by the temperature gradient. As the electrons diffuse away from the hot probe, they leave behind a positively charged zone at the hot probe and the negatively charged mobile electrons tend to build up near the cold probe. This results in the hot probe becoming positive with respect to the cold probe. By a similar set of arguments, if the material were p-type, positively charged holes would be the majority carriers, and the polarity of the induced voltage would be reversed.

Dynamic Mechanical Properties

Interferometry with a stroboscopic LED and laser Doppler vibrometry can both be used to measure

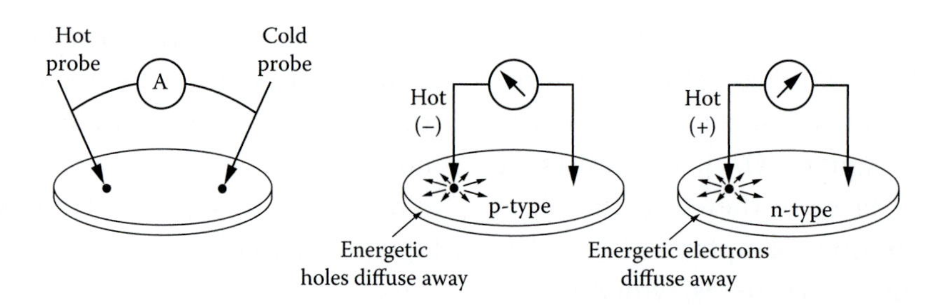

FIGURE 6.39 Using a hot-point probe to determine whether a semiconductor is n- or p-type.

contactless velocity and displacement of vibrating structures. A stroboscopic LED freezes device motion and enables an interferometer, as shown in Figure 6.11, to measure in-plane and out-of-plane motion at frequencies up to 10 MHz (this addition to an interferometer is called a *dynamic MEMS capability*).

A laser Doppler vibrometer is an interferometry system for accurately measuring velocity and displacement of vibrating structures again without contact. The system is based on the principle of the detection of the Doppler shift of coherent laser light scattered from a small area of the moving test structure. This system detects the Doppler shift of a coherent laser and then converts the frequency shift to velocity. As the laser light has a very high frequency, a direct demodulation of the light is impossible; therefore, an optical interferometer is used to mix the scattered light coherently with a reference beam. Such an arrangement can be based on a Michelson interferometer, as shown in Figure 6.10.

Structure Metrology

Introduction

Structural metrology measures crystallinity, defects, surface roughness, etc. In the structural metrology category, we review stylus profilometry for surface roughness and transmission electron microscopy (TEM) for crystallinity.

Stylus Profilometry

Surface texture of a sample is, in general, broken up into three components: roughness, waviness, and form (see also Volume II, Chapter 6, on mechanical energy-based removing). Roughness refers to the marks made on a sample surface by the machining tool itself, e.g., grooves from the tool or from each grinding granule on a grinding wheel. Waviness is the result of the distance between the tool and the workpiece changing during machining. This may be caused by vibration in the tool caused by sources outside the machining tool (e.g., a truck passing by) or by sources inside the tool (e.g., a worn spindle). Form errors are yet larger dimensional errors because the machine tool path is not always straight or it may be worn. This creates surface irregularities, but in a consistent manner, because the machine follow the same wrong path. All three surface finish components exist simultaneously and overlap one another. One often will want to look at each (roughness, waviness, and form) separately, so we make the assumption (a correct one, in most cases) that roughness has a shorter wavelength than waviness, which in turn has a shorter wavelength than form.

A stylus profilometer uses a hard tip (e.g., diamond) dragged over a surface with very light pressure and simultaneously measures form, dimension, texture, and step height. The first such devices for surface roughness measurement came about around 1930. The movement of the tip actuates a linear variable differential transducer (LVDT) that converts movement in an electrical signal. The tip of the profilometer is conical and has a finite rounded shape, which interacts with the sample being scanned. It may measure film thickness (step height) changes of 200 Å to 65 μm with a vertical resolution of about 1 nm. The vertical sensitivity is in the nanometer range, but steep edge profiles are distorted because of the shape of the tip, as indicated in Figure 6.40. The horizontal resolution depends on tip radius, so small-radius tips are better but are more likely to be damaged by mishandling.

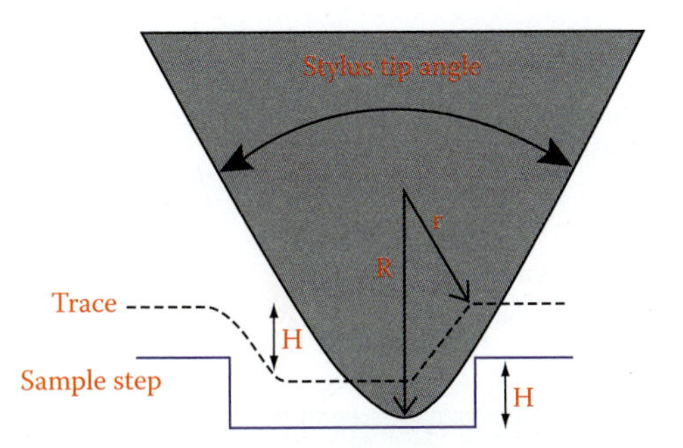

FIGURE 6.40 The stylus of a stylus profilometer.

As we saw earlier, an atomic force microscope (AFM) is a new instrument for measuring roughness of a sample surface. As for its design, an AFM has similarities to a conventional stylus profilometer; however, an AFM can image a sample surface and precisely measure up to nanometer size in three dimensions. Because of the sharper tip and small loading force, the lateral resolution in AFM is greatly improved in comparison with the conventional profilometer. In Table 6.8 we compare the major types of profilometers.

Transmission Electron Microscopy

Ernst Ruska and Max Knoll built the first transmission electron microscope (TEM) in 1931. As in the case of the first optical microscope, this was a very crude machine compared with its modern descendants. It did, however, put some important ideas about how to image with electrons to the test. For example, instead of using glass lenses to bend and focus light, electron microscopes use magnetic coils to do the same for electrons (H. Busch was the first to use a magnetic coil like a lens in 1926). Ruska used his first electron microscope to take pictures of gold and copper surfaces and produced images at about ×10 (about the same magnification as the 1595 Janssen microscope). The development of the electron microscope was much faster, though, than that of the optical microscope, and it was not long before electron microscopes far surpassed the resolution of optical microscopes (Ruska constructed a microscope that surpassed the 200-nm optical limit in 1933). Following the development

Table 6.8 Profilometers Compared

	White Light Interferometry and Phase Shift Technology	Stylus Contact	Atomic Force Microscope
Advantages	Excellent depth resolution, 0.1 nm; very good lateral resolution down, to 0.3 μm	Lowest-price technique for standard systems	Best lateral resolutions for detail imaging
	Use charge-coupled device camera to take roughness measurement in a few seconds	Depending on design can test steep angular surfaces	Can reveal surface structure in details
	Stitching technique allows coverage of large surfaces		
	Ideal for many microelectronic applications		
Disadvantages	Difficulty on high-angular surfaces or surface where line of interference pattern is difficult to see; chromatic aberration technique works better on these surfaces	Tip size often larger than 10 μm but can be down to 1 μm in higher-priced systems	Slow compared with white light systems
	More limited than white light chromatic aberration in the *z*-direction; stitching allows a few millimeters	Depending on size of probe and load applied, deformation may occur on softer materials, such as polymers or plastics	Requires more knowledge for use
		Normally two-dimensional instruments	Relatively small surfaces can be tested (100 μm by 100 μm)
		Slow, especially if three-dimensional imaging is performed	

Source: http://www.microphotonics.com/compare.htm.

of the first commercial available electron microscope by Ruska and the company Siemens in 1939, the electron microscope rapidly became an invaluable tool to science and engineering. Much of what we know about the nanoworld we owe indeed to the electron microscope. Although the technique of x-ray diffraction preceded the invention of the electron microscope and provided compelling evidence for atomic crystal structure, atoms themselves had never been seen until high-resolution TEMs came about.

A TEM maybe compared to a slide projector. In a slide projector, a beam of light passes through (transmits) a slide, and as the light passes through, the structures and objects on the slide modulate it. This transmitted beam is then projected onto the viewing screen, forming an enlarged image of the slide. TEMs work much the same way except that they shine a beam of electrons through a very thin specimen, and whatever electrons are transmitted are projected onto a phosphor screen or recorded on film or by a digital camera for the user to see. TEMs are used to characterize the microstructure of materials with very high spatial resolution. The electron gun, condenser lenses, and vacuum system in a TEM are similar to the ones in an SEM (see above), but in a TEM, a higher acceleration voltage of 120–200 kV is used (the operating voltage in an SEM ranges from 30 to 60 kV). In a scanning transmission electron microscope (STEM), the beam is rastered across the sample to form the image.

With a TEM, information about the morphology, crystal structure, defects, crystal phases, and composition is obtained by a combination of electron-optical imaging, electron diffraction, and small probe (20 Å) capabilities. The trade-off for this diverse range of structural information and high resolution is the challenge of producing very thin samples for electron transmission (Figure 6.41). A TEM sample must be 1000 Å or less in thickness in the area of interest. Sample preparation for TEM is more tedious and time consuming than for almost any other analytical technique.

There are three basic modes of TEM operation: the bright-field imaging mode, the dark-field imaging mode, and the diffraction mode. The way to choose which electrons form the image is by inserting an aperture into the back focal plane of the objective lens.

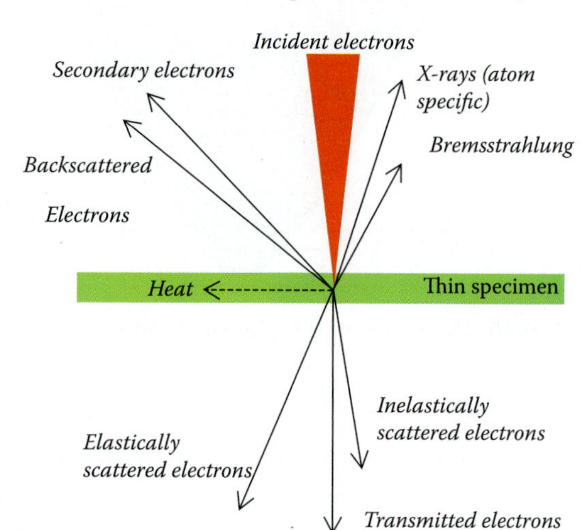

FIGURE 6.41 Electron path through a thin specimen (<1,000 Å) in transmission electron microscopy and scanning transmission electron microscopy.

In the bright-field (BF) mode of the TEM, an aperture is placed in the back focal plane of the objective lens, which allows only the direct beam to pass (Figure 6.42a). In this case, the image results from a weakening of the direct beam by its interaction with the sample. Therefore, mass-thickness and diffraction contrast contribute to image formation: thick areas, areas in which heavy atoms are enriched, and crystalline areas appear with dark contrast.

In dark-field (DF) images, the direct beam is blocked by the aperture while one or more diffracted beams are allowed to pass the objective aperture. Since diffracted beams have strongly interacted with the specimen, very useful information is present in DF images, e.g., about planar defects, stacking faults, or particle size (Figure 6.42b).

We can also think of the TEM as a diffraction camera. To obtain lattice images, a larger objective aperture is selected to allow many beams, including the direct beam, to pass (Figure 6.42c). The image is formed by the interference of the diffracted beams with the direct beam (phase contrast). If the point resolution of the microscope is sufficiently high and a suitable sample is oriented along a zone axis, then high-resolution TEM (HRTEM) images are obtained. In many cases, the atomic structure of a specimen can directly be investigated by HRTEM.

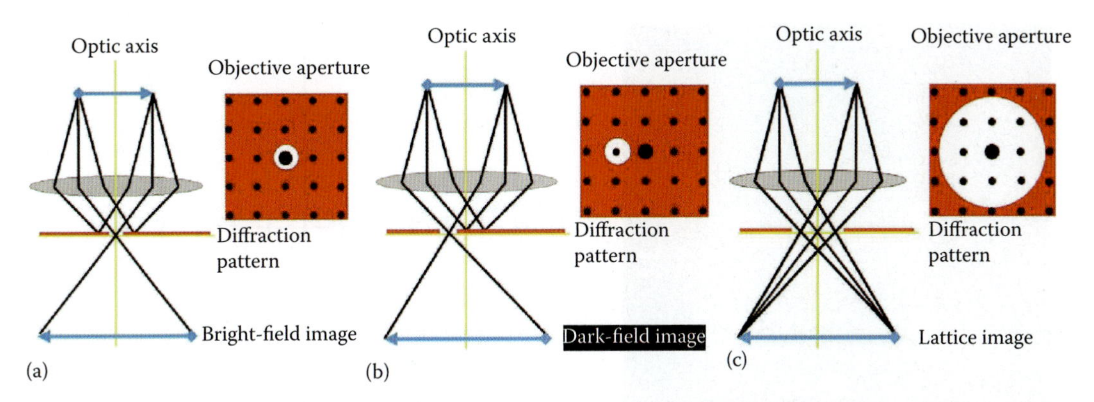

FIGURE 6.42 Bright-field image (a), dark-field image (b), and diffraction (c) in scanning transmission electron microscopy.

The objective lens forms a diffraction pattern in the back focal plane with electrons scattered by the sample and combines them to generate an image in the image plane (first intermediate image) as illustrated in Figure 6.43. Thus, diffraction pattern and image are simultaneously present in the TEM. In imaging mode (left panel in Figure 6.43), an objective aperture can be inserted in the back focal plane to select one or more beams that contribute to the final image (BF, DF, and HRTEM). In selected area electron diffraction (SAED), an aperture in the plane of the first intermediate image defines the region for which the diffraction is obtained (Figure 6.43, right panel). Which of them appears in the plane of the second intermediate image and is magnified by the projective lens on the viewing screen depends on the intermediate lens. Switching from real space (image) to reciprocal space (diffraction pattern) is easily achieved by changing the strength of the intermediate lens. By manipulating the magnetic lenses of the microscope, the diffraction pattern may thus be observed by projecting it onto the screen instead of the image (Figure 6.43).

An example of what a diffraction pattern obtained in this way may look like is shown in Figure 6.44. The center spot in this figure corresponds to the transmitted electron beam, and the other spots are diffracted portions of the initial electron beam. The location of the spots is governed by Bragg's law ($n\lambda = 2d \sin\theta$; see Volume I, Figure 2.17 and Equation 2.20).

Transmission and scanning transmission electron microscopy can produce images of atomic lattices with 2.3-Å point-to-point resolution. In Table 6.9, we compare scanning electron microscopy (SEM) and transmission electron microscopy (TEM) with optical microscopy (OM) and scanning proximal microscopy (SPM) (using both STMs and AFMs).

Compositional Metrology

Introduction

As examples of chemical composition metrology we discuss energy dispersive analysis of x-rays (EDAX) [also known as *energy dispersive x-ray spectroscopy* (EDX) or EDS], a standard procedure for identifying and quantifying elemental composition of specimen surfaces to shallow depths and Auger electron spectroscopy (AES) and scanning Auger

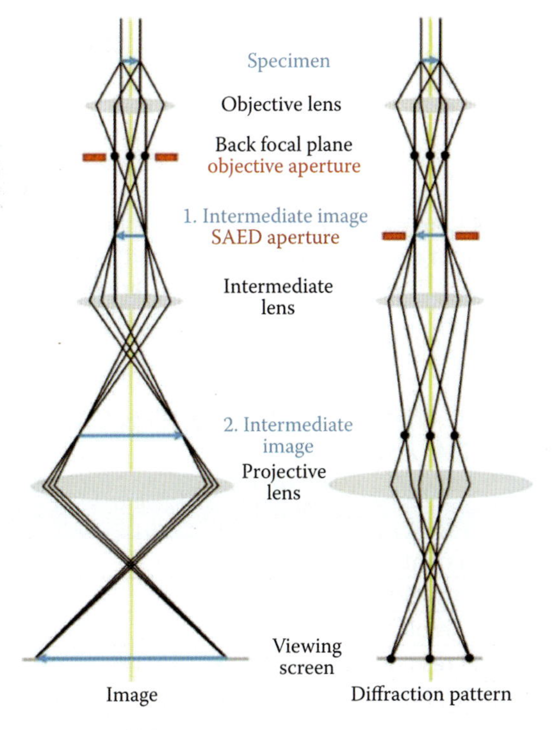

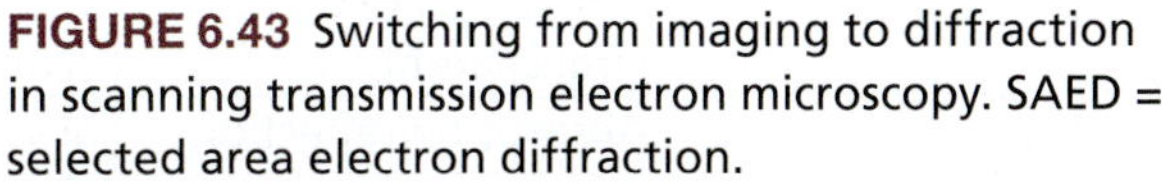
FIGURE 6.43 Switching from imaging to diffraction in scanning transmission electron microscopy. SAED = selected area electron diffraction.

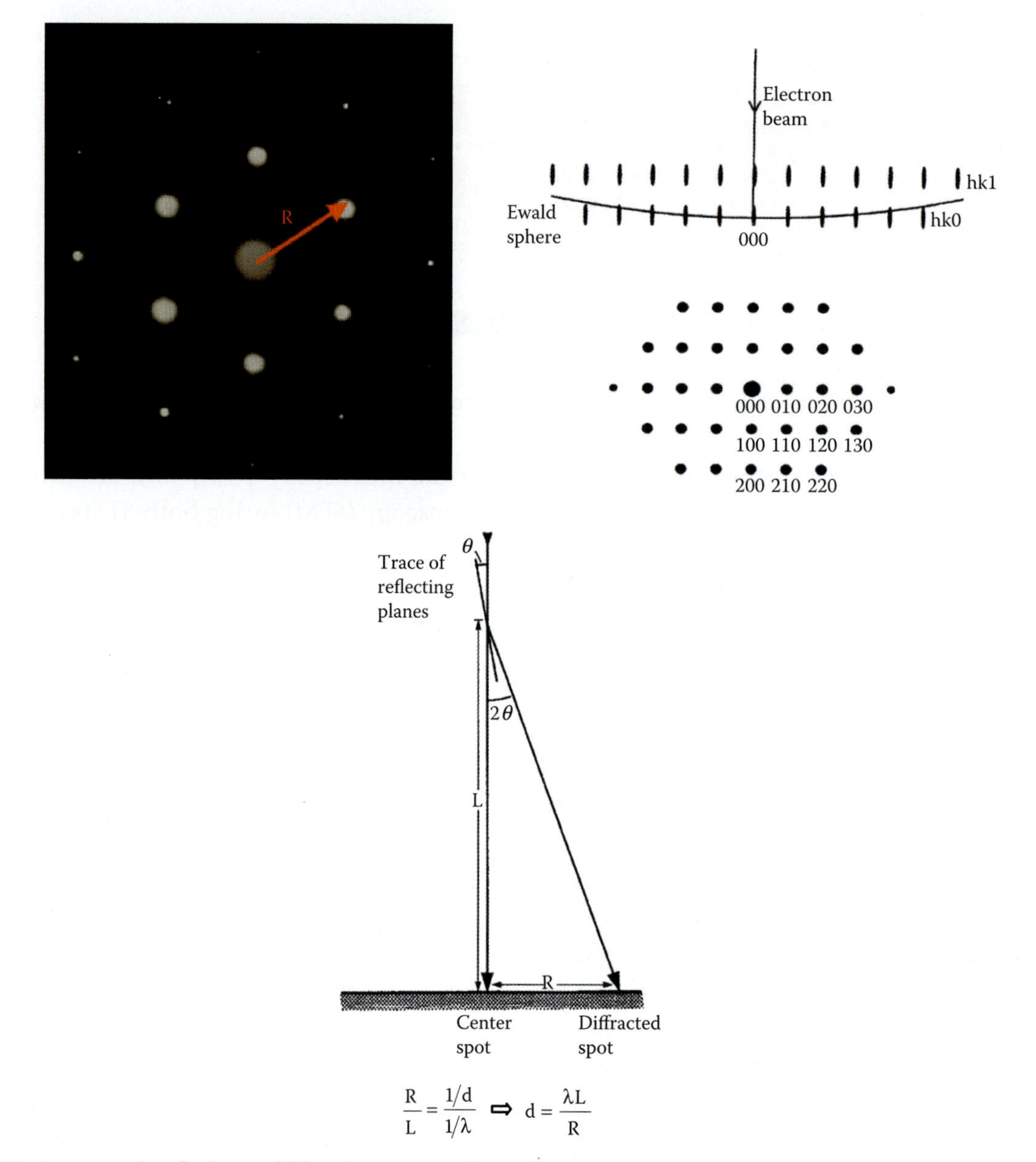

FIGURE 6.44 An example of what a diffraction pattern in a scanning transmission electron microscope may look like. The center spot corresponds to the transmitted electron beam, and the other spots are diffracted portions of the initial electron beam. The locations of the spots are governed by Bragg's law.

Table 6.9 The Use of SEMs and TEMs Compared with OM and Scanning Proximal Microscopy (Using Both STMs and AFMs)

	OM	SEM/TEM	Scanning Proximal Probe Microscopy
Environment	Air, liquid	Vacuum	Air, liquid, UHV
Depth of field	Small	Large	Medium
Lateral resolution	1 μm	1–5 nm: SEM 0.1 nm: TEM	2–10 nm: AFM 0.1 nm: STM
Vertical resolution	N/A	N/A	1.1 nm: AFM 0.2 0.01 nm: STM
Magnification	Up to 2,000	Up to 1,000,000	Up to 100,000,000
Sample	Not completely transparent	Unchargeable, vacuum compatible; thin film: TEM	Surface height <10 mm
Contrast	Absorption/reflection	Scattering/diffraction	Tunneling

microscopy (SAM), which can identify every atom from Li on up.

EDAX (EDX or EDS)

Energy dispersive analysis of x-rays (EDAX) is also called *EDX* and, more and more frequently, *EDS*. This compositional metrology technique is carried out in conjunction with scanning electron microscopy (SEM) and transmission electron microscopy (TEM). A modern trend in electron microscopy is indeed to fit x-ray analysis equipment as a bolt-on accessory to electron microscopes. Bombarding a specimen with electrons causes x-rays of characteristic wavelengths and energies to be emitted from the spot where the beam strikes the sample, and an EDS detector analyzes the x-ray photons by energy rather than wavelength. The EDS detector (sometimes called *EDX detector*) converts the energy of each individual x-ray into a voltage signal of proportional size. The detector involves a large single-crystal Si that either has been treated to approximate an ideal semiconductor or is of high-enough purity to truly be an intrinsic semiconductor. Front and back contacts on the Si piece are kept at several kilovolts potential relative to each other, and x-rays are converted into a charge by ionization of atoms in the crystal. To keep electronic noise at a minimum, the Si detector is cooled in liquid nitrogen. X-rays that pass through the front contact tend to dissipate their energy, creating electron-hole pairs in the intrinsic region. Next, this charge is converted to a voltage signal by a field effect transistor (FET) preamplifier, and the voltage signal is input into a pulse processor for measurement. Since each electron-hole pair has a characteristic creation energy, the total number of charge carriers created is proportional to the energy of the incident x-ray. A computer keeps track of the number of counts within each energy range, and the total collected x-ray spectrum can then be determined. Computer analysis of energy spectra makes it possible to measure accurately the nature and quantity of different elements in the material. This can then be used to make a chemical map of the surface. The technique is of little use to biologists because light elements such as carbon produce too weak an x-ray signal. However, it is of great value in materials science, particularly because an area as small as 1 μm^2 can be analyzed with precision (http://www.edax.com).

Typically, x-ray analysis is a bulk sensitive technique because of the small cross-section [10^{-6} Å^2 compared with 1 Å^2 for LEED (see above)], but grazing incidence (<1°) makes it surface sensitive (total reflection, because the refractive index of x-rays is slightly smaller than unity). The refracted wave becomes an evanescent wave traveling along the surface within a few 10 Å. This is called *gracing incidence x-ray diffraction* (GIXRD).

Auger Electron Spectroscopy and Scanning Auger Microscopy

Upon electron bombardment of metals, Pierre Auger, in 1925, observed (at first in a cloud chamber, and then on photographic plates) the occurrence of electron emissions with precisely determined energies. These electrons were later named Auger electrons and may serve to identify their parent atoms. It was J. J. Lander, who in 1953 suggested the idea of using Auger electron spectroscopy (AES) for surface analysis. AES was implemented as an analytic tool in 1967 (by Larry Harris), after increasing the sensitivity of the method by using differential spectra to discriminate the small Auger peaks in the electronic spectra. AES is now the most commonly used method by which to investigate surface composition.

The Auger process is a three-electron concerted process, as illustrated in Figure 6.45. An incident electron beam (2–10 keV) hits an atom and ejects a core electron, leaving a hole in a highly excited atom. The ionized atom relaxes by emitting either an x-ray photon (x-ray fluorescence) or by ejecting another electron (Auger electron). For lighter elements Auger emission is favored over x-ray fluorescence. In Auger spectroscopy the state of excitation is quenched by another electron falling into the hole left by the ejected core electron. The binding energy of the first electron is always very much greater than that of the second, so that there is still considerable excess energy in the system. This extra energy is shed by the ejection of a third electron from another shell. The residual kinetic energy of the ejected electron is measured and is a unique quantity determined by the binding energies of the three electrons and the work function of the materials involved.

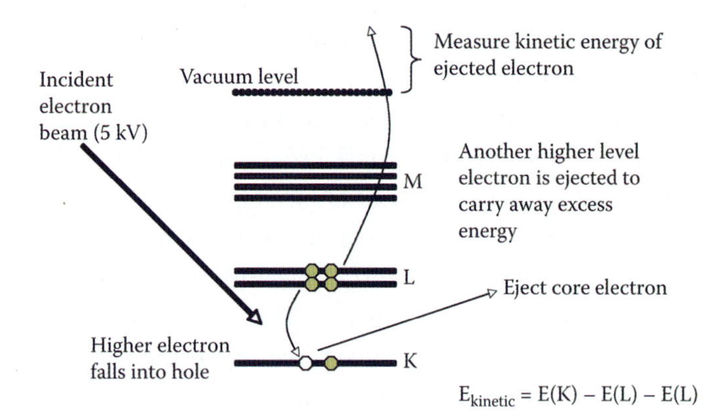

FIGURE 6.45 The Auger process.

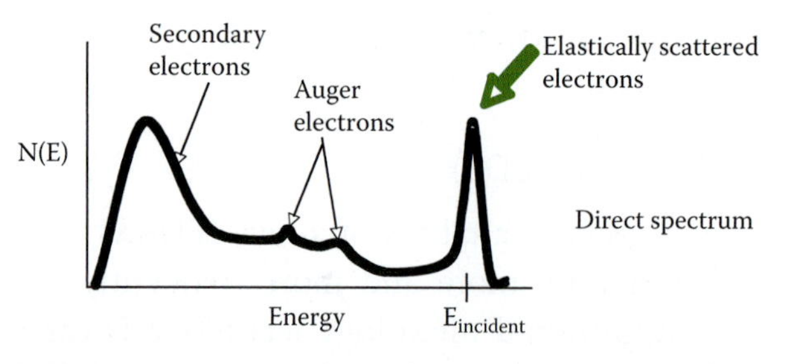

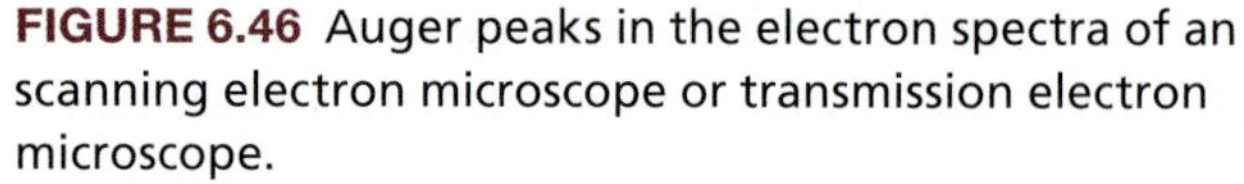

FIGURE 6.46 Auger peaks in the electron spectra of an scanning electron microscope or transmission electron microscope.

An Auger process is identified by the three letters of the atomic shells that participate in the process. The transition in Figure 6.45, for example, is a KLL Auger transition.

In Figure 6.18 above, we showed the relative intensity versus abundance of the various emission phenomena from a sample under e-beam radiation in an SEM and a TEM. Some of the electrons, in the energy region between backscattered and secondary electrons, arise from the Auger process occurring in atoms near the surface. Many Auger electrons will fully lose their energy via inelastic collisions in the solid. Therefore, the analyzer will collect only those Auger electrons originating from the near surface region. The number of emitted electrons N(E) versus energy is redrawn in Figure 6.46, with emphasis on the Auger electron peaks. Electrons emitted in the 100–1000 eV range (higher than secondary, lower than backscattered) are energy analyzed, and a spectrum is obtained that is unique to every atom. Often, hemispherical analyzers are used because of its better geometric properties. Hence, it gives very specific atomic identity information rather than the general small/large *z*-guidance provided by backscattered electrons (BSEs) in an SEM. Auger spectroscopy can identify every atom from Li on up (since three electrons are involved in the Auger process, H and He do not produce Auger electrons). Auger spectra for all elements are compiled in an Auger atlas.

Typically, the relatively small Auger signals N(E) are superimposed on a large background. Therefore the spectra are usually taken in the derivative mode, by applying a modulation voltage on the analyzer and detecting with a lock-in amplifier. This is demonstrated in the AES example of Figure 6.47.

Since the electron beam can be can be focused to a spot of <1 nm, one can carry out SAM. In the excitation process, the backscattered electrons also find their way into neighboring regions of the sample and can excite electrons and induce an Auger process in regions of the sample that were not in the original excitation volume. The Auger signal then arises from an apparently larger region of the sample than the incident electron beam directly excites. This degrades the spatial resolution in the SAM process by a factor of 2–5 compared with the same SEM experiment. As a consequence, the penetration depth and scattering within the solid typically limits the SAM resolution to ~20 nm, but recent instruments claim a resolution of ~6 nm. By correlating a specific signal with the excitation beam's location, a chemical map of a surface can be produced, as illustrated in Figure 6.48. Although chemical shifts lead to changes in the Auger energies, AES is usually not used to get chemical binding information because of the three electrons involved.

Summary: Microscope Comparisons

In Figure 6.49 we compare the lateral resolutions of optical microscopes (OMs), scanning electron microscopes (SEMs), interference microscopes (IMs), and proximal probe techniques such as the use of scanning tunneling microscopes (STMs) and atomic force microscopes (AFMs).

OM is usually what one thinks of when studying microscopy. It is prevalent and widely available. Its resolution is limited by the diffraction of light, which is limited to about ½ of the wavelength of the light being diffracted through the optical lensing elements of the microscope. While this is several

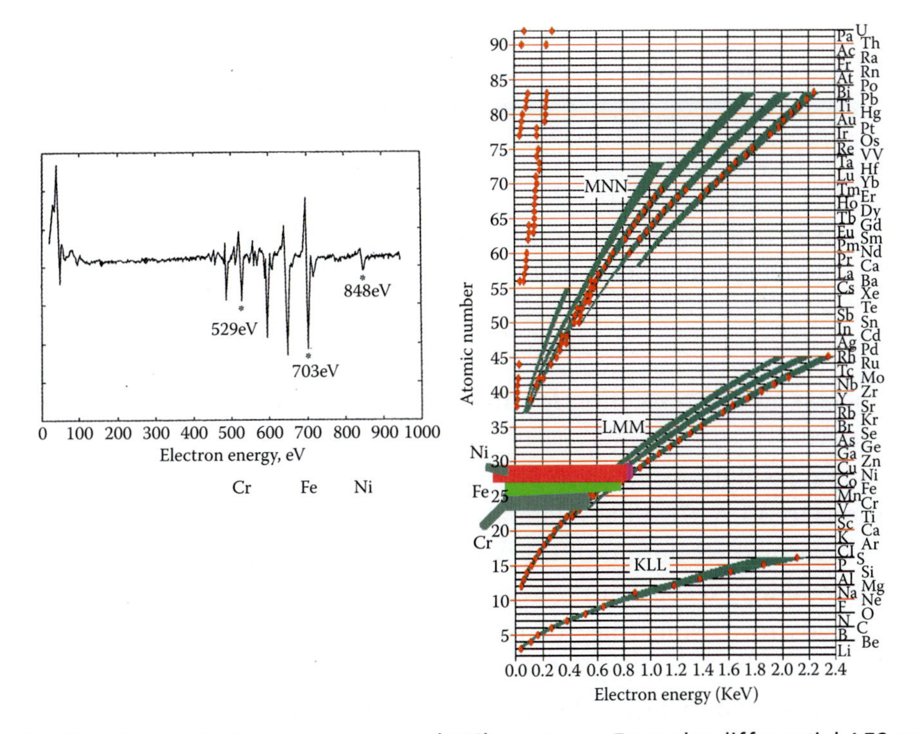

FIGURE 6.47 Example of an Auger electron spectroscopy (AES) spectrum. From the differential AES spectrum, Cr, Fe, and Ni have been identified.

hundred nanometers, in practice it is challenging to get good optical resolution down to the 1-μm range. Interference microscopy is a newer technique that uses optical wavelength light. Its lateral resolution is controlled exactly like that of optical microscopy, and it has the same limitations. As we will see below, its virtue comes from its vertical resolution and its wide dynamic range.

In an SEM, the resolution is limited by the diffraction of electron waves. Since these are in the angstrom range, SEMs with resolution of a few nanometers and a wide magnification range, from × 20 to × 650,000 are quite widely deployed. Recent reports have even described a new instrument able to observe atoms (a few angstroms), but these are not available for use yet. Probe microscopes such as STMs and AFMs have a resolution that is controlled by the sharpness of the probe tip. This can be in the range of a few nanometers, but the differing contrast mechanisms mean that an STM can easily achieve

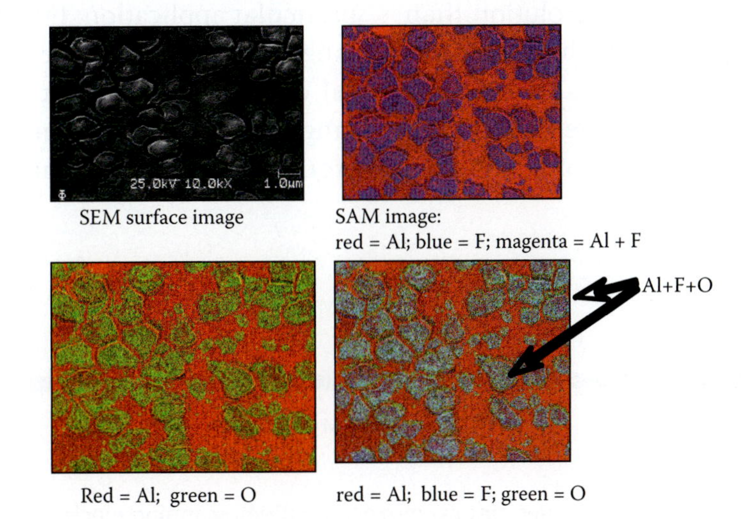

FIGURE 6.48 Scanning electron microscopy and Auger images of an aluminum oxide surface, in the absence and presence of fluorine contamination.

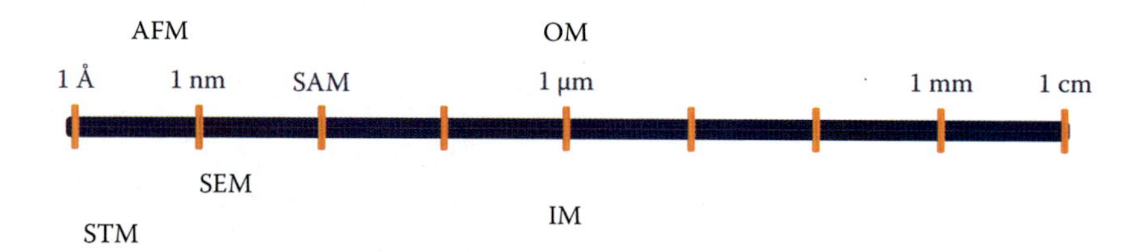

FIGURE 6.49 Lateral resolution of microscopies: optical microscopy (OM), scanning electron microscopy (SEM), scanning Auger microscopy (SAM), interference microscopy (IM), scanning tunneling microscopy (STM), and atomic force microscopy (AFM).

atomic resolution, whereas atomic resolution for an AFM is difficult to achieve.

AFM comes with lateral resolutions down to 30 Å and a vertical resolutions to 0.1 Å. STM has a lateral resolution of 1 Å and a vertical resolution of 0.1 Å. SAM is closely related to SEM, except that the scattering in the sample leads to a "smearing" of the incident beam and broadens the effective spot size. Its resolution general lags behind that of an SEM by a factor of 2–5.

Vertical resolutions for the types of microscopes listed above are quite different from their lateral resolutions, as is evident from Figure 6.50. OMs, SEMs, and SAMs do not give much vertical information at all. While it is rather straightforward to derive quantitative lateral data, these techniques do not provide for quantitation in the vertical direction. To counter this, it is possible to tilt samples and image their sides. However, in a specific image, the vertical information is very poor. Hence, these techniques have much better lateral resolution than vertical resolution. In contrast, probe microscopes and IMs have much better vertical resolution than lateral resolution. The IM, even though it is an optical technique, has a resolution limit approaching 1 nm. The AFM and the STM have vertical resolutions in the subatomic range of tens of picometers. The best instruments can resolve down to a few hundred femtometers.

We need to image large things as well as small. So another important point is, which instrument can provide the largest field of view? What is the range of features, largest to smallest, that can be observed? The resolution limits listed above do define the lower limits of the dynamic range of the microscopes reviewed, and the upper limit is controlled by the size of the field of view that can be imaged at once. OM is able to image from macroscopic features down to its resolution limit. The scanning electron techniques generally have a maximum field of view of a few millimeters. The proximal probe microscopes are quite limited in their maximum field of view. Also, the limitations are different for the lateral and vertical directions. Lateral limits range from a few micrometers to over 100 μm. By contrast, the range in the vertical direction is limited to around a few micrometers. These limits often are determining factors in assessing the applicability of a probe microscope technique for a particular application. IM has a very wide dynamic range. It is able to image features varying in vertical displacement of over a millimeter. The lateral range is set by its maximum field of view, which is around 5 mm. The dynamic ranges of the various

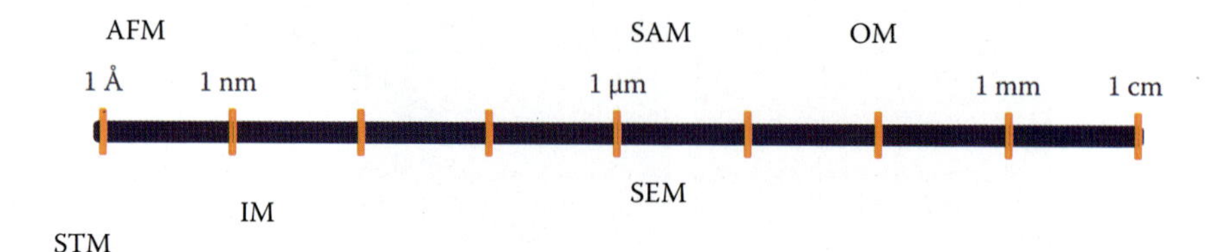

FIGURE 6.50 Vertical resolution of microscopies: optical microscopy (OM), scanning electron microscopy (SEM), scanning Auger microscopy (SAM), interference microscopy (IM), scanning tunneling microscopy (STM), and atomic force microscopy (AFM).

microscopes, considered here, are summarized in Figure 6.51.

MEMS Based Metrology: In Situ Stress-Measuring Techniques

Introduction

A stressed thin film will bend a thin substrate by a measurable degree. A tensile stress bends the surface and makes it concave, while a compressive stress renders the surface convex (see Volume II, Figure 7.58). The most common methods for measuring the stress in a thin film are based on this substrate bending principle. The deformation of a thin substrate due to stress is measured either by observing the displacement of the center of a circular disk or by using a thin cantilevered beam as a substrate and calculating the radius of curvature of the beam and hence the stress from the deflection of the free end. More sophisticated local stress measurements use analytical tools such as x-ray,[14] acoustics, Raman spectroscopy,[15] infrared spectroscopy,[16] and electron-diffraction techniques. Local stress does not necessarily mean the same as the stress measured by substrate bending techniques, since stress is defined microscopically, while deformations are induced mostly macroscopically. The relationship between macroscopic forces and displacements and internal differential deformation, therefore, must be modeled carefully. Local stress measurements may also be made using in situ surface micromachined structures such as strain gauges made directly out of the film to be tested.[17–22] The dimensions and structure of a thin suspended and pressurized micromachined membrane, as we learned above, may be measured by stylus profilometry, light microscopy, or interferometry. Intrinsic stress influences the frequency response of mobile microstructures (see Volume II, Equation 7.29), which can be measured by laser Doppler vibrometer or with an interferometer equipped with a stroboscope (see Figure 6.11: optical WYKO white-light profilometer with dynamic MEMS capability). Whereas residual stress can be determined from wafer curvature and dynamic deflection data, material structure of the film can be studied by x-ray diffraction and transmission electron microscopy (TEM). Krulevitch, among others, attempted to link the material structure of polysilicon and its residual strain.[23] Below, we will review stress-measuring techniques, starting with the more traditional ones and subsequently clarifying the problems and opportunities in stress measuring with in situ surface micromachined devices.

Disk Method: Biaxial Measurements of Mechanical Properties of Thin Films

For all practical purposes, only stresses in the x and y directions are of interest in determining overall thin-film stress, as a film under high stress can only expand or contract by bending the substrate and deforming it in a vertical direction. Vertical deformations will not induce stresses in a substrate because it freely moves in that direction. The latter condition enables us to obtain quite accurate stress values by measuring changes in bow or radius of curvature of a substrate. The residual stresses in thin films are large, and sensitive optical gauges may measure the associated substrate deflections.

The disc method most commonly used is based on a measurement of the deflection in the center of the disc substrate (say a Si wafer) before and after processing. Since any change in wafer shape is directly attributable to the stress in the deposited film, it

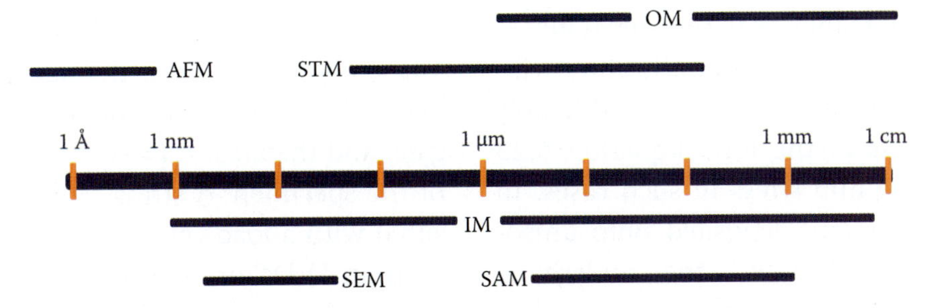

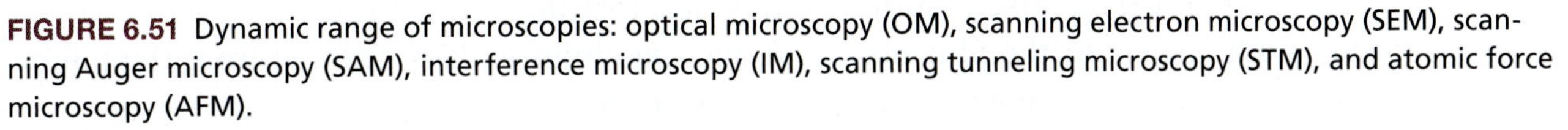

FIGURE 6.51 Dynamic range of microscopies: optical microscopy (OM), scanning electron microscopy (SEM), scanning Auger microscopy (SAM), interference microscopy (IM), scanning tunneling microscopy (STM), and atomic force microscopy (AFM).

is relatively straightforward to calculate stress by measuring these changes. Stress in films using this method is found through the Stoney equation,[24] relating film stress to substrate curvature, as follows:

$$\sigma = \frac{1}{R}\frac{E}{6(1-\nu)}\frac{T^2}{t} \tag{6.23}$$

where R = measured radius of curvature of the bent substrate
E/(1 – ν) = biaxial modulus of the substrate
T = thickness of the substrate
t = thickness of the applied film[25]

The underlying assumptions include the following:

- The disc substrate is thin and has transversely isotropic elastic properties with respect to the film normal.
- The applied film thickness is much less than the substrate thickness.
- The film thickness is uniform.
- The temperature of the disk substrate/film system is uniform.
- The disc substrate/film system is mechanically free.
- The disc substrate without film has no bow.
- Stress is equibiaxial and homogeneous over the entire substrate.
- Film stress is constant through the film thickness.

For most films on Si, we assume that $t \leq T$; for example, t/T measures ~10^{-3} for thin films on Si. The acceptability of the uniform thickness, homogeneous, and equibiaxial stress assumptions depend on the deposition process. Chemical vapor deposition (CVD) is a widely used process, as it produces relatively uniform films; however, sputter-deposited films can vary considerably over the substrate. In regard to the assumption of stress uniformity with film thickness, residual stress can vary considerably through the thickness of the film. Equation 6.23 gives only an average film stress in such cases. In cases where thin films are deposited onto anisotropic single-crystal substrates, the underlying assumption of a substrate with transversely isotropic elastic properties with respect to the film normal is not completely justified. Using single-crystal silicon substrates possessing moderately anisotropic properties such as <100> or <111> oriented wafers (Volume I, Equation 4.87) satisfies the transverse isotropy argument. Any curvature inherent in the substrate must be measured before film deposition and algebraically added to the final measured radius of curvature. To give an idea of the degree of curvature, 1 μm of thermal oxide may cause a 30-μm warp of a 4-inch silicon wafer, corresponding to a radius of curvature of 41.7 m.

The following companies offer practical disc method-based instruments to measure stress on wafers: ADE Corp. (Newton, MA; acquired by KLA-Tencor in 2006), Corning Tropel (Fairport, NY; http://www.corning.com/metrology), Ionic Systems (San Jose, CA; http://www.ionic.com), and KLA Tencor Instruments (Mountain View, CA; http://www.kla-tencor.com).[25] Figure 6.52a illustrates the sample output from KLA Tencor's optical stress analysis system. Figure 6.52b represents the measuring principle of Ionic Systems' optical stress analyzer. None of the above techniques satisfies the need for measuring stress in low modulus materials such as polyimide. For the latter applications, the suspended membrane approach (see below) is more suited.

Dog-Bone Method: Uniaxial Measurements of Mechanical Properties of Thin Films

Many problems associated with handling thin films in stress test equipment may be bypassed by applying micromachining techniques. One simple example of problems encountered with thin films is the measurement of uniaxial tension to establish the Young's modulus. This method, effective for macroscopic samples, proves problematic for small samples. The test formula is illustrated in Figure 6.53. The gauge length, *L*, of the bar-shaped structure in Figure 6.53a represents the region we allow to elongate, and the area, A (= $W \times H$), is the cross-section of the specimen. A stress, *F*/*A*, is applied and measured with a load cell; the strain, $\delta L/L$, is measured with an LVDT or another displacement transducer (a typical instrument used is the Instron 1123). The Young's modulus is then deduced from the following equation.

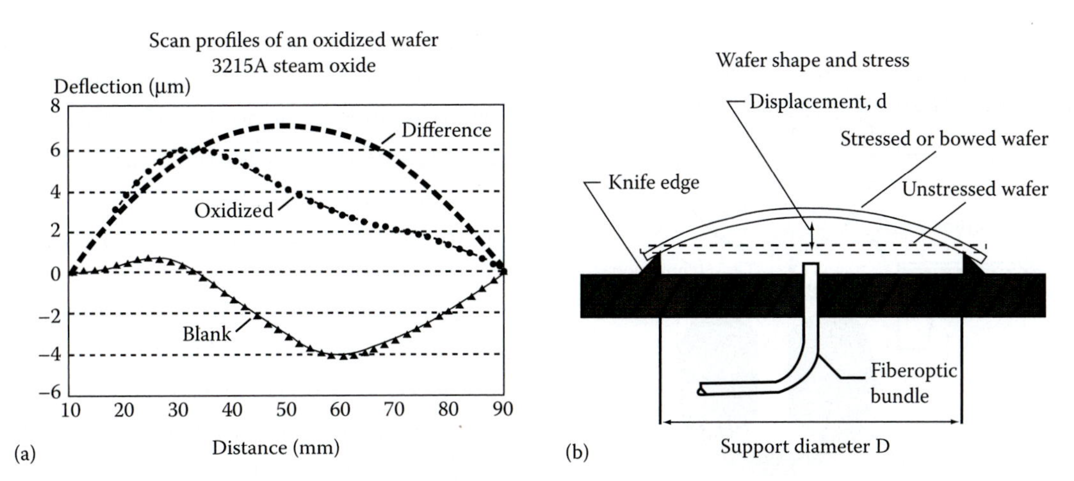

FIGURE 6.52 Curvature measurement for stress analysis. (a) Sample output from KLA Tencor's FLX 2908 stress analysis instrument, showing how stress is derived from changes in wafer curvature. (b) The reflected light technique, used by Ionic Systems to measure wafer curvature. (From Singer, P. 1992. Film stress and how to measure it. *Semicond Int* 15:54–58. With permission.[25])

$$E = \left(\frac{F}{A}\right)\left(\frac{L}{\delta L}\right) \tag{6.24}$$

The obvious problem, for small or large samples, is how to grip the sample without changing A. Under elongation, A will indeed contract by $(W + H)\delta\nu L$. In general, making a dog-bone shaped structure (also called an *Instron specimen*) solves that problem, as shown in Figure 6.53b. Still, the grips introduced in an Instron sample can produce end effects and uncertainties in determining L. Making Instron specimens in thin films is even more of a challenge, since the thin film needs to be removed from the surface, possibly changing the stress state, while the removal itself may modify the film.

As in the case of adhesion (see Volume II, Figure 7.56), some new techniques for testing stress in thin films, based on micromachining, are being explored. These microtechniques prove more advantageous than the whole wafer disc technique in that they are able to make local (in situ) measurements.

The fabrication of micro-Instron specimens of thin polyimide samples is illustrated in Figure 6.54A.[26] Polyimide is deposited on a p+ Si membrane in multiple coats. Each coat is prebaked at 130°C for 15 minutes. After reaching the desired thickness, the film is cured at 400°C in nitrogen for 1 h (Figure 6.54Aa). The polyimide is then covered with a 3,000-Å layer of evaporated aluminum (Figure 6.54Ab). The aluminum layer is patterned by wet etching (in phosphoric-acetic-nitric solution, referred to as the *PAN etch*) into the Instron specimen shape (Figure 6.54Ac). Dry etching transfers the pattern to the polyimide (Figure 6.54Ad). After removing the Al mask by wet etching, the p^+ support is removed by a wet isotropic etch (HNA) or an SF_6 plasma etch (Figure 6.54Ae), and finally the side silicon is removed along four pre-etched scribe lines, releasing the residual stress (Figure 6.54Af). The

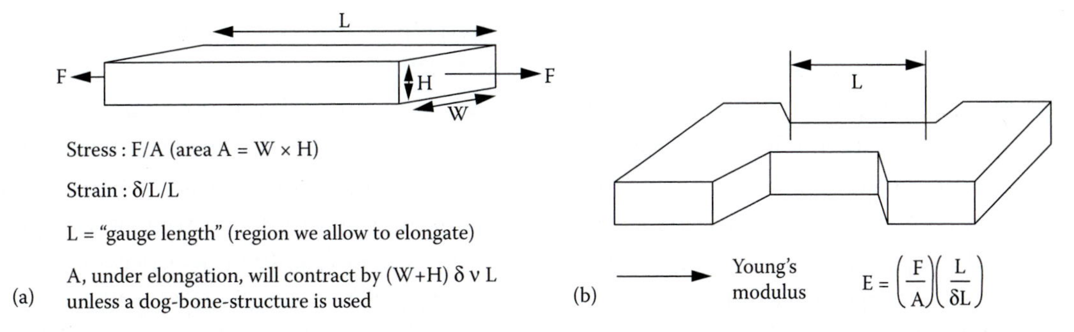

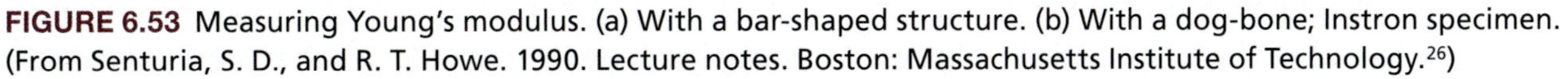

FIGURE 6.53 Measuring Young's modulus. (a) With a bar-shaped structure. (b) With a dog-bone; Instron specimen. (From Senturia, S. D., and R. T. Howe. 1990. Lecture notes. Boston: Massachusetts Institute of Technology.[26])

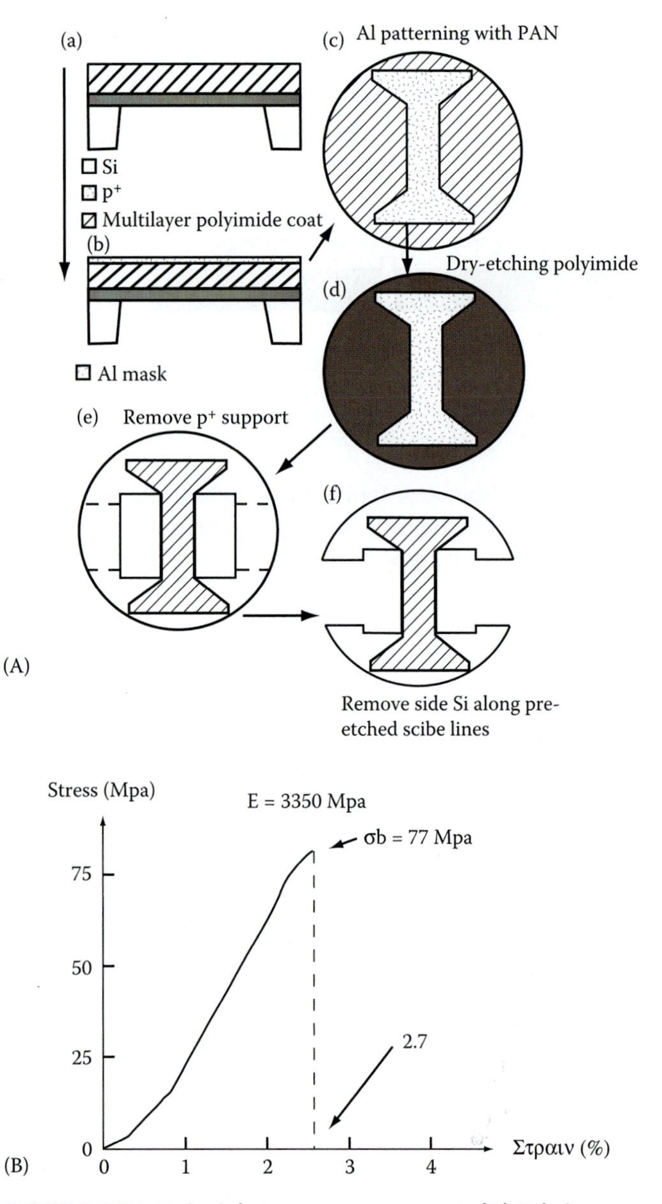

FIGURE 6.54 Uniaxial stress measurement. (A) Fabrication process of a dog-bone sample for measurement of uniaxial strain. (B) Stress versus strain for DuPont's 2525 polyimide. (Courtesy of Dr. F. Maseeh, IntelliSense.)

remaining silicon acts as supports for the grips of the Instron.[27] The resulting structure can be manipulated like any other macrosample without the need for removal of the film from its substrate. This technique enables the gathering of stress/strain data for a variety of commercially available polyimides.[28] A typical measurement result for DuPont's polyimide 2525, illustrated in Figure 6.54B, gives a break stress and strain of 77 MPa (σ_b) and 2.7% (ε_b), respectively, and 3350 MPa for the Young's modulus.

Figure 6.55a illustrates a micromachined test structure able to establish the strain and the ultimate stress of a thin film.[29] A suspended rectangular polymer membrane is patterned into an asymmetric structure before removing the thin supporting Si. Once released, the wide suspended strip (width, W_1) pulls on the thinner necks (total width, W_2), resulting in a deflection, δ, from its original mask position toward the right to its final position after release. The residual tensile stress in the film drives the deformation, δ, as shown in Figure 6.55a. By varying the geometry, it is possible to create structures exhibiting small strain in the thinner sections as opposed to others that exceed the ultimate strain of the film. For structures where the strain is small enough to be modeled with linear elastic behavior, the deflection, δ, can be related to the strain as follows:

$$\varepsilon = \frac{\sigma}{E} = \frac{\delta\left(\frac{W_1}{L_1} + \frac{W_2}{L_2}\right)}{W_1 - W_2} \tag{6.25}$$

where the geometries are defined as illustrated in Figure 6.55a. Figure 6.55b displays a photograph of two released structures, one with thicker necks, the other with necks so thin that they fractured upon release of the film. On the basis of the residual tensile strain of the film and the geometry of the structures that failed, the ultimate strain of the particular polyimide used was determined to be 4.5%.

Using similar micromachined tensile test structures, Biebl et al.[30] measured the fracture strength of undoped and doped polysilicon and found 2.84 ± 0.09 GPa for undoped material and 2.11 ± 0.10 GPa in the case of phosphorus doping, 2.77 ± 0.08 GPa for boron doping, and 2.70 ± 0.09 GPa for arsenic doping. No statistically significant differences were observed between samples released using concentrated HF or buffered HF. However, a 17% decrease of the fracture stress was observed for a 100% increase in etching time. These data contrast with Greek et al.'s[31] in situ tensile strength test result of 768 MPa for an undoped polysilicon film, a mean tensile strength approximately one-tenth that of single-crystal Si (6 GPa).[32] We normally expect polycrystalline films to be stronger than single-crystal films (see Volume II, Chapter 7). Greek et al.[31] explain this discrepancy for polysilicon by pointing out that their polysilicon

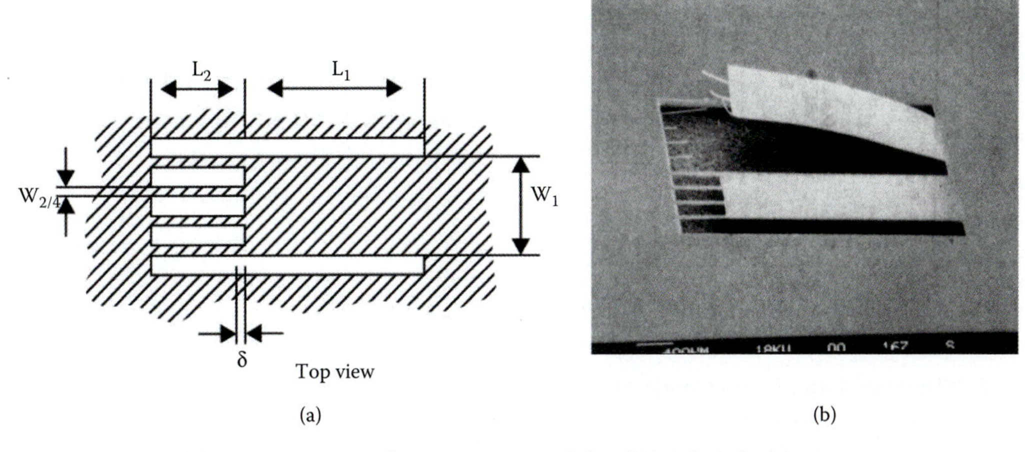

FIGURE 6.55 Ultimate strain. (a) Test structures for stress-to-modulus (strain) and ultimate stress measurements. (b) Two released structures, one of which has exceeded the ultimate strain of the film, resulting in fracture of the necks. (From Senturia, S. 1987. In *Proceedings: IEEE Micro Robots and Teleoperators Workshop* 3/1–5. Hyannis, MA. With permission.[29])

films have a very rough surface compared with single-crystal material, containing many locations of stress concentration where a fracture crack can be initiated.

Biaxial Measurements of Mechanical Properties of Thin Films: Suspended Membrane Methods

Disc stress-measuring techniques are not suitable for measuring stress in low modulus tensile materials such as polyimides, whereas suspended membranes are very convenient for this purpose. The same micromachined test structure used for adhesion testing, sometimes called the *blister test,* as shown in Volume II, Figure 7.56, can measure the tensile stress in low-modulus materials. This type of test structure ensues from shaping a silicon diaphragm by conventional anisotropic etching, followed by applying the coating, and finally removing the supporting silicon from the back with an SF_6 plasma.[29] By pressurizing one side of the membrane and measuring the deflection, one can extract both the residual stress and the biaxial modulus of the membrane. Pressure to the suspended film can be applied by a gas or by a point-load applicator.[33] The load-deflection curve at moderate deflections (strains less than 5%) answers to the following equation:

$$p = C_1 \frac{\sigma t d}{a^2} + C_2 \left(\frac{E}{1-\nu} \right) \frac{t d^3}{a^2} \qquad (6.26)$$

where

p = pressure differential across the film
d = center deflection
a = the initial radius
t = the membrane thickness
σ = the initial film stress

In the simplified Cabrera model for circular membranes, the constants C_1 and C_2 equal 4 and 8/3, respectively. For more rigorous solutions for both circular and rectangular membranes and references to other proposed models, consult Maseeh-Tehrani.[27] The relationship in Equation 6.26 can simultaneously determine σ and the biaxial modulus $E/1 - \nu$; plotting pa^2/dt versus $(d/a)^2$ should yield a straight line. The residual stress can be extracted from the intercept and the biaxial modulus from the slope of the least-squares best-fit line.[29] A typical result obtained via such measurements is represented in Figure 6.56. For the same DuPont polyimide 2525, measuring a Young's modulus of 3350 MPa in the uniaxial test (Figure 6.54B), the measurements give 5540 MPa for the biaxial modulus and 32 MPa for the residual stress. The residual stress-to-biaxial modulus ratio, also referred to as the residual biaxial strain, thus reaches 0.6%. The latter quantity must be compared with the ultimate strain when evaluating potential reliability problems associated with cracking of films. By loading the membranes to the elastic limit point, yield stress and strain can be determined as with the uniaxial test.

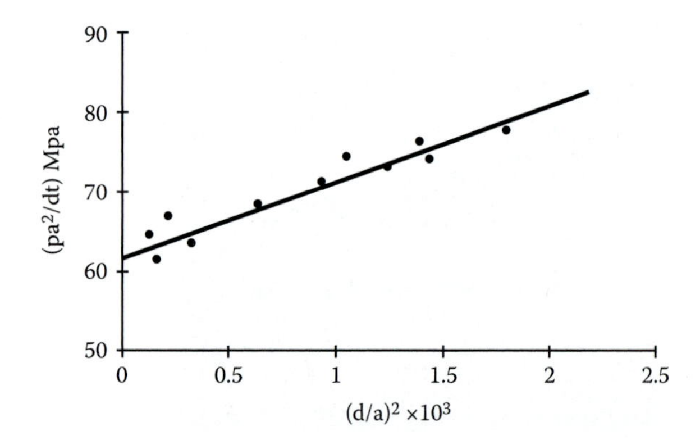

FIGURE 6.56 Load deflection data of a polyimide membrane (DuPont 2525). From the intercept, a residual stress of 32 MPa was calculated, and from the slope, a biaxial modulus of 5540 MPa was calculated. (From Senturia, S. 1987. In *Proceedings: IEEE Micro Robots and Teleoperators Workshop* 3/1–5. Hyannis, MA. With permission.[29])

Poisson Ratio for Thin Films

The Poisson ratio for thin films presents us with more difficulties to measure than the Young's modulus, as thin films tend to bend out of plane in response to in-plane shear. Maseeh and Senturia[28] combine uniaxial and biaxial measurements to calculate the in-plane Poisson ratio of polyimides. For example, for the DuPont polyimide 2525, they determined 3350 MPa for E and 5540 MPa for the biaxial modulus $[E/(1 - \nu)]$, leading to a Poisson ratio (ν) of 0.41 ± 0.1. The errors on both the biaxial and uniaxial measurements need to be reduced in order to develop more confidence in the extracted value of the Poisson ratio. At present, the precision on the Poisson ratio is limited to about 20%.

Other Surface Micromachined Structures to Gauge Intrinsic Stress

Introduction

Various other surface micromachined structures have been used to measure mechanical properties of thin films. We will give a short review here, but the interested reader might want to consult the original references for more details.

Clamped-Clamped Beams

Several groups have used rows of clamped-clamped beams (bridges) with incrementally increasing lengths to determine the critical buckling load and hence deduce the residual compressive stress in polysilicon films (Figure 6.57a).[34,35] The residual strain, $\varepsilon = \sigma/E$, is obtained from the critical length, L_c, at which buckling occurs (Euler's formula for elastic instability of struts):

$$\varepsilon = \frac{4\pi^2}{A}\frac{I}{L_c^2} \qquad (6.27)$$

where A is the beam cross-sectional area and I the moment of inertia. As an example, with a maximum beam length of 500 μm and a film thickness of 1.0 μm, the buckling beam method can detect compressive stress as small as 0.5 MPa. This simple Euler approach does not take into consideration additional effects such as internal moments resulting from gradients in residual stress.

Ring Crossbar Structures

Tensile strain can be measured by a series of rings (Figure 6.57b) constrained to the substrate at two points on a diameter and spanned orthogonally by a clamped-clamped beam. After removal of the sacrificial layer, tensile strain in the ring places the spanning beam in compression; the critical buckling length of the beam can be related to the average strain.[37]

Vernier Gauges

Both clamped-clamped beams and ring structures need to be implemented in entire arrays of structures. They do not allow easy integration with active microstructures because of space constraints. As opposed to proof structures, one might use vernier gauges to measure the displacement of structures induced by residual strain.[36] The idea was first explored by Kim,[38] whose device consisted of two cantilever beams fixed at two opposite points. The end movement of the beams caused by the residual strain is measured by a vernier gauge. This method requires only one structure, but the best resolution for strain measurement reported is only 0.02% for 500-μm beams. Moreover, the vernier gauge device may indicate an erroneous strain when an out-of-plane strain gradient occurs.[36] Other types of direct strain measurement devices are the T- and H-shaped structures from Allen et al.[20] and

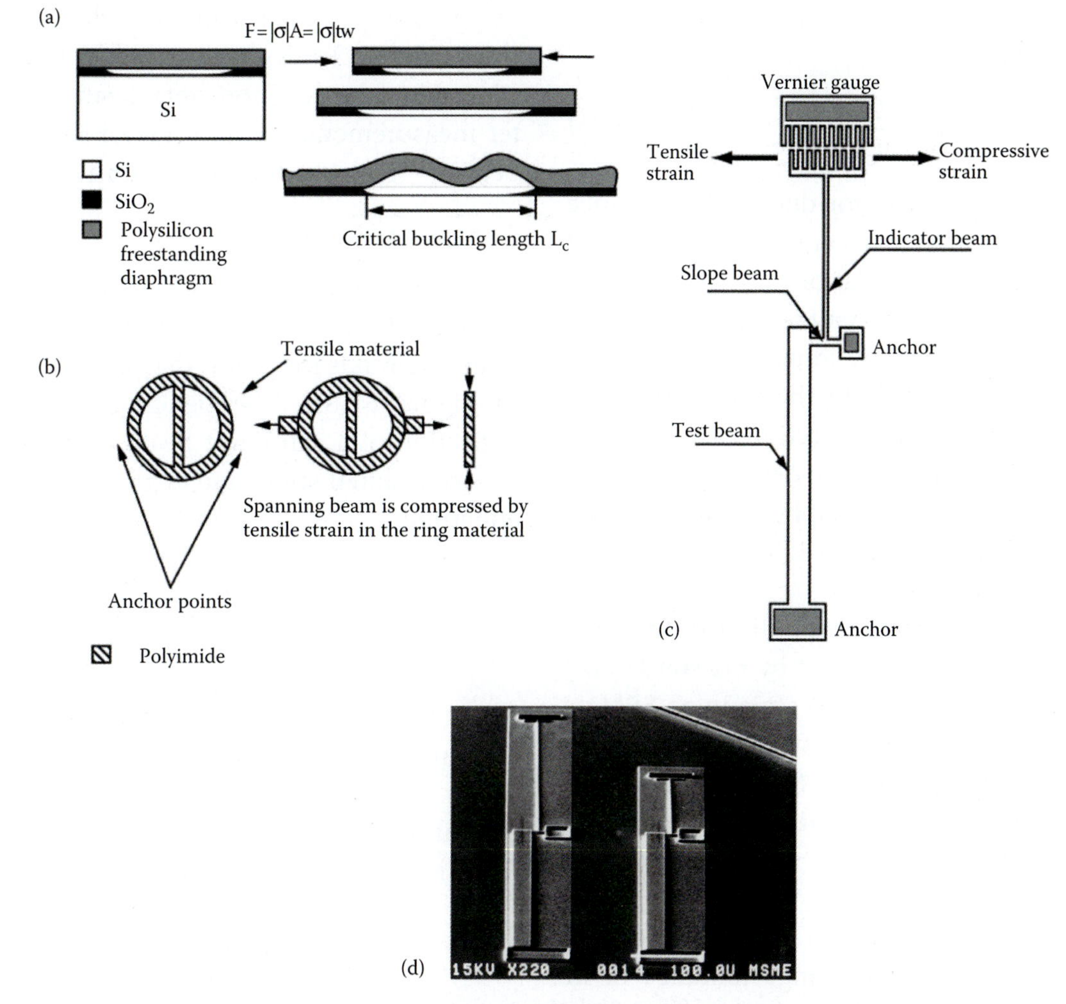

FIGURE 6.57 Some micromachined structures used for stress measurements. (a) Clamped-clamped beams: measuring the critical buckling length of clamped-clamped beams enables measurement of residual stress. (b) Crossbar rings: tensile stress can be measured by buckling induced in the crossbar of a ring structure. (c) A schematic of a strain gauge capable of measuring tensile or compressive stress. (d) Scanning electron microscope microphotograph of two strain gauges. (c and d from Lin, L. 1993. Selective encapsulations of MEMS: Micro channels, needles, resonators, and electromechanical filters. PhD thesis, University of California, Berkeley. With permission.[36])

Mehregany et al.[17] Optical measurement of the movement at the top of the T- or H-shape structures becomes possible only with very long beams (greater than 2.5 mm). They occupy large areas, and their complexity requires finite element methods to analyze their output. The same is true for the strain magnification structure by Goosen et al.[39] This structure measures strain by interconnecting two opposed beams such that the residual strain in the beams causes a third beam to rotate as a gauge needle. The rotation of the gauge needle quantifies the residual strain. A schematic of a micromachined strain gauge capable of measuring tensile or compressive residual stress, as shown in Figure 6.57c, was developed by Lin at University of California, Berkeley.[36] Figure 6.57d represents an SEM photograph of Lin's strain gauge. This gauge by far outranks the various in situ gauges explored. The strain gauge uses only one structure, can be fabricated in situ with active devices, determines tensile or compressive strain under optical microscopes, has a fine resolution of 0.001%, and resists the out-of-plane strain gradient. When the device is released in the sacrificial etch step, the test beam (length L_t) expands or contracts, depending on the sign of the residual stress in the film, causing the compliant slope beam (length L_s) to deflect into an S shape. The indicator beam (length L_i), attached to the deforming beam at its point of inflection, rotates through an angle θ, and the deflection, δ, is read on the vernier scale. The residual strain is calculated as follows:

$$\varepsilon_f = \frac{2L_s\delta}{3L_iL_tC} \tag{6.28}$$

where C is a correction factor due to the presence of the indicator beam.[36] This equation was derived from simple beam theory relationships and assumes that no out-of-plane motion will occur. The accuracy of the strain gauge is greatly improved, because its output is independent of both the thickness of the deposited film and the cross-section of the microstructure. Krulevitch used these devices to measure residual stress in in situ phosphorus-doped polysilicon films,[23] while Lin tested LPCVD silicon-rich silicon nitride films with it.[36]

An improved micromachined indicator structure, inspired by Lin's work, was built by Ericson et al.[40] By reading an integrated nonius scale in an SEM or an optical microscope, internal stress was measured with a resolution better than 0.5 MPa.[40,41] Both thick (10 μm) and thin (2 μm) polysilicon films were characterized this way.

Lateral Resonators

Biebl et al.[42] extracted the Young's modulus of in situ phosphorus-doped polysilicon by measuring the mechanical response of polysilicon linear lateral comb-drive resonators (see Volume II, Figure 7.76). The results reveal a value of 130 ± 5 GPa for the Young's modulus of highly phosphorus-doped films deposited at 610°C with a phosphine-to-silane mole ratio of 1.0×10^{-2} and annealing at 1050°C. For a deposition at 560°C with a phosphine-to-silane ratio of 1.6×10^{-3}, a Young's modulus of 147 ± 6 GPa was extracted.

Stress Nonuniformity Measurement by Cantilever Beams and Cantilever Spirals

Introduction

The uniformity of stress through the depth of a film introduces a property that it is extremely important to control. Variations in the magnitude and direction of the stress in the vertical direction can cause cantilevered structures to curl toward or away from the substrate. Stress gradients present in the polysilicon film must thus be controlled to ensure predictable behavior of designed structures when released from the substrate. To determine the thickness variation in residual stress, noncontact surface profilometer measurements on an array of simple cantilever beams[43, 44] or cantilever spirals can be used.[25]

Cantilever Beams

The deflections resulting from stress variation through the thickness of simple cantilever beams after their release from the substrate are shown in Figure 6.58A. The bending moment causing deflection of a cantilever beam follows out of prerelease residual stress and is given by the following equation:

$$M = \int_{-t/2}^{t/2} zb\sigma(z)dz \tag{6.29}$$

where $\sigma(z)$ represents the residual stress in the film as a function of thickness and b stands for cantilever width. Assuming a linear strain gradient Γ (physical dimensions 1/length) such that $\sigma(z) = E\Gamma z$, Equation 6.29 converts as follows:

$$\Gamma = \frac{12M}{Ebt^3} = \frac{M}{EI} \tag{6.30}$$

where the moment of inertia, I, for a rectangular cross-section is given by $I = bt^3/12$. The measured deflection z, i.e., the vertical deflection of the cantilever's endpoint, from beam theory for a cantilever with an applied end moment, is given as follows.

$$z = \frac{ML^2}{2EI} = \frac{\Gamma L^2}{2} \tag{6.31}$$

Figure 6.58B represents a topographical contour map of an array of polysilicon cantilevers. The cantilevers vary in length from 25 to 300 μm by 25-μm increments. Notice that the tip of the longest cantilever resides at a lower height (approximately 0.9 μm closer to the substrate) than the anchored support, indicating a downward-bending moment.[43] The gradients can be reduced or eliminated with a high-temperature anneal. With integrated electronics on the same chip, long high-temperature processing must be avoided. Therefore, stress gradients can limit the length of cantilevered structures used in surface-micromachined designs.

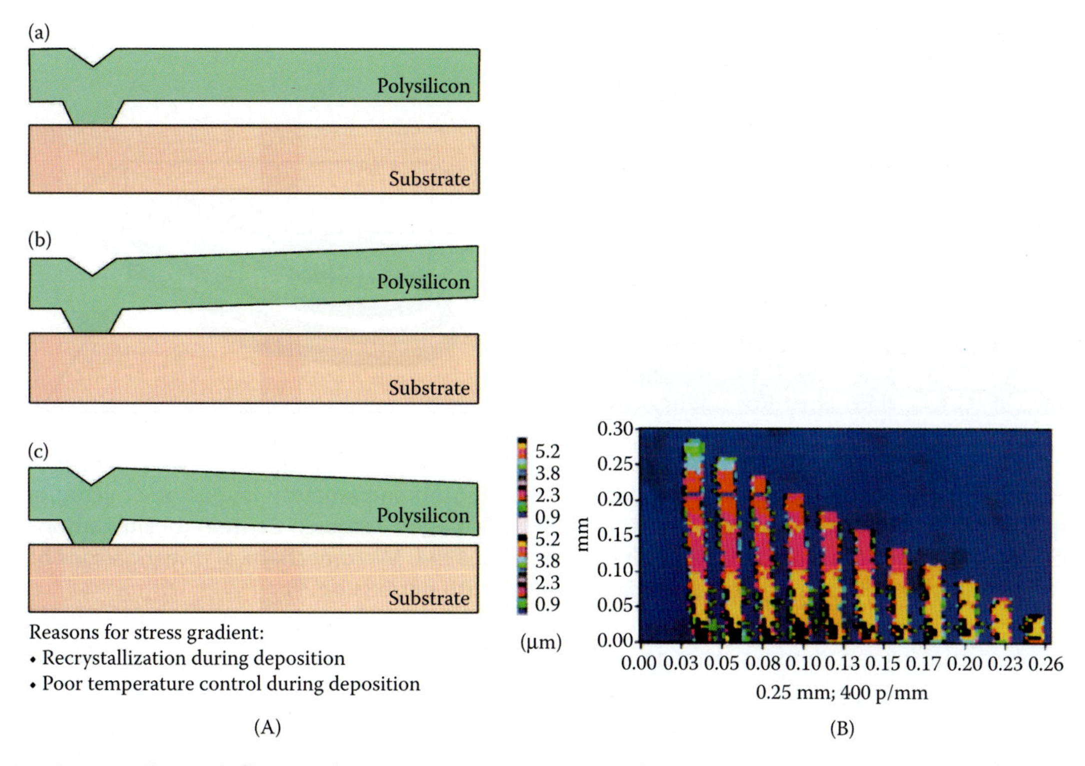

FIGURE 6.58 Microcantilever deflection for measuring stress nonuniformity. (A) Micro-cantilever deflection for measuring stress nonuniformity. (a) No gradient. (b) Higher tensile stress near the surface. (c) Lower tensile stress near the surface. (B) Topographical contour map of polysilicon cantilever array. (From Core, T. A., W. K. Tsang, and S. J. Sherman. 1993. Fabrication technology for an integrated surface-micromachined sensor. *Solid State Technol* 36:39–47. With permission.[43])

The use of silicon-on-insulator (SOI)-based micromachined cantilevers for measuring surface stress induced during adsorption of biomolecules is illustrated in Volume I, Example 4.1.

Cantilever Spirals

Residual stress gradients also can be measured by Fan's cantilever spiral, as shown in Figure 6.59a.[45] Spirals anchored at the inside spring upward, rotate, and contract with positive strain gradient (tending to curl a cantilever upward), whereas spirals anchored at the outside deflect in a similar manner in response to a negative gradient. Theoretically, positive and negative gradients produce spirals with mirror symmetry.[45] The strain gradient can be determined from spiral structures by measuring the amount of lateral contraction, the change in height, or the amount of rotation. Krulevitch presented the computer code for the spiral simulation in his doctoral thesis.[23] Figure 6.59b shows a simulated spiral with a bending moment of $\Gamma = \pm 3.0$ mm^{-1} after release.

Krulevitch compared all the above surface micromachined structures for stress and stress gradient measurements on polysilicon films. His comments are summarized in Table 6.10.[23] He found that the fixed-fixed beam structures for determining compressive stress from the buckling criterion produced remarkably self-consistent and repeatable results. Wafer curvature stress profiling proved reliable for determining average stress and the true stress gradient as compared with micromachined spirals. Measurements of curled cantilevers could not be used a lot, as the strain gradients mainly were negative for polysilicon, leading to cantilevers contacting the substrate. The strain gauge dial structures were useful over a rather limited strain-gradient range. With too large a strain gradient, curling of the long beams overshadows expansion effects and makes the vernier indicator unreadable.

Modeling

Modeling is the use of mathematical equations to simulate and predict real events and processes. In this modeling section we introduce finite element analysis and review software packages helpful in the

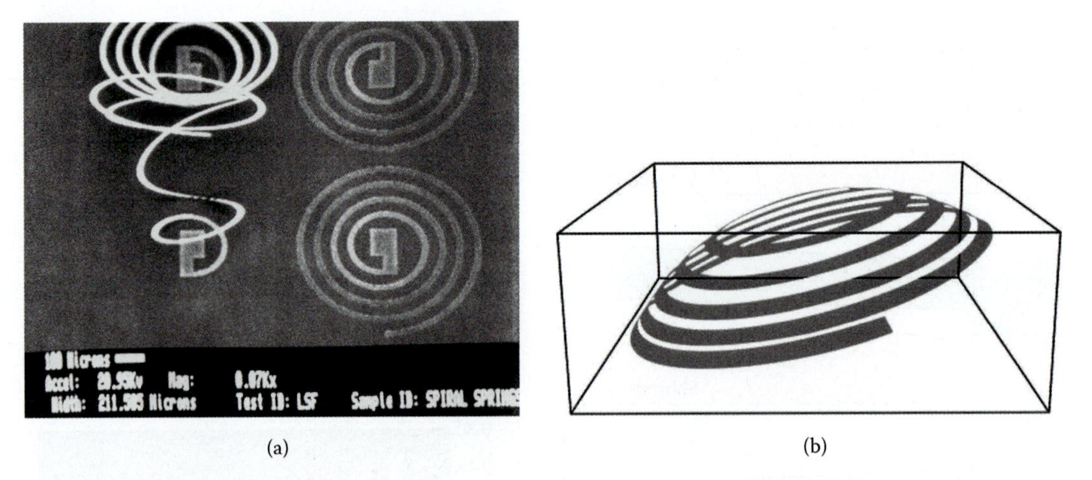

(a) (b)

FIGURE 6.59 Cantilever spirals for stress gradient measurement. (a) Scanning electron microscope micrographs of spirals from an as-deposited polysilicon. (Courtesy of Dr. L.-S. Fan, IBM, Alamaden Research Center.) (b) Simulation of a thin-film micromachined spiral with $G = 3.0\ mm^{-1}$. (From Krulevitch, P. A. 1994. Micromechanical investigations of silicon and Ni–Ti–Cu thin films. PhD thesis, University of California, Berkeley. With permission.[23])

design phase of miniaturized devices, that is, software for computer-aided design (CAD).

After brainstorming about sensor specifications and sensor transduction principles and making preliminary designs on paper and the white board, it is a good idea to initiate a CAD of the overall microsystem to grasp how all the components, including the package, fit together. In Figure 6.60 we show a three-dimensional visualization (construction) of a NiTi-based valve (the principle of this shape memory type valve is discussed in Chapter 8, this volume, on actuators).

More than visualization is needed though for CAD, and a software package in general is structured as sketched in Figure 6.61 with the design aids used to create the design, simulation to develop the technology, and verification to check the design. The final verification is during fabrication itself; the goal here is to avoid wasteful and slow experiments by carrying out less costly computer work and to get the fabrication "right" the first time.[53]

During design, the process may be separated in a conceptual design and simulation phase and a phase of final design of masks and processes. The ideal suite of CAD tools required for each activity is summarized in Table 6.11;[54] besides visualization, simulation, and verification, a good MEMS CAD package typically includes an extensive materials database.

With CAD for MEMS application in mind, IC development software packages have been expanded with finite element analysis (FEA) programs and

Table 6.10 Summary of Various Techniques for Measuring Residual Film Stress

Measurement Technique	Measurable Stress State	Remarks
Wafer curvature	Stress gradient, average stress	Average stress over entire wafer, provides true stress gradient, approximately 5 MPa resolution
Vernier strain gauges	Average stress	Local stress, small dynamic range, resolution = 2 MPa
Spiral cantilevers	Stress gradient	Local stress, provides equivalent linear gradient
Curling beam cantilevers	Large positive stress gradient	Local stress, provides equivalent linear gradient
Fixed-fixed beams	Average compressive stress	Local stress measurement

Source: Krulevitch, P. A. 1994. Micromechanical investigations of silicon and Ni–Ti–Cu thin films. PhD thesis, University of California, Berkeley. With permission.[23]

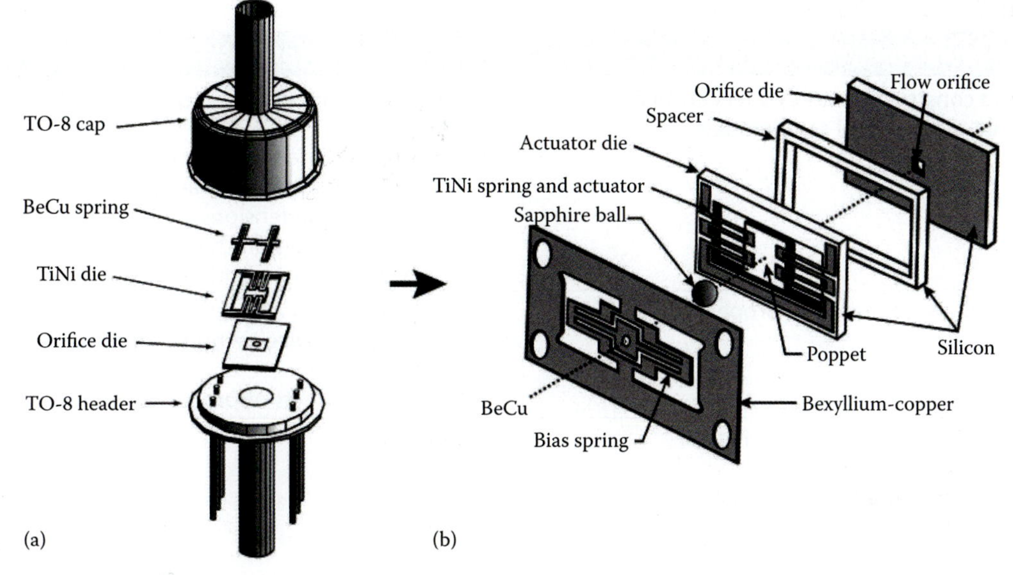

FIGURE 6.60 Microflow Analytical's shape memory alloy valve. Normally closed micro valve. A current through the NiTi die lifts it from the orifice die. (Courtesy of Dr. D. Johnson, TiNi Corporation.)

materials databases. This type of software is paying off, even for IC development where there is a growing need for the ability to perform mechanical analysis of microelectronic devices, both in ensuring structural reliability against failure of thin film layers and in evaluating the effects of various external loads, including the effects of temperature and humidity. An understanding of the required processes and materials properties that ensure reproducibility for commercial MEMS products has emerged slowly, and several CAD systems that have facilitated the wider acceptance of MEMS are now on the market. Most include a materials database that can be updated by the user. Details on CAD packages for MEMS is preceded by a short introduction to finite element analysis. We finish this chapter by referring to software packages of interest in NEMS and bio-NEMS modeling.

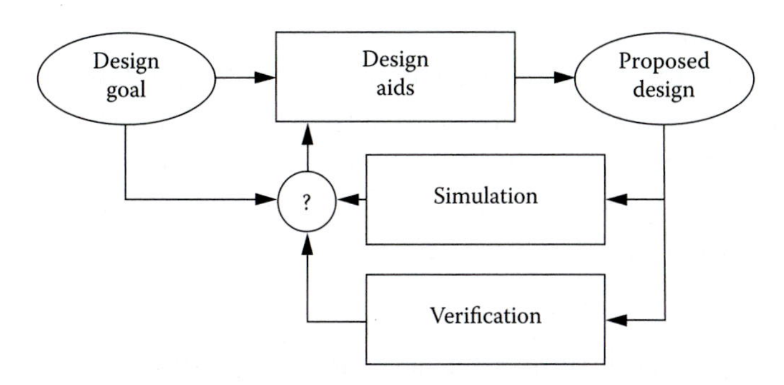

FIGURE 6.61 Design, simulation, and verification.[54]

Finite Element Analysis

Introduction

Finite element analysis is used in almost every engineering discipline. Early FEA packages were limited to simulators using linear static stress analysis. Today, they have been extended to include nonlinear static stress, dynamic stress (vibration), fluid flow, heat transfer, electrostatic forces, etc. These capabilities are frequently combined to perform coupled-force analyses that consider multiple physical phenomena and are tightly integrated within a CAD interface. The best-developed application areas, which also happen to be very important for microsystem design, are electrostatics, stress/strain analysis of solid structures, heat conduction analysis, and fluid dynamics. FEA typically includes the steps outlined below, where we will assume that we are dealing with a stress/strain analysis of a solid structure.

Geometry Development

The essence of FEA is "divide and conquer," and the first step consists of dividing the whole MEMS structure under analysis into a finite number of subdivisions of special shapes or elements that are interconnected at specific points, called *nodes* (Figure 6.62). The nodes are discrete points of the MEMS structure where the analysis will reflect the

Table 6.11 Ideal Tool Suites for MEMS

Conceptual Design and Simulation	Final Design of Masks and Process
Rapid construction and visualization of three-dimensional solid models	Process simulation or process database, including: • Lithographic and etch process biases (the difference between as-drawn and as-fabricated dimensions) • Process tolerances on thicknesses, lateral dimensions, doping and resistivity levels
A database of materials properties	Design optimization and sensitivity analysis: • Variation of device sizing to optimize performance • Analysis of effects of process tolerances
Simulation tools for basic physical phenomena. For example: • Thermal analysis: heat flow • Mechanical and structural analysis: deformation • Electrostatic analysis: capacitance and charge density • Magnetostatic analysis: Inductance and flux density • Fluid analysis: pressures and flow	Mask layout
Coupled-force simulators. For example: • Thermally induced deformation • Electrostatic and magnetostatic actuators • Interaction of fluids with deformable structures	Design verification, including: • Construction of a three-dimensional solid model of the design, using the actual masks and process sequence • Checking the design for violation of any design rules imposed by the process • Simulation of the expected performance of the design, including the construction of macromodels of performance useable in circuit simulators to assess overall system performance
Formulation and use of macromodels. For example: • Lumped mechanical equivalents for complex structures • Equivalent electric circuit of a resonant sensor • Feedback representation for coupled-force problems	

Source: Senturia, S. D., and R. T. Howe. 1990. Lecture notes. Boston: Massachusetts Institute of Technology.[26]

response of the component to an applied loading. This response is defined in terms of nodal degrees of freedom. In the case of stress analysis, up to six degrees of freedom are possible at each node (three components of translation and three components of rotation), depending on the element type selected (e.g., beam, plate, 2D, and 3D elements). Element selection is a function of product geometry and loading conditions. The element selected affects the results as each element has characteristic properties. A model can use more than one type of element.

Material Property Assignment

In the second stress/strain finite element analysis step, the modulus of elasticity or rigidity and Poisson ratio are used to define material properties for each element. For a nonlinear material condition (elastic-plastic), a stress-strain curve represents the material properties.

Mesh Generation

The grid of connecting elements at common nodes is called the *mesh* (Figure 6.62). Based on the element

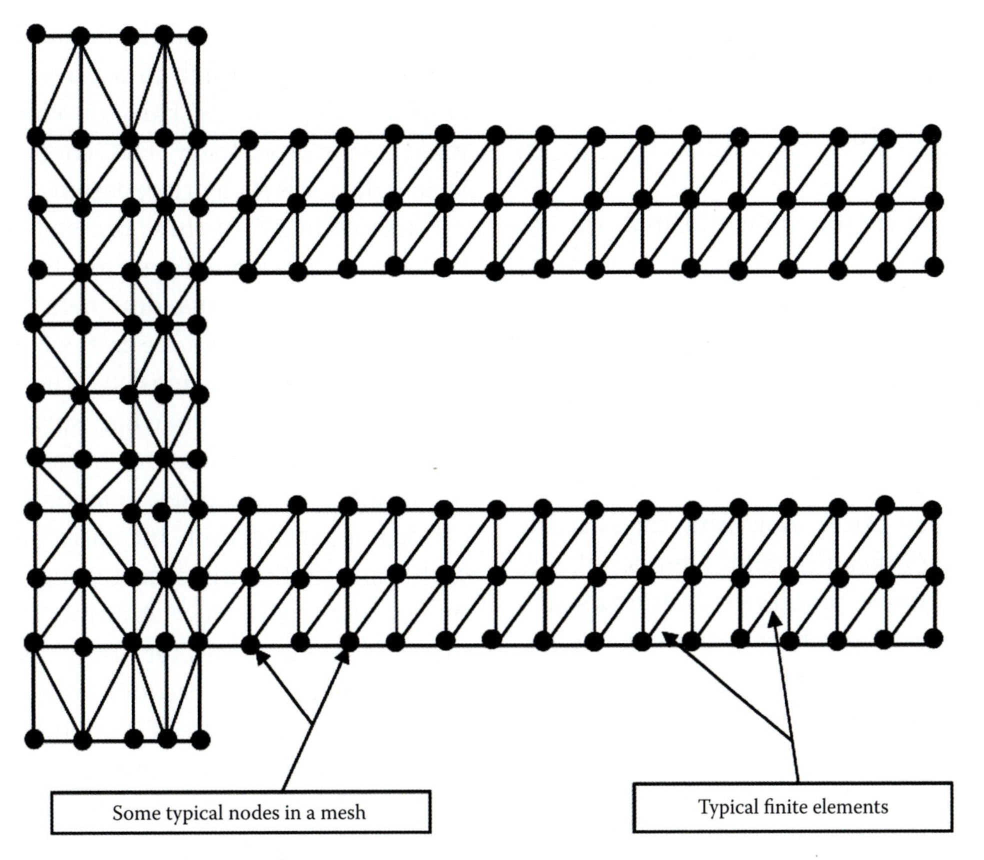

FIGURE 6.62 Finite element analysis mesh.

types selected, automatic mesh generation subdivides the geometry into finite elements in a third step of the FEA process. The element density within each segment of the structure is either chosen or automatically determined. When adjacent elements share nodes, the displacement field is continuous across the shared element boundaries and loads can be transferred between the elements. Analysis is performed on the elements instead of the whole structure, and solutions at the element level are "assembled" to get the corresponding solutions for the whole structure. Mathematically, this is achieved by calculating nodal stiffness properties for each element and arranging these into matrices. The appropriate matrix transformation generates a global stiffness matrix from the existing element matrix.

Boundary and Loading Conditions

In a fourth step, the appropriate boundary and loading conditions apply constraints to the model. In stress analysis, the boundary conditions specify displacement constraints (i.e., in the x-, y-, or z-direction), and the loading conditions apply loads to the model such as concentrated forces at specified nodes or pressure at specified edge surfaces of the element.

Run Analysis

During analysis, in a fifth step, the program processes the equation matrices with the applied loads and boundary conditions to calculate displacements, strain, natural frequencies, or other data specified by the user. This provides a stress distribution across the entire MEMS structure. The high-stress regions should have the highest element density, as a finer mesh increases the accuracy of the model. Various adaptive methods find the critical regions in the model and make the necessary mesh refinement to reduce the error for the next iteration before reaching convergence.

Results

Results obtained in a sixth step include nodal and element information, displacement at nodes, stresses and strains in each element, and various forms of graphical display of the solution. In this regard, a von Mises stress plot often is used. The von

Mises stress (also called *effective* or *equivalent stress*) is defined as follows:

$$\sigma_M = \sqrt{\frac{(\sigma_1 - \sigma_2)^2 + (\sigma_2 - \sigma_3)^2 + (\sigma_1 - \sigma_3)^2}{2}} \quad (6.32)$$

where σ_1, σ_2, and σ_3 are the three principal stresses. The von Mises stress is compared with the yield strength for plastic yielding and for the prediction of rupture of the structure.[55]

Data Correlation and Design Optimization

Finally, in a seventh step, experimental data are collected to correlate the FEA model results and to formulate a baseline. After comparing the baseline results, design modification and remodeling are available. This iterative process constitutes design optimization, namely, combining the engineering requirements, geometric parameters, CAD model, and performance goals into a computer simulation to achieve the optimum design.

As mentioned in the introduction, in addition to considering a part's ability to withstand mechanical stresses, today's FEA software also enables engineers to predict other real-world stresses. These might include the effects of extreme temperatures or temperature change (heat transfer analysis), the flow of fluids through and around objects (fluid flow analysis), or voltage distributions over the surface or throughout the volume of an object (electrostatic analysis). As these effects often are coupled, it is important that the FEA program consider their effects on one another. For instance, a MEMS chip may be heating up over time, cooling down by airflow from a fan, vibrating against other parts, and be electrically charged. A typical approach is to isolate and calculate each variable then feed the results into the FEA program one at a time. But since each variable could also affect the others, either a coupled analysis or tools for relating results is required.

CAD for MEMS

Introduction

CAD tools were essential to the evolution of MEMS from laboratory status to a bona fide and accessible manufacturing process. Without them, fabrication would have remained in the domain of experts, and evolution of the design process would still rely on empirical approaches. Early developments in CAD for MEMS evolved from 2D IC circuit work and often lacked the tools for a priori design of micromechanical devices in 2.5 or three dimensions essential for high-aspect ratio microsystems. Given the freedom to fabricate 3D systems, designers think in terms of the whole system, not a series of 2D reticulations of the system (i.e., masks). The mask-to-model approach is still somewhat suitable for surface micromachined devices but becomes less useful as the number of processes, materials, and mechanical degrees of freedom increases, as in the case of high-aspect ratio micromachining (especially for a process such as LIGA).[56] A number of current commercial CAD packages for MEMS are competing for market share, and the codes have been improving steadily. For the future, CAD programs addressing generic miniaturization problems rather than just micromachining for Si and poly-Si would be highly desirable. The architecture should have both lithographic and nonlithographic machining options, adhere to accepted format standards for data communication, and integrate available design, modeling, and simulation software from both traditional and nontraditional (i.e., IC based manufacturing) industries wherever possible.

Derived CAD Packages for MEMS

For some of the simplest micromachined structures, modest modification of existing IC software packages enabled designing of the micromachine as in regular IC processing. Marshall et al.,[57] for example, modified the standard Magic VLSI mask layout package (http://opencircuitdesign.com/magic/archive/papers/Magic_A_VLSI_Layout_System.pdf) to permit the design of simple micromachined structures using CMOS processes. This modification makes standard foundry work through the MOSIS service (http://www.mosis.org) possible.[58] Other traditional IC mask layout software packages are used by MEMS engineers from companies such as TMA and Cadence. In the public domain, one can turn to KIC from the University of California, Berkeley.[59] For a comparison of various layout design and verification software packages, visit http://www.ee.rochester.edu/users/sde/cad/ LDV.html. Mechanical drafting tools with interfaces to mask making, such as AutoCAD, have full geometric flexibility. Other popular mechanical CAD tools for 3D solid model construction, meshing, application of loads and boundary conditions, and visualization of results include Pro/Engineer, PATRAN, and I-DEAS. Suppliers of CAD systems to the semiconductor industry also offer one- and two-dimensional process simulators. These simulators cover standard semiconductor processes such as oxidation, diffusion, thin film deposition, and plasma etching. An example is SAMPLE[60–62] for simulation of projection lithography, deposition, and etching. Simulation tools for basic physical properties include ABAQUS for structural and thermal finite-element simulation and FASTCAP (with its follow-on MEMCAP) (available for download from http://www.rle.mit.edu/cpg/research_codes.htm) for electrostatic boundary-element simulation.

COSMOS (http://www.cosmosm.com) has a MEMS analysis capability for mechanical, thermal, electrostatic, electromagnetic, and fluid flow modeling. COSMOS sells a version that is tightly integrated with the 3D modeling software SolidWorks™. SolidWorks is a player in the mechanical CAD tools arena for 3D solid model construction. Its tight integration with the COSMOS FEA package also enabled it to enter the MEMS simulation industry. AnSoft (http://www.ansoft.com) has software products for 2D and 3D electromagnetic field and thermal field finite-element simulation capabilities applicable to RF-MEMS analysis and design. AnSoft's products integrate electromagnetic, circuit, and system engineering modules. Field equation solvers such as QuickField (http://www.quickfield.com) are increasingly being used for MEMS modeling. Such software provides a quick analysis tool for 2D field problems such as electromagnetics, heat transfer, and fluid flow.

CAD Specifically Developed for MEMS

Some early examples of CAD programs developed specifically for MEMS include Oyster from IBM,[63] MEMCAD from Massachusetts Institute of Technology (MIT),[64] CAEMEMS from the University of Michigan,[65] and the SESES system from ETH

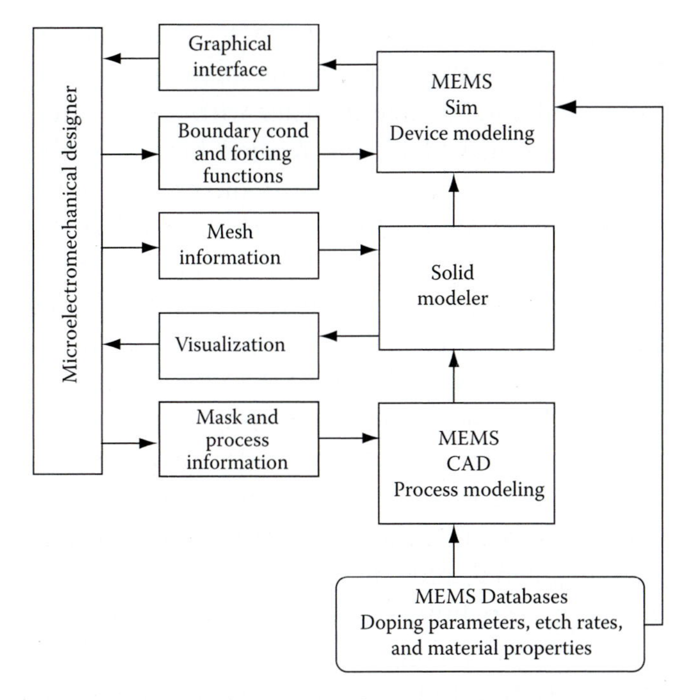

FIGURE 6.63 Block diagram of the CAEMEMS system. [Crary, S., and Y. Zhang. 1990. In *Proceedings: IEEE Micro Electro Mechanical Systems (MEMS '90)*, 113–14. Napa Valley, CA. With permission.[65]]

(Zurich).[66] These programs mainly address bulk Si micromachining and poly-Si surface micromachining applications.

The CAEMEMS program provides a database of material properties and process model parameters, a process modeler that takes process information as input and produces a solid model of the device to be fabricated, a solid modeler that allows for visualization and design verification, and a device modeler that performs finite element simulations of components and systems. The database can be updated by the user as new information becomes available. A block diagram of the CAEMEMS software package is shown in Figure 6.63.[65] ANSYS Multiphysics is the version of ANSYS FEA software that has been used for MEMS modeling. ANSYS Multiphysics has a broad physics capability, such as structural modal, static, transient, electrostatic-structural, high-frequency and low-frequency electromagnetics, electrothermal-structural coupled physics, free surfaces and capillary action, capacitance extraction, Newtonian/non-Newtonian continuum flow, and fluidic structural capability to evaluate damping effects, among others.

The architecture of MEMCAD, illustrated in Figure 6.64a, integrates various simulators, databases, and a solid state modeler with a user interface.[54] As outlined by the dashed blocks in the figure, the CAD system consists of three sections: the microelectronic CAD section, the mechanical CAD section, and the material property simulator. The interactions among these three sections and the flow of information is denoted by the direction of the arrows. In MEMCAD V1.0, the primary interface for mechanical modeling was through PATRAN. MEMCAD V2.0 provides this function from I-DEAS. In the microelectronic section, the mask is created using KIC. SUPREM 3 and SAMPLE are integrated to provide depth and cross-sectional modeling capabilities. Microcosm licensed the MEMCAD technology from MIT and released the first redesigned commercial version in June 1996. Besides MEMCAD V4.0, Microcosm also developed FlumeCAD, geared toward modeling of microfluidic devices. In January 2001 Microcosm Technologies changed its name to Coventor (http://www.coventor.com).

IntelliSense Corp. sells IntelliSuite™ (http:// www.intellisense.com),[67] incorporating the MEMaterial® database and the anistropic etching module AnisE®. The material database contains electrical, mechanical, optical, and physical properties of semiconductor thin films collected from the literature. As in the case of CAEMEMS, the database can be updated by the user. The architecture of IntelliCAD, with a central graphical user interface, is shown in Figure 6.64b. Figure 6.65 shows a typical output of MEMaterial. The selected 3D plot shows the density variation of a plasma-enhanced chemical vapor-deposited Si nitride film as a function of deposition temperature and pressure. AnisE simulates single-crystal silicon anistropic etching. It predicts the effects of etchant's temperature, concentration, and etch time on the final 3D geometry. Users can study etch stops, corner compensation, higher-order etch planes, and process tolerances for single- or double-sided masks. The program added RF-MEMS design tool (2002) and a bio-MEMS design tool (2003).

Other simulators for anistropic etching of Si are available. For example, the Anisotropic Crystalline Etch Simulation (ACES™) is a first personal computer-based 3D etch simulator from the University of Illinois at Urbana-Champaign. The program can simulate silicon etching with different front-surface orientations in different etchants. 2D and 3D simulation programs for anisotropic etching profiles have also been developed by Sequin at the University of California, Berkeley,[68] Koide et al. at Hitachi,[69] and Li et al. at Caltech.[70] These programs enable the prediction of a change in cross-sectional shape of a feature in a Si wafer with arbitrary crystallographic orientation and with a mask including concave and convex edges.

Coventor now markets CoventorWare,™ which consists of a system-level modeling environment, a 2D layout editor, materials property editor and database, a 3D model generator, and a multiphysics numerical analysis framework. It has also added a new module that interfaces with IC design environment by exporting models to IC simulators from Synopsys and Cadence.

CFD Research Corporation (CFDRC) is another provider of advanced software tools for modeling, simulation, and design of MEMS. Their software has been used for the analysis of a wide variety of physical phenomena such as fluid flow, heat transfer, combustion, and fluid-structure interaction. Industrial applications include MEMS, electronics packaging, semiconductor turbomachinery, and automotive engineering (http://www.cfdrc.com/).

MEMS Pro, originally from MEMSCAP, provides a unified MEMS and IC design environment. It includes certain software elements licensed from Tanner Research. MEMS Pro was sold in 2004 to SoftMEMS (http://www.softmems.com/) and is still marketed under the same name. It provides an integrated MEMS and IC- design environment with system-level tools, layout editing, 3D Solid Modeler (3D view generator from layout and fabrication process description), FEA modeling tools, and MEMS verification tool for design rule checks.

SUGAR V3.1 from University of California, Berkeley, is a free simulation program for MEMS devices composed of beams, electrostatic gaps, anchors, circuits elements etc. The approach is based on a modified nodal analysis to solve coupled nonlinear differential equations. The simulation algorithms are implemented in MATLAB®

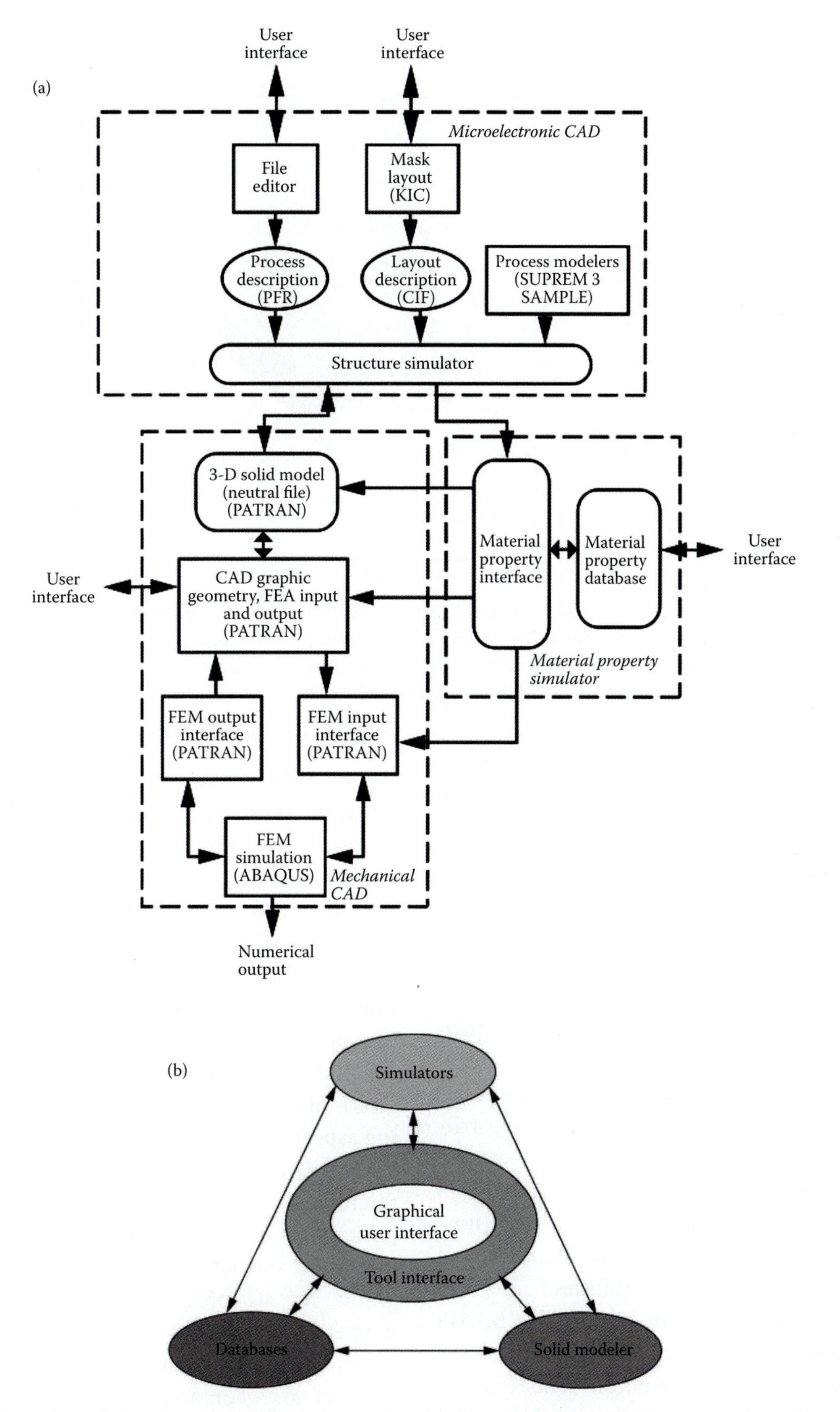

FIGURE 6.64 (a) Outline of the Massachusetts Institute of Technology MEMCAD. (b) Schematic of IntelliCAD architecture. (Courtesy of Dr. F. Maseeh, IntelliSense.)

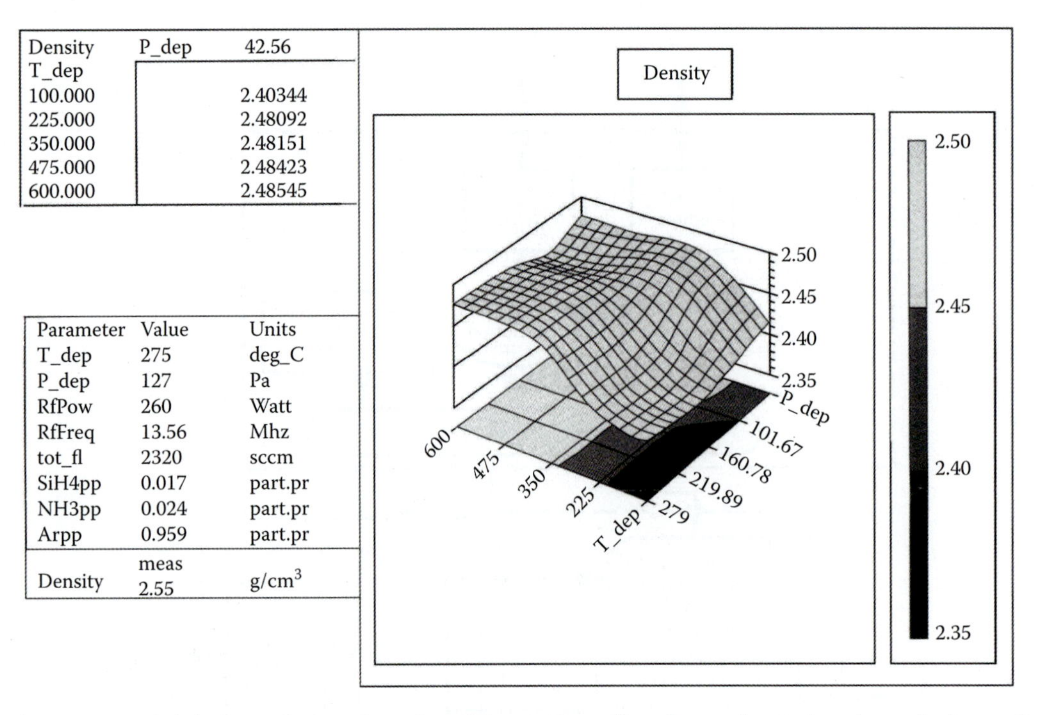

FIGURE 6.65 The MEMaterial design window from IntelliSense Corp. The three-dimensional graph shows the variation of Si nitride density deposited by plasma-enhanced chemical vapor-deposition. The table at the bottom left shows the process parameters, their numerical values, and units. The resulting density is shown at the bottom of the table. The user can alter any of the parameters. (Courtesy of Dr. F. Maseeh, IntelliSense.)

and are portable across all Unix and personal computer platforms.

FEMLAB, marketed by COMSOL (http://www.comsol.com), has entered the MEMS market with its MEMS Module. FEMLAB is a partial differential equation-based FEA modeling tool. It has modules for mechanical, thermal, DC and AC electrical, and fluidic modeling of devices among others. The FEMLAB MEMS Module includes ready-made applications that cover microfluidics plus electromagnetic-structural, thermal-structural, and fluid-structural interactions. Its MEMS-specific application modes are electrostatics, stress and strain, piezoelectric, and electrokinetics.

ALGOR (http://www.algor.com) has a MEMS module for mechanical, thermal, and fluid flow modeling. It also has electrostatic analysis and piezoelectric material model capabilities.

The major improvements in the leading MEMS software have been in a tighter integration with solid modeling as well as IC design environments, system-level modeling for quick analysis, capabilities of modeling microfluidics for the bio-MEMS industry, and capabilities for modeling RF-MEMS. Some FEA software packages have also added MEMS modules.

Future of CAD for MEMS and NEMS

Future MEMS CAD packages should seamlessly integrate 2D IC processes with 3D MEMS and traditional machining in a wide variety of materials. A conspicuous example for the need of such a 3D microdesign tool is in micromolding of LIGA shapes, in which nontraditional (lithography) and traditional (plastic molding) technologies are combined (see Volume II, Chapter 10). As an example of work in this direction, Hill et al.[71] explored the use of the computer software I-DEAS Master Series™ Thermoplastic Molding to model the micromolding aspect of such LIGA parts and have found that it accurately describes the filling characteristics of the microparts. What is needed now is to extend this package to include lithography steps.

A good introduction to the available CAD programs for IC processes comes from Fichtner.[72] A most detailed account of simulation and design issues of microsystems and microstructures is by Adey et al.[73] For access to continuously updated information on MEMS modeling, visit the MEMS Interchange on the Web (http://www-mtl.mit.edu/semisubway/semi subway.html); equally useful is http://www.memsnet.org.

With the advent of NEMS and bio-MEMS, molecular modeling has come to the forefront. Molecular modeling is a collective term that refers to theoretical methods and computational techniques to model or mimic the behavior of molecules. For molecular modeling (say on protein or DNA sensors), software can now successfully model the dynamics of most molecular interactions under numerous static and dynamic conditions. Molecular dynamics (MD) is a specialized discipline of molecular modeling and computes the behavior of a system as a function of time. The main justification of the MD method is that statistical ensemble averages are equal to time averages of the system. Because molecular systems generally consist of a vast number of particles, it is impossible to find the properties of such complex systems analytically; MD simulation circumvents this problem by using numerical methods. Molecular dynamics simulations are playing an increasingly important role in many areas of science and engineering, from biology and pharmacy to nanoelectronics and structural materials. We review only some available software packages for molecular modeling.

A lot of the modeling work is geared toward DNA and proteins; software is available to graphically manipulate 3D images of proteins and nucleic acids. The modules perform complex molecular and quantum mechanical calculation and protein structure homology analysis. Files in standard formats such as that of the Protein Data Bank can be imported.

Some other programs to check out are the following:

- CPMD (Car-Parrinello Molecular Dynamics)—This code can be used to perform *ab initio* molecular dynamics calculations. It allows for time-dependent density functional calculations, wave function optimization, and path-integral molecular dynamics.
- Siesta (Spanish Initiative for Electronic Simulations with Thousands of Atoms)—This code uses numerically truncated orbitals (single- and double-zeta approach) to build an order-N density functional code. This code is ideal for modeling large scale nanostructures (i.e., nanotubes, nanowires, and molecules).
- NWChem—This code is a highly parallel quantum chemistry package capable of self-consistent field, Hartree-Fock (RHF, UHF), and Gaussian density functional calculations, as well as several other techniques.
- Atomistix (ATK and VNL)—This commercial package provides the ability to calculate I-V curves in nanostructures and molecular junctions. The code is based on a nonequilibrium Green's function approach, and it also takes advantage of numerically truncated orbitals to allow for calculations of large systems.
- LAMMPS—This is a general purpose molecular dynamics simulator that has the option to use Leonard Jones potentials, embedded atom potentials, and potentials for biomolecules and proteins. This parallel code can easily handle systems with thousands of atoms. The ability to incorporate the effect of temperature provides an important complement to density functional techniques.
- MPB (MIT Photonic Bands package)—This code can calculate the band structure and electromagnetic modes of periodic dielectric structures.
- UTQUANT—This is a quasi-static CV simulator for one-dimensional silicon MOS structures.
- SEMC-2D (Schrödinger Equation Monte Carlo 2D)—This is a simulator for quantum transport and inelastic scattering effects in nanoscale semiconductor devices, such as nanoscale double-gate MOSFETs and tunnel injection lasers.

Questions

Questions by Chuan Zhang, UC Irvine

6.1: Compare the similarities and differences of scanning near-field optical microscopy (NSOM) and traditional optical microscopy. Explain how NSOM exceeds the resolution limit of traditional optical microscopes.

6.2: Classical far-field microscopy was introduced by Anton van Leeuwenhoek in the 16th century. It was a major breakthrough that played an important role in the advancement of science, engineering, and medicine. Briefly explain what is the underlying physical limitation regarding the magnification/resolution of classical far-field optical microscopy.

6.3: What does the NSOM acronym stand for? What are the principles of NSOM? What is the difference between apertureless NSOM and aperture-based NSOM and what dictates their resolution?

6.4: Scanning tunneling microscopes (STMs) and atomic force microscopes (AFMs) are two so-called *proximal probe techniques* with the power to reveal the atomic world to us.
(a) What is the difference between these two?
(b) A crucial component in an STM is the integrated sharp tip, which has a curvature radius of about 50 Å, an angle of aperture of 5°, and a height of 10 μm. Sketch a process to fabricate this integrated tip on a Si wafer.

6.5: What is the effect of the following changes on quantum yield of fluorescence? Explain the reason for your answer.
(a) Raising the temperature.
(b) Adding a heavy atom to the solvent.
(c) Lowering solvent viscosity.
(d) What is the physical process that makes phosphorescence rare?

6.6: What does AFM stand for and what is the principle of operation of AFM? What kind of information can you obtain by using AFM? Briefly describe an application where AFM is employed.

6.7: Briefly state the difference between fluorescence and phosphorescence.

6.8: Please define the following analytical terms. Use equations when appropriate.
(a) Beer's law
(b) Transmittance
(c) Quantum yield
(d) Sensitivity
(e) Precision
(f) Accuracy

6.9: Suggest the type of system that you would employ for the separation of the following compounds. Justify your choice.
(a) DNA with a size of 500 bases in length and DNA with a size of 2,000 bases in length
(b) A protein of a MW = 15,000 Daltons and another protein of a MW = 100,000 Daltons
(c) A protein with a MW = 50,000 Daltons and a fluorophore with a MW = 300
(d) A mixture of long-chain unsaturated aliphatic compounds

6.10: Compare the merits of an AFM with transmission electron microscopy (TEM) and a scanning electron microscopy (SEM).

6.11: Why do the samples for TEM need to be extremely thin, while those for SEM don't need to be?

6.12: Both energy dispersive spectroscopy (EDS) and Auger electron spectroscopy (AES) can be utilized to conduct the composition analysis of a sample. Which one of them can be used to analyze light elements such as Li, Be, and C?

6.13: How does two-photon confocal microscopy enhance the resolution of fluorescence microscopy?

6.14: List the signals generated when electrons interact with a solid. What information about a material can be obtained from each of those signals?

6.15: Explain the Auger process.

6.16: Review the various steps in a typical finite element analysis (FEA), e.g., a stress-strain problem.

6.17: The identity of a semiconductor can be determined by measuring its intrinsic carrier concentrations at various temperatures. An unknown semiconductor is submerged in ice water (273K) and boiling water (373K). Its intrinsic carrier concentrations, n_i, were found to be 4.6×10^{18} m^{-3} and 3.1×10^{20} m^{-3} at 273K and 373K, respectively. What is the bandgap and identify of the semiconductor?

Further Reading

ASTM Standard Guide for Computed Tomography (CT) Imaging. Designation E 1441–93, Philadelphia: ASTM.

Binnig, G., C. Quate, and C. Gerber. 1986. Atomic force microscopy. *Phys Rev Lett* 56:930–33.

Binnig, G., and H. Rohrer. 2000. Scanning tunneling microscopy. *IBM Journal of Research and Development* 44(1–2):279–93.

Cao, S., U. Brand, T. Kleine-Besten, W. Hoffmann, H. Schwenke, S. Bütefisch, and S. Büttgenbach. 2002. Recent developments in dimensional metrology for microsystem components. *Microsystems Technologies* 8:3–6.

Ceremuga, J. 2003. Obtaining inspection of high aspect ratio microstructures using a programmable optical microscope. Masters thesis, Georgia Institute of Technology, Atlanta, GA.

Chinn, D., P. Ostendorp, L. Garrett, P. Guthrei, M. Haugh, R. Kershmann, and T. Kurfess. 2002. Three dimensional imaging of LIGA-made micromachines using digital volumetric imaging. *Proceedings of the Japan-USA Symposium on Flexible Automation*, Hiroshima, Japan, Vol. 1:565–68.

Chinn, D., P. Ostendorp, M. Haugh, R. Kershmann, T. Kurfess, A. Claudet, and T. Tucker. Three dimensional imaging of LIGA-made microcomponents. *ASME Journal of Manufacturing Science and Engineering* 126:813–21.

Corle, T., and G. Kino. 1996. *Confocal Scanning Optical Microscopy and Related Imaging Systems.* San Diego: Academic Press.

El-Hakim, S. 1990. Some solutions to vision dimensional metrology problems. *Close-Range Photogrammetry Meets Machine Vision*. SPIE 1395:480–87.

Fan, K., C. Chu, and J. Mou. 2001. Development of a low-cost autofocusing probe for profile measurement. *Measurement Sci Tech* 12:2137–46.

Fukiwara, M., A. Yamaguchi, K. Takamasu, and S. Ozono. 2001. Evaluation of stages of nano-CMM. *Initiatives of Precision Engineering at the Beginning of a Millennium*, 634–638, New York: Springer.

Griffith, J., H. Marchman, G. Miller, and L. Hopkins. 1995. Dimensional metrology with scanning probe microscopes. *J Vacuum Sci Tech* 13(3):1100–05.

Groot, P., and L. Deck. 1994. Surface profiling by frequency-domain analysis of white light interferograms. *Proceedings of SPIE—The International Society for Optical Engineering* 2248:101–04.

Hall, A., and L. Degertekin. 2002. Integrated optical interferometric detection method for micromachined capacitive acoustic transducers. *Appl Phys Lett* 80:3859–61.

Heller, A. 2003. Nondestructive characterization at the mesoscale. *Science and Technology Review*, Livermore, CA: Lawrence Livermore National Laboratory.

Hibbard, R., and M. Bono. 2003. Meso-scale metrology tools: A survey of relevant tools and a discussion of their strengths and weaknesses. *Proceedings of Machines and Processes for Micro-scale and Meso-scale Fabrication, Metrology, and Assembly*, 70–72.

Kim, B., A. Razavi, L. Degertekin, and T. Kurfess. 2003. Micromachined interferometer for MEMS metrology. *Proceedings of Machines and Processes for Micro-scale and Meso-scale Fabrication, Metrology, and Assembly*, 73–78.

Kim, B., A. Razavi, F. Degertekin, and T. Kurfess. 2002. Micromachined interferometer for measuring dynamic response of microstructures. *Proceedings of 2002 ASME International Mechanical Engineering Congress, and Exposition, MEMS Symposium*, New Orleans, LA.

Kim, B., Schmittdiel, M. C., Degertekin, F. L., and T. R. Kurfess. 2004. Scanning grating microinterferometer for MEMS metrology. *ASME Journal of Manufacturing Science and Engineering* 126:807–12.

Kirkland, E. 2003. A nano coordinate machine for optical dimensional metrology. Master's thesis, Georgia Institute of Technology, Atlanta GA.

Kramar, J., J. Jun, W. Penzes, F. Scire, E. Teague, and J. Villarrubia. 1999. Grating pitch measurements with the molecular measuring machine. *SPIE 3806, Conference on Recent Advances in Metrology, Characterization, and Standards for Optical Digital Data Disks*, 46–53.

Kramar, J., E. Amatucci, D. Gilsinn, J. Jun, W. Penzes, F. Scire, E. Teague, and J. Villarrubia. 1999. Toward nanometer accuracy measurements. *SPIE 3677, Metrology, Inspection and Process Control for Microlithography XIII*, 1017–28.

Kurfess, T., D. Chinn, P. Ostendorp, A. Claudet, and T. Tucker. 2002. Metrology for micro-components. *Proceedings of the JSME/ASME International Conference on Materials and Processing 2002*, Honolulu, Hawaii, 2:77–82.

Lagerquist, M., W. Bither, and R. Brouillette. 1996. Improving SEM linewidth metrology by two-dimensional scanning force microscopy. *SPIE 2725, Metrology, Inspection and Process Control for Microlithography X*, 494–503.

Larrabee, R., and M. Postek. 1994. Parameters characterizing the measurement of a critical dimension. *Proceedings of SPIE—The International Society for Optical Engineering* CR52:2–24.

Li, X., M. Rahman, K. Liu, K. Neo, and C. Chan. 2003. Nano-precision measurement of diamond tool edge radius for wafer fabrication. *J Materials Process Tech* 140:358–62.

Mack, C., S. Jug, R. Hones, P. Apte, S. Williams, and M. Pochkowski. 2001. Metrology and analysis of two dimensional SEM patterns. *SPIE 4344, Metrology, Inspection and Process Control for Microlithography XV, 169–76.*

Marchman, H. 1996. Nanometer-scale dimensional metrology with noncontact atomic force microscopy. *SPIE 2725, Metrology, Inspection and Process Control for Microlithography X.*

Marchman, H., and N. Dunham. 1998. AFM: A valid reference tool? *SPIE 3332, Metrology, Inspection and Process Control for Microlithography XII*, 2–9.

Marchman, H. 1998. Critical dimension metrology. *Microlithography: Science and Technology*, New York: Marcel Dekker, Inc.

Marschner, T., G. Eytan, and O. Dror. 2001. Determination of best focus and exposure dose using CD-SEM side-wall imaging. *SPIE 4344, Metrology, Inspection and Process Control for Microlithography XV*, 355–65.

Martz, H., and G. Albrecht. 2003. Nondestructive characterization technologies for metrology of micro/mesoscale assemblies. *Proceedings of Machines and Processes for Micro-scale and Meso-scale Fabrication, Metrology, and Assembly*, 131–141.

Meyyappan, A., M. Klos, and S. Muckenhirn. 2001. Foot (bottom corner) measurement of a structure with SPM. *SPIE 4344, Metrology, Inspection and Process Control for Microlithography XV*, 733–38.

Nichols, J. F., and Kurfess, T. R. 2004. Metrology of high aspect ratio MEMS. *J Microsyst Tech* 10:556–59.

O'Mahony, C., M. Hill, M. Brunet, R. Duane, and A. Mathewson. 2003. Characterization of micromechanical structures using white-light interferometry. *Measurement Sci Tech* 14:1807–14.

Ogura, I., and Y. Okazaki. 2003. A study of development of small-cmm probe detecting contact angle. *ASPE Summer Topical Meeting—Coordinate Measuring Machines, 349.*

Opsal, J., H. Chu, Y. Wen, Y. Chang, and G. Li. 2002. Fundamental solutions for real-time optical CD metrology. *SPIE* 4344, *Metrology, Inspection and Process Control for Microlithography XVI*, Santa Clara, CA, March 4–7, 2002, p. 163.

Peggs, G., A. Lewis, and R. Leach. 2003. Measuring in three dimensions at the mesoscopic scale. *Proceedings of Machines and Processes for Micro-scale and Meso-scale Fabrication, Metrology, and Assembly*, 53–57.

Peggs, G., A. Lewis, and S. Oldfield. 1999. Design for a compact high-accuracy CMM. *CIRP Annals—Manufacturing Technology* 48(1):417–20.

Postek, M. 1994. Scanning electron microscope metrology. *Proceedings of SPIE—The International Society for Optical Engineering* CR52:46–90.

Rizivi, S., and A. Meyyappan. 1999. Atomic force microscopy: A diagnostic tool (in) for mask making in the coming years. *SPIE* 3677, *Metrology, Inspection and Process Control for Microlithography XIII*.

Roth, J., E. Felkel, and P. Groot. 2003. Optical metrology. *Proceedings of Machines and Processes for Micro-scale and Meso-scale Fabrication, Metrology, and Assembly*, 87–92.

Schellekens, P., H. Haitjema, and W. Pril. 2001. A silicon-etched probe for 3-D coordinate measurements with an uncertainty below 0.1 μm. *IEEE Transactions on Instrumentation and Measurement* 50(6):1519–23.

Shilling, M. 2003. Two dimensional analysis of mesoscale parts using image processing techniques. Masters thesis, Georgia Institute of Technology, Atlanta, GA.

Shiozawa, H., Y. Fukutomi, T. Ushioda, and S. Yoshimura. 1998. Development of ultra-precision 3D-CMM based on 3-D metrology frame. *Proc ASPE* 18:15–18.

Simon, M., and C. Sauerwein. 2001. Quality control of light metal castings by 3D computed tomography. http://www.ndt.net/article/wcndt00/papers/idn730/idn730.htm.

St. Clair, L., and A. Mirza. 2000. Metrology for MEMS manufacturing. *Sensors Magazine* 17(7):24–33.

Storment, C., D. Borkholder, V. Westerlind, J. Suh, N. Maluf, and G. Kovacs. 1994. Flexible, dry-released process for aluminum electrostatic actuators. *J Microelectromech Syst* 3(3):90–96.

Svetkoff, D., and D. Kilgus. 1991. Influence of object structure on the accuracy of 3D systems for metrology. *SPIE 1614 Optics, Illumination, and Image Sensing for Maching Vision VI*, 218–30.

Takamasu, K., M. Fujiwara, A. Yamaguchi, M. Hiraki, and S. Ozono. 2001. Evaluation of thermal drift of nano-CMM. *Proc. of 2nd EUSPEN International Conference*, Turin, Italy, May 27–31.

Takamasu, K., M. Hiraki, K. Enami, and S. Ozono. 1999. Development of nano-CMM and parallel-CMM. *Proc. of Intl. Dimensional Metrology Workshop*, May 10–13.

Uhlmann, E., D. Oberschmidt, and G. Kunath-Fandri. 2003. 3D-analysis of microstructures with confocal laser scanning microscopy. *Proceedings of Machines and Processes for Micro-scale and Meso-scale Fabrication, Metrology, and Assembly*, 93–97.

Veeco. 2003. *Scanning probe/atomic force microscopy: Technology overview and update.* Santa Barbara, CA: Veeco Instruments, Inc.

VIEW Engineering. Voyager 6x12 specifications. http://www.vieweng.com, 2003.

Walch, K., A. Meyyappan, S. Muckenhirn, and J. Margail. 2001. Measurement of sidewall, line and line-edge roughness with scanning probe microscopy. *SPIE 4344, Metrology, Inspection and Process Control for Microlithography XV*, 726–32.

Wyant, J. 2002. White light interferometry. *Proceedings of SPIE—The International Society for Optical Engineering* 4737:98–107.

Zygo Corporation. 2003. *NewView 5000 specifications.* Middlefield, CT: Zygo Corporation.

References

1. Quirk, M., and J. Serda. 2001. *Semiconductor manufacturing technology*. Upper Saddle River, NJ: Prentice Hall.
2. Nichols, J. F. 2004. *Metrology of high aspect ratio MEMS, PhD diss.*, Georgia Institute of Technology.
3. Ehmann, K., D. Bourell, M. L. Culpepper, T. J. Hodgson, T. R. Kurfess, M. Madou, K. Rajurkar, and R. DeVor. 2007. *Micromanufacturing: International research and development*. New York: Springer.
4. Denk, W., J. Strickler, and W. Webb. 1990. Two-photon laser scanning fluorescence microscopy. *Science* 248:73–76.
5. Tompkins, H. G., and E. A. Irene, eds. 2005. *Handbook of ellipsometry.* Berlin: Springer.
6. Goldstein, J., D. Newbury, D. Joy, C. Lyman, P. Echlin, E. Lifshin, L. Sawyer, and J. Michael. 2003. *Scanning electron microscopy and x-ray microanalysis.* New York: Kluwer Academic/Plenum Publishers.
7. Synge, E. H. 1928. A suggested method for extending microscopic resolution into the ultramicroscopic region. *Phil Mag* 6:356–62.
8. O'Keefe, J. A. 1956. Resolving power of visible light. *J Opt Soc Am* 46:359.
9. Ash, E. A., and G. Nicholls. 1972. Super-resolution aperture scanning microscope. *Nature* 237:510.
10. Pohl, D. W., W. Denk, and M. Lanz. 1984. Optical stethoscopy: Image recording with resolution lambda/20. *Appl Phys Lett* 44:651–53.
11. Harootunian, A., E. Betzig, M. Isaacson, and A. Lewis. 1986. Super-resolution fluorescence near-field scanning microscopy. *Appl Phys Lett* 49:674–76.
12. Vinci, R. P., and J. C. Braveman. 1991. Mechanical testing of thin films. *6th International Conference on Solid-State Sensors and Actuators (Transducers '91)*. San Francisco, June 24–27, 1991, pp. 943–48.
13. Bushan, B., ed. 1995. *Handbook of micro/nanotribology.* Boca Raton, FL: CRC Press.
14. Wong, S. M. 1978. Residual stress measurements on chromium films by x-ray diffraction using the sin2 Y method. *Thin Solid Films* 53:65–71.
15. Nishioka, T., Y. Shinoda, and Y. Ohmachi. 1985. Raman microprobe analysis of stress in Ge and GaAs/Ge on silicon dioxide-coated silicon substrates. *J Appl Phys* 57:276–81.
16. Marco, S., J. Samitier, O. Ruiz, J. R. Morante, J. Esteve-Tinto, and J. Bausells. 1991. Stress measurement of SiO_2-polycrystalline silicon structures for micromechanical devices by means of infrared spectroscopy technique. *6th International Conference on Solid-State Sensors and Actuators (Transducers '91)*. San Francisco, June 24–27, 1991, pp. 209–12.
17. Mehregany, M., R. T. Howe, and S. D. Senturia. 1987. Novel microstructures for the in situ measurement of mechanical properties of thin films. *J Appl Phys* 62:3579–84.
18. Jaccodine, R. J., and W. A. Schlegel. 1966. Measurements of strains at Si-SiO_2 interface. *J Appl Phys* 37:2429–34.
19. Bromley, E. I., J. N. Randall, D. C. Flanders, and R. W. Mountain. 1983. A technique for the determination of stress in thin films. *J Vac Sci Technol* B1:1364–66.
20. Allen, M. G., M. Mehregany, R. T. Howe, and S. D. Senturia. 1987. Microfabricated structures for the in situ measurement of residual stress, Young's modulus and ultimate strain of thin films. *Appl Phys Lett* 51:241–43.
21. Zhang, L. M., D. Uttamchandani, and B. Culshaw. 1991. Measurement of the mechanical properties of silicon microresonators. *Sensors Actuators* A A29:79–84.
22. Pratt, R. I., G. C. Johnson, R. T. Howe, and J. C. Chang. 1991. Micromechanical structures for thin film characterization. *6th International Conference on Solid-State Sensors and Actuators (Transducers '91)*. San Francisco, June 24–27, 1991, pp. 205–08.

23. Krulevitch, P. A. 1994. Micromechanical investigations of silicon and Ni–Ti–Cu thin films. PhD thesis, University of California, Berkeley.
24. Hoffman, R. W. 1976. In *Physics of nonmetallic thin films (NATO advanced study institutes series: Series B, physics)*, ed. C. H. S. Dupuy and A. A. Cachard, 273–353. New York: Plenum Press.
25. Singer, P. 1992. Film stress and how to measure it. *Semicond Int* 15:54–58.
26. Senturia, S. D., and R. T. Howe. 1990. Lecture notes. Boston: Massachusetts Institute of Technology.
27. Maseeh-Tehrani, F. 1990. Characterization of mechanical properties of microelectronic thin films. PhD thesis, Massachusetts Institute of Technology.
28. Maseeh, F., and S. D. Senturia. 1989. In *Polyimides: Materials, chemistry, and characterization*, ed. C. Feger, M. M. Khojasteh, and J. E. McGrath, 575–84. Amsterdam, the Netherlands: Elsevier Science Publishers B.V.
29. Senturia, S. 1987. Can we design microbotic devices without knowing the mechanical properties of materials? *Proceedings: IEEE Micro Robots and Teleoperators Workshop*. Hyannis, MA, November 1987, 3/1–5.
30. Biebl, M., and H. von Philipsborn. 1993. Fracture Strength of Doped and Undoped Panical Filters. PhD thesis, University of California, Berkeley.
31. Greek, S., F. Ericson, S. Johansson, and J.-Å. Schweitz. 1995. In situ tensile strength measurement of thick-film and thin film micromachined structures. *8th International Conference on Solid-State Sensors and Actuators (Transducers '95)*. Stockholm, Sweden, June 25–29, 1995, pp. 56–59.
32. Ericson, F., and J.-Å. Schweitz. 1990. Micromechanical fracture strength of silicon. *J Appl Phys* 68:5840–44.
33. Vinci, R. P., and J. C. Braveman. 1991. Mechanical testing of thin films. *6th International Conference on Solid-State Sensors and Actuators (Transducers '91)*. San Francisco, June 24–27, 1991, pp. 943–48.
34. Sekimoto, M., H. Yoshihara, and T. Ohkubo. 1982. Silicon nitride single-layer x-ray mask. *J Vac Sci Technol* 21:1017–21.
35. Guckel, H., T. Randazzo, and D. W. Burns. 1985. A simple technique for the determination of mechanical strain in thin films with applications to polysilicon. *J Appl Phys* 57:1671–75.
36. Lin, L. 1993. Selective encapsulations of MEMS: Micro channels, needles, resonators, and electromechanical filters. PhD thesis, University of California, Berkeley.
37. Guckel, H., D. W. Burns, C. C. G. Visser, H. A. C. Tilmans, and D. Deroo. 1988. Fine-grained polysilicon films with built-in tensile strain. *IEEE Trans Electron Devices* 35:800–01.
38. Kim, C. J. 1991. Silicon electromechanical microgrippers: Design, fabrication, and testing. PhD thesis, University of California, Berkeley.
39. Goosen, J. F. L., B. P. van Drieenhuizen, P. J. French, and R. F. Wolfenbuttel. 1993. In *7th International Conference on Solid-State Sensors and Actuators (Transducers '93)*. Yokohama, Japan, June 7–10, 1993, pp. 783–86.
40. Ericson, F., S. Greek, J. Soderkvist, and J.-Å. Schweitz. 1995. High sensitive internal film stress measurement by an improved micromachined indicator structure. *8th International Conference on Solid-State Sensors and Actuators (Transducers '95)*. Stockholm, Sweden, June 25–29, 1995, pp. 84–87.
41. Benitez, M. A., J. Esteve, M. S. Benrakkad, J. R. Morante, J. Samitier, and J. Å. Schweitz. 1995. Stress profile characterization and test structures analysis of single and double ion implanted LPCVD polycrystalline silicon. *8th International Conference on Solid-State Sensors and Actuators (Transducers '95)*. Stockholm, Sweden, June 25–29, 1995, pp. 88–91.
42. Biebl, M., G. Brandl, and R. T. Howe. 1995. Young's modulus of in situ phosphorus-doped polysilicon. *8th International Conference on Solid-State Sensors and Actuators (Transducers '95)*. Stockholm, Sweden, June 25–29, 1995, pp. 80–83.
43. Core, T. A., W. K. Tsang, and S. J. Sherman. 1993. Fabrication technology for an integrated surface-micromachined sensor. *Solid State Technol* 36:39–47.
44. Chu, W. H., M. Mehregany, X. Ning, and P. Pirouz. 1992. Measurement of residual stress-induced bending moment of p+ silicon films. *Mat Res Soc Symp* 239:169.
45. Fan, L.-S., R. S. Muller, W. Yun, J. Huang, and R. T. Howe. 1990. Spiral microstructures for the measurement of average strain gradients in thin films. *Proceedings: IEEE Micro Electro Mechanical Systems (MEMS '90)*. Napa Valley, CA, February 11–14, 1990, pp. 177–81.
46. Campbell, D. S. 1970. In *Handbook of thin film technology*, eds. L. I. Maissel and R. Glang. New York: McGraw-Hill.
47. Bushan, B. 1996. In *Proceedings: IEEE Ninth Annual International Workshop on Micro Electro Mechanical Systems*, 91–98. San Diego, CA.
48. Bushan, B., Ed. 1995. *Handbook of micro/nanotribology*. Boca Raton, FL: CRC Press.
49. Howe, R. T., and R. S. Muller. 1983. Polycrystalline silicon micromechanical beams. *J Electrochem Soc* 130:1420–23.
50. Howe, R. T. 1985. In *Micromachining and micropackaging of transducers*, eds. C. D. Fung, P. W. Cheung, W. H. Ko, and D. G. Fleming, 169–87. New York: Elsevier.
51. Yun, W. 1992. A surface micromachined accelerometer with integrated CMOS detection circuitry. PhD thesis, University of California, Berkeley.
52. Adams, A. C. 1988. In *VLSI technology*, ed. S. M. Sze. New York: McGraw-Hill.
53. Maseeh, F., and S. D. Senturia. 1990. In *Technical Digest: 1990 Solid State Sensor and Actuator Workshop*. Hilton Head Island, SC, June 1990, pp. 55–60.
54. Senturia, S. D., and R. T. Howe. 1990. Lecture notes. Boston: Massachusetts Institute of Technology.
55. Cook, R. D. 1995. *Finite element modeling for stress analysis*. New York: Wiley.
56. Murphy, M. 1995. Personal communication.
57. Marshall, J. C., M. Parameswaran, M. E. Zaghloul, and M. Gaitan. 1992. High-level CAD melds micromachine devices with foundries. *IEEE Circuits Devices Mag* 8:10–17.
58. Tomovich, C., ed. 1988. *MOSIS user manual*. Los Angeles, CA: University of Southern California.
59. Billingsley, G. C. 1983. *Report no. UCB/ERL M83/62*. Berkeley, CA: University of California, Berkeley.
60. ERL. 1985. *SAMPLE version 1.6a user's guide*. Berkeley, CA: Electronics Research Laboratory, University of California, Berkeley.
61. Oldham, W. G., S. N. Nandgaonkar, A. R. Neureuther, and M. O'Toole. 1979. A general simulator for VLSI lithography and etching processes—Part I: Application to projection lithography. *IEEE Trans Electron Devices* ED-26:717–22.

62. Oldham, W. G., A. R. Neureuther, J. L. Reynolds, S. N. Nandgaonkar, and C. Sung. 1980. A general simulator for VLSI lithography and etching processes—Part 2: Application to deposition and etching. *IEEE Trans Electron Devices* ED-27:1455–59.
63. Koppelman, G. M. 1989. In *Proceedings: IEEE Micro Electro Mechanical Systems (MEMS '89)*, 88–93. Salt Lake City, UT.
64. Gilbert, J. R., G. K. Ananthasuresh, and S. D. Senturia. 1996. 3-D modeling and simulation of contact problems and hysteresis in coupled electromechanics. *Ninth Annual International Workshop on Micro Electro Mechanical Systems (MEMS '96)*. San Diego, CA, February 11–15, 1996, pp. 127–32.
65. Crary, S., and Y. Zhang. 1990. CAEMEMS: an integrated computer aided engineering workbench for micro electromechanical systems. *Proceedings: IEEE Micro Electro Mechanical Systems (MEMS '90)*. Napa Valley, CA, pp. 113–14.
66. Korvink, J. G., J. Funk, M. Roos, G. Wachutka, and H. Baltes. 1994. SESES: a comprehensive MEMS modeling system. *IEEE International Workshop on Micro Electro Mechanical Systems (MEMS '94)*. Oiso, Japan, January 25–28, 1994, pp. 22–27.
67. Maseeh, F. 1994. A novel multidimensional semiconductor material analysis tool. *Solid State Technol* 37:83–84.
68. Sequin, C. H. 1992. Computer simulation of anisotropic crystal etching. *Sensors Actuators* A A34:225–41.
69. Koide, A., K. Sato, and S. Tanaka. 1991. Simulation of two-dimensional etch profile of silicon during orientation-dependent anisotropic etching. *Proceedings: IEEE Micro Electro Mechanical Systems (MEMS '91)*. Nara, Japan, pp. 216–20.
70. Li, G., T. Hubbard, and E. K. Antonsson. 1998. SEGS: on-line etch simulator. *1998 International Conference on Modeling and Simulation of Microsystems*. Santa Clara, CA.
71. Hill, S. D. J., K. P. Kamper, U. Dasbach, J. Dopper, W. Ehrfeld, and M. Kaupert. 1995. In *Simulation and design of microsystems and microstructures*, eds. R. A. Adey, A. Lahrmann, and C. Lessmollmann, 276–83. Billerica, MA: Computational Mechanics Publications.
72. Fichtner, W. 1988. In *VLSI technology*, ed. S. M. Sze, 422–65. New York: McGraw-Hill.
73. Adey, R. A., A. Lahrmann, and C. Lessmollmann, eds. 1995. *Simulation and design of microsystems and microstructures*. Billerica, MA: Computational Mechanics Publications.

Part II

Scaling Laws, Actuators, and Power and Brains in Miniature Devices

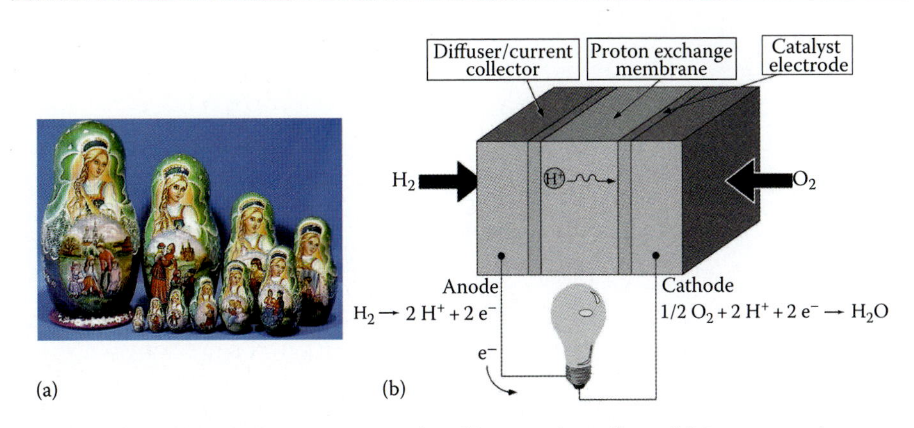

(a) Nested Russian dolls as an example of isometric scaling. (b) Proton exchange membrane fuel cell.

The civilization of abundance is also that of famines in African and other places. The collapse of totalitarian regimes has left intact all the evils of the democratic liberal societies, ruled by the demon of money. Marx's famous phrase about religion as the opiate of the masses can now be applied, and more accurately; to television, which will end up anaesthetizing the human race, sunk in an idiotic beatitude.

Octavio Paz in a passage on the science and technology of the West in *In Light of India*

Napoleon asked: "Monsieur Laplace, why wasn't the Creator mentioned in your book on Celestial Mechanics?"
Laplace replied: "Sir, I have no need for that hypothesis."

Introduction to Part II

Chapter 7 Scaling Laws
Chapter 8 Actuators
Chapter 9 Power and Brains in Miniature Devices

Introduction to Part II

Of the three chapters in Part II, the first deals with scaling laws, the second deals with actuators, and the third discusses surrounding power generation and the implementation of brains in miniaturized devices.

In Chapter 7, on scaling laws, we look at scaling from both intuitive and mathematical points of view. We are especially interested in deviations from linear scaling, where downscaling reveals new unexpected physical and chemical phenomena. The scaling chapter also constitutes an introduction to the subsequent treatises on actuators in Chapter 8 and on power and brains in Chapter 9. An actuator—like a sensor—is a device that converts energy from one form to another. In the case of an actuator, one is interested in the ensuing action; in the case of a sensor, one is interested in the information garnered. Scaling enables one to compare various actuator mechanisms, such as those proposed to propel a rotor blade, bend a thin Si membrane, or move fluids in fluidic channels. Power generation poses quite a challenge for miniaturization science; the smaller the power source, the less its total capacity. We will consider the powering of miniaturized equipment, as well as the miniaturization of power sources themselves, in Chapter 9. In the same chapter, we also discuss the different strategies for making micromachines smarter.

Part II should help the reader to choose a preferred actuator mechanism and/or power source through a better understanding of scaling and give some idea on how to incorporate intelligence in MEMS.

7

Scaling Laws

Our solar system as an example of isometric scaling.

Outline

Introduction

The electronics revolution, which began in the 1960s, drove the miniaturization of radio, television, hard disk drives, camcorders, personal digital assistants, etc., to the point where miniaturization now appears commonplace. More recent is the miniaturization of components that are not based on integrated circuits in hardware such as physical, chemical, and biological sensors and analytical instruments. Progress in this newer area is derived from mature, traditional manufacturing technologies such as mechanical precision machining, as well as from lithography-based nontraditional microelectromechanical system (MEMS) and nanoelectromechanical system (NEMS) techniques. The encroaching limits of our planet's resources and the continuing deterioration of the environment are adding urgency to this trend of scaling things down in size.

In Table 7.1 we list sound reasons for miniaturizing systems, actuators, power sources, sensors, and components. In some cases, the reason is favorable scaling—for example, smaller devices tend to be faster and consume less power. Scaling may also disfavor miniaturization, as often happens with actuators and power sources; smaller actuators exert less force, and smaller power sources harness less power.

In this chapter, we analyze scaling laws and the size regimes where macrotheories start requiring corrections. We aspire to elucidate the unexpected behavior of some of these micromachines and to better understand why, in some cases, it makes sense to miniaturize a device for reasons beyond economics, volume, and weight. Following some comments about human microintuition, we introduce the concepts of

TABLE 7.1 Motivations to Miniaturize Systems, Actuators, Power Sources, Sensors, and Components

Miniaturization Attributes	Motivations
Low energy* and little material consumed	There are limited resources on planet Earth
Arrays of sensors	Redundancy, wider dynamic range, and increased selectivity through pattern recognition
Small	Smaller is lower in cost, minimal invasive
Favorable scaling laws (in some cases)	Forces that scale with a low power become more prominent in the microdomain; if these are positive attributes then miniaturization is favorable, e.g., surface tension becomes more important than gravity in narrower capillaries
Batch and beyond batch techniques (roll-to-roll manufacturing)	Lowers cost
Disposable	Helps avoid contamination
Breakdown of macrolaws in physics and chemistry	New physics and chemistry might be developed
Increased sensitivity (in some cases)	Nonlinear effects can increase a sensor's sensitivity, e.g., amperometric sensors
Smaller building blocks	The smaller the building blocks, the more sophisticated a system that can be built

* There is a significant debate about this point. If the total environmental "cost" for a product is considered, miniaturized products might be worse for the planet. It is similar to the debate about hybrid cars: they save on gas, but the environmental impacts of producing the batteries are often ignored.

isometry, allometry, and dimensional analysis—mathematical techniques that help us predict the influence of downscaling more reliably. The lessons learned are then applied to some scaling applications in both natural and human-made "machines." We end the chapter by introducing some examples where scaling laws break down and where noncontinuum models must be invoked. In a continuum, all quantities are defined everywhere and vary continuously from point to point; in a noncontinuum this continuity breaks down.

Microintuition

As children, we learned to deal with everyday objects having size and scale dimensions that are naturally "sized up" by our senses, and we were mystified and enthralled by anything much larger (Gulliver) or smaller (Tom Thumb) (see Figure 7.1). It is indeed much more difficult to get a feel for or understand the consequences of dimensions of physical systems many orders of magnitude larger or smaller than what we commonly experience. Of all our senses, vision informs our mind the most. The unaided eyes generally give us the ability to discern things as small in size as an ant's eye or the point of a pin (approximate size, 50 μm). However, telescopes and microscopes continue to improve and greatly extend our sensory capabilities in our quest to understand and utilize our material world better, including systems spanning extreme length scales from the astronomical to the atomic. These new sets of eyes have greatly assisted us in assessing the size and scale of diverse natural and artificial material systems. Today telescopes allow us to see galaxies 13 billion light-years away (10^{25} m), and microscopes allow us to see even single atoms (about 10^{-10} m)!

Q: Gulliver was twelve times as tall as the Lilliputians. How much should they feed him? Could it be twelve times their own food ration? Hint: A person's food needs are related to his or her mass (volume), which depends on the cube of the linear dimension.

A: Let l_G and V_G denote Gulliver's linear and volume dimensions. Let l_L and V_L denote a Lilliputian's linear and volume dimensions. Gulliver is twelve times as tall as the Lilliputians, or

FIGURE 7.1 *Tom Thumb* originally by the Brothers Grimm (a) and *Gulliver's Travels* by Jonathan Swift (b).

$l_G = 12\ l_L$. Then, $V_G \propto (l_G)^3$ and $V_L \propto (l_L)^3$, so $V_G/V_L = (l_G)^3/(l_L)^3 = (12\ l_L)^3/(l_L)^3 = 12^3 = 1728$. Gulliver needs to be fed 1728 times as much food each day as a Lilliputian. This may be the real reason Gulliver was driven out of Lilliput. (Courtesy of W. Wilson, University of Central Oklahoma).

Humans are accustomed to thinking in distances and sizes—even microsizes have become quite intuitive, more so than microtimes. To accommodate the human inability to comprehend minute segments of time, Isaac Asimov introduced a measure based on the speed of light. Asimov's *light-meter* is the time required for light (in a vacuum) to cover the distance of a meter. An illustrative case is the half-life of radioactive particles. Saying that the half-life of a first particle is of the order of a hundredth of a millionth of a second and that of a second particle is of a thousandth of a trillionth of a second makes little impression. On the other hand, saying that the half-life of one type of particle is of the order of a light-meter and that of the other is of a tenth of a light-micrometer is more effective. We find linear extrapolation of length easy but are quickly at a loss when considering the implications that shrinking of time segments has or shrinking of surface-to-volume ratios on the relative strength of external forces working on miniaturized structures. In dealing with very small devices, our *microintuition*—as Bill Trimmer calls it—is often misleading (see also http://home.earthlink.net/~trimmerw/mems/tutorials.html).[1]

The effect of downscaling, especially into the nanodomain, is indeed not well understood and leads to some very surprising and sometimes counterintuitive results. For example, scaling shows that NEMS devices feature fundamental frequencies in the microwave range, active masses in the femtogram range, mechanical quality factors (Qs) in the tens of thousands, force sensitivities in the attonewton level, mass sensitivities at the level of individual atoms, and heat capacities far below yoctocalories. NEMS will make for electromechanical components with response times and operating frequencies that are as fast as most of today's electron devices. We now introduce a rigid mathematical formulation for scaling of a system, illustrating the newly gained insights with observations from nature. Our aim is to develop a systematic approach about the likely behavior of downsized systems so we do not need to rely on microintuition alone.

In Figure 7.2 we illustrate some known and not-so-well-known objects and their respective sizes,

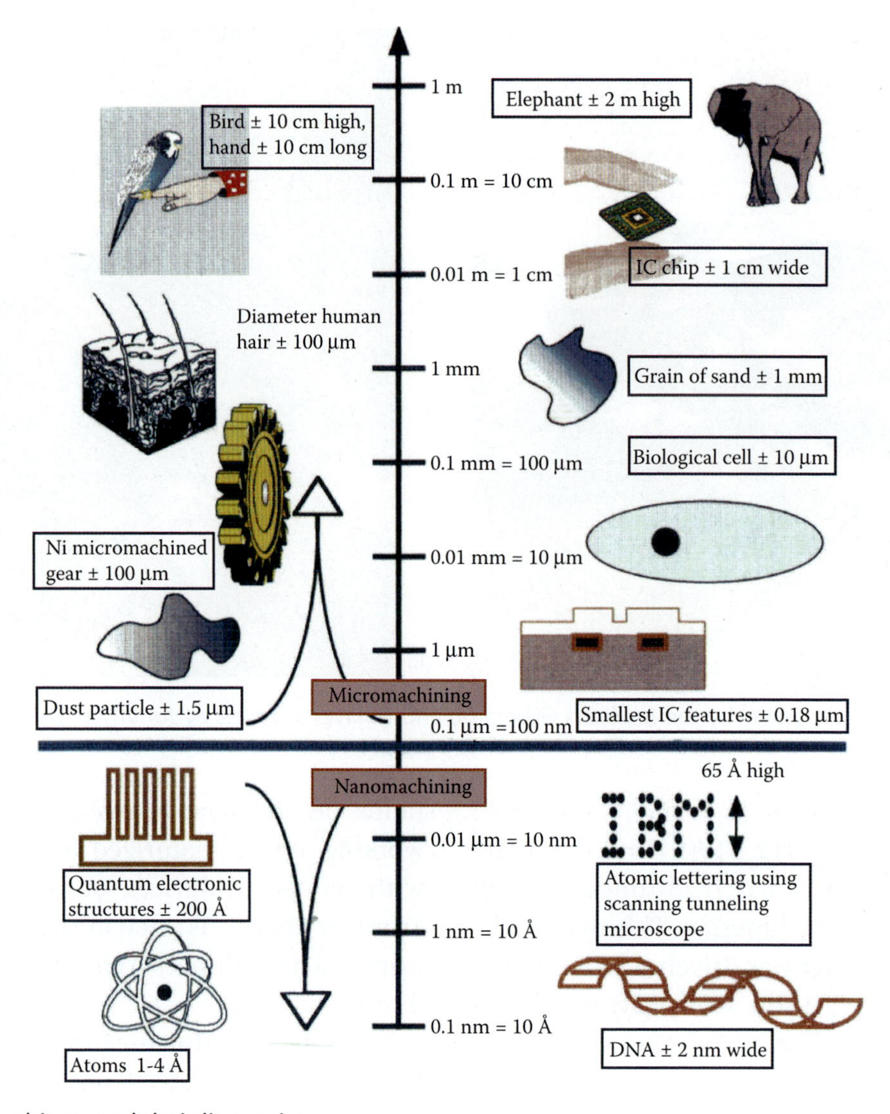

FIGURE 7.2 Various objects and their linear size.

including MEMS and NEMS, organized in powers of ten.

Mathematical Approach to Scaling

Introduction

In his book *On Growth and Form* (first published in 1917), D'Arcy Wentworth Thompson explains the profound effects of scaling in nature. Thompson (1860–1948) was a mathematically trained zoologist who pioneered the field we call *biomathematics* today (scaling laws in nature are a subset of biomathematics). He recognized many strong mathematical patterns in nature (spiral seashells, zebra stripes, etc.) and tried to find an explanation for the physical principles behind them. His book is the work of a true maverick and still inspires those interested in a mathematical understanding of biology. For a more recent account on growth and form in nature, including a more detailed mathematical treatment of scaling in animals, we refer to Knut Schmidt-Nielsen.[2] A popularizing and exquisitely illustrated work on the same topic is Thomas McMahon and John Bonner's *On Size and Life*.[3]

Similarity or Isometry, Allometry, Fractals, and Dimensional Analysis

A scaling law describes the variation of physical quantities with the size of the system. To understand scaling, it is essential to develop a notion for the concept of similarity or isometry. The idea is familiar from geometry: for example, two triangles are said to be similar or isometric if all of their angles are equal, even if the sides of the two triangles are of different lengths. The two triangles have the same shape; the larger one is simply a scaled up version of the smaller one. This notion can be generalized to include complex organisms or human-made artifacts such as engines, boats, and aircraft.

In the most general case, to compare the relative sizes of two structures, *x* and *y*, one may write:

$$y = bx^a \quad (7.1)$$

where *a* and *b* are constants. If the exponent a equals zero, then *y* and *x* are independent. This holds, for example, for the size of a cell (*y*) as a function of animal size (*x*); the cell diameter (*b*) is independent of animal size (about 10 μm, see Chapter 2). In case of a linear relationship between *y* and *x*, *a* equals 1, and *x* and *y* are said to be isometric or exhibit geometric similarity—the proportions of the structures that are being compared do not change with a change in size. Isometric change occurs when each dimension of a structure is scaled up or down regardless of the disproportionate change in such parameters as surface area and volume. If *x* doubles in length, then so does *y*. The proportions remain constant and do not change with a change in size. Besides the aforementioned triangles, the nine planets and the ten nested Russian dolls shown above are all simple examples of isometry, in which the proportions do not change with size.

From the rules of isometry, we expect that for geometrically similar animals of the same mass density, the body mass, *m*, will be proportional to the cube of the body length, l^3, and that the area, *S*, will be proportional to l^2. In a number of species, body surface areas proportional to the 2/3 power of body mass are just what is found. This is illustrated in Figure 7.3 where we have plotted the surface area versus body mass for a species of salamander in a log-log plot. To elucidate the plot in Figure 7.3[4], we first rewrite Equation 7.1 in its logarithmic form and obtain the following.

$$\log y = \log b + a \log x \quad (7.2)$$

In the example case the slope *a* equals 2/3, or 0.67, and fits the isometric prediction exactly. The value of *b* is of significance as well; it fixes the value for *y* when *x* is equal to 1. Comparisons made over a limited size range, say within a given species of plants or animals and types of bridges or boats, are often isometric.

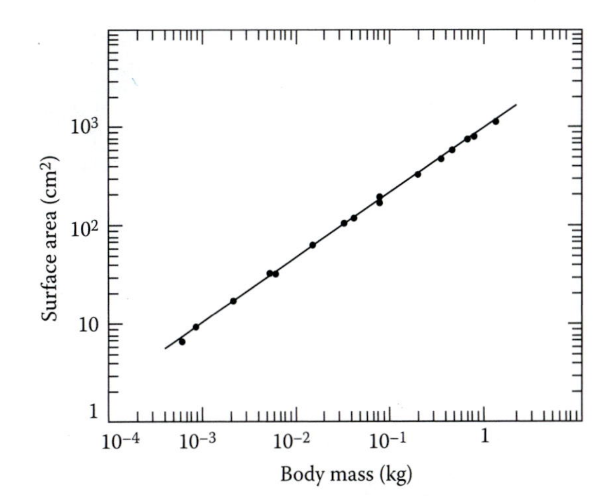

FIGURE 7.3 Surface area versus body mass for a species of salamander, the dwarf siren (*Pseudobranchus striatus*). The surface area, *S*, is proportional to the body mass, *m*, raised to the 0.67 power.

When, on the other hand, comparing animals, plants, engines, or boats of greatly differing size, proportions do often change with size, and one speaks about allometry. Allometric change occurs when the proportions of an organism adjust to the different rates at which surface area, volume, and other physical parameters change with differing sizes. It is sometimes referred to as *differential scaling*. Perhaps the most famous allometric relationship pertains to the metabolic machinery of animals. In 1932, Max Kleiber, an American veterinary scientist, found that the basal metabolic rate (the rate at which the body generates heat when the organism is at rest), as measured by the rate of oxygen consumption or heat production, is proportional to $m^{0.75}$ over a very wide range of animals. In Figure 7.4, we show what has become known as *Kleiber's law*, which says that the metabolic rate for a wide range of different species is proportional to $m^{0.75}$ (the broken line with a slope of 0.67 = 2/3 has been added for comparison). This law has been shown to be approximately true over twenty orders of magnitude (from single cells to elephants). There are a large variety of other phenomena that change with body size according to quarter-power law scaling. For example, a cat,

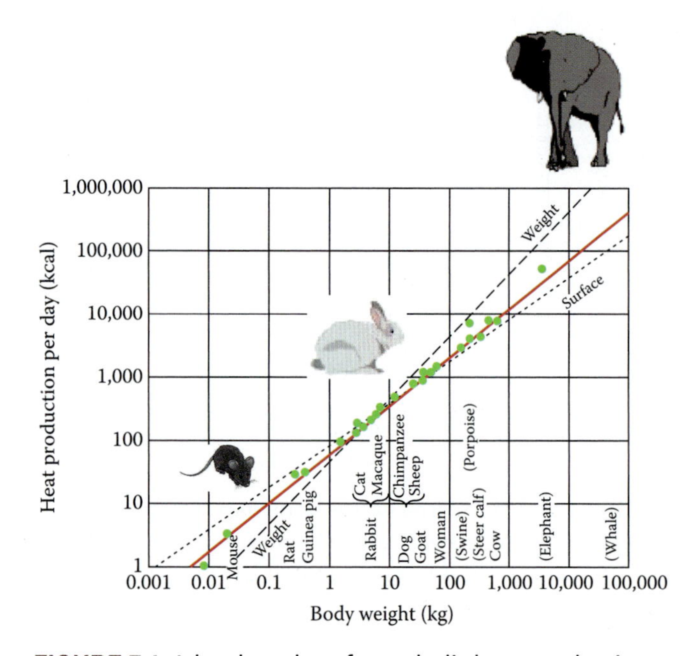

FIGURE 7.4 A log-log plot of metabolic heat production against body weight, taken from Kleiber's 1947 article. It is clear that most points (green) closely fit the red line with a slope of three quarters, rather than the weight line (slope of 1) or the surface line (slope of two thirds). Note that although the differences between the lines seem small, this is a log-log plot, so the deviations are actually vast.

100 times the weight of a mouse lives about $100^{1/4}$, or about three times, as long as a mouse, and heartbeat scales as $m^{-1/4}$, so the cat's heart beats one-third as fast as a mouse's.

The reason for Kleiber's allometric relation has eluded scientists for a long time. For metabolism, molecules inside and outside cells, such as carbohydrates, enzymes, hormones, nucleic acids, must interact and react with each other. If a cell and its organelles were not limited to a size of only a few micrometers in diameter, then it would be difficult for the various molecules to diffuse and come in contact within a reasonable amount of time. The small cell size maximizes surface-to-volume ratio (S/V), allowing for the most efficient metabolism. In larger organisms, individual cells are accessed via distribution networks: in plants, a vascular system of xylem and phloem carries water and nutrients to and from the different parts of the organism, and in animals, the circulatory system of veins plays the same role. If S/V was the main factor involved in controlling metabolic rate, that rate would be expected to scale to the body mass of an organism to the two-thirds power. Plotting the basal rate of oxygen consumption rate within a given species is indeed often found to be proportional to $m^{0.67}$. On this basis, for a cat, 100 times more massive than a mouse, we expect to have a basal metabolism of $100^{0.67}$, or about 21.5 times that of the mouse. But Kleiber's law, which applies when one looks across different species and over a wide range of sizes, means that a cat's metabolic rate is not 21.5 times greater than a mouse's, but 31.6 times. One of the more inspiring explanations for Kleiber's law came from ecologists Enquist and Brown and physicist West in 1997. These authors explain that the generic principle underlying Kleiber's law is that nutrient supply networks in animals and plants form a branching fractal network* to reach all cells in an organism. They showed that they could derive Kleiber's 3/4-power scaling law by invoking these fractal networks. In their analysis, then, the metabolism scales as the dimension of the circulatory system, which is after all what delivers the energy being metabolized.† Allometric changes in proportions are often compensations for competing requirements of surface area, cross-sectional area, and volume. In this case, nature evidently evolved a method to overcome the natural limitation of simple geometric scaling by developing these more efficient fractal-like webs to connect all the similarly sized cells within an organism. Fractals are an optimal geometry for minimizing the work lost due to the transfer network while maximizing the effective surface area (see Figure 7.5, with a tree as an example of a fractal). We believe that this insight might be of great value in human engineering endeavors where reagents and reaction products must be distributed with maximum efficiency throughout the bulk of a device, be it a fuel cell, a battery or a sensor. In electrochemical systems, it is advantageous to have a large S/V when there is the need to transfer the signal or power effectively to an electrical network.

* In a fractal network, each tiny part is a replica of the whole. In the current context, think about blood capillaries in our bodies and their botanical equivalent in trees: a silhouette of the human circulatory system and of roots and branches of a tree actually look remarkably similar.

† The theory is not without is detractors; Dodds at MIT, for example, disputes the very existence of the 3/4 law.[5] A 2005 paper by West and coworkers seems to vindicate this team's work.[6]

FIGURE 7.5 A tree as an example of a fractal.

We will demonstrate in Chapter 9, this volume, in the section on miniaturizing power sources, that fractal geometries for the electrodes might be the best solution for maximizing the current while minimizing the internal resistance.

Having reviewed the basics of isometry and allometry, we now introduce the concept of dimensional analysis, one of the simplest yet most powerful tools in a physicist's bag of tricks. Dimensional analysis is important when modeling physical phenomena, for instance, predicting the performance of a large plane on the basis of observing a small-scale model in a wind tunnel or the reverse—more relevant in miniaturization science—predicting the behavior of a miniaturized machine on the basis of the performance of a larger model. Besides predicting the behavior in scale-up or scale-down experiments, dimensional analysis is of use in the deduction of physics laws (physics demands dimensional consistency in equations), reduction of number of variables by introducing dimensionless groups and order of magnitude analysis (which terms dominate in an equation). Langhaar described dimensional analysis as "a method by which we deduce information about a phenomenon from the single premise that the phenomenon can be described by a dimensionally correct equation among certain variables."[7]

The dimensional specification for a physical variable is its *dimensional formula*. The dimensional formula shows the fundamental or base quantities (denoted by capital letters) raised to the appropriate powers and enclosed in brackets. There are seven fundamental base quantities in these dimensional formulas: time, length, mass, electrical current, temperature, amount of a substance in moles, and luminous intensity. All other quantities, dimensions, and units are known as *derived quantities*. For example, the dimensional formula for acceleration, a, is $[LT^{-2}]$. In the simplest use, dimensional analysis helps one decide whether an equation is correct as the dimensions on both sides of the equality sign must be the same and one can figure out the dimensions of a particular variable if the others are known.

In a more advanced use of dimensional analysis, the object is to substitute a set of dimensionless numbers for the dimensional physical variables that describe the problem—pressure, velocity, density, and so on. As the dimensionless numbers are products or ratios of the physical variables, this process always succeeds in reducing the number of variables in the problem. This reduction in the number of variables is a very worthwhile exercise since it frequently saves a great deal of experimental effort.

We introduce first the two most important theorems of dimensional analysis and follow that by listing a number of detailed steps to carry out a typical dimensional analysis using the exponent method or Buckingham method. Finally, we present a couple of examples and an exercise that will turn out to be of use later on in this chapter.

First Theorem

The dimensional homogeneity theorem states that any physical quantity is dimensionally a power law monomial: [Any Physical Quantity] = $[M^a L^b T^c]$.

Second Theorem

The Buckingham Π theorem (Pi theorem) states that if a system has k physical quantities of relevance that depend on r independent dimensions, then there are a total of $k - r$ mutually independent dimensionless groups $\pi_1, \pi_2, \ldots, \pi_{k-r}$. The dimensionless groups are called *pi-terms*. The behavior of the system is then describable by a dimensionless equation $F(\pi_1, \pi_2, \ldots, \pi_{k-r}) = 0$. The number of pi-terms is thus fewer than the original number of variables by r.

In practice, one proceeds as follows:

1. Make an informed guess about which parameters (variables), k, are likely to be important

in the phenomenon under scrutiny. These are typically all variables required to describe the problem geometrically (e.g., diameter, D), materials properties [e.g., fluid properties (ρ,μ)], physical constants (e.g., gravity, g—would things work differently on the moon?), and external effects (dp/dx, a pressure differential).

2. Express each variable in terms of its fundamental quantities (dimensions), $[M]$, $[L]$, $[T]$, etc.
3. Determine the required number of dimensionless parameters ($k - r$).
4. Select a number of repeating variables equal to the number of dimensions (= r). All fundamental quantities must be included in this set, and each repeating variable must be dimensionless independent of the others.
5. Form a dimensionless parameter π by multiplying one of the nonrepeating variables by the product of the repeating variables, each raised to an unknown exponent.
6. Solve for the unknown exponents.
7. Repeat this process for each nonrepeating variable.
8. Express results as a relationship among the dimensionless parameters: $F(\pi_1, \pi_2, \pi_3, \ldots) = 0$.

Example 7.1

We apply the Buckingham method now to some example problems, starting with a simple one. Say you want to find out the drag force F_D on a body immersed in a flowing fluid (say a fish or a submarine moving underwater—let's call it a swimmer). The drag force F_D is the force opposing the object's velocity and will depend on the following:

$$F_D = f(l, v, \rho, \mu) \qquad (7.3)$$

where k = 5 (number of variables/parameters) and r = 3 number of dimensions (L, T, and M).

Thus, the number of dimensionless numbers is expected to be 5 – 3 = 2. This is summarized in Table 7.2.

Select "repeating" variables: l, v, and ρ. Combine these with the rest of the variables: μ and F_D. So the first pi-term becomes $\pi_1 = \mu(l^a v^b \rho^c)$.

TABLE 7.2 There Are Five Parameters or Variables (k) and Three Fundamental Quantities

F_D	l	v	ρ	μ
MLT^{-2}	L	LT^{-1}	ML^{-3}	$ML^{-1}T^{-1}$

To calculate the exponents, one writes the following equation:

$$\pi_1 = \mu(l^a v^b \rho^c) \qquad (7.4)$$

and since this combination is supposed to be dimensionless:

$$M^0L^0T^0 = (ML^{-1}T^{-1})(L)^a(LT^{-1})^b(ML^{-3})^c$$
$$M: 0 = 1 + c \rightarrow c = -1$$
$$L: 0 = -1 + a + b - 3c \rightarrow a = -1$$
$$T: 0 = -1 - b \rightarrow b = -1$$

The first π term is thus as follows:

$$\pi_1 = \mu/lv\rho,\ \pi_1 = 1/\text{Reynolds number } (1/\text{Re}) \qquad (7.5)$$

usually written as follows:

$$\text{Re} = \rho vl/\mu \qquad (6.71)$$

The Reynolds number, familiar from Volume I, Chapter 6, appears in a large number of fluid flow problems. It is the ratio of inertial forces over viscous forces. A more complete discussion of the Reynolds number is given in Volume I, Chapter 6, on fluidics. A list of other common dimensionless groups is represented in Appendix 7A. The second dimensionless group (pi-term) should be chosen in such a way that it is certain to be independent of the first pi-term. One way to do that is to make sure that at least one physical parameter that was present in the first group is missing from the second and vice versa. In the example case, we replace the viscosity μ by the drag force F_D, as follows:

$$\pi_2 = F_D(\rho^d l^e v^f) \qquad (7.6)$$

$$M^0L^0T^0 = (MLT^{-2})(ML^{-3})^d(L)^e(LT^{-1})^f$$

$$M: 0 = 1 + d \rightarrow d = -1$$

$$L: 0 = 1 - 3d + e + f \rightarrow e = -2$$

$$T: 0 = -2 - f \rightarrow f = -2$$

This way we obtain a second dimensionless group, given as follows.

$$\pi_2 = F_D/\rho\ l^2v^2 \tag{7.7}$$

This second group, $F_D/\rho\ l^2v^2$, may be considered a dimensionless drag force. Using *A* as the wetted surface of the submerged swimmer, this dimensionless group may be rewritten as the drag coefficient C_D. Since $\pi_2 = f(\pi_1)$ we can also write the following.

$$C_D = F_D/\rho l^2v^2 = f(Re) \tag{7.8}$$

For a given Reynolds number, the best design for a race car, submarine, or airplane is the one that minimizes drag coefficient C_D (see below under "Scaling in Flying").

Example 7.2

In a second example, we apply dimensional analysis to the drag force on a rough sphere in a flowing solution. Drag force on a rough sphere (with diameter *D* and roughness *k*) is a function of *D*, ρ, μ, *v*, and *k*. The relative roughness of the sphere is given as *k*/*D*:

k = 6 (number of variables/parameters)

r = 3 number of dimensions (*L*, *T*, and *M*)

Thus, the number of dimensionless numbers is expected to be 6 − 3 = 3. This is summarized in Table 7.3.

Select "repeating" variables: *D*, *v*, and ρ. Combine these with the rest of the variables: F_D, μ, and *k*. So the first pi-term is given as follows:

$$\pi_1 = \mu(D^a v^b \rho^c) \tag{7.9}$$

and since this combination is dimensionless:

$$M^0L^0T^0 = (ML^{-1}T^{-1})(L)^a(LT^{-1})^b(ML^{-3})^c$$

$$M: 0 = 1 + c \rightarrow c = -1$$

TABLE 7.3 There Are Six Parameters or Variables (k) and Three Fundamental Quantities

F_D	D	v	k	ρ	μ
MLT^{-2}	L	LT^{-1}	L	ML^{-3}	$ML^{-1}T^{-1}$

$$L: 0 = -1 + a + b - 3c \rightarrow a = -1$$

$$T: 0 = -1 - b \rightarrow b = -1$$

we again obtain $\pi_1 = \mu/lv\rho$, π_1 = 1/Reynolds number, abbreviated as Re = ρvl/η (Volume I, Equation 6.71). For the second dimensionless group we choose the repeating variables *D*, *v*, and ρ and combine those with the nonrepeating variables F_D, μ, and *k*.

$$\pi_2 = k(D^a v^b \rho^c) \tag{7.10}$$

$$M^0L^0T^0 = (L)(L)^a(LT^{-1})^b(ML^{-3})^c$$

$$M: 0 = c \rightarrow c = 0$$

$$L: 0 = 1 + a + b - 3c \rightarrow a = -1$$

$$T: 0 = -b \rightarrow b = 0$$

$$\pi_2 \equiv k/D \tag{7.11}$$

For the third pi-term, we write a dimensionless group as follows:

$$\pi_3 = F_D(D^a v^b \rho^c) \tag{7.12}$$

$$M^0L^0T^0 = (MLT^{-2})(L)^a(LT^{-1})^b(ML^{-3})^c$$

$$M: 0 = c + 1 \rightarrow c = -1$$

$$L: 0 = 1 + a + b - 3c \rightarrow a = -2$$

$$T: 0 = -2 - b \rightarrow b = -2$$

$$\pi_3 \equiv F_D/D^2v^2\rho \tag{7.13}$$

And since $\pi_3 \equiv f(\pi_2, \pi_1)$, we can also write the following equation:

$$F_D/D^2v^2\rho = f(\rho vl/\mu,\ k/d) \tag{7.14}$$

We encountered this expression when discussing the friction factor for pipe flow in Volume I, Chapter 6 (see Volume I, Equation 6.73). With the math introduced so far, we can express similitude or similarity a bit better now than the way we represented it in Equation 7.1. As we noted before, similarity is very important to be able to predict the behavior of a model with respect to the real life object. In miniaturization science the model is usually very large compared with the actual prototype device. Suppose the physical problem described in terms of dimensionless parameters is given as $\pi_1 = f(\pi_2, \pi_3, \ldots, \pi_n)$; applied to the prototype, this transforms into $\pi_{1p} = f(\pi_{2p},\pi_{3p},\ldots,\pi_{np})$, and for the model of the prototype it reads as $\pi_{1m} = f(\pi_{2m},\pi_{3m},\ldots,\pi_{nm})$. The similarity requirements for modeling then become the following:

$$\begin{aligned} \pi_{2m} &\equiv \pi_{2p} \\ \pi_{3m} &\equiv \pi_{3p} \\ &\cdots\cdots \\ \pi_{nm} &\equiv \pi_{np} \\ \Rightarrow \text{then } \pi_{1m} &\equiv \pi_{1p} \end{aligned} \tag{7.15}$$

In other words, under these conditions the dependent variables for prototype and model will be the same. Similarity comes in different flavors. There is, first of all, geometric similarity (scale factor); i.e., the ratio of lengths in the scale model compared with the full scale case is constant (all angles are identical). Then there is kinematic similarity, in which the ratios of velocity, acceleration, and flow rate between the scale model and the full scale are constant (time-dependent behavior of the model). Finally, there is dynamic similarity, where the ratios of forces on scale model and full-scale case are constant.

To illustrate these distinctive similarities, consider predicting the drag on a thin rectangular plate (*w* × *h*) from a model placed normal to a flow. The drag function in this case will be $F_D = f(w, h, \mu, \rho, \text{and } v)$. Dimensional analysis shows that $\pi_1 = f(\pi_2, \pi_3)$, and this applies to both model and prototype. Working out the problem leads to an expression very similar to the one in Equation 7.14 (drag on rough sphere instead of on a plate):

$$F_D/\,w^2v^2\rho = f(\rho v l/\mu,\ w/h) \tag{7.16}$$

We can design a model to predict the drag on the prototype, since the drag on the model will be $F_{Dm}/w_m^2v_m^2\rho_m = f(\rho_m v_m l_m/\mu_m,\ w_m/h_m)$ as $\pi_{1m} = f(\pi_{2m}, \pi_{3m})$, and for the prototype we will have $F_{Dp}/w_p^2v_p^2\rho_p = f(\rho_p v_p l_p/\mu_p, w_p/h_p)$ as $\pi_{1p} = f(\pi_{2p}, \pi_{3p})$. The geometric similarity conditions give us the size of the model, with $\pi_{2m} \equiv \pi_{2p}$, $w_m/h_m = w_p/h_p$. For the kinematic similarity, we look at the π_3 terms ($\pi_{3m} = \pi_{3p}$), namely $\rho_m v_m l_m/\mu_m = \rho_p v_p l_p/\mu_p$, or the velocity of the model is as follows:

$$v_m = \frac{\mu_m \rho_p l_p}{\mu_p \rho_m l_m} v_p \tag{7.17}$$

For the dynamic similarity—the drag force—on the prototype, we calculate from $\pi_{1m} \equiv \pi_{1p}$.

$$F_{Dp} = \left(\frac{w_p}{w_m}\right)^2 \frac{\rho_p}{\rho_m} \left(\frac{v_p}{v_m}\right)^2 F_{Dm} \tag{7.18}$$

Exercise 7.1

A model of a submarine is 1/20th scale and is moving in fresh water (20°C) at a speed of 2 m/s. Find the speed of the model and the ratio of the drag force on the model to that of the real submarine moving in seawater. The density of seawater ρ is 1015 kg/m³, and $\nu = \mu/\rho = 1.4 \times 10^{-6}$ m²/s.

From the Reynolds number ($Re_m = Re_p$), we can calculate the speed of the model in fresh water (kinematic similarity).

$$\frac{v_m l_m}{\nu_m} = \frac{v_p l_p}{\nu_p} \text{ or } v_m = \frac{\nu_m l_p}{\nu_p l_m} v_p = \frac{1}{1.4} \times \frac{20}{1} \times 2\text{m/s}$$
$$v_m = 28.6 \text{ m/s}$$

To calculate the ratio of the drag forces we rely on the dynamic similarity.

$$\frac{F_m}{\rho_m v_m^2 l_m^2} = \frac{F_p}{\rho_p v_p^2 l_p^2} \text{ or}$$
$$\frac{F_m}{F_p} = \frac{\rho_m v_m^2 l_m^2}{\rho_p v_p^2 l_p^2} = \frac{1000}{1015}\left(\frac{28.6}{2}\right)^2\left(\frac{1}{20}\right)^2$$
$$\frac{F_m}{F_p} = 0.504$$

Trimmer's Vertical Bracket Notation

From the preceding section, we appreciate that as the scale of structures decreases, so does the importance of phenomena that vary with the largest power of the linear dimension *l*. Phenomena that are more weakly dependent on size dominate in small dimensions: electrostatics (l^2), friction (l^2), surface tension (l), diffusion ($l^{1/2}$), and van der Waals forces ($l^{1/4}$). High-power forces include gravity (l^3) (the mass of a system, *m*, scales as l^3), inertia (l^3), magnetism (l^2, l^3, or l^4, depending on the exact configuration), flow (l^4), and thermal emission

(l^2 to l^4). Scaling laws may be positive or negative order. Positive-order laws imply that the property grows with increasing scale, and negative-order laws imply that the property grows with decreasing scale. A zero-order law implies that a property is invariant with scale. In other words, positive orders affect large objects and negative orders small objects.

Trimmer introduced an elegant method to express different scaling laws by using a vertical bracket notation. For different possible forces, he writes:

$$\begin{bmatrix} l^1 \\ l^2 \\ l^3 \\ \cdot \\ \cdot \\ l^n \end{bmatrix} = F \qquad (7.19)$$

The top element in this notation refers to the case where the force scales as l^1. The next one down refers to a case where the force scales as l^2, etc. If the system becomes one-tenth its original size, all the dimensions decrease by a tenth. The mass of a system, m, scales as l^3, and as systems become smaller, the scaling of the force also determines the acceleration a, transit time t, and the amount of power per unit volume (PV^{-1}). For a generalized case with a force F scaling as l^F, we obtain

$$a = \frac{F}{m} = [l^F][l^{-3}] \qquad (7.20)$$

for acceleration a,

$$t = \sqrt{\frac{2xm}{F}} = \left([l^1][l^3][l^{-F}]\right)^{\frac{1}{2}} \qquad (7.21)$$

for transit time t, with x = distance, and

$$\frac{P}{V} = \frac{Fx}{tV} \qquad (7.22)$$

for power per unit volume. Applying Equation 7.20 to Equation 7.22 to calculate a, t, and PV^{-1} for forces scaling with varying power of the linear dimension, we obtain the following:

$$\begin{bmatrix} l^1 \\ l^2 \\ l^3 \\ \cdot \\ \cdot \\ l^n \end{bmatrix} = F \rightarrow \begin{bmatrix} l^{-2} \\ l^{-1} \\ l^0 \\ \cdot \\ \cdot \\ l^{n-3} \end{bmatrix} = a \rightarrow \begin{bmatrix} l^{1.5} \\ l^1 \\ l^{0.5} \\ \cdot \\ \cdot \\ l^{\sqrt{4-n}} \end{bmatrix} = t \rightarrow \begin{bmatrix} l^{-2.5} \\ l^{-1} \\ l^{0.5} \\ \cdot \\ \cdot \\ l^{n-2-\sqrt{4-n}} \end{bmatrix} = \frac{P}{V} \qquad (7.23)$$

We can best appreciate the usage of the vertical bracket representation from an example. In the case of electrostatic actuation, $l^F = l^2$, and from Equation 7.23 one deduces $a = l^{-1}$, $t = l^1$, and for the power density $PV^{-1} = l^{-1}$ (an increase of a factor of 10).

For force laws with a power higher than l^2, the power generated per volume degrades as the scale decreases. Even in a case with $F = l^4$, the time required to perform a task remains constant when the system is scaled down. This is an observation we understand intuitively: small things tend to be quick. We should keep in mind that beneficial effects are easily overshadowed by loss mechanisms, which scale in the same way or become even more important in the microdomain.

In Table 7.4 we list various physical phenomena and their scaling. The highest-order law is that for the moment of inertia (l^5); mass moment of inertia becomes rapidly unimportant for small-scale systems, and consequently small motors, both electric and combustion powered, are able to reach top speed much faster than large motors.

Scaling Examples

The Importance of the Surface-to-Volume Ratio

If a system is reduced isomorphically in size (i.e., scaled down with all dimensions of the system decreased uniformly), the changes in length, area, and volume ratios alter the relative influence of various physical effects that determine the overall operation in unexpected ways. For geometrically similar (isometric) objects of the same mass density, body mass (m), is proportional to the cube of the body length, l^3, and the surface area is proportional to l^2 or the surface-to-volume ratio (S/V) equals $l^{2/3,}$ i.e., $l^{0.67}$.

Table 7.4 Scaling of a Selected Set of Physical Phenomena

Physical Quantity	Scaling Exponent of l	Units
Area	2	m^2
Bending stiffness	1	$N\ m^{-1}$
Buoyant force	3	N
Capacitance	1	F
Capacitor electric field	−1	Vm^{-1}
Deformation	1	m
Drag and lift forces	2 + 2n (n is fluid relative velocity)	N
Electrostatic energy	3	J
Electrostatic force	2	N
Frictional force	2	N
Heat capacity	3	$J\ K^{-1}$
Inductance	1	H
Magnetic force	4	N
Mass (m)	3	kg
Mass moment of inertia	5	$Kg{\cdot}m^2$
Ohmic current	2	A
Resistance	−1	Ω
Resistive power loss	1	$A^2\ \Omega$
Shear stiffness	1	$N\ m^{-1}$
Strength	2	$N\ m^{-2}$
Strength-to weight ratio	−1	m
Surface tension force	1	N
Thermal conductance	1	$W\ K^{-1}$
Thermal time constant	2	s
Viscous forces	1 + v (v is fluid relative velocity)	N
Voltage	1	V
Volume (V)	3	m^3

As isometric objects or organisms shrink in size, the S/V increases. The surface area of a smaller structure is relatively larger compared to its volume than that of a larger object. There are many examples in biology and daily experience to choose from to illustrate the pervasive importance of increasing S/V with decreasing size. The abundance curve of types of animals as a function of their size as shown in Figure 7.6 illustrates the importance of a change in S/V as a function of size. For warm-blooded mammals there are significant problems associated with being too small. The very small pygmy or dwarf shrew (about 4 cm long), for example, must eat continuously or freeze to death. The heat loss from a living creature is roughly proportional to its surface area (l^2), and the rate of compensating heat generation through eating is roughly proportional to its volume (l^3). As animals get smaller, a greater percentage of their energy intake is required to balance the heat loss. A warm-blooded animal smaller than the dwarf shrew or the Cuban bee hummingbird (about 5 cm) becomes improbable; it could neither obtain nor digest the food required to maintain its constant body temperature. Insects circumvent the problem of heat loss by being cold-blooded.

On the large end of the animal size spectrum, there are no land animals larger than the African elephant at 3.80 m. Land animals fighting the resistance of gravity can grow only so big before becoming too clumsy and inefficient. Large animals also find few niches on earth that can accommodate them, and as a consequence there are fewer large animals than small ones. For creatures of the sea, the same physical barrier of gravity to size does not

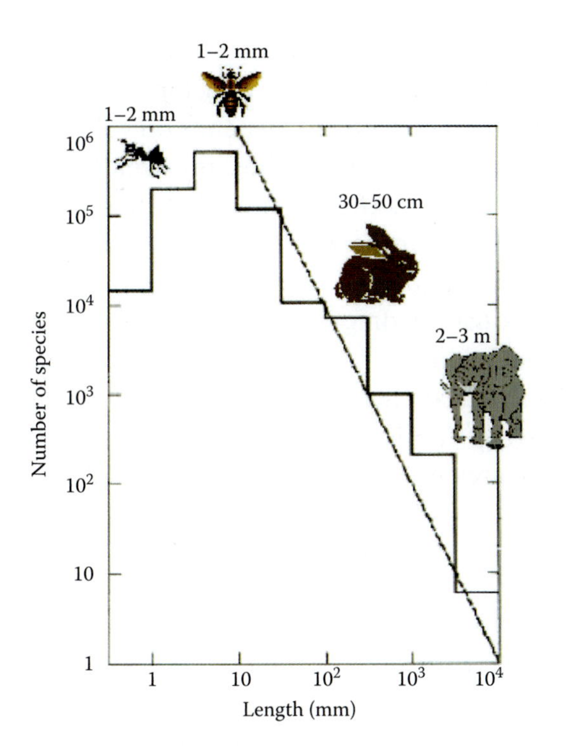

FIGURE 7.6 An abundance of types of animals in various size categories.

exist. The largest sea creature in volume is a blue whale, a little more than 20 m (60 ft.) long, much larger than the elephant. In terms of their length, the range of living things stretches over about 8 orders of magnitude, from the minute prokaryotic bacterium-like *Mycoplasma* (0.3 μm) to the massive blue whale (20 m).

At the small end of the size spectrum, smaller birds and animals are quicker and more agile, and insects are most plentiful. Scaling laws and the many niches in nature where they can hide and survive make insects perhaps the most successful animal species. Morowitz postulates that scaling laws must impose a lower limit for life in a dry environment even for cold-blooded animals, explaining why the abundance curve exhibits a peak.[8] Smaller organisms, because of their very large S/V, cannot retain their vital fluids long enough to survive, and no independent living organism (we exclude viruses) is smaller than the smallest bacteria at 0.2 μm. Water-based life increases its range of sizes both above and below that of terrestrial animals by evading gravity in the large-size range and drying out in the small-size range.

Faster evaporation in dry environments associated with larger surface-to-volume ratios also has important consequences for the retention of liquids in small biosensors, the evaporation of ejected drops in drop delivery equipment, the aqueous cocktails in miniaturized polymerase chain reaction (PCR) chambers, etc. In all these, evaporation quickly becomes a problem with decreasing drop size. Solutions include adding a hygroscopic material; mixing with solvents with lower vapor pressure (e.g., glycerol); topping of the solution with a low-vapor-pressure, nonmixing liquid; and working in a solvent-saturated environment.

Scaling in Surface Swimming

Earlier we mentioned the work on scaling in nature by Knut Schmidt-Nielsen, McMahon and Bonner, and Vogel in the context of Biomimetics (see also Chapter 2). In the same context, Hayashi presented an analysis of the natural laws that govern the relationship between speed and body length in water, on land, and in air.[9] Hayashi also analyzed the effect that miniaturization has on the strength of materials, various surface phenomena, and manufacturing accuracy. The latter is an aspect we covered in Volume II, Chapter 6, on mechanical energy-based removing. Here we will briefly look at scaling in surface swimming followed by an introduction to scaling in flight.

The resistance fish and other swimmers must overcome moving on the surface of a body of water is not gravity (l^3) but "skin friction," which increases only as the square of the linear dimensions (l^2). In this case, a larger size leads to a distinct advantage in that the larger the surface of the swimmer, the greater its swimming speed. This can be understood as follows. The available energy (E) for swimming speed (v) depends on the mass of the creature's muscles or engines (l^3), while its motion through water is opposed by the skin-friction resistance, R (l^2). This leads to $E \sim R \times v^2$ or $v \sim \sqrt{l}$; in other words, the bigger fish or the bigger ship moves faster, but only in the ratio of the square root of the increasing length. The latter is captured in the Froude number. The Froude number Fr is the ratio of inertial to gravitational forces. William Froude (1810–1879), British naval architect and engineer, realized that the factor that most limits the speed of ships or swimmers is the energy involved in the production of waves by the swimmer and by noting that wave speed increases with the square root of the wavelength. Surface swimmers find it hard to go faster than the waves they create, and by making waves of longer length, the bigger swimmer makes waves that travel faster. Wave production depends on two forces, inertia and gravity. The inertia of the moving water keeps it wavy, and gravity tries to flatten the surface; the interaction of these two phenomena sets the speed of the wave. Surface swimmers thus vary in how they produce waves on the basis of their length, so replace the mass in Equation 7.24 by l^3 times density. Inertial acceleration is hard to measure, so we replace it by velocity divided by time and obtain Equation 7.25. Time in Equation 7.25 can be gotten rid of by noting that length over time is velocity, and the Froude number is obtained in Equation 7.26.

$$Fr = \frac{ma}{mg} \tag{7.24}$$

$$Fr = \frac{\rho l^3 vt^{-1}}{\rho l^3 g} \tag{7.25}$$

$$Fr = \frac{\rho l^3 vt^{-1}}{\rho l^3 g} = \frac{\rho l^2 v^2}{\rho l^3 g} = \frac{v^2}{gl} \tag{7.26}$$

Scaling in Flying

Aircraft and ocean liners operate in fluid environments with a range of Reynolds numbers in the millions to hundreds of millions because the viscous forces are small compared with inertial forces (see Volume I, Equation 6.71). At the other end of the spectrum are small insects, birds, and fish, whose viscous forces are of the same order of magnitude as the inertial forces. Obviously, depending on size, different strategies will be required for both flying and swimming. Although the same fluid mechanical rules govern motion through air and water, flying, as Icarus could attest (see the inset heading Chapter 2, this volume), is considerably harder than swimming. The lower density of air helps a flying machine go forward faster (thrust), but it necessitates a larger force to keep it aloft. For an object to hover immobile in air takes an upward force that counterbalances weight, while forward thrust requires a force equal to the drag force (F_D) at its specific flying speed. Weight depends on volume, while drag depends on area. Thus the smaller creature finds weight less troublesome, but drag becomes a real "drag"—in other words, for a tiny insect, staying aloft against gravity's pull is easy but making headway against the drag of the air is tougher than for a bird or a plane.

The drag coefficient C_D, a dimensionless number, is given by $C_D = F_D/\rho\, l^2 v^2 = f\,(Re)$ (Equation 7.8), with surface area A replacing l^2 and a factor 1/2 added:

$$C_D = \frac{F_D}{\frac{1}{2}\rho v^2 A} \tag{7.27}$$

where ρ = density of the fluid
v = free stream velocity
A = characteristic area of the body*
F_D = drag force

The drag coefficient, C_D, is the parameter that best describes the resistance to motion. As we gleaned from Equation 7.8, the drag coefficient varies with the Reynolds number, but the coefficient also depends on the angle of attack (AOA). This is the angle that the flying object penetrates the air with. It is measured in degrees. In the case of an airfoil (Figure 7.7), at zero degrees AOA, the front of the airfoil is pointed directly into the stream. At a 90° AOA, it is pointed straight up.

We learned in Volume I, Chapter 6, on fluidics, that C_D also depends on surface roughness; the drag coefficient decreases with increasing roughness as it delays the boundary layer separation. In Figure 7.8A we show the historical reduction in drag on cars. The C_D of a Porsche 924 Turbo, for example, is 0.34, while that for a Toyota Tercel (liftback) is 0.54. In Figure 7.8B we also show the drag coefficient of some different familiar shapes.

Lift force, F_L, is the force normal to the direction of the object's velocity, v. Lift force, like drag force, depends on surface area and goes up with the object's

* A = frontal area (thick body, cylinder, cars, projectiles) = plan area (wide/flat body, wings, hydrofoils) = wetted area (surface ships, barges). This is a simplification, as both frictional and pressure drag forces should be considered, and factor A has a different meaning in both. For a more rigorous treatment, see Granger.[6]

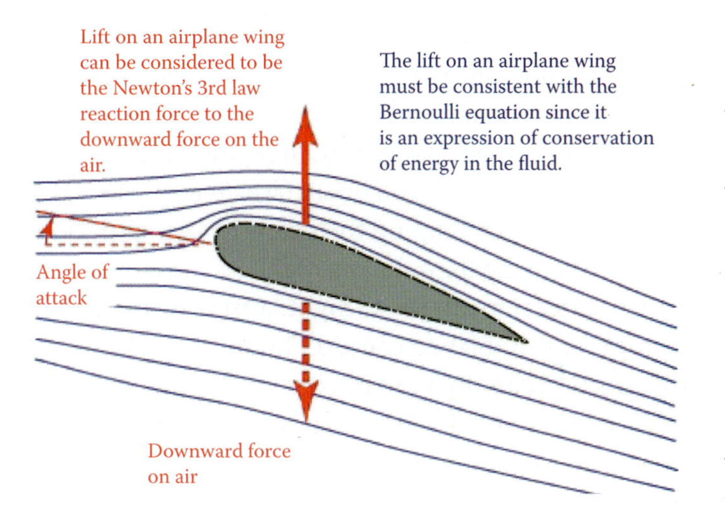

FIGURE 7.7 Angle of attack of an airfoil. For Bernoulli equation, see Volume I, Chapter 6, on fluidics (Equation 6.44). Notice that the Bernoulli interpretation of lift is not entirely correct (see below).

speed through the air. Many of us were taught that airplanes fly as a result of the Bernoulli's principle, which says that if air speeds up, pressure is lowered (Volume I, Equation 6.44). From the shape of the wing shown in Figure 7.7 the path over the top is longer than over the bottom, and in the Bernoulli model a wing generates lift because the air goes faster over the top, creating a region of low pressure, and thus lift. This begs the question of how inverted flight is possible, and naturally one may also ask why the air goes faster over the top of the wing than over the bottom. The usual explanation is that when the air separates at the leading edge of the wing, the part that goes over the top must converge at the trailing edge with the part that goes under the bottom. This is the so-called *principle of equal transit times* (http://www.allstar.fiu.edu/AERO/airflylvl3.htm, http://www.grc.nasa.gov/www/K-12/airplane/

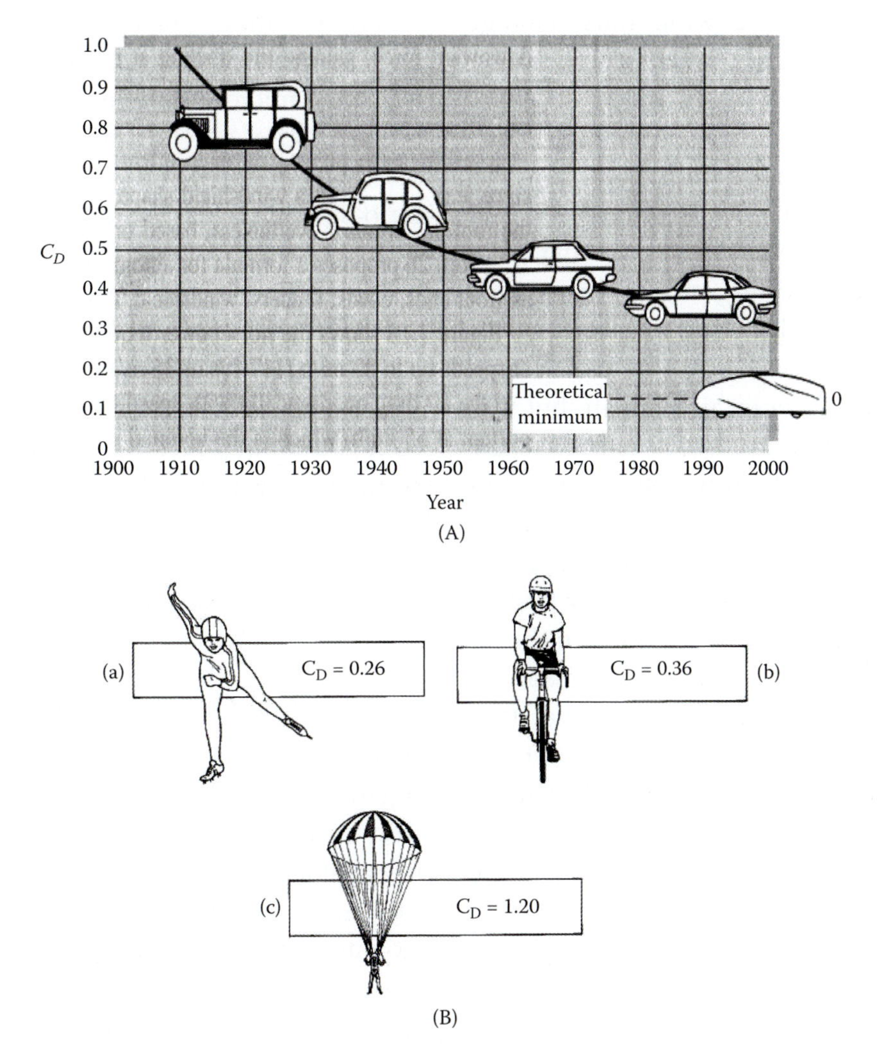

FIGURE 7.8 (A) Historical reduction of drag on cars. (B) Drag on a variety of shapes: (a) skater; (b) cyclist; (c) parachutist.

lift1.html). But in a wind tunnel simulation, where colored smoke is introduced, one sees that the air that goes over the top of the wing gets to the trailing edge before the air that goes under the wing. In fact, closer inspection reveals that the air going under the wing is slowed down from the "free-stream" velocity of the air. A calculation of the average speeds of the air over and under the wing can be made, and from the Bernoulli equation the resulting pressure forces and thus lift can be determined. The problem is that this calculation shows that in order to generate the required lift for a typical small airplane, the distance over the top of the wing must be about 50% longer than under the bottom; this would make for a very odd wing shape. An airplane wing more typically has a top surface that is only 1.5–2.5% longer than the bottom surface. So a Cessna 172 would have to fly at over 400 mph to generate enough lift. Clearly, something in this description of lift is wrong. As Newton's laws suggests, the wing must change something of the air to get lift, and that something, as revealed from wind tunnel tests, is that air passes over the wing and is bent down. The bending of the air is the action, and the reaction is the lift on the wing. The natural question now becomes, how does the wing divert the air down? When a moving fluid, such as air or water, comes into contact with a curved surface, it tries to follow that surface; this is known as the *Coanda effect*. The reason behind the Coanda effect is viscosity; viscosity in air is small, but it is enough for the air molecules to want to stick to the wing surface. Because the fluid near the surface has a change in velocity, the fluid flow is bent toward the surface. When the air is bent around the top of the wing, it pulls on the air above it, accelerating that air down; otherwise there would be voids left in the air above the wing. This pulling causes the pressure to become lower above the wing. It is the acceleration of the air above the wing in the downward direction that gives lift. The lift of a wing is equal to the change in momentum of the air it is diverting down.

The dimensionless lift coefficient C_L is given by the following equation:

$$C_L = \frac{F_L}{\frac{1}{2}\rho v^2 A} \tag{7.28}$$

Both lift and drag coefficients are a function of the Reynolds number, AOA, geometric shape (i.e., the thickness distribution), and speed. The lift and drag forces can be changed by changing (most easily) the AOA or velocity (v). The surface area (A) can also be changed by changing the shape or configuration of, say, the wing flaps during flight. As in swimming, larger objects mean faster flight: a fruit fly might hit 3 mph, a bumblebee can do about 12 mph, and a Boeing 747 can do 920 kph.

Doubling the length of a wing gives a flying machine four times the lift (and drag), but it also increases the weight by a factor of 8. The best-designed wings produce a lot of lift while suffering little drag. The largest flying animals alive today are the soaring birds such as the 10-kg wandering albatross, with a wingspan of 3.5 meters. Lift relative to drag gets worse as wings get smaller, and gliding becomes impossible with wing lengths of less than 1 mm.[10] As a point of reference, ordinary airplanes have a lift/drag ratio of 10–100, whereas that of a micromachined airplane fabricated from Si by Kubo et al. (1.56 mm long and with a mass of 10.8 mg) was measured at 0.4.[11] The low lift/drag ratio of the latter is due to the influence of viscosity, which gets more pernicious the smaller the flying object. When a glider descends, the angle depends entirely on the ratio of lift to drag. The larger this ratio, the more horizontal the glide. Kubo et al.'s micromachined airplane had an AOA less than 45°, so it could not glide. Small flying animals use fast-beating wings to produce both lift and thrust. Their wings not only beat up and down but to some extent also fore and aft. The drag force produced by down beating must exceed that produced by up beating.

Scaling in Surface Tension and Adhesion

Inertia and gravity forces rule the world of large organisms and machines, while the forces of attraction between molecules rule the world of minute organisms. A large and lumbering elephant cares little about viscosity and even less about surface tension and diffusion, but insects are similar in size to many human-made miniature machines, and viscosity, surface tension, and diffusion play an important role for both.

Capillary forces, as we saw above, are caused by surface tension and provide a striking example of scaling laws. The mass of a liquid in a capillary tube, and hence the weight, scales as l^3 and decreases more rapidly than the surface tension, which scales as l, as the system becomes smaller. That is why it is more difficult to empty liquids from a capillary than to spill coffee from a cup.

One easily recognizes how insects take advantage of surface tension and several other scaling phenomena: some insects can walk on water (e.g., a water strider), and most jump farther in proportion to their size than humans can. Walking on water is based on the surface tension at the water surface (l^1), which easily supports an insect's weight (l^3). A 10-mg, mosquito-sized insect needs a mere millimeter of total foot edge to be supported by surface tension; a 60-kg human would need 8000 m of foot edge. The effort for a jump is proportional to the mass (m) and the height (h) to which that weight is raised. In other words, $E \sim m \times h$. The biological force in a muscle available for this work is proportional to the mass of the muscle or to the mass of the animal. It follows that h is, or tends to be, a constant. As a consequence, animals tend to jump to the same actual height independent of their size. Insects have other advantages: to circumvent problems with excessive heat loss, insects are cold-blooded. Having adapted so well to so many niches in nature, it is no wonder insects are so abundant (Figure 7.6).

As we saw in Volume I, Chapter 7, molecular adhesive forces increase in proportion to the surface l^2. The strength of the attraction force between two solid surfaces drops off very quickly, though, as the distance between the two solids increases (see Volume I, Equation 7.26 and Table 7.8). The larger the bodies, the smaller the chance they come in close contact, and intimate contact is greatly aided by small size. Bonner suggested that adhesive forces, just like surface tension, scale as l^1. In Figure 7.9, following Bonner,[4] we plot the forces due to mass (l^3) and molecular cohesion (l^1) (both surface tension and adhesion) as a function of length in a log-log plot. The vertical axis measures force in arbitrary units. The lines for l^3 and l^1 are made to cross arbitrarily at 1 mm, based on the empirical observation that structures of that length have to deal in

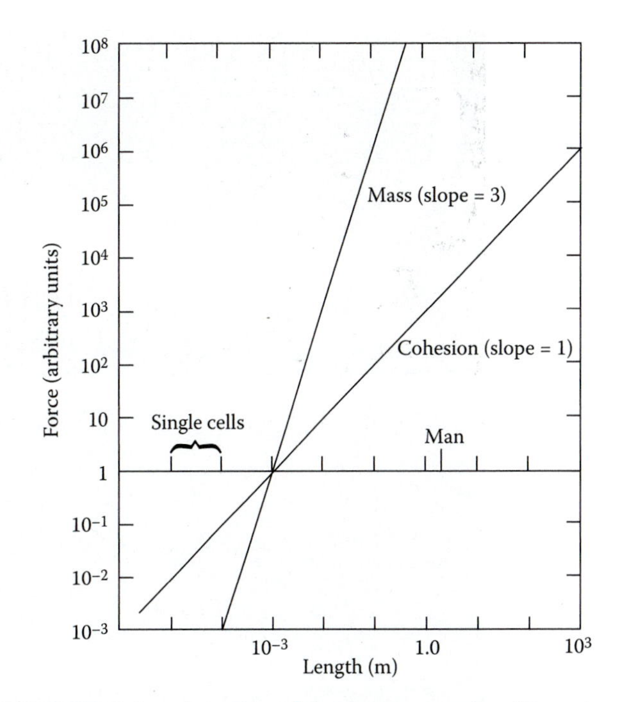

FIGURE 7.9 Log-log plot of forces due to mass (l^3) and molecular cohesion (l^2) (both surface tension and adhesion) as a function of length. The vertical axis measures force in arbitrary units. The lines for l^3 and l^1 are made to cross arbitrarily at 1 mm.[4]

equal amounts with the impact of gravity and cohesion forces. This graph helps explain how creatures below a certain size, can walk on walls and ceilings. Protuberances on the feet of these creatures function to enhance the molecular adhesion. From Figure 7.9, we see that small insects and other wall-walking animals must operate below a critical size at which the line proportional to body weight intersects the line proportional to cohesive force.

Kellar Autumn et al. confirmed, using a MEMS force sensor, that in the case of a gecko, the gecko setae (microhairs) are responsible for the remarkable van der Waals-based dry adhesion to walls.[12] This effect is almost exclusively based on the size and shape of the gecko's foot hair or setae and is quite independent of surface chemistry. Gecko foot hair is made of five microdiameter seta fibers, and each of these fibers is equipped with between 100 and 1000 nanofibers (spatulae), which are 200 nm in diameter (see Figure 7.10 for a picture of a gecko, a gecko foot, setae, and spatulae). In order for this type of dry and clean adhesion (the hairs are hydrophobic in nature) to work, only a small preload force normal to the surface is required to force the compliant

FIGURE 7.10 A gecko (a), gecko foot (b), setae (c), and spatulae (d). (Courtesy of Dr. K. Autumn, http://www.lclark.edu/~autumn/PNAS.)

hairs to configure themselves properly to a rough or smooth surface. The adhesion force can be as high as 10 N per 1 cm^2 area, and the foot will remain on until peeled off. Menon et al. have been working on synthetic gecko hair to enable gecko inspired climbing robots.[13] For an update on gecko adhesion, visit http://www.lclark.edu/~autumn/PNAS.

Scaling and Diffusion

Diffusion plays a particularly significant role in the lives of very small organisms. Diffusion is the consequence of the endless and random wandering of every molecule of every gas or liquid at temperatures above 0 K. Diffusional effects come into play at yet a smaller length scale than surface tension. For a spherical molecule, the Stokes-Einstein diffusion coefficient, D (m^2/s), we saw in Volume I, Chapter 6, is given by (Volume I, Equation 6.29) the following:

$$D = \frac{k_B T}{6\pi\eta r} \quad \text{(Volume I, Equation 6.29)}$$

where k_B = Boltzmann constant ($1.38 \times 10\,m^{-23} J\,K^{-1}$)
T = absolute temperature (K)
η = absolute viscosity (kg/m·s)
r = hydrodynamic radius

A small molecule, with a molecular weight of between 500 and 1000, will have a diffusion coefficient of about $5 \times 10^{-5}\,m^2/s$. According to the random walk equation, the diffusion length, x, of a molecule in solution is given by the Einstein-Smoluchowski expression (Volume I, Equation 6.100):

$$x = \sqrt{2D\tau} \quad \text{(Volume I, Equation 6.100)}$$

TABLE 7.5 A Drop of Saline Solution in a Small Bucket

Volume	1 μL	1 nL (10^{-9} L)	1 pL (10^{-12} L)	1 fL (10^{-15} L)	1 aL (10^{-18} L)
Length of each cube side	1 mm	100 μm	10 μm	1 μm	100 nm
Time to diffuse over a cube side	500 s	5 s	0.050 s	0.5 ms	0.05 ms
Number of molecules in a 1 μM solution	6×10^{11}	6×10^{8}	6×10^{5}	600	6

Sources: Manz, A., J. Harrison, E. Verpoorte, J. Fettinger, A. Paulus, H. Lüdi, and H. Widmer. 1992. Planar chips technology for miniaturization and integration of separation techniques into monitoring systems: Capillary electrophoresis on a chip. *J Chromatogr* 593:253–58;[14] and Brown, P. R., and E. Grushka, eds. 1993. *Advances in chromatography*. New York: Marcel Dekker.[15]

where τ is the time required for a molecule to diffuse over a distance x. From Volume I, Equation 6.100, diffusion of a molecule in the bulk of a liquid over a length of 10 μm is a million times faster than diffusion over 1 cm. Table 7.5[14,15] illustrates how the volume of a drop of liquid (say, of a saline solution, in liters), relates to the linear dimension of a cube containing it. Assuming a diffusion constant D of 10^{-5} cm^2s^{-1}, the table also lists the time it takes for a molecule to diffuse across one side of the containment volume; it will take a molecule about 500 s to diffuse over a distance of 1 mm but only 0.5 ms to cross a distance of 1 μm. Finally, the table lists the total number of molecules the various volumes filled with a 1 μM solution contain.

As we saw in Volume I, Chapter 6, on fluidics, mixing, although only mediated by diffusion, is very fast at the microlevel and should allow for reaction times to be determined by inherent kinetics rather than the time it takes for reactant species to meet in solution. In nature, only the smallest animals rely on diffusion for transport; animals made up of more than a few cells cannot rely on diffusion anymore to move materials within themselves. They augment transport with hearts, blood vessels, pumped lungs, digestive tubes, etc. These distribution networks typically constitute fractals. Fractals are an optimal geometry for minimizing the work lost due to the transfer network while maximizing the effective surface area. Along this line, one can envision making arrays of parallel microchemical reactors. Mixing small amounts of fluid in a large set of these parallel microreactors leads to a much higher mixing and reaction efficiency than when mixing the same total amount of reagents in one big reactor vessel all at once. Based on the latter, it is possible to design chemical reactors where scaling up is perfectly linear, a feat not possible when scaling up chemical reactors in the traditional mode from milliliters to hundreds of liters in big tanks.

Mixing of liquids in small microfluidic manifolds is addressed in Volume I, Chapter 6, on fluidics.

Scaling of Strength-to-Weight Ratio and Inertia

From the previous section, we recognize the important role of gravity for larger systems. Gravity, for example, does not allow large creatures to support themselves with surface tension on the surface of water. It makes large objects fall faster than small ones and necessitates that larger aircraft fly faster to stay aloft. To counteract gravity, large land mammals have stiffer bones than small animals and require thicker legs. The legs of the elephant appear as rather straight, stubby columns, while those of an insect are long and spindly. The diameter of a tall homogeneous body such as a tree must grow as the power 3/2 of its height, which accounts for Goethe's *Es ist dafür gesorgt, dass die Bäume nicht in den Himmel wachsen** in *Dichtung und Wahrheit*.[16] Human design reflects the same influence. If we build two geometrically similar bridges, the larger will be the weaker of the two, and it will be so in the ratio of their linear dimension, l. The strength of the iron girders in the two bridges varies with the square of the linear dimension, that is, l^2, but the weight of the whole structure varies with the cube of its linear dimension,

* Translation: It has been arranged for trees not to grow all the way into heaven.

that is, l^3. Scaling up the entire bridge by a factor of two gives us columns that are four times as strong but must also bear eight times the load—the safety factor is halved. The girders, being thicker, also suffer greater self-loading, which requires them to be thicker yet. To engineer around gravity, stiffer materials and alternative designs are required. The larger the structure is, the more severe the strain. By reducing the size of a device, the structural stiffness generally increases relative to inertia-imposed loads.

The strength-to-weight ratio scales as l^{-1} (area over weight, i.e., l^2/l^3), and as a consequence small things are relatively stronger. Consider the strength comparison between a human and an ant. The human body is 300 times that of an ant, and a human can carry approximately one body weight. The strength-to-weight ratio law predicts that a 1/300th scale human Lilliputian could carry 300 times its body weight. Although not exactly right, an ant can carry objects 10–50 times (but not 300 times) its own weight. The discrepancy is caused by the difference between a scaled human form to that of an ant. Because of the increased strength-to-weight ratio, an ant is proportionately more slender than a human; in other words, an ant scaled to human size would have legs far too thin to support its own weight.

Inertia is the tendency of both fluids and solids to remain at rest or to keep moving. The force involved in starting a movement or stopping it equals the mass of the object times the acceleration or deceleration. The highest-order scaling laws are those involving moments of inertia, that is:

$$I = \int r^2 dm \tag{7.29}$$

with l^5. Small things are easier to start and stop. An example of this is small motors, electric or combustion, which are able to reach top speed in a fraction of a second; large motors may require several seconds to reach full speed.

An important bottleneck in the development of microactuators in general is that resistive forces, such as viscous drag (l^2) and surface tension (l), exceed motive forces [mass (l^3) × surface (l^2) = (l^5)] in the microdomain. Miniaturization engineers look for driving forces that scale more advantageously, such as electrostastics (l^2) and capillary force (l).

The above scaling examples constitute but a few illustrations; many more scaling examples can be found in Chapter 8, this volume, on actuators; Volume I, Chapter 6, on fluidics; Chapter 9, this volume, on power and brains in miniature devices; and throughout the book.

Breakdown of Scaling Laws

Introduction

Scaling laws break down when the continuum theory for the phenomenon under study no longer applies. The latter occurs any time boundary conditions for the macroscale case are violated. Many examples of breakdown of scaling laws can be found throughout the book. A detailed example was worked out in Volume I, Chapter 7, where we saw that electrochemical detectors with electrodes of the size of the diffusion boundary layer thickness display noncontinuum behavior. Here we give as an additional example, namely that of electrostatic devices with electrode spacing of the size of the mean free path of the surrounding gas molecules.

Scaling in Electrostatics

Boundaries and Continuum Theory in Electrostatics

Electrostatic charges, we will see in Chapter 8, on actuators, arise from buildup or deficit of free electrons in a material. An electrically charged material can exert an attractive force on oppositely charged objects or a repulsive force on similarly charged ones. To appreciate scaling issues in electrostatics, we follow Trimmer's analysis of isometric scaling of the maximum stored energy in a simple parallel plate capacitor. The ability of a capacitor, *C*, to store charge is its capacitance, *C*, and the capacitance is related to the applied potential *V* as C = Q/V. From Gauss's law (Volume I, Equation 5.37) applied to two plates of cross-sectional area *A*, containing a dielectric (and hence charge density *Q/A*), one obtains the following:

$$E = \frac{Q/A}{\varepsilon_r \varepsilon_0} \tag{7.30}$$

and since $C = Q/V$ and E is given by V/d, where d is the distance between the two plates of the capacitor, Equation 7.30 can be rewritten as follows.

$$C = \frac{\varepsilon_r \varepsilon_0 A}{d} \quad (7.31)$$

From Equation 7.31, it is clear that the greater the relative permittivity ε_r, of the material between the capacitor plates, the greater the capacitance. This can be understood as follows, since a dielectric is an insulator all a gap material can do in response to a charge on the capacitor plates is to accumulate a countercharge. The greater the relative permittivity, the greater the effect of this charges redistribution; thus the greater the effective "stored charge." Based on Equation 7.30, the insertion of a dielectric reduces the electric field inside the dielectric by a factor ε_r.

The dimensions of the capacitor in Figure 7.11 are w, v, and d, and they all scale with l. We first calculate the incremental work dW to add charge dQ to this capacitor at voltage V:

$$dW = V(Q) \cdot dQ = \left(\frac{Q}{C}\right) \cdot dQ \quad (7.32)$$

The total work W to charge to Q and thus the energy stored in the capacitor is then given as follows:

$$U = W = \frac{1}{C}\int_0^Q Q\,dQ = \frac{1}{2}\frac{Q^2}{C} = \frac{1}{2}CV^2 \quad (7.33)$$

Above a critical voltage, V_b, the electrostatic forces polarizing the molecules in the gap are so strong that electrons are torn free and charge flows and the maximum electrostatic potential energy, U_m, stored in the capacitor in Figure 7.11 is then given by the following:

$$U_m = \frac{1}{2}CV_b^2 = \frac{\varepsilon_r v_0 w\, v V_b^2}{2d} \quad (7.34)$$

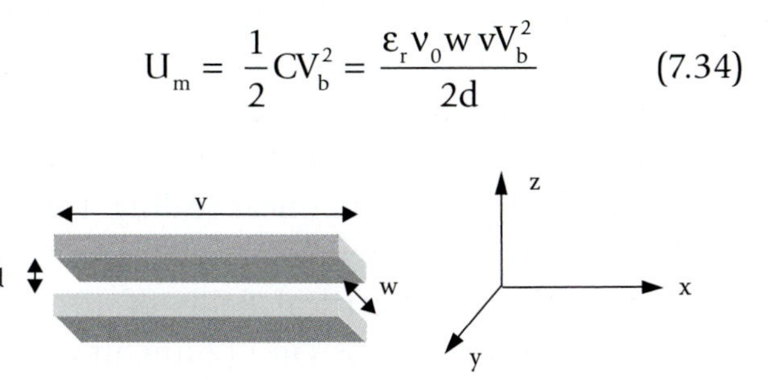

FIGURE 7.11 A parallel plate capacitor of plate size $w \times v$ and separation d stores a maximum potential energy of U_m.

where V_b = electrical breakdown voltage and $\varepsilon_r\varepsilon_0 \times w \times v/d$ = capacitance C. The permittivity of vacuum, ε_0, and the relative permittivity, ε_r, remain unchanged with scaling, so we can assign them an l^0 dimension. We also assume, for now, that V_b scales linearly with d. Intuitively, we expect a smaller gap, d, to result, indeed, in a lower breakdown voltage, V_b. However, as we shall see next (contrary to our macrointuition), at very small capacitor plate separation, the linear relationship between V_b and d breaks down. The breakdown voltage of the capacitor actually starts increasing with decreasing plate separation (see discussion below on continuum breakdown in electrostatics and the discussion on the Paschen curve below, in Figure 7.12).

Writing out all the dimensions in Equation 7.34 in terms of l, we obtain the following:

$$U_m \propto \frac{l^0\, l^0 l^1\, l^1\, (l^1)^2}{l^1} = l^3 \quad (7.35)$$

and the maximum energy stored in the capacitor scales as l^3. It follows that if l decreases by a factor of 10 (i.e., w, v, and d all decrease simultaneously), the stored maximum potential energy in the capacitor decreases by a factor of 1000.

The maximum electrostatic energy density, U'_m (in J/m³), between the capacitor plates is obtained from Equation 7.34 by dividing the maximum potential energy, U_m, by the volume, V_0, of the capacitor gap, $V_0 = w \times v \times d$:

$$U'_m = \frac{\varepsilon_r v_0 E_b^2}{2} \quad (7.36)$$

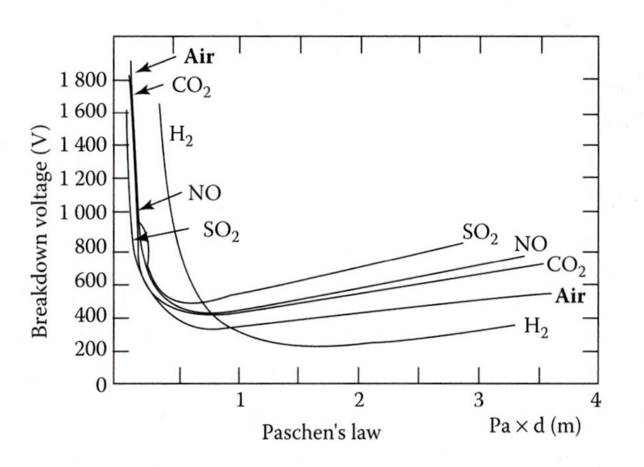

FIGURE 7.12 Breakdown voltage between two conductors, plotted as a function of the product of the distance between them and the surrounding gas pressure (P).

This expression for the energy density is general and is not restricted to the special case of a constant field in a parallel plate capacitor. In the case of an air capacitor, E_b is limited to approximately 3×10^6 Vm^{-1} by the electrical breakdown of air. Thus, the stored energy for an air capacitor is about 40 Jm^{-3}.[17] In power capacitors (see Chapter 9, on power and brains in miniature devices), with special dielectrics and a field of 10×10^6 Vm^{-1}, energy densities of 1 kJm^{-3} have been achieved.

We are interested in using electrostatic force in all types of actuators (see Chapter 8, on actuators), and that force, *F*, is the negative spatial derivative of the energy, *U*. In a case where we move the plate of a capacitor with respect to the other plate, the electric force will be the spatial derivative, in that direction, of the stored energy *U*. A translational movement of one plate with respect to the other in a microactuator in the direction *x* leads to a force, F_x, as follows:

$$F_x = -\frac{\partial U}{\partial x} = \frac{V^2}{2}\frac{\partial C}{\partial x} = \frac{l^3}{l^1} = l^2 \qquad (7.37)$$

where *V* is the applied voltage. The electrostatic force for a constant field is thus found to scale as l^2. This is often an advantage, because the mass and, hence, inertial forces scale as l^3. The electrostatic force gains over inertial forces as the size of the system is decreased. A decrease in size by a factor of 10 leads to a decrease of the inertial forces by a factor of 1000, whereas the electrostatic force decreases by a factor of only 100. It is thus not hard to exert rapid accelerations in microsystems with the aid of electrical fields. These are the type of features we will explore further in different types of electrostatic actuators in Chapter 8.

Breakdown of Continuum Theory in Electrostatics

We will now consider where macroscopic electrostatic laws err in the microscopic regime. The fundamental boundary condition in a gas pertains to the main free path, λ, the average distance traveled by a molecule between two successive collisions. The mean free path, λ, can be derived from the kinetic theory of gases as follows:

$$\lambda = \frac{1}{\sqrt{2}\pi n a^2} = \frac{kT}{\sqrt{2}\pi P a^2} \qquad (7.38)$$

where n = density (the number of molecules per unit volume)
T = temperature
k = Boltzmann constant
P = pressure
a = size of the molecule

Using Avogadro's number, we know that at room temperature and 1 atmosphere, 1 cm^3 contains 2.69×10^{19} molecules. Assuming that oxygen and nitrogen have a molecule size of about 3 Å, one calculates then a mean free path in air of about 60 nm.

Electrostatic charges, we saw earlier, arise from buildup or deficit of free electrons in a material. An electrically charged material can exert an attractive force on oppositely charged objects or a repulsive force on similarly charged ones. At small distances between two conductors in air, the electrical field is not isotropic as assumed above. This can be gleaned from the so-called *Paschen curve* in Figure 7.12. This figure represents the breakdown voltage between two conductors, plotted as a function of the product of the distance between them and the surrounding gas pressure (*P*). In the case of an air capacitor, the field strength required to cause sparks between the two electrodes (E_b), we saw in the section above, is approximately 3×10^6 Vm^{-1}. The general shape of the Paschen curve can be understood more easily by assuming that the pressure, *P*, remains constant at 1 atm, thus making the *x*-axis a simple distance axis. We notice that, on the right side of the Paschen curve, at large electrode distances, the field is constant and the scaling laws introduced above all pertain. But for smaller electrode gaps (below about 5 μm in air), the curve sharply reverses, bending upward and leading to higher electrical breakdown fields for thinner air gaps. Electrical breakdown in these small gaps does occur at much larger voltages than those predicted from linear scaling. From Bart et al.[17] and Busch-Vishniac[18] and references therein, we quote observed electrical fields of $10^8 \times$ Vm^{-1} (1.5-μm air gap), 1.7×10^8 Vm^{-1} (2-μm air gap), and 3.2×10^7 Vm^{-1} (12.5-μm air gap). The upper limit is the electrical field measured for small gaps in vacuum, that is, 3.0×10^8 Vm^{-1}.[19] Surface roughness results in a lower average breakdown voltage, but even then, these fields are significantly higher than the 3×10^6 Vm^{-1} quoted above for breakdown of capacitors with large air gaps. Assuming a field strength of 3×10^8 Vm^{-1}, Equation 7.36 predicts an energy density of about 4×10^5 Jm^{-3}, compared with 40 Jm^{-3} without the Paschen effect. This unexpected result derives from the fact that there are not enough ionizing collisions to induce an avalanche between electrodes with a separation comparable to the mean free path of the surrounding gas. When the gap between two conductors approaches the mean free path of the molecules (λ), statistically fewer molecules are present to be ionized between those closely spaced conductors. Higher fields and higher-energy densities can be achieved before the critical breakdown voltage, V_b, is reached. The Paschen curve also illustrates, how, in terms of influencing the breakdown voltage, reducing the gap between two electrodes is equivalent to reducing the surrounding gas pressure: the x-axis is the product of pressure (P) and distance (d), and they exert the same effect on the breakdown voltage. One can thus actually say that reducing the distance between two conductors in a gas is equivalent to reducing the pressure in a macroscopic set-up.

In addition to the advantage of higher-energy densities in micromachined electrostatic systems, the Paschen effect suggests yet another advantage. The scaling of the electrical field (E_b) in the nonisotropic region is more like $l^{-1/2}$. According to Equation 7.37, in the macroregime, the force on a capacitor scales as l^2, or as a surface force. In the Paschen regime, the force scales with l so that with a factor of 10 size reduction, the inertial force still decreases by a factor of 1000, but the electrostatic force decreases only by a factor of 10. On the basis of Equation 7.23, we further see that acceleration and transit times are also higher than for the isotropic system.

Working on the left side of the Paschen curve renders high-field operation possible for a wide variety of electrostatic microdevices without incurring catastrophic sparking. Examples include the electrostatic motors illustrated in Figure 8.5, the electrostatic switch in Figure 8.19, and the author's volcano ionization sources presented in the example below.

Example: Ion Mobility Spectrometer

In an ion mobility spectrometer (IMS), ions produced at atmospheric pressure in an ionization cell by a ^{63}Ni beta source are accelerated in a drift chamber (uniform field of 150–250 V/cm), where they are separated according to their mobilities, detected as a current on a Faraday plate, and plotted on a time axis in accordance with their time of arrival (see Figure 7.13). About twenty years ago, this author set out to micromachine some critical components of an IMS instrument. We recognized that field uniformity, temperature, and pressure could be controlled with micromachined Si sensors. Our aim, though, was to substitute the ^{63}Ni ionization source

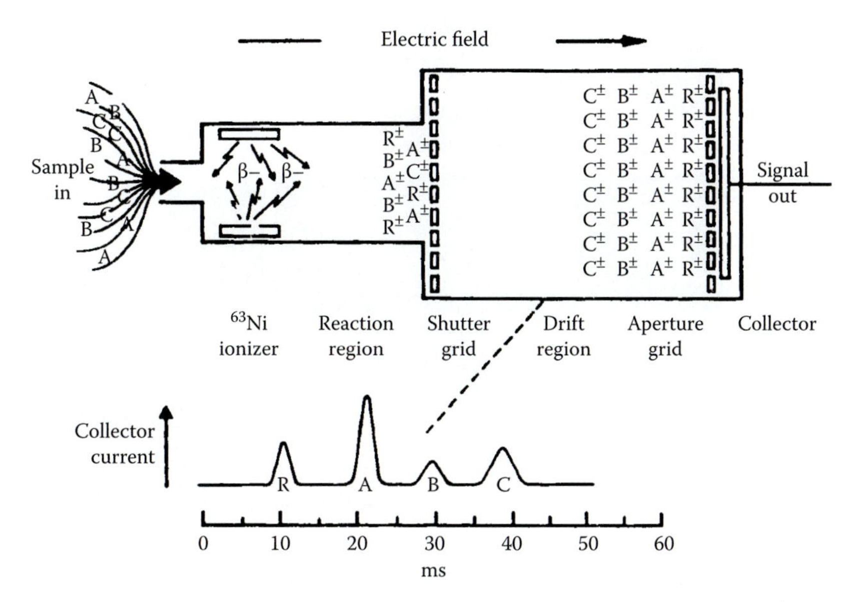

FIGURE 7.13 Ion mobility spectrometry with a Ni ionization source.

with a less fragmenting ionization source employing a "softer" field ionization (FI) technique. In FI, one creates mainly positive parent ions (M^+), and thus simpler spectra for complex environmental gas mixtures result. FI is well known from FI mass spectrometry (FIMS). But mass spectrometry is carried out in vacuum, and considering the extreme electric fields required for FI, one might expect that, under atmospheric pressure conditions, catastrophic electrical breakdown between the ionizing electrodes would occur. The idea about how to avoid breakdown or sparking at atmospheric pressures is based on the peculiar behavior of microelectrodes operating at the left side of the Paschen curve (see Figure 7.12), where much higher voltages are required for breakdown to occur.

For an ionization source operating on the left side of the Paschen curve, we microfabricated an array of microvolcanoes with a typical throat opening of 1 μm and a volcano rim to gate distance of less than 1 μm. Both the volcano rim and gate electrode may be made from a variety of inert metals. A scanning electron microscope (SEM) photomicrograph of a single microvolcano and an array of microvolcanoes is shown in Figure 7.14a (the metal used in this case is Pt). Given the small dimensions of the microvolcanoes, the intense electric fields required for field ionization can be produced with significantly

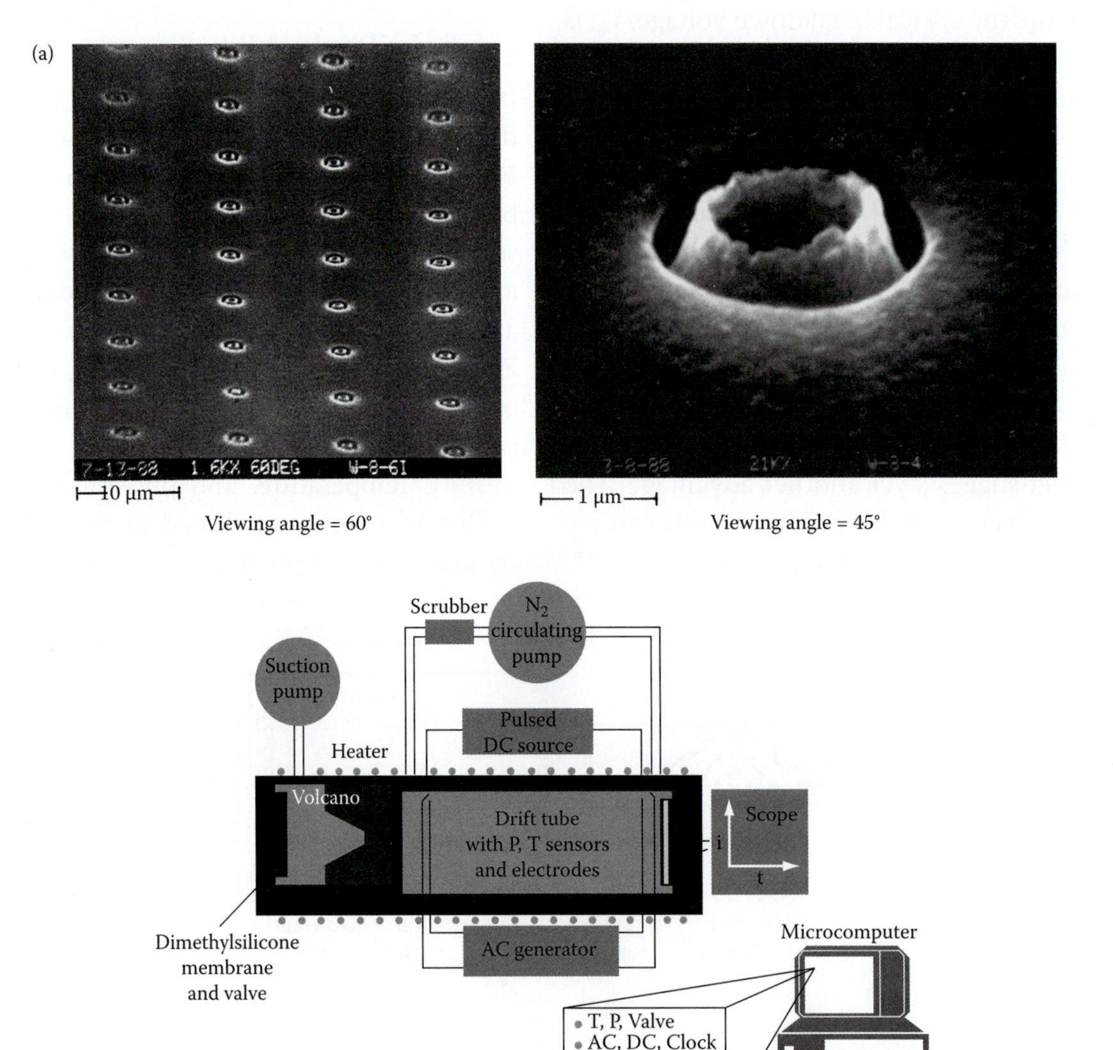

FIGURE 7.14 Ion mobility spectrometry with micromachined volcano sources. (a) Scanning electron microscope image of micromachined volcanoes. (b) Setup for ion mobility spectrometry with microvolcano ionization sources.

lower voltages than other FI sources. Referring back to the Paschen curve, it is clear that the microvolcano sources should operate at atmospheric pressure with voltages up to several hundred volts and perhaps as high as a kilovolt before they induce sparking. In collaboration with Dr. M. Coggiola, at SRI International, we did show that these sources can indeed resist electrical breakdown at atmospheric pressure and produce usable current levels from positive ions formed from gases such as pyridine, butane, and toluene in a setup like that shown in Figure 7.14b.[20]

The fabrication process of the microvolcanoes, which are embedded in a brittle 1-μm-thick Si dioxide layer, was very problematic, and failure rate was very high. Sturdier, more reliable volcano sources are needed to continue this type of development. Today, much better, high-aspect-ratio micromachining techniques are available, and it would be very worthwhile to reinitiate such a study.

Questions

Questions by Omid Rohani, UC Irvine

7.1: What do you expect will be the influence of miniaturization on a potentiometric sensor and on an amperometric sensor?

7.2: Explain in detail the scaling laws for flying and swimming.

7.3: The surface waves in water are characterized by an angular velocity ω and a wave number $k = 2\pi\lambda$ (λ, wavelength). In this problem we intend to find a dispersion relation expressing ω as a function of k.

(a) Using Buckingham's π theorem, find a dispersion relation containing ω and k. Accept that the depth d, the wave height h, the acceleration of gravity g, the waters density ρ, and the water surface tension γ play roles in this problem and neglect the water viscosity.

Appendix 7A: Common Dimensional Numbers

Parameter	Definition	Qualitative Ratio	Importance of Effects
Reynolds number	$R_E = \rho UL/\mu$	Inertia/Viscosity	Always
Mach number	MA = U/A	Flow speed/Sound speed	Compressible flow
Froude number	$Fr = U^2/gL$	Inertia/gravity	Free surface flow
Weber number	$W_e = \rho U^2 L/\gamma$	Inertia/Surface tension	Free surface flow
Cavitation number (Euler number)	$Ca = (p - p_v)/\rho U^2$	Pressure/Inertia	Cavitation
Prandtl number	$Pr = C_p\mu/k$	Dissipation/conduction	Heat convection
Eckert number	$E_c = U^2/c_p T_o$	Kinetic energy/enthalpy	Dissipation
Specific heat ratio	$\gamma = c_p/c_v$	Enthalpy/internal energy	Compressible flow
Strouhal number	$St = \omega L/U$	Oscillation/mean speed	Oscillating flow
Roughness ratio	ε/L	Wall roughness/Body length	Turbulent, rough walls
Grashof number	$Gr = \beta\Delta TgL^3\rho^2/\mu^2$	Buoyancy/viscosity	Natural convection
Temperature ratio	T_w/T_o	Wall temperature/stream temperature	Heat transfer
Pressure coefficient	$C_p = (p-p_o)/0.5\rho U^2$	Static pressure/dynamic pressure	Aerodynamics, hydrodynamics
Lift coefficient	$C_L = L/0.5\rho U^2 A$	Lift force/dynamics force	Aerodynamics, hydrodynamics
Drag coefficient	$C_D = D/0.5\rho U^2 A$	Lift force/dynamics force	Aerodynamics, hydrodynamics
Euler number	$E_u = g_c\Delta P/\rho V^2$	Friction head × velocity head	Momentum transfer, fluid frictions
Weber number	$W_e = DV^2\,\rho/g_c\sigma$	Inertial force/surface tension force	Momentum transfer

(b) Now suppose that the water is deep compared to the wave length and the wave height is small compared to the wave length. Also suppose that the waves are long. Try to write the relation you got in part (a) in a simpler form expressing ω just as a function of k.

7.4: Explain the DC breakdown voltage versus electrode distance curve and how it is relevant to dry etching. How is miniaturization of an electrode set equivalent to creating a local vacuum?

7.5: Why is the range of sizes in water-based life much greater than terrestrial life range? Discuss both above and below limits of these ranges.

7.6: Fill in the blank with the correct number for these scaling laws.

(a) Mass scales with a factor of _______.

(b) Stress scales with a factor of _______.

(c) Natural frequency scales with a factor of _______.

(d) Viscous damping scales with a factor of _______.

(e) Coulomb damping scales with a factor _______.

(f) Elastic coulomb damping scales with a factor of _______.

(g) Surface adhesion scales with a factor of _______.

(h) Power scales with a factor of _______.

7.7: How is the number of steps one animal must take per unit distance related to the animal's weight?

7.8: How does velocity scale with size (ignore drag/damping)? (A short two- or three- equation answer—the solution may be surprising.)

7.9: What methods do you recommend to overcome the problem of retention of liquids caused by large surface-to-volume ratios in small biosensors?

7.10: A simple and accurate viscometer can be made from a length of capillary tubing. If the flow rate and pressure drop are measured, and the tube geometry is known, the viscosity can be computed. A test of a certain liquid in a capillary viscometer gave the following data:

- flow rate: 880 mm^3/s
- tube diameter: 500 μm
- tube length: 1 m
- pressure drop: 1.0 MPa

Determine the viscosity of the liquid

7.11: The accepted transition Reynolds number for flow in a circular pipe is Re ~2300. For flow through a 6 cm dia. pipe, at what velocity will this occur at 20°C for (a) airflow and (b) water flow?

7.12: What are the forces that play an important role for (a) very large and (b) very small devices?

7.13: Explain Kleiber's law. What is the generic principle underlying Kleiber's law?

7.14: How do the following forces scale?

(a) Electrostatic force in a constant electric field

(b) Pneumatic and hydraulic forces caused by pressure

7.15: How does the acceleration, transit time, and produced power for each case scale in problem 7.14?

7.16: The required power in a stirring device in a tank is a function of diameter of the device d, number of rotations of the motor per unit time N, density of liquid ρ and viscosity of the liquid η.

(a) Using Buckingham's π theorem, obtain a relation between power P and d, N, ρ, η.

(b) Supposing that the power P is proportional to the square of N, determine by what factor would P be expected to increase if d is doubled.

7.17: A centrifugal pump for pumping the lubricating oil at 15°C rotates at 1000 rpm. For testing this pump, we make a model pump which works with 20°C air. Assuming the diameter of model is 4 times the diameter of the prototype, find the speed at which the model pump should run.

7.18: Complete similarity between a ship model and a full-sized ship requires the same Reynolds and Froude numbers. Suppose that

$$Re = \rho VL/\mu = 11 \times 10^8 \text{ and } Fr^{-1} = gL/V^2 = 33.25$$

Assuming a model 1/100 the size of the ship, can you design an experiment having full similarity?

7.19: In the miniaturization of analytical instruments, scaling laws and breakdown of scaling laws often determine whether miniaturization will favor sensitivity or not. What will happen to the sensitivity of the following techniques upon miniaturization: (a) the optical path in UV spectrometer; (b) a potentiometric sensor (e.g., a pH sensor); (c) an amperometric sensor (e.g., an oxygen gas sensor); (d) the column in a GC?

7.20: If you miniaturize an absorption-based optical analytical instrument and one based on luminescence, which one scales down more favorably? Explain why.

7. 21: What is the primary assumption of continuum mechanics? Give examples where continuum theory breaks down and how MEMS, in some cases, may take advantage of this breakdown.

7.22: Using dimensional analysis, derive the flow Q of an ideal liquid through an orifice in terms of the density of the liquid ρ, the pressure difference ΔP, and the diameter of the orifice d.

References

1. Trimmer, W. S., ed. 1997. *Micromechanics and MEMS classic and seminal papers to 1990.* New York: IEEE.
2. Schmidt-Nielsen, K. 1984. *Scaling: Why is animal size so important?* Cambridge, United Kingdom: Cambridge University Press.
3. McMahon, T., and J. T. Bonner. 1985. *On size and life.* New York: Scientific American Library.
4. Bonner, J. T. 2006. *Why size matters: from bacteria to blue whales.* Princeton, NJ: Princeton University Press.
5. Dodds, S., D. H. Rothman, and J. S. Weitz. 2000. Re-examination of the "3/4-law" of metabolism. *J Theor Biol* 209:9–27.
6. Gillooly, J. F., A. P. Allen, G. B. West, and J. H. Brown. 2005. The rate of DNA evolution: Effects of body size and temperature on the molecular clock. *Proc Natl Acad Sci U S A* 102:140–45.
7. Langhaar, H. L. 1951. *Dimensional analysis and the theory of models.* New York: Wiley.
8. Morowitz, H. J. 1985. *Mayonnaise and the origin of life.* New York: Berkley Books.
9. Hayashi, T. 1994. Micromechanism and their characteristics. *IEEE International Workshop on Micro Electro Mechanical Systems (MEMS '94).* Oiso, Japan, January 25–28, 1994, pp. 39–44.
10. Fujimasa, I. 1996. *Micromachines: A new era in mechanical engineering.* Oxford, United Kingdom: Oxford University Press.
11. Kubo, Y., I. Shimoyama, and H. Miura. 1993. Study of insect-based flying microrobots. *IEEE International Conference on Robotics and Automation.* Atlanta, GA, May 1993, pp. 386–91.
12. Autumn, K., Y. A. Liang, S. T. Hsieh, W. Zesch, W. P. Chan, T. W. Kenny, R. Fearing, and R. J. Full. 2000. Adhesive force of a single gecko foot-hair. *Nature* 405:681–85.
13. Menon, C., and S. Metin. 2006. A biomimetic climbing robot based on the gecko. *J Bionic Eng* 3:115–25.
14. Manz, A., J. Harrison, E. Verpoorte, J. Fettinger, A. Paulus, H. Lüdi, and H. Widmer. 1992. Planar chips technology for miniaturization and integration of separation techniques into monitoring systems: Capillary electrophoresis on a chip. *J Chromatogr* 593:253–58.
15. Brown, P. R., and E. Grushka, eds. 1993. *Advances in chromatography.* New York: Marcel Dekker.
16. Thompson, D. W. 1992 (reprint of 1917 edition). *On growth and form, Canto ed.* Cambridge, United Kingdom: Cambridge University Press.
17. Bart, S. F., T. A. Lober, R. T. Howe, J. H. Lang, and M. F. Schlecht. 1988. Design considerations for micromachined electric actuators. *Sensors Actuators* 14:269–92.
18. Busch-Vishniac, I. J. 1992. The case for magnetically driven microactuators. *Sensors Actuators A* A33:207–20.
19. Bollee, B. 1969. Electrostatic motors. *Philips Tech Rev* 30:178–94.
20. Madou, M. J., and S. R. Morrison. 1991. High-field operation of submicrometer devices at atmospheric pressure. *1991 International Conference on Solid-State Sensors and Actuators.* San Francisco, June 24–28, 1991, p. 145.

8

Actuators

Outline

MEMS electrostatic comb actuator. (Courtesy of MEMX.)

> Their knowledge did not lead one to the contemplation of divinity or liberation; their science was action: nature obeyed them, and in their cities the power of the rich and strong was less oppressive. The old magic was now in reach of everyone who knew the formula for the spell.
>
> **Octavio Paz in a passage on the science and technology of the West in *In Light of India***

Introduction

An actuator, like a sensor, is a device that converts energy from one form to another (Figure 8.1). In the case of an actuator, one is interested in the ensuing action, while in the case of a sensor what is of interest is the information gained. On the subject of sensor/actuators, we review electrical, magnetic, thermal, chemical, and electrochemical phenomena. From the previous chapter, we know that the dominant physical quantities may change as a function of length scale. In this context, we saw how gravitational (inertial) forces become less effective in the microdomain, whereas van der Waals forces, electrostatic forces, and surface tension forces become dominant. In this chapter we compare different sensor/actuators on the basis of their scaling behavior and other properties,

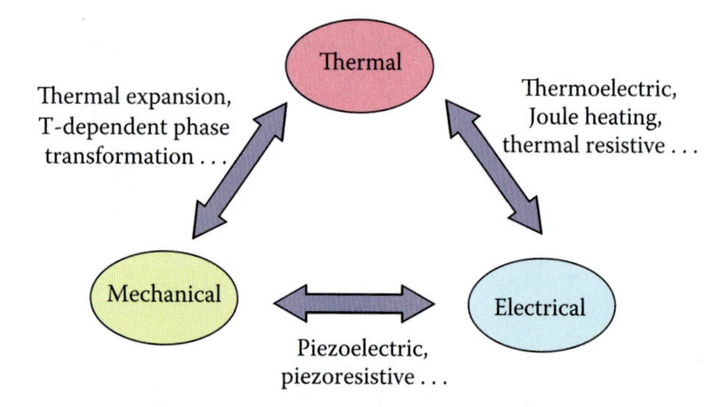

FIGURE 8.1 Energy conversion in sensor and actuator systems.

which, besides size and cost, include dynamic range, displacement (linear or angular), force or torque, response time, and power consumption (see Table 8.1). The results of these comparisons must be seen as guideposts toward a decision only on what actuator to use; yet other considerations (e.g., size of the absolute forces involved, the potential for integration with electronics, materials choice, and manufacturability) may influence the design rules for a particular microsensor or microactuator.

TABLE 8.1 Sensor/Actuator Design Principles and Considerations. Numbers refer to the numbered explanations below the table

• Operational principle (e.g., electrostatic capacitor, shape memory effect, solenoid inductor, thermoelectric effect, piezoelectric, thermal expansion, and themoresistive)
• Scaling behavior
• Size and cost
• Materials choices
• Process feasibility
• Size of absolute forces obtained
• Sensitivity (signal-to-noise ratio) (1)
• Accuracy (3)
• Output force/energy/power
• Dynamic range of frequency response (8)
• Hysteresis (4)
• Resolution (7)
• Nonlinearity (also linearity) (5)
• Noise (6)
• Span or bandwidth range (2)
• Limitations: quality factor, failure mode
• Temperature (or other factors) dependence
• Reliability
• Manufacturability
• Integration and packaging

1. Sensitivity. The sensitivity of a sensor/actuator is defined in terms of the relationship between input signal and output signal. The sensitivity is, in general, the ratio of a small change in output signal to a small change in input signal. As such, it may be expressed as the derivative of the transfer function* with respect to the input signal. A typical unit is volts/kelvin. A thermometer would have high sensitivity if a small temperature change resulted in a large voltage change.
2. Span or bandwidth. The range of input signals that may be converted to output signals by the sensor/actuator. Signals outside of this range are expected to cause unacceptably large inaccuracy. This span or dynamic range is usually specified by the sensor/actuator supplier as the range over which other performance characteristics described in the data sheets are expected to apply.
3. Accuracy. Generally defined as the largest expected error between actual and ideal output signals. Sometimes this is quoted as a fraction of the full-scale output. For example, a thermometer might be guaranteed to be accurate within 5% of full-scale output (FSO).
4. Hysteresis. Some sensors/actuators do not return to the same output value when the input stimulus is cycled up or down. The width of the expected error in terms of the measured quantity is defined as the hysteresis. A typical unit is percentage of full-scale output.
5. Nonlinearity (often called *linearity*). The maximum deviation from a linear transfer function over the specified dynamic range. There are several measures of this error. The most common compares the actual transfer function with the "best straight line," which lies midway between the two parallel lines that encompass the entire transfer function over the specified dynamic

* The functional relationship between input signal and output signal. Usually, this function is represented as a graph showing the relationship between the input and output signal, and the details of this relationship may constitute a complete description of the sensor/actuator characteristics. For expensive sensors/actuators which are individually calibrated, this might take the form of a certified calibration curve.

range of the device. This choice of comparison method is popular because it makes most sensors/actuators look their best.

6. Noise. All sensors/actuators produce some output noise in addition to the output signal. This noise limits the performance of the system. Noise is generally distributed across the frequency spectrum. Many common noise sources produce a white noise distribution, which is to say that the spectral noise density is the same at all frequencies. Since there is an inverse relationship between the bandwidth and measurement time, it can be said that the noise decreases with the square root of the measurement time.
7. Resolution. The resolution of a sensor/actuator is defined as the minimum detectable signal fluctuation. Since fluctuations are temporal phenomena, there is some relationship between the timescale for the fluctuation and the minimum detectable amplitude. Therefore, the definition of resolution must include some information about the nature of the measurement being carried out.
8. Dynamic range or frequency response. All sensors/actuators have finite response times to an instantaneous change in input signal. In addition, many sensors have decay times, which would represent the time after a step change in physical signal for the sensor output to decay to its original value. The reciprocal of these times correspond to the upper and lower cutoff frequencies, respectively. The bandwidth of a sensor is the frequency range between these two frequencies.

Jargon from different science disciplines and various application fields often makes interdisciplinary microelectromechanical system (MEMS) and nanoelectromechanical system (NEMS) discussions challenging. A useful scheme for standardizing the classification of miniaturized sensors and actuators would make communication between researchers from different disciplines (manufacturers, immunologists, mechanical engineers, market researchers, etc.) easier. Several such attempts have been made in that direction. We state our favored approach here. White[1] describes a classification scheme derived from a Hitachi Research Laboratory communication. He distinguishes ten different domains or fields:

1. Acoustic
2. Biological
3. Chemical
4. Electrical
5. Magnetic
6. Mechanical
7. Optical
8. Radiant
9. Thermal
10. Other

Middlehoek and Audet[2] follow Lion[3] and contract this list to six domains:

1. Radiant
2. Mechanical
3. Thermal
4. Electrical
5. Magnetic
6. Chemical

For simplicity, we adopt the latter list here. Based on Gopel et al.[4] and Habekotte,[5] Table 8.2 exemplifies sensing/actuating principles in these six signal domains. Since the listed energy domains can be energy input as well as energy output, a 6×6 matrix results. The table also presents an example device in each field. We may have a mechanical sensor/actuator converting mechanical energy into thermal energy (i.e., a thermomechanical device such as a friction calorimeter), but also a thermal sensor/actuator converting thermal energy into mechanical energy (e.g., a mechanothermal sensor such as a bimetallic strip).

In the literature, no logic seems to be applied concerning the use of "thermo-optical" or "optothermal." Middlehoek and Audet[2] further distinguish between self-generating and modulating sensors/actuators. A device based on a modulating principle requires an auxiliary energy source. An example of a modulating sensor is a fiber-optic magnetic-field sensor in which a magnetostrictive jacket is used to convert a magnetic field into an induced strain in the optical fiber. A sensor/actuator based on a self-generating principle does not require an auxiliary energy source (e.g., a thermocouple in

Table 8.2 Example of Transduction Principles

Input (Primary Signal)	Output (Secondary Signal)					
	Mechanical (Mechano-)	Thermal (Thermo-)	Electrical (Electro-)	Magnetic (Magneto-)	Radiant (Photo- or Radio-)	Chemical (Chemo-)
Mechanical	Acoustics, fluidics	Friction calorimeter, cooling effects	Piezoelectricity, piezoresistivity	Piezomagnetic effect	Photoelasticity, Doppler effect	Motor proteins
Thermal	Thermal expansion, bimetallic strip		Pyroelectricity, Seebeck effect		Radiant emission	Reaction activation
Electrical	Piezoelectricity, electrometer	Joule heating, Peltier effect	Langmuir probe	Biot-Savart's law	Electroluminescence, Kerr effect	Electrolysis
Magnetic	Magnetostriction, magnetometer	Thermomagnetic	Magnetoresistance		Faraday effect, Cotton-Moulton effect	
Radiant	Radiation pressure	Thermopile, bolometer	Photoelectric, dember		Optical bistability	Photoreactions
Chemical	Hygrometer	Calorimeter, thermal conductivity	Amperometry, flame ionization, Volta effect	Nuclear magnetic resonance	Chemiluminescence	

which a change in temperature directly results in an electrical signal that can be measured). The difference between the two is illustrated in Figure 8.2.[6] Since the modulating sensor needs another energy input, one also refers to such sensors as *passive*, whereas the self-generating sensors are *active*.

Irrespective of the fact of dealing with a small sensor, an actuator, or a microsystem, we can always describe the function of the miniaturized device in terms of the matrix in Table 8.2.

There are many reasons for wanting to miniaturize systems, actuators, power sources, sensors, and components. In some cases, the reason is favorable scaling—for example, smaller devices tend to be faster and consume less power. Scaling may also disfavor miniaturization, as often happens with actuators

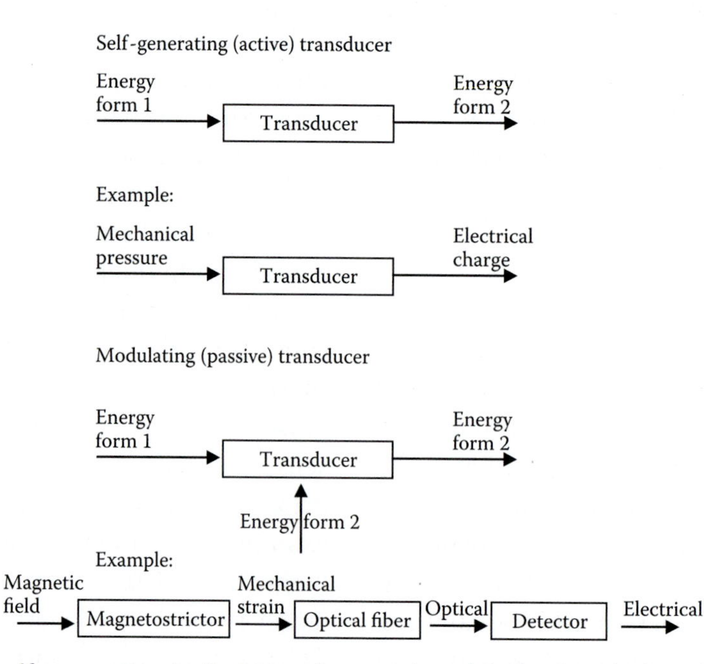

FIGURE 8.2 Comparison of self-generating (active) transducer and modulating (passive) transducer; the piezoelectric sensor represents a self-generating transducer, and the fiber optic magnetic field sensor represents a modulating transducer. (After National Research Council.)

and power sources; smaller actuators exert less force, and smaller power sources harness less power.

Finally, actuator materials are often called *smart materials*, which refers to materials that undergo precise transformations through physical interactions. Examples include piezoelectric materials, shape memory alloys (SMAs), conductive polymers, electrochromic materials, smart gels, rheological materials, biomaterials, electrostrictive materials, magnetostrictive materials, and electrorheological fluids. All these materials have potential uses in actuators that control chatter in precision machine tools, improved robotic parts that move faster and with greater accuracy, smaller microelectronic circuits in machines ranging from computers to photolithography printers, and health-monitoring fibers for bridges, buildings, and utility poles.

Electric Actuators

Under electric actuators, we consider all actuators that involve electrical fields (DC or AC); this includes, foremost, electrostatic and piezoelectric actuators, which we study in the most detail. Besides electrostatic and piezoelectric devices, we consider dielectric induction effects on particles in liquids and on liquids themselves, where one speaks about dielectrophoresis (covered in Volume I, Chapter 6) and electrohydrodynamics (EHD), respectively. Finally, under electric actuators, we also explain how electrostrictive materials, pyroelectrics, thermoelectrics, electrets, and electrorheological fluids work.

Electrostatics

Introduction

In electrostatic sensors and actuators, electrically charged materials are used. These materials can exert an attractive force on oppositely charged objects or a repulsive force on similarly charged ones. We start by introducing the simplest of electrostatic actuators, namely a parallel plate capacitor. This is followed by an introduction to electrical and mechanical equivalent circuits and the correspondence of their respective elements. This gives us the background for considering the dynamic behavior of a parallel plate capacitor. We then further describe the two most important characteristics of electrostatic actuators, i.e., their natural frequency and quality factor. We also introduce some other electrostatic actuators, such as wobble motors and switches, and compare them with other types of actuators. We finish this section with a summary about electrostatic actuators and a more general approach to modeling them.

Parallel Plate Capacitor Based Actuators

A parallel plate capacitor, as shown in Figure 8.3, is an example of a simple electrostatic actuator. We calculate the stored energy density in a capacitor with plate area *A* and distance between the plates *d* as follows:

$$U = \frac{1}{2}\,\varepsilon_0\,\varepsilon_r\,\frac{A}{d}\,V^2 = \frac{1}{2}\,\varepsilon\,(Ad)\,E^2 \tag{8.1}$$

(see also Equation 7.34, with $\varepsilon = \varepsilon_r\,\varepsilon_0$ wv = A and $V_b = V$).

The stored energy density is then obtained by dividing *U* from Equation 8.1 by the volume of the capacitor (*Ad*), or:

$$U' = \frac{1}{2}\varepsilon E^2 = \frac{1}{2}\mathbf{D}\cdot\mathbf{E} \tag{8.2}$$

where we bolded D and E to remind the reader that we are dealing with vectors here (see also Equation 7.36, modified to describe all fields, *E*, not just the breakdown field, E_b, and with $\varepsilon = \varepsilon_r\,\varepsilon_0$):

The electrostatic surface force is given by the following equation (a generalization of Equation 7.37):

$$F = -\nabla U \tag{8.3}$$

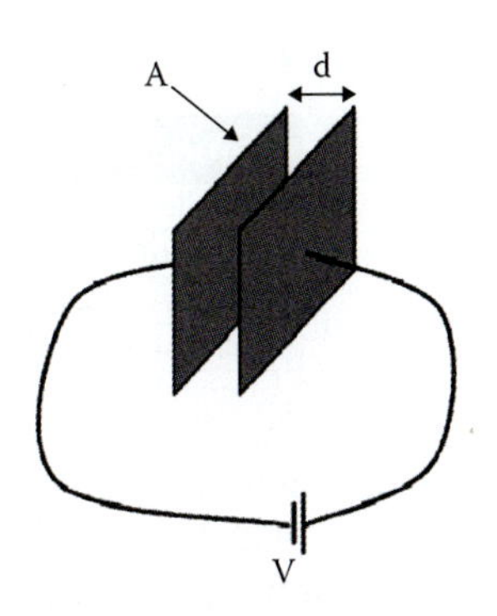

FIGURE 8.3 Parallel plate capacitor.

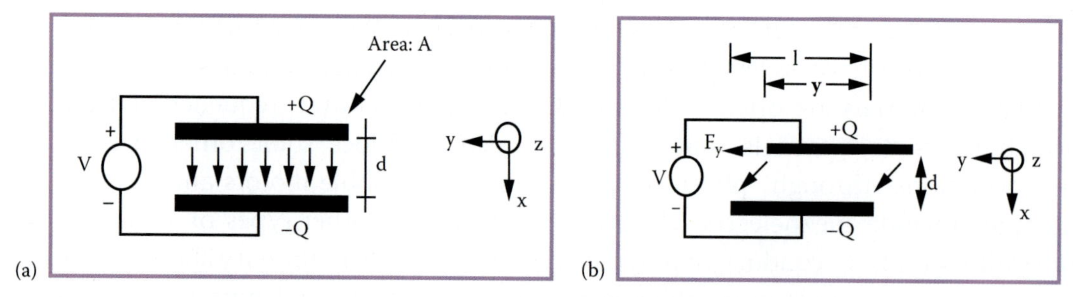

FIGURE 8.4 Vertical-driven (a) and parallel-driven (b) parallel plate capacitor actuators.

Let us now consider parallel- and vertical-driven parallel plate capacitor actuators, as depicted in Figure 8.4.

If the two plates of the parallel capacitor in a microactuator are displaced perpendicular to each other (*d* changes along the *x*-direction), the force of attraction in the *x*-direction, based on Equations 8.1 (with $\varepsilon_r = 1$) and 8.3, is given as follows:

$$F_x = -\frac{\partial U}{\partial x} = \frac{1}{2}\frac{\varepsilon_0 A}{x^2}V^2 \tag{8.4}$$

Vertically driven resonant microstructures operate in this mode (see Figure 8.5a). Vibration in the *x*-direction in this case is excited electrostatically, and motion is detected electrostatically as well; that is, by sensing the change in capacitance.[7]

In Chapter 7, we saw that the electrostatic force, F_x, for a constant field, scales as l^2, since:

$$F_x = -\frac{\partial U}{\partial x} = \frac{V^2}{2}\frac{\partial C}{\partial x} = \frac{l^3}{l^1} = l^2 \tag{7.37}$$

where *V* is the applied voltage. This l^2 scaling of the electrostatic force often constitutes an advantage compared with other forces when miniaturizing an actuator. For example, mass and hence inertial forces scale as l^3, and from Chapter 7, we know that the lower the exponent of *l*, the more important the force becomes in the microdomain (see also below and Figure 8.8).

A large displacement in a micromachined electrostatic actuator element can best be achieved if one plate (the actuator) moves parallel to the other capacitor plate rather than perpendicular to it (see Figures 8.4b and 8.5b). For a parallel movement of the plates (in the *y*-direction or *z*-direction), Equations 8.1 and 8.3 result in a force of attraction parallel to the plates, which for the *y*-direction is given by the following equation:

$$F_y = \frac{1}{2}\frac{\varepsilon_0 z}{d}V^2 \tag{8.5}$$

Notice that the force is constant for a *y* directional translation, i.e., the force is scale-independent in

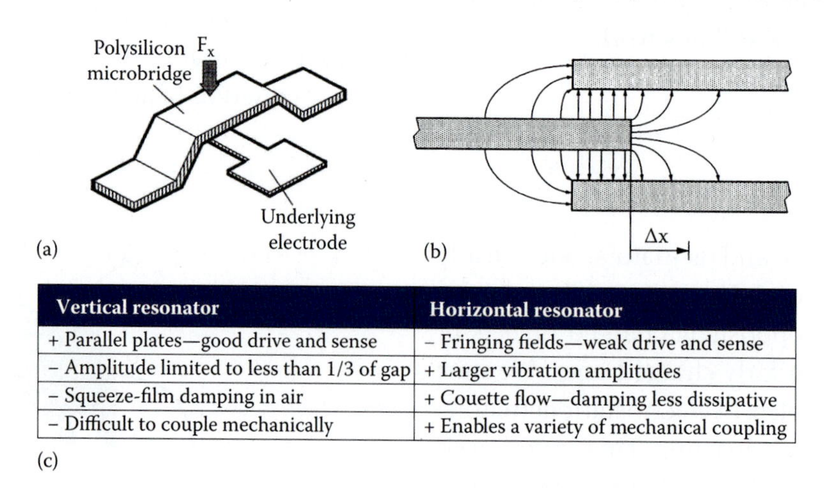

Vertical resonator	Horizontal resonator
+ Parallel plates—good drive and sense	– Fringing fields—weak drive and sense
– Amplitude limited to less than 1/3 of gap	+ Larger vibration amplitudes
– Squeeze-film damping in air	+ Couette flow—damping less dissipative
– Difficult to couple mechanically	+ Enables a variety of mechanical coupling

(c)

FIGURE 8.5 Electrostatic actuation. (a) Vertically driven polysilicon microbridge. (b) Laterally driven electrostatic actuator. (After Tang, W. C.-K. 1990. Electrostatic comb drive for resonant sensor and actuator applications. PhD thesis, University of California, Berkeley.[7]) (c) Summary of the pros and cons of vertical and lateral resonators, from a lecture by Dr. Bill Tang (University of California, Irvine).

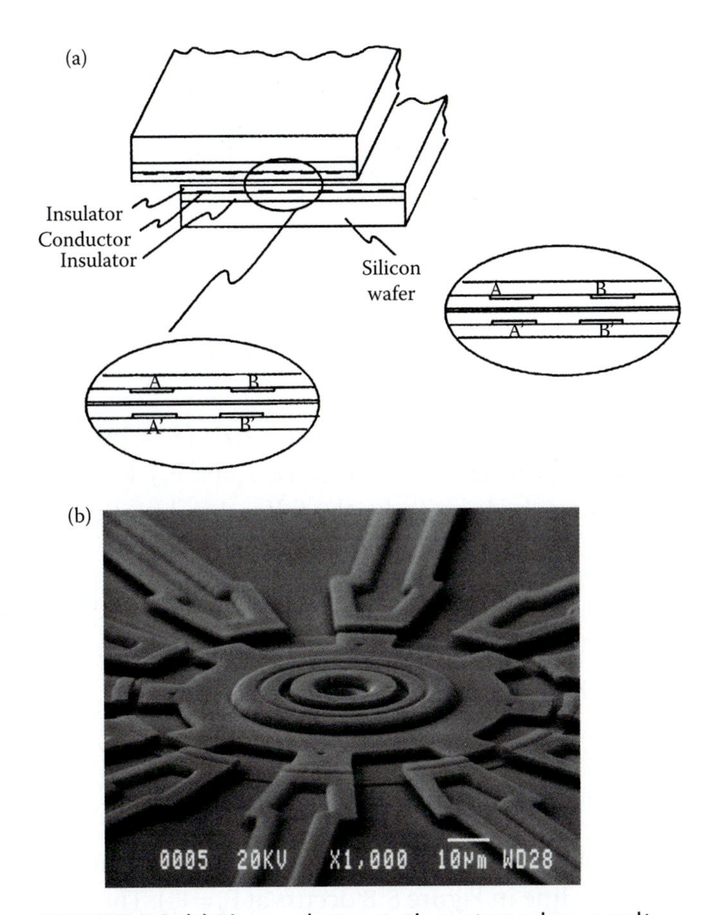

FIGURE 8.6 (a) Linear electrostatic motor: when a voltage is applied to the misaligned plates A-A′, a force is exerted, which aligns plates A-A′. Now, plates B-B′ are misaligned and in a position to be activated to cause a motion. (After Trimmer, W. S. N., and K. J. Gabriel. 1987. Design considerations for a practical electrostatic micromotor. *Sensors Actuators* 11:189.[11]) (b) Rotational electrostatic motor: scanning electron microscope micrograph of a polysilicon 12:8 salient pole micromotor. The rotor sits atop a 0.5-mm-thick layer of polysilicon that acts as an electrostatic shield. Rotor, hub, and stators are formed from 1.5-mm-thick polysilicon. A 2.0-mm-thick polysilicon disk is attached to the rotor. In turn, the hub overlaps this disk to pin the rotor onto the substrate. (Courtesy of D. Koester, MCNC-MEMS Technology Applications Center.) The first polysilicon motors were made at University of California, Berkeley (Fan, Tai, and Muller), Massachusetts Institute of Technology, and AT&T. Typical starting voltages were >100 V, and operating voltages were >50 V.

that direction ($F \sim l^0V^2$). In this case, the force parallel to the plates wants to realign the plates. This is exploited, for example, in laterally driven linear and rotary electrostatic motors or side-drive motors (see Figure 8.6a and b). The lateral force remains constant during the movement as long as the fringing fields can be neglected (see Figure 8.5b). Besides longer strokes, driving and sensing of planar microstructures parallel to the substrate have two other major advantages over a vertical movement of the plates: the forces change linearly with distance (see Equation 8.5), and dissipative squeeze damping is avoided.[8–10] The latter is related to the magnitude of the viscous losses when moving structures displace fluids (usually air) from small separator gaps. Losses are larger for vertical displacement of fluids, leading to squeeze-film damping and a resultant low-quality factor (*Q*) (see below) for resonating elements. Couette flow (see Volume I, Chapter 6, on fluidics) in the gap between the structure and the substrate for lateral motion of the actuator is much less dissipative than squeeze-film damping, and higher *Q* resonators are obtained. Figure 8.5c summarizes the pros and cons of vertical and lateral resonators.

Lateral electrostatic actuation is very important in linear resonators with comb-like structures, such as the device shown in the header of this chapter. These comb drives are used, for example, in accelerometers [see the Analog Devices' ADXL-50 accelerometer, with a surface micromachined capacitive sensor (center), on-chip excitation, self-test, and signal-conditioning circuitry in Volume II, Figure 7.92]. By applying a voltage, a movable shuttle moves toward a fixed part (Volume II, Figures 7.76 and 7.93). For a rectangular design of the resonator beams, the force and the microactuator displacement is proportional to the ratio of structure height, *T*, versus separation gap, *d*, that is, the achievable aspect ratio and the number of fingers, *N*:

$$F = \frac{1}{2}\frac{\varepsilon_0 TN}{d}V^2 \tag{8.6}$$

An efficient comb drive with a large *F* can thus be achieved by designing many high comb fingers with narrow gaps. In polysilicon surface micromachining, the thickness, *T*, is typically limited to about 2–3 µm, whereas with LIGA or pseudo-LIGA methods, a thickness of several hundred micrometers is possible (Volume II, Chapter 10). Clearly, the latter machining methods have the potential to produce the strongest electrostatic actuators. Mohr et al. further showed that by making the comb fingers trapezoidal in shape, the maximum displacement of the interdigitated fingers can be further increased,

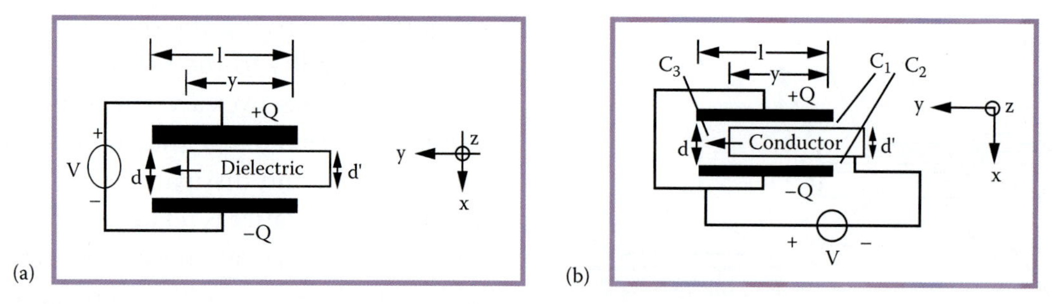

FIGURE 8.7 Dielectric (*a*) or metallic (*b*) slabs of material are moved into the gap of a parallel plate capacitor.

as smaller capacitor gaps are possible this way.[12] For a more rigorous mathematical derivation of Equation 8.6, we refer to the finite-element simulation of comb drives by Tang.[7]

When a dielectric or a metallic slab is moved between the two plates of a parallel plate capacitor as shown in Figure 8.7, Equation 8.5 must be modified as follows:

$$F_y = \frac{1}{2}\frac{(\varepsilon_d - \varepsilon_0)z}{d'}V^2 \qquad (8.7)$$

for a dielectric slab of thickness *d′* and dielectric constant ε_d (Figure 8.7a). For a metal slab of thickness *d′* inserted in the capacitor gap, one derives the following equation (Figure 8.7b):

$$F_y = \frac{1}{2}\frac{\varepsilon_0 z d'}{d(d-d')}V^2 \qquad (8.8)$$

It is important to be able to evaluate the relative importance of different competing forces as one downscales an actuator. In Figure 4.39, we compare gravity, van der Waals, electrostatic, and surface tension forces that are involved when picking and placing small objects during assembly. From this graph it is clear that surface forces become more dominant as structures become smaller. Here we detail the relative importance of gravitational versus electrostatic forces for deciding upon an actuator mechanism in the microworld. In Figure 8.8, we consider competing gravitational and electrostatic forces working on a cube of material with length *L*, density ρ, and charge *q*. The gravitational force, F_g, working on the cube is given as follows:

$$F_g = g\rho V = g\rho L^3 \qquad (8.9)$$

Based on Equation 8.4, with $\sigma = q/A$ ($A = L^2$) and $AE = q/\varepsilon$, the electrostatic force F_e ($\equiv F_x$) working upon the cube is given by the following equation:

$$F_e = \frac{1}{2}\varepsilon AE^2 = \frac{\sigma^2}{2\varepsilon}L^2 \qquad (8.10)$$

Hence, when:

$$L < \frac{\sigma^2}{2g\rho\varepsilon} \qquad (8.11)$$

electrostatics dominates (the intersection of the red and blue line in Figure 8.8 occurs at $F_g = F_e$). The electrostatic force, as we learn in Chapter 7, gains over inertial forces as the size of the system is decreased: the electrostatic force, with the lowest exponent in *l*, becomes the dominant force in the microdomain (surface force is l^2 vs. l^3 for gravitation). A decrease in size by a factor of 10 leads to a decrease of the inertial forces by a factor of 1000, whereas the electrostatic force decreases by a factor of only 100. It is thus not hard to exert rapid accelerations in microsystems with the aid of electrical fields.

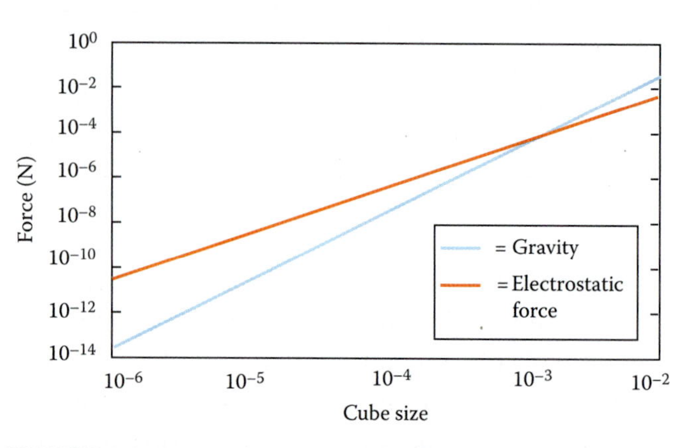

FIGURE 8.8 Competing gravitational and electrostatic forces working on a cube of length *L*, charge *q*, and density ρ (see also Figure 4.39).

TABLE 8.3 Equivalent Electrical and Mechanical Terms

Displacement, g	⇔	Charge, q
Force, F	⇔	Voltage, V
Velocity, dz/dt	⇔	Current, I (= dq/dt)
Mass, m	⇔	Inductance, L
Spring, k	⇔	Capacitance, 1/C
Damping, b (dashpot)	⇔	Resistance, R

Mechanical and electrical equivalent circuits.

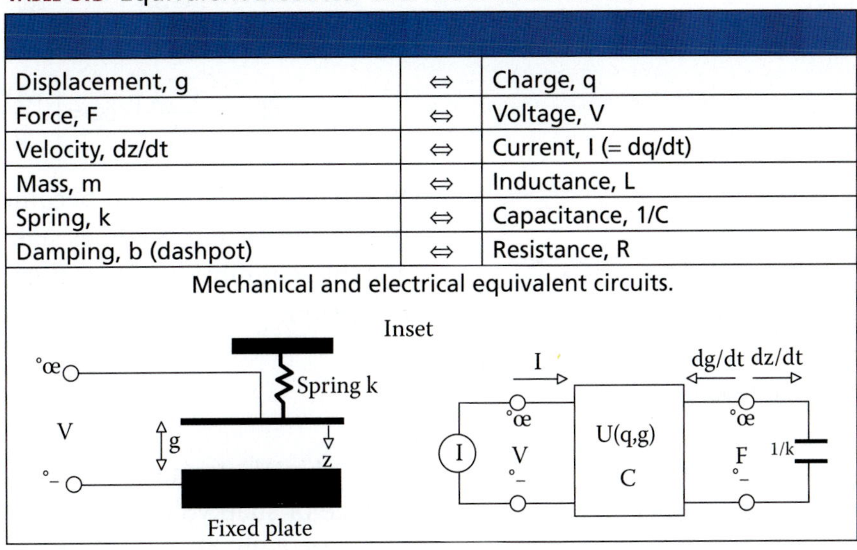

Correspondence between Electrical and Mechanical Circuits

In this section we briefly introduce the correspondence between electrical and mechanical circuits.

For modeling of electrostatic actuators as shown in Figure 8.7, it is useful to make abstractions of the electrical and mechanical system elements as illustrated in Table 8.3. In this table we list the correspondence between the parameters in electrical and mechanical circuits. In the inset in this table, the gap of the electrostatic actuator is marked with the letter *g*, not to be confused with the gravitational constant. Movement of the electrostatically actuated structure with spring constant *k* is in the *z*-direction.

We will come back to modeling electrostatic actuators at the end of "Electrostatics."

Dynamics of Electrostatic Actuators

Dynamics is the branch of classical mechanics that is concerned with the motion of bodies. Here we consider the dynamics of an electrostatic actuator. In the electrostatic actuator shown in Figure 8.9, the electrostatic force, $F_e(V)$ (see Equation 8.4, with x replaced by g, the instantaneous gap spacing, and ε_0 replaced by ε) works against a mechanical restraint $F_s = kz$. The latter is Hooke's law, with k the spring constant and z the displacement of the movable plate, $z = g - g_0$, with g_0 the initial gap spacing, i.e., the gap without applied voltage.

A stable equilibrium results when the electrostatic and mechanical force are equal in absolute magnitude but opposite in sign, i.e., $F_e = -F_s$, or as follows:

$$F_e = \frac{1}{2}\frac{\varepsilon A V^2}{g^2} = -kz = k(g_0 - g) \tag{8.12}$$

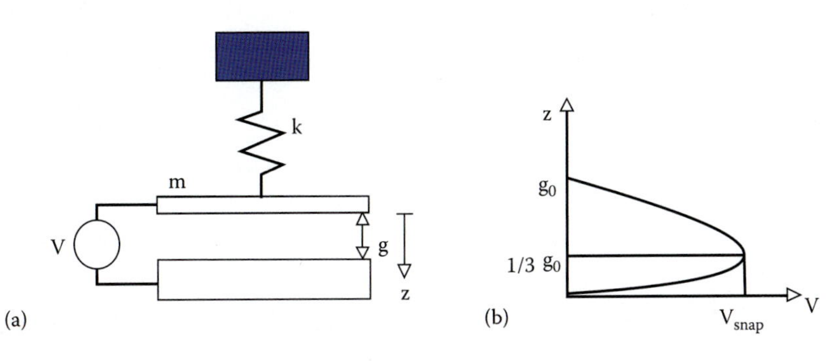

FIGURE 8.9 (a) Schematic of an electrostatic actuator. (b) At the snap voltage, V_{snap}, there is an electrostatic instability or a pull-in of the electrostatic actuator, where g_0 is the initial gap spacing and the snap voltage occurs at 1/3 of the initial gap spacing.

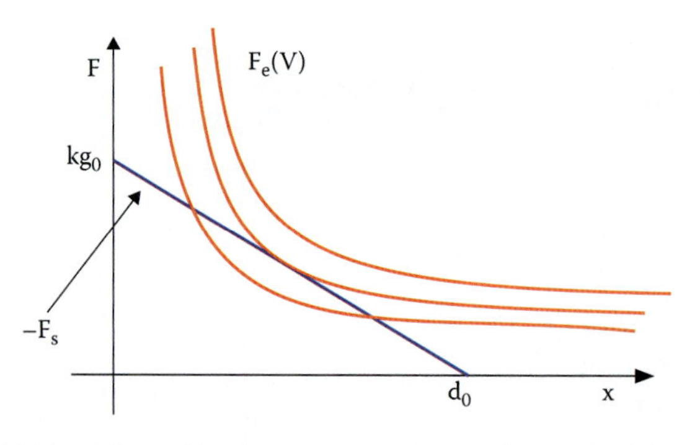

FIGURE 8.10 Positioning of a capacitor plate. The blue line is simply Hooke's law, and the three red lines are the electrostatic force for three different voltages.

This relationship is illustrated in Figure 8.10 for three different voltage values. Notice that the blue line (Hooke's law) intersects the lowest red F_e (V)-curve (electrical force) at two points and touches the next red F_e (V)-curve at only one point. These intersection points and the contact point are now analyzed further.

From Figure 8.10, we recognize that at $g = g_0$ ($z = 0$), when there is no voltage applied, $F_s = 0$. When the electrostatic actuator pull-in has shunted the capacitor plates, the gap $g = 0$ ($z = -g_0$), and the force $F_s = kg_0$.

On the basis of Equation 8.12, we derive the following voltage dependence on z:

$$V = \sqrt{\frac{2kz}{\varepsilon A}}(g_0 - z) \tag{8.13}$$

This is plotted in Figure 8.9b. When the applied voltage, V, is larger than V_{snap} (the snap or pull-in voltage), an unstable situation results (the system loses equilibrium) and the capacitor plates snap together. Let us analyze the conditions for this electrostatic instability in a bit more detail. An equilibrium point is unstable if in the presence of a small perturbation the net force does not tend to return the system back to the equilibrium position. The net force is the sum of the electrostatic and mechanical force, $F_{net} = F_e + F_s$, or, from Equation 8.12:

$$F_{net} = -\frac{\varepsilon AV^2}{2g^2} + k(g_0 - g) \tag{8.14}$$

For a stable equilibrium point, the variation of the net force (δF_{net}) should be of the opposite sign to the perturbation (δg), i.e., $\delta F_{net}/\delta g < 0$. We calculate ΔF_{net} as follows:

$$\delta F_{net} = \left.\frac{\partial F_{net}}{\partial g}\right|_V \delta g = \left(\frac{\varepsilon AV^2}{g^3} - k\right)\delta g \tag{8.15}$$

When $\delta g > 0$, the system becomes unstable if $\delta F_{net} > 0$, and the system becomes stable if $\delta F_{net} < 0$, or for a stable equilibrium, one must have the following condition:

$$k > \frac{\varepsilon AV^2}{g^3} \tag{8.16}$$

At the electrostatic instability voltage, pull-in voltage or snap voltage, V_{snap} (the peak of the curve in Figure 8.9b: $\delta V/\delta z = 0$), the net force $F_{net} = 0$, and one obtains the following:

$$k = -\frac{\varepsilon AV_{snap}^2}{g_{snap}^3} \tag{8.17}$$

and:

$$F_{net} = -\frac{kg_{snap}}{2} + k(g_0 - g_{snap}) = 0 \tag{8.18}$$

From Equation 8.18, we derive for the value of g_{snap}:

$$g_{snap} = \frac{2}{3}g_0 \tag{8.19}$$

The value for V_{snap}, the electrostatic instability or snap voltage, is then as follows:

$$V_{snap} = \sqrt{\frac{8kg_0^3}{27\varepsilon A}} \tag{8.20}$$

Representing a normalized displacement as:

$$\zeta = \frac{z}{g_0} = 1 - \frac{g}{g_0} \tag{8.21}$$

And a normalized voltage as:

$$v = \frac{V}{V_{snap}} \tag{8.22}$$

The equilibrium condition can be represented as:

$$\frac{4}{27}\frac{v^2}{(1-\zeta)^2} = \zeta \tag{8.23}$$

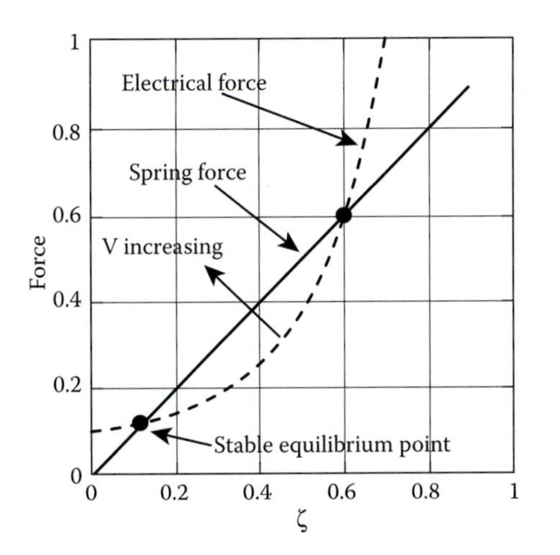

FIGURE 8.11 Force, F, as a function of ζ (with $v = 0.8$). The spring force is marked by a solid line and the electrical force by an arched line (see also Figure 8.10).

The two intersection points merge when $\zeta = 1/3$ ($g_{snap} = 2/3g_0$) and $v = 1$ ($V = V_{snap}$), and this point corresponds to the pull-in point. As shown in Figure 8.10, this is the point where the blue line just touches the red curve. In Figure 8.11, we plot the force, F, as a function of ζ (with $v = 0.8$). The spring force is marked by a solid line and the electrical force by an arched one.

Fundamental Frequency and Quality Factor of Electrostatic Resonators

Introduction

Two very important characteristics for electrostatic actuators are the fundamental frequency and the quality factor of the moving parts in these systems. The fundamental frequency of a resonating part is the frequency at which a free system vibrates. There are two different common definitions of the Q of a resonator. One is via the energy storage, where the Q is given as 2π times the ratio of the stored energy to the energy dissipated per oscillation cycle, or equivalently the ratio of the stored energy to the energy dissipated per radian of the oscillation. The other definition is via the resonance bandwidth in which the Q is the ratio of the resonance frequency f_0 and the full width at half-maximum (FWHM) bandwidth, Δf, of the resonance. Both definitions are equivalent only in the limit of weakly damped oscillations, i.e., for high Q values.

Fundamental Frequency

The fundamental or natural frequency, f_0, of a simple spring-mass system resonator is given by:

$$f_0 = \frac{1}{2\pi}\sqrt{\frac{k}{m}} \quad (8.24)$$

and:

$$\omega_0 = 2\pi f_0 = \sqrt{\frac{k}{m}} \quad (8.25)$$

with k the spring constant (defined as load/deflection), m the mass, and ω_0 the natural circular frequency. The spring constant k in the two preceding equations scales as l, and m scales as l^3; it follows that the resonance frequency scales as $f_0 \sim l^{-1}$, confirming that the smaller the resonator the higher its resonant frequency.

For a more complicated resonator, of the type shown in the header of this chapter, the same fundamental equation holds but with the spring constant k replaced by K, that is, the effective spring constant (also stiffness constant or stiffness matrix), and m replaced by M, that is, the effective mass or mass matrix of the structure. The stiffness matrix varies with geometry as well as with the induced stresses.

With reference to Table 8.3, we recognize that just as a spring-mass combination has a natural (resonant) frequency determined by the stiffness of the spring and the magnitude of the mass, so too do the values of an inductor and capacitor determine the natural oscillating frequency of an electrical inductance-capacitance (LC) circuit. The larger the inductance, the longer it takes a given moving charge to set up a magnetic field. Since a given charge produces only a small voltage across a large capacitor, the larger the capacitance the longer it will take to charge and discharge that capacitor. The natural oscillating frequency of an LC circuit is given by the following expressions:

$$f_0 = \frac{1}{2\pi}\sqrt{\frac{1}{LC}} \quad (8.26)$$

and:

$$\omega_0 = 2\pi f_0 = \sqrt{\frac{1}{LC}} \quad (8.27)$$

Quality Factor

The Q of a resonator, in one of the two prevailing definitions (see introduction above), is given as the ratio of stored over lost energy in one cycle.

$$Q = 2\pi \frac{E}{\Delta E} = 2\pi \frac{\tau}{T} = \tau\omega_0 \tag{8.28}$$

For a mechanical system as sketched in the inset in Table 8.3, the mechanical quality factor is then calculated as follows:

$$Q = \tau\omega_0 = \frac{m}{b}\omega_0 = \frac{\sqrt{km}}{b} \tag{8.29}$$

with b the damping in the system. From the discussion in the introduction above, m and k should be replaced by M and K for more complex resonator structures.

Analogously, for an electrical system, the quality factor is given by the following:

$$Q = \tau\omega_0 = \frac{L}{R}\omega_0 = \frac{1}{R}\sqrt{\frac{L}{C}} \tag{8.30}$$

with L the inductance and C the capacitance (see Table 8.3).

Typical mechanical quality factors for laterally driven polysilicon resonant microstructures (see figure in the header of this chapter) are about 100 or less.[8] A poly-SiC lateral resonator, a material with a high Young's modules and thus a high k, at $<10^{-5}$ Torr, has been micromachined and features Q values >100,000.[13] In a good vacuum, b in Equation 8.29 becomes very small, further improving the factor Q. For NEMS devices, Qs attained today in moderate vacuum are in the range of 10^3 to 10^5. This greatly exceeds typical Qs from electrical resonators. A high Q is linked to small internal dissipation (D = 1/Q), imparting, in turn, low operating power levels and high attainable force sensitivity to nanodevices.[14]

Merits of Electrostatic Devices

To analyze the merits of electrostatic actuators, we start by deriving the maximum energy density one can store in such an actuator. In Chapter 7, on scaling, we saw that the maximum electrostatic energy density, U'_m (in J/m^3), between the capacitor plates of an electrostatic actuator is obtained by dividing the maximum potential energy, U_m, by the volume, V_0, of the capacitor gap, $V_0 = w \times v \times d$ (see Figure 7.11):

$$U'_m = \frac{\varepsilon_r \varepsilon_0 E_b^2}{2} \tag{7.36}$$

where E_b represents the highest attainable electric field in the actuator, i.e., the electric breakdown field strength. For an air capacitor, E_b is limited to approximately 3×10^6 Vm^{-1} by the electrical breakdown of air. Thus, the stored energy for an air capacitor is about 40 Jm^{-3}. Electrostatics is a surface force, and the surface force density, F', is given by the following (derivative of Equation 8.4):

$$F' = \frac{\partial F}{\partial A} = \frac{\varepsilon_r \varepsilon_0 E^2}{2} \tag{8.31}$$

where A stands for the surface area. The surface force density equals the energy density in the field. Hence, one wants to use the maximum field possible for the largest possible force. In Chapter 7, we saw that this force scales as l^2, which is favorable in the microdomain where forces with the lowest exponent win out (e.g., electrostatics with l^2 win out over the l^4 scaling of magnetic actuators).

Our reasoning so far involves isotropic electrical fields only. In Chapter 7 we saw that besides favorable scaling, the Paschen effect represents another advantage for microelectrostatic systems. Working on the left side of the Paschen curve (see Figure 7.12) renders very high-field operation possible for a wide variety of electrostatic microdevices without incurring catastrophic sparking. Examples include the electrostatic motors illustrated in Figure 8.6b, the electrostatic switch in Figure 8.13, and the author's volcano ionization sources presented in Example 7.1 at the end of Chapter 7 (Figure 7.14). The field in the nonisotropic Paschen regime scales more nearly as $l^{-1/2}$, whereas the force scales as l^1. With a factor of 10 reduction in size, the inertial force still decreases by a factor of 1000, but the electrostatic force decreases only by a factor of 10. Acceleration and transit times are also higher than for the isotropic system. Superficially, this seems to put magnetic motors, based on volume forces, at a disadvantage compared with electrostatic micromotors. We shall see, however, that this is not the end of the story. For example, magnetic field energy can be made two orders of magnitude greater than the best one can achieve with electrostatic fields and small air gaps in air or vacuum.[15]

We are interested in using electrostatic force in all types of actuators: not only simple parallel plate capacitors but also linear and rotational motors. As we saw earlier, the force, *F*, is the negative spatial derivative of the energy, *U*, in the system (Equation 8.3). In the case we move the plate of a capacitor with respect to another plate, the electric force will be the spatial derivative, in that direction, of the stored energy *U*. The translational movement of one plate with respect to the other in a linear motor moving in the *x*-direction leads to a force, F_x, as follows:

$$F_x = -\frac{\partial U}{\partial x} = \frac{V^2}{2}\frac{\partial C}{\partial x} = \frac{1^3}{1^1} = 1^2 \quad (7.37)$$

where *V* is the applied voltage. The electrostatic force for a constant field is thus found to scale as l^2. This is often an advantage, because the mass and, hence, inertial forces scale as l^3. The latter point was clearly illustrated in Figure 8.8.

From Equation 7.37 translational motor action is associated with a change in the capacitance: an increase in capacitance for a motor, a decrease in capacitance for a generator. For a rotational motor a rotational force, F_r, by analogy to Equation 7.37 is defined as follows:

$$F_r = -\frac{\partial U}{\partial \theta} \quad (8.32)$$

where $\partial\theta$ is the differential angular displacement in radians. From Equations 7.37 and 8.32, F_r can be expressed in terms of the change in capacitance.

$$F_r = -\frac{V^2}{2}\frac{\partial C}{\partial \theta} = -\frac{q^2}{2C^2}\frac{\partial C}{\partial \theta} \quad (8.33)$$

The second expression on the right is obtained by introducing the relation $V = qC^{-1}$, where *q* is the total charge on the capacitor plates. Detailed calculations of the capacitance of electrostatic motors/generators can be found in Mahadevan[16,17] and in Kumar et al.[18] Torque of a rotational motor is given by the following:

$$T = r \times F_r \quad (8.34)$$

With *r* the radius and F_r perpendicular, Equation 8.34 simply becomes $T = r \cdot F_r$. The power, *P*, generated by the motor can easily be calculated from the torque as follows:

$$P = T\omega = T2\pi f \quad (8.35)$$

where ω represents the angular frequency and *f* the frequency at which the motor rotates. From Equation 8.35, one would expect that as electrostatic motors decrease, the inherent increase in frequency could help to offset the decrease in torque. However, static and dynamic frictional forces (surface forces!) come into play and are a major barrier. Despite these problems, surface-micromachined electrostatic motors of the type shown in Figure 8.6b, with rotational speeds of 15,000 rpm and continuous operation for over a week, have proven to be possible at voltages below 300 V.[19] Friction-reduction methods include deposition of a silicon nitride sliding surface; electrostatic, magnetic, or other types of levitation; and replacing sliding contacts with rolling contacts in wobble or harmonic motors.[20,21]

In harmonic motors, the rotor rolls on the inside of the stator without slipping and rotates slightly in the process (Figure 8.12). It is this slight rotation

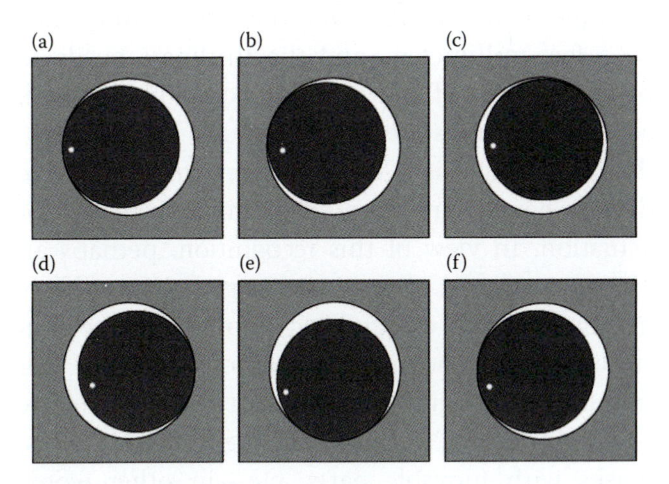

FIGURE 8.12 Wobble or harmonic motors. (a) The rotor is touching the edge of the hole to the left. The white dot marks a point on the rotor; (b)–(f) show the rotor as it progressively rolls around the hole in a clockwise manner. Note the position of the white dot in (f) after one rotation. The rotor has twisted about its axis slightly in a counterclockwise direction. As the rotor rolls repeatedly around the hole, the twisting motion produces an output torque.

of the rotor that produces the output motion of the motor. Applying voltages between the rotor and electrodes in the stator hole electrostatically drives the rotor. Each of the rotations of the rotor about the stator hole produces another small twist of the rotor axis, producing a beneficial gear reduction. Advantages of this type of motor are the rolling motion, which avoids sliding friction, the closeness of the rotor and stator, enabling a higher electrostatic energy density, and the gear reduction, producing a large torque. Harmonic motors do not scale up well, however, and Trimmer compares large wobble motors to an unbalanced washing machine.[22] In the microdomain, inertial effects decrease, and these unbalanced forces are insignificant for microharmonic motors.

To drive a load several hundred micrometers in thickness and width and several millimeters in diameter, the required torque of a motor is of the order of 10^{-5} Nm. However, the torque generated with a typical surface-micromachined electrostatic motor, 100–150 μm in diameter and with an air gap of about 2 μm, is only in the nano-newton-meter or pico-newton-meter range (see Figure 8.6b).[21] Calculation for an outer rotor, surface-micromachined wobble motor,* with copper electroplated structural elements, indicates that to achieve a torque of 10^{-4} to 10^{-5} Nm, the stator radius must be a few millimeters, and the thickness needs to be of the order of 10 μm.[24] As frequently observed, actuators do not scale advantageously in the microdomain. Torque is a volume effect, and small-size devices have a limited range of force available for actuation. In view of this recognition, perhaps an excessive amount of research has gone into surface-micromachined electrostatic motors. These flat surface micromachined micromotors could still find a use, though, as shutters for charged particle or photon beams, memory writing/reading devices, masks with movable parts, etc.—in other words for applications in which the required forces are minimal. Below, we will compare electrostatic motors with the more complicated and typically larger

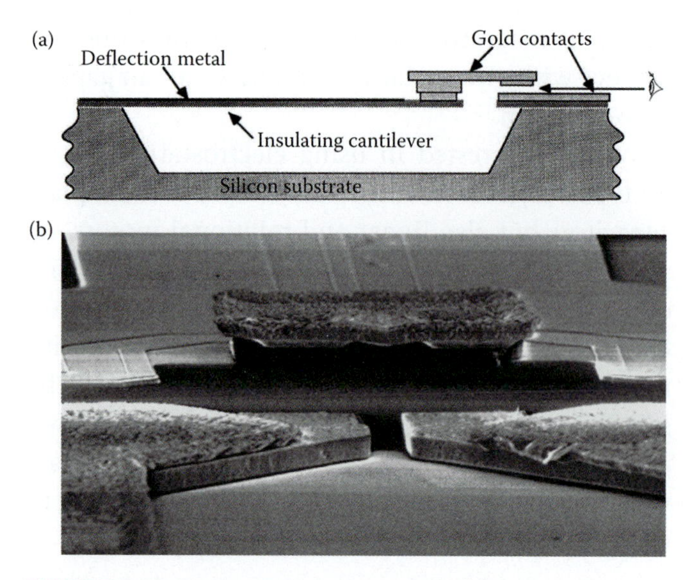

FIGURE 8.13 Electrostatic microswitch. (a) Schematic; (b) Scanning electron microscope micrograph, looking straight into the opened contact. Switch built by the author's research group.

micromachined magnetic motors that, it appears, can perform a broader range of work more readily.

Besides different types of monolithic electrostatic micromotors,[19,25] electrostatic actuators have been used for hybrid mounted micromotors;[26] microvalves;[28] mechanical resonators for use in, for example, gyros and accelerometers;[7] displacement actuators for optical components (e.g., for positioning micromirrors);[28] and switches.[29] An electrostatic microswitch developed by this author is illustrated in Figure 8.13. Upon applying a bias between the cantilever and the substrate, two metal gold pads make physical contact enabling large currents to flow. A mechanical switch as shown has advantages over an electronic switch, as its off-impedance (air gap) is infinite and its on-impedance (metal contact) is very low.

Summary Electrostatic Actuators: Performance and Modeling

Electrostatic fields can exert great forces, but generally only across very short distances (1 μm). The extremely low current consumption associated with electrostatic devices makes for highly efficient actuation and the high speeds (up to 100 kHz) and ease of fabrication of electrostatic devices has enabled many applications (see above). In Figure 8.14, where

* A detailed description of an electrostatic eccentric drive micromotor (wobble motor) can be found in Price et al.[23]

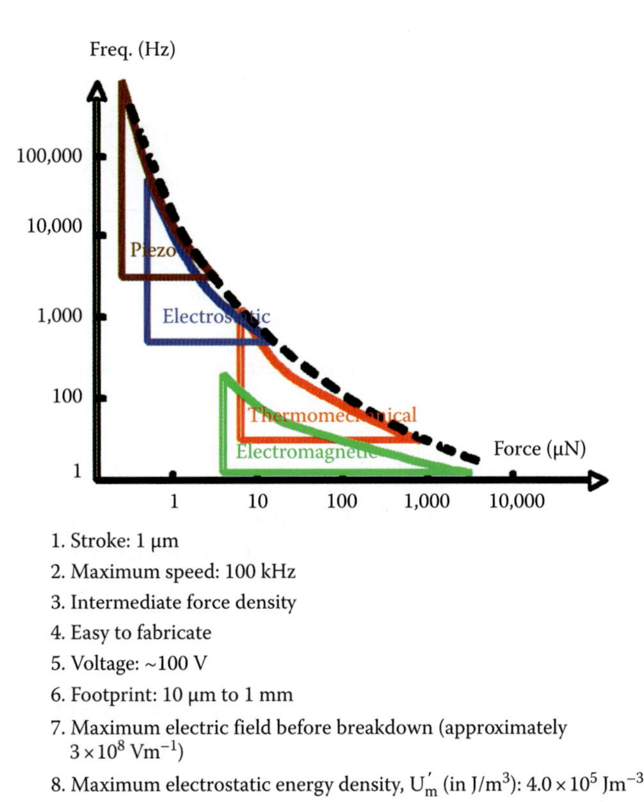

FIGURE 8.14 Summary of electrostatic actuators.

we plot frequency versus force, we summarize some of the most salient electrostatic actuator characteristics in comparison to actuators based on other principles.

In Figure 8.15 we generalize the description of the dynamics of an electromechanical system. We employ x as the direction of displacement (instead of z; see above) and K for a combined stiffness constant (instead of k; see above). The differential equation, describing the electrical equivalent circuit shown here (Figure 8.15b), is as follows:

$$L\frac{d}{dt}\left(\frac{dQ}{dt}\right)+R\left(\frac{dQ}{dt}\right)+\frac{1}{C}Q(t)=V_i$$

and with
$$\frac{d}{dt}\left(\frac{dQ}{dt}\right)=\ddot{Q}(t) \quad \text{and} \quad \frac{dx}{dt}=\dot{Q}(t)$$

$$L\ddot{Q}(t)+R\dot{Q}(t)+\frac{1}{C}Q(t)=V_i \tag{8.36}$$

where V_i is the input voltage and V_0 the output voltage. L is inductance, R is resistance, and C is capacitance. For the mechanical equivalent circuit (Figure 8.15a), the differential equation is as follows:

$$m\frac{d}{dt}\left(\frac{dx}{dt}\right)+b\left(\frac{dx}{dt}\right)+Kx(t)=F$$

with
$$\frac{d}{dt}\left(\frac{dx}{dt}\right)=\ddot{x}(t) \quad \text{and} \quad \left(\frac{dx}{dt}\right)=\dot{x}(t)$$

$$m\ddot{x}(t)+b\dot{x}(t)+Kx(t)=F \tag{8.37}$$

where the external force F is the input, the displacement x is the output, b is damping, m is the mass, and K is the stiffness.

The modeling equations gathered above are in the time domain, but often analysis of a system is easier

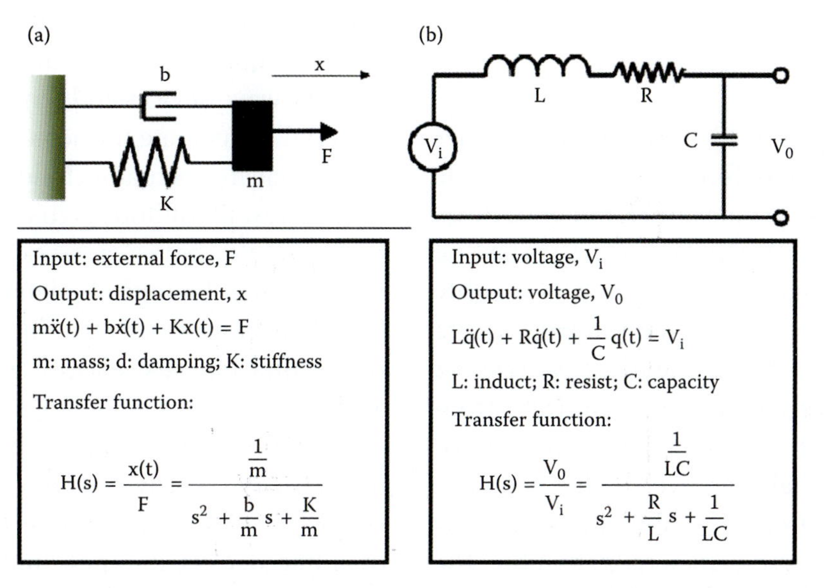

FIGURE 8.15 General description of the dynamics of an electromechanical system.

to perform in the frequency domain. In order to convert these equations to the frequency domain, we apply the Laplace transform to determine the transfer function [$H(s)$] of the system. As a consequence functions with a real dependent variable (such as time) are converted into functions with a complex dependent variable (such as frequency, often represented by s), which are easier to manipulate. The transfer function of the system then is the ratio of the output Laplace transform to the input Laplace transform assuming zero initial conditions. Many important characteristics of dynamic or control systems can be determined from the transfer function. The general procedure to find the transfer function of a linear differential equation from input to output is to take the Laplace transforms of both sides assuming zero conditions and to solve for the ratio of the output Laplace over the input Laplace. For the electrical case the transfer function is as follows:

$$H(s) = \frac{V_0}{V_i} = \frac{\frac{1}{LC}}{s^2 + \frac{R}{L}s + \frac{1}{LC}} \tag{8.38}$$

And for the mechanical case, it is as follows:

$$H(s) = \frac{x(t)}{F} = \frac{\frac{1}{m}}{s^2 + \frac{b}{m}s + \frac{K}{m}} \tag{8.39}$$

Piezoelectricity

Introduction

In 1880, Pierre Curie and his brother Paul-Jacques discovered that external forces applied to single crystals of quartz and some other minerals (e.g., Rochelle salt) generate a charge on the surface of these crystals. The charge, they found, is roughly proportional to the applied mechanical stress (force per unit area). This is called the *direct piezoelectric effect*. The effect received its name in 1881 from Wilhelm Hankel, and it remained a curiosity until 1921, when Walter Cady discovered that a quartz resonator could be used for stabilizing electronic oscillators.[30] Piezoelectric materials exhibit the inverse effects as well: an applied voltage generates

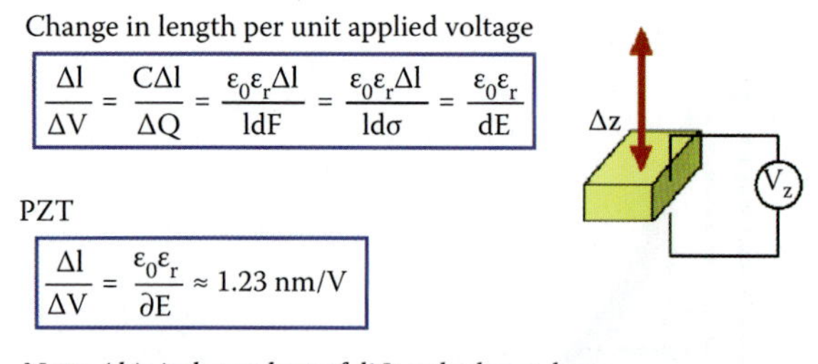

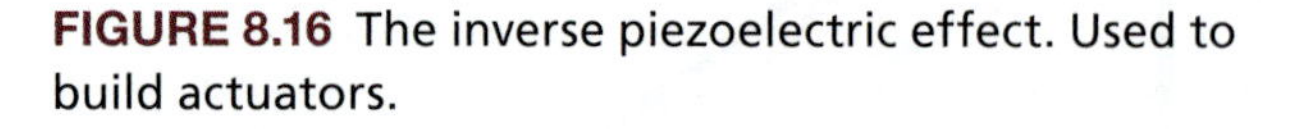

FIGURE 8.16 The inverse piezoelectric effect. Used to build actuators.

a deformation of the crystal (Figure 8.16). Lippman predicted this effect from thermodynamic principles, and the Curie brothers confirmed it experimentally a year later (1881). In the direct case, the crystals work like sensors; in the indirect case, piezoelectrics exhibit an actuation behavior. The mechanical waves generated by piezoelectrics are also called *acoustic waves*. Acoustic wave devices are so named because their detection mechanism is a mechanical, or acoustic, wave. Piezoelectric acoustic wave sensors apply an oscillating electric field to create a mechanical wave, which propagates through the substrate and is then converted back to an electric field for measurement. As the acoustic wave propagates through or on the surface of the piezoelectric material, any changes to the characteristics of the propagation path affect the velocity and/or amplitude of the wave. Changes in velocity can be monitored by measuring the frequency or phase characteristics of the sensor and can then be correlated to the corresponding physical quantity being measured. Virtually all acoustic wave devices and sensors use a piezoelectric material to generate the acoustic waves.

Piezoelectric actuators follow mammalian muscle as the most ubiquitous actuator principle in nature. For example, bones possess piezoelectric properties.[31] Other biological materials that have been found to be piezoelectric include tendon, dentin, ivory, intestine, silk, elastin, wood, and the nucleic acids.

Piezoelectricity is not to be confused with ferroelectricity, discovered in 1920 by J. Valasek. Ferroelectricity is the property of a spontaneous or induced electric dipole moment. In crystals of Rochelle salt ($NaKC_4H_4O_6 \cdot 4H_2O$), the unit cells have a spontaneous asymmetric charge distribution, and the orientation of this polarization can be switched by applying a field strong enough to drag the unsymmetrically placed ions into equivalent positions in the opposite direction. All ferroelectric materials are piezoelectric, but the converse is not always true. Piezoelectricity relates to the crystalline ionic structure; ferroelectricity relates to electron spin. Above the so-called *Curie temperature*, spontaneous polarization of a ferroelectric is lost, because thermal vibrations randomize the dipole orientations. The transition from a nonferroelectric to a ferroelectric state is analogous to the magnetic ordering transition in a ferromagnet below its Curie temperature. Ferroic materials are multifunctional materials, materials having both sensing and actuation functions. A large number of applications of ferroelectric materials exploit an indirect consequence of ferroelectricity, such as dielectric, piezoelectric, pyroelectric, and electrooptic properties. The biggest use of ferroelectric materials is as dielectrics in capacitors, ferroelectrics for thin films for nonvolatile memories, piezoelectric materials for medical ultrasound and imagining and actuators, and electrooptical materials for data storage and displays. Ferroelectric materials exhibit strong electrostriction.

Electrostriction, covered below, is a secondary coupling in which the strain is proportional to square of the electric field. Electrostriction, a property of all dielectrics but pronounced only for ferroelectrics, is similar to the piezoelectric effect in that it involves an increase in length parallel to an applied electric field. In electrostrictive materials, in contrast to piezoelectric materials, the direction of this small change in geometry does not reverse if the direction of the electrical field is reversed.

Mechanism of Piezoelectricity

A simplified model of piezoelectricity entails the notion of anions (–) and cations (+) moving in opposite directions under the influence of an electric field or a mechanical force. The forces generated by this motion cause lattice deformation for noncentrosymmetric crystals because of the presence of both high- and low-stiffness ionic bonds. The effect for quartz is illustrated in Figure 8.17. If the cell shown here is deformed along the *x*- or *y*-axis, the O-ion is displaced, and positive or negative charges are formed. As a result, all piezoelectric materials are necessarily anisotropic; in cases of central symmetry, an applied force does not yield an electric polarization. By applying mechanical deformations to piezoelectric crystals, electric dipoles are generated, and a potential difference develops when changing those mechanical deformations. Silicon is not piezoelectric because it is cubic and covalent rather than

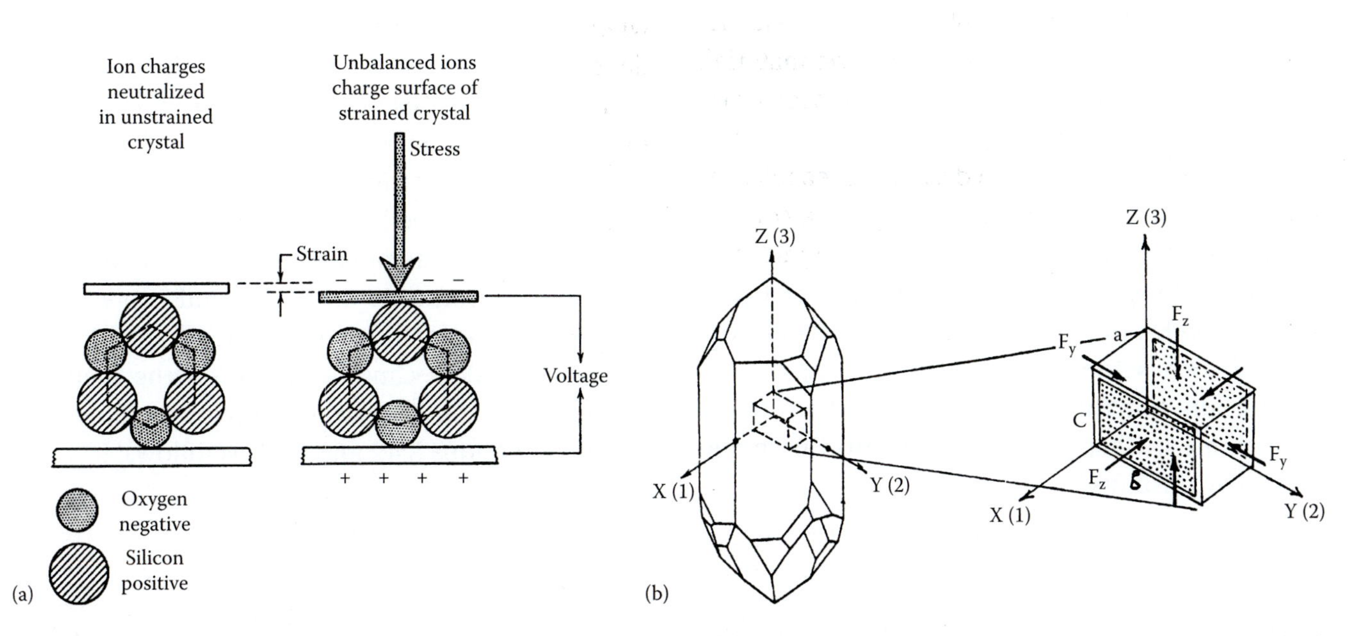

FIGURE 8.17 Piezoelectricity in an ionic crystal such as quartz. (a) Ion position in quartz lattice with and without applied stress. (b) Element cut from quartz crystal under stress.

ionic and noncentrosymmetric. It is important to remember that the potential and the associated currents in piezoelectric materials are a function of the continuously changing mechanical deformation. Therefore, typical and practical uses are in situations involving dynamic strains of an oscillatory nature.[32]

As we saw in Volume I, Chapter 2, the thousands of crystals found in nature can be grouped into 230 space groups on the basis of symmetry elements. Since the relative position of the symmetry elements in a crystal is not important, and only their orientation counts, the 230 space groups can be reduced to just thirty-two point groups. The thirty-two point groups further divide over the seven well-known crystal systems: triclinic, monoclinic, orthorhombic, tertragonal, trigonal, hexagonal, and cubic. Of the thirty-two point groups, twenty-one do not have a center of symmetry (i.e., they are noncentrosymmetric). The latter crystals possess one or more crystallographically unique directional axes, and all, except the 432 point group, show the piezoelectric effect along unique directional axes. Piezoelectric properties are thus present in twenty of the thirty-two different crystallographic point groups, although only a few of them are used. They are found also in amorphous ferroelectric materials. Of the twenty crystallographic classes, only ten display ferroelectric properties.

A more detailed understanding of piezoelectricity is based on an understanding of the piezoelectric equations describing the coupling between electric and mechanical strains in a piezoelectric material. Polarization and stress are vector and tensor properties, respectively, and in general, arbitrary components of each can be related via the piezoelectric effect. For this reason, piezoelectricity is a complicated property, and up to eighteen constants may be required to specify it. When a stress, σ (FA^{-1}), is applied to a slab of material cut from a piezoelectric material such as quartz, the resulting one-dimensional strain, ε, in the elastic range, can be written as (see Volume I, Chapter 4, "Stress-Strain Curve and Elasticity Constants"):

$$\varepsilon = S\sigma \tag{8.40}$$

which represents Hooke's law, where S stands for the compliance ($S^{-1} = E$ = Young's modulus). With a potential difference applied across the faces of the piezoelectric, an electric field, E (Vm^{-1}), is established and we obtain the following (see Volume I, Equation 3.37):

$$D = \varepsilon_r E = \varepsilon_0 E(1+\chi) = P \tag{8.41}$$

where D = electric displacement (or electric flux density)
ε_r = relative permittivity of the material (dimensionless)
E = electric field
ε_0 = permittivity of vacuum in $A \times s\ V^{-1} \times m^{-1}$
χ = electrical susceptibility
P = polarization, that is, the electric dipole moment per unit volume of material

Equation 8.40 contains only mechanical quantities, and Equation 8.41 contains only electrical parameters. Piezoelectric materials possess a special interlocking behavior in which electrical charges are produced by straining the material, and mechanical forces are produced by subjecting the material to an electric field.[33] For a one-dimensional piezoelectric material with electric field and stress in the same direction, according to the principle of energy conservation from thermodynamics, we can describe this situation as follows:[34]

$$D = d\sigma + \varepsilon(d)^{\sigma} E \tag{8.42}$$

$$D = e\varepsilon + \varepsilon(d)^{\varepsilon} E \tag{8.43}$$

where d represents a piezoelectric constant (charge density/applied stress) expressed in coulombs divided by newtons (C/N), and $\varepsilon(d)\sigma$ is the permittivity at constant stress [we added a "(d)" to distinguish the dielectric constant from the strain ε]. To work with d coefficients is useful when charge generators are contemplated, for example, in piezoelectric accelerometers. In the second expression, e stands for a piezoelectric constant (charge density/applied strain) in Cm^{-1}, and $\varepsilon(d)^{\varepsilon}$ is the dielectric constant at a constant strain. Solving for E, we can rewrite Equations 8.42 and 8.43 as follows:

$$E = -g\sigma + \frac{D}{\varepsilon(d)^{\sigma}} \tag{8.44}$$

$$E = -h\varepsilon + \frac{D}{\varepsilon(d)^{\varepsilon}} \tag{8.45}$$

where $g = d/\varepsilon\ (d)^{\sigma}$ is a piezoelectric constant (field/applied stress), also called a *voltage constant*, in V/m/N, and $h = e/\varepsilon\ (d)^{e}$ is a piezoelectric constant (field/applied strain) in V/m. The above set of equations describes the direct piezoelectric effect. The equations for the inverse effect are written as follows:

$$\varepsilon = dE + S^{E}\sigma \tag{8.46}$$

$$\varepsilon = gD + S^{D}\sigma \tag{8.47}$$

where d and g = piezoelectric constants given in mV^{-1} and mC^{-1}, respectively
S^E = compliance at constant field
S^D = the compliance at constant electric flux density

Solving Equations 8.46 and 8.47 for σ results in the following:

$$\sigma = -eE + E(Y)^{E}\varepsilon \tag{8.48}$$

$$\sigma = -hD + E(Y)^{D}\varepsilon \tag{8.49}$$

where $e = d/S^E$ and $h = g/S^D$ = piezoelectric constants given in N/V/m and N/C, respectively
$E(Y)^E$ = Young's modulus [we added a "(Y)" here to distinguish the Young's modulus from the electric field *E*] under constant electrical field
$E(Y)^D$ = Young's modulus under constant electric flux density

Equations 8.42 to 8.49 are known as the *piezoelectric constitutive relations*. They are summarized in Table 8.4, together with the definition of the four piezoelectric constants *d*, *e*, *g*, and *h*.

If no external field is imposed, according to Equations 8.41, 8.42, and 8.43, a stress, σ, will lead to the following polarization, *P*, or direct effect:

$$P = d\sigma \tag{8.50}$$

and, according to Equations 8.46 and 8.47, if there is no external stress applied, a field, *E*, will lead to the following strain and converse effect, ε:

$$\varepsilon = dE \tag{8.51}$$

so that the same constant—the piezoelectric coefficient *d*—is used for the direct and reverse effects.

The ratio of the converted energy of one kind (mechanical or electrical) stored at any instant in a piezoelectric to the input energy of the second kind (electrical or mechanical) is defined as the square of the coupling coefficient:

$$k = \sqrt{\frac{\text{mechanical energy stored}}{\text{electrical energy applied}}} = \sqrt{\frac{\text{electrical energy stored}}{\text{mechanical energy applied}}} \tag{8.52}$$

The electromechanical coupling coefficient, *k*, is thus a measure of the interchange of electrical and mechanical energy. It can be shown that *k* equals the geometric mean of the piezoelectric voltage coefficient (*g*) and the piezoelectric stress coefficient (*e*) and is indicative of the ability of a material to both detect and generate mechanical vibrations:[35]

$$k = \sqrt{ge} \tag{8.53}$$

To activate the maximum piezoelectric strain in a given crystal, the orientation dependence of the piezoelectric effect should be carefully considered.

Table 8.4 Piezoelectric Constitutive Equations and Definitions of Piezoelectric Parameters

Piezoelectric Equations	Equation Number in Text	Definitions of Constants	SI Units
$D = d\sigma + \varepsilon(d)\sigma E$	8.42	d = charge density/applied stress	C/N
$D = e\varepsilon + \varepsilon(d)\varepsilon E$	8.43	e = charge density/applied strain	C/m
$E = -g\sigma + D/\varepsilon(d)\sigma$	8.44	g = field/applied stress	V/m/N
$E = -h\varepsilon + D/\varepsilon(d)\varepsilon$	8.45	h = field/applied strain	V/m
$\varepsilon = dE + S^E\sigma$	8.46	d = strain/applied field	m/V
$\varepsilon = gD + S^D\sigma$	8.47	g = strain/applied charge density	m/C
$\sigma = -eE + E(Y)^E\varepsilon$	8.48	e = stress/applied field	N/V/m
$\sigma = -hD + E(Y)^D\varepsilon$	8.49	h = stress/applied charge density	N/C

For the general case of a piezoelectric crystal, rather than a one-dimensional piezoelectric as treated above, the piezoelectric constitutive relations must be generalized. The generalized elastic response of a piezoelectric crystal to an electric field (see Equation 8.51) may then be expressed as follows:

$$\varepsilon_k = \sum_{i=1}^{3} \mathbf{d}_{ik}\mathbf{E}_i \tag{8.54}$$

and the polarization in the absence of a field, from Equation 8.50, as follows:

$$\mathbf{P}_i = \sum_{k=1}^{6} \mathbf{d}_{ik}\sigma_k \tag{8.55}$$

In these equations, i = 1, 2, and 3 make up the indices of the components of polarization and k = 1, 2, …, 6 the indices of the components of mechanical stress and strain. In the piezoelectric constants, the first subscript refers to the direction of the field; the second subscript refers to the direction of the stress. The subscripts 1, 2, and 3 indicate the *x*-, *y*-, and *z*-axes, respectively. For example, in d_{33}, i = 3 indicates that the polarizing electrodes are perpendicular to the 3-axis (i.e., *z*-axis), and the piezoelectric-induced strain or applied stress is in the 3-direction (along the *z*-axis) as well. For the coupling coefficient, *k* (Equation 8.53), the same notation is used. The above expressions are equivalent to writing (strain tensor) = (d tensor) × (electric field vector) and (polarization vector) = (d tensor) × (stress tensor), respectively. Piezoelectricity, we saw in Volume I, Chapter 2, on crystallography, is an example of a vector-tensor effect; an electric field (vector) causes a mechanical deformation (tensor). Elastic deformation under the influence of a stress tensor is an example of a tensor-tensor effect.

On the basis of this notation, we can now generalize the constitutive relations as:

$$\mathbf{D}_i = \mathbf{d}_{ik}\sigma_\kappa + \varepsilon(\mathbf{d})_{ik}\sigma\mathbf{E}_i \tag{8.56}$$

and:

$$\varepsilon_k = \mathbf{d}_{ik}\mathbf{E}_i + \mathbf{S}^{E}_{ik}\sigma_k \tag{8.57}$$

where $\mathbf{D}_i$ = electric displacement vector (3 × 1), C/m²
$\mathbf{E}_i$ = electric field vector (3 × 1), V/m
σ_i = stress (force/unit area) tensor (6 × 1), N/m²
ε_k = strain (relative displacement) tensor (6 × 1)
$\mathbf{d}_{ik}$ = a matrix of piezoelectric constants (3 × 6), m/V
$\varepsilon(\mathbf{d})_{ik}$ = permittivity measured at constant stress (3 × 3), As/Vm
$\mathbf{S}^{E}_{ik}$ = elastic compliance matrix when subjected to a constant electric field (6 × 6), m²/N

The other equations in Table 8.4 may be generalized in the same fashion.

From Volume I, Equation 4.83, we recall that the strain tensor has six components: σ_x, σ_y, σ_z for compression or tension and τ_{xy}, τ_{xz}, and τ_{yz} for the shear components. The piezoelectric behavior of a crystal can be completely described if the set of data related to the piezoelectric constants (d_{ik}), elastic compliances (S_{ik}), and permittivities [$\varepsilon(d)^{\sigma}_{ik}$] is given. This set can be introduced with a 9 × 9 matrix in which columns are associated with mechanical stresses and field strengths and rows refer to strains and polarizations:[36]

	σ_x	σ_y	σ_z	τ_{xy}	τ_{xz}	τ_{yz}	E_x	E_y	E_z
ε_x	S_{11}	S_{12}	S_{13}	S_{14}	S_{15}	S_{16}	d_{11}	d_{21}	d_{31}
ε_y	S_{21}	S_{22}	S_{23}	S_{24}	S_{25}	S_{26}	d_{12}	d_{22}	d_{32}
ε_z	S_{31}	S_{32}	S_{33}	S_{34}	S_{35}	S_{36}	d_{13}	d_{23}	d_{33}
γ_{xy}	S_{41}	S_{42}	S_{43}	S_{44}	S_{45}	S_{46}	d_{14}	d_{24}	d_{34}
γ_{xz}	S_{51}	S_{52}	S_{53}	S_{54}	S_{55}	S_{56}	d_{15}	d_{25}	d_{35}
γ_{yz}	S_{61}	S_{62}	S_{63}	S_{64}	S_{65}	S_{66}	d_{16}	d_{26}	d_{36}
P_x	d_{11}	d_{12}	d_{13}	d_{14}	d_{15}	d_{16}	ε_{11}	ε_{12}	ε_{13}
P_y	d_{21}	d_{22}	d_{23}	d_{24}	d_{25}	d_{26}	ε_{21}	ε_{22}	ε_{23}
P_z	d_{31}	d_{32}	d_{33}	d_{34}	d_{35}	d_{36}	ε_{31}	ε_{32}	ε_{33}

(8.58)

This matrix is symmetrical ($S_{ik} = S_{ki}$, $d_{ik} = d_{ki}$, and $\varepsilon_{ik} = \varepsilon_{ki}$) and reduces to forty-five different terms, including twenty-one compliances, six permittivities, and eighteen piezoelectric constants. In the top left of this matrix (rectangular box), we recognize Equation 4.83, but because of the presence of piezoelectricity, the matrix is this much larger. For simplicity, we have replaced $\varepsilon(d)^{\sigma}_{ik}$ by ε_{ik} in the above matrix. A further simplified notation with one subscript is often introduced for dielectric constants: $\varepsilon_{11} = \varepsilon_1$, $\varepsilon_{22} = \varepsilon_2$, and $\varepsilon_{33} = \varepsilon_3$. For the defined orientation of the crystallographic axes, $\varepsilon_{ik} = 0$ for i ≠ k.

For crystalline quartz, the eighteen remaining piezoelectric constants further reduce to two:

$$d_{ik} = \begin{bmatrix} d_{11} & d_{12} & 0 & d_{14} & 0 & 0 \\ 0 & 0 & 0 & 0 & d_{25} & d_{26} \\ 0 & 0 & 0 & 0 & 0 & 0 \end{bmatrix} \quad (8.59)$$

with $d_{11} = -d_{12} = -d_{26}/2 = 2.31$ pC/N and $d_{14} = -d_{25} = 0.73$ pC/N. The meaning of d_{14} is as follows: a torsion stress of 1 N/m^2 around the axis 1 (x) of quartz (direction 4, or xz) induces a charge density of 0.73 pC/N in two metal plates on the material in direction 1 (x).

Exercise 8.1:

A piezoelectric ceramic of 1 cm thickness in the 3-direction and 10 cm length in the 1-direction is subjected to a potential of 0.01 volts in the 3-direction, and the resulting change in length is 10^{-11} cm. What is the numerical value and subscript of the *d* constant? Answer: d_{31} = strain developed/field applied = (10^{-11} cm/10 cm)/(0.01 V/0.01 m) = 10^{-12} mV^{-1}.

Exercise 8.2:

If a voltage *V* is applied across the thickness of a piezoelectric slab, the displacements ΔL, ΔW, and Δt along the length, width, and thickness directions, respectively, are given by $\Delta L = d_{31} \times V \times L/t$, $\Delta W = d_{31} \times V \times W/t$, and $\Delta t = d_{33} \times V$, where *L* and *W* are the length and width of the plate, respectively, and *t* is the thickness or separation between the electrodes. When a force *F* is applied along the length, width, or thickness direction of the same slab, a voltage *V* is measured, given in each of the three cases as $V = d_{31} \times F/\varepsilon \times L$, $V = d_{31} \times F/\varepsilon \times W$, and $V = d_{33} \times F \times t/\varepsilon \times L \times W$.

At frequencies far below the resonance, piezoelectric transducers are fundamentally capacitors. Since, in a capacitor, the voltage *V* is related to the charge Q via the capacitance *C* (Q = CV), the charge coefficients d_{ik} are related in the same way to the voltage coefficients g_{ik} via ε_{ik}, or:

$$d_{ik} = \varepsilon(d)^{\sigma}_{ik} \cdot g_{ik} \cdot \varepsilon_0 \quad (8.60)$$

with ε_0 being the permittivity of free space. At resonance, the dielectric constant is reduced by a factor of $(1 - k)$, where k is the coupling coefficient of the mode in question.

For an in-depth mathematical treatise of piezoelectricity, we refer to *The Theory of Piezoelectric Shells and Plates*.[37] Other excellent references are by Cady[38] and Zelenka.[39]

Overview of Types of Acoustic Waves

Introduction

Acoustic wave devices involve piezoelectric substrates or the coupling of piezoelectric materials to nonpiezoelectric structures to (e.g., a Si cantilever equipped with a thin film of piezoelectric ZnO). These devices are described by the mode of sound wave propagation through the piezoelectric substrate or through the linked nonpiezoelectric material. Primarily, their velocities and displacement directions distinguish sound wave types. It is important to remember here that acoustic waves are the result of induced particle vibration that depends on the elastic properties of the solid. In Table 8.5 we summarize several, but not all, of the wave modes possible in solids.

Metal transducers launch acoustic waves at ultrasonic frequencies and the type and resonant frequency of such waves are fixed by the piezoelectric crystal orientation, the thickness (t) of the piezoelectric used, and the geometry of the metal transducers. The resulting mechanical waves are called *transverse* or *shear waves* when the particle displacement is perpendicular to the direction of the wave propagation. Shear waves cannot propagate in fluids, and in a solid they typically travel at half the velocity of a longitudinal wave in the same material. In contrast, compressional or longitudinal acoustic waves are defined when particle displacement is parallel to

Table 8.5 Acoustic Wave Types in Solids

Wave Type	Particle Vibrations
Longitudinal	Parallel to wave direction
Transverse (shear)	Perpendicular to wave direction
Surface (Rayleigh)	Elliptical orbit, symmetrical mode
Plate (Lamb)	Component perpendicular to surface (extensional wave)
Plate (Love)	Parallel to plane layer, perpendicular to wave direction

the direction of the acoustic wave travel, and these waves can be transmitted in solids and liquids.

Bulk Acoustic Wave Devices

While bulk acoustic waves (BAWs) propagate through the volume of the piezoelectric substrate, the propagation of surface acoustic waves (SAWs) or Rayleigh waves is restricted to a distance of about one acoustic wavelength from the piezoelectric substrate surface. Rayleigh waves are somewhat similar to water waves in that the motion of particles is both longitudinal and transverse in a plane containing the direction of propagation and the normal to the surface.

A commonly used BAW device is the thickness shear mode (TSM) resonator shown in Figure 8.18. In TSM mode, bulk transverse waves with particle displacement parallel to the surface of the sensor are generated. Such waves, on propagation into a liquid, suffer severe attenuation of the acoustic energy due to the viscous forces in the liquid. The device pictured in Figure 8.18 is more commonly known as a *quartz crystal microbalance* (QCM) and is used, among other things, to measure deposited metal thickness in an evaporation station (see Volume II, Figure 7.2a). It consists of an AT-cut quartz disc with metal electrodes on opposite sides to effect the application of an oscillating electric field. A quartz AT cut is made normal to the y-axis but rotated 35.25° from the z-axis. In this direction, quartz has a zero temperature coefficient at 40°C, making the AT cut ideal for frequency control applications. Quartz is a hexagonal crystal (see Figure 8.17), and if the electrical field is aligned with the y-axis, the resulting deformation is a shear along the neutral z-axis. To achieve a shear motion, the cut must be normal to the y-axis so that charging electrodes placed on the two faces of the wafer can produce the field. The thickness (t) of the quartz

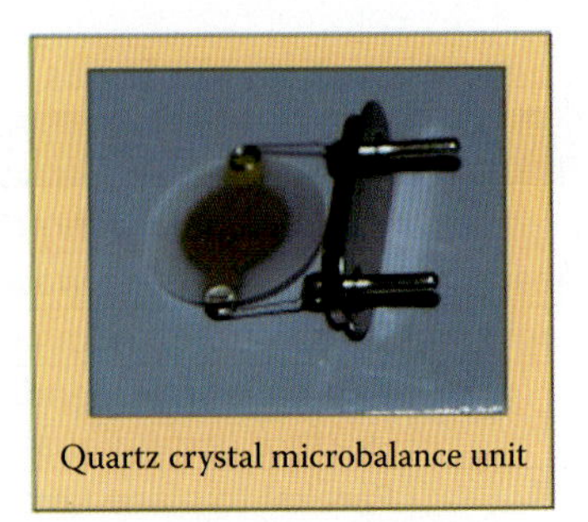

FIGURE 8.18 Quartz crystal microbalance.

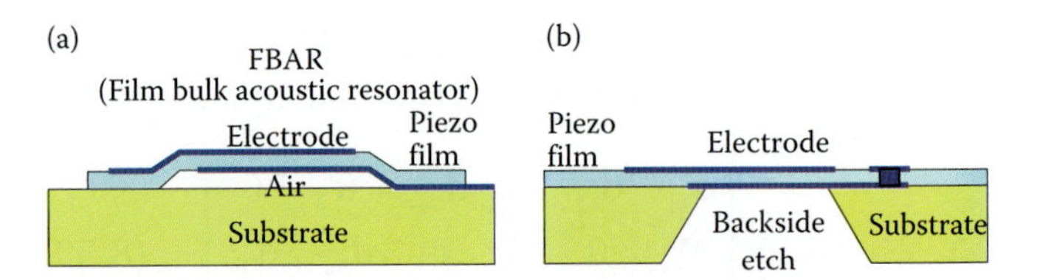

FIGURE 8.19 Two methods to make film bulk acoustic resonators (FBARS). The piezo film is typically aluminum nitride (AlN).

plate determines the wavelengths (λ) of the fundamental ($n = 1$) and harmonic resonances through the expression $\lambda = 2t/n$ ($n = 3, 5, 7$). Consequently, the resonant frequency for the fundamental mode is restricted by the thickness t of the quartz wafers to the range of approximately 5–20 MHz.

A more recent type of BAW device is a film bulk acoustic resonator (FBAR) that is used in mobile telephones. FBARs feature a $Q > 400$ at 1.9 GHz and often use aluminum nitride (AlN) as the piezoelectric material. The major difference between an FBAR and the QCM of Figure 8.19 is that the AlN thin film piezoelectric is much thinner than the quartz cut. Two types of FBAR constructs are shown in Figure 8.19. Figure 8.19a involves the etching of a sacrificial support layer below the piezofilm film. Figure 8.19b consists of etching the substrate from the back of the wafer to the front surface creating a so-called *pothole structure*.

Surface Acoustic Waves

It was Lord Rayleigh, in 1887, who discovered the surface acoustic wave (SAW) mode of sound propagation and predicted their properties. Named for their discoverer, Rayleigh waves have a longitudinal and a vertical shear component that couple with a medium in contact with the device's surface. Rayleigh waves are also of interest to geologists since these surface waves are excited by earthquakes.

A SAW device, as illustrated in Figure 8.20, comprises a thick ST-cut quartz plate and two interdigital metal transducers (IDTs).* In a typical "delay line" configuration, input and output IDTs generate or receive Rayleigh waves on the same side of a piezoelectric substrate. It was White and Voltmer[40]

* An interdigital transducer (IDT), or interdigitated transducer, is a device which consists of two interlocking comb-shaped metallic coatings.

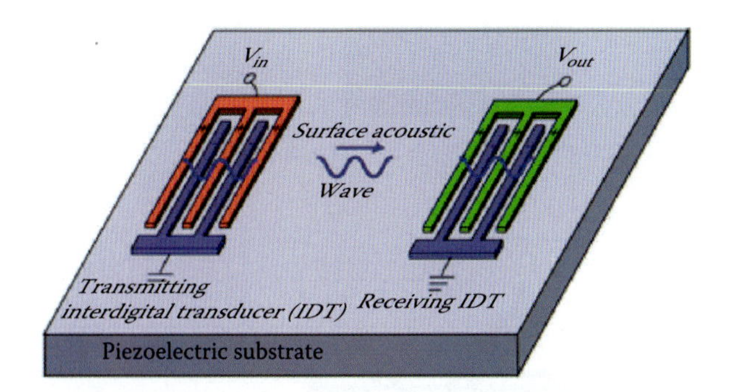

FIGURE 8.20 Surface-acoustic wave device; see also the hygrometer in Figure 8.26.

who pioneered the marriage of piezoelectricity with microelectronics through the demonstration that photolithographic techniques could be employed to deposit IDTs on piezoelectric substrates to excite Rayleigh waves. Rayleigh waves have a velocity that is ~5 orders of magnitude less than the corresponding electromagnetic wave, making Rayleigh surface waves among the slowest to propagate in solids. The wave amplitudes are typically ~10 Å, and the wavelengths range from 1 to 100 μm. The frequency of a SAW device is determined by the "finger" spacing of the IDTs ranging from 30 to 500 MHz (so they can be more than ten times faster than a QCM).

The surface particles in a Rayleigh wave move elliptically, resulting in a wave consisting of both shear and compressional components. The latter provokes an important attenuation effect in liquids that prevents the application of SAW devices in such media. An exception to this loss of sound energy occurs for devices using waves that propagate at a velocity lower than the sound velocity in the liquid. Regardless of the displacement components, such modes do not radiate coherently and are thus relatively undamped by liquids [see next section, on flexural plate wave (FPW) devices].

Flexural Plate Waves

A final important category of surface-launched acoustic wave sensors is those composed of very thin plate devices, where the thickness of a piezoelectric substrate is reduced to a dimension on the order of the acoustic wavelength itself. These devices use a thin piezoelectric substrate, or plate, as an acoustic waveguide that confines the acoustic energy between the upper and lower surfaces of the plate. The devices that generate such waves are termed *FPW devices*, as illustrated in Figure 8.21, and the waves they generate are called *Lamb waves*. Although surface-excited, the generated Lamb waves travel through the bulk of the piezoelectric membrane, and since both surfaces undergo displacement, detection may occur on either side. This is a very important advantage, as one side may feature the IDTs that must be isolated from, say, conducting liquids or corrosive gases, while the other side can be used as a sensing surface. Lamb waves generated in a plate of finite thickness can be of both symmetric and antisymmetric modes when referring to the median plane of the plate. Since the piezoelectric plate is so thin, the operating frequencies of FPW devices are substantially lower than for other surface-launched acoustic wave devices (2–7 MHz). As a consequence of the minimal thickness of the plate, the wave velocity is lower than the compressional velocity of sound in liquids. Accordingly, compressional waves are not coupled in liquids, resulting in less attenuation of acoustic energy. This is a consequence of Snell's law: the acoustic wave is trapped in the plate much as a light wave is trapped in an optical waveguide. This means that FPWs can even be used in liquids.

A practical FPW device, as shown in Figure 8.21, has a design similar to that of a SAW device. The basic elements of an FPW device consist of a composite plate of low-stress thin substrate material (e.g., 0.8–4-μm silicon nitride), a thin zinc oxide film (0.3–1.5 μm) as piezoelectric, and a pair of aluminum IDTs on the ZnO layer. IDTs are positioned in a way to form a delay-line configuration. A metal ground plane (0.2–05 μm thick) is sandwiched between the silicon nitride and piezoelectric layer. A chemical sensitive film may be deposited on the opposite side of the membrane. The IDTs launch and receive the plate waves (e.g., the 0th-order antisymmetric Lamb waves*) and, together with an amplifier, form a feed-

* Lamb waves are associated with Rayleigh waves on a plate whose thickness is smaller than the acoustic wavelength. They can be considered as two Rayleigh waves propagating on both sides of the thin plate. Two kinds of waves can propagate through the plate independently, namely the symmetric and the antisymmetric waves. The A_0 wave is the antisymmetric Lamb wave with the lowest phase velocity. This makes the wave suitable for liquid sensing because the A_0 wave does not excite compressional waves in a loading liquid and thus reduces scattering energy if its phase velocity is lower than the sound velocity of liquid. The A_0 mode Lamb wave sensor is also called the *FPW sensor*.

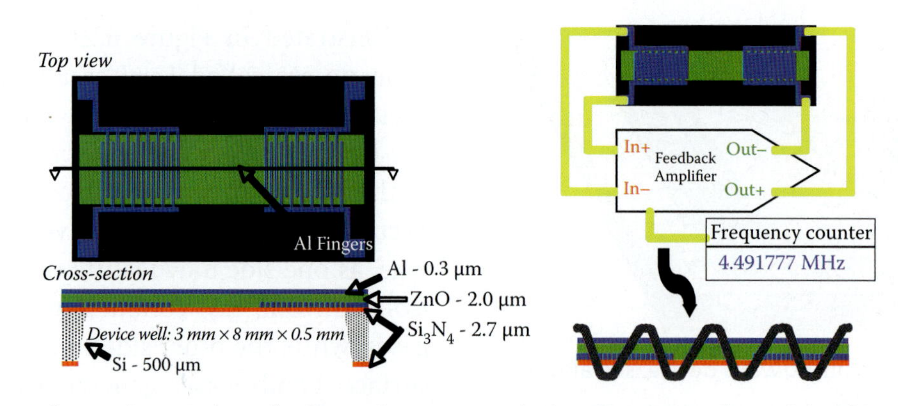

FIGURE 8.21 Construction and operation of a flexural pate wave device. (Courtesy of Dr. Dick White, University of California, Berkeley.)

back oscillator whose output frequency may provide a measure of the mass per unit area of the membrane, including a chemically sensitive film deposited on it. Summarizing the differences between an FPW device and a SAW device: 1) FPW devices operate at lower frequencies (a few megahertz as opposed to hundreds of megahertz for SAW sensors); 2) the IDTs are deposited on the opposite side of the FPW sensing material, and thus the sensor can be easier applied to liquids because the IDT electrodes are insulated from the liquid; and 3) the plate thickness is much smaller than the acoustic wavelength. FPW devices offer high mass sensitivity at much lower operating frequency than SAW devices.

Piezoelectric actuators generally produce very strong forces and very small motions that can be amplified, for example, by making the piezoelectric material part of a bimorph or a stack (Figure 8.22). Stacked structures attain larger displacements but are more difficult to implement in a miniaturized device. Piezo materials operate at high speed and return to a neutral position when the power is switched off. Alternating currents produce oscillations in a piezoelectric, and operation at the specimen's fundamental resonant frequency produces the largest elongation and at the highest power efficiency.

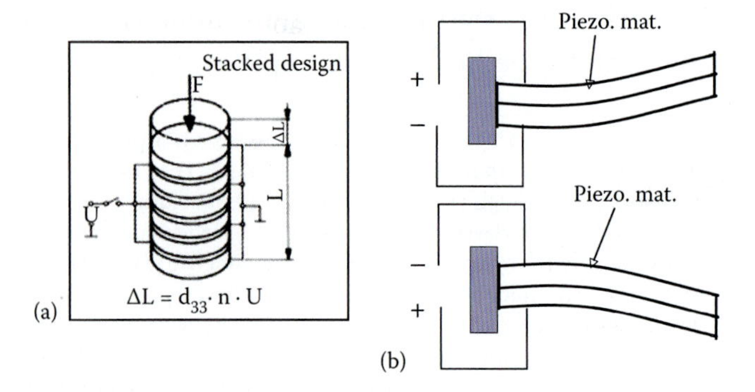

FIGURE 8.22 Piezoelectric stack actuator (a) and piezoelectric bimorph actuator (b).

Overview of Piezoelectric Materials

The first extensively used piezoelectric materials were naturally occurring crystals —quartz and Rochelle salt ($NaKC_4H_4O_6 \cdot 4H_2O$). Quartz has great physical and chemical stability but only a small coupling coefficient, *k*. Applications of quartz includes oscillators and resonators (see, for example, the quartz angular rate sensor based on a tuning fork in Figure 5.1). Rochelle salt has a great piezoelectric effect, but it is soluble in water and has poor temperature characteristics. It has been used in various transducers such as phonograph pickups.

In synthetic piezoelectrics, ceramics formed by many tightly compacted crystallites are most popular. Many piezoelectric materials, such as barium titanate ($BaTiO_3$), lead titanate ($PbTiO_3$), lead zirconate titanate (PZT), lead lanthanum zirconate titanate (PLZT), lead magnesium niobate (PMN), potassium niobate ($KNbO_3$), potassium sodium niobate ($K_xNa_{1-x}NbO_3$), and potassium tantalate niobate [$K(Ta_xNb_{1-x})O_3$], have a perovskite structure, that is, a structure of the type ABO_3. These ceramics are all ferroelectrics. As we saw in Volume I, Chapter 2, this absence of a center of symmetry in the crystal structure gives rise to spontaneous polarization. The crystal structure is cubic above the Curie temperature and tetragonal as it cools down. To align the dipoles in the compacted monocrystals in the same direction (i.e., to polarize or pole them), they are subjected to a strong electric field during their fabrication process. Above the Curie point, the dipole directions in ferroelectric materials have random orientations. To align the dipoles, fields of 10 kVcm^{-1} are common at temperatures slightly above the Curie temperature. The ceramic is then cooled while maintaining the field. When the field is removed, the crystallites cannot reorder in random form because of the mechanical stresses accumulated, resulting in a permanent electric polarization.

Barium titanate was the first material to be developed as a piezoceramic (see also Volume I, Figure 2.10, where the use of this material as a pyroelectric was highlighted). It was discovered independently, in the United States (Waigner and Salomon, 1942), the Soviet Union (Wul and Goldman, 1945), and Japan (Miyake and Ueda, 1946). Applications include detection of mechanical vibration, construction of actuators, and generation of acoustic and ultrasonic vibrations.

PZT is a solid solution of $PbTiO_3$ (tetragonal, with six poling directions) and $PbZrO_3$ (rhombohedral, with eight poling directions). This piezoceramic comes in various compositions such as PZT-4, PZT-5, PZT-5A, and PZT-5H. It was discovered in 1955 by Takagi, Shirane, and Sawagachi in Japan and Jaffe in the United States. The material has a high Curie temperature and is used in devices such as ceramic filters and piezoelectric igniters. Important for MEMS, PZT can be fashioned in thin film form by RF sputtering and rapid thermal annealing and subsequent poling. Alternatively, a sol-gel process is used to fabricate these films. One obtains a piezoelectric material with a relative dielectric constant of more than 1000 this way.[41] Abe and Reed[42] prefer sputtering and argue that the sol-gel technique is less attractive for MEMS applications, given that a 1-μm-thick film would require repeated spin coatings and pyrolysis steps. They prepared PZT films by sputtering from a composite target and found substrate heating during deposition to be crucial to transform the films into the perovskite phase during subsequent annealing. Piezoelectric thin films, such as PZT, zinc oxide (ZnO), and aluminum nitride (AlN), have the cost advantage over crystalline materials. Among these three, the electromechanical coupling effect of PZT is 3 to 9 times larger and the dielectric constant 100 times larger than that of AlN and ZnO, which makes PZT potentially the best candidate for thin film acoustic sensors. However, the polycrystalline structure of PZT and the required heat treatment during the coating process complicate the realization. A problem with all these polycrystalline materials in general relates to their temperature sensitivity and aging (loss of piezo properties) when approaching the Curie temperature.

Polymers such as polyvinylidene fluoride (PVDF), lacking central symmetry, also display piezoelectric properties. This piezopolymer is also pyroelectric, capable of converting changes in temperature to electrical output (see below). Compared with quartz and ceramics, piezofilm is obviously more pliant and lighter in weight. In addition, it is rugged, quite inert, and low cost. This engineering material could provide a sensing solution for vibration sensors, acceleration and shock sensors, passive infrared (IR) sensors, solid state switches, and acoustic and ultrasonic sensors. It should be noted that the peak reversible stress and strain developed by PVDF approximate 3 MPa and 0.1%, respectively. The latter values are rather small, and as for piezoelectric ceramics and magnetostrictive materials (see below), mechanical amplifiers are required to produce the larger displacements needed for many applications. A 200-fold mechanical amplification is required to make PVDF's peak strain similar to that of mammalian muscle.[43] Table 8.6[44,45] summarizes some of the physical properties making PVDF an interesting sensor material.

AlN is a wide-bandgap semiconductor and an excellent candidate for the integration of acoustic wave devices on chips with silicon-based electronics. AlN is a stiff and light material that exhibits excellent piezoelectric properties, including an electromechanical coupling coefficient of 0.088 and a high in-plane acoustic velocity of ~5700 m/s. As a piezoelectric ceramic, aluminum nitride is compatible with standard silicon MEMS processing and has found widespread use in FBARs (see Figure 8.19) and SAW devices (Figure 8.20) AlN may also prove to be useful for the integration of mechanical devices and in the field of packaging of bio-MEMS

Table 8.6 Typical Properties of Piezofilm (Uniaxially Oriented Film)

Film Thickness (μm)	10–100
Piezoelectric strain constants (pC/N)	
d_{31}	28
d_{33}	−35
Young's modulus ($10^9/m^2$)	
E_{33}	5.4
Poisson ratio	
v_{21}	0.25
v_{31}	0.57
v_{32}	0.45
Pyroelectric coefficient (10^{-6} C/m²K)	−30
Mass density (10^3 kg/m³)	1.78
Temperature range (°C)	−40 to 145
Breakdown voltage (V/μm)	80
Maximum operating voltage (V/μm)	30
Yield strength (10^6 N/m²)	45–55
Permittivity (10^{-12} F/m) at 10kHz	106–113
Capacitance (pF/cm²) at 10 kHz	380
Electromechanical coupling factor, 10 Hz	
k_{31}	13%
k_{32}	1.7%
Piezoelectric stress constant (10^{-3} m/mC/m²)	
g_{31}	216
g_{33}	−339
Volume resistivity (Ω · m)	$>10^{13}$

Sources: AMP. 1993. *Piezo film sensors.* Valley Forge, PA: AMP;[44] and Nalwa, H. S., ed. 1995. *Ferroelectric polymers: Chemistry, physics, and applications.* New York: Marcel Dekker.[45]

devices because of its superior thermal conductivity 130–140 (W/mK, room temperature to 100°C) and nontoxicity. The material is often grown by DC magnetron sputtering or by LPCVD on a (001) sapphire substrate (heated to 300–400°C) and can also be epitaxially grown on <111> silicon. Even atomic layer deposition (ALD) has been used to deposit AlN.

The last piezoelectric material we cover here is ZnO, a very versatile piezoelectric that has been used in cantilever-beam accelerometers, gas sensors, FPW devices, IR detectors, anemometers, tactile sensor arrays, etc. A typical change in length per electric field unit (perpendicular to the field E) for ZnO equals 5×10^{-10} C/N compared with 100–200 $\times 10^{-10}$ C/N for PZT-type thin film ceramics. Techniques that have been used for zinc-oxide thin film deposition include RF and dc sputtering, chemical vapor deposition, ion plating, reactive magnetron sputtering, and planar magnetron sputtering (in Polla and Muller[46] and references therein). For thin film deposition of ZnO, planar magnetron sputtering appears to give the best piezoelectric and pyroelectric characteristics; using this technique highly oriented ZnO films have been deposited on SiO_2, polycrystalline silicon, and bare silicon substrates. The compatibility of ZnO with silicon processing is demonstrated, for example, in the commercially available acoustic sensor from Honeywell Corporation, shown in Figure 8.27. The microphone chip illustrated here incorporates ZnO deposited on a silicon substrate, including signal-conditioning circuitry. Muller[47] found the best thin film crystallinity at a sputtering power of 200 W with a 10 mTorr[46] ambient gas mixture consisting of an equal mix of oxygen and argon. The distance between the substrate and target measures about 4 cm, with the substrate temperature maintained at 230°C during deposition. Tjhen et al.[48] characterized the thin film properties of sputtered ZnO and found the electrical properties to be sensitive to substrate material, stress, and surface condition.

The considerations in designing sensors with AlN, ZnO, or PZT films include the following:

1. Value of electromechanical coupling
2. Good adhesion to the substrate
3. Resistance to environmental effects (e.g., humidity, temperature)
4. VLSI process compatibility
5. Cost effectiveness

To further improve the mechanical properties of piezoelectric sensors, piezoelectric composite materials are sometimes used. These are heterogeneous systems consisting of two or more different phases, of which at least one shows piezoelectric properties. Hirata et al.,[49] for example, used LIGA to make arrays of PZT rods with diameters below 20 μm and an aspect ratio of over 5. The PZT rods are embedded in a polymer matrix to make a PZT/polymer composite. The PZT slurry, average particle size 0.4 μm, is injected into a LIGA resist mold (for information on slurry casting of piezoceramics see Preu et al.[50] and Lubitz[51]). After solidification, the resist is removed by plasma etching, and the array of

Table 8.7 Comparison of Some Important Piezoelectric Materials

Material	Piezoelectric Constant d (pC/N)			Dielectric Constant	Young's Modulus E (GPa)	Curie Temperature (°C)	Maximum k (Coupling Coefficient)
AlN	−0.48 (d_{15})	−0.58 (d_{31})	1.55 (d_{33})	9.4	$E_{11} = 410 \pm 10$ $E_{12} = 149 \pm 10$ $E_{13} = 99 \pm 4$ $E_{33} = 389 \pm 10$ $E_{44} = 125 \pm 5$	Not known (>1150 °C)	
Quartz	2.31 (d_{11})	0.73 (d_{14})		$\varepsilon_1 = 4.52$ $\varepsilon_3 = 4.63$	107	550	0.1
PZT (depending on composition)	80–593 (d_{33})	−94 to −274 (d_{31})	494–784 (d_{15})	$\varepsilon_3 = 425–1900$	53	193–490	0.69–0.75
PVDF (Kynar)	23 (d_{31})	4 (d_{32})	−35 (d_{33})	12	3	>MP (150)	0.2
ZnO	−12 (d_{15})	12 (d_{33})	(−4.7) (d_{31})	$\varepsilon_3 = 8.2$	123		
$BaTiO_3$	78 (d_{31})	190 (d_{33})		1,700			

Sources: Based on Pallas-Areny, R., and J.G. Webster. 1991. *Sensors and signal conditioning.* New York: Wiley;[34] Khazan, A.D. 1994. *Transducers and their elements.* Englewood Cliffs, NJ: PTR Prentice Hall;[36] and Flynn, A.M., L.S. Tavrow, S.F. Bart, R.A. Brooks, D.J. Ehrlich, K.R. Udayakumar, and L.E. Cross. 1992. Piezoelectric micromotors. *J Microelectromech Syst* 1:44.[41]

ceramic posts is cast in an epoxy resin. Resolution of acoustic-imaging transducers, for example, for medical applications, should gain resolution as the PZT rods undergo further miniaturization as a result of minimized cross-talk in the array.

Typical piezoelectric materials are compared in Table 8.7. For a survey of inorganic ferroelectrics, refer to *Ferroelectric Transducers and Sensors.*[35] For a detailed review of ferroelectric polymers, refer to *Ferroelectric Polymers: Chemistry, Physics, and Applications.*[45]

Piezoelectric Applications

Overview

When the characteristics of an acoustic path change, a change in acoustic output results. As the acoustic wave propagates through or on the surface of a piezoelectric material, changes to the characteristics of the propagation path affect the velocity and/or amplitude of the wave. Those changes in velocity can be monitored by measuring the frequency or the phase characteristics of the piezoelectric device and can then be correlated to the corresponding physical quantity being measured. In principle, then, all acoustic wave devices are sensors in that they are sensitive to many different physical parameters. The range of phenomena that can be detected by acoustic devices can be further expanded dramatically by coating the piezoelectric material or a structure coupled to it with materials that undergo changes in their mass, elasticity, or conductivity upon exposure to some physical or chemical stimulus. These sensors then may become pressure, torque, shock, and force detectors if under an applied stress the dynamics of the propagating medium change. They become mass, or gravimetric, sensors when particles are allowed to contact the propagation medium, changing the stress on it. They are turned into vapor sensors when a coating is applied that absorbs only specific chemical vapors. These devices work by measuring the mass of the absorbed vapor. A heating element under the chemical film can also be used to desorb chemicals from the device. If the coating absorbs specific biological chemicals in liquids, the detectors become biosensors.

Bulk ceramic piezoelectric devices, say on the basis of PZTs, have been widely used for decades, but thin film applications are more recent arrivals. Telecommunications constitutes the largest market today, accounting for about 3 billion acoustic wave filters annually, primarily for mobile cell phones. This use involves SAW devices, and lately, FBARs. These components are used as bandpass filters. Some of the new emerging applications for acoustic wave devices include all types of sensors for automotive applications (accelerometers and torque and tire pressure sensors), medical applications (chemical sensors), and industrial and commercial applications (vapor, humidity, temperature, and mass sensors). Some of these sensors are capable of being passively and wirelessly interrogated (no sensor power source required). The use of piezoelectrics in actuators is also a growth area and includes micromotors, ignitors, bimorphs, impact printer heads, speakers, microphones, precision positioning stages, microscale surgical tools,[52] pumps, inchworm stepping motors, and tip-positioning stages for scanning tunneling microscope systems.

The small strains, usually less than 0.1%, and high stresses (e.g., 35 MPa) generated by piezoelectric devices have spawned a diverse range of actuator applications, as noted above (see also Figure 8.23 below; see Figure 1.11 for a micropiezopump from Animas-Debiotech's). Bulk piezoelectric transducers have also been used for a long time in medical ultrasound, in both passive and active modes.[53] In the passive mode, the transducer acts as a sound receiver only; it converts mechanical sound energy into an electrical signal. In the active mode (the converse effect), the transducer is an active sound transmitter; electrical energy is transformed into mechanical sound energy. Mostly, piezoelectrics are used in a pulse echo mode in which the transducers are used to simultaneously perform active and passive functions. A sound wave is propagated into the surrounding medium, and a faint echo is received after a small time gap due to the acoustic mismatch between the interface materials. This principle is used for ultrasonic imaging applications. When used in the human body with a sound velocity of 1500 ms^{-1}, the resolution varies from 1 mm to 50 μm for the frequency range of 1.5–3 MHz. The maximum frequency is dictated by the attenuation in the body, that is, 0.5 dB/cm MHz.

In the following we detail four of the applications mentioned above, namely, micromotors, pumps, acoustic-based chemical sensors, and acoustic imaging.

Piezoelectric Motors

Piezoelectric actuators, as Flynn et al.[41] point out, offer significant advantages over both electrostatic (see above) and electromagnetic (see below) actuators. With motors, for example, the greater the energy density that can be stored in the gap between rotor and stator, the greater the potential for converting

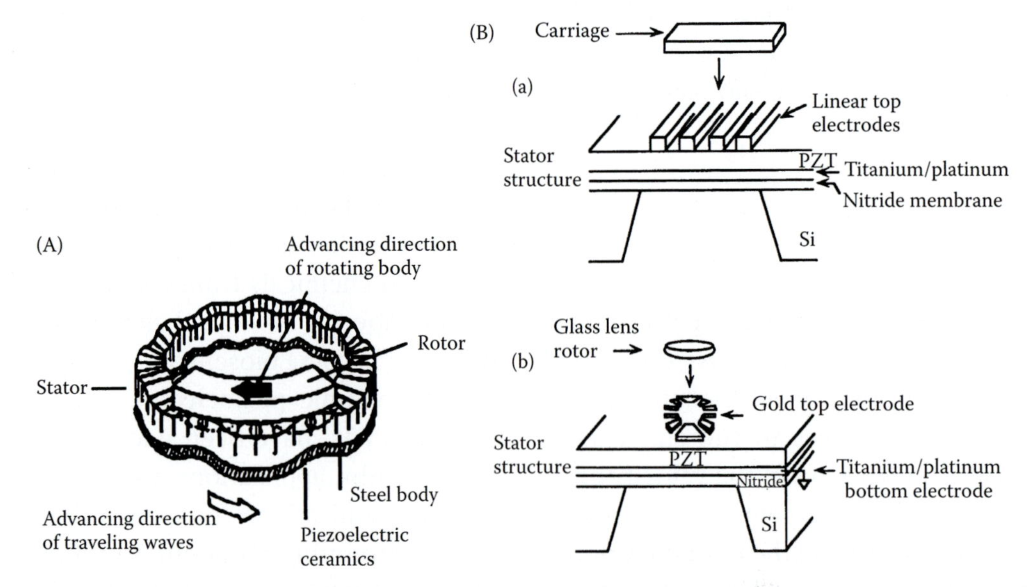

FIGURE 8.23 Piezoelectric motors. (A) Schematic of commercially available Panasonic ultrasonic motor: an electrically induced wave of mechanical deformation travels through a piezoelectric medium in the stator, moving the rotor body along through friction. (B) Thin-film lead zirconate titanate (PZT) motors. (a) Linear motor: the PZT stator is patterned onto a silicon membrane. The stator can deflect more because the nitride membrane is thin; titanium and platinum form the ground electrode, and the linear gold stripes are the top electrodes. A carriage is deposited on the stator by hand. (b) Rotary motor: identical to the linear motor except that the top electrodes are patterned in a circle. (After Flynn, A.M., L.S. Tavrow, S.F. Bart, R.A. Brooks, D.J. Ehrlich, K.R. Udayakumar, and L.E. Cross. 1992. Piezoelectric micromotors. *J Microelectromech Syst* 1:44.[41])

that energy to torque or useful work. In what follows, we compare the energy density one can store in the rotor/stator gap for various motor mechanisms. For electrostatic motors, the maximum energy density storable in the air gap, is determined by E_b, the maximum electric field before breakdown (approximately 3×10^8 Vm^{-1} for 1-μm gaps). From Chapter 7 we recall:

$$U'_m = \frac{\varepsilon_r \varepsilon_0 E_b^2}{2} \quad (7.36)$$

where $\varepsilon_r = \varepsilon_{air}$ represents the permittivity of air (almost equal to that of free space). This results in an energy density of 4.0×10^5 Jm^{-3}. With a piezoelectric motor made from ferroelectric material such as PZT, the energy density again is determined by the maximum electric field before breakdown. Thin film PZT can withstand high electric fields, about the same as an air gap ($E_b = 3 \times 10^8$ Vm^{-1}), but the dielectric constant is 3 orders of magnitude larger ($\varepsilon_{PZT} = 1300\ \varepsilon_{air}$) than air:

$$U'_m = \frac{\varepsilon_{pzt} \varepsilon_0 E_b^2}{2} \quad (8.61)$$

In principle then, the energy density also should be 3 orders of magnitude larger. In case of magnetic motors, the magnetic energy density, U'_m, stored in an air gap is given by:

$$U'_m = B^2/2\mu_0 \quad (8.62)$$

where B is the magnitude of the magnetic flux density in the gap and μ_0 the magnetic permeability of free space (see further below Equation 8.80). With magnetic actuators, the gap energy density, by pushing B into saturation (about 1.5 T), leads to 950,000 Jm^{-3}. The energy density stored in the rotor/stator gap for magnetic, electrostatic, piezoelectric, and SMA actuation (see below) are compared in Table 8.8.

Obviously, piezoelectric ultrasonic motors have an energy advantage. Other advantages, summarized from Flynn et al.,[41] include the following:

- Low voltages: no air gap is needed; mechanical forces are generated by applying a voltage directly across the piezoelectric film. With a 0.3-μm thin film, only a few volts are required, as opposed to hundreds of volts needed in air gap electrostatic motors.
- Geardown: motors can be fabricated without the need of a gearbox. Electrostatic wobble motors are also able to produce an inherent gear reduction but do not have the high dielectric advantage.
- No levitation: with electrostatic motors, levitation and flatness are very important in obtaining good sliding motion of the rotor around the bearing. The piezoelectric motor depends on friction so that no levitation is required, and it can be freely sized.
- Axial coupling: electrostatic motors require axial symmetry around the bearing. Since height is difficult to obtain with most nontraditional micromachining techniques (except for LIGA and pseudo-LIGA methods), a limited area is available for energy transduction. With the piezoelectric traveling-wave motors, linear or rotary motors can be built. As the stator is flat, with the rotor sitting on top of it, planar technologies are very capable of creating extra area to couple power out.
- Rotor material: the rotor can be of any material. A very important factor is that the conductivity of fluids being pumped with such motors does not affect the device, whereas an

Table 8.8 Comparing Energy Densities for Magnetic, Shape Memory Alloy, Electrostatic, and Piezoelectric Actuation

Principle	Maximum Work Energy Density	Equation	Special Drive Condition
Magnetic	9.5×10^5 Jm^{-3}	½ B^2/μ_0	1.5 T
Shape memory alloy	10.4×10^6 Jm^{-3} (from stress-strain isotherms)		1.4 W mm^{-3}
Electrostatic	4×10^5 Jm^{-3}	½ $\varepsilon_0 \varepsilon_{air} Eb^2$	3×10^8 Vm^{-1}, 1-μm gap in air
Piezoelectric	5.2×10^7 Jm^{-3}	½ $\varepsilon_0 \varepsilon_{PZT} Eb^2$	3×10^8 Vm^{-1}, 1-μm-thick lead zirconate titanate film

electrostatic motor would be shunted by a conductive liquid.
- Holding torque: because they are based on friction, piezoelectric ultrasonic motors can maintain holding torque even in the absence of applied power.

A schematic of a commercially available piezoelectric motor from Panasonic, which uses an ultrasonic traveling wave, is shown in Figure 8.23A. Two bulk ceramic PZT layers, each segmented with alternating poled regions, are placed on top of one another. For a given polarity of the applied bias, one segment contracts and the neighboring segment expands. With the two ceramic pieces put on top of each other with their segmented areas out of phase, any point on the stator then moves with the rotor being pulled along the stator surface through frictional coupling.

Thin film PZT linear and rotary motors by Flynn et al.[41] and Udayakumar et al.[54] are illustrated in Figure 8.23B. The stators in both cases are microfabricated using lithography techniques. Carriages and rotors may be made out of any type of material (in the rotational case, a glass lens was used for a rotor). The work demonstrates that, for a 5-V excitation, 1.6×10^{-12} N/m/V^2 normalized torque could be achieved, as opposed to 1.4×10^{-15} N/m/V^2 for electrostatic motors operating at 100 V.

The piezoelectric effect scales down with the bulk of the material; therefore, miniaturization opportunities are limited, and hybrid-type microactuators, where larger pieces of piezoelectric material are glued onto a substrate, are sometimes more reasonable. This has been the approach followed, for example, in the fabrication of piezoelectric micropumps (see below and Figure 1.11),[55,56] linear stepper motor,[57] and microvalves.[58]

Piezoelectric Pumping

In a tube vibrating with acoustic energy, an axially directed, acoustic-streaming force is generated along the inner surface.[59] This driving force can be generated along the entire length of a narrow tube to create a distributed pump that moves fluids without an externally applied pressure. The pump surface area and the driving force per unit volume of enclosed fluid increase as the tube diameter decreases

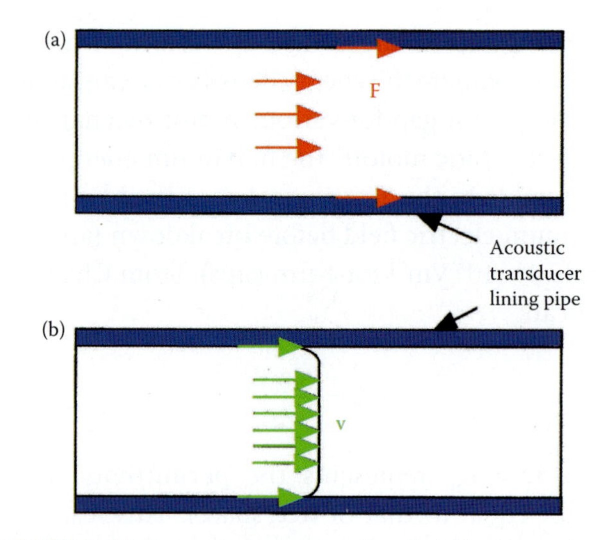

FIGURE 8.24 Acoustic streaming in a fluid inside a pipe. (a) Streaming force generated by an axially directed, traveling acoustic wave that grazes the inside walls of the tube. (b) The resulting steady state velocity distribution (blunt flow).

(pumping force $F \sim l^{-1}$). The axially directed, steady state force acting on the fluid is proportional to the square of the acoustic-wave amplitude and diminishes exponentially with distance from the wall. This force is sketched in Figure 8.24a. Because the force reaches a maximum near the walls of the tube, blunt flow is established in the fluid, as shown in Figure 8.24b, establishing an attractive technology to choose for miniaturization.

Unique properties of acoustic streaming pumps include the ability to pump without directly contacting the fluid so that a wide variety of liquids and gases may be pumped, and a simple compact structure with zero internal dead-volume. Tens of individual acoustic pumps could be operated in coordination to make an active network of interconnecting channels. One implementation of a piezopump is based on a flexural plate wave (FPW) delay line (see also Figure 8.21). An FPW device consists of a thin (~1 μm thick) membrane supported on all sides by a silicon chip. The membrane is coated with a piezoelectric film (say, ZnO) and aluminum interdigital transducers (IDTs). An RF voltage is applied to one IDT, producing mechanical stress in the piezoelectric layer, which generates flexural acoustic waves in the membrane (see Figure 8.25). The first IDT thus provides the electric field necessary to displace the substrate and form an acoustic wave. This wave propagates

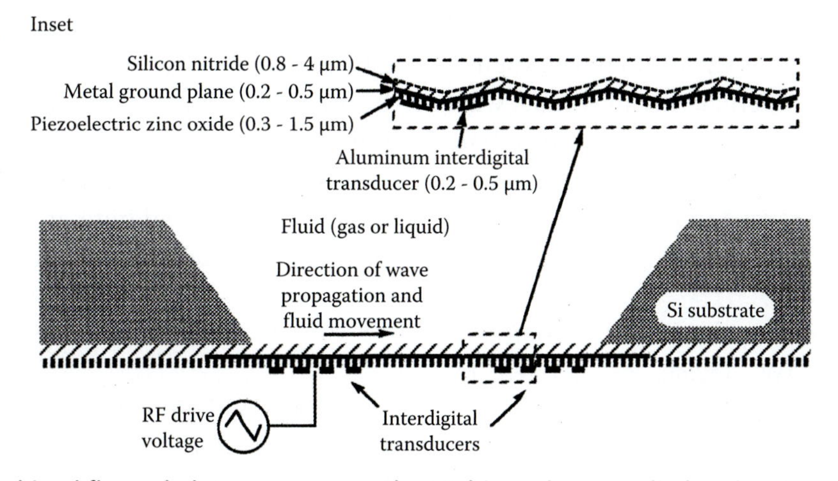

FIGURE 8.25 Micromachined flexural plate wave pump. The RF drive voltage applied to the piezoelectric interdigital transducers induces the flexural wave motion shown in the inset. Fluid motion is induced near the surface by acoustic stress. (Courtesy of Dr. R. White, University of California, Berkeley.)

through the substrate, where it is converted back to an electric field at the second IDT. Magnified, these waves look like the ripples in a flag waving in the breeze, as sketched in the inset of Figure 8.25. This wave motion induces acoustic streaming in the fluid next to the membrane. Wave propagation and fluid velocity both operate from left to right as shown in the figure. The structure generates large acoustic-wave amplitudes and is particularly effective for producing streaming; it has been proven to move air at speeds of 30 mm/s and water at 0.3 mm/s with only 5 V of RF drive.[60,61] Moroney et al. both translated and rotated small polysilicon blocks (2 μm thick and 50 μm square to 250 by 500 μm) using traveling ultrasonic flexural waves.[60] Acoustic streaming has also been established with the related SAW delay line.[62] As shown in Figure 8.20, a SAW device is fabricated by depositing interdigitated electrodes on the surface of a piezoelectric (typically $LiNbO_3$, $LiTaO_3$, and more recently also PZT thin films). An elastic wave generated at the input IDT travels along the surface and is detected by the output IDT. Besides their use for delay lines and filters in mobile phones, television and microwave communications, SAW devices have been used for many chemical sensor applications (see next) and in streaming experiments as discussed here.[63] Miyazaki et al., using a thin metal pipe flattened and bonded to a piezoelectric plate, produced a head pressure of about 10 mmH_2O and a flow rate of 0.02 cm^3/s at 40 V, peak to peak, applied to the piezoelectric ceramic plate.[62]

Acoustic Sensing of Mass

Introduction

In this section we say a few words about chemical and biological sensing by weighing with acoustic devices. Loading it directly onto a piezoelectric material can sense the mass of an analyte; alternatively, a thin layer of piezoelectric material might drive another resonator structure on which the mass accumulates as in microcantilever beams and resonator-comb drives. The main difficulty with mass sensors is achieving selectivity. In these applications, the adsorption of chemicals of interest is usually accomplished through polymer layers as sensitive coatings. One approach to improve selectivity further is to fabricate an array of sensors, each having a different partially selective film and using a pattern recognition algorithm to identify the species present (see artificial nose in Chapter 10). Another approach is to couple a front end to the sensor, such as a gas chromatograph, which separates the species and enables identification based on time of arrival at the sensor. These devices typically offer a signal-to-noise (S/N) ratio on the order of 100, with limits of detection (LOD) around 10 ppm. Mass sensitive piezotransducers used for chemical sensing are categorized by operating principle as bulk quartz resonators (BAWs), FBARs, SAW devices, FPW devices, cantilever beam resonators (see Example 8.1), and comb drive resonators.

As we saw above, all acoustic sensors will function well in gaseous or vacuum environments, but only a subset operates efficiently in contact with liquids. The shear horizontal wave does not radiate appreciable energy into liquids, allowing liquid operation without excessive damping.

Mass-Sensing BAW Devices

The mass sensitivity of a quartz crystal microbalance (QCM) or a BAW device, as derived by Sauerbrey[64] in 1959, is given by:

$$\Delta f = -\frac{2f_0^2}{Av_q\rho_q}\Delta m = -\frac{Cf_0^2}{A}\Delta m \qquad (8.63)$$

with:

$$C = \frac{2}{v_q\rho_q}$$

where Δf = frequency shift due to the added mass in hertz

C = a constant

f_0 = fundamental frequency of the oscillator in Hz

$\Delta m/A$ = surface mass loading in kg/m^2

v_q = shear wave velocity in quartz (= 3340 ms^{-1})

ρ_q = density of quartz crystal (= 2648 kgm^{-3})

Assumptions in the derivation of this equation are that the sensor film is thin and rigidly coupled to the resonator surface, the density and acoustic properties of the film are identical to that of the quartz resonator (for very thin coatings, the influence of the coating layer on the acoustic properties is assumed to be negligible), the film is uniform and evenly coated onto the mass-sensitive electrode surface, no mechanical stress is applied to the crystal by the film, and no interfacial slip occurs at the quartz/film boundary. The equation does not apply to non-Newtonian fluids, viscoelastic films, thick films, and liquid environment. In essence, this treats any added layer on the device as being composed of quartz; that is, the new film is considered as extending the acoustic wavelength of the device. The equation predicts that the sensitivity of the BAW sensor is proportional to the square of the natural frequency. In practice, though, the natural frequency is limited since it cannot increase without decreasing the thickness of the crystal. A practical lower thickness of about 150 μm imposes a frequency upper limit of approximately 10 MHz. This frequency determines the maximum sensitivity, which based on this equation is calculated to be 0.23 kHz/μg.

In order to compare the various gravimetric sensing devices reviewed here, it is convenient to define the mass sensitivity of a sensor as follows:

$$S_m = \lim_{\Delta m \to 0} \frac{1}{f}\frac{\Delta f}{\Delta m} = \frac{1}{f}\frac{df}{dm} \qquad (8.64)$$

where Δm and dm are normalized to the active sensor area of the device. The sensitivity is thus the fractional change of the resonant frequency of the structure with addition of mass to the sensor. Another important characterization is the sensors's minimum detection limits. If the limit of mass resolution is defined as a mass sensitivity resulting in a frequency shift three times larger than the oscillator fluctuation (which is say 0.1 Hz), the minimum detectable mass density (MDMD) may be derived by rearranging Equation 8.64 as follows:

$$\Delta m_{min} = \frac{1}{S_m}\frac{\Delta f_{min}}{f} \qquad (8.65)$$

where Δm_{min} and Δf_{min} (say 0.3 Hz) are the MDMD and minimum detectable frequency change, respectively.

Based on Equation 8.64, for a thickness shear mode (TSM) device (a QCM), the theoretical sensitivity, S_m, varies as $1/\rho_q d$, where d is the device thickness. In the case of a 1-cm by 1-cm device operating at 6 MHz and with a typical thickness (d), an MDMD of 5 ng/cm^2 is calculated (Equation 8.65). This result is shown in Table 8.9.

Thin-film bulk acoustic resonators combine very high frequencies (GHz) with very thin deposited piezoelectrics (~1 μm or less). White and coworkers made an aerosol monitor using an FBAR device (see Figure 8.19) and calculated a theoretical mass limit of detection of 0.001 ng for it.[65] An FBAR is not only much more sensitive than a QCM, it is also significantly smaller.

Mass-Sensing SAW Devices

In the previous section, we saw that thinner BAW devices operate at higher frequencies and increase the mass sensitivity. However, thinning down

Table 8.9 Gravimetric Sensitivities of Acoustic Sensors

Sensor	Device Description	Theoretical Mass Sensitivity, S_m (cm²/g)	Typical Operating frequency (MHz)	Theoretical Mass Limit of Detection (ng)	Typical Size
Thickness shear mode or quartz crystal microbalance	AT-cut quartz	$-1/\rho d$	6	5	1 × 1 cm
Surface acoustic wave	ST-cut quartz	$-K(\sigma)/\rho\lambda$	20–500	0.1	2 × 6 mm
Flexural plate wave	ZnO, SiN, Al	$-1/2\rho d$	2–5	0.002	1 × 6 mm
Film bulk acoustic resonator	AlN, SiN, Al, Au	$-1/\rho d$	1000–2000	0.001	100 × 100 μm
Microcantilever with mass loading distributed evenly over the cantilever surface		$S_m \sim 1/\rho d$	5–0.02	0.02	Micrometer-size (see Figure 8.77, where t is used for thickness)

Sources: Ballantine, D.S., R.M. White, S.L. Martin, A.J. Ricco, E.T. Zellers, G.C. Frye, and H. Wohltjen. 1997. *Acoustic wave sensors: Theory, design, and physico-chemical applications*. San Diego, CA: Academic Press;[63] and Kobrin, P., C. Seabury, A. Harker, and R. O'Toole. 1999. Thin-film resonant chemical sensor with resonant acoustic isolator. US Patent 5936150.[66]

quartz plates beyond the normal range results in fragile devices that are difficult to manufacture and to handle. For this reason, SAW devices have been investigated for gravimetry, typically operating at 100 MHz. Chemical vapor sensors based on SAW devices were first reported in 1979.[67] Most of them rely on the mass sensitivity of the detector, in conjunction with a chemically selective coating that absorbs the vapors of interest and results in an increased mass loading of the device. One disadvantage of the SAW devices is that Rayleigh waves are surface-normal waves, making them poorly suited for liquid sensing. When a SAW sensor is contacted by a liquid, the resulting compressional waves cause an excessive attenuation of the surface wave.

Often one works with two SAW devices, one of which is used as a reference, effectively minimizing the effects of temperature variations. As we saw above, SAWs are sound waves that travel parallel to the surface with their displacement amplitude decaying into the material so that they are confined to within roughly one wavelength of the surface. Interdigitated electrodes deposited on the piezoelectric material generate the acoustic waves and receive the signal. The pitch of the transducer defines the wavelength. SAW detectors, as shown in Figures 8.20 and 8.26, typically have a higher natural frequency than BAW detectors (30–500 MHz vs. a few MHz), and since signal strength is proportional to the fundamental frequency squared (Equation 8.63), they are more sensitive than quartz crystal microbalances. For the detailed derivation of the sensitivity S_m and MDMD for a SAW device, consult Ballantine et al.[63] They calculated $S_m = -K(\sigma)/\rho\lambda$, where λ is the wavelength and $K(\sigma)$ is a function of the Poisson ratio, σ. As reflected in Table 8.9, the mass sensitivity of a SAW device might be up to 50 times higher than that for a QCM, while a SAW device is typically somewhat smaller in size.

The superior sensitivity of a SAW device over a QCM may also be understood by recognizing that the sensitivity of the acoustic mass sensor is proportional to the amount of acoustic energy in the propagation path that is being perturbed. Bulk acoustic wave sensors typically disperse the energy from the sensing surface through the whole bulk of the material all the way to the other surface. This distribution of energy minimizes the energy density on the surface, which is where the sensing is done. SAW sensors, conversely, focus their energy on the surface, tending to make them more sensitive. However, SAW detectors are susceptible to changes in stiffness, dielectric constant, and relative permittivity relative to the surface mass, which can degrade sensitivity. BAW detectors have the principal advantage of negligible influences from conductivity and permittivity.

For both TSM and SAW sensors, to obtain higher sensitivity, one must increase the operating frequency.

SAW Example Application: Dew Point/Humidity Sensor

If a SAW sensor is temperature controlled and exposed to the ambient atmosphere, water will

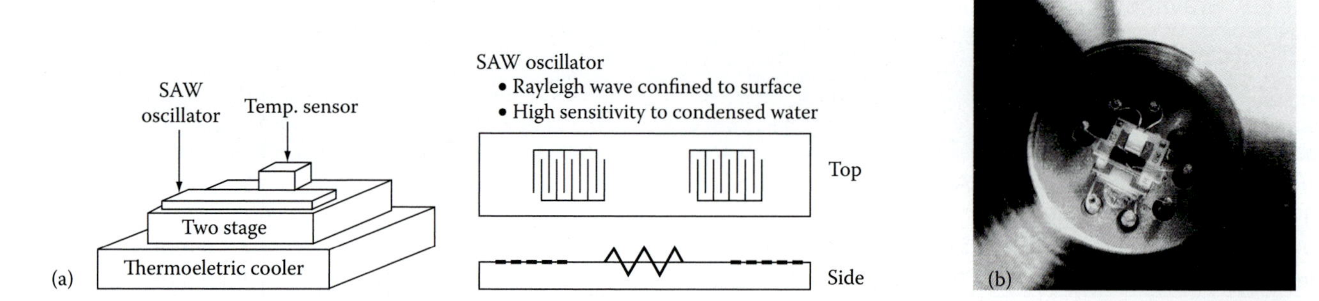

FIGURE 8.26 Surface acoustic wave (SAW) hygrometer. (a) schematic of the direct dew point sensor.[68] (Courtesy of Dr. M. Hoenk, Jet Propulsion Laboratory, Pasadena, CA.) (b) photograph of SAW hygrometer. The SAW device (long, rectangular bar) is mounted next to a platinum temperature sensor (square white package), on top of a miniature two-stage thermoelectric cooler. The device size is approximately 1 × 1 cm.

condense on it at the dew point temperature, making it an effective dew point sensor. Current commercial instruments for high-precision dew point measurements are based on optical techniques, which have cost, contamination, accuracy, sensitivity, and long-term stability issues. At Jet Propulsion Laboratory, as part of a Micro Weather Station development program, a new microhygrometer operating on accurate dew point measuring principles was demonstrated. A schematic illustrating the SAW hygrometer is shown in Figure 8.26. This device precisely measures the dew point/frost point temperature of ambient air. A quartz SAW microsensor detects the presence of thin layers of water on its surface as its temperature is varied by a miniature thermoelectric cooler. A fast digital feedback loop is used to control the sensor's temperature such that the amount of water on the sensor is held constant. This temperature is the dew point or frost point. This approach is functionally similar to the conventional chilled mirror hygrometers. However, the SAW moisture sensor provides much higher sensitivity and faster response for small amounts of water than does the optical detectors used in conventional hygrometers. This advantage, coupled with a high-performance digital feedback loop and precise measurements of the sensor temperature, enables accurate measurements with a response over an order of magnitude faster than conventional airborne hygrometers.

Mass-Sensing FPW Devices

FPW devices with sorbent polymer films respond to vapors in a manner similar to SAW devices coated with the same polymer. The FPW vapor sensor, however, is slightly smaller (see Table 8.9) and offers lower absolute noise levels and hence lower vapor detection limits. Earlier we listed other advantages associated with using an FPW or Lamb wave device for mass sensing. Lamb waves can propagate along a substrate even when immersed in water. Even though particle displacement is in the direction normal to the substrate, the low velocity of the acoustic waves preclude them from leaking into the environment. The accompanying electronics necessary for Lamb wave device applications also work at lower frequencies than those of SAW devices and are thus easier to design and cheaper to implement. Most importantly, as shown in Table 8.9, the mass sensitivity of a Lamb wave sensor is considerably greater than that of a SAW or Rayleigh wave sensor. The sensitivity calculated by Ballantine et al. for an FPW is given as $S_m \sim -1/2\rho d$.[63] In an FPW device, higher sensitivity is reached by reducing the plate thickness, d (this may be made less than a micrometer!). Mass sensitivity increases as membranes thickness decreases, and frequency decreases at the same time. This scaling law is in contrast to SAW devices, where mass sensitivity increases as frequency increases.

Acoustic Imaging

There is substantial interest in two-dimensional arrays of acoustic transducers for three-dimensional acoustic imaging. Such arrays may be used in handheld diver's sonar, medical ultrasound imaging, and

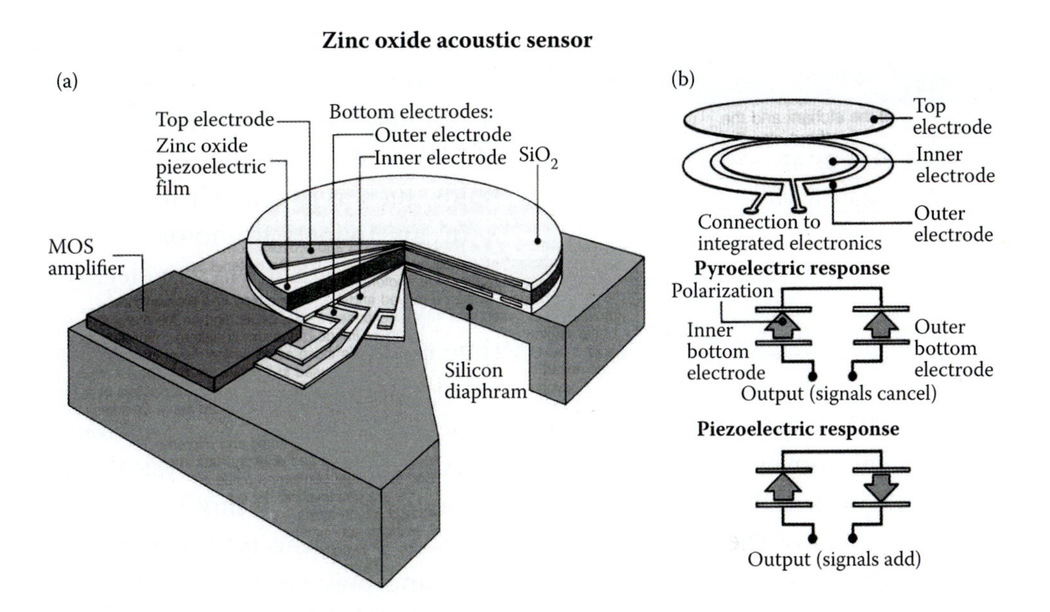

FIGURE 8.27 Zinc oxide acoustic sensor. (a) Two parallel plate electrodes in Honeywell's acoustic sensor act as capacitors. (b) Voltages due to temperature variations (pyroelectric effect) on zinc oxide film cancel, while those due to pressure add, doubling the output. (After Allen, R. 1984. Sensors in silicon. *High Technology,* September: 43–81.[70])

nondestructive testing. Bernstein et al.[69] microfabricated a 16 × 16 array of PZT pixels (3 μm thick and 0.4 × 0.4 mm each) for imaging purposes. The PZT pixels were deposited by sol-gel technology with polyimide in between the pixels. In a novel approach, the ferroelectric film was poled laterally (that is, in the plane of the deposited film) so that much larger interelectrode distances became possible. Instead of using metal electrodes on opposite sides of the thin PZT film, metal electrodes with a 30-μm spacing on the top side of the film resulted in a 30-fold increase in voltage sensitivity. With a stress applied in the same direction as the poling, there was a further gain of a factor of 2 due to a higher ferroelectric coupling constant in that direction. In total, a 30 dB improved sensitivity, compared with conventional polarizing across the thickness of the PZT film, was measured.[69] A commercially available acoustic sensor from Honeywell Corporation is shown in Figure 8.27.[70] The microphone chip illustrated incorporates ZnO deposited on a silicon substrate. A unique annular electrode design permits the parasitic signal in the zinc oxide film to be canceled because of temperature variations while doubling the output because of pressure (see Figure 8.27b).

Cast or spun-on and poled polymeric films of PVDF can also be employed to integrate piezoelectrics inexpensively into silicon micromachines. Gluing of commercial PVDF film with a urethane adhesive also works. Kolesar and Dyson,[71] for example, realized a two-dimensional electrically multiplexed tactile sensor this way. A 40-μm-thick PVDF film was glued onto an 8 × 8 array of taxel electrodes (400 × 400 μm each) electrically coupled to a set of sixty-four MOSFET amplifiers located around the periphery of this array. The response of the tactile sensor was found to be linear for loads between 0.008 and 1.35 N. Crude imaging of the applied loads also was demonstrated.

Piezoelectricity Summary

Some pros for using piezoelectricity as an actuator mechanism are high-energy density/bandwidth and lower operating voltages than electrostatics. Some cons are fabrication complexity and reproducibility of the actuator characteristics, and at the microscale, piezoelectrics provide very limited strain/displacement.

In Table 8.10 (see also Figure 8.14) we summarize some of the salient points associated with piezoelectric actuators.

Dielectric Induction

In the case of dielectric induction effects on particles in liquids and on fluids themselves, one speaks about

TABLE 8.10 Salient Features of Piezoelectric Actuators

1. Stroke: small, 10–2 μm
2. Speed: >100 kHz (see Figure 8.14)
3. Force density: low (see Figure 8.14)
4. Complex to fabricate
5. Voltage: ~10–100 V
6. Footprint: 1–10 μm
7. Maximum energy density, U_m (in J/m^3): 52 × 106 Jm^{-3}

dielectrophoresis and EHD, respectively. The word *dielectrophoresis* implies the movement of particles through induction, while in *electrohydrodynamics* the emphasis is on the movement of the fluid itself. In dielectrophoresis, the induced charges can be attractive or repulsive with respect to the inducing field, depending on the dielectric constants of the particle and the medium in which the particle is embedded. This principle can be used for sorting, moving, and separating small objects such as biological cells.[72–74] The phenomenon is detailed in Volume I, Chapter 6, on fluidics.

In EHD pumping, fluid forces are generated by the interaction of electric fields and charges in the fluid. In such pumps electrodes are regularly spaced along a microchannel and each electrode is phase shifted with respect to the next. In contrast to forces generated by mechanical pumping using an impeller or bellows, EHD pumping requires no moving parts.[75–77] The interaction of electrical fields with induced electrical charge in the fluid yields a force that then transfers momentum to the fluid.[78] The EHD effect has also been put to use in mixing of fluids of different conductivities. Choi mixed DI water and saline solution in a 10-pL volume this way.[79] Surface charges induced at the interface of the two liquid streams react with the applied electric fields to generate shear forces leading to rapid mixing of the streams passing the electrodes. Reasonable electric fields can only be built up with acceptable voltage levels within microstructures; therefore, this principle becomes more and more effective with decreasing dimensions. A requirement for the continued existence of free charge is the presence of a spatial gradient in the conductivity or permittivity. Free charge injected into a region without gradient will relax in a time characterized by the relaxation time of the charge. One way to accomplish a conductivity gradient in the bulk of a slightly conductive fluid is by imposing a temperature gradient. A limitation of EHD is its reduced effectiveness with conductive fluids, such as in biological environments.

In an induction motor, localized charges are induced in the rotor by the electrostatic field generated by the stator electrodes. Since a finite relaxation time is needed for this excess charge to be diffused, a driving torque is generated. An induction motor is driven by a traveling wave of voltage and uses charge relaxation rather than physically salient poles in a variable-capacitance motor as shown in Figure 8.6b.[80] Consequently, the rotor can be a smooth uniform plate and may even be a fluid.[81] In the case of the rotor blades, fabrication involves some rather tricky planarization steps. The conductivity of the induction motor materials, on the other hand, strongly affects performance, so fabrication difficulties associated with variable-capacitance motors with rotor blades are traded for those associated with conductivity control in induction motors.[82]

Electrostriction

Electrostriction is a property of all dielectric materials caused by the presence of randomly aligned electrical domains within the material. With an electric field applied, the opposite sides of the domains become differently charged and attract each other, reducing material thickness in the direction of the applied field (and increasing thickness in the orthogonal directions in accordance with the Poisson's ratio. The resulting strain (ratio of deformation to the original dimension) is proportional to the square of the polarization. Reversal of the electric field does not reverse the direction of the deformation. In other words, electrostrictive materials—basically all dielectrics—develop mechanical deformations when subjected to an external electrical field; that is, they show an increase in length parallel to the field. Even apparently symmetric crystals of nonconducting materials exhibit this effect, but it is usually smaller than the piezoelectric effect.[83] The phenomenon is attributed to the rotation of small electrical domains in the material upon exposure to a field. In the absence of the field,

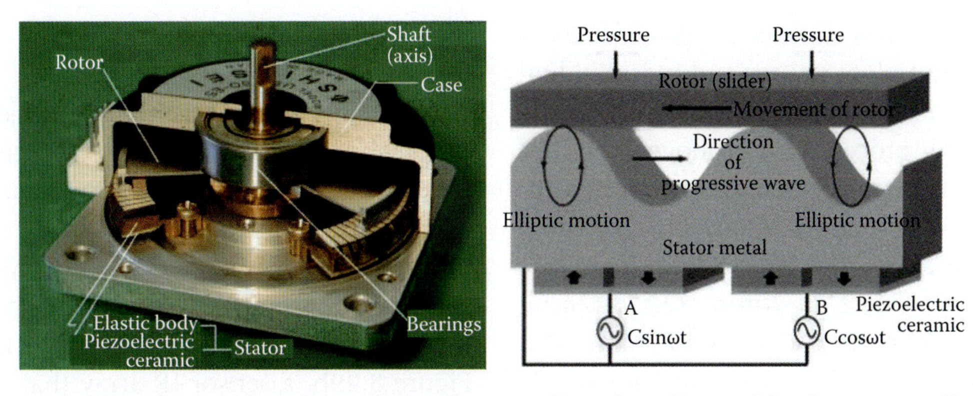

FIGURE 8.28 A ceramic piezoelectrostrictor in an ultrasonic motor. (From http://www.shinsei-motor.com/English/index.html.)

the domains are oriented randomly. The alignment of the electrical domains parallel to the electrical field results in the development of a deformation field in the electrostrictive material. Usually, the effect is small, but a few materials, such as certain titanates and zirconates, show a large effect, in which case they are ferroelectric and exhibit properties analogous to ferromagnetic materials. In both groups of materials, spontaneous polarization occurs within small regions or domains. The electric dipoles in ferroelectric materials or the elementary magnets in ferromagnetic materials are in parallel alignment within each domain. Like piezoelectrics, electrostrictors are often based on the perovskite crystal structure. Two examples of electrostrictive materials are barium titanate and lead magnesium niobate (PMN).[32,33] In the case of PMN, electrostriction gives rise to a mechanical deformation comparable to that induced by the piezoelectric effect in PZT. Fripp and Hagood compared the relative merits of piezoceramics and electrostrictors for vibration suppression.[83] Uchino made astronomy mirrors, the shape of which is continuously adjustable by bonding silvered glass to a series of electrostrictive layers that enable one to make minute adjustments to the mirror's shape in response to an applied field.[84]

A ceramic piezoelectrostrictor is used in the motor shown in Figure 8.28. This ultrasonic motor uses ultrasonic vibration (20 kHz or above) as the driving source and comprises a stator (a piezoelectrostrictor ceramic glued to a metal as the elastic body) and a rotor (the dynamic body). The piezoelectrostrictor ceramic is glued onto one side of the metal stator. The rotor is pressed against the opposite stator metal surface. Comb tooth-shaped grooves are created on the metal stator surface facing the rotor. The rotor is pressed tightly against this comb-toothed side of the stator metal surface. The ceramic element generates ultrasonic vibration when a specific high-frequency voltage is applied and the piezoelectric ceramic itself is expanding and contracting. The ultrasonic wave, i.e., swelling and falling of the stator surface, travels along the surface of the stator in a single direction. The ultrasonic vibration that travels in this manner is called the *progressive wave*. The comb tooth-shaped grooves on the stator surface are devised to make the amplitude of the elliptic motion large and to reduce abrasion. As the progressive wave travels and undulates through the contact, an elliptic motion is generated. The locus of the elliptic motion points in the opposite direction of the progressive wave traveling on the stator surface.

Pyroelectricity

The spontaneous polarization in ferrorelectrics is temperature sensitive. This is called the *pyroelectric effect*, which was discovered in tourmaline [(Na,Ca)(Li,Mg,Al)(Al,Fe,Mn)$_6$ $(BO_3)_3$ $(Si_6O_{18})(OH_4)$], the so-called *Ceylon magnet* (class 3m) by Teophrast in 314 BC and first recorded in the scientific literature in 1824 by Brewster.[32,85] The word is derived from the Greek for "heat electricity" and is the development of electric polarization in classes of noncentrosymmetric crystals subjected to a temperature change. Electrical polarity in a biological material is produced by a change in temperature. Pyroelectricity is probably a basic physical property of all living organisms. First discovered in 1966 in tendon and bone, it has since been shown to exist in most animal and plant tissues and in individual cells. Pyroelectricity appears to play a fundamental part in the growth processes (morphogenesis) and in physiological functions (such as sensory perception) of organisms. From Volume I, Chapter 2, we know that there are twenty-one classes lacking a center of symmetry: 1, 2, m, 2mm, 222, 3, 3mm, 32, 4, –4, 4mm, –42m, 422, 6, –6, 6mm, 62m, 622, 23, 43m, and 432. Among materials electrically polarized because of crystal structure, pyroelectric materials are such that they become electrically polarized with changes in temperature, and the amount of electric charge produced varies according to temperature change. The phenomenon is present in crystallized dielectrics with one or more polar axes of symmetry. In polar crystals, the centroids of positive and negative charges in a unit cell do not coincide. As a result, nonzero polarization, P, of the crystal occurs. Pyroelectricity is thus only possible in crystals lacking a center of symmetry and also have a unique polar axis: 1, 2, m, 2mm, 3, 3mm, 4, 4mm, 6, and 6mm. The polarization, P, of the crystal changes with temperature, and this is the result of a structure change with temperature. While all pyroelectric materials are piezoelectric, the opposite is not true; for example, quartz is piezoelectric but not pyroelectric. Both zinc oxide, with its wurtzite crystal structure, and PZT, with a ferroelectric perovskite structure, feature an intrinsic dipole moment and show pyroelectricity even in thin film form. In Volume I, Figure 2.10, we already illustrated the pyroelectric effect in $BaTiO_3$. The strongest pyroelectric effect for a thin film is measured with the sol-gel processed tetragonal $PbTiO_3$.[48,86] The pyroelectric constant, p, of these materials is defined as the differential change in polarization with temperature in the case of uniform heating, constant stress (σ), and low electric field (E) in the crystal (see also row 1 in Volume I, Table 2.1):

$$p = \left(\frac{\partial P}{\partial T}\right)_{\sigma, E = 0} \tag{8.66}$$

p is expressed in C/cm² K. Thin films of ZnO, PZT, and $PbTiO_3$ all show pyroelectric coefficients similar to the bulk material: $0.95–1.05 \times 10^{-9}$ C/cm² K for ZnO, $50–70 \times 10^{-9}$ for PZT, and $95–125 \times 10^{-9}$ for $PbTiO_3$.[48,86]

In the pyroelectric IR (for 10 μm) sensor configuration shown in Figure 8.29a, two sensing elements are connected in electrical opposition and one is shielded from the radiation source, to cancel out the effect of any drift in ambient temperature. A germanium window is used because it is opaque to visible light but transparent at wavelengths near 10 μm. In Figure 8.29b, a sensor IR array that was developed in the 1980s is shown. Detector elements made of lead strontium titanate (PST) or barium strontium titanate (BST) were commonly used. These materials have Curie temperatures close to room temperature. The devices operate at temperatures above the Curie level (in the so-called *paraelectric phase*), and an applied electric field induces the pyroelectric effect. Laser-assisted chemical etching produces an array with a large number of detector pixels. The elements on that array are soldered (or "bump bonded") to a silicon multiplexer. Contemporary versions have as many as 384 × 288 pixels, and a noise-equivalent temperature difference (NETD) is a measure of the detector-noise-limited sensitivity of single-band IR systems) as low as 75 mK. Micromachined thin-film elements with even lower NETDs are under development.

Single crystals of triglycine sulfate (TSG), $LiTaO_3$, and $(Sr,Ba)Nb_2O_6$ are widely used for heat-sensing applications. Using $LiTaO_3$, Hsieh et al.[87] produced sensitive, wide-range anemometers. In all thermal anemometers, the goal is to measure heat loss. Usually, this is accomplished by measuring the resistance of a temperature-sensitive resistor; but with a pyroelectric sensor, heat loss generates a current directly. Gas flows of 0.1 mL/min to over 20 L/min were detected. It should be remembered that only changing temperatures can be detected this way. For a crystal maintained at a constant temperature, the charges will decay rapidly.

Using a thin (<5 μm), freestanding membrane of poly(vinylidene fluoride-trifluoroethylene) [P(VDF/TrFE)] copolymer in a Si frame, Zavracky et al. made a pyroelectric IR detector array.[88] The freestanding film is provided with top and bottom semitransparent Cr/Au electrodes, and incident IR light produces a voltage between these electrodes. The advantage of

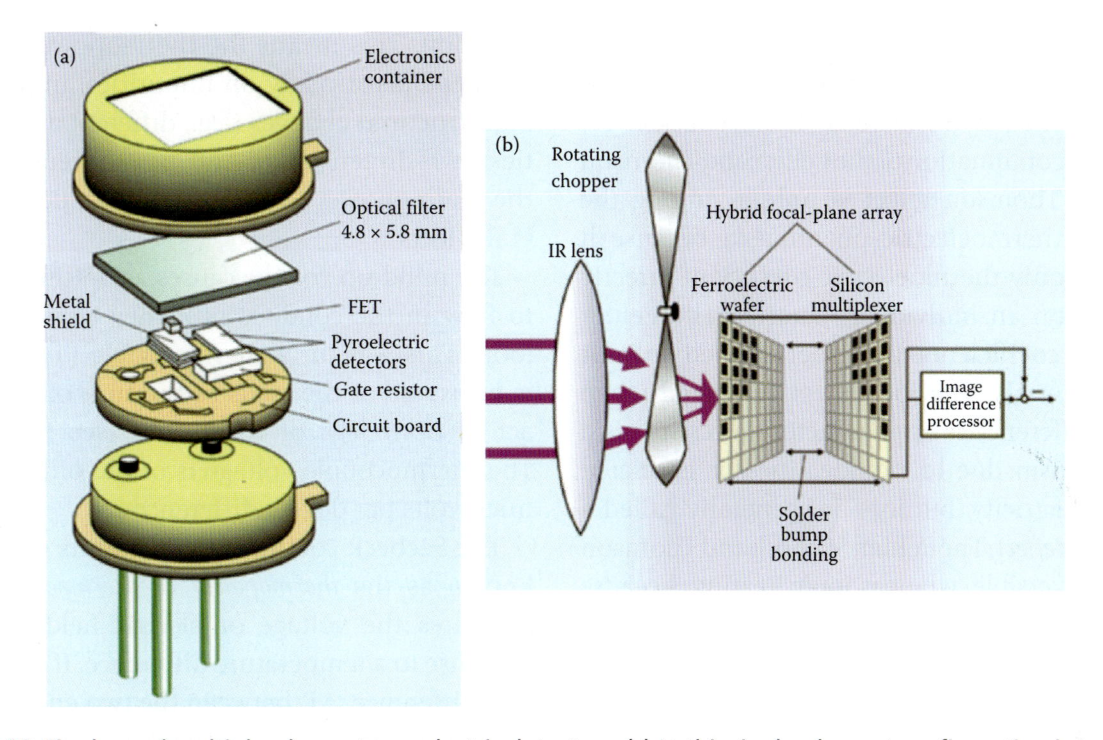

FIGURE 8.29 Single- and multiple-element pyroelectric detectors. (a) In this single-element configuration intended to monitor radiation with a wavelength near 10 μm, two lead titanate pyroelectric elements are used. One detector element is exposed to incoming radiation, another shielded under a metal strip. Both crystals are electrically connected with opposite polarities, to cancel out the effect of any drift in ambient temperature. (b) A ferroelectric-hybrid focal-plane device comprises a lens, usually made of germanium to block visible light, and an array of individual ferroelectric elements, each of which is bonded by tiny balls of solder to elements from a silicon multiplexer. The incident infrared radiation must be periodically blocked, here by a chopper, to ensure that a temperature variation is measured. (From http://www.physicstoday.org/vol-58/iss-8/p31.html.)

this polymeric pyroelectric is that it is well suited for microfabrication, since it can be applied to a substrate by spin on from a solvent-based precursor. Complementary metal-oxide-silicon (CMOS) devices fabricated in the Si frame can be used for amplification and signal processing. To prevent thermal conduction from the sensing film to the substrate, which limits sensitivity, the thin polymer film may be supported by a thin silicon nitride film.

Pyroelectric materials are used in remote control devices, night vision, and thermal sensors.

Thermoelectricity*

Introduction

Thermoelectricity covers three separate phenomena: the Seebeck effect, the Peltier effect, and the Thomson effect. The Seebeck phenomenon, discovered by the Estonian-German physicist Seebeck in 1821, is perhaps the best known, since it is used in thermocouples for the measurement of temperature differences. The effect is that a voltage is created in the presence of a temperature difference between two different metals or semiconductors. The measured voltage is proportional to the temperature difference, and the proportionality constant is known as the *Seebeck coefficient*, *S*.

Independently, the French physicist Peltier discovered the reverse effect, i.e., a current flow across a junction results into heating or cooling. In the absence of a magnetic field, the Seebeck and Peltier effects are the two most important thermoelectric effects.

The Thomson effect is the heating or cooling of a current-carrying conductor with a temperature gradient. Any current-carrying conductor (except for a superconductor) with a temperature difference between two points either absorbs or emits heat,

* Thermoelectric actuators could equally well be discussed under thermal actuators.

depending on the material. The Thomson effect was predicted and experimentally observed by William Thomson (Lord Kelvin) in 1851. The Seebeck effect constitutes a combination of the Peltier and Thomson effects. The Thomson effect is unique among the three main thermoelectric phenomena because it features the only thermoelectric coefficient directly measurable on an individual material. The Peltier and Seebeck coefficients can be determined only for pairs of materials. The Joule heat generated when a voltage difference is applied across a resistor is a loss mechanism due to nonidealities; it is related to thermoelectricity, but it is not generally called a *thermoelectric effect*. The Peltier-Seebeck and Thomson effects are reversible, whereas Joule heating, because of the second law of thermodynamics, cannot be reversible.

Mechanism behind Thermoelectricity

An applied temperature difference between two different metals or semiconductors causes charge carriers, electrons or holes, to diffuse from the hot side to the cold side, similar to the way a classical gas expands when heated. The diffusing charges are scattered by impurities, imperfections, and lattice vibrations (phonons), and since scattering is energy dependent, the hot and cold carriers diffuse at different rates. This creates a higher density of carriers at one end of the material, and the separation of charges creates an electric field. This electric field, however, opposes the uneven scattering of carriers, and an equilibrium is reached with the net number of carriers diffusing in one direction canceled by the net number of carriers moving in the opposite direction. This means the thermopower of a material depends strongly on impurities, imperfections, and structural changes (e.g., different dimensionalities, two-dimensional, and one-dimensional), and the thermopower of a material is a collection of several effects.

The build-up voltage causes a continuous current to flow in the conductors if they form a complete loop. This way, in a thermocouple (TC), a voltage is measured at the leads as the sum of the voltages across both legs of the device (see Figure 8.30). The thermocouple voltage is of the order of several microvolts per degree difference.

The Seebeck coefficient, S (in units of V/K), also known as the *thermopower* or *thermoelectric power*, measures the voltage or electric field induced in response to a temperature difference. If the temperature difference, ΔT, between the two ends of a material is small, then the thermopower of a material is defined as follows:

$$S = \frac{\Delta V}{\Delta T} \tag{8.67}$$

and a thermoelectric voltage ΔV is seen at the terminals. The thermopower S is an important material parameter and determines the efficiency of a thermoelectric material. A larger induced thermoelectric voltage for a given temperature gradient leads to a larger thermoelectric efficiency.

For the thermocouple in Figure 8.30, the measured electromotive force (EMF) is calculated as follows:

$$V = \int_{T_1}^{T_2} [S_A(T) - S_B(T)]dT \tag{8.68}$$

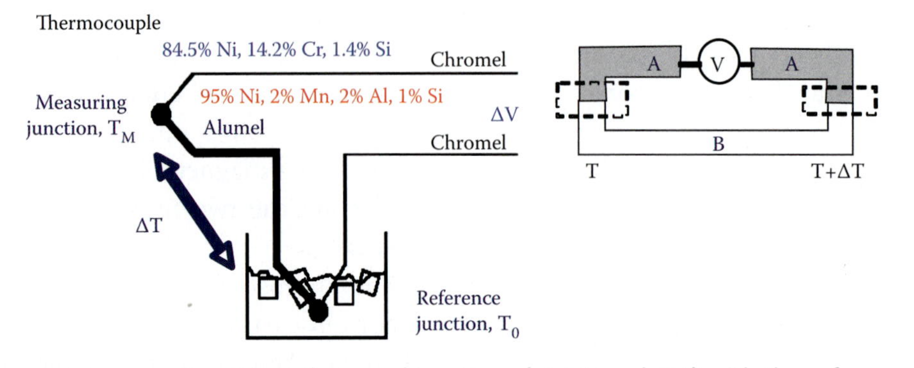

FIGURE 8.30 Thermocouple operation. Dissimilar metal junctions (at *A*/*B* and *B*/*A*) with the reference junction at T_0 (say, an ice bath) and the measuring junction at T_M, lead to a voltage reading *V*. (From a talk by Chien-Neng Liao, NTHU.)

where S_A and S_B are the Seebeck coefficients of the metals *A* and *B*, and $\Delta T = T_1 - T_2$ is the temperature differential between the two junctions. The Seebeck coefficients are nonlinear and depend on the conductors' absolute temperature, material, and molecular structure. A thermocouple can be used to measure a temperature difference directly, or to measure an absolute temperature, by setting one end to a known temperature. Several thermocouples in series are called a *thermopile*. The thermocouple can also be used to extract energy in applications where heat and temperature differences are available "for free" (see energy harvesting in Chapter 9). Thin-film thermocouples have been used, for example, to power a watch on the basis of the temperature difference between its cool front face and the warm skin contact. A disadvantage of this approach is the very low efficiency of the conversion.

Typically, a metal displays a small thermopower because it comes with half-filled bands. In contrast, semiconductors can be doped with an excess amount of electrons or holes and thus can have larger positive or negative values of the thermopower depending on the charge of the excess carriers.

The Peltier effect is illustrated in Figure 8.31. When charge carriers flow from one material into another in which their average transport energy is higher, they absorb thermal energy from the lattice, and this cools the junction between the two materials. One can think of this as the electron gas expanding in the energy space as they enter the second material. The thermoelectric process is reversible, so if the direction of the current is reversed, energy is released to the lattice, and there is heating at the junction.

The definition of the Peltier coefficient, Π_{AB}, is given as the ratio of heat absorption (J_Q) over current flow (J_E) (see Figure 8.31).

$$\Pi_{AB} = \frac{J_Q}{J_E} \tag{8.69}$$

As mentioned above, there is no direct experimental method to determine an absolute Seebeck coefficient or absolute Peltier coefficient for an individual material. However, in 1854, Thomson found two relationships, now called the *Thomson* or *Kelvin relationships*, relating the three thermoelectric coefficients. Therefore, only one can be considered unique. The absolute temperature, *T*; the Peltier coefficient, Π; and the Seebeck coefficient, *S*, are related by the first Thomson relation:

$$\Pi_{AB} = ST \tag{8.70}$$

In turn, *S* and *T* are related to the Thomson coefficient, μ, by the second Thomson relation as follows:

$$\mu = T\frac{dS}{dT} \tag{8.71}$$

The efficiency of a conventional thermoelectric cooler is controlled by a parameter *Z*, given by:

$$Z = \frac{S^2\sigma}{\kappa} \tag{8.72}$$

where σ and κ are the electrical and total thermal conductivities (with both lattice and electrons contributions) of the material, respectively, and *S* is the Seebeck coefficient (in μV/K). More commonly a *dimensionless figure of merit ZT* is defined by multiplying Z with the average temperature T. Larger values of ZT indicate a greater thermodynamic efficiency and *ZT* is a convenient figure for comparing the

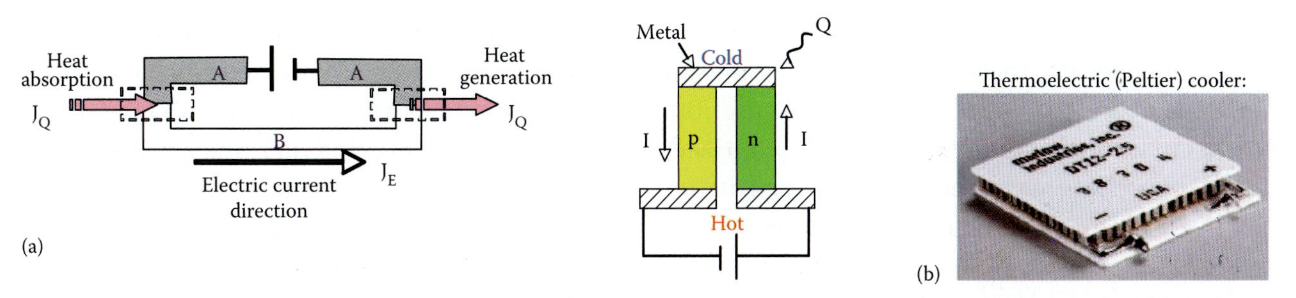

FIGURE 8.31 In a Peltier device, current flow across a junction of dissimilar metals causes a heat flux, thus cooling one side and heating the other. (a) Illustration of the Peltier effect. (b) Picture of a Peltier cooler. (From a talk by Chien-Neng Liao, NTHU.)

efficiency of devices using different materials. One can show that the efficiency for cooling, represented by Φ, and that of heating, symbolized by η, both increase with an increase of:

$$(1 + ZT)^{\frac{1}{2}} = \left(1 + \frac{S^2\sigma}{\kappa}T\right)^{\frac{1}{2}} \quad (8.73)$$

Consequently, requirements for a good thermoelectric (TE) material are a high figure of merit (*ZT*), i.e., a high Seebeck coefficient (S), a high electrical conductivity (σ), and a low thermal conductivity (κ). For good thermoelectric materials one has a *ZT* of ~1, and this remained the upper limit for many years. Thermoelectric compounds include bismuth telluride (Bi_2Te_3), used for refrigeration, bismuth antimony (BiSb), lead telluride (PbTe), and silicon germanium ($Si_{1-x}Ge_x$), used for power generation. These materials today are used for smaller cooling/heating applications; they are not highly efficient but are environmentally friendly and very reliable. On a larger scale, thermoelectric refrigeration would be very interesting as it involves no toxic chlorofluorocarbons (CFCs) and no moving parts and is quick and reliable. The biggest problem is that *ZT* has to be improved from ~0.9 to 3 or higher to become more commercially attractive [*ZT* needs to improve to >1.3 at 300 K for a major impact in electronics cooling and to around 2.5 for a revolutionary impact in air conditioning and power from waste heat (energy scavenging)]. The pioneering work of Hicks and Dresselhaus in 1993 renewed interest in thermoelectrics.[89] Dresselhaus et al. showed that lower-dimensional structures such as nanowires based on bismuth, BiSb, Bi_2Te_3, and SiGe nanotechnology could lead to a higher *ZT*. This group demonstrated that electrons in low-dimensional semiconductors such as quantum wells and wires exhibit an improved thermoelectric power factor. This is due to the fact that electron motion perpendicular to the potential barrier is quantized creating sharp features in the electronic density of states (DOS). The higher density of states at the Fermi level increases σ, and the increased phonon boundary scattering lowers κ (see also Volume I, Chapter 3 and Chapter 3, this volume). Both factors contribute to make *ZT* higher. In Figure 8.32, we show the thermoelectric figure-of-merit *ZT* of a Bi quantum well and quantum wire as a function of size (see also Chapter 3, this volume).

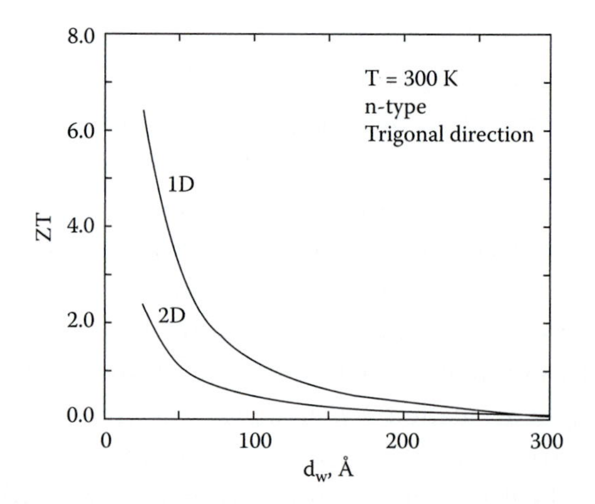

FIGURE 8.32 Thermoelectric figure of merit *ZT* of bismuth quantum well and quantum wire as a function of dimension.[89]

Because of the massive amount of processing know-how and the relatively large Seebeck coefficient, polysilicon is an attractive material for thermocouples. The Seebeck coefficient for 0.38-μm-thick n-polysilicon films with a sheet resistance of 30 Ω/□ is −100 μV/K; for n-polysilicon with a sheet resistance of 2600 Ω/□, it is −450 μV/K. For p-polysilicon with a 400 Ω/□ sheet resistance, it is 270 μV/K. Silicon might also start playing a more and more important role in heating and cooling applications. In 2008, a team from Caltech and a team from University of California, Berkeley, separately reported that they were able to increase silicon's ability to convert heat into electric current by as much as 100 times.[90,91] Bulk silicon is only an average material for thermoelectric conversion, because its thermal conductivity is too high; heat travels across it so well that it is difficult to create the necessary temperature gradient. Bulk silicon at room temperature has a *ZT* of only 0.01, but the Berkeley team increased that to 0.4, and the Caltech team increased it to 0.6. This puts silicon nanowires almost on par with bismuth telluride (see above). Making thermoelectric devices out of silicon, which is much more abundant, cheaper, and easier to handle than bismuth telluride, could help create a large market for these devices.

An excellent 2006 review on nanoscale thermal transport and microrefrigerators on a chip is by Dr. Ali Shakouri.[92]

Electrets

A permanently charged dielectric, an electret, forms an elegant base to improve upon electrostatic actuation by setting up a permanent electric field, thereby reducing or eliminating the need for a large applied bias. There are many different methods to form electrets from high-impedance dielectrics, such as the corona, the liquid contact, and the electron beam methods.[93,94] An electret-based micromachine may incorporate a capacitor charged because of the presence of the electret between the capacitor plates while one of the plates is sensitive to an external force (e.g., pressure). Micromachined electret microphones have been built this way and exhibit an open-circuit sensitivity of about 2.5 mV/μbar at 1 kHz.[95] By charging up SiO_2 with a 300-V corona, electrets in these subminiature microphones exhibit a charge decay time constant amounting to more than 100 years of expected operation. Similarly, an electret-based pressure sensor with a permanently charged polymer foil (commercially available Teflon FEP) as the electret has been manufactured. The foil was charged by electron beam exposure. A maximum sensitivity of 10mA/A/100 mmHg (i.e., the measured relative change of drain current), about ten times higher than the sensitivity of piezoresistive pressure sensors with comparable dimensions ($1 \times 2 \times 0.3$ mm), was determined.

Wolffenbuttel et al.[96] point out that an insulating, electret-implanted rotor circumvents two major limitations of electrostatic motors, namely, the relatively large voltages and the friction between the rotor and the stator during rotation. Replacing the polysilicon rotor in a motor as shown in Figure 8.6b allows propulsion at smaller drive voltages as well as electrostatic levitation and lateral alignment of the rotor, making it virtually contactless. Temperature and humidity sensitivity constitute two of the biggest drawbacks of employing electrets in actuators.

Earlier, we pointed out that bone has piezoelectric properties; Mascarenhas shows that bone can also be made into an electret,[97,98] and Fukada et al.[99] report that plastic electrets applied to bone produce alterations in bone growth.

Electrorheological Fluids

Introduction: Field-Responsive Fluids

Field-responsive fluids refer to colloidal dispersions whose rheological properties are modified under the influence of an external field, typically electric or magnetic. Upon application of such fields, induced dipoles in the dispersion cause the particles to form columns that align with the applied field. Field-responsive fluids are commonly classified as electrorheological (ER), magnetorheological (MR), and ferrofluids (FF). FF are a variant of MR fluids, having as main requirement particle size below 20 nm.

The idea of field-responsive fluids dates back to the eighteenth century. As early as 1779, Gowan Knight attempted to obtain a magnetic fluid by suspending iron fillings in water but the fillings would settle too quickly. In 1896, Duff[100] observed a reversible 0.5% change of viscosity in castor oil and glycerin when applying an electric field of 27 KV/cm. In 1947 Winslow discovered a similar but larger effect when dispersing fine dielectric powders in a nonconducting fluid[101] that he called *electroviscous.*[102] Later, this term was changed to *electrorheological.* A year later, Rabinow[103] published his discovery of a similar effect, but applying a magnetic field instead, giving birth to *magnetorheological fluids.* Since then, a considerable amount of research has been conducted on the topic. Many patents were filed in the 1950s, but the lack of commercial devices and probably the fact that most suspensions suffered from particle sedimentation dampened the initial enthusiasm. In the 1980s, research revived with ER fluids as the main focus since they were thought to be closer to commercial applications. This initial hunch proved to be wrong and MR fluids became the center of attention in the 1990s and eventually led to the application of MR fluids in commercial dampers, loudspeakers, and dynamic seals. Although the interest on ER fluids is still large, they have not yet been applied in commercial devices. The reasons for the switch from ER to MR devices are that the MR effect is an order of magnitude stronger than that of ER fluids and that the required magnetic field levels are easier to achieve.

Given the fact that it is easy to obtain large electric and magnetic fields in the microdomain, research on adapting ER and MR fluids to MEMS should follow soon.

Here we only discuss electrorheological fluids. MR fluids and FF are covered below, under magnetic actuators.

Electrorheological Fluids

ER fluids are a class of colloidal dispersions, usually dielectric particles in a nonconducting fluid, that exhibit large reversible changes in their viscosity or flow rate; that is, in their rheological behavior when subjected to external electrical fields (typically of 1–4 KV/mm). These changes in rheological behavior are typically manifested by a pseudo-phase change from the liquid to a viscous, semisolid, gel-like state and result in a dramatic increase in flow resistance (the viscosity might increase by several orders of magnitude). The total power required is rather low, and the response of ER fluids to an electrical stimulus is typically in the millisecond range. The electrical field induces dipoles in the particulate phase, that is, the fine dielectric powder suspended in nonconducting fluid, and forges an interaction between the individual particles. The induced dipole then aligns with the field to minimize dipole-dipole interaction energy. Eventually, columnar structures like those presented in Figure 8.33 are formed.[104,105]

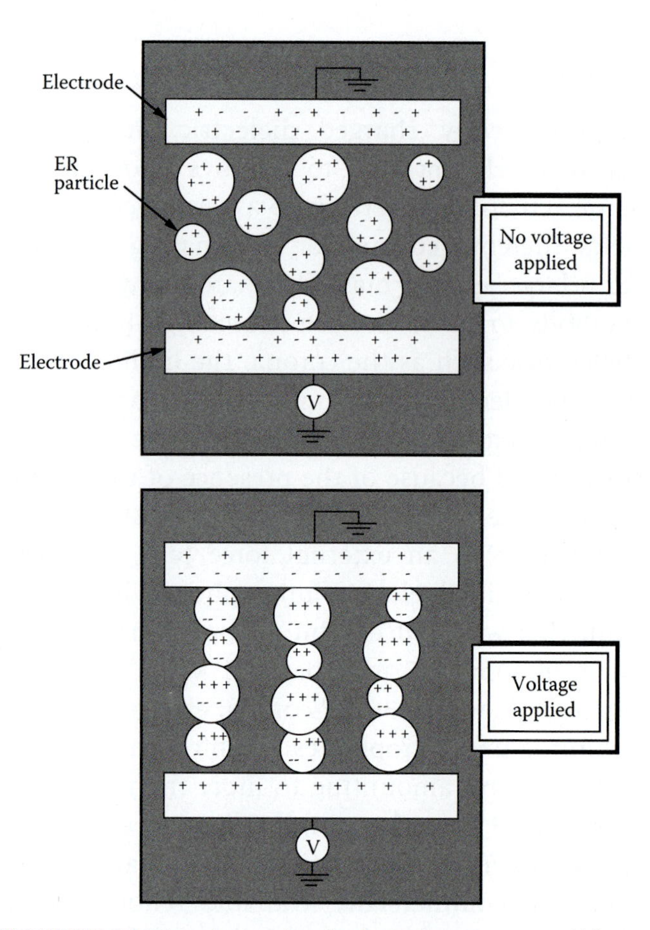

FIGURE 8.33 The electrorheological phenomenon. (After Gandhi, M.V., and B.S. Thompson. 1992. *Smart materials and structures*. London: Chapman & Hall.[32])

Willis Winslow, who discovered and patented[101] the electrorheological phenomenon in 1947, worked with silica gel in mineral oil. More recent formulations are based on suspensions of particles of polymers or ceramic materials in silicone oil. Typical products proposed with these actuators include clutches and transmission systems, hydraulic valves, and vibration isolation systems such as engine mounts and shock absorbers. The bulk conductance of the carrier fluid can present a difficulty to power usage; true nonconducting fluids work best. Problems with separation of fluid and particles due to evaporation, sedimentation, centrifugal forces, and electrophoresis can lead to device failures. Wear due to abrasion from particles can also cause device failure. Electrorheological fluids are very temperature sensitive, and both boiling and freezing are major issues. Furthermore, ER fluids are sensitive to impurities since polarization mechanisms might be changed by them.

In 2003 a "giant electrorheological effect" was reported by Wen et al.,[106] who demonstrated an ER effect of the order of magnitude of an MR one. Electrorheological devices have not yet reached the commercial market, while MR fluids are already used in some applications (see "Magnetorheological Fluids" below).

Magnetic Actuators

Introduction

In the macroworld, electromagnetic (EM) forces dominate the development of actuators such as conventional motors. On the other hand, large electrostatic motors such as the 1889 capacitive motor made by Zipernowsky[107] are rarities rather than commodities. Because of the three-dimensional nature of magnets and solenoids, EM systems are difficult to micromachine using planar integrated circuit (IC) processes. Electrostatic motors are simpler and more compatible with IC fabrication, and electrostatics often scale more favorably into the microdomain. These factors explain the current preponderance of electrostatic actuators in micromachining. We shall learn from a detailed comparison of electrostatic versus magnetic actuation that both present advantages and disadvantages and that both have severe problems when high-power output is demanded from the miniaturized systems in which they are embedded.

Principle behind Magnetic Actuation

A good starting point to understand magnetic actuators is Ampère's circuital law, used to calculate the magnetic field strength **H** (in A/m) around a current-carrying wire (Volume I, Figure 5.25). That law, we saw in Volume I, Chapter 5, states that the line integral of an electric current density *J* running through a straight wire generates a magnetic field strength **H** (in A/m) around that wire. The integral of **H** along the circular path *C* is equal to the current *I*, which in turn can be obtained as the integral of the current density **J** across the cross-section of the wire *S*:

$$\oint_C \mathbf{H} \cdot dl = \int_S \mathbf{J} \cdot ds \qquad \text{(Volume I, Equation 5.28)}$$

When applying this to a solenoid inductor as shown in Figure 8.34 one obtains the following:

$$\oint_C \mathbf{H} \cdot dl = NI \tag{8.74}$$

where *N* is the number of windings in the solenoid. The electric inductance *L* (in *H*) of the solenoid equals *N*Φ/*I* (see Volume I, Table 5.1), where Φ is the

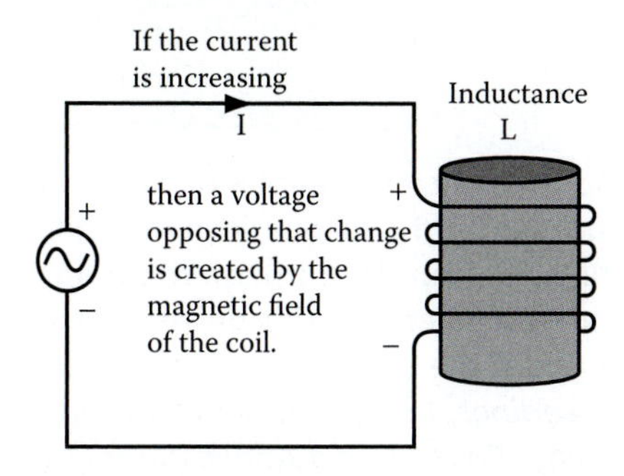

FIGURE 8.34 Solenoid inductor.

magnetic flux. With Φ = BS = μHS (with μ the magnetic permeability and *B* the magnetic induction field) and, using Equation 8.74, we obtain:

$$\Phi = \mu \frac{NI}{1} S \tag{8.75}$$

where *l* is the length of the solenoid, so that for the inductance *L* of the solenoid we get the following:

$$L = \mu \frac{N^2}{1} S \tag{8.76}$$

We can now calculate the stored magnetic energy (U_m) and energy density (U'_m) of the solenoid. The electromotive force, or V_{EMF}, generated by the coil is given by (see also first term on the left in Equation 8.36):

$$V_{EMF} = -L\frac{dI}{dt} = -N\frac{d\Phi}{dt} \tag{8.77}$$

So that for the stored energy:

$$U_m = \int -(V_{EMF})dq = \int_0^1 L\left(\frac{dI}{dt}\right)dq = \frac{1}{2}LI^2 \tag{8.78}$$

or:

$$U_m = \frac{1}{2}\mu\left(\frac{N}{1}\right)^2 S1 \frac{B^2}{\mu^2(N/1)^2} = \frac{B^2}{2\mu} S1 \tag{8.79}$$

and for the energy density we divide by the volume (V = Sl) to obtain the magnetic energy density, U'_m, as follows:

$$U'_m = B^2/2\mu \tag{8.80}$$

which, in the case of an air gap, was introduced earlier as Equation 8.62, where *B* is the magnitude of the magnetic flux density in the gap and μ_0 the magnetic permeability of free space. Earlier, we also mentioned that for magnetic actuators in air, the gap energy density by pushing *B* into saturation (about 1.5 T) leads to 950,000 Jm^{-3}. The energy density stored in the rotor/stator gap for magnetic, electrostatic, piezoelectric, and SMA actuation (see below) was compared in Table 8.8.

By analogy with the electrostatic energy density from Equation 8.2:

$$U' = \frac{1}{2}\varepsilon E^2 = \frac{1}{2}\mathbf{D}\cdot\mathbf{E} \tag{8.2}$$

the magnetic energy density can be written out as follows:

$$U'_m = \frac{B^2}{2\mu} = \frac{1}{2}\mathbf{H}\cdot\mathbf{B} \qquad (8.81)$$

And by analogy to the electrostatic surface force *F*, given earlier as follows:

$$F = -\nabla U \qquad (8.3)$$

we calculate a translational magnetic force, F_m, as (see Figure 8.35, where translation is in the *y* direction):

$$F_m = -\nabla U_m \qquad (8.82)$$

with:

$$U_m = \frac{1}{2}LI^2 = \frac{1}{2}\mu N^2 lSI^2 \qquad (8.79)$$

we obtain for a small movement, *dy*:

$$dU_m = \frac{1}{2}N^2SI^2\left[\mu(1 - dy) + \mu_0 dy\right] \qquad (8.83)$$

Or, with $\mu = \mu_0(1 + \chi)$, we calculate the following for the magnetic restoring force along the *y*-axis:

$$F_m = \frac{dU_m}{dy} = -\frac{1}{2}N^2SI^2\mu_0\chi \qquad (8.84)$$

To garner some magnetic actuator scaling insights, we rewrite Equation 5.28 of Volume I as follows:

$$\oint_c \mathbf{B}\cdot d\mathbf{l} = \mu_0 I = \mu_0\int_s \mathbf{J}\cdot d\mathbf{s} \qquad (8.85)$$

This equation describes the creation of a magnetic field by a current, with *S* defining the area of the surface bounded by *C*, and *J* the current density. We use this expression for an example calculation of scaling of the solenoid shown in Figure 8.36.[108] As the system becomes smaller, area ds decreases more rapidly than *dl*, and *B* decreases unless the current density *J* is increased. We refer to the appendix in Trimmer[109] for a more detailed analysis. His results indicate that different assumptions about current density and heat transfer lead to different scaling laws. When the current density is kept constant and the size of the system is decreased, the force between two EM wires or coils scales as (l^4). In other words, a size reduction of 10 means a magnetic force reduction of 10,000. Electrostatic force scales as l^2, and from Chapter 7, on scaling, we remember that the lower the coefficient in *l* the better the scaling behavior into the microdomain. So from this vantage point, electrostatic actuators scale better than EM ones. The scaling results of the interaction between a coil and a permanent magnet are somewhat better: (l^3). To improve the situation further, we can increase the current density. The resulting heat, as we will see below, is more effectively removed from microstructures, thus avoiding overheating and improving the scaling. Constant heat flow per unit area of the windings results in l^3 ($l^{2.5}$ with a permanent magnet). A constant temperature difference between windings and environment yields l^2 (and also l^2 for the permanent magnet case). For the last magnetic

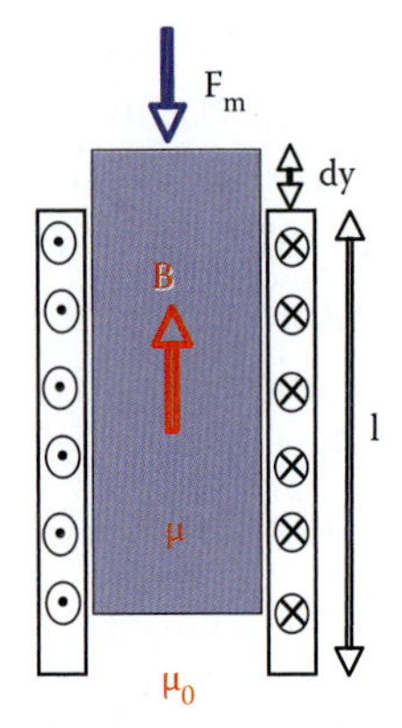

FIGURE 8.35 Magnetic force in a translational motion.

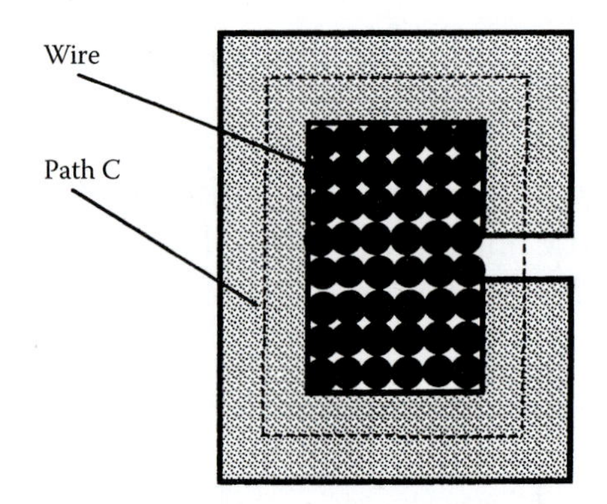

FIGURE 8.36 The line integral of B·dl over a closed curve C is equal to μ_0, the magnetic permeability of free space times the current through this loop C. As the system becomes smaller, the area decreases more rapidly than the length of the dashed line. [After Trimmer, T. 1990. In *Proceedings, Integrated Micro-Motion Systems: Micromachining, Control, and Applications (3rd Toyota Conference)*, 1–15. Aichi, Japan.[108]]

case, where the force scales as (l^2), the power that must be dissipated per unit volume scales as (l^{-1}), or, when the scale is decreased by a factor of 10, 10 times as much power must be dissipated. The use of superconductors could eliminate this problem (see below).

Examples of Magnetic Actuators

Introduction

The magnetic actuation examples we consider here are magnetic levitation for making valves, magnetic micromotors, and a bead array counter (BARC).

Levitation Valve

Friction and wear in miniaturized systems primarily relate to the surface contact between solids, in particular in bearing surfaces supporting the load of the micromachinery. They scale as l^2 and become increasingly important in the microdomain (see also Chapter 7). The friction in early electrostatic micromotors was actually found to be comparable to the friction of brake materials on cast iron.[110] Levitation eliminates wear and friction, two factors affecting reliability and control. Levitation is obtained from magnetic,[111] electrostatic,[112,113] and fluidic forces.[114] Here, we consider magnetic levitation for making a valve.

Levitation with magnetic devices, or *maglev*, can be achieved by various methods, such as with permanent magnets, electromagnets (including superconducting magnets), and diamagnetic bodies. Working with permanent magnets leads to better scaling behavior than working with electromagnets (see above). Wagner et al.[115] levitated a small, rare-earth permanent magnet out of the plane of a silicon substrate equipped with a planar coil as shown in Figure 8.37b. This moving structure may be used to close off a hole in a valve seat, making for a magnetically actuated valve. The permanent magnet was glued onto a silicon micromachined thin plate suspended by a thin silicon spring made from suspending beams parallel to the substrate edge. The monolithically integrated seventeen-turn planar coil was used to generate a magnetic field force, which forces the magnet on its spring suspended platform to move vertically (z-direction).[116] Using planar IC technologies to manufacture the coil ensures an optimum heat flow within the device (see below under Thermal Actuators). If μ_z, the magnetization of the permanent magnet, is independent of the magnetic

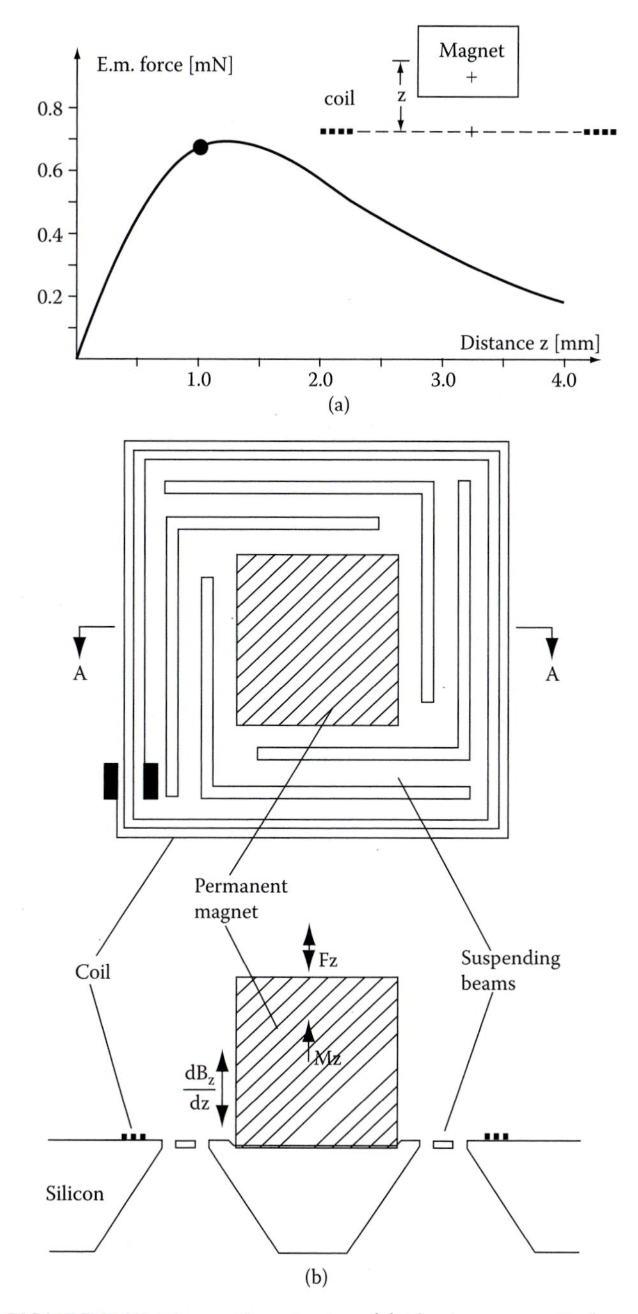

FIGURE 8.37 Magnetic actuator. (a) Electromagnetic force on a magnet. The dot indicates the magnet position in the fabricated device. (b) Schematic of a vertical magnetic microactuator with integrated planar coil and hybrid-mounted permanent magnet. [After Benecke, W. 1991. Silicon-microactuators: Activation mechanisms and scaling problems. In *1991 International Conference on Solid-State Sensors and Actuators (Transducers '91)*, 46–50. San Francisco, CA.[116]]

induction field, B, the vertical force acting on the permanent magnet is given by the following:

$$F_{m,z} = \mu_z \int \frac{dB_z}{dz} dV \tag{8.86}$$

where B_z represents the vertical component of the magnetic field produced by the planar coil. The magnetic force is proportional to the volume of the magnet. Thin magnetic layers will, in general, be insufficient to generate high forces. Problems with commercially available NdFeB permanent magnets include the fact that these magnets, like other rare-earth-based magnets, incorporate on their side (nonpole) surfaces a magnetically reversed layer of about 20 μm thick. This surface demagnetization limits the miniaturization of these permanent magnets to a few hundred micrometers; the smallest commercially available size is about 0.3 mm. The deflection of the highlighted actuator is given by Hooke's law as follows:

$$\Delta z = \frac{F_{m,z}}{k} \tag{8.87}$$

with k symbolizing the spring constant. The magnitude of the magnetic field depends on the current density achieved in the windings generating the field, and this current in turn is limited by heating of the solenoid metal (gold, in this case). Sheet resistance of thin-film conductors, electromigration, and thermal and geometric constraints all conspire to reduce the achievable magnetic energy density. Wagner et al.[115,117] assume constant current and very efficient power dissipation and derive a magnetic force with quadratic scaling and a deflection exhibiting linear scaling. To obtain constant current, this group has the number of coil turns scaling linearly with the cross-section of the coil wire. Quadratic force scaling, also found for electrostatics, lends itself well to the microrange. In practice, the magnetic force in the example increases rapidly at first and then decreases slowly with z, exhibiting a maximum value of 0.68 mN at a distance of 1.25 mm from the coil plane (see Figure 8.37a).[116] Increasing the volume of the permanent magnet thus works only up to a point, as the magnetic field decreases slowly with z. For a driving current range between –300 and +300 mA, an elevation of the permanent magnet of 143 μm is achieved, resulting in a mean slope of 24 μm per 100 mA. This example indicates that at least for linear actuators, magnetic actuation can be designed to scale as well as electrostatics and that these actuators are not too hard to implement when allowing for some final manual assembly.

Along the same lines, a meso (intermediate) scale, electromagnetically actuated, normally closed valve was realized using low-temperature cofired ceramic (LTCC) tape technology by Gongora-Rubio et al.[118] In this hybrid device, LTCC green tape technology is used for the electromagnet [several layers of planar silver spirals on alumina (Dupont 951 series)] and for the fluid manifold (alumina) fabricated at the same time. Anisotropically etched Si is employed for the planar rectangular spring (similar to the Benecke spring in Figure 8.37b) and a high-energy SmCo for the mini-permanent magnet. Using a 900-Gauss SmCo (1-mm diameter) magnet, a 200-μm deflection of the rectangular planar Si spring was obtained. The square spiral spring is covered with a polysiloxane film; the rectangular spring butts up against a polysiloxane valve seat on the ceramic coil stack and is lifted of the valve seat when powered up.

Magnetic Micromotors

Electromagnetic (EM) microactuators with glued magnets of millimeter dimensions have been used for different types of vertical, torsional, and multiaxial actuation. Here we consider EM micromotors and compare them with electrostatic micromotors.

Making micromotors has fascinated the MEMS community from day one. To witness, on November 28, 1960, William McLellan collected $1000 from Richard Feynman for having made the first operating electric motor only 1/64 inch in size.[119] The handmade motor was a more or less standard, two-phase, permanent-magnet motor.

Actuators in general, we remarked earlier, often do not scale advantageously in the microdomain. The torque of both electrostatic and electromagnetic motors depends on the shearing stress on the rotor integrated over its whole area, and force production is proportional to changes in the stored energy in the gap between rotor and stator (see Equation 8.82). The amount of force that may be generated per unit substrate area is thus proportional to the height of the motor. Large-aspect ratio rotor structures are therefore desired for both types of motors (or actuators in general for that matter). Torque is a volume effect, and small, flat-surfaced micromachined micromotors have a only a small force available for actuation. For most practical applications, motors with torques larger than 10^{-6} to 10^{-7} Nm are required,* but, as we saw above, the torque generated with a typical surface-micromachined electrostatic motor is only in the nano-newton-meter or pico-newton-meter range (see Figure 8.6b for a typical surface micromachined motor). In practice, the performance with respect to torque and speed of a traditional miniature magnetic motor with a 1-mm diameter to 2-mm-long permanent magnetic rotor is incomparably better than the surface micromachined motors described here and in Volume II, Chapter 7.[120]

With LIGA (see Volume II, Chapter 10) or a combination of LIGA (or pseudo-LIGA) and precision engineering, large aspect ratio magnetic motors can be built that do produce torques that are in the 10^{-6} to 10^{-7} Nm range. As an example, in Figure 8.38 we show one of the best-performing micromotors, fabricated using the sacrificial LIGA process in combination with traditional precision machining. Essentially, a permanent magnetic rotor follows a rotating magnetic field in the stator coils. Both rotor and stator in this micromotor are made from pure nickel. Such a motor with a diameter of 285 μm operates at speeds slightly above 30,000 rpm, with a current excitation level of 600 mA. No evidence of deterioration was evident even after 50 million rotations.[121] The motor manufacture illustrated here shows how LIGA micromachining and precision engineering constitute complementary techniques for producing individual parts that are then assembled afterward. Only components with the smallest features are produced by LIGA; other parts are produced by traditional precision mechanical methods. The cost of producing monolithic magnetic LIGA micromotors urges one to investigate less expensive LIGA-like technologies or hybrid approaches (LIGA combined with traditional machining) as more accessible and adequate alternatives.

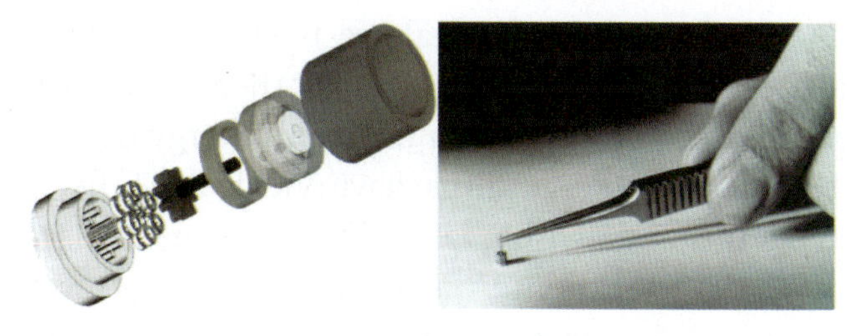

FIGURE 8.38 Large-aspect-ratio magnetic motor. (Courtesy of IMM, Germany.)

A University of Wisconsin team[122] demonstrated a series of magnetic LIGA motors, typically consisting of a plated nickel rotor that is free to rotate about a fixed shaft. Some of the motors made by this group can also be run electrostatically. Given their dramatically increased torque, these motors are again much better than their similar sized surface micromachined cousins. Using a pseudo-LIGA approach, Ahn et al. made a meander-type induction coil involving polyimide and electroplated high-permeability Ni (81%)-Fe (19%).[123] An EM motor based on this approach rotated at speeds up to 500 rpm with applied currents of 300–500 mA and a driving voltage of less than 1 V. The torque predicted for a stator current of 500 mA was 1.2 μN.

High-aspect-ratio techniques, we pointed out, also benefit torques in electrostatic micromotors. Wallrabe et al.,[124] for example, present design rules and tests of electrostatic LIGA micromotors. Minimum driving voltages needed were measured to be about 60 V and optimized designs are expected to deliver torques in the micro-newton-meter range. Suppose now that we use the same high-aspect-ratio machining techniques for both magnetic and electric micromotors, which one is then expected to be the best? For this we turn our attention again to the maximum energy density achievable in the stator-rotor gap. In conventional motors using iron, magnetic induction is limited to 1.5 T because of saturation, producing an energy density of about 9×10^5 Jm^{-3}. This is already more than twice the achievable electrostatic energy in a 1-μm gap (see Table 8.8). Moreover, it is important to remember that, in the case of electrostatic motors, there is little room for improvement, since the electrostatic

* As a reference point, to drive a cylindrical load of several hundred microns in diameter and several mm long, the required torque is of the order of 10^{-5} Nm.

energy density assumed above is close to what is achievable in vacuum. In contrast, there is plenty of room for improvement in magnetic energy density. Thin ferromagnetic films have yielded 2-T fields, and substantially higher fields are generated, at much larger expense, using superconductors (see below).[125] For example, using small-bore 10–15-T superconducting magnets, the achieved energy density of 4–9 × 10^7 J m^{-3} is roughly two orders of magnitude larger than the highest possible electric field density. For now, though, the 10–15-T superconducting magnets fall outside the realm of micromachining.

The efficiency—the ratio of power used in performing a desired task to the total amount of power consumed for a current-driven magnetic motor—will be worse than for a voltage-driven electrostatic device. However, Busch-Vishniac argues, since friction in small gaps, rather than electrical losses, are the major losses in microactuators, mechanical losses render nearly all types equally inefficient.[125] Moreover, since the gap in magnetics may be larger, friction is actually easier to avoid in magnetic actuators. One could add that a wider separation gap would also make the actuator less sensitive to dust and humidity.

In addition to the problem of minuscule amounts of power being produced by minuscule motors of both types, that of heat dissipation in a micromotor below a certain size needs to be solved. Busch-Vishniac calculated that a generic problem of heat dissipation exists if microfabricated microactuators continue to shrink in size without a concomitant reduction in the amount of power used in the system. For instance, a motor dissipating 100 mW should not have a rotor smaller than 20 μm, as the temperature might rise above 250°C, resulting in thermal breakdown of the metals in the microstructure. Classical models for heat convection were used to reach this conclusion. A more accurate model would result in a reduction of the heat lost by convection as the effective viscosity in the gap is greater than that predicted by the classical model (see section on heat transfer in Volume I, Chapter 6). In other words, the results generated with the classical model should be viewed as a conservative approximation; real microdevices might run hotter yet.

The more immediate challenge for both types of motors, however, is in making them in batch fashion. MEMS activities in micromachined motors today have generated plenty of new insights into phenomena such as stiction, friction, and wear of various materials but have not yet produced viable, batch-produced micromotors that can compete with a proven product such as the Panasonic piezoelectric device pictured in Figure 8.23a.

The above analysis seems to put magnetic actuation at an advantage over electrostatic actuation and even suggests that micromachined actuators are not always possible or useful. We compare magnetic and electrostatic actuators in general in a separate section, "Comparing Magnetic with Electrostatic Actuation."

The Bead Array Counter

In the bead array counter (BARC), we describe the use of a magnetic field gradient to levitate and remove paramagnetic microbeads that are not specifically bound to the surface of an array of giant magnetoresistive (GMR) sensors.[126] Recent developments in thin-film magnetic technology has resulted in films that exhibit a large change in resistance in response to a magnetic field. This phenomenon is known as GMR to distinguish it from conventional anisotropic magnetoresistance (AMR). AMR resistors exhibit a change of resistance of <3%, but GMR materials achieve a change of as much as 10–20%. GMR films have two or more magnetic layers separated by a nonmagnetic layer. Because of spin-dependent scattering of the conduction electrons, the resistance is maximum when the magnetic moments of the layers are antiparallel and minimum when they are parallel.

The BARC uses molecular recognition, such as DNA hybridization, magnetic microbeads, and magnetoresistive sensors (such as giant magnetoresistive sensors or GMR sensors), and the application of a magnetic field to detect and identify binding of a molecule with a molecular receptor. In one particular embodiment, DNA probes are patterned directly above an array of GMR sensors, as shown in Figure 8.39. Sample containing complementary DNA hybridizes with the probes on the surface. Magnetic or paramagnetic beads are labeled with a binding material (e.g., streptavidin),

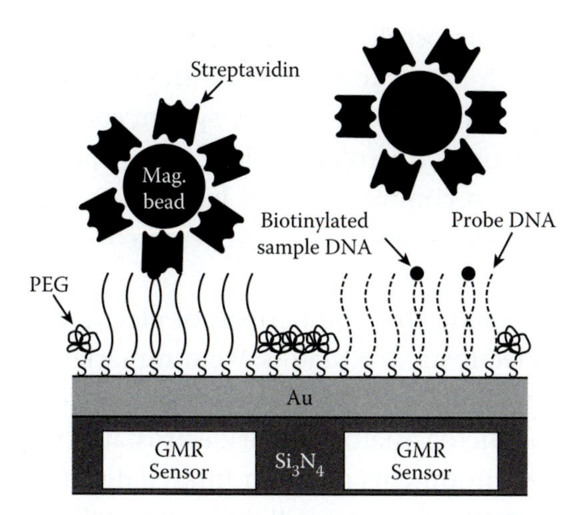

FIGURE 8.39 Schematic diagram of the bead array counter (BARC) chip surface chemistry and hybridization assay. Thiolated DNA probes are patterned onto a gold layer directly above the giant magnetoresistive (GMR) sensors on the BARC chip. Biotinylated sample DNA is then added, and it hybridizes with the DNA probes on the surface when the complementary sequence is present. Unbound sample DNA is washed away. Streptavidin-coated magnetic beads are injected over the chip surface, binding to biotinylated sample DNA hybridized on the BARC chip. Beads that are not specifically bound are removed by applying a magnetic field. Bound beads are detected by the GMR sensors. (Based on Edelstein, R.L., C.R. Tamanaha, P.E. Sheenan, M.M. Miller, D.R. Baselt, L.J. Whitman, and R.J. Colton. 2000. The BARC biosensor applied to the detection of biological warfare agents. *Biosens Bioelectron* 14:805.[126])

and they bind to biotinylated sample DNA hybridized on the BARC chip. Nonspecifically adherent microbeads are removed by means of a controlled magnetic field—applying the magnet removes >95% of these microbeads. A change in the output of the magnetic field sensors indicates the presence of magnetic particles bound to sensors and thereby indicates the presence and concentration of target molecules in the sample. The detection of the presence of beads requires a uniform magnetic field, but removing the beads requires a nonuniform large field gradient. The latter is accomplished by moving a portion of the electromagnetic core closer to the chip with a plunger.

Comparing Magnetic with Electrostatic Actuation

The debate about the relative merits of electrostatic versus magnetic actuation, especially for driving micromotors, has preoccupied many MEMS researchers over the last two decades.[119,127] Fujita and coworkers, for example, have argued that electrostatic actuation is preferred,[128–130] pointing to the following attributes of surface micromachined electrostatic actuators:

1. Thin insulating layers such as SiO_2 or Si_3N_4 exhibit breakdown strengths as high as 2 MV/cm. The power density in this field is 7×10^5 J/m^3; this value is equal to the power density of a 1.3-T magnetic field. The contracting pressure induced by this field is 1.3 MPa. A voltage of about 100 V is sufficient to generate the strong fields mentioned.
2. The electrostatic force is a surface force exhibiting a favorable scaling law. The actuation is simple, as it involves only a pair of electrodes separated by an insulator.
3. The electrostatic actuator is driven by voltage, and voltage switching is far easier and faster than current switching (as in EM actuators). Energy loss through Joule heating is also lower.
4. Weight and power consumption are low.

However, earlier, the comparison between electrostatic and magnetic micromotors demonstrated that many factors besides scaling need to be considered when deciding upon a certain type of actuation principle. Whereas in some cases, the magnetic power might scale disadvantageously into the microscale, the absolute magnetic forces achievable are much larger.

While there is some validity, especially in the case of micromotors, to the arguments favoring magnetic actuators, the criticism of micromachined electrostatic actuators is too harsh where linear micromachined electrostatic devices such as resonators are concerned. In the latter application, power output is of little concern, and these devices have found a major application, for example, in the analog devices accelerometer described in Volume II, Example 7.1. Similarly, for most optical applications such as digital mirror technology from TI (Volume II, Example 7.2), the amount of power required to deflect a laser beam with a micromachined mirror is minimal.

Table 8.11 Salient Points of Electromagnetic Actuators

1. Maximum stroke: 10^{-1} μm
2. Maximum speed: 10 kHz (see Figure 8.14)
3. Intermediate force density: 10^{-2}
4. Complex to fabricate
6. Large footprint: 1–10 mm
7. Special drive conditions: 1.5 T
8. Maximum electrostatic energy density, U'_m (in J/m^3): 9.5×10^5 Jm^{-3}

For many years, micromagnetics have found major use in Hall chips and thin-film heads for magnetic discs (see Figure 1.1d). Besides magnetic micromotors, magnetic valves, and BRAC chips, described above, micromagnetic efforts continue on making denser thin-film magnetic heads;[131] high-performance magnetographic printing heads;[132] electronic switching components;[133] different types of vertical, torsional, and multiaxial actuators;[117] and contactless magnetic transmission of force to ferrofluids (FF) (see "Ferrofluids" below).[115] The pros of magnetic actuation include high force and stroke for the size of the actuator, the ability to attract and repel, multiaxis capability, and the linear current response. The most outstanding advantage of magnetic actuators is the long range of the force. With dimensions above 1 mm and for larger forces, actuators based on permanent magnets become a good choice (see Table 8.11). The cons might be listed as large actuator footprint (approximately a centimeter), requirement of external field sources, complex microscale manufacturing/assembly processes, high-power consumption, and interactions of the magnetic field with nearby system components. Incorporating microfabrication of a permanent magnet into common MEMS fabrication processes remains an active field of research (see also Figure 8.14).

To conclude this comparison of electromagnetic versus electrostatic actuators, we summarize the key points in Table 8.12. Our conclusion regarding the status of microactuators is that electrostatics is useful in dry environments and over limited distances and that electromagnetics is still difficult to collapse into integrated structures.

Magnetostriction

Introduction

The magnetostrictive effect is the contraction or expansion of a material under the influence of a magnetic field and the inverse effect of changes in magnetization due to a stress in the material. It is the magnetic equivalent of electrostriction (see above). This bidirectional effect between the magnetic and mechanical states of a magnetostrictive or piezomagnetic material presents one with another transduction capability that can be used for building actuators and sensors. It was the Scotsman James Joule (1818–1889) who, in 1842, found that when a nickel rod is magnetized it contracts very slightly (the effect is also called the *Joule effect*). A very early use of electrostrictive materials (1861) was for the

Table 8.12 Comparison of Electrostatic versus Magnetic Actuators

	Electrostatic	Magnetic
Field energy density	4×10^5 Jm^{-3}(max)	4×10^7 Jm^{-3} (10 T)
Scaling	(l^2)	(l^3) (constant current)
Gap contamination sensitivity	Very sensitive to humidity and attracts dust	Fairly insensitive
Integrated circuit compatibility	Good	Not very good
Range	Short range	Long range
Power efficiency	Very good in the absence of mechanical friction	More power consuming even in the absence of mechanical friction
Implementation of levitation schemes	Possible	Easier implemented because larger gaps are possible
Miniaturization	Excellent	Difficult
Complexity	Low	High
Control	Voltage switching control is faster, easier to make, and more efficient	Current switching control is less efficient, more complex, and less efficient

generation of sound and ultrasound, and one of the first telephone earpieces was magnetostrictive. The magnetostrictive effect is actually familiar to all of us in the humming of transformers (expanding and contracting of the coils).

How Does Magnetostriction Work?

The magnetostrictive coefficient, ε, is the fractional change in length ($\Delta l/l$) as the magnetization of the material increases from zero to its saturation value ($\Delta l/l = \varepsilon$) (see Figure 8.40). Its value is typically in the parts-per-million range (e.g., 40×10^{-6}). It can be positive or negative in a direction parallel to the magnetic field and is independent of the direction of that field, indicating a square-law type of relationship, such as:

$$\varepsilon = \frac{\Delta 1}{1} = cB_0^2 \qquad (8.88)$$

where ε = static strain

B_0 = flux density

c (m^4Wb^{-2}) = a material constant

For a small flux density ($B \ll B_0$), taking the derivative of Equation 8.88, one derives the basic magnetostrictive strain equation:

$$\varepsilon = \beta B \qquad (8.89)$$

where $\beta = 2cB_0$ defines a magnetostrictive strain constant with dimensions of m^2Wb^{-1}. As the magnetic field, *B*, increases, the length change saturates and increasing the field strength further brings about no additional change. As in the case of piezoelectricity, the inverse effect exists as well; a changing flux density results when modulating the stress on a piezomagnet. This reverse effect is called the *Villari effect.*

The magnetostriction phenomenon is attributed to the rotations of small magnetic domains in the

FIGURE 8.40 Magnetostrictive materials: the Joule effect.

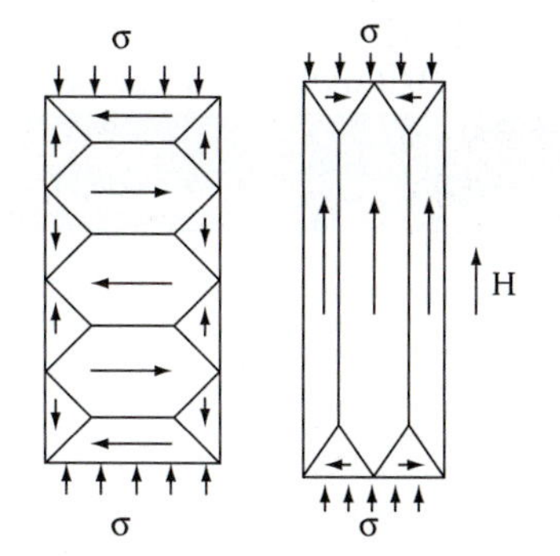

FIGURE 8.41 Magnetostriction. With a compressive load applied to a magnetostrictive material, the domain structure is oriented perpendicular to the applied force. As a magnetic field is introduced, the domain structure rotates, producing the maximum possible strain in the material.

material, randomly orientated in the absence of a magnetic field. The orienting of these small domains by the imposition of a magnetic field results in the development of a strain field (Figure 8.41).

As the intensity of the field increases, more and more domains line up with the field until saturation is reached. Magnetostrictive actuation, like piezoelectric actuation, is important when large forces must be obtained over small distances. Applications for magnetostrictive devices include high-force linear motors, positioners for adaptive optics, active vibration or noise control systems, medical and industrial ultrasonics, pumps, and sonar. Both find applications as sonar sensors, vibration dampers, and tunable-compliance materials.

The Wiedemann effect describes what happens when an axial magnetic field is applied to a magnetostrictive wire, with a current passing through it, i.e., a twisting occurs at the location of the axial magnetic field (Figure 8.42). The twisting is caused by interaction of the axial magnetic field, from an

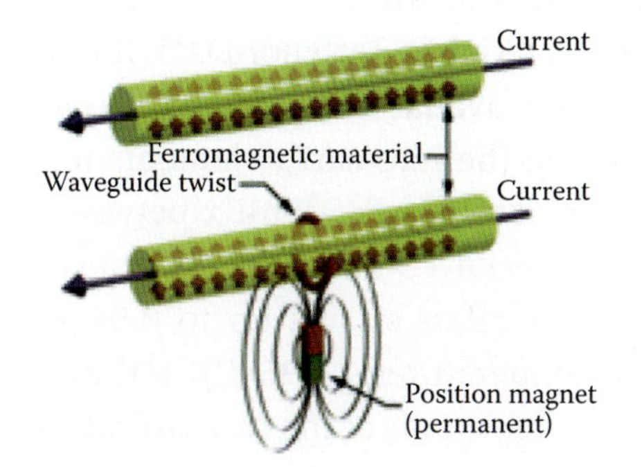

FIGURE 8.42 The Wiedemann effect: the twisting due to an axial magnetic field applied to a ferromagnetic material carrying a current.

Table 8.13 Magnetostrictive Materials with Their Magnetostrictive Coefficients

Material	Saturation Magnetostriction (μm/m)
Nickel	−28
49Co, 49Fe, 2V	−65
Iron	+5
50Ni, 50Fe	+28
87Fe,13Al	+30
95Ni, 5Fe	−35
Cobalt	−50
$CoFe_2O_4$	−250

external permanent magnet, with the magnetic field along the wire, due to the current in the wire. The inverse effect, that of creation of an axial magnetic field by a magnetostrictive material when subjected to a torque, is known as the *Matteucci effect*. Together, these effects are used in torque magnetostrictive sensors.

Magnetostrictive Materials

The amount of magnetostriction in base elements and simple alloys is small, on the order of 10^{-6} m/m. The magnetostrictive effect is exhibited by the transitional metals including iron, cobalt, and nickel and their alloys. The magnetostrictive coefficients of some magnetostrictive materials are shown in Table 8.13. The highest room temperature magnetostriction of a pure element is that of Co, which saturates at 60 microstrains.

"Giant magnetostriction" materials are materials in which the magnetostrictive coefficient exceeds 1000 μm/m (that is about two hundred times larger than that of iron). These materials are today's choice for magnetostrictive sensors and actuators. They were developed in the 1960s for underwater radar; a typical material is Terfenol-D.[104] Terfenol-D is a commercially available magnetostrictive material incorporating the rare-earth elements terbium and dysprosium ($Tb_xDy_{1-x}Fe_y$, with x between 0.27 and 0.3 and y between 1.9 and 1.98) (Etrema Company). This material offers strains up to 0.002—at liquid nitrogen temperatures (−196°C), terbium, dysprosium, and compounds of the two can exhibits strains of 0.01. The Terfenol-D effect is often referred to as the *giant magnetostrictive effect*. It works in the opposite sense of iron magnetostriction; the material expands rather than contracts. An alloy of iron with samarium also displays a giant magnetostrictive effect but, like iron, in the shrinking sense (see below and Honda et al.[134]).

Sputter-deposited magnetostrictive films present an interesting opportunity for actuation in micromachines where contactless, high-frequency operation is desired. Terbium-iron alloys are typical magnetostrictive materials that can be deposited by sputtering. Quandt et al., using an RF-sputtered TbDyFe film (10 μm thick) on a silicon cantilever 2 cm long and 50 μm thick, could deflect the silicon cantilever by more than 200 μm in an external field of only 30 mT.[135] The cantilever was operated at a frequency of 500 Hz, and no degradation could be observed after more than 10^7 operations. The same group is exploring the use of these films for valves and pumps. Honda et al. discovered that amorphous Tb-Fe films exhibit positive magnetostriction, while amorphous Sm-Fe thin films exhibit negative magnetostriction (see above), and this group built magnetostrictive bimorph cantilevers and traveling machines as illustrated in Figure 8.43a and 8.48b.[134] The actuation behavior of the magnetostrictive bimorph cantilever is as follows: a 1-cm-long polyimide beam (thicknesses of 7.5, 50, and 125 μm were experimented with) is sandwiched between an upper Tb-Fe film and a bottom Sm-Fe film, each 1 μm thick. When a magnetic field is applied along the cantilever length direction, the Tb-Fe film expands and the Sm-Fe film contracts, bending the polyimide beam downward. With a magnetic field along the width direction, the Tb-Fe film contracts and the Sm-Fe film expands, deflecting the beam upward. One version of this group's traveling machine, shown in Figure 8.43b, consists of a 7.5-μm-thick polyimide film equipped with magnetostrictive bimorph actuator layers and with two legs inclined so that it travels in one direction. With an alternating magnetic field of 100 Oe at 50 Hz applied along the machine length direction, it vibrates and travels at an average speed of 0.5 mm/s in the arrow direction.

A major application of magnetostriction is in noncontact torque sensing. An integrated silicon micromachined sensor head for torque and force measurements was first proposed in 1994 by

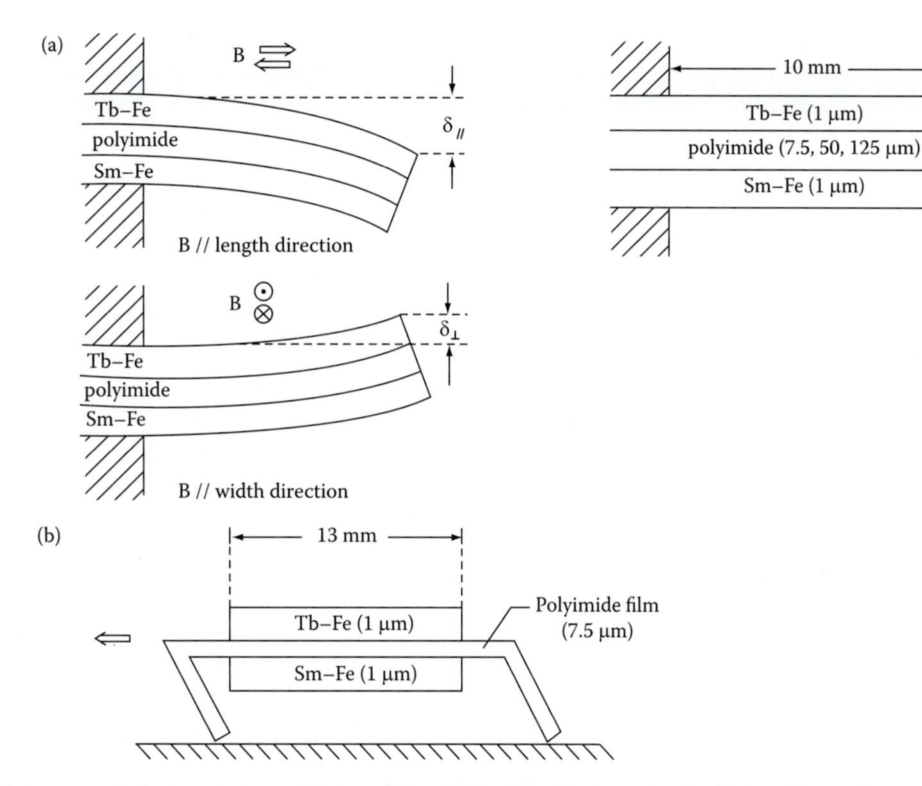

FIGURE 8.43 Magnetostrictive actuators. (a) Actuation behavior of a magnetostrictive bimorph cantilever. (b) Schematic view of traveling machine. [After Honda, T., K.I. Arai, and M. Yamaguchi. 1994. Fabrication of actuators using magnetostrictive thin films. In *IEEE International Workshop on Micro Electro Mechanical Systems (MEMS '94)*, 51–56. Oiso, Japan.[134]]

Rombach and Langheinrich.[136] It is based on a magnetic yoke with an exciting and receiving coil that detects the change of permeability of a $Co_{75}Si_{15}B_{10}$ ribbon fixed to the shaft surface of an automotive vehicle. The $Co_{75}Si_{15}B_{10}$ is an amorphous metal alloy rather than the crystalline alloys discussed so far. The most common of these materials is Metglas 2605SC, an amorphous alloy of iron, with boron, silicon, and carbon [for more on these so-called *bulk metallic glasses* (BMGs), see Chapter 5]. These materials exhibit a smaller magnetostrictive distortion at saturation than do the crystalline iron-terbium alloys, and generate smaller forces, but they provide a larger response for small applied fields and vice versa—small mechanical distortions create a larger change in magnetic properties. These amorphous magnetostrictors are used in sensitive strain gauges and in accelerometers as well.[105,137]

Comparing Magnetostricion with Piezoelectrics

The power requirements for giant magnetostriction are greater than for piezoelectric materials, but the actuation offers a larger displacement, and the ratio of mass per unit stress is greater than with a PZT actuator.[138] Magnetostrictive materials have other advantages; they are metal alloys rather than brittle ceramics, and they do not exhibit dielectric breakdown under high fields as do piezoelectric and ferroelectric materials.[105] There are obviously relative advantages and disadvantages to the two technologies; some of these are listed in Table 8.14.

Magnetohydrodynamic Pumping

Magnetohydrodynamic (MHD) pumping is widely used in nuclear power plant cooling systems and has been applied for pumping and sensing of conductive gases and liquid metals.[139] It has not been exploited much in the microdomain.

MHD pumping, like EHD pumping, involves no moving parts, has a continuous flow, and is compatible with solutions containing biological specimens—both phenomena are attractive as scales go down. A MEMS MHD pump is schematically illustrated in Figure 8.44. The effect is based on the Lorentz force, F_L (see Volume I,

TABLE 8.14 Comparison of Magnetostrictive and Piezoelectric Materials

Parameter	Magnetostrictive	Piezoelectric
Frequency range (ultrasonic range is from 20 to 200 kHz)	<20 kHz More limited because of size	20–200 kHz Most versatile
Audible noise	More noise	Less noisy; higher-frequency piezoelectric equipment is easier brought under Occupational Safety and Health Administration (OSHA) limits by appropriate design and acoustic shielding
Power requirements	Larger	Lower
Displacement	Larger	Lower
Transducer reliability	Both are mature, highly reliable, engineered devices	
Generator reliability	2-year warranty	1-year warranty
Effect of aging; the electrical activity of piezoelectric materials slowly degrades as a half-life function from the time they are "poled" material during manufacture	No depolarization problem	In the case of piezoelectric materials, the reduction in activity after 100 days brings the activity level to within 1% of the ultimate level it will achieve over the next 17 years of aging
Energy efficiency	<50%: magnetic systems are usually less than 50% efficient because of the energy lost in heating of the coils and the effects of magnetic hysteresis	>70%: piezoelectric transducers are extremely efficient because of the direct conversion of electrical to mechanical energy in a single step
Material strengths	Magnetostrictors are metal alloys rather than brittle ceramics, and they do not exhibit dielectric breakdown under high fields as do piezoelectric and ferroelectric materials	Brittle ceramics that do exhibit dielectric breakdown

Equation 3.64), which is the force generated by a current-carrying conductor in a magnetic field. This magnetic force, in the case of a fluid, can be written out as follows:

$$F_L = \sigma EB \tag{8.90}$$

where σ = conductivity of the fluid ($\Omega^{-1}m^{-1}$)
B = permanent magnetic field (Gauss)
E = electric field applied to the pump (V/m), which is $E = V/L$

Heng et al., using LIGA, microfabricated an MHD-type micropump. When operating in the DC mode, they observed significant bubble generation due to electrolysis of the fluid being pumped. The same group later worked on an AC-driven nozzle/diffuser type MHD to eliminate or reduce bubble formation.[140] In an AC MHD micropump, a sinusoidal AC electrical current and a perpendicular sinusoidal AC magnetic field pass through the electrolyte solution and transverse to a microchannel. The time-averaged

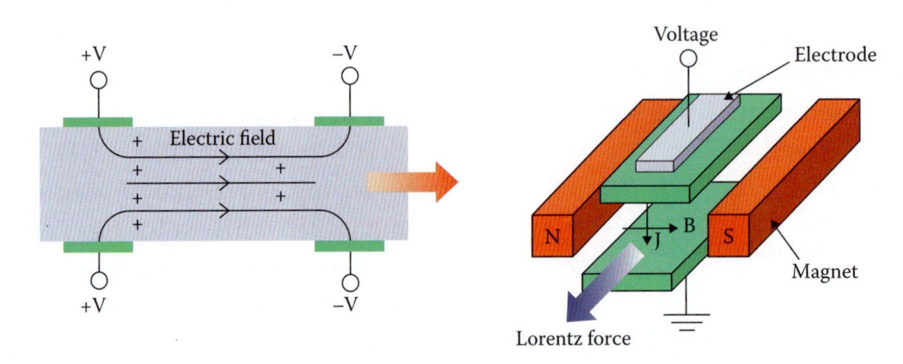

FIGURE 8.44 Schematic of a microelectromechanical system magnetohydrodynamic pump.

Lorentz force in this arrangement is given by the following equation:

$$F_L = IBw\int_0^{2\pi} \sin\omega t \sin(\omega t + \phi) d\omega t \qquad (8.91)$$

where I is the current amplitude and w the width of the microchannel. The ability to control the phase allows for controlling both the flow speed and the flow direction. Such an AC MHD micropump was demonstrated by Lemoff et al.[141] The same group also microfabricated an AC MHD fluidic switch in silicon with a switching speed faster than 0.033 s.[142] The switch is made by integrating two AC MHD pumps into different arms of a fluidic Y-channel. With a 1 M NaCl solution, a flow velocity of a 0.5 to 0.6 mm/s was registered in a 300-μm-deep, 1-mm-wide channel.

Magnetorheological Fluids, Ferrofluids, and Ferrowaxes

Magnetorheological Fluids

Magnetorheological (MR) fluids are colloidal dispersions of mesoscale size (1–10 μm) ferromagnetic particles suspended in polar or nonpolar carrier fluids. The particles are typically high-purity iron powders, but ceramic and alloys are used as well. These particles are multidomain and exhibit low levels of magnetic coercivity. As expected, the clumping of the suspended particles is caused by magnetic attraction.

When unperturbed by a magnetic field, MR fluids appear rather liquid. Upon applying a magnetic field, the viscosity of the fluid increases with the magnetic field magnitude in a nonlinear fashion. Upon reaching very high fields, the MR fluid becomes solid-like (Figure 8.45). As in the case of ER fluids, MR fluids

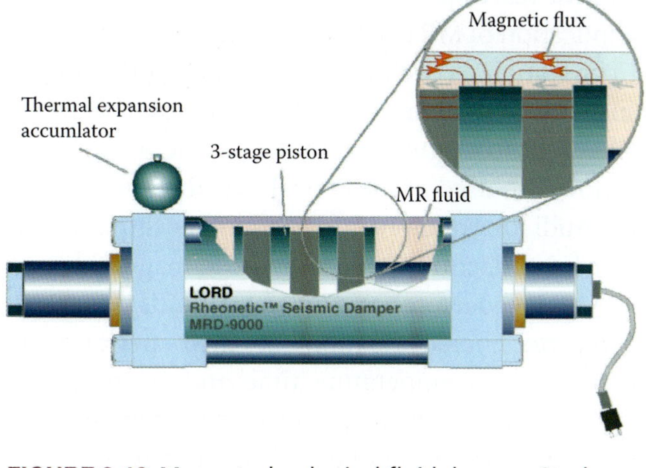

FIGURE 8.46 Magnetorheological fluid damper. Stroke: ±8 cm. Maximum input power: <50 watts. Maximum force (nominal): 200,000 N.

reverse to their original state upon removal of the external field. Overall response time of an MR system is of the order of milliseconds.[143]

MR fluids were discovered and patented in 1948 by Jacob Rabinow,[103] who was then with the National Bureau of Standards (NBS) [now the National Institute of Standards and Technology (NIST)]. After this invention, MR fluids basically disappeared from both the academic and industrial scene until the late 1980s. It is believed that the reason for this decline of interest in MR fluids in the early 1950s was that sedimentation and agglomeration of the magnetic particles were deemed unsolvable. This explains why knowledge of MR fluids is not as mature as that of ER fluids (see above), where the research remained more or less active after their initial discovery. Surprisingly, commercial success came first to MR fluids.

The main applications of MR fluids have been in the development of active vibration control and in the transfer of torque or force systems. In Figure 8.46, an MR fluid damper used in active building

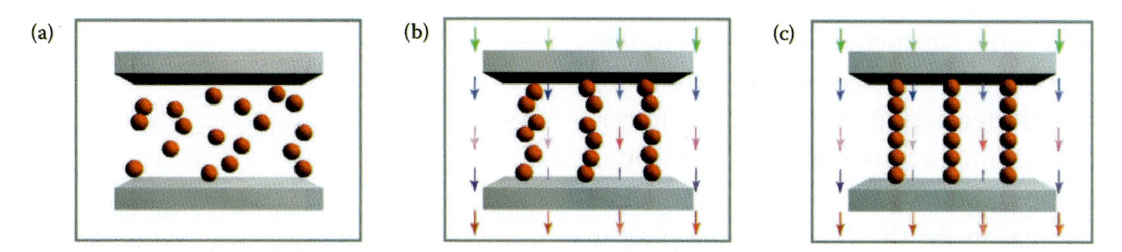

FIGURE 8.45 Magnetorheological fluids: (a) No magnetic field. (b) Intermediate magnetic field. (c) High magnetic field.

control (earthquakes) is shown. Another excellent application of MR fluids is the MagneRide by Delphi Corporation. A well-marketed magnetorheological suspension system, it was first used by GM in the Chevrolet Corvette C5, Cadillac Seville, and STS and is now also used in the Ferrari 599, the Audi TT, and the Audi R8. For more information, see http://delphi.com/manufacturers/auto/other/ride/magneride.

A major goal for the future in MR fluids is the more complete characterization of their performance depending on temperature, time, and shearing stress cycles. A hurdle in the path toward commercialization is the lack of suppliers of high-purity iron powder that makes the cost of MR fluid relatively high, even in large volumes.[143]

Ferrofluids

There are some notable efforts in nanoparticle-based MR fluids. Ferrofluids (FF) are a variant of MR fluids and are differentiated only by the size of the suspended particles used. In FF, one deals with particles smaller than 20 nm in diameter, whereas MR fluids are designed with particles in the micrometer size range (1–10 μm). The magnetic behavior of nanostructured materials has been investigated since the early days of nanotechnology. Some very significant results were first obtained by Herzer and Warlimont in 1992,[143] and an excellent review of the field by Herzer appeared in 1997.[144] When a ferromagnetic material is magnetized in one direction, it will not relax back to zero magnetization when the imposed magnetizing field is removed. It must be driven back to zero by a field in the opposite direction. If an alternating magnetic field is applied to such material, its magnetization will thus trace out a loop called a *hysteresis loop*. It is well known that the microstructure, especially grain size, determines the hysteresis loop of a ferromagnetic material.[144] It is found that the coercivity, H_c, the intensity of the magnetic field needed to reduce the magnetization of a ferromagnetic material to zero after it has reached saturation, increases with decreasing grain size down to values of about 40 nm for a pure metal such as iron or about 700 nm for hard magnets based on rare earth metals.[144] The reason for the H_c increasing proportional to $1/d$ is that as particles decrease in size, the magnetic domain size coincides more and more with the grain size. In this size range, multidomain (M-D) materials change to single-domain (S-D) materials. In the single-domain region, a change of magnetization can be produced only by rotation of the magnetization of a single grain or crystallite as a whole, which requires strong magnetic fields, depending on the shape and anisotropy of the particles. As the size of the magnetic particles is further scaled down, a transformation from ferromagnetic to superparamagnetic behavior occurs and the curve of H_c versus grain size d exhibits a peak at d_s (see Figure 8.47).[144] At particle size d_s, the material behaves as a purely single-domain material, and at d_p, the transition to a superparamagnetic is complete.

Typical particle sizes for the ferrro- to superparamagnetic (SP) phase transformation are between 10 and 20 nm for oxides and 1–3 nm for metals. This range corresponds to the band of lines drawn through the peak in the H_c-d curve. In the nanoparticle regime, between d_s and d_p, the coercivity exhibits a d^6 dependence.[144] In the superparamagnetic state of the material, the room temperature thermal energy, kT, where k is the Boltzmann constant, overcomes the magnetostatic energy of the particle (= domain), resulting in minimal hysteresis. In this case, even when the temperature is below the Curie or Neel temperature (and hence the thermal energy is not sufficient to overcome the coupling forces between neighboring atoms), the thermal energy is sufficient to change the direction of magnetization of the

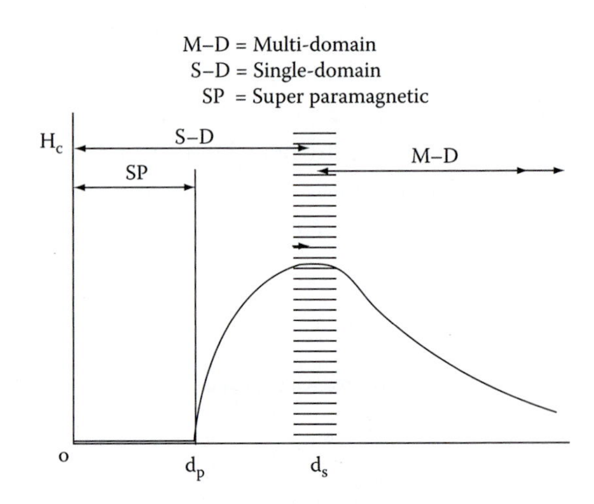

FIGURE 8.47 Coercivity (H_c) versus particle size. (Based on Herzer, G. 1997. In *Handbook of magnetic materials*, ed. K.H.J. Bushow, 417. Elsevier Science B.V.[144])

entire crystallite. The resulting fluctuations in the direction of magnetization cause the magnetic field to average to zero at d_p. Thus, the material behaves in a manner similar to paramagnetism, except that instead of each individual atom being independently influenced by an external magnetic field, the magnetic moment of the entire crystallite tends to align with the magnetic field. Superparamagnetic particles are attracted to a magnetic field but retain no residual magnetism after the field is removed (no hysteresis). These superparamagnetic nanoparticles feature a high-saturation magnetization, high relative permeability, low coercivity, high resistivity, and low hysteresis loss, and all of this benefits a large number of applications,* including in FF.

Superparamagnetic particles can be dispersed in a liquid, e.g., oil, without perceptible agglomeration if treated properly (see below and colloidal stability in Volume I, Chapter 7). Because of the nanometer sized particles in FF, sedimentation is prevented by Brownian motion, and agglomeration is avoided through the use of surfactants or ionic shells that coat the particles. If no coating is used, attractive van der Waals forces would cause the agglomeration of the particles. A typical FF constitutes a stable colloidal suspension of single-domain magnetic nanoparticles dispersed in a liquid carrier in a volume fraction of typically 2–10% (Figure 8.48). As explained above, MR fluids exhibit different viscosities depending on the strength of the magnetic field applied; FF, on the other hand, maintain their viscosity regardless of the magnitude of the magnetic field. Where MR fluids solidify in strong magnetic fields, FF always remain liquid.

In their natural state, with no field applied, FF are subjected to three forces: 1) repulsive forces due to the implemented coating, 2) attractive van der Waals forces, and 3) magnetic dipole-dipole interactions, which can be either repulsive or attractive. A true FF maintains equilibrium between the forces mentioned even in the presence of strong magnetic fields. A stable FF will not conglomerate for several years, until eventually the surfactants or the coatings on the particles break down and gradually cause the irreversible conglomeration of the particles into bigger clusters. At this point, the behavior of the FF tends to simulate that of an MR fluid.

FIGURE 8.48 Ferrofluid fixed with a magnetic field. The bottle is filled with ferrofluid and shows the magnetic force of the ferrofluid acting on the permanent magnet (Institute for Machine Tools and Factory Management, IWF, Germany).

Dispersed nanoparticles in FF are typically magnetite (Fe_3O_4) or maghemite (γ-Fe_2O_3), although cobalt, iron, and manganese zinc ferrite Zn_xMn_{1-x} have been used as well. The fluid carrier might be polar (water, in which case the FF are highly biocompatible) or nonpolar (hydrocarbons, fluorocarbons). Obtaining magnetic nanoparticles itself constitutes an important research field, and different methods have been attempted, with precipitation being the most common one (see Volume I, Chapter 7 and Chapter 3, this volume). Current applications of FF include positioning systems for multiaxle movement (see Figure 8.49), compliant positioning systems, dynamic sealing (e.g., on computer hard disks), bearings, adaptive gripping devices and dampers, and heat dissipation. Suspended superparamagnetic particles are also used to tag biomaterials of interest, the targeted species can be removed from a matrix using a magnetic field, but they do not agglomerate (i.e., they stay suspended).

Although perhaps not immediately or obviously useful, FF present a striking behavior when subjected to a normal magnetic field of strength above a critical value. Such fields induce a normal-field instability, causing the FF to instantly protrude from the surface in spikes aligned to the field lines, as

* On the other hand, it is also because of superparamagnetism that computer hard drive technologies are expected to stop growing once they reach a density of 150 gigabits per square inch.

Technical Data:

Load capacity: F < 10 N (in normal direction)
Accuracy of Positioning: ~ 1 μm
Max. velocity: 0.1 m/sec (uncontrolled); 0.02 m/sec (precise control)
Max. Range: s = 10 mm (limited by the volume of fluid)
Coil current: I < 5 A (for optimized coils and field formers)

FIGURE 8.49 View of a two-axis ferrofluid positioning system with two paired coils for each direction (Institute for Machine Tools and Factory Management, IWF, Germany).

shown in Figure 8.50. The spikes are characteristic of equilibrium of gravitational force, magnetic force, and surface tension.

Many excellent videos have been created demonstrating this interesting behavior. A special mention goes to those obtained by Sachiko Kodama (http://www.kodama.hc.uec.ac.jp/protrudeflow/index.html), where art and technology are cleverly mixed to obtain beautiful FF patterns and behaviors.

Another interesting application along the same artistic lines is the creation of fractal droplets on a surface by confining a drop of FF between two

FIGURE 8.50. A normal-field instability caused by a magnet placed underneath the ferrofluid. (From Sachiko Kodama, http://www.kodama.hc.uec.ac.jp/protrudeflow/index.html.)

FIGURE 8.51 Fractals on a surface obtained by magnetic interaction of a field with a ferrofluid drop enclosed in a Hele-Shaw cell.[146] (From Prof. Markus Zahn, Massachusetts Institute of Technology.)

glass slides in a Hele-Shaw cell* with a 0.9–1.4 mm gap (see Rinaldi and Zahn[145] and http://www.aps.org/units/dfd). When an FF drop is placed in such a cell and is subjected to simultaneous in-plane rotating and DC axial uniform magnetic fields and the magnitude of the fields increases, different, striking patterns are obtained (see Figure 8.51). The fluorocarbon-based FF used for the results shown here has a saturation magnetization of ≅400 Gauss and a low field magnetic susceptibility of $\chi \approx 3$. The FF is surrounded by a 50/50 mixture of isopropyl alcohol and deionized water that prevents the FF from wetting the glass plates. The rotational field strength is up to 100 Gauss rms at frequencies 20–40 Hz while the dc axial field is varied from 0 to ~250 Gauss.[146]

Next we detail an important microfluidic application of FF (i.e., valving and pumping) and highlight the prospect of ferrowaxes in fluidic platforms.

Hartshorne et al. demonstrated a FF-based valve.[147] In their work, a 600-cP FF was used to implement plugs to block fluidic channels. In air, this FF valve could hold pressures of up to 12 kPa (1.74 psi). When tested in water, water leakage at the FF/channel wall interface was observed. By making the channel walls hydrophobic, this water leak could be reduced. In the latter case, the valve could withstand pressures up to 10 kPa (1.45 psi). Besides their low pressure rating, FF plugs require the application of a magnetic

* A Hele-Shaw cell consists of two flat plates that are parallel to each other and are separated by a small distance, with at least one of the two plates being transparent.

field at all times in order to maintain the plug at the desired position.

An elegant approach for pumping with FF was introduced by Hatch and coworkers.[148] This team manipulated two FF plugs in a circular channel using a magnet that moves in a circular motion and maintains the plugs in place using stationary magnets. Using this scheme, one is able to pump defined volumes of liquid, given by the separation between the ferroplugs and the radius of the circular channel containing both pumping liquid and plugs.

Our team at University of California, Irvine, is attempting to use modified FF valves to create an inexpensive, normally closed, reusable valve that does not consume any power to stay closed. This type of valve is essential for all types of microfluidic platforms. For example, diagnostic reagents often must be stored on a chip or platform for months before they actually are used (the shelf-life of a diagnostics lab-on-a-chip device must be at least six months). Unless one has good, inexpensive liquid and vapor barriers/valves, liquids or vapors from the different reagent compartments will contaminate the system and even mix with each other, rendering the platform useless in a matter of hours. In principle, a regular pure wax valve fits the description: inexpensive, normally closed, and a good barrier for both liquids and vapors. Pal and colleagues[149] implemented pure wax valves, using paraffin wax and other waxes with higher melting points. Moving the resistance-heated molten wax by applying air pressure toggled the valve. Similarly, Liu and coworkers[150] used paraffin wax in their single-use valve. In this case, resistive heaters or Peltier elements melted the wax, while air pressure was again used for moving the molten wax. The authors claimed zero leakage when checking sealing against air. They measured a maximum holding pressure of 40 psi, while Pal et al. achieved 250 psi. Both groups succeeded in using a wax valve for sealing a polymerase chain reaction (PCR) chamber (remember that the maximum temperature needed for PCR thermocycling is ~92°C; see Figure 2.53). Liu et al. reported the use of paraffin valves (melting point ~50°C) 10 mm away from the PCR chamber, while Pal et al., using waxes with higher melting points (~85°C), could operate as close as ~5 mm from the PCR chamber. Use of microcrystalline waxes, with melting points up to 95°C, would be better yet for use in PCR applications.

An important drawback of the aforementioned wax valves is the need for an air pump to control the movement of the molten wax. This is where modified FF valves come in. A normally closed, reusable FF valve that provides effective sealing of a channel, for both vapors and liquids, and can be moved magnetically would thus be a very desirable feature. In the case of a regular FF valve, though, that objective is not met: to stay closed, a magnetic field needs to be applied, which is impractical during storage. Ferrowaxes, a modified type of FF valve, might be able to solve this problem.

In a ferrowax, magnetic nanoparticles are dispersed in a molten wax, and phase changes between solid and liquid are temperature controlled. A magnetic wax offers the opportunity of magnetic manipulation at will while the wax is molten and solidification at desired location (e.g., a valve seat) upon removal of the heat source. Additionally the wax, in its liquid phase, conforms to its holding container, and through the use of different kinds of waxes, the melting point can be tailored to different applications. Paraffin waxes appear to be the best choice for melting points of around 70–80°C, whereas microcrystalline waxes can take temperatures of around 95°C. Thanks to its nonpolar nature, the wax is immiscible in most biological samples. Advantages and disadvantages of the use of ferrowax are summarized in Table 8.15.

Kwang W. Oh and colleagues[151] demonstrated a ferrowax valve in a Y-shaped microfluidic junction. The valve is opened and closed by heating (using resistance heating) and magnetically displaced at will. The valve presented no leakage up to pressures of 50 psi.

Park and coworkers[152] implemented a single-use ferrowax valve in a compact disc (CD)-like centrifugal microfluidic device. This research group at Samsung took advantage of the absorption of laser energy by the embedded magnetic nanoparticles. The laser source operates at relatively low intensity to induce heat in the ferrowax causing a quick phase change to liquid. The pressure generated by the phase change causes the plug to move in a direction established a priori by the device's geometry, thus either

Table 8.15 Pros and Cons of Using Ferrowaxes

Pros	Cons
Relatively inexpensive	Heat source is required adding to the cost and complexity of the system
Energy efficient	Finite time of ferrowax phase change limits the time available for manipulation
Absorption of specific electromagnetic energy	Not yet commercially available (as of 2008)
Precision control	Industry scalability, since they require deposition of hot, viscous slurry
Nonpolar carrier used	
Melting point can be tailored	Melting point restricts the temperatures the at which the device might be used and/or stored
Do not interact with sample	
Conform to container geometry and provide an excellent seal after solidifying	

blocking or unblocking a channel or chamber. In this case, the magnetic nanoparticles are used solely to enhance heat absorption of the ferrowax and to induce a quicker phase change than that of paraffin wax alone. Our team at University of California, Irvine, tested the quality of the sealing of a ferrovalve for vapors and liquids qualitatively by immersing ferrowax sealed channels, containing calcium chloride, in a liquid environment. Even after more than twenty months, the calcium chloride did not show any sign of water uptake. A solidified ferrowax plug of around 2 mm in length with a cross-section of 1.75 mm^2 was also tested in a circular channel on the CD platform; the plug was located at a distance of 80 mm from the center of the rotating disc and did not show leakage up to 6000 rpm.

Regarding a real pumping application with ferrowax on the CD, the author's research group is currently exploring the use of these pumps to bring liquid samples back toward the center of the CD.[153] Indeed, a serious drawback in current CD microfluidics, besides the need of the type of valve described above, is the inability, after spinning fluids from the center to the rim, to move them back toward the center of the CD. This prevents the implementation of more complex diagnostic systems on a single standard-size CD, since the whole process of sample preparation and detection cannot be fitted easily onto a single CD platform. An FF plug could pump the sample back to the center of the CD for additional process steps (see also Volume I, Chapter 6, on fluidics, for more details on this CD platform).

The concept that we are implementing is the use of a ferrowax or FF plug as a pulling or pushing plug for the magnetic field to move the liquid inside a microchannel. To minimize water leakage at the interface of the pull/push plug and the CD plastic the walls of the channel are made hydrophobic. The concept is sketched in Figure 8.52. Theoretically, engineering a very soft ferrowax could eliminate the need for a heat source in a ferrowax system. In this case, the torque generated by the magnetic particles under a magnetic field could perhaps be made greater than the cohesion forces within the material, making the material move when a magnetic field is applied. The soft material might be a soft paraffin, gel-like polymer, or gelatinized nonpolar carrier that exhibits a "solid-like" behavior.

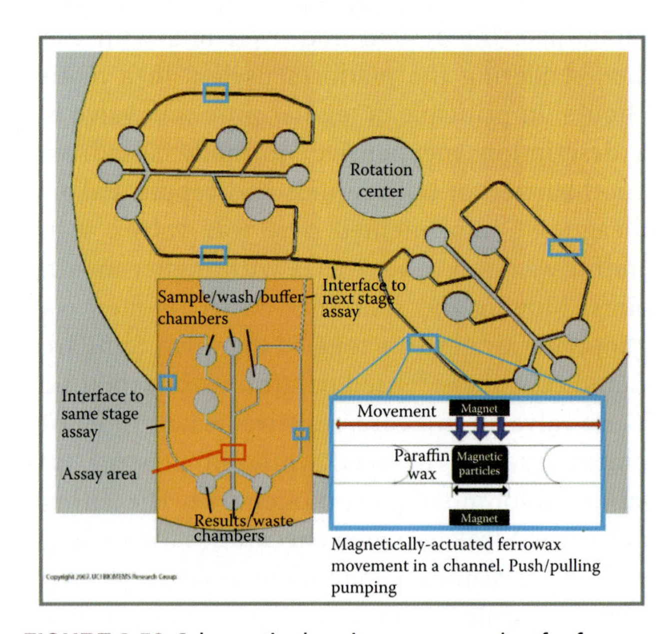

FIGURE 8.52 Schematic showing an example of a ferrowax pump. Ferrowax plugs, magnetically actuated by an external magnetic field, pull/push a sample toward the center of a compact disc (CD)-like centrifugal microfluidics platform. This way a sample might be reprocessed in one single CD platform, enabling a full diagnostic test (i.e., separation, filtration, amplification, and quantification) to be contained in a single platform. (From Zoval, J., and M.J. Madou. 2004. Centrifuge-based fluidic platforms. *Proc IEEE* 92(1):140–53.[153])

Superconductivity

Introduction

Given that actuators and power sources, in general, exhibit poor scaling in the microdomain, superconductors could play a dominant role in the more efficient powering and actuating of micromechanisms. The superconducting state is a new phase state in which, below a certain critical temperature (T_c), zero resistance and perfect diamagnetism is reached. It is well known that superconductors can levitate magnets because of their diamagnetism (Meissner effect). In this section we elaborate on superconductivity as it makes for the ideal Meissner levitators.

The levitating permanent magnets described above are ferromagnets with a positive and very large magnetic susceptibility, χ_m (χ_m usually is only constant in a small range of magnetization). Levitation can also be based on diamagnetic materials for which χ_m is small and negative. In diamagnetism, as a magnet approaches a diamagnetic material, magnetic dipole moments are induced in the diamagnetic material that oppose the applied field. These dipole moments lead to magnetic forces that tend to push away the magnet. Some ordinary diamagnetic materials, such as carbon and bismuth, with a χ_m of about -1×10^{-6} (compared with -0.20×10^{-6} for Au or -0.11×10^{-6} for Cu), can levitate magnets at room temperature. Although only a small force per unit mass, this kind of levitation does not require power, operates at room temperature, and can be used with a variety of materials. Kim et al., for example, levitated a permanent-magnet mover by applying the Meissner effect.[110] The ideal would of course be to have superconductivity at room temperature, but even a material superconducting at liquid nitrogen temperature (77 K) is amazing [liquid nitrogen ($1.04/L) is much cheaper than liquid helium ($3.75/L) and far easier to handle].

The Discoveries

Superconductivity represents a macroscopic quantum effect: one of the few cases where quantum effects become observable on a macroscopic scale. The electrons in a superconductor behave as a single particle. The Dutch physicist, Heike Kammerlingh-Onnes (1853–1926) (Figure 8.53), discovered the phenomenon in mercury in 1911, and in 1913, he received a Nobel Prize for his work.* The superconducting state may be defined as a new phase state in which, below a certain critical temperature (T_c), zero resistance and perfect diamagnetism is reached.

Heike Kamerlingh Onnes
(1853-1926)

FIGURE 8.53 The Dutch physicist Heike Kammerlingh-Onnes (1853–1926).

Most high-purity metals, when cooled to temperatures nearing 0 K, exhibit a gradually decreasing electrical resistivity, approaching some small yet finite value characteristic of the particular metal. For a few superconducting materials, such as mercury and indium, resistivity abruptly plunges from a finite value to one that is virtually zero at very low temperatures and remains there upon further cooling. Loss of resistance to electrical current flow occurs below a critical temperature, T_c. Below that temperature superconductors offer an energy transfer medium with virtually no power loss. In the case of pure elements such as mercury (see Figure 8.54) and indium, the T_c is between a few and 10 K. The change from the normal state to the superconducting state occurs over a temperature range of about a thousandth of a degree. One might wonder then if the resistance, R, really is zero and how one measures a zero resistance. To measure $R = 0$, a loop of a superconductor is made, a current is injected in it, and the magnetic field due to this current is measured as a function of time. Carrying out these experiments

* Kammerlingh-Onnes was also the first to liquefy helium at 4.2 K, using a "fancy" refrigerator, in 1908.

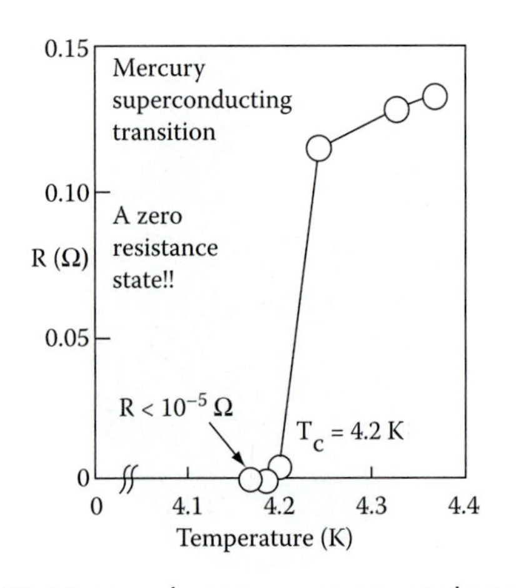

FIGURE 8.54 Mercury becomes a superconductor below 4.2 K.

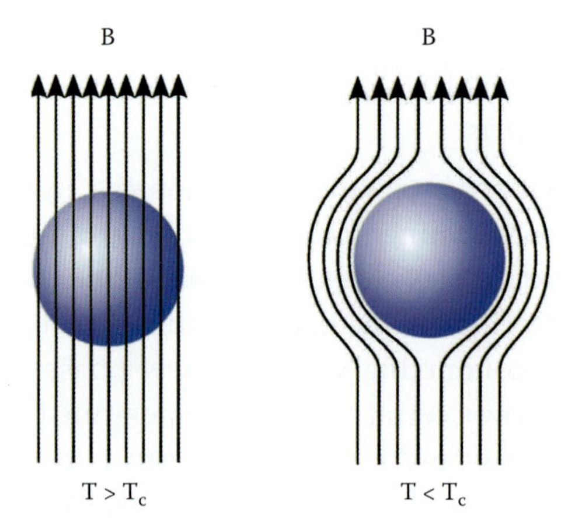

FIGURE 8.55 The Meissner effect for type I superconductors at $T > T_c$ and $T < T_c$.

turns out to be very boring, after monitoring the current via the magnetic field it generates for a number of *years* and observing no change, scientists typically shut the experiment down. We cannot say from this experiment that the resistance is really zero, but we can at least say it is far smaller than any capability or patience of ours to measure (the theory says it really is zero). In other words, once set in motion, electrical current will flow forever in a closed loop of a superconductor (R = 0), and all magnetic fields will be expelled from the same chunk of material (perfect diamagnetism). The Holy Grail for superconductivity researchers is, of course, a room-temperature superconductor, which would exhibit perpetual motion. But even a material superconducting at liquid nitrogen temperature (77 K) is amazing. [Liquid nitrogen ($1.04/L) is much cheaper than liquid helium ($3.75/L) and far easier to handle.]

The curious magnetic property of superconductors in which the material expels all traces of magnetism from its interior is called the *Meissner effect*; it was discovered in Berlin, in 1933, by Walther Meissner and Robert Ochsenfeld. Superconductivity baffled classical physicists. Below T_c there can be no resistance, so there can be no voltage inside the superconductor, and from Faraday's law it follows that the magnetic field inside cannot change. However, we actually observe that the field lines from the external field are "expelled" from the superconductor. How is that possible? The explanation is that a supercurrent (screening current) is induced on the surface in a direction such that it cancels the external field (see Figure 8.55). The magnetic field generated is strong enough to overcome gravity, so that a magnet can be made to levitate over the superconductor (see below, Figure 8.60). A superconductor such as this, which exhibits a complete Meissner effect, is called a *type I superconductor*, a soft superconductor, or a low-critical-temperature superconductor (LTS). Such a superconductor is limited in its current-carrying capability because it can tolerate only very small magnetic fields. As we shall see below, type II semiconductors can carry much larger currents but are not perfectly diamagnetic.

Among pure elements, niobium shows the highest temperature for superconductivity, and there are about a dozen niobium compounds (examples include niobium carbide, niobium titanium, niobium gallium, and niobium germanium) that all show super conductivity up to about 30 K (liquid hydrogen range). Nb_3Ge was discovered in 1973 and was long the record-holder for highest T_c (see Figure 8.56). Some of the niobium compounds have been drawn into wires to produce very compact magnetic fields for use in, for example, magnetic resonance imaging (MRI). In Table 8.16 we present a review of typical metal superconductors.

Until 1986, the T_c was 23 K for the best low-temperature superconductors, requiring liquid helium for maintaining the superconductivity state. In 1986, Bednorz and Muller (Figure 8.57), working at the IBM Zurich Research Laboratory, discovered higher-temperature (35 K) superconductivity in a lanthanium-barium-copper oxide. Their compound was a ceramic material and, ironically, a very poor conductor at room temperature.

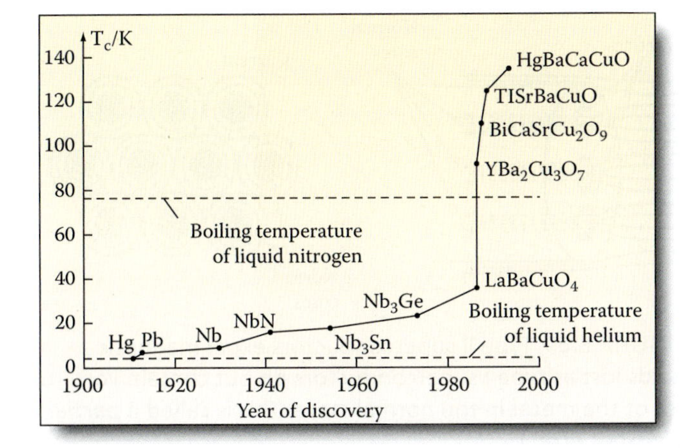

FIGURE 8.56 The critical temperature, T_c, of various superconductors plotted against their discovery dates.

FIGURE 8.57 George Bednorz and Alex Muller, 1987 Noble Prize winners.

Bednorz and Muller's ceramic was the first nonmetal that begins to superconduct above the boiling point of liquid hydrogen. Their invention triggered a race to find superconducting materials with transition temperatures above 77 K, the boiling point of liquid nitrogen. Overnight, anybody who knew about superconductors or ceramics started cooking up new compounds in his or her furnace, and a record number of high-temperature superconductors was announced over the next couple of years (see Figure 8.56). Paul Chu, of the University of Houston, made a stunning breakthrough when he announced the discovery of a material that became superconducting at 93 K, well above liquid nitrogen temperature. Early in 1987, Chu circulated a preprint of a paper describing superconductivity at 93 K in a ceramic containing Yb, Ba, Cu, and O. When the paper appeared in print, every Yb (ytterbium) had been replaced by Y (yttrium)! It was just a typo, Chu* explained. By January 1988, another copper-oxide material was shown to exhibit superconductivity at a high temperature, 110 K. The world record T_c for a superconductor is 138 K, for a compound containing mercury, thallium, barium, calcium, copper, and oxygen. Some superconductor ceramic materials are listed in Table 8.17. The new breed of ceramic superconductors is referred to as *hard superconductors, type II superconductors*, or *high-critical-temperature superconductors* (HTS). The two types of superconductors are compared in Figure 8.58.

In the case of a small applied magnetic field, type II superconductors act like type I superconductors. In the presence of large magnetic fields, a type II "sacrifices" part of itself so that the rest can remain superconducting. These superconductors contain vortices that are filaments of the material in the normal

TABLE 8.16 Type I Superconductors, Soft Superconductors, and Low-Critical-Temperature Superconductors

Compound	T_c (K)
Al	1.17
α-Hg	4.15
In	3.41
Nb	9.25
Ru	0.49
Sn	3.72
Ti	0.40
Zn	0.85
Al_2Y	0.35
$AuPb_2$	3.15
InPb	6.65
Ir_2Th	6.50
Nb_2SnV	9.80
CuS	1.62
Nb_3Sn	18.00
TiO	0.58
SnO	3.81
$(SN)_x$	0.26

* This would not be the last intrigue haunting superconductor research. In May 2002, outside researchers presented evidence to Bell Labs management of possible manipulation of data involving five separate papers published by its scientists in *Science, Nature,* and *Applied Physics Letters* over a two-year period by Jan Hendrik Schön as lead author and his colleagues at Murray Hill and elsewhere as coauthors. Schön is the only researcher who coauthored all five papers in question.

Table 8.17 Examples of Type II Superconductors with Their Critical Temperatures (T_c)

Compound	T_c (K)
$YBa_2Cu_3O_7$	93
$YBa_2Cu_4O_8$	80
$Y_2Ba_4Cu_7O_{15}$	93
$Bi_2CaSr_2Cu_2O_8$	92
$Bi_2Ca_2Sr_2Cu_3O_{10}$	110
$Tl_2CaBa_2Cu_2O_8$	119
$Tl_2Ca_2Ba_2Cu_3O_{10}$	128
$TlCaBa_2Cu_2O_7$	103
$TlCa_2Ba_2Cu_3O_8$	110
$Tl_{0.5}Pb_{0.5}Ca_2Sr_2Cu_3O_9$	120

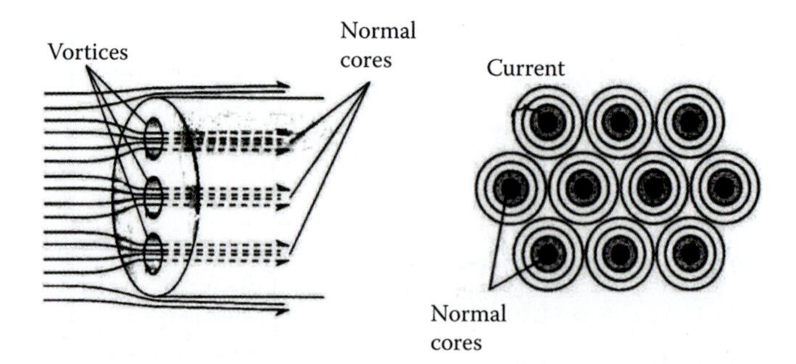

FIGURE 8.59 Type II superconductors expel magnetic fields just as type I superconductors do but contain vortices of the metal in the normal state. This is called a *partial Meissner effect.*

state as shown in Figure 8.59. The magnetic field lines pass through those vortices. High-temperature superconductors thus exhibit only a partial Meissner effect. Compared with type I superconductors, type II superconductors can carry enormous currents (in some cases, superconducting wires can carry currents that are thousands of times larger than those that can be handled by ordinary wires of similar dimensions), but these brittle ceramics are very difficult to fabricate into wires.

In Figure 8.60, we show a magnetic cube, placed on a disc made of a superconductor material. As the disc is cooled down with liquid nitrogen, superconductivity sets in below its critical temperature and the magnetic repulsion levitates the cube.

The Model of John Bardeen, Leon Cooper, and Robert Schrieffer (BCS Model)

The mechanism of superconductivity remained a mystery from 1911 until 1957. The abrupt change from normal state to superconductive state indicated clearly that superconductivity is not merely an extension of ordinary conductivity. Three important observations having to do with superconductors that gave hints about its origins were the following:

1. The Meissner effect exists.
2. The "best" metals do not seem to be superconductors. Among superconducting metals, better metals make poorer superconductors, and vice versa.
3. A lighter isotope of mercury becomes superconducting at a higher temperature (i.e., more readily) than a heavier isotope.

From quantum theory, we know that electrical conductivity arises because of acceleration of electrons near the Fermi surface and that electrons move in conductor-like waves rather than as particles. Because they act as waves, electrons may travel

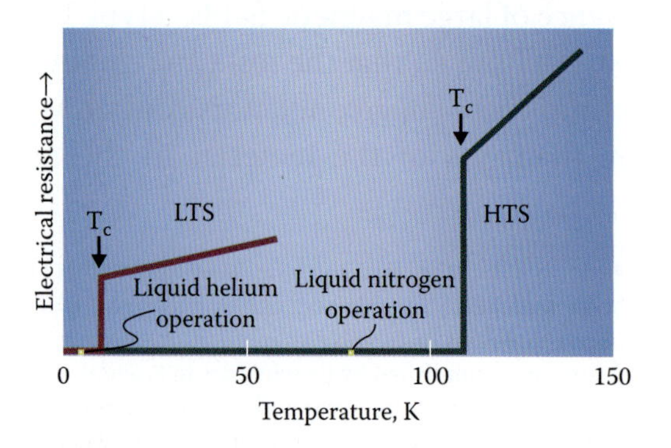

FIGURE 8.58 Low- and high-critical-temperature (T_c) superconductors compared.

FIGURE 8.60 A magnetic cube is placed on a disc made of a superconductor material. As the disc is cooled down with liquid nitrogen, superconductivity sets in below its critical temperature, and the magnetic repulsion levitates the cube.

FIGURE 8.61 The theory of superconductivity by Bardeen, Cooper, and Schrieffer (left to right) earned them a joint Nobel Prize in 1972.

through a metal without colliding with the metal ions, and consequently, in principle at least, one could get zero resistance. If electrons do not collide with metal ions, what do they collide with? After all, metals do have finite resistivities? As we discovered in Volume I, Chapter 3, even in a pure material, thermal oscillations of the atoms, or phonons, will scatter moving electrons, causing finite resistivity; electrons do not bump into ions at the regular crystal sites but scatter from ions that happen to be vibrating and in the "wrong" place when the electron wave passes by. Impurity atoms and defects such as vacancies, dislocations, and twin and grain boundaries will scatter the electrons even more, further increasing the resistivity. When the temperature is lowered, the phonon contribution to resistivity is expected to fall away, but the contribution caused by impurities should remain finite. In the case of superconductivity, all these resistance-causing factors are somehow overcome.

John Bardeen, Leon Cooper, and Robert Schrieffer (Figure 8.61), all at the University of Illinois, first explained superconductivity in 1957 and won the 1971 Nobel Prize for their BCS model (this was Bardeen's second Nobel Prize). The model works for low-temperature superconductivity only; there is no explanation yet for the ceramic HTS. The key observation that led to the BCS model was the isotope effect, which puts one on the trail of phonons; since isotopes are involved, atoms, and perhaps their movement, must somehow play a role in the phenomenon. To start the argument, remember that in a normal metal conductivity (σ) is proportional to scattering time (τ). In superconductors, there are no final states for the scattered electrons to go to, so scattering is simply quantum mechanically not allowed. What this means is that the difference between a normal metal and a superconductor is that in the case of the superconductor there is a bandgap, as shown in Figure 8.62.

Because of this bandgap in momentum space, scattering is forbidden, i.e., there are no final states for scattered electrons to go to. The energy gap is caused by attractive interaction between electrons. Somehow pairing of electrons reduces their energy. This is rather curious; how can electrons attract each other and form pairs with a lower energy content? The answer is that there is a *phonon glue* that causes their interaction to be attractive.

This needs some further detailing. Phonons, as we know, are periodic vibrations of atoms in solids; think of plucking an atom with nanotweezers. The forces between the different atoms pull the plucked

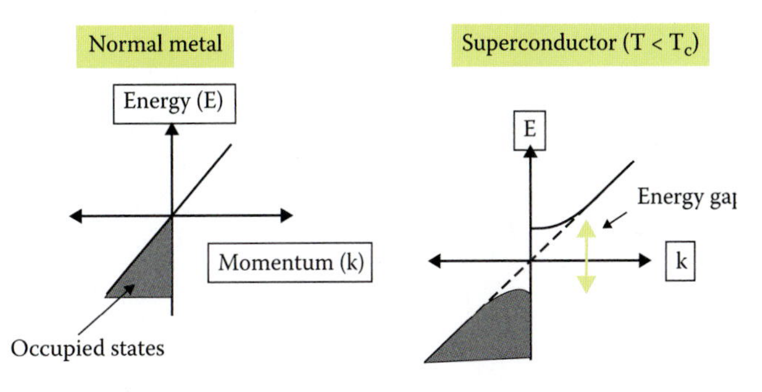

FIGURE 8.62 An energy bandgap in momentum space in a superconductor.

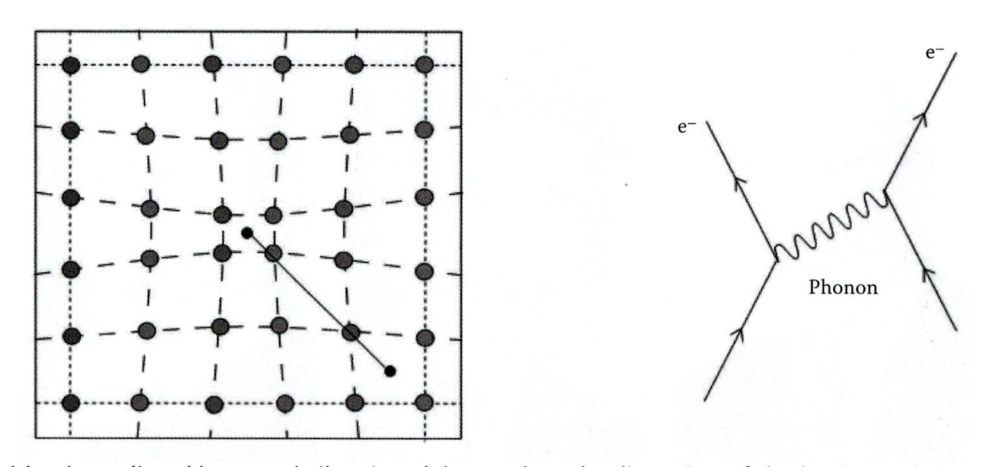

FIGURE 8.63 Pairing is mediated by crystal vibrations (phonons), and a distortion of the lattice as shown causes Cooper pair formation.

atom back into position, setting up a wave of atomic vibrations that travels through the material. Pairing is mediated by these crystal vibrations (phonons), and it is a distortion of the lattice, as shown in Figure 8.63, that allows the formation of the Cooper pairs, as shown in Figure 8.64, to take place.

A close-up of the crystal deformations leading to the formation of Cooper pairs is shown in Figure 8.65. As electrons travel between the ion cores of the metal they exert an attractive force on the positive nuclei, and these move very slightly away from their idealized positions. As a result of this disturbance, the distribution of positive charges is no longer uniform and features regions of small excess of positive charges. This extra charge is sufficient to attract a second electron, and so two electrons are linked, albeit indirectly. If the attraction exceeds the electron-electron repulsion, a "bound pair" is formed. The bound pair is called a *Cooper pair*. This mechanism explains why good conductors such as Cu and Ag are poor superconductors; these materials are good conductors at room temperature exactly because their conduction electrons do not strongly interact with the ion cores, and as a consequence we would not expect such materials to form good superconductors since the mechanism for forming pairs is very weak. This pair formation also explains the abrupt change from normal to superconducting state. Such a sudden transition is known as a *phase change*. Just as ice melts suddenly at 0°C, the thermal motion of the molecules at a critical temperature abruptly destroys the Cooper pairs.

The model also explains why a lighter isotope of mercury becomes superconducting at a higher temperature (i.e., more readily) than a heavier isotope; the smaller the mass of the positive ion core, the greater the distortion an electron can induce (to be more exact, $T_c M^{1/2}$ = constant). But how do those Cooper pairs explain superconductivity? Electrons in a superconductor move in correlated pairs in such a way that the properties of each pair are the same

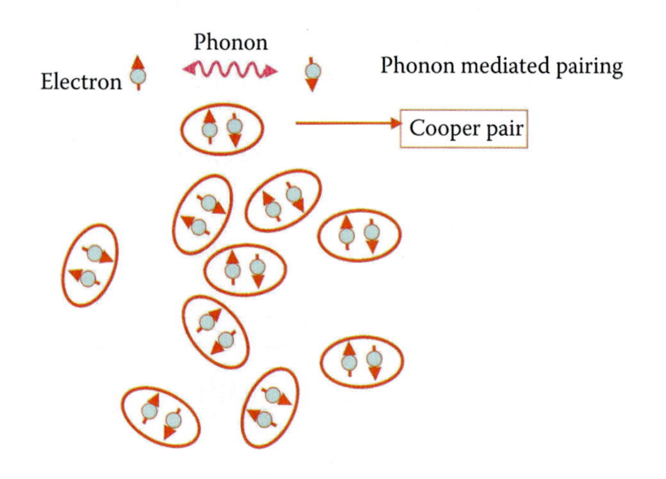

FIGURE 8.64 Cooper pair formation.

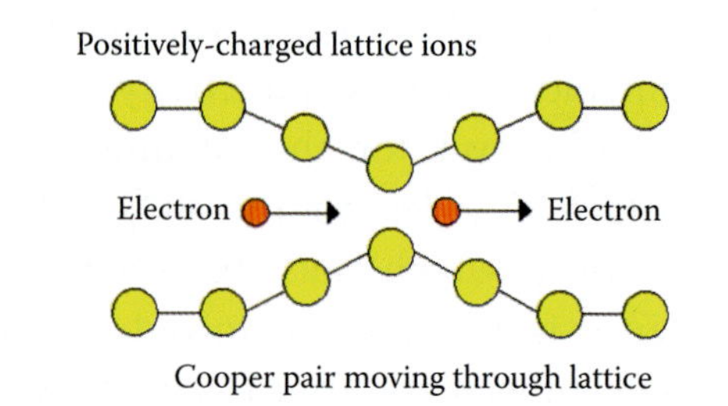

FIGURE 8.65 A lattice deformation forming a Cooper pair.

as those of all the others. These paired electrons are not necessarily close; on average they might be separated by hundreds or even thousands of atom spacings. The maximum distance up to which the states of paired electrons are correlated to produce superconductivity is called the *coherence length*. Coherence length of Cooper pairs is about 10^{-4} cm.

The electrons in Cooper pairs have opposite spins and equal and opposite linear momenta, or the pairs have zero spin and zero momentum, which makes them bosons instead of fermions (see Figure 8.66 and Volume I, Chapter 3).

In the absence of a current there is no net movement of electrons, and all the electron pairs have an average velocity of zero. Zero-spin particles such as Cooper pairs are bosons (like photons and phonons), and they "like" each other so much that they all crowd together in the same state (energy and momentum). Cooper pairs condense into one macroscopic quantum state with $\sim 10^{23}$ particles that are all coherent. If a current flows, all the pairs have the same velocity and move in the same direction, as one collective body. In the superconducting state, the Cooper pairs are constantly scattering each other, but the total momentum remains constant. Remember—the paired electrons have opposite momentum. If one gains momentum, the other loses momentum. Because all Cooper pairs act together, a single lattice ion cannot scatter a single Cooper pair or they do not lose energy by interacting with the lattice. With all the electrons acting as a team, the only resistance possible is for all electron pairs to simultaneously scatter into a different state, an event that is so implausible that all attempts to measure any resistance in a superconductor have failed. This is the same as saying that there are no energy levels for the scattered pairs to go to, or that there is an energy gap. Because all the electrons act together to form a single quantum state, electrons flow around a wire as electrons in an atom do. Zeeman splitting takes place when a material is placed in a high magnetic field. When Zeeman splitting occurs in superconductors, the field is great enough to force a Cooper pair to occupy the same energy level and thus spin in the same direction. This breaks the Cooper pairs and superconductivity.

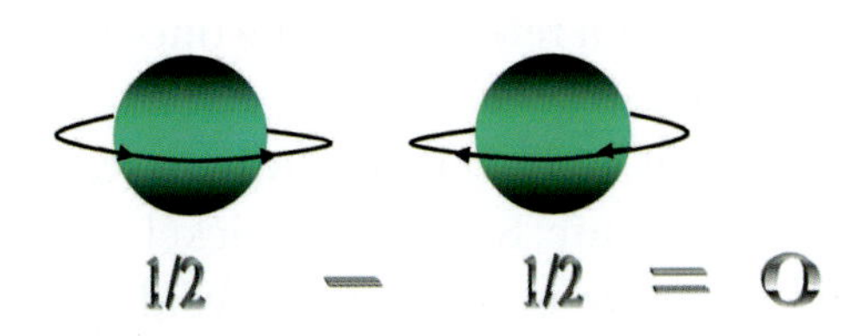

FIGURE 8.66 How to make bosons out of electrons.

In summary, the BCS model is based on a phonon-mediated attractive interaction between electrons, leading to the formation of Cooper pairs; these pairs act like bosons, which fall into a coherent superconducting state. BCS theory beautifully explained all aspects of superconductors, but it seemed to predict that 30 K (give or take a couple of degrees) is the highest possible temperature at which a material could be superconducting, as above that temperature lattice vibrations destroy the Cooper pairing: the BCS model explains the LTS but not the HTS. BCS theory is clearly a good starting point, and just as clearly it is not the final answer. If we understood HTS, it might be possible to propose a room-temperature superconductor.

Thermal Actuators[68]

Introduction

In thermomechanical actuators (TMAs), thermal energy is transformed into mechanical movement. A μ-TMA generates controlled motions via thermal strains. The simplicity of the system is the major reason that many MEMS devices are actuated by thermal means.

In Figure 8.67, we illustrate an example of a TMA, i.e., a parallel beam μ-TMA. By applying a current, the narrow beam shown in this structure is heated

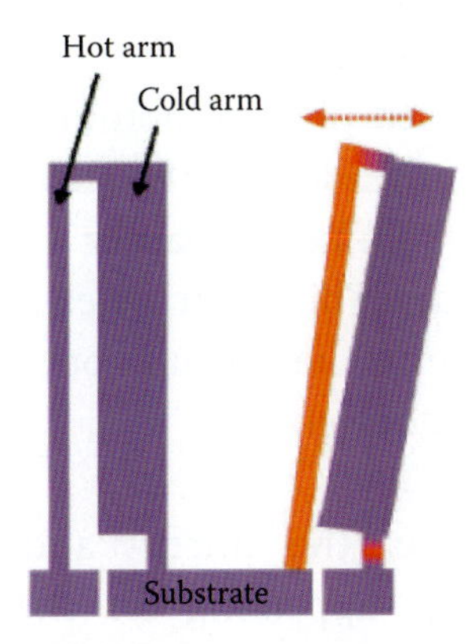

FIGURE 8.67 Parallel beam thermomechanical actuator.

up more and expands also more because of its higher resistance and higher current density than the wider beam, causing a movement as indicated. Removing the current restores the structure to its original position.

Scaling of Thermal Actuators

Scaling laws in heating and cooling, such as the total amount of heat required to start a thermal actuator, its heating and cooling rate, the thermal stresses or distortions expected, and possible damage to delicate MEMS and NEMS components, were detailed in Volume I, Chapter 6. In the same chapter we also analyzed the breakdown of the heat transfer continuum theory in the microdomain, and in Volume I, Chapter 3, we discussed quantized thermal transport in nanoscale devices.

Here we reiterate the most salient points of those scaling insights as they apply to thermal actuators:

1. Small devices, having a smaller mass, heat and cool faster than larger structures (see Volume I, Equation 6.68).
2. The Biot number (Bi) relates the heat transfer resistance inside and at the surface of a body (Volume I, Equation 6.70). This dimensionless number gives an indication of the potential for thermal stresses. For Bi $\ll 1$, internal temperature gradients become small and the microstructure can be treated as having a uniform temperature. In the microscale, the Bi is small, so that conduction accounts for 99% of the energy transport. Consequently, with small devices there is less worry about thermal stresses induced by thermal gradients, and systems with larger internal heat generation capacity by volume can be built.
3. The increased phonon-boundary scattering in nanostructures lowers the thermal conductivity κ (see Volume I, Chapter 3 and Chapter 3, this volume). This contributes, for example, to making better heat insulators and more efficient thermoelectrics.

From the above, the micro- and nanodomain seems especially appealing for thermal actuator applications.

Types of Thermomechanical Actuators

Introduction

In this section we compare pneumatic, bimetallic, and shape memory alloy thermal actuators by considering their use in microvalves. These three types of thermal actuator mechanisms employed in microvalves are illustrated in Figure 8.68. In the section that follows we detail the choice of actuator types in commercially available valves.

Thermopneumatic Actuators

The thermopneumatic valve (see Figure 8.68a) traps a liquid in a sealed cavity containing a thin film heating resistor along one side of the cavity, with a flexible diaphragm wall forming the opposite side.[154] Upon heating, the fluid in the chamber expands and evaporates, raising the pressure in the sealed cavity and causing the flexible wall to bulge outward. In a normally open valve, this bulging diaphragm closes a nearby orifice. In a closed valve, as illustrated in Figure 8.68a, actuation levers a silicon body away from an orifice. Compared with other microvalve technologies, thermopneumatic actuation exerts tremendous force through a long stroke. In a normally open valve, over 20 N through a stroke of 50 μm was demonstrated.[154] This long stroke allows a thermopneumatically actuated valve to control higher flow rates and a wider range of pressures. This type of valve was made at Redwood Microsystems, Inc., a company that is now defunct.

Bimetallic Actuators

The second thermal actuator, illustrated in Figure 8.68b, is a bimetal. A bimetal involves the combination of two materials with distinct thermal expansion coefficients. A bimetal strip will bend when heated or cooled from the initial reference temperature because of incompatible thermal expansions of the materials that are bonded together. The bimetal returns to its initial reference shape once the applied heat is removed.

In the case of a bimetal structure based valve, as illustrated in Figure 8.68b, the nickel and silicon relax and the valve is closed when cold; upon heating, the nickel expands more than the silicon, lifting

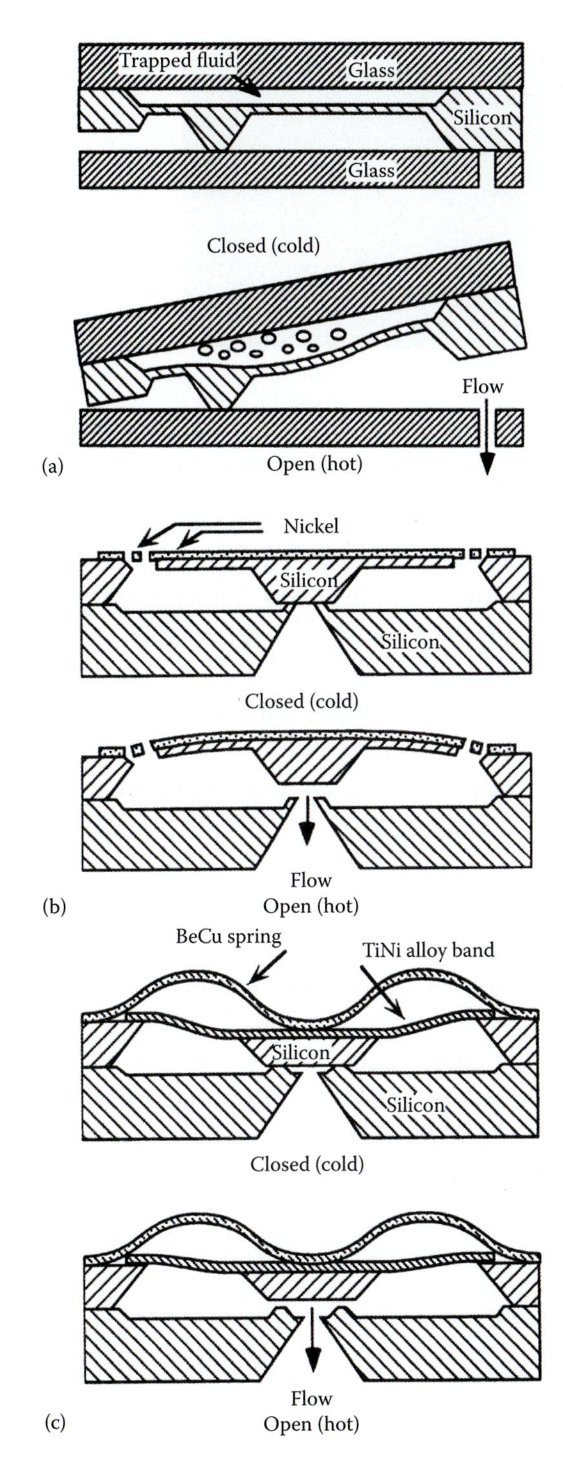

FIGURE 8.68 Schematics of thermally actuated valves. (a) Schematic of a normally closed thermopneumatic valve. (After Zdeblick, M.J., R. Anderson, J. Jankowski, B. Kline-Schoder, L. Christel, R. Miles, and W. Weber. 1994. In *Technical Digest: 1994 Solid State Sensor and Actuator Workshop*, 251. Hilton Head Island, SC.[154]) (b) Schematic of a normally closed nickel/silicon bimetal valve. (c) Schematic of a normally closed shape memory alloy valve. [After Barth, P.W. 1995. In *8th International Conference on Solid-State Sensors and Actuators (Transducers '95)*, 276. Stockholm, Sweden.]

the silicon body from the valve seat. In terms of size, bimetal actuators are smaller than thermopneumatic ones and consume less power, but they lack the longer stroke of the latter.[155]

Shape Memory Alloy (SMA) Actuators

A third thermal actuator is based on shape memory alloys (see Figure 8.68c). Shape memory alloys undergo reversible thermal-mechanical transformations of their atomic structure at certain temperatures. Specifically, SMAs such as NiTi undergo crystalline phase transformations from a weak and easily deformable state at low temperatures to a hard and difficult-to-deform state at higher temperatures. The material is first held in the desired shape and heated to well above the transition temperature (T_{tr}). At these temperatures, the crystal structure is in the austenite or parent phase. Cooling transforms the material into martensite, the low-temperature phase, which plastically deforms (solder-like) by as much as 10% at relatively low stresses (yield strength is only 100 Mpa). When the SMA is heated again above its transition temperature, T_{tr}, it transforms back to the high-temperature phase (austenite) and reverses to its originally high-temperature shape, exerting a substantial force (>100 MN/m^2). The austenite phase has a much higher yield point and can sustain stresses up to 560 MPa without permanent deformation.[156] The large forces make SMAs ideal for actuation purposes; piezoelectric and electrostatic actuators exert only a fraction of the SMA force, but they act much faster. The temperature interval over which the shape change takes place typically is 10–20°C. The lowest energy path to the austenite exactly retraces the atomic movements responsible for the deformation, causing the shape memory. The discovery of the effect in 1951 involved a gold-cadmium alloy, but this was soon extended to a broad range of other alloys, including titanium-nickel, copper-aluminum-nickel, iron-nickel, and iron-platinum alloys. For practical applications, Ni-Ti, Cu-Zn-Al, and Cu-Al-Ni all have been tried. Depending on the type of alloy and the alloy composition, critical temperatures range between −150°C and +150°C.

Besides their use in μ-valves, shape memory alloys have been used in numerous applications (see Figure 8.69) including medical (e.g., microsurgical

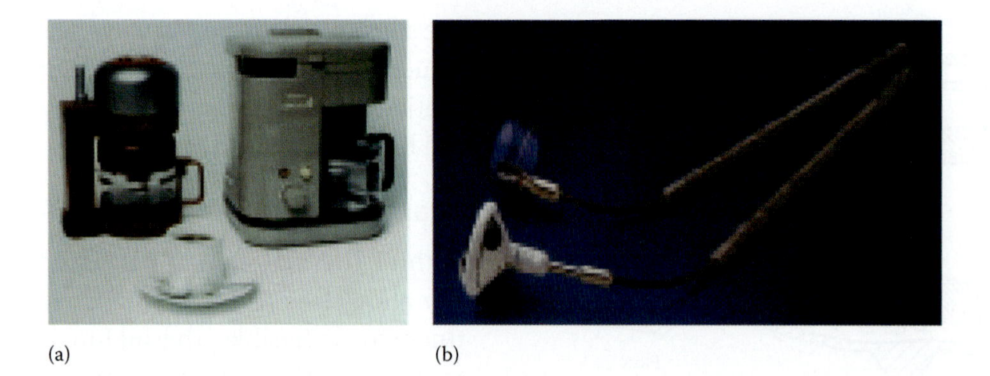

(a) (b)

FIGURE 8.69 Examples of shape memory alloy (SMA) applications. SMA valve opens and releases hot water at the proper temperature (a). NiTi surgical tools: precise bending of the tool to the proper shape in microsurgery (b).

instruments, stents, artificial muscle), aerospace (e.g., connectors, lock rings), automotive (e.g., Ni-Ti thermostat), industrial (e.g., pipe connectors), consumer (e.g., eyeglass frames), and safety (e.g., fire safety valves).

The most extensive work with shape memory alloys involves bulk materials and bulk titanium-nickel alloys in the form of wires and rods, which are commercially available under the name Nitinol™. Nitinol, a composition transforming near room temperature, is traditionally prepared with 50% Ti and 50% Ni. Increasing the Ni content decreases the transition temperature by about 25°C per 0.2 at % Ni. The material is a good electrical conductor, with a resistivity of 80 μΩ·cm, but a relatively poor thermal conductor, with a thermal conductivity about one-tenth that of silicon.

The more IC-compatible sputtered films were made in the late 1980s by Walker et al.[157] at Bell Labs and by Busch and Johnson[156] at TiNi Alloy Company (http://www.sma-mems.com). Thin-film, nickel-titanium shape memory alloys up to 50 μm in thickness have been sputter deposited. These films exhibit memory behavior comparable to bulk material, and the phase transition to and from martensite lies entirely above ambient temperature (see Figure 8.70).[158] When deposited at room temperature, NiTi films are amorphous. Heating to 500°C causes the film to crystallize and acquire its shape memory property.[158]

Some materials and processes are incompatible with such high temperatures, requiring either rapid thermal annealing (RTA) or a special choice of materials and careful selection of the order in which processes are carried out. Another limitation of thin-film NiTi is the upper limit of 70–80°C for transition temperature T_{tr}, which is too low for applications such as valves for fuel injection systems and many military applications. Experimentation with higher-temperature tertiary alloys such as TiNiHf is under way to alleviate this problem. The bandwidth of the NiTi films is slightly improved, to about 5 Hz, compared with the NiTi wire (about 1 Hz) because of the more effective cooling in the microdomain. Another advantage of TiNi thin-film actuators is that they are biocompatible and transistor-transistor logic (TTL) voltage compatible.

We compared the energy density and drive conditions for NiTi with magnetic, electrostatic, and piezoactuators in Table 8.8. NiTi outperforms magnetic and electrostatic actuation in terms of energy density and may cause large motions in the range from 10 μm to 1 mm, operational requirements not well matched by electrostatic or piezoelectric

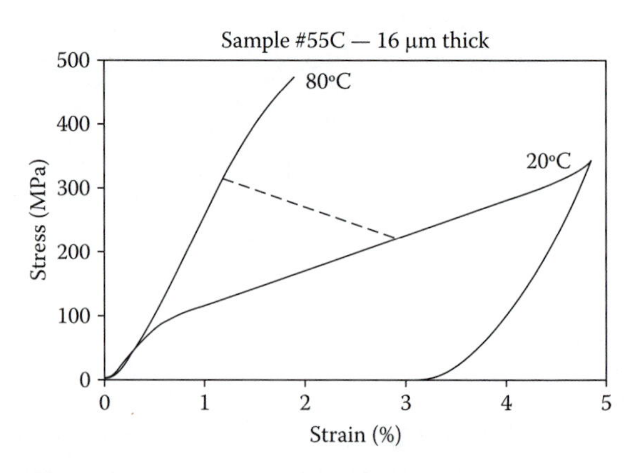

FIGURE 8.70 Stress-strain curve for a TiNi shape memory alloy thin film. The sloping, broken line indicates a load line spring force resisting the actuator, showing a 2% repeatable strain recovery. (After Johnson, A.D. 1991. Vacuum-deposited TiNi shape memory film: Characterization and applications in microdevices. *J Micromech Microeng* 1:34.[158])

technologies. As the required voltage for actuation is low (a couple of volts), it becomes compatible with IC technology. The phase change in SMAs is temperature driven, and as heat must be removed before the next cycle starts, this makes the system inefficient—the cycle rate may be slowed down by the achievable rate of heat transfer. The ultimate efficiency cannot exceed that of the Carnot cycle operating between the same temperature limits.[159] Practical SMA actuators have an efficiency typically less than 10%; for example, 50 J of heat input might be required to obtain 1 J of mechanical energy output.[158] However, in very small devices, heat transfer is rapid, which makes this type of actuator more attractive. Another potential disadvantage is the need, in most applications, of mechanical biasing devices to provide a return force such as the BeCu spring, enabling cyclical behavior in the normally closed valve sketched in Figure 8.68c. The beryllium-copper spring puts bands of SMA in tension over the orifice at low temperature. The SMA bands constitute electrical resistors and are heated by current passage. At high temperature, the SMA element works against the BeCu spring to reach its high-temperature preset form and moves away from the valve seat, opening the valve. At low temperatures, the spring easily pushes the silicon body back onto the valve seat, closing the valve. The spring adds somewhat to the size of the SMA actuator and also further slows down the response time (0.5–2 Hz). If the high-temperature shape does not change upon cooling the specimen and transforming it to the martensic phase, the phenomenon is referred to as a *one-way* shape memory effect, and springs or other mechanical biasing mechanisms are required. Solutions to avoid mechanical biasing might be within reach: SMAs can, under certain circumstances, remember their low-temperature shape as well, enabling cyclic device operation without the need of a bias.[160] In the latter case, one refers to a "reversible" shape memory effect. The low bandwidth of SMA usually is assumed to be determined by the relatively long cooling thermal time constant. Hunter et al. report an interesting unexplained effect in which very large brief current pulses ($>10^9$ A/m^2), imposed during externally shortening and lengthening cycles, alter the subsequent NiTi switching properties.[161] The altered NiTi shortens and lengthens very rapidly (within 40 ms) and generates a maximum extrapolated stress of 230 MN/m^2 and a peak power/mass approaching 50 kW/kg.[161]

Shape memory plastics also are being developed. Norsorex, for example, is the trade name for a polynorbornene polymer with excellent shape recovery properties at shape memory temperatures of over 35°C. Another polymer is Zeon Shable, a polyester-based polymer blend.[162] There are many potential applications for shape memory alloys and patents abound in this area (upwards of 15,000).

Summarizing, the advantages of SMA actuation include high force/weight and force/volume ratios, large deformation, simplicity of implementation via a heating current fed through the SMA material, simple cooling by the ambient environment (especially effective for small structures with a small Biot number), and inexpensive raw materials. Disadvantages include the one-way operation necessitating a bias force, heating/cooling cycles that reduce the bandwidth, the number of cycles reducing the maximal deformation (after $>10^5$ cycles, the maximal deformation is less than 4%), the fact that cycling changes the properties of the alloy, hysteresis (10 ... 30°C), and nonlinearity.

Comparison of Actuator Principles in Commercial Valves

Miniature, highly accurate valves are of importance in many industrial operation processes. In Table 8.18, listing four commercially available thermal valve technologies, we compare the merits of the various underlying thermal actuator mechanism reviewed in this section.[155] Parameters of interest for valve performance include power consumption, actuation speed, force, stroke, cost, and package size.

Hewlett-Packard (HP) has come up with a thermally actuated valve intended for a very small, possibly handheld, gas chromatography (GC) system.[163,164] A schematic of HP's normally closed nickel/silicon bimetal valve was shown earlier in Figure 8.68c (a more detailed view is presented in Figure 8.71b). Resistive heating of the nickel/silicon bimetal causes the membrane deformation and opening of the valve. The competing IC Sensor's

Table 8.18 Comparison of Micromachined Thermal Valves

Parameters	Shape Memory Alloy (Microflow Analytical, Inc.)	Al (IC Sensors, Inc.)	Ni/Si (HP)	Thermopneumatic (Redwood Microsystems)
Pressure (PSIG)	80	25	150	100
Flow (sccm, air or N_2) at T = 25 °C	6000	100	1000	2000
Power (W)	0.29	0.5	1.03	2.0
Response time (ms)	100	100	200	400

Source: After Barth, P.W. 1995. In *8th International Conference on Solid-State Sensors and Actuators (Transducers '95)*, 276. Stockholm, Sweden.[155]

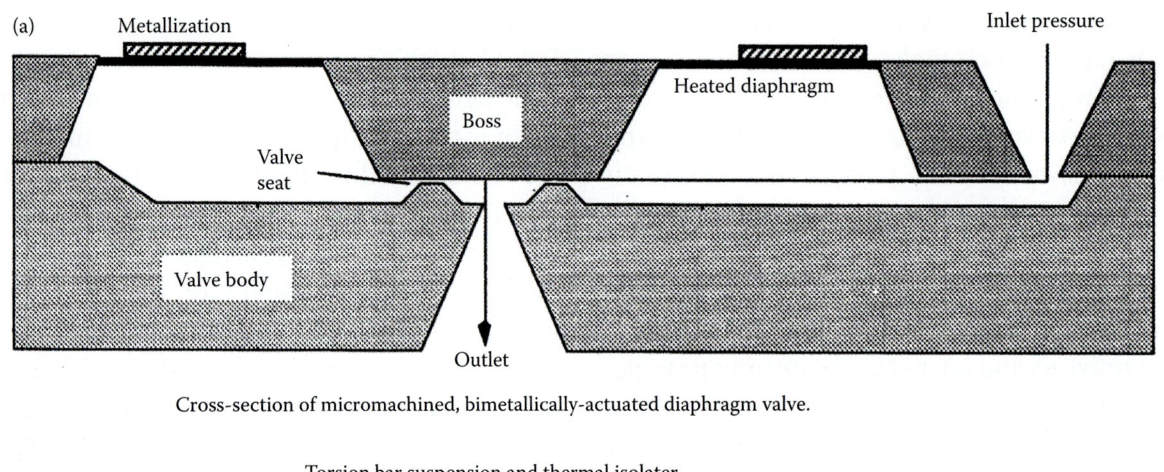

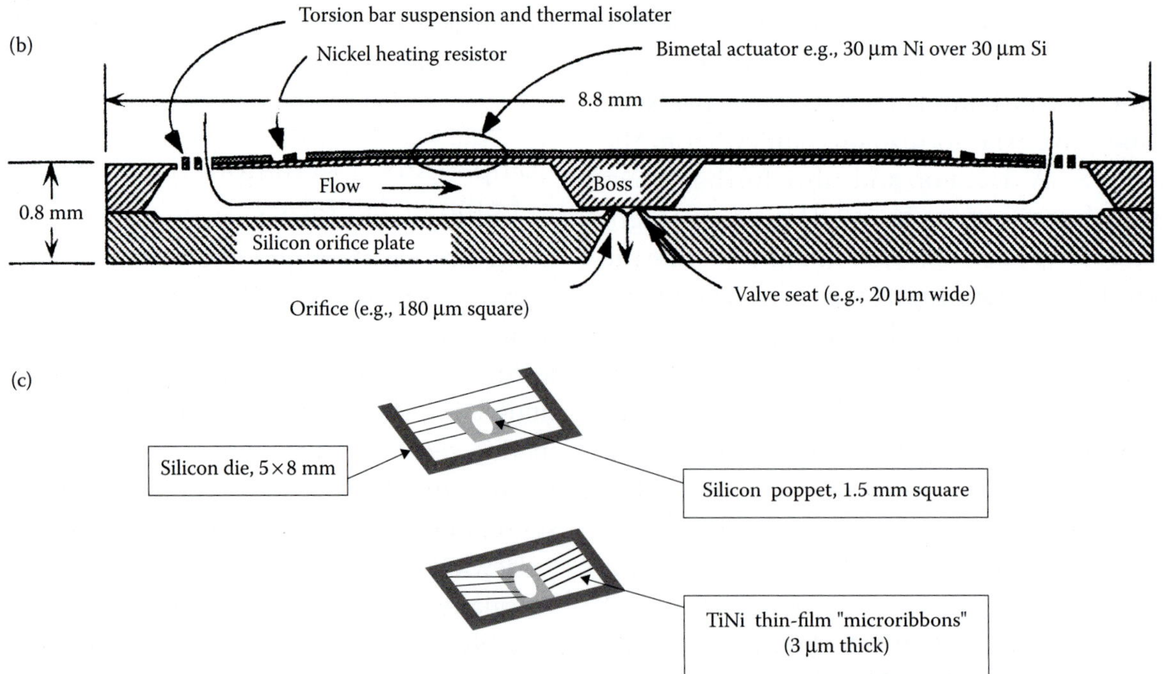

FIGURE 8.71 Commercially available, micromachined, thermally actuated valves. (a) IC Sensor's bimetallic Al/Si valve. (b) Hewlett-Packard's bimetallic Ni/Si valve. (c) The TiNi actuator consists of a rectangle of silicon, 5 mm wide and 8 mm long, that has been partially removed to leave a poppet supported by 8 TiNi thin-film "microribbons." The lower drawing shows these microribbons stretched. When electrical current runs through these microribbons, the shape memory property causes them to contract. The upper view shows the microribbons in their contracted state. The poppet displacement is more than 100 µm, and it can exert a force of as much as 0.5 N (http://www.tinialloy.com/devices.html).

valve (Figure 8.71a) is also a normally closed bimetallic valve but uses an annular aluminum region on a Si diaphragm with integral diffused resistors instead. By varying the electrical power dissipated in the resistors, and thus the temperature of the Si diaphragm, the displacement of a central boss can be controlled.[165] The HP valve exhibits an increase of an order of magnitude in both pressure and flow control capability compared with the IC Sensor's valve. This improvement is made possible by using a torsion bar suspension (see Figure 8.71b) and a nickel-and-silicon bimetal combination, which allows larger valve dimensions and higher stress levels for consequent longer stroke and higher force. For very small structures, this thermal type of actuation becomes an attractive option because of geometric scaling (see comments above and in Volume I, Chapter 6 on the Biot number). The amount of thermal mass decreases as the volume of heated material decreases, and the thermal loss decreases as the thickness of the support structure for the bimetal decreases, while the force per unit area remains high. The normally closed thermopneumatic valve (the Fluistor™) from Redwood Microsystems (now defunct) is shown in Figure 8.68a; in this valve, a liquid-filled cavity, upon heating, forces a silicon diaphragm to lift off the valve seat (a normally open valve is also available from the same company). Although providing a tremendous force (>20 N) through a long stroke (between 50 and 150 μm), this valve is thermally inefficient because of the large volume of heated fluid and is difficult to assemble because of the working fluid in the device. Moreover, it is quite slow (~ 400 ms). A fourth alternative is the normally closed SMA valve from Microflow, sold by TiNi Corporation, which is shown in Figure 8.71c.[166] Sputtered thin-film NiTi is the SMA element and is deformed in tension through an integral BeCu bias spring, pushing it against a valve seat. Passing a small current through the actuator element heats it up, and it regains its original shape, lifting it against the spring from the valve seat and allowing gas to flow. From the point of view of performance, the HP valve stands out; in terms of manufacturability and IC process compatibility, the IC valve has the edge. Redwood Microsystems has gone the furthest in incorporating its valves into higher-level subsystems, including pressure and feedback control, to create a completely electronically programmable pressure regulator. This company has also paid more attention to applying its valves in biomedical instruments carrying fluids.[154,155]

Table 8.19 Salient Points of Thermal Actuators

1. Maximum stroke: 100 μm
2. Maximum speed: 10 s to 1 kHz (low bandwidth; see Figure 8.19)
3. Maximum force: 1–10 mN (high force; see Figure 8.19)
4. Easy to fabricate
6. Footprint: 10 μm to mm
7. Special drive conditions: 1–5 V
8. High power and temperatures
9. Multiaxis operation

Summary: Thermal Actuators

In Table 8.19, we list the salient points associated with thermal actuators. See also Figure 8.14.

Other Actuators and Some Final Actuator Comparisons: Final Comparison

Introduction

In closing this chapter on actuators, we review two more interesting actuator principles, i.e., those based on chemical/electrochemical forces and those based on the quantum mechanical Casimir force, and present some final actuator comparison figures.

Chemical/Electrochemical Actuators

A variety of electrochemical and chemical actuators have been demonstrated. In Volume II, Example 4.3, we reviewed an electrochemical disruptable valve invented by the author (see Volume II, Figure 4.98). Hamberg et al.[167] used electrolysis of water to create pressure on a micromachined membrane. Osmotic pumps use a semipermeable membrane over a pump body to realize a mechanical stroke; the mechanism is clarified in Figure 8.72.[168] A solution in a closed vessel is covered on one side with a semipermeable membrane. A hypertonic solution outside the pump draws solvent from the hypotonic solution inside the pump body, reducing the volume inside and bending the membrane inward. Such an

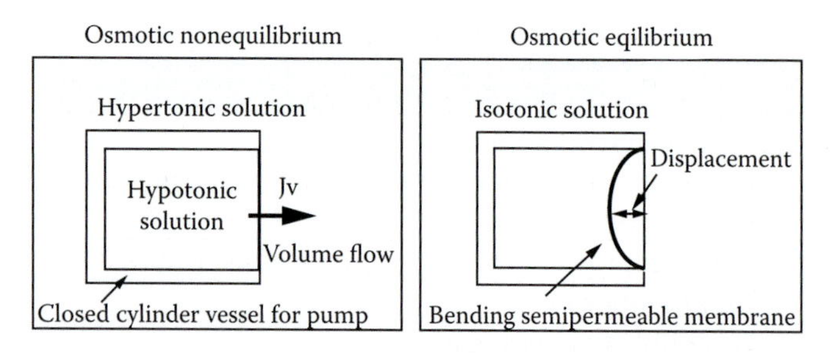

FIGURE 8.72 Mechanism of the osmotic pump. The decrease in volume bends the semipermeable membrane. JV corresponds to the volume flow out of the pump chamber. [After Nagakura, T., K. Ishihara, T. Furukawa, K. Masuda, and T. Tsuda. 1995. In *8th International Conference on Solid-State Sensors and Actuators (Transducers '95)*, 287. Stockholm, Sweden.[168]]

osmotic pump was proposed for delivering insulin to a patient, depending on the patient's changing glucose concentration. In this application, the membrane displacement caused by a high glucose concentration in the patient pushes insulin out of the pump into the patient's body.[168]

Also, hydrogels and redox polymers and blends of the two have been used as chemical or electrochemical actuators. Examples include the actuator hydrogel valve in Volume II, Figure 8.15. One of the most useful features of a polymer actuator is that swelling and shrinking of the material can be triggered by a wide variety of external stimuli. This may involve pH, ionic strength, magnetic field, electrical field, etc. A number of these environmentally sensitive or "inteligent" materials are listed in Table 8.20.[169] Most of these polymers exhibit reversible structural modifications upon repeated changes in the external environment. These polymer systems are of great interest as drug delivery vehicles. A Web site describing progress in the general area of electroactive polymers is the WorldWide Electroactive Polymers Newsletter, or Artificial Muscles Newsletter (http://ndeaa.jpl.nasa.gov/nasa-nde/lommas/eap/EAP-web.htm).

In Volume I, Chapter 6, we described a gold/polypyrrole actuator μ-flap (see Volume I, Figure 6.64) as a means of active mixing in the microdomain. A thin Au film (1 μm), covered with a thin layer of polypyrrole, forms an electrochemical actuator. If submerged in an ionic solution, upon application of a bias small solvated cations in the solution move in and out of the redox polymer matrix, swelling and shrinking, respectively, its apparent volume. The volume changes in the polymer generate stresses on the Au layer surface, bending it upward or downward in a "flapping" motion. Movement frequencies in the 1–20 Hz range can be achieved with this type of actuator.

Casimir Forces: A Quantum Actuator

We normally think of a vacuum as space with nothing in it, and we would then also tend to conclude that its energy is zero. But this is inconsistent with the time-energy uncertainty principle that predicts that the energy of a vacuum fluctuates (Volume I, Equation 3.107). It is therefore possible that empty space locally

TABLE 8.20 Environmentally Sensitive Polymers for Drug Delivery

Stimulus	Hydrogel	Mechanism
pH	Acidic or basic hydrogel	Change in pH—swelling—release of drug
Ionic strength	Ionic hydrogel	Change in ionic strength—change in concentration of ions inside gel—change in swelling—release of drug
Chemical species	Hydrogel containing electron-accepting groups	Electron-donating compounds—formation of charge/transfer complex—change in swelling—release of drug
Enzyme substrate	Hydrogel containing immobilized enzymes	Substrate present—enzymatic conversion—product changes swelling of gel—release of drug
Magnetic	Magnetic particles dispersed in aliginate microspheres	Applied magnetic field—change in pores in gel—change in swelling—release of drug
Thermal	Thermoresponsive hydrogel poly(N-isopropylacrylamide)	Change in temperature—change in polymer-polymer and water-polymer interactions—change in swelling—release of drug
Electrical	Polyelectrolyte hydrogel	Applied electric field—membrane charging—electrophoresis of charged drug—change in swelling—release of drug
Ultrasound irradiation	Ethylene-vinyl alcohol hydrogel	Ultrasound irradiation—temperature increase—release of drug

Source: After Brannon-Peppas, L. 1997. Polymers in controlled drug delivery. *Med Plastics Biomater* 4:34.[169]

does not have zero energy but may have sufficient energy ΔE for a very short time, Δt, to create particles and their antiparticles. This is called the *zero point energy.* Empty space is a seething mass of quantum fluctuations creating virtual particles that recombine and annihilate, just below the threshold of reality (Volume I, Figure 7.29). These fluctuations have been measured as the Casimir effect. The Casimir effect is a small attractive force that acts between two close parallel uncharged conducting plates in a vacuum as illustrated in Volume I, Figures 3.69 and 7.30. Casimir realized that only virtual particles of certain wavelengths can appear between the plates, while those of any wavelength can appear on the outside. Particles other than the photon also contribute a small effect, but only the photon force is measurable. All bosons, such as photons, produce an attractive Casimir force, while fermions make a repulsive contribution.

The attractive Casimir force, F_c, between two plates of area A separated by a distance r can be calculated as follows:

$$F_c = -\frac{\pi^2}{240}\frac{\hbar c}{r^4}A \tag{8.92}$$

where c is the speed of light. Steve K. Lamoreaux[170] measured this tiny force in 1996 with a torsion pendulum (Figure 8.73). One can quickly establish from Equation 8.92 that two mirrors of 1 cm^2 separated by a distance of 1 μm have an attractive Casimir force of 130 nN. Although this force appears small,

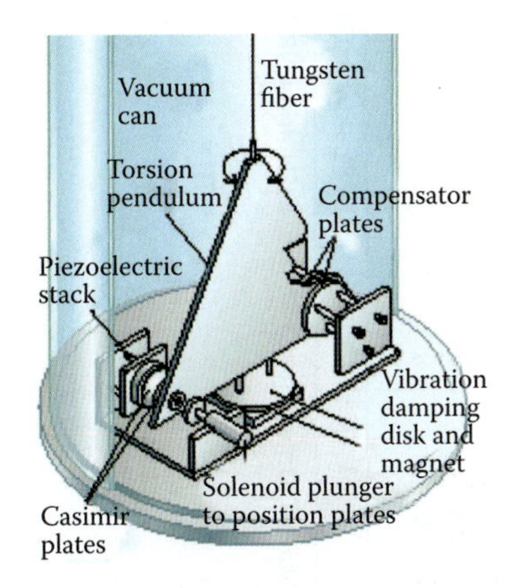

FIGURE 8.73 This experiment by Steve K. Lamoreaux[170] relies on a torsion pendulum. A current applied to the piezoelectric stack tries to move the Casimir plate on the pendulum. The compensator plates hold the pendulum still. The voltage needed to prevent any twisting serves as a measure of the Casimir effect.

at distances below a micrometer, the Casimir force becomes the dominant force between two neutral objects. At separations of 10 nm, the Casimir effect produces the equivalent of 1 atmosphere pressure! Obviously this is a force to reckon with in MEMS and NEMS devices.

As summarized in Table 8.21, Lamoreux's measurement of the Casimir force was followed by many subsequent attempts using MEMS and NEMS actuators to measure this force.

The zero point energy is also connected to Einstein's famous cosmological constant:

$$\Lambda = \frac{8\pi g}{3c^2}\rho \tag{8.93}$$

where g = gravitational constant
c = speed of light
ρ = energy density of vacuum

Einstein included this term in the equations for general relativity because he was dissatisfied that his equations would not allow for a stationary universe, which was the model of the universe at the time. Soon after Einstein developed his theory, observations by Edwin Hubble (1929) proofed that the universe was expanding. Einstein then abandoned the cosmological constant and called it the "biggest blunder" of his life. The cosmological constant remains of interest though, as observations made in the late 1990s of distance-redshift relations suggest that the universe is accelerating. These observations can be explained by assuming a very small positive cosmological constant in Einstein's equations. The current standard model of cosmology, the Lambda-Cold Dark Matter model (CDM), thus includes the cosmological constant, which is measured to be on the order of 10^{-35} s^{-2}, or 10^{-47} GeV^4, or 10^{-29} g/cm^3, or about 10^{-120} in Planck units.

Final Comparison of Actuator Mechanisms

Finally, in Figures 8.74 and 8.75 we show two very complete figures summarizing the state of the art in actuators by Bell.[171] The first is a figure comparing all types of actuators in terms of maximum displacement versus maximum frequency; the second compares their maximum displacement versus maximum force exerted.

Table 8.21 Attempts Using MEMS and NEMS Actuators to Measure Casimir Force

Investigators	Year	Geometry	Method	Distance Scale (nm)	Materials	Pressure (mbar)	Temp. (K)	Accuracy (%)
S.K. Lamoreaux	1997		Torsion pendulum	600–6000	Au (500 nm)	10^{-4}	300	5
U. Mohideen and A. Roy	1998		AFM	100–900	Al (300 nm) + AuPd (20 nm)	5×10^{-2}	300	2
A. Roy and U. Mohideen	1999		AFM	100–900	Al (250 nm) + AuPd (8 nm)	5×10^{-2}	300	2
G.L. Klimthitskaya, A. Roy, U. Mohideen, and V.M. Mostepanenko	1999		AFM	100–900	Al (300 nm) + AuPd (20 nm)	5×10^{-2}	300	1
T. Ederth	2000		Piezo-tube manipulator	20–100	50 μm Au wires coated in thiol SAM	1000	300	1
H.B. Chan, V.A. Aksyuk, R.N. Kleiman, D.J. Bishop, and F. Capasso	2001		MEMS torsion bar capacitance	90–1000	Au (200 nm) + Cr underlayer	1000	300	1
G. Bressi, G. Carugno, R. Onofrio, and G. Ruoso	2002		Interferometry	500–3000	Cr (50 nm) on Si	10^{-5}	300	15
R.S. Decca, D. Lopez, E. Fischbach, and D. E. Krause	2003		MEMS torsion bar capacitance	200–2000	Cu/Au	10^{-4}	300	1
NANOCASE	2005		AFM, MEMS	10–1000	Si, Au	10^{-11}	20–1000	<1

Source: Antezza Mauro, University of Trento, INFN and CNR-INFM BEC Center on Bose-Einstein Condensation, Trento, Italy.

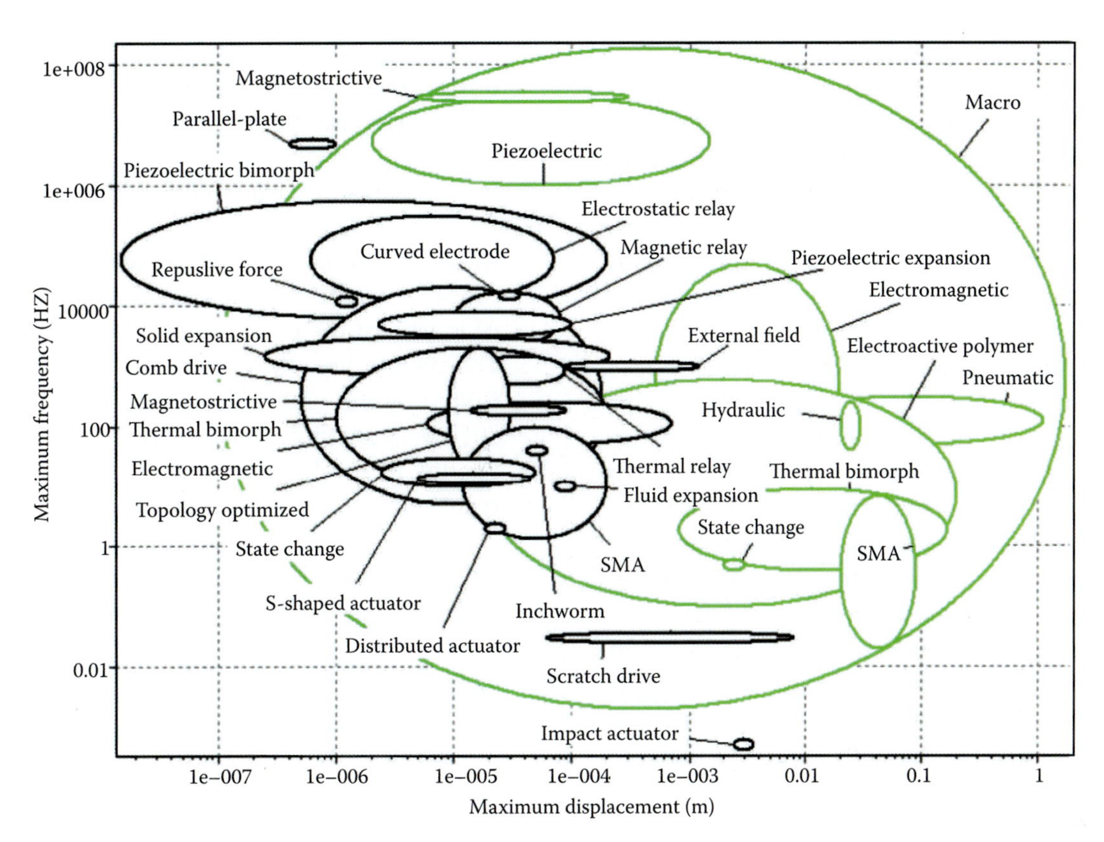

FIGURE 8.74 Comparison of all types of actuators in terms of maximum displacement versus maximum frequency.[171]

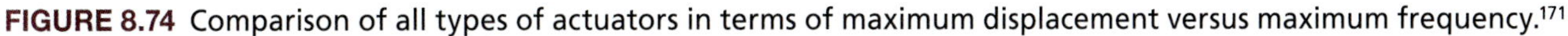

Example 8.1: Mass Sensing Actuators

In Volume I, Example 4.1, we analyzed the design criteria for a static mass sensitive piezoresistive cantilever. In that example, we were interested only in the static bending of the cantilever upon mass loading. Here we consider the mass sensitivity of different types of resonators, including cantilever beams. A mass sensitive resonator transforms an additional mass loading into

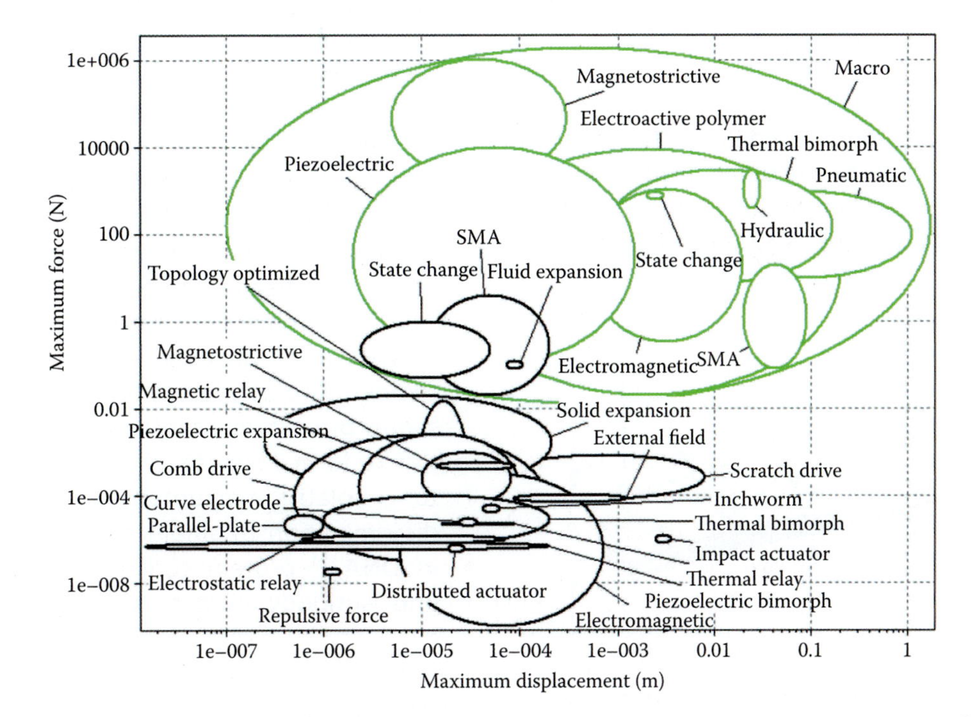

FIGURE 8.75 Comparison of all types of actuators in terms of maximum displacement versus maximum force exerted.[171]

a resonance frequency shift. Resonance frequency shifts of cantilevers can be measured with high precision using optical reflection with a diode laser and a linear position sensitive detector (PSD). The cantilever measuring principle may also be based on the detection of changes in phase, deflection, amplitude, and *Q*.[172] The excitation of MEMS resonators, say a comb drive or a cantilever, is most often realized by electrostatic actuation. The drawbacks of this approach include the relatively high supply voltage and the nonlinearity, linked to the varying distance between the electrodes. An alternative actuation scheme uses the combination of a high-quality piezoelectric thin film with a mechanical resonator. This has several advantages: it avoids the nanometer-size gaps typically required for electrostatic transduction and has a better power handling capability, better impedance handling with electronics, and better potential to reach 10 GHz and above (especially important for very sensitive mass-sensing applications).

Consider as the resonator a uniform beam with a rectangular cross-section, which is fixed at one end and deflected at the other (Figure 8.76). Such a beam experiences lengthwise stress that is compressive below the centerline and tensile above it. The effective spring constant, *K*, of this beam is given by the following:

$$K = EWt^3/4L^3 \tag{8.94}$$

with *E* the Young's modulus and *W*, *t*, *L* the width, thickness, and length of the beam, respectively. Representative MEMS values for *W*, *t*, and *L* are 20, 0.6, and 100 μm, with a resulting value for *K* of 0.1 N/m.

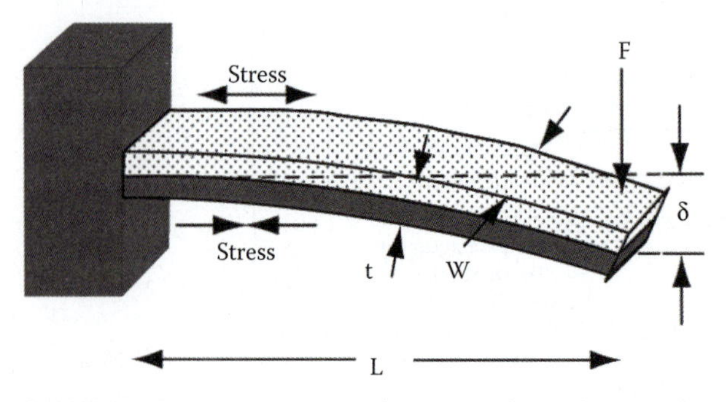

FIGURE 8.76 Stresses in a uniform cantilever beam when deflected.

For this simple cantilever beam the maximum deflection, δ, induced by a weight, W_e, at its free end is given by the following:

$$\delta = \frac{W_e L^3}{3EI} \tag{8.95}$$

in which *E* and *I* are, respectively, the Young's modulus of the beam material and the moment of inertia of the beam, which is given as $W t^3/12$. The weight W_e = Mg, that is, the mass (*M*) multiplied by the gravitational acceleration (*g*). The stiffness constant, *K*, from Equation 8.94, corresponds to the load/deflection, or $W_e/\delta = 3EI/L^3$, resulting in a natural frequency, f_0, of:

$$f_0 = \frac{1}{2\pi}\sqrt{\frac{3EI}{ML^3}} = \frac{1}{2\pi}\sqrt{\frac{EWt^3}{4ML^3}} = \frac{t}{4\pi L^2}\sqrt{\frac{E}{\rho}} \tag{8.96}$$

where ρ is the density of the material.* For a Si-based stress-free MEMS cantilever with dimensions of 100 × 3 × 0.1 (*L* × *W* × *t*, in μm), the calculated resonant frequency is 12 kHz (Young's modulus E = 160 GPa); for a NEMS beam of 0.1 × 0.01 × 0.01 (*L* × *W* × *t*, in μm), the frequency is 1.2 GHz. Using a stiffer material (higher Young's modulus *E*) such as SiC (E = 400 GPa), the corresponding numbers are even higher, that is, 19 kHz and 1.9 GHz, respectively— reflecting the higher $(E/\rho)^{1/2}$ in Equation 8.96. In the NEMS range, mechanical devices are almost as fast as today's electronic devices. Making things yet smaller, the ultimate resonant frequency is reached in the THz range, that is, for molecular vibrations.[174] An intermediate frequency of 1.25 MHz was observed for an 8.3-μm-long, 95-nm-thick SiO_2 beam.[175]

In Figure 8.77 we show an example of a set of micromachined single-crystal silicon cantilevers. These beams can be used for real-time, in situ measurements of physical parameters such as viscosity,[176] pressure, density, flow rate and temperature. The latter, for instance, is achieved simply by coating a cantilever with metal on one side to form a bimetal.[177] Mercury vapor,[177] moisture, volatile mercaptans, DNA hybrization,[178] discrimination of single-nucleotide mismatches in DNA,[179] protein conformational changes,[180] and antibody-antigen binding all have been

* For a rigorous derivation, including the case of a thin cantilever that has been metalized with a layer that contributes to the functioning of the device, see Den Hartog.[173]

FIGURE 8.77 Scanning electron micrograph of a section of a microfabricated silicon cantilever array (eight cantilevers, each 1 μm thick, 500 μm long, and 100 μm wide, with a pitch of 250 μm, spring constant 0.02 Nm^{-1}. (Reprinted with permission from Fritz, J., M.K. Baller, H.P. Lang, H. Rothuizen, P. Vettinger, E. Meyer, H.-J. Güntherodt, C. Gerber, and J.K. Gimzewski. 2000. Translating biomolecular recognition into nanomechanics. *Science* 288:316. Copyright 2000, American Association for the Advancement of Science. Courtesy of Dr. J. Fritz, Massachusetts Institute of Technology Media Lab.[182])

monitored using cantilevers.[181] For chemical and biochemical sensing, the microcantilever surface must be derivatized with chemically selective coatings, perhaps through self-assembled alkanethiols or organosilane films, direct covalent attachment of molecular receptors, dip coating (e.g., polymeric resins), or adsorption/evaporation (e.g., sol-gel matrices).

Under the assumption of negligible variation in spring constant and uniformly distributed mass loading of the beam, Equation 8.96 describes the resonance frequency dependence on mass of the cantilever. Adding a discrete mass, M_d, to the cantilever defines a total effective mass, M_T, of the cantilever-adsorbate system as $M_T = M + M_d$. Equation 8.96 describes the mass dependence of a mass sensing cantilever, in which case the effective mass is that of the cantilever plus an adsorbate. The effective mass of the beam, M, is related to the mass of the beam, M_b, through $M = nM_b$, where n is a geometric parameter, which for a rectangular bar is 0.24.[183] This results in an expression for the resonant frequency given by the following:

$$f_0 = \frac{1}{2\pi}\sqrt{\frac{EWt^3}{4M_TL^3}} = \frac{1}{2\pi}\sqrt{\frac{EWt^3}{4L^3(M_d + 0.24WtL\rho)}} \tag{8.97}$$

where ρ is the density of the cantilever. To compare the mass sensitivity, S_m (given by Equation 8.64), of a cantilever sensor to that of other gravimetric devices, see Table 8.9.

In a case where mass loading is distributed evenly over the cantilever surface, S_m, as noted in Table 8.9, is given by the following (in Table 8.9, *d* was used for thickness instead of *t*).

$$S_m = \frac{1}{\rho t} \tag{8.98}$$

The positive sign indicates that, as mass is added, the resonant frequency increases, corresponding to the increase of the cross-sectional thickness of the beam. For an end-loaded cantilever (mass only allowed to accumulate in a small area at the free end of the beam) this results in:

$$S_m = \frac{-\xi}{2\rho(\xi t_d + 0.24t)} \tag{8.99}$$

where ξ and t_d are the fractional area coverage and thickness of the deposited mass, respectively. The negative sign denotes that the frequency is decreasing with increasing mass piling up.

From the above we conclude that for mass sensing with a microcantilever, a uniform mass loading is preferred. From Equations 8.98 and 8.99, it is further advantageous to make the cantilever as low density (ρ) and as thin as possible (*t*). Dramatic further improvements in analytical figures of merit can be achieved by scaling the cantilever size down to the nanoscale.

In Figure 8.78 we compare the size of a typical quartz crystal microbalance (QCM) transducer, with a minimal volume of 80 mm^3, with an array of ten micromachined cantilevers (e.g., based on silicon nitride) that requires as little as 0.001 mm^3 of volume.

From Table 8.9 cantilevers are much more sensitive than quartz microbalances and are

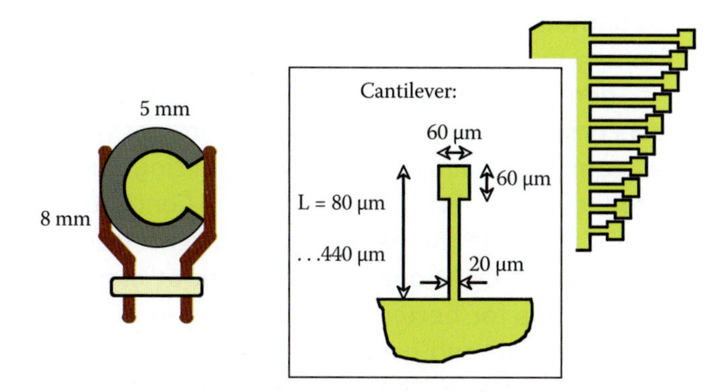

FIGURE 8.78 Size comparison of a quartz crystal microbalance and a micromachined array of cantilevers.

easier to downscale. Nanomechanical resonators enable the measurement of mass with extraordinary sensitivity. Samples as light as 7 zeptograms have been weighed in vacuum with cantilevers.[184] Resolving such small mass changes requires the resonator to be very light with a high *Q*. In solution, viscosity severely degrades both of these characteristics, preventing many sensing applications where fluid is required. One needs to excite with a positive feedback loop of a piezoelectric or measure in air or vacuum. FBARs, we saw earlier, can be downscaled more easily than BAWs, and for similar size they are even more sensitive than cantilevers (see Table 8.9).

An alternative approach, avoiding the viscous damping problem, is to use a static formulation, pioneered by Thundat et al. from Oak Ridge National Laboratory,[177] and involves the measurement of bending induced by the adsorption of molecules onto a thin microcantilever that has two chemically different surfaces. The Thundat et al. approach was covered in Volume I, Chapter 4 (Example 4.1).

Burg et al. pioneered yet a different approach to eliminate viscous damping by placing the solution inside a hollow resonator that is surrounded by vacuum.[185] This group demonstrated that suspended microchannel resonators can weigh single nanoparticles, single bacterial cells, and submonolayers of adsorbed proteins in water with subfemtogram resolution (1 Hz bandwidth). The combination of the low resonator mass (100 ng) and high-quality factor (Q = 15,000) enables an improvement in mass resolution of six orders of magnitude over a high-end commercial quartz crystal microbalance. This development gives access to intriguing applications, such as mass-based flow cytometry, the direct detection of pathogens, or the nonoptical sizing and mass density measurement of colloidal particles.

There are several (near) commercial biosensors based on microcantilever structures, as well as companies aiming at commercializing such products (e.g., Protiveris, Graviton, Concentris and Cantisens® Research). The current trend in SAW and QCM devices is to work at higher frequencies for better sensitivity, and in silicon MEMS, micro- and nanocantilvers are replacing quartz devices, as they are easier to integrate and easier to make into many device arrays. The key strength of all these devices is that they hold the promise of label-free measurements. In Chapter 10, this volume, we will also see that the RF-MEMS community is hard at work trying to replace quartz and AlN resonators with silicon MEMS-based oscillators.

Acknowledgments

With thanks to Ramses Madou, Marleen Madou, and George O'Quinn.

Questions

Questions and answers by Dr. Marc Madou, assisted by Mr. George O'Quinn, UC Berkeley

8.1: Fill in the blank with the correct word.
- (a) Piezoelectric materials produce ______ when they are deformed by a force.
- (b) Silicon is ______ piezoelectric.
- (c) Viscosity of a fluid is the tendency of the fluid to ______ .
- (d) Large field strength can be used to produce large ______ with smaller ______ .
- (e) ______ is the major loss mechanism in microactuators.
- (f) Shape memory alloys recover the shape they are given when they are heated to a temperature that is higher than the ______.

8.2: What is the difference between a sensor and an actuator?

8.3: What is the underlying principle of ferrofluids (FF) and how do they differ from magnetorheological fluids (MR)?

8.4: Describe, using equations, why one might prefer a parallel-driven parallel plate capacitor in an electrostatic actuator over a vertical-driven one?

8.5: Prove that the snap voltage V_{snap} in a parallel plate capacitor is given by 1/3 g_0, where g_0 is the initial gap spacing (with V = 0).

8.6: Make a table listing corresponding mechanical and electrical equivalent circuit components.

8.7: Suggest a normally closed valve technology based on phase-change of a material that works for both vapors and liquids and does not require power to remain closed but only to open.

8.8: What transduction principles are used in biosensors?

8.9: What is the difference between a self-exciting and a modulating sensor? Give at least one example of each used to sense the same quantity.

8.10: Describe different methods to levitate and move a biological cell to a desired spot on a glass slide.

8.11: Why are wobble motors effective when small, yet ineffective when large, and what role does friction play in a wobble motor?

8.12: Explain the Seebeck effect, Peltier effect, Thomson effect, and Joule heating. How are they important to thermoelectric devices?

8.13: What are the characteristics of shape memory alloy (SMA) actuators that cause them to have a slower response time, but higher force than piezoelectric and electrostatic actuators?

8.14: What is the main difference between piezoelectricity and ferroelectricity?

8.15: How are Cooper pairs formed?

8.16: Why do good conductors make poor superconductors?

8.17: Why is the Casimir force an important consideration in MEMS and NEMS?

8.18: Explain magnetostriction and compare it to the piezoelectric effect.

8.19: Name five mechanical measurands.

References

1. White, R. M. 1987. A sensor classification scheme. *IEEE Trans Ultrason Ferroelectr Freq Control* UFFC-34:124–26.
2. Middlehoek, S., and S. Audet. 1989. *Silicon sensors.* London: Academic Press.
3. Lion, K. S. 1969. Transducers: Problems and prospects. *IEEE Trans Ind Electron* 16:2–5.
4. Gopel, W., J. Hesse, and J. N. Zemel, eds. 1989. *Sensors: A comprehensive survey.* New York: VCH.
5. Habekotte, E. 1993. In *Sensoren and Actuatoren in de Werktuigbouw/Machinebouw 15–96.* The Hague, the Netherlands: Centrum voor Micro-Electronica.
6. National Research Council. 1995. NRC.
7. Tang, W. C.-K. 1990. Electrostatic comb drive for resonant sensor and actuator applications. PhD thesis, University of California, Berkeley.
8. Tang, W. C., T.-C. H. Nguyen, and R. T. Howe. 1989. Laterally driven polysilicon resonant microstructures. *Proceedings: IEEE Micro Electro Mechanical Systems (MEMS '89)*. Salt Lake City, UT, February 20–22, 1989, pp. 53–59.
9. Denn, M. M. 1980. *Process fluid mechanics.* Englewood Cliffs, NJ: Prentice Hall.
10. Schmidt, M. A. 1988. Microsensors for the measurement of shear forces in turbulent boundary layers. Ph.D. thesis, Massachusetts Institute of Technology.
11. Trimmer, W. S. N., and K. J. Gabriel. 1987. Design considerations for a practical electrostatic micro-motor. *Sensors Actuators* 11:189–206.
12. Mohr, J., P. Bley, M. Strohrmann, and U. Wallrabe. 1992. Microactuators fabricated by the LIGA process. *J Micromech Microeng* 2:234–41.
13. Roy, S., A. K. McIlwain, R. G. DeAnna, A. J. Fleischman, R. K. Burla, C. A. Zorman, and M. Mehregany. 2000. SiC resonant devices for high Q and high temperature applications. *Solid-State Sensor and Actuator Workshop*, pp. 22–25. Hilton Head Island, SC: Transducers Research Foundation.
14. Roukes, M. L. 2000. Nanoelectromechanical systems. *Solid-State Sensor and Actuator Workshop*, pp. 367–76. Hilton Head Island, SC: Transducers Research Foundation.
15. Busch-Vishniac, I. J. 1992. The case for magnetically driven microactuators. *Sensors Actuators A* A33:207–20.
16. Mahadevan, R. 1987. In *Proceedings: IEEE Micro Robots and Teleoperators Workshop*, 15/1–8. Hyannis, MA.
17. Mahadevan, R. 1990. In *Proceedings: IEEE Micro Electro Mechanical Systems (MEMS '90)*. Napa Valley, CA, February 11–14, 1990, pp. 120–27.
18. Kumar, S., D. Cho, and W. N. Carr. 1992. Experimental study of electric suspension for microbearings. *J Microelectromech Syst* 1:23–30.
19. Mehregany, M., P. Nagarkar, S. D. Senturia, and J. H. Lang. 1990. Operation of microfabricated harmonic and ordinary side-drive motors. *Proceedings: IEEE Micro Electro Mechanical Systems (MEMS '90)*. Napa Valley, CA, February 11–14, 1990, pp. 82–88.
20. Mehregany, M., S. D. Senturia, and J. H. Lang. 1990. Friction and wear in microfabricated harmonic side-drive motor. *Technical Digest: IEEE Solid-State Sensor and Actuator Workshop.* Hilton Head Island, SC, June 4–7, 1990, pp. 17–22.
21. Fujita, H., and K. J. Gabriel. 1991. New opportunities for micro actuators. *6th International Conference on Solid-State Sensors and Actuators (Transducers '91)*. San Francisco, CA, June 1991, pp. 14–20.
22. Trimmer, W. S., ed. 1997. *Micromechanics and MEMS classic and seminal papers to 1990.* New York: IEEE.
23. Price, R. H., S. J. Cunningham, and S. C. Jacobsen. 1992. Field analysis for the electrostatic eccentric drive micromotor (wobble motor). *J Electrost* 28:7–38.
24. Furuhata, T., T. Hirano, L. H. Lane, R. E. Fontana, L. S. Fan, and H. Fujita. 1993. Outer rotor surface-micromachined wobble motor. *Proceedings: IEEE Micro Electro Mechanical Systems (MEMS '93)*. Fort Lauderdale, FL, February 7–10, 1993, pp. 161–66.
25. Tai, Y., L. Fan, and R. Muller. 1989. IC processed micromotors: design, technology, and testing. *Proceedings: IEEE Micro Electro Mechanical Systems (MEMS '89)*. Salt Lake City, UT, February 20–22, 1989, pp. 1–6.
26. Fujita, H. 1990. In *Proceedings: Micro System Technologies '90*. Berlin, Germany, p. 818.

27. Ohnstein, T., T. Fukiura, J. Ridley, and V. Bonne. 1990. Micromachined silicon microvalve. *Proceedings: IEEE Micro Electro Mechanical Systems (MEMS '90)*. Napa Valley, CA, February 11–14, 1990, pp. 95–98.
28. Marxer, C., O. Manzardo, H.-P. Herzig, R. Dändliker, and N. F. de Rooij. 1999. An electrostatic actuator with large dynamic range and linear displacement-voltage behaviour for a miniature spectrometer. *Transducers '99* 1:786–89.
29. Petersen, K. 1979. Micromechanical membrane switches on silicon. *IBM J Res Dev* Report no. 23:376.
30. Cady, W. G. 1922. The piezo-electric resonator. *Proceedings of the Institute of Radio Engineers*, April 1922, pp. 83–114.
31. Fukuda, E., and I. Yasuda. 1957. On the piezoelectric effect in bone. *J Phys Soc Jpn* 12:1158–62.
32. Gandhi, M. V., and B. S. Thompson. 1992. *Smart materials and structures*. London: Chapman and Hall.
33. Hueter, T. F., and R. H. Bolt. 1955. *Sonics*. New York: Wiley.
34. Pallas-Areny, R., and J. G. Webster. 1991. *Sensors and signal conditioning*. New York: Wiley.
35. Herbert, J. M. 1982. *Ferroelectric transducers and sensors*. New York: Gordon and Breach Science Publishers.
36. Khazan, A. D. 1994. *Transducers and their elements*. Englewood Cliffs, NJ: PTR Prentice Hall.
37. Rogacheva, N. N. 1994. The theory of piezoelectric shells and plates. Boca Raton, FL: CRC Press.
38. Cady, W. G. 1964. *Piezoelectricity*. New York: Dover.
39. Zelenka, J. 1986. *Piezoelectric resonators and their applications*. Amsterdam: Elsevier.
40. White, R. M., and F. W. Voltmer. 1965. Direct piezoelectric coupling to surface elastic waves. *Appl Phys Lett* 7:314–316.
41. Flynn, A. M., L. S. Tavrow, S. F. Bart, R. A. Brooks, D. J. Ehrlich, K. R. Udayakumar, and L. E. Cross. 1992. Piezoelectric micromotors. *J Microelectromech Syst* 1:44–52.
42. Abe, T., and M. L. Reed. 1994. In *IEEE International Workshop on Micro Electro Mechanical Systems (MEMS '94)*, 164–69. Oiso, Japan.
43. Hunter, I. W., and S. Lafontaine. 1992. A comparison of muscle with artificial actuators. *Technical Digest: 1992 Solid State Sensor and Actuator Workshop*. Hilton Head Island, SC, June 22–25, 1992, pp. 178–85.
44. AMP. 1993. *Piezo film sensors*. Valley Forge, PA: AMP.
45. Nalwa, H. S., ed. 1995. *Ferroelectric polymers: Chemistry, physics, and applications*. New York: Marcel Dekker.
46. Polla, D., and R. S. Muller. 1986. Zinc-oxide thin films for integrated-sensor applications. *Technical Digest: IEEE Solid-State Sensors Workshop*. Hilton Head Island, SC, June 2–5, 1986.
47. Muller, R. S. 1987. From ICs to microstructures materials and technologies. *Proceedings: IEEE Micro Robots and Teleoperators Workshop*. Hyannis, MA, November 9–11, 1987, IEEE Catalog No. 87TH0204-8.
48. Tjhen, W., T. Tamagawa, C.-P. Ye, C.-C. Hsueh, P. Schiller, and D. L. Polla. 1991. In *Proceedings: IEEE Micro Electro Mechanical Systems (MEMS '91)*. Nara, Japan, pp. 114–19.
49. Hirata, Y., H. Okuyama, S. Ogino, T. Numazawa, and H. Takada. 1995. In *Proceedings: IEEE Micro Electro Mechanical Systems (MEMS '95)*. Amsterdam, the Netherlands, pp. 191–96.
50. Preu, G., A. Wolff, D. Cramer, and U. Bast. 1991. In *Proceedings: Second European Ceramic Society Conference (2nd ECerS '91)*, 2005–9. Augsburg, Germany.
51. Lubitz, K. 1989. *Mikrostrukturierung von piezokeramik*. Berlin: VDI-Tagungsbericht, 796.
52. Son, I.-S., A. Lal, B. Hubbard, and T. Olson. 2000. In *Solid-State Sensor and Actuator Workshop*, 206–9. Hilton Head Island, SC: Transducers Research Foundation.
53. Christensen, D. A. 1988. *Ultrasonic bioinstrumentation*. New York: Wiley.
54. Udayakumar, K. R., S. F. Bart, A. M. Flynn, J. Chen, L. S. Tavrow, L. E. Cross, R. A. Brooks, and D. J. Ehrlich. 1991. Ferroelectric thin films for ultrasonic micromotors. *Proceedings: IEEE Micro Electro Mechanical Systems (MEMS '91)*. Nara, Japan, January 30–February 2, 1991, pp. 109–13.
55. van Lintel, H. T. G., F. C. M. van der Pol, and S. Bouwstra. 1988. A piezoelectric micropump based on micromachining of silicon. *Sensors Actuators* 15:153–67.
56. Smits, J. G. 1990. Piezoelectric micropump with three valves working peristaltically. *Sensors Actuators A* A21:2030–6.
57. Judy, J. W., D. L. Polla, and W. P. Robbins. 1990. Experimental model and IC-process design of a nanometer linear piezoelectric stepper motor. *Microstructures, Sensors, and Actuators*. American Society of Mechanical Engineers, November 1990, pp. 11–17.
58. Shoji, S., S. Nakagawa, and M. Esashi. 1990. Micropump and sample-injector for integrated chemical analyzing systems. *Sensors Actuators A* A21:189–92.
59. Nyborg, W. L. M. 1965. Acoustic streaming. In *Physical acoustics*, ed. W. P. Mason, 265–331. New York: Academic Press.
60. Moroney, R. M., R. M. White, and R. T. Howe. 1990. Ultrasonic micromotors: physics and applications. *IEEE Micro Electro Mechanical Systems*. New York, February 1990, pp. 182–87.
61. Moroney, R. M., R. M. White, and R. T. Howe. 1991. Microtransport induced by ultrasonic waves. *Appl Phys Lett* 59:774–76.
62. Miyazaki, S., T. Kawai, and M. Araragi. 1991. A piezo-electric pump driven by a flexural progressive wave. *Proceedings: IEEE Micro Electro Mechanical Systems (MEMS '91)*. Nara, Japan, January 30–February 2, 1991, pp. 283–88.
63. Ballantine, D. S., R. M. White, S. L. Martin, A. J. Ricco, E. T. Zellers, G. C. Frye, and H. Wohltjen. 1997. *Acoustic wave sensors: Theory, design, and physico-chemical applications*. San Diego, CA: Academic Press.
64. Sauerbrey, G. Z. 1959. The use of an oscillator for weighing thin layers and for microweighing. *Z Phys* 155:206–12.
65. Black, J. P., A. Elium, R. M. White, M. G. Apte, L. A. Gundel, and R. Cambie. 2007. 6D-2 MEMS-enabled miniaturized particulate matter monitor employing 1.6 GHz aluminum nitride thin-film bulk acoustic wave resonator (FBAR) and thermophoretic precipitator. *Ultrasonics symposium, 2007*, New York, October 28–31, 2007, pp. 476–79.
66. Kobrin, P., C. Seabury, A. Harker, and R. O'Toole. 1999. *Thin-film resonant chemical sensor with resonant acoustic isolator*. US Patent 5936150.
67. Wohltjen, H., and R. E. Dessy. 1979. Surface acoustic wave probe for chemical analysis, parts I–III. *Anal Chem* 5:1458–75.
68. Hoenk, M. E., T. R. Van Zandt, D. A. McWatters, R. K. Watson, C. Kukkonen III, W. Kaiser, and D. Cheng. 1995. *Surface acoustic wave hygrometer flight tests on the NASA DC-8 airborne laboratory*. Pasadena, CA: Jet Propulsion Laboratory.
69. Bernstein, J. J., J. Bottari, K. Houston, G. Kirkos, and R. Miller. 2000. In *Solid-State Sensor and Actuator Workshop*, 281–84. Hilton Head Island, SC: Transducers Research Foundation.

70. Allen, R. 1984. Sensors in silicon. *High Technology,* September: 43–81.
71. Kolesar, E. S., and C. S. Dyson. 1995. Object imaging with a piezoelectric robotic tactile sensor. *J Microelectromech Syst* 4:87–96.
72. Glaser, K., and G. Fuhr. 1987. The spin of cells in rotating high frequency electric fields. In *Mechanistic approaches to interactions of electric and electromagnetic fields with living systems,* eds. M. Blank and E. Findl. New York: Plenum Press, 271–89.
73. Fuhr, G., R. Hagedorn, T. Muller, B. Wagner, and W. Benecke. 1991. Linear motion of dielectric particles and living cells in microfabricated structures induced by travelling electric fields. *Proceedings: IEEE Micro Electro Mechanical Systems (MEMS '91).* Nara, Japan, January 30–February 2, 1991, pp. 259–64.
74. Moesner, F. M., and T. Higuchi. 1995. Devices for particle handling by an ac electric field. *Proceedings: IEEE Micro Electro Mechanical Systems (MEMS '95).* Amsterdam, the Netherlands, January 29–February 2, 1995, pp. 66–71.
75. Fuhr, G., R. Hagedorn, T. Muller, W. Benecke, and B. Wagner. 1992. Microfabricated electrohydrodynamic (EHD) pumps for liquids of higher conductivity. *J Microelectromech Syst* 1:95–98.
76. Bart, S. F., L. S. Tavrow, M. Mehregany, and J. H. Lang. 1990. Microfabricated electrohydrodynamic pumps. *Sensors Actuators A* A21:193–97.
77. Shoji, S. 1998. In *Microsystem technology in chemistry and life sciences,* eds. A. Manz and H. Becker, 161–88. New York: Springer.
78. Richter, A., A. Plettner, K. Hoffmann, and H. Sandmaier. 1991. Electrohydrodynamic pumping and flow measurement. *Proceedings: IEEE Micro Electro Mechanical Systems (MEMS '91).* Nara, Japan, January 30–February 2, 1991, pp. 271–76.
79. Choi, J.-W., and C. H. Ahn. 2000. An active micro mixer using electrohydrodynamic (EHD) convection. *Solid-State Sensor and Actuator Workshop.* Hilton Head Island, SC, June 4–8, 2000, pp. 52–55.
80. Bart, S. F., T. A. Lober, R. T. Howe, J. H. Lang, and M. F. Schlecht. 1988. Design considerations for micromachined electric actuators. *Sensors Actuators* 14:269–92.
81. Melcher, J. R. 1966. Traveling-wave induced electroconvection. *Phys Fluids* 9:1548–55.
82. Bart, S., and J. H. Lang. 1989. An analysis of electroquasistatic induction micromotors. *Sensors Actuators* 20:97–106.
83. Fripp, M. L. R., and N. W. Hagood. 1995. Comparison of electrostrictive and piezoceramic actuation for vibration suppression (Proceedings paper). In *Smart structures and materials 1995: Smart structures and integrated systems,* ed. I. Chopra. Bellingham, WA: SPIE, May 8, 1995, pp. 334–48.
84. Uchino, K. 1993. Ceramic actuators: Principles and applications. *MRS Bull* 18:42.
85. Lang, S. B. 1974. Sourcebook of pyroelectricity. New York: Gordon and Breach.
86. Ye, C., T. Tamagawa, and D. L. Polla. 1991. Pyroelectric PbTiO3 thin films for microsensor applications *6th International Conference on Solid-State Sensors and Actuators (Transducers '91).* San Francisco, CA, June 24–27, 1991, pp. 904–07.
87. Hsieh, H. Y., A. Spetz, and J. N. Zemel. 1991. Wide range pyroelectric anemometers for gas flow measurements. *6th International Conference on Solid-State Sensors and Actuators (Transducers '91).* San Francisco, CA, June 24–27, 1991, pp. 38–40.
88. Zavracky, P. M., K. Warner, G. Jenkins, S. Etienne, C. Logan, and R. Grace. 1998. In *Solid-State Sensor and Actuator Workshop.* Hilton Head Island, SC, 209–11.
89. Hicks, L. D., and M. S. Dresselhaus. 1993. Effect of quantum-well structures on the thermoelectric figure of merit. *Phys Rev B Condens Matter* 47:727–31.
90. Yang, P., A. I. Hochbaum, R. Chen, R. D. Delgado, W. Liang, E. C. Garnett, M. Najarian, and A. Majumdar. 2008. Rough silicon nanowires as high performance thermoelectric materials. *Nature* 451:163–67.
91. Heath, J. R., A. Boukai, Y. Bunimovich, J. Tahir-Kheli, J. Yu, and W. A. Goddard. 2008. Silicon nanowires as efficient thermoelectric materials. *Nature* 451:168–71.
92. Shakouri, A. 2006. Nanoscale thermal transport and microrefrigerators on a chip. *Proceedings of the IEEE* 94:1613–38.
93. Voorthuyzen, J. A., and P. Bergveld. 1988. The PRESSFET: An integrated electret-MOSFET based pressure sensor. *Sensors Actuators* 14:349–60.
94. Voorthuyzen, J. A., P. Bergveld, and A. J. Sprenkels. 1989. Semiconductor-based electret sensors for sound and pressure. *IEEE Trans Electr Insul* 24:267–76.
95. Sprenkels, A. J., R. A. Groothengel, A. J. Verloop, and P. Bergveld. 1989. Development of an electret microphone in silicon. *Sensors Actuators* 17:509–12.
96. Wolffenbuttel, R. F., J. F. L. Goosen, and P. M. Sarro. 1991. Design considerations for a permanent-rotor-charge excited micromotor with an electrostatic bearing. *Sensors Actuators A* A25–27:583–90.
97. Mascarenhas, S. 1973. The electret state: A new property of bone. *Electrets,* ed. M. M. Perlman Princeton, NJ: The Electrochemical Society, p. 650.
98. Mascarenhas, S. 1974. The electret effect in bone and polymers and the bound-water problem. *Ann NY Acad Sci* 238:36.
99. Fukada, E., T. Takamaster, and I. Yasuda. 1975. Callus formation by electret. *Jpn J Appl Phys* 14:2079.
100. Duff, A. W. 1896. The viscosity of polarized dielectrics. *Phys Rev* 41:23–38.
101. Winslow, W. M. 1947. *Method and means for translating electrical impulses into mechanical force.* US Patent 2,417,850.
102. Winslow, W. M. 1947. Induced vibration of suspensions. *J Appl Phys* 20:1137–40.
103. Rabinow, J. 1948. The magnetic fluid clutch. *AIEE Trans* 67(II):1308–15.
104. Gandhi, M. V., and B. S. Thompson. 1992. *Smart materials and structures.* London: Chapman & Hall.
105. Ball, P. 1997. *Made to measure: New materials for the 21st century.* Princeton, NJ: Princeton University Press.
106. Wen, W., X. Huang, S. Yang, K. Lu, and P. Sheng. 2003. The giant electrorheolgical effect in suspension of nanoparticles. *Nature Materials* 2:727–30.
107. Zipernowsky. 1889. Zipernowsky electrostatic motor. *Electr World* 14:260.
108. Trimmer, T. 1990. Micromechanical systems. *Proceedings, Integrated Micro-Motion Systems: Micromachining, Control, and Applications (3rd Toyota Conference).* Aichi, Japan, October 1990, pp. 1–15.
109. Trimmer, W. S. N. 1989. Microrobots and micromechanical systems. *Sensors Actuators* 19:267–87.
110. Kim, Y.-K., M. Katsurai, and H. Fujita. 1990. Fabrication and testing of a micro superconducting actuator using the Meissner effect. *Proceedings: IEEE Micro Electro Mechanical Systems (MEMS '90).* Napa Valley, CA, February 11–14, 1990, pp. 61–66.

111. Pelrine, R., and I. Busch-Vishniac. 1987. Magnetically levitated micro-machines. *Proceedings: IEEE Micro Robots and Teleoperators Workshop*. Hyannis, MA, November 1987, 19/1–5.
112. Rousselet, J., G. H. Markx, and R. Pethig. 1998. Separation of erythrocytes and latex beads by dielectrophoretic levitation and hyperlayer field-flow fractionation. *Colloids Surfaces A* 140:209–16.
113. Kumar, S., D. Cho, and W. N. Carr. 1992. Experimental study of electric suspension for microbearings. *J Microelectromech Syst* 1:23–30.
114. Pister, K. S. J., R. S. Fearing, and R. T. Howe. 1990. A planar air levitated electrostatic actuator system. *Proceedings: IEEE Micro Electro Mechanical Systems (MEMS '90)*. Napa Valley, CA, February 11–14, 1990, pp. 67–71.
115. Wagner, B., M. Kreutzer, and W. Benecke. 1993. Permanent magnet micromotors on silicon substrates. *J Microelectromech Syst* 2:23–29.
116. Benecke, W. 1991. Silicon-microactuators: Activation mechanisms and scaling problems. In *1991 International Conference on Solid-State Sensors and Actuators (Transducers '91)*. San Francisco, CA, June 24–27, 1991, pp. 46–50.
117. Wagner, B., M. Kreutzer, and W. Benecke. 1991. Electromagnetic microactuators with multiple degrees of freedom. *6th International Conference on Solid-State Sensors and Actuators (Transducers '91)*. San Francisco, CA, June 24–27, 1991, pp. 614–17.
118. Gongora-Rubio, M., L. Sola-Laguna, M. Smith, and J. J. Santiago-Aviles. 1999. In *Microfluidic devices and systems II (proceedings of the SPIE)*. Santa Clara, CA: SPIE, 230–39.
119. Price, R. H., S. J. Cunningham, and S. C. Jacobsen. 1992. Field analysis for the electrostatic eccentric drive micromotor (wobble motor). *J Electrost* 28:7–38.
120. Goemans, P. A. F. M. 1994. Microsystems and energy: the role of energy. *Microsystem technology: Exploring opportunities*, ed. G. K. Lebbink. Alphen aan de Rijn/Zaventem: Samsom BedrijfsInformatie bv, pp. 50–64.
121. Christenson, T. R., H. Gückel, K. J. Skrobis, and J. Klein. 1992. In *Solid-State Sensors and Actuators Workshop*. Hilton Head Island, SC, June 22–25, 1992, pp. 51–54.
122. Guckel, H., K. J. Skrobis, T. R. Christenson, J. Klein, S. Han, B. Choi, E. G. Lovell, and T. W. Chapma. 1991. Fabrication and testing of the planar magnetic micromotor. *J Micromech Microeng* 1:135–38.
123. Ahn, C. H., Y. J. Kim, and M. G. Allen. 1993. A planar variable reluctance micromotor with fully integrated stator and wrapped coils. *IEEE microelectromechanical systems*. Fort Lauderdale, FL, February 1993, pp. 1–6. New York: IEEE.
124. Wallrabe, U., P. Bley, B. Krevet, W. Menz, and J. Mohr. 1994. Design rules and test of electrostatic micromotors made by the LIGA process. *J Micromech Microeng* 4:40–45.
125. Busch-Vishniac, I. J. 1992. The case for magnetically driven microactuators. *Sensors Actuators A* A33:207–20.
126. Edelstein, R. L., C. R. Tamanaha, P. E. Sheenan, M. M. Miller, D. R. Baselt, L. J. Whitman, and R. J. Colton. 2000. The BARC biosensor applied to the detection of biological warfare agents. *Biosens Bioelectron* 14:805–13.
127. Busch, J. D., and A. D. Johnson. 1990. Shape-memory properties in Ni-Ti sputter-deposited film. *J Appl Phys* 68:6224–28.
128. Fujita, H., and A. Omodaka. 1987. Electrostatic actuators for micromechatronics. *Proceedings: IEEE Micro Robots and Teleoperators Workshop*. Hyannis, MA , November 1987, 14/1–10.
129. Fujita, H. 1990. In *Proceedings: Micro System Technologies '90*, 818. Berlin, Germany.
130. Takeshima, N., K. J. Gabriel, M. Ozaki, J. Takahashi, H. Horiguchi, and H. Fujita. 1991. Electrostatic parallelogram actuators. *6th International Conference on Solid-State Sensors and Actuators (Transducers '91)*. San Francisco, CA, June 24–27, 1991, pp. 63–66.
131. Romankiw, L. T. 1989. In *Proceedings: Symposium on Magnetic Materials, Processes, and Devices*, 39–53. Hollywood, FL.
132. Cardot, F., J. Gobet, M. Bogdanski, and F. Rudolf. 1994. Fabrication of a magnetic transducer composed of a high-density array of microelectromagnets with on-chip electronics. *Sensors Actuators A* A43:11–16.
133. Ahn, C. H., Y. J. Kim, and M. G. Allen. 1993. A fully integrated micromachined toroidal inductor with a nickel-iron magnetic core (the switched dc/dc converter application). *7th International Conference on Solid-State Sensors and Actuators (Transducers '93)*. Yokohama, Japan, June 7–10, 1993, pp. 70–73.
134. Honda, T., K. I. Arai, and M. Yamaguchi. 1994. Fabrication of actuators using magnetostrictive thin films. In *IEEE International Workshop on Micro Electro Mechanical Systems (MEMS '94)*. Oiso, Japan, January 25–28, 1994, pp. 51–56.
135. Quandt, E., and K. Seemann. 1995. Fabrication of giant magnetostrictive thin film actuators. *Proceedings: IEEE Micro Electro Mechanical Systems (MEMS '95)*. Amsterdam, the Netherlands, January 29–February 2, 1995, pp. 273–77.
136. Rombach, P., and W. Langheinrich. 1994. An integrated sensor head in silicon for contactless detection of torque and force. *Sensors Actuators A* A41–42:410–16.
137. Hathaway, K. B., and A. E. Clark. 1993. Magnetostrictive materials. *MRS Bull* April: 34.
138. Fukuda, T., H. Hosokai, H. Ohyama, H. Hashimoto, and F. Arai. 1991. Giant magnetostrictive alloy (GMA) applications to micro-mobile robot as a micro actuator without power supply cables. *Proceedings: IEEE Micro Electro Mechanical Systems (MEMS '91)*. Nara, Japan, January 30–February 2, 1991, pp. 210–15.
139. Panholzer, R. 1963. Electromagnetic pumps. *Elec Eng* 2:128–35.
140. Heng, K.-H., L. Huang, W. Wang, and M. C. Murphy. 1999. Development of a diffuser/nozzle type micropump based on magnetohydrodynamic (MHD) principle. *Microfluidic devices and systems II*, eds. C. H. Ahn and A. B. Frazier. San Jose, CA, September 20–21, 1999, pp. 66–73.
141. Lemoff, A. V., A. P. Lee, R. R. Miles, and C. F. McConaghy. 1999. An ac magnetohydrodynamic micropump: towards a true integrated microfluidic system. *Technical Digest: 10th International Conference on Solid-State Sensors and Actuators*. Sendai, Japan, pp. 1126–29.
142. Lemoff, A. V., and A. P. Lee. 2000. In *Micro total analysis systems*, eds. A. van den Berg, W. Olthuis, and P. Bergveld. Enschede, the Netherlands: Kluwer Academic Publishers.
143. Phule, P., and J. Ginder. 1998. The materials science of field-responsive fluids. *MRS Bull* 23:19–21.

144. Herzer, G. 1997. Nanocrystalline soft magnetic alloys. In *Handbook of magnetic materials,* ed. K. H. J. Buschow, 417–61. Amsterdam: Elsevier Science B.V.
145. Rinaldi, C., and M. Zahn. 2002. Effects of spin viscosity on ferrofluid duct flow profiles in alternating and rotating magnetic fields. *J Magnetism Magnetic Materials* 252: 172–75.
146. Lorenz, C., and M. Zahn. 2003. Hele-Shaw ferrohydrodynamics for rotating and dc axial magnetic fields. 2003 Gallery of Fluid Motion. *Phys Fluids* 15:S4.
147. Hartshorne, H., C. J. Backhouse, and W. E. Lee. 2004. Ferrofluid-based microchip pump and valve. *Sensors Actuators B* 99:592–600.
148. Hatch, A., A. Kamholz, G. Holman, P. Yager, and K. Bohringer. 2001. A ferrofluidic magnetic micropump. *J Microelectromech Syst* 10:215–21.
149. Pal, R., M. Yang, B. N. Johnson, D. T. Burke, and M. A. Burns. 2004. Phase change microvalve for integrated devices. *Anal Chem* 76:3740–48.
150. Liu, R. H., J. Bonanno, J. Yang, R. Lenigk, and P. Grodzinski. 2004. Single-use, thermally actuated paraffin valves for microfluidic applications. *Sens Actuators* B 98:328–36.
151. Oh, K. W., K. Namkoong, and C. Park. 2005. A phase change microvalve using meltable magnetic material: Ferro-wax. In *9th International Conference on Miniaturized Systems for Chemistry and Life Sciences: uTAS*. Boston, MA, October 9–13, 2005, pp. 554–56.
152. Park, J. M., Y. Cho, B.-S. Lee, J.-G. Lee, and C. Ko. 2007. Multifunctional microvalves controlled by optical illumination on nanoheaters and its application in centrifugal microfluidic devices. *Lab Chip* 7:557–64.
153. Zoval, J., and M. J. Madou. 2004. Centrifuge-based fluidic platforms. *Proc IEEE* 92(1):140–53.
154. Zdeblick, M. J., R. Anderson, J. Jankowski, B. Kline-Schoder, L. Christel, R. Miles, and W. Weber. 1994. Thermopneumatically actuated microvalves and integrated electro-fluidic circuits. *Technical Digest: 1994 Solid State Sensor and Actuator Workshop*. Hilton Head Island, SC, June 13–16, 1994, pp. 251–55.
155. Barth, P. W. 1995. Silicon microvalves for gas flow control. *8th International Conference on Solid-State Sensors and Actuators (Transducers '95)*. Stockholm, Sweden, June 25–29, 1995, pp. 276–80.
156. Busch, J. D., and A. D. Johnson. 1990. Shape-memory properties in Ni-Ti sputter-deposited film. *J Appl Phys* 68:6224–28.
157. Walker, J. A., K. J. Gabriel, and M. Mehregany. 1990. Thin-film processing of TiNi shape memory alloy. *Sensors Actuators A* A21:243–46.
158. Johnson, A. D. 1991. Vacuum-deposited TiNi shape memory film: Characterization and applications in microdevices. *J Micromech Microeng* 1:34–41.
159. Dario, P., M. Bergamasco, L. Bernardi, and A. Bicchi. 1987. A shape memory alloy actuating module for fine manipulation. *Proceedings: IEEE Micro Robots and Teleoperators Workshop*. Hyannis, MA, November 1987, 16/1–5.
160. Quandt, E., C. Halene, H. Holleck, K. Feit, M. Kohl, and P. Schlossmacher. 1995. Sputter deposition of TiNi and TiNiPd films displaying the two way shape memory effect. *8th International Conference on Solid-State Sensors and Actuators (Transducers '95)*. Stockholm, Sweden, June 25–29, 1995, pp. 202–05.
161. Hunter, I. W., and S. Lafontaine. 1992. A comparison of muscle with artificial actuators. *Technical Digest: 1992 Solid State Sensor and Actuator Workshop*. Hilton Head Island, SC, June 22–25, 1992, pp. 178–85.
162. Gandhi, M. V., and B. S. Thompson. 1992. *Smart materials and structures.* London: Chapman & Hall.
163. Gordon, G. B., and P. W. Barth. 1991. Thermally-actuated microminiature valve, Hewlett-Packard, Palo Alto, CA. US Patent 5058856.
164. Barth, P. W., C. C. Beatty, and L. A. Field. 1981. Characterization of the silicon electrode. *Surf Sci* 108:135–52.
165. Jerman, H. 1991. Electrically-activated, normally-closed diaphragm valves. *6th International Conference on Solid-State Sensors and Actuators (Transducers '91)*. San Francisco, CA, June 24–27, 1991, pp. 1045–48.
166. Editorial. 1995. *Microflow Analytical.*
167. Hamberg, M. W., C. Neagu, J. G. E. Gardeniers, D. J. Ijntema, and M. Elwenspoek. 1995. Electrochemical micro actuator. *Proceedings: IEEE Micro Electro Mechanical Systems (MEMS '95)*. Amsterdam, the Netherlands, January 29–February 2, 1995, pp. 106–10.
168. Nagakura, T., K. Ishihara, T. Furukawa, K. Masuda, and T. Tsuda. 1995. Auto-regulated medical pump without energy supply. *8th International Conference on Solid-State Sensors and Actuators (Transducers '95)*. Stockholm, Sweden, June 25–29, 1995, pp. 287–90.
169. Brannon-Peppas, L. 1997. Polymers in controlled drug delivery. *Med Plastics Biomater* 4:34–44.
170. Lamoreaux, S. K. 1997. Demonstration of the Casimir force in the 0.6 to 6 μm range. *Phys Rev Lett* 78:5.
171. Bell, D. J., T. J. Lu, N. A. Fleck, and S. M. Spearing. 2005. MEMS actuators and sensors: Observations on their performance and selection. *J Microelectromech Syst* 15:S153–64.
172. Finot, E., A. Passian, and T. Thundat. 2008. Measurement of mechanical properties of cantilever shaped materials. *Sensors* 8:3497–541.
173. Den Hartog, J. P. 1956. *Mechanical vibrations.* New York: McGraw-Hill.
174. Roukes, M. L. 2000. In *Solid-State Sensor and Actuator Workshop*. Hilton Head Island, SC: Transducers Research Foundation, pp. 367–76.
175. Petersen, K. E. 1982. Silicon as a mechanical material. *Proc IEEE* 70:420–57.
176. Oden, P. I., G. Y. Chen, R. A. Steele, R. J. Warmack, and T. Thundat. 1996. Viscous drag measurements utilizing microfabricated cantilevers. *Appl Phys Lett* 68:1465–69.
177. Thundat, T., P. I. Oden, and R. J. Warmack. 1997. Microcantilever sensors. *Nanoscale Microscale Thermophys Eng* 1:185–99.
178. Thundat, T., G. Y. Chen, R. J. Warmack, D. P. Allison, and E. A. Wachter. 1995. Vapor detection using resonating microcantilevers. *Anal Chem* 67:519–21.
179. Hansen, K. M., H.-F. Ji, G. Wu, R. Datar, R. Cote, A. Majumdar, and T. Thundat. 2001. Cantilever-based optical deflection assay for discrimination of DNA single-nucleotide mismatches. *Anal Chem* 73:1567–71.
180. Moulin, A. M. 2000. Microcantilever-based biosensors. *Ultramicroscopy* 82:23–31.
181. Thundat, T., L. A. Bottomley, S. Meller, W. H. Velander, and R. Van Tassell. 2001. In *Immunoassays: Methods and protocols,* eds. A. L. Ghindilis, A. R. Pavlov, and P. B. Atanajov. Totowa, NJ: Humana Press.

182. Fritz, J., M. K. Baller, H. P. Lang, H. Rothuizen, P. Vettinger, E. Meyer, H.-J. Güntherodt, C. Gerber, and J. K. Gimzewski. 2000. Translating biomolecular recognition into nanomechanics. *Science* 288:316–18.
183. Chen, G. Y., T. Thundat, E. A. Wachter, and R. J. Warmack. 1995. Adsorption-induced surface stress and its effects on resonance frequency of cantilevers. *J Appl Phys* 77:3618–22.
184. Yang, Y. T., C. Callegari, X. L. Feng, K. L. Ekinci, and M. L. Roukes. 2006. Zeptogram-scale nanomechanical mass sensing. *Nano Lett* 6:583–86.
185. Burg, T. P., M. Godin, S. M. Knudsen, W. Shen, G. Carlson, J. S. Foster, K. Babcock, and S. R. Manalis. 2007. Weighing of biomolecules, single cells and single nanoparticles in fluid. *Nature* 446:1066–69.

9

Power and Brains in Miniature Devices

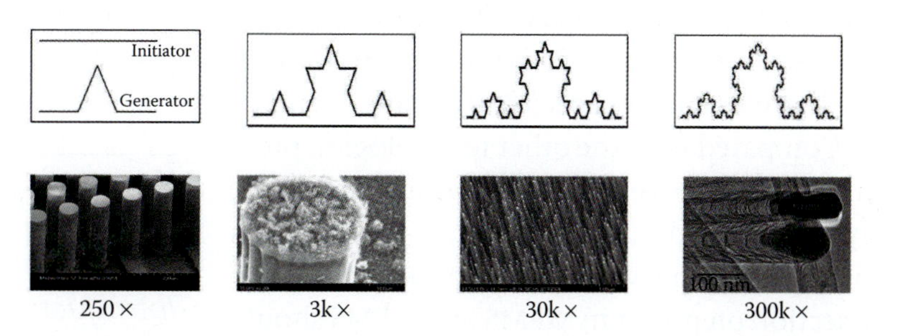

Different orders of a Koch curve seen in parallel with carbon fractal electrodes at different magnifications. Fractal electrodes can minimize internal resistance while maximizing surface-to-volume ratios. Carbon-MEMS shows great promise for use in the fabrication of fractal electrodes for electrochemical applications.

Outline

Introduction

In this chapter we consider power and brains in miniature devices. We analyze the challenges and promises of powering miniaturized devices and of the miniaturization of power sources themselves. The power section starts with a comparison of onboard energy sources, such as batteries, capacitors, fuel cells, internal combustion engines, and solar cells. Next, we consider energy scavenging methods, such as taking advantage of environmental vibrations or thermal gradients.

Miniaturization of power sources scales disadvantageously, as the specific energy (energy per volume unit: l^3) of the power source determines the volume for a given application. That volume, besides the active materials (say, Li in a Li-ion battery), must also include packaging, and packaging efficiency degrades as the power source gets smaller.

As in the case of cell phones, personal digital assistants (PDAs), distributed sensor networks, portable computers, etc., small, lightweight, and long-lasting energy sources are one of the most urgently needed MEMS and NEMS breakthrough technologies. A pertinent quote from *Wired* magazine in April 2004 reads, "If we don't do something about increasing battery life, we're toast. The biggest impediment to our technological future isn't extending Moore's law. No, the biggest challenge to progress is much more ordinary: It's battery life."[1]

Micro- and nanosystems require even smaller power sources, as weight and volume of onboard energy sources are disproportionately large compared with the systems they power. The roles of energy storage and energy dissipation in microsystems differ considerably from the world of practical daily experience. Designing micro- and nanosystems demands close examination of energy budgets and taking advantage of the merits of smallness as well as minimizing its adverse effects. In Figure 9.1 we compare progress made in improving processors, hard disk drive (HDD) capacity, and battery energy stored. Compared with the other technologies, progress in battery improvement is obviously much more sluggish, with a growth of, at most, 2–3% per year: there is obviously no Moore's law for batteries.

The section on power in MEMS and NEMS should help the reader in choosing a preferred power source for his or her MEMS/NEMS device.

We also consider the implementation of computing power or "brains" in miniaturized systems. After a short summary of current and projected computer technology and its comparison with nature's computers, we consider the latest developments in a

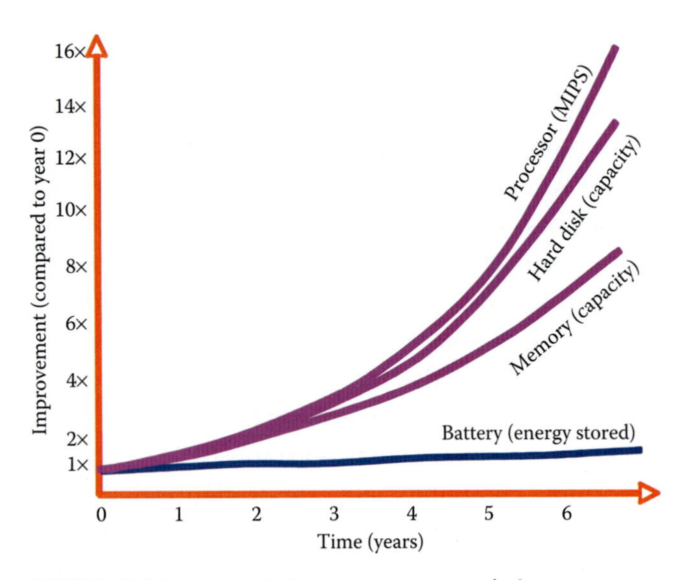

FIGURE 9.1 Progress in improvements made in processors, hard disk capacity, and battery energy stored. No Moore's law in batteries: 2–3% growth per year. MIPS, million instructions per second.

field closely related and important to microrobotics: artificial intelligence (AI). We will discuss how both bottom-up and top-down approaches to AI might impact miniaturization technology. Complexity theory and artificial life are examples of bottom-up approaches to building AI that feature the self-organizing or emergent properties touched upon in Chapter 3. Artificial neural networks represent a bottom-up approach to developing brains heavily dependent on learning cycles. Electronic noses and tongues have already been developed on the basis of such neural nets (see Chapter 10). As in the case of machining, we will learn that bottom-up approaches for embedding intelligence in miniature devices are gaining more and more interest.

Power in MEMS and NEMS

Onboard Energy Sources

Introduction

Onboard energy sources for MEMS devices may include radioactive sources, batteries, capacitors, fuel cells, internal combustion engines, solar cells, etc. Combustion processes for electrical power generation provide advantages over onboard batteries, even when the low conversion efficiency in the combustion process from thermal energy to electrical energy is taken into account. For example, hydrocarbon fuels provide an energy storage density between 40 and 50 MJ/kg, whereas modern nickel metal hydride batteries, commonly used in laptop computers, provide only 0.4 MJ/kg. Thus, even at only 4% conversion efficiency from thermal to electrical energy, hydrocarbon fuels provide about five times higher energy storage density than batteries.

There has been a lot of attention given recently to MEMS and even NEMS power sources. The goal is to put as much energy as possible in the smallest amount of space. To form an idea of what is achievable with known power systems, we list the energy one can store in 1 mm^3 of various power systems in Table 9.1. From this table, the highest energy density available is from a radioactive source, but such sources may never be practical for consumer applications because of environmental and safety concerns (although americium is used as the current source

Table 9.1 Energy Stored in 1 mm^3

Power capacitor	4 μJ/mm^3	1 μW for 4 s
Thick-film battery	1 J/mm^3	270 μW for 1 h
Thin-film battery	2.5 J/mm^3	0.7 mW for l h
Solar cell (1 × 1 × 0.1 mm^3)		0.1 mW
Gasoline	300 J/mm^3	3 mW for 1 day
180 Ta ($T_{1/2}$ = 8 h)	≈1 MJ/mm^3	34 W
178 Hf ($T_{1/2}$ = 31 years)	>10 MJ/mm^3	160 mW

in some fire alarms). Chemical energy in the form of batteries provides the next highest density. Solar light, at noon on a sunny day, gives us 100 mW/cm^2 (office lights are 7.2 mW/cm^2). Collectors of solar light work with varying efficiency; single-crystal Si provides a 14–30% efficient conversion of light to power. Polysilicon provides about 10–15% efficiency, and organic dyes are 5–10% efficient.

Future stand-alone microsystems will almost surely employ hybrid strategies, using low levels of power for low-level activities and quick bursts of energy, perhaps relying on a capacitor, for computing, communication, or maneuvering.

MEMS and NEMS Batteries, Capacitors, Fuel Cells, Combustion Engines, and Solar Cells

Batteries

To judge the benefits and problems associated with different battery technologies, some general properties of batteries must be introduced. It is usually desirable that the amount of energy stored in a given mass or volume of active battery material is as high as possible. To compare the energy content of batteries, the terms *specific energy* (expressed in Wh/kg) and *energy density* (in Wh/L) are used, whereas the rate capability is expressed as *specific power* (in W/kg) and *power density* (in W/L). To reach the goal of a high specific energy and energy density, two fundamental requirements must be met by the battery electrode materials: 1) a high specific charge (in Ah/kg) and charge density (in Ah/L), i.e., a high number of available charge carriers per mass and volume unit of the material; and 2) a high (positive electrode) and low (negative electrode) standard redox potential of the respective electrode redox reactions, leading to a high cell voltage (see also Volume I, Chapter 7). Moreover, in rechargeable batteries (also called *reversible* or *secondary batteries*), reactions at both negative and positive electrode have to be highly reversible to maintain the specific charge for hundreds of charge/discharge cycles. With an assumed cycling efficiency of even 99% in each cycle, the available specific charge after 100 cycles would be only ~37% of the initial value (after 500 cycles, less than 1%), and electrode materials should thus combine high specific charge with good rechargeability.

In Table 9.2 we list primary and secondary batteries for onboard power systems, and for comparison we also included data on microfuel cells and

Table 9.2 Examples of Onboard Power Systems

On Chip			
Power Sources	**Energy Density (Wh/kg)**	**Wh/L**	**Power Density (W/kg)**
Primary batteries			
Zinc air	200	1050–1560	150
Zn-Cl_2	150	100–180	90
Alkaline	125–225	150–440	
Silver Oxide	155–285	250–500	
Lithium	100–200	340–500	>200
Secondary batteries			
Lead-acid	30–40	40	30–40
Ni-Cd	200	170	150–200
NiMH	150–200	250	150
Li–ion	100	300	200
Microfuel cell		4000	
Ultracapacitors		13–28	200

ultracapacitors. The specific energy (energy per volume unit) of the onboard power sources listed in Table 9.2, determines the proper active volume for a given application. If the volume of the packaging is taken into account, energy volume densities of primary lithium batteries may reach 340–500 Wh/L; these batteries generate the highest specific energy of any commercially available battery. If we assume that the average power consumption of a MEMS device is 100 mW, we need slightly more than 1 cm^3 of lithium battery to last for 1 year, assuming we can use 100% of the charge in the battery. The energy density of rechargeable batteries is typically less than half of that of primary batteries, so in that case we need to be recharging every 3–4 months. Today, commercially available batteries are obviously too big for most MEMS devices, and they are even too big for many mobile consumer applications.

An important technical challenge that arises in considering downscaling batteries, besides the obvious fact that their capacity scales as l^3, is that battery packaging efficiency scales poorly in the microdomain: there is relatively more packaging in smaller batteries than in large ones. What is needed, then, in the microbattery arena is a set of new packaging strategies, new materials, and new manufacturing techniques. In order to be able to make an onboard micropower source using MEMS technology, one also expects that the materials involved are compatible with microfabrication processes (e.g., being stable under vacuum conditions; no liquid electrolytes), that the battery is rechargeable (because its life cycle will be short), and that the reaction products are compatible with the MEMS device.

In terms of commercially available batteries for MEMS, Li primary and Li-ion reversible batteries are the two most important choices today. In lithium primary batteries, the anode, consisting of high-purity lithium, may be combined with many different cathode materials, resulting in different voltages ranging from 1.5 to 3.9 V. For electrolyte, batteries contain organic solvents in which lithium salts are dissolved or they feature solid polymers that also may function as electrolytes. Beyond the button Li batteries 4.8 mm (diameter) by 1.4 mm (high) used in watches and cameras, progress in miniaturizing high-energy-density power sources has been limited. Button Li cells are now the best available primary energy sources for microsystems. The basic principle of rechargeable Li-ion batteries is electrochemical intercalation and de-intercalation of lithium in both electrodes. As we noted above, an ideal secondary or reversible battery has a high-energy capacity, fast charging time, and long cycle life. The capacity is determined by the lithium saturation concentration of the electrode materials. Many materials combination have been researched, and an often touted approach is the use of carbon nanotubes (CNTs). The highest Li concentration can be stored in CNTs if all the interstitial sites (inter-shell van der Waals spaces, inter-tube channels and inner cores) are accessible for Li intercalation. Nalimova et al. proved that two Li atoms per C atom can be stored in multiwall carbon nanotubes (MWCNTs) under a high pressure of 6 GPa.[2,3] This is a very high Li loading, but it involves a chemical reaction, not an intercalation, and its usefulness for a battery is not established. Gaucher et al. reported high degrees of irreversible (500 mAh/g) and reversible (450 mAh/g) insertion of lithium ions in MWCNTs in an aprotic medium.[4] They noticed that the structure of the MWCNTs plays an important role in specific capacity and cycle stability. The capacity of less-graphitized MWCNT was higher than that of the well-graphitized MWCNT, although the latter exhibited better cycle stability and charge/discharge rates.[4] In spite of such interesting behaviors of CNTs as the host of lithium intercalation, their large irreversibility capacity and their poor discharge curves relative to those of lithium metal leave the material not satisfactory yet for commercial applications.

Batteries and fuel cell materials deposited with integrated circuit (IC) technologies on the device substrate itself remain in the research stage.[5] Often, the thin-film materials deposited in constructing those batteries, such as Li, TiS_2, V_2O_5, etc., prove incompatible with the IC process, and the prospect of integrating them with ICs remains remote. Ultrathin, solid-state Li cells—"energy paper"—are beginning to emerge. Kanebo introduced the polymeric polyacenic semiconductor (PAS)-based battery as far back as 1993. The polymer PAS film in the battery is only 200 μm thick and has an active surface area of 2200 m^2/g. It serves as the anode, and the cathode is lithium based. The voltage is 3.3 V, corresponding to 3 Ni-Cd elements in series. Unfortunately, the energy density, taking the complete, packaged battery into account, is only 5.5 Wh/L.[6]

If a MEMS device uses 10 μW of power continuously, in a day it will consume ~1 J of energy. Since the energy density for Li-ion thin-film systems is in the range of ~2 J mm^{-3}, matching or exceeding standard commercial lithium-ion systems, it would appear that this battery fits the task. In reality, this is not so: thin-film batteries, as shown in Figure 9.2, are two-dimensional (2D) devices and are necessarily thin in order to prevent power losses typically associated with the slow transport of ions, and this form factor leads to a low capacity per mm^2. The energy per unit area as reported for several lithium thin-film batteries ranges from 0.25 to $\sim 2 \times 10^{-2}$ Jmm^{-2};[7] consequently, with a typical device thickness on the order of 15 μm, a thin-film battery would require a significant amount of area in order to supply 1 Jmm^{-3}.

The real question then becomes whether a thin-film battery can supply the energy that the MEMS device requires within the constraints of the areal "footprint" available for the battery. Unfortunately, a thin-film battery of 1 mm^2, despite having an excellent energy per unit volume, falls far short of being able to power a typical MEMS for more than one day. If the areal footprint is made 100 times larger (at 1 cm^2), the thin-film approach becomes more acceptable,[7,8] These important consequences of the 2D nature of thin-film batteries are frequently overlooked. The calculation by Koeneman et al.,[9] for example, ignored the 2D character of thin-film batteries when they concluded that these batteries could carry out some 60,000 actuations of a "smart bearing."[7] When the actual area available for the power source on the device is considered, only about 1200 actuations are possible. In order to power devices with limited real estate and maintain a small areal footprint, batteries must somehow make better use of their thickness dimension.[7]

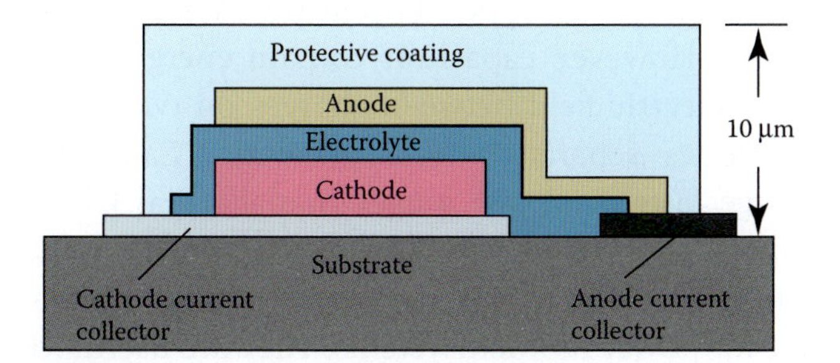

FIGURE 9.2 Example thin-film battery.

In recent years there has been a growing realization that improved microbattery performance can be attained by configuring the electrode materials currently employed in 2D batteries into micro-three-dimensional (μ-3D) architectures (high-rise batteries). In this approach, cell structures are designed to maximize power and energy density while maintaining short ion transport distances.[7] There are several possible architectures that can achieve this same goal, but an underlying principle in all μ-3D batteries is that the transport between the electrodes remains one-dimensional at the microscopic level, while the electrodes are configured in complex non-planar geometries in order to increase the energy density of the cell within the same footprint area.[7,10]

Detailed calculations by Long et al.[7] show that microbatteries configured as μ-3D structures can readily achieve 1 Jmm^{-2}, the amount required to power a typical MEMS device for a day. Perhaps the most obvious μ-3D design is one consisting of interdigitated electrode posts, shown in Figure 9.3a. The anode and cathode consist of arrays of rods separated by a continuous electrolyte phase. The spatial arrangement of the anode and cathode arrays determines the current-potential distribution. The short transport distances lead to a much lower ohmic resistance as compared with traditional planar battery configurations. The molecular diffusion equation indicates that diffusion over a 10-μm distance is 1 million-fold faster than diffusion over a 1-cm distance, since $x = (2D\tau)^{1/2}$ (Volume I, Equation 6.100), where D is the diffusion constant and τ the time required for a molecule to diffuse over distance x. Thus, μ-3D configurations with narrow electrode posts offer a means of maintaining "short" diffusion distances, thereby increasing the power capacity of the battery. For example, Hart et al. first consider a thin-film 2D battery that comprises a 1-cm^2 area for both anode and cathode, each 22.5 μm thick and separated by a 5-μm-thick electrolyte.[10] The total volume of electrodes and electrolyte in this case is 5×10^{-3} cm^3. It is relatively straightforward

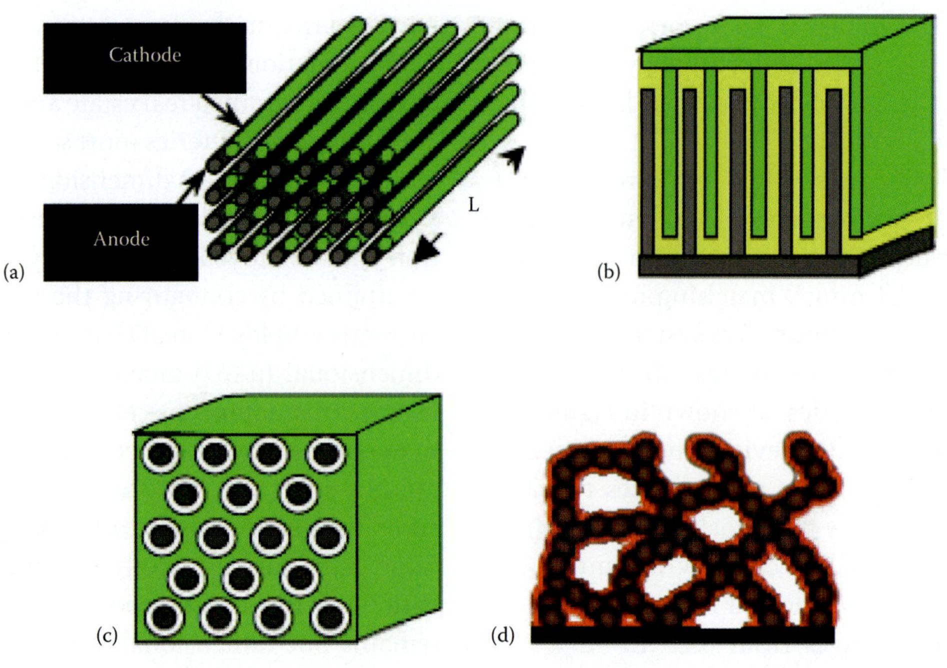

FIGURE 9.3 Examples of prospective three-dimensional architectures for charge-insertion batteries. (a) Array of interdigitated cylindrical cathodes and anodes; (b) Interdigitated plate array of cathodes and anodes; (c) Rod array of cylindrical anodes coated with a thin layer of ion-conducting dielectric (electrolyte), with the remaining free volume filled with the cathode material; (d) Aperiodic "sponge" architectures in which the solid network of the sponge serves as the charge-insertion cathode, which is coated with an ultrathin layer of ion-conducting dielectric (electrolyte), and in which the remaining free volume is filled with an interpenetrating, continuous anode.[7]

to calculate then that a corresponding μ-3D array battery (see Figure 9.3) with a series of 5-μm-radius interdigitated cathodes and anodes, a 5-μm surface-to-surface electrode spacing, and occupying the same total volume (i.e., 5×10^{-3} cm^3) contains only about 39% of the energy capacity of the thin-film 2D design. The lower-energy capacity is due to a higher percentage of the total volume being occupied by the electrolyte. On the other hand, the active cathode and anode surface areas in the three-dimensional (3D) design are 3.5 cm^2 each, significantly larger than the 2D design. There are some other intriguing advantages of the μ-3D design that are not directly reflected in the above numbers. For instance, the transport length scale in the thin-film 2D battery is 350% larger than in the 3D design. This means that the μ-3D design is significantly less susceptible to ohmic losses and other transport limitations. To achieve equal transport length scales in the 2D design would require a 350% increase in the areal footprint in order to maintain equal cell volume, a significant disadvantage in employing these devices in MEMS and microelectronic applications. While the above comparison of 2D and μ-3D designs indicates that the μ-3D cell has an inherently lower-energy capacity per total cell volume, in fact, the capacity of the 3D design can be increased without limit by increasing the battery height, without sacrificing the small area footprint or high-power density.

For an example of this type of μ-3D electrode configuration, see the carbon-MEMS (C-MEMS) battery in Figure 5.18. This type of battery is being developed by Enevate (http://www.enevate.com/).

Capacitors

Rechargeable and disposable (primary) batteries, reviewed above, use a chemical reaction to produce energy. The problem is that after many charges and discharges, even a rechargeable (secondary) battery loses capacity, to the point where the user has to discard it. However, capacitors contain energy stored in an electric field between charges on two metal electrodes separated by a dielectric. Capacitors charge faster and last longer than normal batteries, and their reversibility, absence of polarity, and extended lifetime make them an attractive alternative for power in microsystems. They are useful for on-chip power conversion, but their energy density is too low to be a real secondary storage component. Electrochemical capacitors improve that perspective; with an energy density on the order of 75 J/cm^3, they are a good potential for secondary storage.

Electrochemical capacitors are called by a number of names: *supercapacitors*, *ultracapacitors*, and *electrochemical double-layer* (ECDL) *capacitors*. Typically, supercapacitor devices consist of two electrodes and an insulating material that separates the electrodes. The insulator prevent electrical contact but allows ions from the electrolyte to pass through. At each electrode, there exists a dipole layer, the so-called *ECDL*. Typically ~0.4 nm thick, the ECDL acts as a capacitor if the potential difference across it is less than that needed to dissociate the electrolyte (into hydrogen or oxygen in the case of water). If the electrolyte is water, this restricts the voltage to 1.23 V, but higher voltages of 2.3 V are possible by using a nonaqueous electrolyte, such as tetraethyl ammonium fluoroborate in propylene carbonate. The capacitance is given by $C = \varepsilon A/d$, where ε is the dielectric constant, A is the electrode surface area, and d is the ECDL thickness. The electrodes are made of materials with a high effective surface area, such as porous metal oxides (e.g., IrO_x), porous carbon, or carbon aerogels, in order to maximize the surface area, A, of the double layer and to achieve a small double layer thickness, d. By choosing and fabricating electrodes with extremely high surface area A (e.g., 1000 m^2/g or greater) and a small d, very large capacitances are achievable (e.g., 1000 F). Carbon nanotube filaments can further increase the surface area of the electrodes and allow capacitors to store yet more energy. Single-walled carbon nanotubes (SWNTs) have the largest surface area to volume ratio of any carbon material, 3000 m^2/g, as all their atoms are on the surface, and they make for the ultimate electrode material. This advantage is used for energy storage in nanotube supercapacitors.[7]

The performance of supercapacitors is judged in terms of their energy density (per unit volume) or $\frac{1}{2}CV^2$ and power density or V^2/R, where R is their series discharge resistance. Another advantage of CNTs in this regard is that they have a much lower resistance than the often-used activated carbon, which could greatly increase their power density. The limiting factors are contacting CNTs to the electrode backing and having the ability to do this at low temperatures and at a low cost, as the competing activated carbon is very cheap.

Since it is possible to carbonize photoresist materials, make them porous, and charge them, it is possible to integrate ultracapacitors on ICs (see C-MEMS in Chapter 5). The most challenging technical issue, just as in the case of a battery, is packaging. Supercapacitors and batteries incorporate very corrosive and reactive materials, making the challenge even more daunting.

The performance of supercapacitors can be gleaned from the *Ragone plot*, as shown in Figure 9.4. A Ragone plot is a log-log plot of the specific power (W/kg) versus the specific energy (Wh/kg).

As can be seen from this plot, supercapacitors or ultracapacitors bridge the gap between capacitors and batteries and have the ability to store greater amounts of energy than conventional capacitors, and they are able to deliver more power than batteries. Supercapacitors also possess a number of desirable qualities that make them an attractive energy storage option: they store and release charge completely reversibly, they are extremely efficient and can withstand a large number of charge/discharge cycles, they can store or release energy very quickly, and they can operate over a wide range of temperatures.

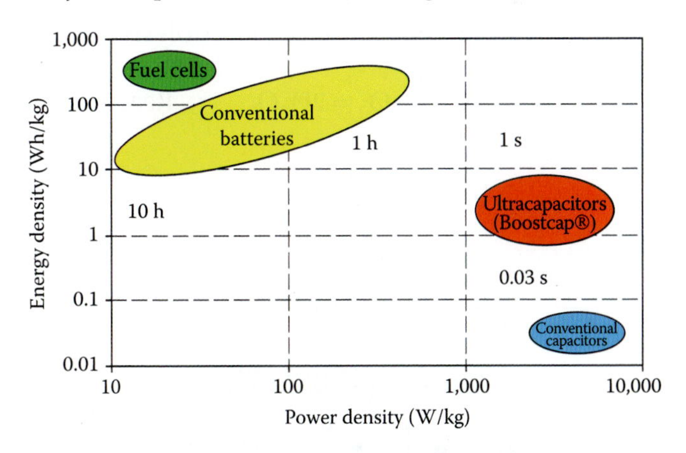

FIGURE 9.4 Ragone plot. The power density versus energy density plot distinguishes various electricity storage systems. Batteries are high-energy density but low-power density, whereas normal capacitors are low-energy density but high-power density. Li^+ batteries have the highest energy density of any battery. Supercapacitors have a high-energy density and high-power density and are intermediate between normal capacitors and batteries. See http://www.tecategroup.com/ultracapacitors/maxwellboostcaps.php.

Fuel Cells

Introduction

Two major challenges of our times are the shrinking of available fossil energy resources and climate change associated with global warming. With humans consuming carbon-based fuels (carbon, natural gas, coal, and tar shale) faster than they can be replaced, yet another "oil war" making the daily news, GM still selling 15–20,000 Hummers a year, and Tata planning to sell millions of its new Nano cars, independence from fossil fuels has finally again become what former President Jimmy Carter called the "moral equivalent of war."

One of the often advanced solutions to the energy crisis is to replace fossil hydrocarbon-based energy with the energy from carbon-free sources such as the sun, nuclear energy, or the hot interior of the earth and use hydrogen as the energy carrier. Hydrogen can be produced by hydrolysis of water using energy from carbon-free sources, and the hydrogen thus produced can serve as fuel in fuel cells to generate electricity, either stationary or mobile.

Fuel cells, discovered in the 1850s, are batteries with renewable energetic materials (fuels) (see inset in the heading of Part II and Figure 9.5). The electrodes in a fuel cell do not take part in the cell reactions they act as reaction sites only. The basic reaction in a fuel cell is reversed electrolysis, or:

$$2H_2 + O_2 = 2H_2O \qquad \text{Reaction 9.1}$$

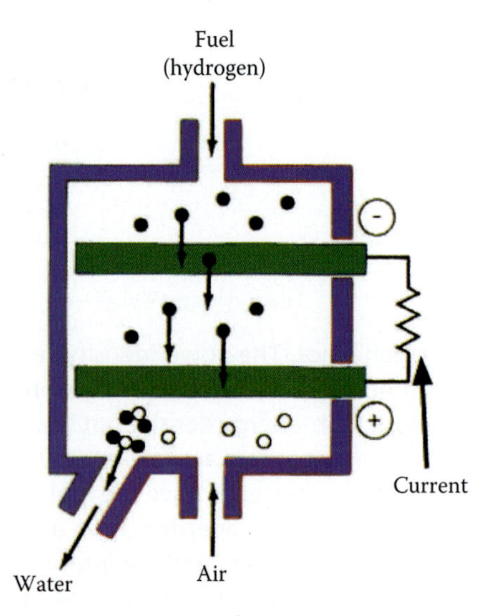

FIGURE 9.5 Schematic of a fuel cell.

Electricity is produced in an anode reaction:

$$H_2 = 2H^+ + 2e^- \qquad \text{Reaction 9.2}$$

and a cathode reaction:

$$\tfrac{1}{2}\, O_2 + 2H^+ + 2e^- = H_2O \qquad \text{Reaction 9.3}$$

Fuel cells can help us generate cleaner power from conventional sources more efficiently and can be viewed as a temporary energy solution until we can harvest more renewable resources (solar, wind) and/or or more powerful sources (fusion) directly.

Differences in electrolytes used determine the type of fuel cell:

- Phosphoric acid fuel cell (PAFC)
- Proton exchange membrane fuel cell (PEM, SPE™, and PEFC)
- Solid oxide fuel cell (SOFC)
- Molten carbonate fuel cell (MCFC)
- Alkaline fuel cell (AFC)

There are quite a few positive fuel cell attributes to enumerate. Power and energy density of a fuel cell are higher than in batteries [e.g., methanol (1550 Wh/L, 1738 Wh/kg) versus lithium-ion (450 Wh/L, 185 Wh/kg); see also the Ragone plot in Figure 9.4]. Fuels used in fuel cells typically have much higher (approximately 10×) energy densities than their battery counterparts, and to increase the power density of a fuel cell, one only needs to increase the surface-to-volume ratio within the cell. The latter is a much simpler task than engineering new material chemistries for batteries. There is no lengthy recharging, as fuel cells are rapidly refueled; instead of charging a battery for an extended amount of time, a fuel cartridge is inserted. The energy conversion process in fuel cells is simpler, and there is an unlimited access to active electrochemical material (fuel), while the amount of active materials accessible for energy conversion in batteries is limited to their volumetric energy density. Fuel cells feature negligible self-discharge, are clean and environment-friendly, and have an unlimited cycle stability. Costs of fuel cells are coming down: through sophisticated engineering the amount of expensive platinum required for catalysts is being lowered, and new manufacturing processes are being introduced. While the cost per Joule ($/J) has gone up for energy-dense batteries (e.g., NiCd weighs 0.5 kg, lasts 1 h, and costs $20, and a comparable Li-ion battery lasts 3 h but costs more than four times more), the cost of a comparable fuel cell that lasts 30 h has gone down (to perhaps <$5).

Disadvantages of fuel cells compared with batteries are lower efficiency (up to 50–60%, compared with about 90% for Li-ion batteries) and the need for peripherals such as pumps, humidifiers, fuel processors, power conditioners, etc. These peripherals have to do with the need to obtain hydrogen from traditional fuels (oil, coal, wood, propane, natural gas, ethanol, etc.), to store fuels and hydrogen (the storage medium), and to manage heat and water (both products of the fuel cell reaction) and power (e.g., convert DC to AC). A typical process to obtain hydrogen is steam reforming, which can process methane into hydrogen (research and development is under way to also reform diesel, gasoline, ethanol, etc.). The basic steam reforming reaction is as follows:

$$CH_4 + 2H_2O = 3H_2 + CO_2 \qquad \text{Reaction 9.4}$$

Intermediate steps in this reaction produce CO, which is reacted to form CO_2. The reaction also requires catalysts to speed up reaction rates, and sulfur compounds must be removed to protect those catalysts.

In Table 9.3 we present a summary of stationary fuel cells (data are from 2003). Fuel cells are still

Table 9.3 Summary of Stationary Fuel Cell Products

200 KW PAFC – UTC Fuel cells

250 KW SOFC – Siemens

Type	Electrolyte	Electrode	Fuel processor	Temp (°C)	Electrical Efficiency (LHV)	Cogeneration potential	Stage of development	Manufacturer
PAFC	Phosphoric acid (liquid/gel)	Noble metal	External	200	35-40%	Hot water Steam (low psi)	Commercial sales 250 units in field	UTC fuel cells
SOFC	Zinconium oxide (solid)	Ceramic	Internal	1000	50-65%	Hot gas Steam optional	Field demonstrations Several units in field	Siemens Westinghouse
MCFC	Alkali carbonates (liquid)	Nickel	Internal	650	45-60%	Hot gas Steam optional	Field demonstrations Expect to have precommercial units in place within one year	Fuel Cell Energy, Inc.
PEMFC	Organic polymer (solid)	Noble metal	External	80	40-45%	NA	Field demonstrations Commercial sales of small residential units	Avista Ballard H Power UTC Fuel cell Nuvera Plug Power

200 KW MCFC – Fuel Cell Energy, Inc.

250 KW PEM – Ballard

too expensive, and in order to introduce them into the huge mobile market, their miniaturization and the integration/simplification of the accompanying peripheral gear is a must (see next section).

Microfuel Cells

Introduction

There are many reasons why one might want to make a miniature fuel cell. To begin with, small fuel cells, especially when they come with less or size-reduced peripheral equipment, will make many more applications possible (e.g., power for mobile phones, iPods, laptops, etc.). The comparison between 2D and μ-3D batteries carried out earlier (see Batteries) can be extended to 2D and μ-3D fuel cells. From simple theoretical considerations, for a specific μ-3D electrode designs, we calculated a 3.5-fold larger surface area than the corresponding 2D design. The surface area of a fuel cell determines its amperage, meaning that fuel cell power is directly proportional to the electrode surface area. Smaller cells have larger surface areas relative to their volume, so they automatically feature increased power density. Microfuel cells might eventually also come at a lower cost because less material is used in their construction. Because of function integration possible in microsystems, the overall complexity may also be reduced (less peripherals), further reducing total weight and volume. Moreover, microfuel cells are expected to have a higher efficiency because of their high surface-to-volume ratio and the corresponding increase in triple points (reaction points where gas, liquid, and electrode meet) and the smaller fluidic channel sizes, which allow more efficient mass transport (for serpentine flow patterns, the flow velocity is inversely proportional to the square of the length scale). Employing microfluidic strategies, gas transport in the cell can be maximized while gas stoichiometry can be maintained and "dead zones" underneath structures minimized. Microfuel cells also may benefit from increased catalyst utilization, because microfabrication and nanostructuring techniques (see e.g., carbon nanohorns, Figure 9.8) allow for more control over catalyst deposition and distribution. Finally, it is easier to maintain a homogeneous environment within a small area, and the internal resistance is lower because of shorter conductive paths.

Example 9.1:

There have been many attempts at miniaturization of fuel cells. A recent example comes from Toshiba, which unveiled a miniature direct methanol fuel cell (DMFC) developed for small electronic devices such as digital audio players and wireless headsets (Figure 9.6). The fuel cell, measures 22 × 56 × 4.5 mm, weighs 8.5 g, and can put out 100 mW of power for as long as 20 h with a 2 cc charge of 95% methanol. The air-breathing cell has only passive components.

DMFCs are a subcategory of polymer exchange membrane fuel cells (PEMFCs) where the fuel, methanol, is not reformed but is fed directly into the fuel cell, avoiding complicated catalytic reforming steps. When a DMFC is operating, the methanol solution and oxygen gas are fed to the surface of the membrane electrode assembly (MEA) through the anode and cathode plates, respectively. Methanol is consumed at the anode catalyst layer, producing protons and electrons and releasing the byproduct CO_2. The protons directly transfer through the PEM to the cathode, while the electrons, blocked by the insulating PEM, have to travel to the cathode via the external circuit, resulting in an electrical current. The protons and electrons combine with O_2 at the cathode catalyst layer to form H_2O. Storage of methanol is easier than that of hydrogen because it does not need to be done at high pressures or low temperatures, since methanol is a liquid. Liquid methanol, as

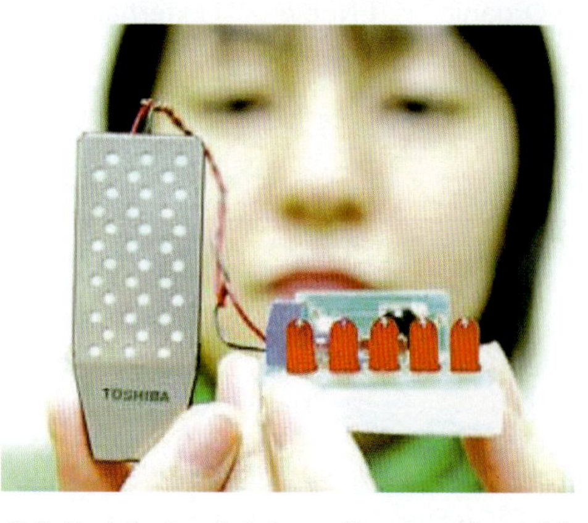

FIGURE 9.6 Toshiba's miniature direct methanol fuel cell.

the fuel of DMFC, has the advantages of easy storage and the ability to refuel instantly with no interruption of operation. The methanol cell can be directly started and operated at room temperature, which is a significant advantage for the applications of the portable electronic devices. Finally, the energy density of methanol is an order of magnitude greater than that of even highly compressed hydrogen. Methanol has an energy density of 17.6 kJ/cm^3, which is about 6 times that of a lithium battery.

Unfortunately, the efficiency of DMFCs is typically low because of the high permeation of methanol through the polymer membrane, a deleterious effect known as methanol crossover.

Fuels and Fuel Storage in Microfuel Cells

Introduction

The advent of microfuel cell-powered gadgets will require progress in the manufacture of fuel storage cartridges (see Figure 9.7).

Fuel sources for fuel cells range from highly flammable to relatively inert. Many miniaturized microfuel cells use methanol to be oxidized directly in DMFCs (see the Toshiba cell described above) or they consume hydrogen that is reformed "on demand" from methanol in reformed methanol fuel cells (RMFCs). Formic acid at a concentration of <85% by weight may be oxidized directly in a formic acid fuel cell. Liquid borohydride is either oxidized directly, in direct borohydride fuel cells (DBFCs), or reformed to hydrogen on demand, in indirect borohydride fuel cells. A butane or a butane/propane mix is oxidized directly by an SOFC system.

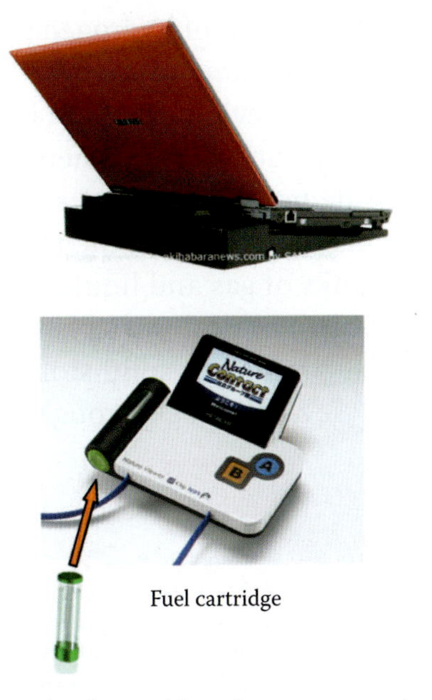

FIGURE 9.7 Fuel cell cartridges in consumer electronics.

Using methanol, Samsung has developed a prototype fuel cell for use in laptops that it claims lasts almost twice as long as other fuel cell systems being readied (see Figure 9.7). The Samsung fuel cell, powered by about 200 cm^3 of liquid methanol, has an energy density of 200 Wh/L. The cell measures 23 cm wide, 8.2 cm in length, and 5.3 cm deep and weighs less than 1 kg. In total, the fuel cell can supply power to a laptop for about 15 h. Such performance substantially improves over the 100–130 Wh/L energy densities of the fuel cells being developed by companies such as Toshiba (see Figure 9.6) and NEC for laptops. While hydrogen-based cells have taken off for home or automobile use, versions based on methanol for use in electronics products have yet to be commercialized. Toshiba and NEC are among the companies that promised methanol fuel cell-based laptops in previous years, but each time, technology launches have been delayed.

In what follows, we analyze in a bit more detail progress made in hydrogen storage using CNTs and carbon nanohorns and the prospect of biomass as fuels in biofuel cells.

Hydrogen Storage

The chemical energy per mass of hydrogen (142 MJ kg^{-1}) is at least three times larger than that of other chemical fuels (for example, the equivalent value for liquid hydrocarbons is 47 MJ kg^{-1}). Also, hydrogen as a fuel leads to pure water as a byproduct, and the hydrogen/oxygen reaction is a reaction for which catalysts have been researched the most. The problem of storing hydrogen safely and effectively is one of the major technological barriers currently preventing the widespread adoption of hydrogen as an energy carrier and the subsequent transition to a so-called "hydrogen economy." Practical problems with storage of hydrogen gas or liquid make solid state storage the most attractive. To appreciate the challenge, consider that 4 kg of hydrogen gas, which is required for a practical driving distance, occupies a volume as large as 49 m^3. Storage solutions must reduce the enormous volume of the hydrogen gas. A solid medium capable of absorbing and releasing large quantities of hydrogen easily and reliably is actively being sought and various types of solids have been proposed, including microporous materials that can store physisorbed molecular hydrogen at low temperatures,[11] intermetallic hydrides that absorb atomic hydrogen as an interstitial,[12] and complex hydrides formed from light elements, such as Li, B, Na, Mg, and Al.[13,14] GE, for example, is working with the Department of Energy (DOE) on $Li_2Mg(NH)_2$, which contains 5.6% hydrogen and has reversible storage capability. As a threshold for economical hydrogen storage, the DOE has set storage requirements for the DOE Freedom CAR program at 6 weight % (wt%) to be considered for technological implementation.

Among the storage methods currently being explored, storing hydrogen by adsorption in metals appears too expensive. However, trapping it in porous materials is not only efficient (all the hydrogen adsorbed can be recovered) but cheap. What's more, the cycle of hydrogen storage and release does not require any reactivation or regeneration of the material. Accordingly, considerable research efforts are going into the evaluation of various nanostructures, such as CNTs, to find the most suitable lightweight hydrogen storage material. Because of their cylindrical and hollow geometry and nanometer-scale diameters, it was predicted that CNTs would store liquids or gases in their inner cores. Claims of storage of hydrogen in CNTs attracted worldwide interest when Dillon et al. published 5–10 wt% hydrogen storage in single-walled carbon nanotubes (SWNTs).[15] Subsequently, claims of high hydrogen storage capacities at room temperature and pressures higher than 100 bar in graphitic nanofibers (GNFs) and lithium- and potassium-doped multiwalled carbon nanotubes (MWCNTs) were made. But in the late 1990s, the identification of many erroneous measurements of gaseous hydrogen uptake by CNTs and CNFs led to a controversy that is still lingering today.*[16] In light of today's knowledge, it is not very likely that at moderate pressures and around room temperature carbon nanostructures can store the amount of hydrogen required for hydrogen-based fuel cells (at least not for fuel cells for cars): using them for storage is possible only at extremely low temperatures (below −196°C) because of the weak interaction between hydrogen and carbon. The future possibility of storing hydrogen inside carbon-based porous materials, as part of a clean energy scheme, depends on the force of the interaction between hydrogen and carbon and on how easily this force can be increased.

Another possibility for hydrogen storage is electrochemical storage in CNTs. In this case, it is not a hydrogen molecule but a hydrogen atom that is physisorbed. Currently, the storage of hydrogen in this absorbed form in carbonaceous materials is considered the most appropriate route to solve the hydrogen storage problem. While most of the studies have focused on the hydrogen storage through physisorption, recent density functional theory (DFT) calculations for single-walled CNTs indicate the potential for up to 7.5 wt% hydrogen storage capacity through chemisorption by saturating the C-C double bonds in the nanotube walls and forming C-H bonds (i.e., hydrogenation; see the articles "Generalized Chemical Reactivity of Curved Surfaces: Carbon Nanotubes"[17] and "Theoretical evaluation of hydrogen storage capacity in pure carbon nanostructures"[18]). However, direct experimental evidence of the high values of the hydrogen capacity through chemisorption has not yet been demonstrated, and so far, the only storage systems that have met the DOE targets are compressed tanks of gas and liquid. A 2004 report by the American Physical Society concluded that even the most promising solid-storage technologies are still several breakthroughs away from practical use.

* Some of the physical properties of hydrogen, including its low molar mass relative to gas phase impurities and its susceptibility to leakage and permeation, together with the sensitivity of the materials that can be used to store it, mean that hydrogen sorption measurements are particularly prone to error.

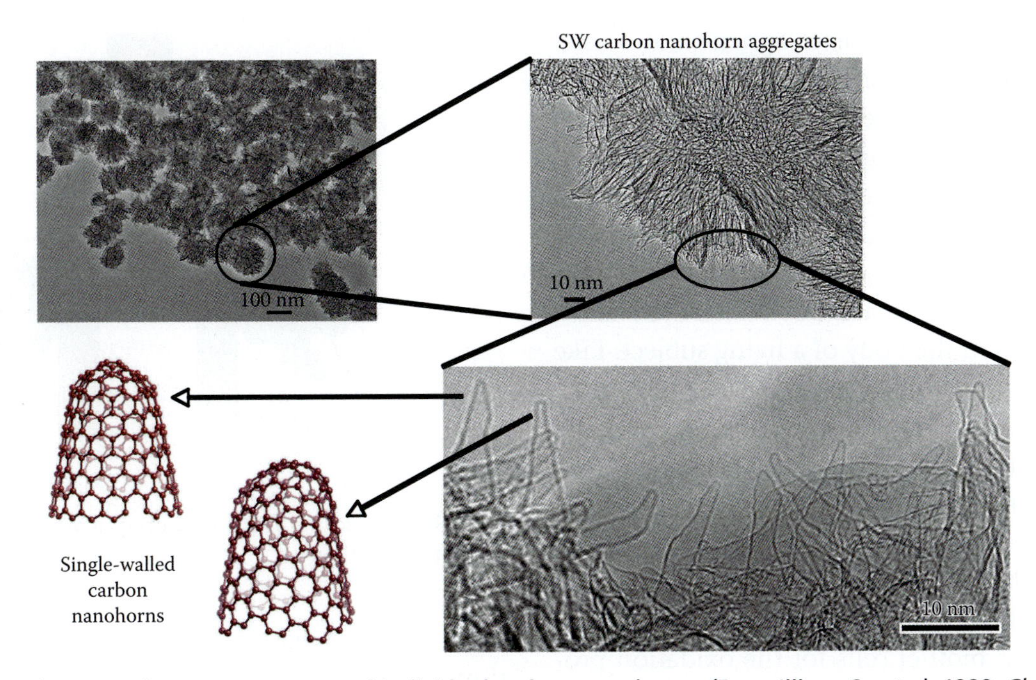

FIGURE 9.8 Carbon nanohorn aggregates and individual carbon nanohorns. (From Iijima, S., et al. 1999. *Chem Phys Lett* 309:165.[20])

Carbon nanohorns offer a better prospect for hydrogen storage in fuel cells than CNTs. Nanohorns are pure carbon structures that are 2–3 nm long, are cone-shaped, and aggregate to form dahlia-shaped structures 80–100 nm in diameter with a large number of horn-shaped, short, single-layered nanotubes that stick out in all directions (Figure 9.8). Since the tips of the cones are pointed, it was suspected that there might be a stronger hydrogen-substrate interaction than in CNTs. Recently, researchers at the Centre National de la Recherche Scientifique (CNRS) confirmed that with carbon nanohorns structures, the hydrogen-carbon interaction is indeed far more stable than with nanotubes.[19] This group demonstrated that nanohorns have remarkable adsorptive behavior because of their large surface area (about 400 m^2/g) and excellent catalytic properties because Pt catalyst is dispersed more homogeneously on the surface of carbon nanohorns. Unlike single-walled CNTs, carbon nanohorns can be made without the use of catalysts. Carbon nanohorn aggregates can be produced with a yield of more than 90% through laser vaporization of carbon at room temperature. Sumio Iijima's group showed that they could also be produced by laser ablation of graphite, and these were the researchers who came up with the name *nanohorns*.[20]

Carbon nanohorns are expected to perform excellent as gas storage material, making use of their high adsorbability.* Carbon nanohorns have actually cleared the DOE threshold for commercial adoption as methane gas storage material.

Biofuels

Not everyone agrees that improved hydrogen storage is the way to further develop microfuel cells. Some researchers point out that proponents of a hydrogen economy claim to address greenhouse gas (GHG) emissions, but in reality, these experts counterargue, the most economic methods of making hydrogen is from natural gas and these processes produce significant amounts of GHGs. Since virtually all living things in the biosphere operate by using biomass as fuel and generate power using fuel cells, these scientists propose that a biomimetic approach would appear to be the most environmentally sustainable method of producing energy. The most common energy currencies on the planet, used by virtually every living thing, are found in landfills, wastewater

* Carbon nanohorns selectively adsorb DNA fractions. Inorganic materials are mostly used today in selecting DNA fractions. However, carbon, with its excellent biocompatibility, may be a better material than inorganic substances. See http://www.nanonet.go.jp/english/mailmag/2004/023a.html.

treatment plants, wood waste, agricultural waste, grasses, etc. The mechanism of biofuel cells involves the oxidation of substrates such as carbohydrates, proteins, or fats by microorganisms or enzymes generating electrons in an external circuit. Biocatalyzed oxidation of organic substances by oxygen or other oxidizers in a two-electrode cell provides a potential means for the conversion of chemical to electrical energy, even in the body of a living subject. Like conventional fuel cells (such as methanol fuel cells), these types of cells can deliver more energy per volume weight than conventional batteries, and the theoretical efficiency of biofuel cells is estimated to be as high as 90% (compared with up to 60% for other fuel cells; see above). Abundant organic raw materials, such as amino acids or glucose, can be used as fuels in biofuel cells for the oxidation process, and molecular oxygen can act as the substrate being reduced. If the biofuel cells use concentrated sources of chemical energy, they can be made safe, small, and light. Moreover, biofuel cells differ from the traditional fuel cells by the material used to catalyze the electrochemical reaction. Instead of using precious metals as catalysts, biofuel cells rely on biological molecules, such as enzymes, to carry out the reaction.

Biofuel cells are categorized into microbial-based and enzymatic biofuel cells, which employ biocatalysts, i.e., either microorganisms or enzymes, respectively, to convert chemical energy into electrical energy. Biocatalysts are attractive compared with transition metal catalysts because 1) they operate at moderate temperatures, 2) they allow for the use of inexpensive fuel cell components, 3) they are renewable, and 4) they can be developed for a variety of cheap fuels. Enzymatic biofuel cells are preferred over microbial ones. Microbial-based biofuel cells have too many disadvantages: complex design, big size, poor reproducibility, and short lifetime based on bacterial survival.

As an example of an enzymatic biofuel cell, consider the glucose cell in Figure 9.9. Here, oxidation of glucose is coupled to the reduction of dissolved O_2. The cell incorporates an anode consisting of sugar-digesting enzymes and mediator, and a cathode comprising oxygen-reducing enzymes and mediator. The anode extracts electrons and hydrogen ions from the sugar (glucose) through enzymatic oxidation as follows.

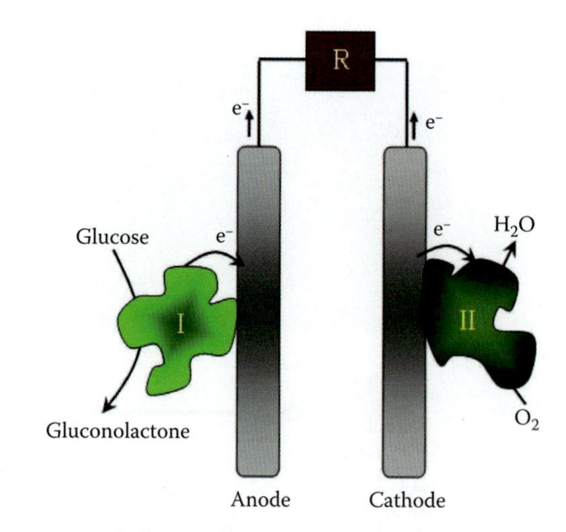

FIGURE 9.9 Biofuel cell based on glucose oxidation and oxygen reduction.

$$\text{Glucose} \rightarrow \text{Gluconolactone} + 2\ H^+ + 2e^- \qquad \text{Reaction 9.4}$$

The hydrogen ion migrates to the cathode, and once it arrives at the cathode, the hydrogen ions and electrons absorb oxygen from the air to produce water.

$$(1/2)O_2 + 2\ H^+ + 2e^- \rightarrow H_2O \qquad \text{Reaction 9.5}$$

Through this electrochemical reaction, electrons pass through the outer circuit to generate electricity. Power output of this cell is the product of the cell voltage, V_{cell}, and the cell current, I_{cell}. The oxidation reaction may be catalyzed by glucose oxidase (I) at the anode, and oxygen reduction may be catalyzed by laccase (II) (or bilirubin oxidase) at the cathode. Since most of the redox enzymes lack direct electrical communication with electrode supports, various electron mediators have to be used to contact the biocatalyst electrically with the electrodes. A redox mediator of appropriate redox potential is required to shuttle electrons between the protein and the electrode surface, because direct electron transfer to buried redox sites within these proteins is generally not possible given the distance of the active site from the electrode surface (for more details, see Volume I, Chapter 7).

For the pictured glucose/O_2 biofuel cell, a V_{cell} higher than 0.5 V can be achieved by the utilization of enzymes and redox mediators of carefully chosen redox potentials. The mediators and enzymes can be immobilized on the electrodes by, among others, codeposition of redox polymer/enzyme and a bifunctional linker, coelectrodeposition and enzyme reconstitution at a functionalized electrode surface to provide membrane-less biofuel cells. The immobilization concept represents an important advancement, as it precludes the need to separate the anode and cathode half-cells from each other using a membrane, provided that no solution redox reaction between fuel and oxygen occurs (as is the case for glucose). See http://www.tappi.org/content/events/10nano/papers/20.2.pdf.

Example 9.2:

Sony, one of the world's largest battery makers, has succeeded in creating a sugar-based biofuel cell. The biofuel cell does not require mixing or convection of the glucose solution or air movement; as it is a passive-type cell, it works simply by supplying sugar solution into the unit. Their test fuel cells measure 39 mm^3 and delivers 50 mW—currently the world's highest level for passive-type biofuel cells of comparable volume. The biobattery casing is made of vegetable-based plastic (polylactate) and is designed in the image of a biological cell. By connecting four cubic cells, it is possible to power a memory-type Walkman (NW-E407), together with a pair of passive-type speakers (no external power source) (see Figure 9.10).

FIGURE 9.10 Sony's biofuel cell based on glucose solutions. Four units are shown to power a Sony Walkman. (Courtesy of Sony.)

Key achievements in the development of the Sony biofuel cell include the following:

1. Enhanced immobilization of enzymes and mediator on the anode. For effective glucose digestion, the anode contains high concentration of enzymes and mediator. The Sony immobilization technology involves two polymers to attach enzyme and mediator to the anode. The polymers feature an opposite charge, so the electrostatic interaction between the two polymers effectively secures the enzymes and the mediator closely together. The ionic balance and immobilization process have been optimized for efficient electron extraction from the glucose.
2. Cathode structure for efficient oxygen absorption. The water content within the cathode environment is vital to ensure optimum conditions for the efficient enzymatic reduction of oxygen. The Sony biofuel cell employs porous carbon electrodes partitioned using a cellophane separator. The optimization of this electrode structure ensures that the appropriate water levels are maintained, enhancing the reactivity of the cathode.
3. Optimization of electrolyte. A phosphate buffer of approximately 0.1 M is generally used within enzymology research; however, an unusually high 1.0 M concentration buffer is used in the Sony biofuel cell. This is based on the discovery that such high concentration levels are effective for maintaining the activity of enzymes immobilized on the electrodes.

Sony continues its development of immobilization systems, electrode composition, and other technologies in order to further enhance power output and durability, with the aim of realizing practical applications for these biobatteries in the near future.[21, 22]

Despite the many advances in enzyme and polymer membrane modified electrode techniques, current enzymatic biofuel cells cannot compete with conventional batteries because of their low cell voltages and power densities. In addition, other key challenges of enzymatic biofuel cells include an uncertain storage and operational stability, tedious pretreatments of special and expensive electrode materials with a subsequent complex chemical, and/or electrochemical functionalization, the adsorption of many interfering species on the electrodes, etc.

Implantable biofuel cells are proposed to convert the energy stored within chemicals present in vivo into usable electrical energy. The microbiofuel cell is potentially an ideal power source for a variety of devices that need to be implanted for short periods of time (i.e., from a few days to a period of a couple of weeks), such as medical sensor/transmitter systems and injectable biosensors, because of its easy integration with the sensor-transmitter system fabricated by MEMS technologies. Current biofuel cells cannot compete with conventional electrochemical cells in powering implantable MEMS devices, such as in situ and in vivo biosensors. Normally these cells would produce continuously a few microwatts, of which less than 1 μW l is required for the operation of the sensor; the transmitter consumes most of the power. With a small ultracapacitor storing ~10 mJ, enough for 1-ms-long 1–10 GHz bursts of 10 μW every 10 s, the transmitted information is easily acquired outside the body. Even though microwave irradiation has been used to charge these devices, if one considers the potential health risks of direct microwave irradiation, the biofuel cell could be more convenient and safe for the patients.

Micromachining of Microfuel Cells

To appreciate the many challenges involved in micromachining a fuel cell, let us start by considering how one is typically constructed (see Figure 9.11). To focus the mind we assume that we are dealing with miniaturization of a passive PEM fuel cell. In general, passive-type fuel systems have a more simple structure and are more suitable for miniaturization, whereas active type systems have a more complicated structure and are suited to higher power devices. A PEMFC contains an electrolyte that is a layer of solid polymer that allows protons to be transmitted from one face to the other. PEMFCs require hydrogen and oxygen (or air) as inputs, and these gases must be humidified in most cases. The operating temperatures are relatively low, at around 60–80°C.

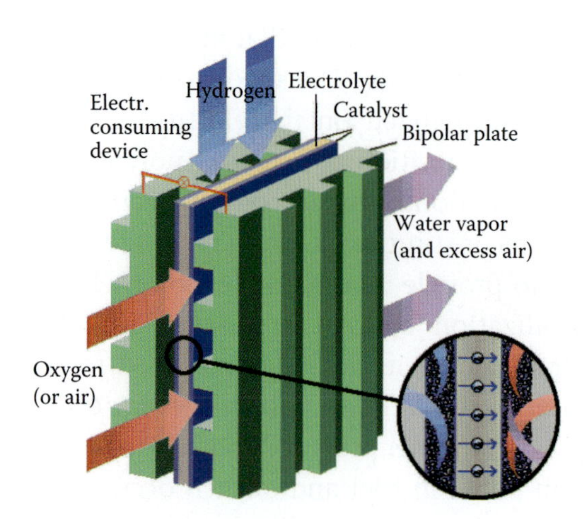

FIGURE 9.11 Construction of a typical fuel cell.

Figure 9.11 shows one fuel cell element only and these elements can be either stacked (Figure 9.12a) or put in an array. A fuel cell array can either be banded or flip-flopped as illustrated in Figure 9.12b.

Bipolar plates have a structural, an electrical, and a fluidic function. The requirements for the bipolar plates in the microfuel cells pictured in Figure 9.11 include high electrical conductivity, resistance to corrosion, chemical compatibility, high thermal conductivity, low leakage of gases, high mechanical robustness, low weight, manufacturability, and low cost. Graphite satisfies all these criteria except for mechanical robustness, manufacturability, and low cost. Bipolar plates are a significant part of the PEMFC stack and account for 80% of the total

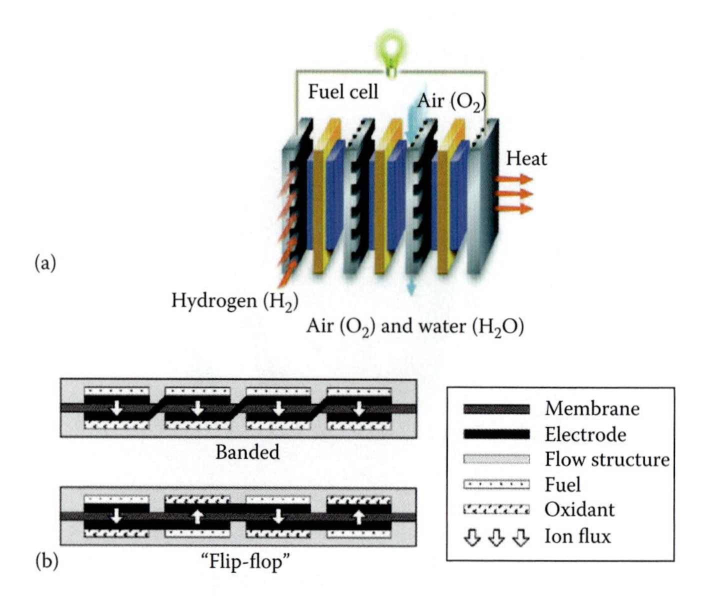

FIGURE 9.12 (a) Fuel cell stack. (b) Two methods for arraying fuel cell elements.

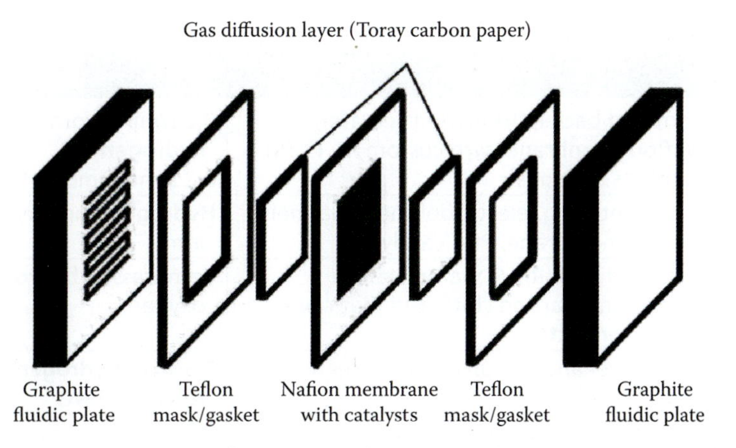

FIGURE 9.13 Bipolar plates (also called graphite fluidic plates) and typical membrane electrode assembly (MEA).

weight and 45% of stack cost. The bipolar plates will obviously be an important first target in miniaturizing a fuel cell. The membrane electrode assembly (MEA) in a PEMFC, illustrated in Figure 9.13, may consist of a Nafion® membrane (Nafion 112, Electrochem, Woburn, MA) coated with catalyst sandwiched between two carbon paper layers (e.g., Toray carbon paper). The carbon papers act as gas diffusion layers, and they are pushed up against the Nafion membrane using two Teflon gaskets.

A wide variety of micromachining techniques have been employed to construct microfuel cells, including Si MEMS-based, printed circuit (PC)-board-based, and carbon MEMS-based. An example of a Si-based MEMS-based fuel cell is shown in Figure 9.14.[23]

In this Si design, electropolished tubs in the highly conductive Si bipolar plates make for gas flow channels. Nafion is used as the membrane, and porous Si plays the role of gas diffusion layer to bring the gas to the Pt/C electrodes on the Nafion surface. Other Si MEMS-based fuel cells are reviewed in Table 9.4.

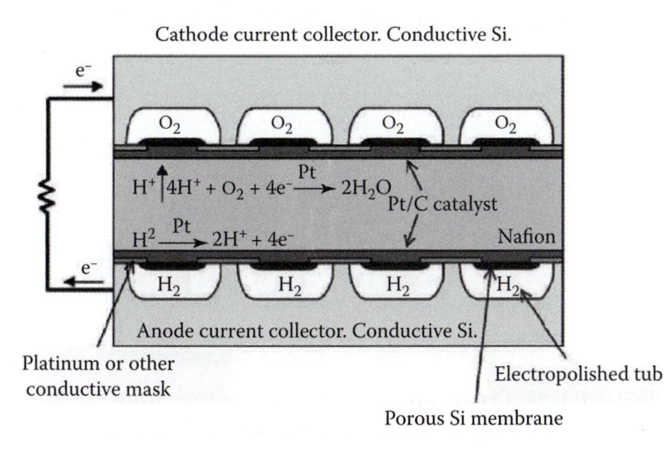

FIGURE 9.14 Silicon-MEMS-based fuel cell.[23]

Silicon is expensive compared with the printed circuit board-based fuel cell depicted in Figure 9.15.

Dr. Ben Park made the world's smallest hydrogen-air fuel cell with carbon-MEMS bipolar plates while working in the author's laboratory.[24] C-MEMS describes a manufacturing technique in which carbon devices are made by exposing a prepatterned organic structure to high temperatures in an inert or reducing environment (i.e., pyrolysis). The C-MEMS process makes the fabrication of intricate carbon shapes as easy as the patterning/machining/molding of a selected polymer material (see Chapter 5 for details).

For the manufacture of C-MEMS bipolar plates, 20-mil (~500-μm)-thick Cirlex® (Fralock, Canoga Park, CA) plastic sheets were machined using a circuit board milling tool (T-Tech, Norcross, GA) with 500-μm diameter end mills to create the fluidics. When starting with a size of the polymer bipolar plates of 1 × 1 cm, shrinkage during the pyrolysis process reduces the plates to 0.8 × 0.8 cm (see Figure 9.16).

For the membrane electrode assembly (MEA), commercial fuel cell electrodes [1 mg/cm^2 loading, 20 wt% Pt/Vulcan XC-72 (EC-20-10-7; Electrochem-Woburn, MA)] are pressed into an activated Nafion membrane* (Nafion 115) (Figure 9.17).

The fluidic channel walls and separators machined from high-temperature polymer sheets are bonded together (using PI-5878G polyimide) to create

* The Nafion membrane is activated as follows: deionized (DI) water for 1 h, 30% H_2O_2 for 1 h, 10 M H_2SO_4 for 1 h, and finally a short rinse in DI water (all at 80°C). The Nafion is stored in DI water until fabrication of the MEA.

TABLE 9.4 Silicon-MEMS-Based Fuel Cells

Source	Design	Fuels	Power Density
S.C. Kelley University of Minnesota	Front and back KOH-etched Si wafer, Nafion Membrane with custom Pt, Pt-Ru catalysts coatings	0.2 L/min hydrated neat **hydrogen** 0.2 L/min ambient air	130 mW/cm²
J.P. Meyers Bell Labs	Porus Si etching, electropolished channels, Nafion membrane, Pt/C catalyst	**Hydrogen** and oxygen at 1 atm	60 mW/cm²
S.J. Lee Stanford University	Front and back RIE Si wafer, Nafion membrane, carbon cloth and platinum, **series-connected**	Compressed **hydrogen** and oxygen	40 mW/cm²
R. Hahn Fraunhofer, IZM	Wafer level and foil processes, screen printing	0.5 sccm **hydrogen (DMFC) NO AIR PUMPING**	80 mW/cm²
S.C. Kelley University of Minnesota	Front and back KOH-etched Si wafer, Nafion Membrane with custom Pt, Pt-Ru catalysts coatings	0.2 L/min 0.5 M **methanol,** H_2O 0.2 L/min ambient air	15 mW/cm²
Y.H. Seo KAIST, Korea	Front and back KOH and RIE Si Wafer, Nafion membrane, Pt catalyst	**Methanol** and ambient air **NO FUEL OR AIR PUMPING**	0.1 mW/cm²

Source: Presentation by Hahn, R., S. Wagner, H. Reichl, M. Krumm, and K. Marquardt. 2003. Fraunhofer IZM and TU-Berlin. Development of micro fuel cells with help of MEMS technologies. MINATEC 2003.

fluidic plates as shown in Figure 9.18. The polymide physical binder also acts as the electrical binder when converted to carbon during pyrolysis. The MEA, fluidic plates, and gas inlet/outlets (cut from conventional syringes; see Figure 9.19b) are brought together, and epoxy is used to seal the entire fuel cell structure.

Making bipolar plates by C-MEMS technology has several benefits. Because one starts from an easy-to-machine polymer, new types of bipolar plates become feasible, and because the bipolar plate fluidics, gas diffusion layer, and catalyst support layer are all made of carbon, they can be integrated and fabricated into a single monolithic structure. This reduces weight, size, complexity, and internal resistance while increasing mechanical robustness. Control over the carbon precursor material allows materials engineering of the carbon itself.

The open-circuit voltage of an individual unoptimized C-MEMS cell has been measured at 871 mV, and the closed-circuit current draw of the fuel cell stabilized at 3.11 mA (with a voltage of 110 mV). The maximum power output was 0.773 mW (1.21 mW/cm²), with an internal resistance of about 210 Ω.[24]

Our team also successfully fabricated a hydrogen-air C-MEMS fuel cell stack, as depicted in Figure 9.19.[24] The gaskets between cell elements are made of silicon rubber, which provides elastic support of the repeating parts and prevents leakage of gases. The end plates are machined in a T-Tech circuit board milling tool to create the screw holes and openings for gas ports. The latter are cut from conventional syringes. Repeating the bipolar plates, the gaskets, and the MEAs and then compressing them with four screws in the four corners of the whole structure results in the completed fuel cell stack, as shown in Figure 9.19.

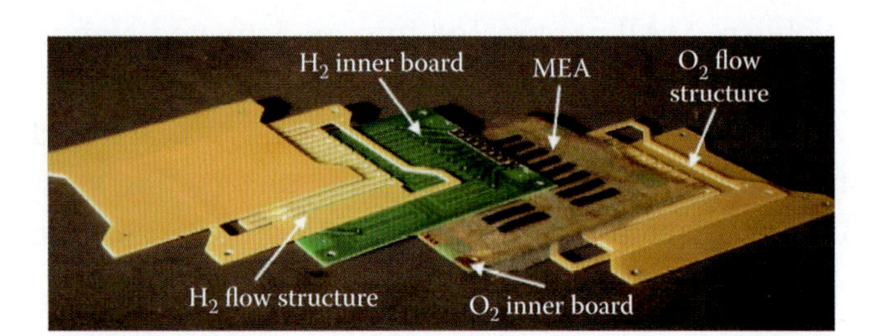

FIGURE 9.15 Printed circuit board fuel cell. (From O'Hayre, R. 2004. Microscale electrochemistry: Application to fuel cells. Thesis, Stanford University.)

FIGURE 9.16 Cirlex bipolar plates of 1 × 1 cm (one structure on the right) shrink during the pyrolysis process to 0.8 × 0.8 cm (two structures on the left).

FIGURE 9.17 The membrane electrode assembly. A Nafion sheet (transparent) is sandwiched between two carbon electrodes.

The C-MEMS fuel cell could be further improved by engineering the carbon electrodes further. This may involve at least three strategies: incorporation of SWNTs and/or multiwall carbon nanotubes (MWNTs), the use of carbon horns, and the use of carbon fractals. Compared with other forms of carbon electrodes (such as graphite and bare glassy carbon electrode), MWNT microelectrodes show better performances because of the dimensions of the MWNTs and their electronic structure.[24] As we pointed out above, fuel cell electrodes, made of carbon nanohorns are expected to help improve the cells' power-generation capacity even further and extend their lifetime, because platinum catalyst nanoparticles disperse better among carbon nanohorns and do not aggregate.[24] The concept of implementing carbon fractal electrodes for both batteries

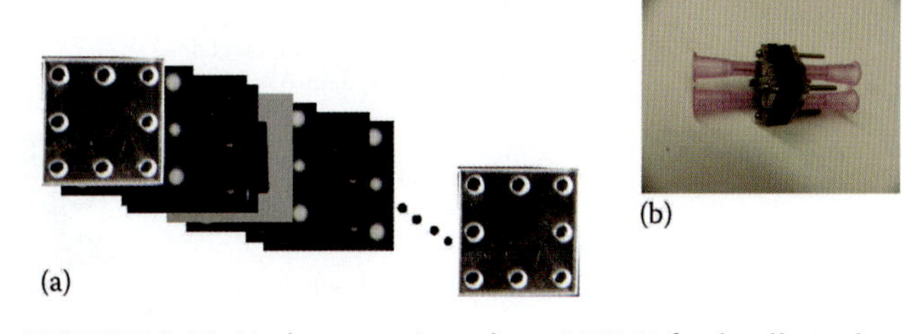

FIGURE 9.19 Hydrogen-air carbon-MEMS fuel cell stack. (a) The MEA, fluidic plates, and gas inlet/outlets (cut from conventional syringes) are brought together, and epoxy is used to seal the entire fuel cell structure (b).

and fuel cells is introduced below (see "Fractals and Fractal Dimensions").

Finally, in the microbiofuel cell context, we briefly highlight hybrid insect-MEMS (HI-MEMS; also called *cyborg bugs*, or *cybugs*), a concept sponsored by the Defense Advanced Research Projects Agency (DARPA). The MEMS on the insect are coupled to the insect anatomy and nervous system and draw power from the insect to control its locomotion by obtaining motion trajectories either from global positioning system (GPS) coordinates or by using radiofrequency (RF)-, optical-, or ultrasonic-based remote control. DARPA has military applications in mind for HI-MEMS, with visions of insect swarms with various embedded MEMS sensors (such as cameras, microphones, and chemical sniffers) penetrating enemy territory (e.g., the Crawford Ranch). The HI-MEMS battalions would perform reconnaissance missions beyond the capabilities of soldiers. But robo-insects, equipped with chemical sensors, might also have civil applications, perhaps to explore a hazardous chemical spill zone and report back about it to humans in a

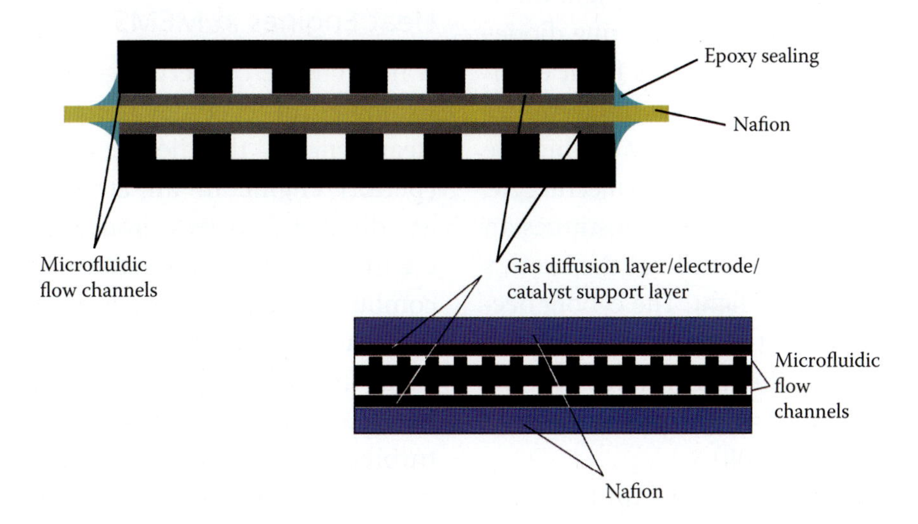

FIGURE 9.18 Assembly of carbon-MEMS fuel cell.

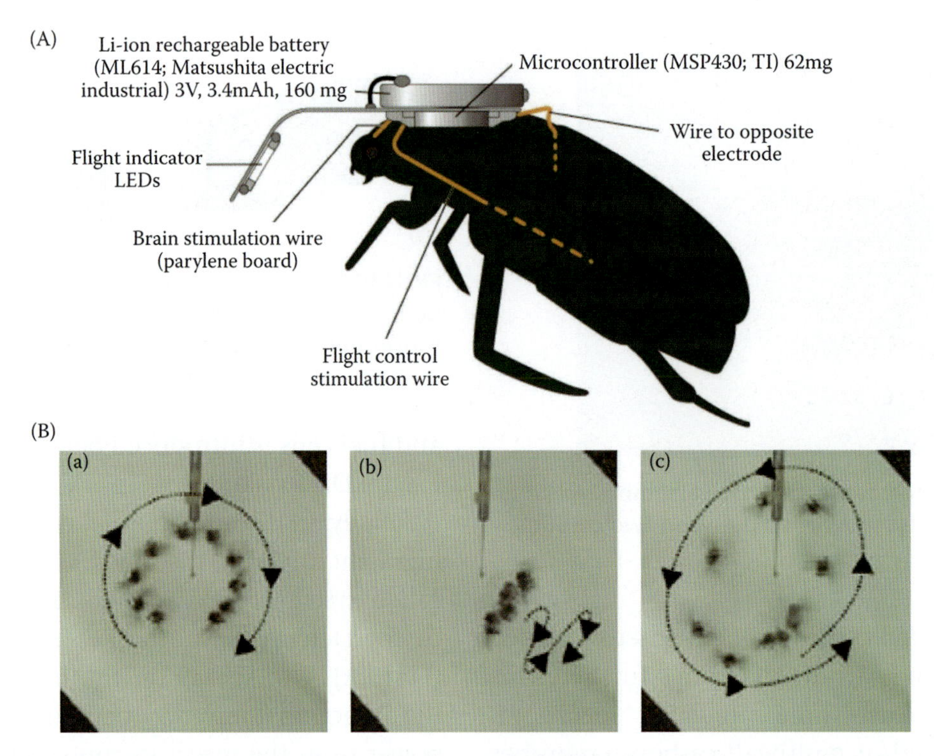

FIGURE 9.20 (A) Cyborg beetle system. The microcontroller controls flight by stimulating brain, wing, and other nerves. An array of ten light-emitting diodes hanging in front of the beetle's eyes is lit to control flight without direct nerve stimulation. The beetle weighs 1–2 g, and the total control system weighs 422 mg. (From Nikkei Electronics Asia, April 2008.) (B) Electrical stimulation of wing muscles on either side initiates a turn. Beetles mounted on a long string (10 cm) were programmed with a continuous sequence of left, pause, right, pause instructions; each instruction lasted 2 s. Left flight muscle stimulation generates a right turn (a), followed by a pause during which the beetle zigs and zags (b), followed by right muscle stimulation that generates a left turn (c). Each photograph consists of ten frames; frames were taken every 0.2 s. (From MEMS 2008, an international academic conference on MEMS that took place January 13–17, 2008, in Tucson, AZ.)

safe zone. A recent embodiment of the HI-MEMS concept comes from a University of Michigan team that successfully created a cyborg unicorn beetle microsystem (Figure 9.20A).[25] The University of Michigan researchers equipped a North American unicorn beetle with an IC, button battery, and a flight direction indicator consisting of ten light-emitting diodes (LEDs) and succeeded in controlling its flight externally. Flight control is handled by inserting electrodes into a portion of the brain immediately behind the beetle's eye and into wing muscles, and injecting AC pulses. The LED directional controller constitutes an independent control system and utilizes the beetle's proclivity to follow a blinking light. The cyborg beetle is able to take off and land, turn left or right, and demonstrate a number of other flight control behaviors. Electrical stimulation of wing muscles on either side initiates a turn (Figure 9.20B).

The DARPA program director, Dr. Amit Lal, got the idea for remote-controlled insects from the 1990 science fiction novel *Sparrowhawk* by Thomas A. Easton. In that novel, Easton writes about genetically engineered animals that are greatly enlarged and then outfitted with implanted control structures.

Heat Engines in MEMS

A heat engine converts heat to mechanical energy. Examples of everyday heat engines include the steam engine, the diesel engine, and the gasoline (petrol) engine in an automobile. Heat engines are divided into two groups: external-combustion engines and internal-combustion engines. External-combustion engines produce hot gases that transfer heat to another fluid, and the energy in this fluid, in turn, is changed into mechanical energy. External-combustion engines include gas and steam turbines and reciprocating steam engines. Internal-combustion engines produce hot gases whose heat energy is changed directly into mechanical energy.

Several groups involved in power MEMS are investigating scaled-down versions of well-established macroscale combustion devices. The idea is that since batteries have low specific energy, and liquid hydrocarbon fuels have a very high specific energy, a miniaturized power generating device, even with a relatively inefficient conversion of hydrocarbon fuels to power would result in increased lifetime and/or reduced weight of an electronic or mechanical system that currently requires batteries for power. A gas turbine powering a generator, for example, could in principle pack thirty times the energy of any battery. Refueling would replace recharging. There are quite a few difficulties with this approach, though, one being that flames extinguish because of heat losses if the dimensions of the combustion chamber are too small. Furthermore, even if flame quenching does not occur, heat and friction losses become increasingly important at smaller scales, since the heat release due to combustion and thus power output scales with the volume of the engine, whereas the heat and friction losses scale with the surface area.

As an example application of microcombustion engines, consider satellite missions that deal with smaller and smaller satellites, so that launch systems and altitude control thrusters of reduced mass and volume are now needed. Thrusters are devices in which propellant is burnt or disintegrates to produce a controlled force in a selected direction. These small rocket engines are expected to produce thrust of the order of tens of newtons at a thrust-to-weight of over 1000, i.e., 10 times the thrust-to-weight of conventional chemical liquid bipropellant engines. A simple MEMS thruster fabricated of a silicon/glass structure is shown in Figure 9.21.[26]

The nozzle, plenum, chamber, and injector of this MEMS thruster are all etched in silicon, and a glass ceiling is bonded to the etched silicon part. Propellants that have been used in this structure include H_2O_2 and N_2H_4 with silver and platinum on Al_2O_3 as catalysts. Hydrogen peroxide is fed through the plenum and injector and reacts in the presence of the catalyst to produce heat and gases in the chamber. The gases expand through the nozzle to produce thrust.

Another postage-stamp-size microthruster, fabricated at Massachusetts Institute of Technology (MIT), was made of six layers of silicon fused together

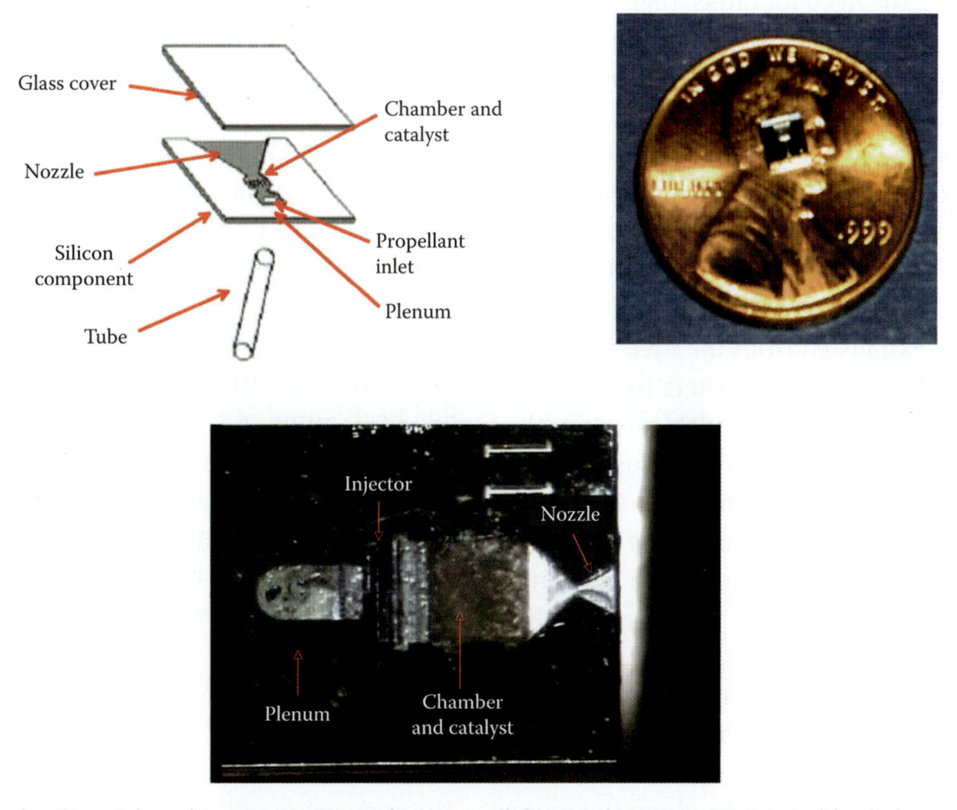

FIGURE 9.21 Simple silicon/glass thruster structure. (From a talk by Apurva Varia, NASA Goddard Thermal Model of MEMS Thruster.)

to make a sandwich 3 mm thick and was projected to produce up to 15 N of thrust by burning oxygen and methane. To prevent melting, ethanol coolant circulates around the tiny flat thrust chamber. The thrust level from these micromachined devices is expected to be very high in relation to their mass. MIT's Adam London thinks that a two-stage, 80-kg rocket might be sufficient to put a Coke-can-sized payload in orbit using 800 of his Si thrusters.[27] A more recent MIT microrocket engine thrust chamber and nozzle design measures 18 × 14 × 3 mm and builds on London's earlier gaseous propellant work, expanding the operating envelope of these motors to higher thrust levels.[28] Directions for improved specific impulse engines include increasing the engine size by a factor of 2–4 and research on hydrogen peroxide as a coolant. Another approach is to use liquid hydrocarbons or solid propellants. With the combustor working at 1500–1600 K, Si with a melting point of 1685 K is out as a building material. SiC and perhaps quartz are alternative candidate materials.

MEMS thruster arrays for satellites offer several advantages over conventional thrusters: they have no moving parts; utilize a variety of propellants; are scalable; eliminate the need for tanks, fuel lines, and valves; and fully integrate the structure of the satellite with the propulsion to power it. The Honeywell-Princeton MEMS Megapixel Microthruster Array (MMMA) consists of a 256 × 256 array of pixels, each one capable of being fired as a separate thruster (Figure 9.22). The thruster operates by first igniting a small amount of lead styphanate, which in turn ignites the main propellant of a nitrocellulose mixture. This is then exhausted from the fired pixel out a channel to produce thrust. Thrusters in this class give very small and precise impulse bits (on the order of tens of mN-s). Another microthruster array, developed by TRW, Caltech, and the Aerospace Corp., is fabricated as a three-layer silicon and glass sandwich, with the middle layer consisting of multiple small propellant cells sealed with a rupturable diaphragm on one side and an ignitor on the other. Each cell is a separate thruster, and when ignited, delivers one impulse bit. Delivering propulsion in discrete increments by igniting thrusters in controlled sequences has lent the technology the name "digital propulsion."

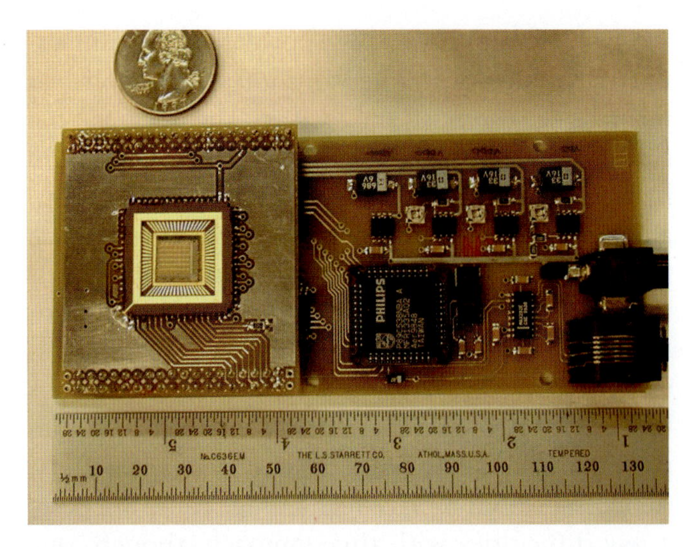

FIGURE 9.22 The MEMS Megapixel Microthruster Array ignitor thruster with its firing control board (http://alfven.princeton.edu/projects/MEMS1.htm).

The MEMS thrusters discussed above are all internal-combustion devices. External-combustion MEMS devices, with power generation by the alternate heating and cooling of a working fluid or a solid (e.g., shape memory alloys) integrated on a chip, have been attempted for driving a load as well. The heating in such an engine results typically from passing a current through a resistor. It would be preferable to use infrared radiation instead, since in that case no leads need to connect to the chip. It has been projected that a gas-based external heat engine of 5 × 5 × 5 mm^3 might provide an output of 10–100 W/kg. With actuators based on shape memory alloys, an output of up to 1 kW/kg is feasible, with an efficiency ten times lower.[6] Problems associated with crafting MEMS external heat engines include the thermal isolation of heating and cooling sections, minimization of friction, and the difficulty of implementing a flywheel. Some of these problems were successfully addressed by Sniegowski and coworkers,[29,30] who demonstrated a surface micromachined microengine capable of delivering torque to a micromechanism. Angular velocities of 600,000 rpm were registered for this engine, driven by steam.

Solar Cells

Introduction

The conversion of light (photons) to electricity (voltage) in a solar cell is called the *photovoltaic* (PV) *effect*. The most iconic form of solar power, the solar cell, consumes no fuel, has a wide power-handling capacity, has a high power-to-weight ratio, produces no pollution, and is more and more often made from inexpensive polycrystalline or amorphous materials, with even polymer-based solar cells being introduced. The operating principle of a solar cell was covered in Volume I, Chapter 4 (Figure 4.42), and in Table 9.5 we present a short historical overview of some important solar cell related milestones since the original observation of the photoeffect by Antoine Becquerel in 1839. As with batteries, progress in improving solar cell efficiency has been rather slow (there is no Moore's law curve for solar cells, either). Several important measures are used to characterize solar cells, the most obvious one being the total amount of electrical power produced for a given amount of solar power input. Expressed as a percentage, this is known as the solar conversion efficiency. Electrical power is the product of current and voltage, so the maximum values for these measurements are important as well: J_{sc} (short circuit current) and V_{oc} (open circuit voltage), respectively. Finally, in order to understand the underlying physics, the *quantum efficiency* is used to compare the chance that one photon (of a particular energy) will create one electron. A solar cell must be capable of producing electricity for at least twenty years without a significant decrease in efficiency, so lifespan is another crucial characteristic.

Table 9.5 Milestones in Solar Cell Development

1839:	Discovery of photovoltaic effect between two electrodes in a liquid (Antoine Cesar Becquerel)
1876:	Photovoltaic effect in Se (Heinrich Hertz)
1883:	First solar electric module (Charles Fritts)
1906:	Pochettino studies the photoconductivity of anthracene
1930:	Research on Cu_2O/Cu solar cell
1941:	Patent on Si solar cell (R. Ohl in Bell Labs)
1954:	Crystalline Si solar cell (Bell Labs); 4% efficiency
1958:	Solar panels as backup and power assist in space (*Vanguard 1*); 5 mW
1970:	Zhores Alferov and his team in the USSR created the first GaAs heterostructure solar cells
1973:	Oil crisis—solar energy is hot, subsequent Republican United States president kills Carter's efforts in favor of oil
1980:	Solar cell using CdTe, $CuInSe_2$, TiO_2, etc.
1991:	Liquid-junction solar cells with photosensitized anode invented by Michael Grätzel and Brian O'Regan at the École Polytechnique Fédérale de Lausanne, known as Grätzel cells
1995:	Yu and Hall make the first bulk polymer/polymer heterojunction photovoltaic cell
2000:	Peters/van Hall used oligomer-C_{60} diads/triads as the active material in photovoltaic cells (see Figure 3.15, for an example coming from the Stoddard-Zinc group)
2001:	Schmidt-Mende made a self-organized liquid crystalline solar cell of hexabenzocoronene and perylene

Because of the soaring oil prices, solar energy is a hot topic again, as it was under President Carter (even Republicans now concede that cheap oil is not a US birthright). The photovoltaic cell is currently the fastest-growing form of alternative energy, increasing by 50% a year, and where a kWh of PV electricity cost 50 cents in 1995, it fell to 20 cents in 2005. BBC Research, which charts technology markets, expects the global solar market to grow to $32 billion by 2012, with thin-film solar cells expanding 45% a year.

At the earth's surface, the average solar energy is ~4×10^{24} J per year, and the 2001 global energy use was ~4×10^{20} J per year (increasing ~2% annually). In the United States alone, the average power requirement is 3.3 TW. With a 10% efficient solar cell, this means that we would need 1.7% of the land area of the United States devoted to PV to quench this thirst; this roughly corresponds to the area occupied by the interstate highways.

As we saw in Volume I, Chapter 4, a traditional semiconductor solar cell is made from an n-type and a p-type semiconductor put in contact with each other. The result is a region at the interface, the p-n junction, where charge carriers are depleted and/or accumulated on each side of the interface. The transfer of charge carriers produces a potential barrier, typically 0.6–0.7 V high. In the case of silicon (bandgap = 1.1 eV), sunlight provides enough energy to promote electrons out of the valence band into the conduction band, and when a load is placed across the cell, these electrons flow from the p-type to the n-type material, move through the external circuit, and then stream back into the p-type material, where they recombine with the valence-band holes they left behind. How sunlight creates an electrical current in a solar cell is detailed a bit further in Figure 9.23.

If the incoming photon energy $E_\gamma < E_g$, then the photon is not absorbed; with $E_\gamma > E_g$, the photon is absorbed and the energy $E_\gamma - E_g$ is lost via thermalization. Additional losses occur through radiative recombination, trap states, and Auger recombination. If the carrier lifetime is less than the recombination lifetime then the charge is swept away by the inbuilt electric field (E). This depends on carrier mobility, defect density, carrier diffusion length, and exciton energy.

The conventional approach to PVs, as sketched in Figure 9.23, comes with several disadvantages. First, the semiconductor bandgap dictates that only photons with the bandgap energy or more contribute to producing a current. Higher-energy photons, at the blue end of the spectrum, have more than enough energy to cross the bandgap, but much of it is wasted as heat.

A cascade or tandem cell can achieve a higher total conversion efficiency by capturing a larger portion of the solar spectrum. In a typical multijunction cell, individual cells with different bandgaps are stacked on top of one another. The cells are stacked in such a way that sunlight falls first on the material having the largest bandgap. Photons not absorbed in the first cell are transmitted to the second cell, which then absorbs the higher-energy portion of the remaining solar radiation while

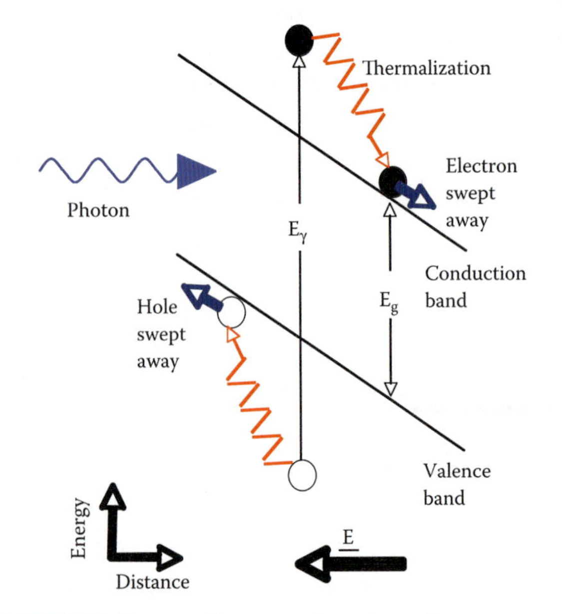

FIGURE 9.23 Energy diagram of a solar cell.

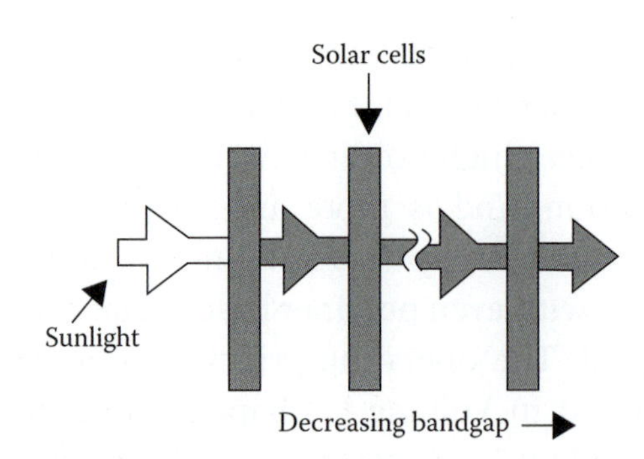

FIGURE 9.24 Multijunction, cascade, or tandem solar cell.

remaining transparent to the lower-energy photons. These selective absorption processes continue through to the final cell, which has the smallest bandgap (Figure 9.24). The ideal bandgap combination for a tandem solar cell is typically 1.75 and 1.1 eV. Such bandgap combinations can be attained using lattice-matched III–V semiconductor material combinations. The best monolithic combination is represented by the GaInP/GaAs tandem cell, possessing an AM 0 efficiency of 26.9%. The ideal efficiency for an infinite stack is 86.8%, and for a GaInP/GaAs/Ge triple-junction cell an efficiency of 32% has been measured.

Multijunction solar cells dramatically improve efficiency, but these cells are still very high-cost and suitable only for large commercial deployments (combinations include GaAs/InGaP, Ge/InGaAsN, InP/InAlAs, InP/InGaAs, and GaN/InGaN). In general terms, the types of cells suitable for rooftop deployment have not changed significantly in efficiency, although costs have dropped somewhat because of increased supply.

A higher light intensity can be focused on the solar cells by the use of mirror and lens systems, perhaps even a system that tracks the sun. This is perhaps too expensive a solution for rooftop implementation but could be useful in MEMS devices.

Another problem associated with conventional solar cells is that in order to capture enough photons in the p-type layer, that part has to be quite "thick." However, a thick p-type layer also increases the chance that electrons and holes will recombine before reaching the p-n junction. These effects combine to produce an upper limit of about 20% on the efficiency of silicon solar cells. The current problem with solar cells is not really the efficiency (20%), but rather the cost of production. Though it has come down by a factor of 100 in fifty years, it still has to go down by a factor of 2–5 to be competitive with fossil plants.

The best known solar cells today remain based on Si, and they are either monocrystalline, polycrystalline, or hydrogenated amorphous (α-Si:H) based. Polycrystalline films are poured and are cheaper and simpler to make than monocrystalline silicon that is Czochralski grown (see Volume I, Chapter 4). The efficiency of polycrystalline cells is, however, somewhat less than that of monocrystalline cells. Hydrogenated amorphous Si (α-Si:H) cells are relatively cheap, but they come with the lowest efficiency (see Table 9.6). Although its semiconducting properties are inferior to those of single-crystal Si, we learned in Chapter 5 that α-Si:H, manufactured by plasma-enhanced chemical vapor deposition (PECVD) from silane, can be doped both n- and p-type (in contrast to α-Si) (see Table 5.8 for state-of-the-art parameters of PECVD α-Si:H). Exposure of α-Si:H to light increases photoconductivity by four to six orders of magnitude, and the absorption coefficient of α-Si:H is more than an order of magnitude larger than that of single-crystal Si near the maximum solar photon energy region of 500 nm. Accordingly, the optimum thickness of the active layer in an α-Si:H solar cell can measure 1 μm, much smaller than that of single-crystal Si solar cells. In Figure 5.15, we compared polycrystalline-based solar cells with ultrathin amorphous Si on glass and ultrathin amorphous Si on flexible substrates.

Solar cell technology represents the most MEMS-compatible technology for power integration. Lee et al.[31] note that hydrogenated amorphous silicon solar cells are a particularly attractive means to realize an onboard MEMS power supply, and they implemented a high-voltage integrated solar cell array[31] as an electrostatic MEMS power supply. By interconnecting 100 individual solar cells in series, this team measured an open circuit potential as high as 150 V under AM 1.5 conditions, a voltage high enough to drive onboard electrostatic actuators. The conversion efficiency in that effort was only 0.2%, however. Sakakibara et al.[32] were able to generate more than 200 V with a similar solar cell on an area of 1 cm^2 and obtained a conversion efficiency of 4.65%. In both cases, amorphous silicon was used in a triple-stacked PV structure generating up to 2.3 V per cell. To obtain a very dense packing of array elements and to make the series connection of the solar cells, the latter group used focused laser beams for patterning electrodes and PV materials.

Table 9.6 Silicon-Based Solar Cell Efficiency in %

Type	Efficiency in Laboratory	Efficiency in Production
Monocrystalline	24	14–17
Polycrystalline	18	13–15
Amorphous	13	5–7

Since solar light is only intermittently available, electric storage elements need to be implemented as well. Along this line, Kimura et al.[33] fabricated a miniature optoelectric transformer consisting of a p-n junction photocell and a multilayer spiral coil transformer.

III/V solar cells are cells with a high efficiency, but they are made of very expensive semiconducting materials. A single junction cell based on GaAs comes with an efficiency of 25.1%. In 1970, Zhores Alferov and his team in the USSR created the first GaAs heterostructure solar cells. Today, the combination of GaAs with germanium and indium gallium phosphide is the basis of a triple-junction solar cell that holds the record for solar light conversion efficiency of over 32% and can operate also with light as concentrated as 2000 Suns.* This type of solar cell powers the robots *Spirit* and *Opportunity*, which are exploring the surface of Mars.

Newer solar cell materials include thin films of copper indium diselenide ($CuInSe_2$), copper indium gallium diselenide (CIGS), and cadmium telluride (CdTe). Commercial leaders in this area are a firm called First Solar, which uses cadmium telluride, and Miasolé, which manufactures a thin-film PV cell based on an ultrathin layer of photoactive CIGS on a stainless steel foil only 50 μm thick. The latter can easily be used in PV modules or incorporated into building materials such as membrane roofing. The efficiencies and bandgaps of GaAs and two of

* Where 1 Sun equals 1000 W/m^2 at the point of arrival. One Sun is a unit of power flux.

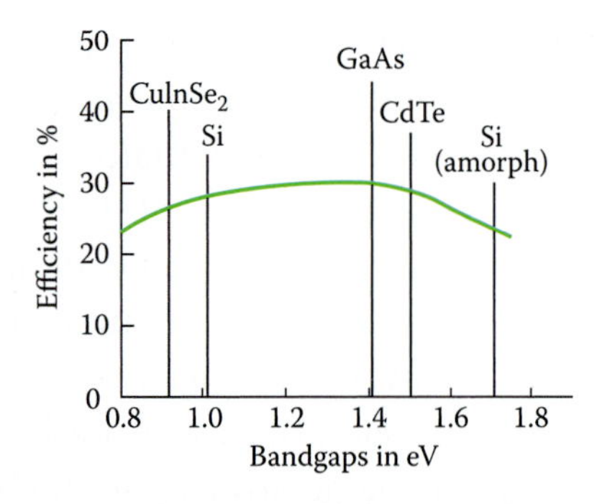

FIGURE 9.25 Bandgaps and efficiencies of the most important solar cell materials. The bars represent theoretical efficiencies, and the smooth line represents practical efficiencies.

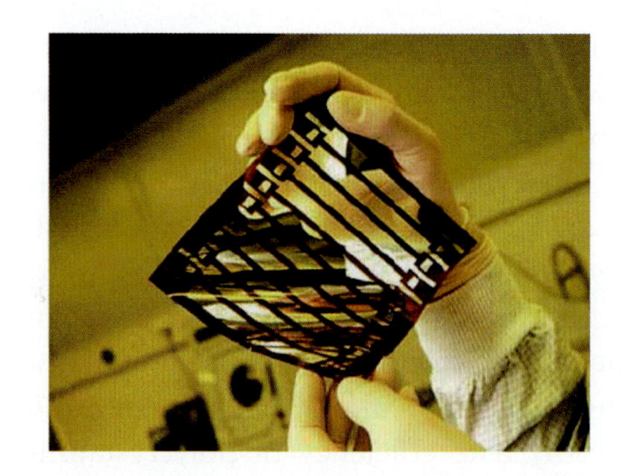

FIGURE 9.26 Polymer or organic photovoltaics.

the newer solar materials are compared with those of single-crystal and amorphous Si in Figure 9.25.

One development that made organic PVs an attractive option is the availability of light-sensitive conjugated polymers: polymers with alternating single and double carbon-carbon (sometimes carbon-nitrogen) bonds. In the 1970s, it was discovered that chemical doping of conjugated polymers increased electronic conductivity several orders of magnitude. Since that time, electronically conducting materials based on conjugated polymers have found many applications, including sensors, LEDs, and solar cells. Advantages of polymer or organic PVs (OPVs) include the fact that they are easily processed over large areas using spin-coating, doctor blade techniques (wet-processing), evaporation through a mask (dry processing), and printing (Figure 9.26). The usage of roll-to-roll printing, especially, guarantees a favorable cost (see Figure 1.5). They also have a low weight, exhibit mechanical flexibility and transparency, and have a bandgap that can easily be tuned chemically by incorporation of different functional groups. Efficiencies stand at 5%, but there are no obstacles identified for 10%. The low cost of plastic PV may enable solar power in applications where it was previously uneconomic (e.g., solar tents, sails, packaging, etc.). For rigid cells, ITO (Indium tin oxide)-coated glass usually serves as the substrate for these organic cells.

Figure 9.27 summarizes the best research solar cell efficiencies for a broad range of materials from universities, research institutes, and private companies. Besides PV converters for solar light and laser

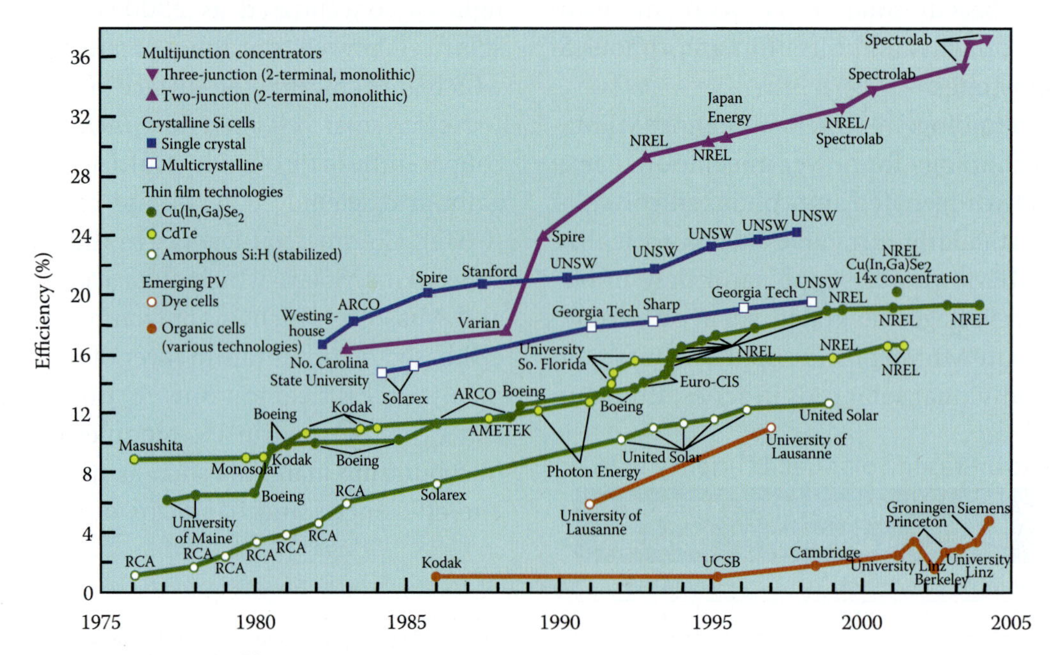

FIGURE 9.27 Research solar cell efficiencies. (From Art Nozik, DOE Solar Energy Workshop, 2005.)

light, microwaves could be used to power microsystems. In the latter, extremely small receivers and converters need to be constructed.

MEMS and NEMS in Solar Cells

Introduction

Micro- and nanotechnologies are crucial to increasing the efficiency of existing solar cells, and NEMS is enabling totally new solar cell designs as well. MEMS techniques allow for the fabrication of lenses (including Fresnel lenses), mirrors, photonic crystals, and cavities and the texturing of solar cell surfaces to enhance the interaction of incoming photons with the solar cell active junctions. As an example of a nanoenhancement of existing solar cells, consider the use of quantum dots (QDs) to replace dyes in luminescent concentrators.

NEMS may make a much bigger impact, though, on future solar cells. QDs, for example, exhibit optical properties that are absent in the bulk material due to the confinement of electron-hole pairs, in a region of a few nanometers and the wavelength at which a QD absorbs or emits radiation can be adjusted at will (see Volume I, Chapters 3 and 5). A first advantage of QDs, then, stems from the fact that an optimum bandgap, corresponding to the highest possible solar-electric energy conversion, can be achieved by using a mixture of QDs of different sizes for harvesting the maximum amount of the incident light (E_γ). A second advantage stems from the increased possibility of multiphoton absorption by nanoparticles. The formation of multiple excitons per absorbed photon takes place only when the energy of the absorbed photon is far greater than the semiconductor bandgap (E_g). Multiple photon absorption does not readily occur in bulk semiconductors, where the excess energy quickly dissipates as heat before it can create other electron-hole pairs. However, in semiconducting QDs, the rate of energy dissipation is reduced dramatically, and with the charge carriers confined within a nanovolume, their interactions are enhanced, increasing the probability for multiple excitons per incoming photon. A third advantage of QDs is that in contrast to traditional crystalline semiconductor materials that are rigid, QDs can be molded into a variety of different forms, such as sheets or three-dimensional arrays. Quantum dots can easily be combined with organic polymers or dyes or made into porous films, and in colloidal form, they can be processed to create junctions on inexpensive substrates such as plastics or thin metal foils. A fourth advantage of QDs is that it is quite easy to decouple the absorption and charge separation process. Decoupling charge separation from light absorption results in many more engineering options to make the solar cell of the future. For example, the charge separation may be achieved in a semiconductor QD, while the light absorption may be in an organic dye adsorbed on its surface.

The broader quantum solar cell strategy is thus to sensitize a first material with bandgap E_g with another material of bandgap (E_q) smaller than E_g (i.e., $E_q < E_\gamma < E_g$), and this second semiconductor material can be a dye (monomeric), a photoactive biopolymer, a QD, or a quantum well. QDs have the advantage over dyes as they degrade less in sunlight. In its simplest form, a quantum well solar cell (QWSC) consists of a p-i-n solar cell with a multiple quantum well (MQW) system added to the intrinsic (*i*) region. The idea, as with a tandem solar cell, is that the quantum wells absorb light of energy below the bulk bandgap, E_g.

MEMS in Solar Cells

MEMS techniques allow for the fabrication of lenses (including Fresnel lenses), mirrors, photonic crystals, cavities, and solar cell surface textures to enhance the interaction of incoming photons with the solar cell. A simple method to put MEMS to work in a conventional solar cell is by surface structuring the cell so that the light hits the PV several times, as illustrated in the pyramid structuring in Figure 9.28.[34]

NEMS in Solar Cells

We present examples of quantum technology benefiting solar cells by highlighting nanoporous semiconductors in the liquid-junction Grätzel solar cell, two-photon excitation in hybrid organic-QD PV devices, and increased efficiency in a QWSCs.

In Volume I, Chapter 7, we learned that in liquid-junction solar cells, to avoid corrosion of the cell, one chooses a redox system where the maximum density

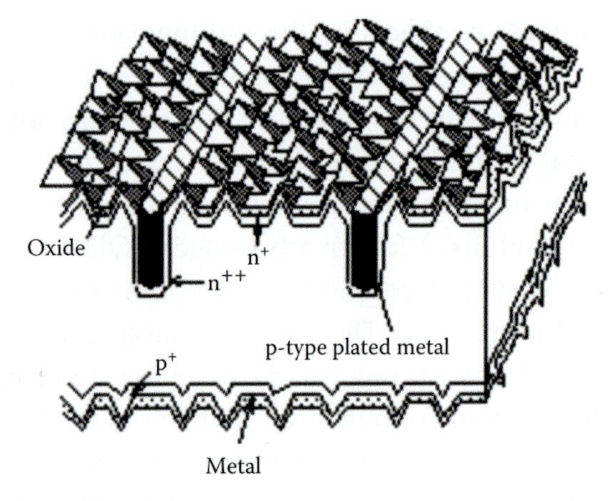

FIGURE 9.28 Surface structuring of a solar cell surface.[34]

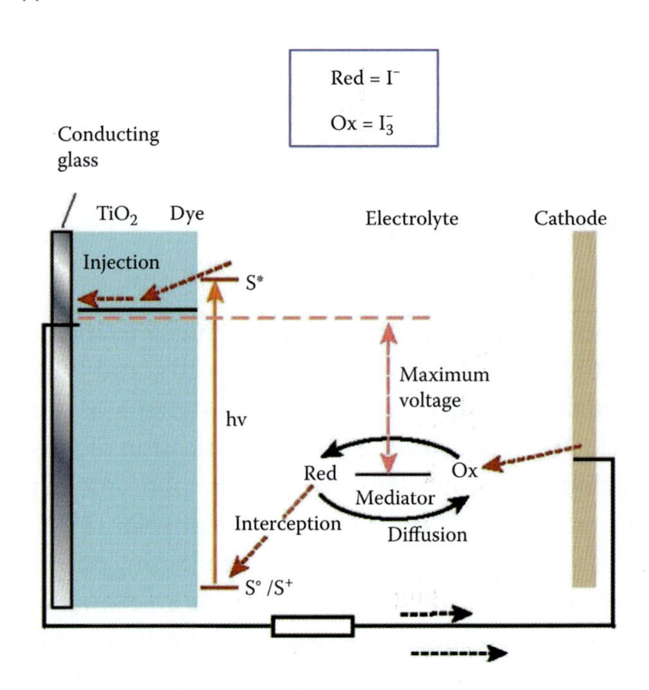

FIGURE 9.29 A liquid-junction solar cell: the dye-sensitized Grätzel design. The *S* energy levels represent the photosensitive dye levels.

of states of the reductant (the peak of the Gaussian in Volume I, Figure 7.62) is at the same level as the valence band edge of the semiconductor.[35] In this configuration, empty levels (holes) in the valence band, generated by shining light on the electrolyte/semiconductor interface, react with the reductant rather than with the semiconductor itself. In such liquid-junction solar cells, appropriate reductants thus compete effectively with the oxidation/corrosion of the semiconductor because they react more avidly with the holes in the valence band than the material itself.[36] The maximum voltage generated by such a cell, in theory, is simply the difference between the Fermi level of the semiconductor and the redox potential of the electrolyte.

The best known liquid-junction solar cell today is the dye-sensitized Grätzel design.[37-39] The Grätzel cell is significantly different from a conventional semiconductor solar cell, in which the semiconductor acts as the source of photoelectrons and also provides the potential barrier to separate the charges to generate the current. In contrast, in the Grätzel dye-sensitized solar cell, the semiconductor is used solely for charge separation, and the photoelectrons are provided from a separate photosensitive dye. As shown in Figure 9.29 the solar cell consists of a TiO_2 layer forming a nanoporous structure with a dye (e.g., ruthenium-polypyridine) spread throughout its surface. The dye molecules are small, and in order to capture a reasonable amount of the incoming light, the nanoporous structure is used as a scaffold holding large numbers of the molecules in a 3D matrix, vastly increasing the number of molecules for a given surface area. The charge separation is provided by the semiconductor-liquid junction contact. The TiO_2 layer sits on a transparent anode made of fluorine-doped tin oxide (SnO_2:F) deposited on the side of the glass plate facing the TiO_2 layer.

The electrolyte solution also holds a redox couple (e.g., I^-/I_3^-) mediator, located in the space between the dye-coated TiO_2 and a cathode, typically a thin film of platinum metal. Photons enter the cell through the transparent SnO_2:F window, and if they have enough energy, they are absorbed by the dye, creating an excited dye state. From this excited state an electron is "injected" into the conduction band of the TiO_2. This way, the dye molecule is oxidized, and it would decompose if it didn't quickly react with iodide in the electrolyte oxidizing it to form triiodide (dye regeneration reaction):

$$3I^- \rightarrow I_3^- + 2e^- \qquad \text{Reaction 9.6}$$

The reaction with iodide occurs very quickly compared with the recombination of the injected electron with the oxidized dye molecule, effectively preventing short-circuiting the solar cell. The injected electron then travels to the cathode via the external circuit, and the triiodide recovers its missing electron by diffusing through the solution to the cathode, where it is reduced back to iodide (redox regeneration reaction):

$$I_3^- + 2e^- \rightarrow 3I^- \qquad \text{Reaction 9.7}$$

The electron injection process does not introduce a hole in the TiO_2, only an extra electron, and recombination of this electron directly from the TiO_2 to the electrolyte is not possible because of differences in energy levels (see Volume I, Chapter 7: if there are no overlapping energy levels, no reaction can occur). This is a very important property of the dye-sensitized Grätzel design: the electron-hole recombination that plagues the efficiency of traditional cells does not exist here. The potential used for external work in this solar cell is given by the following equation:

$$V_{ext} = E_F - V_{redox} \qquad (9.1)$$

The quantum efficiency of a Grätzel-cell is very high because the huge surface area of the TiO_2 nanostructure makes the chance that a photon will be absorbed very high, and the dyes are very effective at converting photons to electrons. Most of the losses are due to conduction losses in the TiO_2 and the clear anode, and the optical losses stem from the front electrode as well. The overall quantum efficiency is about 90%, with the "lost" 10% being largely accounted for by the optical losses in the anode. Although the dye is highly efficient in turning photons into electrons, only those electrons with enough energy to cross the TiO_2 bandgap result in current being produced. The bandgap is larger than that of silicon, which means that fewer of the photons in sunlight are usable for power generation. In addition, the electrolyte limits the speed at which the dye molecules can regain their electrons and become available for photoexcitation again. These factors limit the current generated by a liquid-junction solar cell; for comparison, a traditional silicon-based solar cell offers about 35 mA/cm^2, whereas a Grätzel cell offers about 20 mA/cm^2. The closest competitors in price-performance terms are the various thin-film approaches, which are currently somewhat more developed commercially.

The dyes used in early experimental Grätzel cells were sensitive only in the high-frequency end of the solar spectrum (ultraviolet and blue). Newer versions have a much wider frequency response; notably, the triscarboxy-terpyridine ruthenium complex [Ru(2,2′,2″-$(COOH)_3$-terpy)$(NCS)_3$] is efficient right into the low-frequency range of red and infrared (IR) light. The wide spectral response results in the dye having a deep brown-black color, and it is referred to simply as *black dye*.[39] These polymers also are also sensitive to infrared are called *thermovoltaics*, and they generate electricity from both light and heat. Current Grätzel cells demonstrate a conversion efficiency of about 11%, whereas common low-cost commercial panels operate between 12% and 15%. Flexible thin-film cells are typically around 8%. Perhaps the major disadvantage of a liquid-junction solar cell is the use of liquid electrolytes, which brings leakage and corrosion problems, as well as temperature stability problems. At low temperatures, the electrolyte can freeze, ending power production and potentially leading to physical damage. Higher temperatures cause the liquid to expand, making sealing the cells a serious problem.

To avoid the above problems of liquid-junction solar cells, several research groups aim to produce a hybrid organic-QD PV device as the solid analog to the Grätzel cell. Researchers at the University of Toronto mixed lead sulfide (PbS) QDs with the polymer poly(2-methoxy-5-[2′-ethylhexyloxy-p-phenylenevinylene]) (MEH-PPV). Wrapping MEH-PPV around the QDs shifted the polymer's absorption into the infrared: on its own the polymer absorbs between ~400 and ~600 nm, and the QDs have absorption peaks that can be tuned from ~800 to ~2000 nm (see Figure 9.30).[40] The best efficiency registered thus far with this approach is 1.6%.[41]

In 2000, Nozik from the National Renewable Energy Laboratory in Golden, Colorado, predicted that theoretically, QDs could increase the efficiency of solar cells through multiple exciton generation.[42] "We have shown that solar cells based on QDs theoretically could convert more than 65 percent of the sun's energy into electricity, approximately doubling the efficiency of solar cells," said Nozik.[42]

In 2004, Schaller and Klimov from the Los Alamos National Laboratory in New Mexico were the first to demonstrate this phenomenon experimentally, using QDs made of lead selenide.[43] Multiple excitons

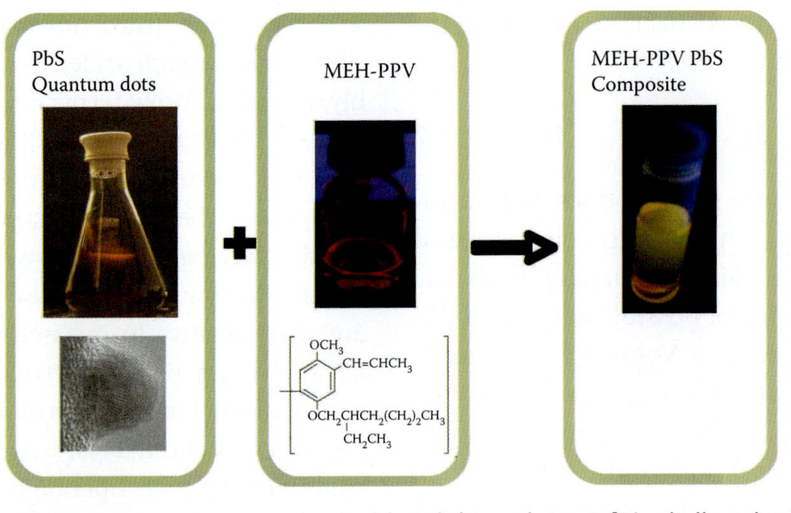

FIGURE 9.30 Lead sulfide (PbS) quantum dots are mixed with poly(2-methoxy-5-[2′-ethylhexyloxy-p-phenylenevinylene]) (MEH-PVV) to make a MEH-PPV PbS composite.[40]

start to form simultaneously as soon as the photon energy reaches twice the bandgap, and a group of researchers led by Nozik demonstrated that the absorption of a single photon by 2.9 nm PbSe QDs yielded not one exciton but three when the energy of the photon absorbed is four times that of the bandgap[43] (a quantum yield of 300%). Quantum dots made of PbS also showed the same phenomenon.

Quantum wells, as we saw in Volume I, Chapters 3 and 5, are ultrathin layers (nanostructures) of a narrower bandgap material (e.g., a GaAs well) sandwiched between regions of wider bandgap (e.g., AlGaAs barrier). They are made possible by the development of modern crystal growth techniques such as molecular beam epitaxy (MBE) and metal-organic vapor phase epitaxy (MOVPE). We also saw how quantum well lasers work at high forward bias and how quantum well photodiodes operate at reverse current and reverse bias. Quantum well solar cells (QWSCs) work in the forward bias and reverse current quadrant of the current-voltage plot, where power can be extracted.[44] In Figure 9.31 a first material with bandgap E_g is contacted with another material of bandgap (E_q) smaller than E_g, i.e., $E_q < E_\gamma < E_g$. The idea, as with a tandem solar cell, is that the quantum wells absorb light of energy below the bulk bandgap, E_g.

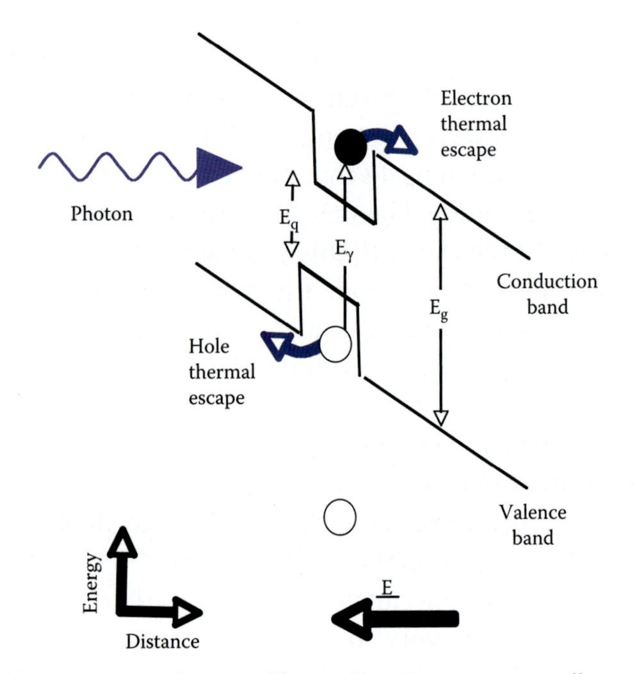

FIGURE 9.31 Schematic illustrating the quantum well solar cell.

In the simplest QWSC, several quantum wells are added to the *i*-region of a conventional p-i-n solar cell of wider bandgap (E_g). The quantum wells do absorb light of energy below the bulk bandgap E_g. If carriers escape from the wells before recombining, they enhance the output current, and short-circuit current (I_{sc}) can be strongly enhanced. The open-circuit voltage (Voc) is reduced, but the cell's efficiency is considerably enhanced.

Energy Scavenging (Harvesting)

MEMS devices either feature onboard power sources such as batteries, capacitors, fuel cells, radioactive elements, etc.; are powered by energy that is beamed in on purpose; or harvest energy from their own environment. Energy harvesting is the process of capturing minute amounts of energy from one or more naturally occurring energy sources and storing that energy for later use. Harvestable energy sources for a MEMS device include mechanical (from sources such as vibration, mechanical stress, and strain, captured by ratchets and springs), thermal (waste energy from furnaces, heaters, and friction, captured by thermocouples, bimorphs, and thermoelectrics), light (captured from sunlight or room light via photosensors, photodiodes, or solar panels), electromagnetic (from inductors, coils, and transformers, captured by coils), natural (wind and ocean currents), the human body (a combination of mechanical and thermal energy naturally generated from bio-organisms or through actions such as walking and sitting), and other energy from chemical and biological sources. All these energy sources are virtually unlimited and essentially free, if they can be captured at or near the system location.

Naturally occurring temperature gradients are one potential source of power. Stark and Stordeur,[45] for example, demonstrated a thermoelectric device that can produce 15 mW of power from a 10°C temperature difference. The thermal gradient principle is employed in satellites, which are hot on the side turned toward the sun but cold on the side facing away from the sun. A satellite can be powered by the energy thus generated for up to thirty years. The principle also operates in reverse: thermoelectric elements can produce cold from electricity and are used these days to cool the interiors of computers. In the 1960s, it was believed that if transistors could replace vacuum tubes for electronics, then surely thermoelectrics could replace refrigerators. Requirements for a good thermoelectric (TE) material are a high figure of merit, *ZT*; this requires a high Seebeck coefficient, *S*, a high electrical conductivity, σ, and a low thermal conductivity, κ (see Equations 8.70–8.73). Unfortunately, *ZT* has to be improved from ~0.9 to 3 or higher to become commercially attractive. (*ZT* needs to improve to over 1.3 at 300 K for a major impact in electronics cooling and to around 2.5 for a revolutionary impact in air conditioning and power from waste heat.)

Some researchers have focused on scavenging power from the human body. Shenck and Paradiso,[46] for example, did create shoe inserts capable of generating 8.4 mW of power under normal walking conditions. [See also MIT Media Lab's Parasitic Power Harvesting project for devices built into a shoe (http://www.media.mit.edu/resenv/power.html).]

Mechanical vibrations are another promising power source. Scavenging the power from commonly occurring vibrations for use by electronics is both feasible and attractive for certain applications. Three popular methods to harvest energy from vibrations are piezoelectric, capacitive, and inductive. In the piezoelectric method, the strain in a piezoelectric material causes a charge separation (voltage across capacitor). Ottman et al.[47] have designed and optimized power circuitry to be used with a piezoelectric vibration-based generator. In a capacitive pick-up, a change in capacitance causes either a voltage or charge increase. Roundy et al.[48] have published electrostatic MEMS designs with simulated power outputs in the range of tens of microwatts per cubic centimeter. With an inductive pick-up, a coil moves through a magnetic field causing current in a wire. Amirtharajah, Chandrakasan, and coworkers[49] demonstrated an electromagnetic vibration-to-electricity converter with a 160-cm^3 volume (4 × 4 × 10 cm) that produces 400 mW of power (or 2.5 mW/cm^3). Table 9.7 compares the features of different vibrational harvesting techniques. Currently, a piezoelectric approach with PZT is the most popular because of its compact configuration and compatibility with MEMS, but its inherent limitations, including aging, depolarization, brittleness, and high-output impedance, limit many applications.[50]

As Goemans points out, the possibility of converting motion into electrical energy can be very attractive for cases where battery replacement is unacceptable, kinetic energy is abundantly available, and space is not too limited.[6] He lists biomedical implants, tire pressure monitoring systems, and electronic locks as potential application areas.[6] Some power harvesting systems are compared with batteries, fuel cells, solar power, and nuclear power in Table 9.8.

Table 9.7 Summary Comparison of the Different Vibrational Types of Harvesting Mechanisms[50]

Type	Advantages	Disadvantages
Electromagnetic	• No need of smart material • No external voltage source	• Bulky size: magnets and pick-up coil • Difficult to integrate with MEMS • Maximum voltage of 0.1 V
Electrostatic	• No need of smart material • Compatible with MEMS • Voltages of 2~10 V	• External voltage (or charge) source • Mechanical constraints needed • Capacitive
Piezoelectric	• No external voltage source • High voltages of 2~10 V • Compact configuration • Compatible with MEMS • High coupling in single crystals	• Depolarization • Brittleness in PZT • Poor coupling in piezo film (PVDF) charge leakage • High-output impedance
Magnetostrictive	• Ultrahigh coupling coefficient >0.9 • No depolarization problem • High flexibility • Suited to high-frequency vibration	• Nonlinear effect • Pick-up coil • May need bias magnets • Difficult to integrate with MEMS

Table 9.8 Energy Harvesting Compared with Batteries, Fuel Cells, and Nuclear Power

	Power (Energy) Density	Source of Estimates
Batteries (zinc-air)	1050–1560 mWh/cm^3 (1.4 V)	Published data from manufacturers
Batteries (lithium ion)	300 mWh/cm^3 (3–4 V)	Published data from manufacturers
Solar (outdoors)	15 mW/cm^2 – direct sun 0.15 mW/cm^2 – cloudy day	Published data and testing
Solar (indoor)	0.006 mW/cm^2 – my desk 0.57 m/Wcm^2 – 12 in. under a 60 W bulb	Testing
Vibrations	0.001–0.1 m/Wcm^3	Simulations and testing
Acoustic noise	3E-6 mW/cm^2 at 75 Db sound level 9.6E-4 mW/cm^2 at 100 Db sound level	Direct calculations from acoustic theory
Passive human powered	1.8 mW (shoe inserts >> 1 cm^2)	Published study
Thermal conversion	0.0018 mW – 10 deg. C gradient	Published study
Nuclear reaction	80 mW/cm^3 1E6 mWh/cm^3	Published data
Fuel cells	300–500 mW/cm^3 ~4000 mWh/cm^3	Published data

Fractal Power

Introduction

In this section, we show how the use of fractal-like electrodes in batteries and fuel cells minimizes the internal resistance while maximizing surface-to-volume ratios. We also demonstrate how C-MEMS techniques show great promise for use in the fabrication of such fractal electrodes for electrochemical applications.

Fractals and Fractal Dimensions

A fractal is a rough or fragmented geometric shape, each part of which is, at least approximately, a smaller copy of the whole. Another definition holds that a fractal is an irregular geometric object with an infinite nesting of structure at all scales. Fractals are said to be self-similar because no matter what the scale of observation, whether we zoom in or out, we appear to be seeing the same object. Fractals came to the public attention through the stunning abstractions generated in computer screen savers by the famous Mandelbrot set. Mandelbrot* introduced the idea of

* As a child in France, Mandelbrot wondered how to use the smooth regularities of Euclidean shapes to model the complexity of the world he saw around him. Where were the circles in nature? Where were the parallel lines and infinite planes? He concluded "clouds are not spheres, mountains are not cones, coastlines are not circles, and bark is not smooth nor does lightning travel in a straight line."[51]

fractals in a paper that asked, how long is the coastline of Great Britain?[51] The coastline is irregular, so a measure with a straight ruler provides an estimate only. The estimated coastal length, *L*, equals the length of the ruler, *s*, multiplied by *N*, the number of such rulers needed to cover the measured object. Mandelbrot realized that the smaller his measuring stick, the longer the coast would turn out to be and that using molecular distances, the coast would have an infinite length. This means, Mandelbrot conjectured, that the Euclidean dimension, *D*, of a coastline is between the one dimension of a straight line and the two of a plane.

If we take an object residing in Euclidean dimension *D* and reduce its linear size by $1/r$ in each spatial direction, its measure (length, area, or volume) would increase by *N* times the original, with *N* given by the following equation:

$$N = r^D \tag{9.2}$$

This is pictured in Figure 9.32.

The German mathematician Felix Hausdorff introduced a more general treatment of dimension *D*. He pointed out that taking the log of both sides of Equation 9.2 one obtains the following:

$$\log(N) = D \log(r) \tag{9.3}$$

And if we solve for *D*, we get the following:

$$D = \log(N)/\log(r) \tag{9.4}$$

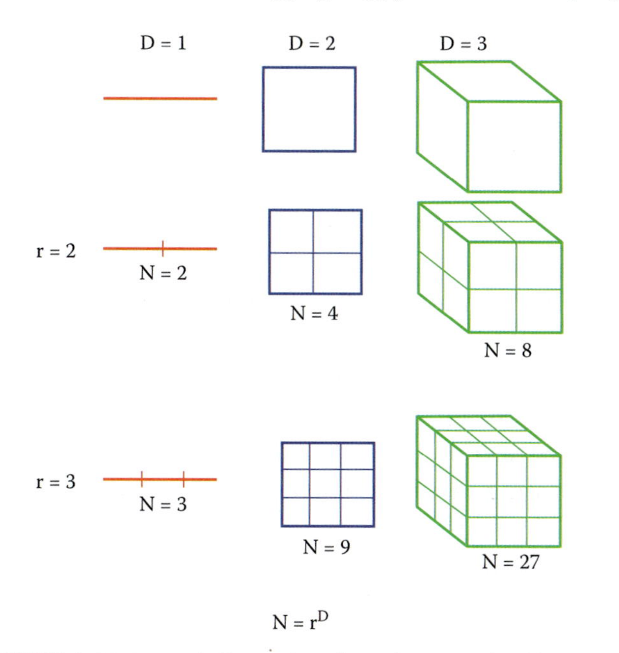

FIGURE 9.32 Fractal dimensionality: the Hausdorff dimension.

The point that emerges is that if dimensionality, *D*, is examined this way, *D* need not be an integer, as it is in Euclidean geometry. It could be a fraction, as it is in fractal geometry. This approach has proved very useful for describing all types of natural objects and for evaluating trajectories of dynamic systems.

It was Lewis Fry Richardson who first noted the regularity between the length *L* of coastlines, rivers, continental boundaries, etc., and the scale size *r*: when plotted in a log-log plot, a linear relation is obtained between log $[L(r)]$ and $\log(r)$. Mandelbrot assigned the term $(1 - D)$ to the slope of such a log-log plot, or:

$$\log[L(r)] = (1 - D)\log(r) + b \tag{9.5}$$

where *D* is the fractal dimension and *b* a constant.

For Great Britain, the slope $1 - D = -0.24$, approximately, or $D = 1.24$, a fractional value. The coastline of South Africa is very smooth, virtually an arc of a circle, and the slope is very near zero, and this gives a fractional dimension $D = 1 - 0 = 1$. This makes sense because the coastline is very nearly a regular Euclidean object, a line, which has dimensionality of 1. In general, the "rougher" the line, the steeper the slope and the larger the fractal dimension (the fractal dimension for the Nile delta is 1.4; that of the Amazon basin is 1.85).

Hausdorffian dimensionality is further illustrated by considering the Koch curve shown in the inset of the header of this chapter. To make a Koch curve, a one-dimensional (1D) line is buckled to take on some fractional dimension between 1D and 2D. We begin with a straight line of length 1, called the *initiator*. We then remove the middle third of the line, and replace it with two lines that each have the same length (1/3) as the remaining lines on each side. This new form is called the *generator*, because it specifies a rule that is used to generate a new form. The rule says to take each line and replace it with four lines, each one-third the length of the original. We can do this iteratively without end; the length of the curve increases with each iteration, and it asymptotically approaches infinite length. If we treat the Koch curve as we did the coastlines and continents, the relation between $\log[L(r)]$ and $\log(r)$ for the Koch curve gives us $D = \log(4)/\log(3) = 1.26$. It is a number that the Koch curve approaches as it becomes ever more finely broken over all possible scales. A huge number of everyday phenomena do seem ruled by such power-laws rather than Gaussian statistics. Some are well-known examples, such as earthquakes and species extinctions. Fractal organization is to be found in all kinds of dissipative structure—structures that "chaotically" branch in order to dump a load of energy as fast as possible (see Chapter 3, "Self-Organization").

Fractals in Nature

Nature always strives for the highest efficiency in all its organisms, and fractal-like networks are nature's answer to this maximization problem, literally "giving life an additional dimension."[52] Fractals in nature represent the geometry of irregular shapes and chaotic systems: trees, rivers, brains, lungs, the cardiovascular system, clouds, and coastlines all can be described by fractals.

In nature, the fundamental equation that describes scaling of various variables with relation to body mass, we saw in Chapter 7, is an allometric equation of the following form:

$$V = aM^b \quad (7.1)$$

where V is the variable in question, a is a constant, and b is the scaling exponent (in Chapter 7 we used $y = bx^a$). The fascinating empirical observation that has perplexed many scientists and has been a source of much debate is the fact that the exponents (b in Equation 7.1) of many variables, including cellular metabolism ($b = -1/4$), heartbeat ($b = -1/4$), maximal population growth ($b = -1/4$), lifespan ($b = 1/4$), blood circulation ($b = 1/4$), embryonic growth/development ($b = 1/4$), metabolic rates of entire organisms ($b = 3/4$), cross-sectional areas of mammalian aortas ($b = 3/4$), and cross-sectional areas of tree trunks ($b = 3/4$) are multiples of 1/4 instead of 1/3.[52] Pure geometric scaling of area (l^2) and volume (l^3) leads to a scaling constant of 2/3, but as Max Kleiber discovered in the early 1930s, the metabolic rates of entire organisms (which should scale with area) scale with respect to body mass (which should scale with volume), with a scaling constant of 3/4. This relationship is now referred to as *Kleiber's law* (see Figure 7.4). We also saw that West et al.[52] proposed a general model for this extra "fourth dimension". By modeling the cardiovascular system as a fractal-like network that ends in terminal units (capillaries) of the same size, they have shown that minimizing the work that the heart performs in pulsatile systems leads to the proper scaling constants that are empirically observed.

Fractals in Electrochemistry

Just as nature has benefited from fractal structures in almost all of its organisms, biomimetic fractal designs in electrochemical devices such as power conversion devices and sensors can also lead to benefits in scaling.

Fractals are an optimal geometry for minimizing the work lost due to the transfer network while maximizing the effective surface area. In many electrochemical systems, it is advantageous to have a large surface-to-volume ratio, while needing to transfer the signal or power effectively to an electrical network. Electrochemical energy conversion devices, such as fuel cells and some types of batteries, as well as sensors, such as glucose sensors, are examples of such applications. Miniaturization of these electrochemical devices to create miniature electrochemical devices poses several problems due to scaling. One issue is that there is a problem of increasing internal resistance in miniature 3D devices. In cases where an electrode is composed of materials that are excellent electrical conductors, such as metals, the internal resistance contribution of the electrode is minimal and can be ignored, but if the electrode is composed of carbon or another material that is a optimal electrochemical material, but not necessarily an optimal electrical material, the internal resistance contribution due to the electrodes can be significant.[53]

The resistance of a bulk or porous network scales inversely to the length scale l if scaled isometrically, as illustrated in Figure 9.33a. Figure 9.33b illustrates the scaling involved when scaling area with a constant thickness. Since the volumes of the structures can be written as follows:

$$V = l^3 \quad \text{for Figure 9.33a} \quad (9.6)$$

$$V = tl^2 \propto l^2 \quad \text{for Figure 9.33b} \quad (9.7)$$

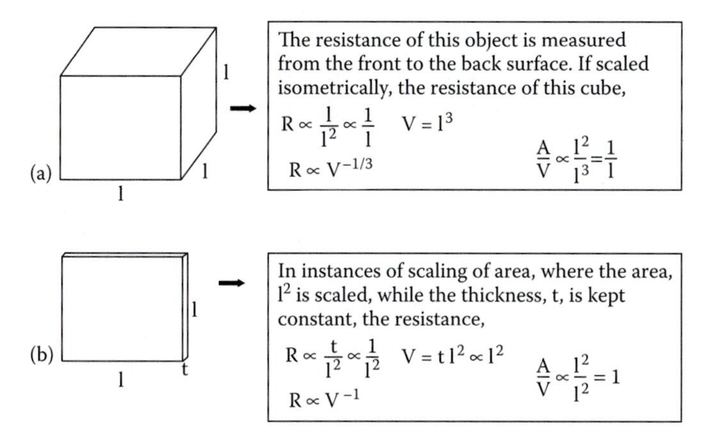

FIGURE 9.33 Scaling of a cube (a) and a sheet (b).

From this figure it can be concluded that resistance of a bulk material scales as follows:

$$R \sim V^{-1/3} \quad \text{for Figure 9.33a} \tag{9.8}$$

when scaling isometrically and as:

$$R \sim V^{-1} \quad \text{for Figure 9.33b} \tag{9.9}$$

when scaling area only.

The surface area and the scaling of the surface-to-volume ratios of the two cases in Figure 9.33 are as follows (assuming $t \ll l$).

$$\frac{A}{V} \propto \frac{l^2}{l^3} = \frac{1}{l} \quad \text{for Figure 9.33a} \tag{9.10}$$

and:

$$A = 4lt + 2l^2 \approx 2l^2 \text{ and } \frac{A}{V} \propto \frac{l^2}{l^2} = 1 \quad \text{for Figure 9.33b} \tag{9.11}$$

The surface-to-volume ratio of a bulk material is inversely proportional to the length scale ($1/l$) when scaling isometrically in three dimensions and is invariant (1) when scaling in two dimensions.

In the case of conventional Li-ion batteries and fuel cells, where the dimensions of the fundamental building blocks (the anode/electrolyte/cathode sandwich) are smaller than the final device dimensions and need to be scaled up, areal scaling is advantageous. As can be seen from Equations 9.8 to 9.11, 2D scaling from a small fundamental building block results in lower internal resistance and higher effective area compared with an isometric three-dimensional scaling approach. Indeed, commercial Li-ion batteries and fuel cells are composed of many layers of anode, cathode, and electrolyte that are rolled, folded, or stacked on top of each other. In the case of miniature power devices, where conventional methods of rolling, folding, and stacking are not applicable, microfabrication deposition techniques have been used to create thin-film batteries. Although applicable in some circumstances, there are instances where a 3D package is needed. As mentioned by Long et al.[7], the limiting factor of increasing the height of 3D miniature batteries is the increased internal resistance of the high-aspect-ratio structures. The use of better electronic and ionic conductors, in addition to the optimal geometries, determines the practical upper limit of these devices. This author believes that fractal networks can prove to be a more favorable architecture for miniature electrochemical devices. Other than the purely electrical advantages, fractal architectures provide a middle ground between high-surface-area electrodes composed of nanomaterials and solid electrodes. Although the high surface area of nanomaterial electrodes does increase performance in many areas, much of the volume of the entire battery can be "wasted" because of the electrolyte around the electrode. In a fractal electrode, much of the volume that would be composed of electrolyte is replaced with electrode material.

The approach used by West et al.[52] to model the fractal nature of the cardiovascular system is mirrored and directly converted to that of a fractal network of electrically conductive material in the model of Park et al.[53] The purpose of this exercise was not to model a specific system but to provide an idealized zeroth-order approximation for a fractal electrochemical system.

There are three general assumptions made by West et al.[52] The first assumption is that a fractal-like branching network is required to supply a volume effectively. In an electrochemical device, it is important to maximize the use of space. For example, fuel cell engineers strive to maximize the density of triple phase boundaries (most of the time by increasing the effective surface area), and batteries and sensors benefit from large surface-to-volume ratios because, in many cases, the chemistry is largely a surface phenomenon (limited by a process near the surface).

In situations where the reaction is limited by mass transfer, the mass transfer problem can be alleviated by reducing the distance the reactants need to travel, i.e., by spreading the reaction sites throughout the entire volume of the device. Just as capillaries are spread out throughout an organism because each capillary can only supply a certain volume of space with oxygen and nutrients, it is beneficial to spread out the reaction surface throughout the volume in an electrochemical device. It is possible to spread out the effective surface area throughout a 3D space using a sponge-like or web-like material composed of fibers or particulates, but in many cases, it becomes difficult to transfer the charge from and to the reaction sites.

Space-filling fractal networks can work to ensure that there is efficient charge transfer from a huge effective surface area to a current collector. The second assumption is that the smallest fractal structure is a size-invariant unit independent of the total volume. There are physical and practical limitations in size for all materials, even though these limits may differ from material to material. Nanoparticles, nanofibers, nanotubes, and other nanoscale materials are all examples of a fundamental building block. The final assumption is that energy used to distribute resources should be minimized.

Figure 9.34 illustrates the fractal electrochemical network with three branches (fractal levels) and a branching factor, n, of two as assumed by our team for our fractal model. N_k is the number of elements at fractal level k, and each level has $N_k = n^k$ elements; s is the fractal level for the smallest elements. J_{limit} is the maximum current density of any element, and R_k, ρ, l_k, A_k, r_k, v_k are the resistance, resistivity, length, area, radius, and service volume, respectively, for a fractal element at level k. A cylindrical geometry for each element is assumed.[53]

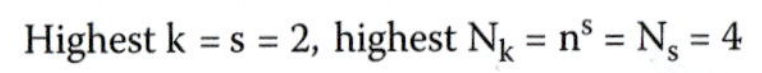

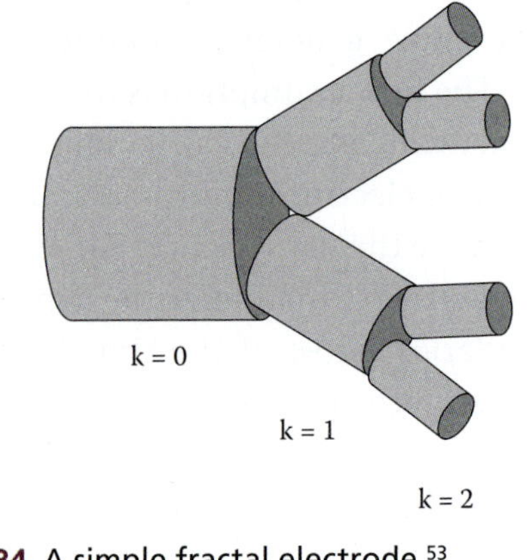

FIGURE 9.34 A simple fractal electrode.[53]

In electrochemical devices, there is loss of energy and an associated voltage drop due to the internal resistance. Minimization of this internal resistance is key to achieving efficient power transfer. This becomes more crucial as the current to the electrical load becomes larger because of Ohm's law:

$$E = IR \tag{9.12}$$

where E is the electrical potential, I is the current, and R is the resistance. In situations where the electrochemical system is limited by surface phenomena, there is a limiting interface current density J_{limit}. The total current, I_{total}, flowing through an electrode with a total surface area (total exchange area) of A_{total} operating in a surface-phenomena-limited regime is as follows:

$$I_{total} = J_{limit} A_{total} \tag{9.13}$$

We define the fractal level with the smallest elements as level s and N_k to be the number of elements at level k. We derived that the total exchange surface area, A_{total}, is proportional to the number of the smallest elements, N_s:

$$I_{total} \propto N_s \tag{9.14}$$

The total current is thus proportional to the number of the smallest elements within the fractal network. Obviously, maximizing the number of small elements within a fractal structure is desired. We also derived the following:

$$R_{total} \propto \frac{1}{N_s} \tag{9.15}$$

Again, as with Equation 9.14, maximizing the number of the smallest elements is desirable. This ensures that current is maximized, while the internal resistance is minimized. From Ohm's law and Equations 9.14 and 9.15, it can be concluded that if a fractal network geometry is used, the voltage drop across the entire structure is as follows:

$$E_{total\ internal\ voltage\ drop} = I_{total} R_{total} = \text{constant} \tag{9.16}$$

The power dissipation, in turn, scales with N_s:

$$P = E_{total} I_{total} = I^2_{total} R_{total} \propto N_s \quad (9.17)$$

We also calculated that V_0, the volume of the liquid where effective exchange occurs with the electrode, the exchange volume can take place as given by the following (for details see Park et al.[53]):

$$V_0 \propto N_s \quad (9.18)$$

Combining Equation 9.18 with Equations 9.14, 9.15, 9.16, and 9.17, one can then summarize the volumetric scaling of the electrical characteristics of a fractal network:

$$I_{total} \propto V_0 \quad (9.19)$$

and:

$$R_{total} \propto \frac{1}{V_0} \quad (9.20)$$

So that $E_{total\ internal\ voltage\ drop} = I_{total} R_{total} = \text{constant}$ again, and:

$$P \propto V_0 \quad (9.21)$$

We also obtained the following:

$$\frac{A_{total}}{V_0} \approx \text{constant} \quad (9.22)$$

In other words the surface-to-volume ratio of a fractal electrode does not change as the volume changes.

Comparing Equation 9.20 with Equation 9.9, it can be concluded that the resistance of fractal electrodes scale similar to a thinly layered structure, not to an isometrically scaled volumetric geometry, and comparing Equation 9.22 with Equation 9.11, the surface-to-volume ratio of a fractal electrode does not change as the volume changes; again the fractal geometry scales more like a layered film than a volumetric electrode. The scaling of the electrical properties of a fractal electrode is summarized and compared with that of a 2D or 3D electrode in Table 9.9.

C-MEMS Fractals

The author believes that C-MEMS[54–57] technology is an attractive path toward the fabrication of 3D fractal-like carbon electrodes. C-MEMS describes a fabrication technique in which carbon devices

Table 9.9 Scaling of a Fractal Compared with Two-Dimensional and Three-Dimensional Scaling

Fractal	3D Scaling	2D Scaling
$I_{total} \propto V_0$		
$R_{total} \propto \frac{1}{V_0}$	$R \propto V^{-1/3}$	$R \propto V^{-1}$
$E_{total\ internal\ voltage\ drop} = \text{constant}$		
$P \propto V_0$		
$\frac{A_{total}}{V_0} \approx \text{constant}$	$\frac{A}{V} \propto \frac{l^2}{l^3} = \frac{1}{l}$.	$\frac{A}{V} \propto \frac{l^2}{l^2} = 1$

are made by treating a prestructured organic material with high temperatures (typically 900°C and higher) in an inert or reducing environment. The details of the fabrication process are described in Chapter 5.

Fabricating a multilevel fractal structure that is idealized by the model that we have described above (and shown in Figure 9.34) is not a trivial issue. One reason behind the difficulty of this task is that the tools that are used for 3D microfabrication do not apply over a large range of length scales. Hybrid 3D fabrication techniques that combine different fabrication methods are therefore required (see Figure 3.11). An example hybrid method is briefly described below for the construction of a 3D fractal-like C-MEMS structure.

Example 9.3: Doped Photoresist Method to Make Fractal-Like Carbon Structures

In this method, standard photolithography is used on a photoresist (SU-8) premixed with carbon nanofibers (CNFs). The CNFs (Hawaii Industrial Labs) are first mixed with the SU-8 100 resin (1% by weight). The doped photoresist is then spin-coated onto a silicon wafer to an approximate 100-μm thickness. The standard photolithographic process to pattern the desired cylindrical photoresist posts then follows. After development, the posts exhibit a few clumps of CNFs on their surfaces that are partially embedded inside the SU-8. Pyrolysis of these structures converts the entire structure into carbon, with a thick coat of clumps of CNFs now covering the side walls of the posts. The CNFs shown can withstand the pyrolysis process and stay attached to the carbon posts. The process of all the

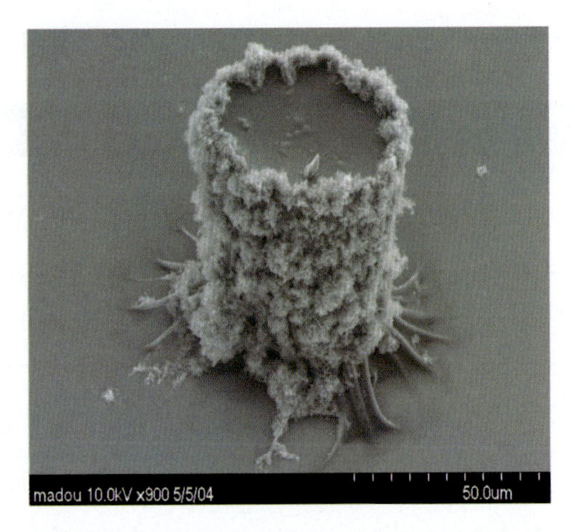

FIGURE 9.35 Scanning electron microscope image of a carbon post with nanofibers.

nanofibers migrating to the cylinder walls during pyrolysis is not yet understood. Figure 9.35 shows a scanning electron microscope image of this fractal-like structure. This is a direct attempt at mimicking the branching that we have mathematically modeled above (see Figure 9.34). Obviously this represents a rather poor fractal-like structure, with only two possible length scales, i.e., 100 μm (for the posts or tree stems defined by lithography) and nanometer-sized CNFs (the "branches").

For a sensor application, a single tree-like carbon microelectrode would already make for a big improvement over current sensor electrodes since it comes with the largest surface area while keeping the internal resistance minimal. Such a tree-like structure could be decorated with catalyst molecules like the decorations on a Christmas tree.

As illustrated in Figure 9.36, plant matter can be converted into carbon, much as charcoal is formed. One could then envision the use of natural occurring organic matter-based fractals to make carbon fractals.

In order to create a truly C-MEMS biomimetic energy conversion device, the anode and cathode, both fractal structures, must be intertwined while maintaining a pinhole-free electrolyte separation between the two in a manner similar to the blood vessels and air pathways in a mammalian lung. This may prove to be the most difficult task in arriving at a fully C-MEMS fractal battery or fuel cell.

Brains in Miniature Devices

Introduction

Before considering the implementation of computing power or brains in miniaturized systems, such as in microrobots, it is a good idea to briefly look at the state of the art in computer development in general and compare it with nature's arsenal using neurons and brains. After this introductory section, we consider the latest developments in artificial intelligence (AI). We discuss how both bottom-up and top-down approaches to AI might affect MEMS. Complexity theory and artificial life are considered next as two examples of bottom-up approaches to building AI that feature the self-organizing or emergent properties touched upon in Chapter 3. Artificial neural networks represent a bottom-up approach heavily dependent on learning cycles. Electronic noses and tongues have already been developed on the basis of such neural nets. As in the case of machining, we will learn that bottom-up approaches for embedding intelligence in MEMS are gaining more and more interest.

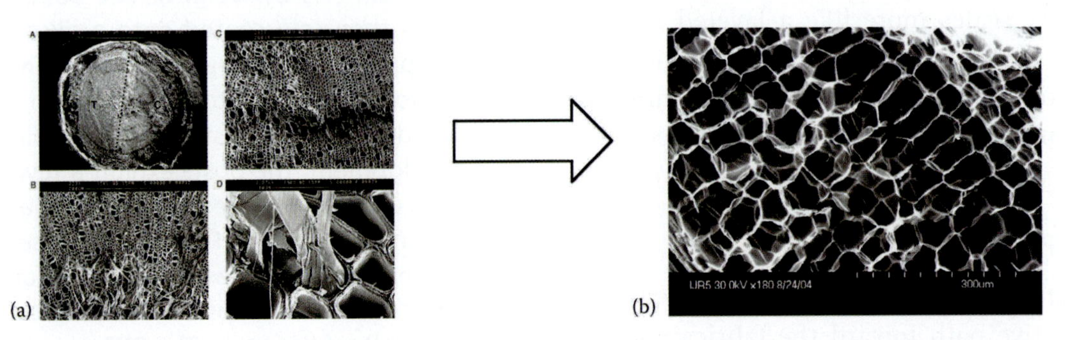

FIGURE 9.36 In the charcoal process, the structure of the plant cells is preserved. The high-temperature process is carried out in an oxygen-starved environment. (a) Twig of a tree with plant cells. (b) Charcoal. The original plant cell walls are preserved in the carbon.

State of the Art in Computer Development

IBM's ASCI White is a 12.3 trillion calculations per second machine the size of two basketball courts; it weighs 106 tons and cost $110 million. The machine runs programs on 8192 central processing units (CPUs) in parallel, resides at the Lawrence Livermore National Laboratory, and is the property of the US Department of Energy. Today, the most ambitious development effort to reach petaflop computers (*flops* is an acronym for floating point operations per second) is being carried out by a consortium headed by Caltech. The new system will have 150-GHz chips, cooled to near absolute zero. Because of their size and cost, such systems obviously remain out of range for MEMS. Closer to home are PCs. Intel's 0.18-μm manufacturing process is used to produce Coppermine Pentium III chips for PCs, which, when working at full speed, are as fast as 1.5 GHz. As we have seen in Volume II, Chapters 1 and 2, Moore's law will continue to hold for the next three to four MOSFET generations, reducing any mainframe to PC status eventually. In Volume I, Chapter 4, we saw how the fundamental thermodynamic limit of a MOSFET switch in a computer is the energy required to produce a binary transition that can be distinguished. The von Neumann-Landauer formula sets that theoretical limit (at 300K) as:

$$E_{bit} > kT \ln 2 \approx 18 \text{ meV}$$

(Volume I, Equation 4.70)

If a MOSFET switch consumes an amount of free energy E and performs N useful operations, the energy efficiency, we saw, is given by the following:

$$\eta = E/N \qquad \text{(Volume I, Equation 4.71)}$$

with η in units of operations per unit energy, or operations/second/watt. Using the thermodynamic energy limit per bit, this, in theory, enables an energy efficiency of $\eta = 3.5 \times 10^{20}$ operations/second/watt. In practice, using the case of 90 nm VLSI technology as an example, E is on the order of 1 fJ (femtojoule) and $\eta = 10^{15}$ operations/second/watt. This means that theoretically we can increase computing efficiency by a factor of more than 100,000. MOSFET downscaling is hampered by energy level quantization, though. The quantum mechanical limits on MOSFET scaling, based on the Heisenberg uncertainty principle, we summarized as follows:

$$x_{min} = \frac{\hbar}{\Delta p_x} = \frac{\hbar}{\sqrt{2m_e E_{bit}}} = \frac{\hbar}{\sqrt{2m_e k_B T \ln 2}} = 1.5 \text{ nm}$$

(Volume I, Equation 4.72)

which puts a fundamental limit on the integration density. For a limit on switching speed, we calculated the following:

$$\Delta E \Delta t \geq \frac{h}{2\pi} = \hbar$$

$$\Rightarrow \tau \sim 0.04 \text{ ps or 25 Tbits/second}$$

(Volume I, Equation 4.73)

Thus, quantum mechanics allows for a very small switch, but other factors, such as a material's temperature tolerance range, will limit before this fundamental limit sets in. Maintaining Moore's exponential increase in transistor count, however, will become progressively more difficult and expensive. Cost may ultimately become more significant than technical boundaries, since it dictates every investment in research and development. Memory storage devices are improving even faster than ICs, but they will also encounter similar obstacles. This has fueled the drive to look beyond silicon and current transistor logic to totally new areas, such as optical, DNA, and quantum computing.

Radical modifications to current computer architecture are under consideration today. For example, it is the electron charge that is used in the conventional semiconductors that comprise the CPUs and dynamic random access memories (DRAMs) in computers today. For the nonvolatile memory, magnetic disk drives are used, and these rely on the spin state of electrons to store large amounts of information and in particular to preserve information when the computer is shut down. There is now the possibility of developing devices that simultaneously use the spin and charge of the electron to vastly increase the capability of information processing, in what are called *spintronics* (spin-based electronics) *devices*. All spintronic devices behave according to this simple scheme: 1) information is stored (written) into spins as a particular spin orientation (up or down), 2) the spins, being attached to mobile electrons, carry the information along a wire, and 3) the information is read at a terminal. Spin orientation of conduction electrons survives for a relatively long time (nanoseconds, compared with tens of femtoseconds during which electron momentum decays), which makes spintronic devices particularly attractive for memory storage, magnetic sensors applications, and quantum computing, where electron spin represents a bit (called a *qubit*) of information. A qubit can exist not only in a state corresponding to the logical state 0 or 1 as in a classical bit, but also in states corresponding to a blend or superposition of these classical states. In other words, a qubit can exist as a zero, a one, or simultaneously as both 0 and 1, with a numerical coefficient representing the probability for each state. Four pairs of classical bits are needed to represent the same information as 1 pair of qubits. With eight qubits we can represent every number between 0 and 255 simultaneously. This allows for the compression of massive amounts of data. Quantum computing thus takes advantage of the concepts of superposition of states and entaglement introduced in Volume I, Chapter 3. This gives rise to many more states of the system, and thus in principle the ability to represent much more information without increasing the number of components—providing that it is possible to devise a way for the states of the system to interact quantum mechanically.[58]

Quantum computers with a handful of qubits have been made, but it seems hard to envision one with more than ten qubits. These early prototypes tend to suffer from decoherence after around 1000 operations. An interesting approach is to work with electrons on nanodots instead of on individual atoms as qubits. Nanowires are used to connect these nanodots and provide a physical entanglement. Providing entanglement through a physical connection between individual atoms is impossible, but between nanodots it can be done and it might avoid decoherence.

Research in these computing alternatives is well under way and even though non-silicon-based processors will not appear for many years to come, it seems quite certain that information technology will not be stopped when Moore's law ceases to be valid. For integrating computing power in MEMS, there are already good solutions today. The evolution of ICs and microprocessors has miniaturized the size of the computing aspect to the point where ubiquitous computing (ubicomp) has become possible, but sensors, output devices, data storage, and power supplies have a lot of catching up to do. It is precisely in this domain—the coupling of computation to the physical world—that micromachining will make its biggest contribution yet. To appreciate how to program memory embedded in MEMS with powerful top-down algorithms or simpler bottom-up learning routines, we will now look at how brains in nature work and consider the latest developments in AI, complexity theory, and artificial life.

The human brain and senses, just like nature's manufacturing methods, evolved in a biological context and are concerned mainly with survival and reproduction. Computing needs in MEMS are different, so we do not want to blindly mimic the brain but study it to learn and adapt strategies that fit our needs.

From Neurons to a Brain

It is important to realize that in nature, the simplest forms of life already exhibit complex reflexive behavior without any nerve cells or a brain at all. Even nature's reflexive behavior is often more complex than what even our "smart" robots can do today. Thus, the topic deserves some attention. A very simple brainless and nerveless organism, such as a paramecium, moves toward food or a comfortable temperature zone, orients itself in gravitational and centrifugal fields, responds to sound waves, swims upstream in a water current and toward a cathode at low applied currents and toward the anode at higher currents, moves away from damaging UV radiation and away from a touch and unfamiliar chemicals. Insects with only a few hundred thousand neurons also manage to fly around in real time and manage to avoid obstacles, although their computers are immensely slow, so they obviously must be able to manage their intelligence in some other ways than our large computers do.

In biology, nerve cells connect sense organs to the brain, and the brain processes incoming signals into perceptions. Nerve cells also run the other way—from the brain to the sense organs, enabling the brain to fine-tune its senses to pick up whatever it considers most important; in other words, perception is a two-way street. The brain itself relies on 10^{10} interconnected neurons (Figure 9.37) in 3.5 pounds or 15 cm^3 of gray matter. These neurons act like armies of switches, interconnected at 10^{14} synapses and collectively firing 10^{16} times a second while consuming less power than an ordinary light bulb, making today's attempts at artificial neural networks (see below) a pale version of their natural counterpart. An individual nerve cell does not posses a smidgen of intelligence; the latter emerges only from the interaction of billions of nerve cells. Nerve cells form long, thin fibers that carry electrical signals. The signals involve ions, not free electrons. The ions do not travel along the nerve cell but across its cell membrane. Electrical activity at one position along the nerve triggers activity in the next position. A charge pulse travels along the axon of a neuron (Figure 9.37) at a speed of 300 feet per second (200 miles per hour). This is slow compared with the speed of an electrical pulse in a wire and a million times slower than that in a Si-based switch (milliseconds versus nanoseconds).[59] Nature obviously does not get its advantage from speed; people are not able to learn, and later remember, more than about 2 bits per second for any extended period. If you could maintain that rate for 12 h every day for 100 years, the total would be about three billion bits–less than what we can store today on a regular 5-in. compact disc (CD).[60] The merits of our brain stem from its 3D nature and interconnections of neurons, as well as their individual complexity. The number advantage of human brains might disappear soon enough. As depicted in Volume II, Figure 2.1, the total number of transistors on a 2D chip will reach the number of neurons in the human brain by the year 2010; computers, some predict, will achieve parity in memory capacity with the human brain shortly after (in Volume II, Figure 2.1, the year 2020 is projected for this event).[61] The area density of the neurons was actually surpassed in the mid-1980s.[62] Neurons are not limited to communicating in an on/off or fire/no-fire mode; they send information in the form of rate of firing, a myriad complicated mechanisms increase or decrease signal strengths at synapses, neurons constantly adapt and change (growth and movement of axons and dendrites), and the brain does not employ a single generic neuron.[59] So not only does nature have an advantage in total number of interconnected switches, the switching behavior of a neuron is also significantly more complex than that of a computer switch.

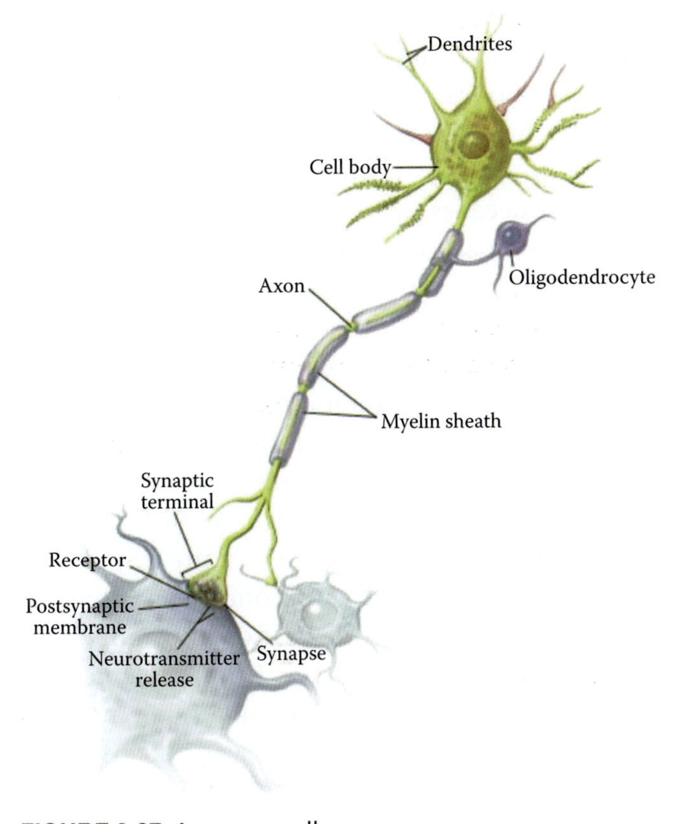

FIGURE 9.37 A neuron cell.

DNA constitutes the long-term read-only memory storage at the heart of life as it evolved on planet earth. A DNA strand with about 2×10^8 nucleotides is tightly packed into a volume of 500 μm^3 (see Figure 4.1). With the assumption that a set of three nucleotides is analogous to a byte, this corresponds to a memory volume density of 1.2 Mb/μm^3 and about 1 Kb/μm in linear density, indicating that memory chips based on DNA as the active elements could have extremely high density. Transcription, or the process of reading the DNA code, base by base, to make proteins, was illustrated in Figure 2.12. Both the packing density and the transcription process are perfect illustrations of the power of nanochemistry at work in nature. The evolutionary development of brains is very slow, though, and that of *Homo sapiens*—evolution's crown jewel (despite Howard Stern and Rush Limbaugh)—emerged only tens of thousands of years ago. According to Kurzweil's law of accelerating returns (Volume II, Chapter 2), the next salient events (e.g., "intelligent machines making other intelligent machines") will emerge much faster than mutations in our DNA can accommodate. For these next salient events, nature's DNA process is too slow: human intelligence has taken over already, and technological innovations will accumulate faster and faster.

DNA computing utilizes the information coding sequence of DNA and the interactions between complimentary strands to perform complex computations. DNA computing was first demonstrated by Adleman of the University of Southern California, in 1994.[63] The computation in Adleman's experiment ran along at 1014 operations per second, a rate of 100 teraflops, or 100 trillion floating point operations per second. Contrast this with the world's fastest supercomputer, NEC's Earth Simulator, running at just 35.8 teraflops. Obviously, computing with DNA has some advantages over current machines. Whereas today's computing technology rests on a highly linear principle of logic and one computation must be completed before the next can begin, the use of DNA means that an enormous number of calculations can take place simultaneously: the largest supercomputers have a few hundred processors run in parallel, but in some first DNA experiments there were 1 trillion processes running in parallel. This parallel power is many times faster than that of traditional machines—a mix of 1,018 strands of DNA could operate at 10,000 times the speed of today's advanced supercomputers.

In such a computer, the hardware consists of enzymes (restriction nuclease and ligase), a double strand of DNA as input data, and a few short DNA molecules as transition rules or software. The solutions come as pieces of DNA that must be isolated to be read. The process uses only one-ten-billionth of a watt versus tens of watts for an electronic microprocessor. There are some very serious limitations though. Although DNA can store a trillion times more information than current storage media, the way in which the information is processed necessitates a massive amount of DNA if larger-scale problems are to be solved. Furthermore, although the computation process takes place at an awesome speed, the "printout" of the result is excruciatingly sluggish, and involves many steps—it took Adleman a week of laboratory work to extract the potential solutions from his DNA cocktail.

Artificial Intelligence

Two Contenders: Top-Down and Bottom-Up

Introduction

Artificial intelligence is the field of research that attempts to embed intelligence into machines; it includes knowledge-based systems, pattern recognition, automatic learning, natural-language understanding, robotics, etc. Two distinct camps within the AI community today are the very ambitious and traditional, analytical top-down approach and the more modest, natural bottom-up approach. It is interesting to note that the same top-down and bottom-up duality showed up in manufacturing a couple of decades later. Traditionalists represent a rational approach to AI, building intelligence from logical first principles and complex algorithms. Research fields influencing this approach are semiconductor technology, cognitive psychology, and information science. Intelligence has proved very difficult (perhaps too difficult) to describe this way, and although the naturalist approach represents a lower level of computing, it has recently been more successful than traditional AI. Nature-inspired AI emphasizes neural network computing, complex system interactions, and self-organization. The natural approach to AI draws from research such as neuroscience, complex adaptive systems, and molecular biology. As in many research fields, rational statements about the relative merits of these two approaches are hard to find. Comparisons are based not so much on proven scientific facts as on the powerful egos, reputations of their institutions, and the old-boy networks propagating their views and influencing government funding agencies; the latter often listen only to the "loudest" voice. From the status of the two contending approaches discussed below, it appears that the natural AI approach has been the more successful. The two AI camps are compared in Table 9.10.

Traditional AI

Traditional AI got its unofficial baptism in 1956 at a small conference in Dartmouth, New Hampshire. Present at the time were luminaries such as Marvin Minsky, John McCarthy, and Edward Feigenbaum, and although they were, as many visionary scientists tend to be, over-the-top optimistic, the fruits of AI are now being realized.[59] An example of the optimism reigning in the early days of AI was the prediction about the early successful realization of the Turing test. Alan Turing, mathematician and computer scientist, had predicted, in a 1950 paper, that by the year 2000 a human judge interviewing a computer and a human foil, both hidden from view, would not be able to differentiate the computer from the human when questioning both via a terminal. Although this has not yet materialized, Ray Kurzweil maintains that Turing was correct and might simply be off by about twenty years.[61]

Surprisingly, breakthroughs in AI did not come from software programs reasoning in a precise, highly ordered, step-by-step fashion, as the logicist originators of AI had predicted but from simpler, so-called *expert systems* that made headway in the early 1980s. The latter sort of AI represents a rather practical or heuristic approach to addressing questions in very specific fields, such as the stock market, geophysics, or diagnosis of trouble with humans or cars. These simpler programs rely on rules learned by experience or training fed into the computer. An expert system will typically have a large number of *if-then* rules, and because the computer can quickly analyze thousands of these rules, it can outperform its human experts easily in more and more fields. One prominent example of AI's accomplishments is IBM's Deep Thought computer. Deep Thought beat grand chess master Brent Larsen in 1988 and sailed past world champion Garry Kasparov in May 1997.[59] The chess program has a human-like ability to zero in on a few most promising pattern of moves but does not really have a deep understanding of chess strategy. This makes it more akin to the natural AI approach we review next. Another example is the huge data and rule base of Lenat's Cyc 9 (*Cyc* from *encyclopedia*).[59] Besides information, Cyc again incorporates many rules; half of its memory content is facts, the other half is rules. The highly touted, ten-year-long Japanese Fifth Generation Project, a program from the 1980s meant to catch Japan up with the United States in software development and based on traditional AI, is widely seen as a failure now. The logical and heuristic approach simply proved to be inadequate. Interestingly, the Sixth Generation Project, also known as the Real-World Computing Project, started in Japan in 1992, aimed to achieve human-brain-like computing capabilities by the year 2002. Also this milestone was not met. Clearly, natural AI is today a more favored approach than traditional AI.

Table 9.10 Two Directions in Artificial Intelligence (AI) Research

<table>
<tr><th></th><th></th><th></th></tr>
<tr><td rowspan="10">AI</td><td>Traditional</td><td rowspan="5">Semiconductor technology
Information sciences
Cognitive psychology</td></tr>
<tr><td>Reductionist</td></tr>
<tr><td>Analytical</td></tr>
<tr><td>Top-down (see nanomachining)</td></tr>
<tr><td>e.g., rule-based expert systems</td></tr>
<tr><td>Natural</td><td rowspan="5">Molecular biology
Neurosciences
Complex adaptive systems</td></tr>
<tr><td>Complex interactions</td></tr>
<tr><td>Bottom-up (see nanochemistry)</td></tr>
<tr><td>Self-organization</td></tr>
<tr><td>e.g., neural networks</td></tr>
</table>

Marvin Minsky remains a strong proponent of the traditional approach to AI, and in *The Society of Mind* he tries to put the field back on track.[64] In a typical broadside against the bottom-up school, Minsky says, "Why bother building a robot that's capable of getting from here to there, if one gets there it can't tell the difference between a table and a cup of coffee?"[65] The book proposes that one should take multiple approaches to solving a problem. Lenat's Cyc, he claims, does not have enough different approaches, and neural networks are just too dumb.

Natural AI

In the early 1980s, some scientists recognized that traditional AI had been trying to solve problems too much from a top-down mode, whereas a bottoms-up mode might be more appropriate. The very complex programs of traditional AI had not been adapting to or learning from the world around them, an innate capability of even the simplest animals. As in manufacturing today, it appeared that nature might provide some guidance. Neuroscientists speculated that the processing of conscious information takes up as little as a thousandth of the human brain's power and that most of the rest goes into lower-level aspects of survival.[61] In other words, there are deep connections between the need to get food and mate (i.e., survival) and intelligence. It is not so surprising, then, that *Homo sapiens'* DNA is 98.6% the same as that of the lowland gorilla and 97.8% the same as that of the orangutan. On the basis of these observations, the neuroscientist Wilson introduced the concept of creating computer simulations or even simple robots that could avoid danger, find food, and deal in general with a more or less complex environment.[66] Wilson called his computer simulations *animats*. Brooks and coworkers at MIT have been implementing similar ideas into a series of small robots that interact and learn from their environment rather than being loaded up front with a blueprint of every possible scenario the robots might encounter. Sensor data from these robots are shared with all of the different microprocessors on the machines. Maes and coworkers enhanced this type of thinking by adding programs that enable the robot to learn from its mistakes by implementing "positive" and "negative" feedback as well as attempting programs for "motivation," such as "aggression," "curiosity," and "hunger."[66] Brooks's robots are a far cry from von Neumann's self-replicating robots but probably closer to what we can expect to accomplish in this decade with smart MEMS.

Complexity Theory

The science of complexity is preoccupied with the investigation of emergent, self-organizing properties in complex systems. The term *emergent* is used for phenomena in which the whole is greater than the sum of the parts; the whole system exhibits collective behavior that seems not to be built into the individual components in an obvious or explicit manner. In Chapter 3 we saw how Ilya Prigogine with his theory of "dissipative structures" was the first to describe, in a rigid mathematical framework, the coexistence of structure and change, balance and flow, stillness and motion, stability and instability, order and disorder, equilibrium and nonequilibrium, and being and becoming characterizing complex systems.[67] The name *dissipative structures* conveys the notion of dissipation on the one hand and order or structure on the other. Prigogine's monumental and revealing theory is exemplified by living organisms maintaining themselves in a stable state far from equilibrium despite continual flow of matter and energy (metabolism) through their system. Systems far from equilibrium cannot be described by classical thermodynamics, which applies only to systems close to equilibrium. The linear mathematics applicable for small flows (also called *fluxes*) characteristic of systems close to equilibrium must be replaced by nonlinear expressions describing the large fluxes associated with nonequilibrium systems. Dissipative structures do not tend toward equilibrium; on the contrary, they exhibit instabilities, bifurcation points, leading to new forms of order that move the system farther and farther away from the equilibrium state with an accompanying increase in complexity. The greater the complexity, the higher the degree of nonlinearity. The behavior of such a nonequilibrium system cannot be described from the individual parts; instead it behaves as an interconnected whole with multiple feedback loops. Nonlinear equations describing dissipative systems usually have more than one solution, and as a consequence, new situations may emerge at any moment. This contrasts with classical thermodynamics, where a system is completely predictable and pinned down by one solution only. From Prigogine we learn that the solution or path taken at an instability or bifurcation point in a dissipative structure depends on the previous history of the system and that its behavior is no longer universal but quite unique to the individual system itself. Moving away from equilibrium, we thus move away from the universal to the unique, from determinacy to indeterminacy, from individual components to an interconnected structure with links between structure and its history and a richness and variety as embodied by life itself.[68] Indeterminacy in dissipative structures is also introduced through small random variations, that is, "noise" from the environment. At bifurcation points, the path taken depends very sensitively on such noise, and the outcome becomes even more unpredictable, with new structures and higher order and complexity emerging spontaneously.

That order increases at bifurcation points does not contradict the second law of thermodynamics, as the increased order comes at the expense of greater disorder of the environment. Contrary to our intuition, chaotic systems are highly organized, exhibiting complex patterns dividing and subdividing again and again at smaller and smaller scales. In fluidic instabilities, turbulence will appear at sufficiently large velocities. In the case of chemical reactions, instabilities appear through repeated self-amplifying feedback. In both cases, self-organization, the spontaneous emergence of order, results from the combined effects of nonequilibrium, irreversibility, feedback loops, and instability. In fluidics, an example is the bathtub vortex, and in chemistry, examples include the so-called *chemical clocks* and *catalytic networks of enzymes*. The latter were shown to not only self-organize but also self-reproduce and evolve (see Chapter 3). In physics, an example is laser light, in which, under special circumstances, far from equilibrium and by "pumping" in energy from the outside, coherence spontaneously emerges from incoherent light. The most spectacular example of a self-organizing systems is the earth itself, as detailed in Lovelock and Margulis' Gaia theory.[69] Although often under attack, this theory is gaining more acceptance.

The reductionist approach of traditional AI fails when problems become extremely complex, and often no understanding is gained by breaking them down into their component parts. Earlier we saw how individual neurons or nerve cells work; not one exhibits any intelligence by itself, but intelligence in our brain emerges as a consequence of the interactions of billions of them. This is true for chemical and biological interactions that go on in living organisms, social structures, the weather, the economy, etc. Life itself, according to Kauffman, is nothing but an emergent property of a certain kind of complex system.[70] Although complexity theory seems to have made impact on market prediction, its impact on understanding life and evolution is still a matter of debate; this debate will probably go on for some time (see also Volume I, Chapter 7). Whereas natural AI is already useful in designing smart MEMS, complexity theory will start becoming important only when we work in the nanodomain and deal with self-organizing molecular systems.

Artificial Life

Artificial life, or a-life, is the study of lifelike creatures built by humans. This new science was officially inaugurated in Los Alamos, New Mexico, at a 1987 gathering organized by Christopher G. Langton.[66] Computer-generated entities or cellular automata behaving as self-operating machines, processing information about their surroundings, proceeding logically, and replicating themselves, were already predicted by the legendary mathematician von Neumann in the late 1940s. Today, many creatures of this type—really nothing more than software code—have been demonstrated, even on laptop computers. Some practitioners believe that by watching these creatures evolve on the computer screen and subjecting them to a variety of stresses, we can learn important lessons about the evolution of real-life systems more quickly and less expensively than with standard experimentation. Early a-life programs include *Boids* (http://www.lifeartificial.info/flocking-boid-behavior), which exhibit flocking behavior uncannily similar to that of real birds. The simple rules in this program control only individual Boids, but somehow flocking group behavior emerges (see complexity theory).[71] *Network Tierra* by Thomas Ray features software simulations of organisms—*creatures* in which each *cell* has its own DNA-like genetic code.[71] *Tierra* can be downloaded from http://www.hip.atr.co.jp/~ray/tierra/tierra.html. *Tierra* organisms, in their simulated world, compete for a limited resource (energy) provided by the computer CPU. The seed electronic organism, the Ancestor in *Tierra*, has only three genes, and besides feeding it is designed to replicate itself and to mutate. Several sources of mutations are programmed in. There is also a *reaper* that kills off creatures that are less well adapted and those that are too old. To everyone's surprise, *Tierra* very quickly developed some lifelike evolution characteristics, such as the development of new species, parasites, defense mechanisms against parasites, symbiotic relationships, competition for food, and even something akin to punctuated equilibrium as proposed by Gould and Eldredge in 1972.[70] The discussion about how increasing complexity arose is an ongoing one. In the case of *Tierra*, there was no tendency toward increasing complexity; on the contrary, simpler creatures evolved over time. Emergence of complexity is a fact of life, though, and Ray is trying to induce it in *Network Tierra*. For evolution to proceed in this artificial world, it has become clear that a lifelike complexity must challenge the organisms. Ray is doing this now by letting his organisms grow on the Internet with its diverse electronic life. The organism, like a screensaver, becomes active only when the host computer is inactive. Ray speculates on the emergence of such intelligent life in *Tierra*-like environments; it will make the Turing test irrelevant. For interested readers, more on artificial life can be found at Avida Artificial Life Group (http://www.krl.caltech.edu/avida), Boids (http://hmt.com/cwr/boids.html), Complexity On-Line (http://www.csu.edu.au/complex), Dawkins (http://www.spacelab.net/~catalj), Primordial Soup Kitchen (http://www.psoup.math.wisc.edu/kitchen.html), Stuart Kauffman (http://www.santafe.edu.edu/People/kauffman), Swarm (http://www.santafe.edu/projects/swarm), and *Tierra* (http://www.hip.atr.co.jp/~ray/tierra/tierra.html). As in the case of complexity theory, artificial life is bound to become more important as we move from MEMS to NEMS.

Artificial Neural Network Software

As we have seen, neurons, to a first-order approximation, work by gathering signals from other neurons, weighing each by varying algebraic amounts, and then internally summing all the inputs. If this cumulative signal exceeds some internally stored threshold value, the neuron fires its own signal that is subsequently gathered in by other downstream neurons. John Hopfield, in 1982, invented a simple mathematical model of a network of nerve cells, called a *neural net*. As a quantum physicist Hopfield asked himself if a neuron in a brain might not act like an atom in a lattice. He reasoned that like an atom can take on discrete states, a neuron might also be found in discrete states: it can fire or not fire. His idea boiled down to the simple idea that just as atoms in a crystal arrange themselves to minimize the energy, so must a neural network minimize its "energy." This led to his celebrated 1982 paper, "Neural Networks and Physical Systems with Emergent Collective Computational Abilities." He showed that if many neural units were hooked together, they acquired computational abilities. The artificial neuron is a mathematical construct that emulates the more salient functions of biological neurons, i.e., the signal integration and threshold firing behavior. The most basic neural software consists of a layer of input and output *nodes*. When the input nodes receive a strong enough signal they "fire," sending their own signals to the output nodes attached to them. The input nodes in such a simple system represent an array of input stimuli, and a response is picked up from the output nodes. To make the neural network produce the right answer, an outside controller, usually a human or a computer programmed with the correct database, must adjust the switches between the input and output till the output produces the right answer. In neural network jargon, these settings are called *weights*. In the simplest case, we make the weights either one or zero; in general, though we set the weights to any of a range of signal pass-through strengths. Increasing the weight between an input and output node means that the input node signal will make a larger contribution to the likelihood that the connected output node will fire. This corresponds to an excitatory signal in a nerve cell. A decrease in weight corresponds to an inhibitory signal and can stop a branch in the net from cascading throughout the net or shut off a nerve cell in the biological equivalent. The intelligence within the neural network is stored within these sundry algebraic connection weights. The process of teaching the network the appropriate weights is a complicated and often lengthy process. Frank Rosenblatt designed the first neural network—a so-called *perceptron*—that could be trained by consecutive small adaptations of the weighing factors.[59] This was achieved by comparing the results of consecutive network attempts at the right answer with the desired results and making small changes to the weighing factors after each comparison; more dramatic adjustments would have wiped out all previous training and give the wrong answers for any other pattern recognition.

Using a Hopfield-style neural net, Sahley and Gelperin worked out a model that faithfully captured the learning a common garden slug (*Limax maximus*) displayed when confronted with smells such as that of quinine (which it dislikes) and carrots (which it likes).[70] Their model neural net is shown in Figure 9.38.

Neural nets have been used in sensor array applications such as artificial noses and tongues.[72] In these instruments, a small number of odor and taste sensors are trained to smell and taste food and drinks containing hundreds of different chemicals. Making selective sensors for each separate chemical would be impossible, but a limited sensor array develops

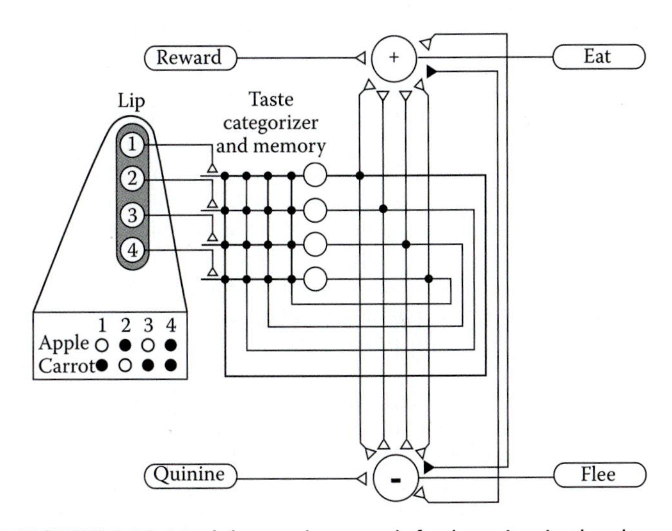

FIGURE 9.38 Model neural network for learning in the slug.

specific patterns (fingerprints) for complex chemical mixtures that correlate well with the human sensory evaluation. This approach has taken a rap for being a "black box"—it gives the right results but does not explain why. Indeed, neural net software packages develop connection traces internally that embody the rules behind the conceptual space on which they are training. For many chemical and biological applications, however, it is not important to know the identity of the individual components of a complex mixture, but it is important to sound an alarm if an unusual pattern is detected; for this application, natural AI seems the appropriate way to embed intelligence.

Acknowledgments

With special thanks to Ramses Madou and Drs. Rabih Zaouk and Benjamin Park.

Questions

9.1: What is meant by bottom-up and top-down artificial intelligence approaches? Which approach makes most sense for the development of smart MEMS?

9.2: Compare the function of a set of neurons in the brain with an artificial neural net. Find some real-life examples where neural nets are used today.

9.3: Explain how μ-3D batteries are better than thin-film batteries for powering a MEMS device.

9.4: Sketch a rough Ragone plot for capacitors, supercapacitors, batteries and fuel cells.

9.5: What MEMS power would you put into a microrobot? A microrocket? A microsubmarine? A microbutterfly?

9.6: Discuss some options for hydrogen storage. Why is it so important for the future of fuel cells?

9.7: As far as fuel cells are concerned what is the merit of biofuel cells?

9.8: Explain for what uses you might prefer a supercapacitor over a Li-ion battery.

9.9: What is the role of a bipolar plate in a fuel cell? How could C-MEMS make for better bipolar plates?

9.10: What are some of the promises and challenges involved with building micromachined combustion engines?

9.11: In what ways can MEMS and NEMS improve solar cells?

9.12: Explain with a diagram how a Grätzel solar cell works.

9.13: What are the fundamental limits in computing power? Use equations.

9.14: Explain how a fractal design of the electrodes might benefit a fuel cell.

9.15: Explain the scaling behavior of a cube and a sheet in terms of resistance and surface to volume ratio.

References

1. Malone, M. S. 2004. Moore's second law. *Wired*, April. http://www.wired.com/wired/archive/12.04/start.html?pg=2.
2. Chesnokov, S. A., V. A. Nalimova, A. G. Rinzler, R. E. Smalley, and J. E. Fischer. 1999. Mechanical energy storage in carbon nanotube springs. *Phys Rev Lett* 82:343–46.
3. Nalimova, V. A., D. E. Sklovsky, G. N. Bondarenko, H. A. Gaucher, S. Bonnamy, and F. Beguin. 1997. Lithium interaction with carbon nanotubes. *Syn Met* 88:89–93.
4. Gaucher, H., E. Frackowiak, S. Gautier, S. Bonnamy, and F. Beguin. 1999. Electrochemical storage of lithium in multiwalled carbon nanotubes. *Carbon* 37:61–69.
5. Bates, J. B., G. R. Gruzalski, and C. F. Luck. 1993. Rechargeable solid-state lithium microbatteries. *Proceedings: IEEE Micro Electro Mechanical Systems (MEMS '93)*. Fort Lauderdale, FL, February 1993, pp. 82–86.
6. Goemans, P. A. F. M. 1994. Microsystems and energy: the role of energy. *Microsystem technology: Exploring opportunities*, ed. G. K. Lebbink. Alphen aan de Rijn/Zaventem: Samsom BedrijfsInformatie bv, pp. 50–64.
7. Long, J. W., B. Dunn, D. R. Rolison, and H. S. White. 2004. Three-dimensional battery architectures. *Chem Rev* 104:4463.
8. LaFollette, R. M., J. N. Harb, and P. Humble. 2001. Microfabricated secondary batteries for remote, autonomous electrical devices. *The Sixteenth Annual Battery Conference on Applications and Advances, 2001*. Long Beach, CA, January 9–12, 2001, pp. 349–54.
9. Koeneman, P. B., I. J. Busch-Vishniac, and K. L. Wood. 1997. Feasibility of micro power supplies for MEMS. *J Microelectromech Syst* 6:355–62.
10. Hart, R. W., H. S. White, B. Dunn, and D. R. Rolison. 2003. 3D microbatteries. *Electrochem Commun* 5:120–23.
11. Thomas, P., Y. Y. Fei, and X. D. Zhu. 2007. Adsorption and desorption of hydrogen on bare and Xe-covered Cu(111). *Appl Phys A* 86:115–21.
12. Sandronk, G. 1999. A panoramic overview of hydrogen storage alloys from a gas reaction point of view. *J Alloys Compounds* 293:887–88.
13. Schuth, F., B. Bogdanovic, and M. Felderhoff. 2004. Light metal hydrides and complex hydrides for hydrogen storage. *Chem Commun* 20:2249–58.
14. Orimo, S., and Y. Nakamori. 2004. Li-N based hydrogen storage materials. *Mater Sci Eng B* 108:48–50.
15. Dillon, A. C., K. M. Jones, T. A. Bekkedahl, and C. H. Kiang. 1997. Storage of hydrogen in single-walled carbon nanotubes. *Nature* 386:377–79.
16. Langmi, H. W., and G. S. McGrady. 2008. Ternary nitrides for hydrogen storage: Li–B–N, Li–Al–N and Li–Ga–N systems. *J Alloys Compounds* 466:287–92.
17. Park, S., D. Srivatava, and K. Cho. 2003. Generalized chemical reactivity of curved surfaces: carbon nanotubes. *Nano Lett.* 3:1273–77.
18. Li, J., T. Furuta, H. Goto, T. Ohashi, Y. Fujiwara, and S. Yip. 2003. Theoretical evaluation of hydrogen storage capacity in pure carbon nanostructures. *J Chem Phys* 119:2376–85.
19. Fernandez-Alonso, F., F. J. Bermejo, C. Cabrillo, R. O. Loutfy, V. Leon, and M. L. Saboungi. 2007. Nature of the bound states of molecular hydrogen in carbon nanohorns. *Phys Rev Lett* 98:215503.
20. Iijima, S. 1999. Nano-aggregates of single-walled graphitic carbon nano-horns. *Chem Phys Lett* 309:13.
21. Sakai, H., Y. Tokita, and T. Hatazawa. 2007. A high-power glucose/oxygen biofuel cell. *Fuel Cell Technology: Biofuel Cells, Enzymatic and Microbial, Division of Fuel Chemistry, 234th ACS National Meeting*. Boston, MA, August 19–23, 2007, p. 196.
22. Admin. 2007. http://www.ecoustics.com.
23. Meyers, J. P., and H. L. Maynard. 2002. Design considerations for miniaturized PEM fuel cells. *J Power Sources* 109:76–88.
24. Park, B., and M. J. Madou. 2006. Design, fabrication, and initial testing of a miniature PEM fuel cell with micro-scale pyrolyzed carbon fluidic plates. *J Power Sources* 162:369–79.
25. Sato, H., C. W. Berry, B. E. Casey, G. Lavella, Y. Yao, J. M. VandenBrooks, and M. M. Maharbiz. 2008. A cyborg beetle: insect flight control through an implantable, tetherless microsystem. *IEEE 21st International Conference on Micro Electro Mechanical Systems 2008*. Tucson, AZ, January 13–17, 2008, pp. 164–67.
26. Varia, A. 2003. Thermal model of MEMS thruster. *FEMCI Workshop 2003*. NASA Goddard Space Flight Center, May 7–8, 2003.
27. London, A. P. 2000. *Development and test of a microfabricated birpropellant rocket engine*, PhD diss. Cambridge, MA: Massachusetts Institute of Technology.
28. Protz, C. S. 2004. *Experimental investigation of microfabricated bipropellant rocket engines*, PhD diss. Cambridge, MA: Massachusetts Institute of Technology.
29. Sniegowski, J., and E. Garcia. 1995. Microfabricated actuators and their application to optics. *Micro-Optics/Micromechanics and Laser Scanning and Shaping (Proceedings of the SPIE)*. San Jose, CA, February 7–9, 1995, pp. 46–64.
30. Garcia, E., and J. Sniegowski. 1995. Surface micromachined microengine. *Sensors Actuators A* A48:203–14.
31. Lee, J. B., Z. Chen, M. G. Allen, A. Rohatgi, and R. Arya. 1995. A miniaturized high-voltage solar cell array as an electrostatic MEMS power supply. *J Microelectromech Syst* 4:102–08.
32. Salalibara, T., H. Izu, T. Kura, W. Shinohara, H. Iwata, S. Kiyama, and S. Tsuda. 1995. High-voltage photovoltaic micro-devices fabricated by a new laser-processing. *Proceedings: IEEE Micro Electro Mechanical Systems (MEMS '95)*. Amsterdam, the Netherlands, January 29–February 2, 1995, pp. 282–87.

33. Kimura, M., N. Miyakoshi, and M. Daibou. 1991. A miniature opto-electric transformer. *Proceedings: IEEE Micro Electro Mechanical Systems (MEMS '91)*. Nara, Japan, January 30–February 2, 1991, pp. 227–32.
34. 2008. School of Photovoltaic and Renewable Energy Engineering, University of New South Wales.
35. Mao, F., N. Mano, and A. Heller. 2003. Long tethers binding redox centers to polymer backbones enhance electron transport in enzyme "wiring" hydrogels. *J Am Chem Soc* 125:4951–57.
36. Madou, M. J., and S. R. Morrison. 1989. *Chemical sensing with solid state devices*. New York: Academic Press.
37. Gratzel, M. 2001. Photoelectrochemical cells. *Nature* 414:15.
38. Gratzel, M. 2005. Solar energy conversion by dye-sensitized photovoltaic cells. *Inorg Chem* 44:6841–51.
39. O'Regan, B., and M. Gratzel. 1991. A low-cost, high-efficiency solar cell based on dye-sensitized colloidal TiO_2 films. *Nature* 353:737–40.
40. Greenham, N. C., X. Peng, and A. P. Alivisatos. 1996. Charge separation and transport in conjugated-polymer/semiconductor-nanocrystal composites studied by photoluminescence quenching and photoconductivity. *Phys Rev B* 54:17628–37.
41. Huynh, W., J. J. Dittmer, and A. P. Alivisatos. 2002. Hybrid nanorod-polymer solar cells. *Science* 295:2425–27.
42. Nozik, A. J. 2002. Quantum dot solar cells. *Pysica E* 14:115–20.
43. Pietryga, J. M., R. D. Schaller, D. Werder, M. H. Stewart, V. I. Klimov, and J. A. Hollinasworth. 2004. Pushing the band gap envelope: Mid-infrared emitting colloidal PbSe quantum dots. *J Am Chem Soc* 126:11752–53.
44. Branham, K., J. L. Marques, J. Hassard, and P. O'Brien. 2000. Quantum-dot concentrator and thermodynamic model for the global redshift. *Appl Phys Lett* 76:1197.
45. Stark, I., and M. Strodeur. 1999. New micro thermoelectric devices based on bismuth telluride-type thin solid films. *Eighteenth International Conference on Thermoelectrics*. Baltimore, MD, Aug. 29–Sept. 2, 1999, pp. 465–72.
46. Shenck, N. S., and J. A. Paradiso. 2001. Energy scavenging with shoe-mounted piezoelectrics. *IEEE Micro* 21:30–42.
47. Ottman, G. K., H. F. Hofmann, A. C. Bhatt, and G. A. Lesieutre. 2002. Adaptive piezoelectric energy harvesting circuit for wireless remote power supply. *IEEE Trans Power Electron* 17:669–76.
48. Roundy, S., B. P. Otis, Y. H. Chee, J. M. Rabaey, and P. Wright. A 1.9-GHz transmit beacon using environmentally scavenged energy. *ISPLED 2003*. Seoul Korea, August 25–27, 2003.
49. Meninger, S., T. O. Mur-Miranda, R. Amirtharajah, A. Chandrakasan, and J. Lang. 1999. Vibration-to-electric energy conversion. *1999 International Symposium on Low Power Electronics and Design*. New York: ACM Press, pp. 48–53.
50. Wang, L., and F. G. Yuan. 2007. Energy harvesting by magnetostrictive material (MsM) for powering wireless sensors in SHM. *SPIE Smart Structures and Materials and NDE and Health Monitoring, 14th International Symposium (SSN07)*, March 18–22, 2007.
51. Mandelbrot, B. B., and J. A. Wheeler. 1983. The fractal geometry of nature. *Am J Phys* 51:286–87.
52. West, G. B., J. H. Brown, and B. J. Enquist. 1997. A general model for the origin of allometric scaling laws in biology. *Science* 276:122–26.
53. Park, B., R. Zaouk, W. Chunlei, and M. J. Madou. 2007. A case for fractal electrodes in electrochemical applications. *J Electrochem Soc* 154:1–5.
54. Park, B. Y., R. Zaouk, C. Wang, and M. J. Madou. 2007. A case for fractal electrodes in electrochemical applications. *J Electrochem Soc* 154:1–5.
55. Wang, C., G. Jia, L. H. Taherabadi, and M. J. Madou. 2005. A novel method for the fabrication of high-aspect ratio C-MEMS structures. *J Microelectromechan Syst* 14: 348–58.
56. Wang, C., L. Taherabadi, G. Jia, M. Madou, Y. Yeh, and B. Dunn. 2004. C-MEMS for the manufacture of 3D microbatteries. *Electrochem Solid-State Lett* 7:A435–38.
57. Wang, C., and M. Madou. 2005. From MEMS to NEMs with carbon. *Biosens Bioelectron* 20:2181–87.
58. Wolf, S. DARPA—DSO. http://www.darpa.mil/dso/thrust/math/quist.htm.
59. Freeman, D. H. 1994. *Brainmakers*. New York: Touchstone.
60. Minsky, M. 1994. Will robots inherit the earth? *Scientific American*, 271 (October):108–13.
61. Kurzweil, R. 1999. *The age of the spiritual machine*. New York: Penguin.
62. Bashir, R. 2001. DNA-mediated artificial nano-biostructures: State of the art, future directions. *Superlattice Microstruct* 29:1–16.
63. Adleman, L. M. 1994. Molecular computation of solutions to combinatorial problems. *Science* 266:1021–24.
64. Minsky, M. 1987. *The society of mind*. New York: Simon and Schuster.
65. Kaku, M. 1997. *Visions: How science will revolutionize the 21st century*. New York: Oxford University Press.
66. Levy, S. 1992. *Artificial life*. New York: Vintage Books.
67. Prigogine, I., and I. Stengers. 1984. *Order out of chaos*. New York: Bantam.
68. Capra, F. 1996. *The web of life*. New York: Anchor Books.
69. Lovelock, J., and L. Margulis. 1974. Biological modulation of the earth's atmosphere. *Icarus* 21:471–89.
70. Stewart, I. 1988. *Life's other secret*. New York: John Wiley & Sons.
71. Morris, R. 1999. *Artificial worlds*. New York: Plenum Trade.
72. Toko, K. 2000. *Biomimetic sensor technology*. Cambridge, United Kingdom: Cambridge University Press.

Part III

Miniaturization Application

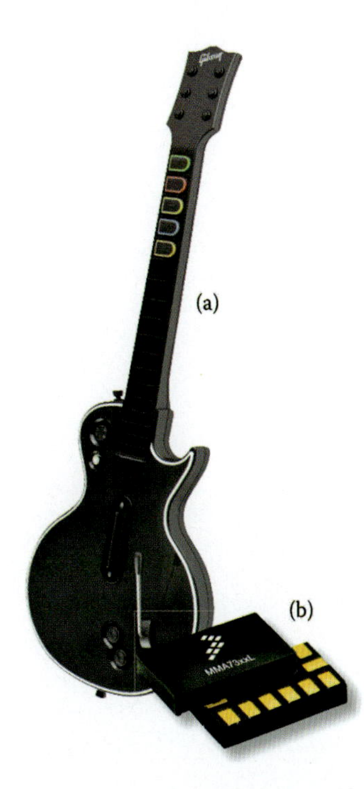

(a) The controller that is included as part of the popular *Guitar Hero* video game relies on (b) a three-axis motion-sensing accelerometer from STMicroelectronics. (From http://archive.electronicdesign.com/files/29/19285/fig_01.gif. With permission.)

Introduction to Part III

Chapter 10 MEMS and NEMS Applications

Introduction to Part III

In Chapter 10 on MEMS and NEMS applications we cover the market dynamics for MEMS and NEMS commercial products. We cover both lithography- and nonlithography-based MEMS. After an analysis of the total world market in MEMS, we consider some selected market categories, i.e., MEMS in the automotive market; MEMS in medical and biotechnology markets (bio-MEMS); MEMS in industrial automation, including environmental monitoring markets; and MEMS in information technology peripherals and telecommunications markets.

The market expectations for NEMS today are much greater than they ever were for MEMS. An important objective in the section on nanotechnology markets is to try to help the reader distinguish between hype and science/facts, perhaps helping to avoid yet another lurking technology bubble.

10

MEMS and NEMS Applications

Outline

Picture of packaged digital micromirror devices (DMD™) from Texas Instruments (TI). These are pixel devices; shown are the VGA (640 × 480), the SVGA (800 × 600), and the XGA (1024 × 768). Digital mirror devices constitute the largest microelectromechanical system (MEMS) product after read-write heads and ink-jet heads.

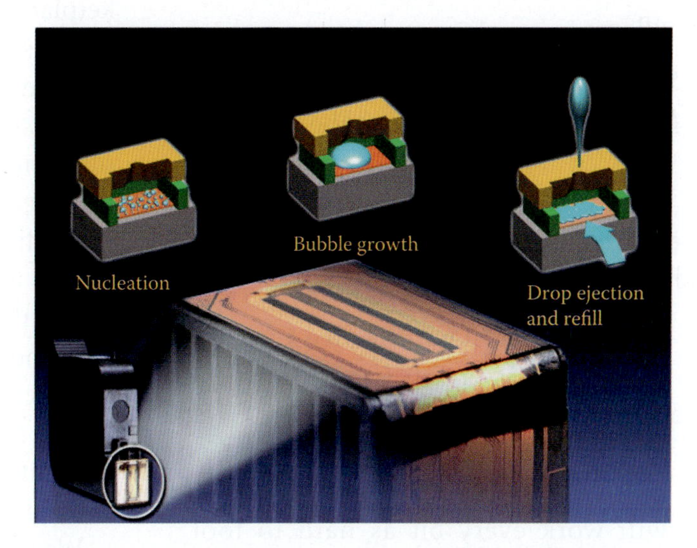

Another major microelectromechanical system (MEMS) application: an ink-jet printer head. These represent the second largest MEMS market, after read-write heads.

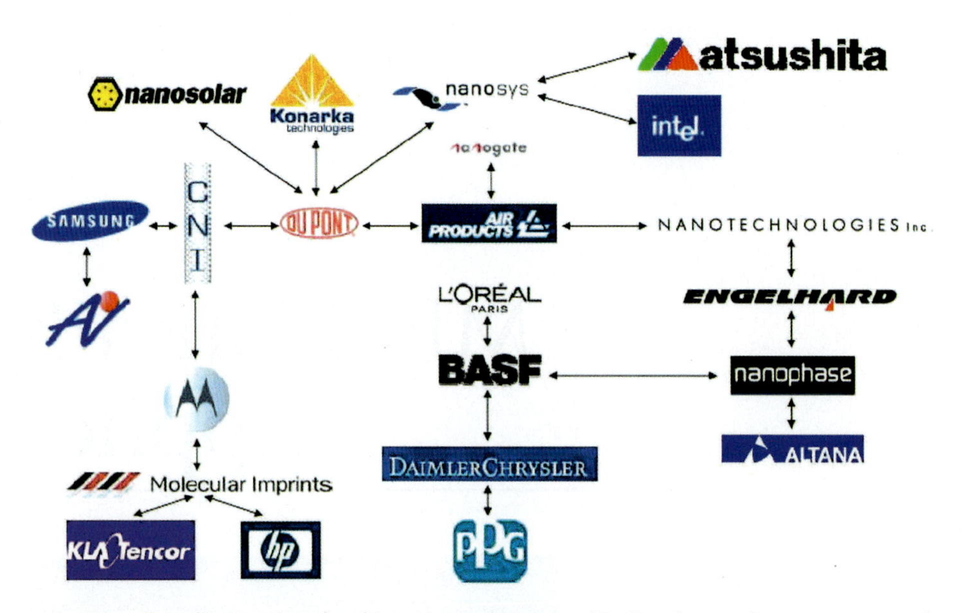

A nanonet of companies, small and large, involved in nanotechnology. Notice that only start-ups use "nano" in their names.

Prediction is extremely difficult. Especially about the future.

Niels Bohr

Cars will cost as little as $200. People will have two-month vacations. They will care little for possessions. The happiest people live in one-factory villages.

Predictions for 1960 by General Motors in a "Futurama"exhibit at the 1939–1940 New York World's Fair

To the electron—may it never be of any use to anybody.

Favorite toast of hardheaded Cavendish scientists in the early 1900s
Michael Riordan and Lillian Hoddeson, Crystal Fire (W. W. Norton, 1998)

People will work every bit as hard to fool themselves as they will to fool others—which makes it very difficult to tell just where the line between foolishness and fraud is located.

Robert Park
Voodoo Science (Oxford University Press, 2000)

Introduction

The field of microelectromechanical systems (MEMS) and nanoelectromechanical systems (NEMS) is currently undergoing a total makeover. Mechanical and optical MEMS (MOEMS) products have made the transition out of the laboratory and into the marketplace with a plethora of new products, including a strong entry into the consumer electronics field (mobile phones, cameras, laptop computers, games, etc.). Biological MEMS (bio-MEMS), on the other hand, is in the doldrums, with little commercial acceptance, while NEMS, especially BIONEMS, is stealing all the research dollars. In commercial terms, if MEMS is entering adolescence, the word *nano* still seems to derive from a verb that means "to seek and get venture capital funding" instead of from the Greek noun for dwarf. Today the market dynamics for MEMS and NEMS commercial products are very different, and we treat MEMS and NEMS applications separately.

We start with a preamble about today's MEMS market character and the current and future position of the United States in MEMS and NEMS in general. We are interested in both lithography-based and non-lithography-based MEMS. Compared with lithography-based MEMS, thenon-lithography MEMS approach is underreported, and it is especially in the latter area that progress in the United States is lacking.

To understand where revenues in MEMS are derived from, and to appreciate where further growth can be expected from, we provide definitions of MEMS, bio-MEMS, microsystems technology (MST), micrototal analysis system (μ-TAS), lab-on-a-chip (LOC), micromechatronics, miniature sensors and actuators, microstructures or microcomponents, and first-level packaged MEMS/MST and the smallest commercialized unit. These terms are familiar to micromachinists but are sometimes confused in marketing reports. An evaluation of MEMS applications better describing the MEMS contribution to the final product and the manufacturing method used to produce the MEMS [e.g., non-lithography-based, such as computer numerical control (CNC) machined, or fabricated using a lithography-based technique, such as LIGA] might be of help in interpreting market studies that are often vague about what the term *microfabricated products* comprises. Companies and research organizations interested in entering the field might then at least be able to judge, based on the underlying study, whether they have the manufacturing expertise or know-how to make them possible contenders.

After an analysis of the total world market in MEMS, we consider some selected market categories, i.e., MEMS in the automotive market; MEMS in medical and biotechnology markets (bio-MEMS); MEMS in industrial automation, including environmental monitoring markets; and MEMS in information technology (IT) peripherals and telecommunications markets.

The market expectations for nanotechnology today are much greater than they ever were for MEMS. An important objective in the section on nanotechnology markets is to try to help the reader distinguish between hype and factual science, perhaps helping to avoid yet another lurking technology bubble.

As highlighted in Chapter 1, which compares traditional manufacturing and MEMS and NEMS, we hope that a better understanding of how to match different manufacturing options with a given application will guide the identification of additional killer MEMS and NEMS applications and encourage more companies and research organizations to innovate faster based on their in-house manufacturing tools and know-how.

Preamble on the MEMS Market

Lithography-Based MEMS

Lithography-based MEMS defines all micromanufacturing involving a lithography step. It was not too long ago that lithography-based MEMS was surrounded with a similar hype that surrounds NEMS today. It was said that MEMS, by using highly automated, parallel-manufacturing techniques, would enjoy the same economy of scale as integrated circuits (ICs) and ride similar cost curves. Using "micromachining" technology (much of it borrowed from the IC industry), pumps, sensors, relays, motors—just about any manufactured gizmo in our daily world—would be shrunk down to the micrometer scale, so that it might be hidden, worn, or implanted and in some cases be inexpensive enough for a disposable application. In those early years of MEMS (the early 1980s), when MEMS was almost 100% silicon-based, it was even projected that long before 2008, its market would be larger than that for ICs. This notion was based on the expectation of many more applications for MEMS and combinations of MEMS and ICs than for ICs alone. This market prediction has not been fulfilled yet: the 2006 worldwide IC market was $211.0 billion (see also Volume I, Appendix 1B); in contrast, the worldwide MEMS market for the same year was estimated between $6.3 and $16.5 billion (depending on which criteria one uses to define MEMS). In other words, after more than twenty years, one could claim a MEMS market of at most 8% of the total IC market, taking into account that this was achieved only by changing the criteria about what qualified as MEMS [e.g., including nonsilicon devices, such as thin-film read-write (RW) heads]. The MEMS market in 2007 consisted principally of RW heads, ink-jet printer heads, digital light processing chips, pressure sensors, accelerometers, flow sensors, gyroscopes, microphones, oscillators, and radiofrequency (RF) switches, which, except for digital light processing (DLP) chips, all are becoming commodity products very swiftly and are now more and more manufactured outside the United States (profit margins have become very thin on most of these products). Early market studies did overstate dramatically what impact Si-based MEMS had on the world economy and how fast that would happen. A realistic number for the market of Si-MEMS products today is less than 3% of the IC market (even the $6.3 billion MEMS sales in 2006, i.e., 3% of the $211 billion IC market for 2006, included nonsilicon devices). The main reason that MEMS sales continue to lag behind IC sales has been the relatively stable price-to-performance (p/p) ratio of sensors and actuators ever since 1960, which is in marked contrast to the p/p ratio of ICs, which has fallen enormously between 1960 and 2007 and is now significantly below that for sensors and actuators. As a consequence of these changes, the cost of a measurement system is, in general, dominated first by the cost of the microactuator and second by the cost of the microsensor.

When including nonsilicon MEMS (which many market studies now do), and especially when also counting non-lithography-based MEMS (e.g., precision-engineered products; see below) (which very few market studies do) MEMS might constitute 10% of all current IC sales.

If some investors still cringe when they hear MEMS mentioned, they might be remembering the 2001 technology bubble around "optical MEMS switches" for telecommunications. Optical channel cross-connecting is a key function in most communication systems. In a *hybrid approach,* one converts optical data streams into electronic data using electronic cross-connection technology. Then, the electronic data streams are converted back into optical signals. In electronic cross-connecting of *N*-by-*N* channels ($N \times N$), *N* may be on the order of thousands. Using arrays of MEMS micromirrors, one can obviate the need for back-and-forth conversion of light to electrical signals. Cross-connecting optical channels directly in the photonic domain is known as *all-optical switching.* These all-optical path switches could, in theory, operate at any wavelength, at any speed, and using any kind of protocol. In 2007, the size of typical commercially available all-optical channel switches may switch between 80 or 128 channels (N = 80 or 128) (see Figure 10.1). But in the late 1990s, all-optical channel cross-connecting switches with N = 1000 were already promised by an army of MOEMS start-up companies (150 worldwide). Unfortunately, the more than $10 billion investment by large telecommunication companies such as Corning, JDF Uniphase, and Nortel in 1999–2000 for those very large arrays of micromirror devices became a textbook example of what former Federal Reserve chairman Alan Greenspan called "irrational exuberance." The all-optical switch mania started in 2000 when Nortel bought up the start-up Xros Inc. for $3.25 billion. By the time Nortel killed off the company in 2002, the market had flooded with competitors, most of them with similar MEMS-based ideas for all-optical switches, and suddenly the 1000-port mania was relegated to the oops file. There were not enough customers for that many start-ups, and

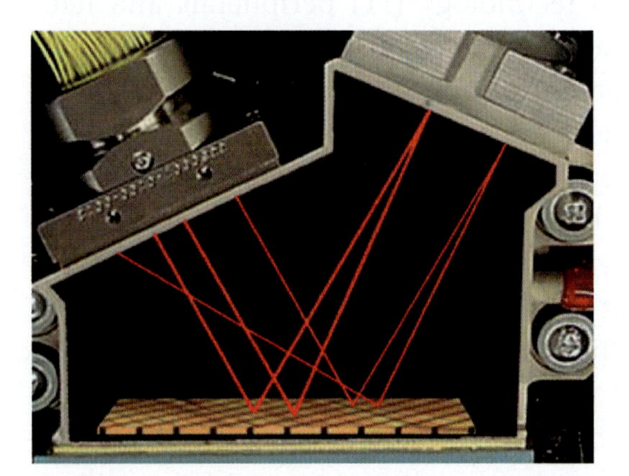

FIGURE 10.1 Glimmerglass Intelligent Optical Switch offers optical to optical switching (http://www.glimmerglass.com/technology.aspx). This switch from Glimmerglass couples light from an input fiber array of up to eighty or more inputs into an output array of the same scale. A microelectromechanical system mirror array directs light from any one input to any of the output fibers. (Images from Glimmerglass.)

the three largest investing companies pulled back, leaving a few individuals very rich, but with neither new jobs nor products to show for their time and with many scientists and engineers out of a job (or working as consultants—often synonymous with jobless in the United States). The rapid growth of the telecommunications industry initially brought great demand for MEMS optical switches. Unfortunately, the expected demand did not meet the actual demand. What was left of the all-optical switching market, in the period between 2002 and 2004, was littered with the broken shells of start-ups that often tried to survive by doing MEMS foundry work. Many of them had never got past the design stage, a few shipped products, and others had nothing to show for but PowerPoint presentations. Surprisingly, even some of the latter managed to cheat investors and large companies out of money as MBAs and scientists alike oversold the technology. The experience embittered many investors, and no wonder.

Today, some surviving MOEMS companies, building all-optical switches, are successfully targeting more modest switch applications; say for apartment buildings or university switchboards and for new emerging applications, such as semiconductor test equipment. Companies such as Glimmerglass claim to have solved the problems previously associated with the MOEMS switches and maintain that their optical switches are more reliable, less complicated, and more cost-effective. The Glimmerglass switch, shown in Figure 10.1, is an 80 × 80 (input × output fibers) switch with promising sales figures. A big challenge for this new generation of optical-switch vendors is overcoming the negative perceptions left over from the boom years.

Instead of switching signals among very large arrays of fiber optic cables for telecommunication switches, Texas Instruments (TI) targeted its digital light processing (DLP) technology to projectors and video displays. Because it focused on the right mass consumer application from the get-go, TI's digital micromirror device (DMD™) turned out to be a big-scale MEMS success (see Example 10.1 below).

A similar case of "irrational exuberance," although on a much smaller scale, played out with the early investments in microfluidics for high-throughput analysis for new drug candidates [called *high-throughput screening* (HTS)] and for high-throughput analysis in diagnostics (Figure 10.2). Early investments for these application came mostly from cash-rich drug companies and huge diagnostic firms, to help them with faster discovery of new drugs and faster diagnostics, respectively, but there was no market beyond some research equipment, and again, too many competitors chased too few customers.

Microfluidic applications in point-of-care (POC) diagnostics—a mass-market application—appear more promising. In diagnostics, one often deals

High Throughput Analysis

Diagnostics

>800 blood tests per hour

Drug discovery

>1000 microtiter plates per day

FIGURE 10.2 High-throughput analysis in diagnostics and high-throughput screening in drug discovery.

with disposable products of considerable size compared with a typical IC, and one must compete with traditional products made in huge batches on paper or plastic substrates or even in roll-to-roll (R2R) continuous processes (say, glucose sensor paper strips and aspirin pills). This leaves little room for batch-type Si technology, and as a consequence, most new biotechnology-related MEMS products are non-silicon-based. Although it failed for HTS applications, nonsilicon microfluidics is poised to succeed in diagnostic applications. This parallels the case of micromirrors that failed for very large communication switches (expensive product/small market) but succeeded for portable projectors (less expensive product/huge market; see next section).

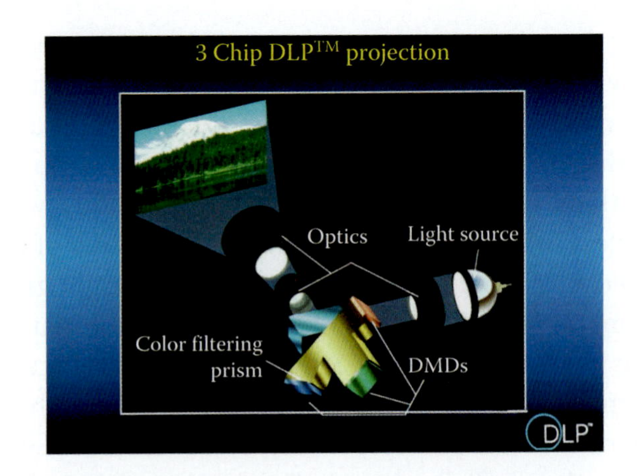

FIGURE 10.3 Texas Instruments DMDs. Digital light processing replaces three conventional LCDs with one to three DMDs, one each for the red, green, and blue components of a video image. Each chip contains approximately 750,000 or more microscopic mirrors or pixels. A single light source (bulb) drives all DMDs, the monochromatic images are converged internally using a prism system, and the resulting full-color image is projected via a lens onto a screen. DLP projectors are lightweight, relatively bright, and simple to set up and operate, and they provide a much better contrast range and higher resolution than LCD projectors. The DLP technology is more expensive at present than LCD.

Example 10.1: Texas Instruments Digital Micromirror Devices

An example of a successful MOEMS application comes from the optical switching arena with the DLP chip from TI. DMDs, invented in 1987 by TI's Larry Hornbeck, consist of an array of hundreds of thousands (from 480,000 to >2,000,000) of electrically actuated micromirrors built as a superstructure over complementary metal-oxide-silicon (CMOS) address circuitry. Two pixels are schematically illustrated in Volume II, Figure 7.98, together with the underlying Si chip and circuitry. Each pixel is made up of a reflective aluminum micromirror supported from a central post. The manufacture of this chip illustrates the integration of a nonsilicon MEMS with a CMOS SRAM 8-inch wafer process line. The surface micromachining process to fabricate DMDs on wafers incorporating SRAM CMOS electronic address and control circuitry is illustrated in Volume II, Figure 7.100. The aluminum alloy DMDs are used in consumer products such as projectors and digital television (Figure 10.3). For a picture of DMD chips, see the illustration in the heading of this chapter.

Millions of those DMDs have now been built into projectors (the first one came out in 1996 from InFocus) and big-screen televisions (the first came out in 2002 from Samsung) (Figure 10.3). It was projected that by 2009, MEMS displays would become the second largest MEMS market, after thin-film read-write (RW) heads (Nexus). A major advantage for TI is that DLP chips are an exception to the high-volume/low-price rule, typically for all ICs and most MEMS devices: the 2004 asking price for a DLP chip was $400; it was estimated at $190 in 2009 (Nexus).[1] There is little question that DMD is the most complex MEMS in the world, and some at TI are bragging that TI has more moving parts on one wafer than Analog Devices Inc. (ADI) builds accelerometers in a year (ADI ships 2 million units a week).

Mechanical and MOEMS devices have been around since the 1980s, and it appears that this market finally has achieved mass-market acceptance in IT peripherals. A primary reason is the demand for read-write heads, ink-jet heads, and microdisplays. Since about 2004–2005, these MEMS devices have been incorporated in consumer products. Read-write heads used to be mostly used in personal computers (PCs) (a market that has slowed down) but can now be found in mp3 players, smart phones, digital video cameras, set-top players, and DVD recorders. Ink-jet heads are migrating from regular printers to printers of digital photos—especially photos from cell phone cameras—and microdisplays found their way from projectors and high-definition, large-screen televisions to microdisplays for mobile handsets [e.g., the iMoD (interferometric modulator display), developed at Qualcomm] (see also "MEMS in IT Peripherals and Telecommunications Markets").

The automotive field, also, continues to be a strong market driver for MEMS, especially for security applications, with the demand for pressure sensors, accelerometers, and gyros. For example, the electronic stability program (ESP) and the tire pressure monitoring systems (TPMS) are currently promoting worldwide yet more use of MEMS in the car industry.

Recently, the market for accelerometers in consumer electronics (CE) (computers, games, phones, etc.) has also started to explode. Consumer electronics have the potential to dwarf the automotive application, not only for accelerometers but also for pressure sensors and gyros. Consequently MEMS growth is expected by some to follow a hockey stick curve, with the value of products rising to $40 billion in 2015 and $200 billion in 2025![2] (see also an estimate by Dr. Marinis from Draper Laboratory: http://www.sem.org/PDF/s01p01.pdf). The problem now is one of profitability rather than the absence of real product opportunities: how to earn money on, say, $0.32 microphones or $1.35 triaxis accelerometers for mobile handsets? Only by becoming more similar to the IC industry can MEMS devices achieve this goal, i.e., through more integration on larger (8-inch) Si wafers. In a sense, as far as mechanical and MOEMS are concerned, the IC industry has now swallowed the MEMS field.

Despite two important "bubble busts" in the MEMS market, we do expect that besides a booming MEMS business in consumer electronics, many new MEMS applications are yet to be realized and that MEMS will be essential in facilitating the handshake between the macro and nano worlds. Furthermore, this author speculates that it is in nanotechnology, especially when considering bottom-up manufacturing (nanochemistry), that a paradigm shift away from IC-type manufacturing will take shape, and that nanotechnology holds the potential for having a much larger impact on society than IC technology ever did. The IC industry will then become subservient to this larger nano industry, just as MEMS is subservient to the IC industry today.

Non-Lithography-Based MEMS

In this book, we also cover non-lithography-based miniaturization techniques and their applications. Compared with lithography-based MEMS, this technology area is underreported, as results often are evolutionary rather than revolutionary. Less widely advertised than lithography-based MEMS advances, which are oversold in public relation hungry universities and government institutes, non-lithography micromachining, practiced mostly in highly competitive private companies of the likes of Sankyo Seiki, Samsung, Philips, Olympus, etc., actually leads to more practical products faster than lithography-based MEMS. Applications include computer hard discs, photocopier drums, mold inserts for compact disc reader heads, high-definition TV projection lenses, and VCR scanning heads. Some examples of non-lithography MEMS products are shown in Figure 10.4. The United States is in a rather difficult position when it comes to non-lithography-based MEMS [such as electrical discharge machining (EDM) and CNC]. Below, we discuss some reasons why the United States is losing out in this type of miniaturization.

Over the last twenty years, the United States' MEMS efforts focused on exploiting silicon planar lithography as the core technology for microstructure fabrication, whereas in Japan and Europe, a much wider variety of miniaturization technologies were pursued, including several non-lithography-based MEMS methods. A prominent example of this, reviewed in Volume II, Chapter 6, is the ten-year Micromachine Technology Project [also called the *Desk Top Factory* (DTF)], initiated in Japan in 1991 and focused on both a wide range of fabrication technologies and a wide range of materials, including extensions to more traditional machining processes (Figure 10.5). More recently, Korea and the European Community have started similar efforts. There are no such concentrated efforts going on in the United States[3] at the moment.

In Japan, where MEMS was driven by mechanical engineering, there was also more focus on the fabrication and assembly of discrete products, whereas MEMS in the United States, being driven by the IC industry, viewed such work as

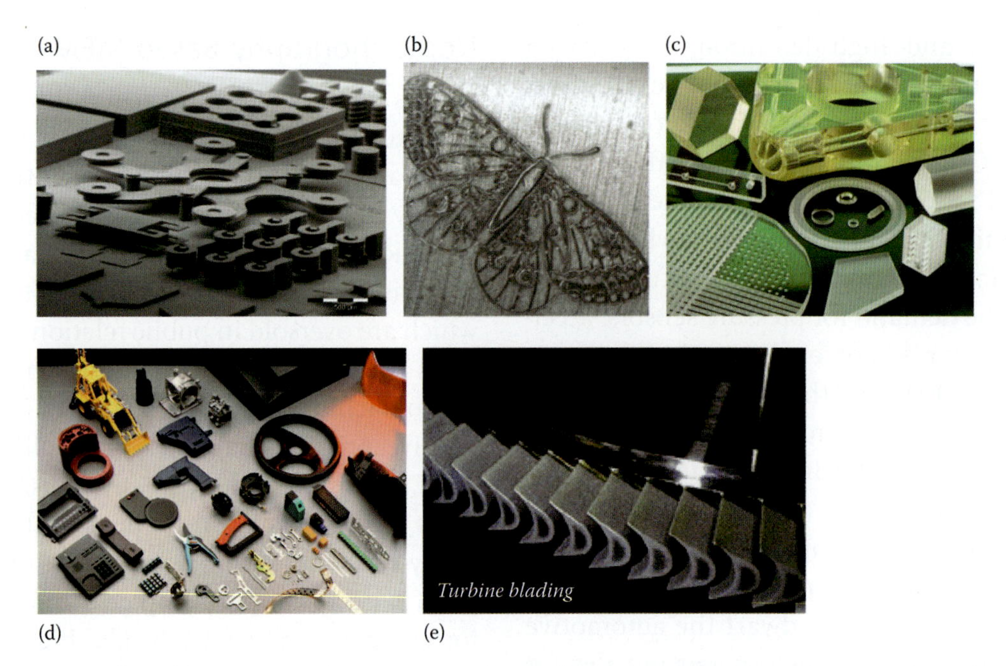

FIGURE 10.4 Examples of nonlithography microelectromechanical system products: products made with precision machining. (a) Electrochemical fabrication (EFAB) is a solid, free-form fabrication technology that can create complex, miniature three-dimensional metal structures that are impossible or impractical to make using other technologies, such as electrical discharge machining (EDM), laser machining, or silicon micromachining. (b) Laser marking and engraving. (c) Precision glass components produced through computer numerical control (CNC) machining, grounding, lapping, and polishing. (d) Electrical discharge-machined products. (e) Turbine blades made with electrochemical machining.

fundamentally violating the paradigm of batch fabrication.[3] Assembly projects are only now gaining vigor in the United States as part of efforts in nanotechnology, embodied, for example, by the Zyvex Nanomanipulator, shown in Figure 10.6 (see also Chapter 4, this volume). The Zyvex Nanomanipulator is able to grab and position objects as small as 10 nm in diameter.

It is difficult to convince U.S. funding agencies to invest in evolutionary-type technology, such as developing better CNC machining tool bits (see Volume II, Chapter 6), but the miniaturization of devices associated with a large number of application fields is today demanding the production of components with manufactured features in the range of a few microns to a few hundred microns, requiring those new tool bits. These fields include optics, electronics, medicine, biotechnology, communications, sensors,

FIGURE 10.5 Example Desk Top Factory (DTF) at AIST (see http://www.dtf.ne.jp/en/achievements.html for other examples).

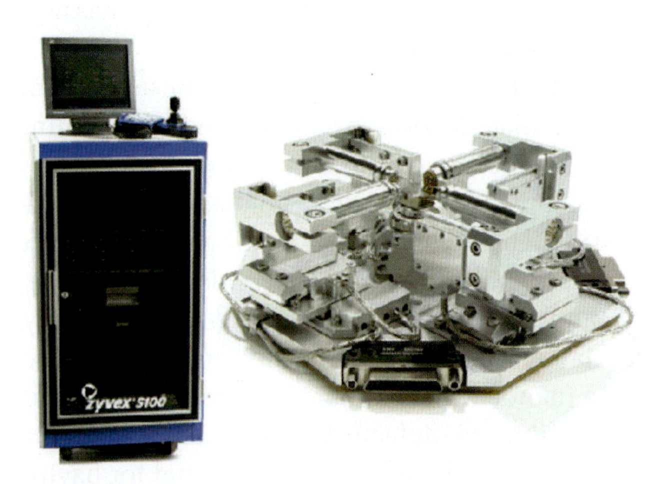

FIGURE 10.6 The S100 Nanomanipulator System from Zyvex is a manipulation and testing tool used with a scanning electron microscope for micro- and nanoscale research, development, and production applications (http://www.zyvex.com).

and avionics, to name just a few. Specific applications are medical implants, diagnostic and remediation devices, microscale batteries and fuel cells, fluidic microchemical reactors requiring microscale pumps, valves and mixing devices, microfluidic systems in general, microholes for fiber optics, micronozzles for high-temperature jets, micromolds and deep x-ray lithography masks, optical lenses, etc. Functional requirements for these applications demand tight tolerances and the use of a wide variety of engineering materials, e.g., steels, titanium, brass, aluminum, platinum, iridium, ceramics, polymers, and composites.

The United States has lost dominance of the tooling industry and the manufacturing skills in many of the precision machining areas needed for the identified applications. The U.S. machine tool industry enjoyed a period of global dominance between the early 1900s and the late 1970s. In the final quarter of the last century, however, almost every major U.S. producer lost domestic market share as a result of import competition. By the early 1990s, several segments of the U.S. machine tool industry appeared poised on the edge of market exit. In a 2007 World Technology Evaluation Center (WTEC) publication, we detailed our observation that countries not manufacturing high-technology goods today are increasingly at a loss, as they do not learn about the newest manufacturing needs from developing the next-generation products.[3]

In the short term, free market forces in a global economy will push mass-product manufacturing of highly accurate miniaturized devices, requiring quality control and multiple assembly steps, toward low-cost, low-wage countries, and the manufacture of small-lot, high-added-value products and items that require intense service assistance will remain in the more developed countries. But the real question is this one: can an advanced country, in the long run, survive without continued input on practical manufacturing, processes, and new materials know-how? The often quoted excuse for exporting manufacturing jobs abroad, usually coming from higher management and MBAs (immune, at least for now, from losing their own jobs), is that the United States and other developed countries will still design, market, and sell those next-generation products. That argument will be rendered wishful thinking, though, when the new materials and processes that go into next-generation products are foreign owned.

MEMS lithography-based and non-lithography-based micromachining will continue to enrich the nations that invest in it most heavily. To stem the hollowing out of the manufacturing base in developed nations, the governments of many countries have made considerable investments in the miniaturization of new products (MEMS and NEMS) and the miniaturization of manufacturing tools (for example, the DTF, mentioned above and detailed in Volume II, Chapter 6). Based on the state-of-the-art and current investment levels (private and government), Europe (especially Germany and Switzerland), and Asia (particularly Japan and Korea) will gain the most from developments in non-lithography-based machining, as they have a long tradition in it and have invested more heavily in this field. As pointed out above, the United States, over the last twenty years, has emphasized lithography-based MEMS with outstanding research results and a dominant market position, but as many MEMS products have become commodity products, Asian countries stand to reap more benefits from them. In more traditional micromanufacturing, the United States is now lagging behind Japan, Germany, Switzerland, and Korea, and as pointed out above, such a lag tends to get worse as time goes on as every new generation of products (from television sets to mobile phones) incorporates some new manufacturing and materials know-how.

If the U.S. government continues to invest in nanotechnology, it will be the United States that will again reap the initial benefits, first in new materials and then in new devices. But to be able to manufacture and gain from those investments in the longer term, it might be wise to start something on the scale of the Manhattan Project in advanced manufacturing techniques; otherwise, the United States will not reap the benefits from its research and development investments itself, as it would no longer be able to implement what it invents.

MEMS Market Studies

As pointed out above, almost all MEMS market projections from before the year 2000 were too optimistic (e.g., $6.3 billion realized vs. a projected more

Table 10.1 From Concept to Full Commercialization

Product	Discovery	Product Development	Cost Reduction	Full Commercialization	Duration (Years)
Pressure sensors	1954–1961	1961–1975 (1961 first prototype at Kulite)	1975–1984	1984	23
Ink-jet printer nozzles	1977 (IBM)	1984–1990	1990–1996	1996	19
Accelerometers	1970 (Kulite piezoresistive)–1979 (Stanford-capacitive)	1985–1990	1990–1995	1995	25 16
Microrelays	1977–1982	1982–1998	1998–2005	2005	28
Photonics/displays	1979 IBM	1986–1998	1998–2005	2001	22
Bio-/chemical sensors	1980–1994	1994–2000	2000–2005	2005	25
Gyros/rate sensors	1982–1990	1990–1996	1996–2004	2004	22
Radiofrequency	1994–1998	1998–2002	2002–2007	2007	13

Source: Grace Associates.

than $200 billion in 2006!). MEMS market studies today have become more judicious, but this author believes that many are still misleading for one or more of the following reasons:

- Many MEMS market studies do not specify whether their market numbers pertain to the MEMS structure alone or whether they include one or more higher system levels. A refreshing change in this regard came from Nexus (http://www.nexus-mems.com) and Yole Développement (http://www.yole.fr), which distinguish in their latest market studies between the first-level packaged MEMS and the smallest commercialized unit. This is illustrated for the case of an ink-jet cartridge in Figure 10.13 (see under Definitions). One cannot buy ink-jet heads alone, and from the point of view of the market (products sold and bought), the smallest commercialized unit is the cartridge or printer head. This decreases the total volume of the market (number of MEMS pieces) but provides a more realistic vision of component-based systems.
- A few market projections still include both MEMS optical switches for telecommunications and MEMS-based microfluidics for high-throughput screening (HTS), despite the fact that both applications have almost disappeared from the commercial scene.
- Most market studies do not specify the machining method used to make the MEMS part. The product could be precision machined, silicon- or non-silicon-based, and involve lithography or not.
- Looking back at the many market reports from the past, it is striking how the development time for a new MEMS product was underestimated over and over again. Table 10.1, based on a report by Grace Associates (http://www.rgrace.com), presents a sobering reality. On average, it still takes about 21 years to go from a new idea to a fully commercialized product. As MEMS products become more integrated and more IC-like, this development time will shrink dramatically.

The lack of more commercially successful applications for micromachines conflicts with the fact that in the evolution of electronic products, miniaturization only follows as a very natural development initiated by a continuing market pull. With radio transceivers, optical pickup systems for compact disc players, hard disks, thin-film read-write (RW) heads, camcorders, etc., features or functions that take up too much space must be compressed; otherwise, they do not survive. Outside the electronics area (such as in automotive sensors and biomedical devices), the market pull is similarly toward miniaturization. So most probably, the real MEMS market has not yet materialized, and product offerings are not meeting the specifications of the market.

There are several reasons why the MEMS market still lags behind the IC market by a factor of 10 or more:

- Micromachining is too often equated with microelectronics, specifically with application-specific integrated circuit (ASIC) manufacture. In reality, MEMS lacks the high degree of generality characteristic of microelectronics. ICs can be grouped in a limited number of classes, within which design and production follow well-defined and common steps. As a consequence, the price/performance (*p*/*p*) ratio allows industry to make profits on complex ICs, and it makes ICs suited for mass production. Silicon-MEMS sensors and actuators, on the other hand, are very specific.*
- Sensor applications also call for complicated, novel packaging schemes and MEMS packaging experience is relatively new. MEMS often have to be in direct contact with their surroundings, and each environment imposes its own constraints. This makes packaging for MEMS relatively more important, in terms of cost and (often) size, than ICs, further illustrating our claim in the Packaging section in Chapter 4, this volume that MEMS design should start from the packaging constraints and work its way to the miniature sensor inside rather than the other way around. Packaging and testing costs as much as 70% of total value MEMS product.
- Successive "bubble busts" [illustrated above in terms of microfluidics for high-throughput screening (HTS) and large optical switch arrays for telecommunications] have scared investors and product planners.
- The market is very fragmented.
- Multidisciplinary knowledge is required.
- There is a lack of capital formation opportunities.
- There is a lack of well-defined direction from roadmaps, industry standards, and industry associations.
- Perhaps equally important has been the lack of marketing skills in the early MEMS companies (and, as Regis McKenna says, "marketing is everything"). A survey suggests that 65% of MEMS gurus entered the field because the technology itself seemed compelling ("technology push"). So, obviously, there was a lack of technology/application matching. Micromachinists initially were too far removed from a detailed understanding of the applications. They were not alert enough to market demands and pushed micromachining technologies they controlled rather than solving industrial problems with the most suitable micromachining technology. In reality a MEMS solution based on a more mature precision engineering technology is as good a solution as one based on, say, surface micromachining.

Lessons Learned

One important MEMS market lesson from the last decade is that only MEMS applications geared toward mass consumption applications are succeeding commercially. As an important consequence, silicon-based MEMS devices (e.g., TI's DMDs, ADI's surface-micromachined accelerometers, and all types of RF-MEMS) must be compatible with existing IC lines to reduce costs. The consumer market should be a strong driving force for the MEMS industry to develop more integrated devices and is pushing the MEMS industry to 8-inch wafer size manufacturing. The need to adhere closer to IC manufacturing rules has become even more urgent, as few U.S. MEMS foundries have survived. As we saw above, the emergence of mass consumer MEMS applications has made MEMS very hot again, and the companies and MEMS foundries that did survive the MOEMS bubble can expect to thrive if they can scale up to compete with traditional IC fabs fast enough. In the author's own experience, MEMS foundries are hard to make profitable in the United States, as they need to keep on accepting a wide variety of jobs to keep the expensive clean-room operation going. In the process, they do not stay long enough on the learning curve of one new product to become world leaders for that particular product. Consequently, these laboratories are moving more and more to Asia, where large companies seem to have more patience in keeping the doors open till the MEMS foundry turns profitable.

* To be fair, for micromachines, the economics dictating profitable IC manufacture do not always apply; a microinstrument—say, a micro-gas chromatograph—may cost considerably more than a typical IC. Moreover, sensors and actuators, often using quite esoteric non-IC type materials, are produced in relatively low volumes and are so very application specific that, in principle, they may command a much higher price tag.

Olympus, in Japan, for example, runs the largest MEMS foundry in the world.[3] Similarly Bosch, in Germany, opened a MEMS foundry in 2005 (Bosch Sensortec). More traditional IC foundries have better survival chances in the United States, but it is worth pointing out that China will control 5% of the IC market by 2010 and the largest number of new 8-inch fabs are in Taiwan and Korea.[3] Although Korea and Taiwan are leading in terms of added capacity in 2007, Japan still holds most of the worldwide capacity, with about 24% followed by Taiwan, the Americas, and South Korea.

In case new non-IC materials and new processes need to be developed for a mass consumer MEMS product, as was the case for TI's digital mirror device, long time commitments (5–10 years) and huge sums of money are required (even with initial government funding) to bring the product to market. Few companies can afford such a large investment of time and money.

Definitions

Introduction

Micromachining marketing studies do not always clarify whether the quoted revenues apply to the micromachine alone or include the micromachine with some ancillary electronics and whether they comprise a higher-level subsystem or describe the whole system. Moreover, the precision machining technology employed often remains unspecified. A good market study should start from a clear definition of the products covered and should specify the manufacturing method.

MEMS

The terms micromachining, microfabrication, micromanufacturing, and MEMS describe the fabrication of devices with at least one of their dimensions in the micrometer range. The acronym MEMS originated in the United States[4] (Figure 10.7) and originally applied exclusively to silicon-based mechanical applications. But just as the twentieth century eventually became the electronics century, the twenty-first century promises to become the age of biotechnology and information/communication technology. This shift has manifested itself so rapidly in miniaturization science that we now see bio-MEMS problems eagerly tackled by electrical and mechanical engineers. Twenty years ago, the MEMS field was dominated by mechanical applications; today, most new applications are either information/communication-related or chemical and biological in nature. As a consequence, today MEMS refers to all subminiaturized systems, including silicon-based mechanical devices, chemical and biological sensors and actuators, and miniature nonsilicon structures (e.g., devices made from plastics or ceramics). The application of MEMS to medical and biotechnology problems is called *bio-MEMS* (which includes microfluidic structures, drug delivery devices, immunosensors, and DNA arrays).

More specific terms, such as MOEMS (e.g., micromirror arrays, fiber optic connectors), RF-MEMS (e.g. inductors, capacitors, antennas), commercial-off-

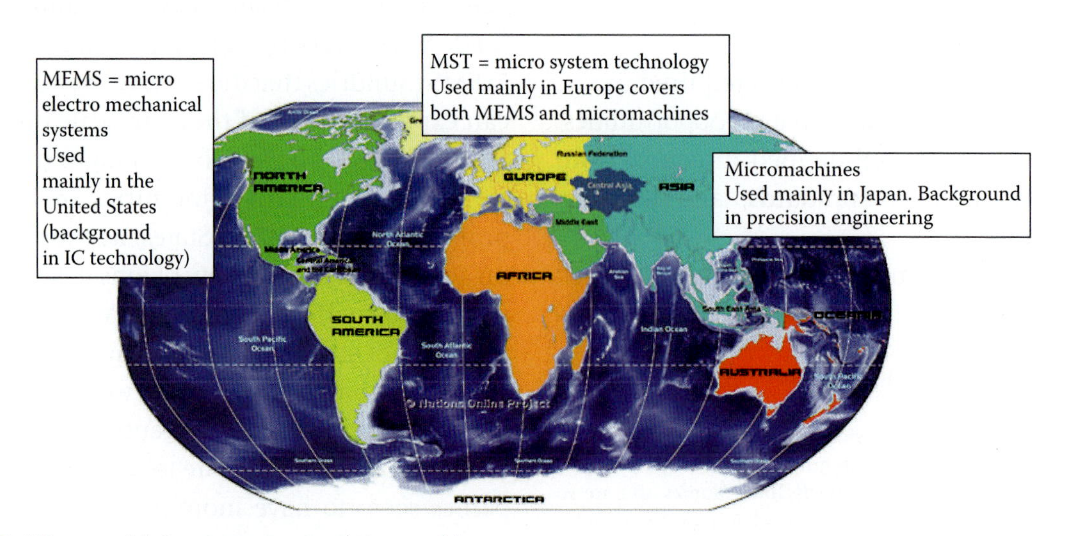

FIGURE 10.7 Micromachining terms around the world.

the-shelf microelectromechanical systems (COTS-MEMS), and high-aspect-ratio MEMS (HARMEMS) have also been introduced, and many more can be expected as many untenured professors want to have at least named something in the MEMS field.

Microsystems Technology

In Europe, the term *microsystems technology* (MST) is more prevalent than MEMS (Figure 10.7). One European definition of MST reads: "A microsystem is an intelligent miniaturized system comprising sensing, processing and/or actuating functions. Normally these would combine two or more of the following: electrical, mechanical, optical, chemical, biological, magnetic or other properties, integrated onto a single multichip hybrid." Alternatively, "A microsystem or microinstrument integrates sensors, actuators, and electronic components on a small footprint, collects and interprets data, makes decisions, and enforces actions upon its environment."[5]

As microsystems are combinations of sensors, actuators, and processing units, they are very application specific. They should be distinguished though from application-specific ICs (ASICs), which can be grouped into a much more limited number of classes within which design and production follow well-defined and common steps.[6]

Micrototal Analysis Systems/Lab-on-a-Chip

Micrototal analysis (μ-TAS) constitutes MST with an analytical function. An alternative term often used is lab-on-a-chip (LOC). In most of the literature today, little distinction is made among MST, μ-TAS, and LOC. Examples of commercial products in this general category are Nanogen's NanoChip™ molecular biology workstation (http://www.nanogen.com) (see Volume II, Example 8.2), TI's Spreeta and handheld reader (http://www.sensata.com/sensors/spreeta-analytical-sensor-highlights.htm) (see Figure 10.47), Caliper's microfluidic LabChip systems (http://www.calipertech.com), RotaPrep's sample preparation stage (Figure 10.51), and Agilent's MicroGC (Figure 10.8). It is important to note that most μ-TAS today still involve sizable instruments. Only some of the components have been miniaturized. Microsystems of this type still carry a price tag similar to that of large instruments ($2,000 to $120,000). The miniaturized components enable some improved

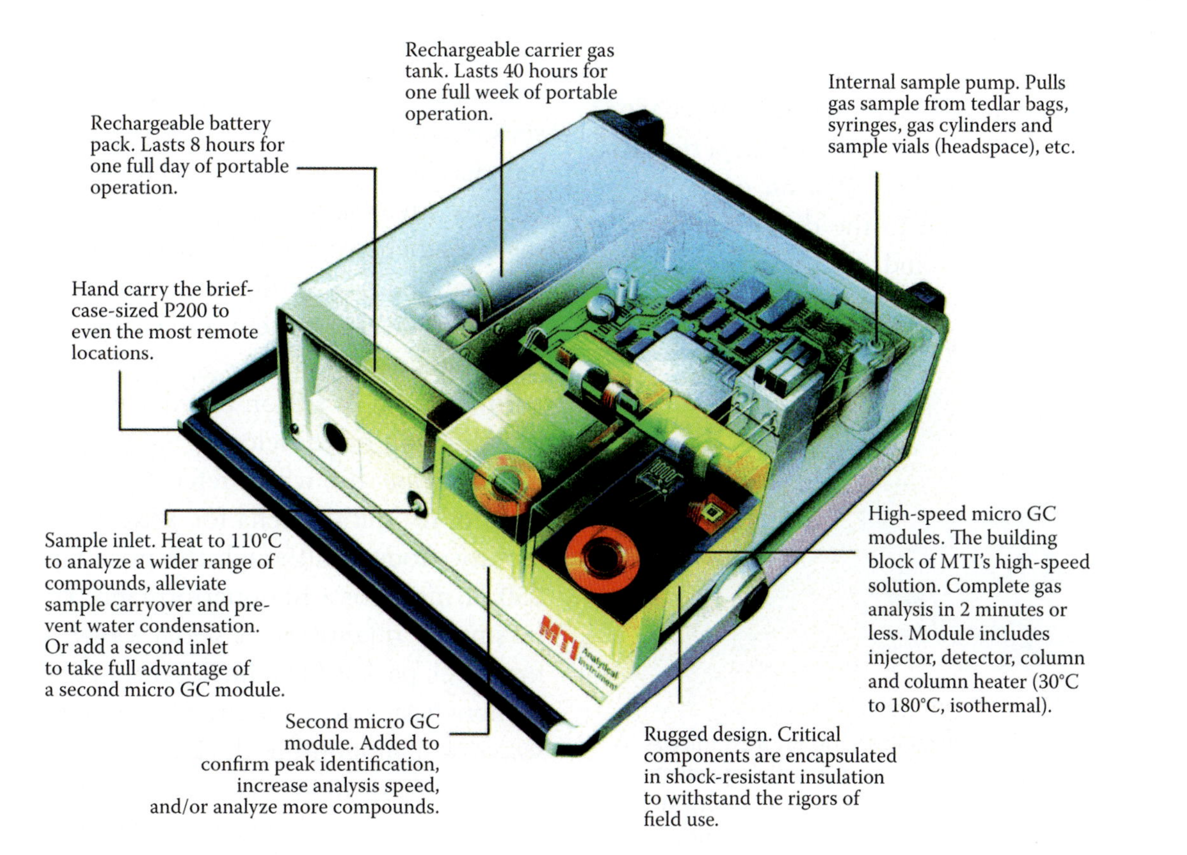

FIGURE 10.8 Agilent's MicroGC (formerly MTI).

functionality, but not yet a lower cost or a smaller, perhaps handheld, instrument. Some of the notable exceptions are the portable surface plasmon resonance sensor (Spreeta) shown in Figure 10.47 (with an instrument of $250 and disposable component of $30) and Johnson & Johnson i-STAT handheld automated blood analyzer, shown in Figure 10.42. The same instrument shown in Figure 10.8, but in a handheld format, would command a price tag somewhere between that of large instruments ($2,000 to $120,000) and sensors ($0.50 to $100), i.e., $200 to $500. To develop more handheld analytical equipment like the Spreeta is one of the μ-TAS challenges that lie ahead. Through the widening of MEMS applications, the word *MEMS* in the United States is more broadly interpreted now, and MST and MEMS have become more or less synonyms.

Micromechatronics

A mechatronics product is defined as a smart device that involves mechanical and electrical principles in design and performs hybrid mechanical and electrical functions. In Japan, the focus has been on micromechatronics or micromachines, defined as follows: "Micromechatronics is the technology that involves the design and manufacture of mechatronics products with dimensions in the range of 1 μm to 1 mm and are capable of performing complex microscopic tasks."[5] (For an example, see Figure 10.9.) In addition to the difference in sizes between mechatronics and micromechanotronics, major differences are in design methodologies, fabrication techniques, and packaging and testing.

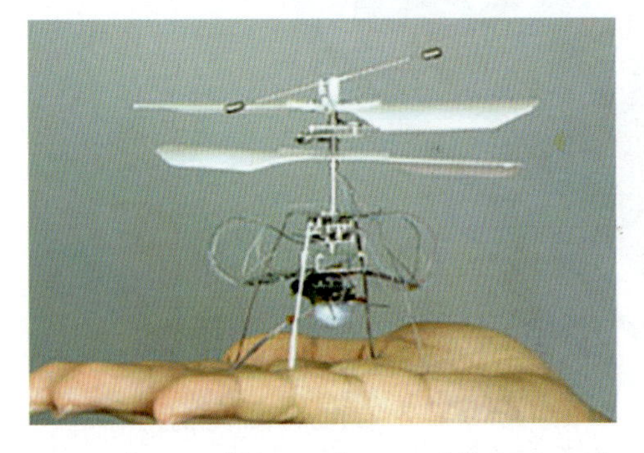

FIGURE 10.9 The world's smallest and lightest micro-flying robot. Epson pioneered the micro-flying robot, based on micromechatronics technology—the world's lightest and most advanced microrobot, which also features Bluetooth wireless control and independent flight.

In micromechatronics, structural components are designed on the base of scaling laws and electromagnetic actuators often are replaced by thermal, piezoelectric, and electrostatic microactuators. Components more often are fabricated by chemical-physical processes than by machine tools. Assembly, packaging, and testing are also different from those in macroscale mechatronics. Whereas MEMS and MST definitions have grown closer together, this Japanese term retains its separate meaning.

Microsensors

A sensor element is a device that converts one form of energy into another (e.g., ZnO, a piezoelectric material, which converts mechanical energy into electricity) and provides the user with a usable energy output in response to a specific measurable input. Measurands may belong to the radiation, thermal, electrical, chemical, mechanical, or magnetic field domains. The sensor element may be built from plastics, semiconductors, metals, ceramics, etc. To qualify as a microsensor, it must have at least one dimension in the micrometer range. Since the early 1980s, the term *sensor element* often invoked the notion of a silicon-built device, but there are many non-silicon-based microsensors.

A sensor includes a sensor element or an array of sensor elements with physical packaging and external electrical or optical connections. Synonyms for "sensor" are *transducer* and *detector*. A sensor system includes the sensor and its assorted signal processing hardware (analog or digital). *Transducer* sometimes refers to a sensor system, especially in the process control industry.

In the case of silicon-based sensors, some additional jargon has developed. A Si sensor element is called a *sensor die*, which refers to a micromachined Si chip. It typically sells for $0.10 to $2 as a commodity product, although the price tag can rise to $50 or more for a high-performance structure sold in smaller quantities. A *silicon sensor* alludes to a first-level, packaged Si die or sensor element with or without basic electronic circuitry. An *integrated silicon sensor* is a monolithic device comprising the sensor and one or more electronic components, which amplify and condition (standardize) the sensor output signal. The standardization of the sensor signal

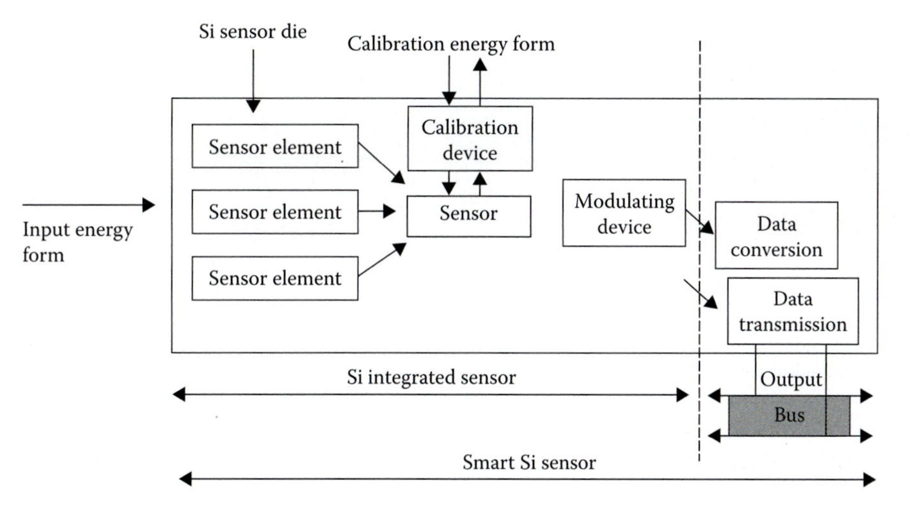

FIGURE 10.10 Definition of a silicon sensor die, a silicon sensor, a silicon integrated sensor, and a smart silicon sensor. (After National Research Council, 1995.)

makes the device bus compatible, enabling efficient communication between the central processor and the sensor. Typical selling prices range from $2.50 for quantities of a million units per year to over $100 for complex sensors in smaller unit quantities. At yet a higher level is a *smart silicon sensor,* which is a packaged integrated sensor containing some part of the signal-processing unit to provide performance enhancement for the user. Signal processing might include autocalibration, interference reduction, compensation for parasitic effects, offset correction, and self-test. The various silicon-based parts defined above are represented schematically in Figure 10.10.

Microactuators

The definition of a sensor element as a device that converts energy from one form into another also applies to an actuator. In the case of an actuator, one is interested in the ensuing action, whereas in the case of a sensor, our interest is in the information garnered. In other words, if the intent is to measure a change, one will refer to a sensor; if the intent is change or action itself, one defines an actuator. The appropriate use of *sensor* or *actuator* is not based on the physics involved but on the intent of the application.[7]

Actuators are components that convert energy into an appropriate action, often dictated by a sensor control unit. They facilitate a function such as opening a valve, positioning a mirror, moving a plug of liquid, etc. Microactuators are actuators with at least some dimension in the micrometer range. Since an actuator "acts," some power is usually needed. Miniaturization of actuators is consequently less obvious than that of sensors, as power does not scale advantageously in the microdomain (power is proportional to the volume, i.e., l^3; see Chapter 8, this volume). To induce micron-scale motion, actuators do not need to be micron-scale themselves. Busch-Vishniac suggests we call the latter *microactuators* in contrast with *microfabricated actuators,* which are micromachined actuators.[8]

The selling price for Si-based actuators in large quantities may range from $5 to $200. A few Si micromachined actuators have reached the commercial market. Examples include Redwood Microsystems' (company now defunct) thermopneumatic valve (Figure 10.11a),[9] TiNi's shape memory alloy microvalve (http://www.sma-mems.com) (Figure 10.11b), and ADI's surface-micromachined accelerometer qualifies (Volume II, Figure 7.92). The latter involves a sensor/actuator combination with a self-test involving electrostatically actuated interdigitated polysilicon fingers (Volume II, Figure 7.93).

Microstructures or Microcomponents

A microstructure or microcomponent refers to a precision machined part that has at least one of its dimensions in the micrometer range but is not a sensor, actuator, or microsystem. Rather, it concerns an item such as a microlens, micromirror, micronozzle, microneedle, etc., that acquires a useful function only when combined with other components. Selling prices of such Si microstructures in large

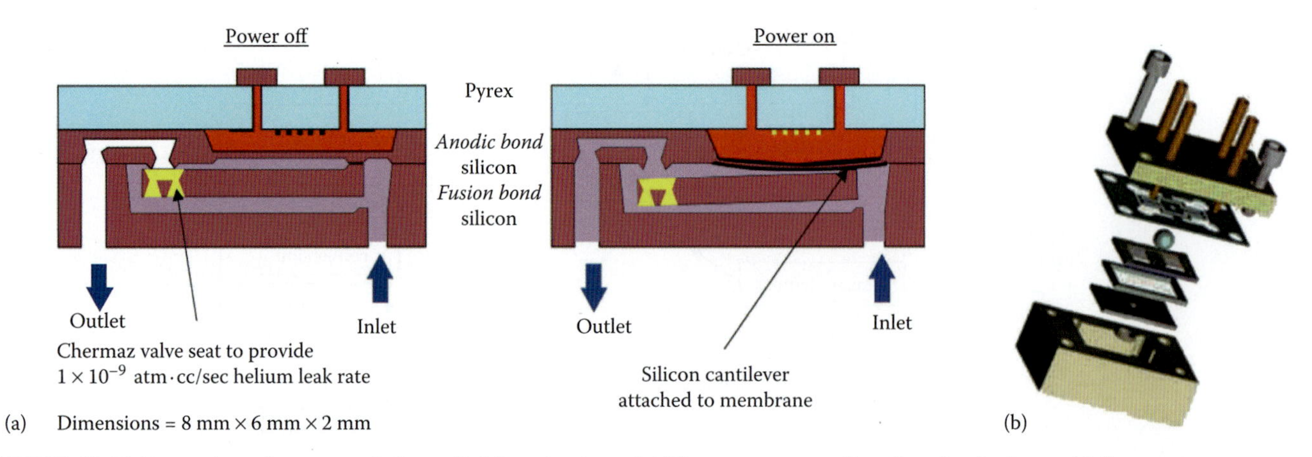

FIGURE 10.11 Examples of commercially available actuators. (a) Thermopneumatic valves by Redwood Microsystems (Fluistor™). Normally closed shut-off microvalve featuring a liquid-filled cavity, which flexes a silicon diaphragm when heated, forcing the valve cover to lift off the valve seat. (After Zdeblick, M. J., R. Anderson, J. Jankowski, B. Kline-Schoder, L. Christel, R. Miles, and W. Weber. 1994. In *Technical Digest: 1994 Solid State Sensor and Actuator Workshop,* 251–55. Hilton Head Island, SC.) (b) TiNi Alloy Company's shape memory alloy valve ($190/valve in small quantities). This is a normally closed microvalve. A current through the NiTi die lifts it from the orifice die. (Courtesy of Dr. D. Johnson, TiNi Alloy Company.)

volumes may range from $0.25 to $100.[10] The latter numbers are hard to confirm, since most microstructures are delivered to clients as part of large development contracts.

An example of a Si microstructure built by the author and his team is the scanning tunneling microscope (STM) cantilever with integrated sharp Si tip shown in Figure 10.12.[11,12] The silicon tip is sharpened by consecutive oxidation sharpening, a process developed by Ravi and Marcus[11] (see also Volume I, Chapter 4, "Orientation and Dopant Dependence of Oxidation Kinetics"). While we delivered the first STM tips to a client company under a best effort-type research and development contract, eventually, these components were sold at a fixed price. Since an STM tip determines the quality of an STM picture, this product is an example of how a small micromachined component might provide a competitive edge to an STM instrument (~$20K to $60K) manufacturer. In this case, it is important that the instrument manufacturer controls the micromachining technology of the crucial microcomponent or, better yet, has it in-house.

First-Level Packaged MEMS and Smallest Commercialized Units

Nexus, a consulting firm on MEMS, makes an important distinction between the first-level packaged MEMS/MST and the smallest commercialized unit.[1] This is illustrated in the case of an ink-jet cartridge

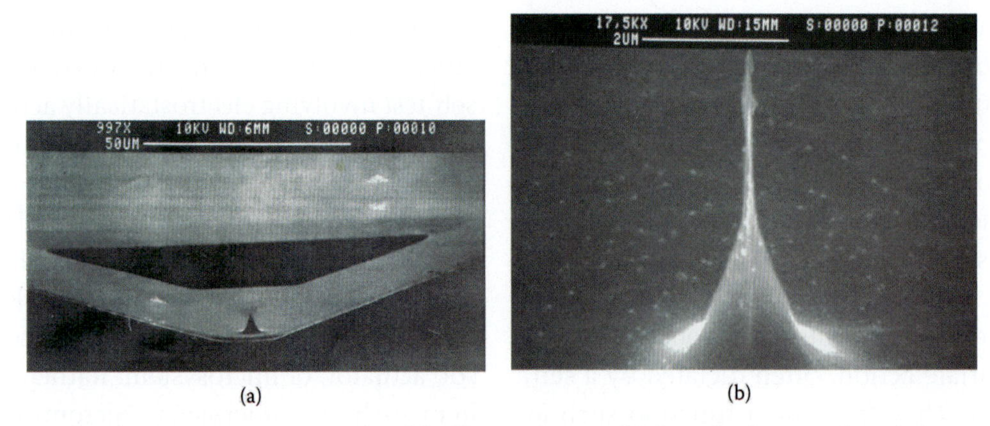

FIGURE 10.12 Example of microcomponent made with silicon micromachining: single-crystal scanning tunneling microscope tip on single-crystal cantilever beam. (a) High-aspect-ratio silicon tip. Photograph taken from a 45° angle. (b) Tip is actually 40% taller than it appears. (From Ravi, T. S., and R. B. Marcus. 1991. Oxidation sharpening of silicon tips. *J Vac Sci Technol B* 9:2733–37; Editorial. 1991. Menlo Park, CA: TSDC [Teknekron]. With permission.)

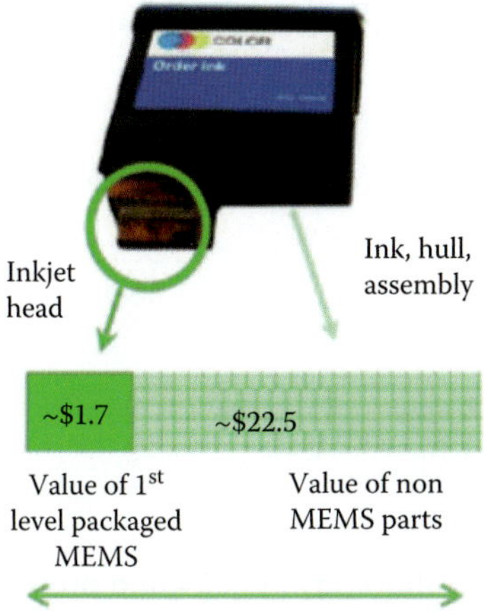

FIGURE 10.13 Example of an ink-jet printer head. First-level packaged MEMS and smallest commercialized unit. (From Nexus. 2006. *Nexus market analysis for MEMS and microsystems III, 2005–2009.*)

in Figure 10.13. One cannot buy ink-jet heads alone, and from the point of view of the market (products sold and bought), the smallest commercialized unit is the cartridge or printer head. The MEMS/MST in the cartridge/printer (~$1.7) is worth much less than the total value of the cartridge/printer (~$22.5).

This simple clarification is very relevant for Si foundries interested in manufacturing the Si die only.

Overall Market for Micromachines

Introduction

Richard Feynman presented two lectures on micromachining: "There's Plenty of Room at the Bottom," at Caltech on December 16, 1959, and "Infinitesimal Machinery," at the Jet Propulsion Labs on February 23, 1983.[13,14] In these now famous lectures, he speculated on the use of micromachining for some esoteric devices such as microfabricated light shutters, probing devices for electronic circuits, and adjustable masks, the first two of which have been realized. He was also convinced of the future of truly three-dimensional circuitry in computers. In general, he doubted the wide applicability of the technology, though. Feynman could not suggest many uses for micromachinery, but he was not worried. He said, "I'm fascinated but I don't know why."

In this section, we investigate applications and overall markets for MEMS and show that the field has matured, with MEMS products now entering consumer electronics so that the market is poised for a major take-off. We conclude with a very positive outlook for MEMS, with the understanding that micromachining be defined as broadly as we have in this book—beyond silicon-based technology and perhaps even beyond batch methods; that is, including continuous manufacturing methods.

Overall MEMS Market Size

The numbers for the overall market size and market projections for miniaturization technology presented here principally are culled from marketing studies and insights by Nexus (http://www.nexus-mems.com), Yole Développement (http://www.yole.fr), Bosch Sensortec (http://www.bosch-sensortec.com), SEMI (http://www.semi.org), Frost & Sullivan (http://www.frost.com), Draper Labs (http://www.draper.com), Wicht Technologie Consulting (WTC) (http://www.wtc-consult.de), Janusz Bryzek (janusz@bryzek.com), and Roger Grace Associates (http://www.rgrace.com).

According to the Nexus Market Analysis for MEMS and Microsystems III 2005–2009,[1] the MEMS/MST market volume was worth $12 billion in 2004 and was expected to reach $25 billion in 2009. The growth was projected to be rapid, with a 16% compound annual growth rate (CAGR) in value in the period 2004–2009 (Figure 10.14). We like to point out that the 1998 Nexus Task Force[5] projected a MEMS/MST market for 2002 of $38 billion already. Obviously, a more detailed analysis about the real MEMS/MST contribution has made for a more careful forecast in this latest report. Considering the traditional Nexus definition used in earlier reports—the market for the smallest unit, incorporating MST component(s), that is commercially available—the market was expected to increase from $36 billion in 2004 to $52 billion in 2009 (Figure 10.14).

In 2005, according to Yole (http://www.yole.fr/), more than half of Si-based MEMS by value, which this consulting firm estimates at $2.6 billion, were produced in North America. Japan was the next largest MEMS producing region, accounting for one-fifth of the total 2005 MEMS revenues ($1.1 billion). About 16% of MEMS sales, over $815 million, originated from European production. Singapore was also a significant manufacturing location for MEMS

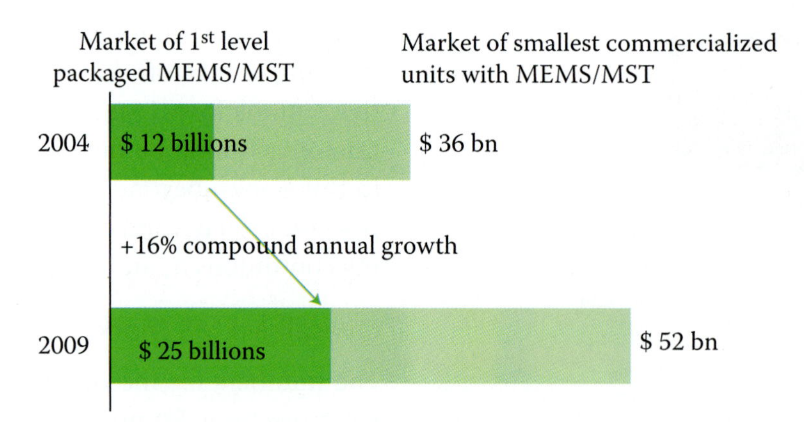

FIGURE 10.14 Total market for microelectromechanical system/microsystems technology in 2004 and projections for 2009. (From Nexus. 2006. *Nexus market analysis for MEMS and microsystems III, 2005–2009.*)

manufacturers (e.g., STMicroelectronics with ink-jet heads), with 11% of MEMS device revenues attributed to Singapore fabs. The MEMS industry is highly fragmented with a large number of device manufacturers worldwide in a wide variety of application fields. Many start-ups and small and medium-sized companies participate in the industry. In 2004, there were 230 manufacturers of MEMS products. Most of these were developing new products, some of which are just starting to significantly affect the market. The top thirty MEMS device manufacturers accounted for about 84% of MEMS device sales in 2005. No single device manufacturer has reached the billion-dollar sales mark as of yet: the average revenue for a top thirty manufacturer is about $147 million (http://www.yole.fr/pagesAn/products/semi.asp).

In Table 10.2 we list 2005 MEMS sales of some important MEMS companies. The top three MEMS companies today are TI, Hewlett-Packard, and Bosch.

In Figure 10.15 we show the Nexus market breakout for sixteen types of first-level packaged MEMS/MST (a seventeenth category, "other," includes microreactors, chip cooler, inclinometers, memories, fingerprints, liquid lenses, microspectrometer, wafer probes, micromirrors for optical processing,* micropumps, micromotors, and chemical analysis systems).[1] According to this breakout, the MEMS/MST market in 2009 was still 70% made up of only three products: read-write (RW) heads, ink-jet heads, and microdisplays. RW heads still represented around 51% of the market in 2009: while traditional application of RW heads in PCs would grow only moderately, the RW market was experiencing a renaissance in consumer electronics, as hard discs were entering music players (e.g., in many iPod models) and smart phones [Samsung introduced the first cell phone with a hard disc drive (HDD) in 2004], as well as digital video cameras, set-top players, and DVD recorders. It is important to note that MEMS will lose a very important market when HDDs are replaced by solid state disc (SSD) memory (a flash drive), as both RW heads and accelerometers will fall away. There are a lot of arguments to suggest that SSDs will replace HDDs in many applications. The SSD market is growing fast, approximately doubling in revenue every year for the foreseeable future. In the long term (5–10 years) the trusted HDD will eventually retire, to be replaced by the flash drive.

Ink-jet heads continued to be one of the most profitable markets for MEMS. MEMS microdisplays are the new MEMS blockbuster, and their revenues were expected to overtake ink-jet heads in 2009. TI was forging ahead with the digital light processing (DLP) chip for projectors and for front projection

TABLE 10.2 Sales (in 2005) of Microelectromechanical System Products by Some of the Largest Players

Company	Sales ($, in Millions)
Texas Instruments	788
Hewlett-Packard	750
Bosch	325
Lexmark	230
ST Microelectronics	200
Seiko	199
Cannon	184
Freescale	182

* In Figure 10.15, Nexus somehow has digital micromirrors categorized under both *other* and *microdisplays*.

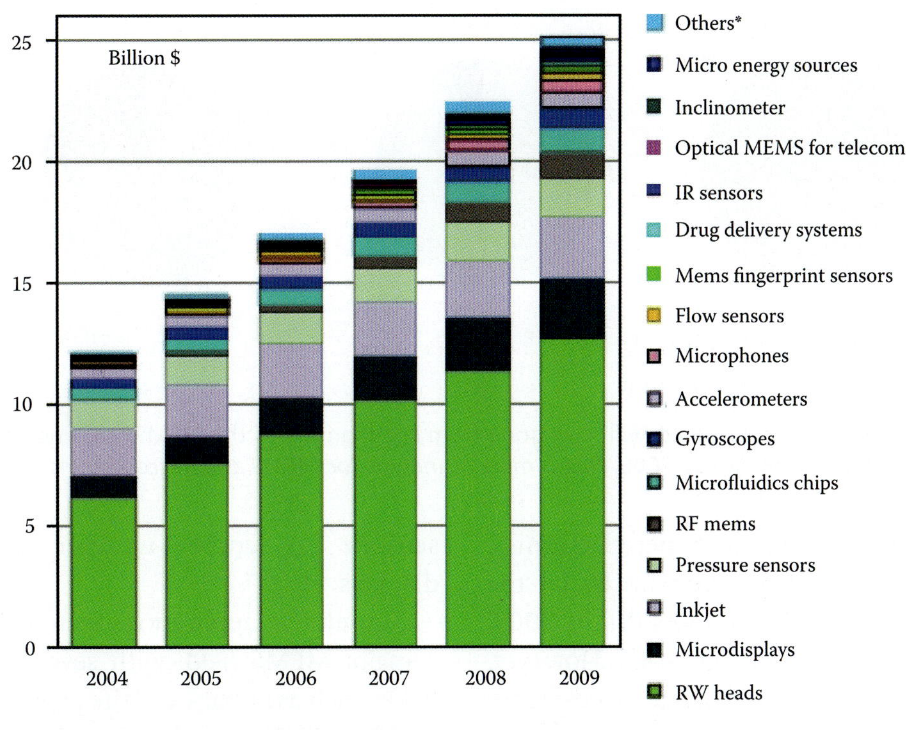

* Others: Microreaction, chip cooler, inclinometers, MEMS memories, MEMS fingerprints, liquid lenses, microspectrometer, wafer probes, micro-mirrors for optical processing, micro-pumps, micromotors, chemical analysis systems

FIGURE 10.15 Nexus market breakout for first-level packaged microelectromechanical system/microsystems technology devices.[1]

and rear projection TVs (home theater). Three other products that, according to Nexus, would make each over $1 billion in sales in 2009 are pressure sensors, RF MEMS, and inertial sensors.[1] Other fast-growing markets were microphones and tire pressure monitors, next to established pressure and motion sensors, which were increasingly being driven by consumer electronics.

The relative contributions to various application fields of these MEMS/MST for the years 2004 and 2009 are illustrated with two pie charts in Figure 10.16. The three dominant application fields were IT

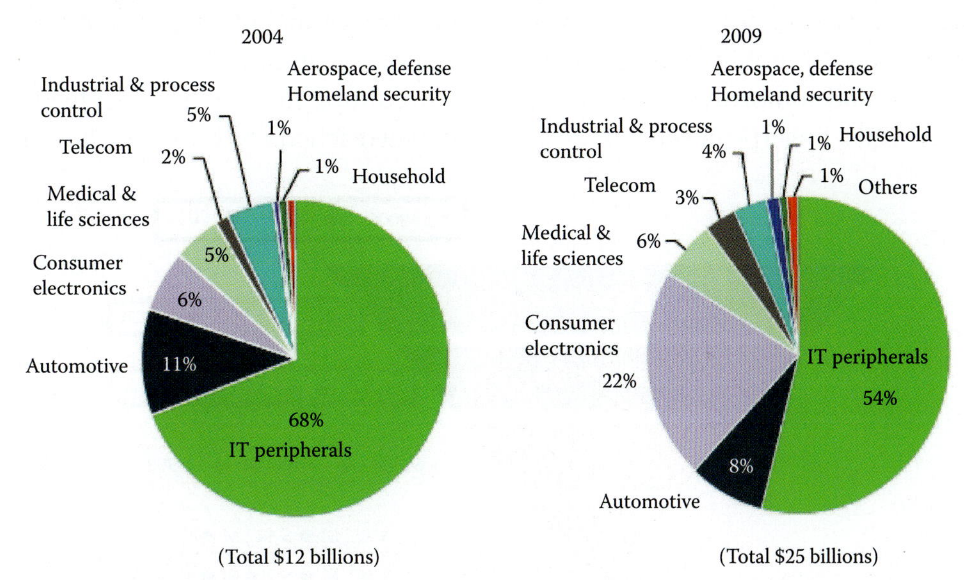

FIGURE 10.16 The relative contributions to various application fields for microelectromechanical system/microsystems technology products for 2004 and projections for 2009. (From Nexus. 2006. *Nexus market analysis for MEMS and microsystems III, 2005–2009.*)

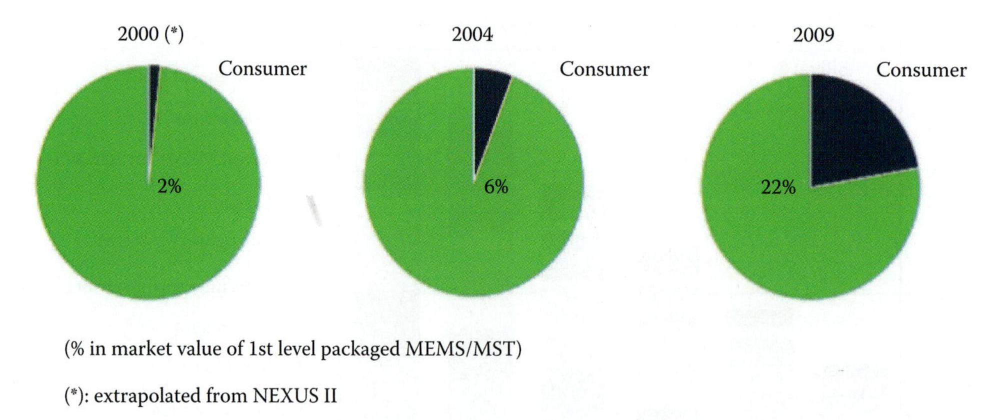

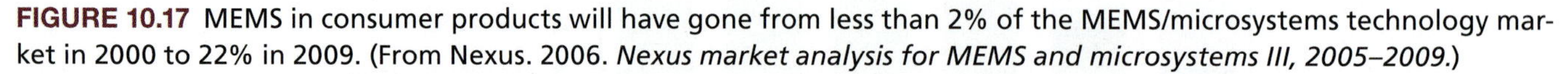

FIGURE 10.17 MEMS in consumer products will have gone from less than 2% of the MEMS/microsystems technology market in 2000 to 22% in 2009. (From Nexus. 2006. *Nexus market analysis for MEMS and microsystems III, 2005–2009.*)

peripherals, automotive, and consumer electronics. IT peripherals were expected to remain by far the first application field of MEMS and MST in 2009, mainly with RW heads and ink-jet heads. However, the share of IT peripherals was expected to decrease from 69% to 54%. The biggest predicted change was in the area of MEMS in consumer electronics (CE) that was predicted to grow from less than 2% of the MEMS/MST market in 2000 to 22% in 2009 (Figure 10.17). Consumer electronics were expected to drive the MEMS market growth for the remainder of this decade. The three drivers for MEMS in consumer electronics were 1) large-screen high-definition television for everybody (rear projection and front projection home theater TVs), 2) more storage in digital equipment (with HDDs that enter digital video cameras, music players, smart phones, set-top players, and DVD recorders), and 3) mobile handsets (with accelerometers today, and tomorrow gyros, microdisplays, micro fuel cells, fingerprint sensors, zoom sensors, gas sensors, weather stations, projection displays, etc.).

Automotive applications were expected to remain a major MEMS field, with several killer application fields, such as airbags and tire pressure monitoring system (TPMS). The numbers of MEMS for this application would still grow at a rapid pace; however, the growth in revenue was expected to be moderate due to a continuous pressure on prices (e.g., gyroscopes increasing 15% per year in units but only 8% in revenue).

Yole analyzes the overall MEMS market structure a bit differently. It introduces a MEMS supply chain, as shown in Figure 10.18, and distinguishes among MEMS-based systems, MEMS alone, front-end equipment, and materials and chemicals.[15] Also, the Yole MEMS numbers refer to Si technology only, so they do not include, for example, RW heads, which are not silicon-based.

According to Yole, the worldwide MEMS systems market reached $40 billion in 2006 and was expected

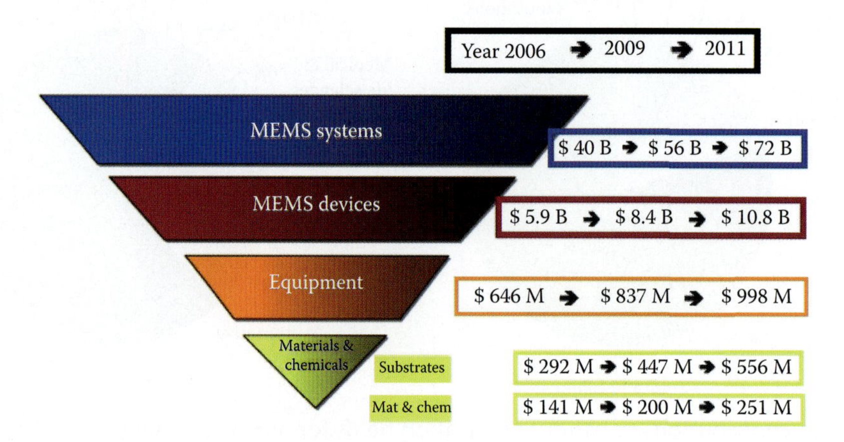

FIGURE 10.18 Yole's microelectromechanical system (MEMS supply chain from 2006 to 2009 to 2011. (From Yole. 2007. Evolution of the MEMS markets and evolution of the MEMS foundry business. *Transducers*.)

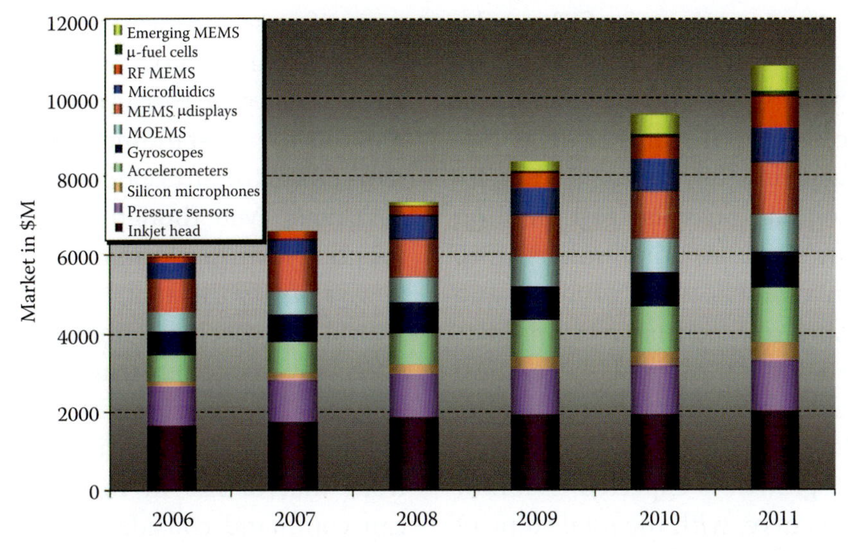

FIGURE 10.19 Global microelectromechanical system market for 2006–2011, according to Yole.

to rise to $56 billion by 2009 and $72 billion by 2011, for a compound annual growth rate (CAGR) of about 15% over the five-year period.[15] Globally, the MEMS market was $5.9 billion in 2006 and was forecast to grow to $8.4 billion in 2009 and to $10.8 billion in 2011, with a 13% CAGR. The MEMS equipment market was $646 million worldwide in 2006 and was expected to expand to $837 million in 2009 and to $998 million in 2011. The five-year CAGR forecast for MEMS equipment was 6%. MEMS materials can be divided into substrates and chemicals and other materials. Together these markets totaled $433 million in 2006 and were forecast to be $807 million in 2011. MEMS materials are expected to grow at a five-year CAGR of 15% through 2011.[15]

The global MEMS market for ten different products and one category called *emerging MEMS*, over the period 2006–2011, according to Yole, is illustrated in Figure 10.19.[15] These numbers are only 1/3 of the Nexus numbers, mostly because Yole does not include RW heads (which are not based on Si technology; Yole's market study is Si-centric) and because under its Emerging MEMS category, Yole is considering some other MEMS products than Nexus lists in its Others category.

Yole puts the ink-jet head market at $1.8 billion in 2006 and expects it to grow to $2 billion in 2011. Although digital imaging is a strong driver for this relatively mature market, Yole expects that the overall trend toward nondisposable ink-jet heads will dampen volume growth for this application.

As in the Nexus report, Yole sees a big change in the MEMS market today in that an important part is now driven by the growth of consumer electronics, which is new in MEMS industry. The period from 2007 to 2011, Yole claims, will certainly see the dominant growth of consumer applications compared with automotive, industrial, and/or medical business. This move toward the consumer applications also is pushing the MEMS industry to 8-inch wafer size manufacturing, and this means more involvement of IC foundries in MEMS business. This switch to 8-inch wafer has a lot to do with the increase of ink-jet head die size and the transition to 8-inch by TI for digital mirror production.

Fast-Changing MEMS Market Characteristics

The very large projected markets for MEMS in market studies published before 2000 are realistic only if nonsilicon approaches are included (e.g., thin-film RW heads) or a first-level system in which the MEMS technology is embedded is counted as well. After the 2001 optical and fluidic MEMS bubbles, major changes in the markets for mechanical and MOEMS devices (RW heads, ink-jet heads, pressure sensors, accelerometers, gyros, digital micromirror arrays, microphones, etc.), are causing upward revisions of MEMS market projections again. These types of MEMS are now experiencing a growth spurt because of their incorporation in consumer products (cars, mobile phones, games, PCs, cameras, etc.). Mechanical and MOEMS largely are made with IC-based techniques, and the cost of ICs can be lowered only when the quantity of parts numbers in the millions per year. Consequently for silicon-based sensors to succeed on the same scale as ICs, the current manufacturing environment dictates that one must concentrate on mass consumption products such as cars, PCs, air conditioners, TVs, cell phones, vacuum cleaners, toys, printers, and dishwashers. This transition to consumer products has finally happened for optical and mechanical MEMS, and for these products the market dynamics have now become similar to those of the IC market. Consequently MEMS growth is expected to follow a hockey stick growth curve, with the total value of MEMS rising, according to one market report, to $40 billion in 2015 and $200 billion in 2025 (Dr. Marinis from Draper Laboratory: http://www.sem.org/PDF/s01p01.pdf). The problem now is more one of profitability rather than the absence of real product opportunities, i.e., sensors will become smaller and smaller, more and more integrated and manufactured on larger (8-inch) Si wafers. In a sense, one might say that the IC industry has swallowed the mechanical and MOEMS field. The more established MEMS such as accelerometers (ADI), ink-jet heads (TI), and RW heads (IBM, Hitachi) are exploited now mostly by large companies. The involvement of MEMS and IC foundries is on the rise as well, and the tremendous cost pressure is driving more and more MEMS manufacturing overseas.

However, one must still make a distinction between ICs and MEMS. There are about 50,000 types of sensors to measure 100 different physical and chemical parameters,[16] so often only 1,000 to 10,000 are needed a year. We can conclude that fragmentation remains the outstanding characteristic in the sensor market. This often excludes integrated semiconductor solutions, even though they might be technically superior. Cost/performance improvement does not always justify the application of IC technology, and hybrid approaches remain an important option especially in microinstrumentation and for chemical sensors. The dynamics of the market for MEMS-based microinstrumentation is, furthermore, different from that of ICs in that MEMS instrumentation can command considerably higher price tags (say, hundreds of dollars rather than $10 or less). In the latter case, smaller numbers of parts may be offset by a higher profit margin. The situation has been compared to that of application-specific integrated circuit (ASIC) manufacture, leading to the acronym ASIS (application-specific integrated system). Hybrid technology will be employed in this case not solely out of cost considerations but because one monolithic manufacturing technique is unlikely to deliver the bandwidth in feature sizes and material types necessary to cover the various needs common in small instrumentation.

The situation for disposable biosensors (e.g., glucose sensors) applied to diagnostics is very different again. In this nascent field one competes with industrial giants and with very efficient established manufacturing processes. In the bio-MEMS field, to succeed similarly to the mechanical and MOEMS

Table 10.3 New MEMS Companies and Products

Company	Product	Where
Silmach	Silicon-based micromotors for the watch industry	France
Perpetuum	Energy harvesting systems using MEMS as building blocks	United Kingdom
Rhevision Varioptic Siimpel (United States)	Liquid lenses for consumer applications	United States France United States
Lilliputian Systems	MEMS-based micro-fuel cells	United States
Qualcomm Liquavista	Innovative microdisplays	United States Netherlands
SiTime Discera Silicon Clocks VTI Technologies	MEMS-based oscillator	United States United States United States Finland
Microvision Philips	New optical microelectromechanical system applications	United States Netherlands

area, one will need to develop mass manufacturing techniques based on very large batches or roll-to-roll manufacturing on paper on plastic substrates.

Most of the risk taking in innovative MEMS concepts is still by U.S. entrepreneurs, although activity in Europe and Asia has picked up, as is clear from the list of new MEMS companies in Table 10.3. There remains about a three- to five-year delay in time between start-up activity in the United States and Europe on a specific subject (e.g., U.S.-based Affymetrix, created in 1992, versus the European OMM in 1996). So far, even successful MEMS start-up companies remain two orders of magnitude smaller than companies in the mainstream IC market.

MEMS Market Segments

Introduction

We consider four MEMS market categories: MEMS in the automotive market; MEMS in medical and biotechnology markets (bio-MEMS); MEMS in industrial automation, including environmental monitoring markets; and MEMS in IT peripherals and telecommunications markets.

MEMS in the Automotive Market

Introduction

Nexus did forecast the number of passenger vehicles, sport-utility vehicles, and light trucks produced globally to reach an estimated 68 million per year by 2009 (add another 15 million for other vehicles, such as buses, trucks, the Popemobile (smile, please), etc.).[1] In a complex electronic/electromechanical system such as the modern car, the need for effective, accurate, reliable, and low-cost electronics and sensors is pressing. While in 1980 electronics accounted for merely 2% of the total cost of a car, today it has reached almost 20%, and it was estimated that by 2010, electronics and software would account for up to 40% of the value of a car (see Figure 10.20). MEMS are making for a larger and larger contribution to this electronic content.

New vehicles feature increasingly sophisticated engine management and safety systems, while a larger number of sensor systems are making their way into lower-priced cars, fueling the overall growth

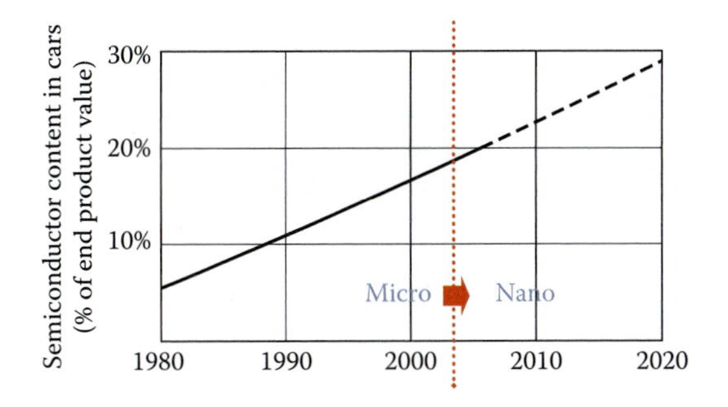

FIGURE 10.20 Semiconductors underpin all business sectors. It was estimated that by 2010, electronics and software would account for up to 40% of the value of a car.

of MEMS in the automotive sensor market. Today, typical mid-sized vehicles contain more than 50 sensors and luxury-class cars more than 100. About one-third of these sensors are MEMS-based, including frontal impact airbags, tire pressure monitoring systems (TPMS), and engine monitors to improve fuel efficiency and lower toxic emissions. This trend will continue as governmental regulations and the needs of the consumer drive auto manufacturers to provide improved safety, quality, and comfort in vehicles. The total market for silicon-based MEMS in safety systems, vehicle navigation, engine management, and cabin comfort, according to a Nexus market report, was \$1.29 billion in 2004, and was expected to grow at 9% CAGR to reach nearly \$2 billion in 2009, as illustrated in Figure 10.21.[1] Between 2000 and 2004, the largest growth in MEMS applications occurred in the automotive market, with applications expanding to the point that a typical automobile contains twenty-five silicon MEMS devices. Pressure, acceleration, and gyros dominate the present Si MEMS automotive market. The technologies involved are mostly Si bulk micromachining and polysilicon surface micromachining.

Even though the quantity of MEMS devices sold into the automotive market is expected to continue climbing steadily, the automotive percentage of the total MEMS market is actually expected to decline, as the consumer electronics category expands.

When considering all automotive sensors (not only MEMS), the worldwide market was more than \$10 billion in 2005, and it was expected to continue to increase at an average annual growth rate of 6% to \$14.2 billion in 2010. From automotive sensors, speed and position (38%) and oxygen (20%) sensors would

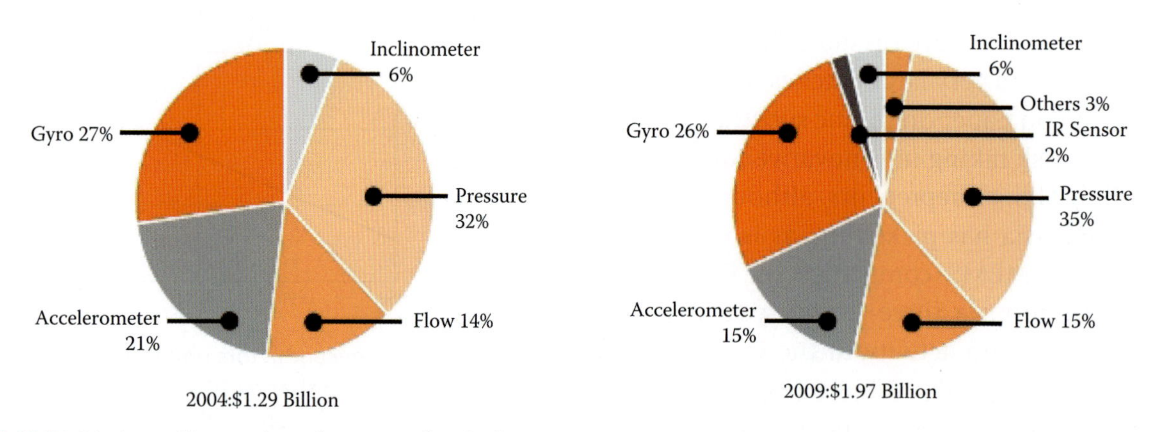

FIGURE 10.21 Various silicon microelectromechanical system sensors in passenger vehicles. (From Nexus. 2006. *Nexus market analysis for MEMS and microsystems III, 2005–2009.*)

remain the largest market segment, rising from $6.2 billion in 2005 to $8.6 billion in 2010 (58%) [*Automotive Sensors*, November 2005 (http://www.theinfoshop.com/study/bc34281-automotive-sensors.html), and Automotive Sensors Market Emerging & Current Sensors Update Now and Future, September 2005 (http://www.researchandmarkets.com/reportinfo.asp?report_id=307920)].

A most important automotive chemical sensor is the so-called *lambda probe*, a solid state oxygen sensor (all new European cars are fitted with a minimum of two lambda sensors). The lambda probe involves traditional ceramics manufacturing, but planar, micromachined oxygen sensors are gaining ground (see Volume II, Figure 9.24).

Many other opportunities for micromachining in automotive sensing exist; a list of sensor needs (met and unmet) in the automotive and transportation industry in general is presented in Table 10.4. There is a continuing migration in the automotive industry from larger electromechanical automotive components to MEMS. In Table 10.5 we summarize this trend. The newest automotive applications for MEMS include better exhaust monitoring devices, wireless sensors for engine management systems, and energy scavengers that harness the energy generated from rotational or vibratory movements within the vehicle as a future power source for TPMS.

Figure 10.22 shows where some of these sensors are positioned in a car. Component cost is a significant factor in the selection criteria of automotive systems designers. Getting your MEMS sensor into a car might take anywhere from seven to ten years. The cost of a MEMS device itself constitutes one-third or less of the total delivered component cost (often $10 or less!). Over 50% of the cost of a typical sensor is in calibration and testing, highlighting the need for innovation. Therefore, a significant challenge to achieve highly efficient design for manufacturability and testing is imposed on all suppliers who wish to be successful in the automotive sector. As a consequence, the use of IC-compatible process lines for MEMS production is growing. There clearly are important benefits in using CMOS or BiCMOS-compatible manufacturing lines to target high-volume MEMS markets, such as the automotive market, and toward adoption of more integrated solutions.

Pressure

Introduction and Market Perspectives

Pressure measurements are absolute, relative, or differential. For most pressure sensors, the common design element is made from a thin, flexible diaphragm that deforms elastically in response to an applied force. Diaphragm thicknesses in these sensors typically range from 20 µm to several millimeters and employ steel, ceramic, or silicon. The ceramic capacitive pressure sensor has a diaphragm cut from relatively thick "cast" sheets of ceramic (e.g., from Kavlico). According to an analysis by Frost & Sullivan (http://www.sensors.frost.com), pressure sensors represent one of the most broad, mature, but very fragmented sensor markets (similar to temperature sensors). In the study "Pressure Sensors and Transmitters Market," the company reveals that the market for all pressure sensors earned revenue of $4 billion in 2004 and projects that this market

Table 10.4 Examples of Sensors and Other Micromachined Components in Use or Needed in Transportation Context

Micromachine	Application	Status
Accelerometers	Airbag release, vehicle dynamics control, antilock braking systems, antitheft systems, active suspension control, headlight leveling	Micromachined products are on the market replacing the old "ball and tube" approach
Air flow sensor	Air/fuel ratio control	On the market but expensive
Alcohol sensor	Detection of alcohol level in blood	No reliable products available
Angle and position sensor	Throttle valve angle, pedal angle or position, and steering angle	All are available
Angular rate sensor	Vehicle dynamics control, navigation, and rollover systems	Available
Azide sensor	Airbag propellant	Does not exist
Battery sensor charge/density	Charge/density of electrolyte	Research
Contraband detector	Customs	No good solutions
Driver identification	Theft prevention	Some solutions available
Head-up display	Information on window	Not affordable yet
Humidity sensor	Cabin climate	Unsatisfying products available
Level sensor	Oil and gas level	Available
Light sensor	Turn on the lights	Available
Micronozzle system	Fuel injection	Available
NOX sensor	Pollution control	Not available
Oil quality monitor	Engine protection	Not available
Oxygen sensor (two are required in all new cars in Europe)	Air/gas ratio	Available, but new technology is needed for more demanding applications
Pressure sensors	Air intake control (manifold absolute pressure), turbocharger pressure, oil pressure, atmospheric pressure, fuel-tank pressure, brake pressure, climate control, fuel pressure, tire pressure (by September 2007, 100% of the vehicles sold on the United States markets were required to have a tire pressure monitoring system)	Micromachined sensors exist and are on the market and replacing LVDT technology
Parking sensors	Collision avoidance	Acoustic technology available
Rain sensors	Automatic	Technology available
Road/bridge condition sensors	Preventive road maintenance	Not available
Road sensors	Roughness, ice, rain	Not available at cost
Rotational speed and phase sensors	Wheel speed for antilock braking system, engine speed, camshaft, and crankshaft phase for motor control, gear shaft speed for transmission control	Available
Smart windows	Defog, clear view	Not available at low cost
Solid state cameras	Looking behind truck or car	Available
Spill detector	Clearing of truck accidents	No good solutions available
Temperature sensors	Inside and outside vehicle	Available
Torque sensors	Work	Under development

will reach $5.6 billion in 2011. The major industry sectors for pressure sensors are automotive, medical, industrial and process control, aerospace, and military. Household items and consumer products represent upcoming market opportunities.

According to Nexus, narrowing the survey down to silicon-MEMS pressure sensors, sales for first level packaged devices in 2004 were $1.084 billion, and they were expected to grow at a 7% CAGR, resulting in a $1.696 billion market in 2009.[1] Narrowing the scope of the survey even further, the automotive Si pressure sensor market was $408 million in 2004, and revenues for 2009 were expected to amount to $677 million (a 10% growth rate). Silicon

TABLE 10.5 Migration Status of Automotive Sensors from Large Electromechanical to MEMS

Application	Previous Approach	MEMS Approach	Status
Coolant pressure	Ceramic capacitive	Bonded silicon strain gage	Available (e.g., Keller, SSI Technologies, Fasco)
Exhaust gas recirculation pressure	Ceramic capacitive	Bulk-micromachined silicon	Available (e.g., Ford, Chrysler)
MAP	LVDT	Bulk-micromachined silicon	Mature (e.g., Delco, Nippondenso, Bosch) and more and more replaced by MAF
Airbag accelerometer	Breed "ball and tube" TRW Teknar "rollamite"	Surface and bulk-micromachined silicon	Mature (Analog Devices, Motorola, SensoNor, Delco, and Nippondenso)
Rate sensor	Piezoelectric	Surface and bulk-micromachined silicon	Available (Daimler Benz, Delco, Bosch)
MAF sensor	Hot wire anemometer	Surface-micromachined silicon	Available (Bosch)
Tire pressure	Not applicable	Bulk-micromachined silicon	Available (Sensonor, Motorola, TRW NovaSensor)
Oxygen sensor	Ceramic casting of YSZ	Green tape laminated planar zirconia sensors	Available but expensive (e.g., NGK's UEGO sensor)

Source: Expanded and updated from http://www.rgrace.com/Papers/auto2.htm.

MEMS costs for pressure sensors have come down from $1000 in the 1960s to $10 in the mid-1980s and, in some cases, less than $0.50 today (varies with application-mostly determined by the required packaging and testing). Pressure sensor applications in the automotive industry are listed in Table 10.4 as manifold absolute pressure (MAP), turbocharger pressure, oil pressure, barometric air pressure (BAP) sensor, fuel-tank vapor pressure, brake pressure, climate control, and tire pressure. According to the pie

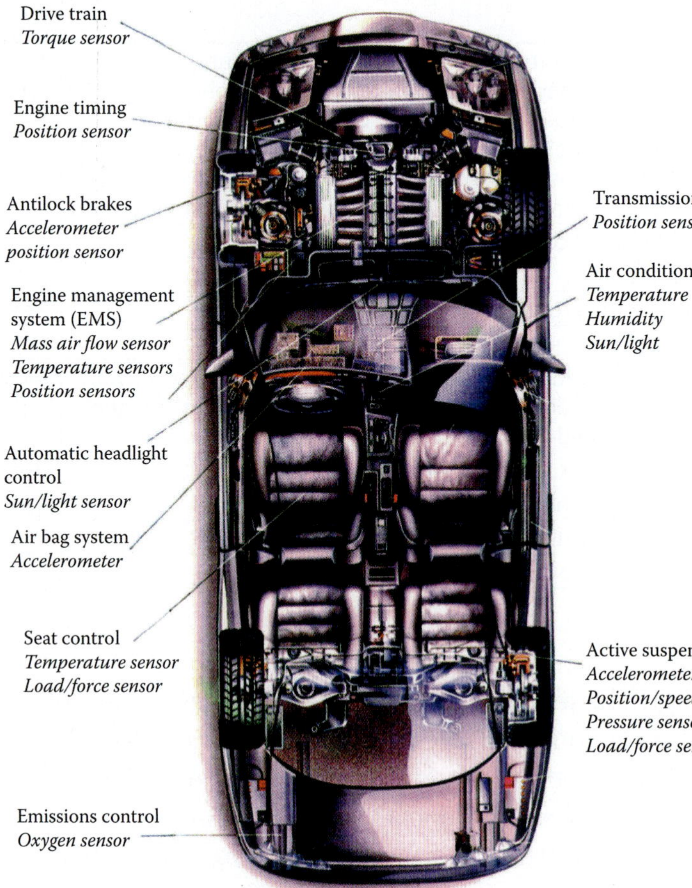

FIGURE 10.22 Automotive sensors.

Table 10.6 Comparison of Pressure Sensor Technology

Parameter	Piezoelectric	Piezoresistive	Capacitive
Self-generating	Yes	No	No
Impedance	High	Low	High
Signal level	High	Low	Moderate
High temperature	Yes	No	No
Accuracy (full scale)	±1%	±1%	±0.2%
Static calibration	No	Yes	Yes
Rugged, high sensitivity	Yes	No	Yes
Damped designs	No	Yes	Yes
Cost	High	Low	High
Electronics required	Yes	No	Yes

(For resonant techniques, see Table 10.7.)

chart in Figure 10.21, pressure sensors will make for 35% of the automotive silicon sensors in 2009 (http://www.wtc-consult.com/cms/cms/upload/PDF/Nexus_Sample_PressureSensors.pdf).

Two main varieties of Si pressure sensors used in cars are piezoresistive and capacitive. In a capacitive design, a thin diaphragm forms one plate of a two-plate device that detects capacitance changes when the diaphragm deflects. In a piezoresistive approach, deflection is measured by the change in the resistance of the silicon strain elements on the diaphragm, or directly in the diaphragm with resistors implanted into the silicon diaphragm. These pressure sensor approaches are compared with a more traditional piezoelectric sensor approach in Table 10.6. A fourth approach is based on resonant structures, as discussed below (see "Resonant Beam Sensors") and compared with piezoresistive and capacitive in Table 10.7. Also, two major manufacturing options are employed for capacitive and piezoresistive pressure sensors: bulk micromachining in single-crystal Si (SCS) and surface micromachining in polysilicon. Recently, Bosch has developed a third pressure sensor manufacturing technology: advanced porous silicon membrane (APSM). This technology makes use of porous silicon to form a cavity underneath the surface of the monocrystalline silicon, which helps to make the wafer manufacturing process fully CMOS compatible and turns the traditional bulk-micromachining process into a surface technology (see Volume II, Figure 4.78). Silicon pressure sensors cover a range spanning vacuum, low pressure (0.02 to 0.1 bar), medium pressure (0.25 to 10 bar), and high pressure (60 to more than 1000 bar).

A capacitive pressure sensor is more sensitive, is less temperature dependent, and holds up better in harsh environments than a piezoresistive one. If one relies on bulk micromachining for the manufacture of a capacitive pressure sensor, it must be carved out of a larger piece of Si to attain the same sensitivity as a piezoresistive sensor, making it more expensive. The electronics for a capacitive pressure sensor are more complicated and expensive. Most of all, in the case of a capacitive sensor, the small signals necessitate integration of sensor and electronics on the same chip. Thus, piezoresistive devices provide higher output signals and are smaller than capacitive sensors, but they require temperature compensation that may be integrated within the same silicon substrate.

The major suppliers of pressure sensors include Honeywell (United States), Bosch (Germany), Infineon (Germany), Freescale (United States), Denso (Japan), Delphi-Delco (United States), All Sensors (United States), Melexis (United States), Integrated

Table 10.7 Performance Features of Resonant, Piezoresistive, and Capacitive Sensing

Feature	Resonant	Piezoresistive	Capacitive
Output form	Frequency	Voltage	Voltage
Resolution	1 part in 10^8	1 part in 10^5	1 part in 10^4–10^5
Accuracy	100–1,000 ppm	500–10,000 ppm	100–100,000 ppm
Power consumption	0.1–10 mW	≈10 mW	<0.1 mW
Temperature cross-sensitivity	-30×10^{-6}/°C	$-1{,}600 \times 10^{-6}$/°C	4×10^{-6}/°C

Source: Greenwood, J. C. 1988. Silicon in mechanical sensors. *J Phys E Sci Instrum* 21:1114–28.

Sensing Systems, Inc. (United States), Silicon Microstructures, Inc. (United States), Kavlico (United States), Dalsa (United States), Merit Sensors (United States), Measurement Specialties, Inc. (United States), GE Nova Sensor (United States), Thales Avionics (France), MEMSCAP AS (Norway), Endevco (United States), and GE Druck (United Kingdom).

Automotive Pressure Sensor Examples

Introduction

Historically, pressure sensors in the automotive industry were used predominantly in engine management, e.g., for the measurement of manifold air pressure (MAP) at up to 5 bar. Silicon MEMS pressure sensors have been used in the automotive industry for many years; Bosch, for example, introduced its first Si pressure sensor for the automotive market in 1993. A MAP sensor determines air intake pressure to provide feedback on the correct air-fuel mixture for pollution control and fuel efficiency. Barometric atmospheric pressure (BAP) sensors compensate the air-fuel mix according to altitude. Fuel vapor pressure monitoring is a mandatory requirement on new U.S. vehicles (since 2006). Oil pressure sensors can be either ceramic capacitive, silicon strain gauges (bonded and fused to membranes), or piezoresistive, with silicon chips in silicone oil-filled reservoirs and capped by stainless steel diaphragms. Also, pressure sensors are used in high-pressure fuel injection systems, coolants, braking, and so forth.

A new market opportunity for Si pressure sensors is the direct tire pressure monitoring system (TPMS), a mandatory requirement on new U.S. vehicles (see below). Side airbags and passenger occupation detection are two additional required applications. The pressure sensor competes against accelerometers in the first case and with an infrared low-resolution imaging array in the latter. Other solutions for passenger detection include iBOLT from Bosch, which uses Hall-effect sensors. Pressure sensors react faster in the event of a side impact and are available on the BMW 5 series cars.

Automotive pressure sensor sales are projected to grow at 10% as a result of increased deployment in engine management applications such as BAP and TPMS. Side airbags and occupancy detection systems, where several different solutions are expected, will also help spur sales of automotive sensors. TPMS is driven by legislation and was expected to exceed $200 million in 2009.

Manifold Absolute Pressure Sensor

A prime example of a Si pressure sensor in the automotive application is the MAP sensor, used to control the air-fuel ratio in the engine's management system. Prices, in 2004, varied from about $3 to $5 in volume. Although MAP feeds into a major market, the application is on the decline. Because of a desire for a more direct and accurate measurement technique, MAP sensors are being replaced by mass air flow (MAF) sensors, which provide a more direct control of the engine. The major drawback to MAF sensors at this point is their high cost (e.g., about $30 for the hot wire anemometer from Hitachi), a significantly higher price than for MAP sensors (<$5). Several micromachining options are being considered in an attempt to reduce the cost of an MAF sensor, but none has yet led to a promising approach. Bosch introduced a thin-film equivalent to a hot wire anemometer in 1995. Several organizations are searching for a MEMS version of such a thin-film anemometer. In MAF-based systems, the MAF sensor replaces the MAP sensor in conjunction with a BAP sensor needed to correct for altitude. These BAP sensors are often identical to MAP sensors. Obviously, losing the MAP sensor application does not mean losing all the pressure sensor applications in engine control. In Figure 10.23, we illustrate the Bosch BAP sensor mounted in a premold housing.

FIGURE 10.23 A barometric air pressure sensor in a premold housing by Bosch Sensortec.

Tire Pressure Monitoring Systems

One of the fastest growing new applications of MEMS pressure sensors is in Tire Pressure Monitoring Systems (TPMS). The U.S. Department of Transportation's National Highway Traffic Safety Administration (NHTSA) has mandated a new safety standard to warn the driver when a tire is significantly underinflated. The mandate required manufacturers to install a four-tire TPMS that is capable of detecting when a tire is more than 25% underinflated and warning the driver (mandated compliance of 100% for passenger cars was in September 2007). This mandate had its origin in 2001, with a series of tire problems experienced by Firestone/Ford (100 accidents). Two TPMS types are available. The first is based on an indirect measurement: slight changes in wheel diameter in response to pressure loss are detected and computed by the changes based on wheel speed measurements provided by the antilock braking system (ABS). The second type of system is based on direct measurement with a pressure sensor, motion switch, and wireless transmitter installed in the tire's valve stem or wheel to make the measurement (Figure 10.24). The sensor sends information on inflation pressure levels by radio signals to a receiving unit inside the vehicle and a dashboard indicator alerts the driver. The direct method is considered to be more accurate than the indirect approach, but the advantage of greater accuracy is offset by a cost of $65 to $80 per vehicle, since a market-friendly price for a system consisting of a pressure sensor, electronics, and a wireless transmitter is $12–$15 per wheel. High-volume manufacturing should reduced the price to below $10. Sixty million TPMS units were expected in 2008 for the U.S. market alone.

Whereas the TPMS technology today involves a unit mounted in the wheel powered by a battery, in the near future we will see the arrival of the intelligent tire. In this case the TPMS unit will be mounted in the tire itself instead of on the wheel. This will represent a tremendous market opportunity: from 2010 onward, 200 million tires per year were expected on new cars and a 1 billion per year replacement tire market (two or three sets in lifetime). By that time, it was anticipated tires would feature an embedded battery-less pressure sensor/wire antenna for continuous monitoring, and additional data on wear and traction [feed data for the electronic stability program (ESP)] will be available. Interestingly, in this scenario the responsibility for safety of the passenger will shift from car to tire manufacturer.

WTC estimated that the market for TPMS systems would grow at 50% in 2008 due to the U.S. mandate described above before leveling off in 2009–2010 to a growth rate of under 10%. Technology options are resistive and capacitive pressure sensors and SAW devices. Infineon/Sensonor and GE Novasensor are the traditional leading MEMS sensor manufacturers for this application. Whereas two companies were competing in this market in 2005 (Infineon/Sensonor and GE Novasensor), as of 2008, fifteen companies had products available. Other players are

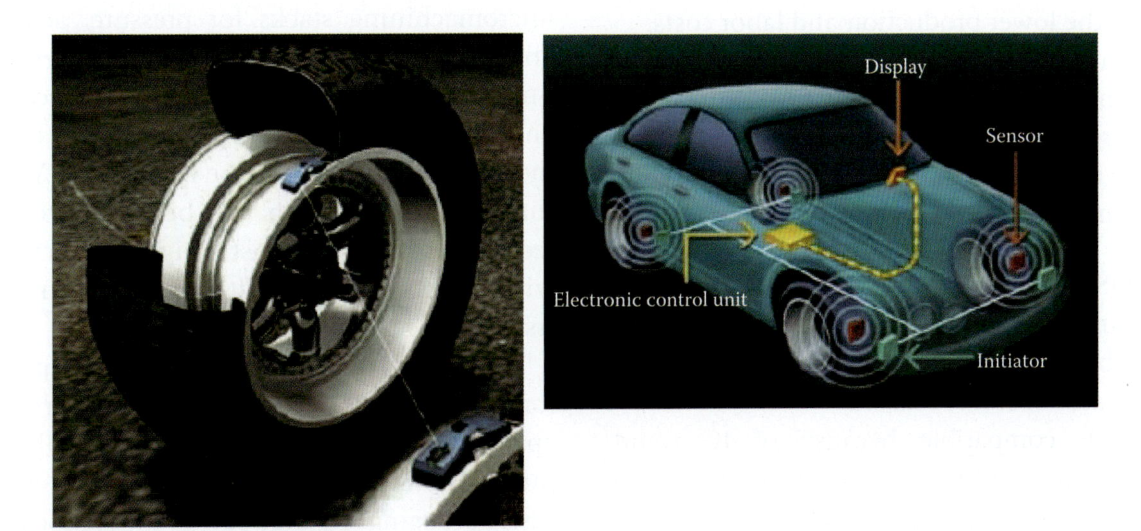

FIGURE 10.24 Microelectromechanical system pressure sensor with wireless data transfer in rim of tire.

TI, Freescale, Bosch, Melexis. Kavlico, Continental, Schraeder, Siemens VDO, and Philips (not the sensor chip). Many other companies have products under development (e.g., at LVsensors, http://www.lvsensors.com).

Important future automotive pressure sensor developments include high-pressure sensors, such as those required for cylinder pressure sensors (one per cylinder). The latter will be required for direct pressure determination inside the combustion chamber, starting with diesel engines. Lower emissions and improved fuel economy are expected to justify the higher cost. The extreme pressure rules out silicon, and in-cylinder sensors may alleviate the need for MAP and air intake MAF devices altogether. Honeywell has licensed fiber-optic pressure sensing technology for this demanding application.

Comparison of Manufacturing Approaches for Pressure Sensing

The majority of Si MEMS pressure sensor processes today are CMOS-compatible. About 55% of production is on 6-inch wafers, and this is expected to increase to 80% in 2010.[15] Manufacturing on 8-inch wafers is becoming more and more popular. It is expected that by 2011, 29% of MEMS manufacturing will be on 8-inch wafers. In 2007, Omron, Freescale, Silex, and DNP all announced new investments in 8-inch manufacturing. The pressure sensors market in the automotive world is under tremendous pressure to reduce cost. Several manufacturers have set up their production centers in Asia Pacific regions because of the lower production and labor costs.

The manufacturing approach for pressure sensing and, to some extent, for all Si micromachined sensors for large industrial market applications where cost is the main issue must incorporate as many of the following attributes as possible:

- CMOS compatibility because of the increasing demand for digital-compatible (see also resonant beam approach, described below), highly integrated Si smart sensors
- Six-inch compatible because of IC trend/availability (and 8-inch silicon MEMS is coming up fast)
- As little fixturing as possible
- Al- (in the future, this might become Cu instead) and IC-line-compatible etchants
- Generic, making it easy to implement a different sensing range or different package or even a different sensor type
- Highest accuracy potential (favors capacitive and resonant devices; see below)
- Wafer-level isolation of electrical connections, making it easier to use in harsh media (removing sensor metallization from the media themselves)
- Dry etching whenever possible
- Single-sided processing if feasible

Polysilicon now holds the cost advantage over a silicon-on-insulator (SOI) approach, but polysilicon will not readily lead to a high accuracy method. Like SOI, bulk micromachining constitutes a more secure option to meet the accuracy demands of future pressure sensor generations but often involves too many glass and/or Si wafers stacked up in one device to allow for an economic solution. Moreover, bulk micromachining is a more expensive technology and up to recently less CMOS compatible than polysilicon. We say up to recently, since Bosch succeeded in making bulk micromachining more CMOS compatible [see the Bosch advanced porous silicon membrane (APSM) process in Volume II, Chapter 4, (Figure 4.77)]. SOI might be able to compete better with polysilicon, as clever designs could obviate the complicated multiple wafer solutions that characterize today's bulk micromachining stacks for pressure, acceleration, and gyros (a three- to five-wafer stack design is not uncommon). SOI technology will also become more and more available and less expensive as it becomes more commonly used in the IC industry. We believe that the SOI approach will become a competitive technology for low-cost, high-accuracy sensors without the reproducibility and long-term stability issues of the polysilicon material.

SOI wafers are finding increasing applications for pressure sensor devices, especially for industrial applications. Most pressure sensors still are bulk micromachined, so wet etching remains a key process. Packaging is a large part of the component cost, and wafer bonding for wafer-level packaging (WLP) is a growing trend (see Chapter 4, this volume, on packaging, assembly, and self-assembly).

We can summarize that, for now, an inexpensive solution for a pressure sensor with moderate accuracy is based on polysilicon. In the long run, a single-crystal approach based on the Bosch APSM process or an SOI approach, perhaps with a resonant design (see next), is more suited for a generic strategy to produce higher accuracy devices.

Piezoresistive, Capacitive, Resonant Beam, and Other Sensor Approaches

Introduction

The arguments we present in this section about what sensor mechanism to choose for a given pressure sensor application pertain equally well to accelerometers and angular rate sensors. The impetus for obtaining higher accuracy and sensitivity in automotive sensors usually hinges on energy efficiency and/or regulatory mandates. This may require a review of the transducing methods, currently still dominated by piezoresistivity. Examples of alternatives include optical techniques such as Honeywell's in-cylinder fiber-optic pressure sensor, the significantly higher sensitivity of both capacitive and resonant approaches, piezoelectric sensing (mainly for high frequency applications), and tunneling current sensing (very sensitive but increased complexity). The inherent digital compatibility of resonant beams may propel them to the foreground for high accuracy and very low-signal sensor applications. Because of the promise held by the resonant sensing application, we detail this technology a bit further here.

Resonant Beam Sensors

Electrical readout in a pressure sensor, an accelerometer, or angular rate sensor may be performed using piezoresistors located in a flexure area, by measuring the change in capacitance of the moving members with respect to the fixed member(s), or by the measurement of frequency change of a small resonant structure. A resonator is a mechanical structure fabricated to vibrate at a particular resonant frequency. The resonant frequencies of such microresonators can be made extremely stable, so they are used as a time base (the quartz tuning fork in watches, for example) or as the sensing element of a resonant sensor. The performance benefits of a well-designed resonant sensor are compared with piezoresistive and capacitive techniques in Table 10.7.[17] A common advantage of resonant beam and piezoresistive is that they do not depend on measurements relative to "fixed" members, as the capacitive approach does. The small signal magnitude (small capacitance change ΔC) of a capacitive sensor means that on-chip electronics are mandatory to avoid errors due to stray capacitance variations. Piezoresistive and capacitive sensing elements must be used with high-precision amplifiers and trimmed analog circuits to accurately sense and condition small changes in the value of the electronic components. These sensitive circuits and small signals can be highly susceptible to fluctuations in temperature, electrical noise, mechanical and electrical drift, temperature and pressure cycling hysteresis, ionizing radiation, and magnetic fields. These factors and others result in measurement inaccuracies, effectively degrading overall precision and resolution. As a result, best-case production performance of solid-state sensors generally has been limited to a long-term accuracy of about 0.1%, or 1000 ppm.

Resonating devices, on the other hand, are affected primarily by the mechanical characteristics of the structure, not by its electrical behavior. In addition, since the system itself measures only frequency, the interface to a digital system vastly is simplified. In practice, resonant sensors can attain measurement accuracies at least 10 times higher than conventional sensing methods. The fabrication of resonant structures, however, is more complex, and the requirement for packaging such devices more demanding than for piezoresistive or capacitive sensors.

The basic building blocks for a resonant beam structure embedded in a pressure-sensitive membrane are shown in Figure 10.25.[18]

The specific structure shown here is one by the Yokogawa Corporation. To operate as a differential pressure sensing device, this chip incorporates two H-shaped bridges, located in the Si substrate and operating at the same identical frequency (90 kHz). Both bridges (resonators) are located at the same strain-free plane (depth) from the chip's surface, but in different locations. A resonant sensor is designed

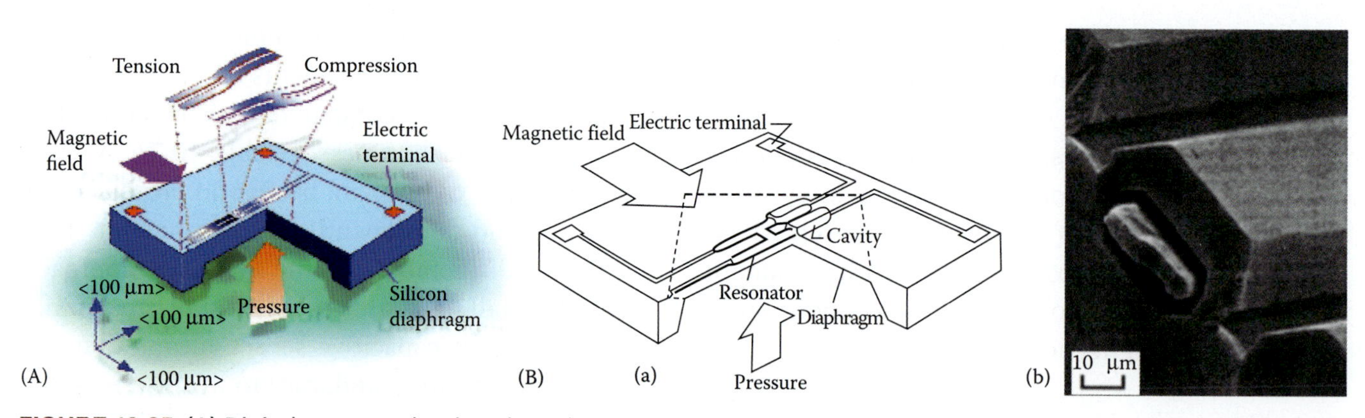

FIGURE 10.25 (A) Digital sensor technology by Yokogawa Corporation. (B) Construction of the sensor (a) and cross-sectional scanning electron microscope photograph of the resonator (b).[18]

such that the resonator's natural frequency is a function of the measurand (pressure in this case), and typically the measurand alters the stiffness, mass, or shape of the resonator, hence causing a change in its resonant frequency. In the current case, when pressure is applied, because of their position and exact depth in the substrate, one bridge goes into tension and the other into compression. The other important features of a resonant sensor are the vibration drive and detection mechanisms. The Yokogawa chip incorporates an electromagnetically driven resonant beam sealed in a vacuum shell, integrated onto a standard silicon pressure sensor diaphragm (an SOI approach is used). The drive mechanism excites the vibrations in the structure while the detection mechanism senses these vibrations. The frequency of the detected vibration forms the output of the resonant device. As pressure is applied the compression bridge changes from 90 to 110 kHz and the tension bridge from 90 to 70 kHz. To obtain a differential output the microprocessor simply counts the change in bridge frequencies (Δ40 kHz). In the case of gauge pressure measurement, both bridges are used. One bridge measures the reference pressure (atmospheric or head pressure), and the other one measures the applied pressure (gauge).

With the digital pressure sensor pictured in Figure 10.26, it was demonstrated that an accuracy of 0.01%, or 100 ppm, is attainable. The device also offers wide dynamic range, high sensitivity, and high stability, along with a frequency output for easy interface to digital compensation circuitry. The chip fabrication is challenging though, comprising four sequential selective epitaxial growth and critical selective etching steps, and the sensor package is extremely complex and therefore expensive to produce. It is, nonetheless, an excellent example of the power of silicon microfabrication technology

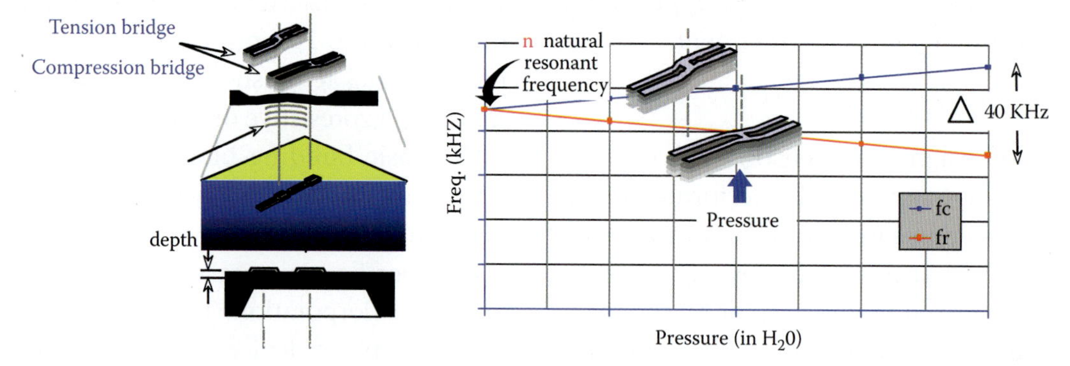

FIGURE 10.26 Operating principle of the digital pressure sensor. The first bridge is located directly in the center of the chip and the other nearer the edge. As pressure is applied, the center bridge will be in a state of "compression" and the outer bridge in a state of "tension." As a result, the natural frequency of the two bridges changes. The compression bridge frequency increases and the tension bridge frequency decreases, thereby effectively creating a differential frequency. The two frequencies are easily managed by a CPU as a time-based calculation directly proportional to the applied differential pressure. (From Yokogawa Corporation of America.)

and represents a major technical achievement. Morover, Yokogawa's application represents the industrial sensing and control business, which is aimed at high-performance sensors that can sell for higher prices in modest volumes.

GE Druck also developed and incorporated a resonant silicon sensor in its pressure transducers (http://www.gesensing.com/druckproducts), as well as GE NovaSensor.[19] Zook et al.[20] fabricated clamped-clamped beams in a polysilicon surface micromachining process and, by locating the polysilicon beam in a high-vacuum polysilicon shell, avoided damping. Surface micromaching simplified the manufacturing process over previous attempts in SOI by Ikeda et al. shown in Figures 10.25 and 10.26[21] and may be able to produce cost-effective resonators as replacements for piezoresistors. The sensitivity to applied axial loads of the beam exceeds the gage factor of single-crystal piezoresistors easily by a factor of 10–100. Devices with sensitivities as high as 700 Hz/g have been reported.[22]

Today, the resonant beam approach remains the least-developed sensing mode, despite the fact that the resonant beam method showed very good potential in the past (although expensive and difficult to implement). So it is interesting that there is not more commercial activity to validate this sensing approach further and making these resonant sensing devices less expensive. It could be that one of the main reasons why direct frequency sensing is not investigated more is the fabulous resolution ADI achieved with its capacitive sensing scheme. Because the signal conditioning is so tightly coupled to the sensing element, the integrated ADI sensors can measure extremely small changes in differential capacitance. As an example, the capacitance resolution of their inexpensive ADXL202E, a surface-micromachined 50-mm^3 accelerometer, is as low as 20 zeptofarads (20×10^{-21} F) (see next section).

Accelerometers

Introduction and Market Perspectives: Victim of Success

WTC estimated that in 2006, the silicon-MEMS market in the automotive sector was $1.6 billion (this firm only counted flow, pressure, acceleration, and rate), making this the second biggest MEMS opportunity after RW-heads and ink-jet print heads. By 2011, the same market was pegged at $2.2 billion, with a CAGR of around 7%. The main applications in revenue terms were, in order, pressure sensors, gyroscopes, accelerometers, and flow sensors, as shown in Figure 10.27.

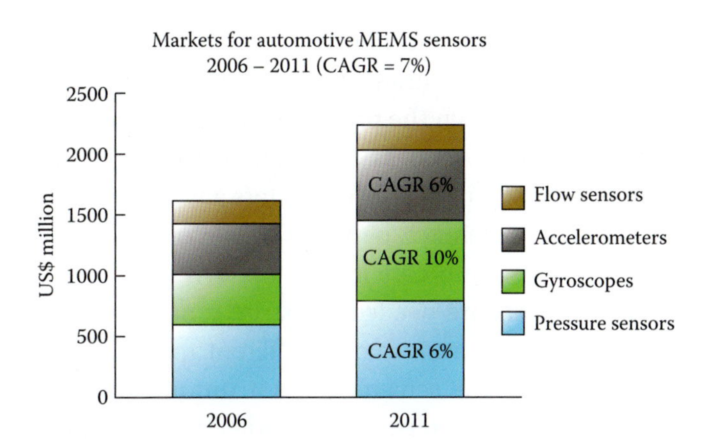

FIGURE 10.27 Markets for automotive microelectromechanical system sensors, according to WTC.

The total number of Si sensors WTC claimed would grow from over 430 million units in 2006 to 780 million in 2011, an annual growth of 13% (for a reference point, consider that by 2006, Bosch had produced 600 million sensors, 130 million units in 2006 alone). This outstrips the dollar growth because of price erosion, running at 4–5% per year.

Yole estimated the 2005 worldwide accelerometer market at $393 million, representing about 8% of the overall silicon-MEMS market. By 2010, they expected the accelerometer market to be $869 million, for a five-year CAGR of 17%.

Major players in this space include Bosch, Infineon, ADI, Denso, and Delphi. Since the early 2000s, manufacturers have been putting accelerometers in consumer electronics such as laptops [it started when IBM (now Lenovo) put a dual-axis accelerometer in its notebook in 2004], cameras, and cell phones. WTC fears that in coming years the accelerometer market will suffer from price pressures reminiscent of that in consumer electronics (10% per year). Accelerometers might become victims of their success!

Accelerometers signified the next big Si micromachining entry in the commercial world after the Si pressure sensor. Today, polysilicon surface-micromachined capacitive and single-crystal Si bulk-micromachined accelerometers have taking over a large share of the automotive acceleration market. Manufacturing volumes of tens of millions a year have brought down the price to less than $2 per sensor. A company such as ADI makes 2 million surface-micromachined accelerometers a week.

A major difference between accelerometers and pressure sensors in cars is that less diverse types of accelerometers are needed. Pressure sensors require various ranges and many different packages to accommodate the sensors in their unique environments. Accelerometers, on the other hand, require little difference in packaging; two main types (i.e., high *g* and low *g*) cover all needs. This explains why the first surface-micromachined commercial Si sensor was an accelerometer rather than a pressure sensor.

The first demonstration of a capacitive Si micromachined accelerometer took place in 1979 at Stanford University.[4] Over the years, a variety of accelerometers based on different physical detection schemes, such as piezoelectric, piezoresistive (thin- and thick-film), capacitive, tunneling, and resonant members, have been attempted in a wide variety of manufacturing schemes. As listed in Table 10.4, in cars, accelerometers are used as crash detection sensors both in the electronic control unit (ECU) and in peripheral sensors such as upfront sensors (UFS) or pedestrian contact sensors close to the bumper, peripheral acceleration sensors (PAS) in the doors, and rear crash sensors.

Automotive Accelerometer Examples

Airbag Accelerometer

The principal mass market for Si acceleration sensors in the automotive industry is in crash detection and airbag release systems. A MEMS-based accelerometer detects with great precision the sudden slowing of a vehicle and fires the airbag when needed. Now produced in the millions per year, these devices have gotten smaller, cheaper, and more capable, riding the learning curve originally promised by MEMS proponents. This learning experience by Bosch-Sensortec is illustrated in Figure 10.28 as a sequence of product generations for automobile airbag systems introduced over time. The MEMS elements are not only getting much smaller than the formerly used fine-mechanical piezoelectric device (shown in Fig. 10.28a) but also much more cost efficient, and yet they include an increasing number of functions (such as self-tests and accuracy check routines).

In 1991, the Intermodal Surface Transportation Efficiency Act (ISTEA) was signed into law. ISTEA airbag requirements ensured that by 1998, 100% of the new U.S. vehicle fleet be equipped with airbags. System integrators like Autoliv expect airbag accelerometer penetration to increase from 40 to 60 million vehicles over the five-year period between 2006 and 2011 (i.e., 80% of cars worldwide). WTC estimated that this penetration equals an insertion of 200 million accelerometers in 2006 going up to 350 million in five years, with as many as eight accelerometers per electric control unit (ECU).

An earlier generation of airbag sensors used in the United States consisted of electromechanical types supplied by Breed and TRW (see Table 10.5). Three electromechanical accelerometers discriminated between zones according to the severity of a vehicle crash. Two sensors were used in the crush zone, usually on frame rails behind the front bumpers or on the lower portion of radiator supports. These sensors determined the severity of impact and activation of the airbag.

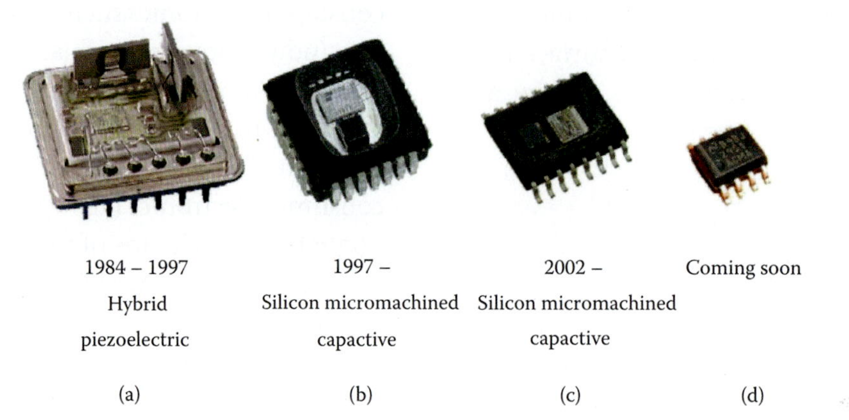

FIGURE 10.28 Generations of high-*g* acceleration sensors used in automotive airbag systems at Bosch.

Another sensor was located outside the crush zone, either in or near the passenger compartment. This sensor, often known as the *safing sensor*, prevented deployment of the airbag in the event someone inadvertently impacted (or tampered with) the front sensors.

Micromachined accelerometers displaced these electromechanical devices, as they made single-point sensing possible. Single-point sensing requires only one accelerometer, located in the passenger compartment, reducing the number of sensors and associated wiring needed in the system. In addition to reducing the number of sensors required in each system, surface-micromachined accelerometers from ADI were costing approximately $4 per unit in volume in 1998 (and below $2 today), significantly less than electromechanical accelerometers at approximately $15 per unit.

The early bulk-micromachined Si accelerometers lacked damping and overload protection until TRW NovaSensor introduced a piezoresistive accelerometer combining a microactuator for reliable self-testing and overload protection on the same chip. This Si bulk-micromachined ±50-*g* accelerometer based on hybrid technology, with signal conditioning electronics on a separate chip, bonded into a TO-5 header, is shown in Chapter 4, this volume, Figure 4.9.[23] In 1991, ADI introduced the first commercial product to use polysilicon surface micromachining, the so-called *ADXL-50*, a capacitive accelerometer with an electrostatic actuator built in for self-testing. The construction of this sensor was studied in Volume II, Example 7.9 (see also Volume II, Figures 7.92–7.96).

More and more vehicles are being equipped with side airbags, ensuring continued growth of the accelerometer market. Primarily ADI, GM/Delco, and Motorola supply silicon accelerometers in the 50-*g* range for U.S. vehicles. Sensonor (Norway) has historically provided a large portion of the European market, while Nippondenso is the major provider in Japan. The ADI accelerometer was the first integrated sensor; most other accelerometers on the market still use a multichip solution to sense and provide the appropriate signal condition.

Suspension Accelerometers

ADI, between 1990 and 2000, branched out from their high-*g* range of airbag (50 *g*) accelerometers, to medium and low-*g* sensors for suspension and stability control devices (say, 0–2 *g*).

Silicon accelerometers in suspension systems face more hurdles than the Si accelerometers in airbag systems. Low *g* is more difficult (signal-to-noise limit), signal conditioning is required with each remotely mounted package (one suspension sensor per wheel), and the sensor is exposed to the harsh automotive environment, i.e., more complex packaging. Many companies, including ADI and Motorola were and still are aggressively pursuing this application. Low-*g* accelerometers are also of utility in antilock braking systems (ABS) featuring traction control. No federal law mandates ABS/traction control, and because of the cost constraints, only high-end and midsize cars are expected to utilize the system.

In 1995, using its patented iMEMS® (integrated MEMS) approach, ADI introduced a 5-*g* accelerometer, pushing the polysilicon surface micromachining technology it pioneered with their 50-*g* device (the ADXL50) a bit further. Also out of ADI, in 1998, came the breakthrough 2-*g* dual-axis ADXL202E accelerometer, an inexpensive, surface-micromachined accelerometer capable of measuring both static (gravitational force) and dynamic (vibration and shock) forces, pushing the surface micromachining sensor technology even further. This sensor detects movements on two axes and outputs digital signals to microcontrollers as simple as 4-bit microcontroller units (MCUs). A single moving "beam" in the ADXL202E has "fingers" that are used to detect beam movement (through capacitance change) in two directions, and this goes through an on-chip algorithm and is sent along as digital data. In Figure 10.29a we show how silicon is used to make the spring and proof mass and how a finger is added to make a variable differential capacitor. The beam is made up of many interdigitated fingers, and each finger can be visualized as shown in this figure. The electronics measure the change in displacement by measuring the change in differential capacitance against a set of fixed fingers (plates). The critical dimensions of fingers and the proof mass (0.7 μg) of this sensor are clarified in Figure 10.29b. The finger height (2 μm for the example product) is fixed by process technology, while the overlap (125 μm here) is adjustable to

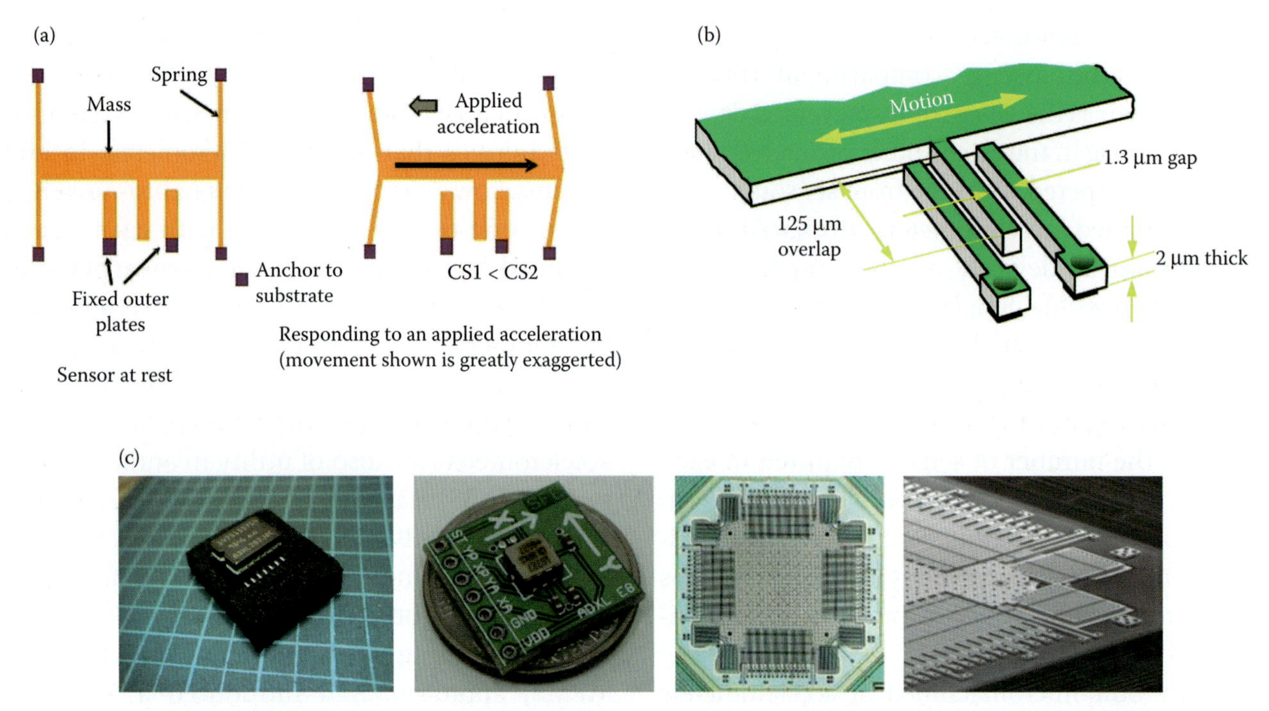

FIGURE 10.29 Analog Devices' ± 2 *g* dual-axis ADXL202E accelerometer is an inexpensive, surface-micromachined accelerometer capable of measuring both static (gravitational force) and dynamic (vibration and shock). (a) Silicon is used to make the spring and proof mass, and fingers are added to make a variable differential capacitor. (b) Critical dimensions and proof mass weight and dimensions for a single finger. The proof mass is 0.7 μg. The accelerometer has some impressive numbers: 0.1 pF per side for the differential capacitor, 20 zF (10^{-21} F) smallest detectable capacitance change, and 2.5 pm minimum detectable beam deflection (one tenth of an atomic diameter). (c) In the ADXL202 beam structure, note the four pairs of orthogonal serpentine springs at the corners of the proof mass. The entire mechanical structure is suspended from the four points where the springs meet.

some extent. Longer fingers are not desirable, because they are harder to manufacture and result in larger beam areas, which in turn translates to more expensive sensors. The fingers are released when the oxide layer between the underlying bipolar IC and the beam is etched away, leaving a 1.6-μm gap with fingers spaced every 1.3 μm. Rather than using two discrete beams placed orthogonally, the fingers that constitute the *x*- and *y*-axes, variable capacitors are integrated along the sides of a single square beam. This results in a reduction of the overall sensor area, yet the larger common beam mass enhances the resolution of the ADXL202E. The differential capacitance of each finger is proportional to the overlapping area between the fixed outer plates and the displacement of the moving finger. Because these are very small capacitors, one needs the largest practical differential capacitance to reduce noise and increase resolution. The fixed outer plates are driven with square waves that are 180 degrees out of phase. When the movable fingers (and hence the beam) are centered between the fixed outer plates, both sides of the differential capacitor have equal capacitance and the AC voltage on the beam is zero. However, if the beam is displaced due to an applied acceleration, the differential capacitance will be unbalanced, and an AC voltage of amplitude proportional to the displacement of the beam results. In this sensor movements as minute as 0.02 Å are detectable. Because of the combination of low cost and high performance acceleration/tilt sensing possible with this device, other applications have become possible. The ADXL202E is currently used in car alarms (for jacking or towing sensing), pen-based handwriting recognition systems, portable electronic games, and wearable sports equipment.

The ADXL202E was an early demonstration that there is a path for MEMS from automotive applications to consumer electronics. The average price of MEMS accelerometers across all applications is decreasing, from an average of $2.50 in 2004 to less than $1.90 in 2009, with consumer applications driving the price erosion. The average selling price (ASP) for a three-axis MEMS accelerometer was still approximately $2.00 in consumer products in 2006, while lowest cost products were already shipping at under $1.50 in 2006. It was expected that by 2009, the average selling price would drop to $1.35, and the least expensive (first-level packaged) three-axis accelerometers would ship for under $1.00.

Comparing Manufacturing Options for Accelerometers

In the automotive market 80% of accelerometers are manufactured using deep reactive ion etch (DRIE) equipment. Growth in the accelerometer market overall actually is driving growth of DRIE equipment sales. The primary wafer size for accelerometers is currently 6-inch (accounting for more than 80% of device production), but 8-inch wafers are expected to make inroads and were projected to account for nearly 10% of production by 2010. Devices are typically processed on silicon, with SOI wafers in use or in development by the major manufacturers.

Obtaining sufficient sensitivity was originally thought to be a potential problem when using polysilicon to manufacture low-*g* accelerometers. To make a large proof mass is especially challenging for polysilicon accelerometers. Process control and handling considerations limit the minimum flexure thickness for suspending the inertial mass to about 2 μm and, given the relatively low density of Si, obtaining deflection at 1 or 2 *g* without making the proof mass large is difficult. Realizing a large proof mass directly affects chip size and thus cost, presenting the developer with an unattractive option. The proof mass of a bulk-micromachined accelerometer may be 0.1 mg, whereas a typical value for a surface-micromachined device is 0.5 μg, 200 times smaller. With a source capacitance in the neighborhood of 150 femtofarads (i.e., 0.15 pF), and orders of magnitude smaller differential capacitance changes, say 100 attofarads (10^{-18} F) for a 1-*g* acceleration, a resolution of 23 zeptofarads (10^{-21} F) is required, and the signal output of these devices easily could be lost in the presence of parasitic capacitance or noise. Despite these early misgivings expressed here, the surface-micromachined ±2-*g* capacitive accelerometer announced in 1995 by ADI demonstrates that low-*g* polysilicon sensors and cost-effective integration of all the electronics is quite possible. It is actually within ADI's abilities to provide on-chip signal conditioning that allows the minute signals to be read with minimal interference. A synchronous demodulation technique ("AC bridge") is used to increase noise immunity and improve the signal to noise ratio. A square-wave carrier drives the sensor, and the modulated signal produced by the change in differential capacitance is amplified and synchronously demodulated, minimizing noise and boosting the signal-to-noise ratio.

In silicon bulk-micromachined accelerometers, the formation of a mesa structure (proof mass) from a single crystal Si is more complicated because of rapid and irregular etching of convex corners by orientation-dependent etchants, such as ethylene diamine pyrocatechol (EDP) and KOH. Corner compensation schemes must be invoked to avoid irregular corners (see Volume II, Chapter 4).

Critics of the capacitive polysilicon accelerometer still maintain that the integration of electronics is too complex and that a bulk-micromachined sensor requires only twelve mask levels, whereas the polysilicon device requires twenty-eight mask levels. Along this line, major Japanese companies have chosen single-crystal solutions for building accelerometers to avoid the difficulties associated with reproducing polysilicon material properties and the possible unknown long-term instabilities.

The pros and cons of surface micromachining (see Volume II, Chapter 7) can then be summarized as follows:

Pros

- A surface-micromachined device has a built-in support, resulting in cost effectiveness.
- Surface micromachining only requires one-sided processing compatible with conventional IC processes.
- Devices made by surface micromachining are smaller in size, have higher yields, and accommodate more on-chip circuitry.

Cons

- The out-of-plane dimensions in surface micromachining in general is limited to ~2 μm, hence the light mass of the proof mass, resulting in a low sensor sensitivity.

- The suspended Si structure is more compliant (due to the high ratio of length to thickness of the device structure) and hence has a tendency to stick to the support substrate, especially in overdrive situations.

As we made clear in the section on pressure sensing, we believe that SOI surface micromachining combines the best features of surface micromachining with the best features of single-crystalline Si. The same benefit befalls the Bosch IMAPS process, which enables electronic integrations just as in the case of ADI (see above). Using dry etching to etch the proof mass from an epimaterial avoids corner compensation issues; the epimaterial ensures larger mass and single-crystal performance. This approach might overtake both surface and bulk micromachining as SOI wafers become less expensive and more available. A generic design of a resonant beam structure in an SOI wafer may be possible. This design could constitute the basis of a wide family of pressure, acceleration, and angular rate sensors.

MEMS Angular Rate Sensors

Introduction

First, some clarifications are in order about the definitions of an angular rate sensor, gyros, and accelerometers. An angular rate sensor is not a gyro, but it is common jargon to refer to an angular rate sensor as an *electronic gyro* or even simply a *gyro*. In reality they are quite different, although one can be made to act like the other. Angular rate sensors measure angular rate (surprise!), i.e., how fast (degrees per second or radians per second) something is rotating. A gyroscope, on the other hand, is used to measure angular position (not rate), using the principle of "rigidity in space" of the gyroscope (see Figure 10.30). For example, this is how the attitude indicator (artificial horizon) of an aircraft works. A gyroscope can be made to measure angular rate (using the principle of precession), but that requires a finely calibrated measuring spring. Conversely, an angular rate sensor can be made to measure angular position by integrating the signal over time. So, an angular rate sensor can be made to act like a gyro, and a gyro can be made to act like an angular rate sensor, but they do not act the same by themselves. Lastly, an accelerometer is nothing like a gyro and cannot be made to act like one. An accelerometer measures linear acceleration. Using calculus, this can be integrated into linear velocity and position. The rotation typically is measured in reference to one of three axes: yaw, pitch, or roll (see Figure 10.30). The difference between an accelerometer and an angular rate sensor is that the gyro measures the angular acceleration (how fast the rate is at which an object rotates changes) and the accelerometer measures how much something is accelerating (not angular).

As an example of how an angular rate sensor could be used, a yaw-axis angular rate sensor that is mounted on a turntable rotating at 33 1/3 rpm (revolutions per minute) would read a constant rotation of 360° times 33 1/3 rpm divided by 60 seconds, or 200°/s. In the case of an electronic angular rate sensor, a voltage proportional to the angular rate, as determined by its sensitivity, measured in millivolts per degree per s (mV/°/s), would be the output. The full-scale voltage range determines how much angular rate can

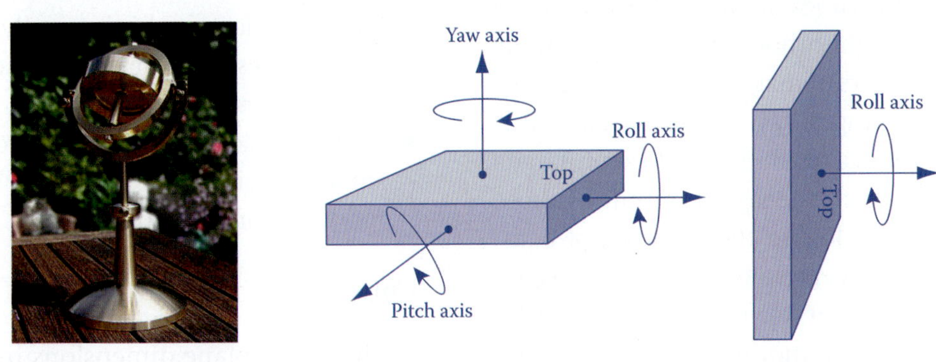

FIGURE 10.30 Illustration of a conventional mechanical gyroscope and the three rotational degrees of freedom it can measure (http://www.brassgoggles.co.uk/images/brassgyroscope.jpg). Shown are gyro axes of rotational sensitivity. Depending on how a gyro normally sits, its primary axis of sensitivity can be one of the three axes of motion: yaw, pitch, or roll (see also Figure 10.31).

be measured. In the turntable example, an angular rate sensor would need to accommodate a full-scale voltage corresponding to at least 200°/s. The full-scale range is limited by the available voltage swing divided by the sensitivity. The ADI MEMS angular rate sensor we will analyze below, the ADXRS300, comes with 1.5 V full-scale and a sensitivity of 5 mV/°/s, so it handles a full-scale of 300°/s. One practical application of an angular rate sensor is to measure how quickly a car is turning. If the angular rate sensor, mounted in the car, senses that the car is spinning out of control, differential braking engages to bring it back into control. As we saw earlier, the angular rate signal can also be integrated over time to determine angular position particularly useful for maintaining continuity of GPS-based navigation when the satellite signal is lost for short periods of time, for instance, when the car is in a tunnel.

Before satellite-based global positioning systems (GPSs), there was the mechanical gyroscope: a reliable rotation sensor, based on the inertial properties of a rapidly spinning rotor. This mechanical contraption, which has been around since the late 1700s, is used for maintaining a fixed orientation with great accuracy regardless of the earth's rotation, as shown in Figure 10.30. Traditional spinning gyroscopes work on the basis that a spinning object that is tilted perpendicularly to the direction of the spin will have a precession. The precession keeps the device oriented in a vertical direction so that the angle relative to the reference surface can be measured. A rotation in the frame thus imparts a torque (rotation) on the spinning disk, which precesses (rotates) as a result (conservation of angular momentum). In other words, the flywheel's large angular momentum counteracts externally applied torques and keeps the orientation of the spin-axle fixed. Practical applications limit the movement to measure only one axis of rotation (roll, pitch, or yaw; see Figure 10.30). Precision-machined mechanical gyros may cost anywhere between \$10,000 and \$100,000.

The ring-laser gyroscope, first discussed by Georges Sagnac (1869–1926) in 1913 and demonstrated in 1962 by Warren Macek of Sperry Corporation, quickly displaced mechanical versions for high-end and high-precision applications such as aviation and is now even used in high-end cars. Sagnac discovered that light sent around a closed loop, in two different directions, would show a phase difference between the two beams when the loop is rotated (refer here to the discussion of the Michelson-Morley experiment in Volume I, Chapter 5, "The Lorentz Transformation"). The Sagnac effect is illustrated in Figure 10.31. The ring-laser gyroscope consists of two laser beams traveling in opposite directions (i.e., counter-propagating) around a closed-loop path. The pulse traveling in the same direction as the rotation of the loop must travel a slightly greater distance than the pulse traveling in the opposite direction. As a result, the counter-rotating pulse arrives at

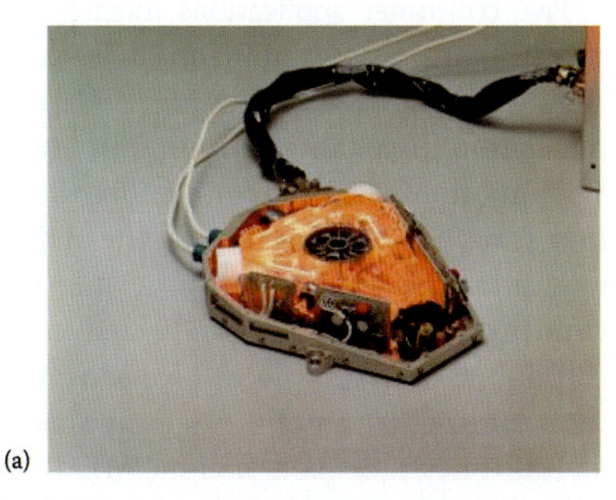

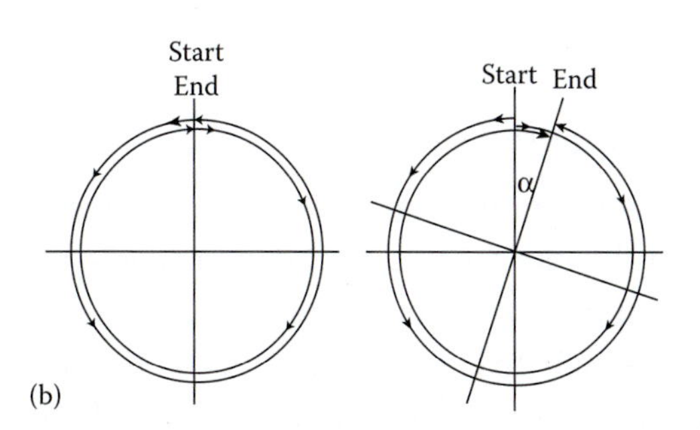

FIGURE 10.31 (a) The RLG. (b) The Sagnac effect: if two pulses of light are sent in opposite directions around a stationary circular loop of radius R, they will travel the same inertial distance at the same speed, so they will arrive at the end point simultaneously. This is illustrated in the left-hand drawing. The drawing on the right indicates what happens if the loop itself is rotating during this procedure. The symbol α denotes the angular displacement of the loop during the time required for the pulses to travel once around the loop. For any positive value of α, the pulse traveling in the same direction as the rotation of the loop must travel a slightly greater distance than the pulse traveling in the opposite direction. As a result, the counter-rotating pulse arrives at the "end" point slightly earlier than the corotating pulse.

the "end" point slightly earlier than the co-rotating pulse. The constructive and destructive interference patterns formed by splitting off and mixing parts of the two beams can be used to determine the rate and direction of rotation of the device itself.

The fiber optic gyroscope (FOG) is based on the Sagnac effect and consists of a loop of single-mode optical fiber (often a polarization-maintaining fiber) and related coupler components, a semiconductor laser or superluminescent LED, and signal-processing electronics. The coupler components generally are fabricated in proton-exchanged $LiNbO_3$ integrated-optic circuits. This material is chosen for its ability to modulate the light beam for improved detection. The FOG, being a simpler device, currently is receiving more attention due to its potential to achieve the required performance at a lower cost than with ring laser gyroscopes (RLG) or mechanical gyroscope technology.

These instruments still cost about $1000. For example, the fiber optic gyro from Andrew Corp. (now KVH Industries) costs $950. It is a cylindrical device that measures 77 mm in diameter and is 88 mm long. The higher-end Honeywell Ring Laser Gyro (RLG) has revolutionized the way the world navigates on land, sea, air, and space. It is highly reliable because it has virtually no moving parts, replacing spinning mass with spinning light.

Interferometric FOGs and MEMS angular rate sensors are changing the commercial landscape of gyros once again. We cover only MEMS gyroscopes (angular rate sensors really) here; they are constructed out of silicon (often SOI) and quartz, or they may be ceramic-based.

Market Perspectives

The requirements for inertial sensors, i.e., accelerometers and angular rate sensors, vary drastically depending on the applications they are intended for. Figures 10.32a and 10.32b give an overview of the main performance requirements for accelerometers and angular rate sensors, respectively, in various applications.

In the late 1990s, inexpensive, solid state, micromachined angular rate sensors became the Holy Grail for many micromachinists around the globe. The same companies that made a lot of headway in micromachined accelerometers, such as ADI, are also at the forefront of Si angular rate sensor development today.

In 2005, according to Yole, global MEMS-based angular rate sensors sales were $558 million, and they were expected to grow by a CAGR of 11% through 2010 to reach $930 million. MEMS gyroscopes are included in three automotive applications: Antirollover systems, global positioning system (GPS) navigation, and electronic stability control (ESC). These automotive applications represent more than 70% of the MEMS gyroscope market. ESC is a rapidly growing market as these systems could potentially reach 100% adoption in new cars shortly after 2010. Current navigation system designs use a combination of global positioning (GPS) and CD-ROM maps in addition to wheel rotation sensors and angular rate sensors, or magnetic compasses. Angular rate sensors enhance the accuracy of navigation instruments in situations where satellite reception is lost, as for example in tunnels, between tall buildings, and in mountainous terrain. The cost of current navigation systems is in the $200 to $800 range, and MEMS solutions are expected to reduce this cost further.

As in the case of accelerometers, once gyros make it big in the automotive markets, they also succeed in consumer electronics (CE). The sensor specifications in consumer electronics are much more relaxed than in the automotive application, but price erosion and product cycles are much faster than in the automotive field.

Five consumer applications integrate MEMS gyroscopes: camcorder stabilization, camera stabilization, cell phone stabilization, video games, and navigation for small electronics. In 2006, Nintendo Wii's motion sensing remote control gave, for the first time, public visibility to inertial MEMS. The new attractive human-machine interface in game controls such as the Wii is expected to penetrate the cell phone market as well. Five sensors are used to monitor orientation of the Segway scooter, sampled at 100 times/s. Sensors include MEMS angular rate sensors and liquid-filled tilt sensors.

According to Yole, quartz and ceramic substrates were estimated to account for 37% by volume of total gyroscope production in 2005. But most new gyroscope development programs were silicon-based, so this figure was expected to decrease to 28% in

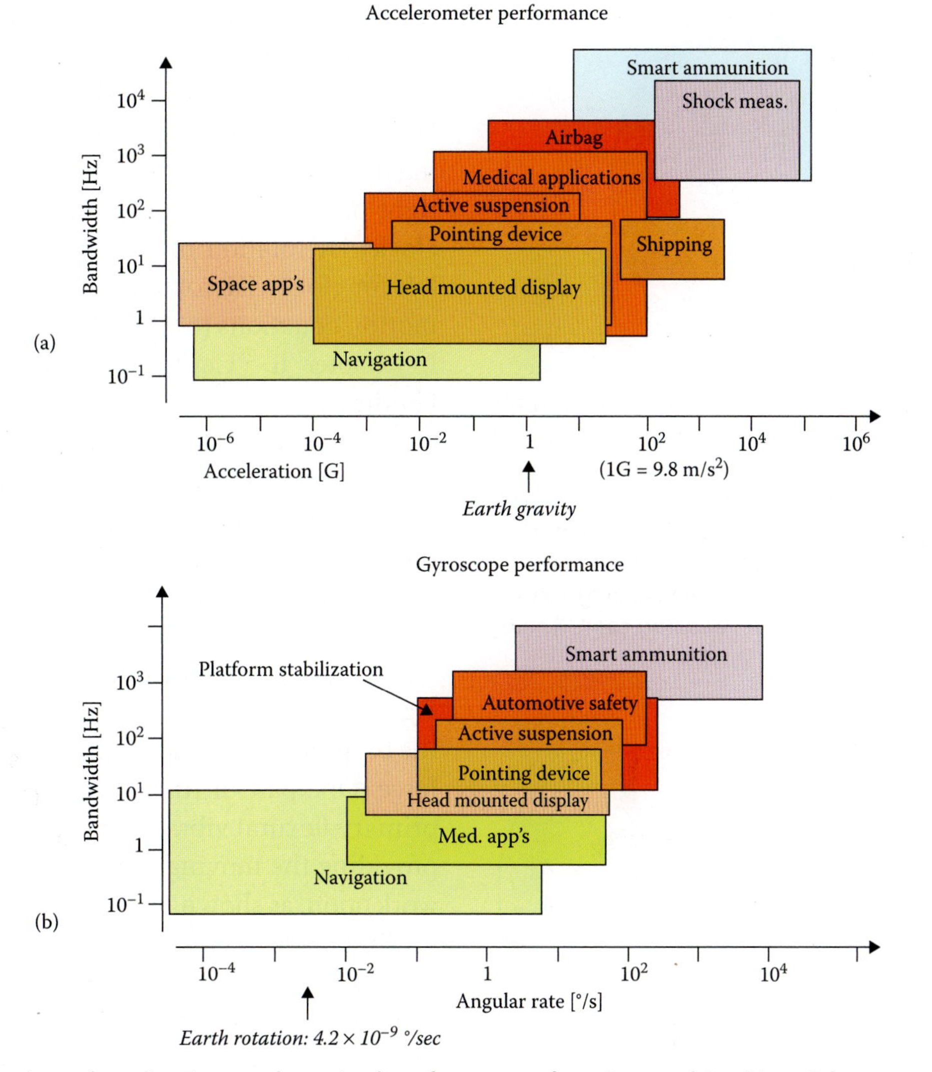

FIGURE 10.32 Overview of applications and required performances for micromachined inertial sensors. (a) Accelerometers; (b) angular rate sensors.

2010. For gyroscopes today, 4-inch wafers are still the dominant technology, but the share of 6-inch wafers is expected to grow to 60% over the next five years. DRIE and SOI increasingly are being adopted in MEMS gyroscope manufacturing.

The average price of a MEMS gyro in defense applications is 4–5 times higher than in automotive applications and more that 15 times the average sales price of consumer applications. Defense and aeronautic applications include missiles guidance, general aviation instruments, platform stabilization, land navigation, and smart munitions in the future. Gyros are made today at ADI, BEI Technologies, Inc., Honeywell, Kionix Inc., Matsushita Electric Industrial Co. Ltd., Motorola Automotive, Murata, Robert Bosch GmbH, SensoNor ASA, Silicon Sensing Systems Ltd, STMicroelectronics, and TRONIC'S Microsystems SA.

MEMS Angular Rate Sensors

Being very small, MEMS gyros obviously cannot operate on the flywheel principle, as that scales with the mass of the flywheel. Even the best MEMS motors still quickly slow down and stop if not externally actuated. The inertial mass of the small MEMS wheels is very, very small (mass moment of inertia, $I = \int r^2 dm$; with a fifth order expression like this, it becomes insignificant very fast when r is made very small). Furthermore, MEMS rotors cannot (thus far) be made so that they precess freely in three dimensions. Instead, MEMS often take advantage of the Coriolis force to detect angular rotation. The Coriolis force, named after Gaspard G. de Coriolis (1792–1843), a French mathematician, is an inertial force that must be applied to maintain a moving body's

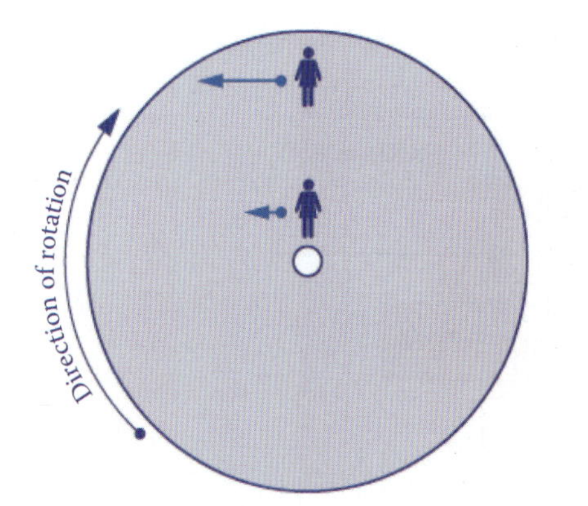

FIGURE 10.33 Illustration of the Coriolis force. A person moving northward toward the outer edge of a rotating platform must increase the westward speed component (blue arrows) to maintain a northbound course. The acceleration required is the Coriolis acceleration. (From 2003. *Analogue Dialogue* 433–37.) New iMEMS angular-rate-sensing gyroscope [John Geen (john.geen@analog.com) and David Krakauer (david.krakauer@analog.com), Analog Devices, Micromachined Products Division.]

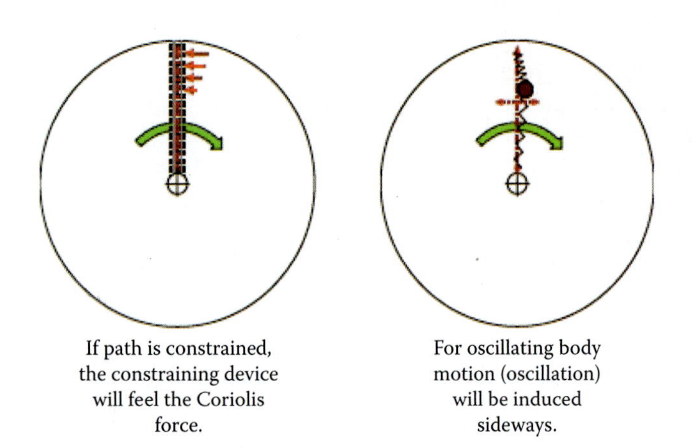

FIGURE 10.34 How to use the Coriolis force in a microelectromechanical system device.

course on a rotating surface. Motion in a rotating reference frame leads to "sideways" movement. Coriolis showed that if the ordinary Newtonian laws of motion of bodies are to be used in a rotating frame of reference, an inertial force, acting to the right of the direction of body motion for counterclockwise rotation of the reference frame or to the left for clockwise rotation, must be included in the motion equations (see also Figure 10.33 and Volume I, Chapter 6, "Centrifugal Fluidic Platform—CD Fluidics").

If the path of the moving body on the rotating surface is constrained, the constraining device will feel the Coriolis force (Figure 10.34). By measuring the imparted force (or its effect on an oscillator), we can measure the angular velocity. Almost all MEMS rate sensors use this feature.

The double-ended tuning fork (DETF), shown in Figure 10.35, is one of several structures that is used as the basis of a MEMS rate sensor based on the Coriolis force. One pair of tines of the DETF is driven into primary flexural vibration by a drive-oscillator, thus providing the moving body for the Coriolis force to work upon, as shown in Figure 10.35. The tines of the fork are made to vibrate in opposite directions but in the plane of the fork, which represents the primary flexural vibration mode. The Coriolis force tends to displace the tips of the fork out of this plane, forcing the tips to describe an ellipitical path as the rotation excites a secondary mode of resonance, a torsional mode coupling with the first flexural mode. A quartz DETF from Systron Donner (Concord, CA; http://systron.com) exploits the piezoelectric properties of

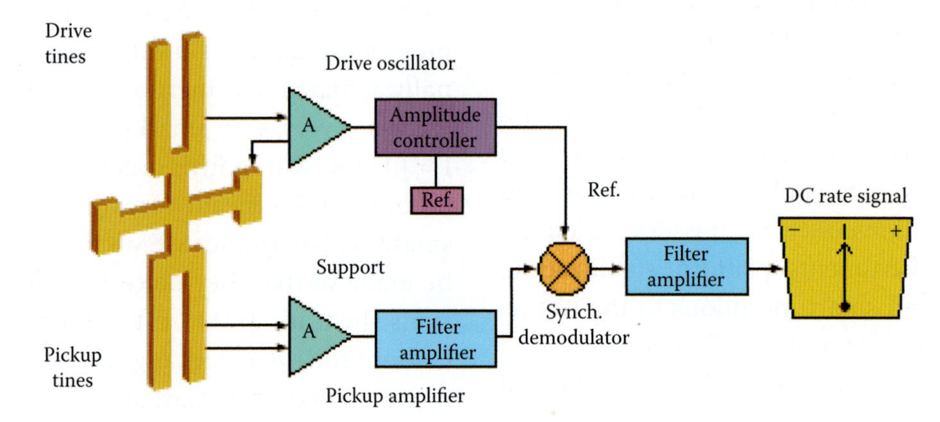

FIGURE 10.35 DETF for a rate sensor. The DETF structure may be etched from quartz or dry etched in the epilayer of a silicon-on-insulator wafer and released by wet etching the oxide layer underneath (from an older version of the http://www.systron.com website).[24]

the material to both excite and sense flexural and torsional vibration modes. The same company made a Si DETF; in this case, the DETF is dry etched out of the episilicon layer of an SOI wafer and under-etched by wet etching. To drive a Si DETF, electrostatic drive electrodes need to be implemented; otherwise, the device works just like the quartz DETF device. Although it gives up some accuracy, the Si device is considerably less expensive and has more than enough accuracy for many applications.

Daimler Benz AG, Stuttgart, Germany, uses a single tuning fork based on SOI micromachining technology for angular rate sensing. The tines of the silicon tuning fork vibrate out of the plane of the die, driven by a thin-film piezoelectric aluminum nitride actuator on top of one of the tines. The Coriolis forces on the tines produce a torque moment about the stem of the tuning fork, giving rise to shear stresses that can be sensed with diffused piezoresistive elements. The shear stress is maximal on the center-line of the stem: this forms the optimal location for the piezoresistive sense elements. The selling price of angular rate gyros such as Bosch's surface-micromachined gyro and Systron Donner's or Matshushita's tuning forks was approximately $25 in 1996, and they were relegated to the top of the line model vehicles.

Other types of angular rate Si sensors, taking advantage of the Coriolis effect, are under development based on comb or ring structures, and both surface and bulk-micromachined sensors have been demonstrated. A comb has the advantage of a larger capacitive area; however, the ring has the advantage of better vibration immunity.

One common comb design, as illustrated in Figure 10.36a, for ADI's ADXRS angular rate sensors, is a proof mass, which can move along the two in-plane

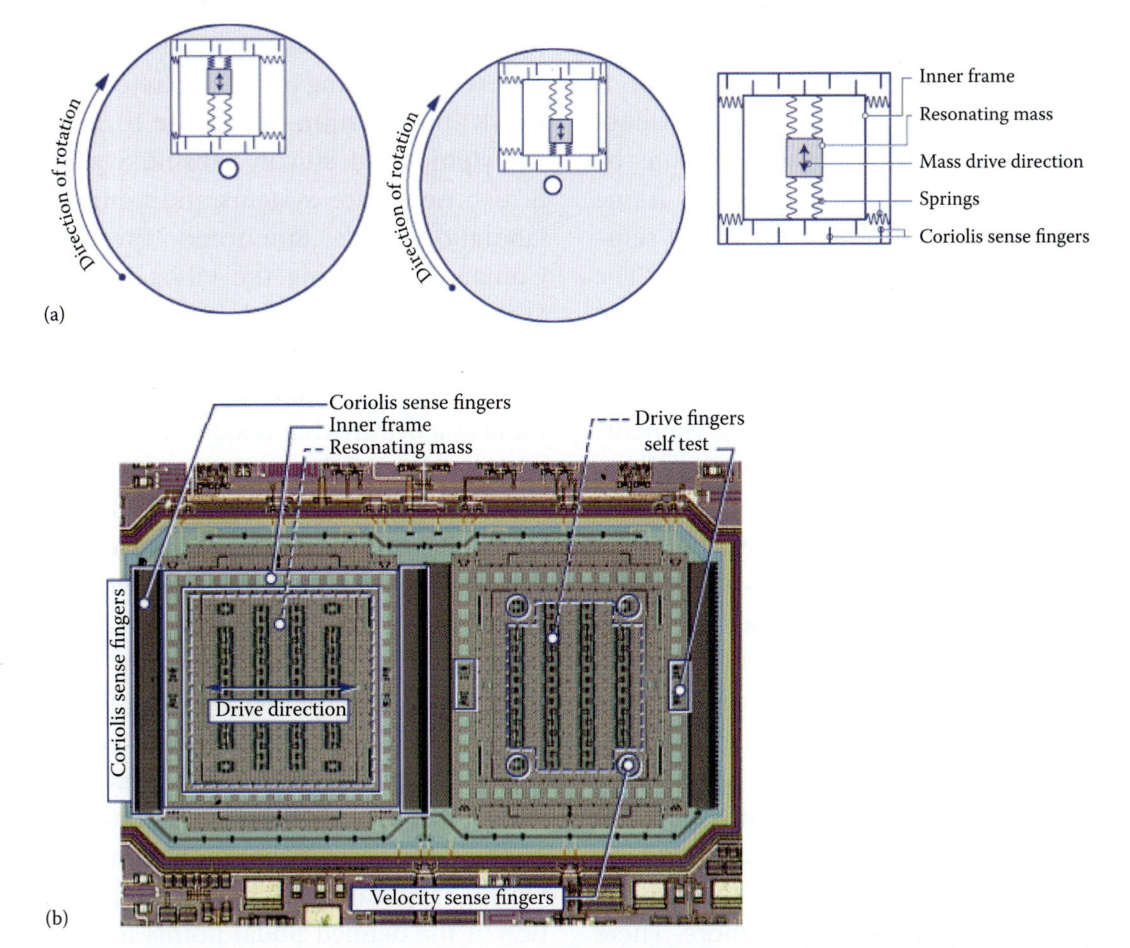

FIGURE 10.36 (a) Design principle behind the ADXRS150 and ADXRS300 from Analog Devices. These yaw-axis gyros have full-scale ranges of 150 and 300°/s, but they can measure rotation about other axes by appropriate mounting orientation. (b) The ADXRS sensors include two structures to enable differential sensing in order to reject environmental shock and vibration. (From 2003. *Analogue Dialogue* 433–37.) New iMEMS Angular-Rate-Sensing Gyroscope. [John Geen (john.geen@analog.com) and David Krakauer (david.krakauer@analog.com), Analog Devices, Micromachined Products Division.]

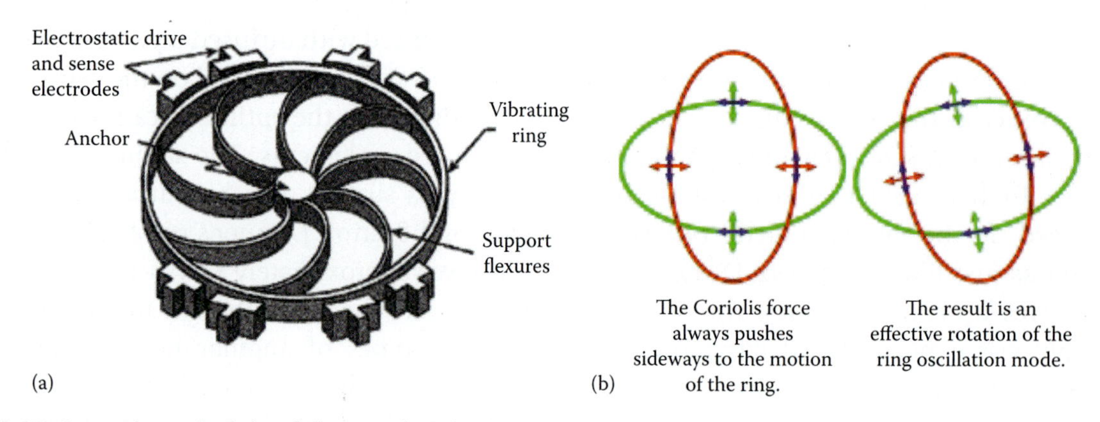

FIGURE 10.37 Operating principle of the rotating ring gyroscope.

axes through a suspension system. The suspended mass is surface-micromachined from polysilicon and tethered to a polysilicon frame so that it can freely resonate only along one of the two in-plane axes. To measure the Coriolis acceleration, the frame containing the resonating mass is in turn tethered to the substrate by springs situated at 90° relative to the resonating motion. With the resonating mass moving, and the surface on which the sensor is mounted rotating, the mass and its frame experience the Coriolis force. ADI's ADXRS sensors measure the displacement of the resonating mass and its frame due to the Coriolis effect through capacitive sensing elements attached to the resonating mass. The comb design in this sensing approach refers to the Coriolis sense fingers that are used to capacitively sense displacement of the frame in response to the force exerted by the mass. These sensing elements are silicon beams interdigitated with two sets of stationary silicon beams attached to the substrate, just as we saw for accelerometers in Figure 10.29b.

The ADXRS gyro electronics can resolve capacitance changes as small as 12×10^{-21} farads (12 zeptofarads) from beam deflections as small as 0.00016 Å (16 femtometers). The only way this can be done in a practical device is by situating the electronics, including amplifiers and filters, on the same die as the mechanical structure. These subatomic displacements are somewhat difficult to comprehend, as the individual atoms on the surface of the moving parts are moving randomly by much more. There are about 10^{12} atoms on the surfaces of the capacitors, so the statistical averaging of their individual motions reduces the uncertainty by a factor of 10^6. So one might wonder whether we can do better by removing the air molecules to prevent their impacting on the sensor. The answer is no. The suspension mass weighs only 4 μg, its flexures are only 1.7 μm wide, and this assembly is suspended over the silicon substrate. It is air that cushions the structure, preventing it from being destroyed by violent shocks and crashes into the substrate. The ADXRS sensors include two identical suspended masses to enable differential sensing in order to reject environmental shock and vibration (see Figure 10.36b).

Delphi[25] and Silicon Sensing Systems (formerly British Aerospace Systems and Equipment), in collaboration with Sumitomo Precision Products Company, are using the vibrating ring gyroscope (VRG) approach to exploit the Coriolis effect.[26] In the vibrating ring approach, a ring is flexured back and forth in resonant mode. The Coriolis effect induces a flexure that is sideways (and out of phase) with the driving flexure (Figure 10.37).

This principle is further detailed in Figure 10.38, which shows a wheel that is driven to vibrate about its axis of symmetry. Rotation about either in-plane axis results in the wheel's tilting, which can be detected with capacitive electrodes under the wheel, as shown in Figure 10.38a. The Delphi-Delco CMOS integrated single-chip surface-micromachined angular rate sensor is shown in Figure 10.38b. The micromachined ring shown here is driven into resonance in the plane of the chip and its vibration forms nodal points. During angular motion, the amplitude of vibration at the defined nodal points increases, and this vibration signal is fed back to reduce the amplitude, thereby maintaining a stationary pattern with nodes at the defined locations. Sacrificial micromachining fabrication techniques are used to allow the ring and

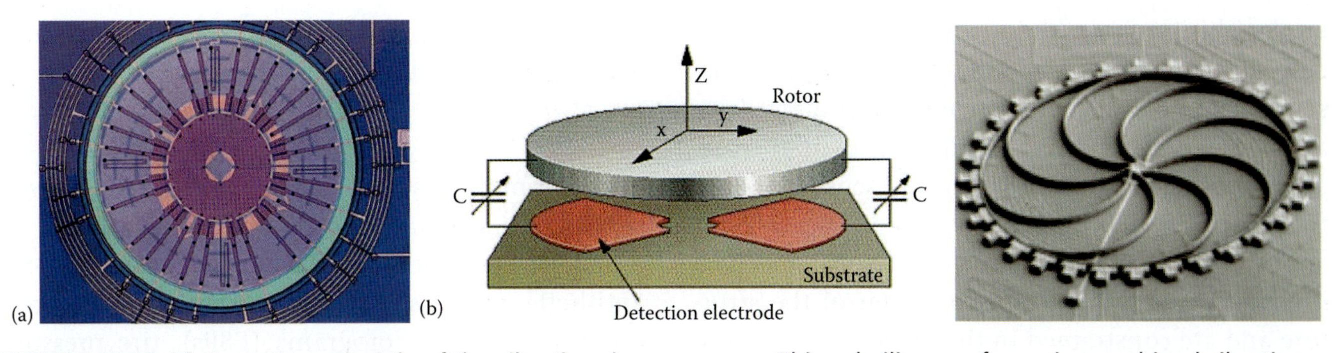

FIGURE 10.38 (a) Operation principle of the vibrating ring gyroscope. This polysilicon surface-micromachined vibrating wheel gyroscope was designed at the Berkeley Sensors and Actuators Center. (b) Delphi-Delco complementary metal-oxide-silicon integrated single-chip surface-micromachined angular rate sensor.

springs to vibrate over the surface of the IC. A scanning electron photomicrograph of the earliest version of the sensor element is shown in Figure 10.38b. The micromachined ring sensor incorporates an electroformed vibrating ring structure fabricated on a silicon IC control chip. Sensors are vacuum-sealed at the chip level using wafer-to-wafer bonding. A vacuum is required since the ring sensor has to be driven into resonance for normal operation with a quality factor >1000. Semicircular springs support the 1-mm-diameter ring and store the vibrational energy. The springs are attached to the substrate with a symmetric post. The post/spring design greatly reduces the effect of packaging stresses on the sensor.

Package-induced stress has been a longstanding problem for many other sensor programs. The Delphi-Delco angular rate sensor is a CMOS integrated single chip and is used for automotive steering assistance, active brake control, and rollover detection.

At the University of California, Irvine, Dr. Shkel's group is developing a new class of micromachined vibratory angular rate sensors, emphasizing structural design concepts to improve robustness, temperature range, and shock survivability. These sensors combine comb and ring mechanical elements. The Shkel group has explored multiple degree-of-freedom (DOF) dynamical systems in order to obtain sensing structures that are inherently insensitive to fluctuations in operation conditions, such as temperature variations, or shock loads.

The Shkel angular rate sensor, shown in Figure 10.39, features a distributed mass architecture utilizing multiple linear oscillators in the drive mode and a single central torsional oscillator in the sense

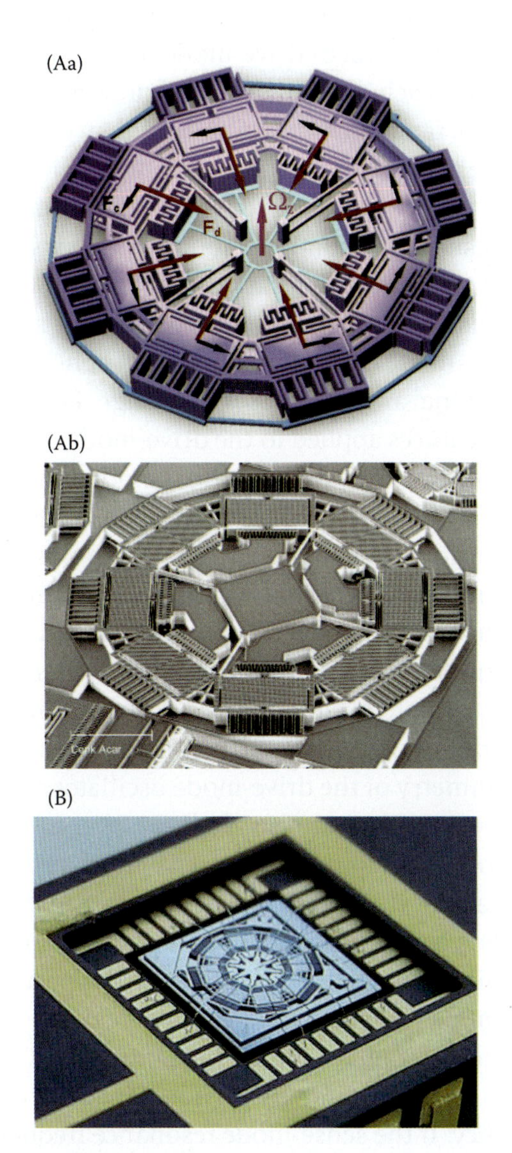

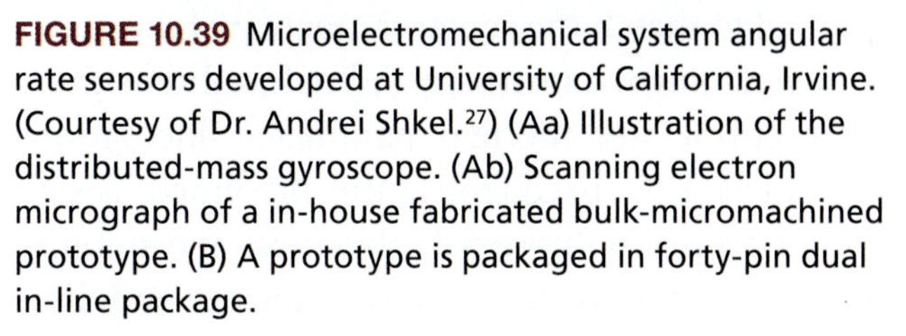

FIGURE 10.39 Microelectromechanical system angular rate sensors developed at University of California, Irvine. (Courtesy of Dr. Andrei Shkel.[27]) (Aa) Illustration of the distributed-mass gyroscope. (Ab) Scanning electron micrograph of a in-house fabricated bulk-micromachined prototype. (B) A prototype is packaged in forty-pin dual in-line package.

mode. This design address a major technical difficulty in the design of MEMS vibratory angular rate sensors, i.e., to keep matched frequencies in the drive and sense directions and eliminate structural cross-axes coupling between the drive and sense modes of the sensor.[27]

The distributed drive-mode oscillators are driven in-phase toward the geometric center of the structure and are constrained in the tangential direction with respect to the supporting frame. In the presence of an angular rotation rate about the *z*-axis, the Coriolis forces are induced on each proof mass orthogonal to each drive-mode oscillation direction. Thus, each of the induced Coriolis force vectors lies in the tangential direction, and they combine to generate a resultant torque on the supporting frame. The net Coriolis torque excites the supporting frame into torsional oscillations about the *z*-axis, which are detected by sensing capacitors.

The multidirectional and axisymmetric nature of the drive-mode oscillators has several benefits over a conventional angular rate sensor design. Firstly, since the drive forces applied to the drive-mode oscillators cancel out in all directions due to the radial symmetry, the net driving force on the structure reduces to zero and eliminates quadrature error in the presence of structural imperfections. Secondly, instability and drift due to mechanical coupling between the drive and sense modes are minimized, since the structure allows complete decoupling of multidirectional linear drive mode and the rotational sense mode. The symmetry of the drive-mode oscillator structure about several axes also cancels the effects of directional residual stresses. The most prominent advantage of the design concept is the capability to provide a wide-bandwidth operation region in the drive-mode frequency response. By designing each drive-mode oscillator to have incrementally spaced resonance frequencies, a constant total Coriolis torque is achieved over a wide range of driving frequency. If the sense-mode resonance frequency is designed to match the center frequency of the drive-frequency range, robustness and reduced sensitivity to structural and thermal parameter fluctuations are achieved. Consequently, the presented design concept results in improved robustness and stability over the operating time of the device, and it is expected to simplify the control architecture and relax tight fabrication and packaging constraints.

Conclusions

The automotive sector has long been a growth market for MEMS sensors with applications in airbag release, vehicle dynamics control, ABSs, antitheft systems, active suspension control, electronic stability programs (ESPs), tire pressure monitoring sytems (TPMSs), headlight leveling, etc. However, the industry continues to change and face new opportunities and challenges—regulations, saturation in some applications, and price erosion issues, to name a few. Today's high-end vehicles feature up to 100 different sensors. About thirty of these are now MEMS based. The market is made up of accelerometers, gyroscopes, and inclinometers, as well as pressure and flow sensors. Emerging applications include infrared (IR) sensors for air quality, microscanners for displays, and, further out, MEMS oscillators and energy scavengers for TPMS.

Consumer electronics applications are growing very rapidly and likely will overtake automotive MEMS applications. There is a convergence of requirements between the automotive and consumer electronics sensors. The automotive market can potentially derive great benefit from the advances in consumer electronics MEMS markets, but automotive accelerometer performance requirements are vastly more stringent than those of consumer electronics.

Although the electronics content of vehicles continually increases, the vehicle market itself is expanding more slowly than in previous years. The net effect is a steady increase in the value of the automotive electronics market. The use of IC-compatible process lines for MEMS production is growing. There are clearly high benefits in using CMOS or BiCMOS-compatible manufacturing lines to target high-volume MEMS markets. Some specific reasons to select monolithic integration as a solution are its cost advantages, commercial imperatives, performance advantages, and pragmatism. Nevertheless, hybrid solutions are still sometimes preferred, as they offer distinct advantages (more flexible manufacturing strategy, shorter development time, a simpler and more easily available manufacturing process, etc.). The integrated MEMS market share is expected to grow in the future. However, there are still bottlenecks to be overcome (process standardization, MEMS is not always a large volume market, etc.).

MEMS in Medical and Biotechnology Markets: Bio-MEMS

Introduction

The per capita annual expenditure for health care is about $3000 in the United States, $3200 in Germany, and $2000 on average in Europe. The total world health market in 2004 was about $793 billion, and large segments include $280 billion for medical instruments and $484 billion for pharmaceuticals. In vitro diagnostics, surprisingly, represented only 3% of the total world health market (i.e., $27.6 billion) (see Figure 10.40; Adams Business Associates).

The biotechnology market covers a large area and has vast and crucial branches among them. Some of these biotechnology branches are genetically modified food (GMO), antibody technologies, antisense technology, biomaterials, biopharmaceuticals, drug discovery, DNA arrays, environmental/industrial biotechnology, enzymes, gene therapy, genetic engineering, genomics, informatics, instrumentation and equipment, molecular biology, pharmaceuticals, proteomics, tissue engineering, and wound care. The biotechnology market had strong growth in recent years, generating a total revenue of $126.3 billion in 2005. It represented a compound annual growth rate (CAGR) of 12.8% for the period 2001–2005. Currently, the United States dominates the market and accounts for 54.3% of the global market revenue.

Medical and biotechnology markets show strong promise for MEMS solutions in a wide variety of applications, as illustrated in Table 10.8. The medical/biotechnology application field of MEMS is referred to as bio-MEMS. In bio-MEMS, one deals with life-saving disposable diagnostic sensors, DNA arrays systems speeding up drug discovery, smart pills improving drug delivery, μ-TAS and LOC, etc., and most of those devices come with disposables. Therefore, the sales volume in bio-MEMS is expected to be eventually significantly larger than that for mechanical MEMS, where sensors are typically in use anywhere from three to ten years.

Progress in the medical and biotechnology area today is only to a minor extent derived from silicon-based bio-MEMS. It has become clear that the straight application of Si technology is very often inadequate in this field compared with its successful application in the automotive world or in mechanical and optical applications in general. Whereas in the automotive application the market for Si micromachined products is mature enough to warrant projection of realistic future revenue numbers, predicted revenues for micromachined sensors and instrumentation for biomedical and biotechnology applications mostly refer to speculation as commercial successes still remain short of realization. In the bio-MEMS field, to succeed as well as mechanical and MOEMS did, one will need to develop mass manufacturing techniques based on very large batches or roll-to-roll manufacturing on paper or plastic substrates. Through a merging of traditional micromanufacturing methods (e.g., electroplating and micromolding) with less traditional methods (e.g., lithography) and continuous manufacturing methods (R2R), we expect more bio-MEMS market breakthroughs.

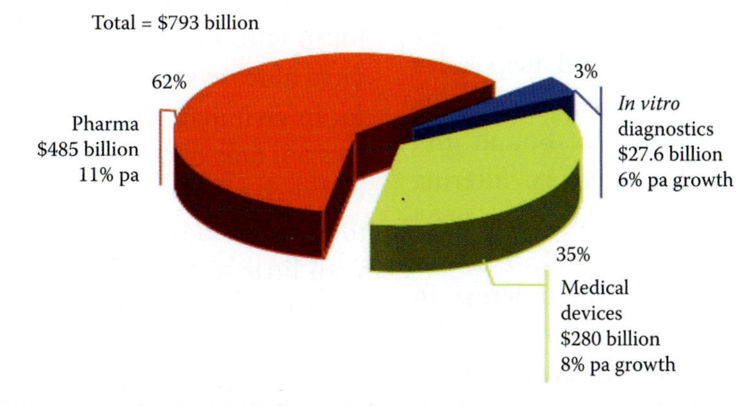

FIGURE 10.40 Global health care market in 2004. (From Adams Business Associates, aba@a-b-a.co.uk.)

TABLE 10.8 Medical/Biotechnology MEMS Applications

Drug Delivery Systems	Typical MEMS Example	Cardiology	Typical MEMS Example
Patches	Cygnus GlucoWatch Biographer (Figure 1.13 this volume)	Pacemakers	One- or two-axis accelerometers for activity monitoring
External and implantable pumps	Animas-Debiotech belt-worn drug delivery pump (Figures 1.10 and 1.11 this volume).	Angioplasty catheters	TRW NovaSensor, Motorola, and Siemens offer disposable pressure sensor (<$1)
Smart pill	Smart pill (Volume II, Figures 4.99 and 4.100)		
Monitoring		**Biotechnologies**	
Point of care testing	i-STAT handheld blood analyzer (Figure 10.42 this volume)	Polymerase chain reaction (PCR)	Polymerase chain reaction (PCR) efforts e.g., at Cepheid
On-line monitoring of blood gases	Dual lumen-based pH, CO_2, and O_2 sensor array (Volume II, Figures 4.92 to 4.96)	Genetic tests and therapy	Nanogen's NanoChip (Volume II, Figure 8.51)
On-line monitoring of pressure	TRW NovaSensor, Motorola, and Siemens offer disposable pressure sensors (<$1)	Electrophoresis	Caliper and Aclara electrophoretic chips
		Liquid handling systems	RotaPrep's CD sample preparation station (Figure 10.51 this volume)
Prostheses/artificial organs		**Minimally invasive surgery**	
Orthopedics, ophthalmology	Nanotexturing of surfaces for enhanced biocompatibility of implants (Chapter 5, this volume)	Cutting tools	Painless ultrasonic cutting tools
Neurology	MEMS neural stimulator probes		

After a write-up on the market character in this diverse field, we look at bio-MEMS applications in in vitro and in vivo diagnostics while highlighting bio-MEMS in biomedical microinstrumentation [micrototal analysis system (μ-TAS) or lab-on-a-chip (LOC)]. Micromachining efforts for building microinstrumentation are seen today as a better approach to solving chemistry and biology problems compared with building more selective chemical sensors. Some underlying reasons are that the resulting instrument may justify significantly higher costs than disposable chemical sensors and help solve the shortcomings of chemical/biological sensors by enabling calibration, separating of bound and unbound species, cleaning the detectors, filtering out unwanted compounds, and obviating the need for very selective chemical sensors.

We conclude with a summary on progress in μ-TAS and LOC in sample preparation for molecular diagnostics, as it is felt that this is the area with the biggest need for bio-MEMS innovation.

Market Character

In terms of the business climate and funding sources in bio-MEMS, the playing ground is uneven and depends very much on the specific medical or biotechnology niche addressed. One overall trend is a move away from the initial application of μ-TAS and LOC to high-throughput screening (HTS) to its current application to point of care (POC) diagnostics instead. Even in the technology boom years (before 2001), it used to be difficult to raise money for in vitro diagnostic bio-MEMS work; for novel in vitro glucose sensors it was especially difficult, and for in vivo diagnostic devices it was harder yet. This was not only because the diagnostics industry is very entrenched and conservative but also because investors had already invested large amounts in that area with little return and because of the uncertainty of health care reimbursement. Moreover, disposable diagnostics were not necessarily perceived positively because of the waste disposal issues. Finally, one had to demonstrate a new diagnostic sensor/instrument 80% toward a finished product in order to even be able to start talking to skeptical investors or conservative large diagnostic companies.

In the case of drug discovery, investors were, up to 2001–2002, more willing to take major risks at an early stage of the technology development. Significantly, from Figure 10.40, medical devices and pharma are a much larger market than in vitro diagnostics. Consequently most microfluidic start-up companies working on μ-TAS or LOC concepts were targeting drug discovery rather than the diagnostic area for early market penetration with bio-MEMS products. The 2001 downturn in the U.S. securities market greatly reduced the value of all of these companies, who must now make money the traditional way—by making products that someone wants to buy.

Today, the situation is reversed with diagnostic applications more attractive than research tools for biotechnology. Diagnostic products are indeed more and more appreciated as important tools in reducing costs of health care. In the past, diagnostic tests were merely used as support data for the real diagnosis, often to confirm the physical diagnosis of the doctor or for legal protection that the correct procedures had been carried out. A major change over the last 5–7 years has been the increased use in true diagnosis and differentiation of conditions in order to achieve the best treatment, i.e., theranostics bringing together the diagnostic and pharmaceutical industries. The driving force for this shift in role for diagnostic tests is the large costs for health care throughout the world at a time when the economies are less robust. The attraction of investing in the development of diagnostic products is being boosted by this change in importance and the recognition that the overall risks are reduced together with an earlier return on modest investments (Adams Business Associates, aba@a-b-a.co.uk).

With this improved business climate, improvements in bio-MEMS fluidics technologies may find now a more sizable market in better diagnostic systems, such as two-point sensor calibration schemes, sensor cleaning devices, sample preparation strategies (see further below), etc., than in HTS, where microfluidics are often only a minuscule part of a large robotic system. Diagnostics is a mass market; HTS is at most a research market (see above in the section Preamble on the MEMS market). We believe that after more significant progress is made in microfluidic platforms, there will be a resurgence of interest in chemical and biological sensors and methods to manufacture them inexpensively for diagnostic applications.

Perhaps this will produce more i-STAT (from Abbott) type products. Abbott's i-STAT, with its planar electrochemical sensor array for in vitro blood electrolyte and blood gas analysis (see below, Figure 10.42), is one of the few survivors of the many companies in the area of miniaturized diagnostics that were started in the 1980s.

The two most promising bio-MEMS areas in the short term are both in molecular diagnostics, and they involve sample preparation for DNA or RNA extraction from a wide range of samples with the same instrument (see Sample Preparation section, below) and the integration of DNA arrays with microfluidics to speed up the hybridization reaction.

In Vitro Diagnostics*

Introduction

Biomedical sensors and instrumentation for in vitro diagnostics represent a huge opportunity for micromachining in the broad sense of the word, through traditional and nontraditional miniaturization methods, continuous manufacturing methods and combinations thereof. To date, the majority of routine clinical tests are performed in large independent clinical laboratories with advanced, automated, analytical equipment. As the amount of resources spent on health care continues to soar, there is an increasing pressure for the decentralization of the POC and toward easy-to-use, miniaturized, inexpensive diagnostic machines and devices. It is in this shift that miniaturization science has so much to offer. With microinstrumentation (μ-TAS, LOC), one envisions that analyses will be performed significantly faster, without the need of

* The diagnostics market (also called *in vitro diagnostics*) includes the instruments and reagents used for the screening, diagnosis, monitoring, or prognosis of disease by laboratory techniques. The products are used in hospitals and private laboratories, in physician offices, and at other point-of-care sites (emergency rooms, intensive care units, outpatient clinics, nursing homes), as well as in home testing (primarily diabetic glucose tests and pregnancy tests).

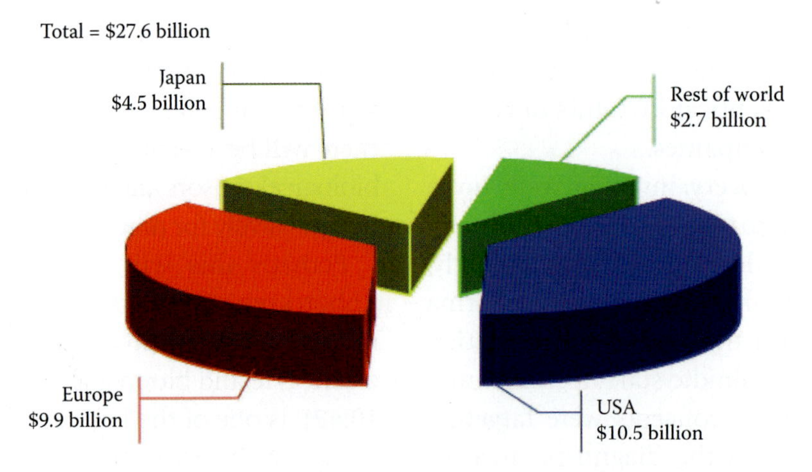

FIGURE 10.41 In-vitro diagnostic markets in 2004. (From Adams Business Associates, aba@a-b-a.co.uk.)

highly trained technicians (e.g., in remote villages, small health stations, etc.), only minute samples will be required, and dead volume in connections will be eliminated. Current in vitro diagnostic tests are categorized as clinical chemistry, blood gas/electrolytes, hematology/flow cytometry, coagulation, DNA tests (molecular diagnostics), microbiology, diabetes, urinalysis, histology/cytology. As shown in Figure 10.41, the worldwide in vitro diagnostics market was about $27.6 billion in sales in 2004, and the annual growth rate was 6% (see also Figure 10.40, which shows that in vitro diagnostics was only 3% of the global health care market for 2004).

The main areas of above-average growth (>6%) in in vitro diagnostics, identified in the 2003 report from the European Diagnostic Manufacturers Association (EDMA; see http://www.edma-ivd.be), are shown in Table 10.9.

As examples of MEMS contributions to in vitro diagnostics, we detail electrochemical sensors for blood electrolyte and blood gas analysis, glucose sensors and DNA arrays. We also cover MEMS contributions to the fabrication of miniaturized well plates, flow cytometry, and surface plasmon resonance (SPR) sensors.

Table 10.9 In Vitro Diagnostics with Above-Average Growth (European Diagnostic Manufacturers Association, http://www.edma-ivd.be)

Glucose tests	>10%
Cardiac tests	>20%
Nucleic acid-based tests	>20%
Genetic testing	>>20%, but from a small base

Other in vitro diagnostic applications are reviewed throughout the book. For example, contributions to the future of immunosensing were reviewed in Volume I, Chapter 7. DNA and protein arrays are also covered in Volume II, Chapter 8 (their fabrication and typical use), and the Nanogen electronic DNA array is covered in Example 8.2 of that chapter. Generic challenges concerning miniaturization in biomedical equipment, such as correct partitioning issues, were addressed in Chapter One of this volume.

Electrochemical Sensors for Blood Electrolytes and Blood Gases

Background: MEMS in Tabletop and Handheld In Vitro Diagnostics

During the past 25 years, the commercialization of MEMS for in vitro diagnostics failed to keep pace with earlier projections. We believe that an overemphasis on integration of chemical sensors with electronic functions on Si hampered the development. We suggest speeding up the progress by applying micromachining tools more correctly. This includes nonsilicon manufacturing, hybrid manufacturing, and roll-to-roll manufacturing, as covered in Chapter 1, this volume.

Progress in sensors for tabletop diagnostic instruments based on nonsilicon manufacturing techniques has been significant. The fully automated Kodak Ektachem is a tabletop clinical blood analyzer based on planar electrochemical sensor slides, and it has been on the market for some time. The Ektachem, the Seralyzer from the Ames Division of Miles Laboratory, the Refletron Plus system from

Roche Diagnostics/Boehringer Mannheim (with a sixteen-test menu), the Analyst from E. I. du Pont de Nemours, the Nucleus from Nova Biomedical, and other, similar instruments all cover a wide range of blood chemistries.[28,29] However, none of these products involve Si micromachining bio-MEMS technology. The progress enabling their development mainly was based on the development of solid-state stabilized reagents for so-called *dry electrochemistry*,[30] on miniaturized electrochemical sensors, and to a minor extent on advances in thin-film technology. Despite the many earlier claims, contributions from micromachining (Si-based or not) are mostly at the component rather than the system level, and many machining technologies are in the experimental stage. Systematic analysis of different machining options as advocated in this book (see Chapter 1, this volume) is only now starting to be applied. In the case of diagnostics, this leaves little room for Si-based technology. As a result of these analyses, one comes to the conclusion that bio-MEMS clearly counts as a fragmented set of supporting technologies rather than a monolithic new industry.

It was anticipated that Si micromachining would play a more prominent role in the development of sensors for the smaller, next-generation, handheld, and portable diagnostic instruments. In reality, micromachining and Si sensor content in the more advanced, handheld and portable, biomedical products is also surprisingly limited. Some such products for analysis of blood electrolytes and gases on the market include the GEM-Premier from Mallinckrodt Sensor Systems, the similar STATPal from PPG, and the Portable Clinical Analyzer from Abbott (the i-STAT system), providing data on sodium, chloride, potassium, urea nitrogen, glucose, and hematocrit in less than 2 min. Another handheld instrument of interest is the AccuMeter cholesterol test from Chemtrak, designed to screen lipids to evaluate cardiovascular disease risk, and the portable blood analysis systems by Diametrics Medical Inc. (electrochemical) and AVL (optical fluorescence). Abaxis introduced the Piccolo™ for general blood chemistry analysis, a small tabletop centrifugal instrument with a spinning reagent disk and optical read-out. Biologix Inc. developed the electrochemistry-based handheld pHacs STAT for blood pH, potassium, and oxygen.[28,31] Of all these products, only the i-STAT sensors are Si based. The Si avenue toward tabletop and handheld instruments has been pursued for more than thirty years with little success to report. Innovation in biomedical sensors for in vitro diagnostics has come mainly from the invention of new chemical detection schemes rather than from clever new micromachining methods. Actually, in biomedical sensors, a reversal toward hybrid and continuous manufacturing methods took place rather than a move toward more integration in Si embodiments.

Blood Electrolyte and Blood Gas Sensors

For blood electrolyte (pH, Na^+, K^+, Ca^{2+}, etc.) and blood gas (O_2 and CO_2) analysis with tabletop or handheld instrumentation, one wants to develop inexpensive, multiuse, or disposable chemical sensors. The principal approach so far has been via electrochemical technology. Microfabrication of electrochemical sensors, using IC technology, has been extremely challenging, mainly due to process incompatibility issues, packaging problems, failure to incorporate a true reference electrode, and the difficulties involving patterning relatively thick organic layers such as ion-selective membranes and hydrogels (see also Volume I, Chapter 7). Most researchers have come to the conclusion that for chemical sensing, especially for in vitro applications, a modular, hybrid approach rather than an integrated Si approach marks the road to progress. Of all the handheld biomedical analyzers on the market, only the i-STAT Portable Clinical Analyzer (Abbott, see http://www.abbottpointofcare.com/) involves Si in the sensor construction itself, primarily because of a historical production mode. Single-crystal Si, it was thought at that time, was the way to manufacture biosensors, and after the equipment investments were made, it was hard to reconsider. In Figure 10.42, we highlight the i-STAT instrument, together with a disposable sensor cartridge.

The Portable Clinical Analyzer from Abbott uses Si in the disposable chemical sensor cartridge only as a substrate and a contacting base; electronics are kept in the handheld reader, clearly a more economical approach than putting the electronics on the disposable (Figure 10.42). The sensor electrodes, gels, and membranes are all deposited with semiconductor-

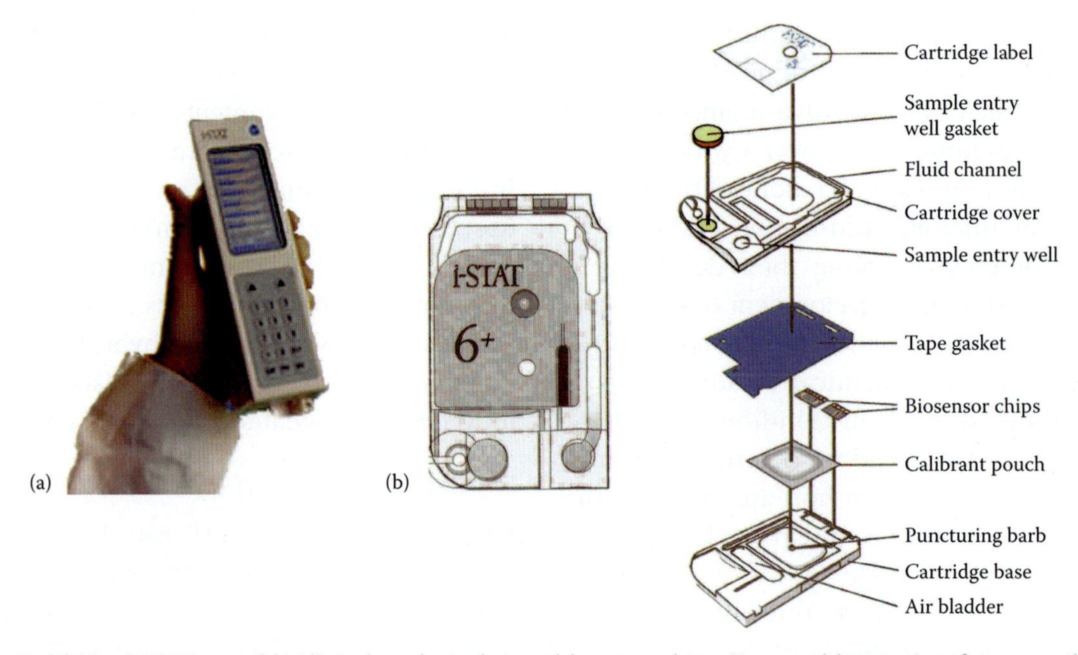

FIGURE 10.42 (a) The i-STAT portable clinical analyzer (now Abbott, see http://www.abbottpointofcare.com/) is used in conjunction with disposable cartridges for determination of a variety of parameters in whole blood. The analyzer stores up to fifty patient records and permits on-screen viewing of test results, as well as transmission of records to a data management system using infrared signals. (b) An i-STAT disposable cartridge.

type processes.[32] Without electronics onboard, the continued use of Si as a support has mostly a historical explanation. It should also be noted that the fluidic channels in the disposable i-STAT cartridge are made from molded plastic and not with Si micromachining. Abbott's manufacturing methods of the Si sensor chip in the i-STAT, after twenty years, still make it difficult to be profitable. This may be related to the manufacturing approach being silicon based and the design being integrated rather than modular. An alternative approach for the manufacture of a microfabricated, modular chemical sensor arrays is introduced in Chapter 1, Example 1.1. Epocal in Toronto uses a very different approach to making biosensor arrays compared to the methods used in manufacturing the i-STAT sensors (http://www.epocal.com). Epocal sensors are fashioned on flex circuits in a single continuous in-line manufacturing process. Flex circuits are adapted from industry standard smart card modules fashioned on 35 mm tape-on-reel format. The Epocal approach is very much the type of progress in biosensor manufacture we promote in Example 1.1.

While the number of new tools available for microfabrication, as gleaned from Chapter 1 of this volume, has grown dramatically, few methods have been gainfully applied to biosensor construction for clinical applications (some of these few methods have been reviewed in Volume II, Chapter 8). In contrast to mechanical sensors, where the excellent mechanical properties of single crystalline Si often tend to favor Si technology, the choice of the optimum manufacturing technology for chemical sensors is far less evident. For chemical sensors, the Si in the sensor often serves no other role than as a passive substrate. Consequently, the following trend emerges: whereas mechanical sensors (pressure, acceleration, temperature, etc.) are moving toward more integration embodied in CMOS-compatible surface micromachining, chemical sensors and biosensors are moving away from integration on Si and toward hybrid technology on plastic or ceramic substrates with silk-screening or drop delivery systems for the application of the organic layers.

Glucose Sensors

In vitro glucose sensing represents the single largest biosensor market, and a significant amount of all research and development in biosensors is geared toward improving this technology. Glucose self-testing by diabetics is one of the fastest growing segments of the world diagnostic market (see Table 10.9).

Diabetes mellitus is a chronic metabolic disorder that is caused by a failure of the body to produce insulin and/or an inability of the body to respond adequately to insulin. Type 1 diabetes occurs most often in children or young adults and accounts for 5–10% of the diagnosed diabetes patient population. Type 2 diabetes, or adult-onset diabetes, accounts for 90–95% of diagnosed diabetes cases worldwide, typically developing in middle-aged adults. Diabetes effects approximately 170 million people worldwide and is increasing, with the World Health Organization (WHO) predicting 300 million diabetics by 2025. The United States alone has 20.8 million people suffering from diabetes. This equates to approximately 6% of the population. Diabetes is the sixth leading cause of death in the United States (based on number of death certificates; National Institute of Diabetes and Digestive and Kidney Diseases). The global diabetes market was worth $18.6 billion in 2005, which was an 11.5% increase from 2004 sales of $16.6 billion. The United States has the dominant share in the global diabetes market, with 49.6% of 2005 global sales. The large U.S. market share is mainly due to the large population and the high prevalence of diabetes, the increasing incidence of diabetes associated with Western diets and an aging population. Kelly Close of Close Concerns Inc. in San Francisco, a consultancy devoted to diabetes and obesity research, estimates that on the basis of industry estimates worldwide sales of blood glucose meters were almost $5.9 billion in 2004. Roche Diagnostics leads, with sales of about $2.3 billion. LifeScan/J&J is second, with $1.7 billion. Together they control two-thirds of the blood glucose marketplace. Bayer, with $800 million, and Abbott Diabetes Care, with $790 million, are neck and neck for third place. The worldwide market for blood glucose test strips and monitors is currently almost $3 billion, and the market is growing rapidly at a sustained rate of 13% a year (for some example blood glucose monitors see Chapter 1, this volume, Figure 1.7). The most prominent method currently available for collecting blood glucose remains the painful and messy finger prick procedure. The recommended sampling rate is four times a day, further discouraging compliance. The need for a painless testing glucose sensor (minimally invasive) is a challenge that many MEMS engineers have worked on. A new method that permits painless blood testing while providing results that meet current standards of accuracy and reliability is key to success (for two examples of blood/interstitial fluid collections devices, see Figure 1.15a with a lancet for blood collections device, and Figure 1.15b for interstitial fluid collection using a laser puncture).

Three generations of glucose sensor technologies are reviewed in Volume I, Chapter 7. A breakthrough came with the 1987 introduction of the amperometric glucose sensor by Medisense (now owned by Abbott), which proved that small, planar electrochemical probes can compete with colorimetric paper strips. The Medisense/Abbott glucose sensor, the Precision QID (shown in Volume I, Figure 7.80), is based on an innovative mediator chemistry and a traditional thick-film manufacturing process. The sales of mediated amperometric biosensors now represents a dominant 65% of the world biosensor market.[33] A comparison of manufacturing techniques for these types of sensors reveals the difficulty of competing using a Si-based approach. One often forgets that in biosensors such as glucose sensors, one is not competing with an expensive serial manufacturing process but with continuous, printing-type processes (see Chapter 1, this volume), making products even cheaper than in a Si type batch production. The cost target for a disposable glucose sensor is about $0.10, about a factor of 10 less than an unpackaged physical sensor die. Unless a significant performance advantage arises, Si technology cannot compete. It is our belief that, besides new chemistries, continuous processes to manufacture disposable electrochemical sensors will be the ultimate answer to the pressing need for disposable, inexpensive diagnostics (see Chapter 1, Figure 1.17, where we used the manufacture of a disposable glucose sensor to illustrate our streamlined decision-making process). A way to go about continuous biosensor manufacture is demonstrated in Chapter 1, Example 1.1: "Proposed Scenario for Continuous Manufacture of Polymer/Metal-Based Biosensors." Besides Abbott's ExacTech blood glucose monitoring system, there are similar products by Johnson & Johnson (Lifescan) and Wampole Laboratories (Answer).[28]

The future of the glucose sensing business is quite intertwined with new sampling methods. Research is emphasizing minimally invasive, noninvasive, and continuous monitoring of blood glucose levels. MEMS microneedles are under development for minimally invasive sampling by several research groups in academia. One example of a minimally invasive method, recently taken off the market, is the GlucoWatch biographer from Cygnus (see Chapter 1, this volume, Figure 1.13). In this device, glucose is extracted through the skin by electro-osmosis, and the glucose is measured amperometrically.[34] Kumetrix (http://www.kumetrix.com) replaces the typical lancet and blood-testing strip with a battery-powered, handheld diagnostic instrument using disposable MEMS cartridges, consisting of a microneedle with a cross-section of about 100 μm, a silicon cuvette (a chamber for the blood sample), and a small window. The device uses capillary action to painlessly draw approximately 100 nL of blood into the cuvette, where it is mixed with a chemical reagent. A laser then measures the glucose and the blood's glucose level is displayed. The MiniMed continuous glucose monitoring system (CGMS; MiniMed Inc., Sylmar, Calif.) is the first commercially available continuous glucose monitor. Each sterile, disposable sensor (again an electrochemical type) used with the CGMS can provide a continuous glucose profile for up to 72 hours. Glucose concentration is measured in the interstitial fluid every 10 seconds, and an average glucose value is stored in memory every 5 minutes. A communications device enables data stored in the monitor to be downloaded and reviewed on a personal computer. A performance evaluation of this very promising technology is reviewed by Gross et al.[35]

The glucose sensor field became relatively stale after the introduction of the MediSense technology. A promising new research direction is touched upon in the section on in vivo diagnostics below and involves glucose binding protein for the detection. Roe and Smoller wrote a very comprehensive review on the state of the art of bloodless glucose measurements as of 1998,[36] and a review of noninvasive glucose monitoring was presented by Klonoff in 1997.[37] Heller and Feldman recently reviewed in vitro electrochemical sensors.[38]

The next generation of electrochemical sensors will be planar, thick-film, hybrid devices, which will have the cost and fabrication advantages of solid state technology but will likely still have some of the same limitations as present electrochemical sensors, that is, repeated calibrations to compensate for long-term drift and the need for periodic maintenance. An important lesson in reviewing this micromachining application is that there is as much need for further optimization of deposition of chemical sensitive layers as for exploration of new manufacturing options. Investigating continuous printing processes for electrochemical sensor manufacture will be most rewarding in the future (see Chapter 1, this volume, Example 1.1). In the meantime, a modular approach with thick-film deposition processes is advisable. Some of the calibration and cleaning problems will be resolved by implementing more sophisticated plastic molded microfluidics.

DNA Arrays

Molecular diagnostics is another one of the fast growing in vitro diagnostic applications (see Table 10.9). With microarrays or DNA chips, thousands of genes or representative oligonucleotide fragments can be arrayed on a single microchip and hybridization with complimentary strands can be sensitively detected (see Volume II, Figures 8.16 and 8.17). The fabrication methods and uses of DNA arrays are reviewed in Volume II, Chapter 8. The two most significant applications for DNA chips are believed to be in gene expression profiling and single-nucleotide polymorphisms (SNPs). In expression profiling, the chip is used to examine the amounts of messenger RNA, which controls how different genes are turned on or off to produce certain types of proteins and cells. If the gene is expressed in one way, it may result in a normal muscle cell, and if expressed in another way it may result is a tumor (see Volume II, Figure 8.23). By profiling the minute variations in a person's DNA called *single-nucleotide polymorphisms* (SNPs) and correlating these to a person's ability to metabolize a drug, personalized drug therapy might become a reality.

When biotechnology started flirting with DNA arrays and microfluidics-based companies, the MEMS market character changed overnight. The opening salvo was perhaps given on March 2, 1995, when Glaxo Wellcome (now GlaxoSmithKline) acquired Affymax, a DNA array company, including its ownership interest in Affymetrix, for $650 million, before there was even a product on the market. This was uncommon for the "traditional" MEMS companies who had always needed to show product revenue and even then got rewarded with only modest initial public offerings (IPOs). Microfluidic companies that were early players in the DNA market fray include Aclara Biosciences, Affymetrix, Nanogen, Caliper, Cepheid, and Orchid. There is some overlap between microfluidics companies and DNA array companies, and there will be much more in the future. The author expects great merit in merging DNA arrays with microfluidics, as it will enable sample preparation for diagnostic DNA arrays and enable working from cruder samples for high-throughput screening or gene expression. All the listed companies did go public and enjoyed large capitalizations due to the explosive stock market and generous evaluations associated with biotechnology companies in general. The "technology bubble" of 2001 took down much of these gains, but it is expected that these array companies will recuperate. The wide acceptance in the United States of genetic testing is not about "if" but about "when," and it will be a burgeoning bio-MEMS application. It is not clear today, though, what is the right format to present molecular diagnostic panels to the market; should it be in small panels and handheld readers for the point-of-care (POC) market with a variety of individual cartridges/test panels (say a chip to screen for population linked genetic diseases, a separate heart disease predeposition chip, etc.), or should it be a high-throughput robotized clinical diagnostics laboratory with hundreds of test panels?

Until large market molecular diagnostic tests, preferably with an instrument including sample preparation, become available, the importance of DNA microarrays in research for disease diagnosis, understanding drug interactions and determining drug candidates is expected to keep the DNA array market growing more slowly. Molecular diagnostics today is less dominated by giant multinationals than traditional diagnostics, and tests are performed in many different specialty labs and are still quite expensive (from a few hundred dollars to several thousands of dollars), so there seems to be a more positive outlook for potential new players.

In 2005, revenue for the total U.S. DNA microarray market totaled approximately $446.8 million, and it is expected to rise to $532.1 million by the end of 2012 (Frost & Sullivan). The compound annual growth rate (CAGR) for 2005 to 2012 is estimated at 10.9%.

Protein chips are also expected to emerge, driven by advances in proteomics. Yet despite their apparent similarity to DNA arrays, protein arrays represent a much smaller market, precisely because they are as finicky as the protein elements spotted/patterned on their surfaces. The biological activity of the capture protein elements, whether antibodies or proteins, must be maintained. One has to be able to print them while preserving their activity, and for a commercial product, the arrays have to be stable for years. As a result, there are relatively few protein arrays on the market. Those that are available come in two basic flavors: general protein arrays and antibody arrays. Protein arrays from Invitrogen, ProtoArray microarrays, are finding utility in various applications, including discovery of autoantibody biomarkers, identification of novel kinase and ubiquitin ligase substrates, and elucidation of protein interactions. Antibody arrays serve a different purpose. Available from such suppliers as Clontech, Thermo Scientific, and Sigma-Aldrich, these products enable protein profiling, examining how levels of several hundred proteins change in response to some treatment. Thermo Scientific's LabVision antibody microarray includes 720 antibodies, and Clontech's Ab Microarray includes 512. Typically, protein samples are labeled with Cy3 and Cy5 dye, and this allows the arrays to be visualized using most standard DNA array scanners.

Miniaturized Well Plates

Almost any new fluidic platform faces the daunting challenge of how to either incorporate or displace the current standard method of dealing with small amounts of chemicals, i.e., the use of miniaturized well plates and associated liquid dispensing systems, as shown in Figure 10.43.

Economic and technological forces are driving the continued push toward further miniaturization

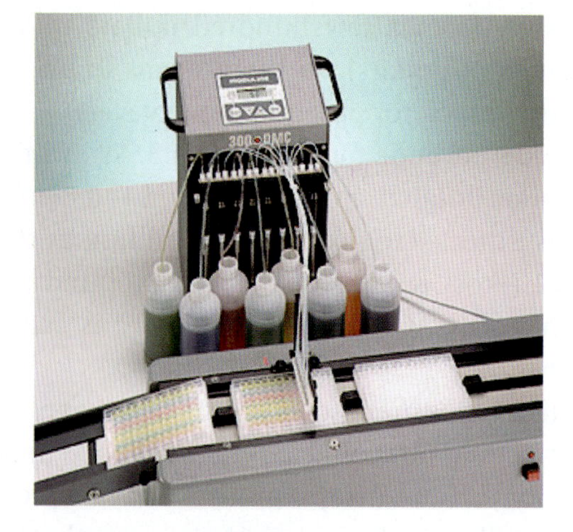

FIGURE 10.43 The 300 DMC dispenser configured for filling 96-well plates using eight different reagents (http://www.varivest.com).

of assays in plate wells for high-throughput screening (HTS) or high-volume diagnostics (Figure 10.2). In HTS, the number of potentially useful molecules is increasing, and the time during which the discovered molecules have a proprietary, strategic advantage is decreasing—in other words, the evaluation of more samples in a shorter amount of time with finite resources compels miniaturization. The number of samples has increased dramatically, primarily because of the concurrent advent of combinatorial and automated synthesis methods together with more precise biochemical or genetic definition of targets using genomic approaches. A first step in the miniaturization process for HTS and high-throughput diagnostics was from test tubes (1-mL range) to 96-well microplates (0.1-mL range) and, more recently, to 384-well (10-μL range) and 1536-well plates (1-μL range). With a 1536-well plate, ultra-high throughput (arbitrarily defined as 10^5 samples per day) becomes feasible. There are some reasons not to miniaturize and push to 1536-well plates and beyond. Certain assays become impossible if wells are too small. Cell-based assays, for example, cannot be miniaturized below 100–1000 cells without cell-to-cell variability becoming the dominant source of sample-to-sample variation, and cells are hard to accommodate in microfluidic systems. There is also substantial capital investment required for laboratories that are not already outfitted with pipetters and automated systems capable of reliably addressing 1536-well plates. For reading, 1536-well compatible readers tend to be more expensive than those limited to 384 wells, and the speed advantage of 1536 is difficult to obtain without expensive charge-coupled device (CCD camera) detectors. Some assays will suffer degradation in signal-to-noise ratio upon miniaturization (e.g., those that employ colorimetry), but many of the assays being developed have sufficient dynamic range to support a smaller format.[39] Engineering challenges associated with handling microliter volumes include: 1) evaporation, which limits the time of exposure to ambient humidity air and can cause sample-to-sample variability across the plate; and 2) fluid handling, which becomes increasingly difficult at smaller and smaller scales. As miniaturization proceeds still further, into the nanoliter-volume assays, evaporation becomes a much more significant issue. At a certain stage, a transition to fully enclosed systems, such as the microfluidic lab-on-a-chip (LOC) systems, becomes a necessity for evaporation and fluid metering. Xeotron developed the XeoChip, a bulk micromachined chip that contains thousands of three-dimensional nanochambers in an area smaller than a postage stamp (http://www.touchbriefings.com/pdf/855/fdd041_applied_tech.pdf). Each chamber serves as a protected reaction vessel for DNA synthesis and subsequent bioassay for use with conventional microarray scanners. The same system can be adapted for the synthesis of RNA, peptides, and other organic molecules.

Flow Cytometry

Flow cytometry is a method for high-throughput measurements of physical and chemical characteristics of cells and other microscopic biological particles. Typically, in commercial cytometers, optical measurements are made as particles are passing, single-file, through a sensing region. The particles are constrained to such a single line-up by hydrodynamic focusing. In this method, the sample flow is a narrow stream formed as the innermost flow of two concentric fluid flows (see Figure 10.44A). The faster the outer sheath flow, the narrower the inner sample flow is focused. Hydrodynamic focusing requires absolute laminar flow in all positions in the flow cell; from some high speed on, the inner stream will defocus through turbulence. Existing commercial

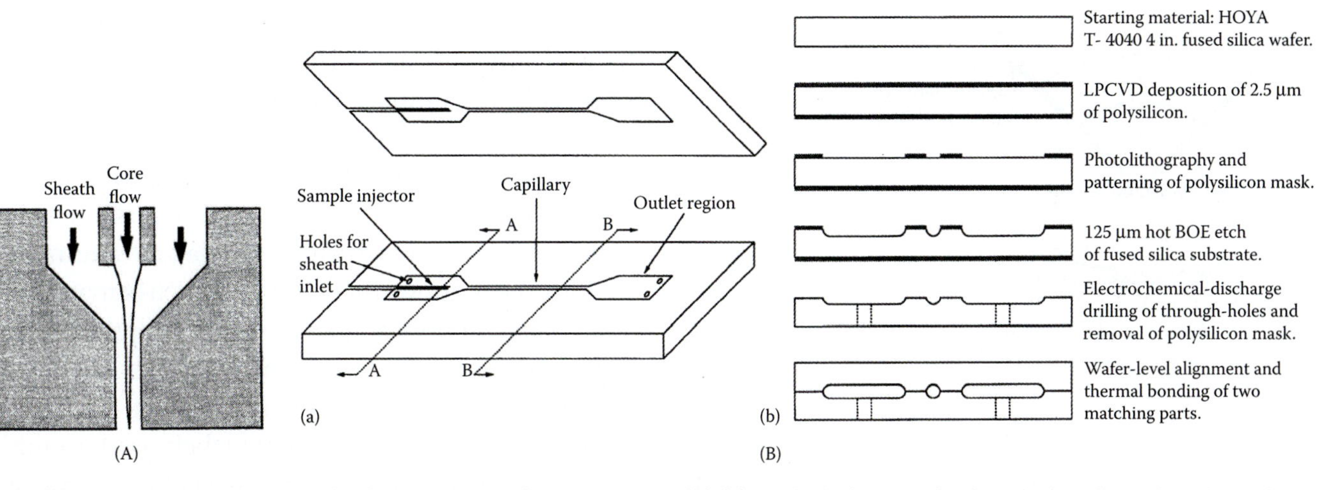

FIGURE 10.44 (A) Schematic of a flow chamber for cytometry. (B) (a) Exploded view of a fused silica flow chamber. The top and bottom have matching injection and outlet areas with a capillary joining them. Through-holes in the bottom plate are for sheath inflow and outflow. The sample is introduced through the injector into the center of the sheath flow. (b) The fabrication sequence shown along cross-section A-A. (After Sobek, D., S. D. Senturia, and M. L. Gray. 1994. In *Technical Digest: 1994 Solid State Sensor and Actuator Workshop,* 260–63. Hilton Head Island, SC.)

equipment requires frequent adjustment of the optical components, but simpler instrumentation is expected through the introduction of optical fibers replacing lenses for sample illumination and light collection. Shapiro and Hercher, in 1986,[40] suggested integration of flow channels, optical elements, and signal conditioning electronics as the next step toward more user-friendly equipment. The first such attempts led to complex micromachined sheath flow cells; for example, a five-layer stainless steel/glass laminate[41] and a four-wafer Si micromachined flow chamber.[42] Sobek et al.[43] have made a much simpler micromachined flow cytometer, which seems to put us back on the right track in terms of correct use of available micromachining tools. Cost consideration did result in a 30 × 15 × 1 mm chip made from two symmetrically machined HOYA T-4040 synthetic quartz (fused silica) parts. These two parts, bonded at 1000°C, define the ovoid cross-section of the sheath flow channel, the circular cross-section of the sample injector, and the ovoid measuring capillary (Figure 10.44B). A smooth, flat surface of the capillary is essential for ensuring good optical measurements in the measuring region. The quartz starting material used here comes in 4-inch wafers, just like Si wafers. This is an important point, as most processing equipment is built to take Si wafers only. The resulting cytometer can focus a sample stream, introduced into the injector with a syringe, to a 10-μm width and can maintain this focusing to a mean velocity up to at least 10 m/s.

The next step in this development would be to find the correct package for the cytometer and an interface to the macroworld. Connectors to the sheath flow ports and the sample injection mechanisms are areas ripe for innovative approaches.

Surface Plasmon Sensor: Spreeta™

Many opportunities for innovative micromachining applications in optical immunosensors present themselves. A prime example can be found in surface plasmon sensors (SPRs). Incident plane polarized light at a specific angle can be almost totally absorbed into a thin metal film (e.g., 500 Å Ag or Au) deposited onto a prism, fiber, or waveguide surface because of resonant coupling of the light energy into a free-electron cloud within the metal (see SPR theory in Volume I, Chapter 5). This effect is known as *surface plasmon resonance* and can be extremely sensitive to the refractive index changes surrounding the metal film. By adsorption of species on the metal film, the specific angle of the light needed to create the resonant absorption changes (Figure 10.45).[44] This adsorption could be from gases or the binding of macromolecules onto the metal film. The main limitation is that the sensitivity depends on the molecular weight, or more correctly, the optical thickness of the absorbed layer, meaning that small

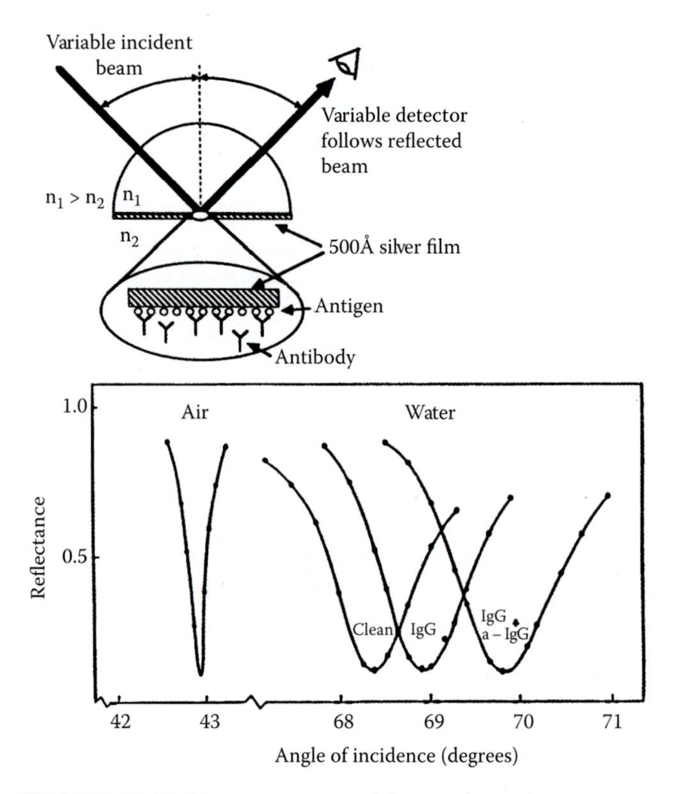

FIGURE 10.45 Measurements with a surface plasmon resonance sensor based on a prism. (From Sutherland, R. M., and C. Dahne. 1987. In *Biosensors: Fundamentals and applications,* eds. A. P. F. Turner, I. Karube, and G. S. Wilson, 655–78. Oxford, United Kingdom: Oxford University Press. With permission.)

molecules will not be good candidates. The SPR-based "immunosensor" from Amersham Pharmacia (Sweden) is a general-purpose, expensive, and rather large instrument incorporating a prism for detection and four parallel flow channels in removable sample chambers (in 2004 Amersham Biosciences, was acquired by GE Healthcare).

A significant amount of research effort is directed toward trying to miniaturize this large optical system. Comparison of diffraction gratings and prism systems clearly indicates that the diffraction grating is preferable for miniaturization. Garabedian et al.[45] attempted to micromachine the whole surface plasmon sensing system by implementing such a grating, but their resulting microinstrument is much too complex for manufacturing. An elegant alternative to both the grating and prism approaches may come from using an optical fiber or a planar optical waveguide. For example, Jorgenson and Yee[46] simplified SPR sensing dramatically by using such a fiber optic approach. As a sensing element, they used a multimode fiber with a section of the cladding removed and a thin layer of highly reflecting metal symmetrically deposited directly on the fiber core (see Figure 10.46a). This design has potential for simplified, robust, remote, and disposable sensing. Unlike traditional SPR measurements, employing a discrete excitation wavelength while modulating the angle of incidence, the SPR fiber optic sensor system uses a fixed angle of incidence and modulates the excitation wavelength. The refractive index can then be determined from the resonance spectrum in the reflected or transmitted spectral intensity distribution. In Figure 10.46b, the SPR sensing surface is enclosed in a flow-cell. Using micromachined planar waveguides in a flow-channel instead of the set-up shown here might result in a promising commercial product.

The most advanced example of a miniaturized SPR instrument is embodied in the miniature SPR sensor from Texas Instruments shown in Figure 10.47.[47–49] This sensor is also an apt illustration of "functional" packaging. As seen in the schematic in Figure 10.47a, light from a light-emitting diode (LED) is reflected from a gold surface in the LED

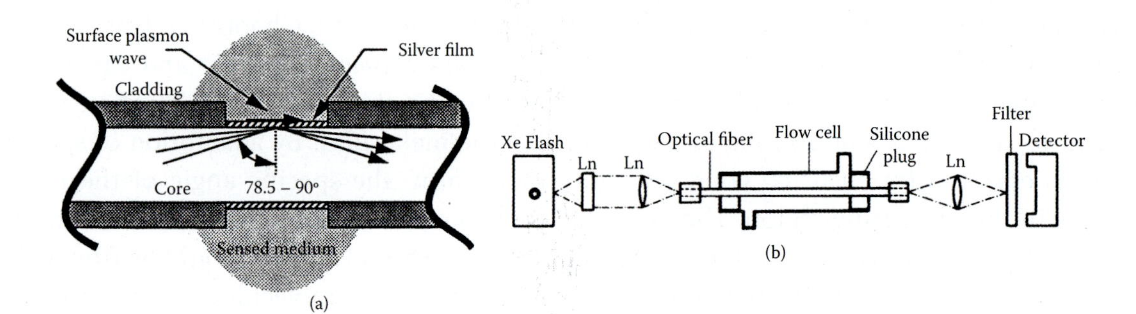

FIGURE 10.46 Simplified surface plasmon resonance (SPR) measurements. (a) Illustration of the SPR fiber. (b) Illustration of the SPR fiber enclosed in a flow-cell. (From Jorgenson, R. C., and S. S. Yee. 1993. A fiber-optic chemical sensor based on surface plasmon resonance. *Sensors Actuators B* B12:213–20. With permission.)

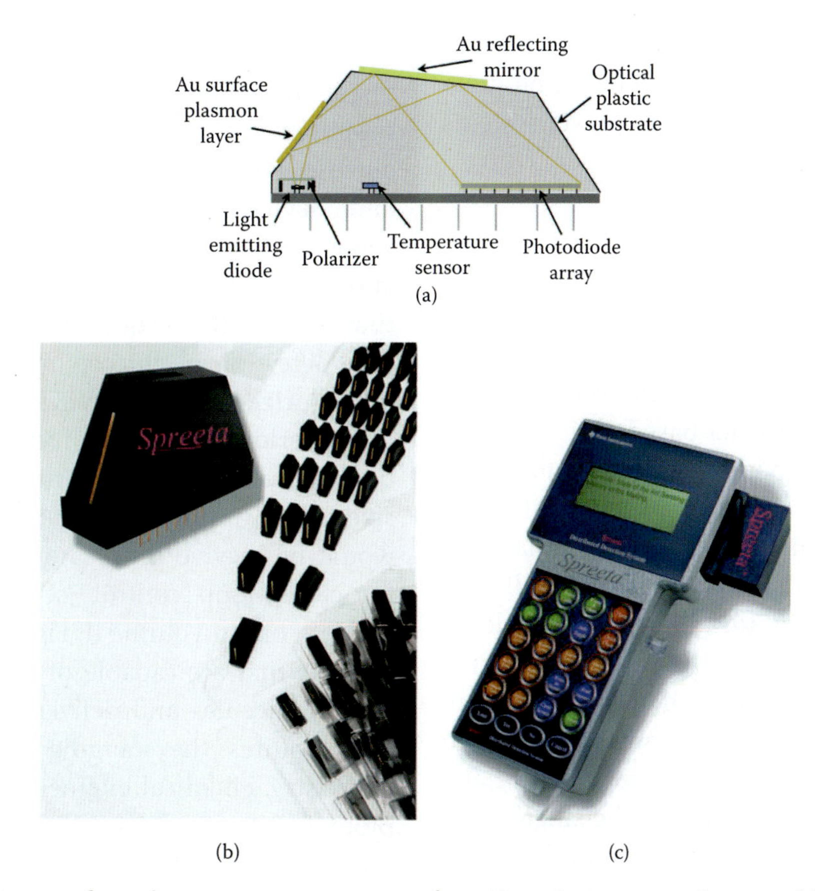

FIGURE 10.47 A miniature surface plasmon resonance sensor from Texas Instruments: Spreeta. (a) Spreeta geometry. (b) Packaged Spreeta unit. (c) Beta version of a cartridge-based Spreeta handheld instrument. (Courtesy of Dr. J. Elkind, Texas Instruments.)

optical plastic substrate/package and is picked up on a photodiode array embedded in the same package. Also contained in the same package is a polarizer for the light emitting diode and a temperature sensor. A plot of reflected light intensity versus angle is read out from the linear diode array as illustrated in Figure 10.48. The current noise floor of the sensor is 7×10^{-7} refractive index units. A beta-version of a cartridge-based Spreeta handheld instrument is shown in Figure 10.47c. Pricing for the disposable today is \$30, and a handheld reader might cost about \$250 (one-tenth as expensive as a Biacore).

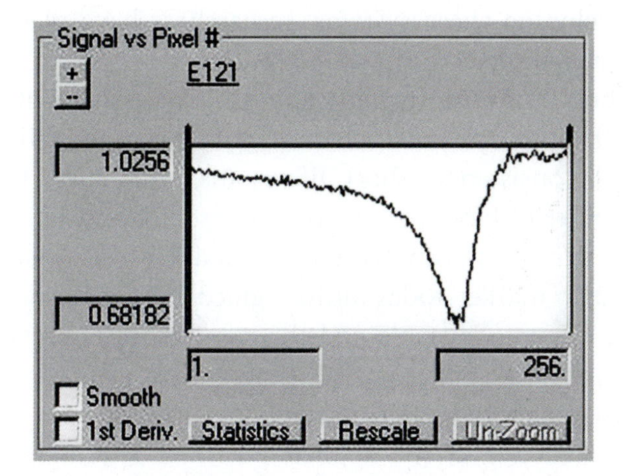

FIGURE 10.48 Surface plasmon resonance curve: reflected light intensity versus pixel number on the photodiode array. (Courtesy of Dr. J. Elkind, Texas Instruments.)

In Vivo MEMS Applications

Introduction

The many problems involving biocompatibility of in vivo devices were addressed in Chapter 5, "Biocompatibility and Surface Modification," where we learned that micromachining might actually provide a way of rendering surfaces more biocompatible by patterning them. Chemical surface modifications are also discussed. After the retrenching from in vivo devices in the 1990s, progress in materials and bio-MEMS has been so swift in the last decade that the problem of in vivo sensing is ready for a new attack. This move back to in vivo sensing is further made easier by the progress in miniaturized telemetry and micropower systems (see Chapter 9, this volume).

In Vivo pH and Blood Gases

The cost of disposable sensors for in vivo use is not as big a concern as for in vitro disposable sensors; size is more critically important because the smaller the sensor, the smaller the intrusion on the patient. It is estimated that a catheter-based pH, CO_2, and O_2 sensor array, as shown in Volume II, Example 4.2, "An Electrochemical Sensor Array Measuring pH, CO_2, and O_2 in a Dual Lumen Catheter," may cost up to $250 and would serve a market of well over $600 million a year in the United States alone. Unfortunately, the warm, wet, saline in vivo environment perhaps puts forth the most severe environment in which Si sensors might be used. No Si-based in vivo chemical sensors are commercially available today. Biocompatibility is the single most complex issue facing in vivo sensor development, and it needs to be addressed up front in the sensor design. This is in line with our philosophy of starting a sensor design from the package. The package for an electrochemical in vivo pH, CO_2, and O_2 sensor typically is a dual lumen catheter modified to incorporate the sensor structure with the required chemically sensitive materials (see Volume II, Example 4.2). In building such a sensor, the author at first neglected to investigate biocompatibility issues up front and instead concentrated on the embedded micromachined electrochemical sensors.[50] The measurements obtained in saline solutions were encouraging, but the funding for continued research dried up before serious biocompatibility testing could be started. The development of the disposable in vivo pH, CO_2, and O_2 probe, the room-temperature micro electronic chemical smart sensor (RT-MECSS) took four years and $6 million. The RT-MECSS puts chemistry and electronics on separate planes while keeping the signal line as short as possible (see Volume II, Figure 4.99).[51] The latter represents a considerable improvement over both ISFET and EGFET (see Volume I, Chapter 7, "Ion-Sensitive Field-Effect Transistors, Extended Gate Field-Effect Transistors, and Isn't FET").

In Vivo Glucose Sensing

There have been many different technologies proposed for in vivo and ex vivo detection of glucose, and a number of ex vivo instruments and devices have been commercialized. While some of these, in particular the ex vivo ones (see above), have been very successful, a technology that meets the requirements of continuous in vivo detection of glucose and can be coupled to a miniaturized responsive delivery of insulin would be superior. Insulin delivery systems that have been developed are external and quite large in size. Ideally, miniaturized drug delivery systems would be in vivo and feature an integrated sensor for closed-loop control (see the responsive drug delivery pill in Volume II, Figure 4.100). Developing systems that are capable of both in vivo detection of biomolecules and delivering a drug is complex and requires the merging of technologies from chemistry, chemical engineering, micromachining, biomaterials engineering, and pharmaceutical science. For example, an ideal in vivo biosensor must be capable of long-term, rapid, reproducible detection of its target biomolecule at physiological levels, must be biocompatible, and must be amenable to miniaturization/microfabrication. Also, the insulin delivery system, controlled by the glucose sensor, should be capable of fast response and hold a sufficient amount of insulin to justify the long-term implantation of the device. The low success level achieved in designing in vivo miniaturized systems so far may partially stem from the fact that when developing these systems, researchers have placed more emphasis on one particular aspect than on others. Some of the persistent difficulties associated with glucose sensors include manufacturing reliability and consistency, short life spans, and cost. The number of biosensors with demonstrated capabilities for in vivo sensing is also limited.[52] Most sensors in the market today utilize glucose oxidase since the enzyme is stable and the cost is low. However, major disadvantages with using glucose oxidase are that a number of in vivo endogenous species are electroactive at the applied potential required for peroxide formation[52] and the production of H_2O_2 renders the eventual disintegration of these sensors.

As we saw in Volume I, Chapter 7, binding proteins associate with a particular analyte in a highly specific manner. Daunert et al. utilized this property in the development of reagentless optical biosensors.[53–56] Specifically, one of the many proteins of interest is the glucose and galactose binding protein (GBP). Glucose binding induces a hinge motion in GBP, and the resulting protein conformational change constitutes the basis of the sensor development. Labeling this protein with a fluorescent probe allows for the monitoring of this change without the addition of any external reagents or substrates. In order to obtain maximal signal versus background fluorescence, it is important that the fluorescent probe be attached to a single site (e.g., a unique cysteine) and at a position where maximum conformational change occurs. This approach has been shown to yield sensors with improved detection limits.[56]

As mentioned above, the most undesirable effect of current daily glucose monitoring is the frequent finger-stick glucose monitoring that the patient needs to perform. Even the more successful implantable devices, such as the MiniMed Continuous Glucose Monitoring System (currently being evaluated in several clinics throughout the United States), need frequent recalibration. This is due to the direct correlation between sensor response and the time-dependent enzyme activity. The use of GBP labeled with fluorescent compounds that have a long lifetime can afford sensors with longer service lives. Lifetime measurements are dependent on the glucose concentration but independent of the concentration of the fluorescently labeled protein present.[57] Thus, since these sensors do not require an active enzyme, a reduction in calibration frequency can be achieved. It is this type of sensor we envision combining with the smart pill in Volume II, Figure 4.100 (pharmacy-on-a-chip).

μ-TAS and LOC in Sample Preparation for Molecular Diagnostics

Introduction

Micromachining and genomics parallel each other in perceived value to society and much is expected from a merging of both fields. Bio-MEMS, in principle, holds the potential to revolutionize molecular diagnostics, as it could lead to much more compact, potentially disposable, and well-integrated molecular diagnostic systems. As we saw in Volume II, Chapter 8, the use of photolithographic techniques has already permeated molecular biology; they are used, for example, to make DNA arrays (e.g., Volume II, Figure 8.16). To enhance DNA hybridization speed, DNA arrays have been embedded in all types of fluidic structures. The principal advantages associated with a bio-MEMS, μ-TAS/LOC approach include the following:

- Small instrument footprint
- Multiplexing so that several analyses can be completed simultaneously
- Microfluidics for fluid manipulation, often obviating the need for traditional macropumps
- Excellent potential for integrating various components onto the device
- Reduced volume requirements for sample and reagents (often expensive)

Here we analyze specifically bio-MEMS contributions for sample preparation in genomics. Like packaging in the semiconductor industry, sample preparation is an area that has been less vigorously pursued than developing the assay itself; perhaps it is perceived as somewhat less glamorous, as is packaging in the IC industry. But, like IC packaging, sample preparation is of overwhelming importance to the end customer. The benefit in developing miniaturized devices for sample preparation in genomics is the low dead volume requirement associated with accuracy, costs of reagents, and small overall footprint, and the difficulty is having enough copies for the analyte to be detectable. The latter implies a flow system reducing the volume during the sample preparation step.

Types of Samples

Procedures for preparing samples for chemical analysis vary dramatically from sample to sample and from analysis to analysis. To reduce the complexity of the discussion, we single out for closer scrutiny the challenges faced in preparing samples for molecular diagnostics. Specifically, we investigate the feasibility of performing sample preparation in a small-footprint integrated instrument capable of performing all the aspects of a common biological DNA assay. The preparation of biomedical samples prior to the diagnostic procedure is often complex, laborious, and time-consuming. It varies, even within the more limited application of DNA testing, depending on the sample type, for example, urine, whole blood, plate bacterial cultures, serum, stool, tissue, etc. Today, the preparation of such samples involves several different instruments and intensive manual operations such as centrifugation, heating and cooling, mixing with reagents, filtration, etc. For most clinical diagnostics, sample preparation is actually the rate-limiting step. In the case of whole blood samples, red blood cells (RBCs) must be separated from leukocytes [white blood cell (WBC) count for 0.2% of human blood cells: 1 μL of whole blood contains 5×10^6 RBCs and 7×10^3 WBCs) and other cells (bacteria, virus particles, etc.). Red blood cells do not have nuclei, the source of nucleic acids, but they do contain contaminants, which interfere with nucleic acid amplification [polymerase chain reaction (PCR)] and other enzymatic manipulations (e.g., both proteins and metal complexes such as hemoglobin bind with nucleic acid). Isolated leukocytes and bacteria samples must be lysed before the extraction of nucleic acid can take place. A traditional procedure for blood DNA isolation comprises separation of the white cell fraction from the whole blood, lysis with detergent, and digestion of proteins with proteinase K (1–6 hours) to purify the target material from contaminating proteins, followed by extraction with phenol-chloroform and precipitation of DNA with alcohol. DNA is not soluble at high ethanol concentrations, so it precipitates out as long strands. Salts, such as sodium chloride, also greatly aid in precipitating DNA. Tissue samples primarily contain cells and connective tissue, which also must be degraded before DNA extraction. The first step in the DNA extraction procedure from tissue is typically to boil a small amount of the tissue in a 5% Chelex solution. The boiling helps to break up the cells in the preparation and denatures proteins. The Chelex protects the DNA sample from DNAases* that might remain after the boiling and that could subsequently contaminate the samples. In the diagnostic approach of the future one envisions a small footprint instrument and a test specific disposable cartridge providing a "sample to answer" solution.

* DNAases are enzymes that occur naturally in all body tissues. They cut DNA, rendering it unsuitable for PCR. Magnesium ions are essential cofactors for DNAases. Chelex resin binds with cations, including Mg^{2+}. By binding with the magnesium ions, the Chelex resin renders DNAases inoperable, thus protecting DNA from their action.

General Sample Preparation Protocols

Sample Size

Miniaturization is a mixed blessing in the case of the amount of sample required to detect a given analyte concentration. That volume is determined by the following:

$$V = \frac{1}{\eta N_A C_i} \tag{10.1}$$

where η = sensor efficiency, with a value between 1 and 0

N_A = Avogadro's number (6.02×10^{23} mol^{-1})

C_i = the concentration of analyte i (moles/L)[58]

The sample volume required is fundamentally dictated by the concentration of the analyte one wants to measure. Manz et al. examined the use of very small sample sizes and concluded that microfluidics is a desirable approach for many diagnostic applications.[59] They summed up their analysis in a graph of target analyte concentration versus required sample volume. This graph shows that many biological chemicals associated with clinical chemistry assays (between 10^{14} and 10^{20} copies/mL) and immunoassays (between 10^7 and 10^{18} copies/mL) might be readily assayed with very small sample volumes, in the range between picoliters and microliters. However, Petersen et al. pointed out that numerous chemicals (and organisms) are routinely present at much lower concentrations, from less than 100 to 10^7 copies/mL.[58] These low-concentration samples include most sources of DNA, which must be detected and analyzed in a growing number of new diagnostic tests. In Figure 10.49, Petersen et al. expand the boundaries of Manz's original graph, which spanned a concentration range from 10^8 to 10^{21} copies/mL and a volume range from 10 mL to 10^{-18} L, and clearly show that the minimum sample volume required for accurate DNA assays is a relatively large volume of 100 μL.

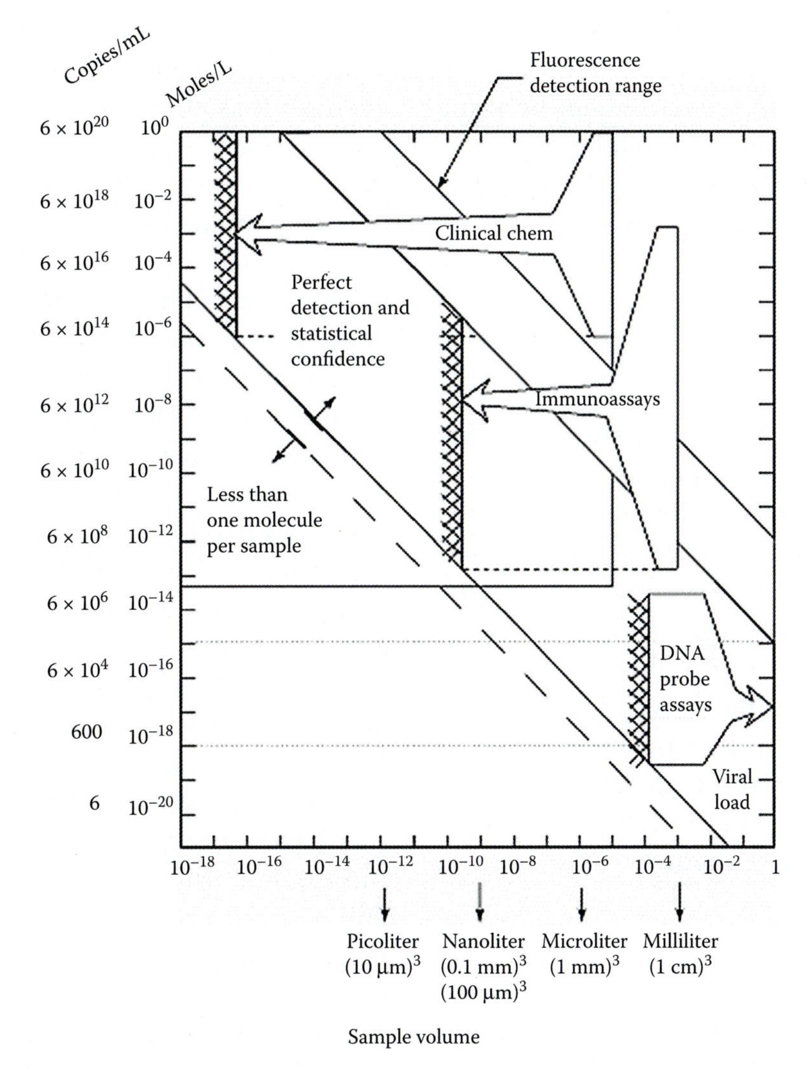

FIGURE 10.49 Sample volume. Scaling of concentrations and volumes.[58] (Courtesy of Dr. Kurt Petersen, Cepheid.)

The amount of original sample required, depending on the analysis method, and the number of DNA copies needed for detection may range from some microliters to several tens of milliliters. Figure 10.49 demonstrates that typical clinical chemistry assays (between 10^{14} and 10^{20} copies/mL) and immunoassays (between 10^7 and 10^{18} copies/mL) are readily carried out with very small sample volumes, in the picoliters and microliters range, but numerous DNA assays deal with less than 100 to 10^7 copies/mL[60] and may require milliliter sample volumes. To summarize, modern DNA diagnostic assays face mounting demands to detect organisms or DNA mutations at very low concentrations, often less than 100 copies per milliliter, in raw biological samples such as blood or urine. Such sensitivity requirements set fundamental physical limitations on the minimum quantities of the starting sample. Using the current nucleic acid hybridization formats and stringency control methods, it remains difficult to detect low copy number nucleic acid targets even with the most sensitive reporter groups (enzyme, fluorophores, radioisotopes) and associated detection systems (fluorometers, luminometers, photon counters, scintillation counters), necessitating a DNA amplification step, for example, using polymerase chain reaction (PCR) (see Chapter 2, "Polymerase Chain Reaction"). PCR is usually considered sensitive enough to detect as few as twenty copies of a target in any given, purified, biological sample. But if the required assay sensitivity is 100 copies per mL, then a minimum of 200 μL of the original sample must be processed to ensure statistical confidence in the result of the assay. If microfluidics and small PCR chambers are used, this means that DNA preconcentration is needed, for example, on silica-type capture surfaces or on arrays of micromachined Si posts.[60] The latter reduces the sample size from macro to micro, which can be accommodated in a microsystem by using some type of a flow system. A modular approach, with a variety of disposable cartridges containing different fluidic options and sensor chemistries in a small fixed instrument, is the most probable solution.

Lysing

Sample preparation for DNA analysis in an integrated device includes releasing the nucleic acid from the sample by cell disruption or lysis, removing contaminants (purification), and recovering/concentrating the nucleic acid. There are many mechanical and nonmechanical cell lysis or disruption methods available. The results of these may be evaluated in terms of the activity level of a cellular enzyme released to the suspending medium measuring the efficiency of the disrupting process. We briefly review both types of lysing approaches and consider their relative merit for integration in microsystems.

Mechanical Methods

Currently, in large nonintegrated tabletop equipment, intracellular products are released from microorganisms or tissue mainly by mechanical disruption of the cells. In this process, the cell envelope is physically broken, releasing all intracellular components into the surrounding medium. There are several types of equipment for mechanical cell disruption commercially available. In a high-pressure homogenizer, a pump forces slurry of sample through a restricted orifice valve. High pressure (up to 1500 bar) is followed by an instant expansion through a special exiting nozzle. Cell disruption is accomplished by three different mechanisms in this system: impingement on the valve, high liquid shear in the orifice, and sudden pressure drop upon discharge, finally causing an explosion of the cell. This approach, although very efficient at cell disruption, is difficult to miniaturize. In a bead mill, cells are agitated in suspension with small abrasive particles. Cells break because of shear forces, grinding between beads, and collisions with beads. This approach is slightly more amenable to miniaturization. Another widely applied method is cell lysis through high-frequency ultrasound. These devices generate intense sonic pressure waves in liquid media. Under the right conditions, the pressure waves cause formation of microbubbles, which grow and collapse violently. Called *cavitation*, the implosion generates a shock wave with enough energy to break cell membranes. Modern ultrasonic processors use piezoelectric generators made of lead zirconate titanate crystals (see Chapter 8, this volume). Vibrations are transmitted down a titanium metal horn or probe tuned to make the processor unit resonate at 15–25 kHz. A relatively new development is a cordless ultrasonic processor supplied by BioSpec Products (Bartlesville, OK). With a tip diameter of 1/8 inch, it easily fits into microtubes and ninety-six-well titer plates. Ultrasonic disintegrators generate considerable heat during processing. For this reason, the sample should be kept ice-cold, and one should be aware that free radicals can be generated during sonication and that these radicals may react with biomolecules. Ultrasonic disruption is an attractive approach for miniaturization of the cell lysis process.

Mechanical disruption methods suffer several drawbacks. Because cells are broken completely, *all* intracellular materials are released. Therefore, the DNA must be separated from a complex mixture of proteins and cell wall fragments. Cell debris produced by mechanical lysis often consists of very small fragments, making the solution difficult to clarify. Moreover, complete product release often requires more than one pass through the membrane disruption device, exacerbating the problem by further reducing the size of the fragments. These fragments are difficult to remove by continuous centrifugation because the throughput of a centrifuge is inversely related to the square of the particle diameter. Filtration is complicated by the gelatinous nature of the homogenate and by its tendency to foul membranes. Because of these drawbacks, several nonmechanical cell lysis approaches have gained favor.

Nonmechanical Methods

Nonmechanical lysis includes thermal, electronic, and chemical methods. Chemical permeabilization of cells can be achieved in numerous ways.[61,62] The most important are enzymatic lysis (e.g., with lysozyme), detergents, and osmotic shock. Chemical methods extract intracellular components from microorganisms by permeabilizing outer cell wall barriers. This can be achieved with organic solvents that act by creation of channels through the cell membrane. Some of the solvents that have been used for this purpose are toluene, ether, phenylethyl alcohol, dimethyl sulfoxide (DMSO), benzene, methanol, and chloroform.

Chemical permeabilization can also be achieved with antibiotics, thionins, surfactants (e.g., NP-40, Triton X-100, Brij, Duponal, CHAPS, SDS, etc.), chaotropic agents, and chelates. An important example in the latter category is EDTA, widely used for permeabilization of Gram-negative microorganisms. Its effectiveness is a result of its ability to bond the divalent cations Ca^{++} and Mg^{++}. These cations stabilize the structure of outer cell membranes, by bonding the lipopolysaccharides to each other. Once these cations are removed by EDTA, the lipopolysaccharides are removed, resulting in increased permeability of the outer cell walls. Chaotropic agents, such as urea and guanidine, are capable of bringing hydrophobic compounds into aqueous solutions. The relative ability of various anions to influence the structural stability of protein (and DNA) in general follows the series $F^- < SO_4^{2-} < Cl^- < Br^- < I^- < ClO_4^- < SCN^- < Cl_3CCOO^-$. This ranking is known as the *Hofmeister series* or *chaotropic series*. Fluoride, chloride, and sulfate salts are structure stabilizers, whereas the salts of other anions are structure destabilizers. Salts that stabilize proteins enhance hydration of proteins and bind weakly, whereas salts that destabilize proteins decrease protein hydration and bind strongly. In other words, the denaturing effect of chaotropic salts is related to destabilization of hydrophobic interactions in proteins. High concentrations (6–8 M) of compounds that tend to break hydrogen bonds, such as urea and guanidine salts, also cause denaturation of proteins. These substances apparently disrupt hydrogen bonds which hold the protein in its unique configuration. However, there also is evidence suggesting that urea and guanidine hydrocholoride may disrupt hydrophobic interactions by promoting the solubility of hydrophobic residues in aqueous solutions. Detergents break open (lyse) cells by destroying the fatty membrane that encloses them. This releases the cell contents, including DNA, into the solution. Detergents also help to strip away proteins that may be associated with the DNA. Enzymes can also be employed to permeabilize cells, but this method is often limited. Enzymes used for enzymatic permeabilization are glycanases, proteases, and mannase. The main drawbacks of employing enzymatic means for recovering intracellular products in large-scale processes are cost and the necessity of removing the lysing enzyme from the product. Basic proteins, such as protamine, or the cationic polysaccharide chitosan can permeabilize yeast cells. Antibiotics that affect bacterial cell-wall biosynthesis, causing loss of viability and often cell lysis (penicillins and cephalosporins, bacitracin, cycloserine, vancomycin), may also be used. Similarly, mammalian cells can be permeabilized by exposure to several natural substances such as streptolysin or even viruses.

While cells exposed to slowly varying extracellular osmotic pressure are usually able to adapt to such changes, cells exposed to rapid changes in external osmolarity can be mechanically injured. This procedure is typically conducted by first allowing the cells to equilibrate internal and external osmotic pressure in a high-sucrose medium and then rapidly diluting away the sucrose. The resulting immediate overpressure of the cytosol is assumed to damage the cell membrane.

Repeated freezing and thawing will thermally lyse a cell, and in some instances, the target DNA may be released by elevating temperatures simultaneously with the denaturing step performed in PCR.[63]

Electronic lysis of cells has been reported by Cheng et al.[64] and Lee and Tai.[65] Of the different types of nonmechanical lysing, the electronic and thermal approach is the most amenable to miniaturization. Chemical approaches should be ranked according to the amounts of chemicals required; the lesser the amount, the more suited the approach to miniaturization

Table 10.10 summarizes some lysis methods and subsequent purification and recovery steps.

TABLE 10.10 Cell Lysis, DNA Purification, and Recovery

Process	Technical Difficulty	Costs	Efficiency	Ease of Miniaturization
Lysis				
Mechanical	High	High	High	Of the mechanical lysis methods reviewed, ultrasonic is most amenable
Chaotropic	Medium	Medium	Medium	Depends strongly on the volume of reagents involved
Thermal	Low	Low	Low	Easy to miniaturize (sometimes this step can be combined with polymerase chain reaction)
Purification				
Ion exchange	Medium	Medium	High	Easy to implement on a miniaturized platform
Beads	Medium	Medium	Medium	Somewhat more cumbersome
Molecular weight cutoff membranes	Low	Low	Low	Easy to implement on a miniaturized platform
Recovery				
Silica elution	Low	Low	High	If a flow-through system is used, miniaturization of low-salt buffer is easy to accommodate
Bead elution	Low	Low	Medium	More cumbersome
Membrane elution	Low	Low	Low	Easiest to miniaturize

Examples of Integrated Sample Preparation

Over the last ten years, several attempts have been made to miniaturize molecular diagnostics, inclusive of sample preparation, in the areas of biological warfare agents and point of care.[58,66–74] These efforts involve microfluidic systems, in which the entire protocol from sample collection through PCR, and in several cases all the way to detection, is embedded in one integrated instrument. In Table 10.11 we review a few of these prototype DNA analysis devices (expanded from Yuen et al.[75]). Extensive efforts pertain to isolating white blood cells from whole blood through the use of micromachined filters (arrays of posts, tortuous channels, comb-shape, and weir-type)[66,76,77] and different designs of silicon microfilters (3–10 μm) across microchannels.[78,79] All these designs feature a high efficiency for separating red from white blood cells. More recently, Yuen et al.[75] CNC-machined a Plexiglas-based module integrating white blood cell separation and nucleic acid amplification reactions. This module comprises a heater-cooler for thermal cycling and a series of 254 × 254 μm microchannels for transporting human blood and reagents in and out of an 8–9-μL dual-purpose (cell separation and PCR) glass-silicon microchip. The filter in this case was of the weir type (3.5-μm filter gap). A region in the human coagulation Factor V gene (226-bp) was directly amplified by the microchip-based PCR from DNA released from white blood cells isolated from the weir filter section.

In the sample preparation approach of the company Tecan Boston (now defunct), a compact disc (CD) (the LabCD; see Volume I, Figure 6.88 and Figure 10.50 below), was used as a sample preparation stage (see also Table 10.11 below). The physics behind this centrifugal microfluidic platform is covered in Volume I, Chapter 6, and its prototyping is detailed in Chapter 5, this volume. The Tecan Boston prototype combined sample preparation and PCR, while detection was performed separately on an ethidium-bromide-stained gel electrophoresis slab. The disposable disc is shown in Figure 10.50. The sample preparation protocol employed is as follows: 1) mixing raw sample (5 μL of whole blood or *Escherichia coli* suspension) with 5 μL of 10 mM NaOH; 2) heating to 95°C for 1–2 minutes to lyse cells, releasing DNA and denaturing proteins inhibitory to PCR; 3) neutralization of lysate by mixing with 5 μL of 16 mM Tris-HCl (pH 7.5); 4) mixing of neutralized lysate with 8–10 μL of liquid PCR reagents and user-selected primers; and 5) thermal cycling. Samples of whole blood or *E. coli* were run on the discs shown in Figure 10.50 using a rotational profile that sequentially drives fluids past capillary valves *d*, through mixing channels *e*, to receiving reservoirs. Reservoirs *a*, *b*, and *c* contain sample, NaOH, and Tris-HCl, respectively. At the lowest rpm,

Table 10.11 Prototype Integrated DNA Analysis Devices

Sample Preparation	Biochemical Reactions	Detection	Fabrication Methods for Device	Source
RNA purification from 1 mL of serum samples spiked with different copy numbers of an RNA transcript that contained a partial HIV sequence	Reverse transcription-PCR, nested PCR, DNase fragmentation and dephosphorylation, and terminal transferase labeling	GeneChip hybridization	Conventional CNC machining	Anderson et al.[67]
White blood cells isolation from <3 μL of human whole blood	PCR using integrated heater-cooler to provide thermal cycling	Not applicable	Conventional CNC machining (standard microfabrication techniques for microchips)	Plexiglas microchip module[75]
5 μL of whole blood or *E. coli* suspension	PCR using thermoelectric devices on a spinning personal computer board platen	Detection is performed separately by electrophoresis	CNC machining of polycarbonate or poly(methylmethacrylate) disc combined with lithography in polydimethylsiloxane for smaller fluidic features	Tecan Boston[74]
Starting from the DEP collection of intact bacteria in a 70-μL sample to DNA hybridization assay (*SLT1* gene of *E. coli*),	Strand displacement amplification, dielectrophoretic collection of *E. coli* bacteria, electronic DNA hybridization	Electric field driven DNA assay	Laminated flexible substrates with electrode arrays (using integrated circuit lithography)	Nanogen[73]
Spores in paper products (0.5 gm in 10 mL of water)	PCR on material eluted from a Si DNA extraction chip (set of 200-μm-high columns, 18 μm in diameter, with a pitch of 34 μm)	Gel electrophoresis in a separate step	Acrylic or polyester plastics and Si chips	Cepheid[60]
30 green-fluorescent protein-transfected THP-1 cells in 25 μm of whole blood	Real-time-PCR	Real-time PCR	Droplets containing surface-functionalized superparamagnetic particles	Institute of Bioengineering and Nanotechnology[80]

Source: Expanded from Yuen et al.[75]
CNC, computer numerical control; PCR, polymerase chain reaction.

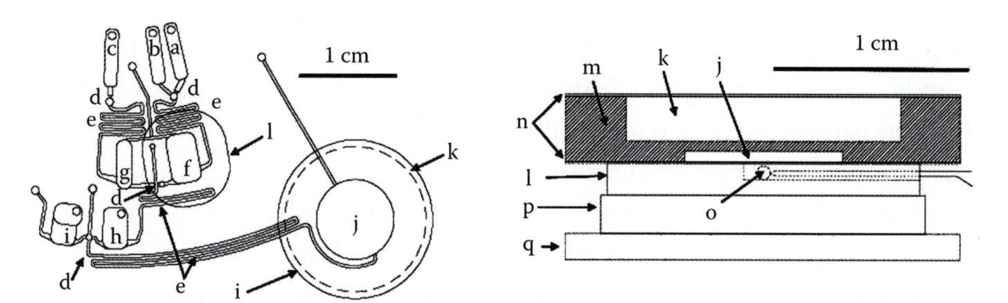

FIGURE 10.50 On the left is the polymerase chain reaction (PCR) structure. The center of the disc is above the figure. The elements are as follows: (a) sample; (b) NaOH; (c) Tris-HCl; (d) capillary valves; (e) mixing channels; (f) lysis chamber; (g) Tris-HCl holding chamber; (h) neutralized lysate holding chamber; (i) PCR reagents; (j) thermal cycling chamber; and (k) air gap. Fluids loaded in (a), (b), and (c) are driven at a first rpm into reservoirs (g) and (f), at which time (f) is heated to 95°C. The rpm is increased and the fluids are driven into (h). The rpm is increased and fluids in (h) and (i) flow into (j). On the right, the cross-section shows disc body (m), air gap (k), sealing layers (n), heat-sink (l), thermoelectric (p), personal computer board (q), and thermistor (o). (Courtesy of Dr. G. Kellogg, Tecan Boston.)

sample and NaOH mix in mixing channel *e* and move to reservoir *f* while Tris-HCl moves to holding reservoir *g*. Chamber *f* is the lysis chamber, which heats the sample/NaOH mixture up to 95°C for 1–2 minutes to lyse the cells and denature the proteins. Heating of the lysis chamber is accomplished by a thermoelectric element mounted on a personal computer-board platen *q* on the spindle of the rotary motor and connected by a slip-ring to a stationary power supply and a temperature controller. The lysis chamber sits on top of a brass heat-sink *l* mounted directly onto the thermoelectric element *p*. A thermistor, *o*, mounted in the heat-sink, enables closed-loop temperature control. By driving up the rpm, the Tris-HCl from the holding chamber *g* and the lysed material in reservoir *f* mix in the serpentine mixing channel neutralizing the lysate and then move into chamber *h*. At yet a higher speed fluids in *h* and *i* flow into the PCR chamber *j*. The PCR reagents used were either reconstituted beads or a solution of necessary reagents held in reservoir *i*. Rapid cycling with small temperature gradients in chamber *j* is achieved by confining the fluid to a thin layer, 0.5 mm thick, adjacent to another thermoelectric element. Above the fluid, an air pocket, *k*, serves as a thermal insulator. As a result, most of the heat transfer occurs via the thermoelectric. Slew rates of ±2°C/s^{-1} have been achieved with fluid volumes of 25 μL; the temperature gradients across the liquid are less than 0.5°C (from simulation and thermocouple measurements). With slight design changes, twelve structures of the type shown in Figure 10.50 may be placed on a 12-cm plastic disc. As we remarked when reviewing fluid propulsion options in Table 1.8 of this volume, a centrifuge platform is an attractive sample preparation stage (wide dynamic range, relatively insensitive to fluid composition, etc.). There is one major issue yet to be resolved, though: unless one works with all dry reagents (and has wetting take place only upon first use of the disc) the liquids held in the sample reservoirs will evaporate and distribute over time as there is no vapor barrier to stop them from expanding within the whole available open volume in the disc. The fluid gating in the disc works only for fluids, not for vapors. Electrokinetic and acoustic fluidic platforms suffer the same problem, so a major remaining challenge for developers of diagnostic platforms based on microfluidics is to implement both fluid and vapor gating. One possible solution is shown in Chapter 1, this volume, Figure 1.12, where reagents of interest are sealed in small blister pouches in a plastic laminate and only when the liquid is needed does a mechanical roller break the liquid and vapor seal. In Figure 8.52 we suggest yet another solution involving ferrowax valves on a CD platform. An external magnet can move these ferrowax plugs to pump or valve fluid plugs on the CD platform.

Kido at RotaPrep, Inc., developed the OrbiPrep system, an integrated, centrifugal sample preparation system. This station does not involve PCR yet, so it is not listed in Table 10.11 but PCR will be integrated in the future. It combines magnetism,[81] centrifugal acceleration, dynamic reagent dispensing,[82] and the Coriolis effect[83] in a system that performs automated, simultaneous milling/homogenization of up to twelve biological samples, followed by the extraction, washing, and elution of purified DNA from each sample. The system consists of a computer-controlled centrifuge-like device that accepts either two-sample rotors or twelve-sample rotors. Figure 10.51a shows a two-sample rotor on the device with its main components labeled. The device is also able to dispense accurate volumes of up to four discrete reagents into the center ring-shaped distribution trough of the rotor while it rotates. This is accomplished by fixing four dispensing needles to the lid of the device. When the lid closes (Figure 10.51b), the tips of the needles enter the trough to rest below the top plane of the rotor. Walls within the trough divide the entering stream of liquid equally among the sample preparation units of the rotor. Figure 10.52 shows the different stages of operation of the OrbiPrep. After loading the samples with buffer into the milling chambers through the sample loading ports, the milling stage involves a slow rotation at 400 rpm that causes the free moving magnet within each milling chamber to oscillate in the radial dimension as the rotor turns through the static magnetic field. The yellow arrows depict the trajectory of the free-moving magnet. The resulting milling action homogenizes biological samples, releasing their contents. The slow rotation prevents the liquid in the milling chamber from moving to the next

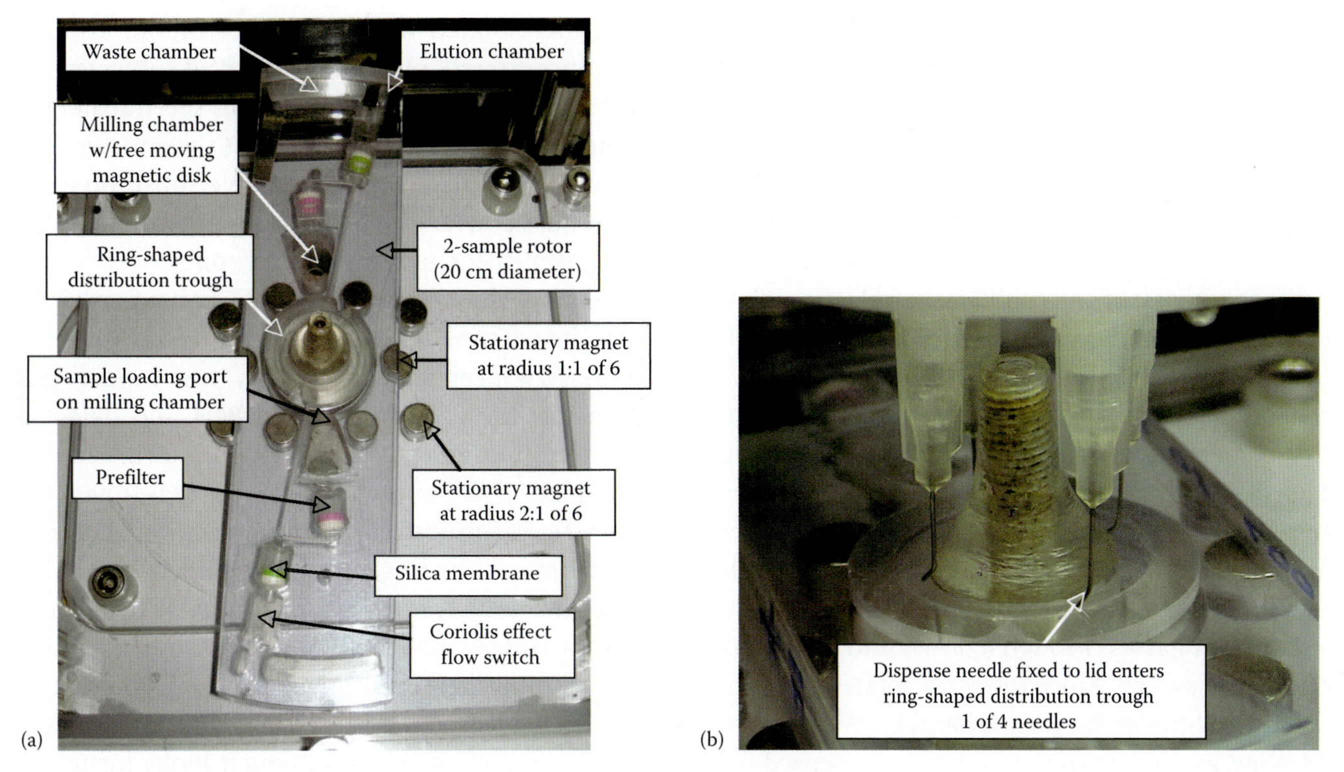

FIGURE 10.51 OrbiPrep system (RotaPrep, Inc.): an integrated centrifugal sample preparation system. (a) Sample preparation station with open lid. (b) Lid closed.

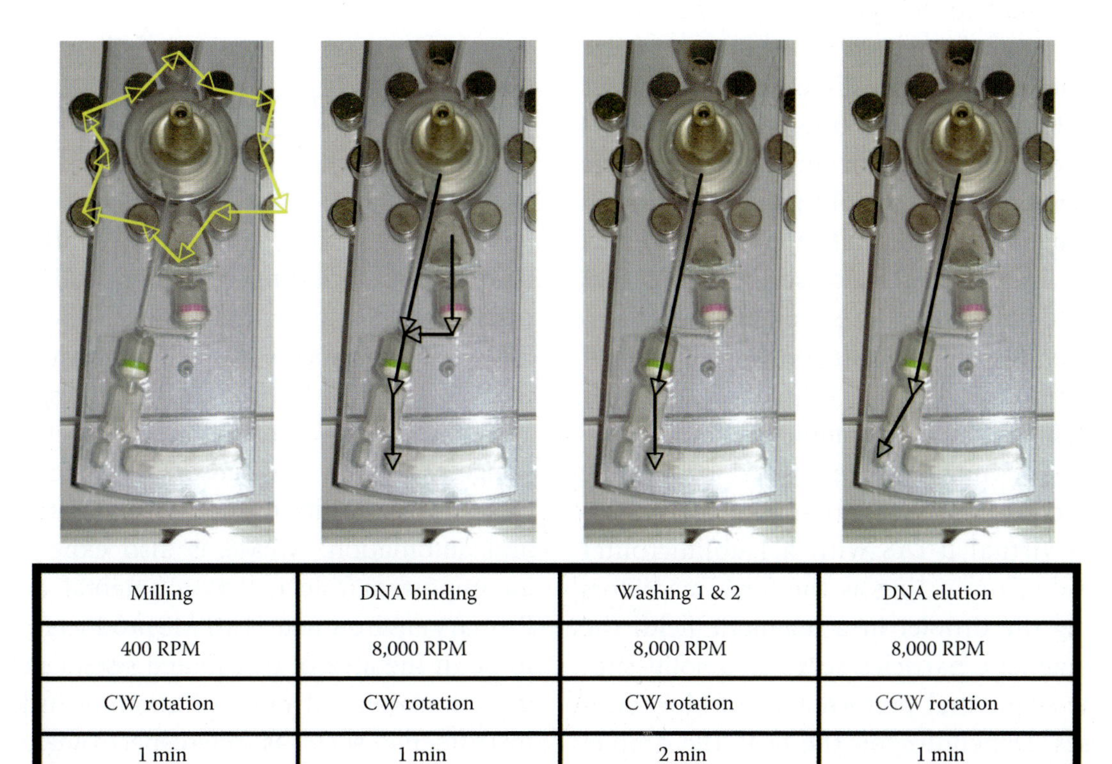

Milling	DNA binding	Washing 1 & 2	DNA elution
400 RPM	8,000 RPM	8,000 RPM	8,000 RPM
CW rotation	CW rotation	CW rotation	CCW rotation
1 min	1 min	2 min	1 min

FIGURE 10.52 OrbiPrep system (RotaPrep, Inc.): an integrated centrifugal sample preparation system. Shown are the different stages of operation of the OrbiPrep.

chamber held there by capillary forces (i.e., the rotation rate is below the burst frequency of the capillary valve separating the two chambers, see Volume I, Equation 6.141). In the DNA binding step, the rate of rotation is increased to 8000 rpm. This causes liquid from the milling chamber to flow downstream (i.e., the rotation speed is now increased above the burst frequency of the capillary valve), through the prefilter, toward the silica membrane. Before the liquid reaches the silica membrane, it is mixed with ethanol that started to be dispensed through the first dispensing needle into the distribution trough. The black arrows depict the flows of the respective liquid streams and the mixing point immediately upstream of the silica membrane. The silica membrane extracts DNA from the combined liquid stream that passes through it down into the Coriolis effect flow switch. Since the rotor rotates clockwise, the stream enters the waste chamber. Washes 1 and 2 follow, as wash buffers 1 and 2 are dispensed from needles 2 and 3 in series. Since the liquid from the milling chamber has been drained, the liquid flow into the silica membrane comes only from the distribution trough, as the black arrows indicate. The rotation continues to be clockwise, so the stream from the silica membrane is directed into the waste chamber. The rotor reverses direction and resumes rotation at 8000 rpm before the elution buffer is dispensed from needle 4. The elution buffer enters the silica membrane and dissolves the DNA from it before flowing downstream into the Coriolis effect flow switch. The counterclockwise rotation causes the stream to flow into the small elution chamber, ready for use in downstream processes such as amplification by PCR. The whole process takes approximately 5 minutes to complete.

Pipper et al.[80] made a free droplet containing surface functionalized superparamagnetic particles into a virtual μ-TAS with a (sub)microliter volume. Aside from acting as the force mediators for actuating the droplet in a magnetic field, the superparamagnetic particles serve as a solid support for the sequential performance of biochemical processes. Depending on the task, the droplet becomes a pump, valve, mixer, extractor, or thermocycler. In an all-automated experiment, thirty green fluorescent protein (GFP)-transfected THP-1 cells are isolated from 25 μm of blood, 100-fold preconcentrated, purified, lysed, and subjected to a real-time PCR (RT-PCR) targeting the transfection vector all within 17 minutes. Fast thermocyles of 7 seconds take place on a disposable substrate under timespace conversion by rotating the droplet clockwise over different temperature zones.

Beyond the prototype stage, Cepheid, in 2001, introduced the GeneXpert for pathogen detection (http://www.cepheid.com). This instrument goes from raw biological sample to PCR and detection within 30 minutes. The detection technology is based on DNA amplification and real-time, fluorometric identification and quantitation. The real-time optical reading of the amplified product is based on four colors (for four targets) per disposable cartridge (the ICORE, which stands for intelligent, cooling/heating, optical reaction). Four ICORE can be loaded at the same time into a GeneXpert instrument. Cepheid's test, which looks for telltale genetic sequences, takes as little as 72 minutes. The GeneXpert machine costs $30,000 to $150,000, depending on how many tests it can run at once. Each individual test cartridge is priced at $42.

MEMS in Industrial Automation, Including Environmental Monitoring Markets

Introduction

According to the Nexus III report, industrial automation, including environmental monitoring, constituted 5% of a $12 billion MEMS market in 2004 and will be a 4% of a $25 billion MEMS market in 2009 (see Figure 10.16). MEMS pressure sensors, gas sensors, flow sensors, accelerometers, infrared imagers, angular rate sensors, temperature probes, etc., all have utility in modern industrial processes and automation. MEMS is also expected to play an important role in environmental applications. Primary environmental MEMS products are expected to be in the area of distributed sensor networks for detection of air and water pollution and in downscaled instruments such as microspectrometers, micro-gas chromatography systems, micro-ion mobility spectrometers, infrared detectors, and perhaps even electronic noses and tongues.

As a wide variety of non-silicon and traditional technologies are involved in the construction of novel industrial and environmental monitors it is harder to come up with accurate MEMS market data in industrial/automation and environmental control. Silicon technology has not penetrated this application as deeply as it has the automotive world. With a plethora of solutions and the unifying theme of silicon-based sensing missing, MEMS market numbers vary much more than for automotive sensing.

Research in industrial and environmental instrumentation miniaturization is especially challenging, as science requirements call for advances in performance with simultaneous large reductions in device mass, volume, and cost. Many conventional instruments operate at or near theoretical limits, and reducing the scale (volume and mass) often leads to sharp reduction of performance. This situation calls for new fundamental principles for instrumentation technology development. In other instances, we shall see, miniaturization is making for better instruments.

From the time a picture of a gas chromatograph (see Volume I, Figure 1.22) the size of a matchbox was printed on the cover of the April 1983 issue of *Scientific American*[84] (Terry and Angell built a first prototype between 1974 and 1975), there has been a continued effort to micromachine analytical instruments. Few MEMS-based microinstruments reached the market. They are still relatively large, tabletop or suitcase size, with only a few key components based on micromachining. One well know example is the Agilent MicroGC (Figure 10.8), looking significantly less integrated than the device suggested in the 1983 *Scientific American* article (the MTI gas chromatograph column itself is not micromachined; only the inlet and detector are). The wisdom of using Si micromachining for microinstrumentation is now in doubt, but looking at the miniaturization job from a precision engineering perspective, one becomes aware of the many alternative manufacturing options available.

MEMS Solid-State Gas Sensors

Most current solid-state gas sensors operate on the principle of modulating the resistance of a metal-oxide thick film by adsorption of gas molecules to its surface at high temperature (see Taguchi sensor, Volume II, Figures 9.11 to 9.13). Typical metal-oxide sensor materials include SnO_2, TiO_2, ZnO, WO_3, and Fe_2O_3. Each metal oxide is sensitive to different gases. For example, tin oxide is most effective at detecting alcohol, hydrogen, hydrogen sulfide, and carbon monoxide, and zinc oxide is useful for detecting halogenated hydrocarbons. Unfortunately, most metal oxides are adversely affected by humidity, which must be controlled at all times. In addition, variations in material properties require that each sensor be individually calibrated.

The market in 2000 for small solid-state gas sensors and sensor systems in the toxic and combustible gas arena was some $120 million in the United States alone. Broken down by product, the U.S. market for these chemical sensors comprises about $25 million for hydrogen sulfide, $20 million for total hydrocarbon monitoring, $12 million each for carbon monoxide and oxygen deficiency, and small percentages for a myriad of other gases. The use of a ceramic tube in a classical Taguchi sensor (see Volume II, Figure 9.11) maximizes the utilization of the power from the coiled Pt wire heater inside the tube, so that most of the power is used to heat the tin oxide covering the tube. The tin-oxide paste is applied on the outside of the ceramic body over thick-film resistance measuring pads by dip coating and is sintered at high temperature. The sintering stabilizes the intergranular contacts where the sensitivity of the gas sensor resides. The structure is not a design suited for mass production, as it involves excessive hand labor. The problems with the thick-film structure are mainly in the area of reproducibility: compressing and sintering a powder, the deposition of the catalyst (e.g., the size of palladium crystallites and how close they are to the intergranular contacts), and the use of binders and other ceramics (e.g., for filtering) are all very difficult to control. It is obvious from our discussion of the Taguchi sensor in Volume II, Chapter 9, that the application of IC techniques to Taguchi sensor manufacture can improve the state of the art dramatically in terms of the required power budget; a power reduction by a factor of ten is quite common. Unfortunately, as explained in Volume II, Chapter 9, no thin-film tin-oxide gas sensor can yet compete with the thick-film-based traditional Taguchi gas sensor in terms of sensitivity. If one could make a thin-film tin-oxide with the same gas sensitivity as a thick film, then micromachining the heater elements, and thin-film deposition of the gas-sensitive material would further improve reproducibility and make for an excellent product. For long-term stability, relatively low temperature semiconductor catalyst should be used (perhaps operating at temperatures <200°C). Figaro Engineering Inc. of Japan pioneered the Taguchi sensor technology in 1962. This company is still one of the leaders today.

A major obstacle to more rapid progress with thin-film micromachined gas sensors remains that the sensitivity of thin films to gases is not yet very well understood. The currently available devices, already quite small and inexpensive, will only be replaced quickly when newer technology improves upon the existing product in terms of reproducibility, cost, sensitivity, response time, concentration range, specificity, reversibility, stability, and power consumption.

At this time, the best course for future progress in the Taguchi-type sensor is to continue to optimize sensitivity of thin-film oxide semiconductors, use planar Si devices (reproducibility and cost), and concentrate on implementing less power-consuming microheater elements. Along this line MicroChemical Systems (MiCS; http://www.microchemical.com) produces micromechanical Si thin-film tin oxide gas sensors. Details about the nature of the thin-film tin oxide are scarce (Figure 10.53).[85] For very low-cost and less demanding applications, the MiCS sensor might succeed, but for more demanding applications, planar thick-film structures on ceramic do present a better option.[86]

FIGURE 10.53 Microelectromechanical system gas sensor by MicroChemical Systems (MiCS; http://www.microchemical.com). Top view of the sensor chip. The key elements of the sensor are a heater that maintains the gas sensitive catalyst layer at a specific temperature (I), a thin supporting thin dielectric layer allowing very low power consumption (II), and a sensitive catalyst layer that changes resistance upon exposure to gases (III). MiCS uses proprietary technology to deposit precise amounts of nanoparticle metal oxide material on the micro hot plate. The very small grain size provides high stability and sensitivity.

An interesting new research direction in the MEMS gas sensor arena might be to take advantage of the micromachined ceramic (Al_2O_3) structures, as illustrated in Chapter 5, Figure 5.9, for both heaters and catalyst support.

The MGS1100 carbon monoxide sensor from Motorola (Schaumburg, IL)[87] is not on the market anymore. The device incorporated a tin-oxide, thin-film sense resistor over a polysilicon resistive heater. The sense resistor and the heater reside over a 2-μm-thick silicon membrane to minimize heat loss through the substrate. Consequently, a mere 47 mW was sufficient to maintain the membrane at 400°C. There is a total of four electrical contacts: two connect to the tin-oxide resistor, and the other two connect to the polysilicon heater. The simplest method to measure resistance is to flow a constant current through the sense element and record the output voltage. Although the electronics were implemented correctly, as you would expect from Motorola, it seems like the fact that a gas sensor is only as good as the selective catalytic nature of its surface was forgotten in this sensor implementation.

The market for humidity sensing in the United States is about the same size as or larger than the total toxic gas sensor market ($120 million). A particularly interesting set of micromachined gas sensors, including humidity sensors, is based on thin-film-coated chemiresistors and acoustic devices.[88] We discussed the Jet Propulsion Laboratory (JPL) microhygrometer in Chapter 8, "SAW Example Application: Dew Point/Humidity Sensor" (Figure 8.26).

Solid State Gas Sensor Arrays: Electronic Noses and Tongues

The selectivity of a chemical sensor is the ability to measure only one chemical in the presence of many other chemicals. Because of the lack of perfect selectivity of chemical sensors, sensor arrays are often implemented. The solid-state gas sensors described above are one of the types of sensors used in arrays in electronic noses to fingerprint complex gas mixtures (other types use conductive polymers, coated quartz microbalances, or carbon black in nonconductive polymer matrix). The sensitivity of Taguchi sensors to different gases depends principally on the choice of catalytic metals and the operating temperature. Although the selectivity of the individual sensors is limited, qualitative and quantitative gas analysis can be performed using pattern-recognition techniques. Optimally, the sensors have a wide dynamic range and a small correlation to the other sensors in the array so that each contributes as much new information as possible. When applying pattern-recognition techniques to chemical analysis one defines chemometrics. The combination of multiple gas sensors and signal analysis using pattern-recognition techniques is the concept behind the *electronic nose* or *E-nose* (Figure 10.54). The sensor array is part of a neural network sensor array (NNSA) consisting of the array, a sampling system, a flow control system, a preprocessor, and a pattern-recognition system.

An electronic nose is an example of biomimetics at work (see also Chapter 2, this volume). In animals, the olfaction system (smell) is perhaps the most mysterious of the senses: even the sense of taste, for example, consists in large part of smell. Two American scientists, Dr. Richard Axel and Dr. Linda B. Buck, who suggested an answer to the enigma of how, with only a limited number of receptors (350), people can smell more than 10,000 different odors, and recall them later, were awarded the 2004 Nobel Prize in Physiology or Medicine.[89] The initial event in the process of olfaction is the recognition of an odorant by olfactory receptors—proteins found in the olfactory epithelium. Olfactory receptors convert the "information" in odorant molecules into electrical signals that are interpreted by the brain. In animals a layer of mucus dissolves the arriving scents and separates out different odor molecules so that they arrive at the receptors at different speeds and times, so there is some type of *snot chromatography* involved as well. The brain is able to interpret this pattern to distinguish a diverse range of smells. In contrast, an artificial nose consists of a much smaller array of chemical sensors, typically between six and twelve, connected to a computer with neural net software capable of recognizing patterns of molecules. (Electronic noses are also touched upon in the context of artificial neural network software in Chapter 9, this volume.)

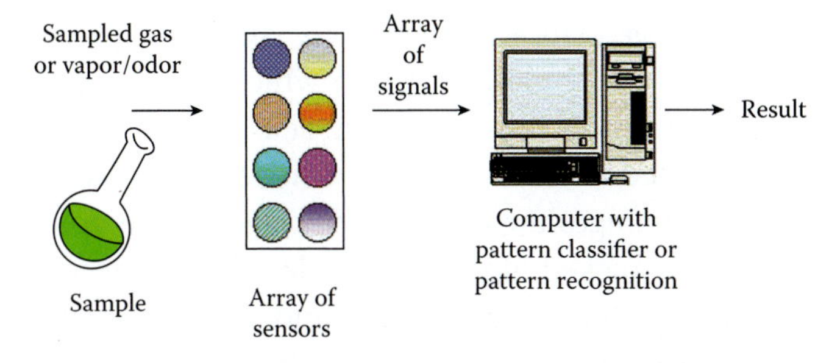

FIGURE 10.54 The electronic nose: an example of biomimetics. Electronic noses were touched upon in the context of neural network software in Chapter 9, this volume.

E-noses were popular at the end of the twentieth century and have been successfully used in a number of applications, e.g., the quality estimation of ground meat, the identification of different paper qualities, the classification of grains with respect to microbial quality, the screening of irradiated tomatoes. NASA used a JPL developed electronic nose on the space shuttle; the manufacturers of Coors beer, Evian water, and Starbucks coffee use electronic noses to test the quality of their products; and E&J Gallo Winery is trying to find out whether they can spot bad corks. E-noses also have been tested to monitor pollution, spot explosives and drugs, and diagnose diseases.

Unfortunately, electronic noses have had a hard time finding acceptance in the marketplace. A lot of training is required of both sensor arrays and the technical staff running the equipment, and the instruments are still large and expensive. The chemical sensors used in the sensor arrays drift too much, so that too-frequent calibration makes the use of the instruments cumbersome. Recognizing this problem, surviving electronic nose companies are now adding mass spectrometry to their sensor arrays (Figure 10.55).

Alpha MOS, Neotronics Scientific Lt., Illumina Inc, AromaScan Inc., and Cyrano Sciences, Inc., were all important players in this field in the 1990s. Alpha MOS, a French company, and AromaScan Inc., a British manufacturer, were the first to develop the electronic nose, or sensor array systems (SAS)

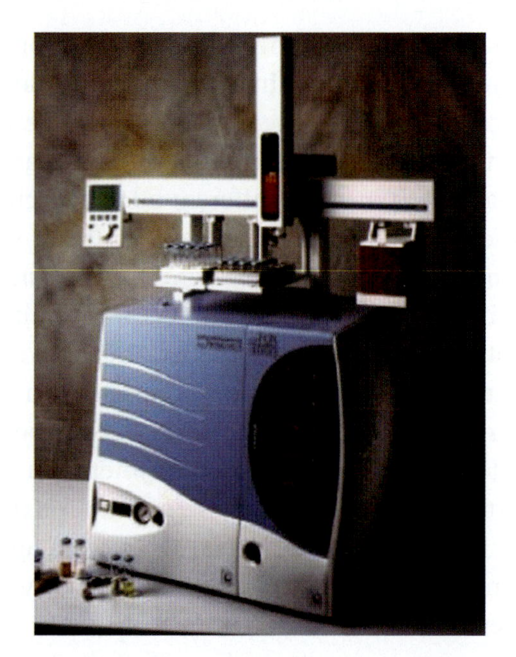

FIGURE 10.55 The Prometheus from Alpha MOS is the world's first odor and VOC analyzer that combines a highly sensitive fingerprint mass spectrometer with a large choice of patented multisensor array technologies to create an extremely flexible analysis platform (http://www.alpha-mos.com/).

technology, for commercial use. At Pittcon 2000, Cyrano showcased a portable electronic nose, the Cyranose 320, which distinguished odors using an array of thirty-two sensors. Each sensor was made of a composite of conductive carbon black and a nonconductive polymer. Today, only Alpha MOS continues as an electronic nose company. There is little left of all the other original contenders. In 2004, the Pasadena-based Cyrano Sciences was acquired by Smiths Detection for $15 million plus earn-out. Neotronics has been amalgamated into the Zellweger Analytics consortium, Illumina Inc. went on to do great things in DNA arrays (see Volume I, Chapter 7, "Bead Arrays for Detection and Analysis of DNA Sequences"), and AromScan became Osmetech, focusing on molecular diagnostics (http:// www.osmetech.com) instead.

The scientific principles of the electronic tongue are similar to the electronic nose but are applied to liquids rather than vapors. An electronic tongue incorporates electrochemical membrane-based sensors (see Volume I, Chapter 7, "Ion-Selective Electrodes") mimicking the five types of taste buds, which sense sourness, saltiness, bitterness, sweetness, and umani (deliciousness). The sensors in the array, like our taste buds, exhibit global selectivity rather than absolute selectivity; in other words, they are not selective to one chemical species, but to a wide variety. The ASTREE Electronic tongue from Alpha MOS is a liquid taste analyzer designed for taste assessment in solutions or solids dissolved in liquids. Like the human tongue, it provides a global "taste fingerprint" of a complex mixture of organic/inorganic compounds. For all tastes, it performs analysis with similar or better detection threshold than the human tongue.

Electrochemical Gas Sensors and Optochemical Gas Sensors

Solid-state semiconductor devices have been heralded as the replacement for electrochemical sensors; however, the electrochemical gas sensor sales are experiencing a growth rate of 30% per year. The latter is probably due to the slow rate of improvement in the selectivity of solid-state gas sensors and superior selectivity of electrochemical sensors brought on the market by companies such as City Technology (http://www.citytech.com) and Neotronics.[90] There is quite some consolidation going on in gas sensor companies. Neotronics and Sieger have been amalgamated into the Zellweger Analytics consortium, and more recently, First Technology has acquired City Technology and Capteur Sensors, while EEV is now part of E2V Technologies. There is plenty of opportunity in the application of MEMS to room-temperature electrochemical sensors, although these sensors are typically small already, they are not made in batches yet.

An optochemical gas sensor approach developed by Quantum Group Inc.[91] was adopted by First Alert as an inexpensive and selective fire alarm (http://www.qginc.com/products.html). This room-temperature sensor comprises a porous, semitransparent substrate (e.g., a glass frit) into which a self-regenerating chemical sensor reagent, selective for CO, is impregnated. Reaction with CO is fast, very selective and changes the color of the reagent. This color change is picked up by a diode photo-detector. Oxygen in the air regenerates and clears the substrate slowly back to its original color (~30 minutes). In this fashion, the sensor is biomimetic, simulating human response to airborne toxins such as carbon monoxide, mercury, ethylene oxide, volatile organic materials, and hydrogen sulfide. This is a very inexpensive and promising approach, and with further fine-tuning of the chemistry of various sensor reactants, we might expect a large family of products based on this optochemical approach, with a selectivity akin to an electrochemical sensor but at a cost similar to that of a solid state gas sensor.

The biomimetic sensor has a functional lifetime of over one year, and the mimicking of the human response to toxins is achieved by the use of a molecular encapsulant that encapsulates at least one gas component. Work is concentrating on extending the sensor's lifetime and speeding up the regeneration of the sensor. The optochemical gas sensor from Quantum Group Inc. could be miniaturized by using nontraditional manufacturing techniques, but finding selective chemistries for other gases, such as NO_2, O_3, and H_2S, is a higher priority.

ISFETs

ISFET technology was covered in Volume I, Chapter 7, "Ion-Sensitive Field-Effect Transistors, Extended Gate Field-Effect Transistors, and Isn't FET." In the United States, ISFETs are manufactured only at the University of Utah, Salt Lake City. Orion Fisher, Corning, Sentron, and Rosemount, as well as UniFet, Inc. (San Diego, CA) sell ISFET-based pH sensors in the United States but do not produce them in-house. Honeywell sells the DuraFET, and in a detailed evaluation study, Sandifer and Voycheck called this sensor remarkable rugged, very stable, and accurate.[92] Orion and Rosemount buy their pH ISFETs in research quantities from Neuchatel (Switzerland) and target the food industry where glass pH electrodes cannot be used for food (pH measurement in dough, for example). Unisensor recently acquired Sentron (www.unisensor.ch). These commercially available ISFETs are for pH only and feature a simple pH sensitive inorganic membrane on the gate. To modify FETs with polymer membranes for detection of other ions is difficult because of the thickness of the membrane and the need for a stable internal reference electrolyte.

As we pointed out in Chapter 1, this volume, usually it is simpler and less expensive to build electrochemical sensors on a nonsilicon substrate (see, for example, the reference electrode in Chapter 1, this volume, Figure 1.18).

Gas Chromatography

Gas chromatography is a technique widely used for the separation and analysis of gaseous samples. A typical gas chromatograph (GC) consists of a carrier gas supply, a sample injection system, a separating fused silica column, an output detector, and a data processing unit, as depicted in the block diagram in Figure 10.56.[93] Separation of the sample vapors is achieved via their differential migration through a capillary column (see Volume I, Chapter 6). A precise and reproducible volume of sample vapor is injected at the input of the column by a valve and swept through it by an inert carrier gas. The

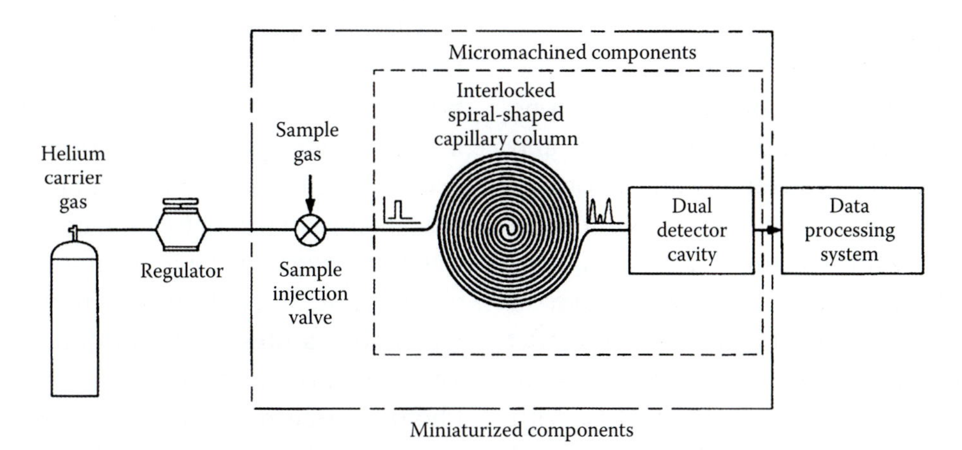

FIGURE 10.56 Functional block diagram of a gas chromatography (GC) system. (After Kolesar, E. S., and R. R. Reston. 1995. Separation and detection of toxic gases with a silicon micromachined gas chromatography system. *Proceedings of the International Conference on Integrated Micro/Nanotechnology for Space Applications, National Aeronautics and Space Administration (NASA) and The Aerospace Corporation*. Houston, TX, October 30–November 3, 1995, pp. 102–14.[93])

column is lined with a liquid stationary phase, a substance capable of absorbing and desorbing each of the component vapors depending upon their unique partition coefficients. The migration rate of each vapor along the column depends on the sample injection pressure, the carrier gas velocity, the temperature, and the degree to which the vapor is absorbed by the stationary phase. The column's output is thus a series of vapor peaks separated by regions of pure carrier gas. To detect those peaks, the output gas stream from the column is passed over a detector that measures a particular property of the gas, such as the thermal conductivity, which can be related to the concentration of sample vapor in the carrier gas. The detector produces a signal, which is amplified and used in the analysis of the sample mixture; the identity and quantity of each vapor in the mixture can be determined from, respectively, its retention time in the column and the area under its output peak.

Gas chromatography systems tend to be large, fragile, and bulky pieces of laboratory equipment, and miniaturization would present many advantages:

- Integration of the instrument into a process
- Portable and handheld units
- Fast response time (1 minute or less) in comparison with a large unit with comparable plate numbers (see Volume I, Chapter 6, "Scaling in Analytical Separation Equipment")
- Less calibration and carrier gases needed
- Low dead volume
- Higher accuracy

The first integrated GC, consisting of a long separating column, a sample injection valve, and a thermal conductivity detector (TCD), fabricated on a single Si wafer, was developed at Stanford as early as 1974 (see Volume I, Figure 1.22). The Agilent MicroGC, based on that original work, uses a traditional fused silica capillary column but retains the other micromachined components (Figure 10.8). By the time MTI produced its first GC, with a 100-μm diameter, fused silica GC columns had become commercially available, and they proved to be superior to the Si micromachined columns. The reasons behind their superiority must be sought in the perfect symmetry of the monolithic silica columns. The smoothness of the stationary phase coating the inside of a capillary column is a function of the profile symmetry. The uniformity of coating thickness is extremely important to maintain very sharp separation peaks. The micromachined designs described in the literature usually consist of a cover sheet of Pyrex glass, which is anodically bonded to a U-shaped, V-groove, or rectangular groove etched in Si.[94] Columns also have been made by joining two U-grooved halves. All these designs have a major potential deficiency: they are bound to accumulate coating on the corners or joints, leading to a less perfect separation.[95] Analytes spend a longer time in areas where the coating is thick than in places where it is thin, leading to a broadening of peaks. Thermal gradients might also be induced by joining dissimilar materials, again causing peak broadening. It also should be realized that the column has to be replaced on a regular basis because of aging and fouling. Therefore, the price of a total integrated system would have to be very low.[96]

Research efforts for making better micromachined GC columns are continuing,[93] but since existing fused silica columns are perfectly adequate, the focus would be more gainfully put on other GC problem arenas. Specifically, micromachining could be of great immediate use in the arena of better, more reproducible, miniaturized valves and fast heating and cooling of thermally isolated column and injector for fast temperature-programmed chromatograms.

A Hewlett-Packard (HP) development of a handheld GC may be the only effort that combines both excellent micromachining capabilities with a long-standing instrumentation knowledge. RVM Scientific, Inc. has also built a novel handheld miniaturized gas chromatograph for detection of contraband drugs in cargo containers and chemical weapons treaty verification. There are several small GC products now being introduced to the market. The SnapShot, for example, is a handheld GC from Photovac International Inc. (Deer Park, NY). The instrument has a battery pack providing 4–6 hours of operation and may be used for personal exposure monitoring. Besides Agilent, Stanford Research Instruments is offering rugged GCs for field use.

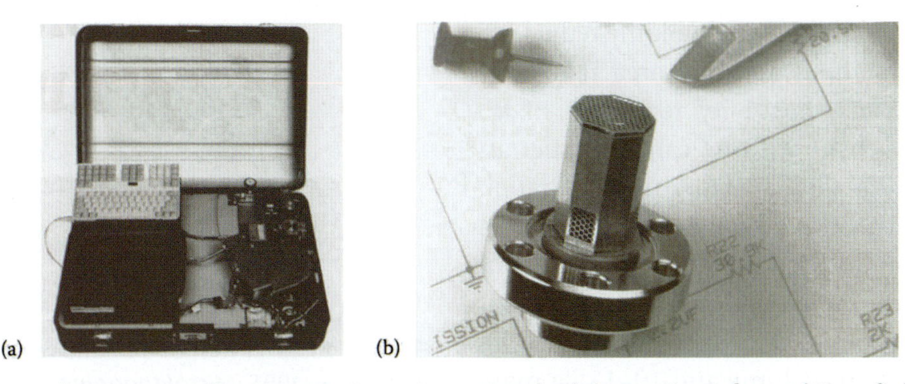

FIGURE 10.57 (a) A gas chromatography-mass spectrometry system built in a suitcase for real-time field use. (Courtesy of Lawrence Livermore National Laboratory, California). (b) A 1-cm scale array of quadrupoles for gas analysis (traditional mechanical precision machining). Built by Ferran Scientific, California.

Mass Spectrometry

Mass spectrometers (MSs) are extremely versatile tools that have found, at least for the moment, their greatest use in laboratory science. The determination of the component atomic and molecular masses and their abundance is indeed essential for the understanding of almost any chemical and/or process analysis. Lack of broader use is attributed to their large mass, volume, and power requirements. Especially powerful would be a miniaturized gas chromatography-mass spectrometry (GC-MS) combination, with the GC for the isolation of the various compounds of a complex mixture and the MS for further identification. Innovative vacuum systems, smaller GC ovens, new sample inlets, and advanced electronics are now making some of this possible. For example, a GC-MS system built in a suitcase by Lawrence Livermore National Laboratory in California for real-time field use is shown in Figure 10.57a.[97] Ferran Scientific has successfully implemented a 1-cm scale array of quadrupoles for gas analysis (Figure 10.57b), and a similar MS is under development at Jet Propulsion Laboratory (JPL) in Pasadena, California. JPL researchers suggest that the Ferran instrument lacks in mechanical precision and hence resolution. In their own work, a miniaturized quadrupole mass filter, one machined with traditional but very precise machining methods (Figure 10.58a) and one dramatically smaller in size fabricated employing LIGA (not shown), was pursued. The JPL research team deemed quadrupole and ion-trap mass spectrometers most suited for miniaturization, as their operation does not require a permanent magnetic field, and they can be fabricated as parallel arrays of spectrometers.[98] The array concept is extremely important considering the following simple scaling insights. Miniaturization of an instrument dimension, l, by a

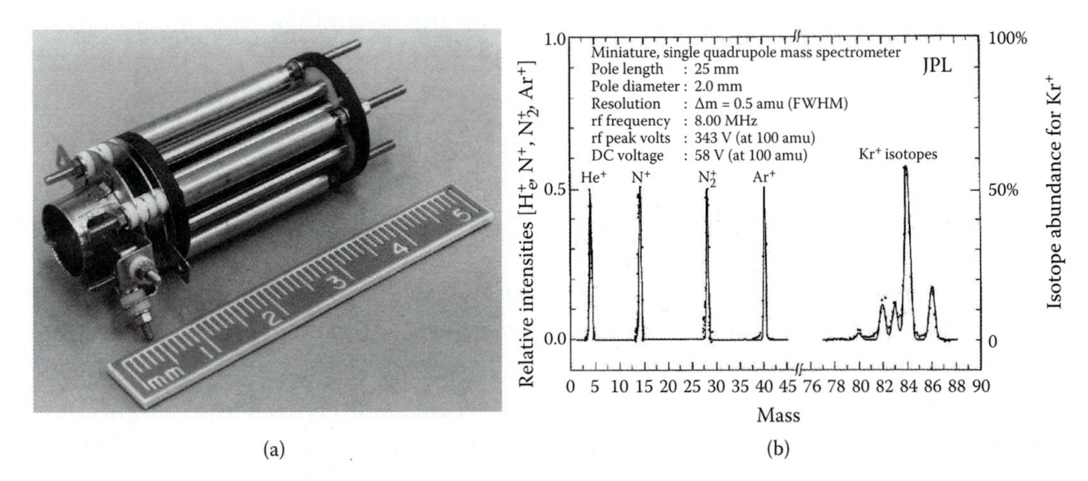

FIGURE 10.58 Miniature mass spectrometry. (a) Jet Propulsion Laboratory's small quadrupole array made by traditional, but very precise, machining. (b) Spectrum generated from the single quadrupole mass spectrometer in (a).

factor k leads to a cubic mass and volume reduction (l^3 scaling) and an input aperture area reduction (l^2 scaling), and thus sensitivity reduction of k^2. To offset the lost sensitivity, one can construct an array of k^2 spectrometers, all working in parallel. One ends up with a spectrometer array with a sensitivity comparable to its larger cousin but with mass and volume reductions of k. A present quadrupole or magnetic-sector instrument may be of the order of 10^4 cm^3 volume and 10–15 kg mass and may require about 20 W power for operation. The JPL traditionally machined microspectrometer, a 4 × 4 array of nine miniature quadrupole mass analyzers (Figure 10.58a), has a mass of less than 0.8 kg (including electronics), is 20 cm^3 in volume, and draws less than 8 W power. The rod alignment, straightness, and roundness accuracy for a mass range of 1 to 250 amu and a mass resolution of 0.5 amu (FWHM) at 200 amu must be 0.15% (see mass spectrum in Figure 10.58b). Initially, the rods were machined at 2 cm long, but they can be further reduced to 1.5 or even 1 cm. The alignment was carried out using a machinable ceramic alignment jig to accurately position the rods. The holes in the jigs are machined using a conventional, high-precision lathe (±0.0002-inch tolerance). With the LIGA technique, the attempt is to make the quadrupole rods 1 mm high and 80 μm in diameter, but these specifications have not yet been met. LIGA might provide more accuracy for the straightness of vertical walls, but the tolerances in the x- and y-directions are still dependent on the aligner. As we may deduce from Volume II, Figure 6.5, a 0.15% tolerance on dimensions below 100 μm is quite a challenge.

Optical Absorption Instruments

The most common analyzers used in analytical chemistry for both gases and liquids are spectroscopic absorption techniques. A considerable amount of effort has been put into miniaturization of such equipment with both traditional and nontraditional machining tools.

Consider, as an example of an environmental application, the detection of methane by means of infrared absorption (based on Beer's law, see Volume I, Equation 7.133). From Volume I, Chapter 7, we know that to obtain a better sensitivity, a longer optical path is required. To accomplish this on a small-footprint instrument of a few square centimeters, a set of LIGA-machined micromirrors may be positioned in a folded configuration, as illustrated in Figure 10.59. The infrared mirrors are preferably made from solid electroplated gold or by evaporating gold on poly(methylmethacrylate) (PMMA) walls. The vertical height of the mirrors is limited by the spot size of the infrared laser beam. The light intensity loss in the miniaturized instrument is due to absorption and scattering by the micromirrors. Gold is the preferred material for the micromirrors because of its high reflectivity in the mid-infrared region. Superpolished gold surfaces can reflect as much as 98% of an incident infrared beam in the specular direction. However, micromachined gold mirrors, electroformed or evaporated on PMMA walls, will likely have an inherent surface roughness, which may cause scattering of incident beams. Working in the infrared range presents an advantage in this case, as the roughness of the mirrors will typically be small compared with the wavelength used. With an angle of incidence close to the normal direction, one can typically accommodate up to 100 reflections before serious losses occur. With a 1.6-cm distance between mirrors, this means 1.6 m of optical path-length, which may not be adequate for a 0.1-ppb sensitivity but would serve well for applications that require only a few parts per million of sensitivity. The conclusion is that one might well be able to miniaturize a gas absorption cell, but a miniaturized infrared analyzer of this sort will be suited only to less demanding applications. No products of this

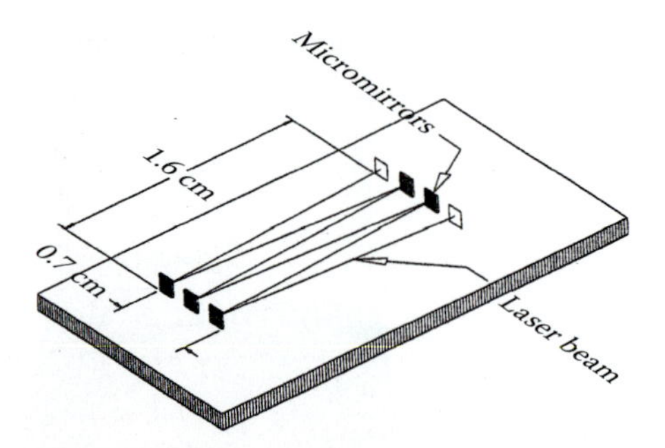

FIGURE 10.59 Layout of a folded optical system. (After Desta et al. 1995. *Microlithography and metrology in micromachining.* Austin, TX: SPIE. With permission.)

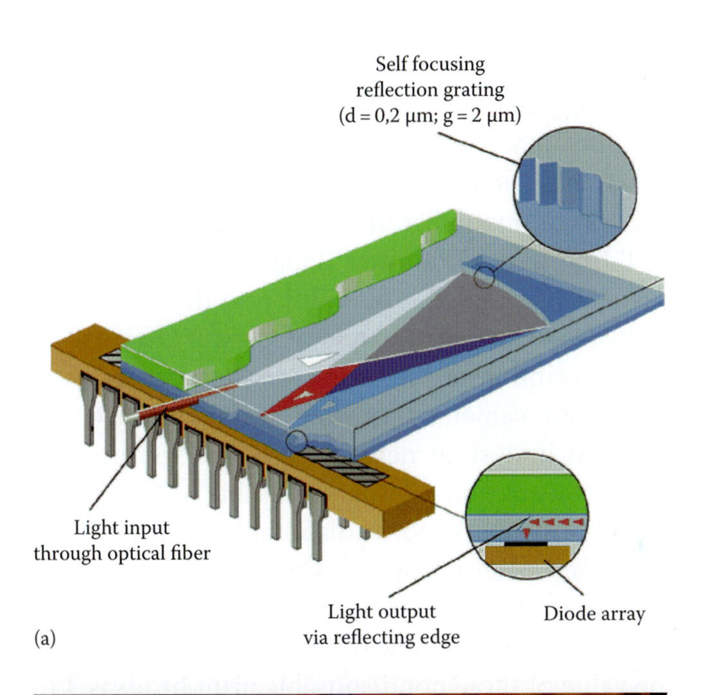

FIGURE 10.60 Microspectrometer from STEAG microParts. Schematic (a) and photograph (b).

sort are on the market yet. In Volume I, Chapter 7, "Scaling in Absorption Spectroscopy and Cavity Ring-Down Spectroscopy," we introduced the cavity ring-down spectroscopy (CRDS) technique, which represents a way around the sensitivity issue when scaling down an optical instrument. As far as we are aware, there is no MEMS version available of the CRDS.

The STEAG microParts VIS microspectrometer is a monolithic dielectric slab waveguide with integrated focusing echelette grating produced by a micromolding technique. The part sells for $100–$300. Light to be analyzed is coupled into the device by an optical fiber. Light reflected from the grating is detected by a diode array. The detector attached at the focal plane is a Hamamatsu photodiode array with 256 pixels. The STEAG microParts microspectrometer is shown in Figure 10.60.

MEMS in IT Peripherals and Telecommunications Markets

Introduction

In the past, automotive applications were setting the pace for increases in manufacturing volumes of MEMS sensors. The future will present quite a different scenario: huge growth rates in MEMS sensors are expected for applications in consumer electronics (CE) products, with a focus on IT and communications, and extensions into health care, life sciences, security systems, domestic appliances, and others. Once a MEMS device succeeds in the automotive application, with its very stringent specifications, the door is wide open to other more relaxed markets, even the consumer market. Analog Devices, for example, is now putting its surface micromachined accelerometers in video games, car and notebook alarms, smart pens, etc.

IT peripherals is a general term used to describe technologies that help produce, manipulate, store, communicate, or disseminate information. From Figure 10.16, this application area today constitutes already the largest MEMS market, dominated by hard disc drive heads, inkjet print heads and microdisplays. Interestingly, at the moment of their market introduction, the first two of these products were not even considered MEMS (see also the Preamble on the MEMS Market section, above).

To illustrate that there are many more MEMS opportunities in this market segment, consider that the mobile phone market crossed the 1 billion unit level in 2006. MEMS applications in phones include microphones/speakers (Knowles Acoustics), accelerometers (Analog Devices) for functions such as scrolling, gyroscopes (Invensense) for in-phone camera-lens stabilization, system timing products/oscillators (SiTime for myriad functions), and RF MEMS for band switching, filters (TeraVicta), and displays (Qualcom/Iridigm). The challenge for mobile phone MEMS manufacturers will be to meet the stringent price constraints demanded by the phone companies while still maintaining necessary reliability and product performance.

Read-Write Heads, Ink-Jet Heads, Microdisplays, Optical Switches, and Accelerometers

Read-Write Heads

Read-write (RW) heads still represented around 51% of the MEMS market in 2009: although the traditional application of RW heads in PCs will grow only moderately, the RW market is experiencing a renaissance in consumer electronics, as hard discs are entering music players (e.g., in many iPod models) and smart phones (Samsung introduced the first cell phone with HDD in 2004), as well as digital video cameras, set-top players, and DVD recorders. Memory read/write technology is an area where MEMS will continue to contribute and astound with its results; an important example we encountered in this regard is the IBM thin film read write head and the millipede (see Volume II, Figure 2.4, and Figure 1.1d, this volume). Major U.S. players in this market segment are Seagate, Conner, IBM, Read-Rite, and Applied Magnetic, in Asia there are companies like Yamaha, TDK, Alps, and a few others. Global revenues from shipments of HDDs were expected to reach $27.7 billion in 2010, up 18.4% from $23.4 billion in 2009. This represents a 2010 total unit shipment forecast of 674.6 million compared to 549.5 million units in 2009 (http://www.emsnow.com/npps/story.cfm?pg=story&id=42187).

It is important to note though that MEMS will lose a very important market when hard disc drives are replaced by solid-state disc (SSD) memory (flash drives), as both RW heads and accelerometers will fall away.

Ink-Jet Heads

At a greater than 25% share, inkjet print heads represent the largest portion of the total MEMS component market. Ink-jet heads continue to be one of the most profitable markets for MEMS. Cartridges for ink-jet printers were introduced in the early 1980s by HP and Canon. In Volume II, Figure 5.30, we show inkjet-printer heads machined with excimer lasers. As shown in Volume II, Figure 8.21b, some of those printer heads are very sophisticated, combining fluidics (at high temperatures!) with integrated electronics, sensors, and actuators. There are two major types of ink-jet print heads. First are the less expensive and disposable bubble-jet printheads, such as the ones made by HP, Canon, Ollivetti, Lexmark, Xerox, etc. Then there are the more expensive piezo print heads, manufactured by companies like Epson, MIT (in Sweden), Brother (Japan), and Frankotyp-Postalia (Germany). The main new driver for ink-jet heads in the next couple of years is expected to be the printing of digital photos—especially from cell phone cameras—growing at 15% to 20% per year. An important new trend in desktop printers is the integration of nondisposable inkjet heads in the printer instead of disposable print heads in cartridges. After Epson and Canon, HP is now also starting to ship printers with integrated print head. Although this will slow down the growth in units, the value of these nondisposable print heads is 3 to 5 times higher.

As competition in the printer marketplace continues to intensify, manufacturers of MEMS inkjet heads must reduce fabrication costs while improving product performance. The vast majority of inkjet head production in 2008 was on 6-inch silicon wafers, but increasing inkjet heads sizes—and therefore die sizes—as well as a general trend toward larger wafer diameters were expected to shift almost 10% of production to 8-inch wafers by 2010. These MEMS devices require high-quality, high-speed slotting of silicon wafers that are diced into the barrier chips, through which ink passes from the reservoir into the nozzles. Currently implemented MEMS techniques for silicon wafer slotting have a number of drawbacks. Wet etching processes require a mask step, making them both slow and costly, while sandblasting creates holes with conical edges, limiting their diameter and density. Dry laser techniques, although somewhat more effective, typically generate debris and microcracks that hamper device quality and performance. Today, key MEMS technologies to fabricate inkjet-heads include increasing use of deep reactive ion etch (DRIE) and SU-8 resins, but new MEMS manufacturing techniques are being explored. The Synova Laser MicroJet (Volume II, Chapter 5, "Water Jet-Guided Laser Machining," Figure 5.23), for example, overcomes some of the machining challenges, ablating the silicon material quickly and efficiently without causing any residual damage. In production environments, the Synova water-driven laser has proven superior to competitive approaches, reducing the cutting time to just 2.5 seconds per slot—faster than any previously implemented slotting technique—with no damage.

It is to be expected that the fluidics and their integration with electronics pioneered in this application field will produce rewarding spin-offs in other application areas such as drop delivery systems, micromachined PCR units, etc., for medical/biomedical applications.

MEMS Microdisplays

MEMS microdisplays are the new MEMS blockbuster, and their revenues were expected to overtake inkjet heads in 2009. Millions of DMDs™ were built into projectors (the first one came out in 1996 from InFocus) and big-screen televisions (the first came out in 2002 from Samsung). Microdisplays found their way from projectors and high-definition, large-screen TVs to microdisplays for mobile handsets (e.g., the iMoD—interferometric modulator display—developed at Qualcomm). It was projected that by 2009 MEMS displays would become the second largest MEMS market, after thin-film RW heads (according to the Nexus report).[1] A major advantage for Texas Instruments (TI) is that DLP chips are an exception to the high-volume/low-price rule, typically for all ICs and most MEMS devices: the 2004 asking price for a DLP chip was $400 in 2004; it was estimated at $190 in 2009 (Nexus).[1]

In an imaginative new application of TI's micromirror technology, NimbleGen Systems developed a set-up that enables researchers to fabricate their own DNA arrays without having to go through the expense of making large sets of expensive photomasks as in the case of the Affymetrix system (see Volume II, Figure 8.19 for a schematic of NimbleGen's virtual mask generation system).

Another elegant MEMS display technology, from Silicon Light Machines, Sunnyvale, California (now owned by Cypress Semiconductor), is the grating light valve, or GLV™ (http://www.siliconlight.com/htmlpgs/homeset/homeframeset.html). In this technology, invented at Stanford, closely spaced parallel rows of reflective ribbons are suspended over a substrate. The gap between the ribbons and substrate is one-quarter wavelength, and in the resting state, the ribbons appear as a continuous reflective surface. With an electrostatic voltage pulling down alternate rows of ribbons, light reflected from the pulled down ribbons travels twice the gap or 1/2 wavelength farther than if reflected from the resting ribbons. The ribbons are arranged such that each element is capable of either reflecting or diffracting light. This allows an array of elements, when appropriately addressed by control signals, to vary the level of light reflected off the surface of the chip. This control of light can be analog (variable control of light level) or digital (switching of light on or off). Because a GLV™ device is reflective, it is highly efficient in its use of light. Besides Silicon Light Machines and TI, Samsung and Sony are also manufacturers of digital mirror technology.

In May 2008, Qualcomm MEMS Technologies, demonstrated the first reflective interferometric modulation (IMOD) color Mirasol™ display for mobile phones. IMOD technology requires no backlighting and reflects light so that wavelengths interfere with each other, creating pure, vivid colors. Mobile devices such as mp3 players and mobile phones stand to benefit from Mirasol displays, which require significantly less power and harness ambient light sources to automatically scale for optimal viewing in virtually any lighting condition. The interferometric modulator element (IMOD) is a simple MEMS device that is composed of two conductive plates. One is a thin-film stack on a glass substrate, and the other is a reflective membrane suspended over the substrate. There is a gap between the two that is filled with air. The IMOD element has two stable states. When no voltage is applied, the plates are separated, and light hitting the substrate is reflected as shown in Figure 10.61. When a small voltage is applied, the plates are pulled together by electrostatic attraction and the light is absorbed, turning the element black. This is the fundamental building block from which Qualcomm Mirasol™ displays are made (Figure 10.61).

MEMS obviously figures central in various designs for next-generation flat panel displays. Illustrative is also the thin-film field emitter work from Capp Spindt (the author's first mentor in MEMS at SRI

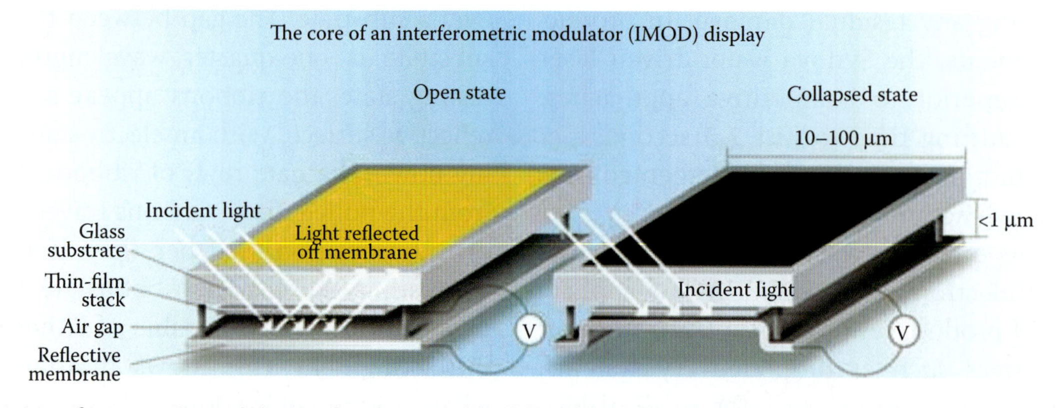

FIGURE 10.61 Interferometric modulation fundamental building block from which Qualcomm Mirasol™ displays are made (http://www.qualcomm.com/qmt/).

International), which is illustrated in Volume II, in Chapter 7, Figure 7.90.

Optical Switches

IC-based processing techniques have been used for a long time in the fabrication of complex opto-electronic ICs, and MEMS has been employed in the fabrication of simple fiber alignment devices. Passive alignment devices such as V-grooves are either fabricated in polymer or Si, with the polymer embodiments enabling high-volume manufacture of low-cost components. The newest developments in optical telecommunications require a new generation of additional components, including the following:

- Wavelength filters
- Micromirror-based all-optical switches
- Configurable add/drop multiplexers to add or drop signals to generate thousands of different frequencies of light
- Tunable filters
- Multiplexers and demultiplexers that direct multiple light beams onto one fiber
- Stabilized optical sources
- Optical polarization controllers
- Integrated optical pumps and MOEMS amplifiers for attenuation, dynamic gain equalization, and dispersion compensation

MEMS is expected to play a role in the manufacture of several of these optical devices. Here we reiterate briefly the role MEMS is playing in just one such application, i.e., large optical switches for telecommunications. As we remember from earlier in this chapter (see under Preamble on the MEMS Market), this application unfortunately led to the so-called *MOEMS bubble* in 2001–2002.

Miniature optical switches, optical cross-connects (OXCs), were expected to change how data are transmitted in everything from communication systems to computers. Miniaturization methods enable the creation of arrays of very small, high-capacity optical switches (see Example 4.4, "Self-Aligned Vertical Mirrors and V-Grooves for a Magnetic Micro-optical Matrix Switch"). These optical switches provide a way of routing light paths from one fiber to another without converting the signals to the electronic domain. OXCs seemed to have arrived at just the right time; optical fiber cables connecting network hubs were indeed quickly outpacing the capacity of electronic switches, and this electronic bottleneck could be solved by this new generation of MOEMS switches. Four example approaches to optical switching are illustrated in Figure 10.62: mechanical (bending of an optical fiber) (a), planar waveguides (where heaters and Peltier devices change the index of refraction to cause the beam path to shift) (b), moving mirror types (c), and the bubble approach (bubbles deflect the light between criss-crossing waveguides) (d).

The router developed at Lucent, for example, could switch 256 input light beams to 256 output fibers without any intermediate optical-electrical conversion (Figure 10.63). Supporting 256 wavelengths, each carrying 40 Gbit/second of traffic, this corresponds to an aggregate throughput of more than 10 Tbit/second. Tiny double-hinged mirrors are used to redirect optical signals by adjusting the mirror tilt.

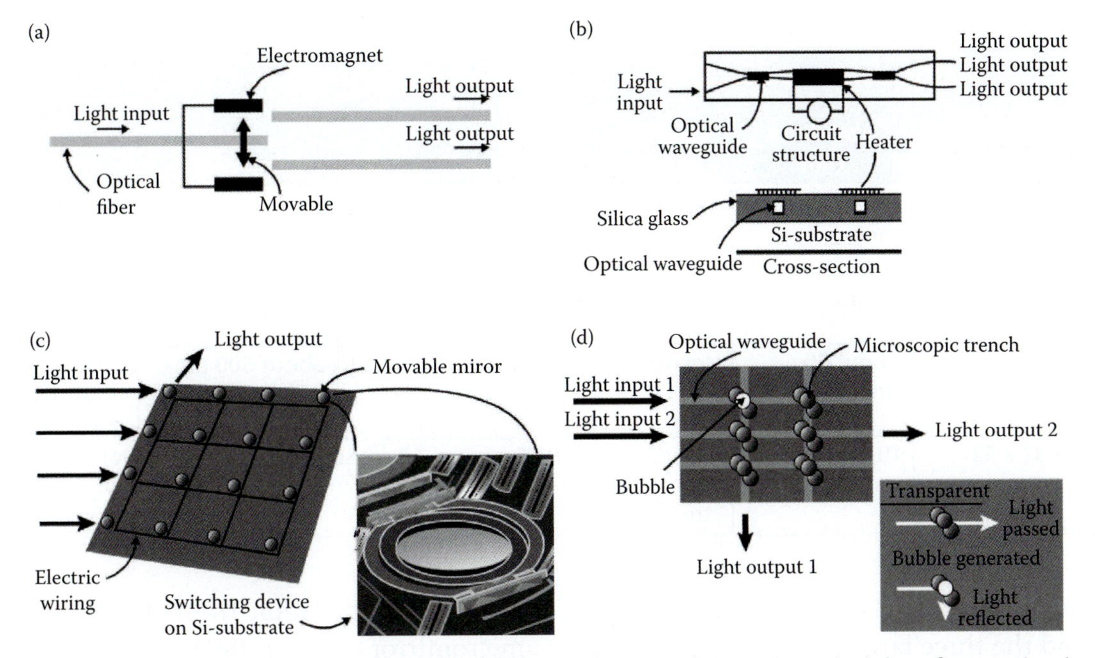

FIGURE 10.62 Example principles of optical switching devices. There are four major principles of operation involved. (a) The mechanical type, where the optical fiber is bent with an electromagnet to change the direction of the light beam. (b) The planar optical waveguide, where heaters and Peltier devices are located near the optical waveguide to change the waveguide temperature. When the temperature changes, so does the index of refraction, causing the beam path to shift. (c) The mirror type. The angles of micromirrors are changed to select different light paths. Mirrors may move on one axis (two-dimensional) or two (three-dimensional). (d) The bubble approach. An oil-filled wall is created along the light path, and bubbles are generated there to reflect the light beam.

A first application for these new switches was expected to be in dense wavelength division multiplexing (DWDM). To squeeze more traffic into a network, one of the strategies is to deploy DWDM. A programmable DWDM add/drop node "drops" one or more wavelengths out of a multiwavelength channel and reroutes them to a different destination while potentially "adding" traffic from another source to the channel.

One of the most impressive optical switches was a 1152 × 1152 OXC by Xros (bought by Nortel and now defunct). The product was projected to be on the market in 2001. Another player was Ilotron a spin-off company from the University of Essex. Most major players in MEMS optical switching were backing the traditional micromirror approach, but Agilent Technologies has a fluidic approach based on its ink-jet technology. In the Agilent switch, bubbles deflect the light between criss-crossing waveguides, and at Corning, liquid crystals are used. In Table 10.12, the four most popular optical switch approaches are compared.

FIGURE 10.63 Lucent free-space microelectromechanical system optical cross-connect mirrors with two-axis, gimbal-mounted micromirrors.

It was expected that optical switches would achieve full commercialization by 2004. The great promise that these truly enabling MOEMS devices seem to offer was at first validated by the acquisition of a series of MOEMS companies. Among these were XROS (a 95-person company), acquired by Nortel Networks for $3.25 billion in stock; IntelliSense, 67% of which was acquired for $500 million by Corning; and Cronos Technologies, acquired for $750 million in stock by JDS Uniphase. However, as we explained in the beginning of this chapter, the more than 10 billion dollar investment by these large telecommunication companies in 1999–2000 became a textbook example of "irrational exuberance." There

TABLE 10.12 Major Characteristics of Optical Switching Devices

Type	Maximum Channels	Switching Time	Insertion Loss	Cross-Talk (Interference with Other Channels)	Device Dimensions	Manufacturers
Mechanical	1 × 2 to 2 × 4	10 milliseconds or less	0.3 dB	–65 dB or less	Two switches in 8.3 mm × 15.6 m × 28 mm	Hitachi Metals, Seiko Instruments, etc.
Planar waveguide	16 × 16	10 milliseconds or less	8 dB for an 8 × 8 switch	–50 dB or less	10 cm square	NTT, etc.
Mirror	256 × 256	About 50 milliseconds	Unknown	Unknown	Mirrors about 500 μm in diameter	Lucent Technologies, A&T, Japan Aviation Electronics, etc.
Bubble	32 × 32	10 milliseconds or less	0.07 dB per trench	–60 dB or less	2 cm square	NTT Electronics, Agilent Technologies, etc.

were not enough customers for the more than eighty start-ups, and the three largest investing companies pulled back. What was left of the all-optical switching market, in the period between 2002 and 2004, was littered with the broken shells of startups that often tried to survive by doing MEMS foundry work. Today, some surviving MOEMS companies, building all-optical switches, are successfully targeting more modest switch applications; say, for apartment buildings or university switchboards, and for new emerging applications such as semiconductor test equipment. Companies such as Glimmerglass claim to have solved the problems previously associated with the MOEMS switches and maintain that their optical switches are more reliable, less complicated, and more cost-effective. The Reflexion eighty-switch pack from Glimerglass is available as either a small core module tailored for communications-systems manufacturers or as a larger evaluation system. The evaluation system includes the Reflexion module, as well as optical-channel power monitors, a fiber patch panel, and an ethernet* control interface. It is all packed into a 4.5- by 9.5- by 7.0-inch rack-mountable case that consumes just 20 W (see Figure 10.1).

As we saw above, instead of switching signals among very large arrays of fiber optic cables for telecommunication switches, TI targeted its digital light processing (DLP) technology to projectors and video displays. Because it focused on the right mass consumer application from the get-go, TI's digital micromirror device (DMD) turned out to be a big-scale MEMS success (see above).

Today, even the large all-optical optical switches are back. Late-stage startup Lambda OpticalSystems (Reston, VA), for example, recently unveiled the LambdaNode 2000, an integrated optical-switching and DWDM transport platform. The system supports 256 wavelengths, each wavelength running at 10 Gbit/s for a total capacity of 2.5 Tbit/s. The device is running live traffic today in a 10-Gbit/s transcontinental network built by the U.S. Naval Research Laboratory. Also, Calient Networks announced details of an interoperability demonstration it conducted with Cisco Systems.

Accelerometers

Recently, the market for accelerometers in consumer electronics (CE) (computers, games, phones, etc.) has also started to explode. Consumer electronics have the potential to dwarf the automotive application, not only of accelerometers but also of pressure sensors and gyros. The first accelerometer application in consumer electronics was for hard drive protection in notebooks. IBM (now Lenovo) puts a dual-axis accelerometer in its notebook: the accelerometer detects falling motion and places the HDD in safe mode prior to impact. Samsung (and others) have experimented with motion-based applications in handsets (gaming, user interfaces, power management, navigation). Early in 2007, Apple announced the iPhone, with motion-based features.

* Ethernet: LAN (local area network) connects your computer to computers close by and around you.

Consequently, MEMS growth is expected by some to follow a hockey stick curve, with the value of products rising to $40 billion in 2015 and $200 billion in 2025.[99] (See also an estimate by Dr. Marinis from Draper Laboratory: http://www.sem.org/PDF/s01p01.pdf.) The problem is now one of profitability rather than the absence of real product opportunities: how to earn money on, say $0.32 microphones or $1.35 triaxis accelerometers for mobile handsets? Only by becoming more similar to the IC industry can MEMS devices achieve this goal, i.e., through more integration on larger (8-inch) Si wafers. In a sense, as far as mechanical and MOEMS are concerned, the IC industry has now swallowed the MEMS field.

RF-MEMS

Wireless communications is showing an explosive growth of emerging commercial and consumer applications of radiofrequency (RF), microwave, and millimeter-wave circuits and systems. Just think that the number of mobile phones was predicted to reach 4.6 billion in 2010 (http://www.cbsnews.com/stories/2010/02/15/business/main6209772.shtml). Several of the new and improved functionalities in mobile phones will be enabled by the use of MEMS products, such as microphones, lenses, and miniaturized data and energy storage devices. Analyst Bourne predicts that by 2010, three-quarters of all cell phones will benefit from MEMS (http://www.eetimes.com/electronics-news/4071288/MEMS-exec-sees-billion-dollar-markets) in 2009. This is making the consumer industry more and more the driving force behind many technology developments in MEMS. RF-MEMS will help further reduce the mobile phone size, power consumption, and cost. According to a market study from WTC covering the 2005–2009 period, the market for RF MEMS components was $126 million in 2004, confirming WTC's forecast from 2002. It was forecasted to grow to over $1.1 billion in 2009 (Figure 10.64).

In this application area, a situation has been reached where the presence of the expensive, off-chip passive RF components, such as fixed capacitors and tunable capacitors (varicaps), high-Q inductors, RF switches and filters, plays a limiting role. MEMS promises to provide higher levels of integration at lower costs and reduce the power consumption. In RF-MEMS the compatibility of the MEMS process with active circuitry processing is critical, and the ideal MEMS solution must have little if any impact on the wafer/electrical yield of the active circuitry. This capability will enable component suppliers to offer increased performance designs while maintaining or reducing device cost. An additional benefit of the totally integrated approach is the inherent reliability of the IC-based processing. One approach is the innovative "above-IC" (AIC) technology that enables the placement of RF-MEMS devices directly on top of the IC by using a thick copper technology compatible with CMOS, BiCMOS and gallium arsenide processes.[100] Over time, integration of RF-MEMS devices may be the means to replace all passive RF chips with on-chip devices, achieving the long-awaited single-chip RF solution. Current technology and process limitations prevent replacement of all passive components with on-chip MEMS components. But placing even some components on-chip offers significant space and cost savings, allowing smaller form factors, benefiting cell phones for example, or added functionality such as internet connectivity.

Commercially available MEMS-based switches (e.g., from Raytheon), with low-loss metallic structures for lower insertion losses and higher linearity, achieve actuation by using either electrostatic or electromagnetic techniques. Infineon, for example, is aiming for electrostatic switches with 10- to 100-microsecond switching times. Electrostatic actuation consumes no current but requires higher actuation voltage (>10 V), whereas electromagnetic actuation requires a lower actuation voltage, but its current consumption is significantly higher (see Chapter 8, this volume). The use of magnetic thin films introduces slightly more processing complexity. Electrostatic switches offer the most promise as reconfiguration switches (e.g., antenna switches or frequency band selection switches), where the key factor is low power consumption.

MEMS manufacturers are also fabricating inductors using a thick copper-on-insulator process to achieve higher quality factors and resonant frequencies. Similarly, vendors are fabricating MEMS-based

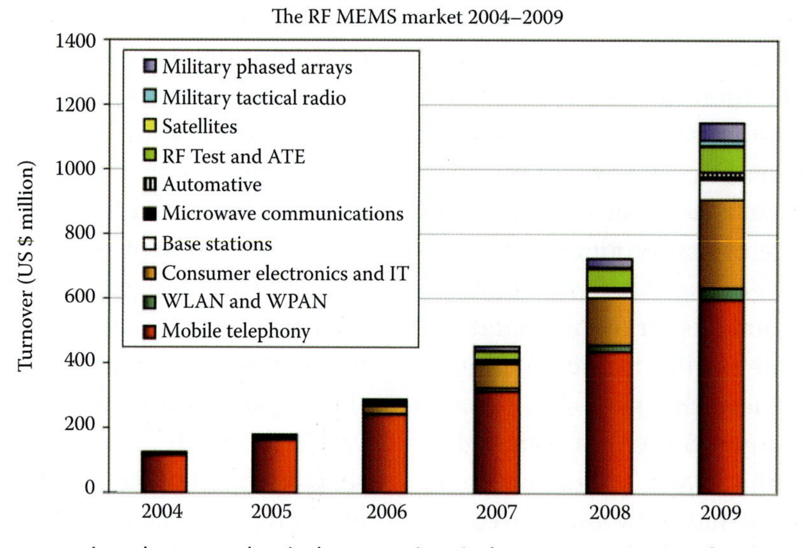

FIGURE 10.64 Radiofrequency microelectromechanical systems in wireless communication for the period 2004–2009 (WTC).

variable capacitors by using a thick copper-on-insulator process to minimize the ohmic losses. MEMS inductors and capacitors are typically suspended with a small air gap insulating them from the lossy silicon or GaAs substrates. The manufacturers achieve tunability of these components by using movable metallic structures with actuation processes similar to the ones used for switches. Typically, the quality factor of inductors using silicon technology have been less than 10 at 2 GHz, so far; this means that most applications have still required off-chip inductors to realize high-performance RF circuits. The objective remains to improve the quality of these devices and the tunability range of the circuits. MEMSCAP has designed high-quality inductors from thick copper on insulator; they range from 1.5 to 15 nH.[101] A quality factor of between 50 and 80 at 2 GHz was obtained. Variable capacitors from the same group offer a high tuning range of more than 2.5, with values ranging between 1 and 10 pF. The combination of inductor and capacitor capabilities optimized for low loss and flexible tunability range is important for the design of voltage-controlled oscillators (VCOs) and filters. Currently, foundries making on-chip RF-MEMS devices are aiming toward fabrication of all the RF-MEMS device types at the same time.

In Volume I, Chapter 4, we saw that for wireless networks such as Wi-Fi and Bluetooth, SiGe is becoming an important choice substrate (see also Chapter 5, this volume where we compare GaAs with SiGe). An example SiGe process is the 130-nm SiGe BiCMOS technology from Jazz Semiconductor. This process combines industry standard 130-nm CMOS with 200 GHz silicon-germanium (SiGe) heterojunction bipolar transistors (HBT) NPN transistors for high-performance RF and millimeter-wave ICs. This technology enables the design of the highest performance circuits in advanced high-speed optical, wireless, and millimeter-wave applications.

Currently, the bulk of the RF-MEMS market services mobile telephony applications, with bulk acoustic wave (BAW) (see Chapter 8, Figure 8.18) duplexers and filters supplied largely by Agilent and Infineon (this product was predicted to continue to dominate and constitute around 40% of the total market in 2009). The RF-MEMS market earned $330 million in 2006, with three companies shipping the newer FBAR (Figure 8.19) duplexers and filters: Avago, Infineon, and EPCOS.

Knowles Acoustics shipped 200 million surface-mount MEMS microphones (a two-chip solution in a single package) and confirmed its clear leadership in this sector, doubling its estimated revenues to $92 million for 2006. Knowles has been selling to cell phone makers for five years and is by far the biggest player in MEMS cell phones. MEMS Tech of Malaysia is a close second. Akustica, Sonion MEMS, and Infineon are other entrants in this area. Today, Akustica is the only maker of a single-chip MEMS microphone, and its rival Sonion MEMS A/S (Roskilde, Denmark) was set to begin volume shipments of MEMS microphones from its new production facility in Ho Chi Minh City, Vietnam, in 2008. Akustica puts all the functions of a miniature electret condenser microphone—including a discrete FET, a separate operational preamplifier chip, and an A/D converter chip—onto a single CMOS MEMS microphone chip.

Another mass market emerged for consumer and IT applications in 2006–2007 with micromechanical resonators. Silicon MEMS start-up companies Discera, Silicon Clocks, and SiTime are currently trying to displace quartz from the oscillator market. They are building packaged silicon MEMS oscillators, consisting of a MEMS resonator and the electrical circuits for the oscillator function and for frequency and temperature compensation. Here, micromechanical resonators are used not as filters but rather as timing devices. MEMS oscillators sales started at the end of 2006 with the first million shipments by SiTime. Seiko Epson also started to ship its QMEMS oscillators. RF-MEMS switches were only in the sampling phase in 2006 (e.g., at Teravicta and Panasonic) but were jolting the market, with Teravicta shipping 10,000 units per month and plans to reach the million mark by the end of 2007. In Chapter 5, "Quartz Properties and Quartz Machining," we made a detailed comparison of Si oscillators with quartz oscillators. More than sixty companies are currently involved in RF MEMS manufacturing, and around a quarter of them were shipping commercial products or samples in 2005. Outside of mass markets, RF test and automated test equipment (ATE) offer the best opportunities, followed by military applications (see Figure 10.64).

MEMS Foundries

As we remarked earlier, the emerging MEMS consumer market is a strong driving force for the MEMS industry to develop MEMS devices integrated with ICs and is pushing the MEMS industry to even adapt 8-inch wafer size manufacturing. Capital requirements for a MEMS production line can be very, very large. And, as with semiconductors, the smaller the features, the more expensive the production line is. Many MEMS startups expended plenty of investor dollars wrestling with exotic materials and getting pilot production lines in place, exhausting their funds before their markets could develop. With this knowledge and the current MEMS upswing, even foundries are now becoming profitable. Along this line, Bosch, a MEMS pioneering company, founded a subsidiary named Bosch Sensortec in early 2005. This subsidiary is dedicated to engineering, developing, and marketing micromechanical sensors with foundry services for consumer electronics, medical technology, security systems, and logistics. The firm is backed by Bosch automotive MEMS development and production and owns a 4000-m^2 CMOS fab and a 4200-m^2 MEMS back-end fab. A new 8-inch fab was planned for 2009. The need to adhere closer to IC manufacturing rules has become even more urgent as few U.S. MEMS foundries have survived. As mentioned earlier, the MEMS foundries that did survive the MOEMS bubble can expect to be doing good business if they can scale up to compete with traditional IC fabs fast enough. In the author's own experience, MEMS foundries are hard to make profitable in the United States, as they need to keep on accepting a wide variety of jobs to keep the expensive clean-room operation going. In the process they do not stay long enough on the learning curve of one new product to become world leaders for that particular product. As a consequence, these labs are moving more and more to Asia, where a large company might have more patience to keep the doors open till the MEMS foundry turns profitable. Olympus, in Japan, for example, already runs the largest MEMS foundry in the world.[102]

Nanotechnology Markets

Introduction

As is clear from the preceding section, although not as large a market as projected, MEMS has made a successful transition out of the laboratory into the marketplace. If one excludes the contributions to nanotechnology markets from genetic engineering (e.g., genetically engineered proteins and DNA probes) and the latest ICs (e.g., chips with sub-100-nm critical dimensions that started shipping in 2003) and next-generation lithography equipment (e.g., extreme UV lithography exposure stations), nanotechnology (NEMS) is still in its infancy, and when a new field emerges it is often the tool-makers and the conference organizers/consultants who make the most money. In his entertaining treatise on Nanotech Pioneers, Steven Edwards[103] compares this to the California gold rush, when the hardware stores and whorehouses made more money than the miners themselves. Nanotechnology quickly became a buzzword and a catchall term to attract funding, and as in the case of MEMS, many existing activities have been relabeled to raise their stock market value.

Whereas in MEMS, products/markets do exist (Bosch makes >100 million MEMS devices a year), in NEMS we deal mostly with manufacturing gear (e.g., carbon-nanotube chemical vapor deposition, dip-pen lithography, nanoimprint lithography, etc.), metrology equipment (Veeco's SPM, Zyvex Nanomanipulators, etc.), software (e.g., Atomostix, MEMulator™, etc.), and products based on nanomaterials (e.g., nano skin cream, suntan lotions, nano-enhanced tennis balls that bounce longer, etc.). Some market studies do also count nanodevices and next-generation lithography equipment in their market estimates. We advice against this practice, as co-opting the IC industry in nanotechnology market projections is bound to be misleading. The emphasis should be put on innovative nanofabrication methodologies making new nanomaterials and nanoproducts and not co-opting efforts that have been going on for a long time. Some will argue that, today at least, besides nanotools research tools, this leaves only low-tech nanomaterials-based products as a new realization. These are products with little technology content and scant manufacturing innovation, and cynics would say that a new transparent sunscreen hardly justifies the current level of investment in nanotechnology.

As with any exciting topic, there is a lot of speculation and exaggerated estimates about the impact nanotechnology will make over time. This author does believe that the ultimate impact of nanotechnology will be broader and deeper than the IC revolution ever was. This is because of the potential of nanotechnology to apply material innovations to so many different situations and challenges. Nanotechnology will profoundly change major industries such as electronics, medicine, transportation, defense, and consumer goods through the development of advanced materials that are lighter, stronger, and more versatile than existing materials. The current nanomaterials boom will be followed by a nanodevices upsurge—it just might be much slower than current market studies suggest.

Overall Market and Market Character

The National Science Foundation (NSF) projects that nanotech will be a $1 trillion industry by 2015; the accuracy of this claim is difficult to assess given the doubts expressed above about what is included in this estimate. Many policy makers have bought into this projection, though; governments across the world poured $4 billion into nanotech research in 2004, and in 2006, the worldwide investment in nanotech research and development was said to be $12.4 billion. There is even a stock index of public companies working on nanotechnology (launched by Lux Research in 2005).

According to a technical market research report, "Nanotechnology: A Realistic Market Assessment" from BCC Research (http://www.bccresearch.com), the global market for nanotechnology was worth $11.6 billion in 2007. This was expected to increase to $12.7 billion in 2008 and $27.0 billion by the end of 2013, with a compound annual growth rate (CAGR) of $16.3% (see Figure 10.65) This makes the NSF market projection for 2015 ($1 trillion) seem unrealistically high. However, as a counterpoint, Lux Research estimates that by 2014, $2.6 trillion in manufactured goods will incorporate nanotechnology—or about 15% of the total global

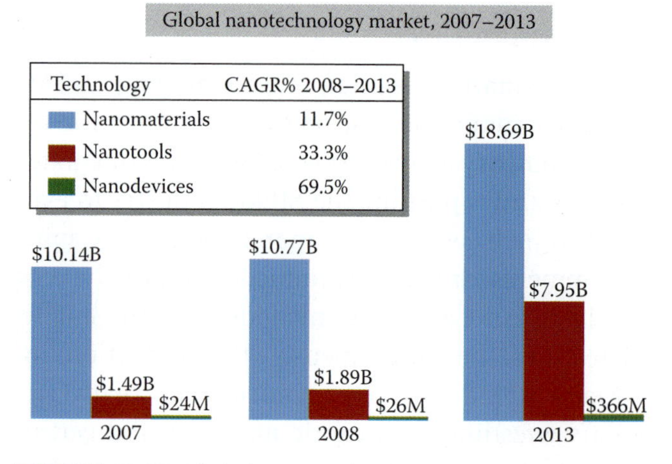

FIGURE 10.65 Global nanotechnology market, 2007–2013. (From BCC Research.)

output will be nanotechnology based! It is hard to figure out what this Lux Research number really means though in terms of the real nanotechnology market size (e.g., TiO_2 and ZnO nanopowders are cheap and would not really constitute a big market share or put many people to work). We actually believe that even the BCC Research numbers are too high. This research firm breaks down the market into nanomaterials, nanotools, and nanodevices. As stated above we are uncertain about the merit of including nanodevices in a nanotechnology market study as today the only nanodevices on the market are the next-generation ICs. These are based on the traditional top-down manufacturing methodologies that were ongoing long before and independent of the current nanofocus. More critical, in terms of coming up with acceptable numbers, is the unfortunate inclusion of current next-generation lithography (NGL) equipment in a market study that should be projecting the impact that new nanotechnologies are bound to make. By including NGL equipment one runs the risk of making projected numbers a self-fulfilling prophecy by broadening the definition of what passes for nanotechnology (this is what happened in the MEMS field). Such work will also not provide sufficient information in order to guide corporate or individual investment decisions. Investors require additional data, such as the size of specific nanotechnology markets, novelty of the manufacturing method (e.g., is it evolutionary or revolutionary), prices, and competition, as well as potential regulation. Nanotechnology market reports should point out specifically whether one is just dealing with an evolutionary extension of IC fabrication or with a revolutionary new nanofabrication method (there are plenty of sophisticated market reports covering the future IC market separately).

Of the three BCC Research nanomarket categories, nanomaterials dominated in 2007, accounting for 87% of the market. Worth an estimated $10.8 billion in 2008, this segment was projected to reach $18.7 billion in 2013, for a CAGR of 11.7%. Nanotools accounted for 12.8% of the market in 2007. Worth an estimated $1.9 billion in 2008, the segment was predicted to reach almost $8.0 billion by the end of 2013, for a CAGR of 33.3%. Nanotools, in which BBC Research includes the nanolithographic tools used to produce the next generation of semiconductors, are projected to grow at a much faster rate than nanomaterials. As a result, that market share was predicted to increase to 29.4% in 2013, while the nanomaterials' share would fall to 69.2% of the total market.

The nanodevices segment, according to BCC Research, will enjoy a CAGR of 69.5% between 2008 and 2013. It was expected to increase from a $26.2 million segment in 2008 to $366.2 million by the end of 2013.

The largest end-user markets for nanotechnology in 2007, according to BCC Research, were environmental remediation (56% of the total market), electronics (20.8%), and energy (14.1%). Electronics, biomedical and consumer applications have much higher projected growth rates than other applications over the next five years (i.e., 30.3%, 56.2%, and 45.9%, respectively.) In contrast, energy applications are projected to grow at a CAGR of only 12.6%, and environmental applications should actually decline by an average of 1.5% per year. It is not clear to this author why this should be the case. Depending on government policies on green technologies, the real numbers might again be quite different.

While it appears inevitable that nanotechnology will have a broad and fundamental impact on many sectors of the world economy, various technical, marketing, and other hurdles need to be overcome before nanotechnology fulfills this promise. These challenges and differences of opinion regarding the speed of the commercial implementation of new nanoproducts are reflected in the wide divergence among existing estimates of the U.S. and global nanotechnology markets. Global market estimates for 2004–2005, for example, range from $8 billion to $68 billion.

Polls show that most Americans know little or nothing about nanotechnology, despite the fact that, according to some estimates, nanotechnology was already incorporated into more than $50 billion in manufactured goods in 2006. Public perceptions about risks—real and perceived—can have large economic consequences. How consumers respond to these early products—in food, electronics, health care, clothing, and cars—is a litmus test for broader market acceptance of nanotechnologies in the future. Along with high expectations, nanotechnology also evokes fear in some people. Bill Joy, cofounder of Sun Microsystems, published an article in 2000 in which he called nanotechnology (along with biotechnology and robotics) a "Faustian bargain" and a "Pandora's box," and called on mankind to relinquish all three technologies. In 2002, Michael Crichton's novel *Prey*, a fictional treatment of the dangers of nanotechnology, made the *New York Times* bestseller list. Interestingly, a recent nationwide survey conducted by the University of Wisconsin–Madison indicates that scientists are more concerned than the general public about the potential health and environmental risks of nanotechnology. Over 30% of scientists interviewed expressed concern that nanotechnology may pose risks to human health, whereas only 20% of ordinary people shared their fears (because they do not understand the technology?).

An important event that helps us appreciate the nanotechnology market dilemma today is what happened to Nanosys in 2004. On April 23 of 2004, Nanosys Inc., a Palo Alto-based developer of inorganic semiconductor nanocrystals, filed for an initial public offering (IPO). Merrill Lynch, Lehman Brothers, CIBC World Markets, and Needham & Co. were underwriting the offering. The company was seeking to raise up to $115 million on the Nasdaq market. But on August 4, 2004, Nanosys Inc. withdrew its IPO. "Based on adverse market conditions," the company explained "Nanosys has determined that it is not advisable at this time to proceed with the proposed offering." Press reports leading up to the IPO had lambasted Nanosys for having only $3 million in revenues in 2003. It was also pointed out that one of the few hard metrics available to value the firm—a price-to-revenue ratio over 120—suggested it was a speculative investment, and that if it performed poorly, it would undermine the ability of other nanotechnology firms to access the public markets. One must also realize that in 2004, the MEMS technology bubbles of 2001–2002 were still vividly remembered.

Nanosys later did raise approximately $40 million in a private equity financing. This financing was led by El Dorado Ventures and includes new investors Masters Capital, Medtronic, Inc., Wasatch Advisors, and others. In addition, there was strong participation from existing investors, including Alexandria Equities, ARCH Venture Partners, CDIB BioScience Ventures, CW Group, Harris & Harris Group, Inc., In-Q-Tel, Intel Capital, H. B. Fuller Company, Lux Capital, Polaris Venture Partners, Prospect Venture Partners, UOB Hermes Asia Technology Fund, and Venrock Associates. Nanosys is using this funding for the ongoing development and manufacturing scale-up of products that incorporate its proprietary, inorganic nanostructures with integrated functionality for multiple industries. Current product development programs include chemical analysis chips for pharmaceutical drug research, fuel cells for portable electronics, nanostructures for displays and phased array antennas, nonvolatile memory for electronic devices, and solid-state lighting products.

The type and number of California-based MEMS and NEMS start-up companies (see Table 10.13) has historically played an important role in what happens in these areas worldwide. In 2000, according to Janusz Bryzeck, there were 100 MEMS/NEMS-based companies in California alone, 33 of which were fabricating MOEMS devices. Decimated by the MOEMS bubble of 2001, "only" sixty-five California MEMS/NEMS companies survived in 2005. In the early years (say, from 1972 on), companies making mechanical MEMS devices (using wet bulk micromachining) played a pivotal role in the development of the market for these mechanical Si sensor products. Petersen's 1982 paper extolling the excellent mechanical properties of single-crystalline silicon helped galvanize academia's involvement in Si micromachining in a major way.[104] Before that time, timid efforts had played out in industry, and practical needs were driving the technology (market pull). Academics often explored micromachines that constituted gadgetry only, and, as a consequence, the field was perceived by some as a technology looking for applications (technology push). Polycrystalline surface micromachining-based (see Volume II, Chapter 7) MEMS companies started emerging around 1991. However, from the mid-1990s on, many of the MEMS companies became bio-MEMS oriented (application of MEMS to biotechnology), and single-crystalline and polycrystalline Si were often not the

Table 10.13 Some California Microelectromechanical System/Nanoelectromechanical System Companies

Company	Location	Date	Overview of Activity and Manufacturing Method
Sensym	Milpitas	1972	Made sensors from silicon and piezoresistive materials. Some cavities in products made by anisotropic etching. *Source:* U.S. Patents.
Endevco	San Juan Capistrano	1975	Produces sensors for vibration, shock, inertial motion, and dynamic pressure measurement using semiconductor technology and piezoceramic processing techniques. *Source:* Endevco.
Irvine Sensors Co.	Costa Mesa	1980	Manufactures some products in a batch process. Products are layered silicon and connected with gold. *Source:* Irvine Sensors Co.
Komag	San Jose	1983	Komag sputters thin-film disks using a vacuum deposition process. During the sputtering process, microscopic magnetic layers are successively deposited on the disk—these layers ultimately provide the magnetic storage capacity for data. *Source:* Komag.
Nanostructures	Santa Clara	1988	Produced thin films such as silicon membranes by electrochemical means or mask production with RIE. *Source:* U.S. Patents.
Abaxis	Sunnyvale	1989	Produces point-of-care blood analyzers from a bottom- and mid-layer of polymethymethacrylate plastic with a top layer of ABS made in a molding process. Some products made from joined thermoplastics through melting. *Source:* U.S. Patents, Abaxis.
Sentir	Santa Clara	1991	Produced sensors, including a pressure sensor made on a ceramic substrate with piezoresistive diaphragm and gel as the transfer medium to transducer. *Source:* U.S. Patents.
Affymetrix	Santa Clara	1992	The company's microarrays were made by a process that begins by coating a 5" by 5" quartz wafer. *Source:* Affymetrix.
Unisun	Newbury Park	1993	Unisun's techniques provide substantial cost advantages to traditional PV products based on silicon wafers and to vacuum deposition techniques typically used to deposit PV thin films. *Source:* Unisun.
Nanogen	San Diego	1993	Makes microarray cartridges the only Si based DNA arrays ever made. *Source:* Nanogen patents.
Front Edge Technology	Baldwin Park	1994	FET develops, manufactures, and markets next-generation, ultra-thin rechargeable batteries for card-type applications. *Source:* Front Edge Technology.
Aclara	Mountain View	1995	Some products made by thermally bonding plastics. *Source:* U.S. Patents.

(Continued)

Table 10.13 Some California Microelectromechanical System/Nanoelectromechanical System Companies (Continued)

Company	Location	Date	Overview of Activity and Manufacturing Method
Nanochip	Fremont	1996	Nanochip is developing the next generation of removable, portable memory chips for consumer applications. Using MEMs technology along with arrays of atomic force probe tips, Nanochip is developing storage chips that will store tens of gigabytes per chip. Our technology is not bound by lithography constraints. *Source:* Nanochip.
Caliper	Mountain View	1996	Microfluidic lab-on-a-chip made of quartz, glass, or plastic-bonded to each other. *Source:* Caliper.
Cepheid	Sunnyvale	1996	Made sample preparation products with the process of electrodeposition, LPVD, glass bonding, and RIE out of materials from the semiconductor industry. *Source:* U.S. Patents.
Fluidigm	South San Francisco	1997	Fluidgm's NanoFlex valve consists of a membrane that deflects under pressure to pinch off the flow of fluids in a microchannel. The valve is made from two separate layers of elastomeric rubber. *Source:* Fluidigm.
Carbon Solutions	Riverside	1998	Manufactures single-walled carbon nanotubes (SWNTs) in bulk quantities. *Source:* Carbon Solutions.
Zyomex	Hayward	1998	Products included fluidic technologies using an etching process. *Source:* U.S. Patents.
PolyFuel	Mountain View	1999	PolyFuel's membranes are based upon hydrocarbon polymers rather than fluorocarbon. *Source:* PolyFuel.
GeneFluidics	Monterey Park	2000	GeneFluidics disposable cartridges and sensor chips are injection-molded, using only plastic substrates and other low-cost materials; the GeneFluidics platform will dramatically reduce the cost of molecular analysis.
Molecular Nanosystems, Inc.	Palo Alto	2001	The company synthesizes carbon nanotubes on predefined locations with the desired orientation on different substrates, manipulates the properties of carbon nanotubes via functionalization, and integrates them into next-generation gas and biological sensors and other high-end electronics devices. *Source:* Nanovip.
Discera	San Jose	2001	Discera offers PureSilicon resonators, which is displacing quartz crystal solutions with systems-on-a-chip alternatives. *Source:* Discera.

Ahwahnee Inc.	San Jose	2002	Ahwahnee's industry-leading CNTs are revolutionizing a number of markets, including fuel cells, energy, and semiconductors. *Source:* Ahwahnee.
Apex Nanomaterials	San Diego	2003	Apex Nanomaterials's world first-class SWNT production technologies using both arcing and CVD methods are generating SWNTs with the best quality and most cost-effective at an industry scale. *Source:* Apex Nanomaterials.
Atomate	Santa Barbara	2003	Atomate Corporation develops complete systems, components, and materials that are optimized for the synthesis of nanostructures using CVD.
NanoSensors, Inc.	Santa Clara	2003	Develops and markets sensors and instruments to detect explosive (X), chemical (C), and biological (B) agents ("XCB") *Source:* Nanosensors.
InvenSense	Santa Clara	2003	InvenSense's gyroscopes takes advantage of major innovations in MEMS mechanical design and fabrication, mixed-signal ASIC design, and wafer-level packaging, which is based on the IC industry. *Source:* Invensense.
Intelligent Optical Systems, Inc.	Torrance	2004	Develops cutting-edge technologies in optical sensing and instrumentation. Current research and development is being conducted in various areas, including Nanoprobes and Materials for Sensing. *Source:* Nanovip.
Unidym, Inc.	Menlo Park	2006	Unidym's initial product is a transparent, conductive film of CNTs that replaces the approximately $1B of expensive and brittle metal oxide films currently employed in touch screens, flat-panel displays, OLEDs, and thin-film solar cells. *Source:* Unidym.
Biodent	Menlo Park	2006	Biodent is the first company to combine printed optoelectronic components with microfluidic systems. This combination is enabled by the fluid processability of organic semiconductor materials. Biodent allows for the first time to print and integrate full optical readout systems directly onto microfluidic devices. *Source:* Biodent.

preferred MEMS substrates for these applications. Later yet, from 1997 on, nanotechnology companies, mostly working on biotechnology type applications (bio-NEMS), started emerging, and here Si was often not involved at all.

For further reading on the history of MEMS and NEMS up to today see Volume I, Chapter 1.

Nanotechnology Tools

To estimate the size and to forecast the market for nanotechnology tools over the next five years, nanotechnology analysis firm Lux Research interviewed CEO's and marketing executives at 21 leading tool vendors worldwide and research and development leaders at forty-nine companies, universities, and national laboratories active in nanotechnology. They correctly excluded established uses of manufacturing and metrology tools in fields such as semiconductors and data storage. This firm believes that the market for tools used to fabricate emerging nanotechnology could grow from less than $20 million in 2004 to nearly $235 million by 2010. At the same time, they claim, the market for metrology tools to inspect matter on the nanometer scale, which currently dominates the $580 million nanotechnology tools market, is expected to see only tepid growth, although still command more than $750 million in 2010. These numbers are lower, and more realistic, than the BCC Research findings (see above) as they exclude the traditional IC tools.

Nanotechnologists basically use three kinds of tools to experiment with matter on the nanometer scale. First, metrology tools, reviewed in Chapter 6, this volume, such as atomic force microscopes (AFMs) and scanning electron microscopes (SEMs) visualize the nanometer-sized objects and represent the most mature and widely used category of tools for emerging nanotechnology. Second, fabrication tools such as nano-imprint lithography and dip-pen nanolithography create structures at the nanometer scale. Thirdly, modeling tools predict nanostructure properties to circumvent the costs and time associated with experiments. Lux Research predicts that the nanotechnology-tools market will see a CAGR of 11% to nearly $1.1 billion in 2010. However, growth will not be even across each of the three market segments. The current market for nanotechnology tools is dominated by inspection tools, which accounted for 95% of the 2004 revenues. The inspection-tools market saw dramatic growth during the early 2000s nanotech explosion, when many university nano centers were being built and equipped. However, the growth outlook for inspection tools looks slow because the university and lab markets are now saturated.

Nanofabrication tools have not yet found commercial use. Still, Lux Research expects these instruments to gain traction over the next five years and move beyond research and development labs to manufacturing floors. In particular, Lux Research sees nano-imprint lithography as the favorite candidate. Lux Research felt dip-pen nanolithography facing sharp resistance, as in many cases researchers would rather have a kit to modify an AFM they already own than purchasing a whole new instrument just to do dip-pen. No wide use for dip-pen nanolithography in production is anticipated before 2010.

Modeling tools claimed less than $10 million in 2004 in the face of entrenched skepticism from potential customers. In five years, Lux Research expects that market to expand to only roughly $40 million, a small fraction of the total nanotechnology-tools market. Still, in a decade the analysts predicted it could break $100 million as power, accuracy, and ease of use of modeling tools continues to improve and more customer success stories emerge.

Nanomaterials

Even though much of the research is still in the initial stages, nanoproducts based on nanomaterials have entered the commercial market. According to an update to the nanotechnology consumer product inventory maintained by the Project on Emerging Nanotechnologies (PEN) (*ScienceDaily*, April 25, 2008), new nanotechnology consumer products are coming on the market at the rate of three or four per week. The number of consumer products using nanotechnology has grown from 212 to 609 since PEN launched the world's first online inventory of manufacturer-identified nanotech goods in March 2006. Health and fitness items, which includes cosmetics and sunscreens, represent 60% of these products. There are thirty-five automotive products in the

PEN inventory, including the Hummer H2. General Motors Corporation bills the H2 as having a cargo bed that "uses about seven pounds of molded in color nanocomposite parts for its trim, center bridge, sail panel and box rail protector." Nanoscale silver is the most cited nanomaterial used. It is found in 143 products, or over 20% of the inventory. Carbon, including carbon nanotubes and fullerenes, is the second highest nanoscale material cited. Other nanoscale materials explicitly referenced in products are zinc (including zinc oxide), titanium (including titanium dioxide), silica, and gold. A list of nanotechnology merchandise—containing everything from nanotech diamonds to cooking oil to golf clubs to iPhones—is available at http://www.nanotechproject.org/consumerproducts. One of the new items among the more than six hundred products now in the inventory is Swissdent Nanowhitening Toothpaste, with "calcium peroxides, in the form of nano-particles." PEN Project Director David Rejeski cites Ace Silver Plus—another of the nine nano toothpastes in the inventory—as an example of the upsurge in nanotechnology consumer products in stores.

There are more than five hundred nanotechnology companies globally, and one-third are in nanoparticles. Of that one-third, more than fifty-five are in carbon nanotubes. Of the present carbon nanotube companies, about twenty are into large-scale production. Some representative example companies in the last category include the following:

1. Nanoledge—(France). Large-scale production (4 ounces per day)—tennis racquets, fibers.
2. Mitsui CNRI (Akashima, Tokyo)—new nanotube plant (120 tons per year at 10% of the present cost).
3. Frontier Carbon (Kitakyusyu City)—Mitsubishi large-scale manufacturing (400 kg per year and 1500 tons in 2007).

Another group of carbon nanotubes companies sells modified carbon nanotubes (e.g., for sensor applications):

1. Carbon Nanotechnologies, the late Richard Smalley's company ($15 million—from angel investors)—single-walled and functionalized tubes.
2. Nantero (>$20 million—DFJ and other venture capital firms)—NRAM™, a high-density, non-volatile random access memory chip.
3. Molecular Nanosystems Stanford ($2 million—band of angel investors)—nanotube sensors.
4. C Sixty ($4 million—CNI angel investors)—protease inhibitor, anti-HIV (merged with CNI).

Carbon nanotubes and fullerenes have very broad applications, and from the above it is seen that large manufacturing companies are stepping in, with the price of product dropping rapidly. There is also a clear trend of specializations developing (e.g., modified carbon nanotubes). There have not been any IPOs in this area yet. Applications of fullerenes include clothes (carbon nanotubes form waterproof tear-resistant cloth fibers), concrete (carbon fibers increase the tensile strength, and halt crack propagation), sports equipment (stronger and lighter tennis rackets, bike parts, golf balls, golf clubs, golf shaft, and baseball bats), ultrahigh-speed flywheels (the high strength/weight ratio enables very high speeds to be achieved), and bridges.

Other nanoparticles are used in catalysis, ceramics, dug delivery and functional materials, and coatings. Existing companies are Degussa, Cabot Corp, Johnson Matthey BASF, and Altana; start-up companies include the following:

1. NanoPhase Technologies (Nasdaq). Patents licensed from Argonne in 1990. This was the first Nanotech IPO (1998). It was overhyped and had to pay compensation of $4.8 (January 2002). Altana Chemie AG acquired a 7% share for $10 million in 2004. Products include ZnO nanoparticles for sunscreens and catalysts. Losing $1.3 million per quarter.
2. Ntera ($30 million)—was lithium-ion batteries, now nanochromic displays.
3. Nanogram ($6.7 million)—nanomaterials.
4. Aveka (buyout from 3M)—particle processing.
5. Oxonica ($4 million)—sunscreens, bio tags, catalysts. Floated on Aim 2005 (overdiluted).
6. Quantum Dot Corp ($37 million)—bio-assays. Acquired by Invitrogen in October 2005.
7. Nanosphere ($10 million). Gold nanoparticle DNA probes for biothreat detection (anthrax).

8. Qinetiq—nanomaterials.
9. Nanomagnetics—was data storage, now water purification, Apaclara.
10. Nanosys—the hyped nanotech IPO (oversold; see above). A Larry Bock company. Sells as much the technology as brand name scientists (Charles Lieber, Hongkun Park, Paul Alivisatos, and many others over the years). Products include biosensors (functionalized inorganic semiconductor nanowires, 1D electron flow; highly sensitive and selective), photovoltaics, nanocrystals in host matrix, solar cells as roof tiles (Matsushita contract), macroelectronics; flexible, low-cost substrate, nanostructured surfaces, hydrophobic and antimicrobial surfaces.

A worthwhile book to consult is *Nanotechnology Applications and Markets* by L. Gasman.[105]

Example 10.2: Cosmetics

The market for cosmetics using nanotechnology is expected to grow significantly over the coming years despite continued concern about the safety of the technology. Nanotechnology is being used increasingly in the personal care industry to develop sunscreens and antiaging formulations. The global market for cosmetics using nanotechnology was valued at $62 million as of 2007 and is forecast to grow annually by 16.6%, reaching $155.8 million by 2012, according to a BCC Research report (NANO17D). This market research company listed sunscreens as one of three applications in the biomedical, pharmaceutical, and cosmetics industries that are expected to account for 95% of the market in 2012. Nanoparticles of titanium dioxide and zinc dioxide are used in sunscreens to increase UV light absorption and make formulations transparent—ultimately making them more appealing to the consumer.

Example 10.3: Textiles

Nanotechnologies are providing incremental improvements to existing textile sectors, but they offer the highest growth in the nontraditional sectors. In the near term, the opportunity for nanotechnology in the textiles industry is in product innovation, not process innovation. Nanotechnology is more likely to be used to produce new materials or enhance the properties of existing materials than to reduce the production cost or improve quality. A number of nanotechnology innovations are already commercially available. These include:

- Stain-, wrinkle-, and liquid-resistant fabrics (see NanoTex)
- Clothing that can absorb body odors
- Clothing that emits deodorant by slow release
- Clothing that changes color with change in light
- Clothing that changes color with external or body heat

There are a number of opportunities that will be explored as nanotechnology develops further:

- New blended fabrics for specific applications (e.g., sportswear, mountainwear, military applications), including the incorporation of carbon nanotubes (CNTs) into fabrics. It is expected that composites with CNTs or interweaving with extruded CNT fibers will introduce higher conductivity as well as high strength.
- Property enhancement or alterations (UV blocking, durability, breathability, flexibility, recyclability, color retention, self-repair etc.); included in this is the introduction of electronic properties into fabrics by treatment with inherently conducting polymers. These provide capabilities such as sensing (chemical and mechanical), energy generation (photovoltaics), energy storage (batteries, supercapacitors), and charge dissipation (antistatic).
- In addition, controlled-release polymers may replenish and/or trigger release of antifungal, surface finish, or medical growth aids to the polymer surface. The triggered release systems may be made to be responsive to stimuli such as changes in temperature, humidity pH, and/or dissolved oxygen.
- Development of specific aesthetic properties (e.g., glow in the dark, color change with angle of light, color change with applied electric field).

The global market for nanotechnology in the textiles industry was around $480 million in 2007. The forecast is that nanobased products and processes will be worth $4.9 billion to the textiles industry by 2015. It is estimated that 24.6% of all textiles products available in 2015 will incorporate some form of nanotechnology, with the greatest level of penetration in the hygiene area, which covers numerous markets from household care to leisure and consumer goods to health care. Nanotechnology-based applications such self-cleaning textiles as will start to make a major impact in 2011. There will likely be a significant revenue growth in nanotechnology products in the sector from 2011 onward.

Questions

10.1: Suggest an array of sensors that could be used for an electronic tongue (5 tastes).

10.2: Why do you think that chemical microsensors have not enjoyed the same commercial success as physical microsensors? How might these deficiencies be overcome in the future?

10.3: Instrumentation miniaturization is a fast developing field. The Agilent 2100 Bioanalyzer fabricated by Agilent & Caliper Technologies Co. has been awarded the Pittcon Editors' Gold Award for 2000. How does this instrument benefit from miniaturization? What are its drawbacks? In miniaturized GC, MS, and IMS, what components have been micromachined?

10.4: At Pittcon 2000, Cyrano showcased a portable electronic nose, the Cyranose 320, which distinguishes odors using an array of 32 sensors. Each sensor is made of a composite of conductive carbon black and a non-conductive polymer. Describe how the sensors work and explain how one "trains" an electronic nose to qualitatively and quantitatively analyze unknown samples.

10.5: *Nanotechnology* is currently the buzzword in the scientific community. What does the term mean to you? Provide examples of nanoscale building blocks. What are the characterization tools (metrology) appropriate for this technology? What strategy would you consider for assembling a nanodevice and why?

10.6: Over the last five years mechanical MEMS applications saw a major change in market acceptance. In what applications and why did this occur?

10.7: The United States seems ahead in lithography-based MEMS and NEMS but is perhaps behind in non-lithography-based MEMS. Explain.

10.8: What are the three market components of NEMS today? How do they compare in size?

10.9: What are the largest MEMS applications in the automotive industry?

10.10: What was behind the infamous optical MEMS bubble bust?

10.11: Explain the concept behind the 1st level packaged MEMS and smallest commercialized unit.

10.12: How does the MEMS market compare in size to the IC market?

10.13: What is the difference between an angular rate sensor, a gyro, and an accelerometer?

10.14: Compare manufacturing options for Si-based pressure sensors and accelerometers. Give pros and cons.

10.15: Give a list of reasons why the ISFET did not become a commercial success.

References

1. NEXUS. 2006. *NEXUS market analysis for MEMS and microsystems III, 2005–2009*. Neuchâtel, Switzerland: NEXUS.
2. Bryzek, J., S. Roundy, B. Bircumshaw, C. Chung, K. Castellino, J. R. Stetter, and M. Vestel. 2006. Marvelous MEMS: Advanced IC sensors and microstructures for high volume applications. *IEEE Circuits Devices Mag* 22:8–28.
3. Ehmann, K. F., D. Bourell, M. I. Culpepper, T. J. Hodgson, T. Kurfess, M. Madou, K. Rajurkar, and R. DeVor. 2007. *International assessment of research and development in micromanufacturing*. New York: Springer.
4. Maluf, N. 2000. *An introduction to microelectromechanical systems engineering*. Boston: Artech House, Boston.
5. Wechsung, R., N. Ünal, J. C. Eloy, and H. Wicht. 1998. Market analysis for micro systems 1996–2002. NEXUS Task Force Report. Berlin, p. 136.
6. Fluitman, J. H. J. 1994. Basic technologies and functions; microsystem technology and silicon micromachining. In *Microsystem technology: Exploring opportunities*. The Hague, The Netherlands: Stichting Toekomst der Techniek, pp. 35–40.

7. National Research Council. 1995. National Research Council.
8. Busch-Vishniac, I. J. 1992. The case for magnetically driven microactuators. *Sensors Actuators A* A33:207–20.
9. Zdeblick, M. J., R. Anderson, J. Jankowski, B. Kline-Schoder, L. Christel, R. Miles, and W. Weber. 1994. Thermopneumatically actuated microvalves and integrated electro-fluidic circuits. *Technical Digest: 1994 Solid State Sensor and Actuator Workshop*. Hilton Head Island, SC , June 13–16, 1994, pp. 251–55.
10. Bryzek, J., K. Petersen, J. R. Mallon, L. Christel, and F. Pourahmadi. 1990. *Silicon sensors and microstructures.* Fremont, CA: Novasensor.
11. Ravi, T. S., and R. B. Marcus. 1991. Oxidation sharpening of silicon tips. *J Vac Sci Technol B* 9:2733–37.
12. Editorial. 1991. Menlo Park, CA: TSDC (Teknekron).
13. Feynman, R. P. 1992. There's plenty of room at the bottom. *J Microelectromech Syst* 1:60–66.
14. Feynman, R. P. 1993. Infinitesimal machinery. *J Microelectromech Syst* 2:4–14.
15. Yole. 2007. Evolution of the MEMS markets and evolution of the MEMS foundry business. *Transducers*.
16. Collins, A. J. 1992. Problems associated with bringing a sensor technology to the market place. *Sensors Actuators A* A31:77–80.
17. Greenwood, J. C. 1988. Silicon in mechanical sensors. *J Phys E Sci Instrum* 21:1114–28.
18. Ikeda, K., H. Kuwayama, T. Kobayashi, T. Watanabe, T. Nishikawa, T. Yoshida, and K. Harada. 1990. Silicon pressure sensor integrates resonant strain gauge on diaphragm. *Sensors Actuators A* A21-A23:146–50.
19. Petersen, K., F. Pourahmadi, J. Brown, P. Parsons, M. Skinner, and J. Tudor. 1991. Resonant beam pressure sensor fabricated with silicon fusion bonding. *6th International Conference on Solid-State Sensors and Actuators (Transducers '91)*. San Francisco, CA, June 24–27, 1991, pp. 664–67.
20. Zook, J. D., D. W. Burns, H. Guckel, J. J. Sniegowski, R. L. Engelstad, and Z. Feng. 1992. Characteristics of polysilicon resonant microbeams. *Sensors Actuators A* 35:51.
21. Ikeda, K., H. Kuwayama, T. Kobayashi, T. Watanabe, T. Nishikawa, and T. Yoshida. 1988. Three dimensional micromachining of silicon resonant strain gauge. *7th Sensor Symposium*. Tokyo, Japan, May 30–31, 1988, pp. 193–96.
22. Burns, D. W., R. D. Horning, W. R. Herb, J. K. Zook, and H. Guckel. 1996. Sealed-cavity resonant microbeam accelerometer. *Sensors Actuators A* 53:249–55.
23. Bryzek, J., K. Petersen, L. Christel, and F. Pourahmadi. 1992. New technologies for silicon accelerometers enable automotive applications. *SAE Technical Papers Series* 920474:25–32.
24. Editorial. 1995. *Commercial brochure*.
25. Chang, S. 1998. An electroformed CMOS integrated angular rate sensor. *Sensors Actuators A* 66:138–43.
26. Lutz, M. G., W. Golderer, J. Gerstenmeier, J. Marek, B. Maihofer, S. Mahler, H. Munzel, and U. Bischof. 1997. A precision yaw rate sensor in silicon micromachining. *Proceedings: 1997 Int. Conf. on Solid-State Sensors and Actuators*. Chicago, June 1997, pp. 847–50.
27. Shkel, A., C. Acar, and A. Shkel. 2005. An approach for increasing drive-mode bandwidth of MEMS vibratory gyroscopes. *J Microelectromechan Syst* 14:520–28.
28. Roe, J. N. 1992. Biosensor development. *Pharm Res* 9:835–44.
29. Editorial. 1993. *Sensor Business Digest*, 1–11.
30. Battaglia, E. A. 1980. *Ion-selective electrode*. US Patent 4214968.
31. Editorial. 1994. Microminiature tools affecting medical products. *BBI Newsletter* 17:88–91.
32. Lauks, I. 1983. Multielement thin film chemical microsensors. *SPIE* 138–50.
33. Turner, A. P. F. 1994. *Third World Congress on Biosensors.* New Orleans, LA, June 1–3, 1994. Amsterdam: Elsevier Advanced Technology.
34. Tierney, M. J., H. L. Kim, M. D. Burns, J. A. Tamada, and R. O. Potts. 2000. Electroanalysis of glucose in transcutaneously extracted samples. *Electroanalysis* 12:1–6.
35. Gross, T. M., B. W. Bode, D. Einhorn, D. M. Kayne, J. H. Reed, N. H. White, and J. J. Mastrototaro. 2000. Performance evaluation of the MiniMed continuous glucose monitoring system during patient home use. *Diabetes Technology and Therapeutics* 2:49–56.
36. Roe, J. N., and B. R. Smoller. 1998. Bloodless glucose measurements. *Therapeutic Drug Carrier Systems* 15:199–241.
37. Klonoff, D. C. 1997. Noninvasive blood glucose monitoring. *Diabetes Care* 20:433–37.
38. Heller, A., and B. Feldman. 2008. Electrochemical glucose sensors and their applications in diabetes management. *Chem Rev* 108:2482–505.
39. Burbaum, J. J. 2000. The evolution of miniaturized well plates. *J Biomol Screen* 5:5–8.
40. Shapiro, H. M., and M. Hercher. 1986. Flow cytometers using optical waveguides in place of lenses for spectroscopic illumination and light collection. *Cytometry* 7:221–23.
41. Miyake, R., H. Ohki, I. Yamazaki, and R. Yabe. 1991. A development of micro sheath flow chamber. *Proceedings: IEEE Micro Electro Mechanical Systems (MEMS '91)*. Nara, Japan, pp. 259–64.
42. Sobek, D., A. M. Young, M. L. Gray, and S. D. Senturia. 1993. An investigation of micro structures, sensors, actuators, machines, and systems. *Proceedings: IEEE Micro Electro Mechanical Systems (MEMS '93)*. Fort Lauderdale, FL, pp. 219–24.
43. Sobek, D., S. D. Senturia, and M. L. Gray. 1994. Microfabricated fused silica flow chambers for flow cytometry. *Technical Digest: 1994 Solid State Sensor and Actuator Workshop*. Hilton Head Island, SC, June 13–16, 1994, pp. 260–63.
44. Sutherland, R. M., and C. Dahne. 1987. In *Biosensors: Fundamentals and applications*, eds. A. P. F. Turner, I. Karube, and G. S. Wilson, 655–78. Oxford, United Kingdom: Oxford University Press.
45. Garabedian, R., C. Gonzalez, J. Richards, A. Knoesen, R. Spencer, S. D. Collins, and R. L. Smith. 1995. Microfabricated surface plasmon sensing system. *Sensors Actuators A* A43:202–07.
46. Jorgenson, R. C., and S. S. Yee. 1993. A fiber-optic chemical sensor based on surface plasmon resonance. *Sensors Actuators B* B12:213–20.
47. Melendez, J., R. Carr, D. U. Bartholomew, K. Kukanskis, J. Elkind, S. Yee, C. Furlong, and R. Woodbury. 1996. A commercial solution for surface plasmon sensing. *Sensors Actuators B* 35:1–5.
48. Melendez, J., R. Carr, D. U. Bartholomew, H. Taneja, S. Yee, C. Jung, and C. Furlong. 1997. Development of a surface plasmon resonance sensor for commercial applications. *Sensors Actuators B* 38–39:375–79.

49. Ballerstadt, R., and J. S. Schultz. 1998. Kinetics of dissolution of concanavalin a/dextran sols in response to glucose measured by surface plasmon resistance. *Sensors Actuators B* 46:50–55.
50. Madou, M. J., and T. Otagawa. 1989. Micron and submicron electrochemical sensors. *AICHE Symposium Series* 267:7–14.
51. Madou, M. J., and T. Otagawa. 1989. *Microelectrochemical sensor and sensor array*. US Patent 4874500.
52. Wilson, G. S., and Y. Hu. 2000. Enzyme-based biosensors for in-vivo measurements. *Chem Rev* 100:2693–704.
53. Schauer-Vukasinovic, V., L. Cullen, and S. Daunert. 1997. Rational design of a calcium sensing system based on induced conformational changes of calmodulin. *J Am Chem Soc* 119:11102–03.
54. Salins, L. L. E., V. Schauer-Vukasinovic, and S. Daunert. 1998. Optical sensing systems based on biomolecular recognition of recombinant proteins. *SPIE* 3270:16–24.
55. Salins, L. L. E., J. Lundgren, and S. Daunert. 1999. A dynamical investigation of acrylodan-labeled mutant phosphate binding protein. *Anal Chem* 71:589–95.
56. Salins, L. L. E., R. A. Ware, C. Ensor, and S. Daunert. 2001. A novel reagentless sensing system for measuring glucose based on the galactose/glucose-binding protein. *Anal Biochem* 294:19–26.
57. Tolosa, L., I. Gryczynski, L. R. Eichhorn, J. D. Dattelbaum, F. N. Castellano, Rao, and J. R. Lakowicz. 1999. Glucose sensor for low-cost lifetime-based sensing using a genetically engineered protein. *Anal Biochem* 267:114–20.
58. Petersen, K. E., W. A. McMillan, G. T. A. Kovacs, M. A. Northrup, L. A. Christel, and F. Pourahmadi. 1998. Towards next generation clinical diagnostic instruments: Scaling and new processing paradigms. *J Biomed Microdevices* 1:71–79.
59. Manz, A., N. Graber, and H. M. Widmer. 1990. Miniaturized total chemical analysis systems: A novel concept for chemical sensing. *Sensors Actuators B* B1:244–48.
60. McMillan, W. A., K. E. Petersen, L. A. Christel, and M. A. Northrup. 1999. Application of advanced microfluidics and rapid PCR to analysis of microbial targets. *Proceedings of the 8th International Symposium on Microbial Ecology*, eds. C. R. Bell, M. Brylinsky, and P. Johnson-Green. Halifax, NS, Canada: Atlantic Canada Society for Microbial Ecology.
61. Li, P. C. H., and D. J. Harrison. 1997. Transport, manipulation, and reaction of biological cells on-chip using electrokinetic effects. *Anal Chem* 69:1564–68.
62. Kuske, C. R., K. L. Banton, D. L. Adorada, P. C. Stark, K. K. Hill, and P. J. Jackson. 1998. Small-scale DNA sample preparation method for field PCR detection of microbial cell and spores in soil. *App Environ Microbiol* 64:2463–72.
63. Belgrader, P., W. Benette, D. Hadley, J. Richards, P. Stratton, R. Mariella, Jr., and F. Milanovich. 1999. PCR detection of bacteria in seven minutes. *Science* 284:449–50.
64. Cheng, J., E. L. Sheldon, L. Wu, A. Uribe, L. O. Gerrue, J. Carrino, M. J. Heller, and J. P. O'Connell. 1998. Preparation and hybridization analysis of DNA/RNA from *E. coli* on microfabricated bioelectronic chips. *Nat Biotechnol* 16:541–45.
65. Lee, S. W., and Y.-C. Tai. 1999. A micro cell lysis device. *Sensors Actuators A* 73:74–79.
66. Wilding, P., L. J. Kricka, J. Cheng, G. Hvichia, M. A. Shoffner, and P. Fortina. 1998. Integrated cell isolation and polymerase chain reaction analysis using silicone microfilter chambers. *Anal Chem* 257:95–100.
67. Anderson, R. C., X. Su, G. J. Bogdan, and J. Fenton. 2000. A miniature integrated device for automated multistep genetic assays. *Nucleic Acids Res* 28:e60.
68. Han, F., and S. J. Lilard. 2000. *In situ* sampling and separation of RNA from individual mammalian cells. *Anal Chem* 72:4073–79.
69. Khandurina, J., T. E. McKnight, S. C. Jacobson, L. C. Waters, R. S. Foote, and J. M. Ramsey. 2000. Integrated system for rapid PCR-based DNA analysis in microfluidic devices. *Anal Chem* 72:2995–3000.
70. Pourahmadi, F., M. Taylor, G. Kovacs, K. Lloyd, S. Sakai, T. Schafer, B. Helton, L. Western, S. Zaner, J. Ching, B. McMillan, P. Belgrader, and M. A. Northrup. 2000. Toward a rapid, integrated, and fully automated DNA diagnostic assay for *Chlamydia tranchomatis* and *Neisseria gonorrhoeae*. *Clin Chem* 46:1511–13.
71. Lagally, E. T., I. Medintz, and R. A. Mathies. 2001. Single-molecule DNA amplification and analysis in an integrated microfluidic device. *Anal Chem* 73:565–70.
72. Krishnan, M., V. Namasivayam, R. Lin, R. Pal, and M. A. Burns. 2001. Microfabricated reaction and separation systems. *Anal Biotechnol* 12:92–98.
73. Yang, J. M., J. Bell, Y. Huang, M. Tirado, D. Thomas, A. H. Forster, R. W. Haigis, P. D. Swanson, R. B. Wallace, B. Martinsons, and M. Krihak. 2001. An integrated, stacked microlaboratory for biological agent detection with DNA and immunoassays. *Biosens Bioelectron* 17:605–18.
74. Kellogg, G. J., T. E. Arnold, B. L. Carvalho, D. C. Duffy, and N. F. Sheppard. 2000. Centrifugal microfluidics: Applications. *Micro total analysis systems,* eds. A. van den Berg, W. Olthuis, and P. Bergveld. Enschede, The Netherlands: Kluwer Academic Publishers, pp. 239–42.
75. Yuen, P. K., L. J. Kricka, P. Fortina, N. J. Panaro, T. Sakazume, and P. Wilding. 2001. Microchip module for blood sample preparation and nucleic acid amplification reactions. *Genome Res* 405–12.
76. Wilding, P., M. A. Shoffner, and L. J. Kricka. 1994. PCR in a silicon microstructure. *Clin Chem* 40:1815–18.
77. Cheng, J., M. A. Shoffner, G. E. Hvichia, L. J. Kricka, and P. Wilding. 1996. Chip PCR: Investigation of different PCR amplification systems in micro-fabricated silicon-glass chips. *Nucleic Acids Res* 24:380–85.
78. Brody, J. P., T. D. Osborn, F. K. Foster, and P. Yager. 1995. A planar microfabricated fluid filter. *8th International Conference on Solid State Sensors and Actuators, and Eurosensors IX*. Stockholm, Sweden, June 1995, pp. 779–82.
79. Fluitman, J. H., A. van den Berg, and T. S. Lammerink. 1995. In *Micro total analysis systems,* eds. A. van den Berg, and P. Bergveld, 73–83. Norwell, MA: Kluwer Academic Publishers.
80. Pipper, J., Y. Zhang, P. Neuzil, and T-M. Hsieh. 2008. Clockwork PCR including sample preparation. *Angew Chem Int Ed* 47:3900–04.
81. Kido, H., M. Micic, D. Smith, J. Zoval, J. Norton, and M. Madou. 2007. A novel, compact disk-like centrifugal microfluidics system for cell lysis and sample homogenization. *Colloids Surf B: Biointerfaces* 58:44–51.
82. Scott, C. D., and J. C.Mailen. 1972. Dynamic introduction of whole-blood samples into fast analyzers. *Clin Chem* 18:749–52.
83. Brenner, T., R. Zengerle, and J. Ducrée. 2003. A flow-switch based on coriolis force. In *Proceedings of μTAS.* Lake Tahoe, CA, October 2003. Norwell, MA: Kluwer Academic.
84. Angell, J. B., S. C. Terry, and P. W. Barth. 1983. Silicon micromechanical devices. *Sci Am* 248:44–57.

85. Editor. 1998. MicroChemical Systems.
86. Madou, M. J. 1994. Compatibility and incompatibility of chemical sensors and analytical equipment with micromachining. *Technical Digest: 1994 Solid State Sensor and Actuator Workshop*. Hilton Head Island, SC, June 13–16, 1994, pp. 164–71.
87. Lyle, R. P., and D. Walters. 1997. In *1997 International Conference on Solid-State Sensors and Actuators*, 975–78. Chicago.
88. Madou, M. J., and S. R. Morrison. 1989. *Chemical sensing with solid state devices*. New York: Academic Press.
89. Buck, L., and R. Axel. 1991. A novel multigene family may encode odorant receptors: A molecular basis for odor recognition. *Cell* 65:175–87.
90. Madou, M. 1994. *NIST Special Publication 865*. Gaithersburg, MD: National Institute of Standards and Technology.
91. Goldstein, M. K. 1991. *Biomimetic sensor that simulates human response to airborne toxins, Quantum Group*. US Patent 5063164.
92. Sandifer, J. R., and J. J. Voycheck. 1999. A review of biosensor and industrial applications of pH-ISFETs and an evaluation of Honeywell's"DuraFET". *Mikrochim Acta* 131: 91–98.
93. Kolesar, E. S., and R. R. Reston. 1995. Separation and detection of toxic gases with a silicon micromachined gas chromatography system. *Proceedings of the International Conference on Integrated Micro/Nanotechnology for Space Applications, National Aeronautics and Space Administration (NASA) and The Aerospace Corporation*. Houston, TX, October 30–November 3, 1995, pp. 102–14.
94. Terry, S. C., J. H. Jerman, and J. B. Angell. 1979. A gas chromatographic air analyzer fabricated on a silicon wafer. *IEEE Trans Electron Dev* ED-26:1880–86.
95. Overton, E. B. Ohio State University. Personal communication.
96. Leeuwis, H., and J. a. v. V. 1994. In *Microsystem technology*, ed. G. K. Lebbink. Alphen aan den Rijn/Zaventem, The Netherlands: Samsom BedrijfsInformatie, pp. 90–94.
97. Goldner, H. 1993. 'Honey, they shrunk the GC!' Gas chromatographs go mini. *Res Dev* June: 45.
98. Stalder, R. E., S. Boumsellek, T. R. Van Zandt, T. W. Kenny, M. H. Hecht, and F. E. Grunthaner. 1993. Micromachined array of electrostatic energy analyzers for charged particles. *J Vac Sci Technol A* 12:2554–58.
99. Bryzek, J., S. Roundy, B. Bircumshaw, C. Chung, K. Castellino, J. R. Stetter, and M. Vestel. 2006. Marvelous MEMS: Advanced IC sensors and microstructures for high volume applications. *IEEE Circuits Devices Mag* 22: 8–28.
100. Leroy, C., M. B. Pisani, C. Hibert, D. Bouvet, M. Puech, and A. M. Ionescu. 2006. High Quality Factor Copper Inductors Integrated in Deep Dry-Etched Quartz Substates. *Symposium on Design, Test, Integration and Packaging of MEMS/MOEMS*, Stresa, Italy, pp. 26–28.
101. Albert, P., J. Costello, and P. Salomon. 2000. RF/microwave applications of MEMS integrated in CMOS technology. *MST News* June: 12–13.
102. Ehmann, K. F., D. Bourell, M. I. Culpepper, T. J. Hodgson, T. Kurfess, M. Madou, K. Rajurkar, and R. DeVor. 2007. *International assessment of research and development in micromanufacturing*. New York: Springer.
103. Edwards, S. A. 2006. *The nanotech pioneers: Where are they taking us?* Weinheim, Berlin: Wiley-VCH.
104. Petersen, K. E. 1982. Silicon as a mechanical material. *Proc IEEE* 70:420–57.
105. Gasman, L. 2006. *Nanotechnology applications and markets*. Norwood, MA: Artech House Publishers.

Index

A

B

C

D

E

N

Q

R

S

T

U

V

W

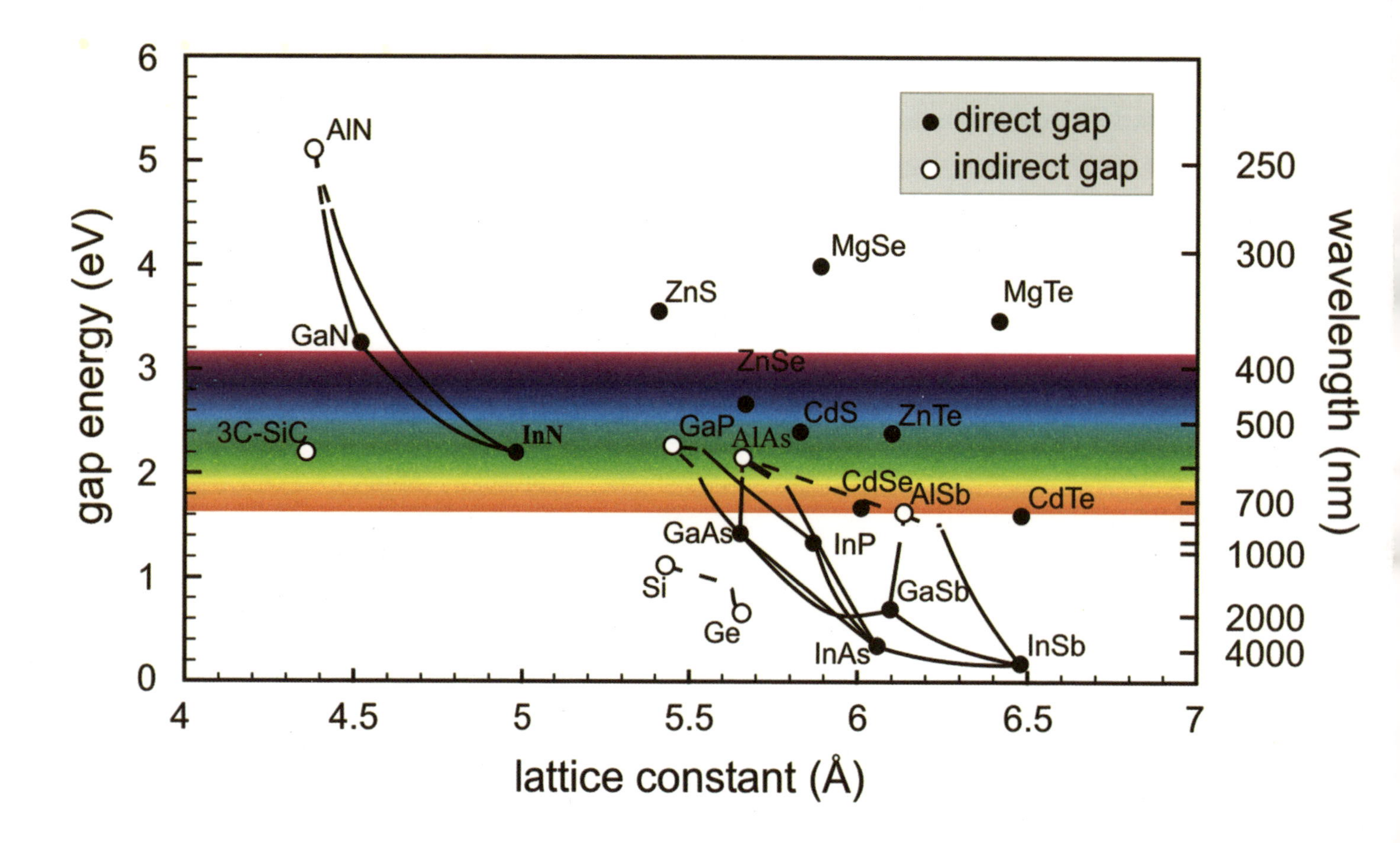

direct gap
indirect gap
gap energy (eV)
wavelength (nm)
lattice constant (Å)
AlN
GaN
InN
3C-SiC
ZnS
MgSe
MgTe
ZnSe
CdS
ZnTe
GaP
AlAs
CdSe
AlSb
CdTe
GaAs
InP
Si
Ge
GaSb
InAs
InSb
0
1
2
3
4
5
6
4
4.5
5
5.5
6
6.5
7
250
300
400
500
700
1000
2000
4000